First Mediterranean Conference on Classical and Quantum Gravity

(MCCQG 2009)

Journal of Physics: Conference Series Volume 222

Crete, Greece
14-18 September 2009

ISBN: 978-1-61738-581-0
ISSN: 1742-6588

Printed from e-media with permission by:

Curran Associates, Inc.
57 Morehouse Lane
Red Hook, NY 12571

Printed by Curran Associates, Inc. (2010)

For permission requests, please contact the Institute of Physics
at the address below.

Institute of Physics
Dirac House, Temple Back
Bristol BS1 6BE UK

Phone: 44 1 17 929 7481
Fax: 44 1 17 920 0979

techtracking@iop.org

Additional copies of this publication are available from:

Curran Associates, Inc.
57 Morehouse Lane
Red Hook, NY 12571 USA
Phone: 845-758-0400
Fax: 845-758-2634
Email: curran@proceedings.com
Web: www.proceedings.com

TABLE OF CONTENTS

STRINGS AND BRANES

CLASSICAL GRAVITY AND ALTERNATIVE THEORIES

COSMOLOGY

GRAVITATIONAL WAVES AND EXPERIMENTS

BLACK HOLES

QUANTUM GRAVITY

PREFACE

This is the fourth volume[1] of articles on the theory of Nonequilibrium Green's functions (NEGF) and their modern application in various fields such as plasma physics, semiconductor physics, molecular electronics and high energy physics. It contains 23 articles written by experts in many-body theory and quantum transport who summarize recent progress in their respective area of research.

The articles are based on talks given at the interdisciplinary conference *Progress in Nonequilibrium Green's functions IV* which was held 17–21 August 2009 at the University of Glasgow, Scotland. This conference continues the tradition of the previous meetings which started in 1999 and which aimed at an informal exchange across field boundaries. The previous meetings and the earlier proceedings proved to be very stimulating not only for young researchers but also for experienced scientists, and we are convinced that this fourth volume will be as successful as the previous ones. As before, this volume includes only extended review-type papers which are written in a way that they are understandable to a broad interdisciplinary audience.

All papers published in this volume of *Journal of Physics: Conference Series* have been peer reviewed through processes administrated by the Editors assuring highest scientific standards. In the review process some papers were substantially revised and improved and some were rejected.

This conference would not have been possible without the remarkable work of the local organizing team around John Barker and Scott Roy and the generous financial support from the University of Glasgow and the Deutsche Forschungsgemeinschaft via SFB-Transregio 24.

Michael Bonitz and Karsten Balzer
Kiel, February 2010

[1] The first two volumes are *Progress in Nonequilibrium Green's functions*, M Bonitz (ed) and *Progress in Nonequilibrium Green's functions II*, M Bonitz and D Semkat (eds), which were published by World Scientific (Singapore), in 2000 and 2003, respectively (ISBN 981-02-4218-2 and ISBN 981-238-271-2). The third volume *Progress in Nonequilibrium Green's functions III* M Bonitz and A Filinov (eds) was published in *Journal of Physics: Conference Series*, volume 35 (2006).

Conference program and participants

Workshop participants

Carl-Olof Almbladh, Lund University, Sweden
Frithjof Anders, Technische Universität Dortmund, Germany
Karsten Balzer, Kiel University, Germany
John Barker, University of Glasgow, Scotland
Sebastian Bauch, Kiel University, Germany
Gordon Baym, University of Illinois at Urbana-Champaign, USA
Michael Bonitz, Kiel University, Germany
Andrew Brown, University of Glasgow, Scotland
Pawel Danielewicz, Michigan State University, USA
Barry Dunietz, University of Michigan, USA
Holger Fehske, Greifswald University, Germany
David Ferry, Arizona State University, USA
Dirk Gericke, University of Warwick, United Kingdom
Matthew Gilbert, University of Illinois at Urbana-Champaign, USA
Maria Hellgren, Lund University, Sweden
David Hochstuhl, Kiel University, Germany
Frank Jahnke, University Bremen, Germany
Antti-Pekka Jauho, Technical University of Denmark, and Helsinki University of Technology
Kimmo Kainulainen, University of Jyväskylä, Finland
Andele Kalvova, Charles University, Prague, Czech Republic
Peter Kleinert, Paul-Drude Institute, Germany
Wolf-Dietrich Kraeft, Rostock University, Germany
Dietrich Kremp, Rostock University, Germany
Antonio Martinez, University of Glasgow, Scotland
John McGrady, University of Glasgow, Scotland
Petri Myohanen, University of Jyväskylä, Finland
Marc Puig von Friesen, Lund University, Sweden
Felix Richter, Rostock University, Germany
Arnau Rios, University of Surrey, United Kingdom
Scott Roy, University of Glasgow, Scotland
Tomi Ruokola, Technical University of Denmark
Stefano Sanvito, Trinity College, Dublin, Ireland
Sebastian Schmitt, Technische Universität Dortmund, Germany
Peter Schmitteckert, Karlsruhe Institute of Technology, Germany

Dirk Semkat, Rostock University, Germany
Vaclav Spicka, Charles University, Prague, Czech Republic
Adrian Stan, University of Jyväskylä, Finland
Gianluca Stefanucci, University of Rome, Italy
Anna-Maija Uimonen, University of Jyväskylä, Finland
Kris van Houcke, Ghent University, Belgium
Robert van Leeuwen, University of Jyväskylä, Finland
Claudio Verdozzi, Lund University, Sweden
Jan Vorberger, University of Warwick, United Kingdom
Jeremy Watling, University of Glasgow, Scotland

List of Talks at the conference

(R)—Review Talk; (I)—Invited Talk; (C)—Contributed Talk.

F. Anders
University of Dortmund
(I) *Quantum transport through nano-devices: A scattering-states numerical renormalization group approach to open quantum systems*

K. Balzer, D. Hochstuhl, S. Bauch, and M. Bonitz
Christian-Albrechts-Universität Kiel
(I) *Efficient grid-based method in non-equilibrium Green's function calculations for inhomogeneous systems*

J. Barker
University of Glasgow
(I) *NEGF quantum transport phase space formalism with compact support*

S. Bauch, D. Hochstuhl, K. Balzer, and M. Bonitz
Christian-Albrechts-Universität Kiel
(C) *Quantum kinetic approach to time-resolved photoionization of multi-electron atoms*

G. Baym
University of Illinois at Urbana-Champaign
(R) *Connections between quark-gluon plasmas and cold atom physics*

P. Danielewicz[1], and A. Rios[2]
[1]Michigan State University, [2]University of Surrey
(I) *Towards Application of Green's Functions to Nonequilibrium Processes in Nuclear Systems*

B. Dunietz
University of Michigan
(I) *A spectroscopic analysis of electron transport dynamics through molecular junctions*

D.K. Ferry[1], L. Huang[1], R. Akis[1], Y.-C. Lai[1], J.P. Bird[2], N. Aoki[3], Y. Ujie[3], S. Motooka[3], T. Morimoto[3], Y. Ochiai[3], R. Brunner[4], R. Meisels[4], and F. Kuchar[4]
[1]Arizona State University, [2]University at Buffalo, [3]Chiba University, [4]Montanuniversität Leoben
(R) *Open Quantum Dots in Bilayer Graphene: Study of Conductance Fluctuations with Recursive Green's Functions*

D. Gericke, K. Wünsch, and J. Vorberger

University of Warwick
(I) *Structural Properties of Warm Dense Matter*

C. Gies, S. Ritter, P. Gartner, and F. Jahnke
University of Bremen
(I) *Photon statistics of semiconductor microcavity lasers and its connection to carrier correlations*

M. Gilbert
University of Illinois at Urbana-Champaign
(I) *Probing the existence and the stability of the Superfluid Phase of Bilayer Graphene*

D. Hochstuhl, S. Bauch, K. Balzer, and M. Bonitz
Christian-Albrechts-Universität Kiel
(C) *Non-equilibrium Green function approach to photoionization processes in atoms and molecules*

A.-P. Jauho
Technical University of Denmark, and Helsinki University of Technology
(I) *Ab initio calculations of phonons and vibrations in nonequilibrium nanowire*

K. Kainulainen
University of Jyväskylä
(I) *Extended Quasiparticle Approximation and nonlocal coherence*

P. Kleinert
Paul-Drude-Institut für Festkörperelektronik
(C) *Unified theory of double-time quantum transport and quantum diffusion*

W.-D. Kraeft
University of Rostock
(C) *Thermodynamics of Strongly Coupled Plasmas*

D. Kremp, D. Semkat, and K.Henneberger
Universität Rostock
(I) *BEC-BCS transition in excitonic systems*

A. Martinez, N. Seoane, A. Brown, J. Barker, and A. Asenov
University of Glasgow
(I) *Dopants and Roughness induced Resonances in thin Si Nanowire transistors: A self-consistent NEGF-Poisson study*

J. McGrady, and V. Georgiev
University of Glasgow
(I) *NEGF for molecular chemistry: a users perspective*

P. Myöhänen[1] A. Stan[1], G. Stefanucci[2,3], and R. van Leeuwen[1,3]
[1]University of Jyväskylä, [2]Università di Roma, [3]European Theoretical Spectroscopy Facility (ETSF)
(I) *Kadanoff-Baym approach to quantum transport through interacting nanoscale systems: From the transient to the steady-state regime*

M. Puig von Friesen, C. Verdozzi, and C.-O. Almbladh
Lund University
(I) *Off-equilibrium dynamics of Hubbard Clusters*

F. Richter, D. Semkat, and K. Henneberger

Journal of Physics: Conference Series **220** (2010) 011002 doi:10.1088/1742-6596/220/1/011002

University of Rostock
(C) *The photon Green's function for bounded media: Splitting property and nonequilibrium radiation laws*

A. Rios[1], P. Danielewicz[2], and B. Barker[2]
[1]University of Surrey, [2]Michigan State University
(C) *Time-dependent Green's Functions Approach to Nuclear Reactions*

T. Ruokola[1], T. Ojanen[2], and A.-P. Jauho[1,3]
[1]Helsinki University of Technology, [2]Freie Universität Berlin, [3]Technical University of Denmark
(C) *Radiative heat transport in quantum circuits*

S. Sanvito
Trinity College, Dublin
(I) *Ab initio electronic transport in macro-molecules from first principles*

P. Schmitteckert
Intitute of Technology Karlsruhe
(I) *Non-equilibrium transport calculations within DMRG*

P. Schmitteckert
Karlsruhe Institute of Technology
(C) *Calculating Green's function from finite systems*

G. Stefanucci
University of Rome
(C) *Time-dependent quantum transport in nanoscopic Josephson junctions: A NEGF approach*

A.-M. Uimonen, P. Myohanen, A. Stan, and R. van Leeuwen
University of Jyväskylä
(C) *Kadanoff-Baym approach to density and current bistability in quantum transport*

K. van Houcke[1], and N. Prokofev[2]
[1]Universiteit Gent, [2] University of Massachusetts
(I) *Diagrammatic Monte Carlo: what happens to the sign-problem?*

B. Velicky, A. Kalvova, and V. Spicka
Charles University of Prague
(I) *Fast transients in mesoscopic systems*

J. Vorberger[1], D.O. Gericke[1], Th. Bornath[2], and M. Schlanges[3]
[1]University of Warwick, [2]Universität Rostock, [3]E.M. Arndt Universität Greifswald
(C) *Energy Transfer Rates and Relaxation in Dense Two-Temperature Plasmas*

List of Posters

A. Alvermann[1], D.M. Edwards[2], and H. Fehske[1]
[1]Ernst-Moritz-Arndt Universität, [2]Imperial College
Calculation of Drude weight for a correlated quantum transport process

S. Bauch, K. Balzer, and M. Bonitz
Christian-Albrechts-Universität Kiel
Quantum Breathing Mode for Fermions and Bosons at Arbitrary Coupling

V.S. Filinov[1], M. Bonitz[2], Y.B. Ivanov, V.V. Skokov, P.R. Levashov[1] and V.E. Fortov[1]
[1]Russian Academy of Sciences, [2]Christian-Albrechts-Universität Kiel, [3]Gesellschaft für

Schwerionenforschung, Darmstadt, [4]Russian Research Center "Kurchatov Institute", Moscow, [5]Joint Institute for Nuclear Research, Dubna
Equation of state of strongly coupled quark-gluon plasma—Path integral Monte Carlo results

M. Heimsoth, D. Hochstuhl, and M. Bonitz
Kiel University
Time-dependent unrestricted Hartree-Fock approximation for strongly correlated mesoscopic Bose systems

T. Koch[1], J. Loos[2], A. Alvermann[1], and H. Fehske[1]
[1]Ernst-Moritz-Arndt Universität, Greifswald, [2]Academy of Sciences of the Czech Republic, Prague
Transport through a vibrating quantum dot: Polaronic effects

P. Myohanen[1], A.-M. Uimonen[1], G. Stefanucci[2], and R. van Leeuwen[1]
[1]University of Jyväskylä, [2]University of Rome
Time-dependent quantum transport with the Kadanoff-Baym equations

A. Rios
University of Surrey
Self-Consistent Green's Functions in equilibrium: the ladder approximation and nuclear matter

S. Schmitt
Technische Universität Dortmund
Single impurity Anderson model out of equilibrium—a comparitive study

D. Semkat, F. Richter, D. Kremp, G. Manzke, W.-D. Kraeft, and K. Henneberger
Universität Rostock,
Ionization equilibrium in electron-hole plasmas: Mott effect vs. Bose-Einstein condensation

C. Verdozzi
Lund University
TDDFT and the Hubbard model

All papers published in this volume of *Journal of Physics: Conference Series* have been peer reviewed through processes administered by the proceedings Editors. Reviews were conducted by expert referees to the professional and scientific standards expected of a proceedings journal published by IOP Publishing.

Possible low energy manifestations of strings and gravity

I. Antoniadis[1]

Department of Physics, CERN - Theory Division, 1211 Geneva 23, Switzerland

E-mail: ignatios.antoniadis@cern.ch

Abstract. Lowering the string scale in the TeV region provides a theoretical framework for solving the mass hierarchy problem and unifying all interactions. The apparent weakness of gravity can then be accounted by the existence of large internal dimensions, in the submillimeter region, and transverse to a braneworld where our universe must be confined. I review the main properties of this scenario and its implications for observations at both particle colliders, and in non-accelerator gravity experiments. I also discuss the warped case and localization of gravity in the presence of infinite size extra dimensions.

1. Introduction

During the last few decades, physics beyond the Standard Model (SM) was guided from the problem of mass hierarchy. This can be formulated as the question of why gravity appears to us so weak compared to the other three known fundamental interactions corresponding to the electromagnetic, weak and strong nuclear forces. Indeed, gravitational interactions are suppressed by a very high energy scale, the Planck mass $M_P \sim 10^{19}$ GeV, associated to a length $l_P \sim 10^{-35}$ m, where they are expected to become important. In a quantum theory, the hierarchy implies a severe fine tuning of the fundamental parameters in more than 30 decimal places in order to keep the masses of elementary particles at their observed values. The reason is that quantum radiative corrections to all masses generated by the Higgs vacuum expectation value (VEV) are proportional to the ultraviolet cutoff which in the presence of gravity is fixed by the Planck mass. As a result, all masses are "attracted" to become about 10^{16} times heavier than their observed values.

Besides compositeness, there are two main theories that have been proposed and studied extensively during the last years, corresponding to different approaches of dealing with the mass hierarchy problem. (1) Low energy supersymmetry with all superparticle masses in the TeV region. Indeed, in the limit of exact supersymmetry, quadratically divergent corrections to the Higgs self-energy are exactly cancelled, while in the softly broken case, they are cutoff by the supersymmetry breaking mass splittings. (2) TeV scale strings, in which quadratic divergences are cutoff by the string scale and low energy supersymmetry is not needed. Both ideas are experimentally testable at high-energy particle colliders and in particular at LHC. Below, I discuss their implementation in string theory.

[1] On leave from CPHT (UMR CNRS 7644) Ecole Polytechnique, F-91128 Palaiseau.

Journal of Physics: Conference Series **222** (2010) 012001 doi:10.1088/1742-6596/222/1/012001

2. Strings and extra dimensions

The appropriate and most convenient framework for low energy supersymmetry and grand unification is the perturbative heterotic string. Indeed, in this theory, gravity and gauge interactions have the same origin, as massless modes of the closed heterotic string, and they are unified at the string scale M_s. As a result, the Planck mass M_P is predicted to be proportional to M_s:

$$M_P = M_s/g\,, \tag{1}$$

where g is the gauge coupling. In the simplest constructions all gauge couplings are the same at the string scale, given by the four-dimensional (4d) string coupling, and thus no grand unified group is needed for unification. In our conventions $\alpha_{\rm GUT} = g^2 \simeq 0.04$, leading to a discrepancy between the string and grand unification scale $M_{\rm GUT}$ by almost two orders of magnitude. Explaining this gap introduces in general new parameters or a new scale, and the predictive power is essentially lost. This is the main defect of this framework, which remains though an open and interesting possibility [1].

The other other perturbative framework that has been studied extensively in the more recent years is type I string theory with D-branes. Unlike in the heterotic string, gauge and gravitational interactions have now different origin. The latter are described again by closed strings, while the former emerge as excitations of open strings with endpoints confined on D-branes [2]. This leads to a braneworld description of our universe, which should be localized on a hypersurface, i.e. a membrane extended in p spatial dimensions, called p-brane (see Fig. 1). Closed strings propagate in all nine dimensions of string theory: in those extended along the p-brane, called parallel, as well as in the transverse ones. On the contrary, open strings are attached on the p-brane. Obviously, our p-brane world must have at least the three known dimensions of space.

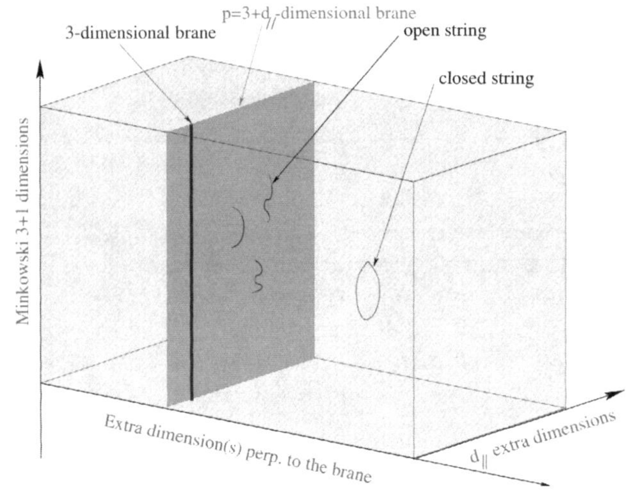

Figure 1. In type I string framework, our Universe contains, besides our 3 spatial dimensions (denoted by the blue line), some extra dimensions ($d_{\|} = p - 3$) parallel to our world p-brane (green plane) where endpoints of open strings are confined, as well as some transverse dimensions (yellow space) where only gravity (described by closed strings) propagate.

But it may contain more: the extra $d_{\|} = p - 3$ parallel dimensions must have a finite size, in order to be unobservable at present energies, and can be as large as $\text{TeV}^{-1} \sim 10^{-18}$ m [3]. On the other hand, transverse dimensions interact with us only gravitationally and experimental bounds are much weaker: their size should be less than about 0.1 mm [4]. In the following, I review the main properties and experimental signatures of low string scale models [5, 6].

2.1. Framework of low scale strings

In type I theory, the different origin of gauge and gravitational interactions implies that the relation between the Planck and string scales is not linear as (1) of the heterotic string. The requirement that string theory should be weakly coupled, constrain the size of all parallel dimensions to be of order of the string length, while transverse dimensions remain unrestricted. Assuming an isotropic transverse space of $n = 9 - p$ compact dimensions of common radius $R_\perp$, one finds:

$$M_P^2 = \frac{1}{g^4} M_s^{2+n} R_\perp^n \,, \qquad g_s \simeq g^2 \,. \tag{2}$$

where g_s is the string coupling. It follows that the type I string scale can be chosen hierarchically smaller than the Planck mass [7, 5] at the expense of introducing extra large transverse dimensions felt only by gravity, while keeping the string coupling small [5]. The weakness of 4d gravity compared to gauge interactions (ratio M_W/M_P) is then attributed to the largeness of the transverse space $R_\perp$ compared to the string length $l_s = M_s^{-1}$.

An important property of these models is that gravity becomes effectively $(4+n)$-dimensional with a strength comparable to those of gauge interactions at the string scale. The first relation of Eq. (2) can be understood as a consequence of the $(4+n)$-dimensional Gauss law for gravity, with

$$M_*^{(4+n)} = M_s^{2+n}/g^4 \tag{3}$$

the effective scale of gravity in $4+n$ dimensions. Taking $M_s \simeq 1$ TeV, one finds a size for the extra dimensions $R_\perp$ varying from 10^8 km, .1 mm, down to a Fermi for $n = 1, 2$, or 6 large dimensions, respectively. This shows that while $n = 1$ is excluded, $n \geq 2$ is allowed by present experimental bounds on gravitational forces [4, 8]. Thus, in these models, gravity appears to us very weak at macroscopic scales because its intensity is spread in the "hidden" extra dimensions. At distances shorter than $R_\perp$, it should deviate from Newton's law, which may be possible to explore in laboratory experiments (see Fig. 2).

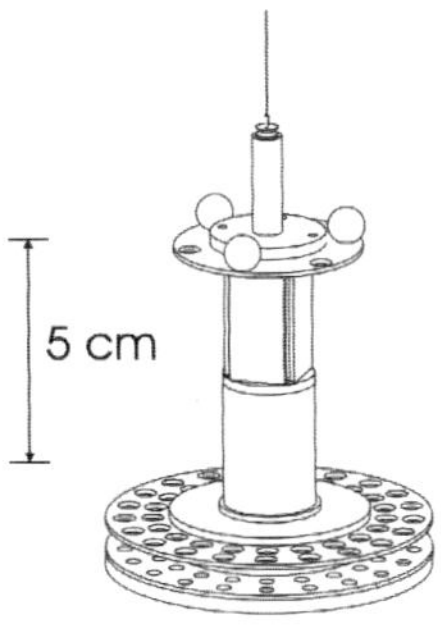

Figure 2. Torsion pendulum that tested Newton's law at 55 μm.

The main experimental implications of TeV scale strings in particle accelerators are of three types, in correspondence with the three different sectors that are generally present: (i) new compactified parallel dimensions, (ii) new extra large transverse dimensions and low scale quantum gravity, and (iii) genuine string and quantum gravity effects. On the other hand, there exist interesting implications in non accelerator table-top experiments due to the exchange of gravitons or other possible states living in the bulk.

3. Experimental implications in accelerators

3.1. World-brane extra dimensions

In this case $RM_s \gtrsim 1$, and the associated compactification scale $R_{\|}^{-1}$ would be the first scale of new physics that should be found increasing the beam energy [3, 9, 10]. The main consequence is the existence of Kaluza-Klein (KK) excitations for all SM particles that propagate along the extra parallel dimensions. Their masses are given by:

$$M_m^2 = M_0^2 + \frac{m^2}{R_{\|}^2} \quad ; \quad m = 0, \pm 1, \pm 2, \dots \tag{4}$$

where we used $d_{\|} = 1$, and M_0 is the higher dimensional mass. The zero-mode $m = 0$ is identified with the 4d state, while the higher modes have the same quantum numbers with the lowest one, except for their mass given in (4). There are two types of experimental signatures of such dimensions [9, 11, 12]: (i) virtual exchange of KK excitations, leading to deviations in cross-sections compared to the SM prediction, that can be used to extract bounds on the compactification scale; (ii) direct production of KK modes.

On general grounds, there can be two different kinds of models with qualitatively different signatures depending on the localization properties of matter fermion fields. If the latter are localized in 3d brane intersections, they do not have excitations and KK momentum is not conserved because of the breaking of translation invariance in the extra dimension(s). KK modes of gauge bosons are then singly produced giving rise to generally strong bounds on the compactification scale and new resonances that can be observed in experiments. Otherwise, they can be produced only in pairs due to the KK momentum conservation, making the bounds weaker but the resonances difficult to observe.

In addition to virtual effects, KK excitations can be produced on-shell at LHC as new resonances [11] (see Fig. 3). There are two different channels, neutral Drell–Yan processes

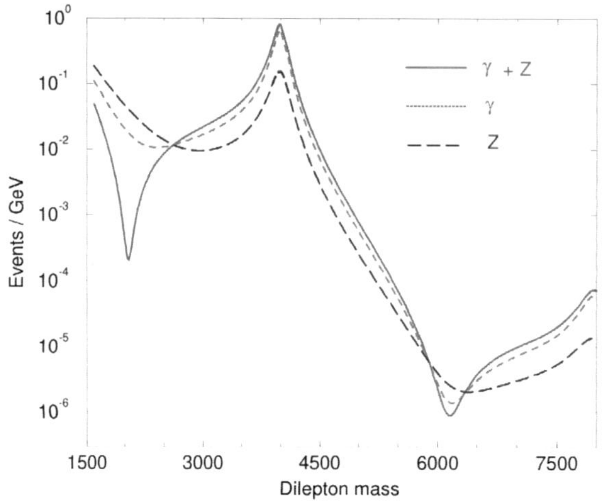

Figure 3. Production of the first KK modes of the photon and of the Z boson at LHC, decaying to electron-positron pairs. The number of expected events is plotted as a function of the energy of the pair in GeV.

$pp \to l^+l^-X$ and the charged channel $l^{\pm}\nu$, corresponding to the production of the KK modes

Table 1. Limits on $R_\perp$ in mm.

Experiment	$n = 2$	$n = 4$	$n = 6$
Collider bounds			
LEP 2	5×10^{-1}	2×10^{-8}	7×10^{-11}
Tevatron	5×10^{-1}	10^{-8}	4×10^{-11}
LHC	4×10^{-3}	6×10^{-10}	3×10^{-12}
NLC	10^{-2}	10^{-9}	6×10^{-12}
Present non-collider bounds			
SN1987A	3×10^{-4}	10^{-8}	6×10^{-10}
COMPTEL	5×10^{-5}	-	-

$\gamma^{(1)}, Z^{(1)}$ and $W_\pm^{(1)}$, respectively. The discovery limits are about 6 TeV, while the exclusion bounds 15 TeV.

On the other hand, if all SM particles propagate in the extra dimension (called universal), KK modes can only be produced in pairs and the lower bound on the compactification scale becomes weaker, of order of 300-500 GeV. Moreover, no resonances can be observed at LHC, so that this scenario appears very similar to low energy supersymmetry. In fact, KK parity can even play the role of R-parity, implying that the lightest KK mode is stable and can be a dark matter candidate in analogy to the LSP [13].

3.2. Extra large transverse dimensions

The main experimental signal is gravitational radiation in the bulk from any physical process on the world-brane. In fact, the very existence of branes breaks translation invariance in the transverse dimensions and gravitons can be emitted from the brane into the bulk. During a collision of center of mass energy $\sqrt{s}$, there are $\sim (\sqrt{s}R_\perp)^n$ KK excitations of gravitons with tiny masses, that can be emitted. Each of these states looks from the 4d point of view as a massive, quasi-stable, extremely weakly coupled (s/M_P^2 suppressed) particle that escapes from the detector. The total effect is a missing-energy cross-section roughly of order:

$$\frac{(\sqrt{s}R_\perp)^n}{M_P^2} \sim \frac{1}{s}\left(\frac{\sqrt{s}}{M_s}\right)^{n+2} . \tag{5}$$

Explicit computation of these effects leads to the bounds given in Table 1.

Fig. 4 shows the cross-section for graviton emission in the bulk, corresponding to the process $pp \to jet+graviton$ at LHC, together with the SM background [14]. There is a particular energy and angular distribution of the produced gravitons that arise from the distribution in mass of KK states of spin-2. This can be contrasted to other sources of missing energy and might be a smoking gun for the extra dimensional nature of such a signal.

In Table 1, there are also included astrophysical and cosmological bounds. Astrophysical bounds [15, 16] arise from the requirement that the radiation of gravitons should not carry on too much of the gravitational binding energy released during core collapse of supernovae. The best cosmological bound [17] is obtained from requiring that decay of bulk gravitons to photons do not generate a spike in the energy spectrum of the photon background measured by the COMPTEL instrument. Bulk gravitons are expected to be produced just before nucleosynthesis due to thermal radiation from the brane. The limits assume that the temperature was at most 1 MeV as nucleosynthesis begins, and become stronger if temperature is increased.

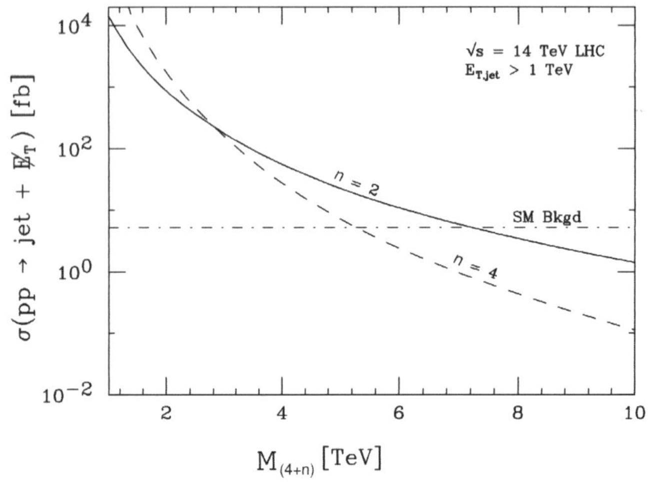

Figure 4. Missing energy due to graviton emission at LHC, as a function of the higher-dimensional gravity scale M_*, produced together with a hadronic jet. The expected cross-section is shown for $n = 2$ and $n = 4$ extra dimensions, together with the SM background.

3.3. String effects

At low energies, the interaction of light (string) states is described by an effective field theory. Their exchange generates in particular four-fermion operators that can be used to extract independent bounds on the string scale. In analogy with the bounds on longitudinal extra dimensions, there are two cases depending on the localization properties of matter fermions. If they come from open strings with both ends on the same stack of branes, exchange of massive open string modes gives rise to dimension eight effective operators, involving four fermions and two space-time derivatives [18, 19]. The corresponding bounds on the string scale are then around 500 GeV. On the other hand, if matter fermions are localized on non-trivial brane intersections, one obtains dimension six four-fermion operators and the bounds become stronger: $M_s \gtrsim 2-3$ TeV [19, 6]. At energies higher than the string scale, new spectacular phenomena are expected to occur, related to string physics and quantum gravity effects, such as possible micro-black hole production [20, 21, 22]. Particle accelerators would then become the best tools for studying quantum gravity and string theory.

Direct production of string resonances in hadron colliders leads generically to a universal deviation from Standard Model in jet distribution [23]. In particular, the first Regge excitation of the gluon has spin 2 and a width an order of magnitude lower than the string scale, leading to a characteristic peak in dijet production; similarly, the first excitations of quarks have spin 3/2. The dijet cross-section is shown in Fig. 5 for LHC energies. The reason for the universal behavior is that tree N-point open superstring amplitudes involving at most two fermions and gluons are completely model independent from the details of the compactification, including the number of supersymmetries that are left unbroken in four dimensions (even if all are broken). Such tree-level amplitudes do not receive contributions from KK, string winding or closed string graviton modes, but are given as a universal sum over exchanges of open string Regge excitations lying in Regge trajectories with masses

$$M_n^2 = M_s^2 n \quad ; \quad n = 0, 1, \ldots \tag{6}$$

and maximal spin $n + 1$.

The relevant partonic cross-sections for dijet production, involving at most two quarks

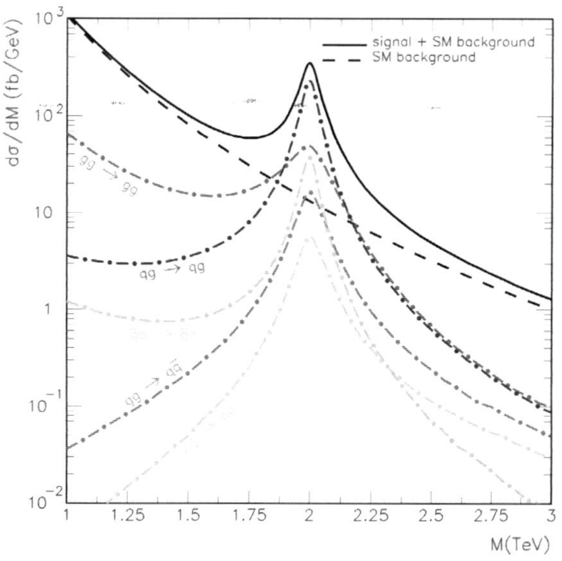

Figure 5. Production of the first Regge excitations at LHC in the dijet channel, for $M_s = 2$ TeV. The cross-section is plotted as a function of the dijet invariant mass M.

are $|\mathcal{M}(gg \to gg)|^2$, $|\mathcal{M}(gg \to q\bar{q})|^2$, $|\mathcal{M}(q\bar{q} \to gg)|^2$ and $|\mathcal{M}(qg \to qg)|^2$, where g and q denote gluons and quarks, respectively. They can be obtained from the first two, up to crossing symmetries, at the full (tree) string level and they are model independent [24]. The low energy expansion of these amplitudes reproduce the usual QCD expressions, while higher order terms describe the string corrections due to the exchange of Regge excitations. Besides these amplitudes, there are also those involving four quarks, such as $|\mathcal{M}(q\bar{q} \to q\bar{q})|^2$ and $|\mathcal{M}(qq \to qq)|^2$. These are model dependent, because the details of the compactification do not decouple. The reason is that they involve four vertices containing twist fields that describe quark states arising from open strings stretched in brane intersections. However, taking into account QCD color factors, their contribution is suppressed because parton luminosities in proton-proton collisions at TeV energies favor at least one gluon in the initial state. As a result, the dominant contribution comes from the model independent cross-sections described above, leading to the effect of Fig. 5.

We finish this section with some comments related to the possible micro-black hole production. Independently on the unresolved issue of the convergence of string perturbation theory in the kinematic region relevant to micro-black hole formation, there is a simple argument showing that at least within the perturbative TeV string framework, the energy threshold for black hole production is far above the LHC reach. Indeed, a string size black hole has a horizon radius $r_H \sim 1$ in string units, while the d-dimensional Newton's constant behaves as $G_N \sim g_s^2$. It follows that the mass of a d-dimensional black hole is [25]:

$$M_{\rm BH} \sim r_H^{d/2-1}/G_N \simeq 1/g_s^2\,. \tag{7}$$

Thus, for a weakly coupled theory, this energy threshold is much higher than the string and the higher dimensional Planck scales M_s and M_* of eq. (3). Comparing this energy threshold with the mass of Regge excitations (6), one finds $n \sim 1/g_s^4$ which is actually compatible with the relation one obtains by identifying the black hole entropy $S_{\rm BH} \sim 1/G_N \sim 1/g_s^2$ with the perturbative string entropy $S_{\rm string} \sim \sqrt{n}$. Using now relation (2), and the value of the Standard

Model gauge couplings $g_s \simeq g^2 \sim 0.1$, one finds that the energy threshold $M_{\rm BH}$ of micro-black hole production is about four orders of magnitude higher than the string scale, implying that one would produce 10^4 string states before reaching $M_{\rm BH}$.

4. Supersymmetry in the bulk and short range forces

4.1. Sub-millimeter forces

Besides the spectacular predictions in accelerators, there are also modifications of gravitation in the sub-millimeter range, which can be tested in "table-top" experiments that measure gravity at short distances. There are three categories of such predictions:

(i) Deviations from the Newton's law $1/r^2$ behavior to $1/r^{2+n}$, which can be observable for $n = 2$ large transverse dimensions of sub-millimeter size.

(ii) New scalar forces in the sub-millimeter range, related to the mechanism of supersymmetry breaking, and mediated by light scalar fields φ with masses [26, 5]:

$$m_\varphi \simeq \frac{m_{susy}^2}{M_P} \simeq 10^{-4} - 10^{-6}\ {\rm eV}\,, \tag{8}$$

for a supersymmetry breaking scale $m_{susy} \simeq 1 - 10$ TeV. They correspond to Compton wavelengths of 1 mm to 10 μm. m_{susy} can be either $1/R_\parallel$ if supersymmetry is broken by compactification [26], or the string scale if it is broken "maximally" on our world-brane [5]. A universal attractive scalar force is mediated by the radion modulus $\varphi \equiv M_P \ln R$, with R the radius of the longitudinal or transverse dimension(s). For $n = 2$, there may be an enhancement factor of the radion mass by $\ln R_\perp M_s \simeq 30$ decreasing its wavelength by an order of magnitude [27].

The coupling of the radius modulus to matter relative to gravity can be easily computed and is given by:

$$\sqrt{\alpha_\varphi} = \frac{1}{M}\frac{\partial M}{\partial \varphi}\ ;\ \ \alpha_\varphi = \begin{cases} \frac{\partial \ln \Lambda_{\rm QCD}}{\partial \ln R} \simeq \frac{1}{3} \ \text{ for } R_\parallel \\ \\ \frac{2n}{n+2} = 1 - 1.5 \text{ for } R_\perp \end{cases} \tag{9}$$

where M denotes a generic physical mass. Such a force can be tested in microgravity experiments and should be contrasted with the change of Newton's law due the presence of extra dimensions that is observable only for $n = 2$ [4, 8]. The resulting bounds from an analysis of the radion effects are [28]: $M_* \gtrsim 6\,{\rm TeV}$.

(iii) Non universal repulsive forces much stronger than gravity, mediated by possible abelian gauge fields in the bulk [15, 29]. Such fields acquire tiny masses of the order of M_s^2/M_P, as in (8), due to brane localized anomalies [29]. Although their gauge coupling is infinitesimally small, $g_A \sim M_s/M_P \simeq 10^{-16}$, it is still bigger that the gravitational coupling E/M_P for typical energies $E \sim 1$ GeV, and the strength of the new force would be $10^6 - 10^8$ stronger than gravity.

In Fig. 6 we depict the actual information from previous, present and upcoming experiments [8, 27]. The solid lines indicate the present limits from the experiments indicated. The excluded regions lie above these solid lines. Measuring gravitational strength forces at short distances is challenging. The horizontal lines correspond to theoretical predictions, in particular for the graviton in the case $n = 2$ and for the radion in the transverse case. These limits are compared to those obtained from particle accelerator experiments in Table 1.

5. Standard Model on D-branes

The gauge group closest to the Standard Model one can easily obtain with D-branes is $U(3) \times U(2) \times U(1)$. The first factor arises from three coincident "color" D-branes. An open string with one end on them is a triplet under $SU(3)$ and carries the same $U(1)$ charge for all three components. Thus, the $U(1)$ factor of $U(3)$ has to be identified with *gauged* baryon number.

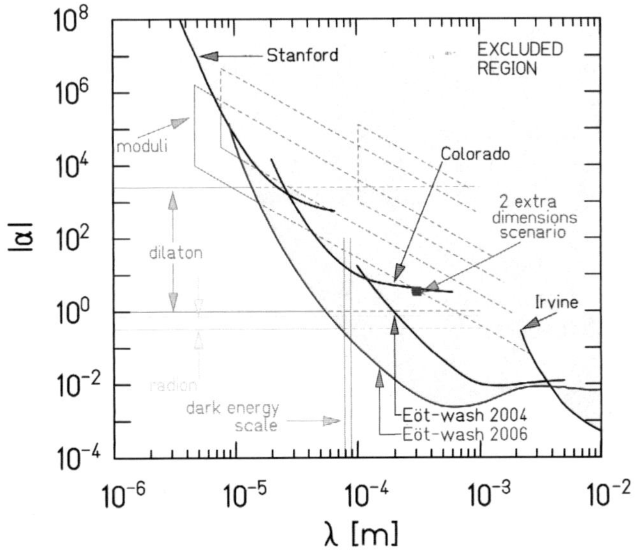

Figure 6. Present limits on new short-range forces (yellow regions), as a function of their range λ and their strength relative to gravity α. The limits are compared to new forces mediated by the graviton in the case of two large extra dimensions, and by the radion.

Similarly, $U(2)$ arises from two coincident "weak" D-branes and the corresponding abelian factor is identified with *gauged* weak-doublet number. Finally, an extra $U(1)$ D-brane is necessary in order to accommodate the Standard Model without breaking the baryon number [30].

It turns out that there are two possible ways of embedding the Standard Model particle spectrum on these stacks of branes [30], which are shown pictorially in Fig. 7. The quark

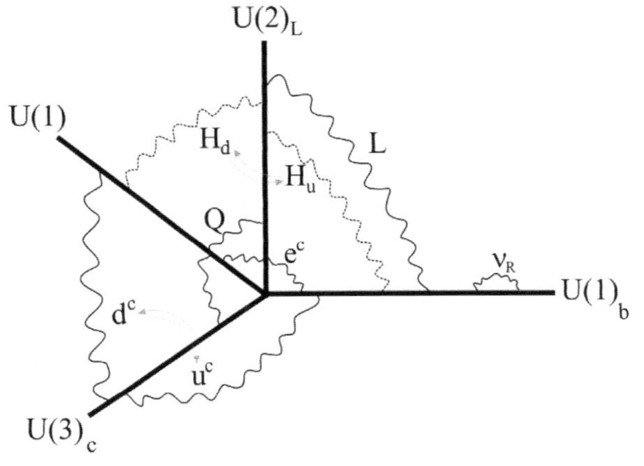

Figure 7. A minimal Standard Model embedding on D-branes.

doublet Q corresponds necessarily to a massless excitation of an open string with its two ends on the two different collections of branes (color and weak). As seen from the figure, a fourth brane stack is needed for a complete embedding, which is chosen to be a $U(1)_b$ extended in the bulk. This is welcome since one can accommodate right handed neutrinos as open string states on the bulk with sufficiently small Yukawa couplings suppressed by the large volume of the bulk [31]. The two models are obtained by an exchange of the up and down antiquarks, u^c and d^c, which correspond to open strings with one end on the color branes and the other either on the $U(1)$ brane, or on the $U(1)_b$ in the bulk. The lepton doublet L arises from an open string stretched between the weak branes and $U(1)_b$, while the antilepton l^c corresponds to a string with one end on the $U(1)$ brane and the other in the bulk. For completeness, we also show the two possible Higgs states H_u and H_d that are both necessary in order to give tree-level masses to all quarks and leptons of the heaviest generation.

6. Non-compact extra dimensions and localized gravity

There are several motivations to study localization of gravity in non-compact extra dimensions: (i) it avoids the problem of fixing the moduli associated to the size of the compactification manifold; (ii) it provides a new approach to the mass hierarchy problem; (iii) there are modifications of gravity at large distances that may have interesting observational consequences. Two types of models have been studied: warped metrics in curved space [32], and infinite size extra dimensions in flat space [33]. The former, although largely inspired by stringy developments and having used many string-theoretic techniques, have not yet a clear and calculable string theory realization [34]. In any case, since curved space is always difficult to handle in string theory, in the following we concentrate mainly on the latter, formulated in flat space with gravity localized on a subspace of the bulk. It turns out that these models of induced gravity have an interesting string theory realization [35] that we describe below, after presenting first a brief overview of the warped case.

6.1. Warped spaces

In these models, space-time is a slice of anti de Sitter space (AdS) in $d = 5$ dimensions while our universe forms a four-dimensional (4d) flat boundary [32]. The corresponding line element is:

$$ds^2 = e^{-2k|y|}\eta_{\mu\nu}dx^\mu dx^\nu + dy^2 \quad ; \quad \Lambda = -24M^3k^2\,, \tag{10}$$

where M, Λ are the 5d Planck mass and cosmological constant, respectively, and the parameter k is the curvature of AdS_5. The fifth coordinate y is restricted on the interval $[0, \pi r_c]$. Thus, this model requires two 'branes', a UV and an IR, located at the two end-points of the interval, $y = 0$ and $y = \pi r_c$, respectively. The vanishing of the 4d cosmological constant requires to fine tune the two tensions: $T = -T' = 24M^3k^2$. The 4d Planck mass is given by:

$$M_P^2 = \frac{1}{k}(1 - e^{-2\pi k r_c})M^3\,. \tag{11}$$

On the other hand, there is an exponential hierarchy of the mass scales in the IR relative to the UV brane due to the warped factor in the metric (10). One can then identify the electroweak scale $M_W \simeq M_P e^{-2\pi k r_c}$.

By studying the corresponding wave functions, one can show that 4d gravity is localized on the UV-brane, while KK gravitons on the IR. It follows that the main prediction of this model is the existence of spin-2 resonances at the TeV scale coupled with (TeV)−1 strength. Their masses are given by:

$$m_n = c_n k e^{-2\pi k r_c} \quad ; \quad c_n \simeq (n + \frac{1}{4})\text{for large}n \tag{12}$$

and they are weakly coupled for $m_n < Me^{-2\pi k r_c}$, *i.e.* for $k < M$. Viable models can be constructed when Standard Model gauge bosons propagate in the 5d bulk, fermions are localized near the UV brane, while the Higgs near the IR [36]. Using AdS/CFT correspondence, they are dual to strongly coupled 4d field theories with composite Higgs; their gauge coupling $g_{YM} \sim M/k$ is strong when spin-2 KK modes are weakly coupled in the gravity dual description.

6.2. *The induced gravity model*

The DGP model and its generalizations are specified by a bulk Einstein-Hilbert (EH) term and a four-dimensional EH term [33]:

$$M^{2+n} \int_{\mathcal{M}_{4+n}} d^4x d^n y \sqrt{G}\, \mathcal{R}_{(4+n)} + M_P^2 \int_{\mathcal{M}_4} d^4x \sqrt{g}\, \mathcal{R}_{(4)}\, ; \quad M_P^2 \equiv r_c^n M^{2+n} \tag{13}$$

with M and M_P the (possibly independent) respective Planck scales. The scale $M \geq 1$ TeV would be related to the short-distance scale below which UV quantum gravity or stringy effects are important. The four-dimensional metric is the restriction of the bulk metric $g_{\mu\nu} = G_{\mu\nu}|$ and we assume the WORLD[2] rigid, allowing the gauge $G_{i\mu}| = 0$ with $i \geq 5$.

6.2.1. Co-dimension one In the case of co-dimension one bulk ($n = 1$) and δ-function localization, it is easy to see that r_c is a crossover scale where gravity changes behavior on the WORLD. Indeed, by Fourrier transform the quadratic part of the action (13) with respect to the 4d position x, at the WORLD position $y = 0$, one obtains $M^{2+n}(p^{2-n} + r_c^n p^2)$, where p is the 4d momentum. It follows that for distances smaller than r_c (large momenta), the first term becomes irrelevant and the graviton propagator on the "brane" exhibits four-dimensional behavior ($1/p^2$) with Planck constant $M_P = M^3 r_c$. On the contrary, at large distances, the first term becomes dominant and the graviton propagator acquires a five-dimensional fall-off ($1/p$) with Planck constant M.

An important consequence of this model is the existence of self-accelerating solutions for the Universe at late times [37]. Indeed, the Friedman equations become:

$$H^2 \pm \frac{H}{r_c} = \frac{8\pi G_N}{3}\rho(t)\,, \tag{14}$$

where H is the Hubble constant and ρ the 4d energy density. For the upper sign (+), one finds usual Robertson-Walker cosmologies extrapolating between 4d to 5d at large scales, while for the lower (negative) sign, one finds a de Sitter solution at late times. This can account for the present dark energy if $r_c \simeq 10^{28}$ cm. On the other hand, this model suffers from the van Dan-Veltman-Zakharov discontinuity of massive gravity, implying a strong coupling regime at distances shorter than the Vainshtein radius $r_V = (r_c^2 w)^{1/3}$, where w is the UV cutoff, corresponding to the 'brane' thickness [38]. Moreover, imposing in general r_c to be larger than the size of the universe, $r_c \gtrsim 10^{28}$ cm, one finds $M \lesssim 100$ MeV, which is in conflict with experimental bounds.

6.2.2. Higher co-dimension The situation changes drastically for more than one non-compact bulk dimensions, $n > 1$, due to the ultraviolet properties of the higher-dimensional theories. Indeed, from the action (13), the effective potential between two test masses in four dimensions

$$\int [d^3x]\, e^{-ip\cdot x}\, V(x) = \frac{D(p)}{1 + r_c^n\, p^2\, D(p)} \left[\tilde{T}_{\mu\nu} T^{\mu\nu} - \frac{1}{2+n}\tilde{T}^\mu_\mu\, T^\nu_\nu\right] \tag{15}$$

$$D(p) = \int [d^n q]\, \frac{f_w(q)}{p^2 + q^2} \tag{16}$$

[2] We avoid calling $\mathcal{M}_4$ a brane because, as we will see below, gravity localizes on singularities of the internal manifold, such as orbifold fixed points. Branes with localized matter can be introduced independently of gravity localization.

is a function of the bulk graviton retarded Green's function $G(x,0;0,0) = \int [d^4p]\, e^{ip\cdot x}\, D(p)$ evaluated for two points localized on the WORLD ($y = y' = 0$). The integral (16) is UV-divergent for $n > 1$ unless there is a non-trivial brane thickness profile $f_w(q)$ of width w. If the four-dimensional WORLD has zero thickness, $f_w(q) \sim 1$, the bulk graviton does not have a normalizable wave function. It therefore cannot contribute to the induced potential, which always takes the form $V(p) \sim 1/p^2$ and Newton's law remains four-dimensional at all distances.

For a non-zero thickness w, there is only one crossover length scale, R_c:

$$R_c = w\left(\frac{r_c}{w}\right)^{\frac{n}{2}}, \tag{17}$$

above which one obtains a higher-dimensional behaviour [39]. Therefore the effective potential presents two regimes: (i) at short distances ($w \ll r \ll R_c$) the gravitational interactions are mediated by the localized four-dimensional graviton and Newton's potential on the WORLD is given by $V(r) \sim 1/r$ and, (ii) at large distances ($r \gg R_c$) the modes of the bulk graviton dominate, changing the potential. Note that for $n = 1$ the expressions (15) and (16) are finite and unambiguously give $V(r) \sim 1/r$ for $r \gg r_c$. For a co-dimension bigger than 1, the precise behavior for large-distance interactions depends *crucially* on the UV completion of the theory. At this point we stress a fundamental difference with the *finite extra dimensions* scenarios. In these cases Newton's law gets higher-dimensional at distances smaller than the characteristic size of the extra dimensions. This is precisely the opposite of the case of infinite volume extra dimensions that we discuss here.

As mentioned above, for higher co-dimension, there is an interplay between UV regularization and IR behavior of the theory. Indeed, several works in the literature raised unitarity [40] and strong coupling problems [41] which depend crucially on the UV completion of the theory. A unitary UV regularization for the higher co-dimension version of the model has been proposed in [42]. It would be interesting to address these questions in a precise string theory context. Actually, using for UV cutoff on the "brane" the 4d Planck length $w \sim l_P$, one gets for the crossover scale (17): $R_c \sim M^{-1}(M_P/M)^{n/2}$. Putting $M \gtrsim 1$ TeV leads to $R_c \lesssim 10^{8(n-2)}$ cm. Imposing $R_c \gtrsim 10^{28}$ cm, one then finds that the number of extra dimensions must be at least six, $n \geq 6$, which is realized nicely in string theory and provides an additional motivation for studying possible string theory realizations.

6.3. String theory realization

The gravity induced model (13) with $n = 6$ is realized as the low-energy effective action of type II string theory on a non-compact six-dimensional manifold $\mathcal{M}_6$ [35], preserving $\mathcal{N} = 2$ supersymmetry in four dimensions. The resulting expressions for the Planck masses M and M_p are:

$$M^2 \sim M_s^2/g_s^{1/2} \quad ; \quad M_P^2 \sim \chi(\frac{c_0}{g_s^2} + c_1)M_s^2\,, \tag{18}$$

with $c_0 = -2\zeta(3)$, $c_1 = \pm 4\zeta(2) = \pm 2\pi^2/3$ and χ is the Euler number of $\mathcal{M}_6$.

It is interesting that the appearance of the induced 4d localized term preserves $\mathcal{N} = 2$ supersymmetry and is independent of the localization mechanism of matter fields (for instance on D-branes). Localization requires the internal space M_6 to have a non-zero Euler characteristic $\chi \neq 0$. Actually, in type IIA/B compactified on a Calabi-Yau manifold, χ counts the difference between the numbers of $\mathcal{N} = 2$ vector multiplets and hypermultiplets: $\chi = \pm 4(n_V - n_H)$ (where the graviton multiplet counts as one vector). Moreover, in the non-compact limit, the Euler number can in general split in different singular points of the internal space, $\chi = \sum_I \chi_I$, giving rise to different localized terms at various points y_I of the internal space. A number of conclusions (confirmed by string calculations in [35]) can be reached by looking closely at (18):

▷ $M_p \gg M$ requires a large non-zero Euler characteristic for M_6, and/or a weak string coupling constant $g_s \to 0$.

▷ Since χ is a topological invariant the localized $\mathcal{R}_{(4)}$ term coming from the closed string sector is universal, independent of the background geometry and dependent only on the internal topology. It is a matter of simple inspection to see that if one wants to have a localized EH term in less than ten dimensions, namely something linear in curvature, with non-compact internal space in all directions, *the only possible dimension is four* (or five in the strong coupling M-theory limit).

▷ In order to find the width w of the localized term, one has to do a separate analysis. On general grounds, using dimensional analysis in the limit $M_P \to \infty$, one expects the effective width to vanish as a power of $l_P \equiv M_P^{-1}$: $w \sim l_P^{\nu}/l_s^{\nu-1}$ with $\nu > 0$. The computation of ν for a general Calabi-Yau space, besides its technical difficulty, presents an additional important complication: from the expression (18), $l_P \sim g_s l_s$ in the weak coupling limit. Thus, w vanishes in perturbation theory and one has to perform a non-perturbative analysis to extract its behavior. Alternatively, one can examine the case of orbifolds. In this limit, $c_0 = 0$, $l_P \sim l_s$, and the hierarchy $M_P > M$ is achieved only in the limit of large χ. In this case, one finds the power $\nu = 1$.

6.3.1. Summary of the results Using $w \sim l_P$ and the relations (18) in the weak coupling limit (with $c_0 \neq 0$), the crossover radius of eq. (17) is given by the string parameters ($n = 6$)

$$R_c = \frac{r_c^3}{w^2} \sim g_s \frac{l_s^4}{l_P^3} \simeq g_s \times 10^{32}\ \text{cm}\,, \tag{19}$$

for $M_s \simeq 1$ TeV. Because R_c has to be of cosmological size, the string coupling can be relatively small, and the Euler number $|\chi| \simeq g_s^2\, l_P \sim g_s^2 \times 10^{32}$ must be very large. The hierarchy is obtained mainly thanks to the large value of χ, so that lowering the bound on R_c lowers the value of χ. Our actual knowledge of gravity at very large distances indicates [43] that R_c should be of the order of the Hubble radius $R_c \simeq 10^{28}$ cm, which implies $g_s \geq 10^{-4}$ and $|\chi| \gtrsim 10^{24}$. A large Euler number implies only a large number of closed string massless particles with no a-priori constraint on the observable gauge and matter sectors, which can be introduced for instance on D3-branes placed at the position where gravity localization occurs. All these particles are localized at the orbifold fixed points (or where the Euler number is concentrated in the general case), and should have sufficiently suppressed gravitational-type couplings, so that their presence with such a huge multiplicity does not contradict observations.

The explicit string realization of localized induced gravity models offers a consistent framework that allows to address a certain number of interesting physics problems. In particular, the effective UV cutoff and the study of the gravity force among matter sources localized on D-branes. It would be also interesting to perform explicit model building and study in detail the phenomenological consequences of these models and compare to other realizations of TeV strings with compact dimensions.

Acknowledgments
Work supported in part by the European Commission under the ERC Advanced Grant 226371 and in part by the CNRS grant GRC APIC PICS 3747.

References

[1] Fore a review, see e.g. K. R. Dienes, Phys. Rept. **287** (1997) 447 [arXiv:hep-th/9602045]; and references therein.
[2] C. Angelantonj and A. Sagnotti, *Phys. Rept.* **371** (2002) 1 [Erratum-ibid. **376** (2003) 339] [arXiv:hep-th/0204089].

[3] I. Antoniadis, *Phys. Lett.* B **246** (1990) 377.
[4] D. J. Kapner, T. S. Cook, E. G. Adelberger, J. H. Gundlach, B. R. Heckel, C. D. Hoyle and H. E. Swanson, Phys. Rev. Lett. **98** (2007) 021101.
[5] N. Arkani-Hamed, S. Dimopoulos and G. R. Dvali, *Phys. Lett.* B **429** (1998) 263 [arXiv:hep-ph/9803315]; I. Antoniadis, N. Arkani-Hamed, S. Dimopoulos and G. R. Dvali, *Phys. Lett.* B **436** (1998) 257 [arXiv:hep-ph/9804398].
[6] For a review see e.g. I. Antoniadis, *Prepared for NATO Advanced Study Institute and EC Summer School on Progress in String, Field and Particle Theory, Cargese, Corsica, France (2002)*; and references therein.
[7] J. D. Lykken, *Phys. Rev.* D **54** (1996) 3693 [arXiv:hep-th/9603133].
[8] J. C. Long and J. C. Price, *Comptes Rendus Physique* **4** (2003) 337; R. S. Decca, D. Lopez, H. B. Chan, E. Fischbach, D. E. Krause and C. R. Jamell, *Phys. Rev. Lett.* **94** (2005) 240401; R. S. Decca et al., arXiv:0706.3283 [hep-ph]; S. J. Smullin, A. A. Geraci, D. M. Weld, J. Chiaverini, S. Holmes and A. Kapitulnik, arXiv:hep-ph/0508204; H. Abele, S. Haeßler and A. Westphal, in 271th WE-Heraeus-Seminar, Bad Honnef (2002).
[9] I. Antoniadis and K. Benakli, *Phys. Lett.* B **326** (1994) 69.
[10] K. R. Dienes, E. Dudas and T. Gherghetta, Phys. Lett. B **436** (1998) 55 [arXiv:hep-ph/9803466]; Nucl. Phys. B **537** (1999) 47 [arXiv:hep-ph/9806292].
[11] I. Antoniadis, K. Benakli and M. Quirós, *Phys. Lett.* B **331** (1994) 313 and *Phys. Lett.* B **460** (1999) 176; P. Nath, Y. Yamada and M. Yamaguchi, *Phys. Lett.* B **466** (1999) 100 T. G. Rizzo and J. D. Wells, *Phys. Rev.* D **61** (2000) 016007; T. G. Rizzo, *Phys. Rev.* D **61** (2000) 055005; A. De Rujula, A. Donini, M. B. Gavela and S. Rigolin, *Phys. Lett.* B **482** (2000) 195;
[12] E. Accomando, I. Antoniadis and K. Benakli, *Nucl. Phys.* B **579** (2000) 3.
[13] G. Servant and T. M. P. Tait, *Nucl. Phys.* B **650** (2003) 391.
[14] G. F. Giudice, R. Rattazzi and J. D. Wells, *Nucl. Phys.* B **544** (1999) 3; E. A. Mirabelli, M. Perelstein and M. E. Peskin, *Phys. Rev. Lett.* **82** (1999) 2236; T. Han, J. D. Lykken and R. Zhang, *Phys. Rev.* D **59** (1999) 105006; K. Cheung and W.-Y. Keung, *Phys. Rev.* D **60** (1999) 112003; C. Balázs *et al.*, *Phys. Rev. Lett.* **83** (1999) 2112; L3 Collaboration (M. Acciarri *et al.*), *Phys. Lett.* B **464** (1999) 135 and **470** (1999) 281: J. L. Hewett, *Phys. Rev. Lett.* **82** (1999) 4765.
[15] N. Arkani-Hamed, S. Dimopoulos and G. Dvali, *Phys. Rev.* D **59** (1999) 086004.
[16] S. Cullen and M. Perelstein, *Phys. Rev. Lett.* **83** (1999) 268; V. Barger, T. Han, C. Kao and R. J. Zhang, *Phys. Lett.* B **461** (1999) 34.
[17] K. Benakli and S. Davidson, *Phys. Rev.* D **60** (1999) 025004; L. J. Hall and D. Smith, *Phys. Rev.* D **60** (1999) 085008.
[18] E. Dudas and J. Mourad, Nucl. Phys. B **575** (2000) 3 [arXiv:hep-th/9911019]; S. Cullen, M. Perelstein and M. E. Peskin, *Phys. Rev.* D **62** (2000) 055012; D. Bourilkov, *Phys. Rev.* D **62** (2000) 076005; L3 Collaboration (M. Acciarri *et al.*), *Phys. Lett.* B **489** (2000) 81.
[19] I. Antoniadis, K. Benakli and A. Laugier, *JHEP* **0105** (2001) 044.
[20] P. C. Argyres, S. Dimopoulos and J. March-Russell, Phys. Lett. B **441** (1998) 96 [arXiv:hep-th/9808138]; T. Banks and W. Fischler, arXiv:hep-th/9906038.
[21] S. B. Giddings and S. Thomas, *Phys. Rev.* D **65** (2002) 056010 [arXiv:hep-ph/0106219]; S. Dimopoulos and G. Landsberg, *Phys. Rev. Lett.* **87** (2001) 161602 [arXiv:hep-ph/0106295].
[22] P. Meade and L. Randall, arXiv:0708.3017 [hep-ph].
[23] L. A. Anchordoqui, H. Goldberg, D. Lust, S. Nawata, S. Stieberger and T. R. Taylor, Phys. Rev. Lett. **101** (2008) 241803 [arXiv:0808.0497 [hep-ph]].
[24] D. Lust, S. Stieberger and T. R. Taylor, Nucl. Phys. B **808** (2009) 1 [arXiv:0807.3333 [hep-th]].
[25] G. T. Horowitz and J. Polchinski, Phys. Rev. D **55** (1997) 6189 [arXiv:hep-th/9612146].
[26] I. Antoniadis, S. Dimopoulos and G. Dvali, *Nucl. Phys.* B **516** (1998) 70; S. Ferrara, C. Kounnas and F. Zwirner, *Nucl. Phys.* B **429** (1994) 589.
[27] I. Antoniadis, K. Benakli, A. Laugier and T. Maillard, *Nucl. Phys.* B **662** (2003) 40 [arXiv:hep-ph/0211409].
[28] E. G. Adelberger, B. R. Heckel, S. Hoedl, C. D. Hoyle, D. J. Kapner and A. Upadhye, Phys. Rev. Lett. **98** (2007) 131104.
[29] I. Antoniadis, E. Kiritsis and J. Rizos, *Nucl. Phys.* B **637** (2002) 92.
[30] I. Antoniadis, E. Kiritsis and T. N. Tomaras, *Phys. Lett.* B **486** (2000) 186; I. Antoniadis, E. Kiritsis, J. Rizos and T. N. Tomaras, *Nucl. Phys.* B **660** (2003) 81.
[31] K. R. Dienes, E. Dudas and T. Gherghetta, *Nucl. Phys.* B **557** (1999) 25 [arXiv:hep-ph/9811428]; N. Arkani-Hamed, S. Dimopoulos, G. R. Dvali and J. March-Russell, *Phys. Rev.* D **65** (2002) 024032 [arXiv:hep-ph/9811448]; G. R. Dvali and A. Y. Smirnov, *Nucl. Phys.* B **563** (1999) 63.
[32] L. Randall and R. Sundrum, *Phys. Rev. Lett.* **83** (1999) 4690 and *Phys. Rev. Lett.* **83** (1999) 3370.
[33] G. R. Dvali, G. Gabadadze and M. Porrati, *Phys. Lett.* B **485** (2000) 208.

[34] H. Verlinde, *Nucl. Phys.* B **580** (2000) 264; S. B. Giddings, S. Kachru and J. Polchinski, *Phys. Rev.* D **66** (2002) 106006.
[35] I. Antoniadis, R. Minasian and P. Vanhove, *Nucl. Phys.* B **648** (2003) 69 [arXiv:hep-th/0209030].
[36] For a review see e.g. R. Contino and A. Pomarol, Comptes Rendus Physique **8** (2007) 1058; and references therein.
[37] C. Deffayet, G. R. Dvali and G. Gabadadze, Phys. Rev. D **65** (2002) 044023 [arXiv:astro-ph/0105068].
[38] M. A. Luty, M. Porrati and R. Rattazzi, arXiv:hep-th/0303116.
[39] G. R. Dvali and G. Gabadadze, *Phys. Rev.* D **63** (2001) 065007; G. R. Dvali, G. Gabadadze, M. Kolanovic and F. Nitti, *Phys. Rev.* D **64** (2001) 084004.
[40] S. L. Dubovsky and V. A. Rubakov, *Phys. Rev.* D **67** (2003) 104014 [arXiv:hep-th/0212222].
[41] V. A. Rubakov, arXiv:hep-th/0303125.
[42] M. Kolanovic, M. Porrati and J. W. Rombouts, *Phys. Rev.* D **68** (2003) 064018 [arXiv:hep-th/0304148].
[43] A. Lue and G. Starkman, *Phys. Rev.* D **67** (2003) 064002.

First Mediterranean Conference on Classical and Quantum Gravity (MCCQG 2009) IOP Publishing
Journal of Physics: Conference Series **222** (2010) 012002 doi:10.1088/1742-6596/222/1/012002

On the relation between local and geometric Lagrangians for higher spins

D. Francia

AstroParticule et Cosmologie (APC) Université Paris VII - Campus Paris Rive Gauche 10, rue Alice Domon et Léonie Duquet F-75205 Paris Cedex 13

E-mail: `francia@apc.univ-paris7.fr`

Abstract. Equations of motion for free higher-spin gauge fields of any symmetry can be formulated in terms of linearised curvatures. On the other hand, gauge invariance alone does not fix the form of the corresponding actions which, in addition, either contain higher derivatives or involve inverse powers of the d'Alembertian operator, thus introducing possible subtleties in degrees of freedom count. We suggest a path to avoid ambiguities, starting from local, unconstrained Lagrangians previously proposed, and integrating out the auxiliary fields from the functional integral, thus generating a unique non-local theory expressed in terms of curvatures.

1. Introduction

One of the unconventional features of higher-spin gauge theories [1, 2, 3, 4, 5, 6, 7, 8, 9, 10] is the peculiar role played by the corresponding generalised curvatures in the formulation of the dynamics.

Higher-spin linearised curvatures were introduced in [11, 12] for symmetric (spinor-) tensors, as proper extensions of the Maxwell field-strength and of the linearised Riemann curvature[1]. They represent the *simplest* gauge invariant tensors that do not vanish on-shell, unless the field itself is pure gauge, with reference to an appropriate generalisation of the abelian gauge transformations of the photon and of the graviton.

The neat systematics underlying their construction, together with the relevance of their lower-spin counterparts, provide basic motivations for the idea that they might play some definite role in a theory of higher-spin gauge fields. On the other hand, due to the increasing number of derivatives needed in their definition, differently from the cases of spin 1 and spin 2, for spin $s \geq 3$ it is not possible to derive from them standard kinetic tensors. For this reason, it was long assumed that one should restrict to the non-geometric formulation of Fronsdal [13], involving only conventional, second-order, differential operators for the construction of free equations of motion for gauge fields of any spin (for an alternative approach see [14]). The price to pay with this choice is that, while no algebraic restrictions (other than symmetry properties of the gauge potential itself) are involved in the definition of higher-spin curvatures, the second-order formulation of [13] requires gauge fields $\varphi_{\mu_1 \cdots \mu_s}$ constrained to be doubly traceless, and subject

[1] In the so-called "metric like" formalism, to be distinguished from the higher-spin generalisation of the Einstein-Cartan formulation of General Relativity, usually termed "frame-like approach" [1, 5].

to an abelian gauge transformation involving a rank-$(s-1)$, traceless gauge parameter:

$$\delta\,\varphi_{\,\mu_1\,\cdots\,\mu_s} \,=\, \partial_{\,\mu_1}\,\Lambda_{\,\mu_2\,\cdots\,\mu_s} \,+\,\cdots\,, \tag{1}$$

$$\begin{aligned} \varphi^{\,\alpha\,\beta}{}_{\alpha\,\beta\,\mu_5\,\cdots\,\mu_s} &\equiv 0\,, \\ \Lambda^{\,\alpha}{}_{\alpha\,\mu_3\,\cdots\,\mu_{s-1}} &\equiv 0\,. \end{aligned} \tag{2}$$

Under these conditions, it is indeed possible to show that the Fronsdal equation

$$\mathcal{F}_{\mu_1\,\ldots\,\mu_s} \,\equiv\, \Box\,\varphi_{\mu_1\,\ldots\mu_s} \,-\, (\partial_{\mu_1}\,\partial^{\,\alpha}\,\varphi_{\alpha\,\mu_2\,\ldots\,\mu_s} \,+\,\ldots) + (\partial_{\mu_1}\partial_{\mu_2}\,\varphi^{\,\alpha}{}_{\alpha\,\mu_3\,\ldots\mu_s} \,+\,\ldots) \,=\, 0\,, \tag{3}$$

propagates the correct polarisations pertaining to a symmetric, massless, spin-s representation of the Poincaré group in D-dimensions[2]. Generalisations of (3) to the case of tensors of mixed-symmetry were found by Labastida to also involve a proper set of algebraic conditions, on the traces of gauge fields and of their gauge parameters [16].

Whereas unusual, the Fronsdal constraints are instrumental in defining a Lagrangian theory in which the number of off-shell components is kept to a minimum. Moreover, the very existence [17] of consistent, non-linear generalisations of (3), makes it is fair to say that there are no compelling reasons suggesting that one should try and get rid of them. On the other hand, simplifications might be expected in a framework where (2) are not to be assumed from the beginning, and in any case it would be rather unsatisfactory if the linearised geometry of [11, 12] were found not to admit deformations relevant for higher-spin non-abelian interactions[3].

Indeed, after [12], the study of higher-spin geometry for symmetric tensors was further pursued in [21] and then extended to the case of mixed-symmetry tensor fields in [22]. Dynamical use of higher-spin curvatures was then proposed in [23, 24], where non-local Lagrangians and equations of motion for symmetric bosons and fermions were investigated, with no *a priori* reference to the Fronsdal formulation. The latter was then shown to be recovered performing the same partial gauge-fixing required in order to remove all non-localities. In a similar spirit, higher-derivative equations of motion for bosons of mixed-symmetry were formulated in [25, 26, 27] in terms of generalised field-strengths (antecedents of non-Lagrangian equations formulated via field-strengths can be found in [11, 14]) while proposals for corresponding higher-derivative or non-local actions were given in [26, 28]. Geometric equations of motion for mixed-symmetry fermions were discussed along similar lines in [4], while in the symmetric case, quadratic deformations of geometric Lagrangians were also constructed [29], providing direct generalisations of the Proca and Fierz-Pauli theories for the description of *massive* fields of any spin.

The basic indication obtained from these results is that, notwithstanding the presence of higher derivatives or non-localities, kinetic tensors built out of curvatures can still be used to describe freely propagating waves of any spin. On the other hand, allowing for non-local operators to be present leads to the consequence that gauge invariance alone is no more a sufficient criterion for the Lagrangian to be unique, and indeed in [30] infinitely many non-local kinetic tensors were shown to exist, for the case of symmetric bosons, besides the basic ones introduced in [23]. As a further selection rule allowing to distinguish among the various options, in [30] it was checked whether those theories reproduced the correct current exchanges between conserved sources, mediated by massless bosons of spin s. Under this requirement it was found

[2] To be precise, only tracelessness of the gauge parameter is needed in the counting of polarisations propagating in (3). Double-tracelessness of φ is then postulated to construct a suitable gauge-invariant Lagrangian. See also related discussions in [15, 12].

[3] Clearly, linearised curvatures can be used to define abelian, Born-Infeld type vertices. They also appear naturally in the quantization of spinning particle models [18], as well as in the description of conformal higher spins [19] (for recent results and more references see [20]).

that, out of the infinitely many geometric Lagrangians consistent with gauge invariance, *only one* non-local theory actually possessed the correct propagator.

On the one hand, this result of uniqueness can be considered satisfactory, since it allows to define a true candidate linear limit for a theory of interacting higher-spins, possibly involving non-linearly deformed curvatures. However, it still leaves unanswered a few questions about the meaning of the whole procedure, given that the check of the propagator is only an *a posteriori* criterion of validity. In particular, the very fact that in the absence of sources infinitely many different equations appear to be consistent, calls for a better understanding of the rationale behind the non-localities of [23, 24, 30], in the spirit of clarifying the meaning of the manipulations involved.

With this purpose in mind, in this note we would like to suggest a different path for the definition of non-local Lagrangians expressed in terms of curvatures, and compare the outcome with the results of [23, 30]. The idea is very simple: we consider a framework where Fronsdal constraints are evaded by means of the introduction of auxiliary fields, so that the theory is still local, and its physical content can be determined using standard techniques. In particular, we resort to the Lagrangians proposed in [31], representing the simplest possible unconstrained ones for the case of symmetric (spinor-) tensors[4]. We then consider the gaussian functional integral for the corresponding theories, and perform the integration over the auxiliary fields. In this way we obtain "effective" non-local Lagrangians involving the physical field φ alone, whose unconstrained gauge invariance implies that they must be expressible in terms of higher-spin curvatures.

The advantage of this procedure is twofold: first, we define in this way an *a priori* criterion to select one member in the class of all possible non-local, geometric Lagrangians. In addition, starting from a theory whose spectrum is known by conventional analysis, we are able to unambiguously relate non-localities to the presence of non-physical fields in the initial local Lagrangians.

We investigate along these lines a few specific cases. After recalling basic facts about higher-spin curvatures and minimal local Lagrangians, which we do in Section 2, we study in Section 3 the form of the non-local effective action for symmetric bosons of spin 3 and spin 4 on flat space-time. We thus show that the integration over the auxiliary fields produces effective, non-local Lagrangians, coinciding with those selected in [30] as the only ones leading to the correct propagators. In addition, in Section 3.3 we broaden our analysis to include the case of (A)dS backgrounds, where we provide the first example of a non-local Lagrangian, for a spin-3 field. Finally, in Section 4 we move our attention to half-integer spins, studying the case of fermions of spin $\frac{5}{2}$. We find in this way the form of the non-local fermionic Lagrangian giving rise to the proper current exchange.

Similar ideas will be exploited in a forthcoming paper [35] with the purpose of analysing the geometrical content of higher-spin *triplets* ([24, 36, 9], and references therein). In the case of those systems, unconstrained gauge-invariance is related to the propagation of several irreducible representations[5], so that, with certain qualifications, a direct correspondence with a sum of constrained Fronsdal Lagrangians can indeed be established [38]. Still, once the auxiliary fields are integrated away, the resulting actions must be expressible in geometrical terms. This suggests in particular that curvatures might play a role in the formulation of higher-spin theories, regardless of whether the Fronsdal-Labastida constraints are assumed or not.

[4] Previous results leading to non-minimal unconstrained Lagrangians can be found in [32]. The generalisation of the local Lagrangians of [31] to the mixed-symmetry case is given in [33]. Constrained descriptions of mixed-symmetry massless fields in the frame-like approach on maximally symmetric backgrounds can be found in [34].
[5] The unconstrained reduction of triplet Lagrangians to the case of irreducible spin s is discussed in [37]. In the frame-like approach, a discussion of the triplets and of their geometrical meaning can be found in [40].

2. Geometric Lagrangians and the issue of uniqueness

Here and in the next section we recall some basic facts, in order to fix the notation and to stress the conceptual issues at stake.

Following the construction of [12], for a rank-3 tensor subject to the abelian gauge transformation (1) the corresponding curvature is

$$\mathcal{R}_{\mu\mu\mu,\,\nu\nu\nu} \,=\, \partial_{\mu}^{3}\,\varphi_{\,\nu\nu\nu} \,-\, \frac{1}{3}\,\partial_{\mu}^{2}\,\partial_{\nu}\,\varphi_{\,\mu\nu\nu} \,+\, \frac{1}{3}\,\partial_{\mu}\,\partial_{\nu}^{2}\,\varphi_{\,\mu\mu\nu} \,-\, \partial_{\nu}^{3}\,\varphi_{\,\mu\mu\mu}\,, \tag{4}$$

in a notation where indices denoted with the same letter are to be understood as being completely symmetrised, without normalization factors, with the minimum numbers of terms required[6], and where in particular $\partial_{\rho}^{k} = \underbrace{\partial_{\rho}\cdots\partial_{\rho}}_{\text{k times}}$ is to be understood as the product of k gradients. For spin 4 the proper generalisation of (4) is

$$\mathcal{R}_{\mu\mu\mu\mu,\,\nu\nu\nu\nu} \,=\, \partial_{\mu}^{4}\,\varphi_{\,\nu\nu\nu\nu} \,-\, \frac{1}{4}\,\partial_{\mu}^{3}\,\partial_{\nu}\,\varphi_{\,\mu\nu\nu\nu} \,+\, \frac{1}{6}\,\partial_{\mu}^{2}\,\partial_{\nu}^{2}\,\varphi_{\,\mu\mu\nu\nu} \,-\, \frac{1}{4}\,\partial_{\mu}\,\partial_{\nu}^{3}\,\varphi_{\,\mu\mu\mu\nu} \,+\, \partial_{\nu}^{4}\,\varphi_{\,\mu\mu\mu\mu}\,, \tag{5}$$

while the general formula for spin s in the present notation looks

$$\mathcal{R}_{\,\mu_{s},\,\nu_{s}} \,=\, \sum_{k=0}^{s} \frac{(-1)^{k}}{\binom{s}{k}}\; \partial_{\mu}^{s-k}\partial_{\nu}^{k}\varphi_{\,\mu_{k},\,\nu_{s-k}}\,, \tag{6}$$

with obvious meaning for subscripts. As previously recalled, the basic property of the tensors (6) is their unconstrained gauge-invariance under the transformation

$$\delta\varphi_{\,\mu_{1}\cdots\mu_{s}} = \partial_{\mu_{1}}\,\Lambda_{\,\mu_{2}\cdots\mu_{s}} + \cdots\,, \tag{7}$$

that in our notation we would simply write as $\delta\varphi = \partial\Lambda$. In addition, they satisfy cyclic and Bianchi identities, reflecting the fact they define irreducible, two-row Young tableaux for $GL(D,\mathbb{R})$. All these properties remain valid if the field φ carries a spinor index as well, so that (6) are also suitable for the definition of a linearised fermionic geometry [12].

In [30] the curvature (4) was used to define a *one-parameter class* of candidate, non-local "Ricci tensors" for a spin-3 field[7]

$$\mathcal{A}_{\varphi}\,(a) \,=\, \frac{1}{\Box}\,\partial\cdot\mathcal{R}^{\prime} \,+\, a\,\frac{\partial^{2}}{\Box^{2}}\,\partial\cdot\mathcal{R}^{\prime\prime}\,, \tag{8}$$

[6] We use the "mostly-plus" space-time metric in d dimensions, denoted with η. Apart from the case of curvatures, whenever there is no risk of confusion all symmetrised indices are left implicit. Lorentz traces are denoted by "primes" or by numbers in square brackets, while divergences are denoted by "$\partial\cdot$". Combinatorial factors can be computed following the rules of [23, 24]. In the product of different tensors full symmetrization of indices is always understood, with no weight factors. Useful combinatorial identities are

$$(\partial^{p}\varphi)^{\prime} \,=\, \Box\partial^{p-2}\varphi + 2\partial^{p-1}\partial\cdot\varphi + \partial^{p}\varphi^{\prime}\,, \qquad\qquad \partial^{p}\partial^{q} \,=\, \binom{p+q}{p}\partial^{p+q}\,,$$

$$\left(\eta^{k}\varphi\right)^{\prime} \,=\, \left[D + 2(s+k-1)\right]\eta^{\,k-1}\varphi + \eta^{k}\varphi^{\prime}\,, \qquad\qquad \eta\eta^{\,n-1} \,=\, n\eta^{\,n}\,.$$

[7] The subscript "φ" in $\mathcal{A}_{\varphi}$ is used to distinguish the non-local, Ricci-like tensors from their local analogues, to be introduced in the next section. Those will be indicated with the symbol $\mathcal{A}$, without subscripts, and will depend on the field φ and on an auxiliary field α.

Journal of Physics: Conference Series **222** (2010) 012002 doi:10.1088/1742-6596/222/1/012002

assumed to define basic equations of motion of the form

$$\mathcal{A}_\varphi(a) = 0 . \tag{9}$$

A first investigation of the consistency of (9) led to the observation that, for almost any value of a, they can be shown to imply the "compensator" equations

$$\mathcal{F} - 3\partial^3 \alpha_\varphi(a) = 0 , \tag{10}$$

where $\alpha_\varphi(a)$ is a non-local tensor whose explicit form depends on a, transforming however always with the trace of the gauge parameter:

$$\delta \alpha_\varphi(a) = \Lambda' . \tag{11}$$

Thus, after a suitable gauge-fixing, infinitely many distinct non-local equations can be reduced to the Fronsdal form. In particular, the especially simple choice $a = 0$ reduces (9) to the equation

$$\frac{1}{\Box}\partial \cdot \mathcal{R}' = 0 , \tag{12}$$

a prototype for the results of the works [23, 24], where free higher-spin Lagrangians formulated in terms of curvatures were first proposed.

On the other hand, as shown in [30], for almost any value of a, the actions associated to the Ricci tensor $\mathcal{A}_\varphi(a)$ turn out to give the *wrong* propagator, thus suggesting that the correct counting of degrees of freedom might involve some further subtleties. Moreover, as recalled in the Introduction, *only one* Lagrangian was proven to provide the correct inverse kinetic operator, defined by the requirement that the current exchange between distant sources effectively projects them onto their transverse-traceless parts in $D-2$ dimensions. For the spin-3 case the "correct" geometric theory is completely characterised by the following quantities:

$$\begin{aligned}
\mathcal{L} &= \frac{1}{2}\varphi\{\mathcal{A}_\varphi - \frac{1}{2}\eta\mathcal{A}'_\varphi\} ,\\
\mathcal{A}_\varphi &= \frac{1}{\Box}\partial\cdot\mathcal{R}' + \frac{\partial^2}{2\Box^2}\partial\cdot\mathcal{R}'' = \mathcal{F} - 3\partial^3\alpha_\varphi ,\\
\alpha_\varphi &= \frac{1}{3\Box^2}\partial\cdot\mathcal{F}' ,
\end{aligned} \tag{13}$$

where in particular second and third of (13) provide the expression of $\mathcal{A}_\varphi$ in terms of curvatures as well as its compensator form. For spin 4 the proper Lagrangian and corresponding Ricci tensor are instead given by

$$\begin{aligned}
\mathcal{L} &= \frac{1}{2}\varphi\{\mathcal{A}_\varphi - \frac{1}{2}\eta\mathcal{A}'_\varphi + \eta^2\mathcal{B}_\varphi\} ,\\
\mathcal{A}_\varphi &= \frac{1}{\Box}\mathcal{R}'' + \frac{1}{2}\frac{\partial^2}{\Box^2}\mathcal{R}''' - 3\frac{\partial^4}{\Box^3}\mathcal{R}^{[4]} = \mathcal{F} - 3\partial^3\alpha_\varphi ,\\
\mathcal{B}_\varphi &= -\frac{3}{8}\frac{1}{\Box}\mathcal{R}^{[4]} = \frac{1}{2}\{\frac{1}{\Box}\partial\cdot\partial\cdot\mathcal{F}' - \mathcal{F}''\} ,\\
\alpha_\varphi &= \frac{1}{3\Box^2}\partial\cdot\mathcal{F}' - \frac{1}{3}\frac{\partial}{\Box^3}\partial\cdot\partial\cdot\mathcal{F}' + \frac{1}{12}\frac{\partial}{\Box^2}\mathcal{F}'' .
\end{aligned} \tag{14}$$

It is clear that the structure of the non-local operators involved in (13) and (14) must retain some special meaning, quite beyond the fact that they guarantee unconstrained gauge invariance of the corresponding Lagrangians, since, as we recalled, the latter can be achieved in several other ways. As a clue to unconver the rationale behind the solutions (13) and (14) we first recall the results of [31, 30], showing how it is possible to obtain full gauge invariance in the simplest way, within the framework of conventional local theories.

Journal of Physics: Conference Series **222** (2010) 012002 doi:10.1088/1742-6596/222/1/012002

2.1. *Minimal local theory*

For a rank-s fully symmetric tensor the first step is to consider the unconstrained variation of the Fronsdal tensor $\mathcal{F}$,

$$\delta\,\mathcal{F}\,=\,3\,\partial^3\,\Lambda'\,, \tag{15}$$

and introduce a spin-$(s-3)$ *compensator* field α, transforming as

$$\delta\,\alpha\,=\,\Lambda'\,, \tag{16}$$

so that the local kinetic tensor

$$\mathcal{A}\,=\,\mathcal{F}\,-\,3\,\partial^3\,\alpha\,, \tag{17}$$

be identically gauge-invariant [24, 36]. Then, exploiting the Bianchi identity for $\mathcal{A}$,

$$\partial\cdot\mathcal{A}\,-\,\frac{1}{2}\,\partial\,\mathcal{A}'\,=\,-\,\frac{3}{2}\,\partial^3\,\{\varphi''\,-\,4\,\partial\cdot\alpha\,-\,\partial\,\alpha'\}\,, \tag{18}$$

it is not difficult to show that a gauge-invariant local Lagrangian can be written in the compact form [31, 30]

$$\mathcal{L}\,=\,\frac{1}{2}\,\varphi\,\{\mathcal{A}\,-\,\frac{1}{2}\,\eta\,\mathcal{A}'\}\,-\,\frac{3}{4}\,\binom{s}{3}\,\alpha\,\partial\cdot\mathcal{A}'\,+\,3\,\binom{s}{4}\,\beta\,\{\varphi''\,-\,4\,\partial\cdot\alpha\,-\,\partial\,\alpha'\}\,, \tag{19}$$

where the Lagrange multiplier β transforms as $\delta\beta\,=\,\partial\cdot\partial\cdot\partial\cdot\Lambda$, while the tensor $\mathcal{C}\,\equiv\,\varphi''\,-\,4\,\partial\cdot\alpha\,-\,\partial\,\alpha'$ is the gauge-invariant completion of the double trace of φ. It is also manifest that Lagrangian (19) possesses the same physical content[8] as Fronsdal's one.

3. Effective non-local Lagrangians for bosons

We would like to establish a direct link between (19) and the geometric Lagrangians of [30], recalled in the previous section. The basic observation is that the integration of the auxiliary fields α and β must define *non-local, effective Lagrangians* for the physical field φ, possessing the same physical content as (19). The unconstrained gauge invariance of the latter, on the other hand, implies that it should be possible to express the resulting effective Lagrangians in terms of curvatures.

Here we would like to perform explicitly this computation for the first few cases, to clarify the mechanism at work in some relatively simple examples. In particular we discuss spin 3 and spin 4 on flat space-time, and show that the corresponding non-local Lagrangians actually coincide with those providing the correct expression for the current exchange, eqs. (13) and (14). In addition, we apply our procedure to the study of a spin-3 field on (A)dS background, to then pass to the fermionic case, where we concentrate on the example of spin $\frac{5}{2}$ [9].

[8] In particular, higher derivatives present in the kinetic operator of the compensator α are harmless. At any rate, it is possible to get rid of them slightly enlarging the field content of (19) [29, 33].

[9] Since we are going to write functional integrals for gauge theories, we should in principle discuss the issue of gauge fixing, following for instance the Faddeev-Popov procedure. We shall not be concerned with this issue here, given that all we want to achieve is an effective Lagrangian for the field φ alone, whose proper quantization would require a separate discussion. For similar reasons, we will not discuss the Wick rotation of our theory to Euclidean space, where in principle functional integrals of gaussian theories can be given a rigorous definition (see e.g. [39]). Rather, we will limit ourselves to the formal use of the rules of gaussian integration in the Minkowskian region. This is tantamount to solving the equations for the auxiliary fields and substituting back in $\mathcal{L}$, with some attention to be paid on the field β, for which in the initial Lagrangian there is no full quadratic form.

3.1. Integrating out auxiliary fields: spin 3 on flat background
Let us denote with $\mathcal{E}_\varphi$ the following combination of the Fronsdal tensor and its trace:

$$\mathcal{E}_\varphi = \mathcal{F} - \frac{1}{2}\eta \mathcal{F}'; \tag{20}$$

the functional integral for the unconstrained spin-3 theory, in the presence of an external source, can thus be written

$$\mathcal{Z}[\mathcal{J}] = \mathcal{N} \int \mathcal{D}\varphi \mathcal{D}\alpha \, e^{i\int d^dx\{\frac{1}{2}\varphi \mathcal{E}_\varphi + \frac{9}{4}\alpha \Box^2 \alpha - \frac{3}{2}\alpha \partial\cdot \mathcal{F}' - \varphi\cdot \mathcal{J}\}}, \tag{21}$$

where $\mathcal{N}$ is an overall normalization. Performing the integration over α one obtains

$$\begin{aligned} \mathcal{Z}[\mathcal{J}] &= \mathcal{N}_\alpha \int \mathcal{D}\varphi \, e^{i\int d^dx\{\mathcal{L}_{eff}(\varphi) - \varphi\cdot\mathcal{J}\}} \\ &= \mathcal{N}_\alpha \int \mathcal{D}\varphi \, e^{i\int d^dx\{\frac{1}{2}\varphi\mathcal{E}_\varphi - \frac{1}{4}\partial\cdot\mathcal{F}' \frac{1}{\Box^2}\partial\cdot\mathcal{F}' - \varphi\cdot\mathcal{J}\}}, \end{aligned} \tag{22}$$

where $\mathcal{N}_\alpha$ indicates a properly modified, field-independent, normalization factor, taking into account the gaussian integration over α. The corresponding non-local, effective Lagrangian is

$$\mathcal{L}_{eff}(\varphi) = \frac{1}{2}\varphi\mathcal{E}_\varphi - \frac{1}{4}\partial\cdot\mathcal{F}' \frac{1}{\Box^2}\partial\cdot\mathcal{F}', \tag{23}$$

and can be shown to coincide, up to total derivatives, with the one defined in (13).

3.2. Integrating out auxiliary fields: spin 4 on flat background
For the unconstrained spin-4 case, rearranging terms in (19), we can write the functional integral as

$$\mathcal{Z}[\mathcal{J}] = \mathcal{N} \int \mathcal{D}\varphi \mathcal{D}\beta \mathcal{D}\alpha \, e^{i\int d^dx\{\frac{1}{2}\varphi\mathcal{E}_\varphi + 3\beta\varphi'' + \frac{1}{2}\alpha A_\alpha \alpha + \alpha B_\alpha(\varphi,\beta) - \varphi\cdot\mathcal{J}\}}, \tag{24}$$

where

$$\begin{aligned} A_\alpha &= 18\Box(\Box + 3\partial\partial\cdot), \\ B_\alpha &= -6\{\partial\cdot\mathcal{F}' + \partial\partial\cdot\partial\cdot\varphi' - \Box\partial\varphi'' - 4\partial\beta\}, \end{aligned} \tag{25}$$

are the relevant operators entering the gaussian integration over α.

Technically, the main issue consists in the inversion of the operator A_α, whose form depends crucially on the rank of the tensors on which it is supposed to act. In the present case what we need is A_α^{-1} when acting on the space of rank-1 tensors (the rank of the compensator α associated to the rank-4 field φ), whose explicit expression is

$$A_\alpha^{-1} = \frac{1}{18\Box^2}(1 - \frac{3}{4}\frac{\partial}{\Box}\partial\cdot). \tag{26}$$

Performing the integration over α in (24), and using (26), we thus obtain

$$\mathcal{Z}[\mathcal{J}] = \mathcal{N}_\alpha \int \mathcal{D}\varphi\mathcal{D}\beta \, e^{i\int d^dx\{\frac{1}{2}\varphi\mathcal{E}_\varphi + 3\beta\varphi'' + B_\alpha(\varphi,\beta)\frac{1}{18\Box^2}(1 - \frac{3}{4}\frac{\partial}{\Box}\partial\cdot)B_\alpha(\varphi,\beta) - \varphi\cdot\mathcal{J}\}}. \tag{27}$$

In the latter expression we can reorganize the various terms, so as to make it explicit the quadratic form in the field β, allowing to perform the second gaussian integration:

$$\mathcal{Z}[\mathcal{J}] = \mathcal{N}_\alpha \int \mathcal{D}\varphi\mathcal{D}\beta \, e^{i\int d^dx\{\frac{1}{2}\varphi\mathcal{E}_\varphi - C_\varphi + \beta\frac{1}{\Box}\beta + \beta(3\varphi'' - \frac{1}{\Box^2}\partial\cdot D_\varphi) - \varphi\cdot\mathcal{J}\}}, \tag{28}$$

where

$$C_{\varphi} = D_{\varphi} \frac{1}{\Box^2} D_{\varphi} + \frac{3}{4} \partial \cdot D_{\varphi} \frac{1}{\Box^3} \partial \cdot D_{\varphi} ,$$
$$D_{\varphi} = \partial \cdot \mathcal{F}' + \partial \partial \cdot \partial \cdot \varphi' - \frac{1}{2} \Box \partial \varphi'' . \tag{29}$$

Integrating over β we are formally left with a theory involving the field φ alone

$$\mathcal{Z}[\mathcal{J}] = \mathcal{N}_{\alpha, \beta} \int \mathcal{D}\varphi \, e^{i \int d^d x \{\mathcal{L}_{eff}(\varphi) - \varphi \cdot \mathcal{J}\}} , \tag{30}$$

where again the resulting non-local, effective Lagrangian

$$\mathcal{L}_{eff}(\varphi) = \frac{1}{2} \varphi \mathcal{E}_{\varphi} - D_{\varphi} \frac{1}{\Box^2} D_{\varphi} - \frac{3}{4} \partial \cdot D_{\varphi} \frac{1}{\Box^3} \partial \cdot D_{\varphi} - \frac{3}{2} \varphi'' \frac{1}{\Box} \partial \cdot D_{\varphi} + \frac{9}{4} \varphi'' \Box \varphi'' , \tag{31}$$

can be shown to be equivalent to the corresponding expression in (14), for instance computing the equations of motion of (31) and verifying that they can be rearranged in the form

$$\mathcal{A}_{\varphi} - \frac{1}{2} \eta \mathcal{A}'_{\varphi} + \eta^2 \mathcal{B}_{\varphi} = 0 , \tag{32}$$

with the various quantities in this expression defined in (14).

3.3. Integrating out auxiliary fields: spin 3 *on (A)dS background*

In [30] the minimal local Lagrangians (19), together with their fermionic counterparts, to be recalled in Section 4, were also generalised to the case of (A)dS backgrounds. On the other hand, the analysis of the non-local theory was considered only for the bosonic case on flat space-time. Here we would like to perform a first step towards a more general discussion, describing the non-local, unconstrained theory on (A)dS background, in the case of a spin-3 field. To this end, we follow two independent procedures.

First, we look for a non-local compensator α_{φ}, in the spirit of [30], and we make sure that the corresponding Lagrangian define the correct propagator. Then we consider the functional integral for the *local* theory, and we show that the integration of the compensator α produces indeed the same effective, non-local Lagrangian. In this fashion, we stress once again, the latter is given a clear interpretation, in terms of a theory whose dynamical content can be analysed using conventional methods.

3.3.1. Construction of the non-local compensator If we were to follow the very same procedure we went through in [30], the first step would be to resort to some proper, covariantised versions of the flat curvatures of [12] (see e.g. [41]). On the other hand, what we really need is a fully gauge invariant candidate "Ricci" tensor on (A)dS, that in the spin-3 case can be more simply constructed resorting to the gauge transformation of the covariantised Fronsdal tensor, when no assumptions are made on the traces of the gauge parameter[10]. To begin with, it might be convenient to recall the deformed (A)dS Fronsdal operator for a spin-s field [42]:

$$\mathcal{F}_L = \mathcal{F} - \frac{1}{L^2} \left\{ [(3 - D - s)(2 - s) - s] \varphi + 2 g \varphi' \right\} , \tag{33}$$

[10] In this way, looking for the kinetic tensor with the highest degree of singularity, we will be directly led to the solution possessing the correct propagator. If we were to start from curvatures, we would find proper (A)dS generalisations of the full family of Ricci-like tensors (8). This simplification, however, is special of the spin-3 case.

where D denotes the space-time dimension and

$$\mathcal{F} = \Box \varphi - \nabla \nabla \cdot \varphi + \nabla^2 \varphi' \tag{34}$$

is the (A)dS-covariantized Fronsdal operator. The unconstrained variation of (33) under $\delta\varphi = \nabla\Lambda$, is

$$\delta \mathcal{F}_L = 3 \nabla^3 \Lambda' - \frac{4}{L^2} g \nabla \Lambda' . \tag{35}$$

Thus, considering the variation of $\nabla \cdot \mathcal{F}_L'$,

$$\delta \nabla \cdot \mathcal{F}_L' = 3 \Box \{\Box - \frac{2}{L^2} (D + 1)\} \Lambda' . \tag{36}$$

it is relatively simple to identify a candidate non-local compensator

$$\alpha_{\varphi, L} = \frac{1}{3 \Box \{\Box - \frac{2}{L^2} (D + 1)\}} \nabla \cdot \mathcal{F}_L' , \tag{37}$$

allowing to define a gauge invariant tensor in the form

$$\mathcal{A}_{\varphi, L} = \mathcal{F}_L - \{3 \nabla^3 - \frac{4}{L^2} g \nabla\} \alpha_{\varphi, L} . \tag{38}$$

The Bianchi identity satisfied by $\mathcal{A}_{\varphi, L}$,

$$\nabla \cdot \mathcal{A}_{\varphi, L} - \frac{1}{2} \nabla \mathcal{A}_{\varphi, L}' \equiv 0 , \tag{39}$$

together with its further property

$$\nabla \cdot \mathcal{A}_{\varphi, L}' \equiv 0 , \tag{40}$$

easily verified from

$$\mathcal{A}_{\varphi, L}' = \mathcal{F}_L' - \{3 \nabla \Box - \frac{6}{L^2} (D + 1) \nabla\} \alpha_{\varphi, L} , \tag{41}$$

imply the existence of a non-local, gauge-invariant Lagrangian of the form

$$\mathcal{L} = \frac{1}{2} \varphi \{\mathcal{A}_{\varphi, L} - \frac{1}{2} g \mathcal{A}_{\varphi, L}'\} - \varphi \cdot \mathcal{J} . \tag{42}$$

It is then straightforward to evaluate the propagator, once the field is coupled to a conserved source. Indeed, following the procedure described in [30, 43], from the Lagrangian equation

$$\mathcal{A}_{\varphi, L} - \frac{1}{2} g \mathcal{A}_{\varphi, L}' = \mathcal{J} , \tag{43}$$

we obtain

$$\mathcal{A}_{\varphi, L} = \mathcal{J} - \frac{1}{D} g \mathcal{J}' . \tag{44}$$

Then, introducing the Lichnerowicz operator $\Box_L$ defined by

$$\Box_L \varphi = \Box \varphi + \frac{1}{L^2} \left[s (D + s - 2) \varphi - 2 g \varphi' \right] , \tag{45}$$

we can rewrite $\mathcal{A}_{\varphi, L}$ for spin 3 as

$$\mathcal{A}_{\varphi, L} = (\Box_L - \frac{4}{L^2}) \varphi + \nabla (\nabla \cdot \varphi - \frac{1}{2} \nabla \varphi') - (3 \nabla^3 - \frac{4}{L^2} g \nabla) \alpha_L . \tag{46}$$

From this expression we can compute the interaction between conserved currents, finding

$$\mathcal{J} \cdot \varphi = \mathcal{J} \frac{1}{\Box_L - \frac{4}{L^2}} \mathcal{J} - \frac{3}{D} \mathcal{J}' \frac{1}{\Box_L - \frac{4}{L^2}} \mathcal{J}' , \tag{47}$$

in agreement with the result found in [30] for the local counterpart of (42).

3.3.2. Integration of the compensator from the local theory While the computation of the propagator already represents a strong consistency check for the Lagrangian (42), we can further clarify its interpretation if we can show that it defines the effective Lagrangian of a more conventional local theory, once the compensator α is integrated away from the functional integral. To this end, following [30], let us write explicitly, for spin 3, the extension to (A)dS of the local unconstrained Lagrangian (19):

$$\mathcal{L} = \frac{e}{2}\varphi\{\mathcal{A}_L - \frac{1}{2}\, g\,\mathcal{A}_L'\} - \frac{3\,e}{4}\alpha\,\nabla\cdot\mathcal{A}_L' , \tag{48}$$

where e denotes the determinant of the vielbein, g is the (A)dS metric, while the tensor $\mathcal{A}_L$ has the same form as in (38), with α being in this case an independent Stueckelberg field, s.t. $\delta\,\alpha = \Lambda'$. In order to consistently eliminate the auxiliary field from the Lagrangian, we can start from

$$\mathcal{Z}\,[\mathcal{J}] = \mathcal{N}\int \mathcal{D}\varphi\,\mathcal{D}\alpha\, e^{\,i\int d^dx\, e\,\{\frac{1}{2}\,\varphi\,\mathcal{E}_\varphi^L + \frac{9}{4}\,\alpha\,\Box\,[\Box - \frac{2}{L^2}\,(D+1)]\,\alpha - \frac{3}{2}\,\alpha\,\nabla\cdot\mathcal{F}_L' - \varphi\cdot\mathcal{J}\}} , \tag{49}$$

where we defined

$$\mathcal{E}_\varphi^L = \mathcal{F}_L - \frac{1}{2}\,g\,\mathcal{F}'_L , \tag{50}$$

and perform the gaussian integration over α, obtaining the effective Lagrangian

$$\mathcal{L}_{eff}\,(\varphi) = \frac{1}{2}\,\varphi\,\mathcal{E}_\varphi^L - \frac{1}{4}\,\nabla\cdot\mathcal{F}_L'\,\frac{1}{\Box\,[\Box - \frac{2}{L^2}\,(D+1)]}\,\nabla\cdot\mathcal{F}_L' , \tag{51}$$

which, in its turn, can be shown to coincide with (42), up to total derivatives.

4. Effective non-local Lagrangians for fermions

Now we would like to extend our considerations to symmetric fermions. We start recalling the construction of the corresponding minimal local Lagrangians [31, 30], to then pass to the integration of the auxiliary field for the case of spin $\frac{5}{2}$, followed by the geometrical interpretation of the result.

4.1. Minimal local theory

The construction of local, unconstrained Lagrangians for symmetric spinor-tensors[11] ψ of rank s (and spin $s+\frac{1}{2}$) closely resembles the corresponding one for bosons, here sketched in Section 2.1. Under the transformation $\delta\,\psi = \partial\,\epsilon$ the unconstrained variation of the Fang-Fronsdal tensor [44],

$$\mathcal{S} = i\,(\not\partial\,\psi - \partial\,\not\psi) , \tag{52}$$

is $\delta\,\mathcal{S} = -\,2\,i\,\partial^2\,\not\epsilon$. Thus, we can build from $\mathcal{S}$ the *fully* gauge invariant operator

$$\mathcal{W} \equiv \mathcal{S} + 2\,i\,\partial^2\xi , \tag{53}$$

where the rank-$(s-2)$ *compensator* ξ transforms as $\delta\,\xi = \not\epsilon$. The Bianchi identity for $\mathcal{W}$,

$$\partial\cdot\mathcal{W} - \frac{1}{2}\,\partial\,\mathcal{W}' - \frac{1}{2}\,\not\partial\,\not{\mathcal{W}} = i\,\partial^2\{\not\psi' - 2\,\partial\cdot\xi - \partial\,\xi' - \not\partial\,\not\xi\} , \tag{54}$$

[11] In the conventions here followed γ_0 is antihermitian, while the γ_i, $i = 1,2,3$ are hermitian. Moreover $\gamma_0\,\gamma_\mu^\dagger\,\gamma_0 = \gamma_\mu$. An additional useful combinatorial rule is $\gamma\cdot(\gamma\,\psi) = (D+2\,s)\,\psi - \gamma\,\not\psi$.

leads naturally to a second gauge-invariant spinor-tensor,

$$\mathcal{Z} \equiv i\left\{\not{\psi}' - 2\partial\cdot\xi - \partial\xi' - \not{\partial}\,\not{\xi}\right\}, \tag{55}$$

directly related to the triple γ-trace constraint on the fermionic gauge field ψ, absent in the Fang-Fronsdal formulation. The minimal flat-space Lagrangians of [31, 30] can then be recovered starting from the trial Lagrangians

$$\mathcal{L}_0 = \frac{1}{2}\bar{\psi}\,\{\mathcal{W} - \frac{1}{2}\eta\,\mathcal{W}' - \frac{1}{2}\gamma\,\not{\mathcal{W}}\} + h.c., \tag{56}$$

and compensating the remainders in their gauge transformations with new terms involving the field ξ and the tensor $\mathcal{Z}$. The complete Lagrangian is finally

$$\begin{aligned}\mathcal{L} =& \frac{1}{2}\bar{\psi}\,\{\mathcal{W} - \frac{1}{2}\gamma\not{\mathcal{W}} - \frac{1}{2}\eta\,\mathcal{W}'\} - \frac{3}{4}\binom{n}{3}\bar{\not{\xi}}\,\partial\cdot\mathcal{W}'\\ &+ \frac{1}{2}\binom{n}{2}\bar{\xi}\,\partial\cdot\not{\mathcal{W}} + \frac{3}{2}\binom{n}{3}\bar{\lambda}\,\mathcal{Z} + h.c.,\end{aligned} \tag{57}$$

where the Lagrange multiplier λ transforms according to $\delta\lambda = \partial\cdot\partial\cdot\epsilon$, in order for $\mathcal{L}$ to be gauge invariant. Lagrangians for constrained spinor-tensors of *any* symmetry, together with their unconstrained extensions, were first presented in the second of [33].

4.2. Integrating out auxiliary fields: flat background

We would like to perform the integration over the auxiliary fields, and investigate the form of the corresponding effective non-local Lagrangian. For simplicity, we limit ourselves to the example of spin $s = \frac{5}{2}$, in which case (57) reduces to[12]

$$\mathcal{L} = i\bar{\psi}\,\mathcal{E}_\psi + 2i\bar{\xi}\,\Box\,\not{\partial}\xi + i(\bar{\xi}\,\partial\cdot\not{\mathcal{S}} - \partial\cdot\bar{\not{\mathcal{S}}}\,\xi), \tag{58}$$

where we defined

$$\mathcal{E}_\psi = \mathcal{S} - \frac{1}{2}\gamma\,\not{\mathcal{S}} - \frac{1}{2}\eta\,\mathcal{S}'. \tag{59}$$

Thus, considering the fermionic partition function in the presence of two grassmannian sources $\bar{\theta}$ and θ,

$$\mathcal{Z}\,[\bar{\theta},\,\theta] = \mathcal{N}\int\mathcal{D}\bar{\psi}\,\mathcal{D}\psi\,\mathcal{D}\bar{\xi}\,\mathcal{D}\xi\, e^{\,i\int d^dx\,\{\mathcal{L} + \bar{\theta}\,\psi + \bar{\psi}\,\theta\}}, \tag{60}$$

and performing the fermionic gaussian integration over $\bar{\xi}$ and ξ we obtain the non-local effective Lagrangian

$$\mathcal{L}_{eff}\,(\bar{\psi},\,\psi) = i\bar{\psi}\,\mathcal{E}_\psi + i\,\partial\cdot\bar{\not{\mathcal{S}}}\,\frac{\not{\partial}}{2\,\Box^2}\,\partial\cdot\not{\mathcal{S}}, \tag{61}$$

which is equivalent, up to total derivatives, to the form

$$\mathcal{L}_{eff}\,(\bar{\psi},\,\psi) = i\,\bar{\psi}\,\{\mathcal{W}_\psi - \frac{1}{2}\gamma\not{\mathcal{W}}_\psi - \frac{1}{2}\eta\,\mathcal{W}'_\psi\}, \tag{62}$$

with the non-local, Dirac-like, kinetic tensor $\mathcal{W}_\psi$ defined as

$$\mathcal{W}_\psi = \mathcal{S} - \partial^2\,\frac{\not{\partial}}{\Box^2}\,\partial\cdot\not{\mathcal{S}}, \tag{63}$$

and satisfying

$$\partial\cdot\not{\mathcal{W}}_\psi \equiv 0, \tag{64}$$

which is relevant to stress in view of the following discussion. Now we would like to make it explicit the content of (62) in terms of curvatures, and compare these results with those obtained in [23, 24, 29] on the geometry of fermionic theories.

[12] To avoid confusion with signs: the last term in (58) is to be interpreted as $\bar{\mathcal{S}}_\mu\,\gamma^\mu$, and not as $\not{\mathcal{S}}^\dagger\,\gamma^0$.

4.3. Geometric interpretation and propagator

Non-local candidate Dirac-Rarita-Schwinger tensors were first proposed in [23, 24], in analogy with the corresponding bosonic quantities computed from the curvatures of [12]. The issue of fermionic geometry was then reconsidered in [29]; there it was shown that, even keeping to a minimum the degree of singularity of the non-local operators involved, in the fermionic case infinitely many independent tensors can actually be constructed.

In particular, for the case of spin $s = \frac{5}{2}$ of interest in this section, starting from the fermionic curvature

$$\mathcal{R}_{\mu\mu,\,\nu\nu} \,=\, \partial_{\mu}^{2}\,\psi_{\nu\,\nu} \,-\, \frac{1}{2}\,\partial_{\mu}\,\partial_{\nu}\,\psi_{\mu\,\nu} \,+\, \partial_{\nu}^{2}\,\psi_{\mu\,\mu}\,, \tag{65}$$

two independent, gauge-invariant, kinetic tensors can be constructed:

$$\begin{aligned} &\frac{\not\partial}{\Box}\,\mathcal{R}_{2}' \equiv D_{2}\,, \\ &\frac{1}{\Box}\,\partial\cdot\not{\mathcal{R}} \equiv \hat{D}_{2} \end{aligned} \tag{66}$$

whose expression in terms of the Fang-Fronsdal operator (52) is

$$\begin{aligned} i\,D_{2} \,&=\, \mathcal{S} \,+\, \frac{\partial^{2}}{\Box}\,\mathcal{S}' \,-\, \frac{\partial}{\Box}\,\partial\cdot\mathcal{S}\,, \\ i\,\hat{D}_{2} \,&=\, \mathcal{S} \,-\, \frac{1}{2}\,\frac{\partial}{\Box}\,\partial\cdot\mathcal{S}\,, \end{aligned} \tag{67}$$

while the corresponding candidate Dirac tensor proposed in [23, 24],

$$\mathcal{S}_{2} \,=\, \mathcal{S} \,+\, \frac{1}{3}\,\frac{\partial^{2}}{\Box}\,\mathcal{S}' \,-\, \frac{2}{3}\,\frac{\partial}{\Box}\,\partial\cdot\mathcal{S}\,, \tag{68}$$

can be shown to be equivalent to the linear combination

$$\mathcal{S}_{2} \,=\, \frac{1}{3}\,i\,D_{2} \,+\, \frac{2}{3}\,i\,\hat{D}_{2}\,. \tag{69}$$

Moreover, in analogy with what already recalled for bosons in the previous sections, additional gauge invariant tensors can be constructed allowing for higher degrees of singularity in the corresponding non-local expressions. On the other hand, the analysis of the current exchange performed for bosons in [30] leads to expect that only one of these theories would display the correct propagator.

For spin $\frac{5}{2}$, in the local formulation of [31, 30], the field equation for ψ in the presence of a source takes the form

$$\mathcal{W} \,-\, \frac{1}{2}\,\gamma\,\not{\mathcal{W}} \,-\, \frac{1}{2}\,\eta\,\mathcal{W}' \,=\, \mathcal{J}\,, \tag{70}$$

where the gauge invariant spinor-tensor $\mathcal{W}$, defined in (53), satisfies the Bianchi identity (54) (where it is to be recalled that $\mathcal{Z} = 0$, due to the equation for $\bar{\lambda}$), implying in particular that, on-shell

$$\partial\cdot\not{\mathcal{W}} \,=\, 0\,, \tag{71}$$

thus accounting for the conservation of $\mathcal{J}$. Therefore, in order to reproduce the correct propagator, we should select a non-local Dirac tensor satisfying the same relations (54) and (71), as well as an equation of the same form as (70).

It is not difficult to check that the Lagrangian theories associated to Dirac-type kinetic tensors (67) (as well as any of their linear combinations) do *not* reproduce (70) nor (71) . Differently, the non-local theory obtained after integration of the compensators $\bar{\xi}$ and ξ, possesses exactly the correct properties, as manifest from (62) and (64). Moreover, while other, independent, non-local compensators could be constructed, such as

$$\hat{\xi}_\psi = -\frac{1}{2}\frac{1}{i\,\Box}\,\mathcal{S}', \tag{72}$$

having the same gauge transformation as $\xi_\psi = -\frac{\not\partial}{2\,i\,\Box^2}\,\partial\cdot\not\mathcal{S}$, nonetheless the corresponding kinetic tensor

$$\hat{\mathcal{W}}_\psi = \mathcal{S} - \frac{\partial^2}{\Box}\,\mathcal{S}' \tag{73}$$

would *not* satisfy (54) and (71). This is another manifestation of the fact that, in the non-local case, it is actually possible to modify quantities by the addition of non-local gauge invariant tensors in such a way that, while gauge transformation properties are obviously preserved, the meaning of the whole construction gets modified in a crucial way. Indeed, there is *only one* geometric theory, for fermions of spin $\frac{5}{2}$, which propagates the correct number of degrees of freedom, and the procedure we followed in this paper can be seen as a safe way to derive it. Finally, its geometric interpretation is made manifest expressing the kinetic tensor (63) in terms of the curvature (65), according to

$$\mathcal{W}_\psi = \mathcal{S} - \partial^2\,\frac{\not\partial}{\Box^2}\,\partial\cdot\not\mathcal{S} = 2\,i\,\frac{1}{\Box}\,\partial\cdot\not\mathcal{R} - i\,\frac{\not\partial}{\Box}\,\mathcal{R}' + i\,\frac{\partial^2}{\Box^2}\,\not\partial\,\mathcal{R}''\,. \tag{74}$$

5. Summary and outlook

We computed effective, non-local Lagrangians for some specific examples of higher-spin gauge fields on flat and (A)dS backgrounds, performing the integration over non-physical fields in the corresponding local theories. The main goal of our computations has been to produce a *definition* of non-local, geometric Lagrangians, possibly devoid of ambiguities on their physical content, clarifying in particular the rationale behind the presence of inverse powers of the d'Alembertian operator in their kinetic tensors.

In this fashion, we were able to provide further support for the particular forms of geometric Lagrangians for symmetric bosons on flat backgrounds first given in [30]. Here we also proposed examples of geometric theories with correct propagators for fermions of spin $\frac{5}{2}$ on flat space-time, and for bosons of spin 3 on (A)dS backgrounds. These results point towards the possibility that a geometric description of interacting higher-spins, involving proper deformations of linearised curvature tensors, could emerge from a conventional, local theory, once all auxiliary fields (or at least a proper subset of them) are integrated away. In particular, in view of the geometrical interpretation of triplets to be given elsewhere [35], it might not be necessary to this end that the local theory be formulated in terms of *unconstrained* fields.

As for what concerns the latter, we leave for future work the generalisation of the results here presented to irreducible, unconstrained bosons and fermions of any spin (and symmetry), as well as the corresponding analysis of massive higher-spin fields.

Acknowledgments

I would like to thank the Organizers of the 1st Mediterranean Conference on Classical and Quantum Gravity for their kind invitation to present my lecture. I am grateful to A. Campoleoni, J. Mourad and A. Sagnotti for useful exchanges and for collaboration on several topics reconsidered in this paper. I also wish to thank the Scuola Normale Superiore of Pisa and the Institute of Physics of the ASCR in Prague for their hospitality while part of this work was being prepared. The present research was supported by APC-Paris VII and by the CNRS through the P2I program, and also in part by the MIUR-PRIN contract 2007-5ATT78 and by the EURYI grant EYI/07/E010 from EUROHORC and ESF.

References

[1] Vasiliev M A 2004 *Fortsch. Phys.* **52** 702 (*Preprint* hep-th/0401177)
[2] Sorokin D 2005 *AIP Conf. Proc.* **767** 172 (*Preprint* hep-th/0405069)
[3] Bouatta N, Compere G and Sagnotti A 2004 (*Preprint* hep-th/0409068)
[4] Bandos I, Bekaert X, de Azcarraga J A, Sorokin D and Tsulaia M 2005 *JHEP* **0505** 031 (*Preprint* hep-th/0501113)
[5] Sagnotti A, Sezgin E and Sundell P (*Preprint* hep-th/0501156)
[6] Bekaert X, Cnockaert S, Iazeolla C and Vasiliev M A 2005 (*Preprint* hep-th/0503128)
[7] Metsaev R R 2006 *Nucl. Phys. B* **759** 147 (*Preprint* hep-th/0512342)
[8] Francia D and Sagnotti A 2006 *J. Phys. Conf. Ser.* **33** 57 (*Preprint* hep-th/0601199)
[9] Fotopoulos A and Tsulaia M (*Preprint* 0805.1346 [hep-th])
[10] Iazeolla C (*Preprint* 0807.0406 [hep-th])
[11] Weinberg S 1965 *Phys. Rev.* **138** B988
[12] de Wit B and Freedman D Z 1980 *Phys. Rev. D* **21** 358
[13] Fronsdal C 1978 *Phys. Rev. D* **18** 3624
[14] Siegel W and Zwiebach B 1987 *Nucl. Phys. B* **282** 125
[15] Curtright T 1979 *Phys. Lett. B* **85** 219.
[16] J. M. F. Labastida, Nucl. Phys. B **322** (1989) 185.
[17] Vasiliev M A 1990 *Phys. Lett. B* **243** 378.
Vasiliev M A 2003 *Phys. Lett. B* **567**139 (*Preprint* hep-th/0304049)
[18] Howe P S, Penati S, Pernici M and Townsend P K 1988 *Phys. Lett. B* **215** 555.
[19] Fradkin E S and Tseytlin A A 1985 *Phys. Rept.* **119** 233.
[20] Bastianelli F, Corradini O and Latini E 2008 *JHEP* **0811** 054 (*Preprint* 0810.0188 [hep-th])
Marnelius R 2009 (*Preprint* 0906.2084 [hep-th])
Vasiliev M A 2009 (*Preprint* 0909.5226 [hep-th])
[21] Damour T and Deser S 1987 *Annales Poincare Phys. Theor.* **47** 277
[22] Dubois-Violette M and Henneaux M 2002 *Commun. Math. Phys.* **226** 393 (*Preprint* math/0110088)
[23] Francia D and Sagnotti A 2002 *Phys. Lett. B* **543** 303 (*Preprint* hep-th/0207002)
[24] Francia D and Sagnotti A 2003 *Class. Quant. Grav.* **20** S473 (*Preprint* hep-th/0212185)
[25] Bekaert X and Boulanger N 2002 *Commun. Math. Phys.* **245** 27 (*Preprint* hep-th/0208058)
[26] de Medeiros P and Hull C 2003 *Commun. Math. Phys.* **235** 255 (*Preprint* hep-th/0208155)
[27] Bekaert X and Boulanger N 2003 *Phys. Lett. B* **561** 183 (*Preprint* hep-th/0301243),
[28] de Medeiros P and Hull C 2003 *JHEP* **0305** 019 (*Preprint* hep-th/0303036)
Bekaert X and Boulanger N 2007 *Commun. Math. Phys.* **271** 723 (*Preprint* hep-th/0606198)
[29] Francia D 2008 *Nucl. Phys. B* **796** 77 (*Preprint* 0710.5378 [hep-th])
Francia D 2008 *Fortsch. Phys.* **56** 800 (*Preprint* 0804.2857 [hep-th])
[30] Francia D, Mourad J and Sagnotti A 2007 *Nucl. Phys. B* **773** 203 (*Preprint* hep-th/0701163),
[31] Francia D and Sagnotti A 2005 *Phys. Lett. B* **624** 93 (*Preprint* hep-th/0507144)
[32] Pashnev A and Tsulaia M 1998 *Mod. Phys. Lett. A* **13** 1853 (*Preprint* hep-th/9803207),
Buchbinder I L, Pashnev A and Tsulaia M 2001 *Phys. Lett. B* **523** 338 (*Preprint* hep-th/0109067)
[33] Campoleoni A, Francia D, Mourad J and Sagnotti A 2009 *Nucl. Phys. B* **815** 289 (*Preprint* 0810.4350 [hep-th])
Campoleoni A, Francia D, Mourad J and Sagnotti A 2010 *Nucl. Phys. B* **828** 425 (*Preprint* 0904.4447 [hep-th])
Campoleoni A (*Preprint* 0910.3155 [hep-th])
[34] Alkalaev K B, Shaynkman O V and Vasiliev M A 2004 *Nucl. Phys. B* **692** 363 (*Preprint* hep-th/0311164)
Skvortsov E D 2008 *JHEP* **0807** 004 (*Preprint* 0801.2268 [hep-th])
Boulanger N, Iazeolla C and Sundell P 2009 *JHEP* **0907** 013 (*Preprint* 0812.3615 [hep-th])
Zinoviev Yu M 2010 *Nucl. Phys. B* **826** 490 (*Preprint* 0907.2140 [hep-th])
[35] Francia D (*to appear*)
[36] Sagnotti A and Tsulaia M 2004 *Nucl. Phys. B* **682** 83 (*Preprint* hep-th/0311257)
[37] Buchbinder I L, Galajinsky A V and Krykhtin V A 2007 *Nucl. Phys. B* **779** 155 (*Preprint* hep-th/0702161)
[38] Fotopoulos A and Tsulaia M 2009 *JHEP* **0910** 050 (*Preprint* [arXiv:0907.4061 [hep-th]])
[39] Glimm J and Jaffe A 1981 *Quantum Physics: A Functional Integral Point of View* (NY: Springer-Verlag)
[40] Sorokin D P and Vasiliev M A 2009 *Nucl. Phys. B* **809** 110 (*Preprint* 0807.0206 [hep-th])
[41] Manvelyan R and Rühl W 2008 *Nucl. Phys. B* **797** 371 (*Preprint* 0705.3528 [hep-th])
[42] Fronsdal C 1979 *Phys. Rev. D* **20** 848.
[43] Francia D, Mourad J and Sagnotti A 2008 *Nucl. Phys. B* **804** 383 (*Preprint* 0803.3832 [hep-th])
[44] Fang J and Fronsdal C 1978 *Phys. Rev. D* **18** 3630

Journal of Physics: Conference Series **222** (2010) 012003 doi:10.1088/1742-6596/222/1/012003

Interrelated aspects of thick braneworlds: 4D gravity localization, smoothness of geometry and mass gap in the graviton spectrum

A Herrera–Aguilar[1], D Malagón–Morejón[2], R R Mora–Luna[3] and U Nucamendi[4]

Instituto de Física y Matemáticas, Universidad Michoacana de San Nicolás de Hidalgo. Edificio C–3, Ciudad Universitaria, C.P. 58040, Morelia, Michoacán, México.

E-mail: herrera@ifm.umich.mx, malagon@ifm.umich.mx, topc@ifm.umich.mx, ulises@ifm.umich.mx

Abstract. We review some interrelated aspects of thick braneworlds which arise within the framework of 5D gravity coupled to a bulk scalar field. When studying localization of 4D gravity in this smooth version of the thin brane Randall-Sundrum model, a kind of dichotomy emerges: if one requires the geometry of the system to be completely smooth, then there is no mass gap in the spectrum of metric fluctuations since it contains just a single massless bound state, interpreted as a 4D graviton, and a tower of massive states starting from zero mass; on the other side, as far as one demands the existence of a mass gap in the graviton spectrum of the theory (in this case there are two bound states, a massless 4D graviton and a massive excitation separated from a continuous spectrum of massive Kaluza-Klein excitations) naked singularities arise at the boundaries of the Riemannian manifold, a fact that obligates one to impose unitary boundary conditions, which are equivalent to eliminating the continuous spectrum of gravitational massive modes, in order to render these singularities harmless from the physical point of view, providing in this way a viable model.

1. Introduction

In recent years, several results have been obtained in the framework of thick braneworld scenarios with a single extra dimension [1]–[38]. These models represent more realistic generalizations of several pioneer thin brane configurations [39]–[52], since if there are extra dimensions, they must be accessible at least at some hypothetic energy scale.

One way to smooth out the singular brane configurations is to replace the delta functions in the action of the system by a self–interacting scalar field. Once we have a smooth brane configuration, it is important from the phenomenological point of view to look for configurations that allow for the existence of a mass gap in the spectrum of gravitational Kaluza–Klein (KK) excitations. This gap fixes the energy scale at which massive modes can be excited and it is relevant for experimentally distinguishing the imprints of the massless zero mode, identified with a stable 4D graviton, from those coming from the massive modes, either discrete or continuous. When there is no mass gap in the spectrum, there are several modes with very small masses that cannot be experimentally distinguished from the massless one.

The relevance of braneworld models is related to the fact that we could live in a higher dimensional world without contradiction with up to day experimentally tested four–dimensional gravitational effects.

An interesting issue is related to the fact that naked singularities arise when a mass gap is present in the five–dimensional gravitational spectrum of the fluctuations of the system when studying four–dimensional gravity localization on thick brane configurations.

In Sec. 2 it is shown that if one considers localization of 4D gravity with a generic warped ansatz for the metric, the presence of a mass gap in the spectrum of metric fluctuations necessarily implies the existence of naked singularities located at the boundaries (or spatial infinity, depending on the used coordinate system) of the fifth dimension or, alternatively, the five–dimensional manifold can be completely smooth iff the massive spectrum of KK excitations is gapless. This analysis is made in terms of a simple, but generic relationship existing between the 5D curvature scalar, the warp factor and the quantum mechanical potential which governs the dynamics of the linearized transverse traceless modes of metric fluctuations[1].

We follow a traditional point of view which states that spaces with naked singularities are physically acceptable only if one imposes unitary boundary conditions that guarantee four–dimensional energy and momentum conservation [2, 3], [53, 54]. By analyzing the spectrum of KK modes of an explicit solution, one can see that after imposing unitary boundary conditions, the continuum part of the spectrum is projected out. Thus, the theory of the discrete part of the spectrum remains unitary because these modes die off rapidly enough as we approach the singularity, a fact that provides a viable model.

We finally summarize our conclusions and give a brief discussion about the physical meaning of imposing the unitary boundary conditions on the system considered in the paper in Section 3.

2. 4D gravity localization: smoothness vs. mass gap

Let us consider the following five–dimensional action [1]–[3]

$$S_5 = \int d^5x\sqrt{|G|}\left[-\frac{1}{4}R_5 + \frac{1}{2}(\nabla\phi)^2 - V(\phi)\right], \tag{1}$$

where ϕ is a bulk scalar field and $V(\phi)$ is a self–interacting potential for the scalar field. We shall study a solution which preserves four–dimensional Poincaré invariance with the metric

$$ds_5^2 = e^{2A(y)}\eta_{nm}dx^n dx^m - dy^2, \tag{2}$$

where $e^{2A(y)}$ is the warp factor, $m, n = 0, 1, 2, 3$.

Let us study the metric fluctuations h_{mn} of (2) given by

$$ds_5^2 = e^{2A(y)}\left[\eta_{mn} + h_{mn}(x, y)\right]dx^m dx^n - dy^2. \tag{3}$$

In the general case one must consider the fluctuations of the scalar field when treating fluctuations of the classical background metric since they are coupled, however, following [1], we shall just study the transverse traceless modes of the background fluctuations h^T_{mn} since they decouple from the sector in which the scalar perturbations vanish.

In order to get a conformally flat metric we perform the coordinate transformation

$$dz = e^{-A}dy. \tag{4}$$

[1] It is important to stress that these computations are done for the sector in which the fluctuations of the scalar field vanish.

Thus, the equation that governs the dynamics of the transverse traceless modes of the metric fluctuations h^T_{mn} becomes [1]–[3]

$$\left(\partial_z^2 + 3A'\partial_z - \Box\right) h^T_{mn}(x,z) = 0, \tag{5}$$

where $A' = dA/dz$ and $\Box$ stands for the 4D D'Alambertian. This equation supports as a particular solution a massless and normalizable 4D graviton given by $h^T_{mn} = K_{mn}e^{ipx}$, where K_{mn} are constant parameters and $p^2 = m^2 = 0$.

We adopt the following ansatz

$$h^T_{mn} = e^{ipx}e^{-3A/2}\Psi_{mn}(z)$$

in order to recast (5) into a Schrödinger's equation form (here we have dropped the subscripts in Ψ for simplicity):

$$[\partial_z^2 - V_{QM}(z) + m^2]\Psi(z) = 0, \tag{6}$$

where m is the 4D mass of the KK excitation modes, and the analog quantum mechanical potential has the form

$$V_{QM}(z) = \frac{3}{2}\left[\partial_z^2 A + \frac{3}{2}(\partial_z A)^2\right]. \tag{7}$$

As it is shown in [4], from (6) it follows that the zero energy wave function is given by

$$\Psi_0(z) = e^{\frac{3}{2}A(z)}. \tag{8}$$

The condition for localizing four–dimensional gravity demands $\Psi_0(z)$ to be normalizable; in other words, the zero energy wave function must satisfy the following relationship

$$\int dz|\Psi_0|^2 < \infty. \tag{9}$$

Finally, it is worth noticing that the curvature scalar corresponding to the ansatz (2) in the z coordinate adopts the form

$$R_5 = 8e^{-2A}\left[\partial_z^2 A + \frac{3}{2}(\partial_z A)^2\right]. \tag{10}$$

As it was pointed out in [4, 55] in order to have a localized massless mode of the 5D graviton, V_{QM} must be a well potential with a negative minimum which approaches a positive value $V_{QM_\infty} \geq 0$ as $|z| \to \infty$. This property ensures the fulfilment of the requirement (9).

Moreover, if we wish to have a well defined effective field theory, it is desirable to obtain solutions where the massless graviton is separated from the massive modes by a mass gap [55]. This phenomenological aspect ensures the lack of arbitrary light KK excitations in the mass spectrum. If the quantum mechanical potential asymptotically approaches a positive value $V_{QM_\infty} > 0$, the existence of a mass gap is guaranteed.

Let us determine how these two aspects are related to the smooth character of the curvature scalar. Thus, we must study the relation among the following aspects of the theory:

a) Localization of 4D gravity in the 5D braneworld.

b) Smoothness of the curvature scalar R_5.

c) Existence of a mass gap in the spectrum of gravitational excitations of the system.

From the expressions for Ψ_0 (8), V_{QM} (7) and R_5 (10) we find that the curvature scalar can be written in terms of the zero energy wave function and the quantum mechanical potential as follows:

$$R_5 = \frac{16}{3}\Psi_0^{-\frac{4}{3}} V_{QM}. \tag{11}$$

Since the warp factor is real, (8) and (9) imply that Ψ_0 must tend to zero asymptotically

$$\Psi_0|_{z\to\infty} \to 0 \tag{12}$$

in order to ensure 4D gravity localization. If we indeed require the existence of a mass gap, V_{QM_∞} must adopt a positive value asymptotically.

Recall that we are considering thick brane configurations that are regular at the position of the brane.

Thus, if a) and c) are fulfilled, (11) implies that R_5 will necessary possess naked singularities as $|z| \to \infty$.

On the other side, if we assume that conditions a) and b) are satisfied, then V_{QM} must vanish asymptotically at least with the same rapidity of $\Psi_0^{-\frac{4}{3}}$, implying, in turn, that there is no mass gap for such a solution of the model.

Thus, if one desires to construct a completely regular thick brane configuration in the framework of the model (1) with the ansatz (2) in which 4D gravity is localized, then the corresponding mass spectrum will have no mass gap between the zero mode bound state, identified with a stable 4D graviton, and the continuous spectrum of massive KK excitations.

Conversely, if we require the existence of a mass gap in such a model restricted by (2), then the scalar curvature will necessary develop naked singularities asymptotically. This singularities can be made harmless if one imposes unitary boundary conditions in the spirit of [2, 3], [53, 54], providing a viable model from the physical point of view.

The fact that the manifold develops naked singularities at the boundaries in this case does not matter if no conserved quantities are allowed to leak out through the boundaries. The Poincaré isometries of metric (2) corresponds to 4D conservation laws. Thus, by considering the translations generated by the Killing vectors $\xi^\mu_m = \delta^\mu_m$, where μ is a five–dimensional index, one can construct currents by contracting with the stress tensor

$$J^\mu = T^{\mu\nu}\xi^m_\nu. \tag{13}$$

These currents satisfy a covariant conservation law of 4D energy and momentum

$$\frac{1}{\sqrt{g}}\partial_\mu\left(\sqrt{g}J^\mu\right) = 0. \tag{14}$$

Therefore we demand that the flux through the singular boundary of spacetime (along the transverse direction) must vanish for all currents in order to ensure that these quantities are conserved in the presence of a singularity:

$$\lim_{z\to\infty}\sqrt{g}J^z = \lim_{z\to\infty}\sqrt{g}g^{zz}\frac{1}{2}\left(\partial_m h^T_{pq}\right)\left(\partial_z h^T_{pq}\right) = 0. \tag{15}$$

We shall see below that these unitary boundary conditions project out all continuum modes from the spectrum of KK graviton excitations, leading to a unitary spectrum consisting of discrete bound states in the theory.

2.1. Solution with a mass gap

The system under consideration allows for the following solution [1]–[3]

$$e^{2A} = [\cos(ay - y_0)]^b, \qquad \phi = \frac{\sqrt{3b}}{2}\ln[\sec(ay - y_0) + \tan(ay - y_0)], \tag{16}$$
$$V(\phi) = \frac{3a^2b^2}{4}\left[1 + \frac{1-2b}{b}\cosh^2\left(\frac{2\phi}{\sqrt{3b}}\right)\right].$$

where a, b and y_0 are arbitrary constants.

Such a solution describes a thick brane located at y_0 when the range of the fifth dimension is $-\pi/2 \le ay - y_0 \le \pi/2$ [27, 28] and involves naked singularities at the boundaries of the manifold [3, 29]; a represents the inverse of the brane's width ($\Delta \sim 1/a$) when $b > 0$.

If $b = 2$ one can invert the coordinate transformation (4) yielding the following relation:

$$\cos[a(y - y_0)] = \mathrm{sech}(az).$$

This transformation decompactifies the fifth dimension and sends the naked singularities that were present at $-\pi/2a$ and $\pi/2a$ to spatial infinity [3, 29]. In terms of z, the warp factor becomes

$$A(z) = \ln \mathrm{sech}(az),$$

and the analog quantum mechanical potential adopts the form [29]

$$V_{QM}(z) = \frac{3a^2}{4}\{3 - 5\mathrm{sech}^2(az)\}. \tag{17}$$

Recall that the eigenvalue spectrum denoted by m^2 parameterizes the spectrum of graviton masses that a four–dimensional observer standing at $z = 0$ sees.

The analog quantum mechanical potential asymptotically approaches a positive value $V_{QM}(\infty) = 9a^2/4$, leading to a mass gap between the zero mode bound state and the first KK massive mode [4].

By setting $u = az$ the Schrödinger equation can be transformed into the standard form possessing a modified Pöschl–Teller potential with $n = 3/2$ (see [56],[29])

$$\left[-\partial_u^2 - n(n+1)\mathrm{sech}^2 u\right]\Psi(u) = \left[\frac{m^2}{a^2} - \frac{9}{4}\right]\Psi(u) = E\Psi(u). \tag{18}$$

Thus, the wave function possesses two bound states: the ground state Ψ_0 with energy E_0 and an excited massive state Ψ_1 with energy E_1. In this case the Schrödinger equation (6) can be solved analytically for arbitrary m and the general solution is a linear combination of associated Legendre functions of first and second kind of degree 3/2 and order $\mu = \sqrt{\frac{9}{4} - \frac{m^2}{a^2}}$:

$$\Psi_m = k_1\, P^{\mu}_{3/2}[\tanh(az)] + k_2\, Q^{\mu}_{3/2}[\tanh(az)], \tag{19}$$

where k_1 and k_2 are integration constants. The ground state corresponds to the massless state $m = 0$ ($\mu = 3/2$) and possesses energy $E_0 = -n^2 = -9/4$, whereas the excited state has mass $m = \sqrt{2}a$ ($\mu = 1/2$) and energy $E_1 = -(n-1)^2 = -1/4$.

Since for the special values $\mu = 3/2$ and $\mu = 1/2$ the associated Legendre function's series are finite, one obtains the following eigenfunction for the massless mode

$$\Psi_0 = C_0\, \mathrm{sech}^{3/2}(az), \tag{20}$$

whereas the excited massive mode adopts the form

$$\Psi_1 = C_1 \sinh(az)\text{sech}^{3/2}(az), \tag{21}$$

where C_0 and C_1 are arbitrary normalization constants.

Since (20) represents the lowest energy eigenfunction of the Schrödinger equation (6) and it has no zeros, it can be interpreted as a massless 4D graviton with no tachyonic instabilities from modes with $m^2 < 0$. On the other side, the eigenfunction (21) represents a normalizable massive graviton localized on the center of the thick brane. The mass gap existing between these two bound states eliminates the potentially dangerous arbitrarily light KK excitations, sending them to energies of the scale of a.

In the spectrum there is also a continuous part of massive modes with $2m > 3a$ (the order becomes purely imaginary $\mu = i\rho$, where $\rho = \sqrt{(m/a)^2 - 9/4}$) that describe plane waves as they approach spatial infinity [29] and give rise to small corrections to Newton's law in 4D flat spacetime in the thin brane limit [4, 30].

However, by imposing the unitary boundary conditions mentioned above, i.e., by demanding the vanishing of the flux through the singular boundary of our spacetime:

$$\lim_{z\to\infty} e^{3A/2}\Psi(z)\partial_z\left[e^{-3A/2}\Psi(z)\right] \sim \Psi(z)\left[(3B - 2A\rho)\cos(a\rho z) + (3A + 2B\rho)\sin(a\rho z)\right] = 0, \tag{22}$$

where we have made use of the asymptotic form of $\Psi(z)$:

$$\Psi(z) \sim A\sin(a\rho z) + B\cos(a\rho z), \tag{23}$$

we get a result implying that A and B must vanish, just eliminating all the continuum modes from the spectrum of KK excitations. Thus, the unitary spectrum consists of two discrete modes because they do not generate any flux into the singularity.

3. Conclusions and discussion

We have a 5D thick braneworld arising in a scalar tensor theory and have investigated the relationship existing among the localization of 4D gravity, the smooth or singular character of the curvature scalar of the classical background, and the existence of a mass gap gap which separates the massless graviton bound state from the massive KK excitations.

We have shown that when considering 4D gravity localization in the 5D thick braneworld, the smooth character of the scalar curvature of the manifold excludes the possibility of the existence of a mass gap in the graviton spectrum of fluctuations; conversely, by demanding the existence of a mass gap between the massless and massive modes of KK excitations, the curvature scalar necessarily develops naked singularities at the boundaries of the manifold.

We further considered an explicit solution which gives rise to a mass gap in the spectrum of graviton fluctuations. This solution contains a massless graviton, one massive excited state with $m = \sqrt{2}a$, and a continuum of modes with a mass gap of size $m = 3a/2$. At very low energies, none of these massive modes can be excited, and an observer at $z = 0$ sees pure four–dimensional gravity. At higher energies the massive state can be excited, giving some corrections to Newton's law and, finally, at energies larger than the gap the whole continuum of modes can be excited. Violations of unitarity occur only when modes that can travel out to the singularities can be excited, i.e. only at energies above the mass of the lightest continuum mode.

For this reason we imposed unitary boundary conditions on the described solution, a fact that eliminates all the continuous modes from the spectrum of KK excitations of the classical background and renders the naked singularities harmless (in the same spirit as in [2, 3], [53, 54]).

Acknowledgments
One of the authors (AHA) is grateful to the organizers of the First Mediterranean Conference on Classical and Quantum Gravity for providing a warm atmosphere during the event. This research was supported by grants CIC–4.16 and CONACYT 60060–J. DMM and RRML acknowledge a PhD grant from CONACYT. AHA and UN thanks SNI for support.

References

[1] De Wolfe O, Freedman D Z, Gubser S S and Karch A 2000 Modeling the fifth dimension with scalars and gravity *Phys. Rev.* D **62** 046008
[2] Gremm M 2000 Four–dimensional gravity on a thick domain wall *Phys. Lett.* B **478** 434
[3] Gremm M 2000 Thick domain walls and singular spaces *Phys. Rev.* D **62** 044017
[4] Csaki C, Erlich J, Hollowood T and Shirman Y 2000 Universal Aspects of gravity localized on thick branes *Nucl. Phys.* B **581** 309
[5] Parikh M K and Solodukhin S N 2001 De Sitter brane gravity: from close–up to panorama *Phys. Lett.* B **503** 384
[6] Wang A 2002 Thick de Sitter 3 branes, dynamic black holes and localization of gravity *Phys. Rev.* D **66** 024024
[7] Minamitsuji M, Naylor W and Sasaki M 2006 Quantum fluctuations on a thick de Sitter brane *Nucl. Phys.* B **737** 121
[8] Bazeia D, Brito F A and Nascimento J R 2003 Supergravity brane worlds and tachyon potentials *Phys. Rev.* D **68** 085007
[9] Bazeia D and Losano L 2006 Deformed defects with applications to braneworlds *Phys. Rev.* D **73** 025016
[10] Ghassemi S, Khakshournia S and Mansouri R 2007 Thick planar domain wall: its thin wall limit and dynamics *Int. J. Mod. Phys.* D **16** 629
[11] Bogdanos C, Dimitriadis A and Tamvakis K 2006 Brane models with a Ricci–coupled scalar field *Phys. Rev.* D **74** 045003
[12] Farakos K and Pasipoularides P 2007 Brane world scenario in the presence of a non–minimally coupled bulk scalar field *J. Phys.: Conf. Ser.* **68** 012041
[13] Bogdanos C and Tamvakis K 2007 Brane cosmological evolution with bulk matter *Phys. Lett.* B **646** 39
[14] Bogdanos C 2007 Exact solutions in 5-D brane models with scalar fields *J. Phys.: Conf. Ser.* **68** 012045
[15] Guerrero R, Melfo A and Pantoja N 2002 Selfgravitating domain walls and the thin wall limit *Phys. Rev.* D **65** 125010
[16] Melfo A, Pantoja N and Skirzewski A 2003 Thick domain wall space–time with and without reflection symmetry *Phys. Rev.* D **67** 105003
[17] Castillo–Felisola O, Melfo A, Pantoja N and Ramirez A 2004 Localizing gravity on exotic thick three–branes *Phys. Rev.* D **70** 104029
[18] Barbosa–Cendejas N and Herrera–Aguilar A 2007 Localizing gravity on thick branes: a solution for massive KK modes of the Schroedinger equation *J. Phys.: Conf. Ser.* **68** 012021
[19] Bazeia D, Gomes A R and Losano L 2009 Gravity localization on thick branes: a numerical approach *Int. J. Mod. Phys.* A **24** (2009) 1135
[20] Dzhunushaliev D, Folomeev V and Minamitsuji M 2009 Thick brane solutions *Preprint* arXiv:0904.1775 [gr–qc]
[21] Liu Y–X, Yang J, Zhao Z–H, Fu C–E, and Duan Y–S 2009 Fermion localization and resonances on a de Sitter thick brane *Preprint* arXiv:0904.1785 [hep-th]
[22] Liu Y–X, Fu C–E, Zhao L and Duan Y–S 2009 Localization and mass spectra of fermions on symmetric and asymmetric thick branes *Preprint* arXiv:0907.0910 [hep-th]
[23] Zhang L–J and Yang G–H 2009 Zero modes of matter fields on scalar flat thick branes *Preprint* arXiv:0907.1178 [hep-th]
[24] Liu Y–X, Zhong Y and Yang K 2009 Warped thick brane solutions of scalar fields with generalized dynamics *Preprint* arXiv:0907.1952 [hep-th]
[25] Liu Y–X, Guo H, Fu C–E and Ren J–R 2009 Localization of matters on anti-de Sitter thick branes *Preprint* arXiv:0907.4424 [hep-th]
[26] Arias O, Cardenas R and Quiros I 2002 Thick brane worlds arising from pure geometry *Nucl. Phys.* B **643** 187
[27] Barbosa–Cendejas N and Herrera–Aguilar A 2005 4D gravity localized in non Z_2–symmetric thick branes *J. High Energy Phys.* JHEP10(2005)101
[28] Barbosa–Cendejas N and Herrera–Aguilar A 2006 Localization of 4D gravity on pure geometrical thick branes *Phys. Rev.* D **73** (2006) 084022. Erratum-ibid. 2008 **77** 049901

[29] Barbosa–Cendejas N, Herrera–Aguilar A, Reyes M A and Schubert C 2008 Mass gap for gravity lozalized on Weyl thick branes *Phys. Rev.* D **77** 126013

[30] Barbosa–Cendejas N, Herrera–Aguilar A, Nucamendi U and Quiros I 2007 Mass hierarchy and mass gap on thick branes with Poincaré symmetry *Preprint* arXiv:0712.3098 [hep-th]

[31] Kehagias A and Tamvakis K 2001 Localized gravitons, gauge bosons and chiral fermions in smooth spaces generated by a bounce *Phys. Lett.* B **504** 38

[32] Liu Y–X, Zhang X–H, Zhang L–D and Duan Y–S 2008 Localization of matters on pure geometrical thick branes *J. High Energy Phys.* JHEP02(2008)067

[33] Zhang X–H, Liu Y–X and Duan Y–S 2008 Localization of fermionic fields on braneworlds with bulk tachyon matter *Mod. Phys. Lett.* A **23** 2093

[34] Liu Y–X, Zhang L–D, Wei S–W and Duan L–D 2008 Localization and mass spectrum of matters on Weyl thick branes *J. High Energy Phys.* JHEP08(2008)041

[35] Liu Y–X, Zhang L–D, Zhang L–J and Duan Y–S 2008 Fermions on thick branes in background of Sine–Gordon kinks *Phys. Rev.* D **78** 065025

[36] Liu Y–X, Zhao Z–H, Wei S–W and Duan Y–S 2009 Bulk matters on symmetric and asymmetric de Sitter thick branes *J. Cosmol. Astropart. Phys.* JCAP02(2009)003

[37] Almeida C A S, Ferreira Jr M M, Gomes A R and Casana R 2009 Fermion localization and resonances on two–field thick branes *Preprint* arXiv:0901.3543 [hep-th]

[38] Liu Y–X, Yang J, Zhao Z–H, Fu C–E and Duan Y–S 2009 Fermion localization and resonances on a de Sitter thick brane *Preprint* arXiv:0904.1785 [hep-th]

[39] Akama K 1982 Pregeometry *Lec. Notes Phys.* **176** 267

[40] Rubakov V A and Shaposhnikov M E 1983 Extra space–time dimensions: towards a solution of the cosmological constant problem *Phys. Lett.* B **125** 139

[41] Antoniadis I 1990 A Possible new dimension at a few TeV *Phys. Lett.* B **246** 317

[42] Gogberashvili M 1999 Four dimensionality in noncompact Kaluza-Klein model *Mod. Phys. Lett.* A **14** 2025

[43] Gogberashvili M 2002 Hierarchy problem in the shell universe model *Int. J. Mod. Phys.* D **11** 1635

[44] Gogberashvili M 2006 Localizing gravity in extra dimensions *Eur. Phys. J.* C **47** 857

[45] Randall L and Sundrum R 1999 An alternative to compactification *Phys. Rev. Lett.* **83** 4690

[46] Randall L and Sundrum R 1999 A large mass hierarchy from a small extra dimension *Phys. Rev Lett.* **83** 3370

[47] Lykken J and Randall L 2000 The shape of gravity *J. High Energy Phys.* JHEP06(2000)014

[48] Garriga J and Tanaka T 2000 Gravity in the brane world *Phys. Rev. Lett.* **84** 2778

[49] Kobayashi S, Koyama K and Soda J 2002 Thick brane worlds and their stability *Phys. Rev.* D **65** 064014

[50] Maartens R 2004 Brane world gravity *Living Rev. Rel.* **7** 07

[51] Gergely L A and Maartens R 2005 Asymmetric brane–worlds with induced gravity *Phys. Rev.* D **71** 024032

[52] Farakos K, Koutsoumbas G and Pasipoularides P 2007 Graviton localization and Newton's law for brane models with a non-minimally coupled bulk scalar field *Phys. Rev.* D **76** 064025

[53] Gell–Mann M and Zwiebach B 1985 Dimensional reduction of space-time induced by nonlinear scalar dynamics and noncompact extra dimensions *Nucl. Phys.* B **260** 569

[54] Cohen A G and Kaplan D B 1999 Solving the hierarchy problem with noncompact extra dimensions *Phys. Lett.* B **470** 52

[55] Brandhuber A and Sfetsos K 1999 Nonstandard compactifications with mass gaps and Newton's law *J. High Energy Phys.* JHEP10(1999)013

[56] Diaz J I, Negro J, Nieto L M and Rosas–Ortiz O 1999 The supersymmetric modified Poschl–Teller and delta–well potentials *J. Phys. A: Math. Gen.* **32** 8447

First Mediterranean Conference on Classical and Quantum Gravity (MCCQG 2009) IOP Publishing
Journal of Physics: Conference Series **222** (2010) 012004 doi:10.1088/1742-6596/222/1/012004

Codimension-2 brane cosmology

Antonios Papazoglou
Institute of Cosmology and Gravitation,
University of Portsmouth,
Portsmouth PO1 3FX, UK

E-mail: antonios.papazoglou@port.ac.uk

Abstract. We study cosmological purely codimension-2 braneworlds in a 6d Lovelock theory, with topological matching conditions where the extrinsic curvature on the brane has no jump. For an arbitrary smooth regularisation of the defect, we show that the brane equations of motion do not close, as there is energy exchange between the brane and the 6d bulk spacetime. In the limit that the deficit angle of the defect is approximately constant, the system closes and there is no leakage of energy in the bulk. We analyse the modified Friedmann equation and we see that there are certain corrections coming from the non-zero extrinsic curvature on the brane. We establish the presence of geometric self-acceleration and a possible curvature domination wedged in between the period of matter and self-acceleration eras as signatures of codimension-2 cosmology.

1. Introduction

Codimension-2 distributional defects are particularly interesting because of the purely topological effect of their self-gravity. A straight distributional string carrying only tension is described by a codimension-2 conical singularity [1], while the ambient bulk spacetime remains flat. Each point of the string worldsheet generates a 2d cone with a non-trivial deficit angle related to the string tension. This topological property of codimension-2 defects generated an activity on this subject mostly in relation to the self-tuning paradigm [2] in brane worlds.

When considering however, non-trivial matter on the defect (as *e.g.*, a cosmological fluid), one cannot find in the context of General Relativity (GR) a distributional solution [3]. This is a old paradox already noticed in 4d GR [4] (see [5] for a mathematical discussion on distributional sources in GR). One had to introduce finite thickness effects [6], which made the problem considerably more difficult, and more importantly, plagued the generality of the result and, hence, its physical relevance. Indeed, the key point when studying distributional sources is that they describe the important main features of brane dynamics for an arbitrary family of finite width regularisations, at the limit of infinitesimal thickness.

The way out of this problem can be found when one understands that it was not the defect construction which is problematic, but rather the gravity theory itself does not have the relevant differential complexity in order to describe complicated distributional solutions. Although 4d GR is the unique tensor theory to have at most second derivative (and not higher than second derivative) field equations, in 5d or 6d one has to add an extra term, the Gauss-Bonnet term, to the action. This generic second order derivative tensor gravity theory is in fact well-known to be Lovelock's theory [7].

The clear-cut hint that this completion of GR in higher dimensions may be relevant to codimension-2 braneworlds came with the work of [8, 9], where it was noticed that upon considering a Gauss-Bonnet bulk theory, one could have, at least in principle, a non trivial energy momentum tensor fuelling geometric junction conditions for a codimension-2 conical defect. This activity came to a halt when it was claimed that the full set of junction conditions plus bulk field equations at the location of the brane led to an inconsistent system [10] or introduced important constraints on the admissible matter on the brane [11], under certain restrictive initial conditions on the braneworld surface [8]. Relaxing these initial conditions in the recent publication [12], revealed that the above problems disappear. Furthermore, it was shown in [12] that the brane matching conditions of [8] inevitably lead to extra codimension-1 distributional matter, whereas the the brane matching conditions of [9] (referred to as topological) lead to a purely codimension-2 defect.

The aim of this talk is the study the cosmological evolution of a purely codimension-2 defect, studied in detail in [12]. For this purpose, we will use the topological matching conditions of [9]. After a brief introduction to the brane equations, *i.e.,* the brane junction conditins and the leading bulk equations evaluated at the brane position, we will go on with investigating the cosmological equations on the brane. We will show that they have the form of modified Lemaître-Friedmann-Robertson-Walker (LFRW) equations, with several interesting new features: a geometric self-acceleration (in agreement with the maximally symmetric solutions of [13]), brane bending effects depending on the equation of state of matter, and energy exchange between the bulk and the brane triggered by a dynamical conical deficit angle.

2. Setup and brane equations of motion for the topological matching conditions

A general system of 6d Lovelock gravity coupled to localised brane sources is described by the total action

$$S = \frac{1}{2\kappa_6^2}\int d^6x\sqrt{-|\mathrm{g}|}\left(\mathcal{R} + \frac{\alpha}{6}\,\mathcal{L}_{GB}\right) + S_{bulk}[\mathrm{g}] + S_{brane}[g]\ , \tag{1}$$

where $S_{bulk}[\mathrm{g}]$ is the action of regular bulk matter, $S_{brane}[g]$ is the action of distributional brane matter and α is the coupling in front of the Gauss-Bonnet Lagrangian density

$$\mathcal{L}_{GB} = \mathcal{R}^2 - 4\mathcal{R}_{MN}\,\mathcal{R}^{MN} + \mathcal{R}_{MNK\Lambda}\,\mathcal{R}^{MNK\Lambda}\ . \tag{2}$$

In the above, the calligraphic quantities refer to the bulk metric tensor g, while the regular ones to the brane metric tensor g. The field equations arising from the action (1) are

$$\mathcal{G}_M^N - \frac{\alpha}{6}\left[\frac{1}{2}\mathcal{L}_{GB}\delta_M^N - 2\mathcal{R}\mathcal{R}_M^N + 4\mathcal{R}_{MK}\mathcal{R}^{NK} + 4\mathcal{R}_{MK}{}^{N}{}_{\Lambda}\mathcal{R}^{K\Lambda} - 2\mathcal{R}_{MK\Lambda\Xi}\mathcal{R}^{NK\Lambda\Xi}\right] = \kappa_6^2(\mathcal{T}_M^N + T_M^N)\ , \tag{3}$$

where $\mathcal{T}_{MN}$ is a regular bulk energy-momentum tensor and T_{MN} is the distributional brane energy-momentum tensor. In the following, we will fix the notation to $3\kappa_6^2 = 4\pi$ to simplify the equations.

A crucial assumption about the dynamics of the system has to do with the matching conditions that one assumes at the brane position. This ambiguity is always present in defects of codimension greater than one. In particular, in the general case one may chose that the extrinsic curvature has a jump when passing from infinitesimally outside the defect to the core of the defect. Such a condition was imposed for example in [8]. The simplest case is the one

Journal of Physics: Conference Series **222** (2010) 012004 doi:10.1088/1742-6596/222/1/012004

where the extrisic curvature is continuous from the core of the defect to outside. These matching conditions were studied in [9] and were referred to as topological.

It has been shown in [12], that the general matching conditions combined with the other bulk equations evaluated the brane position require always some codimension-1 distributional matter and therefore do not describe a pure codimension-2 brane. For this reason in the following we will focus on the *topological matching conditions.*

Assuming that there is axial symmetry in the bulk, the bulk metric ansatz can be written in the brane Gaussian-Normal coordinates as

$$ds_6^2 = dr^2 + L^2(x,r)d\theta^2 + g_{\mu\nu}(x,r)dx^\mu dx^\nu \ , \tag{4}$$

As usually done when dealing with distributional sources, the matching conditions are derived by integrating around the singular space. Here, we seek to integrate over the conical space, say $\mathcal{C}$, with angular metric function

$$L(x,r) = \beta(x)r + \mathcal{O}(r^2) \ , \tag{5}$$

where $\delta = 2\pi(1-\beta(x))$ is the conical deficit. Notice, the coefficients are allowed to depend on the "brane" coordinates x^μ. The angle θ varies in the interval $[0,2\pi)$.

Since we are interested in the pure codimension-2 case, the distributional brane energy-momentum tensor should have a Dirac singularity as

$$T_{MN} \equiv T_{\mu\nu}\frac{\delta(r)}{2\pi L}\delta^\mu_M\delta^\nu_N \ . \tag{6}$$

Then, by integrating the equations of motion, the only non-zero contribution yielding the codimension-2 matching conditions reads [9],

$$\alpha\,\frac{1}{4\pi}\int d^2y\sqrt{g_2}\,R_2\left[G_{\mu\nu} + W_{\mu\nu} - \frac{3}{2\alpha}g_{\mu\nu}\right] = T_{\mu\nu} \ , \tag{7}$$

where $R_2 = -2L''/L$ is the curvature of the 2d internal space, and a $'$ denotes ∂_r. The Einstein equation is corrected with the following contribution from the extrinsic curvature $K_{\mu\nu} \equiv \frac{1}{2}\partial_r g_{\mu\nu}$ on the brane

$$W_{\mu\nu} = K^\lambda_\mu K_{\nu\lambda} - KK_{\mu\nu} + \frac{1}{2}g_{\mu\nu}(K^2 - K^2_{\kappa\lambda}) \ , \tag{8}$$

which is identical to the one in [8].

In order to evaluate the integral of (7) which is singular at the tip, we assume a family of *smooth* regularisations C_λ of the cone, parametrised by λ, with smooth C^2 everywhere caps (for details see [14]), pasted smoothly to the rest of the cone at a boundary ∂C, which we assume does not depend on the parameter λ (see left part of Fig.1). The distributional limit is obtained at the $\lambda \longrightarrow \lambda_c$ limit, where the cap tends to the singular conical space. The width of the regularised defect is supposed to be small enough so that the r-dependent brackets can be approximated with the leading term in their r-expansion, which is just the value of the brackets at $r = 0$. Then the integral has to be performed over only the internal curvature. The Chern-Gauss-Bonnet theorem relates the geometric curvature for arbitrary λ, to the Euler characteristic χ of the regularised cap and the integral of the geodesic curvature k_g of the boundary ∂C. Then, for every cap labeled by λ, the following holds

$$\frac{1}{4\pi}\int_{C_\lambda} d^2y\sqrt{g_\lambda}\,R_\lambda = \chi - \frac{1}{2\pi}\int_{\partial C_\lambda} k_g\,d\theta = 1-\beta \ , \tag{9}$$

Journal of Physics: Conference Series **222** (2010) 012004 doi:10.1088/1742-6596/222/1/012004

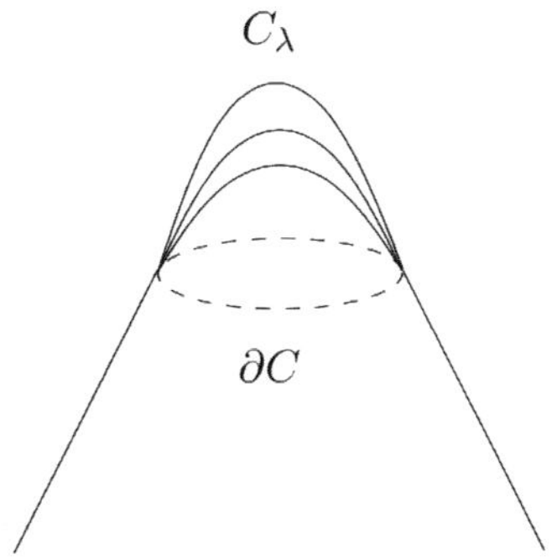

Figure 1. The conical singularity is regularised by a family of regular interiors C_λ smoothly connected to the exterior solution of the cone at the boundary ∂C.

since the λ-labeled caps are topologically equivalent to a disk which has $\chi = 1$ and since $k_g = L'$. The above result only used the fact that the expansion of L close to the conical defect has the expansion (5) and is independent of the smooth regularisation of the interior of the defect. Finally, with the use of the above theorem, the codimension-2 matching conditions for the topological case is

$$G_{\mu\nu} + W_{\mu\nu} - \frac{3}{2\alpha} g_{\mu\nu} = \frac{1}{\alpha(1-\beta)} T_{\mu\nu} \ . \tag{10}$$

Let us now look at the leading $\mathcal{O}(1/r)$ terms of the equations of motion. These come from the contributions of terms of the equations of motion which are multiplied by $L'/L = 1/r + \mathcal{O}(1)$, or $\nabla_\mu L'/L = (\nabla_\mu \beta/\beta) \cdot 1/r + \mathcal{O}(1)$, with ∇_μ the covariant differentiation with respect to the metric $g_{\mu\nu}$.

With the natural assumption that the energy momentum tensor $\mathcal{T}_{MN}$ does not blow up close to the distributional singularity (otherwise the singularity would not be distributional), we obtain that the $\mathcal{O}(1/r)$ terms of the (rr) and the $(r\mu)$ equations yield respectively

$$K^{\mu\nu}\left(G_{\mu\nu} + W_{\mu\nu} - \frac{3}{2\alpha} g_{\mu\nu}\right) = 0 \ , \tag{11}$$

$$\frac{\nabla^\nu \beta}{\beta}\left(G_{\mu\nu} + W_{\mu\nu} - \frac{3}{2\alpha} g_{\mu\nu}\right) + \nabla^\nu W_{\mu\nu} = 0 \ . \tag{12}$$

These two equations act as constraint equations for the two quantities $K_{\mu\nu}$, β, which have to be determined and substituted back to the brane Einstein equation (10). In the general case, there are 11 independent unknown functions, but there are 5 independent constraint equations in (11), (12). Therefore, there will in general be functions which cannot be determined by the local equations around the brane, but need to be determined by the bulk solution. This, in fact, is clearly visible in the case of isotropic cosmology on the brane, as we will see in the next section.

Finally, if we differentiate the brane Einstein equation (10) and use the constraint (12), we can derive the energy conservation equation

$$\nabla^\nu T_{\mu\nu} = -\frac{\nabla^\nu \beta}{\beta(1-\beta)} T_{\mu\nu} \ . \tag{13}$$

From the above, we see that the energy is not strictly conserved on the brane, but can radiate in the bulk if the deficit angle changes. We can therefore anticipate that cosmological evolutions, close to the standard 4d one, will have β almost constant. We will come to this point in the next section.

3. Codimension-2 Cosmological Evolution

Let us apply now the brane equations that we found in the previous section to a brane with cosmological symmetries (homogeneity and isotropy along the brane spatial directions). For this purpose we adopt the following metric ansatz for the 4d metric $g_{\mu\nu}$ in (4) describing LFRW cosmology

$$ds^2 = -n^2(t,r)dt^2 + a^2(t,r)ds_\kappa^2 + dr^2 + L^2(t,r)d\theta^2 \ , \tag{14}$$

with usual maximally symmetric spatial line element for $\kappa = \pm 1, 0$ curvatures. The expansion of n and a around the brane is

$$n(t,r) = 1 + Nr + \mathcal{O}(r^2) \quad , \quad a(t,r) = a\left[1 + Ar + \mathcal{O}(r^2)\right] \ , \tag{15}$$

where with $a = a(t)$ we denote the scale factor on the brane. We will assume that the energy-momentum tensor on the brane is the one of perfect fluid $T_\mu^\nu = (-\rho, P, P, P)$. The energy density and pressure may contain contributions from the vacuum energy of the brane.

Then for the smoothly regularised brane with topological boundary conditions we obtain that the equations (10), (11), (12) give the following system

$$\frac{\rho}{3\alpha(1-\beta)} = H^2 + \frac{\kappa}{a^2} + \frac{1}{2\alpha} - A^2 \ , \tag{16}$$

$$-\frac{\rho + 3P}{6\alpha(1-\beta)} = \frac{\ddot{a}}{a} + \frac{1}{2\alpha} - fA^2 \ , \tag{17}$$

$$f \equiv \frac{N}{A} = \frac{3P}{\rho} \ , \tag{18}$$

$$2A\dot{A} + 2H(1-f)A^2 = \frac{\dot{\beta}}{\beta}\frac{\rho}{3\alpha(1-\beta)} \ . \tag{19}$$

In order to derive the equations (18) and (19), we have used the first one (16). Here, we recognize the first two equations as a modified version of the 4d Friedmann and acceleration equations. Note that β is generically a function of time which can make the effective 4d Planck scale time varying. A second important point is the appearance of an effective cosmological constant in the face of α, which is of geometric origin. A last difference from the standard 4d equations is the presence of the extrinsic curvature correction parametrised by A.

The above system of equations has evidently one free function, which we can take to be $\beta(t)$. This function is expected to be fixed by the asymptotic dynamics of the bulk and cannot be determined by the local equations of motion. As mentioned previously, the time dependence of $\beta(t)$ is translated to energy exchange between the brane and the bulk, since the brane conservation equation (13) reads

$$\dot{\rho} + 3H(\rho + P) + \frac{\dot{\beta}}{\beta(1-\beta)}\rho = 0 \ . \tag{20}$$

This variation of β induces a variation of the effective 4d gravitational constant read in (10) and is constrained during the early cosmology by the primordial abundances at the nucleosynthesis epoch [15]. Even more stringent constraints expected from solar system observations from the fact that the theory with varying β is similar to a scalar-tensor theory. Not knowing the full family of solutions in the bulk, we choose to consider the case where β is *approximately constant.*

This, in fact, will be a consequence of the equations of motion, had we considered a ring regularisation of the topological matching condition case [12], instead of the smooth one that we used in this talk. This case is bound to give us, at least seemingly, acceptable 4d cosmology, and has the merit that the system of equations is closed and does not depend on undetermined (by the matching conditions) functions.

When β is constant, we can solve for the extrinsic curvature A for a given, constant equation of state for the brane matter. For that purpose, we split the energy density and pressure to a part which has the form of brane vacuum energy λ and part which describes brane matter

$$\rho = \lambda + \rho_m \quad , \quad P = -\lambda + P_m \; , \tag{21}$$

with the brane matter equation of state defined as $w = P_m/\rho_m$. Then, for $w \neq -1$, we can integrate (19) as

$$A^2 = \frac{C^2}{(\lambda + \rho_m)^2}\, \rho_m^{\frac{8}{3(1+w)}} \; , \tag{22}$$

with C^2 a positive integration constant. With this expression, the Friedmann and acceleration equations (16), (17) become

$$H^2 + \frac{\kappa}{a^2} = \frac{\kappa_4^2}{3}\rho_m + \left(\frac{\kappa_4^2}{3}\lambda - \frac{1}{2\alpha}\right) + \frac{C^2}{(\lambda+\rho_m)^2}\,\rho_m^{\frac{8}{3(1+w)}} \; , \tag{23}$$

$$\frac{\ddot{a}}{a} = -\frac{\kappa_4^2}{6}(1+3w)\rho_m + \left(\frac{\kappa_4^2}{3}\lambda - \frac{1}{2\alpha}\right) + 3\frac{C^2}{(\lambda+\rho_m)^3}(w\rho_m - \lambda)\rho_m^{\frac{8}{3(1+w)}} \; , \tag{24}$$

where $\kappa_4^2 = 1/\alpha(1-\beta)$, which show a non-trivial correction to the cosmological equations as a function of w. The magnitude of this correction depends on the integration constant C^2. In the following, we will try to see whether it is possible for large enough C^2, consistent with observations, to have some interesting modification to cosmology. From the integration of the continuity equation (20) for β =const. we obtain

$$\rho_m = \frac{\rho_0(1-\beta)}{\beta a^{3(1+w)}} \; , \tag{25}$$

which holds also for the special $w = -1$ case.

For $C^2 \neq 0$, our 6d bulk geometry has a genuine curvature singularity at $r = 0$ (apart from the distributional one). In fact, this is to be expected from purely geometrical considerations (see [16]). Higher codimension defects, when considered in their zero width limit, will develop curvature singularities. These are expected to be smoothable once we take finite width corrections into account. Therefore, we expect finite width effects to be important at the UV sector and our distributional approximation to break down even though this will not show up necessarily in the field equations themselves!

Let us now analyse some particular cases of interest.

3.1. Self-accelerating branes

The first interesting case is the one in which the tension of the brane is vanishing $\lambda = 0$. In this case, we see from (23) that there is an *asymptotic accelerating phase* for $\alpha < 0$. This acceleration is purely due to the geometric Gauss-Bonnet term and therefore is a case of self-acceleration.

w	ρ_m	A^2
-1	const.	a^{-8}
0	a^{-3}	a^{-2}
$1/3$	a^{-4}	const.

Figure 2. The functional dependence on the scale factor of the energy density and the extrinsic curvature correction to the Friedmann equation for different equations of state of the brane matter in the self-accelerating case. From top to bottom we have the equations of state for cosmological constant, dust and radiation.

This corresponds to generalisation of the self-accelerating solutions of [13] with the addition of matter. The correction to the Friedmann equation is given by

$$A^2 = C^2 \, \rho_m^{\frac{2(1-3w)}{3(1+w)}} = \frac{\tilde{C}^2}{a^{2(1-3w)}} \, . \tag{26}$$

Note that the second equality holds also for $w = -1$. In the table in Fig.2 we list the form of corrections that we obtain for different constant equations of state. It is worth noting three interesting limits of the cosmological evolution in this case.

First is the one where the brane is dominated by cosmological constant. This may go against the initial assumption that the tension of the brane vanishes, because of the ambiguity to separate the vacuum part of the energy density from the matter energy density. If a matter component behaving as inflaton or dark energy dominates at some period of the history of the Universe, it will have the behaviour that we note here. So, for this case the extrinsic curvature plays the role of a dark radiation squared term a^{-8} that will dominate the cosmology at early times.

A second interesting limit is when the matter equation of state is that of radiation $w = 1/3$. Then, the extrinsic curvature correction is of the form of constant vacuum energy. One would be tempted to use this vacuum energy to drive early Universe inflation. This seems to be possible, but around matter-radiation equality the Universe will be dominated by this new vacuum energy, making the phenomenology of the model problematic.

Finally, when the matter equation of state is that of dust $w = 0$, the extrinsic curvature correction is of the form of curvature (a^{-2}). In this case, we can see a possibility of having an observable signature of the codimension-2 cosmology, if the constant C^2 is chosen so that there is a brief period of curvature domination around the vacuum energy domination period. This scenario is depicted in Fig. 3. In order to obtain the standard epochs, *i.e.*, matter domination and then cosmological constant domination, we need to have that $\rho_m(a_{eq}) > 1/(2|\alpha|)$ and $(\kappa_4^2/3)\rho_m(a_{eq}) > A^2(a_{eq})$, where a_{eq} is the scale factor at radiation-matter equality and a_{acc} the one at the moment of cosmological constant domination. In order that the brief period of curvature domination is observable, we need $A^2(a_{acc}) > 1/(2|\alpha|)$. Therefore combining these inequalities we see that we need

$$\frac{1}{2|\alpha|} < A^2(a_{acc}) < \frac{\kappa_4^2}{3}\rho_m(a_{eq}) \, . \tag{27}$$

Then, a straightforward manipulation with the help of (25), (26) provides us with the following constraint on C^2

$$\left(\frac{\rho_0(1-\beta)}{\beta}\right)^{-2/3} \frac{a_{acc}^2}{2|\alpha|} < C^2 < \frac{\kappa_4^2}{3}\left(\frac{\rho_0(1-\beta)}{\beta}\right)^{1/3}\left(\frac{a_{acc}^2}{a_{eq}^3}\right) \, . \tag{28}$$

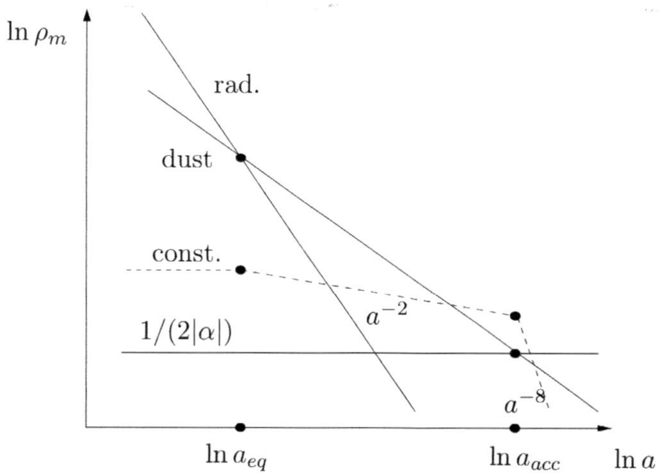

Figure 3. The log-log evolution of the energy densities as a function of the scale factor. With solid lines, there are the contributions of the standard matter and the geometrically induced $1/(2|\alpha|)$ cosmological constant. With dashed lines are the evolutions of the extra component ($\propto A^2$) in the Friedmann equation due to the extrinsic curvature of the brane. The dependence of this extra component on the scale factor is noted. This dependence changes whenever a new matter component becomes dominant, *i.e.*, at matter domination at a_{eq} and at the cosmological constant domination at a_{acc}.

It is not yet clear if such a brief period of curvature domination is phenomenologically viable and it requires certainly further study to see the potential observational signatures of such a case.

3.2. Self-tuning branes

A second interesting case is when self-tuning can be realised. For that possibility, the topological quantity β is used to accurately cancel the vacuum energy contribution of the brane to the effective cosmological constant. In more details, as discussed in [13], by tuning β one can make the two vacuum energy contributions, the geometrical one $1/2|\alpha|$ and the brane tension one λ, cancel

$$\frac{\kappa_4^2}{3}\lambda - \frac{1}{2\alpha} = 0 \ . \tag{29}$$

This limit is orthogonal to the self-accelerating case that we studied before. The effective Friedmann equation for the present case is given by the expression

$$H^2 + \frac{\kappa}{a^2} = \frac{\kappa_4^2}{3}\rho_m + \frac{C^2}{(\frac{3}{2\alpha\kappa_4^2} + \rho_m)^2}\,\rho_m^{\frac{8}{3(1+w)}} \ , \tag{30}$$

which includes a non-trivial correction beyond the standard linear to energy density term. Let us note here that the actual self-tuning is not visible from this Friedmann equation. This is because the present is valid for $\beta = \text{const}$. Instead, one should consider the time-varying deficit angle case in order to see how (29) can be dynamically achieved at the early Universe evolution. Since before nucleosynthesis there are no constraints on the variation of Newton's constant, one could have an acceptable cosmology with varying β. However, then, the brane equations do not close and the mechanism of self-tuning should be dictated by bulk boundary conditions.

4. Conclusions

In this talk we have have dicussed the cosmological equations for a codimension-2 brane in 6d bulk Lovelock gravity. These were obtained by considering the junction conditions on the codimension-2 brane and the leading parts of the bulk equations evaluated at the brane position. For that we used the, so called, topological matching conditions [9], where the extrinsic curvature is continuous outside the defect and at the core of the defect, since these are the ones that do not require any codimension-1 extra sources [12].

Studying the cosmology of these conical branes, we have found several interesting features. Firstly, quite generically the higher order terms give ordinary LFRW equations, as though one included an induced gravity term on the brane [17]. This is quite different from codimension-1 cosmology [18], where the ordinary behaviour is recovered only at late times. Corrections to this standard evolution are threefold: there is a geometric acceleration scale related to $1/\alpha$, which restates the problem of a minute cosmological constant as a gravitational hierarchy between the bulk Gauss-Bonnet and Einstein-Hilbert term. Secondly, extrinsic curvature corrections are apparent and they are dependent of the equation of state of the perfect-fluid matter. In other words, as the brane evolves in the bulk, a dust equation of state can lead to an extra component behaving as curvature, or a radiation equation of state to a cosmological constant term, and so on. To put it in a nutshell, each matter fluid introduces two differing fluid components in the modified LFRW equations. Last, but not least, as the brane evolves it can generically radiate in the bulk. Again, this is unlike the vacuum bulk codimension-1 braneworlds.

The above characteristics can be used to constrain the model in question with respect to cosmological observations. Furthermore, on the theoretical side it would be interesting to have particular bulk solutions manifesting the cosmology evolution we have found, and in particular setting the boundary conditions in order to fix β. The most important information we have given here in this direction is that it is pure 6d Gauss-Bonnet gravity that gives the dynamical LFRW-like evolution on the brane. The 6d Einstein term gives the possibility for a phase of geometric acceleration. Furthermore, a varying β could have interesting consequences for a cosmological self-tuning scenario following equation (29). However, note that the evolution in β would then leave an imprint on the cosmological evolution equations. These are amongst the open, interesting questions that this work puts forward and which we hope will be answered in the near future.

Acknowledgments

I would like to thank Christos Charmousis and Georgios Kofinas for a very enjoyable collaboration. A.P. is supported by the Marie Curie Intra-European Fellowship EIF-039189.

References

[1] A. Vilenkin, Phys. Rev. D **23** (1981) 852.

[2] J. W. Chen, M. A. Luty and E. Ponton, JHEP **0009** (2000) 012 [arXiv:hep-th/0003067]; S. M. Carroll and M. M. Guica, arXiv:hep-th/0302067; I. Navarro, JCAP **0309** (2003) 004 [arXiv:hep-th/0302129]; Y. Aghababaie, C. P. Burgess, S. L. Parameswaran and F. Quevedo, Nucl. Phys. B **680** (2004) 389 [arXiv:hep-th/0304256]; A. Kehagias, Phys. Lett. B **600** (2004) 133 [arXiv:hep-th/0406025]; S. Randjbar-Daemi and V. A. Rubakov, JHEP **0410** (2004) 054 [arXiv:hep-th/0407176]; H. M. Lee and A. Papazoglou, Nucl. Phys. B **705** (2005) 152 [arXiv:hep-th/0407208].

[3] J. M. Cline, J. Descheneau, M. Giovannini and J. Vinet, JHEP **0306** (2003) 048 [arXiv:hep-th/0304147].

[4] W. Israel, Nuovo Cim. B **44S10** (1966) 1 [Erratum-ibid. B **48**, 463 (1967 NUCIA,B44,1.1966)].

[5] R. P. Geroch and J. H. Traschen, Phys. Rev. D **36** (1987) 1017.

[6] M. Peloso, L. Sorbo and G. Tasinato, Phys. Rev. D **73** (2006) 104025 [arXiv:hep-th/0603026]; E. Papantonopoulos, A. Papazoglou and V. Zamarias, JHEP **0703** (2007) 002 [arXiv:hep-th/0611311];

T. Kobayashi and M. Minamitsuji, Phys. Rev. D **75** (2007) 104013 [arXiv:hep-th/0703029]. C. P. Burgess, D. Hoover and G. Tasinato, JHEP **0709** (2007) 124 [arXiv:0705.3212 [hep-th]].
[7] D. Lovelock, J. Math. Phys. **12** (1971) 498; B. Zumino, Phys. Rept. **137** (1986) 109.
[8] P. Bostock, R. Gregory, I. Navarro and J. Santiago, Phys. Rev. Lett. **92** (2004) 221601 [arXiv:hep-th/0311074].
[9] C. Charmousis and R. Zegers, JHEP **0508** (2005) 075 [arXiv:hep-th/0502170].
[10] G. Kofinas, Phys. Lett. B **633** (2006) 141 [arXiv:hep-th/0506035].
[11] E. Papantonopoulos and A. Papazoglou, JCAP **0507** (2005) 004 [arXiv:hep-th/0501112].
[12] C. Charmousis, G. Kofinas and A. Papazoglou, JCAP **1001** (2010) 022 [arXiv:0907.1640 [hep-th]].
[13] C. Charmousis and A. Papazoglou, JHEP **0807** (2008) 062 [arXiv:0804.2121 [hep-th]].
[14] B. Boisseau, C. Charmousis and B. Linet, Phys. Rev. D **55** (1997) 616 [arXiv:gr-qc/9607029].
[15] C. Bambi, M. Giannotti and F. L. Villante, Phys. Rev. D **71** (2005) 123524 [arXiv:astro-ph/0503502].
[16] R. Gregory, Nucl. Phys. B **467** (1996) 159 [arXiv:hep-th/9510202].
[17] G. R. Dvali, G. Gabadadze and M. Porrati, Phys. Lett. B **485** (2000) 208 [arXiv:hep-th/0005016]; C. Deffayet, Phys. Lett. B **502** (2001) 199 [arXiv:hep-th/0010186].
[18] P. Binetruy, C. Deffayet, U. Ellwanger and D. Langlois, Phys. Lett. B **477** (2000) 285 [arXiv:hep-th/9910219]; P. Bowcock, C. Charmousis and R. Gregory, Class. Quant. Grav. **17** (2000) 4745 [arXiv:hep-th/0007177].

First Mediterranean Conference on Classical and Quantum Gravity (MCCQG 2009) IOP Publishing
Journal of Physics: Conference Series **222** (2010) 012005 doi:10.1088/1742-6596/222/1/012005

Evaporation of large black holes in AdS

Jorge V Rocha

Centro Multidisciplinar de Astrofísica - CENTRA, Dept. de Física, Instituto Superior Técnico, Av. Rovisco Pais 1, 1049-001 Lisboa, Portugal

E-mail: jorge.v.rocha@ist.utl.pt

Abstract. The AdS/CFT correspondence offers a new perspective on the long-standing black hole information paradox. However, to be able to use the available gauge/gravity machinery one is forced to consider so-called 'large' black holes in AdS, and these objects are thermodynamically stable – they do not evaporate. We describe a simple toy model that allows large AdS black holes to decay, by coupling the emitted radiation to an external scalar field propagating in an auxiliary space. This effectively changes the properties of the boundary of AdS, making it partly absorbing. We demonstrate that the evaporation process never ceases by explicitly presenting (a) the transmission coefficient for a wave scattering from the bulk into auxiliary space and (b) the greybody factor for a black 3-brane in an AdS background. Therefore, the model provides an interesting framework to address the information paradox using AdS/CFT techniques.

1. Introduction

Despite many efforts over more than thirty years, the black hole information paradox still remains to be elucidated. This long-standing issue is directly related to foundational concepts in physics, namely *unitarity* and *locality*, and its resolution is expected to hold the key to a formulation of quantum gravity.

The famous information loss problem was posed by Hawking in 1976 [1] shortly after his revolutionary discovery that black holes radiate [2]. Furthermore, according to Hawking's original computation the nature of the emitted radiation is that of a featureless blackbody spectrum [2]. It was soon realized that information seemed to be lost in the course of formation and consequent evaporation of black holes. This becomes evident when one considers the formation of a black hole by gravitational collapse of an initial matter distribution in a pure quantum state. Assuming a black hole does form and that it decays completely according to Hawking's law, the final state would be a thermal (i.e., maximally mixed) state, leading to an obvious violation of unitarity. It should be noted that one does not need a black hole to evaporate completely to arrive at a paradox: information must start being recovered in the Hawking particles as soon as the black hole radiates half its initial entropy if the evolution is unitary [3]. So one just needs to start with a large enough black hole.

The AdS/CFT correspondence [4, 5, 6], holographically relating gravitational dynamics in an AdS background with a gauge theory living on the boundary of AdS, offers a new perspective on this old problem. Shortly after the discovery of this duality it was pointed out that the manifest unitarity of the dual field theory should prevent any information loss of occurring [7]. Of course, this argument by itself does not resolve the paradox; one would still need to explain what went wrong in Hawking's argument. But the power of the gauge/gravity duality may well shed some light and provide indications for its resolution.

In order to take advantage of the AdS/CFT correspondence we should consider a black hole in AdS. These objects come in two classes: *small* and *large*. As in the asymptotically flat background case, small AdS black holes are thermodynamically unstable (they have negative specific heat) and do not have a clear interpretation in terms of the dual CFT [8, 9]. On the other hand, large AdS black holes have positive specific heat and are classically stable [10]. These objects are dual to a high temperature thermal state in the CFT [8]. To make progress exploiting the AdS/CFT correspondence we should then consider a large black hole in AdS. However, as it stands there is a clear obstacle to the formulation of the information paradox using large AdS black holes: these objects are stable and they do not evaporate!

The main purpose of the evaporon model [11, 12] is to provide a *simple* setting in which large black holes in AdS can decay, allowing to monitor the evaporation process employing the AdS/CFT duality and ultimately addressing the information paradox.

2. The evaporon model

The thermodynamical stability of large AdS black holes can be traced back to the total reflectivity of the time-like boundary of AdS. All the radiation emitted by the black hole is reflected at the boundary and is reabsorbed before the black hole evaporates. The system approaches an equilibrium state consisting of a black hole with temperature T_H and the surrounding Hawking radiation at the same temperature. The rate at which the black hole emits is compensated by the rate of absorption.

In order to allow large AdS black holes to evaporate, a simple model in the context of the AdS_5/CFT_4 correspondence was constructed in [11]. Recall that the gauge/gravity correspondence relates the CFT generating functional for a gauge invariant single trace operator $\mathcal{O}$ to extrema of the type IIB supergravity action with prescribed boundary conditions for the bulk field Φ dual to the operator $\mathcal{O}$. The non-normalizable solution for the supergravity field determines the source term for the dual operator on the CFT side. In the proposed model, we work at the level of the supergravity action and in the large N limit in which all the bulk fields decouple so we can consider Φ independently.

- First, one attaches a $(1+1)$-dimensional flat spacetime $H = \{(t,z)|\, z \geq 0\}$ to the boundary of AdS_5 and inserts there a scalar field $\sigma(t,z)$ with a standard kinetic term. This field was dubbed the *evaporon*.
- Next, we add an interaction term that couples the evaporon to the supergravity field Φ (hereafter referred to as the dilaton, for simplicity). The interaction is localized on the boundary of AdS_5 (which is 'identified' with the boundary of the additional space H). This is a four-dimensional space parametrized by the timelike coordinate t and the non-holographic spatial coordinates $\mathbf{x}$. Typically, to regulate the computations one cuts off the AdS space[1] at a holographic coordinate $y = \varepsilon$ and the limit $\varepsilon \to 0$ is taken in the end. The interaction is chosen to take place at $y = \varepsilon$ and in such a way as to enforce fall-off conditions for the dilaton in accordance with the non-normalizable solution for a massless scalar field in AdS_5.
- Finally, one is required to add specific counterterms in the evaporon action to renormalize the theory. These counterterms were judicially computed in [11] and their inclusion is crucial to obtain finite results when one takes the cutoff $\varepsilon \to 0$.

[1] We shall adopt coordinates suited for a Poincaré patch, for which the AdS metric can be expressed as $ds^2 = R^2/y^2\left[-dt^2 + d\mathbf{x}^2 + dy^2\right]$. The boundary of AdS lies at $y = 0$.

According to the above discussion, our model is described by the following action for the evaporon-dilaton system (see reference [11] for more details):

$$S[\Phi,\sigma] = S_\Phi[\Phi] + S_\sigma[\sigma] + S_{int}[\Phi,\sigma] + S_{c.t.}[\sigma] , \tag{1}$$

$$S_\Phi[\Phi] = -\frac{1}{2\kappa^2}\int_{\mathrm{AdS}} dt\, d^3\mathbf{x}\, dy\, \sqrt{-g}\frac{1}{2}g^{ab}\partial_a\Phi\partial_b\Phi ,$$

$$S_\sigma[\sigma] = -\int_H dt\, dz\, \frac{1}{2}\left[-(\partial_t\sigma)^2 + (\partial_z\sigma)^2\right] ,$$

$$S_{int}[\Phi,\sigma] = \lambda\int_{\partial\mathrm{AdS}} dt\, d^3\mathbf{x}\, \sqrt{-h}\, \Phi(t,\mathbf{x},\varepsilon)\, \sigma(t,0) ,$$

where g_{ab} denotes the bulk metric, g is its determinant and h represents the determinant of the metric induced on the interaction surface $\partial\mathrm{AdS} \equiv \{(t,\mathbf{x},y)|\, y=\varepsilon\}$. Here, λ plays the role of a coupling constant and κ^2 is related to the 5-dimensional Newton constant G_5 through $\kappa^2 = 8\pi G_5$. The necessary counterterms mentioned previously, which are also localized on the interaction surface $\partial H = \{(t,z)|\, z=0\}$, turn out to be

$$S_{c.t.}[\sigma] = \kappa^2\lambda^2 R^5 V_3 \int_{\partial H} dt \left[-\frac{1}{4\varepsilon^4}\sigma^2 - \frac{1}{24\varepsilon^2}\dot\sigma^2 + \frac{1}{128}\ln\left(\frac{\varepsilon^2}{\mu^2}\right)\ddot\sigma^2\right] , \tag{2}$$

where $V_3 = \int d^3\mathbf{x}$ is the volume of the transverse space and μ is an IR cutoff needed to regulate the expression for the counterterm. Had we been considering AdS in global coordinates, the transverse space would have been S^3 and presumably the cutoff would be related to the AdS scale R.

2.1. The transmission coefficient

The equations of motion for the evaporon-dilaton system can be easily derived from the action (1), supplemented with the counterterms (2):

$$y^2\partial_y^2\Phi - 3y\partial_y\Phi - y^2\partial_t^2\Phi = -2\kappa^2\lambda\, R\,\varepsilon\,\delta(y-\varepsilon)\,\sigma(t,0) , \tag{3}$$

$$-\partial_t^2\sigma + \partial_z^2\sigma = -\delta(z)\lambda V_3\, R^4\varepsilon^{-4}\Phi(t,\varepsilon) + \delta(z)\, 2\kappa^2\lambda^2 R^5 V_3 \left[\frac{\sigma}{4\varepsilon^4} - \frac{\partial_t^2\sigma}{24\varepsilon^2} - \frac{\partial_t^4\sigma}{128}\ln\left(\frac{\varepsilon^2}{\mu^2}\right)\right] . \tag{4}$$

By construction, the evaporon only couples to dilaton zero-modes in transverse space, i.e., Φ is independent of the coordinates $\mathbf{x}$ in the above equations. This is one of the simple features of the model.

The solutions of equations (3) and (4) can be immediately written. Imposing normalizability restricts the form of the dilaton wavefunction in the region near the boundary ($y < \varepsilon$), whereas the solutions for the evaporon are simply plane waves:

$$\Phi(t,y) = e^{-i\omega t}\left[\beta\, y^2 J_2(\omega y) + \gamma\, y^2 Y_2(\omega y)\right] + h.c. \qquad \text{for } y > \varepsilon , \tag{5}$$

$$\Phi(t,y) = \alpha\, e^{-i\omega t} y^2 J_2(\omega y) + h.c. \qquad \text{for } y < \varepsilon , \tag{6}$$

$$\sigma(t,z) = Ae^{-i\omega(t+z)} + Be^{-i\omega(t-z)} + h.c. \, . \tag{7}$$

The problem is now reduced to the scattering of waves in one dimension. Assuming an outgoing dilaton wave is incident on the interaction surface and $A = 0$, i.e., there is no evaporon wave coming from infinity and therefore it is purely outgoing, what fraction of energy is transmitted to the evaporon in the auxiliary space? The continuity of the dilaton at $y = \varepsilon$ and the discontinuity of the first derivatives of Φ and σ (determined by the right hand sides of equations (3) and (4)) can be used to relate the coefficients α, β, γ and B. The final result of this calculation yields the following formula for the transmission coefficient:

$$|\mathcal{T}|^2 = \frac{4\,\mathrm{Im}(\beta\gamma^*)}{|\beta|^2 + |\gamma|^2 + 2\,\mathrm{Im}(\beta\gamma^*)} = \frac{2}{1 + \frac{1}{4\pi}\left(\frac{\omega_4}{\omega}\right)^3 + \pi\left(\frac{\omega}{\omega_4}\right)^3\left[1 + \frac{4}{\pi^2}(\ln(\mu\,\omega))^2\right]}\,, \tag{8}$$

where we have defined an effective coupling constant $\omega_4 \equiv 8(2\lambda^2\kappa^2 R^5 V_3)^{-1/3}$. The result (see Figure 1) depends logarithmically on the scale μ but all the dependence on the cutoff ε has canceled out, as necessary, after having taken the limit $\varepsilon \to 0$. Nevertheless, for this to be possible it was vital to include the counterterms in the action: the transmission coefficient would not be finite without their addition.

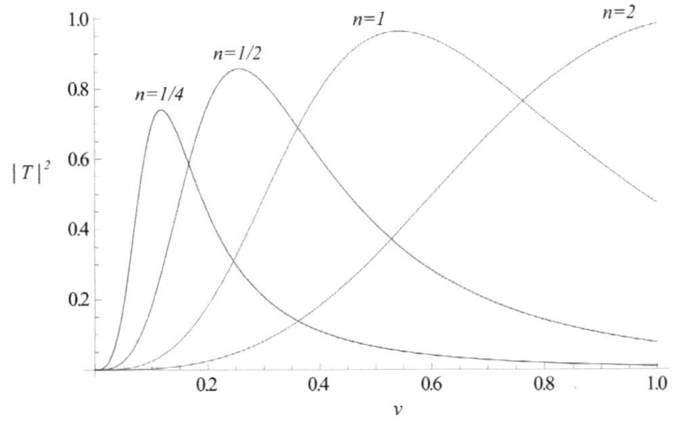

Figure 1. The transmission coefficient as a function of the frequency in units of the IR cutoff scale $\nu \equiv \mu\,\omega$. The result is shown for four different values of the effective coupling constant, equally in units of the IR cutoff, $n \equiv \mu\,\omega_4$.

3. Large black holes in AdS and their evaporation

The evaporon model, and in particular the previous computation of the transmission coefficient, were presented in the planar limit of AdS, in which the boundary becomes $\mathbb{R} \times \mathbb{R}^3$. To preserve the transverse translational symmetry it is natural to consider a black hole possessing a horizon with three flat directions, i.e., a black 3-brane in AdS_5. In the following subsection we shall outline the computation of the greybody factors for this geometry, originally performed in [12], to which the reader is referred for more details. An adaptation of the techniques used in [13] can be employed to derive an expression for the greybody factor in the low frequency regime.

The metric describing a black 3-brane with $\mathrm{AdS}_5 \times S^5$ asymptotics was found in [14] as the decoupling limit of a stack of D3-branes:

$$ds^2 = H^{-1/2}(r)\left[-f(r)dt^2 + d\mathbf{x}^2\right] + H^{1/2}(r)\left[f^{-1}(r)dr^2 + r^2 d\Omega_5^2\right]\,, \tag{9}$$

where

$$H(r) = \left(\frac{R}{r}\right)^4 \quad , \quad f(r) = 1 - \left(\frac{r_H}{r}\right)^4 . \tag{10}$$

This is a solution of type IIB string theory in 10 dimensions of the form $X \times S^5$, where the five-sphere has constant radius R. The five dimensional manifold X features a regular horizon at $r = r_H$ and in the limit $r \to \infty$ it approaches the metric of AdS_5 in Poincaré coordinates[2]. Since the S^5 plays no role in the following we shall drop it from now on.

[2] The relation between the radial coordinate r and the previously used coordinate y is simply $y = R^2/r$.

The Hawking temperature T_H, as well as the energy and entropy per unit 3-volume of the black 3-brane are given by

$$T_H = \frac{r_H}{\pi R^2} , \qquad \frac{E}{V_3} = \frac{3r_H^4}{2\kappa^2 R^5} , \qquad \frac{S}{V_3} = \frac{2\pi r_H^3}{\kappa^2 R^3} . \tag{11}$$

These black objects have positive specific heat [10] and therefore do not evaporate if the boundary of AdS is totally reflective. In this sense, these black branes fall in the category of the *large* black holes in AdS.

3.1. *The greybody factor*

Recall that greybody factors correct the otherwise perfect blackbody emission spectrum of a black hole, by accounting for the fact that the Hawking radiation needs to traverse frequency-dependent potential barriers originated by the background geometry in order to reach asymptotic infinity.

Consider then a massless, minimally coupled scalar field Φ propagating in the geometry of the black brane, obtained from the metric (9) by dropping the five-sphere. It is convenient to Fourier transform and study modes with definite frequency ω and wavevector $\mathbf{k}$ separately, so take $\Phi(t, \mathbf{x}, r) = e^{i\omega t - i\mathbf{k}\cdot\mathbf{x}}\phi_{\omega,\mathbf{k}}(r)$. The equation of motion for the dilaton wavefunction in the above geometry can be cast in Schrödinger-like form:

$$\left[\partial_x^2 + \omega^2 - V(r)\right]\left(r^{3/2}\phi\right) = 0 , \tag{12}$$

where the potential V is given by

$$V(r) = \frac{15}{4}\frac{r^2}{R^4} f(r)^2 + 6\frac{r_H^4}{r^2 R^4} f(r) + m^2\omega^2 f(r) , \tag{13}$$

m is defined by $\mathbf{k}^2 = m^2\omega^2$ (taking values between 0 and 1) and the tortoise coordinate x is determined (up to an additive constant) by $dx = H^{1/2}(r) f^{-1}(r) dr$.

Focusing on the near-horizon region ($r \simeq r_H$ and $V(r) \ll \omega^2$) the solution reduces to

$$\phi(r) = Ae^{i\omega x} , \tag{14}$$

where we chose the linear combination that is purely ingoing at the horizon. In the asymptotic region the solutions may be expressed in terms of Hankel functions of degree two:

$$\phi(r) = C_1(1-m^2)\frac{\omega^2 R^4}{r^2} H_2^{(1)}\left(\sqrt{1-m^2}\frac{\omega R^2}{r}\right) + C_2(1-m^2)\frac{\omega^2 R^4}{r^2} H_2^{(2)}\left(\sqrt{1-m^2}\frac{\omega R^2}{r}\right) . \tag{15}$$

The first and second terms are associated with the outgoing and ingoing parts of the wavefunction ϕ, respectively.

The flux per unit physical transverse area near the horizon can be shown to be

$$J_{hor} = \frac{r_H^3}{R^3}\omega|A|^2 , \tag{16}$$

and the asymptotic flux per unit physical area may be decomposed as $J_{asy} = J_{in} - J_{out}$, where

$$J_{in} = \frac{2}{\pi}(1-m^2)^2\omega^4 R^3 |C_2|^2 \quad , \quad J_{out} = \frac{2}{\pi}(1-m^2)^2\omega^4 R^3 |C_1|^2 . \tag{17}$$

The greybody factor is then defined [13] by the ratio of the (ingoing) flux at the horizon by the ingoing part of the asymptotic flux:

$$\Gamma_m(\omega) \equiv \frac{J_{hor}}{J_{in}} \ . \tag{18}$$

When the frequency is small compared to the temperature of the black brane $\omega \ll T_H$ an analytic expression can be obtained for the greybody factor by matching solutions valid in overlapping regions. The result obtained in [12] was

$$\Gamma_m(\omega) \simeq \frac{\pi}{2}(1-m^2)^2\widetilde{\omega}^3 \ , \qquad \text{for} \quad \widetilde{\omega} \ll 1/\pi \ . \tag{19}$$

Here we have used a dimensionless combination of the frequency, the AdS scale and the horizon radius:

$$\widetilde{\omega} \equiv \omega \frac{R^2}{r_H} \ . \tag{20}$$

Besides m, this is the *only* other dimensionless parameter that enters the equation of motion.

A global analytic solution of equation (12) interpolating between the horizon and the boundary of AdS appears to be out of reach but one can proceed resorting to numerical methods. Essentially, one imposes the ingoing boundary condition at the horizon and then integrates the equation of motion all the way out to the boundary. We know that the solution has the behavior displayed in equations (14) and (15) when $r \to r_H$ and $r \to \infty$, respectively. To compute the greybody factor we just need to express the constants C_1 and C_2 characterizing the asymptotic solution in terms of the constant A which determines the solution near the horizon and employ expression (18).

The results are presented in figures 2 and 3 for three values of m. The greybody factor smoothly interpolates between 0 at low frequencies and its expected asymptotic value of 1 at high energies. In the low frequency regime the numerical results are in very good agreement with the analytic approximation (19). Reference [12] also analyzed the high frequency regime and, although not completely conclusive, there are clear indications that the greybody factor approaches the asymptotic value in a power-law fashion, specifically $1 - \Gamma_0(\widetilde{\omega}) \sim \widetilde{\omega}^{-8}$ for large frequencies (and $m = 0$). The dependence on m was also considered: the greybody factor flattens out for larger m and for $m = 1$ it vanishes identically, since this situation corresponds to propagation of the wave parallel to the brane.

3.2. The decay rate

Once the transmission coefficient and the greybody factor are known one can proceed to determine the rate of evaporation of the considered black brane in the context of the evaporon model. The asymptotic flux of evaporons per unit frequency interval is weighted by three terms: the blackbody spectrum, $\left[e^{\omega/T_H} - 1\right]^{-1}$, the greybody factor, $\Gamma(\omega)$, and the transmission coefficient at the dilaton-evaporon interface, $|\mathcal{T}|^2$. The energy emitted per unit time is given by

$$\frac{dE}{dt} = \int \frac{d\omega}{2\pi} \frac{\omega \, \Gamma_0(\omega)}{e^{\omega/T_H} - 1} |\mathcal{T}|^2(\omega) \ . \tag{21}$$

Note that, *by construction*, the evaporon only couples to the dilaton mode independent of the transverse directions $\mathbf{x}$. Thus, for this calculation only the greybody factor for $m = 0$ is needed.

The decay rate as written in (21) is the overlap of two functions: $|\mathcal{T}|^2$, which peaks roughly at $\omega \approx \omega_4$, and the remainder, peaking roughly at $\omega \approx T_H$. As the black hole decays the temperature will decrease and so one of the peaks will shift, whereas the other peak remains

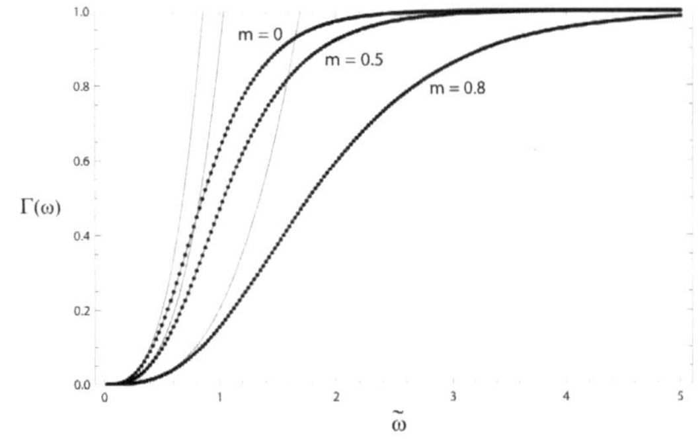

Figure 2. The greybody factor as a function of the dimensionless frequency $\widetilde{\omega} = \omega R^2/r_H$ for three values of m. The dots correspond to the numerical results and interpolating lines have been added to guide the eye. The thin red lines represent to the analytic results valid in the limit of low frequencies.

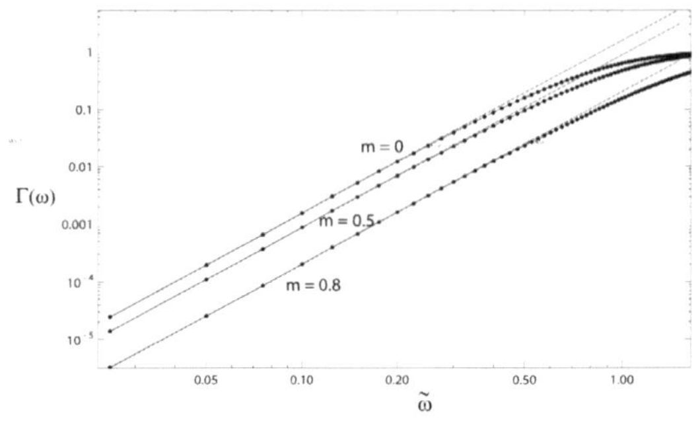

Figure 3. A close-up of the greybody factor around small $\widetilde{\omega}$ is displayed in a log-log plot. The top curves correspond to $m = 0$, the middle ones to $m = 0.5$ and the lower ones to $m = 0.8$. The agreement between the numerical result and the analytic approximation valid in the regime low frequencies is manifest.

fixed at some value dependent on the choice of the coupling constant λ. Nevertheless, the overlap of the two functions is always *finite* and therefore the evaporation proceeds indefinitely.

Performing the integration (21) that would yield the *exact* rate of evaporation is not an easy task for two reasons: first, the greybody factor is only know numerically and for a limited range of frequencies and, secondly, due to our ignorance regarding the IR cutoff μ. However, we can make progress by obtaining a lower bound on the decay rate. This can be accomplished by truncating the domain of integration at a maximum frequency, ω_{max}. In particular, if $\widetilde{\omega}_{max} \ll 1, \widetilde{\omega}_4$ we can use the expression (19) valid at low frequencies and simultaneously the term in the transmission coefficient containing μ can be neglected[3]. One finds that

$$\frac{dE}{dt} > \frac{\pi r_H^5}{\omega_4^3 R^{10}} \int_0^{\widetilde{\omega}_{max}} d\widetilde{\omega} \frac{\widetilde{\omega}^7}{e^{\pi\widetilde{\omega}} - 1} \,. \tag{22}$$

This integral can be evaluated exacly and it is given in terms of polylogarithms. For $\widetilde{\omega}_{max} \simeq 0.1$ its numerical value is 4×10^{-9}. In any event, the lower bound (22) allows us to obtain an upper limit for the evaporation time. For illustration purposes, using the relations (11) with $V_3 = \mathrm{Vol}(S^3)R^3 = 2\pi^2 R^3$, we find that the time it takes for the horizon length to drop from $r_H = \infty$ down to $r_H = R$ is bounded by $\Delta t_{evap} \lesssim 10^{10} \omega_4^3 R^7 \kappa^{-2}$.

4. Conclusions

In conclusion, we have presented a toy model in which large black holes in AdS are allowed to evaporate in finite time. This was accomplished by effectively relaxing the boundary conditions on an AdS bulk field, coupling it to an external scalar field living in an auxiliary (asymptotically flat) space.

The main virtue of the evaporon model is that it provides a framework in which one may address the information paradox with AdS/CFT tools and the evaporation of the black hole can be followed with the CFT dual frame in a controlled way. Indeed, from the dual field theory point of view the interpretation is clear: we are adding an extra sector to the CFT and the

[3] Here, we have changed variables from ω to $\widetilde{\omega}$ as given by the definition (20).

evaporation of a large black hole in AdS corresponds to the transfer of energy to this infinite external reservoir.

Thus the expectation, as advocated since the early days of the AdS/CFT correspondence, is that there is no information loss in the process of formation and evaporation of black holes. Nevertheless, the issue of how the paradox is resolved is still unanswered. An interesting possibility for further exploration of the evaporon model would be to analyze also the formation of a black hole by collapsing a shell of matter in this context. It is our hope that this model may be useful to develop intuition and perhaps pinpoint the elusive ingredient necessary to settle the black hole information paradox.

Acknowledgments
I would like to thank the participants and the organizers of the *First Mediterranean Conference on Classical and Quantum Gravity*, and specially Elias Vagenas. This work was partially supported by NSF grant PHY04-56556 and by Fundação para a Ciência e Tecnologia (FCT) - Portugal through project PTDC/FIS/64175/2006. I also acknowledge financial support from FCT fellowship SFRH/BPD/47332/2008.

References

[1] Hawking S W 1976 Breakdown of predictability in gravitational collapse *Phys. Rev.* D **14** 2460
[2] Hawking S W 1975 Particle creation by black holes *Commun. Math. Phys.* **43** 199
[3] Page D N 1993 Information in black hole radiation *Phys. Rev. Lett.* **71** 3743 (*Preprint* hep-th/9306083)
[4] Maldacena J M 1998 The large N limit of superconformal field theories and supergravity *Adv. Theor. Math. Phys.* **2** 231 (*Preprint* hep-th/9711200)
[5] Gubser S S, Klebanov I R and Polyakov A M 1998 Gauge theory correlators from non-critical string theory *Phys. Lett.* B **428** 105 (*Preprint* hep-th/9802109)
[6] Witten E 1998 Anti-de Sitter space and holography *Adv. Theor. Math. Phys.* **2** 253 (*Preprint* hep-th/9802150)
[7] Lowe D A and Thorlacius L 1999 AdS/CFT and the information paradox *Phys. Rev.* D **60** 104012 (*Preprint* hep-th/9903237)
[8] Witten E 1998 Anti-de Sitter space, thermal phase transition, and confinement in gauge theories *Adv. Theor. Math. Phys.* **2** 505 (*Preprint* hep-th/9803131)
[9] Horowitz G T and Hubeny V E 2000 Quasinormal modes of AdS black holes and the approach to thermal equilibrium *Phys. Rev.* D **62** 024027 (*Preprint* hep-th/9909056)
[10] Hawking S W and Page D N 1983 Thermodynamics of black holes in anti-de Sitter space *Commun. Math. Phys.* **87** 577
[11] Rocha J V 2008 Evaporation of large black holes in AdS: coupling to the evaporon *J. High Energy Phys.* JHEP08(2008)075 (*Preprint* 0804.0055 [hep-th])
[12] Rocha J V 2009 Evaporation of large black holes in AdS: greybody factor and decay rate *J. High Energy Phys.* JHEP08(2009)027 (*Preprint* 0905.4373 [hep-th])
[13] Harmark T, Natário J and Schiappa R 2007 Greybody Factors for d-Dimensional Black Holes (*Preprint* 0708.0017 [hep-th])
[14] Horowitz G T and Ross S F 1998 Possible resolution of black hole singularities from large N gauge theory *J. High Energy Phys.* JHEP04(1998)015 (*Preprint* hep-th/9803085)

Journal of Physics: Conference Series **222** (2010) 012006 doi:10.1088/1742-6596/222/1/012006

Domain walls with non-Abelian orientational moduli

Minoru Eto[a], **Toshiaki Fujimori**[b], **Muneto Nitta**[c], **Keisuke Ohashi**[d], **and Norisuke Sakai**[e]

[a] Theoretical Physics Laboratory, RIKEN, Saitama, 351-0198, Japan,
[b] Department of Physics, Tokyo Institute of Technology, Tokyo 152-8551, Japan,
[c] Department of Physics, and Research and Education center for Natural Sciences, Keio University, Hiyoshi 4-1-1, Yokohama, Kanagawa 223-8521, Japan,
[d] Department of Physics, Kyoto University, Kyoto 606-8502, Japan,
[e] Department of Mathematics, Tokyo Woman's Christian University, Tokyo 167-8585, Japan

E-mail: [a] `meto(at)riken.jp`; [b] `fujimori(at)th.phys.titech.ac.jp`;
[c] `nitta(at)phys-h.keio.ac.jp`; [d] `ohashi(at)gauge.scphys.kyoto-u.ac.jp`;
[e] `Speaker, sakai(at)lab.twcu.ac.jp`

Abstract. Domain walls with non-Abelian orientational moduli are constructed in $U(N)$ gauge theories coupled to Higgs scalar fields with degenerate masses. The associated global symmetry is broken by the domain walls, resulting in the Nambu-Goldstone (and quasi-Nambu-Goldstone) bosons, which form the non-Abelian orientational moduli. As walls separate, the wave functions of the non-Abelian orientational moduli spread between domain walls. By taking the limit of Higgs mass differences to vanish, we clarify the convertion of wall position moduli into the non-Abelian orientational moduli. The moduli space metric and its Kähler potential of the effective field theory on the domain walls are constructed. We consider two models: a $U(1)$ gauge theory with several charged Higgs fields, and a $U(N)$ gauge theory with $2N$ Higgs fields in the fundamental representation. More details are found in our paper[1].

1. BPS solitons

Solitons are useful to build unified models with extra dimensions, and to provide all or part of nonperturbative effects. If a global symmetry of the theory is spontaneously broken by the presence of solitons, Nambu-Goldstone (NG) bosons come out and form (a part of) the moduli space of the soliton. In the case of non-Abelian global symmetry, the resulting massless modes can have non-Abelian orientational moduli. Quite often solitons have parameters, which are called moduli. If we promote these parameters to fields on the world volume of the soliton, they become massless fields in the low-energy effective field theory on the soliton.

Simplest of solitons is the domain wall which depends only on one spatial dimension, namely co-dimension one soliton. In order to have a domain wall, we need to have discrete vacua. As the simplest theory with two discrete degenerate vacua, let us consider a $1+1$ dimensional theory of real scalar field φ with a double well potential

$$\mathcal{L} = \partial_\mu \varphi \partial^\mu \varphi - \lambda(\varphi^2 - v^2)^2, \qquad \lambda > 0. \tag{1}$$

If there is a field configuration connecting the two discrete degenerate vacua, $\varphi_+ \equiv v$ and $\varphi_- \equiv -v$, we obtain a domain wall separating two vacua, as a kink, whose energy density is localized, resulting in the domain wall. The nontrivial boundary condition at infinity assures the topological stability of the configuration: topological charge is characterized by $\pi_0(\mathcal{M})$. To obtain such a solution, let us assume a static configuration depending only on one spatial direction y, and form the following complete square of the energy density

$$\begin{aligned}\mathcal{E} &= (\partial_y\varphi)^2 + \lambda(\varphi^2 - v^2)^2 \\ &= (\partial_y\varphi + \sqrt{\lambda}(\varphi^2 - v^2))^2 + \partial_y\left[2\sqrt{\lambda}\left(v^2\varphi - \frac{\varphi^3}{3}\right)\right].\end{aligned} \tag{2}$$

This is called the Bogomol'nyi completion giving the lower bound of energy which is called the Bogomol'nyi bound

$$\int_{-\infty}^{\infty} dy\mathcal{E} \geq \left[2\sqrt{\lambda}\left(v^2\varphi - \frac{\varphi^3}{3}\right)\right]_{-\infty}^{\infty}. \tag{3}$$

The Bogomol'nyi bound is saturated if and only if the following first order differential equation is satisfied:

$$\partial_y\varphi + \sqrt{\lambda}(\varphi^2 - v^2) = 0, \tag{4}$$

which is called the Bogomol'nyi-Prasad-Sommerfield (BPS) equation for the domain wall. The BPS equation is easily solved to give the BPS domain wall solution

$$\varphi = v\tanh(\sqrt{\lambda}v(y - y_0)). \tag{5}$$

The BPS domain wall solution has a single modulus y_0, whose physical meaning is the position of the wall. This modulus can also be understood as a NG mode corresponding to the spontaneously broken translational symmetry. We can promote the real scalar to complex scalar, and add a fermion with appropriate interactions to make the system supersymmetric. Namely we can embed the theory (1) into a supersymmetric theory with four supercharges. The BPS solution preserves the half of the supersymmetry in this embeded theory. This is a typical example of the BPS solitons which can be understood as the BPS state preserving a part of the supersymmetry in a supersymmetric theory.

As another example of solitons, we can consider vortex in a theory with the Abelian gauge field W_μ coupled to a charged complex scalar field ϕ

$$\mathcal{L} = -\frac{1}{4e^2}F_{\mu\nu}F^{\mu\nu} + \mathcal{D}^\mu\phi(\mathcal{D}_\mu\phi)^\dagger - \frac{\lambda}{4}\left(\phi\phi^\dagger - v^2\right)^2, \tag{6}$$

where $\mathcal{D}_\mu\phi = (\partial_\mu + iW_\mu)\phi$ denotes the covariant derivative, and $F_{\mu\nu} = \partial_\mu W_\nu - \partial_\nu W_\mu$ the field strength. The mapping from the infinity in the x, y plane (S^1) to the vacuum manifold $|\phi| = v, S^1$ gives a topological charge $\pi_1(S^1)$ assuring the stability of the vortex configuration.

$$k = -\frac{1}{2\pi}\int d^2x\ F_{12} \tag{7}$$

The position of a single vortex is again a modulus, since the vortex can be formed anywhere in the x, y plane. If there are two or more vortices, the gauge field induces repulsive force between the vortices, and the scalar field induces attractive interactions[1]. When $e^2 = \lambda$, the two interactions cancel each other and there is no static force between vortices. Therefore vortices can be placed at anywhere relative to each other, and the relative positions become additional moduli. This critical case corresponds precisely to the case where the theory can be embedded into supersymmetric theory by adding fermions appropriately. The vortex solutions can then

[1] This applies to vortices with vorticity of the same sign. If the signs of vorticities are opposite, both gauge field interactions and scalar interactions become attractive.

be understood as the BPS solitons preserving the half of the supersymmetry. If there are internal global symmetry which are broken by the presence of solitons, the NG modes emerge. In particular, if the global symmetry is non-Abelian, we can have non-Abelian orientational moduli for the soliton. These non-Abelian orientational moduli exhibits interesting properties similar to the D-branes in string theory.

2. BPS equations for $U(N)$ gauge theories and the moduli matrix

We consider $U(N_{\rm C})$ gauge theory in space-time dimension $d = 4 + 1$ with a real scalar field Σ in the adjoint representation and $N_{\rm F}\,(> N_{\rm C})$ flavors of massive Higgs scalar fields in the fundamental representation, denoted as an $N_{\rm C} \times N_{\rm F}$ matrix H. Choosing the minimal kinetic term, we obtain

$$\mathcal{L} = \mathcal{L}_{\rm kin} - V, \tag{8}$$

$$\mathcal{L}_{\rm kin} = {\rm Tr}\left(-\frac{1}{2g^2}F_{\mu\nu}F^{\mu\nu} + \frac{1}{g^2}\mathcal{D}_\mu\Sigma\,\mathcal{D}^\mu\Sigma + \mathcal{D}^\mu H\,(\mathcal{D}_\mu H)^\dagger\right), \tag{9}$$

where the covariant derivatives and field strengths are defined as $\mathcal{D}_\mu\Sigma = \partial_\mu\Sigma + i[W_\mu,\Sigma]$, $\mathcal{D}_\mu H = (\partial_\mu + iW_\mu)H$, $F_{\mu\nu} = -i[\mathcal{D}_\mu, \mathcal{D}_\nu]$. Our convention for the space-time metric is $\eta_{\mu\nu} = {\rm diag}(+,-,\cdots,-)$. The scalar potential V is given in terms of a diagonal mass matrix M and a real parameter c as

$$V = {\rm Tr}\Big[\frac{g^2}{4}\left(c\mathbf{1} - HH^\dagger\right)^2 + (\Sigma H - HM)(\Sigma H - HM)^\dagger\Big]. \tag{10}$$

The 1/2 BPS equations for domain walls interpolating the discrete vacua can be obtained by usual Bogomol'nyi completion of the energy

$$E = \int_{-\infty}^{\infty} dy\,{\rm Tr}\left[(\mathcal{D}_y H - HM + \Sigma H)^2 + \frac{1}{g^2}\left(\mathcal{D}_y\Sigma - \frac{g^2}{2}\left(c\mathbf{1} - HH^\dagger\right)\right)^2 + c\,\mathcal{D}_y\Sigma\right] \tag{11}$$

$$\geq c\Big[{\rm Tr}\,\Sigma(\infty) - {\rm Tr}\,\Sigma(-\infty)\Big]. \tag{12}$$

The first order differential equations for the configurations saturating this energy bound are of the form [7]

$$\mathcal{D}_y H = HM - \Sigma H, \qquad \mathcal{D}_y\Sigma = \frac{g^2}{2}\left(c\mathbf{1} - HH^\dagger\right). \tag{13}$$

Here we consider static configurations depending only on the y-direction.

Let us solve these 1/2 BPS equations. Firstly the first equation can be solved by [7]

$$H = S^{-1}(y)H_0 e^{My}, \quad \Sigma + iW_y = S^{-1}(y)\partial_y S(y). \tag{14}$$

Here H_0, called the moduli matrix, is an $N_{\rm C} \times N_{\rm F}$ constant complex matrix of rank $N_{\rm C}$, and contains all the moduli parameters of solutions. The matrix valued quantity $S(y) \in GL(N_{\rm C}, \mathbf{C})$ is determined by the second equation in (13) which can be converted to the following equation for $\Omega \equiv SS^\dagger$:

$$\frac{1}{cg^2}[\partial_y(\Omega^{-1}\partial_y\Omega)] = \mathbf{1}_{N_{\rm C}} - \Omega^{-1}\Omega_0, \qquad \Omega_0 \equiv \frac{1}{c}H_0 e^{2My}H_0^\dagger. \tag{15}$$

This equation is called the master equation for domain walls. From the vacuum conditions at spatial infinities $y \to \pm\infty$, we can see that the solution Ω of the master equation should satisfy the boundary condition $\Omega \to \Omega_0$ as $y \to \pm\infty$. It determines S for a given moduli matrix H_0 up to the gauge transformations $S^{-1} \to US^{-1}$, $U \in U(N_{\rm C})$ and then the physical fields can be obtained through (14). Note that the master equation is symmetric under the following V-transformations

$$H_0 \to VH_0 \quad \text{and} \quad S(y) \to VS(y) \quad \text{with} \quad V \in GL(N_{\rm C}, \mathbf{C}), \tag{16}$$

and if the moduli matrices are related by the V-transformations $H_0' = VH_0$, they give physically equivalent configurations. We call this equivalence relation as the V-equivalence relation and denote it as $H_0 \sim VH_0$. The solution of the master equation exists and unique for any given H_0 at least for the $U(1)$ gauge theory [5]. For $U(N)$ gauge theory the number of moduli parameters agrees with the the result of index theorem [6].

3. Non-Abelian orientational moduli of walls in a $U(1)$ gauge theory

So far, domain walls with eight supercharges have been mostly considered in gauge theories with $U(1)$ gauge field [2]–[5], $U(1) \times U(1)$ gauge field [11], or $U(N)$ gauge fields [7]–[10] coupled to Higgs scalar fields with *non-degenerate* masses except for [12, 13]. In the case of non-degenerate Higgs masses, the flavor symmetry is Abelian: $U(1)^{N_{\rm F}-1}$ and the symmetry of the vacua is also Abelian. As a result each domain wall carries a $U(1)$ orientational modulus [2, 7]; The moduli space of a single domain wall is

$$\mathcal{W}^{k=1} \simeq \mathbf{R} \times S^1. \tag{17}$$

From this viewpoint, these domain walls should be called Abelian domain walls even when the gauge symmetry of the Lagrangian is non-Abelian [7]–[11].

Let us see the non-Abelian orientational moduli in a simple example of the Abelian gauge theory coupled with the $N_{\rm F} = 4$ Higgs fields.

3.1. Vacua of the $U(1)$ gauge theory

The massless vacuum manifold is $T^\star\mathbf{C}P^3$ where the base manifold is parametrized by

$$\mathbf{C}P^3 = \left\{HH^\dagger = c\right\}/U(1), \quad H = \sqrt{c}\,(h_1,\ h_2,\ h_3,\ h_4), \tag{18}$$

where the quotient is the overall $U(1)$. The vacuum manifold is expressed as (the inside and the surface of) a triangular pyramid in the 3 dimensional space $(|h_1|^2, |h_2|^2, |h_3|^3)$, as shown in Fig. 1 (c). When the mass matrix containing a small parameter ϵ $(0 \le \epsilon \in \mathbf{R})$

$$M = \mathrm{diag}\left(m,\ \frac{m\epsilon}{2},\ -\frac{m\epsilon}{2},\ -m\right) \tag{19}$$

is turned on, the vacuum manifold is lifted except for four points and the flavor symmetry breaks from $SU(4)$ to $U(1)^3$. These discrete vacua are the four vertices of the pyramid shown in Fig. 1 (a). We label those vacua as $\langle A\rangle$ $(A = 1, 2, 3, 4)$. The vacuum expectation value (VEV) of the vacuum $\langle A\rangle$ is $h_B = \delta_{AB}$. Taking a limit of $\epsilon \to 0$, the second and the third Higgs fields become degenerate so that the flavor symmetry enhances from $U(1)^3$ to $U(1)^2 \times SU(2) \in SU(4)$. There are two isolated vacua and one degenerate vacuum $\mathbf{C}P^1 \simeq SU(2)/U(1)$ represented by a line connecting $\langle 2\rangle$ and $\langle 3\rangle$ as shown by a thick line in Fig.1(b). We denote this degenerate vacuum as $\langle$2-3$\rangle$.

3.2. Domain walls in the $U(1)$ gauge theory

There exist domain wall solutions interpolating vacua in the model with fully or partially non-degenerate Higgs masses. In the case of $N_{\rm C} = 1$, the moduli matrix and the V-equivalence (16) take the form of

$$H_0 = (\phi_1,\ \phi_2,\ \phi_3,\ \phi_4) \sim \lambda\,(\phi_1,\ \phi_2,\ \phi_3,\ \phi_4)\,, \quad \lambda \in \mathbf{C}^*. \tag{20}$$

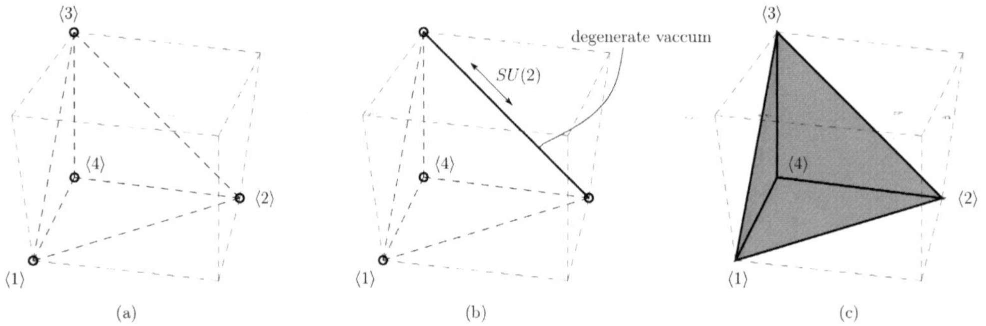

Figure 1. Vacua for various cases of mass configurations plotted in the three-dimensional space of Higgs fields $h_i^2, i = 1,2,3$ with $\sum_{i=1}^4 h_i^2 = 1$. (a) non-degenerate massive vacua (b) massive degenerate and non-degenerate vacua (c) massless vacuum.

In terms of the moduli matrix the vacua $\langle A \rangle$ are described by $\phi_B = \delta_{BA}$ for $B = 1,2,3,4$. Since we want to consider the domain wall interpolating the vacua $\langle 1 \rangle$ and $\langle 4 \rangle$ (passing by $\langle 2 \rangle, \langle 3 \rangle$ on the way), the parameter ϕ_1 and ϕ_4 should not be zero while ϕ_2, ϕ_3 can become zero. So the moduli space corresponding to the multiple domain walls which connect $\langle 1 \rangle$ and $\langle 4 \rangle$ is

$$\mathcal{M} \simeq \left(\mathbf{C}^2 \times (\mathbf{C}^*)^2\right) // \mathbf{C}^* \simeq \mathbf{C}^* \times \mathbf{C}^2, \tag{21}$$

where double slash denotes identification by the V-transformation. Here the part $\mathbf{C}^* \simeq \mathbf{R} \times U(1)$ represents the translational modulus and the associated phase modulus.

When we take the gauge coupling g to infinity, the model reduces to a nonlinear sigma model whose target space is the Higgs branch of the vacua $T^*\mathbf{C}P^3$ in the original theory. To make the discussion simple, we take this limit for a while. One benefit to consider the nonlinear sigma model is that the BPS equations can be analytically solved. In fact the solutions are expressed as [8]

$$H = \frac{1}{\sqrt{\Omega_0}} H_0 e^{My} \quad \text{with} \quad \Omega_0 \equiv H_0 e^{2My} H_0^\dagger. \tag{22}$$

A domain wall solution corresponds to a trajectory connecting the vertex $\langle 1 \rangle$ and $\langle 4 \rangle$. Flows from $\langle 1 \rangle$ to $\langle 4 \rangle$ inside the pyramid are shown in Fig. 2.

Physical meaning of the moduli parameters becomes much clearer by using the V-equivalence relation (20) to fix the form of the moduli matrix as

$$H_0 = \left(1,\ e^{\varphi_1},\ e^{\varphi_1+\varphi_2},\ e^{\varphi_1+\varphi_2+\varphi_3}\right). \tag{23}$$

Furthermore, one may be visually able to see the "kink" configuration in the profile of the field $\Sigma = (1/2)\partial_y \log \Omega_0$. In the vacuum region $\langle A \rangle$ the function $\Sigma(y)$ takes the value $\Sigma = m_A$. Several solutions are shown in Fig. 3. The domain wall positions can be roughly read from the moduli matrix in Eq. (23) as

$$\frac{y_+}{L_+} = \varphi_1 + \varphi_1^*, \quad \frac{y_0}{L_0} = \varphi_2 + \varphi_2^*, \quad \frac{y_-}{L_-} = \varphi_3 + \varphi_3^*, \tag{24}$$

where y_+ is the position of the right wall and y_0, y_- are the positions for the middle and the left walls, respectively. Here $L_{\pm,0}$ stands for the width of each wall

$$L_+ \equiv \frac{2}{m(2-\epsilon)}, \quad L_0 \equiv \frac{1}{m\epsilon}, \quad L_- \equiv \frac{2}{m(2-\epsilon)}. \tag{25}$$

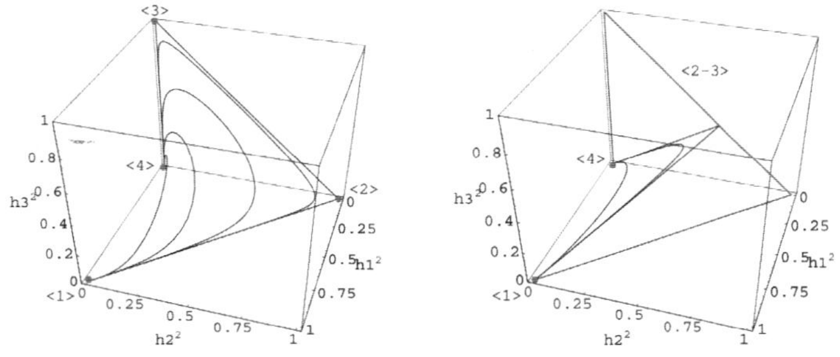

(a) non-degenerate mass (b) degenerate mass

Figure 2. Domain wall trajectories in the target space $\mathbf{C}P^3$ for non-degenerate mass (a) and for degenerate mass (b).

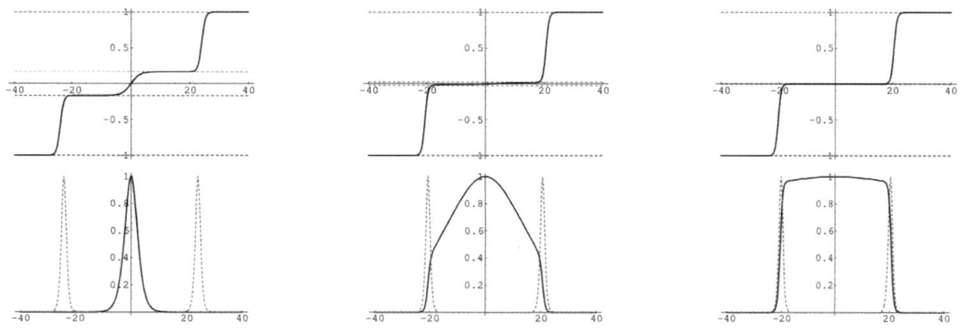

(a) $\epsilon = 1/3$ (b) $\epsilon = 1/20$ (c) $\epsilon = 1/100$

Figure 3. Configuration of Σ (first row) and density of the Kähler metric of φ_1, φ_2 and φ_3 (second row). Moduli parameters are $(\varphi_1, \varphi_2, \varphi_3) = (20, 0, -20)$ and $m = 1$.

This rough estimation is, of course, valid only for well separated walls whose positions are aligned as $y_- \ll y_0 \ll y_+$, see Fig. 3 (a). Each domain wall is accompanied by a complex moduli parameter φ_i whose real part is related to the wall position and imaginary part is the $U(1)$ internal symmetry (the NG mode associated with the broken $U(1)$ flavor symmetry).

To argue symmetry aspects of the moduli parameters, first let us consider a model which has completely non-degenerate masses and domain walls interpolating between those vacua. The global symmetry explicitly breaks from $SU(4)$ to $U(1)^3 \subset SU(4)$. We take, as the unbroken global symmetries, $U_1(1), U_2(1)$ and $U_3(1)$ with generators $\mathrm{diag}\,(1,-1,-1,1)$, $\mathrm{diag}\,(1,0,0,-1)$, and $\mathrm{diag}\,(0,1,-1,0)$ respectively. Each vacuum $\langle A\rangle$ preserves all of these symmetries. However, once domain walls connecting those vacua appear, they break all or a part of these symmetries. For example, the moduli matrix $H_0 = (1, 0, 0, \phi_4)$ corresponding to a domain wall connecting two vacua $\langle 1\rangle$ and $\langle 4\rangle$ breaks $U_2(1)$ but still preserves $U_1(1)$ and $U_3(1)$. Here note that overall phase can be absorbed by the V-transformation (20). Therefore the phase of the moduli parameter ϕ_4 corresponds to nothing but the broken global symmetry $U_2(1)$. This implies the NG mode localizes around the domain wall as we saw above. For the moduli matrix $H_0 = (1, \phi_2, 0, \phi_4)$, which corresponds to two domain walls connecting three vacua $\langle 1\rangle \to \langle 2\rangle \to \langle 4\rangle$, the symmetry $U_3(1)$ in addition to $U_2(1)$ breaks while a combination of $U_1(1)$ and $U_3(1)$ is still preserved. Moreover, when we turn on the third element in the moduli matrix as $H_0 = (1, \phi_2, \phi_3, \phi_4)$, the third vacuum region appears and then the configuration has three domain walls connecting four

Journal of Physics: Conference Series **222** (2010) 012006 doi:10.1088/1742-6596/222/1/012006

vacua $\langle 1\rangle \to \langle 2\rangle \to \langle 3\rangle \to \langle 4\rangle$. In this case all of $U(1)^3$ are broken by the domain walls, so that corresponding three NG modes appear. These three NG modes are described by imaginary parts of $\log\phi$, which are combined with the three positions (24), to form three complex coordinates of the moduli space $\mathbf{C}^2 \times \mathbf{C}^*$.

Next we consider a limit where the second and the third masses are degenerate ($\epsilon \to 0$ in the mass matrix (19)). In the limit the global symmetry $U_1(1) \times U_2(1) \times U_3(1)$ is enhanced to $U_1(1) \times U_2(1) \times SU(2)$. At the same time, the degenerate vacuum $\langle$2-3$\rangle$ appear instead of the two isolated vacua $\langle 2\rangle$ and $\langle 3\rangle$ as shown in Fig. 1 (b). At the degenerate vacuum, $U_1(1), U_2(1)$ are preserved but $SU(2)$ is broken to $U_3(1)$. Therefore the degenerate vacuum $\langle 2{-}3\rangle$ is $SU(2)/U_3(1) = \mathbf{C}P^1$. Non-vanishing $\phi_4 \neq 0$ causes the wall interpolating two vacua $\langle 1\rangle \to \langle 4\rangle$ and breaks only $U_2(1)$ again. Once the degenerate vacuum appears in the configuration such as two domain walls connecting vacua like $\langle 1\rangle \to \langle 2{-}3\rangle \to \langle 4\rangle$, the breaking pattern of the global symmetry becomes different from that in the case of fully non-degenerate masses. The moduli matrix $H_0 = (1, \phi_2, \phi_3, \phi_4)$ describes such domain walls. Note that the second and the third elements break $SU(2)$ completely. The global symmetry $U_1(1) \times U_2(1) \times SU(2)$ are broken to $U(1)$ which is a mixture of $U_1(1)$ and $H \in SU(2)$. Emergence of the second wall and further $U(1)$-symmetry breaking are related to the facts that $|\phi_2|^2 + |\phi_3|^2 \neq 0$ and $\phi_4 \neq 0$. These facts imply that the modes corresponding to the two broken $U(1)$'s localize around the walls accompanied by the two position moduli and the mode corresponding to $SU(2)/H$ have support in a region around the degenerate vacuum $\langle 2{-}3\rangle$. We can count the number of the moduli parameters as follows. Two real parameters $\{|\phi_2|^2 + |\phi_3|^2, |\phi_4|^2\}$ correspond to the positions of the two walls whereas remaining four parameters correspond to the broken global symmetry $U_1(1) \times U_2(1) \times SU(2)/U(1)$. This is again consistent with $\dim_{\mathbf{R}}(\mathbf{C}^2 \times \mathbf{C}^*)$.

In the Fig. 3 we showed domain wall configurations of the three domain walls connecting the four vacua. As the parameter ϵ decreases, the width of the middle domain wall connecting the vacua $\langle 2\rangle$ and $\langle 3\rangle$ becomes broad and the tension of the wall becomes small since they are proportional to $1/\epsilon$ and ϵ, respectively. When the width of the middle wall becomes larger than the separation of two outside walls, $L_0 > y_+ - y_-$, we can no longer see the middle wall. The density of the Kähler metric for the moduli parameters φ_1, φ_2 and φ_3 in the strong gauge coupling limit are shown in the second row of Fig. 3. The Kähler potential in the strong coupling limit is given by $K = c\int dy\ \log\Omega_0$ [14]. When three walls are well isolated as Fig. 3 (a), three modes corresponding to the moduli parameters φ_1, φ_2 and φ_3 are localized on the respective domain walls. As ϵ decreases, the density of the Kähler metric of φ_2 is no longer localized but is stretched between two outside domain walls. In the limit where $\epsilon \to 0$ the physical meaning of φ_2 as the position and the internal phase associated with the middle domain wall should be completely discarded. Instead, φ_2 gives the non-Abelian orientational moduli which comes from the $\mathbf{C}P^1$ zero modes of the degenerate vacua $\langle 2{-}3\rangle$. For fixed moduli parameters $\varphi_1, \varphi_2, \varphi_3$, the domain wall solution as a function of y sweeps out a trajectory in the target space $\mathbf{C}P^3$. These domain wall trajectories are shown for various values of moduli parameters in Fig. 2: the non-degenerate mass case (a) and the degenerate mass case (b). For the degenerate mass case, the trajectories do not go out from the triangular plane whose vertices are $\langle 1\rangle, \langle 4\rangle$ and one point on the edge between $\langle 2\rangle$ and $\langle 3\rangle$.

4. The Generalized Shifman-Yung (GSY) Model

4.1. U(N) gauge theory

Let us now consider non-Abelian gauge theory with degenerate masses of the Higgs fields. Previously considered model is the $U(2)$ gauge theory with four Higgs fields in the fundamental representation with the mass matrix $M = \mathrm{diag}(m, m, -m, -m)$ [12, 13], which we call the Shifman-Yung Model. The model enjoys a flavor symmetry $SU(2)_{\mathrm{L}} \times SU(2)_{\mathrm{R}} \times U(1)_{\mathrm{A}}$. It has been demonstrated that the coincident domain wall configurations break the flavor symmetry to $SU(2)_{\rm V}$ and the NG bosons corresponding to $[SU(2)_{\rm L} \times SU(2)_{\rm R} \times U(1)_{\rm A}]/SU(2)_{\rm V} \simeq U(2)$ appear in the effective action on the walls.

By generalizing the Shifman-Yung model, we consider the $U(N)$ gauge theory with $N_{\rm F} = 2N$ Higgs fields in the fundamental representation whose mass matrix is given by

$$M = m\frac{\sigma_3}{2} \otimes \mathbf{1}_N = \frac{1}{2}{\rm diag}(\overbrace{m, \cdots, m}^{N}, \overbrace{-m, \cdots, -m}^{N}). \tag{26}$$

This system has a non-Abelian flavor symmetry $SU(N)_{\rm L} \times SU(N)_{\rm R} \times U(1)_{\rm A}$. Since we have only two mass parameters m and $-m$, possible vacua are classified by an integer $0 \leq k \leq N$: in the k-th vacua, there is a configuration in which k flavors of the first half and $N-k$ flavors of the latter half take non-vanishing values and then Σ and H are

$$\begin{aligned} \Sigma|_{\rm vacuum} &= \frac{1}{2}{\rm diag}(\overbrace{m, \cdots, m}^{k}, \overbrace{-m, \cdots, -m}^{N-k}), \\ H|_{\rm vacuum} &= \sqrt{c}\begin{pmatrix} \mathbf{1}_k & \mathbf{0} & \mathbf{0}_k & \mathbf{0} \\ \mathbf{0} & \mathbf{0}_{N-k} & \mathbf{0} & \mathbf{1}_{N-k} \end{pmatrix}. \end{aligned} \tag{27}$$

This vacuum is labeled as $(k, N-k)$. The flavor symmetry $SU(N)_{\rm L}$ is broken down to $SU(k)_{\rm C+L} \times SU(N-k)_{\rm L} \times U(1)_{\rm C+L}$, and $SU(N)_{\rm R}$ is broken down to $SU(k)_{\rm R} \times SU(N-k)_{\rm C+R} \times U(1)_{\rm C+R}$, where $C+L, C+R$ denote the locking of the L, R flavor symmetry with the $U(N)_{\rm C}$ color symmetry. Consequently the number of the discrete components of the vacua is $N+1$ in this system. Therefore in this vacua there emerge $4k(N-k)$ NG modes, which parametrize the direct product of two Grassmann manifolds,

$$G^{\rm L}_{N,k} \times G^{\rm R}_{N,k}. \tag{28}$$

Walls are obtained by interpolating between a vacuum at $y = -\infty$ and another vacuum at $y = \infty$. The boundary conditions at both infinities define topological sectors. For a given topological sector, we may find several walls. The maximal number of walls in this system is N, which are obtained for the following maximal topological sector

$$H = \begin{cases} \sqrt{c}(\mathbf{1}_N, \mathbf{0}_N) & \text{at } y = +\infty \\ \sqrt{c}(\mathbf{0}_N, \mathbf{1}_N) & \text{at } y = -\infty \end{cases}. \tag{29}$$

The unbroken symmetries of the vacua $(N, 0)$ and $(0, N)$ which we consider as the boundary condition of domain walls here, are $SU(N)_{\rm C+L} \times U(1)_{\rm C+L} \times SU(N)_{\rm R}$ and $SU(N)_{\rm L} \times SU(N)_{\rm C+R} \times U(1)_{\rm C+R}$, respectively. In this case, the moduli matrix H_0 can be set into the following form without loss of generality:

$$H_0 = \sqrt{c}(\mathbf{1}_N, e^{\phi}) \sim \sqrt{c}(e^{-\phi}, \mathbf{1}_N), \tag{30}$$

where e^{ϕ} is an element of $GL(N, \mathbf{C})$ and ϕ describes the moduli space of walls of this system, and the two forms are related by the V-transformation (16). Therefore the moduli space of domain walls in the GSY model is

$$\mathcal{M} \simeq GL(N, \mathbf{C})[= U(N)^{\mathbf{C}}] \simeq [\mathbf{C}^* \times SL(N, \mathbf{C})]/Z_N. \tag{31}$$

This moduli space admits the isometry

$$e^{\phi} \to e^{i\alpha} g_{\rm L} e^{\phi} g_{\rm R}^{\dagger} \tag{32}$$

with $(g_{\rm L}, g_{\rm R}) \in SU(N)_{\rm L} \times SU(N)_{\rm R}$ and $e^{i\alpha} \in U(1)_{\rm A}$. This is because the domain wall solutions break the symmetry of the two vacua $(N, 0)$ and $(0, N)$, $G = SU(N)_{\rm C+L} \times SU(N)_{\rm C+R} \times U(1)_{\rm C+L-R}$, down to its subgroup.

4.2. Nambu-Goldstone (NG) modes and quasi-NG modes

Note the fact that the global symmetry $G = SU(N)_{\rm L} \times SU(N)_{\rm R} \times U(1)_{\rm A}$ in (32) acts on the moduli space metric as an isometry whereas the complexified group $G^{\bf C} = SL(N, {\bf C})_{\rm L} \times SL(N, {\bf C})_{\rm R} \times {\bf C}^*$ acts on it transitively but not as an isometry. Therefore $G^{\bf C}$ action may change the point in moduli space to another with a different symmetry structure. Since the symmetry of Lagrangian is G but not $G^{\bf C}$ we can use only G when we discuss the symmetry structure at each point in moduli space. General ϕ can be transformed by G to

$$e^{\phi} = {\rm diag.}(v_1, v_2, \cdots, v_N) \tag{33}$$

with v_i real parameters. When all v_i's coincide (coincident walls), the unbroken symmetry is the maximal $H_{\rm max} = SU(N)_{\rm V}$, so we call it the symmetric point. There are N^2 massless NG bosons and N^2 quasi-NG bosons. When all v_i's are different from each other (separated walls), $H_{\rm max} = SU(N)_{\rm V}$ is further broken down to $H_{\rm min} = U(1)_{\rm V}^{N-1}$. Here the numbers of NG bosons and quasi-NG bosons are $2N^2 - (N-1)$ and $N-1$, respectively. These $N-1$ quasi-NG bosons correspond to the $N-1$ parameters v_i without the overall factor. Therefore some of quasi-NG bosons at the symmetric point in the moduli space change to the NG bosons parametrizing $H_{\rm max}/H_{\rm min} = SU(N)_{\rm V}/U(1)_{\rm V}^{N-1}$ reflecting this further symmetry breaking. When some v_i's coincide, some non-Abelian groups are recovered: $H = U(1)_{\rm V}^r \times \prod U(n_i)_{\rm V}$. Then the NG modes $H_{\rm max}/H$ are supplied from quasi-NG modes.

The diagonal moduli parameters v_i (quasi-NG bosons) in Eq. (33) correspond to the positions of N domain walls. When all domain walls are separated, the unbroken symmetry is $U(1)_{\rm V}^{N-1}$. When positions of n domain walls coincide, $U(n)_{\rm V}$ symmetry is recovered. This phenomenon has a resemblance to the case of D-branes. However, there is a crucial difference: the symmetry in our case of domain walls is a global symmetry, whereas that of D-branes is a local gauge symmetry. However in the case of the $d = 2+1$ wall world-volume, massless scalars can be dualized to gauge fields. Shifman and Yung [12] expected that the off-diagonal gauge bosons of $U(N)$ (which are originally the off-diagonal NG bosons of $U(N)$ before taking a duality) will become massive when domain walls are separated, in order to interpret domain walls as D-branes. However, our analysis shows that the off-diagonal NG bosons of $U(N)$ remain massless, and instead some of the quasi-NG bosons become NG bosons for further symmetry breaking with the total number of massless bosons unchanged as explained.

We have obtained the low-energy effective Lagrangian explicitly for this model. We found the following. The wave functions of the NG boson for translation, for $U(1)^N$, and quasi-NG bosons are localized, and that other massless modes are extended between two domain walls, if walls are separated. Wave functions of all the massless modes become identical in the limit of coincident walls. These behaviors are different from D-branes, although there are some similarlities. More precise details of our results can be found in Ref.[1].

When we introduce complex masses for Higgs fields, domain wall junction or network emerge as 1/4 BPS states [15]-[17]. In this case too, non-Abelain NG modes appear in the effective action, when some masses are degenerate [18], [19].

Acknowledgments

This work is supported in part by Grant-in-Aid for Scientific Research from the Ministry of Education, Culture, Sports, Science and Technology, Japan No.21540279, No.21244036 (N.S.).

Appendix

We will briefly give the derivation of our master equation (15). Let us first note an identity for variations $\delta_i, i = 1, 2$ of arbitary regular matrix A

$$\delta_1(A^{-1}\delta_2 A) = A^{-1}\delta_1\delta_2 A - A^{-1}\delta_1 A A^{-1}\delta_2 A = A^{-1}(\delta_2\delta_1 A A^{-1} - \delta_1 A A^{-1}\delta_2 A A^{-1})A$$

$$= A^{-1}\delta_2(\delta_1 A A^{-1})A \tag{A.1}$$

For gauge invariant quantity $\Omega = SS^\dagger$, we obtain

$$\begin{aligned}
&\delta_1(\Omega\delta_2\Omega^{-1}) = \delta_1(SS^\dagger\delta_2(S^{-1\dagger}S^{-1})) = \delta_1(SS^\dagger(\delta_2 S^{-1\dagger})S^{-1} + S\delta_2 S^{-1})\\
&= S\left(\delta_1(S^\dagger\delta_2 S^{-1\dagger}) + S^{-1}\delta_1 SS^\dagger\delta_2 S^{-1\dagger} + S^\dagger\delta_2 S^{-1\dagger}\delta_1 S^{-1}S + S^{-1}\delta_1(S\delta_2 S^{-1})S\right)S^{-1}\\
&= S\left(\delta_1(S^\dagger\delta_2 S^{-1\dagger}) - \delta_1 S^{-1}SS^\dagger\delta_2 S^{-1\dagger} + S^\dagger\delta_2 S^{-1\dagger}\delta_1 S^{-1}S + \delta_2(\delta_1 S^{-1}S)\right)S^{-1}
\end{aligned} \tag{A.2}$$

By choosing both δ_1 and δ_2 to be dirivative in y, and noting that

$$\Sigma = \frac{1}{2}(S^{-1}\partial_y S + \partial_y S^\dagger S^{-1\dagger}), \quad W_y = \frac{1}{2i}(S^{-1}\partial_y S - \partial_y S^\dagger S^{-1\dagger}), \tag{A.3}$$

we obtain the master equation (15) for the domain walls. Choosing δ_1 and δ_2 as ∂_z and $\partial_{\bar{z}}$ respectively, we can obtain the master equation for vortex.

References

[1] M. Eto, T. Fujimori, M. Nitta, K. Ohashi and N. Sakai, Phys. Rev. D **77** (2008) 125008 [arXiv:0802.3135 [hep-th]].

[2] E. R. C. Abraham and P. K. Townsend, Phys. Lett. B **291**, 85 (1992); Phys. Lett. B **295**, 225 (1992).

[3] J. P. Gauntlett, D. Tong and P. K. Townsend, Phys. Rev. D **64**, 025010 (2001) [arXiv:hep-th/0012178]; D. Tong, Phys. Rev. D **66**, 025013 (2002) [arXiv:hep-th/0202012]; JHEP **0304**, 031 (2003) [arXiv:hep-th/0303151]; K. S. M. Lee, Phys. Rev. D **67**, 045009 (2003) [arXiv:hep-th/0211058]; M. Arai, M. Naganuma, M. Nitta, and N. Sakai, Nucl. Phys. B **652**, 35 (2003) [arXiv:hep-th/0211103]; "BPS Wall in N=2 SUSY Nonlinear Sigma Model with Eguchi-Hanson Manifold" in Garden of Quanta - In honor of Hiroshi Ezawa, Eds. by J. Arafune et al. (World Scientific Publishing Co. Pte. Ltd. Singapore, 2003) pp 299-325, [arXiv:hep-th/0302028]; M. Arai, E. Ivanov and J. Niederle, Nucl. Phys. B **680**, 23 (2004) [arXiv:hep-th/0312037].

[4] Y. Isozumi, K. Ohashi and N. Sakai, JHEP **0311**, 061 (2003) [arXiv:hep-th/0310130]; JHEP **0311**, 060 (2003) [arXiv:hep-th/0310189].

[5] N. Sakai and Y. Yang, Commun. Math. Phys. **267**, 783 (2006) [arXiv:hep-th/0505136].

[6] N. Sakai and D. Tong, JHEP **0503** (2005) 019 [arXiv:hep-th/0501207].

[7] Y. Isozumi, M. Nitta, K. Ohashi and N. Sakai, Phys. Rev. Lett. **93**, 161601 (2004) [arXiv:hep-th/0404198].

[8] Y. Isozumi, M. Nitta, K. Ohashi and N. Sakai, Phys. Rev. D **70**, 125014 (2004) [arXiv:hep-th/0405194].

[9] Y. Isozumi, M. Nitta, K. Ohashi and N. Sakai, Phys. Rev. D **71**, 065018 (2005) [arXiv:hep-th/0405129].

[10] M. Eto, Y. Isozumi, M. Nitta, K. Ohashi, K. Ohta and N. Sakai, Phys. Rev. D **71**, 125006 (2005) [arXiv:hep-th/0412024].

[11] M. Eto, Y. Isozumi, M. Nitta, K. Ohashi, K. Ohta, N. Sakai and Y. Tachikawa, Phys. Rev. D **71**, 105009 (2005) [arXiv:hep-th/0503033].

[12] M. Shifman and A. Yung, Phys. Rev. D **70**, 025013 (2004) [arXiv:hep-th/0312257].

[13] M. Eto, M. Nitta, K. Ohashi and D. Tong, Phys. Rev. Lett. **95**, 252003 (2005) [arXiv:hep-th/0508130].

[14] M. Eto, Y. Isozumi, M. Nitta, K. Ohashi and N. Sakai, Phys. Rev. D **73**, 125008 (2006) [arXiv:hep-th/0602289].

[15] M. Eto, Y. Isozumi, M. Nitta, K. Ohashi and N. Sakai, Phys. Rev. D **72** (2005) 085004 [arXiv:hep-th/0506135].

[16] M. Eto, Y. Isozumi, M. Nitta, K. Ohashi and N. Sakai, Phys. Lett. B **632** (2006) 384 [arXiv:hep-th/0508241].

[17] M. Eto, Y. Isozumi, M. Nitta, K. Ohashi, K. Ohta and N. Sakai, AIP Conf. Proc. **805** (2006) 354 [arXiv:hep-th/0509127].

[18] M. Eto, T. Fujimori, T. Nagashima, M. Nitta, K. Ohashi and N. Sakai, Phys. Rev. D **75** (2007) 045010 [arXiv:hep-th/0612003].

[19] M. Eto, T. Fujimori, T. Nagashima, M. Nitta, K. Ohashi and N. Sakai, Phys. Rev. D **76** (2007) 125025 [arXiv:0707.3267 [hep-th]].

Journal of Physics: Conference Series **222** (2010) 012007 doi:10.1088/1742-6596/222/1/012007

The Holographic Universe

P L McFadden and K Skenderis

Institute for Theoretical Physics, Valckenierstraat 65, 1018XE Amsterdam, the Netherlands.

E-mail: P.L.McFadden@uva.nl, K.Skenderis@uva.nl

Abstract.

We present a holographic description of four-dimensional single-scalar inflationary universes in terms of a three-dimensional quantum field theory (QFT). The holographic description correctly reproduces standard inflationary predictions in their regime of applicability. In the opposite case, wherein gravity is strongly coupled at early times, we propose a holographic description in terms of perturbative QFT and present models capable of satisfying the current observational constraints while exhibiting a phenomenology distinct from standard inflation. This provides a qualitatively new method for generating a nearly scale-invariant spectrum of primordial cosmological perturbations.

1. Introduction.

The notion of holography emerged from black hole physics as an answer to the question: why is the entropy of a black hole proportional to the area of its horizon? Since entropy is an extensive quantity, one would have expected the entropy to be instead proportional to the volume the black hole occupies. Typically, entropy is a measure of the number of degrees of freedom. The scaling of the gravitational entropy has thus been taken as an indication of a new fundamental principle, the holographic principle, that underlies any quantum theory of gravity. More precisely, holography states that any quantum gravitational system in $(d+1)$ dimensions should have a dual description in terms of a QFT without gravity in one dimension less [1]. Indeed, if gravity is holographic this would explain the scaling behavior of the black hole entropy, since the entropy of a QFT scales like the volume, which is the same as the area in one higher dimension.

Holography provides a new paradigm for physical reality, the consequences of which we are only just beginning to comprehend. According to this picture one of the macroscopic dimensions of spacetime and one of the forces in the universe, namely gravity, are emergent phenomena in an underlying lower-dimensional QFT. Concrete realizations of holography have been found in string theory and a precise holographic dictionary was established shortly thereafter [2, 3, 4]. To date, almost all such realizations involve spacetimes with a negative cosmological constant. The arguments that led to the holographic principle apply more generally, however, and suggest that one should be able to establish a holographic dictionary that applies to our own universe. The purpose of the work presented in [5] and further discussed here is to propose a concrete holographic framework that applies to our own universe, and in particular to its cosmological evolution. More precisely, we will describe how to set up holography for inflationary cosmology.

Any holographic proposal for cosmology should specify (i) what the dual QFT is, and (ii) how it can be used to compute cosmological observables. Having defined such a duality, the new

Journal of Physics: Conference Series **222** (2010) 012007 doi:10.1088/1742-6596/222/1/012007

description should recover established results in their regime of applicability. Indeed, we will see that our holographic models correctly reproduce standard inflationary results when standard inflation is applicable, namely, under the assumption that gravity was weakly coupled at early times. Yet perhaps more importantly, this approach also gives a qualitatively new method for generating a nearly scale-invariant spectrum when gravity was *strongly coupled* at early times. As we will discuss later, there exist holographic models that are capable of satisfying all current observational constraints, while exhibiting a phenomenology distinct from standard inflation.

Over the last two decades striking new observations have transformed cosmology from a qualitative to quantitative science [6]. These observations reveal a spatially flat universe, endowed with small-amplitude primordial perturbations that are approximately Gaussian and adiabatic with a nearly scale-invariant spectrum. This data is consistent with the generic predictions of inflationary cosmology and set inflation as the leading theoretical paradigm for the initial conditions of Big Bang cosmology. Yet inflation, despite these successes, is still unsatisfactory in a number of ways: it generically requires fine tuning and there are trans-Planckian issues and questions about the initial conditions for inflation [7].

With future observations promising an unprecedented era of precision cosmology, it becomes imperative that inflation is embedded in a UV complete theory (indeed there is increasing amount of effort devoted in embedding inflation in string theory) and it is also important that alternative scenarios are developed. The holographic approach that we undertake provides both. Firstly, the holographic models we will discuss here are three dimensional super-renormalizable theories and thus are UV complete. Secondly, holographic dualities are strong/weak coupling dualities meaning that in the regime where the one description is weakly coupled the other is strongly coupled. This provides an arena for constructing new models with intrinsic strong-coupling gravitational dynamics at early times that have only a weakly coupled three-dimensional QFT description and are thus outside the class of model described by standard inflation. As we will see such models can lead to qualitatively different predictions for the cosmological observables that will be measured in the near future. Furthermore, quite apart from the fresh perspective on early universe cosmology such an approach offers, there are also more pragmatic reasons for developing a holographic framework for cosmology: uncovering the structure of three-dimensional QFT in cosmological observables brings in new intuition about their structure and may lead to more efficient computational techniques, *cf.* the computation of non-Gaussianities in [8].

The holographic description we propose uses the one-to-one correspondence between cosmologies and domain-wall spacetimes discussed in [9, 10] and assumes that the standard gauge/gravity duality is valid. More precisely, the steps involved are illustrated in Fig. 1. The first step is to map any given inflationary model to a domain-wall spacetime. For inflationary cosmologies that at late times approach either a de Sitter spacetime or a power-law scaling solution[1], the corresponding domain-wall solutions describe holographic renormalization group flows. For these cases there is an operational gauge/gravity duality, namely one has a dual description in terms of a three-dimensional QFT. Now, the map between cosmologies and domain-walls can equivalently be expressed entirely in terms of QFT variables, and amounts to a certain analytic continuation of parameters and momenta. Applying this analytic continuation we obtain the QFT dual of the original cosmological spacetime.

We shall call the resulting theory a 'pseudo'-QFT because we currently only have an operational definition of this theory. Namely, we do the computations in the QFT theory dual to the corresponding domain-wall and then apply the analytic continuation. Perhaps a more fundamental perspective is to consider the QFT action, with complex parameters and complex fields as the fundamental objects, and then to consider the results on different real domains as

[1] This era should then be followed by a hot big bang cosmology, as in standard discussions. Here we only discuss the very early universe, *i.e.*, the times when the primordial cosmological perturbations were generated (the inflationary epoch).

Journal of Physics: Conference Series **222** (2010) 012007 doi:10.1088/1742-6596/222/1/012007

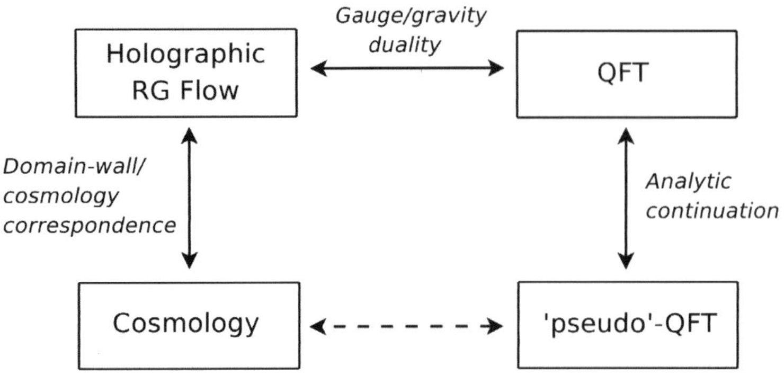

Figure 1. The 'pseudo'-QFT dual to inflationary cosmology is operationally defined using the correspondence of cosmologies to domain-walls and standard gauge/gravity duality.

applicable to either domain-walls or cosmologies. Note that the supergravity embedding of the domain-wall/cosmology correspondence discussed in [11] works in precisely this way.

This paper is organized as follows. In the next section we discuss the domain-wall/cosmology correspondence. Then in Section 3 we discuss the cosmological observables that we would like to compute holographically and in Section 4 we present the holographic analysis. In Section 5 we discuss the analytic continuation to pseudo-QFT and Sections 6 and 7 discuss the new models that are strongly coupled at early times but have a weakly coupled QFT description.

2. Domain-wall/cosmology correspondence.

We explain in this section the lefthand vertical line in Fig. 1, namely, the correspondence between cosmologies and domain-wall spacetimes. For simplicity, we focus on spatially flat universes equipped with a single minimally coupled scalar field, but the results can be extended to more general cases (*eg.*, non-flat, multi-scalar, non-canonical kinetic terms, *etc*). The metric and scalar field for the unperturbed solution are given by

$$\mathrm{d}s^2 = \eta \mathrm{d}z^2 + a^2(z)\mathrm{d}\vec{x}^2, \qquad \Phi = \varphi(z), \tag{1}$$

where $\eta = -1$ in the case of cosmology, in which case z is the time coordinate, and $\eta = +1$ in the case of domain-wall solutions[2] in which case z is the radial coordinate. We take the domain-wall to be Euclidean for later convenience. A Lorentzian domain-wall may be obtained by continuing one of the x^i coordinates to become the time coordinate [10]. The continuation to a Euclidean domain-wall is convenient, however, because the QFT vacuum state implicit in the Euclidean formulation maps to the Bunch-Davies vacuum on the cosmology side. Other choices of cosmological vacuum require considering the boundary QFT in different states, as may be accomplished using the real-time formalism of [12].

With the appropriate choice of $a(z)$ and $\varphi(z)$, the configuration in (1) solves the field equations that follow from the action

$$S = \frac{\eta}{2\kappa^2}\int \mathrm{d}^4x\sqrt{|g|}\,[-R + (\partial\Phi)^2 + 2\kappa^2 V(\Phi)], \tag{2}$$

where $\kappa^2 = 8\pi G$ and we are taking the scalar field Φ to be dimensionless. Note that when $\eta = -1$, the kinetic terms have the appropriate signs for a mostly plus Lorentzian signature metric and when $\eta = +1$, the kinetic terms have the correct sign for Euclidean signature. Had

[2] The name 'domain-wall spacetime' dates back to earlier work featuring solutions of this form that interpolate between two stationary points of the scalar field potential, one at $z = +\infty$ and another at $z = -\infty$. In the present context the name is somewhat misleading, however, since we consider only the $z > 0$ part of the geometry. We will nevertheless stick with this terminology as it is standard usage in high-energy physics.

we expressed both actions in the same signature metric, they would differ only in the sign of the potential. It follows that for every flat FRW solution of a model with potential V there is a corresponding domain-wall solution of a model with potential $-V$ [9, 10].

For background solutions in which the evolution of the scalar field is (piece-wise) *monotonic*, $\varphi(z)$ can be inverted to give $z(\varphi)$ permitting the Hubble rate $H = \dot{a}/a$ to be re-expressed as $H(z) = -(1/2)W(\varphi)$, where $W(\varphi)$ is known as the 'fake superpotential'. In this case, the complete equations of motion for the background take the simple form

$$\frac{\dot{a}}{a} = -\frac{1}{2}W, \qquad \dot{\varphi} = W_{,\varphi}, \qquad 2\eta\kappa^2 V = (W_{,\varphi})^2 - \frac{3}{2}W^2. \tag{3}$$

This first-order formalism goes back to the work of [13] (for cosmology), where it was obtained by application of the Hamilton-Jacobi method. In [10] this formalism was linked to the notion of (fake) (pseudo-) supersymmetry.

We will now extend the correspondence to encompass linear perturbations around the background solution. The linearly perturbed metric takes the general form

$$\begin{aligned} ds^2 &= \eta[1+2\phi(z,\vec{x})]dz^2 + 2a^2(z)[\partial_i\nu(z,\vec{x}) + \nu_i(z,\vec{x})]dz dx^i + a^2(z)[\delta_{ij} + h_{ij}(z,\vec{x})]dx^i dx^j, \\ \Phi &= \varphi(z) + \delta\varphi(z,\vec{x}), \end{aligned} \tag{4}$$

where ν_i is transverse. The spatial metric perturbation h_{ij} may be decomposed as

$$h_{ij}(z,\vec{x}) = -2\psi(z,\vec{x})\delta_{ij} + 2\partial_i\partial_j\chi(z,\vec{x}) + 2\partial_{(i}w_{j)}(z,\vec{x}) + \gamma_{ij}(z,\vec{x}), \tag{5}$$

where ω_i is transverse and $\gamma_{ij}(z,x)$ is transverse traceless. The metric perturbations may then be combined into the gauge-invariant combinations

$$\zeta = \psi + (H/\dot{\varphi})\delta\varphi, \tag{6}$$

$$\hat{\phi} = \phi - (\delta\varphi/\dot{\varphi})^{\dot{}}, \tag{7}$$

$$\hat{\nu} = \nu - \dot{\chi} - \eta(\delta\varphi/a^2\dot{\varphi}), \tag{8}$$

$$\hat{\nu}_i = \nu_i - \dot{\omega}_i. \tag{9}$$

Physically, ζ represents the curvature perturbation on comoving hypersurfaces and has the useful property that it tends to a constant on superhorizon scales[3].

The equations of motion for cosmological perturbations have been worked out long ago, see [15] and references therein, while the corresponding analysis for domain-walls may be found in [16, 17, 18]. In the present case, there are two independent perturbations represented by ζ and γ_{ij}, since the Hamiltonian and momentum constraints are equivalent to

$$\hat{\phi} = -\frac{\dot{\zeta}}{H}, \qquad \hat{\nu} = -\frac{\eta\zeta}{a^2 H} + \frac{\epsilon\dot{\zeta}}{q^2}, \qquad \hat{\nu}_i = 0, \tag{10}$$

where $\vec{q}$ is the comoving wavevector of the perturbations, and the background quantity $\epsilon(z)$ is defined as $\epsilon = -\dot{H}/H^2 = 2(W_{,\varphi}/W)^2$. (In standard inflation ϵ would be the usual slow-roll parameter; however, we do not assume slow roll here). From the remaining Einstein equations, one then finds the following equations of motion for ζ and γ_{ij}:

$$\begin{aligned} 0 &= \ddot{\zeta} + (3H + \dot{\epsilon}/\epsilon)\dot{\zeta} - \eta a^{-2}q^2\zeta, \\ 0 &= \ddot{\gamma}_{ij} + 3H\dot{\gamma}_{ij} - \eta a^{-2}q^2\gamma_{ij}. \end{aligned} \tag{11}$$

[3] Following the end of inflation, the superhorizon value of ζ then remains constant until horizon re-entry, irrespective of the dynamics of the intervening evolution, provided that no entropy perturbations are produced [14, 15].

Journal of Physics: Conference Series **222** (2010) 012007 doi:10.1088/1742-6596/222/1/012007

Defining now the analytically continued variables $\bar{\kappa}$ and $\bar{q}$ according to

$$\bar{\kappa}^2 = -\kappa^2, \qquad \bar{q} = -iq\,, \tag{12}$$

it is easy to see that a perturbed cosmological solution written in terms of the variables κ and q continues to a perturbed Euclidean domain-wall solution expressed in terms of the variables $\bar{\kappa}$ and $\bar{q}$. Note that (11) only requires $\bar{q}^2 = -q^2$ and in (12) we made a choice of a branch cut in the function $q = \sqrt{q^2}$ (for reasons to be explained in the next section). As it is clear from (3), continuing κ is equivalent to continuing V. Here we prefer to continue κ since, as we will see, the former has a clear interpretation in terms of dual QFT variables.

We have thus established that the correspondence between cosmologies and domain-walls holds, not only for the background solutions, but also for linear perturbations around them. This is the basis for the relation between power spectra and holographic 2-point functions, to be discussed momentarily. The argument can be generalized to arbitrary order to relate non-Gaussianities to holographic higher-point functions [19].

3. Cosmological observables.

In the inflationary paradigm, cosmological perturbations originate on sub-horizon scales as quantum fluctuations of the vacuum. Quantizing the perturbations in the usual manner, one finds the scalar and tensor superhorizon power spectra

$$\begin{aligned}\Delta_S^2(q) &= \frac{q^3}{2\pi^2}\langle\zeta(q)\zeta(-q)\rangle = \frac{q^3}{2\pi^2}|\zeta_{q(0)}|^2,\\ \Delta_T^2(q) &= \frac{q^3}{2\pi^2}\langle\gamma_{ij}(q)\gamma_{ij}(-q)\rangle = \frac{2q^3}{\pi^2}|\gamma_{q(0)}|^2,\end{aligned} \tag{13}$$

where $\gamma_{q(0)}$ and $\zeta_{q(0)}$ are the constant late-time values of the cosmological mode functions $\gamma_q(z)$ and $\zeta_q(z)$.

The mode functions are themselves solutions of the classical equations of motion (11) (setting $\gamma_{ij} = \gamma_q e_{ij}$, for some time-independent polarization tensor e_{ij}). To select a unique solution for each mode function we impose the Bunch-Davies vacuum condition $\zeta_q, \gamma_q \sim \exp(-iq\tau)$ as $\tau \to -\infty$, where the conformal time $\tau = \int^z \mathrm{d}z'/a(z')$. The normalization of each solution (up to an overall phase) may then be fixed by imposing the canonical commutation relations for the corresponding quantum fields. This leads to the Wronskian conditions,

$$i = \zeta_q \Pi_q^{(\zeta)*} - \Pi_q^{(\zeta)}\zeta_q^*, \qquad i/2 = \gamma_q \Pi_q^{(\gamma)*} - \Pi_q^{(\gamma)}\gamma_q^*, \tag{14}$$

where $\Pi_q^{(\zeta)} = 2\epsilon a^3\kappa^{-2}\dot{\zeta}_q$ and $\Pi_q^{(\gamma)} = (1/4)a^3\kappa^{-2}\dot{\gamma}_q$ are the canonical momenta associated with each mode function, and we have set $\hbar$ to unity.

To make connection with the holographic analysis to follow, we introduce the linear response functions E and Ω satisfying

$$\Pi_q^{(\zeta)} = \Omega\,\zeta_q, \qquad \Pi_q^{(\gamma)} = E\,\gamma_q. \tag{15}$$

These quantities are well-defined since we have already selected a unique solution for each mode function. Substituting these definitions into the Wronskian conditions, which are valid at all times, the cosmological power spectra may be re-expressed as

$$\Delta_S^2(q) = \frac{-q^3}{4\pi^2 \mathrm{Im}\Omega_{(0)}(q)}, \quad \Delta_T^2(q) = \frac{-q^3}{2\pi^2 \mathrm{Im}E_{(0)}(q)}, \tag{16}$$

Journal of Physics: Conference Series **222** (2010) 012007 doi:10.1088/1742-6596/222/1/012007

where $\mathrm{Im}\Omega_{(0)}$ and $\mathrm{Im}E_{(0)}$ are the constant late-time values of the imaginary part of the response functions (more precisely, the subscript indicates that this is the part of the response function that is invariant under dilatations, see the discussion at the end of subsection 4.1). We will see shortly how the response functions also give the 2-point function of the pseudo-QFT.

Let us now consider the corresponding domain-wall solution obtained by the applying the continuation (12). The early-time behavior $\sim \exp(-iq\tau)$ of the cosmological perturbations maps to the exponentially decaying behavior $\sim \exp(\bar{q}\tau)$ in the interior of the domain-wall ($\tau \to -\infty$). Such regularity in the interior is a prerequisite for holography, explaining our choice of sign in the continuation of q.

The domain-wall response functions $\bar{E}$ and $\bar{\Omega}$ [18] are defined analogously to (15), namely

$$\bar{\Pi}^{(\zeta)}_{\bar{q}} = -\bar{\Omega}\,\zeta_{\bar{q}} \qquad \bar{\Pi}^{(\gamma)}_{\bar{q}} = -\bar{E}\,\gamma_{\bar{q}}, \tag{17}$$

where $\bar{\Pi}^{(\zeta)}_{\bar{q}} = 2\epsilon a^3 \bar{\kappa}^{-2}\dot{\zeta}_{\bar{q}}$ and $\bar{\Pi}^{(\gamma)}_{\bar{q}} = (1/4)a^3\bar{\kappa}^{-2}\dot{\gamma}_{\bar{q}}$ are the radial canonical momenta. The minus sign in (17) are inserted so that

$$\bar{\Omega}(-iq) = \Omega(q), \qquad \bar{E}(-iq) = E(q). \tag{18}$$

By choosing the arbitrary overall phase of the cosmological perturbations appropriately, we may ensure that the domain-wall perturbations are everywhere real. The domain-wall response functions are then purely real, while their cosmological counterparts are complex.

4. Holographic analysis.

In this section we briefly review relevant material from gauge/gravity duality, corresponding to the upper horizontal line in Fig. 1. There are two classes of domain-wall solutions for which holography is well understood:

Asymptotically AdS domain-walls. In this case the solution behaves asymptotically as

$$a(z) \sim e^z, \qquad \varphi \sim 0 \qquad \text{as} \quad z \to \infty. \tag{19}$$

The boundary theory has a UV fixed point which corresponds to the bulk AdS critical point. Depending on the rate at which φ approaches zero as $z \to \infty$, the QFT is either a deformation of the conformal field theory (CFT), or else the CFT in a state in which the dual scalar operator acquires a nonvanishing vacuum expectation value (see [20] for details). Under the domain-wall/cosmology correspondence, these solutions are mapped to cosmologies that are asymptotically de Sitter at late times.

Asymptotically power-law solutions. In this case the solution behaves asymptotically as

$$a(z) \sim (z/z_0)^n, \quad \varphi \sim \sqrt{2n}\log(z/z_0) \quad \text{as} \quad z \to \infty, \tag{20}$$

where $z_0 = n-1$. This case has only very recently been understood [21]. For $n = 7$ the asymptotic geometry is the near-horizon limit of a stack of D2 brane solutions. In general, these solutions describe QFTs with a dimensionful coupling constant in the regime where the dimensionality of the coupling constant drives the dynamics. Under the domain-wall/cosmology correspondence, these solutions are mapped to cosmologies that are asymptotically power-law at late times.

4.1. Basics of holography.
Gauge/gravity duality is an exact equivalence between a bulk gravitational theory and a boundary QFT. Typically, the boundary QFT is a gauge theory that admits a large N expansion. The N here denotes the rank of the gauge group: an example of such theory, with gauge group $SU(N)$, is discussed in section 7.1. The large N limit consists of taking $N \to \infty$ while keeping fixed the 't Hooft coupling constant $\lambda = g_{YM}^2 N$ [22]. One can show that in this limit only planar diagrams survive. On the bulk side, taking the large N limit means that one suppresses loop effects. The value of λ then controls whether the supergravity approximation is valid or not.

The duality relates bulk fields with gauge-invariant operators of the boundary theory. For example, the bulk metric corresponds to the boundary stress-energy tensor, T_{ij}, while bulk scalar fields correspond to boundary scalar operators, such as $\operatorname{tr} F_{ij}F^{ij}$, where F_{ij} is the field strength of the gauge field and the trace is over the gauge group indices. Correlation functions of such gauge invariant operators can then be extracted from the asymptotics of bulk solutions and conversely, given correlation functions of the dual operators, one can reconstruct asymptotic solutions.

To understand the holographic computations we need to know a few things about the structure of asymptotic solutions of the field equations. The discuss below refers specifically to a four dimensional bulk spacetime (the general features are the same in any dimension but the details are different). The results for the case of asymptotically power-law spacetimes can be obtained from the results for asymptotically $AdS_{2\sigma+1}$ spacetimes via a dimensional reduction on a $T^{2\sigma-3}$ torus and analytic continuation in σ [23]. The most general asymptotic solution for both cases can be shown to be of the form [24, 21]

$$\begin{aligned} ds^2 &= dr^2 + g_{ij}(r,x)dx^i dx^j, \\ g_{ij}(r,x) &= e^{2r}\left(g_{(0)ij}(x) + e^{-2r}g_{(2)ij}(x) + \cdots + e^{-2\sigma r}g_{(2\sigma)ij}(x) + \cdots\right), \end{aligned} \tag{21}$$

where $\sigma = 3/2$ for solutions of Einstein's equations with negative cosmological constant and $\sigma = (3n-1)/2(n-1) > 3/2$ for the case of asymptotically power-law solutions. In the latter case, the metric above is given in the so-called dual frame [25]: the Einstein frame metric g^E_{ij} in (2) is related to the metric g^D_{ij} in (21) via the Weyl transformation $g^E_{ij} = \exp(\lambda\Phi)g^D_{ij}$, where $\lambda = \sqrt{2/n}$. In (21), $g_{(0)ij}(x)$ is an arbitrary (non-degenerate) three-dimensional metric, from which the $g_{(2k)ij}(x)$ with $k < \sigma$ are locally determined, while $g_{(2\sigma)ij}(x)$ is only partially constrained by the asymptotic analysis of the field equations. One can show, however, that this coefficient is directly related to the expectation value of the boundary stress-energy tensor [24, 21]:

$$\langle T_{ij}\rangle = \frac{1}{2\bar{\kappa}^2}(2\sigma g_{(2\sigma)ij}). \tag{22}$$

Consider now a scalar field[4] Ψ that is dual to an operator $\mathcal{O}$ of dimension Δ. The conformal dimension depends on the mass of Ψ via $m^2 = \Delta(3-\Delta)$ in the asymptotically AdS case and is equal to $\Delta = 4$ in the asymptotically power-law case (for appropriately normalized $\mathcal{O}$, see [21]). The asymptotic expansion for Ψ has a form analogous to (21),

$$\Psi(x,r) = e^{(\Delta-3)r}(\Psi_{(0)} + e^{-2r}\Psi_{(2)} + \cdots + e^{-2\tilde{\sigma}r}\Psi_{(2\tilde{\sigma})}(x) + \cdots), \tag{23}$$

Here $\Psi_{(0)}(x)$ is unconstrained and is the source for the dual operator $\mathcal{O}$, with all subleading terms, up to order $\exp(-2\tilde{\sigma}r)$, then being locally determined in terms of the sources. The following term in the series, $\Psi_{(2\tilde{\sigma})}$, is undetermined and is related to the expectation value of

[4] In the asymptotically AdS case, $\Psi = \Phi$, while in the asymptotically power-law case, $\Psi = \exp((n-1)\Phi/\sqrt{2n})$. To match with the discussion in [21] for the D2-brane case ($n = 7$), note that $\Phi_{\text{here}} = -(\sqrt{14}/5)\phi_{\text{there}}$, where ϕ_{there} is the scalar field in [21] for $p = 2$.

Journal of Physics: Conference Series **222** (2010) 012007 doi:10.1088/1742-6596/222/1/012007

dual operator $\langle \mathcal{O} \rangle$. In the asymptotically AdS case, $2\tilde{\sigma} = (2\Delta - 3)$, while in the asymptotically power-law case, $\tilde{\sigma} = \sigma = (3n-1)/2(n-1)$. (For details, see [20, 21]).

The constraints on $g_{(2\sigma)ij}$ due to the Einstein equations imply

$$\nabla^i \langle T_{ij} \rangle + \langle \mathcal{O} \rangle \nabla_i \Psi_{(0)} = 0, \qquad (24)$$
$$\langle T^i_i \rangle + (3 - \Delta) \Psi_{(0)} \langle \mathcal{O} \rangle = 0,$$

and these are precisely the expected Ward identities[5].

The relation (21)-(22) can be read in two ways: (i) given a supergravity solution it allows us to read off the QFT data encoded by the solution; (ii) given QFT data it provides a reconstruction of the bulk spacetime in the neighborhood of the boundary. Note that the latter is true even when the supergravity approximation is not valid in the interior of spacetime (because the curvature is large there). The terms exhibited in (21), apart from $g_{(2\sigma)ij}$, are non-normalizable terms and are not affected by dynamics. The first term that is affected by dynamics is $g_{(2\sigma)ij}$ and this is indeed unconstrained asymptotically, except for the constraints due to symmetries (namely the Ward identities (24)). The gauge/gravity duality provides a dual description of dynamics, so the statement of the duality is that $g_{(2\sigma)ij}$ determined from QFT via (22) should agree with the value obtained from string dynamics in a spacetime with these asymptotics. When the gravity approximation is valid throughout, the asymptotics yield sufficient information to uniquely reconstruct a regular bulk solution: the pair $(g_{(0)ij}, g_{(2\sigma)ij})$ (and similar for the scalar field) are coordinates in the covariant phase space of the gravitational theory [26].

An alternative way to express these results that we will use in the next subsection is to use the radial Hamiltonian formulation of [18]. This is a Hamiltonian formulation where the radial direction plays the role of time. In this formalism the expectation value of the stress-energy tensor is given by

$$\langle T_{ij} \rangle = \left(\frac{-2}{\sqrt{g}} \bar{\Pi}_{ij} \right)_{(3)}, \qquad (25)$$

where $\bar{\Pi}_{ij}$ is the radial canonical momentum and we will momentarily explain the meaning of subscript (3). There is a similar formula for the expectation value of $\mathcal{O}$ that we will not need here. A fundamental property of the spacetimes (21) is that the radial derivative is, to leading order as $r \to \infty$, proportional to the dilatation operator δ_D,

$$\partial_r = \delta_D (1 + O[e^{-2r}]), \qquad (26)$$

where

$$\delta_D g_{ij}(x, r) = 2 g_{ij}(x, r), \qquad \delta_D \Psi(x, r) = (\Delta - 3) \Psi(x, r) \qquad (27)$$

(one can verify (26) by inspection of (21) and (23)). The expressions in (27) are the standard dilatation transformation rules for a metric and a source that couples to an operator of dimension Δ in three spacetime dimensions. Equation (26) is a precise version of the often quoted relation between the radial direction and the energy scale of the dual QFT. In our context, equation (26) implies that one can trade the radial expansion for an expansion in terms of eigenfunctions of dilatation operator. An eigenfunction $A_{(n)}$ of weight n is by definition,

$$\delta_D A_{(n)} = -n A_{(n)}. \qquad (28)$$

As follows from (26), $A_{(n)} \sim e^{-nr}(1 + O[e^{-2r}])$, so the radial expansion and the expansion is eigenfunctions of the dilatation operator are closely related. The chief advantage of the radial

[5] Note there is no conformal anomaly in three dimensions.

Hamiltonian reformulation is that the expansion is now manifestly covariant, whereas expanding in a particular coordinate is not a covariant operation. Now, the radial canonical momentum can be decomposed in eigenfunctions of the dilatation operator and the subscript (3) in (25) indicates that we should pick the part with dilatation eigenvalue[6] 3. One may have expected this on general grounds, since the dimension of stress energy tensor is three dimensions is 3. More generally, the expectation value of an operator of dimension Δ is given by the piece of the corresponding radial canonical momentum of weight Δ. Note that in general the radial canonical momentum contains also pieces with weight less than Δ (*i.e.*, less than 3 in the case of T_{ij}): these pieces diverge as $r \to \infty$. One of the advantages of radial Hamiltonian formalism [18] is that renormalization is simple: one simply discards all pieces with weight less than Δ (plus additional logarithmic divergences at weight Δ, when present). One can indeed show that removing these pieces is equivalent to adding local covariant counterterms to the on-shell action.

The radial canonical momentum for the asymptotically AdS case is equal to

$$\bar{\Pi}_{ij} = \frac{1}{2\bar{\kappa}^2}\sqrt{g}(K_{ij} - Kg_{ij}), \tag{29}$$

with $K_{ij} = (1/2)\partial_r g_{ij}$ the extrinsic curvature of constant-r slices. In the case of asymptotically power-law backgrounds, the relevant canonical momentum is that of the dual frame [21], namely

$$\bar{\Pi}^D_{ij} = \frac{1}{2\bar{\kappa}^2}\sqrt{g}e^{\lambda\Phi}\left(K_{ij} - (K + \lambda\partial_r\Phi)\delta^i_j\right), \tag{30}$$

where $\lambda = \sqrt{2/n}$.

4.2. 2-point functions.

To linear order in the sources, the variation of the 1-point function for the stress-energy tensor is

$$\delta\langle T_{ij}(x)\rangle = -\int \mathrm{d}^3y\sqrt{g_{(0)}}\left(\frac{1}{2}\langle T_{ij}(x)T_{kl}(y)\rangle\delta g^{kl}_{(0)}(y) + \langle T_{ij}(x)\mathcal{O}(y)\rangle\delta\varphi_{(0)}(y)\right). \tag{31}$$

It follows that in order to obtain the holographic 2-point functions we need to solve the the linearized equations of motion about the domain-wall solution with Dirichlet boundary conditions at infinity and imposing regularity in the interior. Transforming to momentum space, on general grounds, the 2-point function takes the form

$$\langle T_{ij}(\bar{q})T_{kl}(-\bar{q})\rangle = A(\bar{q})\Pi_{ijkl} + B(\bar{q})\pi_{ij}\pi_{kl}, \tag{32}$$

where Π_{ijkl} is the three-dimensional transverse traceless projection operator defined by

$$\Pi_{ijkl} = \frac{1}{2}(\pi_{ik}\pi_{lj} + \pi_{il}\pi_{kj} - \pi_{ij}\pi_{kl}), \qquad \pi_{ij} = \delta_{ij} - \frac{\bar{q}_i\bar{q}_j}{\bar{q}^2}. \tag{33}$$

Decomposing the boundary metric as in (5), we then find

$$\delta\langle T_{ij}\rangle = \frac{1}{2}A(\bar{q})\gamma_{(0)ij} - 2B(\bar{q})\psi_{(0)}\pi_{ij} - \langle T_{ij}(\bar{q})\mathcal{O}(-\bar{q})\rangle\delta\varphi_{(0)}. \tag{34}$$

From (25), this expression is to be compared with the bulk radial canonical momentum expanded to linear order. We begin by writing the perturbed metric as in (4) and (5), with the lapse and

[6] In odd (bulk) dimensions, the transformation rule of this specific coefficient has also an additional anomalous contribution due to the conformal anomaly. There is no conformal anomaly in our case, and this coefficient is a true eigenfunction of the dilatation operator.

shift perturbations gauged to zero ($\phi = \nu = \nu_i = 0$), and we additionally set ω_i to zero using the constraints (10) and a spatial gauge transformation. Then, for the case of asymptotically AdS backgrounds, using (29) we find

$$\delta\langle T^i_j\rangle = \frac{1}{\bar{\kappa}^2}[\delta K\delta^i_j - \delta K^i_j]_{(3)} = -\frac{1}{\bar{\kappa}^2}\left[2\dot{\psi}\delta^i_j + \bar{q}^2\dot{\chi}\pi^i_j + \frac{1}{2}\dot{\gamma}^i_j\right]_{(3)}. \tag{35}$$

Expressing (10) in the present gauge, it follows that

$$2\dot{\psi} = \dot{\varphi}\delta\varphi, \tag{36}$$
$$\bar{q}^2\dot{\chi} = \frac{\bar{q}^2\psi}{a^2H} - \epsilon\dot{\zeta} = \frac{\bar{q}^2\psi}{a^2H} + \frac{\bar{\kappa}^2\bar{\Omega}\zeta}{2a^3} = \left(\frac{\bar{q}^2}{a^2H} + \frac{\bar{\kappa}^2\bar{\Omega}}{2a^3}\right)\psi + \frac{\bar{\kappa}^2\bar{\Omega}H}{2a^3\dot{\varphi}}\delta\varphi,$$

where in the last line we have used the definition of the domain-wall response function $\bar{\Omega}$ in (17) and expanded ζ according to (6). Then, using the definition of the response function $\bar{E}$ also in (17), we find

$$\delta\langle T^i_j\rangle = \left[\frac{2\bar{E}}{a^3}\gamma^i_j - \left(\frac{\bar{q}^2}{\bar{\kappa}^2a^2H} + \frac{\bar{\Omega}}{2a^3}\right)\psi\pi^i_j - \left(\frac{H\bar{\Omega}}{2a^3\dot{\varphi}}\pi^i_j + \frac{\dot{\varphi}}{\bar{\kappa}^2}\delta^i_j\right)\delta\varphi\right]_{(3)}. \tag{37}$$

Since the scale factor a has dilatation weight -1 (as follows from (27)), comparison with (34) yields[7]

$$A(\bar{q}) = 4\bar{E}_{(0)}(\bar{q}), \qquad B(\bar{q}) = \frac{1}{4}\bar{\Omega}_{(0)}(\bar{q}), \tag{38}$$

where the zero subscript indicates the pieces of the response functions that have zero weight under dilatations, so in particular they are independent of r as $r \to \infty$. Note that, in general, $\bar{E}$ and $\bar{\Omega}$ diverge as $r \to \infty$. Extracting $\bar{E}_{(0)}$ and $\bar{\Omega}_{(0)}$ correctly then requires first determining the terms with eigenvalue less than zero and subtracting these from $\bar{E}$ and $\bar{\Omega}$, before taking the limit $r \to \infty$ (see [18] and the example in the next subsection). The issue here is that the subtraction of the infinite pieces may induce a change in the finite part as well. This can happen if the local covariant counterterms needed to cancel the infinities necessarily have a finite part as well.

Considering now the case of backgrounds that are asymptotically power-law, as before we decompose the metric perturbations as in (4) and (5), and gauge the shift and vector perturbations to zero ($\nu = \nu_i = \omega_i = 0$). This time, however, we choose the lapse perturbation to be $\phi = (\lambda/2)\delta\varphi$. Transforming to the dual frame, we then find

$$\mathrm{d}\tilde{s}^2 = e^{-\lambda\Phi}\mathrm{d}s^2 = \mathrm{d}r^2 + \tilde{a}^2[\delta_{ij} + \tilde{h}_{ij}]\mathrm{d}x^i\mathrm{d}x^j,$$
$$\tilde{h}_{ij} = -2\tilde{\psi}\delta_{ij} + 2\partial_i\partial_j\tilde{\chi} + \tilde{\gamma}_{ij}, \tag{39}$$

where $\tilde{a} = \exp(-\lambda\varphi/2)a$, with $\lambda = \sqrt{2/n}$ and the radial coordinate $\mathrm{d}r = \exp(-\lambda\varphi/2)\mathrm{d}z$. The dual frame perturbations are related to their Einstein frame counterparts by

$$\tilde{\psi} = \psi + (\lambda/2)\delta\varphi, \qquad \tilde{\chi} = \chi, \qquad \tilde{\gamma}_{ij} = \gamma_{ij}. \tag{40}$$

From (25) and (30), we have

$$\delta\langle T^i_j\rangle = \frac{1}{\bar{\kappa}^2}\left[e^{\lambda\varphi}\left((\delta\tilde{K} + \lambda\delta\varphi_{,r})\delta^i_j - \delta\tilde{K}^i_j + (\ldots)\delta\varphi\right)\right]_{(3)}$$
$$= -\frac{1}{\bar{\kappa}^2}\left[e^{\lambda\varphi}\left(2\tilde{\psi}_{,r}\delta^i_j - \lambda\delta\varphi_{,r}\delta^i_j + \bar{q}^2\tilde{\chi}_{,r}\pi^i_j + \frac{1}{2}\tilde{\gamma}^i_{j,r} + (\ldots)\delta\varphi\right)\right]_{(3)}$$
$$= -\frac{1}{\bar{\kappa}^2}\left[e^{3\lambda\varphi/2}\left(2\dot{\psi}\delta^i_j + \bar{q}^2\dot{\chi}\pi^i_j + \frac{1}{2}\dot{\gamma}^i_j + (\ldots)\delta\varphi\right)\right]_{(3)}. \tag{41}$$

[7] The term proportional to $\bar{q}^2$ in the coefficient of ψ in (37) only contributes a contact term to $B(\bar{q})$ which we drop.

It follows from (34) that the terms proportional to $\delta\varphi$ do not contribute to the stress tensor 2-point function that we are interested in, so in the above (and below) these terms have been suppressed. The Einstein frame constraints (10), when expressed in the present gauge, yield

$$2\dot{\psi} = (\ldots)\delta\varphi, \qquad \bar{q}^2\dot{\chi} = \left(\frac{\bar{q}^2}{a^2H} + \frac{\bar{\kappa}^2\bar{\Omega}}{2a^3}\right)\psi + (\ldots)\delta\varphi. \tag{42}$$

Thus we have

$$\delta\langle T^i_j\rangle = \left[\frac{2\bar{E}}{\tilde{a}^3}\tilde{\gamma}^i_j - \left(\frac{\bar{q}^2 e^{\lambda\varphi/2}}{\bar{\kappa}^2\tilde{a}^2 H} + \frac{\bar{\Omega}}{2\tilde{a}^3}\right)\tilde{\psi}\pi^i_j + (\ldots)\delta\varphi\right]_{(3)}, \tag{43}$$

and since the dilatation weight of $\tilde{a}$ is -1, we again recover (38) modulo contact terms.

4.3. Example: Power-law inflation.

To illustrate the above discussion, let us consider the domain-wall backgrounds equal (rather than asymptotic) to (20) discussed in [21], namely

$$a = (z/z_0)^n, \qquad \varphi = \sqrt{2n}\ln(z/z_0), \qquad z_0 = n-1 > 0. \tag{44}$$

Under the domain-wall/cosmology correspondence these solutions are mapped to cosmologies undergoing exact power-law inflation. While this model is strongly constrained by the WMAP data [6], this need not concern us here since our purpose is simply to illustrate the steps involved in the holographic computation. Furthermore, in Section 7 we will see that the strong coupling version of these models (*i.e.*, where gravity is strongly coupled at early times but the dual three-dimensional QFT is weakly coupled) are compatible with observations.

Following [10], one can obtain the fake superpotential from the solution yielding $W = -(2n/z_0)\exp(-\varphi/\sqrt{2n})$. It follows that $\epsilon = 1/n$ and both γ_{ij} and ζ obey the same equation of motion, which for the domain-wall spacetime reads

$$0 = \ddot{\zeta}_{\bar{q}} + (3n/z)\dot{\zeta}_{\bar{q}} - (z/z_0)^{-2n}\bar{q}^2\zeta_{\bar{q}}. \tag{45}$$

Imposing regularity in the interior, the solution is

$$\zeta_{\bar{q}} = C_{\bar{q}}\rho^\sigma K_\sigma(\rho), \tag{46}$$

where K_σ is a modified Bessel function of the second kind of order $\sigma = (3n-1)/2(n-1) > 3/2$, $C_{\bar{q}}$ is an arbitrary function of $\bar{q}$ and the radial coordinate $\rho = \bar{q}(z/z_0)^{1-n}$. The boundary $z \to \infty$ corresponds to $\rho = 0$ while the domain-wall interior corresponds to $\rho \to \infty$. The corresponding radial canonical momentum is equal to

$$\bar{\Pi}^{(\zeta)}_{\bar{q}} = \frac{2\epsilon}{\bar{\kappa}^2}a^3\dot{\zeta}_{\bar{q}} = -\frac{2C_{\bar{q}}}{n\bar{\kappa}^2}\left(\frac{\rho}{\bar{q}}\right)^{-2\sigma}\rho\partial_\rho(\rho^\sigma K_\sigma(\rho)). \tag{47}$$

Expanding about $\rho = 0$, we find

$$\begin{aligned}\zeta_{\bar{q}} &= C_{\bar{q}}\left(1 + \frac{1}{4(1-\sigma)}\rho^2 + \ldots - \frac{\Gamma(1-\sigma)}{4^\sigma\Gamma(1+\sigma)}\rho^{2\sigma} + \ldots\right), \\ \bar{\Pi}^{(\zeta)}_{\bar{q}} &= -C_{\bar{q}}\frac{2\bar{q}^{2\sigma}}{n\bar{\kappa}^2}\left(\frac{1}{2(1-\sigma)}\rho^{2(1-\sigma)} + \ldots - \frac{2\sigma\Gamma(1-\sigma)}{4^\sigma\Gamma(1+\sigma)} + \ldots\right),\end{aligned} \tag{48}$$

and thus

$$\Omega(\bar{q}) = -\frac{\bar{\Pi}^{(\zeta)}_{\bar{q}}}{\zeta_{\bar{q}}} = \frac{2\bar{q}^{2\sigma}}{n\bar{\kappa}^2}\left(\frac{1}{2(1-\sigma)}\rho^{2(1-\sigma)} + \ldots - \frac{2\sigma\Gamma(1-\sigma)}{4^\sigma\Gamma(1+\sigma)} + \ldots\right). \tag{49}$$

As expected, this diverges as $\rho \to 0$. To compute the 2-point function we need to identify the parts that have negative dilatation eigenvalue, subtract them from (49) and then take $\rho \to 0$.

To do this, we first transform to the dual frame via $g^D_{ij} = e^{-\sqrt{(2/n)}\varphi} g^E_{ij}$ and then change radial variable, $r = z_0 \ln(z/z_0) = -\ln(\rho/\bar{q})$. The metric is now that of AdS,

$$\mathrm{d}s^2 = \mathrm{d}r^2 + e^{2r}\mathrm{d}\vec{x}^2, \tag{50}$$

and the dilatation operator is exactly equal to the radial derivative,

$$\delta_D = \partial_r = -\rho\partial_\rho. \tag{51}$$

(This reflects the fact that the AdS isometry group is the same as the conformal group in one dimension less). It follows that any monomial in ρ is an eigenfunction of δ_D,

$$\delta_D \rho^n = -n\rho^n, \tag{52}$$

and one can simply identify in (49) all terms with negative eigenvalue; for example, $\bar{\Omega}_{(-2\sigma+2)} = \bar{q}^{2\sigma}/(n\bar{\kappa}^2(1-\sigma))\rho^{-2(\sigma-1)}$. We then have

$$\bar{\Omega}_{(0)} = -\frac{4\sigma\Gamma(1-\sigma)}{n4^\sigma\Gamma(1+\sigma)}\bar{\kappa}^{-2}\bar{q}^{2\sigma}. \tag{53}$$

In this example, the identification of the terms with negative eigenvalues could be accomplished by inspection. In more complicated examples, however, this is no longer the case, so we briefly indicate here how one could compute them (see [18] for a more complete discussion). Starting from (45), and using the definition of the corresponding canonical momentum (the first equality in (47)), one obtains

$$\dot{\bar{\Pi}}_{\bar{q}} - \frac{2\bar{q}^2}{n\bar{\kappa}^2} a\zeta_{\bar{q}} = 0. \tag{54}$$

Inserting in this equation the definition of the response function, and changing the radial coordinate from z to r, one obtains

$$\partial_r\bar{\Omega} - \alpha^2\bar{\Omega}^2 e^{-2\sigma r} + (\bar{q}^2/\alpha^2)e^{2(\sigma-1)r} = 0, \tag{55}$$

where $\alpha^2 = n\bar{\kappa}^2/2$. This may now be solved asymptotically by expanding Ω in dilatation eigenvalues,

$$\bar{\Omega} = \sum_{k\geq 1} \bar{\Omega}_{(-2\sigma+2k)}, \tag{56}$$

making use[8] of $\partial_r = \delta_D$ and collecting all terms with the same weight. For example, to leading order, at weight $(-2\sigma+2)$, only the first and last term in (55) can have this weight, and one obtains $\bar{\Omega}_{(-2\sigma+2)} = (\bar{q}^2/\alpha^2 2(1-\sigma))\exp(2(\sigma-1)r)$ in agreement with our earlier result. Through iteration, one may obtain all coefficients with negative eigenvalue.

Having obtained $\bar{\Omega}_{(0)}$, we finally compute $B(\bar{q})$:

$$B(\bar{q}) = \frac{1}{4}\bar{\Omega}_{(0)} = -\frac{\sigma\Gamma(1-\sigma)}{n4^\sigma\Gamma(1+\sigma)}\bar{\kappa}^{-2}\bar{q}^{2\sigma} = -\frac{\pi}{4^\sigma\Gamma^2(\sigma)\, n\sin\pi\sigma}\bar{\kappa}^{-2}\bar{q}^{2\sigma}. \tag{57}$$

[8] In examples where the background solution is only asymptotically AdS, the relation between the dilatation operator and the radial derivative contains subleading terms (see (26)) and these that must be taken into account in this computation, see [18] for a complete discussion.

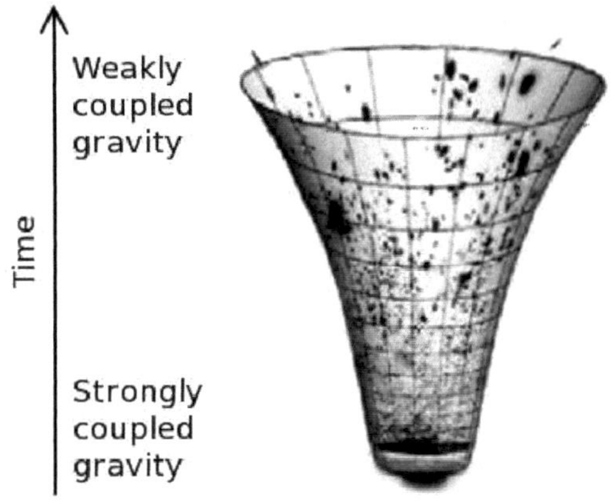

Figure 2. Using holography it is possible to describe the generation of primordial cosmological perturbations during an early time epoch in which the gravitational description is *strongly coupled.* At later times we envision a smooth transition to a conventional hot big bang description in which gravity is weakly coupled. (Figure adapted from [27]).

A near-identical argument holds for the tensors γ_{ij} yielding $\bar{\Omega}_{(0)} = (8/n)\bar{E}_{(0)}$, and hence $A(\bar{q}) = 2nB(\bar{q})$. Via the domain-wall/cosmology correspondence, applying the continuations (12), the imaginary parts of the cosmological response functions are

$$\mathrm{Im}\Omega_{(0)} = (8/n)\mathrm{Im}E_{(0)} = -\frac{4\pi}{n4^{\sigma}\Gamma^2(\sigma)}\kappa^{-2}q^{2\sigma}. \tag{58}$$

From (16), we then recover the expected cosmological power spectra:

$$\Delta_S^2(q) = \frac{n}{16}\Delta_T^2(q) = \frac{n4^{\sigma-2}\Gamma^2(\sigma)}{\pi^3}\kappa^2 q^{3-2\sigma}. \tag{59}$$

5. Continuation to the pseudo-QFT.

We now wish to re-express the bulk analytic continuation (12) in terms of QFT variables, corresponding to the vertical line on the right-hand side of Fig. 1. First, the dimensionful coupling of the theory, corresponding to the deformation by the operator $\mathcal{O}$ can be read off from the asymptotics of Φ. Since Φ does not continue, neither does the coupling constant. Second, $\bar{\kappa}^{-2}$ is proportional to the square of the number of colors, $\bar{N}^2$. It follows that the continuation (12) amounts to

$$\bar{N}^2 = -N^2, \qquad \bar{q} = -iq, \tag{60}$$

where the barred quantities are associated with the QFT dual to the domain-wall and the unbarred quantities are associated with the pseudo-QFT dual to the cosmology. (Thus $\bar{N}$ is the rank of the gauge group of the QFT dual to the domain-wall spacetime, while N is the rank of gauge group of the pseudo-QFT dual to the cosmological spacetime). We therefore find that the power spectrum for any inflationary cosmology that is asymptotically de Sitter or asymptotically power-law can be directly computed from the 2-point function of a three-dimensional QFT via the formulae:

$$\Delta_S^2(q) = \frac{-q^3}{16\pi^2\mathrm{Im}B(-iq)}, \qquad \Delta_T^2(q) = \frac{-2q^3}{\pi^2\mathrm{Im}A(-iq)}. \tag{61}$$

This is one of our principal results.

6. Beyond the weak gravitational description.

In the discussion so far we have assumed that the description in terms of gravity coupled to a scalar field is valid at early times, and that the perturbative quantization of fluctuations can be justified. The holographic description also allows us to obtain results when these assumptions do not hold. At early times, the theory may be strongly coupled with no useful description in terms of low-energy fields (such as the metric and the scalar field). The holographic set-up allows us to extract the late-time behavior of the system, which can be expressed in terms of low-energy fields, from QFT correlators. This is the counterpart of the discussion in Section 4 where we saw that in gauge/gravity duality the asymptotic behavior of bulk fields near the boundary of spacetime is reconstructed by the correlators of the dual QFT. This late-time behavior is precisely the information we need to compute the primordial power spectra and other cosmological observables. We will assume that this postulated early-time phase in which gravity is strongly coupled is subsequently followed by a smooth transition to the usual hot big bang cosmology in which gravity is weakly coupled, see Fig. 2.

7. Holographic phenomenology for cosmology.

Ideally one would deduce from a string/M-theoretic construction what the dual QFT is. Instead we initiate here a holographic phenomenological approach. The dual QFT would involve scalars, fermions and gauge fields and it should admit a large N limit. The question is then whether one can find a theory which is compatible with current observations. In particular, one might consider either deformations of CFTs or theories with a single dimensionful parameter in the regime where the dimensionality of the coupling constant drives the dynamics, as these QFTs have already featured in our discussion above.

7.1. A prototype dual QFT.

We will discuss here super-renormalizable theories that contain one dimensionful coupling constant. A prototype example is three-dimensional $SU(\bar{N})$ Yang-Mills theory[9] coupled to a number of scalars and fermions, all transforming in the adjoint of $SU(\bar{N})$. Theories of this type are typical in AdS/CFT where they appear as the worldvolume theories of D-branes. A general such model that admits a large $\bar{N}$ limit is

$$S = \frac{1}{g_{\mathrm{YM}}^2}\int \mathrm{d}^3x\, \mathrm{tr}\Big[\frac{1}{2}F_{ij}^I F^{Iij} + \frac{1}{2}(D\phi^J)^2 + \frac{1}{2}(D\chi^K)^2 + \bar{\psi}^L \not{D}\psi^L \\ + \lambda_{M_1M_2M_3M_4}\Phi^{M_1}\Phi^{M_2}\Phi^{M_3}\Phi^{M_4} + \mu^{\alpha\beta}_{ML_1L_2}\Phi^M\psi^{L_1}_\alpha\psi^{L_2}_\beta\Big], \tag{62}$$

where we consider $\mathcal{N}_A$ gauge fields A^I ($I = 1, \ldots, \mathcal{N}_A$); $\mathcal{N}_\phi$ minimal scalars ϕ^J ($J = 1, \ldots, \mathcal{N}_\phi$); $\mathcal{N}_\chi$ conformal scalars χ^K ($K = 1, \ldots, \mathcal{N}_\chi$) and $\mathcal{N}_\psi$ fermions ψ^L ($L = 1, \ldots, \mathcal{N}_\psi$). The couplings $\lambda_{M_1M_2M_3M_4}$ and $\mu^{\alpha\beta}_{ML_1L_2}$ (where α and β are spinor indices) are dimensionless, and we have grouped the scalars appearing in the interaction terms as $\Phi^M = (\{\phi^J\},\{\chi^K\})$. Note that all terms in the Lagrangian have dimension 4 so this QFT is indeed of the same type as the QFTs dual to asymptotically power law solutions. The operator $\mathcal{O}$ that featured in our earlier discussion of holography is closely related to the Lagrangian (see the discussion in Section 4 of [21]).

The conformally coupled scalars have an $R\chi^2$ coupling when we couple the theory to gravity; on a flat background this means the conformally coupled scalars have a different stress energy tensor from their minimally coupled counterparts. Specifically, on a flat background, the stress

[9] We write the rank of the QFT gauge group here as $\bar{N}$ since we will first be performing calculations using the QFT dual to the domain-wall spacetime before analytically continuing to the pseudo-QFT.

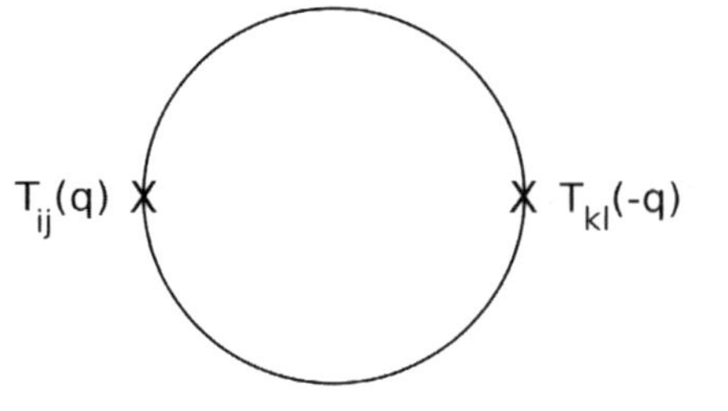

Figure 3. 1-loop contribution to $\langle T_{ij}(\bar{q})T_{kl}(-\bar{q})\rangle$. We sum over the contributions from gauge fields, scalars and fermions, with each diagram yielding a contribution of order $\sim \bar{N}^2\bar{q}^3$.

tensor is given by

$$T_{ij} = \frac{1}{g_{\rm YM}^2}\,{\rm tr}\Big[2F^I_{ik}F^{I\,k}_j + D_i\phi^J D_j\phi^J + D_i\chi^K D_j\chi^K - \frac{1}{8}D_iD_j(\chi^K)^2 + \frac{1}{2}\bar{\psi}^L\gamma_{(i}\overleftrightarrow{D}_{j)}\psi^L \\ - \delta_{ij}\Big(\frac{1}{2}F^I_{kl}F^{I\,kl} + \frac{1}{2}(D\phi^J)^2 + \frac{1}{2}(D\chi^K)^2 - \frac{1}{8}D^2(\chi^K)^2 \\ + \lambda_{M_1M_2M_3M_4}\Phi^{M_1}\Phi^{M_2}\Phi^{M_3}\Phi^{M_4} + \mu^{\alpha\beta}_{ML_1L_2}\Phi^M\psi^{L_1}_\alpha\psi^{L_2}_\beta\Big)\Big]. \tag{63}$$

To extract predictions, we need to compute the coefficients $A(\bar{q})$ and $B(\bar{q})$ appearing in the general decomposition (32) of the stress tensor 2-point function, analytically continue the results, and then insert them in the holographic formulae (61) for the power spectra. To facilitate comparison with the observational data, we will additionally make use of the standard cosmological parameterizations

$$\Delta^2_S(q) = \Delta^2_S(q_0)\left(\frac{q}{q_0}\right)^{n_S(q)-1}, \qquad \Delta^2_T(q) = \Delta^2_T(q_0)\left(\frac{q}{q_0}\right)^{n_T(q)}, \tag{64}$$

where $\Delta^2_{S/T}(q_0)$ is the scalar/tensor amplitude at some chosen pivot scale q_0, and $n_{S/T}(q)$ is the scalar/tensor spectral tilt.

7.2. 1-loop calculation.

The leading contribution to the 2-point function of the stress tensor is at one loop (see Fig. 3). Since the stress tensor has dimension three, and the only dimensionful quantity that can appear to this order is $\bar{q}$ (1-loop amplitudes are independent of g^2_{YM}), it follows that

$$A(\bar{q}) = C_A\bar{N}^2\bar{q}^3 + O(g^2_{\rm YM}), \qquad B(\bar{q}) = C_B\bar{N}^2\bar{q}^3 + O(g^2_{\rm YM}), \tag{65}$$

where C_A and C_B are numerical coefficients whose value depends only on the field content. Explicit calculation then reveals that

$$C_A = (\mathcal{N}_A + \mathcal{N}_\phi + \mathcal{N}_\chi + 2\mathcal{N}_\psi)/256, \qquad C_B = (\mathcal{N}_A + \mathcal{N}_\phi)/256. \tag{66}$$

Inserting (65) into our holographic formulae (61), we find

$$\Delta^2_S(q) = \frac{1}{16\pi^2N^2C_B} + O(g^2_{\rm YM}), \qquad \Delta^2_T(q) = \frac{2}{\pi^2N^2C_A} + O(g^2_{\rm YM}). \tag{67}$$

Comparing with (64), we immediately see that the power spectra are *scale-invariant* to leading order (*i.e.* $n_S = 1 + O(g^2_{\rm YM})$, $n_T = O(g^2_{\rm YM})$), regardless of the precise field content of the model. To estimate the value of N we may compare with the observed amplitude of the scalar power spectrum. From the WMAP data [6] we have $\Delta^2_S(q_0) \sim O(10^{-9})$, hence $N \sim O(10^4)$, justifying our use of the large N limit.

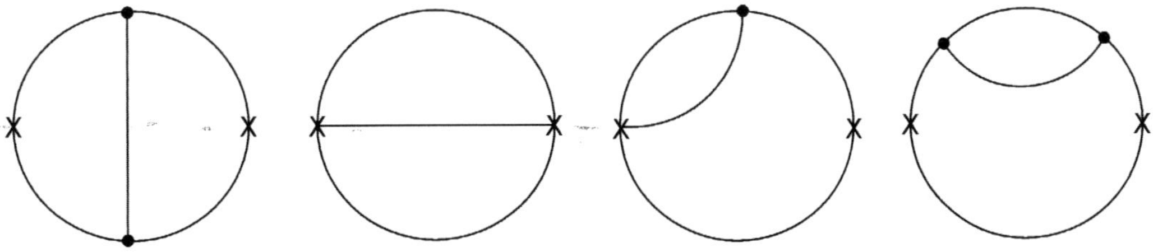

Figure 4. Diagram topologies contributing at 2-loop order. Each diagram consists of an overall factor of $\bar{N}^3 g_{\rm YM}^2$ multiplying an integral with superficial degree of divergence two. After dimensional regularization and renormalization, the integrals evaluate to $\sim \bar{q}^2 \ln(\bar{q}/\bar{q}_*)$, and so overall each diagram yields a contribution to the stress tensor 2-point function of order $\sim \bar{N}^3 g_{\rm YM}^2 \bar{q}^2 \ln(\bar{q}/\bar{q}_*)$, or equivalently $\sim \bar{N}^2 \bar{q}^3 g_{\rm eff}^2 \ln(\bar{q}/\bar{q}_*)$.

The observational data also serve to provide an upper bound on the ratio of tensor to scalar power spectra. From (67), we find

$$r = \Delta_T^2/\Delta_S^2 = 32 C_B/C_A, \tag{68}$$

and hence an upper bound on r translates into a constraint on the field content of the dual QFT. A smaller upper bound on r requires increasing the number of conformal scalars and massless fermions and/or decreasing the number of gauge fields and minimal scalars.

7.3. 2-loop corrections.

Corrections to the stress tensor 2-point function at 2-loop order[10] give rise to small deviations from scale invariance. The full result will be reported elsewhere [19], however, it is easy to obtain an order of magnitude estimate on general grounds. The perturbative expansion depends on the effective dimensionless coupling constant $g_{\rm eff}^2 = g_{\rm YM}^2 \bar{N}/\bar{q}$. Either from inspection or from direct calculation of some of the diagrams contributing at $O(g_{\rm eff}^2)$ (see Fig. 4), one finds

$$\begin{aligned} A(\bar{q}) &= C_A \bar{N}^2 \bar{q}^3 [1 + D_A g_{\rm eff}^2 \ln(\bar{q}/\bar{q}_*) + O(g_{\rm eff}^4)], \\ B(\bar{q}) &= C_B \bar{N}^2 \bar{q}^3 [1 + D_B g_{\rm eff}^2 \ln(\bar{q}/\bar{q}_*) + O(g_{\rm eff}^4)], \end{aligned} \tag{69}$$

where D_A and D_B are numerical coefficients of order one whose value depends only on the field content. To compute D_A and D_B precisely requires summing all the relevant 2-loop diagrams.

Inserting these two-loop corrected results into the holographic formulae (61), we find

$$\begin{aligned} \Delta_S^2(q) &= \frac{1}{16\pi^2 N^2 C_B}[1 - D_B g_{\rm eff}^2 \ln(q/q_*) + O(g_{\rm eff}^4)], \\ \Delta_T^2(q) &= \frac{2}{\pi^2 N^2 C_A}[1 - D_A g_{\rm eff}^2 \ln(q/q_*) + O(g_{\rm eff}^4)], \end{aligned} \tag{70}$$

where the analytically continued effective coupling $g_{\rm eff}^2 = g_{\rm YM}^2 N/q$. In comparison, expanding (64) yields

$$\begin{aligned} \Delta_S^2(q) &= \Delta_S^2(q_0)[1 + (n_S(q) - 1)\ln(q/q_0) + O\left((n_S(q)-1)^2\right)], \\ \Delta_T^2(q) &= \Delta_T^2(q_0)[1 + n_T(q)\ln(q/q_0) + O\left(n_T(q)^2\right)]. \end{aligned} \tag{71}$$

[10] Super-renormalizable theories have infrared divergences, but large N resummation leads to well-defined expressions with $g_{\rm YM}^2$ effectively playing the role of an infrared regulator. The exact amplitudes are nonanalytic functions of the coupling constant [28]. Note that our analytic continuation to pseudo-QFT does not involve the coupling constant.

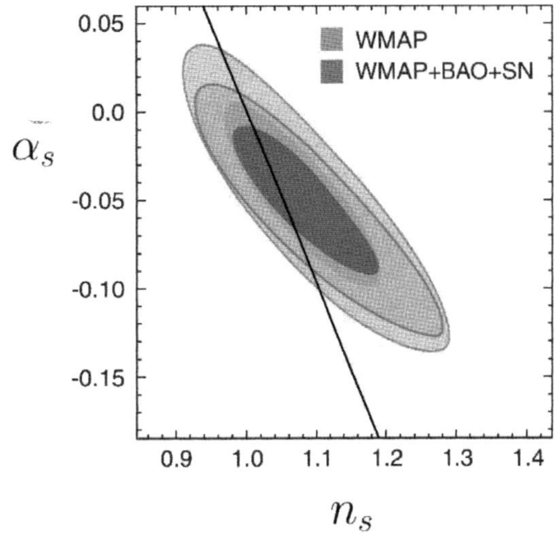

Figure 5. The straight line is the leading order prediction of holographic models with a single dimensionful coupling constant for the correlation of the running α_s and the scalar tilt n_s. The data show the 68% and 95% CL constraints (marginalizing over tensors) at $q_0 = 0.002\,\mathrm{Mpc}^{-1}$, and are taken from Fig. 4 of [6]. As new data appear the allowed region should shrink to a point, which is predicted to lie close to the line.

Identifying the renormalization scale q_* with the pivot scale q_0, we then see that the spectral amplitudes given in (67) are correct to $O(g_{\rm eff}^4)$, and that the corresponding spectral tilts are

$$n_S(q)-1 = -D_B g_{\rm eff}^2 + O(g_{\rm eff}^4), \qquad n_T(q) = -D_A g_{\rm eff}^2 + O(g_{\rm eff}^4). \tag{72}$$

Comparing with the WMAP data, from Table 4 of [6] we find that $(n_s-1) \sim O(10^{-2})$ at $q = 0.002\,\mathrm{Mpc}^{-1}$, and hence $g_{\rm eff}^2 \sim O(10^{-2})$ also, justifying our perturbative treatment of the QFT.

To determine whether the spectral tilts are red or blue requires evaluating the signs of D_A and D_B, which will in general depend on the field content of the QFT. It is nonetheless still possible to extract predictions which are independent of the field content: for example, in these models, the scalar spectral index runs as

$$\alpha_s = \mathrm{d}n_s/\mathrm{d}\ln q = -(n_s-1) + O(g_{\rm eff}^4). \tag{73}$$

This prediction is qualitatively different from slow-roll inflation, for which $\alpha_s/(n_s-1)$ is of first-order in slow-roll [29], yet is nonetheless consistent with the WMAP observational constraints on n_s and α_s given in [6] for a wide range of values of n_s and α_s, as illustrated in Fig. 5.

7.4. Non-Gaussianities.
Once N, $g_{\rm YM}^2$ and the field content are fixed, all other cosmological observables (such as non-Gaussianities, *etc.*) follow uniquely from straightforward computations. We will present details of the correspondence between higher-order QFT correlation functions and non-Gaussian cosmological observables elsewhere [19]. Our results indicate, however, that the non-Gaussianity parameter $f_{NL}^{\rm local}$ [30] is independent of N to leading order, consistent with current observational evidence [6].

8. Conclusions.
Let us summarize the main results. We have presented a holographic framework for early-universe cosmology describing the period of time corresponding to the inflationary epoch. In particular, we have shown how to compute cosmological observables by performing calculations with a three-dimensional dual QFT. This procedure was discussed explicitly for the case of the primordial power spectrum, which is related to the 2-point functions of the dual QFT.

Higher-point functions are related to non-Gaussianities, as will be discussed elsewhere. When gravity is weakly coupled at early times, holography correctly reproduces standard inflationary predictions for cosmological observables. When gravity is instead strongly coupled at early times, one finds new models that have a weakly coupled QFT description. We saw how models of this type exist that are compatible with current observations yet nevertheless have a distinct phenomenology from standard inflation. The proposed holographic approach thus provides a qualitatively new method for generating a nearly scale-invariant spectrum of primordial cosmological perturbations.

A special case to consider is de Sitter spacetime. A correspondence between de Sitter (dS) and CFT has been proposed in the past [31], and it is natural to wish to understand the relation of the present work to dS/CFT. Under the domain-wall/cosmology correspondence, dS is mapped to AdS, and our results for dS follow via suitable analytic continuation from the results for AdS. It is well known that quantum correlators, such as, for example, the Feynman propagators of massive scalar fields in AdS and dS spacetime, do not map to each other under analytic continuation, see for example [32]. This has been one of the obstacles in trying to establish a dS/CFT correspondence using analytic continuation. In our case, however, we do not map quantum de Sitter correlators to quantum AdS correlators. Instead, we map dS correlators directly to correlation functions of the dual QFT, and (as we have seen) one can successfully establish such a map. An approach to dS/CFT that has more in common with the present work is that of Maldacena [8], although our analytic continuation is different: we analytically continue both the momenta and Newton's constant, whereas in [8] it was the dS radius that was continued. Furthermore, compared to previous works [33, 34, 35, 36] that focused mostly on the computation of the scalar power spectrum for asymptotically dS geometries, we gave a complete discussion of both the scalar and tensor power spectra and our work includes the case of asymptotically power-law inflation. Finally, we proposed a precise definition of the dual QFT.

While uncovering an underlying holographic structure in inflationary cosmology is conceptually important, this in itself would lead to no new hard results were one to remain in the regime where inflationary computations are well-justified. The reason is that in this regime, the holographic computations simply reproduce well-known results. Rather, the power of the holographic approach is that it leads to new models in which the gravitational dynamics were strongly coupled at early times; for these models the standard inflationary computations do not apply. This is precisely the distinguishing feature of the present work: because we give an explicit definition of the dual QFT, we are able to obtain just such models with strongly coupled gravitational dynamics at early times. These models have a weakly coupled dual QFT description, permitting the analysis of a scenario that would otherwise be quite intractable.

We have seen how the dual QFT may be defined in an operational sense by first performing all computations with the ordinary QFT dual to a holographic RG flow, and then continuing the number of colors N and the momenta appropriately. In the large N limit the correlators are given as a series in $1/N^2$, and so the analytic continuation simply amounts to inserting minus signs. Furthermore, the continuation in momenta is such that the effective coupling $g^2_{\text{eff}} = g^2_{\text{YM}}N/q$, where q is a momentum, remains real. It would be very interesting to understand such 'pseudo'-QFTs from first principles. This would allow for a non-perturbative definition of the theory and would presumably elucidate many puzzling features of quantum gravity such as, for example, the entropy of de Sitter spacetime. Note also that the dual QFT provides a complete smooth description of the system, including that of the initial singularity that is (generically) present in the background FRW solution. Our focus here was to extract the late-time behavior of the system, but it would be very interesting to understand the implications for singularity resolution.

For the present, however, the main question is whether or not holographic models exist that are compatible with current observations but have a distinct phenomenology from standard inflation. We have shown that the answer to this question is affirmative. Initiating a holographic phenomenological approach, we found that it is straightforward to satisfy the current observational constraints using simple QFT models containing only a few parameters. The near scale invariance of the cosmological power spectra follows immediately from simple dimensional considerations. A number of parameters in the QFT model may then be estimated or constrained using the present observational data; once the remaining parameters have been fixed, all cosmological observables (including non-Gaussianities) then follow from direct computation. Note that these are complete models: there are no UV issues as these theories are super-renormalizable and furthermore the dimensionful coupling constant acts as an infrared regulator. Even without knowing the values of all the parameters in the QFT model, it is still perfectly possible to obtain concrete predictions, since not all cosmological observables depend on the full set of QFT parameters. Examples include the scale invariance of the power spectra at leading order, as well as the running of the spectral index discussed above. In general, we expect to obtain predictions that are qualitatively different from those of standard inflationary scenarios based on weakly coupled gravity. These expectations are borne out by the form of the running we found for the scalar spectral index.

Clearly, the proposed phenomenological approach to holographic cosmology is worthy of further development. Over the next few years, forthcoming experiments (in particular the Planck satellite) promise to dramatically improve the observational constraints on many important cosmological parameters. It may well be that future observations confirm the predictions of holographic models of the type advocated here. The success of such an endeavor might then provide the first observational evidence for the holographic nature of our universe.

Acknowledgments.
We thank I. Kanitscheider, B. van Rees and M. Taylor for useful discussions. The authors acknowledge the support of NWO; KS through a VICI grant and PLM through a VENI grant.

References

[1] 't Hooft G 1993 Dimensional reduction in quantum gravity *Preprint* gr-qc/9310026; Susskind L 1995 The world as a hologram *J. Math. Phys.* **36** 6377 (*Preprint* hep-th/9409089)

[2] Maldacena J M 1998 The large N limit of superconformal field theories and supergravity *Adv. Theor. Math. Phys.* **2** 231–52 (*Preprint* hep-th/9711200)

[3] Gubser S S, Klebanov I R, and Polyakov A M 1998 Gauge theory correlators from noncritical string theory *Phys. Lett.* B **428** 105–14 (*Preprint* hep-th/9802109)

[4] Witten E 1998 Anti-de Sitter space and holography *Adv. Theor. Math. Phys.* **2** 253–91 (*Preprint* hep-th/9802150)

[5] McFadden P L and Skenderis K 2009 Holography for cosmology *Preprint* 0907.5542

[6] Komatsu E *et al.* 2009 Five-year Wilkinson Microwave Anisotropy Probe (WMAP) observations: cosmological interpretation *Astrophys. J. Suppl.* **180** 330 (*Preprint* 0803.0547)

[7] Turok N 2002 A critical review of inflation *Class. Quant. Grav.* **19** 3449-67; Brandenberger R H 1999 Inflationary cosmology: progress and problems *Preprint* hep-ph/9910410

[8] Maldacena J M 2003 Non-Gaussian features of primordial fluctuations in single field inflationary models *JHEP* **0205** 013 (*Preprint* astro-ph/0210603)

[9] Cvetic M and Soleng H H 1995 Naked singularities in dilatonic domain-wall spacetimes *Phys. Rev.* D **51** 5768 (*Preprint* hep-th/9411170)

[10] Skenderis K and Townsend P K 2006 Hidden supersymmetry of domain walls and cosmologies *Phys. Rev. Lett.* **96** 191301 (*Preprint* hep-th/0602260); 2007 Pseudo-supersymmetry and the domain-wall/cosmology correspondence *J. Phys.* A **40** 6733 (*Preprint* hep-th/0610253)

[11] Bergshoeff E A, Hartong J, Ploegh A, Rosseel J and van den Bleeken D 2007 Pseudo-supersymmetry and a tale of alternate realities JHEP **0707** 067 (*Preprint* 0704.3559); Skenderis K, Townsend P K and van Proeyen A 2007 Domain-wall/cosmology correspondence in AdS/dS supergravity JHEP **0708** 036 (*Preprint* 0704.3918)

[12] Skenderis K and van Rees B C 2008 Real-time gauge/gravity duality *Phys. Rev. Lett.* **101** 081601 (*Preprint* 0805.0150); 2009 Real-time gauge/gravity duality: prescription, renormalization and examples *JHEP* **0805** 085 (*Preprint* 0812.2909)

[13] Salopek D S and Bond J R 1990 Nonlinear evolution of long wavelength metric fluctuations in inflationary models *Phys. Rev.* D **42** 3936
[14] Bardeen J M, Steinhardt P J and Turner M S 1983 Spontaneous creation of almost scale-free density perturbations in an inflationary universe *Phys. Rev.* D **28** 679
[15] Mukhanov V F, Feldman H A and Brandenberger R H 1992 Theory of cosmological perturbations *Phys. Rept.* **215** 203
[16] DeWolfe O and Freedman D Z 2000 Notes on fluctuations and correlation functions in holographic renormalization group flows *Preprint* hep-th/0002226
[17] Bianchi M, Freedman D Z and Skenderis K 2001 How to go with an RG flow *JHEP* **0108** 041 (*Preprint* hep-th/0105276)
[18] Papadimitriou I and Skenderis K 2004 AdS/CFT correspondence and geometry *Preprint* hep-th/0404176; 2004 Correlation functions in holographic RG flows *JHEP* **0410** 075 (*Preprint* hep-th/0407071)
[19] McFadden P L and Skenderis K to appear
[20] Skenderis K 2002 Lecture notes on holographic renormalization *Class. Quant. Grav.* **19** 5849 (*Preprint* hep-th/0209067)
[21] Kanitscheider I, Skenderis K and Taylor M 2008 Precision holography for non-conformal branes *JHEP* **0809** 094 (*Preprint* 0807.3324)
[22] 't Hooft G 1974 A planar diagram theory for strong interactions *Nucl. Phys.* B **72** 461
[23] Kanitscheider I and Skenderis K 2009 Universal hydrodynamics of non-conformal branes *JHEP* **0904** 062 (*Preprint* 0901.1487)
[24] de Haro S, Solodukhin S N and Skenderis K 2001 Holographic reconstruction of spacetime and renormalization in the AdS/CFT correspondence *Commun. Math. Phys.* **217** 595 (*Preprint* hep-th/0002230)
[25] Boonstra H J, Skenderis K and Townsend P K 1999 The domain wall/QFT correspondence *JHEP* **9801** 003 (*Preprint* hep-th/9807137)
[26] Papadimitriou I and Skenderis K 2005 Thermodynamics of asymptotically locally AdS spacetimes *JHEP* **0508** 004 (*Preprint* hep-th/0505190)
[27] http://x-journals.com/2009/the-search-for-dark-energy-from-the-ground-up/
[28] Jackiw R and Templeton S 1981 How super-renormalizable theories cure their infrared divergences *Phys. Rev.* D **23** 2291; Appelquist T and Pisarski R D 1981 High-temperature Yang-Mills theories and three-dimensional quantum chromodynamics *Phys. Rev.* D **23** 2305
[29] Kosowsky A and Turner M 1995 CBR anisotropy and the running of the scalar spectral index *Phys. Rev.* D **52** 1739 (*Preprint* astro-ph/9504071)
[30] Komatsu E and Spergel D N 2001 Acoustic signatures in the primary microwave background bispectrum *Phys. Rev.* D **63** 063002 (*Preprint* astro-ph/0005036)
[31] Strominger A 2001 The dS/CFT correspondence *JHEP* **0110** 034 (*Preprint* hep-th/0106113)
[32] Das A K and Dunne G V 2006 Large-order perturbation theory and de Sitter/anti-de Sitter effective actions *Phys. Rev.* D **74** 044029 (*Preprint* hep-th/0607168)
[33] Larsen F, van der Schaar J P and Leigh R G 2002 de Sitter holography and the cosmic microwave background *JHEP* **0204** 047 (*Preprint* hep-th/0202127)
[34] Larsen F and McNees R 2003 Inflation and de Sitter holography *JHEP* **0307** 051 (*Preprint* hep-th/0307026)
[35] van der Schaar J P 2004 Inflationary perturbations from deformed CFT *JHEP* **0401** 070 (*Preprint* hep-th/0307271)
[36] Larsen F and McNees R 2004 Holography, diffeomorphisms, and scaling violations in the CMB *JHEP* **0407** 062 (*Preprint* hep-th/0402050)

First Mediterranean Conference on Classical and Quantum Gravity (MCCQG 2009) IOP Publishing
Journal of Physics: Conference Series **222** (2010) 012008 doi:10.1088/1742-6596/222/1/012008

Naked singularities, branes and Chern-Simons couplings: The dark side of the 2+1 black hole

Talk given at the *Mediterranean Conference on Classical and Quantum Gravity*, Kolymbari, Greece, September 14-18, 2009.

Jorge Zanelli

Centro de Estudios Científicos (CECS), Valdivia, Chile

E-mail: z@cecs.cl

Abstract. Branes are naked singularities, analogous to linear or planar defects in crystals. Zero-branes in AdS spacetimes are "negative mass black holes", which can be generalized to higher-dimensional branes. When these solutions are endowed with angular momentum, the extremal spinning branes correspond to BPS states. On the other hand, the $2p$-branes, spanning a $(2p+1)$-dimensional worldsheet, provide a naturally coupling to CS field theories defined on a D-dimensional spacetime, with $D > 2p+1$. In this picture, the field that lives in the D-dimensional spacetime, as well as the sources that couple to it are made out of the same stuff –an $SO(D-1,2)$ connection. The fact that on the brane the AdS group is necessarily broken down to $SO(2p,2)$, brings in a number of tensor fields that play the role of charged matter living on the brane.

1. Introduction

Naked singularities are often viewed as unphysical, dreadful solutions that violate basic laws of physics. These structures would be some sort of Pandora's box, from which anything can emerge, including "green slime, lost socks and TV sets" [1]. In view of this outrageous behavior, it has been suggested that nature abhors naked singularities, and a *cosmic censorship* mechanism, supposedly built-in in gravitation theory, would provide a forceful clothing for singularities. Clearly, an elegant way out of uncomfortable paradoxes.

Black hole solutions with mass and angular momentum parameters M and J exhibit a naked singularity at the center if [1] $M < |J|$. Hence, the existence of a cosmic censorship mechanism would imply that nature favors $M \geq |J|$, and in particular, $M \geq 0$. This agrees with the expectation that localized distributions of matter must have positive energy/mass. Not only elegant, but useful as well.

Numerous attempts to rigorously prove the cosmic conjecture have failed, and some numerical experiments of gravitational collapse under reasonable initial conditions, inevitably lead to the formation of naked singularities [2]. However, in particular cases it can be verified that nature doesn't accept naked singularities quietly. It is a simple exercise to check that a charged

[1] Here we have used the units in which M and J are dimensionless. The right dimensions will be restored in the next section.

extremal black hole repels a charged particle whose charge is larger than its mass [3]. Other numerical experiments also provide convincing evidence that extremal naked singularities are generically unstable under linearized perturbations [4]: a small localized perturbation around a naked singularity (e.g., a black hole with negative mass parameter), grows exponentially in finite time for generic initial data in the linearized approximation. Although the unbounded growth of the perturbation means that the validity of the linear approximation eventually breaks down, it is a strong indication that something smells rotten in the negative mass black hole.

The standard four-dimensional Schwarzschild black hole has a curvature that diverges for $r \to 0$, and if the mass parameter $M < 0$ there is no horizon around the singularity. This is a problem because a divergent Riemann tensor corresponds to an unbounded energy-momentum density. A naked singularity where the curvature tensor diverges implies an infinite accumulation of energy in a finite region and that would naturally lead to a breakdown in the laws of physics.

Possibly the simplest case of naked singularity is the static three-dimensional black hole with $M < 0$. The geometry of the $2 + 1$ black hole, unlike its four dimensional counterpart, has constant negative curvature for any sign of M [5]. Thus, even in the absence of a horizon surrounding the singularity, an external observer would never have access to a region of infinite curvature. The naked singularity of a negative mass 2+1 black hole is similar to the singularity at the tip of a cone, a point where the manifold structure breaks down because differentiability fails. Nevertheless, this type of singularity is sufficiently mild to allow unambiguous interpretations. The curvature singularity at the tip of a cone the has the form of a Dirac delta, which means that its integral cannot represent an infinite amount of energy concentrated in a finite region. Thus, a conical singularity shouldn't be expected to be the source of infinitely growing excitations like those o! bserved in the instabilities mentioned above.

A naked singularity defined by the set of points where the manifold structure ends, such as a boundary, a domain wall, or a topological defect, is not necessarily a source of paradoxes, but determines the boundary conditions for the dynamical fields, including those that determine the spacetime geometry itself. In the presence of a boundary of this type it is necessary to specify the conditions on the fields, which must be in agreement with the topological obstruction produced by the singularity. In the action principle, this is afforded by the presence of an appropriate source term, that represents the coupling of the geometry to a point-like source or, more generically, to a $2p$-brane.

2. Review of the 2+1 Black Hole

The geometry of the 2+1 black hole with mass M and angular momentum J is given by the metric

$$ds^2 = -f(r)dt^2 + [f(r)]^{-1}dr^2 + r^2(Ndt + d\phi)^2, \tag{1}$$

where $f(r) = -M + \frac{r^2}{\ell^2} - \frac{J^2}{4r^2}$, $N = -\frac{J}{2r^2}$, and ℓ is a yardstick that defines the unit of length (AdS radius). The physical spectrum is the region of the $M - J$ plane defined by $M \geq |J|\ell^{-1} \geq 0$, that corresponds to black hole states. Additionally, the point $M = -1$, $J = 0$ represents anti-de Sitter space. The states $M = |J|\ell^{-1} > 0$ are extremal black holes, and the state $M = |J|\ell^{-1} = 0$ is the lowest energy (ground state) configuration. As noted in [6], this locally AdS geometry can be obtained by a finite isometric identification in the AdS covering space. That is, the black hole manifold is a quotient of AdS_3 by a Killing vector of AdS_3.

$$\Gamma = H^{2,2}/\xi(M, J), \tag{2}$$

where the Killing vector $\xi(M, J)$ is a boost. The AdS_3 geometry is defined as a pseudosphere embedded in a four-dimensional flat spacetime $\mathbb{R}^{2,2}$,

$$-x_0^2 + x_1^2 + x_3^2 - x_4^2 = -\ell^2 \tag{3}$$

The explicit form of the identification in the coordinates of the embedded pseudosphere is

$$\xi_{Generic} = r_+ J_{12} - r_- J_{03}, \tag{4}$$

where

$$r_\pm^2 = \frac{M}{2}\left[1 \pm \sqrt{1 - \frac{J^2}{M^2\ell^2}}\right]\ell^2, \tag{5}$$

and J_{ab} is the generator of rotations/boosts in the $a-b$ plane.

The ground state ($M = 0 = J$) and the extremal ($M = J$) black holes are obtained by means of the following identifications:

$$\xi_{Vac} = \frac{1}{2}(J_{12} + J_{03} + J_{02} - J_{13}), and \tag{6}$$

$$\xi_{Ext} = r_+(J_{01} - J_{23}) + \frac{1}{2}(J_{12} - J_{20} + J_{03} - J_{13}), \tag{7}$$

respectively. Notice that the Killing vector for the extremal and the vacuum solutions are not obtained by just taking the appropriate limits in the space of parameters of $\xi_{Generic}$. For a good review of the geometry and physics of the 2+1 black hole, see, e.g., [7].

3. Conical defect in AdS

It was observed by Izquierdo and Townsend, that a static naked singularity solution ($M < 0, J = 0$) corresponds to conical defect [8]. To illustrate the point, consider a two-dimensional cone, described by locally flat coordinates $(x^1, x^2) = (r\cos\phi_{12}, r\sin\phi_{12})$, where the radial coordinate takes values $0 \leq r < \infty$, and the azimuthal angle ϕ_{12} has a deficit, $0 \leq \phi_{12} \leq 2\pi(1-\alpha)$. The angular deficit produces a naked singularity at the center, $r = 0$, where the curvature has a Dirac δ. This cannot be seen just from reading the metric, which in these coordinates has the standard flat form, $ds^2 = dr^2 + r^2 d\phi_{12}^2$. It is the identification $\phi_{12} \simeq \phi_{12} + 2\pi(1-\alpha)$, with $0 \leq \alpha < 1$, that is responsible for the singularity at $r = 0$. A standard azimuthal angle ϕ of period 2π can be introduced by rescaling ϕ_{12} ($\phi_{12} = [1-\alpha]\phi$), making the topological defect manifest through a factor in the angular component of the metric. The resulting Riemann curvature two-form in two dimensions is $R^1{}_2 = d\omega^1{}_2$, where $\omega^1{}_2 = -d\phi_{12}$. Then, as pointed out in [9], the identity $dd\phi_{12} = -2\pi\alpha\,\delta(T_{12})\,d\Omega_{12}$ is valid in the sense of Stokes' theorem upon integration, and the curvature exhibits the singularity at the origin,

$$R^1{}_2 = 2\pi\alpha\,\delta(T_{12})\,d\Omega^{12}. \tag{8}$$

The two-form Dirac delta $\delta(T_{12})d\Omega_{12}$ has support at $r = 0$ on the two-dimensional plane T_{12}. It can also be checked that the torsion tensor vanishes, thanks to the property of the Dirac distribution, $r\delta(r) = 0$.

The result can be re-interpreted as an identification by the Killing vector $\xi = \xi^i\partial_i = -2\pi\alpha\,\partial_{\phi_{12}}$, where $\partial_{\phi_{12}} = x_1\partial_2 - x_2\partial_1$ is the generator of rotations around the origin in the 1-2 plane. The angular defect results from the identification $x^i \simeq x^i + \xi^i$,

$$\begin{pmatrix} x^1 \\ x^2 \end{pmatrix} \simeq \begin{pmatrix} \cos 2\pi\alpha & \sin 2\pi\alpha \\ -\sin 2\pi\alpha & \cos 2\pi\alpha \end{pmatrix} \begin{pmatrix} x^1 \\ x^2 \end{pmatrix}. \tag{9}$$

Note that the strength of the curvature singularity in the identified geometry is the magnitude of the angular deficit, $2\pi\alpha$, times the two-form delta distribution, $\delta(T_{12})d\Omega_{12}$, which can be identified as the source of the conical singularity. This happens whenever a curvature singularity is the result of an identification by a spacelike Killing vector that leaves fixed points [10]. The

Killing vector ξ is everywhere spacelike, and hence it does not produce closed timelike curves. This geometry has constant negative curvature everywhere, except at the point $r = 0$. Therefore, this is a solution of the Einstein equations with a source, obtained from the action

$$I[g] = \int_{M^3} \left\langle \frac{1}{2} A \wedge dA + \frac{1}{3} A \wedge A \wedge A \right\rangle - \int_{\Gamma^1} \langle j \wedge A \rangle \,, \tag{10}$$

where M^3 is the three-dimensional spacetime manifold, Γ^1 is the worldline defined by the point $r = 0$, and j is the two-form source defined by the right hand side of (8).

4. The "negative mass" black hole

Consider now a conical topological defect in the AdS spacetime. In polar coordinates (here we follow the notation and conventions of [11]),

$$ds^2 = -\left(1 + \frac{r^2}{\ell^2}\right) dt^2 + \left(1 + \frac{r^2}{\ell^2}\right)^{-1} dr^2 + r^2 (1-\alpha)^2 d\phi^2, \tag{11}$$

where the azimuthal angle has the range $0 \leq \phi \leq 2\pi$, and $0 \leq \alpha \leq 1$ is the angular deficit. With a coordinate rescaling, it is possible to rewrite this metric as a black hole of mass M,

$$ds^2 = -\left(-M + \frac{r^2}{\ell^2}\right) dt^2 + \left(-M + \frac{r^2}{\ell^2}\right)^{-1} dr^2 + r^2 d\phi^2, \tag{12}$$

where the mass parameter is related to the angular defect by

$$M = -\,(1-\alpha)^2 \,. \tag{13}$$

Clearly this means that the conical defects belong to the negative part of the black hole energy spectrum. The configuration without angular deficit ($\alpha = 0$) corresponds to the AdS spacetime ($M = -1$), while the maximal deficit, $\alpha = 1$, represents the zero-mass black hole.

A natural question arises here: Are these solutions stable configurations? In order to answer this, one could analyze the perturbative spectrum. However, as is well known, 2+1 gravity has no local degrees of freedom, and hence, the corresponding spectrum is rather trivial: the excitations are all pure gauge transformations connected to the identity. The only nontrivial excitations are of topological character, resulting from identifications on AdS_3: black holes ($M \geq 0$) are obtained by identifying with a timelike Killing vector, corresponding to a boost; the topological defect states ($M < 0$) result from identifying with a spatial rotation. There is nothing to prevent these states from decaying –e. g., via Hawking radiation– into a zero-mass black hole or into global AdS_3, the only two stable configurations continuously connected with $-1 \leq M \leq 0$, $J = 0$.

5. BPS States

The only stable states are those for which a Bogomol'nyi bound is saturated (BPS states). In those cases, the existence of a globally defined Killing spinor (covariantly constant spinor), is sufficient to guarantee that a residual supersymmetry is preserved. The presence of some residual unbroken supersymmetries implies the existence of a lower bound for the sum of charges, the Bogomol'nyi bound. This leads to the positivity of energy in standard supergravity [12, 13, 14, 15], and ensures the stability of the configurations that saturate the energy bound (BPS states). These BPS configurations define good vacua for a perturbative expansion and, in order to establish that a given geometry is a stable ground states, one should investigate whether the geometry admits a globally defined covariantly constant spinor or not [16].

A globally defined Killing spinor is a solution of the equation $D\psi = 0$, in the spacetime region surrounding the topological defect, where D is the covariant derivative operator in (2+1)-dimensional, locally AdS spacetime. In the present case, this means

$$D\psi = \left(d + \frac{1}{4}\omega^{ab}\Gamma_{ab} + \frac{1}{2\ell}e^{a}\Gamma_{a}\right)\psi = 0, \tag{14}$$

where the spacetime region is defined by $r > 0$, $-\infty < t < +\infty$, and $0 \leq \phi \leq 2\pi$. The spin connection and vielbein for the locally AdS and torsion-free spacetime determined by the metric (12) are

$$\begin{array}{ll} e^0 = \ell^{-1}\left(-M + \frac{r^2}{\ell^2}\right)^{1/2} dt, & \omega^{01} = \frac{r}{\ell^2}dt \\ e^1 = \left(-M + \frac{r^2}{\ell^2}\right)^{-1/2} \ell dr, & \omega^{12} = -\left(-M + \frac{r^2}{\ell^2}\right)^{1/2} d\phi \\ e^2 = r d\phi\,, & \omega^{02} = 0 \end{array} \cdot \tag{15}$$

In addition, the spinor might be coupled to a connection field (electromagnetism or some other), but let us limit the discussion to the simplest possibility.

Direct computation shows that the solution takes the form,[2]

$$\psi = \exp[f(r)\Gamma_1] \exp[\frac{i}{2}\alpha\left(\phi + \frac{t}{\ell}\right)]\psi_0\ , \tag{16}$$

where ψ_0 is a constant spinor such that $\Gamma_0\psi_0 = -\psi_0$. It is clear that the only acceptable conditions for a globally defined ψ are: $\psi(\phi+2\pi) = \psi(\phi)$, (single-valued), or $\psi(\phi+2\pi) = -\psi(\phi)$ (double-valued). The first case implies $\alpha = 0$, while the second requires $\alpha = 1$. These two cases are precisely the conditions for the globally AdS spacetime and the massless black hole, respectively. This means that these naked singularities are not BPS states and therefore are not protected from disintegration into more stable configurations (either the $M = 0$ or the $M = -1$ states).

This situation could have bee anticipated: the $M > 0$, $J = 0$ black holes are not BPS either [17]. It is possible, however, to turn a positive mass black hole into a BPS state by boosting it with the right amount of angular momentum. The question then comes to mind: would it be possible to stabilize a naked singularity as well? The answer is yes, and the situation reflects in every detail the positive mass case: one can boost a black hole into a spinning state, and the same can be done with a topological defect. The result is a geometry that resembles the spinning BTZ solution,

$$ds^2 = -f^{\,2}(r)dt^2 + \frac{dr^2}{f^{\,2}(r)} + r^2\left(Ndt + d\phi\right)^2, \tag{17}$$

where, as in the standard case,

$$f^{\,2}(r) = -M + \frac{r^2}{\ell^2} + \frac{J^2}{4r^2}\ , and\ N = -\frac{J}{2r^2}\ , \tag{18}$$

but now $M\ell^2 < -|J|$ implies that both horizons are absent

$$r_{\pm}^2 = \frac{M\ell^2}{2}\left(1 \pm \sqrt{1 - \frac{J^2}{M^2\ell^2}}\right),\ \ imaginary. \tag{19}$$

[2] See, e.g., Refs. [11, 18] for details.

For the extreme case $M\ell^2 = -|J|$ there is one globally defined Killing spinor given by [17]

$$\psi = h(r)\begin{pmatrix} 1 \\ \pm 1 \end{pmatrix}, \tag{20}$$

where $h(r) = \sqrt{\frac{r}{\ell} - \frac{\ell M}{r}}$.

6. Generalizations

A similar result is obtained if the spinor is coupled to an electromagnetic potential [18]. In that case, the spinor must be complex and the Killing equation (14) has an additional term $ieA\psi$. An interesting electromagnetic source in this spacetime is the static magnetic pole at $r = 0$. Its electromagnetic potential is $A = gd\phi$, which looks like an exact form, but it is not: The zero-form ϕ is not a well defined (single-valued) function in the $r - \phi$ plane, as can be easily checked by computing the integral of A around a closed loop around the origin ($\int A = 2\pi g \neq 0$).[3] In the presence of this magnetic pole, the Killing spinor gets an extra contribution to the phase from the electromagnetic flux through the! center of the constant time slices. Now, if the flux exactly matches the angular deficit of the conical defect, a Killing spinor can be globally defined everywhere in the 3D spacetime for $r \neq 0$. The matching condition is [18],

$$(1 - \alpha)(1 + 2eg) \in \mathbb{Z}. \tag{21}$$

Note that there are multiple options to satisfy this condition for $2eg \in \mathbb{Z}$ and $0 < \alpha < 1$. For instance, for $eg = 1$, α can take the values 1/3 and 2/3, for $eg = 8$, $\alpha = 1/17, 2/17, \ldots, 16/17$, etc.

These results can be generalized to other gauge groups, to higher dimensions, or both. The example of a zero brane in three spacetime dimensions analyzed above can be easily generalized to a 0-brane in higher dimensions, or to branes whose worldvolumes are codimension 2 manifolds. More generically, one can consider $2p$-branes embedded in a $(2n+1)$-dimensional spacetime, with $n > p$. These possibilities have been discussed in [10] for the case of gravitational Chern-Simons theories, as well as generic couplings between CS forms and $2p$-branes.

The general result can be summarized as follows:

• A $2p$-brane can be seen as a charged object that couples to a connection through the Chern-Simons form. The world history of the brane, Γ^{2p+1}, is the manifold on which the CS form is defined. This world history is a submanifold that can be isometrically embedded in a $(2n + 1)$-dimensional spacetime ($n \geq p$), where the brane is a topological defect.

• Zero-branes in spacetimes of $D > 3$, obtained by elimination of and angular sector are not constant curvature manifolds . This is because a solid angle deficit cannot be produced by an identification with a Killing vector, as is the case of the $2 + 1$ black hole –with either positive or negative mass. The result is a naked curvature singularity which is likely to be unstable.

• Branes whose world history is a manifold of codimension 2, can be obtained by identification with a single Killing vector. The resulting spacetime has the same local curvature as the space before the identification, but the topology and the causal structure depend on the nature of the Killing vector used. These are topological defects in the same class with the conical singularities discussed earlier. Several identifications can be made simultaneously, provided the corresponding Killing vectors commute, as for example in a two-dimensional torus.

• Identifications can also be performed even if the resulting manifold is not a quotient space. For example, one could make a polyhedron by removing several angular sectors centered around different points. The resulting solution is a multi-centered collection of conical defects. The

[3] This is consistent with the notation $F = dA = 2\pi g\delta(r)$, although this notation might be confusing since it suggests that $r = 0$ is a regular point in the manifold, while it is actually an excluded point in the $r - \phi$ plane.

only constraint that seems to exist on such structure is that the sum of defects cannot exceed the value for the zero-mass black hole,

$$\sum_i \alpha_i \leq 1. \tag{22}$$

7. Discussion

Action: The branes discussed here are geometrical structures obtained from the application of certain boundary conditions that introduce peculiar topological features, but they are essentially made from the same stuff as the rest of the spacetime geometry. These structures do not require the introduction of additional matter: the same connection 1-form that defines the dynamics of the bulk geometry can be used to couple to the brane.

For instance, the spacetime geometry in 2+1 dimensions is determined by a Chern-Simons action for the $SO(2,2)$ algebra [19]. This has been generalized to higher dimensions and even to the supersymmetric case [20]. What is remarkable in the theory that emerges when one allows for conical defects or their higher dimensional generalizations, is that these new structures don't require additional fields to implement them: The dynamics of the intrinsic brane geometry is governed by the CS action living on the brane history, as in (10) [21, 22]. For an observer living in the bulk, however, the brane looks as some source that could even be charged, the charge being related to the magnitude of the angular deficit at the brane.

This seems like a consistent picture –at least classically–, in which the distinction between bulk degrees of freedom (connection fields $A|_M$) and brane degrees of freedom (connection fields restricted to the brane world $A|_\Gamma$) is metaphysical. Naturally, the components of the connection 1-form, when restricted to Γ, split into the components of a connection for a smaller symmetry group, and some 0-forms fields transforming in some tensor representations of the reduced symmetry group. To an observer in the bulk, these tensor fields would look like "matter" fields, charged under the reduced gauge group, in a standard field-theoretical interpretation [23].

Solutions: The "negative energy black hole states" identified as topological defects, are clearly not black holes but naked singularities. In 2+1 dimensions these singularities are topological defects, and hence quite harmless. These solutions are similar to ordinary (classical) elementary particles: They are point-like objects, devoid of a horizon and therefore temperature and entropy cannot be defined for them. An interesting question is whether these states belong to the microscopic states that contribute to the black hole entropy count. It seems that in principle they should, but it is not obvious how to count these states.

The interpretation of these states as classical particles seems more natural, and is also supported by the fact that some of these states, such as the extreme case $M = -|J|$, or in the electrically charged solution with (21), are BPS. These states are supersymmetric invariant ground states that exhibit a positive energy spectrum of excitations, and therefore, perturbatively stable. This corresponds exactly to the idea of an elementary particle: a classically stable state.

Acknowledgments The author wishes to thank J. D. Edelstein, A. Garbarz, G. Giribet, C. Martínez, O. Mišković, and P. Mora, for many enlightening discussions over the past year. He also thanks Theodosios Christodoulakis, Elias Vagenas and the entire staff of the Kolymbari meeting for their warm hospitality and enjoyable atmosphere. This work was supported in part by FONDECYT Grants 11070146, 1061291, 7070117 and 7080201. The Centro de Estudios Científicos (CECS) is funded by the Chilean Government through the Millennium Science Initiative and the Centers of Excellence Base Financing Program of Conicyt. CECS is also supported by a group of private companies which at present includes Antofagasta Minerals, Arauco, Empresas CMPC, Indura, Naviera Ultragas and Telefónica del Sur.

[1] See J. Earman, *Bangs, crunches, whimpers, and shrieks: Singularities and acausalities in relativistic space-times,* Oxford Univ. Pr., NY, 1995.

[2] D. M. Eardley and L. Smarr, *Time function in numerical relativity. Marginally bound dust collapse*, Phys. Rev. **D19**, 2239 (1979); D. Christodoulou, *Violation of cosmic censorship in the gravitational collapse of a dust cloud*, Commun. Math. Phys. **93**, 171 (1984).

[3] See, for example, T. Jacobson, *Introductory Lectures on Black Hole Thermodynamics*, lectures given at the Unversity of Utrecht in 1996; available at http://www.glue.umd.edu/ tajac/BHTlectures/lectures.ps.

[4] G. Dotti, R. Gleiser and J. Pullin, *Instability of charged and rotating naked singularities*, Phys. Lett. **B644**, 289 (2007) [arXiv:gr-qc/0607052].

[5] M. Bañados, C. Teitelboim and J. Zanelli, *The Black hole in three-dimensional space-time*, Phys. Rev. Lett. **69** (1992) 1849 [arXiv:hep-th/9204099].

[6] M. Bañados, M. Henneaux, C. Teitelboim and J. Zanelli, *Geometry of the (2+1) black hole*, Phys. Rev. **D48** (1993) 1506 [arXiv:gr-qc/9302012].

[7] S. Carlip, *The (2+1)-Dimensional black hole*, Class. Quant. Grav. **12**, 2853 (1995) [arXiv:gr-qc/9506079]; *Gravity in (2+1)-Dimensions*, Cambridge University Press (1998).

[8] J. M. Izquierdo and P. K. Townsend, *Supersymmetric space-times in (2+1) adS supergravity models*, Class. Quant. Grav. **12**, 895 (1995) . [arXiv:gr-qc/9501018].

[9] R. Jackiw, *Five lectures on planar gravity*, Lectures given at SILARG VII, Cocoyoc, Mexico, Dec 1990.

[10] O. Mišković and J. Zanelli, *Couplings between Chern-Simons gravities and 2p-branes,* Phys. Rev. **D80**, 044003 (2009) [arXiv: 0901.0737 (hep-th)].

[11] O. Mišković and J. Zanelli, *On the negative spectrum of the 2+1 black hole*, Phys. Rev. **D79** (2009) 105011. [arXiv:0904.0475 [hep-th]].

[12] L. F. Abbott and S. Deser, *Stability Of Gravity With A Cosmological Constant*, Nucl. Phys. **B195**, 76 (1982); *Charge Definition In Nonabelian Gauge Theories*, Phys. Lett. **B116**, 259 (1982).

[13] E. Witten, *A Simple Proof Of The Positive Energy Theorem*, Commun. Math. Phys. **80**, 381 (1981).

[14] S. Deser and C. Teitelboim, *Supergravity Has Positive Energy*, Phys. Rev. Lett. **39**, 249 (1977).

[15] M. T. Grisaru, *Positivity Of The Energy In Einstein Theory*, Phys. Lett. **B73**, 207 (1978).

[16] O. Mišković, R. Troncoso and J. Zanelli, *Dynamics and BPS states of AdS(5) supergravity with a Gauss-Bonnet term*, Phys. Lett. **B637**, 317 (2006) [arXiv:hep-th/0603183].

[17] O. Coussaert and M. Henneaux, *Supersymmetry of the (2+1) black holes*, Phys. Rev. Lett. **72**, 183 (1994) [arXiv:hep-th/9310194].

[18] J. D. Edelstein, A. Garbarz, O. Mišković and J. Zanelli, *BPS Branes in CS AdS Super Gravity*, (to appear).

[19] A. Achúcarro and P. K. Townsend, *A Chern-Simons Action for Three-Dimensional anti-De Sitter Supergravity Theories*, Phys. Lett. **B180**, 89 (1986). E. Witten, *(2+1)-Dimensional Gravity as an Exactly Soluble System*, Nucl. Phys. **B311**, 46 (1988).

[20] A. H. Chamseddine, *Topological gravity and supergravity in various dimensions*, Nucl. Phys. **B346**, 213 (1990). R. Troncoso and J. Zanelli, *New gauge supergravity in seven and eleven dimensions*, Phys. Rev. **D58**, 101703 (1998) [arXiv:hep-th/9710180]. M. Bañados, R. Troncoso and J. Zanelli, *Higher dimensional Chern-Simons supergravity*, Phys. Rev. **D54**, 2605 (1996) [arXiv:gr-qc/9601003].

[21] J. Zanelli, *Uses of Chern-Simons actions*, in "Ten Years of the AdS/CFT Conjecture", AIP Conf. Proc. **1031**, 115 (2008) [arXiv:0805.1778 [hep-th]].

[22] J. D. Edelstein and J. Zanelli, *Sources for Chern-Simons theories*, in "Festschrift on the occasion of Claudio Bunster's 60th birthday" [arXiv:0807.4217 (hep-th)]; *Topologically induced gravity*, J. Phys. **A41**, 304002 (2008).

[23] A. Anabalon, S. Willison and J. Zanelli, *The Universe as a topological defect*, Phys. Rev. **D77**, 044019 (2008) [arXiv:hep-th/0702192]; *Topologically induced gravity*, J. Phys. **A41**, 304002 (2008); *Topological roots of four dimensional gravity*, PoS CARGESE **2007** (2007) 013.

Massive 3d gravity and Bigravity

Máximo Bañados [1]
P.Universidad Católica de Chile, Av. Vicuna Mackenna 4860, Santiago, Chile, and
Astrophysics, University of Oxford, DWB, Keble Road, Oxford, OX1 3RH, UK

Abstract. We review some results concerning bigravity and multigravity. Bigravity was considered for the first time by Isham, Salam and Strathdee almost 40 years ago. We update some results and in particular consider the three-dimensional version which has some interesting features.

1. Introduction

In this contribution we shall discuss several properties of multigravity. The first example of a multigravity theory was considered almost 40 years ago by Isham-Salam and Stradthdee [1]. Their action has two metrics $g_{(1)}$ and $g_{(2)}$ and has the form

$$I[g_1, g_2] = \int [\sqrt{g_1}R_1 + \sqrt{g_2}R_2 - V(g_1, g_2) + L(g_1, g_2, \text{matter})] \tag{1}$$

Let us discuss some properties and motivations to consider this action and its generalizations with N metrics.

- The multigravity version, $g_1, g_2, g_3, ..., g_N$, is the natural (but incomplete) "spin 2 Yang-Mills" theory.
- For the particular action (1), the 4d linear spectrum contains a massive and a massless gravitons. Thus, this action provides a covariant action for a massive graviton.
- The action (1) can be written in any spacetime dimension d. At $d = 3$, the spectrum contains a single massive graviton (the massless particle is trivial). This provides more challenges for AdS_3/CFT_2. At the same time provides yet another action for a massive graviton in three dimensions besides TMG [2] and NMG [3, 4] (see also [5]).
- As other alternative theories, bigravity has interesting cosmological applications in cosmology. The second metric can be seen as a 'matter field' providing an alternative to dark matter [6, 7, 8].

Bigravity has received intermittent but consistent attention since it was first presented. For some recent work see [9, 10, 11, 12, 13] (an incomplete list!).

The original work presented here was developed in several collaborations, in particular, [14, 8, 7, 15].

[1] JS Guggenheim Memorial Foundation Fellow

An important property of multigravity is the fact that an interaction potential preserving the full $diff_d^N$ invariance does not exist [16]. Naively, one could motivate multigravity as an analogy with Yang-Mills theory extending the action of one Maxwell field to N fields,

$$\int F_{\mu\nu}F^{\mu\nu} \quad \rightarrow \quad \int F_{1\,\mu\nu}F_1^{\mu\nu} + F_{2\,\mu\nu}F_2^{\mu\nu} + ... + \textbf{interactions}$$

It is extremely tempting to do the same with the spin two field:

$$\int \sqrt{g}(R-2\Lambda) \quad \rightarrow \quad \int \sqrt{g_1}(R_1 - 2\Lambda_1) + \sqrt{g_2}(R_2 - 2\Lambda_2) + ... \\ + \textbf{interactions}$$

However, for spin 2, the interactions do not have the same nice structure. The N-Gravity action functional

$$I[\overset{(a)}{g}_{\mu\nu}] = \sum_{a=1}^{N} \int \sqrt{g^{(a)}}\, R(g^{(a)}) \tag{2}$$

is invariant under independent transformations on each metric

$$\delta \overset{(a)}{g}_{\mu\nu} = \partial_\mu \xi^\alpha_{(a)} \overset{(a)}{g}_{\alpha\nu} + \partial_\nu \xi^\alpha_{(a)} \overset{(a)}{g}_{\mu\alpha} + \xi^\alpha_{(a)} \partial_\alpha \overset{(a)}{g}_{\mu\nu}, \qquad a = 1...N. \tag{3}$$

However, no interaction between the different metrics exists that preserves this full diff invariance. Only the diagonal subgroup can be preserved. This is a perturbative theorem. For weak fields,

$$\overset{(a)}{g}_{\mu\nu} = \eta_{\mu\nu} + \overset{(a)}{h}_{\mu\nu} \tag{4}$$

no cross potential $V(\overset{(a)}{h}_{\mu\nu})$ preserving the full N dimensional symmetry exists.

Now, the no-go theorem says nothing about interactions preserving the diagonal subgroup. We shall consider the Isham-Salam-Strathdee family ($g_1 = g, g_2 = f$),

$$\begin{aligned} I[g_{\mu\nu}, f_{\mu\nu}] \;=\; & \frac{1}{16\pi G}\int \Big[\sqrt{g}(R_g - 2\Lambda) + \sigma\sqrt{f}(R_f - 2\Lambda) \\ & + \frac{\nu}{4}\sqrt{f}(g_{\mu\nu} - f_{\mu\nu})(g_{\alpha\beta} - f_{\alpha\beta})(f^{\mu\alpha}f^{\nu\beta} - f^{\mu\nu}f^{\alpha\beta}) \\ & + \mathcal{L}_m(g_{\mu\nu} + \sigma f_{\mu\nu}, \Phi) \Big] \end{aligned} \tag{5}$$

Observe the particular coupling to the matter Lagrangian. We shall not consider a matter lagrangian for each metric, as often done in the literature. We shall argue below that the coupling of normal matter to $g_{\mu\nu} + \sigma f_{\mu\nu}$ is the mot natural choice. The point is that $g_{\mu\nu} + \sigma f_{\mu\nu}$ describes a massless graviton (see below) and therefore masses are still attracted by the action of spin 2 massless particle in this theory, just as in ordinary gravity.

Any modification of general relativity must have a regime where it reproduces Einstein gravity. The theory described by the action (5) has a phase in which is identical to GR. The equations of motion that extremize (5) ($8\pi G = 1$) are,

$$\begin{aligned} G_{\mu\nu}(g) + \Lambda g_{\mu\nu} &= \nu(g-f)_{\mu\nu} + T_{\mu\nu} \\ \sigma\left[G_{\mu\nu}(f) + \Lambda f_{\mu\nu}\right] &= -\nu(g-f)_{\mu\nu} + \sigma T_{\mu\nu} \end{aligned}$$

where $(g-f)_{\mu\nu}$ is a complicated combination of $g_{\mu\nu}, f_{\mu\nu}$. The important point here is that this term vanishes at $f_{\mu\nu} = g_{\mu\nu}$.

Thanks to the particular matter coupling to $g_{\mu\nu} + \sigma f_{\mu\nu}$ (massless mode), these equations admit a class of solutions with

$$f_{\mu\nu} = g_{\mu\nu} \tag{6}$$

where $g_{\mu\nu}$ is any solution to Einstein equations ($8\pi G = 1$),

$$G_{\mu\nu} = T_{\mu\nu}. \tag{7}$$

We conclude that any solution $g_{\mu\nu}$ to Einstein equations can be uplifted to be a solution of bigravity by choosing $f_{\mu\nu} = g_{\mu\nu}$. This is perhaps not too relevant. What is more important is the converse statement. All solutions to bigravity are of the form $f_{\mu\nu} = g_{\mu\nu}$ plus a short range massive fluctuation.

This can be seen as follows. Consider the linear fluctuations around $f_{\mu\nu} = g_{\mu\nu} = \bar{g}_{\mu\nu}$

$$g_{\mu\nu} = \bar{g}_{\mu\nu} + h'_{\mu\nu}, \qquad f_{\mu\nu} = \bar{g}_{\mu\nu} + k'_{\mu\nu}. \tag{8}$$

The Isham-Salam-Strathdee action (5) becomes

$$I_{\text{linear}} \simeq \int [h^{\mu\nu}\mathcal{G}(h)_{\mu\nu} + h^{\mu\nu}T_{\mu\nu}] + \left[k^{\mu\nu}\mathcal{G}(k)_{\mu\nu} - \frac{\nu}{4\sigma}(k^{\mu\nu}k_{\mu\nu} - k^2)\right] \tag{9}$$

where $h_{\mu\nu}$ and $k_{\mu\nu}$ are linear combinations of the fluctuations in (8)

$$h_{\mu\nu} = \frac{1}{1+\sigma}(h'_{\mu\nu} + \sigma k'_{\mu\nu}), \qquad k_{\mu\nu} = \frac{1}{1+\sigma}(h'_{\mu\nu} - k'_{\mu\nu}). \tag{10}$$

and $\mathcal{G}$ is the Pauli-Fierz operator. $h_{\mu\nu}$ describes a massless graviton and is coupled to matter. On the other hand $k_{\mu\nu}$ is the massive mode. Both are decoupled. Both actions are unitary and stable. Summarizing:

- $f_{\mu\nu} - g_{\mu\nu} \simeq k_{\mu\nu}$=massive mode, short range. Thus, $f_{\mu\nu} = g_{\mu\nu}$ to any desired scale, and we recover Einstein gravity coupled to matter.

- $g_{\mu\nu} + \sigma f_{\mu\nu} = \bar{g}_{\mu\nu} + h_{\mu\nu}$=massless mode coupled to matter.

2. Bigravity and Cosmology

Linear bigravity is equivalent to Einstein gravity, plus a massive short-range decoupled graviton. Let us explore now other non-linear aspects of this theory. We shall see that an alternative to dark matter emerges.

In cosmology, the Universe goes through several phases: Radiation, matter, acceleration. In a FRW model,

$$ds^2 = -dt^2 + a(t)^2 d\vec{x} \cdot d\vec{x} \tag{11}$$

the dynamics of $a(t)$ is given by the Friedmann equation,

$$\frac{\dot{a}^2}{a^2} = \frac{\Omega_r}{a^4} + \frac{\Omega_m}{a^3} + \Omega_\Lambda \tag{12}$$

The dark matter mystery is that fact that the observed baryonic Ω_m is smaller than expected from Universe dynamics and CMB observations.

Journal of Physics: Conference Series **222** (2010) 012009 doi:10.1088/1742-6596/222/1/012009

Let us show that bigravity contributes exactly at order $\frac{1}{a^3}$ to the Friedman equation. Consider the homogeneous, isotropic ansatz in the bigravity theory,

$$\begin{aligned} ds^2 &= -dt^2 + a(t)^2 d\vec{x}^2, \\ df^2 &= -X(t)^2 dt^2 + Y(t)^2 d\vec{x}^2. \end{aligned}$$

Note that while $g_{tt} = -1$ $X(t)$ must be left arbitrary in order to have a solution

The equation for $a(t)$ (Friedmann equation) is

$$\frac{\dot{a}^2}{a^2} = \frac{\nu Y^3/X}{a^3} + \frac{\Omega_{baryons}}{a^3} + \frac{\Omega_r}{a^4} + \Omega_\Lambda \tag{13}$$

plus two equations for $X(t)$ and $Y(t)$ that we omit. See [6, 7, 8] for details. A remarkable property of bigravity is that near $a = 0$ the combination $\nu Y^3/X$ is constant. Thus we can set

$$\nu Y^3/X = \Omega_{dark\ matter}$$

and the contribution from $f_{\mu\nu}$ to the Friedmann equation is just like pressureless matter; dark matter!

The dark matter interpretation can be taken much further. Fluctuations around Friedmann Universe can be introduced as

$$\begin{aligned} ds^2 &= -(1+2\Psi)dt^2 + a(t)^2(1-2\Phi)\delta_{ij}dx^i dx^j \\ df^2 &= -X(t)^2(1+2\Xi)dt^2 - Y^2\partial_i\beta\, dt dx^i \\ &\quad + Y(t)^2[(1-2\chi)\delta_{ij} + \partial_i\partial_j\mu]dx^i dx^j \end{aligned}$$

and have the expected properties for dark matter. See [8] for details (the action considered in this reference is slightly different from (5)).

3. Bigravity and black holes

The Schwarzschild-dS metric is a solution to the full set of equations

$$ds^2 = -\left(1 - \frac{2M}{r} - \Lambda r^2\right) dt^2 + \frac{dr^2}{1 - \frac{2M}{r} - \Lambda r^2} + r^2 d\Omega^2$$

provided we choose the second metric to be equal to $g_{\mu\nu}$. Thus, the action (5) does have black holes. A natural question is whether these are the most general set of black holes. The answer, first discussed in [17], is negative and an interesting family of other black holes does exist. The following metric is an exact solution to the action (5) [17]

$$\begin{aligned} ds^2 &= -\left(1 - \frac{2M}{r} + \Lambda_1 r^2\right) dt^2 + \frac{dr^2}{1 - \frac{2M}{r} + \Lambda_1 r^2} + r^2 d\Omega^2 \\ df^2 &= -\left(1 - \frac{2\mu}{r} + \Lambda_2 r^2\right) dt'^{\,2} + \frac{dr^2}{1 - \frac{2\mu}{r} + \Lambda_2 r^2} + r^2 d\Omega^2 \end{aligned}$$

with $t' = t + \xi(r)$ where $\xi(r)$ is a function of r only (see [17] for the full metric).

- Λ_1 and Λ_2 are integration constants. Only one combination of them is related to action couplings.
- The mass parameters M and μ are independent integration constants.

Journal of Physics: Conference Series **222** (2010) 012009 doi:10.1088/1742-6596/222/1/012009

The thermodynamical properties, the Kerr and Kerr-Newman extensions of these black holes are not known.

Some (modest) progress can be achieved in three dimensions. The full rotating metric analogous to the Isham-Storey black hole is given by [15]

$$ds^2 = -(-\Lambda r^2 - M)dt^2 + \frac{dr^2}{-\Lambda r^2 - M + \frac{J^2}{4r^2}} + Jdtd\phi + r^2 d\phi^2 \tag{14}$$

and $df^2 = f_{\mu\nu}dx^\mu dx^\nu$ is given by:

$$\begin{aligned} df^2 = & -\left(\tfrac{2r^2}{\lambda^2\ell^2} - M_f\right)dt^2 + \frac{8r^2\ell^2(2r^4\lambda^2+2r^2(M_f\lambda^2\ell^2-\ell^2M_g\lambda^2-2M_g\ell^2)+\ell^2J_g^2)dr^2}{\lambda^2(4r^4-4r^2M_g\ell^2+\ell^2J_g^2)^2} + r^2d\varphi^2 \\ & + J_g dtd\varphi - \frac{4\ell r\sqrt{(r^2(2-\lambda^2)+\lambda^2\ell^2(M_g-M_f))(4r^2(2M_g-M_f\lambda^2)+J_g^2(\lambda^2-2))}dtdr}{\lambda^2(4r^4-4r^2M_g\ell^2+J_g^2\ell^2)} \end{aligned}$$

where

$$2\nu\lambda^2 + 4\sigma\lambda\nu + \nu + \sigma = 0 \tag{15}$$

Both $g_{\mu\nu}$ and $f_{\mu\nu}$ have constant curvature (just like in NMG),

$$\begin{aligned} R(g)^{\mu\nu}{}_{\alpha\beta} &= \Lambda_g \delta^{[\mu\nu]}_{[\alpha\beta]} \\ R(f)^{\mu\nu}{}_{\alpha\beta} &= \Lambda_f \delta^{[\mu\nu]}_{[\alpha\beta]} \end{aligned}$$

Both black holes differ from AdS only in their global properties. Again, the thermodynamical properties of these black holes are not known. Since the curvature of black holes is constant, this seems to the indicate that the full solution can be build as identifications along some Killing vector. Isolating the particular vector is an interesting problem. One may also wonder whether the Cardy formula works in this context [18].

The asymptotic structure of the bigravity black holes present some interesting features. The $r \to \infty$ leading terms in both metrics are

$$\begin{aligned} ds^2 &\simeq -\frac{r^2}{\ell^2}dt^2 + \frac{\ell^2 dr^2}{r^2} + r^2 d\varphi^2 \\ df^2 &\simeq \frac{1}{2}\left(-\frac{2}{\lambda^2}\frac{r^2}{\ell^2}dt^2 + \frac{\ell^2 dr^2}{r^2} + r^2 d\varphi^2\right). \end{aligned}$$

Thus, both metrics are asymptotically AdS, but with different "speeds of light". Each metric has 6 Killing vectors but 4 are common to both: time translations $t \to t + \epsilon$, and SL(2,$\Re$) isometries of Euclidean 2-dimensional AdS_2: SL(2,$\Re$) $\times$ $\Re$. This is quite similar to the solutions discussed in [19]. A critical case occurs for

$$\lambda^2 = 2$$

where the full asymptotic $SO(2,2)$ symmetry is restored. A natural question in the context of 3d gravity is whether this group can be extended to the conformal group.

4. $\lambda^2 = 2$ and Brown-Henneaux symmetry in 3d bigravity

Brown and Henneaux [20] discovered that the phase space of 3d AdS gravity is classified by an infinite number of charges $L_n, \bar{L}_n$ satisfying the Virasoro algebra with central charge

$$c = \frac{3\ell}{2G}. \tag{16}$$

The goal of this section is to extend the results of Brown-Henneaux to bigravity. Let us first briefly review the results of [20].

All solutions to 3d gravity are diffeomorphic to AdS space

$$ds^2 = -\left(1+r^2\right)dt^2 + \frac{dr^2}{1+r^2} + r^2 d\varphi^2. \tag{17}$$

This statement seems to contradict the existence of an infinite number of charges classifying solutions. What happens is that, on open spaces, not all diffeomorphisms are un-physical. The Brown-Henneaux charges are asymptotic charges associated to non-trivial coordinate changes.

The distinction between trivial "un-physical" and non-trivial "physical" changes of coordinates was systematized by Regge and Teitelboim [21]. The transformations $x'^\mu = x^\mu + \epsilon^\mu(x)$ are generated by the Hamiltonian constraint plus a boundary term

$$H[\epsilon] = \int_\Sigma \epsilon^\mu \mathcal{H}_\mu + \int_{\partial\Sigma} \epsilon^\mu q_\mu; \qquad \mathcal{H}_\mu = 0, \;\; q_\mu \neq 0 \tag{18}$$

(i) If the boundary term vanishes, $\int_{\partial\Sigma} \epsilon^\mu q_\mu = 0$, then one has a true "gauge transformation" generated by a constraint.

(ii) Conversely, if $\int_{\partial\Sigma} \epsilon^\mu q_\mu \neq 0$ one has a fake "gauge transformation" not generated by a constraint. It does change the physical state. $Q \equiv \int_{\partial\Sigma} \epsilon^\mu q_\mu$ is the Noether charge associated to the non-trivial transformation.

The set of "physical" coordinate changes depends on boundary conditions and is in general a non-trivial task to isolate them. We shall consider here a simple example that exhibits the richness of the Brown-Henneaux calculation.

Consider the following three-dimensional metric

$$ds^2 = r^2 dz d\bar{z} + \frac{dr^2}{r^2} + \underbrace{T(z)dz^2 + \bar{T}(\bar{z})d\bar{z}^2} + \cdots,$$

T and $\bar{T}$ are 'fluctuations' on the background given by the first two terms. The background is just AdS_3 space (17) in Poincare (light-like) coordinates. The asymptotic Einstein equations impose on the fluctuations the conditions

$$\bar{\partial} T = 0, \qquad \partial \bar{T} = 0.$$

Consider the set of coordinate changes that preserve that background (but allowing the fluctuations T and $\bar{T}$ to transform). The non-trivial, unphysical, coordinate changes are conformal transformations

$$z \to f(z), \qquad \bar{z} \to g(\bar{z}).$$

As a concrete example, consider the chiral case ($\bar{T} = 0$) and act with the coordinate change $z = f(z')$

$$\begin{aligned} ds^2 &= r^2 dz d\bar{z} + \frac{dr^2}{r^2} + T(z)dz^2 \\ &= r^2 \partial' f \, dz' d\bar{z} + \frac{dr^2}{r^2} + T(z)(\partial' f)^2 dz'^2 \end{aligned}$$

We can eliminate the Jacobian $\partial' f$ by doing

$$r^2 \partial' f = r'^2, \qquad \bar{z} = \bar{z}' - \frac{1}{2r'^2}\frac{\partial'^2 f}{\partial' f}.$$

Journal of Physics: Conference Series **222** (2010) 012009 doi:10.1088/1742-6596/222/1/012009

and go back to exactly the metric we started from

$$ds^2 = r'^2 dz' d\bar{z}' + \frac{dr'^2}{r'^2} + T'(z')dz'^2,$$

where –note the Schwarz derivative–

$$T'(z') = T(z)\,(\partial' f)^2 - \frac{1}{2}\{f, z'\}.$$

This would all be meaningless –just a coordinate redefinition– if it wasn't for the fact that $z = f(z')$ is not trivial. The actual value of the central charge needs a bit more work. The formula $\delta T = [T, T]$, also proved by Brown-Henneaux [22], normalizes T and one finds,

$$c = \frac{3\ell}{2G}. \tag{19}$$

This result became strongly entangled with black hole thermodynamics after the important papers [18, 23]. The idea is as follows. The black hole is asymptotically anti-de Sitter with $\ell MG = L_0 + \bar{L}_0, JG = L_0 - \bar{L}_0$.. Due to the AdS/CFT correspondence one expects a CFT_2 with a central charge equal to (19) (up to sub-leading quantum corrections). Next, for **any** unitary modular invariant 2d CFT, the number of states consistent with given (large) $L_0, \bar{L}_0$ is (Cardy formula)

$$\rho(T, \bar{T}) = e^{2\pi\sqrt{\frac{c}{6}L_0} + 2\pi\sqrt{\frac{c}{6}\bar{L}_0}}.$$

Plugging the black hole values for $L_0, \bar{L}_0$ and obtain exactly Bekenstein-Hawking entropy [18, 23].

$$\rho(M, J) = e^{\frac{2\pi r_+}{4G}}.$$

We would like now to extend the Brown-Henneaux analysis to bigravity. The details of this problem can be found in [15]. The bigravity Brown-Henneaux asymptotic solution is,

$$\begin{aligned} ds^2 &\sim \frac{dr^2}{r^2} + r^2 dz d\bar{z} + T(z)dz^2 + \bar{T}(\bar{z})dz^2 \\ df^2 &\sim \frac{1}{2}\left(\frac{dr^2}{r^2} + r^2 dz d\bar{z} + Q(z)dz^2 + \bar{Q}(\bar{z})dz^2 + \frac{dr}{r}(P(z)dz + \bar{P}(\bar{z})d\bar{z})\right) \end{aligned}$$

where the Einstein × Einstein equations imply

$$\bar{\partial}T = \bar{\partial}Q = \bar{\partial}P = 0, \qquad \partial\bar{T} = \partial\bar{Q} = \partial\bar{P} = 0,$$

Note that $P(z)$ and $\bar{P}(\bar{z})$ can be set to zero by a *trivial* coordinate redefinition, but not simultaneously in both metrics. In this sense $P, \bar{P}$ are physical. Furthermore, due to the $1/r$ slow fall-off they appear in the charges.

The set of non-trivial coordinate transformations that preserve the asymptotic conditions and have non-zero charges are conformal maps $z \longrightarrow f(z) \simeq z + \epsilon(z)$. The metric fluctuations transform as

$$\begin{aligned} \delta Q &= -\epsilon\partial Q - 2\partial\epsilon Q + \frac{1}{2}\partial^3\epsilon + P\partial^2\epsilon \\ \delta P &= -\epsilon\partial P - \partial\epsilon P \end{aligned}$$

and the same for the $\bar{z}-$ sector.

Note that P is (1,0) conformal field and ∂P is (2,0), as expected. On the other hand Q is almost a $(2,0)$ field. Interestingly the combination $Q + \partial P$ is a $(2,0)$ field.

We can now compute the charges associated to these transformations and confirm that they are finite. Since the potential in the action does not have derivatives, the charges are the sum of two ADM functionals, one from each metric. The "holomorphic" charge is

$$\begin{aligned} Q &= \overbrace{Q_g}^{} + \overbrace{\sigma Q_f}^{} \\ &= T(z) + Q(z) + \partial P(z), \end{aligned}$$

The total central charge is (computed by usual methods)

$$c = \frac{3\ell}{2G} \frac{1}{\sqrt{1+2\sqrt{2}\nu}} \left(1 + \frac{\sigma}{\sqrt{2}}\right) \tag{20}$$

Now, recall that the action (5) had another AdS background with $g_{\mu\nu} = f_{\mu\nu}$. (Observe the factor $\frac{1}{2}$ in $f_{\mu\nu}$ above: the black holes displayed here are not fluctuations on the background with $f_{\mu\nu} = g_{\mu\nu}$.) The central charge on the "strongly coupled" background with $f_{\mu\nu} = g_{\mu\nu}$ is

$$c_0 = \frac{3\ell}{2G}(1+\sigma). \tag{21}$$

In the whole range of (allowed) parameters it is observed that

$$c_0 > c. \tag{22}$$

Thus, there might be a flow interpolating these two CFT's.

5. Conclusions and open problems

Multi-gravity is an interesting theory highly unexplored and full of open problems. I hope I have succeed in this Contribution to motivate this theory and explain its main features. I would like to end by mentioning some open problems.

(i) Does the Cardy formula apply in three-dimensional bigravity?
(ii) What is the entropy and temperature for bigravity black holes?
(iii) Can one classify solutions with spherical symmetry? Birkhoff theorem?
(iv) How different is New Massive gravity from bigravity ($d = 3$)
(v) "Chiral" gravity. At $d = 3$ the choice

$$\int \sqrt{g}R(g) - \sqrt{f}R(f)$$

gives a massive vector. The central charges are zero, and the dual theory should be a log CFT. Note that an action of this type has recently been considered in [24].

The author was partially supported by Alma-Conicyt Grant # 31080001; Fondecyt Grant #1060648; and the J.S. Guggenheim Memorial Foundation.

[1] Isham C J, Salam A and Strathdee J A 1971 *Phys. Rev.* **D3** 867–873
[2] Deser S, Jackiw R and Templeton S 1982 *Ann. Phys.* **140** 372–411
[3] Bergshoeff E A, Hohm O and Townsend P K 2009 *Phys. Rev. Lett.* **102** 201301 (*Preprint* 0901.1766)
[4] Bergshoeff E A, Hohm O and Townsend P K 2009 *Phys. Rev.* **D79** 124042 (*Preprint* 0905.1259)

[5] Bergshoeff E, Hohm O and Townsend P 2009 (*Preprint* 0912.2944)
[6] Bañados M 2008 *Phys. Rev.* **D77** 123534 (*Preprint* 0801.4103)
[7] Bañados M, Gomberoff A, Rodrigues D C and Skordis C 2009 *Phys. Rev.* **D79** 063515 (*Preprint* 0811.1270)
[8] Bañados M, Ferreira P G and Skordis C 2009 *Phys. Rev.* **D79** 063511 (*Preprint* 0811.1272)
[9] Damour T and Kogan I I 2002 *Phys. Rev.* **D66** 104024 (*Preprint* hep-th/0206042)
[10] Arkani-Hamed N, Georgi H and Schwartz M D 2003 *Annals Phys.* **305** 96–118 (*Preprint* hep-th/0210184)
[11] Blas D, Deffayet C and Garriga J 2007 *Phys. Rev.* **D76** 104036 (*Preprint* 0705.1982)
[12] Berezhiani Z, Comelli D, Nesti F and Pilo L 2007 *Phys. Rev. Lett.* **99** 131101 (*Preprint* hep-th/0703264)
[13] Blas D 2008 *Aspects of Infrared Modifications of Gravity* Ph.D. thesis (*Preprint* 0809.3744)
[14] Li W, Song W and Strominger A 2008 *JHEP* **04** 082 (*Preprint* 0801.4566)
[15] Bañados M and Theisen S 2009 *JHEP* **11** 033 (*Preprint* 0909.1163)
[16] Boulanger N, Damour T, Gualtieri L and Henneaux M 2000 (*Preprint* hep-th/0009109)
[17] Isham C J and Storey D 1978 *Phys. Rev. D* **18** 1047–1052
[18] Strominger A 1998 *JHEP* **02** 009 (*Preprint* hep-th/9712251)
[19] Anninos D, Li W, Padi M, Song W and Strominger A 2009 *JHEP* **03** 130 (*Preprint* 0807.3040)
[20] Brown J D and Henneaux M 1986 *Commun. Math. Phys.* **104** 207–226
[21] Regge T and Teitelboim C 1974 *Ann. Phys.* **88** 286
[22] Brown J D and Henneaux M 1986 *J. Math. Phys.* **27** 489–491
[23] Birmingham D, Sachs I and Sen S 1998 *Phys. Lett.* **B424** 275–280 (*Preprint* hep-th/9801019)
[24] Milgrom M 2009 *Phys. Rev.* **D80** 123536 (*Preprint* 0912.0790)

First Mediterranean Conference on Classical and Quantum Gravity (MCCQG 2009) IOP Publishing
Journal of Physics: Conference Series **222** (2010) 012010 doi:10.1088/1742-6596/222/1/012010

Dark matter as a dynamic effect due to a non-minimal gravitational coupling with matter (I): Analytical results

O Bertolami[1], J. Páramos

Instituto de Plasmas e Fusão Nuclear, Instituto Superior Técnico,
Avenida Rovisco Pais 1, 1049-001 Lisboa, Portugal

E-mail: `orfeu@cosmos.ist.utl.pt, paramos@ist.edu`

Abstract. In this work the phenomenology of models possessing a non-minimal coupling between matter and geometry is discussed, with a particular focus on the possibility of describing the flattening of the galactic rotation curves as a dynamically generated effect derived from this modification to General Relativity. Two possibilities are discussed: firstly, that the observed discrepancy between the measured rotation velocity and the classical prediction is due to a deviation from geodesic motion, due to a non-(covariant) conservation of the energy-momentum tensor; secondly, that even if the principle of energy conservation holds, the dynamical effects arising due to the non-trivial terms in the Einstein equations of motion can give rise to an extra density contribution that may be interpreted as dark matter. In this work, The mechanism of the latter alternative is detailed; a numerical session ascertaining the order of magnitude of the relevant parameters is undertaken in another contribution to this volume, with possible cosmological implications discussed.

1. Introduction

Many modifications of the theories of gravity are strongly motivated by the issue of an alternative solution to the problem of the flattening of the galaxy rotation curves — one which does not require dark matter in the strictest sense, *i.e.* weakly or non-interacting matter fields [1, 2, 3, 4]. With this goal in mind, standard $f(R)$ models [5, 6, 7] have been studied endowed with power-law curvature terms $f_1(R) \propto R^n$ in the action, instead of the linear curvature depicted in the General Relativity (GR) action. For $n > 1$, an addition to the Newtonian potential $\Delta\Phi(r) = -GM(r/r_c)^\beta/2r$ is found, with β a function of n and r_c being a parameter characteristic of each galaxy. A fit of several rotation curves indicates that $n = 3.5$ (corresponding to $\beta = 0.817$) yields the best agreement with observations [8].

In another approach, it has been shown that an asymptotically flat velocity dispersion curve can be obtained if the curvature term in the Einstein-Hilbert action includes a logarithmic factor, $f_1(R) = f_0 R(1 + v \log R)$ [9]; given the non-relativistic condition $v \ll 1$, this may be approached by a power-law with an exponent very close to unity, $f_1(R) \approx f_0 R^{1+v^2}$.

Since this description does not take into consideration any galaxy-dependent parameters, an universal asymptotic velocity v is obtained; this is in stark contrast with the Tully-Fisher and

[1] Also at Departamento de Física, Instituto Superior Técnico.

Faber-Jackson relations, which posit a dependence of the visible mass of a galaxy $M \propto v^m$, for spiral ($m = 4$) and elliptical galaxies ($m = 6$), respectively.

This work reviews the results obtained in Ref. [10], to which the reader is directed for a more technical discussion and the full set of computations. It comprises a small introduction to the non-minimal coupling model under scrutiny, proceeds with the implied non-geodesic motion already hinted above, and then proposes the already mentioned mechanism for dynamically mimicking dark matter; a numerical session and discussion follows in another contribution to this volume.

2. The model

The non-minimal coupling between matter and geometry is embodied in the following action functional [11],

$$S = \int \left[\frac{1}{2} f_1(R) + [1 + \lambda f_2(R)] \mathcal{L}_m \right] \sqrt{-g} d^4 x, \tag{1}$$

where $f_i(R)$ (with $i = 1, 2$) are arbitrary functions of the scalar curvature R, $\mathcal{L}_m$ is the Lagrangian density of matter and g is the metric determinant. The non-minimal coupling f_2 is gauged via the coupling constant λ (with dimensions $[\lambda] = [f_2]^{-1}$). The standard Einstein-Hilbert action is recovered by taking $f_2 = 0$ and $f_1 = 2\kappa(R - 2\Lambda)$, where $\kappa = c^4/16\pi G$ and Λ is the cosmological constant (from now on, one works in a unit system where $c = 1$).

The modified Einstein equations of motion arise from variation of the action with respect to the metric $g_{\mu\nu}$,

$$\left(F_1 + 2\lambda F_2 \mathcal{L}_m\right) R_{\mu\nu} - \frac{1}{2} f_1 g_{\mu\nu} = \left(\square_{\mu\nu} - g_{\mu\nu}\square\right)\left(F_1 + 2\lambda F_2 \mathcal{L}_m\right) + \left(1 + \lambda f_2\right) T_{\mu\nu}, \tag{2}$$

defining $\square_{\mu\nu} \equiv \nabla_\mu \nabla_\nu$ for convenience, and $F_i(R) \equiv f_i'(R)$. The matter energy-momentum tensor is defined as

$$T_{\mu\nu} = -\frac{2}{\sqrt{-g}} \frac{\delta\left(\sqrt{-g}\mathcal{L}_m\right)}{\delta g^{\mu\nu}}, \tag{3}$$

so that the trace of Eq. (2) reads

$$\left(F_1 + 2\lambda F_2 \mathcal{L}_m\right) R - 2 f_1 = -3\square\left(F_1 + 2\lambda F_2 \mathcal{L}_m\right) + \left(1 + \lambda f_2\right) T. \tag{4}$$

The Bianchi identities, $\nabla^\mu G_{\mu\nu} = 0$ imply the covariant "non-conservation law"

$$\nabla^\mu T_{\mu\nu} = \frac{\lambda F_2}{1 + \lambda f_2} \left(g_{\mu\nu}\mathcal{L}_m - T_{\mu\nu}\right) \nabla^\mu R, \tag{5}$$

interpreted as a signal of energy-momentum exchange between matter and geometry due to the non-trivial terms [12].

For simplicity, and since one aims studying the phenomenology of the non-minimal coupling $f_2(R)$, one sets $f_1(R) = 2\kappa R$ (the GR form discarding the cosmological constant Λ); Eq. (2) then reduces to

$$\left(1 + \frac{\lambda}{\kappa} F_2 \mathcal{L}_m\right) R_{\mu\nu} - \frac{1}{2} R g_{\mu\nu} = \frac{\lambda}{\kappa}\left(\square_{\mu\nu} - g_{\mu\nu}\square\right)\left(F_2 \mathcal{L}_m\right) + \frac{1}{2\kappa}\left(1 + \lambda f_2\right) T_{\mu\nu}, \tag{6}$$

and, equivalently, Eq. (4) becomes

$$\left(1 - \frac{\lambda}{\kappa} F_2 \mathcal{L}_m\right) R = 3\frac{\lambda}{\kappa}\square\left(F_2 \mathcal{L}_m\right) - \frac{1}{2\kappa}\left(1 + \lambda f_2\right) T. \tag{7}$$

3. Flattening of the galaxy rotation curves due to non-geodesical motion

From a fundamental standpoint, the most striking characteristic of the model here considered is the non-conservation of the energy-momentum tensor of matter; assuming that matter is described by a perfect fluid distribution with the usual energy-momentum tensor

$$T_{\mu\nu} = (\rho + p)\, U_\mu U_\nu + p g_{\mu\nu}, \tag{8}$$

where ρ is the energy density, $p(\rho)$ is the pressure (given by a suitable equation of state) and U_μ is the four-velocity (satisfying the normalization condition $U_\mu U^\mu = -1$), Eq. (5) leads to an extra force imparted upon a test particle,

$$f^\mu = \frac{1}{\rho + p}\left[\frac{\lambda F_2}{1+\lambda f_2}\left(\mathcal{L}_m + p\right)\nabla_\nu R + \nabla_\nu p\right] h^{\mu\nu}, \tag{9}$$

with the projection operator $h_{\mu\nu} = g_{\mu\nu} - U_\mu U_\nu$, so that $h_{\mu\nu}U^\mu = 0$ and the extra force is orthogonal to the four-velocity.

In GR, the specific form of the Lagrangian density does not appear explicitly in the equations of motion, but only via the derived energy-momentum tensor Eq. (3). For this reason, and particularly if one is not interested in describing the thermodynamics of the perfect fluid, this functional is usually ignored. This, of course, changes in the present situation, since $\mathcal{L}_m$ appears explicitly in Eqs. (6) and (9).

In GR, the Lagrangian density $\mathcal{L}_m$ of a perfect fluid several equivalent on-shell expressions are admitted: for instance, $\mathcal{L}_{m0} = -\rho$, $\mathcal{L}_{m1} = p$ and $\mathcal{L}_{m2} = -na$, where n is the particle number density, and a is the physical free energy, defined as $a = \rho/n - Ts$, with T the fluid temperature and s the entropy per particle. However, in the presence of a non-minimal gravitational coupling with matter, this degeneracy is lifted and the original, bare Lagrangian density $\mathcal{L}_m = -\rho$ should be considered [13, 14].

One may further simplify computations by noticing that as matter in galaxies is non-relativistic (asides from the inner core region), and it may be described approximately by a pressureless dust distribution, $p \ll \rho c^2$.

With the above consideration in mind, one now assumes a static, spherically or cylindrically symmetric *Ansatz* for the problem and that the velocity has only a tangential component, so that $h^{\mu r} = g^{\mu r}$ and so the extra-force Eq. (9) reads

$$f^\mu = -\frac{\lambda F_2}{1+\lambda f_2} R'(r) h^{\mu r}, \tag{10}$$

where the prime denotes differentiation with respect to the radial coordinate r.

Endowed with the expression above, one may now tackle the issue of describing the flattening of the galaxy rotation curves; instead of an asymptotic velocity dispersion $v(r) \rightarrow v_\infty$, one assumes that the reported flattening allows for a shallow decrease, *i.e.*, a profile of the form

$$v^2(r) = v_\infty^2 \left(\frac{r_*}{r}\right)^\beta, \tag{11}$$

where r_* is an unknown normalization constant and $0 \leq \beta < 1$ is the outer slope of the rotation curve. Identifying the extra-force with the observed deviation leads to the expression

$$f^r \simeq -\lambda F_2 R'(r) - \frac{v_\infty^2}{r}\left(\frac{r_*}{r}\right)^\beta, \tag{12}$$

which one may use to ascertain the form of the non-minimal coupling $f_2(R)$; in order to do so, one requires some bearing on the dependence of the scalar curvature on the radial distance $R = R(r)$; to do so, one recalls that it is implicitly assumed that the effect of the non-trivial terms appearing in Eq. (6) is perturbative — so that the geodesics derived from the metric are similar to those found in GR — and the non-minimal coupling manifests itself simply through an extra-force deviating test particles from geodesic motion.

As will be seen below, this approach is self-consistent; for now, one assumes that the scalar curvature is given by its classical expression $R = -8\pi GT$, with $T = -\rho$ the trace of the energy-momentum tensor $T_{\mu\nu}$. Thus, the issue of obtaining a dependence $R(r)$ simplifies to ascribing a particular profile for the density distribution $\rho(r)$ of matter. For this, one takes a power-law profile of the form

$$\rho(r) \approx \rho_0 (a/r)^m, \tag{13}$$

with $m \neq 0$ and ρ_0, a characteristic scales.

With the differential equation Eq. (12) in a closed form, one may solve for $f_2(r)$ and, reversing the dependences above, find the aimed coupling $f_2(R)$. After some algebraic steps (detailed in Ref. [10]), one finds that

$$f_2(R) = \begin{cases} -\dfrac{1}{\beta}\left(\dfrac{R}{R_*}\right)^{\beta/m} & , \beta \neq 0 \\ -\dfrac{1}{m}\log\left(\dfrac{R}{R_*}\right) & , \beta = 0 \end{cases} , \tag{14}$$

defining the characteristic scale $R_* = 8\pi G\rho_0(a/r_*)^m$.

With the above solution, it is trivial to show that the conditions for a negligible effect of the non-minimal coupling $f_2(R)$ on the Einstein field Eq. (6),

$$|\lambda f_2(R)| \ll 1 \quad , \qquad \lambda\left[F_2(R)\rho\right]'' \ll \rho, \tag{15}$$

(using that $T = \mathcal{L}_m = -\rho$) are satisfied for a wide domain, $1\ kpc \lesssim r \lesssim 50\ kpc$, where the flattening of the rotation curves is measured for a wide range of values of r_* (mainly due to the smallness of the gauging constant $\lambda = v_\infty^2 \approx 10^{-12}$): one could set, e.g., $r_* = 1\ kpc$.

As a result, the assumption of a perturbative regime for the equations of motion Eq. (6) is indeed self-consistent — although the deviation from geodesical motion is relevant, dominating the Newtonian acceleration and leading to the flattening of the galaxy rotation curve.

Before continuing, one remarks that a similar exercise would ensue if one was to assume that the velocity dispersion curve follows a logarithmically decaying profile,

$$v^2(r) = v_\infty^2 \frac{r_*}{r}\log\left(\frac{r}{r_*}\right), \tag{16}$$

instead of the simpler power-law assumed above. Indeed, and in the context of standard galactic dark matter descriptions, this radial dependence leads to a dark matter density distribution given by the so-called Navarro-Frenk-White (NFW) density profile, which is widely used in the literature [15] (see Appendix A of Ref. [10] for a discussion of different density profiles). Assuming this decay law for $v^2(r)$ leads to the non-minimal coupling

$$f_2(R) = -\left(\frac{R}{R_*}\right)^{1/m}\left[1 + \frac{1}{m}\log\left(\frac{R_*}{R}\right)\right]. \tag{17}$$

and, as before, the perturbative conditions Eq. (15) are trivially satisfied within the relevant range.

To close this section, one first remarks that the obtained expression Eq. (14) for the non-minimal coupling $f_2(R)$ in the flattened rotation curve case $\beta = 0$ is strikingly similar to that of Ref. [9], where the non-trivial curvature term is given by $f_1(R) = R[1 + v^2 \log(R/R_*)] \approx R(R/R_*)^{v_\infty^2}$. Following that proposal, one may suggestively rewrite the action Eq. (1) as

$$S = \int \left[8\pi G R + \left(\frac{R_*}{R} \right)^\alpha \mathcal{L}_m \right] \sqrt{-g} d^4 x, \tag{18}$$

defining $\alpha = v_\infty^2 / m$.

However, this similarity with Ref. [9] leads to the same caveat, namely an undesired universality of the asymptotic value of the predicted galaxy rotation curves: assuming similar outer slopes m for the visible matter density profile ρ of galaxies, one concludes that v_∞ and β should be fixed parameters of the model, unadjustable for different galaxies. This is not observed for the available rotation curves, in which the asymptotic value v_∞ may vary somewhat (although of the order of magnitude $v_\infty \sim 100 km/s$). One may assume that for different density profiles, spatial symmetry and morphology considerations would contribute to alleviate this issue, but the obtained universality remains a noticeable disadvantage of this approach — thus prompting the search for alternatives within the ample phenomenological arsenal provided by the model under scrutiny.

4. Pressureless dust with non-minimal gravitational coupling

Motivated by the considerations put forward in the last paragraph, one now assumes the opposite stance: instead of considering a perturbative effect of the non-minimal coupling, but a large deviation from geodesic motion, one posits that the energy-momentum tensor of matter is approximately conserved, but the geodesics thus followed by test particles differ significantly from those of GR. As in the previous section, this approach also proves to be self- consistent.

Given the above, the problem becomes similar to those faced by standard $f(R)$ theories (that is, with $f_2(R) = 1$): solve the Einstein field equations so that the obtained metric gives rise to the observed galaxy rotation curves. Since one ultimately aims to describe dark matter as a dynamical effect stemming from the action functional Eq. (1), it is productive to try to express the effect of the non-minimal coupling in terms of an equivalent, mimicked dark matter density. This also allows for a direct comparison with existing profiles for the distribution of dark matter, arising both from simulations as well as observational evidence.

For generality, a power law coupling of the form

$$f_2(R) = \left(\frac{R}{R_0} \right)^n , , \tag{19}$$

is assumed, with n a yet unspecified exponent and R_0 a characteristic scale (which amounts to a rewriting of the gauging constant λ).

It should be highlighted that Eq. (19) does not exclude a more elaborate form for the non-minimal coupling, which could then be written as a series

$$f_2(R) = \sum_p \left(\frac{R}{R_{0p}} \right)^p , \tag{20}$$

summed over the rational exponents p of the "full" model (integer or not); if this is the case, the choice of Eq. (19) is to be viewed as a valid approximation within the galactic context (that is, the dominant term of the series for the corresponding range of R). For this reason, this assumption is not in contradiction with the previous work by the authors [16], which approached the possibility of including a linear coupling $f_2 = R/R_1$ (assumed relevant in a high density environment such as the Sun).

Again, one models matter within a galaxy as a pressureless perfect fluid, *i.e.*, dust, characterized by a density ρ; the corresponding energy-momentum tensor thus reads

$$T_{\mu\nu} = \rho U_\mu U_\nu \rightarrow T = -\rho. \tag{21}$$

Using the Lagrangian density $\mathcal{L}_m = -\rho$ [13], Eq. (6) reads

$$\left[1 - \frac{n}{\kappa}\left(\frac{R}{R_0}\right)^n \frac{\rho}{R}\right] R_{\mu\nu} - \frac{1}{2} R g_{\mu\nu} = \frac{n}{\kappa}\left(g_{\mu\nu}\Box - \Box_{\mu\nu}\right)\left[\left(\frac{R}{R_0}\right)^n \frac{\rho}{R}\right] + \frac{1}{2\kappa}\left[1 + \left(\frac{R}{R_0}\right)^n\right] \rho U_\mu U_\nu, \tag{22}$$

and Eq. (7) becomes

$$R = \frac{1}{2\kappa}\left[1 + (1-2n)\left(\frac{R}{R_0}\right)^n\right]\rho - \frac{3n}{\kappa}\Box\left[\left(\frac{R}{R_0}\right)^n \frac{\rho}{R}\right], \tag{23}$$

As stated at the beginning of this section, the additional terms induced by the non-minimal gravitational coupling $f_2(R)$ are interpreted as contributing to the dark matter halo density profile ρ_{DM}. This can be better ascertained by ignoring the matter contribution arising from GR, so that the non-minimal terms dominate far away the galactic core (where the rotation curve flattens). In this exterior region, one may rewrite Eq. (23) as

$$R \simeq \frac{1-2n}{2\kappa}\left(\frac{R}{R_0}\right)^n \rho - 3\frac{n}{\kappa}\Box\left[\left(\frac{R}{R_0}\right)^n \frac{\rho}{R}\right]. \tag{24}$$

4.1. Static solution

Assuming a known visible matter density ρ, solving the above equation amounts to obtaining the relation $R = R(\rho)$; this is somewhat analog to the previous exercise, where one assumed $\rho(r)$ and computed $R(r)$. Naturally, the classical identification $R = 2\kappa\rho$ fails, since it would lead to a differential equation for the visible matter density which might not be satisfied by the considered profile $\rho(r)$. Inspection of Eq. (24) shows that a solution defined implicitly by the expression

$$R = \frac{1-2n}{2\kappa}\left(\frac{R}{R_0}\right)^n \rho, \tag{25}$$

is self-consistent, since the gradient term vanishes everywhere; accordingly, this is dubbed as a "static" solution. This yields

$$R = R_0\left[\frac{1-2n}{2\kappa}\frac{\rho}{R_0}\right]^{1/(1-n)} = R_0\left[(1-2n)\frac{\rho}{\rho_0}\right]^{1/(1-n)}, \tag{26}$$

introducing the characteristic density $\rho_0 \equiv R_0/2\kappa$.

In order to interpret the additional terms shown in the Einstein field Eqs. (22), one must first show that the above "static" solution provides a suitable description of dark matter, *i.e.* the full tensorial aspect of the problem must be taken into account, not merely the algebraic identification stemming from the trace Eq. (24). Indeed, replacing Eq. (26) in Eq. (22) allows one to conclude that the dynamic effect due to the non-minimal coupling behaves as a perfect fluid characterized by an energy-momentum tensor analog to that of matter, so that this mimicked "dark matter" shares the same 4-velocity as visible matter, but has a pressure and density given by

$$\rho_{dm} = \frac{1-n}{1-4n}\rho_0\left[(1-2n)\frac{\rho}{\rho_0}\right]^{1/(1-n)}, \tag{27}$$

thus yielding a non-vanishing pressure and an equation of state of the form

Journal of Physics: Conference Series **222** (2010) 012010 doi:10.1088/1742-6596/222/1/012010

$$p_{dm} = \frac{n}{1-n}\rho_{dm}. \tag{28}$$

Hence, at a sufficiently large distance from the galactic core, where the rotation curve has flattened, the non-minimal gravitational coupling with ordinary matter mimics a dark matter component which is dragged by ordinary matter, and may have a negative density $\rho_{dm} < 0$, if $1/4 < n < 1$; although strange, this is not a pathology of the model, as the curvature in the considered outer region will remain positive, $R > 0$. As will be seen below, relevant models for the density profile of dark matter lead to negative exponents n, so that the density is indeed positive defined.

4.2. Mimicking dark matter density profiles

Many dark matter models resort to the assumption that, at the region where the galaxy rotation curve has flattened, both visible and dark matter might be modelled by a power law density profile, characterized by the outer slopes m and m', respectively (see Appendix A of Ref. [10] for a thorough discussion),

$$\rho \approx \rho_v \left(\frac{a}{r}\right)^m \quad , \qquad \rho_{dm} \approx \rho_d \left(\frac{a'}{r}\right)^{m'} . \tag{29}$$

Thus, one may establish the algebraic relation $\rho_{dm} \propto \rho^p$, defining the slope ratio $p \equiv m'/m$. The constant ρ_v is specific to each visible matter distribution ρ, and ρ_d will be related to this quantity and to the previously defined ρ_0. Since a similar power law was obtained in the previous paragraph, this relation allows for a direct translation between the parameters R_0, n of the non-minimal coupling model and the mimicked dark matter density profile quantities a', m' and ρ_d, assuming a given visible matter density profile parameters,

$$\frac{1}{1-n} = \frac{m'}{m} \equiv p \rightarrow n = 1 - \frac{m}{m'} = 1 - \frac{1}{p}, \tag{30}$$

and

$$\rho_d = \frac{1}{4+3p}\left(\frac{2}{p} - 1\right)^p \rho_0^{1-p}\rho_v^p. \tag{31}$$

One may assume that the length scale a' is inherited from the visible matter density, so that $a' \sim a$.

Clearly, the long range dominance of the mimicked dark matter over visible matter indicates that the density profile of the former should be shallower than that of the former, so that $m' < m$. Hence, one posits that $p \equiv m'/m < 1$, thus a negative exponent n is expected. This range is of physical significance: since dark matter is relevant at large distances, but the scalar curvature increases towards the galactic core, the non-minimal coupling $f_2(R)$ should scale with an inverse power of the curvature.

For concreteness, one ascertains what is the relevant range for the exponent n; to do so, one has to assume a particular visible matter density profile. In this study, the Hernquist distribution is considered [17],

$$\rho(r) = \frac{M}{2\pi}\frac{a}{r}\frac{1}{(r+a)^3}, \tag{32}$$

so that $m = 4$, and Eq. (30) leads to $n = 1 - 4/m'$. Bearing in mind that one aims replicating the dark matter distribution, it is useful to resort to two widely used candidates, namely the NFW profile

$$\rho_{dm}(r) = \frac{\rho_{cp}}{\left(\frac{r}{a}\right)\left(1+\frac{r}{a}\right)^2} \tag{33}$$

with $m' = 3 \rightarrow n = -1/3$, and the isothermal sphere profile

$$\rho(r) = \frac{v_\infty^2}{2\pi G r^2}, \tag{34}$$

with $m' = 2 \rightarrow n = -1$.

Both dark matter models are characterized by negative exponents n, as indicated above. The obtained values $n = -1/3$ and $n = -1$ should be taken into consideration in the following developments, although the general exponent n is retained.

4.3. Background matching

The elegant "static" solution obtained before can be fruitfully used to determine several characteristics of the model, as has been extensively discussed in Ref. [10]; most calculations are hence omitted, with the focus shifted to a comprehensive discussion of the physical significance of these results.

Firstly, one notices that a negative power-law coupling $f_2 = (R/R_0)^n$ (with $n < 0$) does not admit an asymptotically flat Minkowski space-time, since it would lead to a singular $R = 0$ scenario. This problem is naturally avoided by noticing that the assumed coupling is valid at the galactic scale, and that the asymptotic behaviour should be identified with the large scale, cosmological background. Hence, the notion of a matching distance r_∞ natural appears, defined so that the density of the mimicked dark matter approaches the order of the critical density, $\rho_{dm}(r_\infty) \sim \rho_c$ (the latter being is defined as $\rho_\infty = 3H_0^2/(8\pi G)$, with $H_0 \simeq 70\ (km/s)/Mpc$ the Hubble constant). Using Eq. (26) yields

$$r_\infty = \left[\frac{2}{3} \times 3^n(1-2n)\right]^{1/4} \left(\frac{c}{r_0 H}\right)^{(1-n)/2} \left(r_s r_0^2 a\right)^{1/4}, \tag{35}$$

having defined the characteristic lengthscale $r_0 = 1/\sqrt{R_0}$. The obtained matching distance which may be regarded as an indicator of the radius of the mimicked "dark" matter halo; for consistency, it should vanish when the non-minimal coupling Eq. (19) is negligible, *i.e.* $R_0 \rightarrow 0$ if the exponent n is negative or $R_0 \rightarrow \infty$ for $n > 0$; this is trivially satisfied, since $r_\infty \propto r_0^{n/2} = R_0^{-n} \rightarrow 0$ (excluding the constant coupling $n = 0$ case).

One may also define the characteristic density

$$\rho_0 = 2\kappa R_0 = \frac{c^2}{8\pi G r_0^2} = \left(\frac{c}{r_0 H}\right)^2 \frac{\rho_\infty}{3} = \left(\frac{2.5\ Gpc}{r_0}\right)^2 \rho_\infty. \tag{36}$$

Notice that the equation of state previously found Eq. (28) enables a negative pressure for any negative exponent n; furthermore, in a cosmological setting, the non-trivial coupling $f_2(R) = (R/R_0)^n$ is given by $(\rho_\infty/\rho_0)^n$. Hence, the obtained relation Eq. (36) does hint a putative unified description of dark matter and dark energy, responsible for the acceleration of the expansion rate of the Universe. Other models of unification of dark energy and dark matter include the Chaplygin gas model [18, 19, 20, 21] and some field theory constructions [22].

4.4. Crossover between visible and dark matter dominance

Proceeding with the analysis of the obtained "static" solution, one remarks that the model parameter R_0 introduces both the length scale r_0 as well as a density scale ρ_0; the former should have a clear physical meaning. Indeed, r_0 is associated with the crossover distance r_c at which the density of the mimicked dark matter component becomes similar to that of visible matter, thus signalling the onset of the flattening of the galaxy rotation curves. By identifying $\rho_{dm}(r_c) = \rho(r_c)$ and using the obtained profiles 27 and Eq. (32), one obtains

$$r_c^4 = 2(1-2n)^{1/n} r_s r_0^2 a, \tag{37}$$

with $r_s = 2GM/c^2$ being the Schwarzschild radius of the galaxy. Hence, the crossover radius r_c is proportional to a sort of "geometric mean" of all the length scales present.

4.5. "Gradient" solution

One ends this section by discussing the relevance of the gradient term on the *r.h.s* of Eq. (23); indeed, although the "static" solution makes it vanish identically, it was obtained by neglecting the contribution from visible matter, since one is assuming that dark matter is dominant at large distances. Conversely, at smaller distances one expects the solution to Eq. (23) to deviate significantly from the "static" solution Eq. (26), interpolating between the latter and the classical identification $R = 8\pi G\rho$. Thus, in the regime before the onset of dark matter dominance $r < r_c$, the contribution from the mentioned gradient terms may be highly relevant.

This may be justified by inserting the GR identification $R = 8\pi G\rho$ into Eq. (23), and reading the gradient terms. To do so, one assumes that the D'Alembertian operator is given by its expression in a flat spacetime (in spherical coordinates), which amounts to neglecting relativistic corrections. This requirement translates into the condition $r_s < r_0$, which is easily satisfied by the mechanism under scrutiny, as shall be seen.

A somewhat evolved analysis depicted in Ref. [10] indeed shows that the gradient term may become dominant below a distance $r_k < a < r_c$,

$$r_k \equiv \left[6n^2(1-n) \left(\frac{a^2}{2r_s} \right)^{1-n} r_0^{2n} \right]^{1/(n+1)} . \tag{38}$$

Notice that, at even shorter distances, relativistic corrections of the form r_s/r should be considered. It can be shown that this is not related to the particular form of the visible matter density distribution (in this case, the Hernquist profile), and that assuming another form for $\rho(r)$ would still lead to an eventual dominance of the gradient term of Eq. (23) below a threshold r_k.

This concern may be attenuated by noticing that one is assuming the validity of the Newtonian regime: if the lengthscales a, r_s and r_0 yield a value for r_k smaller than the Schwarzschild radius, then the need to include relativistic corrections to the computation of the gradient term makes the whole approach inconsistent. As a numerical session shall show, the condition $r_k > r_s$ is not always satisfied, so that the above expression for r_k should be considered with due care.

Skipping over the technicalities, it may be shown that it is impossible to simultaneously ensure that the gradient term on the *r.h.s* of Eq. (23) is negligible for all distances: it dominates at either $r < a$ or $r > a$. This is not unexpected, given that the lengthscale a signals the shift of behaviour of the Hernquist profile Eq. (32). Hence, one is forced to conclude that the "static" solution Eq. (26) is not valid after all! In particular, the discussion of the dominance of the gradient term in the outer region $a < r < r_c$ is flawed, as the crossover distance r_c assumed for its validity.

If the gradient term indeed dominates the *r.h.s* of Eq. (23) at short ranges, it turns out that the corresponding "gradient" solution to this equation is the relevant quantity to assess, and not the "static" one already described. By the same token, it is legitimate to assume that this "gradient" solution will play the role of dark matter, so that it should dominate the contribution of visible matter at long ranges. Thus, Eq. (23) may be approximated by

$$R \approx -\frac{3n}{\kappa}\left(\frac{d^2}{dr^2}+\frac{2}{r}\frac{d}{dr}\right)\left[\left(\frac{R}{R_0}\right)^n \frac{\rho}{R}\right]. \tag{39}$$

One now introduces the redefined field

$$\varrho = \left(\frac{R}{R_0}\right)^{-n/(1-n)}, \tag{40}$$

and coordinate $y = r_0/r$; the dominance of this "gradient" solution over the visible matter contribution occurs in the range $r > r_k$, that is, $y < y_k$.

In order to better understand the evolution of this "gradient" solution, one may compute the order of magnitude of the quantities involved at $y = y_k$ (by recalling that, for smaller physical distances $y > y_k$, the classical solution $R = 8\pi G\rho$ is valid and inserting the Hernquist profile for $\rho(r)$). One finds that

$$\left(\frac{d^2\varrho^{n-1}}{dy^2}\right)_{y=y_k} = -\frac{1}{3n}\left(\frac{r_s a}{r_0^2}\right). \tag{41}$$

Inserting typical values for galaxies $a = (1-10)$ *kpc* and $r_s = (10^{-3}-10^{-1})$ *pc* (corresponding to galaxies of approximately $(10^{10}-10^{12})$ solar masses) and recalling the validity condition for the Newtonian regime, $r_0 > r_s$, one concludes that the *r.h.s.* of the above equation is less than unity; it turns out that the values involved are extremely small (of the order 10^{-14} or lower), so that one may safely consider that Eq. (39) is essentially given by

$$\frac{d^2\varrho^{n-1}}{dy^2} \approx 0 \tag{42}$$

at large distances, *i.e.* small values of y — the full form of Eq. (39) with the above redefinitions would be that of a generalized Emden-Fowler differential equation. Hence, the "gradient" solution is very well approximated by $\varrho =$ const.; recalling Eq. (40) leads to

$$R \propto R_0\left(\frac{\rho}{\rho_0}\right)^{1/(1-n)}, \tag{43}$$

that is, the "gradient" solution tracks the "static" solution! Hence, the previous results are partially vindicated, and the identification of the induced effect due to the non-minimal coupling $f_2(R)$ with a perfect fluid with equation of state Eq. (28) is still valid.

4.6. Tully-Fisher law

The Tully-Fisher law is an empirical relationship between the intrinsic luminosity L and a power of the rotation velocity v_∞ of a spiral galaxy. An analogous relationship exists for elliptical galaxies, the Faber-Jackson relation. Since the luminosity is proportional to the visible mass, both may be written as $M \propto v_\infty^\sigma$, with $3 < \sigma \lesssim 4$.

If, for simplicity, one assumes that $M_{dm}(r) \propto r$ at large distances, thus producing a flat rotation curve, the familiar $v_\infty^2 \propto M_{dm}$ relation is recovered; resorting to the scaling Eq. (27), this yields

$$M \propto v^{2(1-n)}, \tag{44}$$

so that the exponent of the Tully-Fisher or Faber-Jackson relations is related to that of the non-minimal coupling $f_2(R)$. For the scenarios already discussed, one obtains $M \propto v^{8/3}$, for $n = -1/3$, mimicking the NFW dark matter density profile, and $M \propto v^4$, for $n = -1$, mimicking a dark matter isothermal sphere.

4.7. Conservation of energy-momentum tensor

Section 3 depicted a first attempt to describe the flattening of the galactic rotation curves as due to the non-conservation of the energy-momentum tensor of matter, *i.e.*, through the non-geodesical motion experienced by fluid particles. Conversely, one now explores the possibility of mimicking "dark matter" as a consequence of the dynamics of Eq. (23). As explicitly stated in the beginning of section 4, one is assuming that matter follows geodesics, so that the solution of Eq. (23) gives rise to the observed rotation curves.

This assertion may now be verified, by rewriting Eq. (9), with $p = 0$ for pressureless matter, the Lagrangian density $\mathcal{L} = -\rho$ and the adopted non-minimal coupling $\lambda f_2(R) = (R/R_0)^n$:

$$f^\mu = -n \frac{\left(\frac{R}{R_0}\right)^n}{1 + \left(\frac{R}{R_0}\right)^n} \left[\log\left(\frac{R}{R_0}\right)\right]_{,\nu} h^{\mu\nu}, \tag{45}$$

with radial component

$$f^r = -n \frac{\left(\frac{R}{R_0}\right)^n}{1 + \left(\frac{R}{R_0}\right)^n} \left[\log\left(\frac{R(r)}{R_0}\right)\right]', \tag{46}$$

assuming spherical symmetry and the Newtonian approximation $h_{rr} = g_{rr} - (v_r/c)^2 \sim 1$.

As discussed in Ref. [10], a clear distinction between the "static" and the "gradient" solutions is that, although both are essentially the same (up to a multiplicative factor), each arises from widely different assumptions: the former assumes that the gradient terms on the *r.h.s.* of Eq. (23) may be disregarded, so that the dominance of dark matter translates into the condition $(R/R_0)^n > 1/(1-2n)$; the latter, however, is based upon the dominance of these gradient terms, which implies the converse relation, $(R/R_0)^n < 1/(1-2n) < 1$, for negative exponent n.

Thus, one concludes that the issue of the covariant conservation of the energy-momentum tensor of matter branches between both solutions: from Eq. (46), one sees that the "static" solution would lead to

$$f^r = -n \left[\log\left(\frac{R}{R_0}\right)\right]' = -n\frac{R'}{R} = \frac{nm'}{r}. \tag{47}$$

since $R \propto \rho_{dm} \propto r^{m'}$ in the outer region of galaxies; given that n is negative, this force correctly points towards the galactic center, thus enabling for a flattening of the rotation curve with a constant asymptotic velocity. However, for small $|n|$ and m', this quantity would be given by $v_\infty/c = \sqrt{-nm'} \sim 1$, in blatant violation of the observed value $v_\infty \sim 10^2\ km/s \sim 10^{-3}c$! Not only is this clearly unphysical (and would undermine the Newtonian approach taken throughout the text), but it also negates the whole purpose of this section, namely to assume geodesic motion.

Contrariwise, the verified dominance of the "gradient" solution and related condition $(R/R_0)^n < 1/(1-2n) < 1$ leads to a radial force

$$f^r = -n \left(\frac{R}{R_0}\right)^{n-1} \left(\frac{R(r)}{R_0}\right)'. \tag{48}$$

and the pathological result stemming from Eq. (47) is avoided.

For the already considered values of $a = (1 - 10)\ kpc$ and $r_s = (10^{-6} - 10^{-4})\ kpc$, it suffices that $r_0 > 200\ kpc$ for the dominance of this extra force over the Newtonian one to occur only for distances greater than 1 Mpc. Hence, this extra force is negligible at galactic scale, and given the actual values for r_0 that will be used afterwards, much greater than this lower limit. Moreover, the energy-momentum tensor is conserved to a very good approximation.

4.8. Energy conditions

One now checks if the strong, null, weak and dominant energy conditions (SEC, NEC, WEC and DEC, respectively) are verified; following Ref. [23], a little algebra yields the condition

$$\left(1+\frac{E_n}{1+\left(\frac{R}{R_0}\right)^{-n}}\right)\rho \geq 0, \tag{49}$$

with

$$E_n = \begin{cases} -2n & , \text{SEC} \\ 0 & , \text{NEC} \\ 2n & , \text{DEC} \\ n & , \text{WEC} \end{cases} . \tag{50}$$

Clearly, both the NEC and SEC are always satisfied, for a negative power law $n < 0$. The WEC and DEC are also fulfilled for the ranges $-1 \leq n \leq 0$ and $-1/2 \leq n \leq 0$, respectively. Since the condition $R > R_0$ is always valid at galactic scales (due to the dominance of the gradient term of Eq. (23)) for the considered models $n = -1$ (isothermal sphere profile) and $n = -1/3$ (NFW profile), one has that $\left(1+(R/R_0)^{-n}\right)^{-1} \approx 0$; hence, the WEC and DEC are also satisfied for these scenarios.

A final check is in order, namely whether the considered mechanism for the mimicking of dark matter does not give rise to the so-called Dolgov-Kawasaki instabilities, whereas small curvature perturbations could arise and expand uncontrollably [24]. Refs. [23, 25] allow one to write the mass scale m_{DK} for these as

$$\begin{aligned} m_{DK}^2 &= \frac{2\kappa - 2\mathcal{L}_m\left[f_2'(R)+f_2''(R)R\right]+f_2'(R)T}{2\mathcal{L}_m f_2''(R)} = \\ &= -\frac{2\kappa R\left(\frac{R}{R_0}\right)^{-n}+n(2n-1)\rho}{2n(n-1)\rho}R, \end{aligned} \tag{51}$$

which are given approximately by

$$m_{DK}^2 \approx \left(\frac{R}{R_0}\right)^{2-n}\frac{\rho_0}{\rho}\frac{r_0^{-2}}{2n(n-1)}. \tag{52}$$

Since m_{DK}^2 is negative, the possibility of instabilities at a length scale r_{DK} is worrisome, with

$$r_{DK} = m_{DK}^{-1} = \sqrt{2n(n-1)}\left(\frac{R_0}{R}\right)^{1-n/2}\sqrt{\frac{\rho}{\rho_0}}r_0. \tag{53}$$

At large distances, where the dark matter-like component dominates and the scalar curvature is greater than the matter density, one finds that this typical length scale r_{DK} grows linearly for the $m' = 2$ isothermal sphere scenario, and with a power-law $r^{3/2}$ for the $m' = 3/2$, corresponding to the NFW density profile.

It should be noted that the generation of these instabilities assumes an initial small perturbation in a constant background curvature; this is best suited for an analysis in cosmology, but should be approached with due care in the present context, where there is a spatial variation of $R(r)$. Indeed, one must instead resort to the characteristic lengthscale over which one may assume a constant curvature, given by $L \equiv -R(r)/R'(r) = r/m'$. One may compare this constant curvature domain L with the instability length scale r_{DK} to ascertain if the assumed perturbative expansion, as embodied in the inequality $r_{DK} < L$, is valid. Equating both quantities yields

$$\left(\frac{r}{a}\right)^{\frac{1+n}{1-n}} < \varrho^{-n+1/2}(2\delta\bar{a})^{\frac{1}{2(1-n)}}\frac{a}{4r_0}\sqrt{\frac{n-1}{2n}}. \tag{54}$$

This relationship immediately shows that the perturbative expansion is invalid if $n = -1$, when dark matter assumes a isothermal sphere distribution; if it takes the form of a NFW profile, $n = -1/3$, one finds that the perturbative expansion is valid only below distances $r < r_s$, where relativistic corrections become relevant. Hence, the Dolgov-Kawasaki instabilities are not physically meaningful in the context of this work.

Acknowledgments

The authors thank E. Vagena for fruitful discussions and the hospitality shown during the *First Mediterranean Conference on Classical and Quantum Gravity*, Crete, September 2009.

[1] Trimble V 1987 *Annual review of astronomy and astrophysics* **25** 425
[2] Bergström L 2000 *Reports on Progress in Physics* **63** 793
[3] Hoekstra H, Yee H K C and Gladders M D 2004 *The Astrophysical Journal* **606** 67
[4] Bertone G, Hooper D and Silk J 2005 *Physics Reports* **405** 279
[5] Capozziello S, Cardone V F and Troisi A 2006 *Journal of Cosmology and Astroparticle Physics* **08** 001
[6] Borowiec A, Godlowski W and Szydlowski M 2007 *Int. J. Geom. Meth. Mod. Phys.* **4** 183
[7] Martins C F and Salucci P 2007 *Monthly Notices of the Royal Astronomical Society* **381** 1103
[8] Capozziello S, Cardone V F and Troisi A 2007 *Monthly Notices of the Royal Astronomical Society* **375** 1423
[9] Böhmer C G, Harko T and Lobo F S N 2008 *Astroparticle Physics* **29** 386
[10] Bertolami O and Páramos J 2009 (*Preprint* arXiv:0906.4757 [astro-ph.GA])
[11] Bertolami O, Böhmer C G, Harko T and Lobo F S N 2007 *Phys. Rev. D* **75** 104016
[12] Bertolami O and Páramos J 2008 *Class. Quantum Grav.* **25** 5017
[13] Bertolami O, Lobo F S N and Páramos J 2008 *Phys. Rev. D* **78** 64036
[14] Faraoni V 2009 (*Preprint* 0912.1249)
[15] Navarro J F, Frenk C S and White S D M 1995 *Monthly Notices of the Royal Astronomical Society* **275** 56
[16] Bertolami O and Páramos J 2008 *Phys. Rev. D* **77** 84018
[17] Hernquist L 1990 *Astrophysical Journal* **356** 359
[18] Kamenshchik A, Moschella U and Pasquier V 2001 *Physics Letters B* **511** 265
[19] Bilić N, Tupper G B and Viollier R D 2002 *Physics Letters B* **535** 17
[20] Bento M C, Bertolami O and Sen A A 2002 *Phys. Rev. D* **66** 43507
[21] Barreiro T, Bertolami O and Torres P 2008 *Phys. Rev. D* **78** 43530
[22] Bertolami O and Rosenfeld R 2008 *Int. J. Mod. Phys. A* **23** 4817
[23] Bertolami O and Sequeira M C 2009 *Phys. Rev. D* **79** 104010
[24] Dolgov A D and Kawasaki M 2003 *Physics Letters B* **573** 1
[25] Faraoni V 2007 *Phys. Rev. D* **76** 127501

First Mediterranean Conference on Classical and Quantum Gravity (MCCQG 2009) IOP Publishing
Journal of Physics: Conference Series **222** (2010) 012011 doi:10.1088/1742-6596/222/1/012011

Dark matter as a dynamic effect due to a non-minimal gravitational coupling with matter (II): Numerical results

J. Páramos, O Bertolami[1]

Instituto de Plasmas e Fusão Nuclear, Instituto Superior Técnico,
Avenida Rovisco Pais 1, 1049-001 Lisboa, Portugal

E-mail: paramos@ist.edu, orfeu@cosmos.ist.utl.pt

Abstract. Following the previous contribution discussing the rich phenomenology of models possessing a non-minimal coupling between matter and geometry, with emphasis on its characteristics and analytical results, the obtained "dark matter" mimicking mechanism is numerically studied. This allows for ascertaining the order of magnitude of the relevant parameters, leading to a validation of the analytical results and the discussion of possible cosmological implications and deviation from universality.

1. Introduction

This work aims to confront the available observational evidence, namely the rotation curves of galaxies, with the analytical results of the "dark matter" mimicking mechanism described in a previous contribution to this volume (see also Ref. [1] for the original discussion and extended results). This will allow for the determination of the relevant model parameter R_0, via the associated lengthscale r_0, and the prediction of possible cosmological implications of the assumed power-law non-minimal coupling.

One chooses to fit numerically several rotation curves of galaxies of E0 type — these are selected since they exhibit an approximate spherical symmetry adequate for comparison with the analytical results obtained in the previous contribution. Seven galaxy rotation curves, are used, with the division into visible and dark matter components as reported in Refs. [2, 3, 4, 5]: NGC 2434, NGC 5846, NGC 6703, NGC 7145, NGC 7192, NGC 7507 and NGC 7626 galaxies.

As stated in the preceding section, particular interest is assigned for $n = -1$ and $n = -1/3$, which correspond to the isothermal sphere or NFW dark matter density profiles. Furthermore, one may recall that the power-law profile $f_2(R) = (R/R_0)^n$ was discussed in the context of a possible series expansion of a more general non-minimal coupling; hence, there is no reason to select *a priori* either case $n = -1$ or $n = -1/3$: instead, one can successfully take advantage of a composite coupling given by

$$f_2(R) = \sqrt[3]{\frac{R_3}{R}} + \frac{R_1}{R}, \tag{1}$$

[1] Also at Departamento de Física, Instituto Superior Técnico.

Journal of Physics: Conference Series **222** (2010) 012011 doi:10.1088/1742-6596/222/1/012011

so that both profiles may be added in order to describe the dark matter density profile and the related galaxy rotation curve.

The proposed non-minimal coupling leads to a more complex expression for the trace of the field equations

$$R = \frac{1}{2\kappa}\left[1 + 3\frac{R_1}{R} + \frac{5}{3}\left(\frac{R_3}{R}\right)^{1/3}\right]\rho + \frac{1}{\kappa}\Box\left(\left[3\frac{R_1}{R} + \left(\frac{R_3}{R}\right)^{1/3}\right]\frac{\rho}{R}\right). \quad (2)$$

One can no longer resort to the identification of a "static" solution, as before, since the non-linearity of the above equation indicates that the solution for this composite scenario is not given by the mere addition of individual solutions; by the same token, no clear equation of state or identification of the mimicked dark matter profile with a perfect fluid arises. Nevertheless, one expects that the main feature of the model are kept, namely the overall mimicking of dark matter density profiles according to the scaling relation between visible matter ρ and "dark matter" ρ_{dm} densities,

$$\rho_{dm} = \frac{1-n}{1-4n}\rho_0\left[(1-2n)\frac{\rho}{\rho_0}\right]^{1/(1-n)}, \quad (3)$$

and the dominance of the gradient term on the *r.h.s.* of the above equation.

2. Numerical study

This stated, one proceeds as follows: each visible matter density profile $\rho(r)$ is derived from the visible component of the corresponding galaxy's rotation curve; two added Hernquist profiles are used, one for the extended gas and another for the core matter distribution. This density profile is then used as an input to Eq. (2), and a best fit scenario for the dark matter component of the rotation curve is obtained by varying the parameters $r_1 = 1/\sqrt{R_1}$ and $r_3 = 1/\sqrt{R_3}$. Although these should be universal quantities, one resorts to individual fits of each galaxy, allowing for a posterior discussion of order of magnitude, deviation between obtained values, any abnormal case or possible trends, *etc.*

The results derived from the numerical solution of Eq. (2) are depicted in Table 1, and shown in Fig.1: as can be seen, the composite non-minimal coupling Eq. (1) provides close fits for all rotation curves in the outer region — of much higher quality than those obtainable if one assumed individual couplings $f_2(R) = R/R_1$ or $f_2(R) = \sqrt[3]{R/R_3}$ (see additional graphs in Ref. [1]). For some galaxies, the overall quality of fit is disturbed by a discrepancy in the inner galactic region. This could result from an uncertainty in the original derivation of the rotation curves, or perhaps due to the deviation from purely spherical symmetry.

From a fundamental standpoint, this deviation could also indicate that the non-minimal coupling should include an additional term that acquires relevance in a higher density environment: recalling an earlier work in the high density context of the Sun, one could admit a linear addition $\delta f_2(R) \propto R$, for instance. Of course, the considered model could be degenerate to the approximation of the pure curvature term given by the GR prescription $f_1(R) = 2\kappa R$, and a more complex picture involving both non-trivial $f_1(R)$ and $f_2(R)$ couplings is in order.

Although details are here suppressed, for brevity, after obtaining the solution to Eq. (2) it is straightforward to compute the gradient term on its *r.h.s.*: as discussed before, this is indeed dominant, so that the mimicked dark matter density profile is almost completely overlapped with it.

Journal of Physics: Conference Series **222** (2010) 012011 doi:10.1088/1742-6596/222/1/012011

Galaxy	Composite				Single	
(NGC)	r_1	r_3	$r_{\infty\ 1}$	$r_{\infty\ 3}$	r_1	r_3
2434	∞	0.9	0	33.1	4.1	0.9
5846	37	∞	138	0	37	34.9
6703	22	∞	61.2	0	22	26.2
7145	22.3	47.3	60.9	14.2	19.9	23.8
7192	14.8	24	86.0	18.3	14.5	7.3
7507	4.9	2.9	178	31.1	4.3	1.1
7626	28	9.6	124	42.5	16.0	7.1

Table 1. Best fit values for the characteristic lengthscales r_1 and r_3 for the composite and separate fits of the galaxy rotation curves with the $n = -1$ isothermal sphere and $n = -1/3$ NFW mimicked "dark matter" scenarios, together with background matching distances $r_{\infty\ 1}$ ($n = -1$) and $r_{\infty\ 3}$ ($n = -1/3$) for the composite non-minimal coupling. The limit $r_i = \infty$ for the composite scenario indicates that the corresponding scale $R_i = 0$. The units used are r_1 (Gpc), r_3 ($10^5\ Gpc$), $r_{\infty\ 1}$ and $r_{\infty\ 3}$ (kpc).

3. Characteristic density, background matching and mimicked "dark matter" inter-dominance

The selected galaxies have their rotation curves fitted by characteristic lengthscales of the order $r_1 \sim 10\ Gpc$ and $r_3 \sim (10^5 - 10^6)\ Gpc$. Although the analytic results derived in the preceding sections do not hold strictly in the assumed composite coupling scenario of Eq. (1), the expression

$$r_\infty = \left[\frac{2}{3} \times 3^n (1-2n)\right]^{1/4} \left(\frac{c}{r_0 H}\right)^{(1-n)/2} \left(r_s r_0^2 a\right)^{1/4} , \tag{4}$$

should nevertheless provide a quantitative figure for the background matching distances,

$$\begin{aligned} r_\infty &= \left[\frac{2}{3} \times 3^n (1-2n)\right]^{1/4} \left(\frac{c}{r_0 H}\right)^{(1-n)/2} \left(r_s r_0^2 a\right)^{1/4} \approx \\ &\simeq \begin{cases} \left(\dfrac{3.8\ Gpc}{r_3}\right)^{2/3} \left(r_s r_3^2 a\right)^{1/4} & , n = -1/3 \\ \left(\dfrac{3.8\ Gpc}{r_1}\right) \left(r_s r_1^2 a\right)^{1/4} & , n = -1 \end{cases} , \end{aligned} \tag{5}$$

after inserting the Hubble constant $H_0 = 70.8\ km\ s^{-1}\ Mpc^{-1}$. Notice that this expression was obtained with the assumption of a single Hernquist profile for the visible matter density, and not for the two component model used in the current fitting session; using the dominant contribution stemming from the extended gas distribution, one obtains the values displayed in the last columns of Table 1.

Since r_∞ may be regarded as signaling the radius of the mimicked dark matter haloes, Table 1 shows that the $n = -1/3$ NFW haloes are always smaller than those with the $n = -1$ isothermal sphere profile — with the exception of NGC 2434 that, however, stands out from the set since it is best fitted by a single NFW "dark matter" component. Also, $r_{\infty\ 1}$ is always larger than the range of the rotation curves, so that the isothermal sphere "dark matter" extends further than the observationally inferred; on the contrary, the value of $r_{\infty\ 3}$ goes from about $0.57L$ (NGC 7145) to $1.73L$ (NGC 7507), L being the full distance considered for the rotation curves.

Journal of Physics: Conference Series **222** (2010) 012011 doi:10.1088/1742-6596/222/1/012011

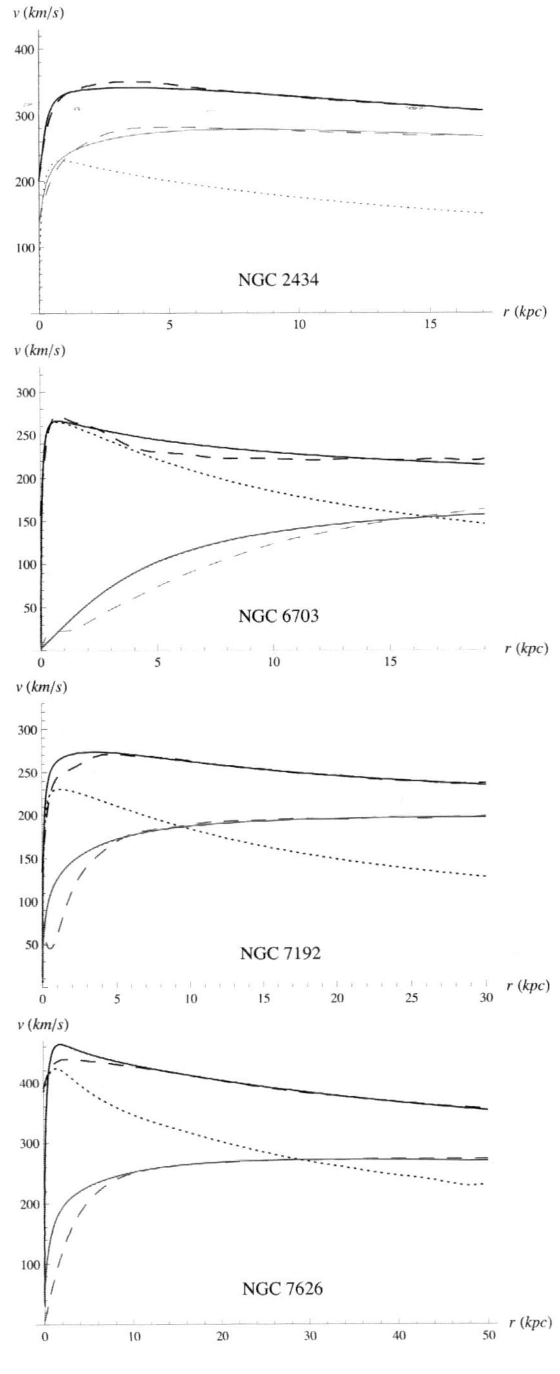

Figure 1. Observed rotation curve (dashed full), decomposed into visible (dotted) and dark matter (dashed grey) contributions [2], superimposed with the mimicked dark matter profile (full grey) arising from the composite non-minimal coupling and resulting full rotation curve (full).

This indicates that the isothermal sphere halo is more significant that its NFW equivalent:

even the exceptional case of NGC 2434, where no isothermal sphere component appears in the best fit scenario, yields a matching distance $r_{\infty\ 3} = 33.1\ kpc$ — only about twice the endpoint of the corresponding rotation curve, $L \simeq 17\ kpc$.

This is confirmed by a simple algebraic computation: since the NFW profile falls as r^{-3}, while the isothermal sphere one displays a shallower r^{-2} long-range dependence, and the respective characteristic densities obey $\rho_3 \ll \rho_1$, it is easy to conclude that the isothermal sphere eventually becomes the dominant dark matter contribution. Indeed, equating both contributions yields

$$\frac{\rho_{dm\ 3}}{\rho_{dm\ 1}} = \left(\frac{r_1^2}{ar_3}\right)^{1/2} \left(\frac{r_s}{a}\right)^{1/4} \left(\frac{a}{r}\right) \sim \frac{1}{10}\frac{a}{r}, \tag{6}$$

so that this dominance occurs at distances larger than $r = a/10 \sim 1\ kpc$. Thus, the considered galactic rotation curves are found to be "mostly flat".

4. Cosmological relevance

As discussed already, the results obtained above are naturally tuned towards the corresponding astrophysical setting, if one assumes that the chosen form for the (composite) non-minimal coupling represents the leading terms of a series expansion of a more evolved expression for $f_2(R)$, suitable in the mentioned range for the scalar curvature R. Hence, there is nothing mandatory concerning a possible cosmological relevance of the composite coupling Eq. (1), with the obtained values for R_1 and R_3: it might be the case that a suitable mechanism applicable in cosmology corresponds to another term of the hypothetical series expansion of $f_2(R)$.

Nevertheless, it is natural to assess if the considered cases $n = -1$ and $n = -1/3$ give rise to implications for cosmology, since this would present a natural candidate for a unification model of dark matter and dark energy. From the discussion of paragraph **??**, one may write the non-minimal coupling Eq. (1), in a cosmological context, as

$$f_2 = \sqrt[3]{\frac{R_3}{R}} + \frac{R_1}{R} \simeq \left(\frac{r_H}{r_3}\right)^{2/3} + \left(\frac{r_H}{r_1}\right)^2 \sim 10^{-4} + 10^{-1}, \tag{7}$$

having used $R = 3/r_H^2$, with $r_H = c/H = 4.2\ Gpc$ the Hubble radius, and inserting the best fit orders of magnitude $r_1 \sim 10\ Gpc$ and $r_3 \sim (10^5 - 10^6)\ Gpc$.

The small value of the $n = -1/3$ non-minimal coupling indicates that this should not be significant at a cosmological level; the higher value of the $n = -1$ power-law coupling (when compared to unity) hints that a simple R_1/R coupling could prove interesting for cosmology.

5. Universality

The obtained values for r_1 and r_3, shown in Table 1, display an undesired variation between galaxies, since these parameters should be universal. Indeed, r_1 averages $\bar{r}_1 = 21.5\ Gpc$ with a standard deviation $\sigma_1 = 10.0\ Gpc$, while r_3 presents $\bar{r}_3 = 1.69 \times 10^6\ Gpc$ and $\sigma_3 = 1.72 \times 10^6\ Gpc$. This deviation from universality is not unexpected, and several causes could be hinted as culprits: although small, the existing assymetry in the selected type E0 galaxies due to the $1 - b/a \lesssim 0.1$ semi-axis deviation from pure sphericity could translate into the difference in obtained values for r_1 and r_3. This shift from the assumed symmetry could be enhanced by localized features and inhomogeneities, which could also act as "seeds" for the dynamical generation of the mimicked "dark matter" component. Furthermore, the reported dominance of the gradient term in the *r.h.s.* of Eq. (2) could boost this effect, if these perturbations vary significantly on a short lengthscale.

By the same token, the relevance of the gradient term could lead to the deviation from universality if the chosen profiles (Hernquist for visible matter, NFW and isothermal sphere for mimicked dark matter) prove not to be so suitable as hoped: this mismatch between the real density profiles and the fitted ones does not need to be large, if a large deviation of its first and second derivatives lies in very localized regions: this could act as a sort of "fitting inhomogeneities" with the same effect as of those discussed above.

Aside from the symmetry and fitting functions selected, more fundamental issues could be behind the variation of the obtained values for r_1 and r_3: in fact, the non-minimal coupling Eq. (1) could be too simplistic, and a more complex model could provide a more universal fit of the galaxy rotation curves.

Also (as discussed with respect to the relatively less successful fits of the inner galactic regions), the simplifying assumption of a linear curvature term $f_1(R) = 2\kappa R$ might also contribute to this deviation from universality, since one knows that power-law terms $f_1(R) \propto R^n$ can also induce a dynamical "dark matter" [6].

Finally, a more interesting idea could lie behind the obtained discrepancy, namely that the visible matter Lagrangian density itself may be different from the assumed perfect fluid form, $\mathcal{L}_m = -\rho$ [7, 8]: this functional dependence is inadequate in the context of the model here addressed, and a more adequate form $\mathcal{L}_m = \mathcal{L}_m(\rho)$ should be considered — perhaps even including other variables related to the thermodynamical description of the matter distribution (see Ref. [9] for an in-depth discussion based upon a similar consideration).

6. Conclusions

In the two contributions to this volume, the possibility of obtaining a solution to the dark matter puzzle, embodied by the flattening of galaxy rotation curves, was approached by resorting to the main phenomenological implications of models possessing a non-minimal coupling of matter to curvature, following results reported in Ref. [1].

As a first attempt, one first examined the non-conservation of the energy-momentum tensor of matter and the implied deviation from geodesic motion, and ascertained what coupling $f_2(R)$ should be so that the derived extra force would lead to the reported flattening of the rotation curves. This requires a logarithmic coupling of the form $\lambda f_2(R) = -v^2/m \log(R/R_*)$ (where m is the outer slope of the visible matter density ρ), which can be approximated by a simpler power-law $\lambda f_2(R) \approx (R_*/R)^\alpha$, with the asymptotic velocity given by $v_\infty^2 = \alpha m$. However, this solution yields an almost universal v_∞, prompting for another possible mechanism for the dynamical mimicking of "dark matter".

This need led to the second, more evolved approach, which is rooted empirically in the phenomenological Tully-Fisher relation: one may instead assume that geodesical motion is preserved, $\nabla^\mu T_{\mu\nu} = 0$ (a condition proved to be self-consistent), but that the metric itself is perturbed: the mimicked dark matter density is then given by the difference $R/2\kappa - \rho$. By resorting to a power-law coupling with matter $f_2(R) = (R/R_0)^n$ (with a negative exponent n yielding the desired effect at low curvatures and long range).

The proposed power-law directly leads to a "dark matter" density that depends on a power of the visible matter density, thus accounting for the Tully-Fisher law in a natural way. The obtained "dark matter" component has a negative pressure, as commonly found in cosmology in dark energy models: this hints the possibility that the non-minimal coupling model might unify dark matter and dark energy.

Given their relevance in the literature, two different scenarios were considered: the NFW and isothermal sphere dark matter profiles ($n = -1/3$ and $n = -1$, respectively). Although separate fits of the selected galaxy rotation curves to each coupling do not yield satisfactory results, a composite coupling of both power-laws produced a much improved adjustment.

The characteristic lengthscales r_1 and r_3 were taken as fitting parameters for each individual galaxy, yielding the order of magnitudes $r_1 \sim 10$ *Gpc* and $r_3 \sim 10^5$ *Gpc*. This enabled the computation of the cosmological background matching distances for each galaxy (which depend on their characteristic lengthscale a) and the obtained astrophysical range showed that the

$n = -1$ isothermal sphere "dark matter" halo dominates the $n = -1/3$ NFW component. Furthermore, the $n = -1$ scenario was shown to have possible relevance in a cosmological context, as $r_1 \sim r_H$, the latter being the Hubble radius — again hinting at a possible unification of the dark components of the Universe.

The lack of the desirable universality in the model parameters R_1 and R_3 was duely noticed, although these quantities are characterized by the same order of magnitude for the best fits obtained; several possible causes for this variation were put forward, from deviation from spherical symmetry to a more complex form for the non-minimal coupling between curvature and matter $f_2(R)$ or the curvature term $f_1(R)$, or the need for a more evolved Lagrangian description of the latter.

As a final remark, one concludes that the rich phenomenology that springs from the model with a non-minimal coupling between matter and curvature enables a direct and elegant alternative to standard dark matter scenarios and many modifications to GR, which usually resort to extensive use of additional fields and other *ad-hoc* features. As an example of the latter, one points out the MOdified Newtonian Dynamics (MOND) hypothesis, which is by itself purely phenomenological, and whose underlying Tensor-Vector-Scalar theory is based upon an extensive paraphernalia of vector and scalar fields [10, 11, 12, 13] (see Ref. [14] for a critical assessment). As a final remark, we point out that the considered model may also be translated into a multi-scalar theory [15], with two scalar fields given by

$$\varphi^1 = \frac{\sqrt{3}}{2} \log \left[1 + n \left(\frac{R}{R_0} \right)^n \right] \qquad \varphi^2 = R, \tag{8}$$

with dynamics driven by a potential

$$U(\varphi^1, \varphi^2) = \frac{1}{4} \exp\left(-\frac{2\sqrt{3}}{3}\varphi^1 \right) \left[\varphi^2 - \frac{f_1(\varphi^2)}{2\kappa} \exp\left(-\frac{2\sqrt{3}}{3}\varphi^1 \right) \right]. \tag{9}$$

Acknowledgments
The authors thank E. Vagena for fruitful discussions and the hospitality shown during the *First Mediterranean Conference on Classical and Quantum Gravity*, Crete, September 2009.

[1] Bertolami O and Páramos J 2009 (*Preprint* arXiv:0906.4757 [astro-ph.GA])
[2] Kronawitter A, Saglia R P, Gerhard O and Bender R 2000 *Astronomy and Astrophysics Supplement* **144** 53
[3] Gerhard O, Kronawitter A, Saglia R P and Bender R 2001 *The Astronomical Journal* **121** 1936
[4] Faber S M, Wegner G, Burstein D, Davies R L, Dressler A, Lynden-Bell D and Terlevich R J 1989 *Astrophysical Journal Supplement Series* **69** 763
[5] Rix H W, de Zeeuw P T, Cretton N, van der Marel R P and Carollo C M 1997 *Astrophysical Journal* **488** 702
[6] Capozziello S, Cardone V F and Troisi A 2007 *Monthly Notices of the Royal Astronomical Society* **375** 1423
[7] Bertolami O, Lobo F S N and Páramos J 2008 *Phys. Rev. D* **78** 64036
[8] Faraoni V 2009 (*Preprint* 0912.1249)
[9] Harko T 2008 *Phys. Lett.* **B669** 376–379 (*Preprint* arXiv:0810.0742 [gr-qc])
[10] Milgrom M 1983 *Astrophysical Journal* **270** 365
[11] Bekenstein J and Milgrom M 1984 *Astrophysical Journal* **286** 7
[12] Bekenstein J D 2004 *Phys. Rev. D* **70** 83509
[13] Angus G W, Famaey B and Zhao H S 2006 *Monthly Notices of the Royal Astronomical Society* **371** 138
[14] Bertolami O and Páramos J 2006 Arxiv: 0611025[gr-qc]
[15] Bertolami O and Páramos J 2008 *Class. Quantum Grav.* **25** 5017

First Mediterranean Conference on Classical and Quantum Gravity (MCCQG 2009) IOP Publishing
Journal of Physics: Conference Series **222** (2010) 012012 doi:10.1088/1742-6596/222/1/012012

The possibility of an accelerating cosmology in Rastall's theory

M Capone[1,4], V F Cardone[2,4] and M L Ruggiero[3,4]

[1] Dipartimento di Matematica, Università di Torino, Via Carlo Alberto 10, 10125 - Torino, Italy

[2] Dipartimento di Fisica Generale "Amedeo Avogadro", Università di Torino, Via Pietro Giuria 1, 10125 - Torino, Italy

[3] UTIU, Università Telematica Internazionale Uninettuno, Corso Vittorio Emanuele II 39, 00186 - Roma, Italy

[4] INFN, Sezione di Torino, Via Pietro Giuria 1, 10125 - Torino, Italy

E-mail: `monica.capone@unito.it`

Abstract. In an attempt to look for a viable mechanism leading to a present day accelerated expansion, we investigate the possibility that the observed cosmic speed up may be recovered in the framework of the Rastall's theory, relying on the non-conservativity of the stress-energy tensor, i.e. $T^{\mu}_{\nu;\mu} \neq 0$. We derive the modified Friedmann equations and show that they correspond to Cardassian-like equations. We also show that, under suitable assumptions on the equation of state of the matter term sourcing the gravitational field, it is indeed possible to get an accelerated expansion, in agreement with the Hubble diagram of both Type Ia Supernovae (SNeIa) and Gamma Ray Bursts (GRBs). Unfortunately, to achieve such a result one has to postulate a matter density parameter larger than the typical $\Omega_M \simeq 0.3$ value inferred from cluster gas mass fraction data. As a further issue, we discuss the possibility to retrieve the Rastall's theory from a Palatini variational principle approach to $f(R)$ gravity. However, such an attempt turns out to be unsuccessful.

1. Introduction

The observed cosmic speed up [1] [2] [3] questions the validity of General Relativity (GR) on large scales. In fact, if on one hand the model of gravitational interaction as described by Einstein's theory is in agreement with many observational tests on relatively small scales, as Solar System and binary pulsars observations show [4], it is well known that in order to make GR agree with the observed acceleration of the Universe the existence of dark energy, a cosmic fluid having exotic properties, has been postulated. Actually, many candidates for explaining the nature of dark energy have been proposed (see e.g. [5], [6] and references therein), some of them relying on the modification of the geometrical structure of the theory, some others on the introduction of physically (up to day) unknown fluids into the equations governing the behaviour of our universe. In this context, we want to consider here a generalization of Einstein's theory based on the requirement that the stress-energy tensor for the matter/energy content is not conserved, i.e. $T^{\mu}_{\nu;\mu} \neq 0$, to check if it is able to describe the accelerated expansion of the universe without evoking other fluids but ordinary matter. In particular, we focus on the so called Rastall's model [7], and we investigate the conditions that the parameters of the theory have to fulfill in order to reproduce the observed accelerated expansion of the Universe. The field equations of this model were deduced by Rastall as ad hoc modifications of GR ones, so as to have a non conserved stress-energy tensor; however, there have been subsequent attempts to deduce Rastall's field equations from a variational principle, but none of them have succeeded [8, 9]. Since these field equations are similar to those that in $f(R)$ gravity are obtained by a Palatini variational principle from a Lagrangian depending on an arbitrary function f of the scalar curvature R [10, 11], we aim at verifying whether Rastall's model can be considered a particular case of $f(R)$ gravity. Furthermore, prompted by a recent paper [12], we check the possibility that Cardassian-like equations [13, 14] can be really derived from Rastall-like theories.

The plan of the paper is as follows: we will firstly give an introduction to Rastall's model in Sec. 2, while the corresponding cosmological scenario and the analogy with the Cardassians expansion model is worked out in Sec. 3. In order to check the possible viability of the Rastall's proposal, we fit the model to the SNeIa and GRBs Hubble diagram as detailed in Sec. 4, while Sec. 5 focuses on a more theoretical aspect investigating the possible connection with $f(R)$ theories in the Palatini formalism. Conclusions are finally presented in Sec. 5.

2. Rastall's Model

In 1972 P. Rastall [7] explored a model in which the stress-energy tensor of the source of the gravitational field, $T_{\mu\nu}$, was not conserved, i.e. the condition $T^{\mu}_{\nu;\mu} \neq 0$ is imposed a priori.

Indeed, Einstein equations[1] read

$$G_{\mu\nu} \doteq R_{\mu\nu} - \frac{1}{2} g_{\mu\nu} R = \kappa_{GR} T_{\mu\nu} \,, \tag{1}$$

where the Ricci tensor is obtained from a metric connection, so that $R_{\mu\nu} = R_{\mu\nu}(g)$ and the scalar curvature R has to be intended as $R \equiv R(g) = g^{\alpha\beta} R_{\alpha\beta}(g)$. Furthermore, we have set $\kappa_{GR} = \frac{8\pi G}{c^4}$. These equations naturally imply the stress-energy tensor conservation as a consequence of the contracted Bianchi identities,

$$G^{\mu}_{\nu;\mu} = 0 \,. \tag{2}$$

It is therefore worth wondering whether it is possible to fulfill the requirement $T^{\mu}_{\nu;\mu} \neq 0$ without violating Eqs.(2). A possible way out could be introducing further geometrical terms on the right hand side of Einstein equations, but one should ask whether this makes sense. Actually, if we insist in deriving these relations from a metric variational approach, the sudden answer would be of course negative: in this case the stress-energy tensor would be surely conserved by construction, so no way to escape the conditions $T^{\mu}_{\nu;\mu} = 0$.

Another remark against the non-conservativity of $T_{\mu\nu}$ focuses on the equivalence principle: as a matter of fact, the conservation of the stress-energy tensor is tested with high accuracy in the realm of Special Relativity (SR). However, one has to go easy with such an approach, as this principle could be misleading [15]. To give an example, when passing from GR to SR, we completely miss the information provided by terms explicitly depending on the curvature tensor, $R^{\alpha}_{\mu\beta\nu}$, as it becomes identically zero in Minkowski spacetime. This means that the two sets of equations

$$\nabla_{\alpha} j^{\alpha} = 0 \tag{3}$$

and

$$\nabla_{\sigma} j^{\sigma} + R^{\alpha}_{\mu\beta\nu} \nabla_{\alpha} j^{\mu} \nabla^{\beta} j^{\nu} = 0 \,, \tag{4}$$

give exactly the same equations, i.e. $\partial_{\alpha} j^{\alpha} = 0$, in SR. So, the straightforward application of the equivalence principle in writing conservation laws should be carefully considered.

[1] Throughout the paper, spacetime is assumed to have the signature $(+, -, -, -)$, and Greek indices run from 0 to 3.

The question is now how to pick up a proper geometrical term such that the Bianchi identities are still not violated, but nevertheless the conservation of the stress-energy tensor of the gravity source is violated. To resume, we ask for a four-vector, say a_ν, such that (i) $T^\mu_{\nu;\mu} = a_\nu$; (ii) $a_\nu \neq 0$ on curved spacetime, but $a_\mu = 0$ on flat spacetime in order not to conflict with the validity of SR. Both these properties hold for the Rastall's proposal, that is

$$T^\mu_{\nu;\mu} = \lambda R_{,\nu} \ , \tag{5}$$

λ being a suitable non-null dimensional constant.

Because of the assumption (5), the field equations are obviously modified and now read

$$R_{\mu\nu} - \frac{1}{2}\left(1 - 2\kappa_r\lambda\right) R g_{\mu\nu} = \kappa_r T_{\mu\nu} \ , \tag{6}$$

where κ_r is a dimensional constant to be determined in order to give the right Poisson equation in the static weak-field limit. It is manifest that in vacuum, where $T_{\mu\nu} = 0$, Rastall's field equations (6) are equivalent to GR ones.

As a matter of fact, the same set of equations can be obtained as the result of a guesswork, that is assuming the left hand side of the sought after equations to be a symmetric tensor consisting only of terms that are linear in the second derivative and/or quadratic in the first derivatives of the metric [16]. Moreover, the time-time component of such equations must give the Poisson equations back for a stationary weak-field. Accordingly, the only requirement we drop with respect to the derivation of the Einstein equations is the one concerning the conservation of $T_{\mu\nu}$. Hence, starting from $G_{\mu\nu} = C_1 R_{\mu\nu} + C_2 R g_{\mu\nu}$, with C_1 and C_2 appropriate constants, we end up with Eqs.(6) again provided that we set:

$$C_2 = \frac{C_1\left(C_1 - 2\right)}{2\left(3 - 2C_1\right)} \ , \tag{7}$$

$$\kappa_r = \frac{8\pi G}{C_1 c^4} \equiv \frac{\kappa_{GR}}{C_1} \ , \tag{8}$$

$$\kappa_r\lambda = \frac{C_1\left(1 - C_1\right)}{2\left(3 - 2C_1\right)} \ , \tag{9}$$

where we have chosen to rewrite all of them in terms of the constant C_1. Note, in particular, that the coupling constant between matter and geometry, κ_r, is not the same as in GR, unless $C_1 = 1$, that is $\lambda = 0$ (i.e., we consistently go back to GR).

Taking the trace of Eqs.(6) gives us the structural or master equation [17]:

$$\left(4\kappa_r\lambda - 1\right) R = \kappa_r T \ . \tag{10}$$

For a traceless stress-energy tensor, $T = 0$ (as for the electromagnetic or a conformally invariant tensor) and two possibilities arise. The first is that $R = 0$ so that we get no difference with standard GR. On the other hand, one could also solve Eq.(10) setting $\kappa_r\lambda = 1/4$, whatever the value of R is. However, inserting this condition in Eq.(9), we get a complex value for C_1 which is clearly meaningless. Therefore, we hereafter assume that $\kappa_r\lambda \neq 1/4$.

Another fundamental question concerns geodesic motions. As it is well known, the equations $T^\mu_{\nu;\mu} = 0$ are nothing but the equations of motion of the fluid we are dealing with. The problem is then, what sort of curves are described in a curved spacetime by a fluid whose stress-energy tensor is not conserved. Following the calculations made by Rastall, we find that in his model the geodesics are those curves characterized by the fact that the scalar curvature R is constant along them. Moreover, it is still possible to speak of conservation of energy for an ideal fluid [7], but again provided that R is constant along the time-like four-velocity vector of the fluid, u_μ. The question remains whether particles creation takes place in the regions where this condition does not hold.

It is worth mentioning that the Rastall's equations (6) can be recast into the same form as the usual Einstein ones. Indeed, one can immediately write

$$G_{\mu\nu} = \kappa_r S_{\mu\nu} \ , \tag{11}$$

where

$$S_{\mu\nu} = T_{\mu\nu} - \frac{\kappa_r \lambda}{4\kappa_r\lambda - 1} g_{\mu\nu} T \ . \tag{12}$$

By construction, this new stress-energy tensor is conserved, $S^{\mu}_{\nu;\mu} = 0$. On introducing $S_{\mu\nu}$, we can recover all the known solutions of Einstein GR by simply taking care of the difference between $S_{\mu\nu}$ and $T_{\mu\nu}$. Furthermore, if we assume $T_{\mu\nu} = (\rho + p)u_\mu u_\nu - p g_{\mu\nu}$, i.e. the source is a perfect fluid with energy density ρ and pressure p, we can explicitely work out an expression for $S_{\mu\nu}$. This turns out to be still a perfect fluid provided we redefine its energy density and pressure as

$$\rho_S = \frac{(3\kappa_r\lambda - 1)\rho + 3\kappa_r\lambda p}{4\kappa_r\lambda - 1} \ , \tag{13}$$

$$p_S = \frac{\kappa_r\lambda\rho + (\kappa_r\lambda - 1)p}{4\kappa_r\lambda - 1} \ . \tag{14}$$

In order to obtain the value of the coupling constant κ_r, we remember that the time-time component of the modified equations should recover the Poisson equation in the static weak-field limit. One thus gets:

$$\frac{\kappa_r}{4\kappa_r\lambda - 1}\left(3\kappa_r\lambda - \frac{1}{2}\right) = \kappa_{GR} \ , \tag{15}$$

whence it is immediate to derive exactly the same coupling as in Einstein gravity only when $\lambda = 0$, that is when the conservation of $T_{\mu\nu}$ is granted.

3. A Cardassian analog and the cosmic speed up

It has been recently claimed [12] that a Cardassian-like [13, 14] modification of the Friedman equation in the form

$$H^2 = \frac{8\pi G}{3c^2}\left[\rho + B(t)(\rho - 3p)^n\right] \ , \tag{16}$$

can be obtained from Rastall-like equations, where $B(t)$ is a function of the cosmic time t. We would now like to show that, although it is indeed possible to recast the Rastall's theory equations in such a way that a Cardassian-like model is recovered, the parameter B in (16) must be a constant.

To this aim, we derive the cosmological equations for the Rastall's theory. We first remember that, when the isotropic and homogenous Robertson-Walker (RW) metric is adopted, in GR one gets the usual Friedmann equations

$$H^2 = \frac{\kappa_{GR} c^2}{3}\rho \ , \tag{17}$$

$$\dot{H} + H^2 = -\frac{\kappa_{GR} c^2}{6}(\rho + 3p) \ , \tag{18}$$

where $H = \dot{a}/a$ is the Hubble parameter, a the scale factor and a dot denotes the derivative with respect to cosmic time t. To get the corresponding equations for the Rastall's theory, one has to insert the RW metric into Eqs.(6) and consider the only independent equations that can be obtained, that is

$$
\begin{aligned}
3\dot{a}^2 - 6\kappa_r\lambda\left(\dot{a}^2 + a\ddot{a}\right) &= \kappa_r a^2 \rho \ , &(19)\\
\dot{a}^2 + 2a\ddot{a} + 6\kappa_r\lambda\left(\dot{a}^2 + a\ddot{a}\right) &= -\kappa_r a^2 p \ , &(20)
\end{aligned}
$$

respectively. The master equation thus becomes

$$6\left(\dot{a}^2 + a\ddot{a}\right) = -\kappa_r a^2 (\rho - 3p) \ , \tag{21}$$

so that, multiplying Eq.(20) by -3 and then adding to Eq.(19), we finally get the first modified Friedmann's Equation:

$$\frac{\ddot{a}}{a} = \frac{\kappa_r}{6} \frac{\rho + 3p - 6\kappa_r\lambda(\rho + p)}{4\kappa_r\lambda - 1} \ , \tag{22}$$

which makes it possible to directly infer the sign of the acceleration. To obtain the second modified Friedmann's equation, it is easier to proceed in a slightly different way. Let us first take the Rastall's equations in the form

$$R_{\mu\nu} = \kappa_r\left(T_{\mu\nu} - \frac{1}{2}\frac{2\kappa_r\lambda - 1}{4\kappa_r\lambda - 1} T g_{\mu\nu}\right) \ . \tag{23}$$

By inserting the RW metric and adding up Eq.(19) with three times Eq.(20), we eventually obtain

$$H^2 = \frac{\kappa_r}{6}\left[(\rho + 3p) + \frac{2\kappa_r\lambda - 1}{4\kappa_r\lambda - 1}(\rho - 3p)\right] \ . \tag{24}$$

It is then only a matter of algebra to rearrange Eqs.(22) and (24) to write them as

$$H^2 = \frac{\kappa_r}{3}\left[\rho - \frac{\kappa_r\lambda}{4\kappa_r\lambda - 1}(\rho - 3p)\right] \ , \tag{25}$$

$$\dot{H} = \frac{\kappa_r}{2(4\kappa_r\lambda - 1)}\left[\rho + p - 4\kappa_r\lambda(\rho - p)\right] \ , \tag{26}$$

which reduce to the standard Friedmann equations (17) and (18) when the parameter λ is switched off. Also note that Eq.(25) has indeed the same expression as the Cardassian-like Eq.(16) provided we set $n = 1$ and accordingly redefine the parameter B. However, it is straightforward to show that B must be a constant. Indeed, from the condition

$$S^{\mu}_{\nu;\mu} = 0 \ , \tag{27}$$

it is immediate to demonstrate that $\kappa_r\lambda$ must be a constant by simply inserting the master equation (21) into Eq.(27) and using the Rastall's requirement $T^{\mu}_{\nu;\mu} = \lambda R_{,\nu}$. Thus, an equation like (16), cannot be self-consistently obtained in the Rastall's model. Moreover, since $T_{\mu\nu}$ is a perfect fluid and remembering the definition of $S_{\mu\nu}$, Eq.(27), we get

$$\dot{\rho} + 3H(\rho + p) = \frac{\kappa_r\lambda}{4\kappa_r\lambda - 1}(\dot{\rho} - 3\dot{p}) \ , \tag{28}$$

which generalizes the continuity equation for the Rastall's theory.

It is worth noticing that, even without integrating the equations, one can immediately predict whether the universe expansion is accelerating or not by simply studying the sign of the right hand side of Eq.(22). Assuming for simplicity that the equation of state of the perfect fluid is a constant, i.e. setting $p = w\rho$, the condition $\ddot{a} > 0$ selects two possible regimes, the first one being

$$w > \frac{6\kappa_r\lambda - 1}{3(1 - 2\kappa_r\lambda)} \,, \tag{29}$$

provided $\kappa_r\lambda > 1/4$. When $\lambda = 0$, however, the above relation reduces to $w > -1/3$ in contrast with the GR result. We have therefore to choose the other solution, namely

$$w < \frac{6\kappa_r\lambda - 1}{3(1 - 2\kappa_r\lambda)} \,, \tag{30}$$

with $\kappa_r\lambda < 1/4$. The right hand side of (30) may be positive or negative depending on the value of $\kappa_r\lambda$. More precisely, if $1/6 < \kappa_r\lambda < 1/4$, then the rhs is positive, while it is negative for $0 < \kappa_r\lambda < 1/6$. It is worth stressing that the model always gives a monotonic behaviour: always decelerated or, as in the above analyzed case, always accelerated.

4. Rastall's model confronted with the data

Neglecting the radiation component, the only fluid sourcing the gravitational field is the standard matter, which can be modeled as dust, i.e. $p = 0$. In such a case, the continuity equation (28) is straightforwardly integrated giving:

$$\rho \propto \rho_0 (1 + z)^{3w_{eff}} \,, \tag{31}$$

with a 0 subscript denoting present day quantities, $z - 1 = \frac{1}{a}$ (having set $a_0 = 1$) the redshift, and

$$w_{eff} = 1 - \frac{\kappa_r\lambda}{4\kappa_r\lambda - 1} \,, \tag{32}$$

an effective equation of state (EoS) for the dust matter, from which a relation between w_{eff} and $\kappa_r\lambda$ is easily deduced. Note that, for $\lambda = 0$, one recovers the usual matter scaling $\rho \propto (1 + z)^3$, while deviation from the standard behaviour occurs in $\lambda \neq 0$ Rastall's theory. Such a different scaling is not surprising at all being an expected consequence of the non-conservativity of the stress-energy tensor. Inserting back Eq.(31) into Eq.(25), we get

$$E^2 \doteq H^2/H_0^2 = (1 + z)^{3w_{eff}} \,, \tag{33}$$

which is all what we need to compute the luminosity distance

$$D_L(z, w_{eff}, h) = d_H(1 + z) \int_0^z \frac{1}{E(z')} dz' \,, \tag{34}$$

with the Hubble radius $d_H = c/H_0 \simeq 3h^{-1}$ Gpc and h the Hubble constant in units of 100 km/sMpc. We have now all the main ingredients to test the viability of the Rastall's model by fitting the predicted luminosity distance to the data on the combined Hubble diagram of SNeIa and GRBs. To this aim, we maximize the following likelihood function:

$$\mathcal{L}(w_{eff}, h) \propto \exp\left(-\frac{\chi^2_{SNeIa} + \chi^2_{GRB}}{2}\right) \times \exp\left[-\left(\frac{h_{HST} - h}{\sigma_{HST}}\right)^2\right] \,, \tag{35}$$

with

$$\chi^2_{SNeIa} = \sum_{i=1}^{\mathcal{N}_{SNeIa}} \left[\frac{\mu_{obs}(z_i) - \mu_{th}(z_i, w_{eff}, h)}{\sigma_i}\right]^2 \,, \tag{36}$$

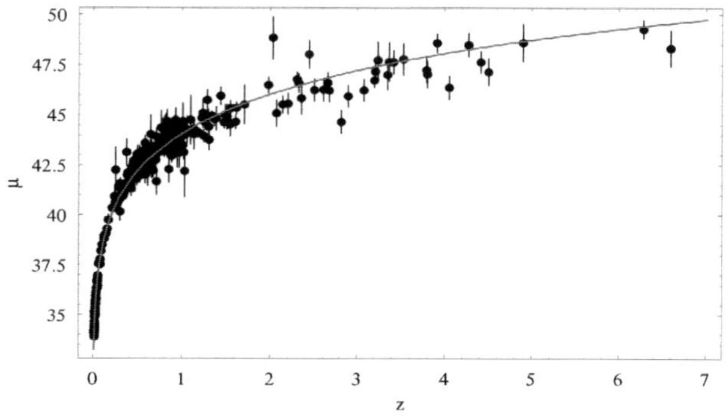

Figure 1. Comparison among predicted and observed SNeIa and GRBs Hubble diagram.

$$\chi^2_{GRB} = \sum_{i=1}^{\mathcal{N}_{GRB}} \left[\frac{\mu_{obs}(z_i) - \mu_{th}(z_i, w_{eff}, h)}{\sigma_i}\right]^2 . \tag{37}$$

The χ^2 terms in (35) take care of the Hubble diagram of SNeIa and GRBs, respectively, and rely on the distance modulus defined as

$$\mu_{th}(z, w_{eff}, h) = 25 + 5 \log D_L(z, w_{eff}, h) . \tag{38}$$

We use the Union [18] dataset for SNeIa and the GRBs sample assembled in Cardone et al. [19] to set the observed quantities (μ_{obs}, σ_i) for the SNeIa and GRBs, respectively. Since the Hubble constant h is degenerate with the (unconstrained) absolute magnitude of a SN, we have added a Gaussian prior on h using the results from the HST Key Project [20] thus setting $(h_{HST}, \sigma_{HST}) = (0.72, 0.08)$.

The best fit model turns out to be

$$(w_{eff}, h) = (0.55, 0.68) ,$$

giving

$$\chi^2_{SNeIa}/d.o.f. = 1.08 , \quad \chi^2_{GRB}/d.o.f. = 2.07 ,$$

where $d.o.f. = N_{SNeIa} + N_{GRB} - N_p$ is the number of degree of freedom of the model, with $N_{SNeIa} = 307$ the number of SNeIa in the Union sample, $N_{GRB} = 69$ the number of GRBs, and $N_p = 2$ the number of parameters of the Rastall's model. While for SNeIa we get a very good reduced χ^2, this is not the case for GRBs so that one could be tempted to deem as unsuccessfull the fit. Actually, Fig.1 shows that the model is indeed fitting quite well both the SNeIa and GRB data so that the large value of $\chi^2_{GRB}/d.o.f.$ should be imputed to the large scatter of the high redshift data around the best fit line, not taken into account by the statistical error on the GRBs distance modulus. In order to further test the model, one can consider the constraints on the matter density parameter. Since our model only contains matter, one could naively think that $\Omega_M = 1$. Actually, one must also take into account that Ω_M is defined using the GR coupling constant κ_{GR} which is related to the Rastall's coupuling κ_r through Eq.(15). It is then a matter of algebra to show that $\Omega_M = w_{eff}$ so that, after marginalizing over h, we get the following constraints:

$$\Omega_M = 0.55^{+0.02}_{-0.03}{}^{+0.05}_{-0.05} ,$$

where we have used the notation $x^{+x_1+x_2}_{-y_1-y_2}$ to mean that x is the median value of the parameter and $(x + x_1, x - y_1), (x + x_2, x - y_2)$ are the 68% and 95% confidence ranges respectively. Note that the value thus obtained is in strong disagreement with the typical $\Omega_M \simeq 0.3$ obtained from both the cosmic microwave background radiation data and cluster gas mass fractions. Such a large matter density parameter is clearly unacceptable and represents a strong evidence against the Rastall's model. It is worth noting that such a result could be qualitatively foreseen considering that, because of the non-conservativity of the matter stress-energy tensor, a sort of matter creation takes place thus increasing Ω_M and leading to the final disagreement.

5. Rastall's field equations from a Palatini approach

Rastall's model has been initially motivated by the need for a theory able to allow a non conservativity of the source stress-energy tensor without violating the Bianchi identities. As such, the original theory was based on purely phenomenological motivations starting directly from the field equations without any attempt to derive them from a variational principle. We intend here to readdress this issue looking for a connection to $f(R)$ theories of gravity, which are obtained by a variational approach, starting from a gravitational Lagrangian depending on a function f of the scalar curvature R (see [10, 11] and references therein). These theories can be studied in the metric formalism, where the action is varied with respect to metric tensor, and in the Palatini formalism, where the action is varied with respect to the metric and the affine connection, which are supposed to be independent from one another. In particular, we want to verify here if Rastall's theory can be considered a particular case of Palatini $f(R)$ gravity, which would suggest how to obtain it from a variational principle.

To this aim, let us first rewrite the field equations in Rastall's theory as follows:

$$R_{\mu\nu} - \frac{1}{2} R g_{\mu\nu} = \kappa_r \left(T_{\mu\nu} - \lambda R g_{\mu\nu} \right) \ , \tag{39}$$

where we stress that $R_{\mu\nu} = R_{\mu\nu}(g)$ and $R \equiv R(g) = g^{\alpha\beta} R_{\alpha\beta}(g)$ for reasons that will be clear soon. Assuming that $f'(\bar{R}) = df(\bar{R})/d\bar{R} \neq 0$, the corresponding field equations [17, 21] for a $f(R)$ theory in the Palatini approach can be written as

$$\bar{R}_{(\mu\nu)}(\Gamma) + \frac{1}{2}\bar{R}(\Gamma, g) g_{\mu\nu} = \frac{\kappa_{GR}}{f'(\bar{R})} T_{\mu\nu} + \frac{1}{2}\left[\frac{f(\bar{R})}{f'(\bar{R})} - \bar{R}(\Gamma, g) \right] g_{\mu\nu} \ , \tag{40}$$

$$\bar{\nabla}_\alpha \left[\sqrt{g} f'(\bar{R}) g^{\mu\nu} \right] = 0 \ , \tag{41}$$

where $\bar{R}_{\mu\nu}(\Gamma)$ is the Ricci tensor of a torsionless connection Γ independent of the metric g, $\bar{R}(\Gamma, g) = g^{\mu\nu}\bar{R}_{\mu\nu}$ is the scalar curvature built with the metric tensor and we have denoted with $\bar{R}_{(\mu\nu)}$ the symmetric part of $\bar{R}_{\mu\nu}$.

To compare Eqs.(39) and (40) - (41), we first rewrite $\bar{R}_{\mu\nu}(\Gamma)$ in terms of the Ricci tensor of the metric, $R_{\mu\nu}(g)$, which gives [22]:

$$\bar{R}_{\mu\nu} = R_{\mu\nu} + \bar{\nabla}_{(\mu} b_{\nu)} - \frac{1}{2} b_\mu b_\nu + g_{\mu\nu} b_\alpha b^\alpha + \frac{1}{2} g_{\mu\nu} \bar{\nabla}_\alpha b^\alpha \doteq R_{\mu\nu} + B_{\mu\nu} \ , \tag{42}$$

where we have dropped the various dependences for sake of clarity. Note that we have here introduced the vector field

$$b_\alpha \doteq -\bar{\nabla}_\alpha \left[\ln f'(\bar{R}) \right] \ , \tag{43}$$

which allows to rewrite Eqs.(41) as

$$\bar{\nabla}_\alpha g_{\mu\nu} = b_\alpha g_{\mu\nu} \ , \tag{44}$$

and to define the rank two tensor field

$$B_{\mu\nu} \doteq \bar{\nabla}_{(\mu} b_{\nu)} - \frac{1}{2} b_\mu b_\nu + g_{\mu\nu} b_\alpha b^\alpha + \frac{1}{2} \bar{\nabla}_\alpha b^\alpha g_{\mu\nu} \ . \tag{45}$$

By saturation of indices in Eqs.(42), one naturally obtains the corrensponding relations between the scalar curvature $\bar{R}(\Gamma, g)$ of the affine connection and that of the metric, $R(g)$, reading:

$$\bar{R} = R + 2\bar{\nabla}_\alpha b^\alpha + \bar{\nabla}^\alpha b_\alpha + \frac{7}{2} b_\alpha b^\alpha \, . \tag{46}$$

Then, inverting Eqs.(42) and (46) and introducing into the Rastall's equations (39) the Ricci tensor and scalar curvature of the metric in terms of the corresponding barred quantities, we obtain the following relations:

$$\bar{R}_{\mu\nu} - \frac{1}{2} g_{\mu\nu} \bar{R} = \kappa_r T_{\mu\nu} + \kappa_r \lambda g_{\mu\nu} \left[2\bar{\nabla}_\alpha b^\alpha + \bar{\nabla}^\alpha b_\alpha + \frac{7}{2} b_\alpha b^\alpha - \bar{R} \right] + \\ + \bar{\nabla}_{(\mu} b_{\nu)} - \frac{1}{2} \left[b_\mu b_\nu + g_{\mu\nu} \left(\frac{3}{2} b_\alpha b^\alpha + \bar{\nabla}_\alpha b^\alpha + \bar{\nabla}^\alpha b_\alpha \right) \right] \, . \tag{47}$$

Note that the structural equation from (47) reads

$$\bar{R} = \frac{\kappa_r}{4\kappa_r \lambda - 1} T + 2\bar{\nabla}_\alpha b^\alpha + \bar{\nabla}^\alpha b_\alpha + \frac{7}{2} b_\alpha b^\alpha \, . \tag{48}$$

Considering Eq.(46), it is immediate to see that Eq.(48) is the same as Eq.(10) thus showing us that the same relation between the metric scalar curvature and the trace of the source stress-energy tensor holds both for the $f(R)$ and Rastall's theories.

From the very definition of b_α and a manipulation of Eq.(40), it is easy to show that, for a general $f(R)$ theory, the following result is valid

$$b_\alpha = -\kappa_{GR} \frac{f''(\bar{R})}{f'(\bar{R})} \frac{T_{,\alpha}}{f''(\bar{R})\bar{R} - f'(\bar{R})} \, . \tag{49}$$

This means that in empty space, or in case of a traceless stress-energy tensor, the vector field b_α is null and, according to (41), the connection Γ is metric, so that $\bar{R}_{\mu\nu}(\Gamma) = R_{\mu\nu}$, and $\bar{R}(\Gamma, g) = R$. As a consequence, the field equations (40) become

$$R_{(\mu\nu)} - \frac{1}{2} R g_{\mu\nu} R = \frac{\kappa_{GR}}{f'(R)} T_{\mu\nu} + \frac{1}{2} \left[\frac{f(R)}{f'(R)} - R \right] g_{\mu\nu} \, . \tag{50}$$

In such a case, we can easily equate the rhs of Eq.(50) with that of Eq.(39) and then work out an analytical expression for $f(R)$ such that the two sets of field equations are identical. Provided $\kappa_r \lambda \neq 1/2$ (since it would correspond to $f(R) \equiv 0$) and setting $\alpha = 2\kappa_r \lambda - 1$, we finally get

$$f(R) = f_0 R^{-\frac{1}{\alpha}} \, . \tag{51}$$

Note that $\lambda = 0$ gives $\alpha = -1$ hence $f(R) = R$ and we correctly recover the GR $f(R)$. On the contrary, for $\lambda \neq 0$, inserting Eq.(51) into Eq.(50) makes it possible to write the field equations as

$$R_{\mu\nu} - \frac{1}{2} R g_{\mu\nu} = \frac{\kappa_{GR}}{f'(R)} T_{\mu\nu} - \kappa_r \lambda R g_{\mu\nu} \, . \tag{52}$$

Comparing Eqs.(39) and (52) shows that, although having the same structure, they are different because of the diverse coupling, being $\kappa_{GR} \neq \kappa_r$. Remembering Eq.(8), we see that the only way to make Eqs.(39) and (52) to coincide is setting

$$f'(R) = C_1 \tag{53}$$

which is clearly incompatible with the solution (51), unless R is a constant. We therefore conclude that the Rastall's theory may be recovered from a $f(R)$ gravity model in the Palatini formalism only when it reduces to GR. Put in other words, this means that Palatini $f(R)$ gravity can not provide a viable Lagrangian from which the Rastall's field equations may be derived.

6. Conclusions

In this paper we have focused on Rastall's theory of gravity, whose field equations are based on the request of the non-conservativity of the stress-energy tensor, i.e. $T^{\mu}_{\nu;\mu} \neq 0$. In particular, we have reexamined this model of gravity to investigate the possibility that it could reproduce the observed cosmic speed up. To this end, we have explicitly worked out the modified Friedmann equations and confronted the model predictions with the available data concerning type Ia Supernovae (SNeIa) and Gamma Ray Bursts (GRBs) : what we have showed is that it is possible to get an accelerated expansion that is in agreement with the Hubble diagram of both SNeIa and GRBs, even if there is unfortunately no possibility to reproduce an accelerated-decelerated-accelerated expansion for our universe as it seems to be requested. These results have also a major drawback: indeed, to get them it is necessary to postulate a matter density parameter much larger than the typical $\Omega_M \simeq 0.3$ value inferred from cluster gas mass fraction data. As a consequence, Rastall's theory is not in agreement with current cosmological observations. We have also shown that Cardassian-like modifications of Friedmann equations are obtained in Rastall's model but, contrary to recent claims, they cannot contain time-dependent parameters. Finally, in an attempt to derive Rastall's field equations from a variational principle, we have studied the connection with $f(R)$ gravity in the Palatini formalism, and we have concluded that the Rastall's theory may be recovered from a $f(R)$ gravity model in the Palatini formalism only when it reduces to GR; thus $f(R)$ gravity cannot provide a variational principle for deriving Rastall's field equations.

Acknowledgments

The authors warmly thank the attendants of the Journal Club on Extended and Alternative Theories of Gravity for useful discussions. MC and VFC are supported by University of Torino and Regione Piemonte. Partial support from INFN projects PD51 and NA12 is acknowledged too.

References

[1] Riess A G *et al* 1998 *Astron. J.* **116** 1009
[2] Perlmutter S *et al* 1999 *Astrophys. J.* **517** 565
[3] Bennet C L *et al* 2003 *Astrophys. J. Suppl.* **148** 1
[4] Will C M 2006 The confrontation between General Relativity and experiment *Living Rev. Relativity* **9** http://www.livingreviews.org/lrr-2006-3
[5] Peebles P J and Ratra B 2003 *Rev. Mod. Phys.* **75** 559
[6] Kamionkowski M 2007 Dark matter and dark energy *Preprint* arXiv:0706.2986
[7] Rastall P 1972 *Phys. Rev.* D **6** 3357
[8] Smalley L L 1993 *Class. Quantum Grav.* **10** 1179
[9] Lindblom L and Hiscock W A 1982 *J. Phys.* A **15** 1827
[10] Sotiriou T and Faraoni V 2008 $f(R)$ theories of gravity *Preprint* arXiv:0805.1726
[11] Capozziello S and Francaviglia M 2007 Extended theories of gravity and their cosmological and astrophysical applications *Preprint* arXiv:0706.1146
[12] Al-Rawaf A S 2008 *Mod. Phys. Lett.* A **23** 2691
[13] Freese K and Lewis M 2002 *Phys. Lett.* B **540** 1
[14] Fay S and Amarzguioui M 2006 *Astronomy and Astrophysics* **460** 37
[15] Trautman A 1965 *Lectures on General Relativity* ed S Deser and K W Ford (Englewood Cliffs, NJ: Prentice-Hall)
[16] Weinberg S 1972 *Gravitation and Cosmology* (New York: J. Wiley and Sons)
[17] Ferraris M, Francaviglia M and Volovich I 1994 *Class. Quant. Grav.* **11** 1505
[18] Kowalski M *et al* 2008 *Astrophys. J.* **686** 749
[19] Cardone V F, Capozziello S and Dainotti M G 2009 An updated gamma ray bursts Hubble diagram *Preprint* arXiv:0901.3194
[20] Freedman W L *et al* 2001 *Astrophys. J.* **553** 47
[21] Ferraris M, Francaviglia M and Volovich I 1993 *Nuovo Cim.* B **108** 1313
[22] Barraco D, Hamity V H and Vucetich H 2002 *Gen. Rel. Grav.* **34** 533

First Mediterranean Conference on Classical and Quantum Gravity (MCCQG 2009) IOP Publishing
Journal of Physics: Conference Series **222** (2010) 012013 doi:10.1088/1742-6596/222/1/012013

Dissipation in Non-equilibrium Spacetime Thermodynamics

G Chirco, S Liberati

SISSA, Via Beirut 2, 34151 Trieste, Italy and INFN sezione di Trieste

E-mail: chirco@sissa.it, liberati@sissa.it

Abstract. It has been recently realized [1, 2] that the thermodynamical derivation of the Einstein equation introduced by Jacobson [3] needs a generalization to a non-equilibrium thermodynamical setting. Here we show that the non-equilibrium character of spacetime thermodynamics — both in GR as well as in $f(R)$ gravity — is generally related to non-local heat fluxes associated with the purely gravitational/internal degrees of freedom of the theory. In particular, we show that the allowed gravitational degrees of freedom can be fixed by the kinematics of the local spacetime causal structure, through the specific Equivalence Principle formulation. In this sense, the thermodynamical description seems to go beyond Einstein's theory as an intrinsic property of gravitation.

1. Introduction

The basic idea of a spacetime thermodynamics [3] bases on the relation between the thermal character of quantum fields vacuum, as perceived by an accelerated Rindler observer, and the stationarity properties of the respective Killing horizon. By making use of the Equivalence Principle [4], the notion of Rindler frame can be used at a local level as an experimental setting for studying the local spacetime dynamics in terms of the way the variation of the spacetime geometry follows from the energy variation of the matter fields [1, 2, 3, 5].

Actually, for a given general spacetime, an approximate local Rindler frame can always be introduced by invoking local flatness (Local Lorentz Invariance) around a point p. By further assuming that the ground state of the fields living in the original spacetime is locally approximated by the Minkowski vacuum, the vacuum state associated with the set of Rindler wedge observers, can be interpreted as an approximate thermal state, with a temperature $T \approx T_{un}\sqrt{-g_{00}}$, where $T_{un} = \hbar/2\pi\chi$ is the the Unruh temperature [6], while $\sqrt{-g_{00}} = \kappa\chi$ is the gravitational Doppler factor defined with the Rindler metric $ds^2 = -\kappa^2\chi^2 d\eta^2 + d\chi^2 + dy^2 + dz^2$. The temperature T stays constant throughout the Rindler wedge and it is well defined on the horizon. Therefore, the thermal character of the Rindler vacuum is effectively extended from the single Rindler observer to the whole wedge.

Moreover, since the accelerated observer in the wedge can only access information on spacelike slices bounded by the bifurcation plane and the vacuum fluctuations between the inside and the outside of the wedge are correlated, she/he will perceive an entanglement entropy. This entropy will scale with the area of the local boundary and diverge with the density of field states in the UV limit. However, with the introduction of an UV cut-off one can make this entropy proportional to the area, that is $S = \alpha A$, where the proportionality factor α can *a priori* depend on the nature of the quantum fields as well as be some complicate function of the position in spacetime [7, 8]. Together with the temperature T, this notion of entropy allows to consider the local Rindler wedge with its Killing horizon as an analogue of an equilibrium canonical ensemble (Gibbs state) bounded by a diathermic wall.

2. Spacetime Thermodynamics

The notion thermal equilibrium for the wedge is ultimately associated to the stationarity of the Killing horizon. In this sense the equilibrium is realized at the horizon bifurcation surface at p, where $dS = \delta(\alpha A) = 0$ by definition.

Moving away from the bifurcation surface along the approximate Kiling horizon, the spacetime becomes dynamical: the presence of matter distorts the local Rindler causal structure and perturbs the thermal state. Thereby, one can describe the slight perturbation of the causal structure by looking at the variation of the affine quantities of the null geodesic comprising the horizon, provided that some condition to relate the spacetime geometry deformation to the variation of the fields energy content can be found.

This condition can be borrowed from classical equilibrium thermodynamics. For a slightly perturbed Gibbs state at temperature T, variations of entropy and internal energy are related by the Clausius relation, $dS = \delta Q/T$, where the change in the mean energy δQ is due to the fluxes into/from an unobservable region of spacetime. For a system undergoing a quasi-static process of energy exchange with the surroundings, the Clausius relation is nothing but a local thermodynamical equilibrium condition.

Now, since the thermal Rindler wedge behaves like a Gibbs state, one can actually use the Clausius relation to relate the horizon entropy and the boost energy across the Rindler horizon, where T is the wedge temperature [3]. As a second fundamental postulate, one can assume a universal entropy density α per unit horizon area δA, such that

$$dS = \alpha \delta A, \tag{1}$$

then implicitly considering a constant UV cut-off for the fields. With this choice, the changes of the entanglement entropy of the fields in the wedge can be effectively described in terms of geometrical variations of the horizon cross-section[3].

Let us highlight here that the assumption $\alpha =$ constant, made in the GR derivation of [1, 3], can be indeed recast as an explicit choice of a specific formulation of the Equivalence principle. As said, generally the UV cut-off α is fixed at the quantum gravity scale. This can be identified as the scale at which the gravitational action is of the order of the quantum of action $\hbar$. For GR this is the standard Planck length $l_p = \sqrt{G\hbar/c^3}$ and hence it is directly related to the Newton constant. However, for a general scalar-tensor theory, i.e. a theory compatible just with the Einstein Equivalence principle (EEP) [4], G is promoted to a spacetime field. As a consequence of this, one should expect that the cut-off will be generically position dependent. In this sense, assuming $dS = \alpha\delta A$ is equivalent to assume the strong formulation of the Equivalence Principle (SEP) [4], hence effectively to allow for the only two SEP-compatible gravity theories: GR and Nordström gravity [9].

Given the assumption in (1), a quantitative expression for the system entropy variation is obtained just by applying the definition for the change of the horizon area in terms of the expansion rate of the null geodesics comprising it, that is

$$\delta A = \int_H \widetilde{\epsilon}\, \theta \, d\lambda \,, \tag{2}$$

with $\widetilde{\epsilon}$ indicating the 2-surface area element of the horizon cross-section, while λ is the affine parameter for the null geodesic congruence comprising the horizon.

Moving away from the equilibrium surface at $\lambda = 0$, along the null congruence, the infinitesimal evolution of θ is given by a linear expansion around its equilibrium value at p, up to the first order in λ, say $\theta \approx \theta_p + \lambda\ d\theta/d\lambda|_p + \mathcal{O}(\lambda^2)$. This first order coefficient will be determined as usual by the Raychaudhuri equation, $d\theta/d\lambda = -1/2\,\theta^2 - \|\sigma\|^2 - R_{ab}\ell^a\ell^b$, where $\|\sigma\|^2$ stands for the squared congruence shear $\sigma^{ab}\sigma_{ab}$, while ℓ^a is the null vector tangent to the null congruence comprising the horizon. In this way, the entropy variation, is directly related to the purely geormtrical degrees of freedom of the horizon congruence, up to $\mathcal{O}(\lambda^2)$

$$dS = \alpha \int_H \widetilde{\epsilon}\, d\lambda \left[\theta - \lambda\left(\frac{1}{2}\theta^2 + \|\sigma\|^2 + R_{ab}\ell^a\ell^b\right)\right]_p . \tag{3}$$

On the other hand, the mean energy variation of the thermal system, given by the boosted energy current flux of matter, is then described by the heat flux across the horizon as

$$\delta Q = \int_H T_{ab}\chi^a d\Sigma^b = \int_H \widetilde{\epsilon}\, d\lambda\ (-\lambda\kappa)\ T_{ab}\ell^a\ell^b\,, \tag{4}$$

where T_{ab} is the matter stress energy tensor, while the volume element is given by $d\Sigma^b = \widetilde{\epsilon}\, d\lambda\, \ell^b$. In the last passage the heat is expressed with respect to the null vector ℓ^a, given the relation $\lambda = -e^{-\kappa v}$ between the λ affine and the v Killing parameter along the approximated Killing horizon, from which $\chi^a = (d\lambda/dv)\,\ell^a$.

At this stage, asking for the Clausius relation to hold for all null vectors ℓ^a, one can equate the $\mathcal{O}(\lambda)$ integrands in (3) and (4). At the zeroth order in λ, the value of heat flux at p is zero, hence one necessarily gets $\theta_p = 0$. Then, to the first order, one finds the constitutive relation

$$\frac{2\pi}{\hbar\alpha}\, T_{ab}\ell^a\ell^b = (\|\sigma\|^2 + R_{ab}\ell^a\ell^b)_p\,. \tag{5}$$

Now, if one further postulates that $\sigma_p = 0$, as originally done in [3], one is left with $2\pi/\hbar\alpha\, T_{ab} = R_{ab} + \Phi\, g_{ab}$, where Φ is an undetermined integration function. Then, by assuming the local energy conservation, that is $\nabla^b T_{ab} = 0$, applying the divergence operator on both sides of (5), and using the contracted Bianchi identity $\nabla^b R_{ab} = \frac{1}{2}\nabla_a R$, one eventually gets $\Phi = -\frac{1}{2}R - \Lambda$, hence $2\pi/\hbar\alpha\, T_{ab} = R_{ab} - \frac{1}{2}R\, g_{ab} - \Lambda\, g_{ab}$, where Λ is some arbitrary integration constant. Once the condition $\alpha = 1/4\hbar G$ is imposed, one can easily recognize the familiar Einstein equations and, noticeably, the latter condition implies that the entropy density of the local Rindler horizon is the same as the one of a black hole.

On the other hand, by allowing for some shear at p, that is for $\sigma_p \neq 0$, the Einstein equation is no more recovered [1, 2]. This argument forces the thermodynamical derivation to be recasted in a non-equilibrium setting, where $dS > \delta Q/T$. In this new context, the Clausius law is replaced by the entropy balance law, $dS = \delta Q/T + d_i S$ and the extra shear term in (5) is then associated with the internal entropy production $d_i S$, generated by the system out of equilibrium. The internal entropy contribution, $\mathcal{O}(\lambda)$, has the form

$$d_i S = -\frac{4\pi\eta}{\hbar}\int_H \widetilde{\epsilon}\,\lambda\, d\lambda\, \|\sigma\|_p^2 \tag{6}$$

and it is interpreted as an internal entropy production term [10] due to some internal spacetime viscosity, with $\eta = \hbar\alpha/4\pi$.

3. Some Remarks

The geometrical properties of the horizon null congruence and consequently the value of affine expansion and shear at p are given with the local spacetime curvature in that point. In order to define an equilibrium surface one just need to require a suitably smooth curvature for the spacetime patch around p, without fixing *a priori* the values of affine expansion and shear. This means that the local Rindler frame is not sensitive to the exact value of the affine expansion and shear at p. Therefore, setting $\sigma_p = 0$ is actually an unjustified arbitrary choice in [3].

Thereby, the assumption that the affine congruence orthogonal to the local spacetime patch around p has zero expansion and shear is equivalent to require that the chosen local patch (and hence its associated null congruence) is less general than the one allowed by the assumed entropy-area relation (or alternatively by the SEP).

On the one hand, allowing for non zero affine shear at p give rise to further interesting clues. In some way, the shear contribution in (5) brings into the entropy balance process a new degree of freedom, which is not fixed by the Ricci tensor and so has nothing to do with the local matter energy sources. Actually, the surface shear is generally related to the Weyl tensor and usually associated with the distortion on the geodesics congruence due to a gravitational perturbation.

In fact, this argument opens an issue about the absence so far of any role for gravitational fluxes in the system energy perturbation mechanism. Due to their non-local nature, the gravitational energy fluxes cannot be taken into account with a proper stress-energy tensor (SET). However, allowing for non-local terms, as the one in $\|\sigma\|^2$ in (5), seems at odds with neglecting the role of these non-local energy contributions.

In this sense, such a spacetime viscosity seems naturally related to the distorsive effect of a gravitational flux, to be intended as a local curvature perturbation which is independent from the Einstein equation. This suggest that gravitational energy fluxes can possibly play a role into the total entropy balance of the system without entering into the Einstein equilibrium relation.

4. Internal entropy in non-equilibrium thermodynamics

In classical non-equilibrium thermodynamics, the rate of change of the entropy is generally written as the sum of two contributions: $dS = d_eS + d_iS$, where d_eS is the rate of entropy exchange with the surroundings, while d_iS comes from the process occurring inside the system and is a non-negative quantity, accordingly to the second law of thermodynamics. In particular, d_iS is zero for reversible (quasi-static) processes and positive for irreversible processes.

However, as some irreversible processes come into play, also the Clausius relation must be generalized to the expression

$$dS = \frac{\delta Q}{T} + \delta N, \tag{7}$$

where δQ is generally referred to as *compensated heat*, that is the heat transferred between the system and its surroundings, while δN, the so called *uncompensated heat*, indicates the amount of entropy associated with the heat which is intrinsic to the system when it undergoes an irreversible process. The above definition generalizes the notion of local equilibrium by extending the entropy balance to the unbalanced contributions related to the irreversible processes, like dissipation (see e.g. [11]).

The generalized Clausius relation (7) is helpful in order to clarify the nature of the equilibrium and non-equilibrium contributions defining our system entropy. In fact, by using the definition of the non-equilibrium entropy given above, we can write $d_eS + d_iS = \delta Q/T + \delta N$ and identify the external and internal entropy contributions in terms of the compensated and uncompensated heat, respectively $d_eS = \delta Q/T$, at the reversible level, and $d_iS = \delta N$, at the irreversible level.

With the notion of generalized Clausius relation, the argument described in Sec.2 acquires a clearer interpretation. Indeed, the extra contribution (6) introduced by the non vanishing horizon shear is an internal entropy production term allowed by the most general choice of the null congruence associated with the starting local spacetime patch compatible with the area-entropy relation for GR (that we linked to the choice of the EP formulation). Therefore, it has to be seen as a by-product of the presence of internal/purely gravitational degrees of freedom of the theory which can be responsible for irreversible dissipative processes.

A possible way to capture non-equilibrium features of the thermal system is to use the analogy between the congruence bundle comprising the horizon and a classical fluid [2]. Indeed, given the Price and Thorne *membrane* approach [12, 13], the local Rindler horizon can be effectively approximated by a timelike hypersurface living just inside the true Rindler wedge, i.e. a timelike *stretched horizon.*

With a 2+1 decomposition, the timelike congruence comprising the stretched horizon is formally equivalent to a 2+1 continuous medium (fluid) living on the spacelike two dimensional cross section of the hypersurface and moving with velocity $\boldsymbol{v}^i$, defined by the unit timelike vector tangent to the hypersurface [2, 12].

This analogy can be used to get an insight on the possible expression for the horizon internal entropy. Indeed, for a slightly anisotropic fluid (viscous medium), the internal entropy is defined by

$$d_i s = -\frac{1}{T}\boldsymbol{P}^{\text{vis}}_{ij}\,\boldsymbol{V}^{ij} = -\frac{1}{T}\,(\boldsymbol{p}^{\text{vis}}\,\nabla^\rho \boldsymbol{v}_\rho + \tilde{\boldsymbol{P}}^{\text{vis}}_{ij}\tilde{\boldsymbol{V}}^{ij}) = \frac{\zeta}{T}\,(\nabla^\rho \boldsymbol{v}_\rho)^2 + \frac{2\eta}{T}\,\|\tilde{\boldsymbol{V}}\|^2, \tag{8}$$

where $\boldsymbol{P}^{\text{vis}}_{ij} = \boldsymbol{p}^{\text{vis}}\boldsymbol{\delta}_{ij} + \tilde{\boldsymbol{P}}^{\text{vis}}_{ij}$ is the viscous pressure acting on the medium, respectively decomposed in its bulk and trace-less components and the linear constitutive equation $\boldsymbol{p}^{\text{vis}} = -\zeta\,\nabla^\rho \boldsymbol{v}_\rho$ and $\tilde{\boldsymbol{P}}^{\text{vis}}_{ij} = -2\eta\,\tilde{\boldsymbol{V}}_{ij}$ have been used, with ζ and η respectively the medium bulk and shear viscosity [11]. Now, from a purely kinematical point of view, we can consider the velocity gradient of the medium, $\nabla_j \boldsymbol{v}_i$, as the equivalent of the deformation tensor field of the horizon congruence (extrinsic curvature), in the Killing frame, $\hat{B}_{ab}$. In both cases, the tensor can be decomposed into trace and symmetric trace-free part, $\boldsymbol{V}_{ij} = 1/3(\nabla^\rho \boldsymbol{v}_\rho)\boldsymbol{\delta}_{ij} + \tilde{\boldsymbol{V}}_{ij}$, where $\boldsymbol{V}_{ij}$ represents the symmetric component of $\nabla_j \boldsymbol{v}_i$, $\boldsymbol{\delta}_{ij}$ is the identity tensor and $\tilde{\boldsymbol{V}}_{ij} = \frac{1}{2}(\nabla_i \boldsymbol{v}_j + \nabla_j \boldsymbol{v}_i)$ is the deviatoric traceless tensor. Similarly, $\hat{B}_{ab} = 1/2\hat{\theta}h_{ab} + \hat{\sigma}_{ab}$, where $\hat{\theta}$ and $\hat{\sigma}_{ab}$, are the usual scalar expansion and shear of the null congruence.

Therefore, given the stretched horizon approximation and by associating the bulk term $(\nabla^\rho \boldsymbol{v}_\rho)$ to $\hat{\theta}$ and the deviatoric traceless tensor $\tilde{\boldsymbol{V}}_{ij}$ to $\hat{\sigma}_{ab}$, one can expect a dissipative internal entropy production term for the horizon congruence, of the form

$$(d_i S)^{\text{vis}} = \frac{1}{T}\int_H \tilde{\boldsymbol{\epsilon}}\, dv\, \zeta\, \hat{\theta}^2 + 2\eta\,\|\hat{\sigma}\|^2. \tag{9}$$

The above expression identifies the congruence kinematical quantities which are responsible for the irreversible viscous transfer of momentum into the system, and for the consequent production of internal entropy.

5. Non-equilibrium Spacetime Thermodynamics: GR

We can now reproduce the thermodynamical derivation of the Einstein equations with a non-equilibrium irreversible thermodynamics setting in a quite general way, by starting from a generic spacelike 2-surface patch comprising p, with non vanishing θ_p and σ_p.

As for the previous derivation, the strong equivalence principle will allow us to use the entropy area relation as in (1). However, in the new setting, the Clausius relation will be generalized to the expression (7), where δN will now encode all the information about both microscopic properties and irreversible perturbations of the system.

The new argument starts from the same definition of entropy, given in (3). Since we are now dealing with a non-equilibrium setting, we expect that the entropy can be expressed as a sum of two different contributions $dS = d_eS + d_iS$. Moreover, we can separate the reversible and irreversible levels, as previously argued,

$$d_eS = \alpha \int_H \widetilde{\epsilon}\, d\lambda\, (\theta - \lambda R_{ab}\ell^a\ell^b)_p = \delta Q/T \;\text{ at the reversible level,} \tag{10}$$

$$d_iS = -\alpha \int_H \widetilde{\epsilon}\, d\lambda\, \lambda \Big(\frac{1}{2}\theta^2 + \|\sigma\|^2\Big)_p = \delta N \;\text{ at the irreversible level.} \tag{11}$$

From the first expression above, one can see that the heat flux is still defined by the expression in (4). Even for the non-equilibrium setting the reversible heat will vanish at $\lambda = 0$. Thereby, at the zero order in λ, one deduces again $\theta_p = 0$, while the relation $R_{ab} + \Phi g_{ab} = (2\pi/\hbar\alpha)T_{ab}$, is recovered for all null vectors ℓ^a at the first order. Following the previous discussion this leads to the Einstein equations if $\alpha = (4\hbar G)^{-1}$.

On the other hand, for the irreversible level, we identify

$$\delta N = d_iS = -\,\alpha \int_H \widetilde{\epsilon}\, d\lambda\, \lambda\, \|\sigma\|_p^2 . \tag{12}$$

This again identifies the shear contribution as an internal entropy term, associating it to some irreversible dissipative process occurring in the thermal Rindler wedge, in accordance with (6).

To get a physical interpretation of δN with respect to the thermal properties of the Rindler wedge, it is helpful to express equation (12) in terms of the Killing horizon parameters. In the new frame,

$$\delta N = d_iS = \frac{\alpha}{\kappa} \int_H \widetilde{\epsilon}\, dv\, \|\hat{\sigma}\|_p^2 \geq 0, \tag{13}$$

in accordance with the second law of thermodynamics.

By a comparison with expression (9), one can actually interpret the expression in (13) as the standard entropy production term for a fluid with shear viscosity η, defined by $2\eta/T = \alpha/\kappa$, that is $\eta = \hbar\alpha/4\pi$, in agreement with the universal relation for the viscosity to entropy density ratio found in the AdS/CFT context [14].

While the previous discussion shows that the spacetime thermodynamics nicely fits into a non-equilibrium setting, we now want to take this arguments a step further and ask whether the expression in (13) can be effectively related to some gravitational energy flux.

The expression for the *uncompensated heat* given in (13) quantifies the energy of the system which is effectively dissipated by the viscous process,

$$T\,\delta N = \frac{\alpha T}{\kappa} \int_H \widetilde{\epsilon}\, dv\, \|\hat{\sigma}\|_p^2 = \frac{1}{8\pi G} \int_H \widetilde{\epsilon}\, dv\, \|\hat{\sigma}\|_p^2, \tag{14}$$

which, by substituting $\alpha = (4\hbar G)^{-1}$, from the reversible sector of the thermodynamical approach, coincides with the Hartle-Hawking formula for the tidal heating of a classical black hole [15, 16, 17, 18, 19].

This is a striking result as it defines the internal entropy production as a purely gravitational effect. Indeed, it can be associated with the work done on the horizon by the perturbative tidal field which is described by the electric part of the Weyl curvature tensor. The horizon viscosity implies that such a work will be converted into internal heat. Hence, the presence of the internal entropy term can be directly related to the process of dissipation via gravitational/internal degrees of freedom. In this sense, the irreversible sector contains the information about the possible activation/propagation of such degrees of freedom of the theory.

6. Non-equilibrium Spacetime Thermodynamics: $f(R)$

In the previous derivation, a crucial assumption lies in the validity of the SEP which allowed to consider the entropy density α as a constant. One might wonder what are the consequences of relaxing such an assumption in favor of the less restrictive Einstein Equivalence Principle (EEP). In this case, one might generically expect that the entropy density is promoted to a spacetime function (basically because the EEP implies a spacetime dependent Newton constant). However, in the definition for the entanglement entropy of the Rindler wedge, this implies a possibly very complicated spacetime dependence for the UV cut-off.

Following the treatment of [1], we consider here the specific case of $f(R)$ gravity, which is known to be equivalent to a single field scalar-tensor theory (more precisely a Brans-Dicke theory with $\omega = 0$ and a specific potential for the scalar field [20]). In this case, the UV cut-off is known to be proportional to some function of the curvature $\boldsymbol{f}(R) \equiv f'(R)$ (where the prime indicates the derivative with respect to R), playing the role of the inverse of the gravitational coupling. In this case, the area entropy relation is known to be given by $S = \alpha\, \boldsymbol{f}(R)$, where α is still a constant (albeit *a priori* different from the one considered in the previous section).

It is easy to see that in this case the entropy variation along the null congruence will be $dS/d\lambda = \alpha(d\boldsymbol{f}/d\lambda\widetilde{\boldsymbol{\epsilon}} + \boldsymbol{f} d\widetilde{\boldsymbol{\epsilon}}/d\lambda)$, where, by definition $\widetilde{\boldsymbol{\epsilon}}^{-1} d\widetilde{\boldsymbol{\epsilon}}/d\lambda = \theta$. Consequently, the entropy change along the horizon will read [1]

$$dS = \alpha \int_H \widetilde{\boldsymbol{\epsilon}}\, d\lambda\, (\dot{\boldsymbol{f}} + \boldsymbol{f}\,\theta), \tag{15}$$

then acquiring, with respect to the previous argument, an extra contribution $\dot{\boldsymbol{f}}$ coupled to the dynamics of the scalar function $\boldsymbol{f}$. (Here the dot stays for differentiation with respect to λ.)

For this reason, in order to set the instantaneous stationarity condition at p, that is $dS = 0$, the affine expansion is no more a good dynamical variable. In this sense, it is helpful to define the quantity $\tilde{\theta} \equiv (\theta\, \boldsymbol{f} + \dot{\boldsymbol{f}})$ as a sort of effective expansion for the congruence. Consequently, the equilibrium surface for the system will be fixed by the condition $\tilde{\theta}_p = 0$, that is $\theta_p = -\dot{\boldsymbol{f}}/\boldsymbol{f}$, where $\dot{\boldsymbol{f}} = \boldsymbol{f}'(R)\, \ell^a\, R_{,a}$ is generally nonzero. In particular, this actually provides an example of local Rindler frame equilibrium surface, for which θ_p is always non-vanishing, apart from the trivial case where $\boldsymbol{f}$ is constant, for which the theory will be equivalent to GR. From the fluid analogy, we could already expect that the presence of the non-vanishing affine expansion would produce a non-equilibrium contribution to the system entropy.

In order to get a quantitative expression for the entropy change in the neighborhood of p, one again can consider an infinitesimal deviation of the entropy from its equilibrium value. Let us then Taylor expand the integrand in (15) around p up to the first order in λ, that is $\tilde{\theta} = \tilde{\theta}_p + \lambda\, (\ddot{\boldsymbol{f}} - \boldsymbol{f}^{-1}\dot{\boldsymbol{f}}^2 + \boldsymbol{f}\,\dot{\theta})_p + \mathcal{O}(\lambda^2)$. Therefore, one can use the Raychaudhuri equation and the geodesic equation $\ell^a \ell^b{}_{;a} = 0$, to obtain the $\mathcal{O}(\lambda)$ expression for the entropy change

$$dS = \alpha \int_H \widetilde{\boldsymbol{\epsilon}}\, d\lambda\, \lambda\, \left[(\boldsymbol{f}_{;ab} - \boldsymbol{f}\, R_{ab})\, \ell^a \ell^b - -3/2\, \boldsymbol{f}\, \theta^2 - \boldsymbol{f}\, \|\sigma\|^2\right]_p, \tag{16}$$

where relation $\tilde{\theta}_p = 0$ is used to substitute $\boldsymbol{f}^{-2}\dot{\boldsymbol{f}}^2 = \theta^2$ at p. Now, keeping the expression in (4) for the heat flux, one can finally reproduce the same approach used in Sec.5.

At the reversible level, the generalized Clausius relation gives the constitutive equation $\boldsymbol{f}\, R_{ab} - \boldsymbol{f}_{;ab} + \Psi\, g_{ab} = (2\pi/\hbar\alpha)\, T_{ab}$ where Ψ is an undetermined function. With the same argument given in Sec.2, one then requires the conservation of the matter stress-energy tensor and use the contracted Bianchi identity to write the commutator of the covariant derivative as $2v^c{}_{;[ab]} = R_{abd}{}^c\, v^d$. In this way, one finds

$$(\boldsymbol{f}\, R_{ab} - \boldsymbol{f}_{;ab})^{;a} = \left(\frac{1}{2} f - \Box \boldsymbol{f}\right)_{,b} \quad \text{and thereby} \quad \Psi = \left(\Box \boldsymbol{f} - \frac{1}{2} f\right). \tag{17}$$

Eventually, equation (17), together with the constitutive equation found leads to the field equations of $f(R)$ gravity

$$\boldsymbol{f}\, R_{ab} - \boldsymbol{f}_{;ab} + \left(\Box \boldsymbol{f} - \frac{1}{2} f\right) g_{ab} = \frac{2\pi}{\hbar\alpha} T_{ab}. \tag{18}$$

with the identification $\alpha = (4\hbar G)^{-1}$. In [1], the same result was obtained starting from the entropy balance relation, assuming $\sigma_p = 0$, and then identifying the extra entropy term in θ in the second line of (16) with a suitable internal entropy term.

Indeed, following the previous discussion, the above term is expected (together with a shear dependent term) as an unavoidable contribution related to the irreversible sector of the generalized Clausius relation (7)

$$\delta N = - \int_H \tilde{\epsilon}\, d\lambda\, \lambda\, (\alpha\, \boldsymbol{f}) \left[\frac{3}{2}\,\theta^2 + \|\sigma\|^2\right]_p, \tag{19}$$

which, as explained in Sec.4, identifies the internal entropy production terms of the system.

As expected, the internal entropy in (19) now shows contributions both from scalar and tensorial degrees of freedom. Indeed, by using the same argument as in the GR case, we again have a natural interpretation for the expression in (19) as the dissipative function of the system.

The shear squared contribution is equivalent to the one found for GR, with a shear viscosity coefficient which now takes a factor $\boldsymbol{f}$, $\eta = \hbar\, \alpha\, \boldsymbol{f}/4\pi$, as a consequence of the UV cut-off chosen for the area entropy relation.

On the other hand, the internal entropy contribution due to the scalar degree of freedom is now given by

$$d_i S_\theta = - \int_H \tilde{\epsilon}\, d\lambda\, \lambda\, (\alpha\, \boldsymbol{f})\, \frac{3}{2}\, \theta_p^2. \tag{20}$$

By making use of the kinematical analogy described in Sec.4 and by expressing the above equation in the Killing frame, one is naturally led to define the bulk viscosity ζ as $\zeta/T = 3\,\alpha\, \boldsymbol{f}/2\kappa$, that is $\zeta = 3\,\hbar\,\alpha\, \boldsymbol{f}/4\pi$, as already found in [1].

Furthermore, in order to give a physical interpretation to (20), one can use the equivalence between $f(R)$ and scalar-tensor gravity, thereby interpreting $\boldsymbol{f}$ as an effective massive dilaton.

The action for $f(R)$ gravity is given by

$$\mathcal{S} = \frac{\hbar\alpha}{4\pi} \int d^4x \sqrt{-g}\, f(R) + \mathcal{S}_{mat}. \tag{21}$$

By introducing an auxiliary field $\varphi \equiv \mathbf{f}(R)$ and assuming $f''(R) \neq 0$ for all R, one can take $V(\varphi)$ as the Legendre transform of $f(R)$ so that $R = V'(\varphi)$, thereby rewriting the expression in (21) as

$$\mathcal{S} = \frac{\hbar\alpha}{4\pi} \int d^4x \sqrt{-g}\, \left[\varphi\, R + V(\varphi)\right] + \mathcal{S}_{mat}. \tag{22}$$

The Euler-Lagrange equations, in the Jordan frame, take the form

$$\varphi \left(R_{ab} - \frac{1}{2} g_{ab} R\right) + \left(g_{ab}\Box - \nabla_a \nabla_b\right) \varphi + \frac{1}{2} g_{ab} V(\varphi) = \frac{2\pi}{\hbar\alpha} T_{ab}, \tag{23}$$

equivalent to field equations given in (18).

In this frame, by using the relation $\tilde{\theta}_p = 0$, one can express the dissipated energy coupled to the bulk and shear viscosity in (20), in terms of the auxiliary scalar field φ

$$T\delta N = - \int_H \tilde{\epsilon}\, d\lambda\, \lambda\, (\alpha\, \varphi)\, T \left[\frac{3}{2}\, \varphi^{-2}\, \dot{\varphi}^2 + \|\sigma\|^2\right]_p. \tag{24}$$

The interpretation of the term related to the shear is straightforward as it is clearly the generalization to a scalar-tensor theory of the tidal heating already obtained for the GR case. More problematic is the interpretation of the bulk viscosity (purely scalar) contribution.

In this direction, as a first step, one can look for the effective field source terms which drive the local deformation of the null horizon congruence. Again, we consider the effective expansion $\tilde{\theta}$ as the suitable quantity to describe the scalar perturbations of the horizon given that $\tilde{\theta} = 0$ is the condition for equilibrium.

Let us then use $\tilde{\theta}$ as the effective expansion rate for the horizon bundle at p. Starting from this equation, where now $\boldsymbol{f}$ is substituted by φ, then using the Raychaudhuri equation and the Jordan frame field equation given in (23), one gets

$$\dot{\tilde{\theta}}_p = -\frac{\theta_p^2}{2} - \|\sigma\|_p^2 - (2\pi/\hbar\alpha)\,\varphi^{-1}\,T_{ab}\ell^a\ell^b - \varphi^{-2}\,\nabla_a\varphi\nabla_b\varphi\,\ell^a\ell^b. \tag{25}$$

Now, since we are interested only in the scalar contributions to $\dot{\tilde{\theta}}_p$, we can set $T_{ab} = 0$ and $\|\sigma\|_p^2 = 0$. Then, by using the equilibrium relation in the new frame $\theta = -\varphi^{-1}\dot{\varphi}$, we are left with $\dot{\tilde{\theta}}_p = -3/2\,\varphi^{-2}\,\nabla_a\varphi\nabla_b\varphi\,\ell^a\ell^b = -3/2\,\varphi^{-2}\,\dot{\varphi}^2$. This equation identifies the quantity $(3/2)\,\varphi^{-2}\,\dot{\varphi}^2$ as what one might define as the gravitational energy flux associated with the solely scalar field degrees of freedom.

In conclusion, we can now provide a clean interpretation of the viscous terms in (24) as those representing the thermal system internal energy loss due to both scalar and tensor gravitational energy fluxes through the horizon. In particular, by moving to the Killing frame, (24) can be rewritten as

$$T\delta N = \frac{1}{8\pi G}\int_H \tilde{\epsilon}\,dv\,\left[\frac{3}{2}\varphi^{-1}\,\dot{\varphi}^2 + \varphi\|\sigma\|^2\right]_p, \tag{26}$$

where we have set $\alpha = (4\hbar G)^{-1}$ as required by the equations of motion (18). The striking similarity of the above expression with the energy loss rate due to the gravitational radiation in scalar-tensor gravity (see e.g. equation (10.135) of [21]) further reinforces the above suggested interpretation. Furthermore, we now see that also for the case of $f(R)$ gravity the relation in (6), the expression for ζ and the above interpretation of (26) provide a correlation between the transmission coefficient of the gravitational energy through the horizon and the horizon congruence viscosity.

7. Discussion

In non-equilibrium spacetime thermodynamics the viscous dissipative effects appear to be naturally associated with purely gravitational energy fluxes. Their association to the irreversible/dissipative sector of the theory strongly suggests an interpretation of their nature as, non-local, internal heat flows associated with the internal spacetime degrees of freedom and clarifies why in GR a local, background independent, description of gravitational waves is precluded.

However, a different issue is the interpretation of the internal entropy production terms with respect to a particular spacetime solution. While the association between internal entropy and allowed form of gravitational fluxes seems quite clear, it might seem however puzzling that the arbitrariness in the choice of initial spacetime patch around p allows for non-zero shear and expansion of the null congruence (and hence for internal entropy production terms) even, for example, if one imagine to have performed the local Rindler wedge construction in a Minkowski spacetime. In fact, the thermodynamical approach is providing us just with the constitutive equations of the thermal system associated with local Rindler wedge, not of the spacetime in which the latter is constructed. The arbitrariness of the choice of the starting local spacetime patch implies that such equations will at most characterize the structure of the gravitational theory selected by the entropy-area relation (the EP formulation). In this sense they will not be associated with physical fluxes or curvatures of the spacetime as a whole.

Another important aspect of this approach highlighted by the work done has to do with the role that the different Equivalence Principle formulations play in selecting the possible gravitational dynamics. The *Strong* Equivalence Principle implies, for a generic choice of the local spacetime patch, $\theta_p = 0$, thereby leading to the equations of motion of GR (with irreversible level only corresponding to tensorial gravitational fluxes). The *Einstein* Equivalence Principle, does not fix either θ nor σ, thereby allowing for a generalized theory of gravity. For the simple case of $f(R)$, we showed how the scalar degree of freedom in fact produces a dissipative contribution which is actually associated with some purely gravitational scalar energy flux through the system boundary (and which is not part of the SET of the scalar field).

The presence of an unavoidable purely scalar gravitational energy flux seems to indicate that GR and scalar-tensor are truly separated theories. However, the very same thermodynamical approach, might also suggest to look at such gravitational theories as different regimes of some more general effective description of gravity. This seems an intriguing possibility worth further investigation as it might lead to a unified framework which associates different gravitational theories with different hydrodynamical regimes of the analogue flow associated with the horizon null congruence.

For what regards the spacetime viscosity, the quantities η and ζ are found to be always related to the UV cut-off scale of the theory through the entropy density. This might suggest an underlying microscopic interpretation of gravity[22] along the ideas of induced gravity (see e.g. [23] for a related discussion).

References

[1] C. Eling, R. Guedens, T. Jacobson, Phys. Rev. Lett. **96**, 121301 (2006).
[2] C. Eling, JEHP **11**, 048 (2008).
[3] T. Jacobson, Phys. Rev. Lett. **75**, 1260 (1995).
[4] C. M. Will, Living Rev. Relativity **9**, (2006), 3. URL: http://www.livingreviews.org/lrr-2006-3
[5] G. Chirco and S. Liberati, "Non-equilibrium Thermodynamics of Spacetime: the Role of Gravitational Dissipation," arXiv:0909.4194 [gr-qc].
[6] W. G. Unruh, Phys. Rev. D **14**, 870 (1976); W.G. Unruh and N. Weiss, Phys. Rev. D **29**, 1656 (1984).
[7] G. 't Hooft, Nucl. Phys. B **256**, 727 (1985).
[8] L. Bombelli, R. K. Koul, J. H. Lee and R. D. Sorkin, Phys. Rev. D **34**, 373 (1986).
[9] J. M. Gerard, Class. Quant. Grav. **24**, 1867 (2007) [arXiv:gr-qc/0607019].
[10] S. R. de Groot and P. Mazur, *Non-equilibrium Thermodynamics* (North-Holland Pub. Co., Amsterdam, 1962).
[11] D. Jou, J. Casas-Vázquez, G. Lebon, *Extended Irreversible Thermodynamics*, 3rd ed. (Springer-Verlag, Berlin Heidelberg. 2001).
[12] R. H. Price and K. S. Thorne, Phys. Rev. D **33**, 915 (1986).
[13] K. S. Thorne, R. H. Price and D. A. Macdonald, *"Black holes: the membrane paradigm"*, (Yale University Press, New Haven USA, 1986).
[14] J. M. Maldacena, Adv. Theor. Math. Phys. **2**, 231 (1998).
[15] S. Chandrasekhar, *The Mathematical Theory of Black Holes* (Clarendon Press, Oxford, 1983).
[16] E. Poisson, Phys. Rev. D **70**, 084044 (2004).
[17] S. W. Hawking, J. B. Hartle, Comm. Math. Phys. **27**, 283 (1972).
[18] E. Poisson, Phys. Rev. Lett. **94**, 161103 (2005).
[19] S. A. Teukolsky, W. H. Press, ApJ. **193**, 443-461 (1974).
[20] T. P. Sotiriou and V. Faraoni, "f(R) Theories Of Gravity," arXiv:0805.1726v3 [gr-qc].
[21] C. M. Will, *Theory and experiment in gravitational physics* (Univeristy Press, Cambridge, Revised edition, 1993).
[22] P. Candelas, D. W. Sciama, Phys. Rev. Lett. **38**, 1372 (1977).
[23] T. Jacobson, "Black hole entropy and induced gravity," arXiv:gr-qc/9404039v1.

Thermodynamic structure of Gravitational field equations

Dawood Kothawala [1]

Inter-University Centre for Astronomy and Astrophysics, Post Bag 04, Ganeshkhind, Pune 411007, INDIA

E-mail: tdawood@iucaa.ernet.in

Abstract. There is an intriguing analogy between the gravitational dynamics of horizons and thermodynamics, which is not yet understood at a deeper level. In fact, it has been shown for several cases that the on-horizon structure of the gravitational field equations in Einstein theory have the structure of *first law of thermodynamics*. In this talk, we discuss how such a structure arises and show that the field equations near any static horizon can be written as: $T\mathrm{d}S - \mathrm{d}E = P\mathrm{d}V$. Moreover, the result extends beyond Einstein theory and holds for Lanczos-Lovelock lagrangians as well. The entropy S we obtain is precisely the Noether charge entropy of Wald, and E provides a natural generalization of *quasi-local energy* of the horizon. We comment on several implications of this result, particularly the notion of gravitational entropy [treated as the Noether charge of diffeomorphism invariance] associated with horizons and it's role in gravitational dynamics arising out of virtual displacements of the horizon.

1. Introduction

There exists a strong mathematical resemblance between the on-horizon structure of the gravitational field equations and the *first law of thermodynamics*. This intriguing structure was first noted by Padmanabhan [1] in the context static spherically symmetric horizon in Einstein gravity. In [1] it was shown that the differential form of the on-horizon Einstein equation [more specifically, the G_0^0 and G_r^r components] can be interpreted as virtual displacement of the horizon, and is, in fact, equivalent to the first law fo thermodynamics: $T\mathrm{d}S = \mathrm{d}E + P_{\perp}\mathrm{d}V$. In this talk, we describe how such a result arises out of near-horizon symmetries of the gravitational field tensor and holds not only for Einstein theory, but arbitrary Lanczos-Lovelock theory as well. In particular, our discussion brings out the key ingredients responsible for the thermodynamic structure of field equations as well as the generality of the result. Moreover, one can identify the expression for horizon entropy with the Noether charge entropy defined by Wald. Therefore, the question of how "gravity knows about thermodynamics at all", finds a natural explanation in the analysis we present. [2]

Notation: The metric signature is $(-,+,+,+)$; latin indices go from 0 to 3, and greek indices from 1 to 3. Capitalized latin indices go over the $(D-2)$ transverse coordinates.

[1] Work done in collaboration with T. Padmanabhan, IUCAA, India.

[2] Only the references most directly related to the present talk have been cited here; for further details and an exhaustive list of references, see [2].

2. Gravitational field equations as a thermodynamic identity

2.1. Background

The coordinate system best suited for our discussion is given by the metric:

$$ds^2 = -N^2 dt^2 + dn^2 + \sigma_{AB} dy^A dy^B \tag{1}$$

where $\sigma_{AB}(n, y^A)$ is the transverse metric, and the Killing horizon, generated by the timelike Killing vector field $\boldsymbol{\xi} = \boldsymbol{\partial}_t$, is approached as $N^2 \to 0$. Demanding the finiteness of curvature invariants on the horizon leads to the following Taylor series expansion for the lapse $N(n, x, y)$ and the transverse metric σ_{AB} [3]:

$$\begin{aligned} N(n,y) &= \kappa n \left[1 - \frac{1}{2} R_\perp(y; n=0) n^2 + O(n^3)\right] && (2) \\ \sigma_{AB} &= \left[\sigma_H(y)\right]_{AB} + \frac{1}{2}\left[\sigma_2(y)\right]_{AB} n^2 + \mathcal{O}(n^3) && (3) \end{aligned}$$

where we have collectively called the transverse coordinates as y, and $R_\perp$ is the Ricciscalar corresponding to y = constant surface. We shall be interested in a small region in the neighbourhood of the spacelike $(D-2)$-surface; more precisely, we shall assume $n \ll R_\perp^{-1/2}$, and would be interested in the $\kappa \to \infty$. That is, we shall require the length scale set by κ to be the smallest of all length scales. The t = constant part of the metric is written by employing *Gaussian normal coordinates* for the spatial part of the metric spanned by (n, y^A), n being the normal distance to the horizon. Consider the (y = constant) null vectors given by:

$$\begin{aligned} l_i &= \left(-1, +N^{-1}, 0, 0\right) \\ k_i &= \left(-1, -N^{-1}, 0, 0\right) \end{aligned} \tag{4}$$

In the limit we are interested in, it is easy to check that these vectors satisfy the geodesic equation in affinely parametrized form, that is; $\nabla_{\boldsymbol{l}} \boldsymbol{l} = 0 = \nabla_{\boldsymbol{k}} \boldsymbol{k}$. [Explicit calculation shows that $\nabla_{\boldsymbol{l}} \boldsymbol{l}$ has only "y"-components, given by:

$$\nabla_{\boldsymbol{l}} \boldsymbol{l} = -\left(2\kappa^{-2}\right)\left[\sigma^{AB}\, \partial_B R_\perp\right]_{n=0} \boldsymbol{\partial}_A + O(n^2) \tag{5}$$

which go to zero in the limit $\partial R_\perp / \kappa^2 \to 0$, which is the limit we are interested in.] The affine parameter λ, defined by $\boldsymbol{l} \cdot \nabla \lambda = 1$ can be found by using the above form of $N(n, y)$; to the leading order, we find that, $\lambda \sim \lambda_H + (1/2)\kappa n^2$ where $\lambda = \lambda_H$ is the location of the horizon. Note that, $N^2 \boldsymbol{l} \to \boldsymbol{\xi}|_H$, which implies, $2\kappa\,(\lambda - \lambda_H)\, \boldsymbol{l} \to \boldsymbol{\xi}|_H$. In subsequent analysis, the differentials of various geometric quantities (such as entropy) defined on the horizon, which are directly involved in the statement of the first law of thermodynamics, are to be interpreted as variations with respect to the affine parameter along the outgoing null geodesics, i.e., λ. This, of course, is the most natural variation that can be chosen on a *null surface*. All throughout, we shall take the *on the horizon* limit by considering a foliation defined by n = constant surfaces and then taking the limit $n \to 0$.

Aside: Before proceeding to the calculation, we should mention the connection of above metric with local coordinate system of an accelerated observer. The most common form in which the latter is written are the so called Fermi normal coordinates, which takes the form (in self-evident notation):

$$\begin{aligned} g_{00} &= -\left[(1 + a_\mu y^\mu)^2 + R_{0\mu 0\nu} y^\mu y^\nu\right] + \mathcal{O}(y^3) \\ g_{0\mu} &= -\frac{2}{3} R_{0\rho\mu\sigma} y^\rho y^\sigma + \mathcal{O}(y^3) \\ g_{\mu\nu} &= \delta_{\mu\nu} - \frac{1}{3} R_{\mu\rho\nu\sigma}\, y^\rho\, y^\sigma + \mathcal{O}(y^3) \end{aligned} \tag{6}$$

It is worth noting that the metric we considered above has a very similar form. Of course, it is simpler than the above metric since while constructing Fermi coordinates, no restriction such as staticity or constant acceleration is imposed. However, it is worth pursuing this point further since one ultimately wants to connect the dynamics of gravity with local horizons of accelerated observers in curved spacetime [see section 3], and so it is important to make precise notions such as local boost Killing vector field, local Rindler temperature etc. Work towards this is currently in progress.

2.2. Thermodynamic structure of Einstein tensor

We will now use the near horizon symmetries of the Einstein tensor to prove that the field equations near the horizon have a thermodynamic interpretation. We begin with the following expression for the on-horizon structure of the Einstein tensor (which can be found in [3];

$$G^{\widehat{\xi}}_{\widehat{\xi}} \,|_H = G^{\widehat{n}}_{\widehat{n}} \,|_H = \frac{1}{2}\mathrm{tr}\,[\sigma_2] - \frac{1}{2}R_{\|} \tag{7}$$

where $R_{\|}$ is the Ricci scalar of the on-horizon transverse metric, $[\sigma_H]_{AB}$. The Einstein tensor components given above are evaluated in an orthonormal tetrad appropriate for a timelike observer moving along the orbit of the Killing vector field generating the Killing horizon. [3]

It is easy to show that, on the horizon: $\delta_\lambda\sqrt{\sigma} = (2\kappa)^{-1}\sqrt{\sigma}\ \mathrm{tr}\,[\sigma_2]\ \delta\lambda$. We can use this to express $G^{\widehat{\xi}}_{\widehat{\xi}}$ in terms of variation of the transverse area with respect to affine parameter λ. Further, using the equality $G^{\widehat{\xi}}_{\widehat{\xi}} \,|_H = G^{\widehat{n}}_{\widehat{n}} \,|_H$ arising out of near-horizon symmetries alongwith the field equations, give

$$T\frac{\partial}{\partial\lambda}\left[\int \frac{1}{4}\sqrt{\sigma}\,\mathrm{d}^2y\right]_H \delta\lambda - \left[\int_H \frac{1}{8\pi}R_{\|}\sqrt{\sigma}\,\mathrm{d}^2y\right]\frac{\delta\lambda}{2} = \int_H P_{\perp}\sqrt{\sigma}\,\mathrm{d}^2y\,\delta\lambda = \overline{F}\,\delta\lambda \tag{8}$$

where we have identified $T = \kappa/2\pi$ as the horizon temperature and used the interpretation of $T^{\widehat{n}}_{\widehat{n}}$ as normal pressure, $P_{\perp}$, *on the horizon.* The last equality defines $\overline{F}$ as the average normal force over the horizon "surface" (in the spirit of membrane paradigm) so that $\overline{F}\,\delta\lambda$ is the virtual work done in displacing the horizon by an affine distance $\delta\lambda$. The above equation can now be recognized as

$$T\delta_\lambda S - \delta_\lambda E = \overline{F}\,\delta\lambda \tag{9}$$

where

$$S = \frac{1}{4}\int_y \sqrt{\sigma}\,\mathrm{d}^2y \tag{10}$$

$$E = \int_\lambda \frac{1}{16\pi}\left[\int_y R_{\|}\sqrt{\sigma}\,\mathrm{d}^2y\right]\delta\lambda \tag{11}$$

give entropy and energy when evaluated at $\lambda = \lambda_H$. While the expression for entropy is familiar, the expression for energy is not, but can be shown to reduce to standard expressions in specific cases.

We have therefore shown that Einstein equations near a static horizon has a highly symmetric structure which has the form of the first law of thermodynamics. It is trivial to show that the above expression reproduces the original result of Padmanabhan for static spherical symmetric case.

[3] The validity of expression (7), with the given Taylor series for $N(n, y^A)$ and $\sigma_{AB}(n, y^A)$, can be easily checked using a symbolic package such as `MAPLE`.

2.3. Thermodynamic structure of Lanczos-Lovelock tensor

A natural generalization of the above result would be to look at Lanczos-Lovelock lagrangians, which are the unique generalizations of Einstein tensor to higher dimensions, and yield equations of motion which are well behaved. We shall simply outline the derivation here; details can be found in [2].

The Lanczos-Lovelock (LL) lagrangians are given by

$$\mathcal{L}_m^{(D)} = \left(\frac{1}{16\pi}\frac{1}{2^m}\right) \boldsymbol{\delta}^{a_1 b_1 \ldots a_m b_m}_{c_1 d_1 \ldots c_m d_m} \boldsymbol{R}^{c_1 d_1}_{ a_1 b_1} \cdots \boldsymbol{R}^{c_m d_m}_{ a_m b_m} \tag{12}$$

The corresponding equations of motion are $2E^i_j = T^i_j$, where

$$E^i_{j(m)} = \left(\frac{1}{16\pi}\frac{m}{2^m}\right) \boldsymbol{\delta}^{a_1 b_1 \ldots a_m b_m}_{j d_1 \ldots c_m d_m} \boldsymbol{R}^{i d_1}_{ a_1 b_1} \cdots \boldsymbol{R}^{c_m d_m}_{ a_m b_m} - \frac{1}{2}\delta^i_j \mathcal{L}_m \tag{13}$$

where m is an integer, and Einstein theory corresponds to $m = 1$. These lagrangians have various special properties which have been discussed extensively in the literature. In particular, these lagrangians satisfy: $\nabla_a \left(\partial\mathcal{L}/\partial R_{abcd}\right) = 0$.

A detailed analysis using Gauss-Codazzi decomposition and Combinatorics leads to

$$E^{\hat{a}}_{\hat{b}}|_H = \left[\begin{array}{cc|c} E_\perp & 0 & 0 \\ 0 & E_\perp & 0 \\ \hline 0 & 0 & E_{\parallel}{}_{[(D-2)\times(D-2)]} \end{array}\right] \tag{14}$$

which generalizes the result of Einstein theory. The object of interest, viz. $E_\perp$, turns out to be

$$E_\perp = \underbrace{\left(\frac{1}{16\pi}\frac{m}{2^m}\right) \sigma^{CA}\ \mathcal{E}^B_A\ [\sigma_2]_{CB}}_{\text{corr to variation of a lower dim. action!}} - \left(\frac{1}{2}\right) \mathcal{L}_m^{(D-2)} \tag{15}$$

Once again, some simple algebraic manipulations similar to those done in the Einstein case puts this in the form (valid on the horizon):

$$2E_\perp \sqrt{\sigma}\,\delta\lambda = \underbrace{\frac{\kappa}{2\pi}\left(\frac{1}{8}\frac{m}{2^{m-1}}\right) \mathcal{E}^{BC}\ \delta_\lambda \sigma_{BC} \sqrt{\sigma}}_{\delta_\lambda S} - \underbrace{\mathcal{L}_m^{(D-2)} \sqrt{\sigma}\,\delta\lambda}_{\delta_\lambda E} \tag{16}$$

where

$$S = 4\pi m \int \mathrm{d}\Sigma\, \mathcal{L}_{m-1}^{(D-2)} \qquad \mapsto \frac{1}{4}\int \mathrm{d}\Sigma \tag{17}$$

$$E = \int^\lambda \delta\lambda \int \mathrm{d}\Sigma\, \mathcal{L}_m^{(D-2)} \qquad \mapsto \int_\lambda \frac{1}{16\pi}\left[\int_y R_\parallel \sqrt{\sigma}\,\mathrm{d}^2 y\right] \delta\lambda \tag{18}$$

where the arrows give corresponding expressions for $D = 4$ Einstein case. The thing to note is that both terms on RHS are expressible as "variations" of quantities locally defined on the horizon! In fact, S is precisely the Noether charge entropy as defined by Wald [4]. The quantity E as defined above gives an expression for energy which matches with known expressions for specific cases [for example, it gives $(1/2)r_H$ for spherically symmetric solutions in $D = 4$ Einstein theory, and reproduces the correct mass for spherically symmetric black hole solutions in Lovelock theory as calculated by others]. In fact, this expression for energy deserves a closer look, since it provides a very natural generalization of quasilocal energy for aspheric black holes in Einstein as well as Lovelock gravity.

Following exactly the same steps as in Einstein case, we conclude that the field equations can be written as: $T\delta_\lambda S - \delta_\lambda E = \overline{F}\delta\lambda$ on the horizon.

3. Comments and Discussion

To put the result in appropriate context [5], begin by noting that what we have shown is that the following relations

$$\underbrace{\boldsymbol{E}(\widehat{\boldsymbol{n}},\widehat{\boldsymbol{n}}) \;=-\boldsymbol{E}(\boldsymbol{u},\boldsymbol{u})}_{near-horizon\ symmetry} \;\;\&\;\; \underbrace{\boldsymbol{E}(\boldsymbol{u},\boldsymbol{u})=(1/2)\;\boldsymbol{T}(\boldsymbol{u},\boldsymbol{u})}_{field\ eq}$$

have a thermodynamic structure. Democracy of all observers then implies that $\boldsymbol{E}=(1/2)\boldsymbol{T}$ are thermodynamic:

$$\boldsymbol{E}(\boldsymbol{u},\boldsymbol{u})=\frac{1}{2}\;\boldsymbol{T}(\boldsymbol{u},\boldsymbol{u}) \underset{all\ \boldsymbol{u}}{\longmapsto} \boldsymbol{E}=\frac{1}{2}\;\boldsymbol{T}$$

Further, note that, in the limit $N\to 0$, we have $N^2\boldsymbol{l}\to\boldsymbol{\xi}$. Using this and $\boldsymbol{u}=\boldsymbol{\xi}/N$, we see that $\boldsymbol{T}(\boldsymbol{u},\boldsymbol{u})\to\boldsymbol{T}(\boldsymbol{\xi},\boldsymbol{l})$, which is precisely the flow of energy when the horizon undergoes a virtual displacement along the outgoing null geodesics. Incidentally, there is another way to look at this which is worth mentioning: the object $g^{\perp}_{ab}=2\xi_{(a}l_{b)}$ is the transverse part of the induced metric, and hence $T_{ab}\xi^a l^b=(1/2)\ \mathrm{tr}_2[T_{ab}]$, where tr_2 is the trace with respect to the metric $2\xi_{(a}l_{b)}$. For spherically symmetric spacetimes, since $T^0_0=T^r_r=P$, we have $(1/2)\ \mathrm{tr}_2[T_{ab}]=P$, which is the result given in [1].

Finally, let us briefly comment on lagrangians which depend on metric and Riemann tensor (but not it's derivatives); i.e., $L=L\left[g_{ab},R_{abcd}\right]$. Equations of motion are given by:

$$P_a^{\ cde}R_{bcde}-2\nabla^c\nabla^d P_{acdb}-\frac{1}{2}Lg_{ab}=8\pi T_{ab} \tag{19}$$

where P_{abcd} is defined as $\partial L/\partial R_{abcd}$, and inherit all the symmetries of the Riemann tensor. It is obvious that analyzing the near-horizon symmetries is not going to be easy, particularly due to the presence of the middle term on LHS; in fact, no such symmetries might exist at all for general lagrangians without imposing additional restrictions. One might concentrate [taking hint from the Lovelock case] on $T_{ab}\xi^a l^b$ without worrying about the symmetries of the field tensor, but then one risks loosing some crucial link between such symmetries and black hole entropy. Indeed, that such a link exists is most clearly evident from the work of Carlip [6], and it would in fact be worthwhile to connect Carlip's analysis with our result. Nonetheless, we observe that the thermodynamic structure of field equations are crucially linked to the near-horizon symmetries and select out a particular class of lagrangians, the Lanczos-Lovelock lagrangians. This seems reasonable since it is only for this class of lagrangians that one obtains second order equations of motion and the initial value problem is well defined. Since we do not have any criterion other than symmetry principles to analyze notions such as gravitational entropy in arbitrary theories of gravity, such a restriction to lagrangians is important since it gives us a handle on the sort of low-energy effective actions we may expect from a full theory of quantum gravity. It must be remembered that two apparently different types of symmetries [definitely connected] are evident in our result: the appearance of Wald entropy which arises as a Noether charge of diffeomorphism invariance (restricted to diffeomorphisms generating isometries) and the near-horizon symmetry of the field tensor, which in fact enforces, in the 2-dimensional subspace orthogonal to the horizon, conformal invariance of matter fields near the Killing horizon. Moreover, our result above already clearly provides a direct connection of entropy so obtained with gravitational dynamics.

4. Acknowledgments

I would like to thank the organizers of MCCQG for a wonderful conference at the beautiful location of Kolymbari.

References

[1] T. Padmanabhan: *Classical and Quantum Thermodynamics of horizons in spherically symmetric spacetimes*, *Class. Quantum Grav.* **19**, 5387 (2002); *Gravity and the Thermodynamics of Horizons*, *Phys. Rept.*, **406**, 49, (2005).

[2] Dawood Kothawala, T. Padmanabhan, *Thermodynamic structure of Lanczos-Lovelock field equations from near-horizon symmetries*, *Phys. Rev.* **D 79**, 104020, (2009).

[3] A. J. M. Medved, D. Martin, M. Visser: *Dirty black holes: Spacetime geometry and near-horizon symmetries*, *Class. Quantum Grav.* **21**, 3111 (2004); *Dirty black holes: Symmetries at stationary non-static horizons*, *Phys. Rev.*, **D. 70**, 024009 (2004).

[4] R. M. Wald, *Black Hole Entropy is Noether Charge*, *Phys.Rev. D* **48** 3427 (1993); V. Iyer, R. M. Wald, *Some Properties of Noether Charge and a Proposal for Dynamical Black Hole Entropy*, *Phys.Rev. D* **50** 846 (1994).

[5] For a recent review of these ideas, see: T. Padmanabhan, *Thermodynamical Aspects of Gravity: New insights*, arXiv:0911.5004v1.

[6] S. Carlip, *Entropy from Conformal Field Theory at Killing Horizons*, *Class. Quantum Grav.* **16** 3327 (1999); *Black Hole Entropy from Horizon Conformal Field Theory*, *Nucl. Phys. B (Proc. Suppl.)* **88** 10 (2000).

First Mediterranean Conference on Classical and Quantum Gravity (MCCQG 2009) IOP Publishing
Journal of Physics: Conference Series **222** (2010) 012015 doi:10.1088/1742-6596/222/1/012015

Observational Constraints on Transverse Gravity: a Generalization of Unimodular Gravity.

J.J. Lopez-Villarejo

E-mail: jj.lopezvillarejo@uam.es
Departamento de Fisica Teorica.
Universidad Autonoma de Madrid
Campus de Cantoblanco, 28049, Madrid (Spain).

Abstract. We explore the hypothesis that the set of symmetries enjoyed by the theory that describes gravity is not the full group of diffeomorphisms (*Diff(M)*), as in General Relativity, but a maximal subgroup of it (*TransverseDiff(M)*), with its elements having a jacobian equal to unity; at the infinitesimal level, the parameter describing the coordinate change $x^\mu \to x^\mu + \xi^\mu(x)$ is transverse, i.e., $\partial_\mu \xi^\mu = 0$. Incidentally, this is the smaller symmetry one needs to propagate consistently a graviton, which is a great theoretical motivation for considering these theories. Also, the determinant of the metric, g, behaves as a "transverse scalar", so that these theories can be seen as a generalization of the better-known unimodular gravity. We present our results on the observational constraints on transverse gravity, in close relation with the claim of equivalence with general scalar-tensor theory. We also comment on the structure of the divergences of the quantum theory to the one-loop order.

1. Introduction & Formalism

The gravitational theory of General Relativity presents as a remarkable feature the property of (active) diffeomorphism gauge invariance. The Einstein-Hilbert action

$$S_{EH} = -\frac{1}{2\kappa}\int d^4x\sqrt{g}R \tag{1}$$

is symmetric[1] under the "Diff" transformation $g_{\mu\nu}(x) \to g_{\mu\nu}(x) + \nabla_\mu\xi_\nu(x) + \nabla_\nu\xi_\mu(x)$, which represents the active version of the infinitesimal coordinate transformation $x^\mu \to x^\mu + \xi^\mu(x)$.

When attempting to quantize this theory in a perturbative expansion, one can take the ansatz $g_{\mu\nu} = \widetilde{g}_{\mu\nu} + h_{\mu\nu}$, where $\widetilde{g}_{\mu\nu}$ is a classical background that fulfills the classical equations of motion (e.o.m), and $h_{\mu\nu}$ is the quantum perturbation. In this way, one gets

$$S = -\frac{1}{2\kappa}\int d^4x[const + o(h_{\mu\nu}^2) + o(h_{\mu\nu}^3) + ...] \tag{2}$$

where the quadratic part in $h_{\mu\nu}$ constitutes the free-theory. In the case of a flat space, with a Minkowski classical background, the free-theory is given by the Fierz-Pauli Lagrangian:

$$\mathcal{L}_{FP} = \mathcal{L}^I + \mathcal{L}^{II} + a\mathcal{L}^{III} + b\,\mathcal{L}^{IV} \quad with\, b = \left(1 - 2a + 3a^2\right)/2 \tag{3}$$

[1] Unlike internal gauge transformations which do not involve the coordinates (e.g., U(1) of electromagnetism), this gauge invariance is not attained at the level of the Lagrangian, but only at the level of the action.

where $\mathcal{L}^I \equiv \frac{1}{4}\partial_\mu h^{\nu\rho}\partial^\mu h_{\nu\rho}$, $\mathcal{L}^{II} = -\frac{1}{2}\partial_\mu h^{\mu\rho}\partial_\nu h^\nu_\rho$, $\mathcal{L}^{III} = \frac{1}{2}\partial^\mu h\partial^\rho h_{\mu\rho}$, $\mathcal{L}^{IV} = -\frac{1}{4}\partial_\mu h\partial^\mu h$. Note that we have allowed for field redefinitions $h_{\mu\nu} \to \phi h_{\mu\nu}$ with respect to the usual presentation of the Fierz-Pauli Lagrangian, when $a = b = 1$. The corresponding action is symmetric under the "linear Diff" transformation

$$h_{\mu\nu} \to h_{\mu\nu} + \partial_\mu\xi_\nu + \partial_\nu\xi_\mu \tag{4}$$

and represents the action for a free massless spin-two particle (graviton). Here, gauge invariance is essential to eliminate the unphysical propagating modes, much like in electromagnetism, where the bad-behaved longitudinal mode is made a gauge artifact through the symmetry $A_\mu \to A_\mu + \partial_\mu\alpha$.

The authors of [1] undertook the task of finding out the most general conditions for a consistent description of the free massless spin-two particle. This involved considerations regarding the little group ISO(2) of the masless spin-two particle and the necessity of triviality of the generators of translations. Their conclusion is that, actually, one only needs the somehow smaller gauge symmetry

$$h_{\mu\nu} \to h_{\mu\nu} + \partial_\mu\xi_\nu + \partial_\nu\xi_\mu, \quad with\ \partial_\mu\xi^\mu = 0, \tag{5}$$

which we will denote as the "Linear TDiff" transformation, to indicate the transverse condition on the vector parameter. Hence, one can think of alternative theories of gravity to General Relativity, based on a slightly different symmetry principle, which however contain the graviton in the linear regime as well. From this observation, there are basically two ways to proceed forward.

The first route was already explored in the aforementioned seminal paper [1]. One should note that the transversality condition in (5) is directly affecting the transformation rule of the trace of $h_{\mu\nu}$: $h^\mu_\mu \to h^\mu_\mu + \partial_\mu\xi^\mu$, such that this quantity is an invariant. Therefore, one can decide to restrict the value of the trace in the theory to have a fixed value, tipically $h^\mu_\mu = 0$. In the full non-linear regime, this naturally translates into the condition $g \equiv -det(g_{\mu\nu}) = 1$. This approach has been named *unimodular gravity.*

In this work we are concerned with the second way to proceed. One can ask himself about the most general theory involving an unconstrained second-rank tensor $h_{\mu\nu}$ and which is compatible with the TDiff symmetry. Since we will be using less symmetry than the standard Diff case, we expect to get a *more general* action also in the full non-linear regime:

$$S_{TDiff} = S_{TDiff}[g_{\mu\nu}, \psi] = S_{Diff}[g_{\mu\nu}, \psi] + S'[g_{\mu\nu}(x), \psi]. \tag{6}$$

We wondered whether this fact could be useful in addressing such problems as the dark energy of dark matter puzzles.

The concrete answer in the linear regime was found in [2]. In general, one gets an extra scalar-like mode propagating as well as the two tensor modes corresponding to the massless graviton — the only exception is when the TDiff symmetry is enhanced with an additional Weyl conformal symmetry, in which case the unimodular approach is recovered. Specifically:

$$\mathcal{L}_{L-TDiff} = \mathcal{L}^I + \mathcal{L}^{II} + a\mathcal{L}^{III} + b\,\mathcal{L}^{IV} \quad with\, a,\ b\ arbitrary \tag{7}$$

where the form of the different pieces in the Lagrangian is given below (3). It should be stressed that this TDiff Langrangian, as well as the following ones, are not covariant with respect to general coordinate transformations[2], so that one has to formulate them in a specific set of coordinates. This is the reason why we are not talking of a *proper* additional scalar mode.

In the non-linear regime, the symmetry (5) *naturally* generalizes[3] to

$$g_{\mu\nu} \to g_{\mu\nu} + \nabla_\mu\xi_\nu + \nabla_\nu\xi_\mu, \quad with\ \partial_\mu\xi^\mu = 0 \tag{8}$$

[2] In passing, one could render the theory covariant by introducing extra objects in it.

[3] This is, however, not a unique answer. The other most "sensible" generalization, where ξ^μ is restricted with $\nabla_\mu\xi^\mu = 0$ (and $\sqrt{g} \to \sqrt{g}$) is doomed since one cannot accommodate a kinetic term for g: $\frac{1}{2}f_k(g)g^{\mu\nu}\partial_\mu g\partial_\nu g$.

$$\Longrightarrow \quad \sqrt{g} \rightarrow \sqrt{g} + \xi^{\mu}\partial_{\mu}\sqrt{g} \tag{9}$$

Hence, the determinant of the metric acts as a scalar under this restricted transformation (it is not, however, a scalar under coordinate transformations). What this entails is that one can have arbitrary powers of g appearing everywhere in the action of the full TDiff theory:

$$\begin{aligned} S_{TDiff} &= -\frac{1}{2\kappa^2}\int_{\mathcal{M}} d^4x\sqrt{g}\left(f(g)R + 2f_{\lambda}(g)\Lambda + \frac{1}{2}f_k(g)g^{\mu\nu}\partial_{\mu}g\partial_{\nu}g\right) \\ &+ \int_{\mathcal{M}} d^4x\sqrt{g}\mathcal{L}_{Matter}[\psi, g_{\mu\nu};\, g], \end{aligned} \tag{10}$$

formulated in a specific set of coordinates.

2. Results

The first question that was examined was the ultraviolet behaviour of these theories. Less symmetry translates in this case into more divergences, as compared with the template of General Relativity. Specifically, a calculation employing the background field method was undertaken in order to extract the 1-loop order divergences [3]. We made use of a correspondence between TDiff theories and scalar-tensor theories in the so-called "unitary gauge", which helped simplify the computations. To 1-loop order, General Relativity is UV (on-shell) finite [4]. These theories are not, due to extra terms arising from the additional g-mode; in fact, the result is very similar to a scalar-tensor theory with the scalar replaced by g. The only exceptions are two already-mentioned limiting cases: General Relativity and the presence of an additional conformal Weyl symmetry (unimodular case).

Then, the observational signatures of these kind of theories were inspected [5]. A general remark is that an almost exact correspondence can be found between these theories and scalar-tensor theories where the g is promoted to a true scalar field ϕ, and a lagrangian multiplier is added to the action to force the constraint $\phi = g$ (the precise form of this correspondence remains to be stablished [in preparation]). In practice, we can thus apply all existing experimental bounds on scalar-tensor gravity to constrain these theories. One should note that, in general, the presence of g "alone" in the matter action will translate into a *non-metric coupling* of the scalar ϕ, when viewed in terms of the scalar-tensor language; hence, we will get wild violations of the (Weak) Equivalence Principle. A mechanism in the spirit of the metric postulate [6], or the one usually advocated in string theory [7] will be generically necessary to render the theory compatible with observations.

3. Conclusions

- TDiff theories are well motivated alternative theories of gravity from a quantum perspective.
- To the 1-loop level, these theories present UV divergences — while General Relativity is UV finite —, with only one significant exception that corresponds, arguably, to Unimodular Gravity.
- The experimental bounds on scalar-tensor theories of gravity are generically applicable to these theories. A mechanism to avoid "excessive" violations of the (Weak) Equivalence Principle is thus necessary to render the theory compatible with experiments.

[1] J. J. van der Bij, H. van Dam and Y. J. Ng, "The Exchange Of Massless Spin Two Particles," Physica **116A**, 307 (1982).
[2] E. Alvarez, D. Blas, J. Garriga and E. Verdaguer, "Transverse Fierz-Pauli symmetry," Nucl. Phys. B **756** (2006) 148 [arXiv:hep-th/0606019].
[3] E. Alvarez, A. F. Faedo and J. J. Lopez-Villarejo, "Ultraviolet behavior of transverse gravity," JHEP **0810** (2008) 023 [arXiv:0807.1293 [hep-th]].

[4] G. 't Hooft and M. J. G. Veltman, "One loop divergencies in the theory of gravitation," Annales Poincare Phys. Theor. A **20** (1974) 69.

[5] E. Alvarez, A. F. Faedo and J. J. Lopez-Villarejo, "Observational constraints on transverse gravity," JCAP **0907** (2009) 002 [arXiv:0904.3298 [hep-th]].

[6] C. M. Will, "The confrontation between general relativity and experiment," Living Rev. Rel. **9**, 3 (2005) [arXiv:gr-qc/0510072].

[7] T. Damour, "Questioning the equivalence principle," arXiv:gr-qc/0109063.

Journal of Physics: Conference Series **222** (2010) 012016 doi:10.1088/1742-6596/222/1/012016

Post-Minkowskian Gravity: Dark Matter as a Relativistic Inertial Effect?

Luca Lusanna
Sezione INFN di Firenze and ACES Topical Team of ESA
Polo Scientifico
Via Sansone 1
50019 Sesto Fiorentino (FI), Italy
Phone: 0039-055-4572334
FAX: 0039-055-4572364

E-mail: lusanna@fi.infn.it

Abstract.
Talk at the *1st Mediterranean Conference in Classical and Quantum Gravity*, held in the Orthodox Academy of Crete in Kolymbari (Greece) from Monday, September 14th to Friday, September 18th, 2009.

A review is given of the theory of non-inertial frames (with the associated inertial effects and the study of the non-relativistic limit) in Minkowski space-time, of parametrized Minkowski theories and of the rest-frame instant form of dynamics for isolated systems admitting a Lagrangian description. The relevance and gauge equivalence of the clock synchronization conventions for the identification of the instantaneous 3-spaces (Euclidean only in inertial frames) are described.

Then this formalism is applied to tetrad gravity in globally hyperbolic, asymptotically Minkowskian space-times without super-translations, where the equivalence principle implies the absence of global inertial frames. The recently discovered York canonical basis, diagonalizing the York-Lichnerowicz approach, allows to identify the gauge variables (inertial effects in general relativity) and the tidal ones (the gravitational waves of the linearized theory) and to clarify the meaning of the Hamilton equations. The role of the gauge variable ${}^3K$, the trace of the extrinsic curvature of the non-Euclidean 3-space (the York time not existing in Newton theory), as a source of inertial effects is emphasized. After the presentation of preliminary results on the linearization of tetrad gravity in the family of non-harmonic 3-orthogonal gauges with a free value of ${}^3K$, we define post-Minkowskian gravitational waves (without post-Newtonian approximations on the matter sources) propagating in a non-Euclidean 3-space, emphasizing the non-graviton-like aspects of gravity. It is conjectured that dark matter may be explained as a relativistic inertial effect induced by ${}^3K$: it would simulate the need to choose a privileged gauge connected with the observational conventions for the description of matter.

As shown in Refs. [1, 2] the recent developments in atom interferometry and in space physics around the Earth require not only a consistent special relativistic treatment of atomic physics in inertial frames in Minkowski space-time and its extension to non-inertial frames, but also the inclusion of the weak field limit of gravity at least in the framework of Einstein's general relativity applied to asymptotically Minkowskian space-times. Once this description is under control, we must quantize the matter both in inertial and non-inertial frames in Minkowski space-time. In this way it is possible to have light rays propagating along flat null geodesics (then

Journal of Physics: Conference Series **222** (2010) 012016 doi:10.1088/1742-6596/222/1/012016

replaced by general relativistic null geodesics in the weak field limit), avoiding to have "photons" reduced to states with two polarizations in a 2-dimensional Hilbert space without a carrier as it is done in the non-relativistic theory of entanglement and in the associated experiments. The existing inclusion of electro-magnetism at the order $1/c$ made by atomic physics destroys the Galilei group and does not allow a consistent definition of the Poincare' one, namely a consistent special relativistic formulation of atomic physics in Minkowski space-time.

While in Galilei space-time both Newtonian time and the Euclidean 3-space, with the associated notion of spatial distance, are *absolute* and Maxwell equations do not exist, in Minkowski space-time only the space-time is *absolute* and there is no intrinsic definition of an instantaneous 3-space where to formulate the Cauchy problem for such equations (the fields cannot be put into a box, because no relativistically covariant box exists). The only intrinsic notion of special relativity is the fixed Minkowski light-cone, describing the locus of incoming and outgoing radiation. The situation becomes more complex in general relativity, where we have a deformed light-cone varying from a point to the other of space-time due to gravity. As a consequence, the pre-requisite to put control on the predictability for future times (a well posed Cauchy problem for the relevant partial differential equations) is the solution of the problem of *clock synchronization*, i.e. of the convention needed to introduce a notion of simultaneity identifying an instantaneous 3-space. Moreover the solution must be such that the transition from a simultaneity convention to another one has to be formulated as a gauge transformation (so that physical results are not influenced by the convention, which only modifies the appearances of phenomena) and it must be possible to extend it to the general relativistic framework, where also the space-time becomes dynamical and the equivalence principle implies the existence (at best) of only global non-inertial frames (the inertial ones exist only locally near an observer in free fall).

Regarding the problem of clock synchronization in presence of gravity near the Earth let us underline the relevance of the ACES mission of ESA [3], programmed for 2013. It will make possible a measurement of the gravitational redshift of the Earth from the two-way link among a microwave clock (PHARAO) on the Space Station and similar clocks on the ground: the proposed microwave link should make possible the control of effects on the scale of 5 picoseconds. This will be a test of post-Newtonian gravity in the framework of Einstein's geometrical view of gravitation: the redshift is a measure of the $1/c^2$ deviation of post-Newtonian null geodesics from Minkowski ones. This is a non-perturbative effect (requiring a re-summation of the whole perturbative expansion) for every quantum field theory, which has to fix the background (and therefore the associated light-cones) to be able to define the quantum Fock space of the theory.

See Ref.[4] for the effects of the general relativistic description of gravity upon atom interferometry.

As already said, in special relativity there is no notion of simultaneity, of instantaneous 3-spaces and of spatial distance. The light postulates say that the two-way (or round-trip; only one clock is involved) velocity of light is a) isotropic and b) constant (a standard constant c replaces the standard of length in existing relativistic metrology). The one-way velocity of light between two observers depends on how their clocks are synchronized (in general is not isotropic and point-dependent). Usually one uses Einstein's convention for clock synchronization: an inertial observer A send a ray of light at x^o_i towards the (in general accelerated) observer B; the ray is reflected towards A at a point P of B world-line and then reabsorbed by A at x^o_f; by convention P is synchronous with the mid-point between emission and absorption on A's world-line, i.e. $x^o_P = x^o_i + \frac{1}{2}(x^o_f - x^o_i)$. This convention selects the Euclidean instantaneous 3-spaces $x^o = ct = const.$ of the inertial frames centered on A. Only in this case the one-way velocity of light between A and B coincides with the two-way one, c. However, if the observer A is accelerated, the convention breaks down, because if *only* the world-line of the accelerated observer A (the *1+3 point of view*) is given, then the only way for defining instantaneous 3-spaces is to identify them with the Euclidean tangent planes orthogonal to the 4-velocity of the observer (the local rest frames). But these planes (they are tangent spaces not 3-spaces!) will intersect each other at a distance from A's world-line of the order of the acceleration lengths of A, so that all the either linearly or rotationally accelerated frames, centered on accelerated observers, based either on Fermi coordinates or on rotating ones, will develop *coordinate singularities*. Therefore their approximated notion of instantaneous 3-spaces cannot be used for a well-posed Cauchy problem for Maxwell equations. See Refs[5, 6, 7] for more details and a rich bibliography on these topics.

Parametrized Minkowski theories [9], [6], [5], with the associated *inertial and non-inertial rest-frame instant form of dynamics*, solve these problems and, together with the results of Refs.[10, 11, 12], allow to get a formulation of relativistic atomic physics [13], [14], [7], both in inertial and non-inertial frames of Minkowski space-time.

To formulate this theory without the coordinate singularities of the 1+3 point of view, we need the *3+1 point of view* [6], in which we assign: a) the world-line of an arbitrary time-like observer; b) an admissible 3+1 splitting of Minkowski space-time, namely a nice foliation with space-like instantaneous 3-spaces (i.e. a clock synchronization convention). This allows to define a *global non-inertial frame* centered on the observer and to use observer-dependent Lorentz-scalar *radar 4-coordinates* $\sigma^A = (\tau; \sigma^r)$, where τ is a monotonically increasing function of the proper time of the observer and σ^r are curvilinear 3-coordinates on the 3-space Σ_τ having the observer as origin. If $x^\mu \mapsto \sigma^A(x)$ is the coordinate transformation from the inertial Cartesian 4-coordinates x^μ to radar coordinates, its inverse $\sigma^A \mapsto x^\mu = z^\mu(\tau, \sigma^r)$ defines the *embedding* functions $z^\mu(\tau, \sigma^r)$ describing the 3-spaces Σ_τ as embedded 3-manifold into Minkowski space-time. The induced 4-metric on Σ_τ is the following functional of the embedding ${}^4g_{AB}(\tau, \sigma^r) = [z^\mu_A\, \eta_{\mu\nu}\, z^\nu_B](\tau, \sigma^r)$, where $z^\mu_A = \partial\, z^\mu / \partial\, \sigma^A$ and ${}^4\eta_{\mu\nu} = \epsilon\, (+---)$ is the flat metric ($\epsilon = \pm 1$ according to either the particle physics $\epsilon = 1$ or the general relativity $\epsilon = -1$ convention). While the 4-vectors $z^\mu_r(\tau, \sigma^u)$ are tangent to Σ_τ, so that the unit normal $l^\mu(\tau, \sigma^u)$ is proportional to $\epsilon^\mu{}_{\alpha\beta\gamma}\, [z^\alpha_1\, z^\beta_2\, z^\gamma_3](\tau, \sigma^u)$, we have $z^\mu_\tau(\tau, \sigma^r) = [N\, l^\mu + N^r\, z^\mu_r](\tau, \sigma^r)$ ($N(\tau, \sigma^r) = \epsilon\, [z^\mu_\tau\, l_\mu](\tau, \sigma^r)$ and $N_r(\tau, \sigma^r) = -\epsilon\, g_{\tau r}(\tau, \sigma^r)$ are the lapse and shift functions).

The foliation is nice and admissible if it satisfies the conditions:
1) $N(\tau, \sigma^r) > 0$ in every point of Σ_τ (the 3-spaces never intersect, avoiding the coordinate singularity of Fermi coordinates);
2) $\epsilon\, {}^4g_{\tau\tau}(\tau, \sigma^r) > 0$, so to avoid the coordinate singularity of the rotating disk, and with the positive-definite 3-metric ${}^3g_{rs}(\tau, \sigma^u) = -\epsilon\, {}^4g_{rs}(\tau, \sigma^u)$ having three positive eigenvalues (these are the Møller conditions [5, 7]);
3) all the 3-spaces Σ_τ must tend to the same space-like hyper-plane at spatial infinity (so that there are always asymptotic inertial observers to be identified with the fixed stars).

These conditions imply that global *rigid* rotations are forbidden in relativistic theories [5]. In Ref.[7] there is the expression of the admissible embedding corresponding to a 3+1 splitting of Minkowski space-time with parallel space-like hyper-planes (not equally spaced due to a linear acceleration) carrying differentially rotating 3-coordinates without the coordinate singularity of the rotating disk. It is the first consistent global non-inertial frame of this type.

In the 3+1 point of view the 4-metric ${}^4g_{AB}(\tau, \vec{\sigma})$ on Σ_τ has the components $\epsilon\, {}^4g_{\tau\tau} = N^2 - N_r\, N^r$, $-\epsilon\, {}^4g_{\tau r} = N_r = {}^3g_{rs}\, N^s$, ${}^3g_{rs} = -\epsilon\, {}^4g_{rs} = \sum_{a=1}^3 {}^3e_{(a)r}\, {}^3e_{(a)s} = \tilde{\phi}^{2/3} \sum_{a=1}^3 e^{2 \sum_{\bar{b}=1}^2 \gamma_{\bar{b}a} R_{\bar{b}}}\, V_{ra}(\theta^i)\, V_{sa}(\theta^i))$, where ${}^3e_{(a)r}(\tau, \sigma^u)$ are cotriads on Σ_τ, $\tilde{\phi}^2(\tau, \sigma^r) = det\, {}^3g_{rs}(\tau, \sigma^r)$ is the 3-volume element on Σ_τ, $\lambda_a(\tau, \sigma^r) = [\tilde{\phi}^{1/3}\, e^{\sum_{\bar{b}=1}^2 \gamma_{\bar{b}a} R_{\bar{b}}}](\tau, \sigma^r)$ are the positive eigenvalues of the 3-metric ($\gamma_{\bar{a}a}$ are suitable numerical constants) and $V(\theta^i(\tau, \sigma^r))$ are diagonalizing rotation matrices depending on three Euler angles.

The components ${}^4g_{AB}$ or the quantities N, N_r, γ, $R_{\bar{a}}$, θ^i, play the role of the *inertial potentials* generating the relativistic apparent forces in the non-inertial frame. It can be shown [7] that the Newtonian inertial potentials are hidden in the functions N, N_r and θ^i. The extrinsic curvature ${}^3K_{rs}(\tau,\sigma^u) = [\frac{1}{2N}(N_{r|s} + N_{s|r} - \partial_\tau\, {}^3g_{rs})](\tau,\sigma^u)$, describing the *shape* of the instantaneous 3-spaces of the non-inertial frame as embedded 3-manifolds of Minkowski space-time, is a functional of the independent inertial potentials ${}^4g_{AB}$.

In parametrized Minkowski theories one considers any isolated system (particles, strings, fields, fluids) admitting a Lagrangian description, because it allows, through the coupling to an external gravitational field, the determination of the matter energy-momentum tensor and of the ten conserved Poincare' generators P^μ and $J^{\mu\nu}$ (assumed finite) of every configuration of the system. Then one replaces the external gravitational 4-metric in the coupled Lagrangian with the 4-metric $g_{AB}(\tau,\sigma^r)$ of an admissible 3+1 splitting of Minkowski space-time and the matter fields with new ones knowing the instantaneous 3-spaces Σ_τ. For instance a Klein-Gordon field $\tilde{\phi}(x)$ will be replaced with $\phi(\tau,\sigma^r) = \tilde{\phi}(z(\tau,\sigma^r))$; the same for every other field. Instead for a relativistic particle with world-line $x^\mu(\tau)$ we must make a choice of its energy sign: then it will be described by 3-coordinates $\eta^r(\tau)$ defined by the intersection of the world-line with Σ_τ: $x^\mu(\tau) = z^\mu(\tau,\eta^r(\tau))$. Differently from all the previous approaches to relativistic mechanics, the dynamical configuration variables are the 3-coordinates $\eta^r_i(\tau)$ and not the world-lines $x^\mu_i(\tau)$ (to rebuild them in an arbitrary frame we need the embedding defining that frame!).

With this procedure we get a Lagrangian depending on the given matter and on the embedding $z^\mu(\tau,\sigma^r)$, which is invariant under frame-preserving diffeomorphisms. As a consequence, there are four first-class constraints (an analogue of the super-Hamiltonian and super-momentum constraints of canonical gravity) implying that the embeddings $z^\mu(\tau,\sigma^r)$ are *gauge variables*, so that all the admissible non-inertial or inertial frames are gauge equivalent, namely physics does *not* depend on the clock synchronization convention and on the choice of the 3-coordinates σ^r: only the appearances of phenomena change by changing the notion of instantaneous 3-space.

As already said, in general relativity the space-time is no more absolute: it becomes dynamical and it is described by ten fields, i.e. by the 4-metric tensor ${}^4g_{\mu\nu}(x)$. To get its Hamiltonian description we will use the same 3+1 point of view and the radar 4-coordinates employed in special relativity. But now the admissible embeddings $x^\mu = z^\mu(\tau,\sigma^r)$ are not dynamical variables: instead their gradients $z^\mu_A(\tau,\sigma^r)$ give the transition coefficient from radar to world 4-coordinates, ${}^4g_{AB}(\tau,\sigma^r) = [z^\mu_A\, z^\nu_B](\tau,\sigma^r)\, {}^4g_{\mu\nu}(z(\tau,\sigma^r))$. As shown in Ref. [8], the dynamical nature of space-time implies that each solution of Einstein's equations dynamically selects a preferred 3+1 splitting of the space-time, namely in general relativity the instantaneous 3-spaces (and therefore the associated clock synchronization convention) are dynamically determined.

If we restrict ourselves to inertial frames, we can define the *inertial rest-frame instant form of dynamics for isolated systems* by choosing the 3+1 splitting corresponding to the intrinsic inertial rest frame of the isolated system centered on an inertial observer: the instantaneous 3-spaces, named Wigner 3-space due to the fact that the 3-vectors inside them are Wigner spin-1 3-vectors [6, 9], are orthogonal to the conserved 4-momentum P^μ of the configuration. In Ref.[7] there is the extension to admissible non-inertial rest frames, where P^μ is orthogonal to the asymptotic space-like hyper-planes to which the instantaneous 3-spaces tend at spatial infinity. This non-inertial family of 3+1 splittings is the only one admitted by the asymptotically Minkowskian space-times covered by canonical gravity formulation discussed below.

In the inertial rest frames we can get the explicit form of the Poincare' generators (in particular of the Lorentz boosts, which, differently from the Galilei ones, are interaction dependent). We can also give the final solution to the old problem of the relativistic extension of the Newtonian center of mass of an isolated system. In its rest frame there are *only* three notions of collective variables, which can be built by using *only* the Poincare' generators (they are *non-local* quantities knowing the whole Σ_τ) [10]: the canonical non-covariant Newton-Wigner center of mass (or center of spin), the non-canonical covariant Fokker-Pryce center of inertia and the non-canonical non-covariant Møller center of energy. All of them tend to the Newtonian center of mass in the non-relativistic limit. See Ref.[6] for the Møller non-covariance world-tube around the Fokker-Pryce 4-vector identified by these collective variables. As shown in Refs.[10, 11, 12] these three variables can be expressed as known functions of the rest time τ, of the canonically conjugate Jacobi data (frozen Cauchy data) $\vec{z} = Mc\,\vec{x}_{NW}(0)$ ($\vec{x}_{NW}(\tau)$ is the standard Newton-Wigner 3-position) and $\vec{h} = \vec{P}/Mc$, of the invariant mass $Mc = \sqrt{\epsilon P^2}$ of the system and of its rest spin $\vec{\bar{S}}$. It is convenient to center the inertial rest frame on the Fokker-Pryce inertial observer. As a consequence, every isolated system (i.e. a closed universe) can be visualized as a decoupled non-covariant collective (non-local) pseudo-particle described by the frozen Jacobi data $\vec{z}$, $\vec{h}$ carrying a *pole-dipole structure*, namely the invariant mass and the rest spin of the system, and with an associated *external* realization of the Poincare' group. The universal breaking of Lorentz covariance is connected to this decoupled non-local collective variable and is irrelevant because all the dynamics of the isolated system leaves inside the Wigner 3-spaces and is Wigner-covariant. In each Wigner 3-space Σ_τ there is a *unfaithful internal* realization of the Poincare' algebra, whose generators are built by using the energy-momentum tensor of the isolated system. While the internal energy and angular momentum are Mc and $\vec{\bar{S}}$ respectively, the internal 3-momentum vanishes: it is the rest frame condition. Also the internal Lorentz boost (whose expression in presence of interactions is given for the first time) vanishes: this condition identifies the covariant non-canonical Fokker-Pryce center of inertia as the natural inertial observer origin of the 3-coordinates σ^r in each Wigner 3-space. As a consequence [13] there are three pairs of second class (interaction-dependent) constraints eliminating the internal 3-center of mass and its conjugate momentum inside the Wigner 3-spaces [14]: this avoids a double counting of the collective variables and allows to re-express the dynamics only in terms of internal Wigner-covariant relative variables. As a consequence, we find that disregarding the unobservable center of mass all the dynamics is described only by relative variables: this is a form of *weak relationism* without the heavy foundational problem of approaches like the one in Ref.[15].

In the case of relativistic particles the reconstruction of their world-lines requires a complex interaction-dependent procedure delineated in Ref.[12]. See Ref.[13] for the comparison with the other formulations of relativistic mechanics developed for the study of the problem of *relativistic bound states* and Ref.[7] for the extension to non-inertial frames, especially to the rest-frame ones.

In Ref.[1] there is a review of the formulation of relativistic atomic physics with the electro-magnetic field in the radiation gauge (so that Coulomb potential governs the mutual interaction of the particles; in Ref.[7] there is its expression in non-inertial frames) given in Refs.[11, 13, 14]. The main result is that, given a suitable regularization of self-energies by means of the use of Grassmann-valued electric charges, there exist a canonical transformation leading to particles mutually interacting with the sum of Coulomb and Darwin potentials plus a *decoupled* transverse radiation field in the rest frame. Therefore at the classical level for the first time we get the identification of the Darwin potential (till now obtainable only from the instantaneous approximations to the Bethe-Salpeter equation in the theory of relativistic bound states) and a way out from the Haag theorem.

Moreover in Ref.[16] we are able to give a consistent quantization of relativistic mechanics in absence of the electro-magnetic field in the inertial rest frame. In it we quantize the frozen

Jacobi data of the canonical non-covariant decoupled center of mass and the Wigner-covariant relative variables on the Wigner hyper-plane. In this way we avoid the causality problems of the Hegerfeldt theorem [17] (the instantaneous spreading of wave packets). Since the center of mass is decoupled, its non-covariance is irrelevant: like for the wave function of the universe, who will observe it?

Due to the need of clock synchronization for the definition of the instantaneous 3-spaces, this Hilbert space $H = H_{com,HJ} \otimes H_{rel}$ ($H_{com,HJ}$ is the Hilbert space of the external center of mass in the Hamilton-Jacobi formulation, while H_{rel} is the Hilbert space of the internal relative variables) is not unitarily equivalent to $H_1 \otimes H_2 \otimes ...$, where H_i are the Hilbert spaces of the individual particles. As a consequence, at the relativistic level the zeroth postulate of non-relativistic quantum mechanics does not hold: the Hilbert space of composite systems is not the tensor product of the Hilbert spaces of the sub-systems.

This quantization can be extended to the class of global non-inertial frames with space-like hyper-planes as 3-spaces and differentially rotating 3-coordinates defined in Ref.[7]. As shown in Ref.[18], in this *multi-temporal quantization* we quantize only the 3-coordinates $\eta_i^r(\tau)$ of the particles and *not* the inertial effects (like the Coriolis and centrifugal ones): they remain c-numbers describing the appearances of phenomena! We get results compatible with atomic spectra.

Instead the quantization of *fields* in non-inertial frames is an *open* problem due to the no-go theorem of Ref.[18] showing the existence of obstructions to the unitary evolution of a massive quantum Klein-Gordon field between two space-like surfaces of Minkowski space-time. Its solution, i.e. the identification of all the 3+1 splittings allowing unitary evolution, will be a prerequisite to any attempt to quantize canonical gravity taking into account the equivalence principle (global inertial frames do not exist!). We have already found global non-inertial frames where the quantization leads to unitary evolution, but we do not yet know the full class of non-inertial frames evading the no-go theorem.

As reviewed in Ref.[1], the quantization defined in Ref.[16] leads to a first formulation of a theory for *relativistic entanglement*. The non validity of the zeroth postulate and the *non-locality* of Poincare' generators imply a *kinematical non-locality* and a *kinematical spatial non-separability* introduced by special relativity, which reduce the relevance of *quantum non-locality* in the study of the foundational problems of quantum mechanics which have to be rephrased in terms of relative variables.

Let us now consider Einstein's general relativity where space-time is dynamical, gravity is described by the 4-metric tensor and the equivalence principle says that global inertial frames do not exist. In it, differently from every other field theory, the 4-metric tensor has a double role: a) like in electro-magnetism and Yang-Mills theory it is a (pre)potential for the gravitational field; b) it also determines the chrono-geometrical structure of space-time through the line element $ds^2 = {}^4g_{\mu\nu}\, dx^\mu\, dx^\nu$. Therefore it teaches relativistic causality to the other fields. In particular it says to massless particles which are the allowed trajectories (null geodesics) in each point of space-time. As already said and shown in Ref.[2], the ACES mission of ESA [3] will give the first precision measurement of the gravitational redshift of the geoid, namely of the $1/c^2$ deformation of Minkowski light-cone caused by the geo-potential. In every quantum field theory, where the definition of the Fock space requires the use of the fixed light-cone of the background, so that property b) is lost and gravity is reduced to the spin-2 massless graviton, this is a non-perturbative effect requiring the resummation of the perturbative expansion.

Since all the properties of the standard model of elementary particles are connected with properties of the representations of the Poincare' group in inertial frames of Minkowski spacetime, we look for a family of space-times admitting the presence of a Poincare' algebra. As a consequence, we shall restrict ourselves to globally hyperbolic, asymptotically Minkowskian at spatial infinity, topologically trivial space-times, for which a well defined Hamiltonian formulation of gravity is possible if we replace the Hilbert action with the ADM one. The 4-metric tends in a suitable way to the flat Minkowski 4-metric ${}^4\eta_{\mu\nu}$ at spatial infinity: having an *asymptotic* Minkowskian background we can avoid to split the 4-metric in the bulk in a background plus perturbations in the weak field limit.

In developing the Hamiltonian formulation, as already said, we use the same 3+1 formalism previously introduced for parametrized Minkowski theories: the basic dynamical variable is now the 4-metric ${}^4g_{AB}(\tau, \sigma^r)$ and not the embedding $z^\mu(\tau, \sigma^r)$. Since tetrad gravity is more natural for the coupling of gravity to the fermions, the 4-metric is decomposed in terms of cotetrads, ${}^4g_{AB} = E_A^{(\alpha)}\, {}^4\eta_{(\alpha)(\beta)}\, E_B^{(\beta)}$ ((α) are flat indices; the cotetrads $E_A^{(\alpha)}$ are the inverse of the tetrads $E^A_{(\alpha)}$ connected to the world tetrads by $E^\mu_{(\alpha)}(x) = z^\mu_A(\tau, \sigma^r)\, E^A_{(\alpha)}(z(\tau, \sigma^r))$), and the ADM action, now a functional of the 16 fields $E_A^{(\alpha)}(\tau, \sigma^r)$, is taken as the action for ADM tetrad gravity. This leads to an interpretation of gravity based on a congruence of time-like observers endowed with orthonormal tetrads: in each point of space-time the time-like axis is the unit 4-velocity of the observer, while the spatial axes are a (gauge) convention for observer's gyroscopes.

The kinematical Poincare' group connecting inertial frames in special relativity and its enlargement to the group of frame-preserving diffeomorphisms required for the treatment of non-inertial frames in parametrized Minkowski theories are now replaced by the full spatio-temporal diffeonorphism group enlarged with the O(3,1) gauge group of the Newman-Penrose approach (the extra gauge freedom acting on the tetrads in the tangent space of each point of space-time and reducing from 16 to 10 the number of independent fields like in metric gravity). The relativity principle of special relativity is replaced with the principle of general covariance (invariance in form of physical laws). Let us remark that the problem of pseudo-tensors in the attempt to describe the gravitational energy, is already present in the non-inertial frames of Minkowski space-time, since they are connected to inertial frames by frame-preserving diffeomorphisms.

See the papers of Ref.[20] for this reformulation of canonical gravity.

If the direction-independent boundary conditions on the 4-metric and its conjugate momenta are such to kill *super-translations* [21], the SPI group of asymptotic symmetries [22] is reduced to the ADM Poincare' group. At the Hamiltonian level a well posed definition of Poisson brackets and of variational principles requires the addition of the DeWitt surface term at spatial infinity to the Dirac Hamiltonian. As shown in Refs.[20], with the previous boundary conditions this term turns out to be the *strong* ADM energy (a flux through a 2-surface at spatial infinity), which is equal to the *weak* ADM energy (expressed as a volume integral over the 3-space) plus constraints. Therefore in this family of space-times there is *not a frozen picture*, like in the family of spatially compact without boundary space-times considered in loop quantum gravity, where the Dirac Hamiltonian is a combination of constraints.

Moreover the absence of super-translations implies that the non-inertial rest frames are the only family of 3+1 splittings admitted by these asymptotically Minkowskian space-times, since the asymptotic Euclidean 3-spaces turn out to be orthogonal to the ADM 4-momentum. Therefore the instantaneous 3-spaces are non-inertial rest frames of the 3-universe and admit asymptotic inertial observers (to be identified with the fixed stars of star catalogues). If ϵ^μ_A are a set of asymptotic flat tetrads, the simplest embedding adapted to the 3+1 splitting of space-time is $x^\mu = z^\mu(\tau, \sigma^r) = x^\mu(\tau) + \epsilon^\mu_r\, \sigma^r$; if the time-like observer origin of the spatial radar coordinates is the inertial, with respect to the asymptotic 4-metric, observer $x^\mu(\tau) = x^\mu_o + \epsilon^\mu_A\, \sigma^A$, then we have

${}^4g_{AB}(\tau,\sigma^r) = \epsilon^\mu_A\,\epsilon^\nu_B\,{}^4g_{\mu\nu}(x)$. As a consequence, the 3-universe contained in an instantaneous 3-space can be described as a decoupled non-covariant non-observable pseudo-particle carrying a pole-dipole structure, whose mass and spin are those identifying the configuration of the "gravitational field plus matter" isolated system present in the 3-universe. As a consequence, the ADM Poincare' algebra with weak generators $\check{P}^A_{ADM}$, $\check{J}^{AB}_{ADM}$ expressed in radar 4-coordinates has to be considered as the unfaithful *internal* Poincare' algebra of special relativity: a) $\check{P}^r_{ADM} \approx 0$ are the rest-frame conditions; b) $\check{J}^{\tau r}_{ADM} \approx 0$ are the gauge-fixings eliminating the internal center of mass inside the 3-space. The weak ADM energy and angular momentum define the rest mass and spin of the 3-universe. In absence of matter Christodoulou - Klainermann space-times [23] are compatible with this description.

With this kind of formalism we can get a deparametrization of general relativity: if we switch off the Newton constant and we choose the flat Minkowski 4-metric in Cartesian coordinates as solution of Einstein's equations, we get the description of the matter present in the 3-universe in the inertial rest frames of Minkowski space-time with the weak ADM Poincare' group collapsing in the Poincare' group of particle physics.

This framework was developed in the works in Refs.[20]. The cotetrads were connected to cotetrads adapted to the 3+1 splitting of space-time (so that the time-like tetrad is boosted to coincide with the unit normal to the instantaneous 3-space Σ_τ, as it is done in Schwinger time gauges) by standard Wigner boosts for time-like vectors of parameters $\varphi_{(a)}(\tau,\sigma^r)$, $a=1,2,3$: $E^{\alpha)}_A = L^{(\alpha)}{}_{(\beta)}(\varphi_{(a)})\,\overset{o}{E}{}^{(\beta)}_A$. The adapted cotetrads have the following expression in terms of cotriads ${}^3e_{(a)r}$ on Σ_τ and of the lapse $N=1+n$ and shift $n_{(a)} = N^r\,{}^3e_{(a)r}$ functions: $\overset{o}{E}{}^{(o)}_\tau = 1+n$, $\overset{o}{E}{}^{(o)}_r = 0$, $\overset{o}{E}{}^{(a)}_\tau = n_{(a)}$, $\overset{o}{E}{}^{(a)}_r = {}^3e_{(a)r}$. The 4-metric becomes ${}^4g_{\tau\tau} = \epsilon\,[(1+n)^2 - \sum_a n^2_{(a)}]$, ${}^4g_{\tau r} = -\epsilon\sum_a n_{(a)}\,{}^3e_{(a)r}$, ${}^4g_{rs} = -\epsilon\,{}^3g_{rs} = -\epsilon\sum_a {}^3e_{(a)r}\,{}^3e_{(a)s}$. The 16 configurational variables in the ADM action are $\varphi_{(a)}$, $1+n$, $n_{(a)}$, ${}^3e_{(a)r}$. There are ten primary constraints (the vanishing of the 7 momenta of boosts, lapse and shift variables plus three constraints describing the rotation on the flat indices (a) of the cotriads) and four secondary ones (the super-Hamiltonian and super-momentum constraints): all of them are first class in the phase space spanned by 16+16 fields. This implies that there are 14 gauge variables describing *inertial effects* and 2 canonical pairs of physical degrees of freedom describing the *tidal effects* of the gravitational field (namely gravitational waves in the weak field limit). In this canonical basis only the momenta ${}^3\pi^r_{(a)}$ conjugated to the cotriads are not vanishing.

Then in Ref.[24] we have found a canonical transformation to a canonical basis adapted to ten of the first class constraints. It implementes the York map of Ref.[25] and diagonalizes the York-Lichnerowicz approach [26]. Its final form is ($\alpha_{(a)}(\tau,\sigma^r)$ are angles)

$\varphi_{(a)}$	$\alpha_{(a)}$	n	$\bar{n}_{(a)}$	θ^r	$\tilde{\phi}$	$R_{\bar{a}}$
$\pi_{\varphi_{(a)}} \approx 0$	$\pi^{(\alpha)}_{(a)} \approx 0$	$\pi_n \approx 0$	$\pi_{\bar{n}_{(a)}} \approx 0$	$\pi^{(\theta)}_r$	$\pi_{\tilde{\phi}} = \frac{c^3}{12\pi G}\,{}^3K$	$\Pi_{\bar{a}}$

$$
\begin{aligned}
{}^3e_{(a)r} &= R_{(a)(b)}(\alpha_{(c)})\,{}^3\bar{e}_{(b)r} = R_{(a)(b)}(\alpha_{(c)})\,V_{rb}(\theta^i)\,\tilde{\phi}^{1/3}\,e^{\sum_{\bar{a}}^{1,2}\gamma_{\bar{a}a}R_{\bar{a}}},\\
{}^4g_{\tau\tau} &= \epsilon\,[(1+n)^2 - \sum_a \bar{n}^2_{(a)}], \qquad {}^4g_{\tau r} = -\epsilon\,\bar{n}_{(a)}\,{}^3\bar{e}_{(a)r},\\
{}^4g_{rs} &= -\epsilon\,{}^3g_{rs} = -\epsilon\,\tilde{\phi}^{2/3}\sum_a V_{ra}(\theta^i)\,V_{sa}(\theta^i)\,e^{2\sum_{\bar{a}}^{1,2}\gamma_{\bar{a}a}R_{\bar{a}}},
\end{aligned}
$$

(1)

In this York canonical basis the *inertial effects* are described by the arbitrary gauge variables $\alpha_{(a)}$, $\varphi_{(a)}$, $1+n$, $\bar{n}_{(a)}$, θ^i, ${}^3K$, while the *tidal effects*, i.e. the physical degrees of freedom of the gravitational field, by the two canonical pairs $R_{\bar{a}}$, $\Pi_{\bar{a}}$, $\bar{a}=1,2$. The momenta $\pi_r^{(\theta)}$ and the 3-volume element $\tilde{\phi}=\sqrt{det\,{}^3g_{rs}}$ have to be found as solutions of the super-momentum and super-hamiltonian (i.e. the Lichmerowicz equation) constraints, respectively.

The gauge variables $\alpha_{(a)}$, $\varphi_{(a)}$ parametrize the extra O(3,1) gauge freedom of the tetrads (the gauge freedom for each observer to choose three gyroscopes as spatial axes and to choose the law for their transport along the world-line). The gauge angles θ^i (i.e. the director cosines of the tangents to the three coordinate lines in each point of Σ_τ) describe the freedom in the choice of the 3-coordinates σ^r on each 3-space: their fixation implies the determination of the shift gauge variables $\bar{n}_{(a)}$, namely the appearances of gravito-magnetism in the chosen 3-coordinate system.

The final basic gauge variable is a momentum, namely the trace ${}^3K(\tau,\sigma^r)$ of the extrinsic curvature (also named the *York time*) of the non-Euclidean 3-space Σ_τ. While in Yang-Mills theory all the gauge variables are configuration ones, the Lorentz signature of space-time implies that ${}^3K$ is a momentum variable: it is a time coordinate, while θ^i are spatial coordinates. While in special relativity the extrinsic curvature of the 3-space of non-inertial frames is a derived gauge quantity (the basic inertial potentials are ${}^4g_{AB}[z]$), here it is an independent gauge variable describing the shape of 3-space as an embedded 3-manifold in the space-time. This is what remains of the special relativistic gauge freedom in the choice of the clock synchronization convention: the other five components of ${}^3K_{rs}$ are determined by $R_{\bar{a}}$, $\Pi_{\bar{a}}$, $\pi_r^{(\theta)}$, $\tilde{\phi}$ and θ^i. This gauge variable has no Newtonian counterpart (the Euclidean 3-space is absolute), because its fixation determines the final shape of the non-Euclidean 3-space and then the lapse gauge variable (i.e. the proper time in each point of 3-space). Moreover this gauge variable gives rise to a negative kinetic term in the weak ADM energy, vanishing only in the gauges ${}^3K(\tau,\vec{\sigma})=0$.

In the York canonical basis the Hamilton equations generated by the Dirac Hamiltonian $H_D=\hat{E}_{ADM}+(constraints)$ are divided in four groups: A) the contracted Bianchi identities, namely the evolution equations for $\tilde{\phi}$ and $\pi_i^{(\theta)}$ (they say that given a solution of the constraints on a Cauchy surface, it remains a solution also at later times); B) the evolution equation for the four basic gauge variables θ^i and ${}^3K$: these equations determine the lapse and the shift functions once the basic gauge variables are fixed; C) the evolution equations for the tidal variables $R_{\bar{a}}$, $\Pi_{\bar{a}}$; D) the Hamilton equations for matter, when present.

Once a gauge is completely fixed, the Hamilton equations become deterministic. Given a solution of the super-momentum and super-Hamiltonian constraints and the Cauchy data for the tidal variables on an initial 3-space, we can find a solution of Einstein's equations in radar 4-coordinates adapted to a time-like observer. To it there is associated a special 3+1 splitting of space-time with dynamically selected instantaneous 3-spaces in accord with Ref.[8]. Then we can get pass to adapted world 4-coordinates ($x^\mu=x^\mu_o+\epsilon^\mu_A\,\sigma^A$) and we can describe the solution in every 4-coordinate system by means of 4-diffeomorphisms.

In Ref.[27] we study the coupling of N charged scalar particles plus the electro-magnetic field to ADM tetrad gravity in this class of asymptotically Minkowskian space-times without super-translations. To regularize the self-energies both the electric charge and the sign of the energy of the particles are Grassmann-valued. The introduction of the non-covariant radiation gauge allows to reformulate the theory in terms of transverse electro-magnetic fields and to extract the generalization of the Coulomb interaction among the particles in the Riemannian instantaneous 3-spaces of global non-inertial frames.

Journal of Physics: Conference Series **222** (2010) 012016 doi:10.1088/1742-6596/222/1/012016

After the reformulation of the whole system in the York canonical basis, we give the restriction of the Hamilton equations and of the constraints to the family of *non-harmonic 3-orthogonal* Schwinger time gauges, in which the instantaneous Riemannian 3-spaces have a non-fixed trace ${}^3K$ of the extrinsic curvature but a diagonal 3-metric. This family of gauges is determined by the gauge fixings $\theta^i(\tau, \sigma^r) \approx 0$ and ${}^3K(\tau, \sigma^r) \approx (arbitrary\, numerical\, function)$.

Starting from the results obtained in Ref.[27] for this family of non-harmonic 3-orthogonal Schwinger gauges, it is possible to define a consistent *linearization* of ADM canonical tetrad gravity plus matter in the weak field approximation [28], to obtain a formulation of Hamiltonian Post-Minkowskian gravity with non-flat Riemannian 3-spaces and asymptotic Minkowski background. This means that the 4-metric tends to the asymptotic Minkowski metric at spatial infinity, ${}^4g_{AB} \rightarrow {}^4\eta_{AB}$. The decomposition ${}^4g_{AB} = {}^4\eta_{AB} + {}^4h_{(1)AB}$, with a first order perturbation ${}^4h_{(1)AB}$ vanishing at spatial infinity, is only used for calculations, but has no intrinsic meaning. Moreover, due to the presence of a ultra-violet cutoff for matter, we can avoid to make Post-Newtonian expansions, namely we get fully relativistic expressions. We have found solutions for the first order quantities $\pi^{(\theta)}_{(1)r}$, $\tilde{\phi} = 1 + 6\,\phi_{(1)}$, $1 + n_{(1)}$, $\bar{n}_{(1)(a)}$. Then we can show that the tidal variables $R_{\bar{a}}$ satisfy a wave equation $R_{\bar{a}} = (known\, functional\, of\, matter)$ with the D'Alambertian associated to the asymptotic Minkowski 4-metric. Therefore, by using a no-incoming radiation condition based on the asymptotic Minkowski light-cone, we get a description of gravitational waves in these non-harmonic gauges, which can be connected to generalized TT(transverse traceless) gauges, as retarded functions of the matter. These gravitational waves do not propagate in inertial frames of the background (like it happens in the standard harmonic gauge description), but in non-Euclidean instantaneous 3-spaces differing from Euclidean 3-spaces at the first order (their intrinsic 3-curvature is detemined by the gravitational waves) and dynamically determined by the linearized solution of Einstein equations. These 3-spaces have a first order extrinsic curvature (with ${}^3K_{(1)}(\tau, \sigma^r)$ describing the clock synchronization convention) and a first order modification of Minkowski light-cone.

We can write explicitly the linearized Hamilton equations for the particles and for the electro-magnetic field: among the forces there are both the inertial potentials and the gravitational waves. If we disregard electro-magnetism, we can study the non-relativistic limit of the particle equations. The preliminary result [28] is that the particle 3-coordinates $\eta^r_i(\tau = ct) = \tilde{\eta}^r_i(t)$ satisfy the equation ($\vec{F}_{Newton}$ is the Newton gravitational force) $m\, \frac{d^2\tilde{\eta}^r_i(t)}{dt^2} = \sum_{j\neq i} F^r_{Newton}(\vec{\tilde{\eta}}_i(t) - \vec{\tilde{\eta}}_j(t)) + \frac{1}{c}\frac{d\tilde{\eta}^r_i(t)}{dt}\Big(\frac{1}{\triangle}\, c^2\, \partial^2_\tau\, {}^3K_{(1)}(\tau = ct, \vec{\sigma})\Big)|_{\vec{\sigma}=\vec{\tilde{\eta}}_i(t)}$. Therefore the (arbitrary in these gauges) double rate of change in time of the trace of the extrinsic curvature creates a post-Newtonian damping (or anti-damping since the sign of ${}^3K_{(1)}$ is not fixed) effect on the motion of particles. This is a inertial effect not existing in Newton theory where the Euclidean 3-space is absolute.

As a consequence there is the possibility of describing part (or maybe all) dark matter as a *relativistic inertial effect* determined by the gauge variable ${}^3K(\tau, \sigma^r)$: the rotation curves of galaxies would then experimentally determine a preferred choice of the instantaneous 3-spaces by using the freedom in ${}^3K_{(1)}$ to fit them. This option would differ both from the non-relativistic MOND approach (where one modifies Newton equations) and from modified gravity theories like the $f(R)$ ones (where one gets a modification of Newton potential) and is under investigation.

Since the implication of this approach to the dark matter problem is the existence of privileged gauges, let us add some remarks on the observables and the gauge problem in general relativity. Since no one is able to evaluate the Dirac observables of the gravitational field and of matter in general relativity (see Pons' talk for the status of the art), let us look at what are the measurable quantities used for observations when gravity is present. Gravitational waves are searched by existing, like LIGO and VIRGO on the Earth surface, or planned, LISA in space, detectors. To study these detectors one must introduce conventional 4-coordinate systems and the transformations rules among them: the terrestrial one ITRS (IERS2003) on the Earth surface, the geodetic celestial one GCRS (IAU2000) near the Earth and the barycentric celestial one BCRS (IAU2000) for the Solar System (see the bibliography of Ref.[2]). While the barycentric frame is considered as an approximate inertial Minkowski frame, the geocentric one is obtained from it by means of a rotation-free Lorentz boost plus post-Newtonian corrections. Therefore the post-Newtonian description of matter extended systems in the Solar System is intrinsically coordinate dependent since it requires the identification of the trajectory of the object. Satellites (think to GPS) are described with NASA coordinates first in ITRS ad then in GCRS. The planets in the Solar System are described in BCRS. The description of stars and galaxies done by astronomers with their star catalogues are based on an extension of BCRS, i.e. on a celestial frame which is an inertial frame of Minkowski space-time reduced to an inertial frame of Galilei space-time. This is due to the fact that the reconstruction of a 4-dimensional space-time from the 2-dimensional observations (light and angles) uses as input the standard cosmological model (the homogeneous and isotropic FRW solution of Einstein's equations) with the constant intrinsic 3-curvature put equal to zero, so that the 3-space is an Euclidean 3-space. As a consequence all the theory of galactic dynamics is strictly non-relativistic. However this set of conventions and assumptions lead to the open problems of dark matter and dark energy and the validity of the standard cosmological model is under investigation. Since the existence of dark matter is deduced from non-relativistic Newton theory in Euclidean 3-space, the conjecture is that we must choose a gauge in general relativity (where the gauge freedom is the choice of 4-coordinates) consistent with the observational conventions. Since all the observational frames have diagonal 3-metric, our family of 3-orthogonal gauges is reasonable. The conjecture coming from our approach is that all the data on what is called dark matter should be used to try to find which value of the York time ${}^3K$ explains the observations if we accept a notion of non-Euclidean 3-spaces. In this way a privileged gauge connected to observations would begin to emerge, to be used to describe the classical physics after the recombination surface.

More information on the role played by the York time will derive from how the following quantities depend upon it (at least in the linearized theory): a) the local proper time of a time-like observer; b) the redshift of rays of light along null geodesics; c) the luminosity distance based on the geodesic deviation equation along null geodesics. The future investigation of these quantities will also give information on dark energy.

Finally, if we will replace the matter with a perfect fluid (for instance dust), this will alow us to try to see whether the York canonical basis can help in developing the back-reaction approach [29] to dark energy, according to which dark energy is a byproduct of the non-linearities of general relativity when one considers spatial mean values on large scales to get a cosmological description of the universe taking into account the inhomogeneity of the observed universe. Since in the 3-orthogonal gauges all the relevant quantities are 3-scalars, it is possible to study the mean value of all the Hamilton equations and not only of the two considered by Buchert and to evaluate them explicitly in the linearized formulation. This will be done in the previous privileged gauge with the recombination surface as Cauchy surface.

[1] L.Lusanna, *Towards Relativistic Atomic Physics and Post-Minkowskian Gravitational Waves*, talk at the Workshop on *Gravitational Waves Detection with Atom Interferometry*, held in Firenze at the Galileo Galilei Institute for Theoretical Physics, February 23-24, 2009 (0908.0209).

[2] L.Lusanna, *The Chrono-Geometrical Structure of General Relativity and Clock Synchronization*, talk at the *First Colloquium Scientific and Fundamental Aspects of the Galileo Programme*, Toulouse, October 1-4, 2007 (0708.0490).

[3] L.Cacciapuoti and C.Salomon, *ACES: Mission Concept and Scientific Objective*, 28/03/2007, ESA document, Estec (ACES_Science_v1_printout.doc).
L.Blanchet, C.Salomon, P.Teyssandier and P.Wolf, *Relativistic Theory for Time and Frequency Transfer to Order* $1/c^3$, Astron.Astrophys. **370**, 320 (2000).
L.Lusanna, *Dynamical Emergence of 3-Space in General Relativity: Implications for the ACES Mission*, in Proc. of the 42th Rencontres de Moriond *Gravitational Waves and Experimental Gravity*, La Thuile (Italy), 11-18 March 2007.
See also the talks at the *SIGRAV Graduate School on Experimental Gravitation in Space*(Firenze, September 25-27, 2006) (http://www.fi.infn.it/GGI-grav-space/egs_s.html); at the Workshop *Advances in Precision Tests and Experimental Gravitation in Space* (Firenze, September 28/30, 2006) (http://www.fi.infn.it/GGI-grav-space/egs_w.html); at the Workshop "Theoretical Aspects of the ACES Mission" (Firenze, April 29-30, 2008) (ftp://cacciapuoti:In73rn0@ftp.estec.esa.int/); at the Workshop on "ACES and Future GNSS-based Earth Observation and Navigation" (Muenchen, May 26-27, 2008) (http://www.iapg.bv.tum.de/12735–aces programme.html).

[4] S.Dimopoulous, P.W.Graham, J.M.Hogan and M.A.Kasevich, *General Relativistic Effects in Atom Interferometry* (arXiv: 0802.4098).
S.Dimopoulous, P.W.Graham, J.M.Hogan, M.A.Kasevich and S.Rajendran, *An Atomic Gravitational Wave Interferometric Sensor (AGIA)* (arXiv: 0806.2125).

[5] D.Alba and L.Lusanna, *Generalized Radar 4-Coordinates and Equal-Time Cauchy Surfaces for Arbitrary Accelerated Observers* (2005), Int.J.Mod.Phys. **D16**, 1149 (2007) (gr-qc/0501090).

[6] L.Lusanna, *The Chrono-Geometrical Structure of Special and General Relativity: A Re-Visitation of Canonical Geometrodynamics*, lectures at 42nd Karpacz Winter School of Theoretical Physics: Current Mathematical Topics in Gravitation and Cosmology, Ladek, Poland, 6-11 Feb 2006, Int.J.Geom.Methods in Mod.Phys. **4**, 79 (2007). (gr-qc/0604120).
L.Lusanna, *The Chronogeometrical Structure of Special and General Relativity: towards a Background-Independent Description of the Gravitational Field and Elementary Particles* (2004), in *General Relativity Research Trends*, ed. A.Reiner, Horizon in World Physics vol. 249 (Nova Science, New York, 2005) (gr-qc/0404122).

[7] D.Alba and L.Lusanna, *Charged Particles and the Electro-Magnetic Field in Non-Inertial Frames*, (arXiv 0812.3057), to appear in Int.J.Geom.Methods in Physics in the form of the following two papers *Charged Particles and the Electro-Magnetic Field in Non-Inertial Frames: I. Admissible 3+1 Splittings of Minkowski Spacetime and the Non-Inertial Rest Frames (0908.0213) and II. Applications: Rotating Frames, Sagnac Effect, Faraday Rotation, Wrap-up Effect (0908.0215).*

[8] L.Lusanna and M.Pauri, *Explaining Leibniz equivalence as difference of non-inertial Appearances: Dis-solution of the Hole Argument and physical individuation of point-events*, History and Philosophy of Modern Physics **37**, 692 (2006) (gr-qc/0604087); *The Physical Role of Gravitational and Gauge Degrees of Freedom in General Relativity. I: Dynamical Synchronization and Generalized Inertial Effects; II: Dirac versus Bergmann Observables and the Objectivity of Space-Time*, Gen.Rel.Grav. **38**, 187 and 229 (2006) (gr-qc/0403081 and 0407007); *Dynamical Emergence of Instantaneous 3-Spaces in a Class of Models of General Relativity*, to appear in the book *Relativity and the Dimensionality of the World*, ed. A. van der Merwe (Springer Series Fundamental Theories of Physics) (gr-qc/0611045).

[9] L. Lusanna, *The N- and 1-Time Classical Descriptions of N-Body Relativistic Kinematics and the Electromagnetic Interaction*, Int. J. Mod. Phys. **A12**, 645 (1997).

[10] D.Alba, L.Lusanna and M.Pauri, *New Directions in Non-Relativistic and Relativistic Rotational and Multipole Kinematics for N-Body and Continuous Systems* (2005), in *Atomic and Molecular Clusters: New Research*, ed.Y.L.Ping (Nova Science, New York, 2006) (hep-th/0505005).
D.Alba, L.Lusanna and M.Pauri, *Centers of Mass and Rotational Kinematics for the Relativistic N-Body Problem in the Rest-Frame Instant Form*, J.Math.Phys. **43**, 1677-1727 (2002) (hep-th/0102087).
D.Alba, L.Lusanna and M.Pauri, *Multipolar Expansions for Closed and Open Systems of Relativistic Particles* , J. Math.Phys. **46**, 062505, 1-36 (2004) (hep-th/0402181).

[11] H.Crater and L.Lusanna, *The Rest-Frame Darwin Potential from the Lienard-Wiechert Solution in the Radiation Gauge*, Ann.Phys.(N.Y.) **289**, 87 (2001)(hep-th/0001046).
D.Alba, H.Crater and L.Lusanna, *The Semiclassical Relativistic Darwin Potential for Spinning Particles in*

the Rest-Frame Instant Form: Two-Body Bound States with Spin 1/2 Constituents, Int.J.Mod.Phys. **A16**, 3365 (2001) (hep-th/0103109).

[12] D.Alba, H.W.Crater and L.Lusanna, *Hamiltonian Relativistic Two-Body Problem: Center of Mass and Orbit Reconstruction*, J.Phys. **A40**, 9585 (2007) (gr-qc/0610200).

[13] D.Alba, H.W.Crater and L.Lusanna, *Towards Relativistic Atom Physics. I. The Rest-Frame Instant Form of Dynamics and a Canonical Transformation for a system of Charged Particles plus the Electro-Magnetic Field*, to appear in Canad.J.Phys. (arXiv: 0806.2383).

[14] D.Alba, H.W.Crater and L.Lusanna, *Towards Relativistic Atom Physics. II. Collective and Relative Relativistic Variables for a System of Charged Particles plus the Electro-Magnetic Field*, to appear in Canad.J.Phys.(0811.0715).

[15] C.Rovelli, *Relational Quantum Mechanics*, Int.J.Theor.Phys. **35**, 1637 (1996) (quant-ph/9609002).
C.Rovelli and M.Smerlak, *Relational EPR*, Found.Phys. **37**, 427 (2007)(quant-ph/0604064).

[16] D.Alba, H.W.Crater and L.Lusanna, *Relativistic Quantum Mechanics and Relativistic Entanglement in the Rest-Frame Instant Form of Dynamics*, (arXiv 0907.1816).

[17] G.C.Hegerfeldt, *Remark on Causality and Particle Localization*, Phys.Rev. **D10**, 3320 (1974); *Instantaneous Spreading and Einstein Causality in Quantum Theory*, Ann.Phys.Lpz. **7**, 716 (1998) (quant-ph/9809030); *Causality, Particle Localization and Positivity of the Energy*, in *Irreversibility and Causality in Quantum Theory - Semigroups and Rigged Hilbert Spaces*, eds. A.Bohm, H.D.Doebner and P.Kielanowski, Lecture Notes in Physics 504, p.238 (Springer , NewYork, 1998) (quant-ph/9806036).

[18] D.Alba and L.Lusanna, *Quantum Mechanics in Non-Inertial Frames with a Multi-Temporal Quantization Scheme: I) Relativistic Particles*, Int.J.Mod.Phys. **A21**, 2781 (2006) (hep-th/0502060)
D.Alba, *Quantum Mechanics in Non-Inertial Frames with a Multi-Temporal Quantization Scheme: II) Non-Relativistic Particles*, Int.J.Mod.Phys. **A21**, 3917 (2006) (hep-th/0504060).

[19] Torre, C.G. and Varadarajan, M. *Functional Evolution of Free Quantum Fields*, Clas. Quantum Grav. **16**, 2651-2668 (1999).
A.Arageorgis, J.Earman and L.Ruetsche, *Weyling the Time Away: the Non-Unitary Implementability of Quantum Field Dynamics o n Curved Spacetimes*, Studies in History and Philosophy of Modern Physics **33**, 151 (2002).

[20] L.Lusanna, *The Rest-Frame Instant Form of Metric Gravity*, Gen.Rel.Grav. **33**, 1579 (2001)(gr-qc/0101048).
L.Lusanna and S.Russo, *A New Parametrization for Tetrad Gravity*, Gen.Rel.Grav. **34**, 189 (2002)(gr-qc/0102074).
R.DePietri, L.Lusanna, L.Martucci and S.Russo, *Dirac's Observables for the Rest-Frame Instant Form of Tetrad Gravity in a Completely Fixed 3-Orthogonal Gauge*, Gen.Rel.Grav. **34**, 877 (2002) (gr-qc/0105084).
J.Agresti, R.De Pietri, L.Lusanna and L.Martucci, *Hamiltonian Linearization of the Rest-Frame Instant Form of Tetrad Gravity in a Completely Fixed 3-Orthogonal Gauge: a Radiation Gauge for Background-Independent Gravitational Waves in a Post-Minkowskian Einstein Spacetime*, Gen.Rel.Grav. **36**, 1055 (2004) (gr-qc/0302084).
J.Agresti, R.De Pietri, L.Lusanna and L.Martucci, *Hamiltonian Linearization of the Rest-Frame Instant Form of Tetrad Gravity in a Completely Fixed 3-Orthogonal Gauge: a Radiation Gauge for Background-Independent Gravitational Waves in a Post-Minkowskian Einstein Spacetime*, Gen.Rel.Grav. **36**, 1055 (2004) (gr-qc/0302084).

[21] T.Regge and C.Teitelboim, *Role of Surface Integrals in the Hamiltonian Formulation of General Relativity*, Ann.Phys.(N.Y.) **88**, 286 (1974).
R.Beig and Ó Murchadha, *The Poincar Group as the Symmetry Group of Canonical General Relativity*, Ann.Phys.(N.Y.) **174**, 463 (1987).
L.B.Szabados, *On the Roots of Poincare' Structure of Asymptotically Flat Space-Times*, Class. Quantum Grav. **20**, 2627 (2003) (gr-qc/0302033).

[22] R.M.Wald, *General Relativity* (Chicago Univ. Press, Chicago, 1984).

[23] D.Christodoulou and S.Klainerman, *The Global Nonlinear Stability of the Minkowski Space* (Princeton, Princeton, 1993).

[24] D.Alba and L.Lusanna, *The York Map as a Shanmugadhasan Canonical Transformationn in Tetrad Gravity and the Role of Non-Inertial Frames in the Geometrical View of the Gravitational Field*, Gen.Rel.Grav. **39**, 2149 (2007) (gr-qc/0604086, v2).

[25] J.Isenberg and J.E.Marsden, *The York Map is a Canonical Transformation*, J.Geom.Phys. **1**, 85 (1984).

[26] I.Ciufolini and J.A.Wheeler, *Gravitation and Inertia* (Princeton Univ.Press, Princeton, 1995).

[27] D.Alba and L.Lusanna, *The Einstein-Maxwell-Particle System in the York Canonical Basis of ADM Tetrad Gravity: I) The Equations of Motion in Arbitrary Schwinger Time Gauges.*, 2009 (arXiv 0907.4087).

[28] D.Alba and L.Lusanna, *The Einstein-Maxwell-Particle System in the York Canonical Basis of ADM Tetrad Gravity: II) The Weak Field Approximation in the 3-Orthogonal Gauges and Hamiltonian Post-Minkowskian*

Gravity: the N-Body Problem and Gravitational Waves with Asymptotic Background., in preparation.
[29] T.Buchert, *Dark Energy from Structure: a Status Report*, Gen.Rel.Grav. **40**, 467 (2008) (0707.2153).

First Mediterranean Conference on Classical and Quantum Gravity (MCCQG 2009) IOP Publishing
Journal of Physics: Conference Series **222** (2010) 012017 doi:10.1088/1742-6596/222/1/012017

Towards the Unification of Gravity and other Interactions: What has been Missed?

Matej Pavšič

Jožef Stefan Institute, Jamova 39, 1000 Ljubljana, Slovenia

E-mail: matej.pavsic@ijs.si

Abstract.
Faced with the persisting problem of the unification of gravity with other fundamental interactions we investigate the possibility of a new paradigm, according to which the basic space of physics is a multidimensional space $\mathcal{C}$ associated with matter configurations. We consider general relativity in $\mathcal{C}$. In spacetime, which is a 4-dimensional subspace of $\mathcal{C}$, we have not only the 4-dimensional gravity, but also other interactions, just as in Kaluza-Klein theories. We then consider a finite dimensional description of extended objects in terms of the center of mass, area, and volume degrees of freedom, which altogether form a 16-dimensional manifold whose tangent space at any point is Clifford algebra Cl(1,3). The latter algebra is very promising for the unification, and it provides description of fermions.

1. Introduction

Our current physical theories are mostly formulated in terms of spacetime, a 4-dimensional manifold whose points correspond to 'events' associated with, e.g., collisions of particles. General relativity is based on the concept of curved spacetime which enables description of gravity. One possible generalization is to consider higher dimensional spacetimes which, besides 4D gravity, include other fundamental interactions. Another possible generalization is to assume that fundamental objects are not point particles but strings, whose excitations contain fundamental particles and interactions. Though very promising, those theories have encountered numerous difficulties, and currently we still do not have a generally accepted 'unified theory'.

When considering extended objects, there is a possibility of formulating the theory in terms of the corresponding configuration space. Such idea has been considered before in numerous illuminating works by Barbour [1]. For some other approaches see refs. [2]. Here I will adopt the view that general relativity should be extended to configuration space $\mathcal{C}$, a multidimensional manifold equipped with metric, connection and curvature. In this picture a 4-dimensional spacetime is just a subspace of $\mathcal{C}$, associated with the degrees of freedom of a single point particle. In order to describe not only 4D gravity, but also other interactions, we can employ the extra dimensions of $\mathcal{C}$.

Journal of Physics: Conference Series **222** (2010) 012017 doi:10.1088/1742-6596/222/1/012017

2. Generalizing relativity

2.1. Configuration space replaces spacetime

Let us consider a system of point particles whose coordinates are X_i^μ, where $\mu = 0, 1, 2, 3$, $i = 1, 2, ..., N$, N being the number of particles in the configuration. We can consider X_i^μ as coordinates in a $4 \times N$-dimensional configuration space $\mathcal{C}$, therefore it is convenient to introduce a more compact notation:

$$\dot{X}_i^\mu \equiv \dot{X}^{(i\mu)} \equiv \dot{X}^M , \qquad M = (i\mu) \tag{1}$$

Let us assume that the action for such system is

$$I[X^M] = M \int \mathrm{d}\tau \, [\dot{X}^M \dot{X}^N G_{MN}(X^M)]^{1/2} \tag{2}$$

which is proportional to the length of a worldline in $\mathcal{C}$. Constant[1] M has the role of *mass* in $\mathcal{C}$, whereas τ is an arbitrary monotonically increasing parameter.

An action which is equivalent to (2) is the Schild action

$$I[X^M] = \int \mathrm{d}\tau \, \dot{X}^M \dot{X}^N \frac{M}{K} G_{MN}(X^M) \tag{3}$$

From the equations of motion derived from (3) it follows that

$$\dot{X}^M \dot{X}^M G_{MN} = K^2 \tag{4}$$

Therefore (3) is a gauge fixed form of the action (2).

We will assume that configuration space $\mathcal{C}$ is a manifold equipped with metric G_{MN}, connection and curvature (that in general does not vanish).

In particular, for the block diagonal metric

$$G_{MN} \equiv G_{(i\mu)(j\nu)} = g_{(i\mu)(j\nu)} = \begin{pmatrix} g_{\mu\nu}(x_1) & 0 & 0 & \cdots \\ 0 & g_{\mu\nu}(x_2) & 0 & \cdots \\ 0 & 0 & g_{\mu\nu}(x_3) & \cdots \\ \vdots & \vdots & \vdots & \vdots \end{pmatrix} \tag{5}$$

the action (3) becomes

$$I[\dot{X}_i^\mu] = \int \mathrm{d}\tau \sum_i \dot{X}_i^\mu \dot{X}_i^\nu \frac{M}{K} g_{\mu\nu}(X_i^\mu) \tag{6}$$

Rewriting eq. (4) as

$$\frac{\dot{X}_1^2}{K^2} = 1 - \frac{\dot{X}_2^2}{K^2} - \frac{\dot{X}_3^2}{K^2} - ... - \frac{\dot{X}_N^2}{K^2} \tag{7}$$

and introducing a new constant $m_1^2 \equiv p_1^2 = M^2 - p_2^2 - p_3^2 - ... - p_N^2$, we find

$$\frac{M^2}{K^2} = \frac{m_1^2}{k_1^2} \quad \text{or} \quad \frac{M}{K} = \frac{m_1}{k_1} \tag{8}$$

Here $k_i^2 = X_i^2 = g_{\mu\nu} \dot{X}_i^\mu \dot{X}_i^\nu$, and $p_i^2 \equiv p_i^\mu p_i^\nu g_{\mu\nu}$. Since the above derivation can be repeated for any $i = 1, 2, ..., N$, we have

$$\frac{M}{K} = \frac{m_i}{k_i} \tag{9}$$

[1] Symbol M, of course, has a different meaning when occurring as an index.

Inserting the latter relation into the action (6) we obtain the Schild action for a system of relativistic particles moving in a gravitational field $g_{\mu\nu}$, in fact the sum of Schild action for individual particles. Thus the ordinary relativistic theory for a many particle system is just a special case of the general action (2) or (3) for a particular, block diagonal metric (5).

By allowing for a more general metric, that cannot be transformed into the form (5) by a choice of coordinates in $\mathcal{C}$, we go beyond the ordinary theory.

Configuration space $\mathcal{C}$ is the space of possible "instantaneous" configurations in M_4. Its points are described by coordinates $x^M \equiv x_i^\mu$. A given configuration traces a worldline $x^M = X^M(\tau)$ in $\mathcal{C}$.

A dynamically possible worldline in $\mathcal{C}$ is a geodesic in $\mathcal{C}$, and it satisfies the variation principle based on the action (2).

Instead of considering a fixed metric of $\mathcal{C}$, let us assume that the metric G_{MN} is dynamical, so that the total action contains a kinetic term for G_{MN}:

$$I[X^M, G_{MN}] = I_m + I_g \tag{10}$$

where

$$I_m = \int \mathrm{d}\tau \, M \, (G_{MN} \, \dot{X}^M \dot{X}^N)^{(1/2)} = \int \mathrm{d}\tau \, M \, (G_{MN} \, \dot{X}^M \dot{X}^N)^{(1/2)} \, \delta^D \, (x - X(\tau)) \, \mathrm{d}^D x \tag{11}$$

and

$$I_g = \frac{1}{16\pi G_D} \int d^D x \, \sqrt{|G|} \, \mathcal{R} \tag{12}$$

Here $\mathcal{R}$ is the curvature scalar in $\mathcal{C}$. So we have *general relativity in configuration space* $\mathcal{C}$. We have arrived at a theory which is analogous to Kaluza-Klein theory. Configuration space is a higher dimensional space, whereas spacetime M_4 is a 4-dimensional subspace of $\mathcal{C}$, associated with a chosen particle.

The concept of configuration space can be used either in macrophysics or in microphysics. Configuration space associated with a system of point particles is finite dimensional. Later we will discuss infinite dimensional configuration spaces associated with strings and branes.

2.2. Equations of motion for a configuration of point particles

The equations of motion derived from the action (10) are the Einstein equations in configuration space $\mathcal{C}$. Let us now split the coordinates of $\mathcal{C}$ into 4-coordinates $X^\mu \equiv X^{1\mu}$, $\mu = 0, 1, 2, 3$ associated with position of a chosen particle, labeled by 1, and the remaining coordinates $X^{\bar{M}}$:

$$X^M = (X^\mu, X^{\bar{M}}) \tag{13}$$

The quadratic form occurring in the action (2) can then be split—according to the well known procedure of Kaluza-Klein theories—into a 4-dimensional part plus the part due to the extra dimensions of configuration space $\mathcal{C}$:

$$\dot{X}^M \dot{X}^N G_{MN} = \dot{X}^\mu \dot{X}^\nu g_{\mu\nu} + \text{extra terms} \tag{14}$$

More precisely, if for the metric of $\mathcal{C}$ we take the ansatz

$$G_{MN} = \begin{pmatrix} g_{\mu\nu} + A_\mu^{\bar{M}} A_\nu^{\bar{N}} \phi_{\bar{M}\bar{N}} \, , & A_\mu^{\bar{N}} \phi_{\bar{M}\bar{N}} \\ A_\nu^{\bar{N}} \phi_{\bar{M}\bar{N}} \, , & \phi_{\bar{M}\bar{N}} \end{pmatrix} \tag{15}$$

then we obtain

$$\dot{X}^M \dot{X}^N G_{MN} = \dot{X}^\mu \dot{X}^\nu g_{\mu\nu} + \dot{X}_{\bar{M}} \dot{X}_{\bar{N}} \, \phi^{\bar{M}\bar{N}} \tag{16}$$

where

$$\dot{X}_{\bar{M}} = G_{\bar{M}N}\dot{X}^N = A_{\bar{M}\mu}\dot{X}^\mu + \phi_{\bar{M}\bar{N}}\dot{X}^{\bar{N}} \tag{17}$$

Inserting expression (16) into the action (11), we have

$$I_m[X^\mu, X^{\bar{M}}] = M \int d\tau \left[\dot{X}^\mu \dot{X}^\nu g_{\mu\nu} + \phi^{\bar{M}\bar{N}}(A_{\bar{M}\mu}\dot{X}^\mu + \phi_{\bar{M}\bar{J}}\dot{X}^{\bar{J}})(A_{\bar{N}\nu}\dot{X}^\nu + \phi_{\bar{N}\bar{K}}\dot{X}^{\bar{K}})\right]^{1/2} \tag{18}$$

Let us now assume that the "internal" subspace of $\mathcal{C}$ admits isometries given by the Killing vector fields $k_\alpha^{\bar{J}}$. Index α runs over the independent Killing vectors, whereas $\bar{J}$, like $\bar{M}$, $\bar{N}$, runs over the "internal" coordinates. Then, as it is customary in Kaluza-Klein theories, we write

$$A_\mu^{\bar{J}} = k_\alpha^{\bar{J}} A_\mu^\alpha \tag{19}$$

The metric $\phi^{\bar{M}\bar{N}}$ of the internal space can be rewritten in terms of a metric $\varphi^{\alpha\beta}$ in the space of isometries:

$$\phi^{\bar{M}\bar{N}} = \varphi^{\alpha\beta} k_\alpha^{\bar{M}} k_\beta^{\bar{N}} + \phi_{\text{extra}}^{\bar{M}\bar{N}} \tag{20}$$

Here $\phi_{\text{extra}}^{\bar{M}\bar{N}}$ are additional terms due to the directions that are orthogonal to ismotries. For particular internal spaces $\bar{\mathcal{C}}$, those additional terms may vanish.

The projections of momenta onto Killing vectors are

$$p_\alpha \equiv k_\alpha^{\bar{J}} P_{\bar{J}} \tag{21}$$

We may chose a coordinate system in $\mathcal{C}$ in which

$$k_\alpha^M = \left(k_\alpha^\mu, k_\alpha^{\bar{M}}\right), \quad k_\alpha^\mu = 0, \quad k_\alpha^{\bar{M}} \neq 0 \tag{22}$$

Variation of the action (18 gives [3]

$$\frac{1}{\lambda}\frac{d}{d\tau}\left(\frac{\dot{X}^\mu}{\lambda}\right) + {}^{(4)}\Gamma^\mu_{\rho\sigma}\frac{\dot{X}^\rho \dot{X}^\sigma}{\lambda^2} + \frac{p_\alpha}{m}F^\alpha_{\mu\nu}\frac{\dot{X}^\nu}{\lambda}$$
$$+ \frac{1}{2m^2}\left(\varphi^{\alpha\beta}_{,\mu} - \varphi^{\alpha\beta}_{,\bar{J}} k^{\bar{J}}_{\alpha'} A^{\alpha'}_\mu\right) p_\alpha p_\beta + \frac{1}{\lambda m}\frac{dm}{d\tau} = 0 \tag{23}$$

where $\lambda = \left(\dot{X}^\mu \dot{X}^\nu g_{\mu\nu}\right)^{1/2}$.

In the above derivation we have used the following relation between the 4-dimensional and the higher dimensional mass

$$\frac{m}{M} = \left(\frac{\dot{X}^\mu X^\nu g_{\mu\nu}}{\dot{X}^M \dot{X}^N G_{MN}}\right)^{1/2} \tag{24}$$

which, for the special case of the block diagonal metric G_{MN} has been already given in eq. (8).

From eq. (23), in which p_α have the role of gauge charges, we see that m has the role of *inertial mass* in 4-dimensions. The 4-dimensional mass m is given by higher dimensional mass M and the contribution due to the extra components of momentum $P_{\bar{M}}$:

$$m^2 = g^{\mu\nu} p_\mu p_\nu = M^2 - \phi^{\bar{M}\bar{N}} p_{\bar{M}} p_{\bar{N}} = M^2 - \varphi^{\alpha\beta} p_\alpha p_\beta \tag{25}$$

These extra components $P_{\bar{M}}$ are in fact momenta of all other particles within the configuration. In general m is not constant, but in configuration spaces with suitable isometries it may be constant. In the last step in eq. (25), for simplicity, we have considered the case in which the terms with $\phi_{\rm extra}^{\bar{M}\bar{N}}$, occurring in eq. (20), vanish.

A configuration under consideration can be the universe. Then, according to this theory, the motion of a subsystem, approximated as a point particle, obeys the law of motion (23). Besides the usual 4-dimensional gravity, there are extra forces. They come from the generalized metric, i.e., the metric of configuration space. Since the inertial mass of a given particle depends on momenta of other particles and their states of motion (their momenta), the Mach principle is automatically incorporated in this theory. Such approach opens a Pandora's box of possibilities to revise our current views on the universe. Persisting problems, such as the horizon problem, dark matter, dark energy, the Pioneer effect, etc., can be examined afresh within this theoretical framework based on the concept of configuration space.

Locality, as we know it in the usual 4-dimensional relativity, no longer holds in this new theory, at least not in general. But in particular, when the metric of $\mathcal{C}$ assumes the block diagonal form (5), we recover the ordinary relativity (special and general), together with locality. However, it is reasonable to expect that metric (5) may not be a solution of the Einstein equations in $\mathcal{C}$. Then the ordinary relativity, i.e., the relativity in M_4, could be recovered as an approximation only. Even before going into the intricate work of solving the equations of general relativity in $\mathcal{C}$, we already have a crucial prediction, namely that locality in spacetime holds only approximately. When considering the universe within this theory, we have to bear in mind that the concept of spacetime has to be replaced by the concept of configuration space $\mathcal{C}$. Locality in M_4 has thus to be replaced by locality in $\mathcal{C}$. More technically this means that, instead of differential equations in M_4 (e.g., the Einstein equations), we have differential equations in $\mathcal{C}$: a given configuration (a point in $\mathcal{C}$) can only influence a nearby configuration (a nearby point in $\mathcal{C}$). Only in certain special cases this translates into the usual notion of locality in M_4 (a subspace of $\mathcal{C}$). The so called 'horizon problem' does not arise in this theory.

3. Configuration spaces associated with strings and branes

Instead of a system of point particles we can consider extended objects such as strings and, in general, branes. A brane configuration is described by the set of functions $X^\mu(\xi^a)$, where ξ^a, $a = 1, 2, ..., n$ is a set of parameters on the brane. We will consider a brane configuration as a point in an infinite dimensional configuration space, called brane space $\mathcal{M}$. Following refs. [4, 5, 3], we will therefore use a condensed notation

$$X^\mu(\xi^a) \equiv X^{\mu(\xi)} \equiv X^M \tag{26}$$

We assume that the branes within classes of tangentially deformed branes are in principle physically distinct objects. All such objects are represented by different points of $\mathcal{M}$-space.

Instead of one brane we can take a 1-parameter family of branes $X^\mu(\tau, \xi^a) \equiv X^{\mu(\xi)}(\tau) \equiv X^M(\tau)$, i.e., a curve (trajectory) in $\mathcal{M}$. In principle every trajectory is kinematically possible. A particular dynamical theory then selects which amongst those kinematically possible branes and trajectories are dynamically possible. We assume that dynamically possible trajectories are *geodesics* in $\mathcal{M}$ determined by the minimal length action [4, 5]:

$$I[X^M] = \int \mathrm{d}\tau \ (\rho_{MN} \dot{X}^M \dot{X}^N)^{(1/2)} \tag{27}$$

Here ρ_{MN} is the metric of $\mathcal{M}$.

In particular, if metric is

$$\rho_{MN} \equiv \rho_{\mu(\xi')\nu(\xi'')} = \kappa \frac{\sqrt{|f(\xi')|}}{\sqrt{\dot{X}^2(\xi')}} \delta(\xi' - \xi'') \eta_{\mu\nu} \tag{28}$$

where $f_{ab} \equiv \partial_a X^\mu \partial_b X^\nu \eta_{\mu\nu}$ is the induced metric on the brane, $f \equiv \det f_{ab}$, $\dot{X}^2 \equiv \dot{X}^\mu \dot{X}^\nu g_{\mu\nu}$, ($\eta_{\mu\nu}$ being the Minkowski metric of the embedding spacetime), then the equations of motion derived from (30) are precisely those of a Dirac-Nambu-Goto brane [4, 5]. Although we started from a brane configuration space in which tangetially deformed branes are considered as distinct objects, the dynamical theory based on the action (27) and the particular choice of metric (28) has for solutions the branes which satisfy such constraints which imply that only the transversal excitations are physical, whereas the tangential excitations are nothing but reparametrizations of ξ^a and τ. For more details see ref. [4].

In this theory we assume that metric (28) is just one particular choice amongst many other possible metrics of $\mathcal{M}$. But dynamically possible metrics are not arbitrary. We assume that they must be solutions of the Einstein equations in $\mathcal{M}$ [4, 5].

We take the brane space $\mathcal{M}$ as an arena for physics. The arena itself is a part of the dynamical system, it is not prescribed in advance. The theory is thus background independent. It is based on the geometric principle which has its roots in the brane space $\mathcal{M}$.

To sum up, the infinite dimensional brane space $\mathcal{M}$ has in principle any metric that is a solution to the Einstein's equations in $\mathcal{M}$. For the particular diagonal metric (28) we obtain the ordinary branes, including strings. But it remains to be checked whether such particular metric is a solution of this generalized dynamical system at all. If not, then this would mean that the ordinary string and brane theory is not exactly embedded into the theory based on dynamical $\mathcal{M}$-space. The proposed theory goes beyond that of the usual strings and branes. It resolves the problem of background independence and the geometric principle behind the string theory . Geometric principle behind the string theory is based on the concept of brane space $\mathcal{M}$, i.e., the configuration space for branes. Occurrence of gauge and gravitational fields in string theories is also elucidated. Such fields are due to string configurations. They occur in the expansion of a string state functional in terms of the Fock space basis. A novel insight is that they occur even within the classical string theory based on the action (27) with $\mathcal{M}$-space metric ρ_{MN}, which is dynamical and satisfies the Einstein equations in $\mathcal{M}$. Multidimensionality of ρ_{MN} allows for extra gauge interactions, besides gravity. In the following we will discuss how in the infinite dimensional space $\mathcal{M}$ one can factor out a finite dimensional subspace.

4. Finite dimensional description of extended objects

As an example let us consider a closed string. It has infinitely many degrees of freedom encoded in $X^\mu(\xi^a)$. But in first approximation we can describe it by four coodinates X^μ of the center of mass only. In the next approximation we can describe it in terms of the coordinates $X^{\mu_1\mu_2}$ of the oriented area enclosed by the string. If the string has finite thickness, and thus it actually is not a string but a 2-brane, then we can also consider the corresponding volume degrees of freedom $X^{\mu_1\mu_2\mu_3}$.

In general, an extended object in 4-dimensional spacetime can be described by 16 coordinates

$$X^M \equiv X^{\mu_1 ... \mu_r}, \quad r = 0,1,2,3,4 \tag{29}$$

They denote a point in a 16-dimensional space C, which is a subspace of the full infinite dimensional space $\mathcal{M}$, the configuration space of the considered extended object. The oriented r-volumes can be elegantly described [7]–[9] by Clifford algebra [6]. We thus consider a space C as a manifold whose tangent space at any point is Clifford algebra. The line element in C is

$$\mathrm{d}S^2 = G_{MN}\, \mathrm{d}x^M \mathrm{d}x^N \tag{30}$$

where G_{MN} now denotes the metric of C. The latter space is called 'Clifford space'.

A worldine $x^M = X^M(\tau)$ in C is associated with the motion of the extended object, e.g., a brane. Let the action be

$$I = \int \mathrm{d}\tau\, (G_{MN}\dot{X}^M \dot{X}^N)^{1/2} \tag{31}$$

If $G_{MN} = \eta_{MN}$ is Minkowski metric, then the equations of motion are

$$\ddot{X}^M \equiv \frac{\mathrm{d}^2 X^M}{\mathrm{d}\tau^2} = 0 \tag{32}$$

They hold for tensionless branes. For the branes with tension one has to replace η_{MN} with a generic metric G_{MN} with non vanishing curvature. Eq. (32) then generalizes to the corresponding geodesic equation

$$\frac{1}{\sqrt{\dot{X}^2}} \left(\frac{\dot{X}^M}{\sqrt{\dot{X}^2}} \right) + \Gamma^M_{JK} \frac{\dot{X}^J \dot{X}^K}{\dot{X}^2} = 0 \tag{33}$$

Such higher dimensional configuration space, associated with branes, enables unification of fundamental interactions à la Kaluza-Klein [10, 8].

Functions $X^M(\tau)$ are just like functions describing the worldline of a point particle. The four values of the index $M = \mu = 0, 1, 2, 3$ in fact correspond to the motion of a point particle in 4-dimensional spacetime M_4. The other values of the index $M = \mu_1 ... \mu_r$ for $r = 0$ (a scalar), $r = 2$ (bivector), $r = 3$ (3-vector) and $r = 4$ (4-vector) are associated with the particle's 'thickness'. From the point of view of spacetime M_4 we have thus a *thick particle*, and not a point particle. The thickness is encoded in 16 coordinates X^M. The elegance of this approach is in the fact that a thick particle in 4-dimensional spacetime M_4 can be described as a point particle in 16-dimensional Clifford space C.

Besides a point particle in Clifford space C that sweeps a wordline $X^M(\tau)$ in C, we can envisage a string in C that sweeps a worldsheet $X^M(\tau, \sigma)$ in C. The four values of the index $M = \mu = 0, 1, 2, 3$ correspond to a string in spacetime M_4. The rest of the indices $M = \mu_1 ... \mu_r$, $r = 0, 2, 3, 4$ label those coordinates which encode the object's thickness. From the point of view of spacetime M_4 the object is a *thick string*. Usual strings are infinitely thin objects. Despite being called 'extended objects', they are in fact not fully extended. Instead of infinitely thin strings we thus consider thick strings. Their thickness is encoded in coordinates $X^M \equiv X^{\mu_1 ... \mu_r}$.

We thus have objects which are strings in a 16-dimensional space C. They are described by the Nambu-Goto action, or, equivalently, by the Polyakov action which, in the conformal gauge, takes the form

$$I = \frac{\kappa}{2} \int \mathrm{d}\tau\, \mathrm{d}\sigma (\dot{X}^M \dot{X}^N - X'^M X'^N) G_{MN} \tag{34}$$

where $\dot{X}^M \equiv \frac{\mathrm{d}X^M}{\mathrm{d}\tau}$ and $X'^M \equiv \frac{\mathrm{d}X^M}{\mathrm{d}\sigma}$.

If the signature of the underlying space time is $+ - - -$, i.e., $(1, 3)$, then the signature of C is $(8, 8)$ [11]. By taking the Jackiw-Kim-Noz definition of vacuum [12, 14], one finds [11] that there are no central terms in the Virasoro algebra, if the D-dimensional space in which the string lives has signature $(D/2, D/2)$.

Instead of adding extra dimensions to spacetime, we can thus start from 4-dimensional spacetime M_4 with signature $(+ - - -)$ and consider the Clifford space C whose dimension is 16, and signature $(8, 8)$. *The necessary extra dimensions for consistency of string theory are in C-space.* This also automatically brings *spinors* into the game. It is an old observation that spinors are the elements of left or right ideals of Clifford algebras [15]–[17] (see also, e.g, refs. [18]–[20]). In other words, spinors are particular sort of Clifford numbers [4]. With the string coordinates $X^M = (X, X^\mu, X^{\mu\nu}, ...)$ we can associate the basis of Clifford algebra $\gamma_M = (\underline{1}, \gamma_\mu, \gamma_{\mu\nu}, ...)$, and consider the Clifford numbers $X^M \gamma_M$ that we call 'polyvectors'[2].

[2] It custummary to describe position in a flat space by vectors. If the space is a"curved" D-dimensional manifold, then its points can be described in terms of vectors belonging to a vector space $\mathbb{R}^D$ which, in particular can be the tangent space at a chosen point of the manifold. In the case of a curved Clifford space C its points are described in terms of 'polyvectors' which are the elements of a Clifford algebra Cl(8,8), a tangent space at a chosen point—"an origin"—of C.

Therefore, the string coordinate polyvectors $X^M\gamma_M$ contain spinors. This is an alternative way of introducing spinors into the string theory [4, 11]. Attempts to achieve a deeper understanding of the structure of supersymmetry within the context of Clifford algebras have been undertaken in refs. [20, 21].

5. Conclusion

In this paper we have considered the concept of the space of all possible matter configurations. If one assumes that positions of all particles are fixed and only the position of one particle is variable, then one has the space of all possible positions of the single particle. This is just the 4-dimensional spacetime. But the latter space is not the whole story, since it is only a subspace of a more general configuration space. What we have missed so far is to employ this more general space in our theoretical considerations. It is true that certain researchers have considered configuration spaces, but the idea has not yet been generally accepted. So far we have been stuck by the fact that we as observers, when moving around, explore only four dimensional spacetime. Such, intuitively reasonable notion, that physical space is associated with the degrees of freedom of a single point particle, has to be revised. There are many particles around, and all their degrees of freedom count; moreover, the particles are actually not point-like, but extended. Thus we have a multidimensional space with the prospect for the unification of gravity with other interactions.

Acknowledgments

This work was supported by the Ministry of High Education, Science, and Technology of Slovenia.

References

[1] Barbour J B 1994 *Class. Quant. Grav.* **11** 2853
Barbour J B 1994 *Class. Quant. Grav.* **11** 2875
Barbour J B 2002 *Class. Quant. Grav.* **19** 3217
Barbour J B 2000 *The End Of Time: The Next Revolution In Physics* (Oxford, UK: Univ. Pr.)
Barbour J B and B. Z. Foster B Z 2008 Constraints and gauge transformations: Dirac's theorem is not always valid *Preprint* 0808.1223 [gr-qc]
[2] Anderson E 2007 *Class. Quant. Grav.* **24** 5317
Anderson E 2008 Triangleland. I. Classical dynamics with exchange of relative angular momentum *Preprint* 0809.1168 [gr-qc]]
Anderson E Triangleland. II. Quantum Mechanics of Pure Shape *Preprint* 0809.3523 [gr-qc]
[3] Pavšič M 2007 Towards a New Paradigm: Relativity in Configuration Space *Preprint* 0712.3660 [gr-qc], published in *Proceedings of the 2nd International Confeerence on Time and Matter*, 26–31 August 2007, Bled Slovenia (Eds. O'Loughlin M, Stanič S, Veberič D)
[4] Pavšič M 2001 *The Landscape of Theoretical Physics: A Global View; From Point Particle to the Brane World and Beyond, in Search of Unifying Principle* (Dordrecht: Kluwer)
[5] Pavšič M 2003 General Principles of Brane Kinematics and Dynamics *Preprint* hep-th/0311060, published in *Proceedings of the EURESCO conference 'What comes beyond the Standard Model'*, 12-17 July 2003, Portorož, Slovenia (Eds. Mankoč Borštnik N, Nielsen H B, Froggatt C D, Lukman D)
[6] Hestenes D 1966 *Space-Time Algebra* (New York: Gordon and Breach)
Hestenes D and Sobcyk G 1984 *Clifford Algebra to Geometric Calculus* (Dordrecht: D. Reidel)
[7] Pavšič M 2003 *Found. Phys.* **33** 1277 (*Preprint* gr-qc/0211085)
[8] Pavšič M 2007 *Found. Phys.* **37** 1197 (*Preprint* hep-th/0605126)
[9] Castro C and Pavšič M 2005 *Prog. Phys.* **1** 31
[10] Pavšič M 2005 *Phys. Lett. B* **614** 85 (*Preprint* hep-th/0412255)
Pavšič M 2006 *Int. J. Mod. Phys.* A **21** 5905 (*Preprint* gr-qc/0507053)
[11] Pavšič M 2005 *Found. Phys.* **35** 1617 (*Preprint* hep-th/0501222)
[12] Cangemi D, Jackiw R and Zwiebach B 1996 *Annals of Physics* **245** 408
Benedict E, Jackiw R and Lee H J 1996 *Phys. Rev.* D **54** 6213
[13] Kim Y S and Noz M E 1973 *Phys. Rev.* D **8** 3521

Kim Y S and Noz M E 1975 *Phys. Rev.* D **12** 122
[14] Pavšič M 1999 *Phys. Lett.* A **254** 119 (*Preprint* hep-th/9812123)
[15] Teitler S 1965 *Supplemento al Nuovo Cimento* **III** 1
Teitler S 1965 *Supplemento al Nuovo Cimento* **III** 15
[16] Budinich P and Trautman A 1989 *Journal of Mathematical Physics* **30** 2125
[17] Lounesto P 2001 *Cifford Algebras and Spinors* (Cambridge: University Press)
[18] Mankoč Borštnik N and Nielsen H B 2003 *J. Math. Phys.* **44** 4817 (*Preprint* hep-th/0303224)
[19] Winnberg J O 1977 J. Math. Phys. **18** 625
[20] de Traubenberg M R 2005 Clifford Algebras in Physics *Preprint* hep-th/0506011
[21] Gates S, Jr., Linch W D and Phillips J 2003 When Superspace Is Not Enough *Preprint* hep-th/0211034
[22] Corwin L 1975 *Rev. Mod. Phys.* **47** 573

Journal of Physics: Conference Series **222** (2010) 012018 doi:10.1088/1742-6596/222/1/012018

Observables in classical canonical gravity: folklore demystified

J. M. Pons

Departament d'Estructura i Constituents de la Matèria and Institut de Ciències del Cosmos, Universitat de Barcelona, Diagonal 647, 08028 Barcelona, Catalonia, Spain

E-mail: pons@ecm.ub.es

D. C. Salisbury

Department of Physics, Austin College, Sherman, Texas 75090-4440, USA
Max-Planck-Institut für Wissenschaftsgeschichte, Boltzmannstrasse 22, 14195 Berlin, Germany

E-mail: dsalisbury@austincollege.edu

K. A. Sundermeyer

Freie Universität Berlin, Fachbereich Physik, Institute for Theoretical Physics, Arnimallee 14, 14195 Berlin, Germany

E-mail: ksun@gmx.de

Abstract. We give an overview of some conceptual difficulties, sometimes called paradoxes, that have puzzled for years the physical interpetation of classical canonical gravity and, by extension, the canonical formulation of generally covariant theories. We identify these difficulties as stemming form some terminological misunderstandings as to what is meant by "gauge invariance", or what is understood classically by a "physical state". We make a thorough analysis of the issue and show that all purported paradoxes disappear when the right terminology is in place. Since this issue is connected with the search of observables - gauge invariant quantities - for these theories, we formally show that time evolving observables can be constructed for every observer. This construction relies on the fixation of the gauge freedom of diffeomorphism invariance by means of a scalar coordinatization. We stress the condition that the coordinatization must be made with scalars. As an example of our method for obtaining observables we discuss the case of the massive particle in AdS spacetime.

1. Introduction

A menacing paradox has been haunting for years the canonical formulation of generally covariant theories. Briefly, the paradox is the following: since the Hamiltonian is nothing but a particular case of the gauge generator, one can - or must - be led to conclude that there is no physical time evolution in such theories, hence "time is frozen". A way of rephrasing this paradox is to argue that the observables, being by definition gauge invariant, must be constants of motion, hence "nothing happens".

So the question is: is there something wrong in the canonical formulation of generally covariant theories? Our answer - which we believe is *the* answer - is: certainly not!. In

[1, 2], building on previous work in [3], we have shown that observables evolving in time can be constructed[1]. We have argued that there is nothing wrong with the canonical formulation, and we have located the origin of the problems in some misunderstandings, basically terminological, that have over the course of time morphed into conceptual ones. We do not want to repeat verbatim here the arguments spelled out in the previous publications, but we will rather try to give an overview of the subject, emphasizing the conceptual aspects as well as other issues not covered in these publications, and refer to [2] for technical details when necessary. The - new - example of the massive particle in an anti-DeSitter (AdS) background will serve to illustrate our results.

We begin in section 2 with some very general aspects of the canonical formulation for general relativity theories and its gauge group (generally covariant theories can be similarly dealt with). In section 3 we introduce and address the purported paradoxes sometimes claimed to plague the formalism. In section 4 we consider the coordinate dependent gauge fixings that will pave the way to the construction of observables given in section 5. Section 6 is devoted to the example, and section 7 to the conclusions.

2. Diffeomorphism in canonical gravity

It was clarified long ago [4] that the gauge group of general relativity (GR) - and of other generally covariant theories - is much larger than the diffeomorphism group. In fact, the scalar density character of the Lagrangian $\mathcal{L}$ ensures the Noether symmetry condition

$$\delta\mathcal{L} = \partial_\mu(\epsilon^\mu \mathcal{L}) \tag{1}$$

for active variations of the fields induced by any *field-dependent* infinitesimal coordinate transformation $x^\mu \to x^\mu - \epsilon^\mu(x, \phi(x))$, with ϕ representing any field or field component. We call this gauge group the diffeomorphism-induced gauge group. Theories satifying (1) are commonly called generally covariant or diffeomorphism invariant. Of course the diffeomorphism group, with ϵ^μ depending on the coordinates only, is just a subgroup of the gauge group.

It was also clarified in [4] that the diffeomorphism group can not be completely realized in phase space. What can be realized is a proper subgroup of the diffeomorphism-induced gauge group - which is sometimes called the Bergmann-Komar group, as we denote it here - in which the field dependent infinitesimal diffeomorphisms ϵ^μ are of the specific form

$$\epsilon^\mu(x, \phi(x)) = n^\mu(x)\xi^0 + \delta^\mu_a \xi^a \tag{2}$$

with $n^\mu = (N^{-1}, -N^{-1}N^a)$, where N is the lapse and N^a the shift, and ξ^μ are arbitrary infinitesimal functions of the coordinates as well as of the fields other than the lapse and shift. The argument given in [4] for this formula (2), see also [5], relies on the requirement of consistency between the algebra of diffeomorphisms and the Poisson bracket structure in phase space. It was realized later [6] that an argument of projectability from configuration-velocity (or tangent space) structures to phase space leads to the same result in a simpler way.

Note that although the gauge group of canonical gravity, the Bergmann-Komar group, is not as large as the gauge group in configuration-velocity space, every infinitesimal diffeomorphism between two given configurations in tangent space can be matched in phase space by an element belonging to the Bergmann-Komar group. In this sense the Bergmann-Komar group is large enough in order to to fully describe general covariance in phase space.

Using the Rosenfeld-Dirac-Bergmann formalism[2] for gauge theories in phase space, the Dirac Hamiltonian of general relativity becomes [9] (here and henceforth sum of repeated indices

[1] The construction is formal, but solves the conceptual issue. See the conclusions for more comments on this point.

[2] See [7] and also [8] in these proceedings for an evaluation of the long forgotten Rosenfled's contribution.

Journal of Physics: Conference Series **222** (2010) 012018 doi:10.1088/1742-6596/222/1/012018

includes spatial integration as well)

$$H_D = N^\mu \mathcal{H}_\mu + \lambda^\mu P_\mu \,, \tag{3}$$

where P_ν are the canonical momenta of the lapse $N = N^0$ and shift N^i, $\mathcal{H}_\mu$ are the Hamiltonian and momentum constraints, and λ^μ are arbitrary functions. The dynamics imposes $\dot{N}^\mu = \lambda^\mu$ and thus one deduces that the lapse and shift are basically arbitrary functions. This is the reason why one usually assigns them values as part of a gauge fixing; for instance one can take the popular $N = 1$ and $N^i = 0$. This partial gauge fixing results in cutting down the Bergmann-Komar gauge group and in particular the infinitesimal gauge generators. The common use of partially gauge fixed actions in the phase space formulation explains why it took so long since the pioneering work of [4] until the entire gauge generator of the Bergmann-Komar group was formulated. It takes the form [6]

$$G_\xi(t) = (\mathcal{H}_\mu + N^\rho C^\nu_{\mu\rho} P_\nu)\xi^\mu + P_\mu \dot{\xi}^\mu \,, \tag{4}$$

(the dot stands for time derivative) where ξ^μ are the arbitrary functions - henceforth called descriptors - associated with an infinitesimal spacetime diffeomorphism through (2), and $C^\nu_{\mu\rho}$ are the structure functions resulting form the algebra of the Hamiltonian and momentum constraints under the Poisson bracket. Note that (4) is the generator of gauge transformations for all phase space variables, including the lapse and shift. One can then explicitely check (2) for all the variables. The Poisson brackets of G_ξ with different descriptors exhibit a soft algebra structure, with structure functions instead of structure constants. The reason can be traced [10, 5, 11] to the relations (2).

This gauge generator G_ξ realizes infinitesimally the full active diffeomorphism invariance of general relativity (GR) in phase space (for diffeomorphisms connected with the identity). It has the four arbitrary functions - in four dimensions - corresponding to four-diffeomorphism invariance. A good gauge fixing - more on this later - would yield the standard counting of degrees of freedom of GR. In conclusion, as regards the gauge symmetries, the canonical formalism is equivalent to the tangent space formalism, with the caveat concerning the aforementioned projectability issue.

3. A debate in canonical gravity

The classical canonical formalism of general GR and, by extension, of any generally covariant theory, has been the subject of a long debate regarding its physical intepretation. This debate has not only affected the physicists practitioners in the field but also the philosophers' community, see for instance [12, 13] and references therein. Let us briefly present the two sides of the debate.

The position on one side is that there ought to be no debate at all because the phase space formalism is equivalent to the formalism in configuration-velocity space, and no one has claimed that any interpretational problem exists in the latter framework. Entire books have been devoted to the experimental tests of GR, and this very language implies that observables exist - alive and kicking. Thus the entire debate must be a consequence of misunderstandings.

On the other side of this debate it has been claimed that the phase space formalism suffers from some difficulties regarding the physical intepretation of what constitutes an observable in these theories. Here we tend to be confronted with paradoxical slogans like "frozen time" or the equally mysterious "nothing happens" picture. According to this view there can no be real physical time evolution for these theories. In another contribution to these proceedings [8] it is shown that this problem already long ago attracted the interest of eminent physicists like Peter Bergmann and P.A.M. Dirac. Without mathematical details, which will be given later, let us briefly recall here the two basic arguments that are wielded on this side of the debate.

3.1. The issue of gauge versus dynamics

The gauge generator, which depends on some arbitrary functions (the descriptors), generates active diffeomorphism-induced variations of phase space variables.But this generator becomes exactly the Dirac Hamiltonian when the descriptors take some specific vaules, actually the values of the lapse and shift. Then the argument goes: if the Dirac Hamiltonian is just a particular case of the gauge generator, this means that the time evolution is just gauge, hence "nothing happens". "Time is frozen". (Incidentally, boundary terms in the Dirac Hamiltonian, or in the gauge generators, leave this puzzling situation unaffected, because such terms have no effect either on the gauge transformations or on the dynamics.)

A variant of this argument proceeds by saying that if one quotients out the gauge-equivalent points in phase space so as to eliminate the gauge freedom, points that represent future - or past - evolution with respect to some initial point, belong to the same equivalence class and are thus identified. Hence "time is frozen".

3.2. The issue of observables as constants of motion

Obervables in phase space must be functionals of the fields having vanishing Poisson brackets with the gauge generator. Consequently, as we have seen, they must have vanishing Poisson brackets with the Dirac Hamiltonian as well; therefore the observables can not have implicit time dependence. On the other hand, because of general covariance, gauge-invariant quantities must be independent of the coordinates and thus in particular they can not have explicit time dependence. Both considerations together imply that the observables must be constants of motion without explicit dependence on the coordinates. Hence we encounter again the paradoxical "nothing happens".

Taken at face value, the case for the "nothing happens" picture seems quite strong, perhaps even insurmountable. We argue that this picture arises in fact from some regrettable misunderstandings, particularly on the use of some basic terms like "gauge invariance", or even the more basic notions of "gauge symmetry" or of what is understood classically by a "physical state". Once we agree on the terminology, clarifications immediately follow and the paradoxical picture quietly fades away. The case is closed.

Now we will address and resolve the two paradoxes just mentioned.

3.3. addressing the issue of gauge versus dynamics

The teminology we propose is the standard one. A symmetry is a map sending solutions (entire on-shell field configurations) of the equations of motion (EOM) to solutions. The symmetry will be gauge if it depends on arbitrary functions of the spacetime coordinates and the fields as well. This terminology is standard, as we say, but let us immediately warn the reader that an important source of confusion can be traced to the influential book by Dirac [14], in which gauge tranformations at a fixed time t_0 are considered. Of course there is nothing wrong with it as an alternative definition, but it is a concept rather different from that of mapping solutions to solutions [15].

Since gauge tranformations map solutions to solutions, our natural arena will be the space $\mathcal{S}$ of on-shell field configurations, i.e., fields obeying the EOM - which we take in phase space. This space is a subset of the much bigger space of general field configurations. An infinitesimal gauge transfomation acts on this bigger space with the ordinary Poisson bracket[3], and its action can be restricted to $\mathcal{S}$ because the generators of gauge transformations define an action which

[3] Note that the Poisson bracket is an equal time action. One way of defining its action is to employ a one-parameter family of phase space variables, with parameter t.

is tangent to $\mathcal{S}$. A point p in $\mathcal{S}$ is an entire spacetime[4] with the fields - solution of the EOM - described in a particular coordinatization. For practical purposes, though, it will be enough to work in a coordinate patch of a given chart. To every point p there is associated an "observer", or "user", who is using such a coordinatization to describe the fields in spacetime.

The gauge generator, acting through the equal time Poisson bracket, produces a symmetry - mapping solutions to solutions - when all times t are considered in (4). We need $G_\xi(t)$, for all times t, to move from a point p to a point p' within the gauge orbit. The gauge generator is used to construct finite gauge transformations, thereby realizing active diffeomorphism-induced transformations. These gauge transformations define equivalence classes within $\mathcal{S}$, which we call gauge orbits. A gauge orbit represents a unique physical state[5], and its different points correspond to different coordinatizations. Certainly, one can also pass from one coordinatization to another by a passive diffeomorphism, but in the active view one moves the description from a point p to a point p' by changing the configuration of the fields without touching the coordinatization. This active view is the one that gets naturally realized in phase space with the use of the Poisson bracket.

We are ready to address the purported paradox anticipated in subsection 3.1. If we take the descriptors $\xi^\mu = N^\mu$ (which, according to (2), give $\epsilon^\mu = \delta^\mu_0$, the rigid time translation diffeomorphism), and take into account that the dynamics dictates $\dot{N}^\mu = \lambda^\mu$, then the gauge generator (4) becomes exactly the Hamiltonian (3). This is the *gauge versus dynamics* puzzle: the Hamiltonian is a particular case of the gauge generator. But, is there really any problem? Should this result really be a suprise?

To see that there is no problem at all with such a coincidence, and more than that, that this accordance is as it must be, just observe that in the space of on-shell field configurations the gauge generator moves from one point p to another p', whereas the Hamiltonian works within every point p, which already represents an entire spacetime. The gauge generator and the Hamiltonian can be mathematically the same object, but their duties are distinct.

To illustrate, suppose we are at the point p and we make an active rigid time translation δt of all our fields - which in the passive view is nothing but a specific change of coordinates - we will move to another point, say p', in the gauge orbit, in which all the fields $\Phi_{p'}$ are related to the fields at p according to $\Phi_{p'}(\mathbf{x}, t) = \Phi_p(\mathbf{x}, t + \delta t)$. The gauge transformation preserves the time coordinate and sets $\Phi_p(\mathbf{x}, t) \rightarrow \Phi_{p'}(\mathbf{x}, t)$. This transformation is indeed generated by the Dirac Hamiltonian, obtained from the gauge generator through the replacement (for all t), $\xi^\mu(t) \rightarrow N^\mu(t)$ and $\dot{\xi}^\mu(t) \rightarrow \dot{N}^\mu(t)$. Thus we see that the Dirac Hamiltonian does precisely what it is designed to do - it effects a global rigid translation in time on solutions of the equations of motion. But clearly this transformation does not begin to exhaust the full range of gauge transformations. In particular it does not effect a change in the functions N^μ other than global translation in time, nor the corresponding changes in the remaining phase space variables.

In connection with the analysis above, we can address also the quotienting issue mentioned in the last paragraph in 3.1. We note again that an imprecise terminology is at the root of the problem. Let us momentarily sit in phase space, instead of within the space of on-shell field configurations. A point in phase space is just a configuration of the fields at a given time[6], and it may be intepreted as the setting of initial conditions for the dynamics in order to build an entire spacetime.

To assert, à la Dirac, that there are gauge transformations connecting points in phase space is misleading to say the least and, should such terminology be used, it must be done under very

[4] Spacetimes will always be understood as encompassing also their on-shell field configurations.
[5] Note that this state is the whole spacetime.
[6] This "time" is simply the time coordinate for an arbitrary observer.

strict and cautious qualifications[7]. These transformations are not gauge symmetries in the first place, because they do not map solutions to solutions. Thus the world "gauge" is not used here in the usual sense. What is true is that these transformations relate possible equivalent sets of initial conditions, equivalent in the sense that they build, by way of the dynamics, gauge equivalent spacetimes. It is obvious that, among these related points in phase space, there will be points representing future or past configurations of other points. They all are associated with a single spacetime physics, i.e., a gauge orbit in $\mathcal{S}$. Indeed, after the quotienting procedure, *each point in this quotient space of phase space represents precisely a whole gauge orbit in* $\mathcal{S}$, and it is nonsense to try to implement the dynamics in such space, because the dynamics takes place at each point in the gauge orbit within S.

In addition, one object is the spacetime physics as a whole, represented by the gauge orbit, and another the physics locally experienced by an observer at a given time according to his or her coordinatization, which we may call the "timeslice" physics. For an observer, this "timeslice" physics changes along his or her own time and when all observers agree on using the same intrinsic coordinatization - see section (4) - they all describe the same changes, which become observable. The "frozen time" picture claimed in subsection 3.1 disappears. More details in the appendix of [2].

3.4. Addressing the issue of observables as constants of motion

In subsection (3.2) the claim is made that observables must have vanishing Poisson brackets with the the gauge generator and hence with the Dirac Hamiltonian. This is entirely correct and indisputable because, by its very definition, observables are gauge invariant. But the next claim in subsection 3.2, asserting that an observable can not depend explicitely on the coordinates is wrong. It originates from a confusion between the active and passive view of diffeomorphism invariance. In the passive view the fields are considered always the same, their mathematical description changing according to the use of different coordinatizations. Instead, in the active view, which is the view taken in phase space, a gauge transformation moves from one configuration to a different one without changing the coordinates. Although both views are equivalent, the coordinates themselves, not being variables in phase space, are gauge invariant in the active view. This explains why an observable may depend explicitly on the coordinates. Requiring the obervables to be independent of the coordinates and at the same time to have vanishing Poisson brackets with the gauge generators is to mix the two pictures of passive and active diffeomorphisms. It is too much of a requirement and therefore it is little wonder that paradoxes occur.

In conclusion, the only acceptable requirement on the observables, as a matter of their definition, is that they must have vanishing Poisson brackets with the the gauge generator. But they are not necessarily constants of motion because of their possible explicit dependences on the coordinates. Again, the "nothing happens", "frozen time" picture fades away.

4. Gauge Fixing and scalar coordinatizations

We will focus on the GR case, though our considerations can be easily extended to other generally covariant metric theories. In a striking difference with the case of gauge theories associated with internal symmetries, a mandatory characteristics of the gauge fixing (GF) in generally covariant theories is that it must have explicit dependence on the coordinates. In particular, the explicit dependence on the time coordinate is a necessary condition [16, 17] for the GF to select a unique point in the gauge orbit and yet not to freeze the dynamics. Our GF constraints will be of the

[7] Dirac's incomplete analysis of gauge transformations, and the subsequent misunderstandings generated by his work, is discussed in [15]

type

$$x^\mu - X^\mu(x) = 0\,.$$

In any gauge theory, a GF must comply with these two conditions:

- *Non-degeneracy* (or *uniqueness*), that is, it must single out a unique point in the gauge orbit, and
- *Completeness*, that is, it must completely fix the dynamics.

We examine these two issues separately for our gauge choice.

4.1. Non-degeneracy

4.1.1. Non-degeneracy and scalar coordinatizations From the physical point of view the non-degeneracy of the GF amounts to the condition that all observers obtain exactly the same description for the fields - the same mathematical functions - when all them agree to use the intrinsic coordinatization provided by the GF. We will show that this physical condition of uniqueness in the description is equivalent to the condition that the fields X^μ involved in the GF must be spacetime scalars. We interpret a choice of intrinsic coordinates as a coordinate transformation from the coordinates x^μ to $X^\mu_x(x)$, where we momentarily label with a subscript x the explicit functions that define the fields[8] X^μ as described by the observer who uses the coordinates x^μ (the x-observer). Suppose that instead of starting with coordinates x^μ we start instead with coordinates $y^\mu = f^\mu(x)$, for which the very same fields defining the intrinsic coordinatization are given by the functions $X^\mu_y(y)$. As seen by the y-observer, taking the intrinsic coordinatization amounts to the coordinate transformation from the coordinates y^μ to $X^\mu_y(y)$. Then the demand that the intrinsic coordinatization is the same for all observers is the demand that the coordinate transformation from X^μ_x to X^μ_y must be the identity transformation, i.e.,

$$X^\mu_x(x) = X^\mu_y(y)\,,$$

which is the condition that the fields X^μ - which take the functional from X^μ_x for the x-observer and X^μ_y for the y-observer - are spacetime scalars.

4.1.2. More on non-degeneracy In addition to the argument above, one can easily show that there is no degeneracy, at least locally - that is, in a given coordinate patch - when we use scalars satisfying the completeness condition. Indeed, suppose that the GF constraints $x^\mu - X^\mu_x(x) = 0$ are satisfied for coordinates x^μ, with X^μ_x being, as before, the functions that express the scalar fields X^μ in the x coordinates. Let $x \to y$ be a change of coordinates $y = f(x)$. Let us check whether the GF is satisfied for these new coordinates. In the y-coordinates the scalar fields are $\tilde{X}^\mu_y(y) = X^\mu_x(x)$. Thus, $y^\mu - \tilde{X}^\mu_y(y) = f^\mu(x) - X^\mu_x(x) = f^\mu(x) - x^\mu$, which can only vanish if $y^\mu = x^\mu$.

We now give some examples of degeneracy that arises as a consequence of attempting to use intrinsic coordinates that do not transform as spacetime scalars. Let us start with a "vector coordinatization". The GF constraints are $x^\mu - J^\mu(x) = 0$, where $J^\mu(x)$ are components of a vector $J = J^\mu(x)\partial_{x^\mu}$. Let us proceed in full detail to show the degeneracy of this GF. Consider a linear transformation of the coordinate system, $x \to y$, so that $y^\mu := M^\mu_\nu x^\nu$. Then

$$\partial_{x^\nu} = \frac{\partial y^\mu}{\partial x^\nu}\partial_{y^\mu} = M^\mu_\nu \partial_{y^\mu}\,,$$

and therefore

$$J = J^\nu(x)\partial_{x^\nu} = J^\nu(x)M^\mu_\nu\partial_{y^\mu} =: \tilde{J}^\mu(y)\partial_{y^\mu}\,.$$

[8] We have no preconception as to whether the X^μ are fields, or field components, or local functionals of the fields. We simply call them "fields".

We thus infer the obvious result that in the new coordinates y^μ the components of the vector field are $\tilde{J}^\mu(y) = M^\mu_\nu J^\nu(x)$. Notice then that

$$y^\mu - \tilde{J}^\mu(y) = M^\mu_\nu(x^\nu - J^\nu(x)) = 0\,.$$

Thus if the GF constraints are satisfied for the x^μ coordinates, they will also be satisfied for the y^μ coordinates. In consequence the GF is degenerate and hence it is not a good GF because it does not select a unique coordinatization. Note that this is a $GL(4,R)$ degeneracy, because matrices M are only required to be invertible.

The following example uses a "tensor coordinatization". We take the GF to be $x^\mu - T^{0\mu}(x) = 0$ with $T^{0\mu}$ components of a tensor field. Considering again a linear transformation of coordinates $y^\mu := M^\mu_\nu x^\nu$, we now will have $\tilde{T}^{\mu\nu}(y) = M^\mu_\rho M^\nu_\sigma T^{\rho\sigma}(x)$, and in particular $\tilde{T}^{0\nu}(y) = M^0_\rho M^\nu_\sigma T^{\rho\sigma}(x)$. Let us choose M such that $M^0_\rho = \delta^0_\rho$. Then,

$$\tilde{T}^{0\nu}(y) = M^\nu_\sigma T^{0\sigma}(x)\,.$$

Consequently, $y^\mu - \tilde{T}^{0\mu}(y) = M^\nu_\sigma(x^\sigma - T^{0\sigma}(x)) = 0$, which explicitely shows the degeneracy of the GF for matrices M satisfying $M^0_\rho = \delta^0_\rho$. The GF $x^\mu - T^{\mu\mu}(x) = 0$ can also easily be shown to be degenerate.

Although we do not claim to have obtained a general proof that all non-scalar coordinatizations are degenerate, it is clear from these examples that the procedure to show such a degeneracy is generalizable to many other cases, though it seems that one has to work them on a case by case basis. In agreement with the physical picture obtained in the previous subsection 4.1.1, we will assume from now on that only a GF made with scalar fields is acceptable.

4.1.3. More on scalars in phase space Consider general relativity coupled with matter with a non derivative coupling and with no additional gauge freedom. Here we prove that a scalar field in phase space - perhaps made with a local functional of the fields - can not depend on the lapse and shift. In fact, under an infinitesimal diffeomorphism ϵ^μ, a generic scalar Φ transforms as $\delta\Phi = \epsilon^\mu \partial_\mu \Phi$, which, using (2), can be written

$$\delta\Phi = \xi^0 n^\mu \partial_\mu \Phi + \xi^i \partial_i \Phi\,. \tag{5}$$

On the other hand, we should be able to find the same transformation by acting with the gauge generator (4), that is,

$$\delta\Phi = \{\Phi,\, G_\xi\} = \xi^\mu\{\Phi,\, \mathcal{H}_\mu + N^\rho C^\nu_{\mu\rho} P_\nu\} + \dot{\xi}^\mu\{\Phi,\, P_\mu\}\,. \tag{6}$$

Due to the arbitrariness of the descriptors ξ^μ, the equality between (5) and (6) implies $\{\Phi,\, P_\mu\} = 0$, thus the scalar Φ can not functionally depend on the lapse and shift.

As a simple example consider the scalar field matter Lagrangian $\mathcal{L}_m = \frac{\sqrt{-g}}{2} g^{\mu\nu}\partial_\mu\phi\partial_\nu\phi$. The ADM decomposition

$$g_{\mu\nu} = \left(\begin{array}{c|c} -N^2 + N^a N_a & N_a \\ \hline N_a & \gamma_{ab} \end{array}\right). \tag{7}$$

implies $\sqrt{-g} = N\sqrt{\gamma}$. The momentum $\pi = \frac{\partial \mathcal{L}_m}{\partial \dot{\phi}}$ permits us to solve for $\dot{\phi}$ in terms of canonical variables as $\dot{\phi} = N^a \partial_a \phi - \frac{N}{\sqrt{\gamma}}\pi$. Then, the scalar $g^{\mu\nu}\partial_\mu\phi\partial_\nu\phi$ becomes, in canonical variables, $\gamma^{abj}\partial_a\phi\partial_b\phi - \frac{\pi^2}{\gamma}$, which is free of dependences on the lapse and shift.

4.2. Completeness

The completeness condition is

$$|\{X^\mu, \mathcal{H}_\nu\}| \neq 0 \,, \tag{8}$$

which essentially indicates that the intrinsic coordinates X^μ evolve under motions generated by the Hamiltonian and momentum constraints. Completeness is then proven as follows. The stabilization of the GF constraints gives new secondary GF constraints which allow for the determination of the lapse and shift; this precisely requires that the matrix function $\{X^\mu, \mathcal{H}_\nu\}$ be invertible. Next, the stabilization of these secondary GF constraints determines the arbitrary functions present in the Dirac Hamiltonian. Thus the dynamics is totally fixed.

In addition to fixing the dynamics, condition (8) guarantees that the fields X^μ are indeed a good local coordinatization, as we now show. One has, specializing the gauge generator (4) to spacetime translations ($\epsilon^\mu = \delta^\mu_0$ for time translations and $\epsilon^\mu = \delta^\mu_i$ for spatial ones),

$$\partial_\nu X^\mu = \mathcal{A}^\mu{}_\rho \mathcal{M}^\rho{}_\nu \,,$$

with $\mathcal{A}$ being the invertible matrix (with discrete as well as continuous indices)

$$\mathcal{A}^\mu{}_\rho := \{X^\mu, \mathcal{H}_\rho\} \,, \tag{9}$$

and

$$\mathcal{M} = \begin{pmatrix} N & \vec{0} \\ \vec{N} & I \end{pmatrix} .$$

Since both $\mathcal{A}$ and $\mathcal{M}$ are invertible (notice that $|\mathcal{M}| = N > 0$), we infer that $\partial_\nu X^\mu$ is invertible. This is the Jacobian of the transformation from coordinates x (those used by an oberver at some arbitrary point p in the gauge orbit) to the intrinsic coordinates defined by the scalars X^μ. Thus, thanks to the completeness condition of the GF, that is, that $\mathcal{A}$ is invertible, we see that the scalars X^μ do indeed define a good local coordinatization.

Note that throughout our analysis global issues concerning the use of a set of four scalar fields to provide for a coordinatization of the whole spacetime remain unaddressed. In general one should expect the condition $|\{X^\mu, \mathcal{H}_\nu\}| \neq 0$ to be satisfied locally, but perhaps not globally. This means that, in different coordinate patches of the manifold's chart, one may have to use different scalars to bring about an intrinsic coordinatization.

4.3. Gauge fixings in GR

Experimental tests of general relativity abound. One may ask how it is accomplished that these tests are "gauge invariant", that is, that they have a true phyiscal meaning? There are two possible answers. On one hand the result of the measurement might be expressible in a gauge invariant way, for instance when measuring the proper time for a particle trajectory. On the other hand, it might be that all observers agree to use the same scalar coordinatization (intrinsic coordinates) to express their results. This is exactly the GF procedure discussed in section 4. We will see in the next section that, at least at the formal level, it is easy to construct observables out of the GF. Once the connection between observables and GF is understood, it is clear that it is enough to focus on measurements made in the framework of an intrinsic coordinate system. These GF procedures are in accordance with the use of special coordinate systems as is customary in the parametrized post-Newtonian formalism [18, 19] and reflected in the IAU 2000 resolutions [20] for astrometry, celestial mechanics and metrology in the relativistic framework.

We think it is worth mentioning in this context the theoretical realization of an intrinsic coordinate system by Rovelli in [21]. A very similar proposal is the Galactic Positioning System, advocated among others by Charles Misner [22], based on four extremely stable pulsars

- extremely stable meaning that its pulse count differs from its proper time by an affine transformation. For use in the neighborhood of the solar system, they should form the vertices of a topological tetrahedron - which is the configurations also considered by Rovelli - with the Earth near its center.

5. Constructing observables

As a first step towards the explicit form of the functional invariants we impose an intrinsic coordinate-dependent gauge condition of the form considered in section 4, $\chi^{(1)\mu} := x^\mu - X^\mu(x) = 0$, where the X^μ are spacetime scalar functions of the canonical fields. Our task is then to find the canonical transformation that moves the field variables to that location on the gauge orbit where the gauge conditions are satisfied. Once we are in possession of this finite transformation we may employ it to transform all the remaining fields. These actively transformed fields, when considered as functionals of the original fields, are the invariants that we seek.

We note that preservation of the gauge conditions under temporal evolution leads to additional constraints $\chi^{(2)\mu} := \delta^\mu_0 - \mathcal{A}^\mu{}_\nu N^\nu \approx 0$, where the matrix $\mathcal{A}^\mu{}_\nu$ has been introduced in (9). Following the lead of Henneaux and Teitelboim[23], further exploited by Dittrich[24] and Thiemann[25], we find it convenient to work with linear combinations $\overline{\zeta}_{(i)\nu}$ of the first class constraints $\zeta_{(1)\mu} := \mathcal{H}_\mu, \zeta_{(2)\mu} := P_\mu$ having the property that $\left\{\chi^{(i)\mu}, \overline{\zeta}_{(j)\nu}\right\} \approx -\delta^i_j\delta^\mu_\nu$. For this purpose we need the inverse of the matrix $\left\{\chi^{(i)\mu}, \zeta_{(j)\nu}\right\}$. The appropriate linear combinations are therefore $\overline{P}_\mu = \mathcal{B}^\nu{}_\mu P_\nu$, where $\mathcal{B}^\alpha{}_\beta$ is the inverse of $\mathcal{A}^\mu{}_\nu$, and $\overline{\mathcal{H}}_\nu = \mathcal{B}^\rho{}_\nu\left(\mathcal{H}_\rho - \mathcal{B}^\mu{}_\lambda N^\sigma\{A^\lambda_\sigma, \mathcal{H}_\rho\}P_\mu\right)$. This new basis exists in general only locally. Letting $\xi = \mathcal{B}^\mu{}_\sigma\overline{\xi}^\sigma$, the gauge generator (4) expressed in terms of this new basis is, up to quadratic terms in the constraints that can be discarded "on-shell" [2],

$$G_{\overline{\xi}}(t) = \overline{P}_\nu\dot{\overline{\xi}}^\nu + \overline{\mathcal{H}}_\nu\overline{\xi}^\nu. \tag{10}$$

The finite active gauge transformation of any dynamical field Φ takes the form

$$exp\left(\{-, G_{\overline{\xi}}\}\right)\Phi = \Phi + \left\{\Phi, G_{\overline{\xi}}\right\} + \frac{1}{2}\left\{\left\{\Phi, G_{\overline{\xi}}\right\}, G_{\overline{\xi}}\right\} + \cdots \tag{11}$$

Since the scalars X^μ can not depend on the lapse and shift (see subsection 4.1.3) the descriptors for the finite gauge transformation that transforms X^μ to x^μ are easily seen to satisfy, on shell,

$$x^\mu = exp\left(\{-, G_{\overline{\xi}}\}\right)X^\mu = X^\mu + \overline{\xi}^\mu, \tag{12}$$

where $\overline{\xi}^\mu := \mathcal{A}^\nu{}_\sigma\xi^\sigma$ is considered a function of the spacetime coordinates and we have made use of the fact that the Poisson brackets $\overline{\mathcal{H}}_\mu$ with themselves vanish "strongly", i.e., they are proportional to terms at least quadratic in the $\overline{\mathcal{H}}_\nu$. The descriptors $\overline{\xi}^\mu = \chi^{(1)\mu}$ and $\dot{\overline{\xi}}^\mu = \chi^{(2)\mu}$ may therefore be substituted into (11), after the Poisson brackets have been computed, to obtain all invariant functionals $\mathcal{I}_\Phi$ associated with the fields Φ.

Although the following results hold for all fields [2] we focus here on fields other than the lapse and shift. Then the explicit expressions for the invariants are

$$\begin{aligned}\mathcal{I}_\Phi &\approx \Phi + \chi^{(1)\mu}\{\Phi, \overline{\mathcal{H}}_\mu\} + \frac{1}{2!}\chi^{(1)\mu}\chi^{(1)\nu}\{\{\Phi, \overline{\mathcal{H}}_\mu\}, \overline{\mathcal{H}}_\nu\} + \cdots \\ &=: \sum_{n=0}^{\infty}\frac{1}{n!}(\chi^{(1)})^n\{\Phi, \overline{\mathcal{H}}\}_{(n)},\end{aligned} \tag{13}$$

where $\approx$ is the symbol for Dirac's weak equality, that is, an equality which holds "on-shell".

With different notation, this expression (13) appeared in the literature in [25] as his equation (2.8) and in [24] as her equation (5.23). Here we have arrived at (13) by a symmetry-inspired procedure, as the effect of the finite gauge transformation that sends any point in the space of on shell field configurations to the specific point satisfying the GF condition. This specific gauge transformation is determined once the set of scalar fields associated with the gauge fixing has been selected. An advantage of the present formulation is that one can send all the fields from the coordinate description of any observer to the description given at the point of the gauge orbit where the gauge conditions are satisfied, and this includes the lapse and shift.

The time rate of change of these invariant functionals satisfies [2]

$$\frac{d}{dt}\mathcal{I}_\Phi \approx \mathcal{I}_{\{\Phi,\,\overline{\mathcal{H}}_0\}}\,. \tag{14}$$

Recognizing that the invariant fields are simply the fields expressed in intrinsic coordinates, that is, the fields as seen by the observer sitting at the point in the gauge orbit where the GF constraints are satisfied, we note that $\mathcal{I}_{\{\Phi,\overline{\mathcal{H}}_0\}} = \{\Phi,\,\overline{\mathcal{H}}_0\}_{\mathcal{P}_G} = \{\Phi,\,N^\mu \mathcal{H}_\mu\}_{\mathcal{P}_G}$. Therefore the invariants satisfy the equations of motion of the gauge fixed fields. We note also that repeated time derivatives of the invariant fields yield the simple expression $\frac{\partial^n}{\partial t^n}\mathcal{I}_\Phi \approx \mathcal{I}_{\{\Phi,\overline{\mathcal{H}}_0\}_{(n)}}$, and that we may therefore express the invariants as follows as a Taylor series in t:

$$\mathcal{I}_\Phi \approx \sum_{n=0}^{\infty} \frac{t^n}{n!}\mathcal{I}_{\{\Phi,\,\overline{\mathcal{H}}_0\}_{(n)}}\Big|_{t=0}\,. \tag{15}$$

This expression displays in a striking manner the notion of *evolving constant of the motion* introduced originally by Rovelli[26]. We stress however, that the evolution we obtain here is nothing other than the evolution determined by Einstein's equations in the gauge-fixed coordinate system. In addition, the potentially infinite set of invariants $\mathcal{I}_{\{\Phi,\overline{\mathcal{H}}_0\}_{(n)}}$ will correspond to Noether symmetries [27, 2]. In generic general relativity the conserved quantities must necessarily be spatially non-local, as pointed out by Torre[28].

Notice that the coefficients of the expansion (15) in the time parameter are constants of motion, because they are invariants without explicit time dependence and, since they have vanishing Poisson brackets with the Hamiltonians, they have no implicit time dependence either. In fact, fixing the explicit time parameter at any arbitrary value, the invariants $\mathcal{I}_\Phi$ become constants of motion.

To close this section we mention two relevant results form [2].

Clearly, the map $\Phi \to \mathcal{I}_\Phi$ is not a canonical transformation because it sends all the gauge equivalent configurations to the same configuration, the one satisfying the gauge fixing conditions. It can nevertheless be understood as a limit of a one-parameter family of canonical maps. Let us consider the functionals, obtained from canonical maps, $\mathcal{K}_\Phi^{(\Lambda)} = exp\left(\{-,\Lambda\,\chi^{(1)\nu}\overline{\mathcal{H}}_\nu\}\right)\Phi$. One can prove that, on shell, [2]

$$\mathcal{K}_\Phi^{(\Lambda)} \approx \sum_{n=0}^{\infty}\frac{1}{n!}(1-e^{-\Lambda})^n(\chi^{(1)})^n\{\Phi,\,\overline{\mathcal{H}}\}_{(n)}\,, \tag{16}$$

from which it follows that

$$\lim_{\Lambda\to\infty}\mathcal{K}_\Phi^{(\Lambda)} \approx \mathcal{I}_\Phi\,.$$

The second result is the observation that the Poisson brackets of the invariants are simply the invariants associated with the Dirac brackets of the fields. This result, already obtained in

[25] for fields other than lapse and shift, holds in fact for for all fields,

$$\{\mathcal{I}_{\Phi^A},\, \mathcal{I}_{\Phi^B}\} \approx \mathcal{I}_{\{\Phi^A,\, \Phi^B\}^*}\,. \tag{17}$$

6. An example: observables for a massive particle in an AdS background

Consider the massive free particle in AdS spacetime. In Poincaré coordinates (actually Poincaré coordinates use $u = \frac{1}{r}$),

$$ds^2 = g_{\mu\nu}dq^\mu dq^\nu = \frac{R^2}{r^2}(\eta_{\mu\nu}dq^\mu dq^\nu) = \frac{R^2}{r^2}(\eta_{\alpha\beta}dq^\alpha dq^\beta + dr^2)$$

with $\mu = (\alpha, n-1)$, $\quad r = q^{n-1}$, $\quad \alpha = (0, \ldots, n-2)$, $\quad \eta_{\alpha\beta} = (-, +, \ldots, +)$. These coordinates only describe a patch of the whole AdS spacetime. Now $r \geq 0$ and $r = 0$ is the boundary.

The Lagrangian for the massive particle in the AdS background is

$$L = \frac{1}{2\,N} g_{\mu\nu}\dot{q}^\mu \dot{q}^\nu - \frac{1}{2}m^2 N\,,$$

with N an auxilary variable. The Dirac Hamiltonian is

$$H_D = \frac{1}{2}N(g^{\mu\nu}p_\mu p_\nu + m^2) + \lambda\pi,$$

where π, the momentum canonically conjugate to N, is the primary constraint, and λ an arbitrary function of time. There is a secondary constraint, namely $\mathcal{H} = \frac{1}{2}(g^{\mu\nu}p_\mu p_\nu + m^2)$. The gauge generator has the form $G = \xi\mathcal{H} + \dot{\xi}\pi$. We choose as a gauge fixing constraint $\chi = t - q^0$. Next, following the usual procedure, we define $A := \{q^0, \mathcal{H}\} = \dfrac{r^2}{R^2}p^0$ (for our purposes, here and henceforth, indices are raised and lowered exclusively with $\eta^{\mu\nu}$, $\eta_{\mu\nu}$.), and

$$\overline{\mathcal{H}} := \frac{1}{A}\mathcal{H} = \frac{1}{2\,p^0}(\eta^{\mu\nu}p_\mu p_\nu + \frac{R^2 m^2}{r^2})\,.$$

Now we are ready to compute the invariants associated with the coordinates and the momenta. Note that we do not write the implicit time dependence in the variables, which is the same as if we were working just in phase space instead of working in the space of trajectories - i.e., field configurations.

Most of the series expansions are trivial. So we get (for $\alpha = (0, i)$)

$$\begin{aligned} \mathcal{I}_{p_\alpha} &= p_\alpha\,, \\ \mathcal{I}_{q^i} &= q^i + \chi\{q^i, \overline{\mathcal{H}}\} = q^i + (t - q^0)\frac{p^i}{p^0} \\ &= (q^i - \frac{p^i}{p^0}q^0) + \frac{p^i}{p^0}t\,. \end{aligned} \tag{18}$$

Thus we extract from $\mathcal{I}_{p_\alpha}$ the constants of motion (CM) p_α. On the other hand, the expansion in the t parameter for $\mathcal{I}_{q^i}$ gives the CM $q^i - \frac{p^i}{p^0}q^0$ and $\frac{p^i}{p^0}$. Using the CM p_α we can identify the CM $C^{0\alpha} := q^0 p^\alpha - p^0 q^\alpha$. To determine the rest of the basic invariants it is convenient to compute $\mathcal{I}_{r^2}$ instead of $\mathcal{I}_r$ (In fact $\mathcal{I}_{r^2} = (\mathcal{I}_r)^2$). We get the finite series

$$\mathcal{I}_{r^2} = r^2 + 2\,\frac{t - q^0}{p^0}\,r\,p_r + (\frac{t - q^0}{p^0})^2(p_r^2 + \frac{R^2\,m^2}{r^2})\,,$$

out of which we identify, as coefficients of the t-expansion, the constants of motion

$$\begin{aligned} C_{(1)} &:= p_r^2 + \frac{R^2\,m^2}{r^2} \\ C_{(2)} &:= r\,p_r - \frac{q^0}{p^0}(p_r^2 + \frac{R^2\,m^2}{r^2}) \\ C_{(3)} &:= r^2 - 2\,\frac{q^0}{p^0}\,r\,p_r + (\frac{q^0}{p^0})^2(p_r^2 + \frac{R^2\,m^2}{r^2})\,. \end{aligned} \tag{19}$$

Note that, due to the Hamiltonian constraint, $C_{(1)} \simeq -\eta^{\alpha\beta}p_\alpha p_\beta$, that is, $C_{(1)}$ is already a combination of other CM. On the other hand, $\frac{1}{2\,p^0}C_{(1)}\,C_{(3)} = C^2_{(2)} + R^2\,m^2$; therefore we only get $C_{(2)}$ as a new constant of motion. Note also that, using the Hamiltonian constraint,

$$\begin{aligned} C_{(2)} &\simeq r\,p_r + \frac{q^0}{p^0}(\eta^{\alpha\beta}p_\alpha p_\beta) = r\,p_r + \frac{1}{p^0}p_\alpha(q^0p^\alpha) \\ &= r\,p_r + \frac{1}{p^0}p_\alpha(C^{0\alpha} + p^0q^\alpha) = \frac{p_\alpha}{p^0}C^{0\alpha} + p_\mu q^\mu\,, \end{aligned} \tag{20}$$

thus we have obtained the CM $S := p_\mu q^\mu$, which is the generator of the scale transformations.

Finally, due to the Hamiltonian constraint, the computation of $\mathcal{I}_{p_r}$, or better $\mathcal{I}_{p_r^2}$, does not yield any new CM's. Nothing new appears either for $\mathcal{I}_N$.

Summarizing, we have obtained $n-1$ momenta, p_α, $n-2$ boost generators, $C^{0\,i}$, and the generator S of scale transformations. These constants of motion are functionally independent and they number $2n-2$. Fixing the values of these CM - such that they satisfy the Hamitonian constraint[9] - does not determine completely the trajectory of the particle. For this we need another independent CM. Actually, it is provided by one of the generators of special conformal transformations. We may take, for instance: $C^0 := 2q^0S - p^0\eta_{\mu\nu}q^\mu q^\nu$. Once we are in possession of $2n-1$ independent CM, since they fix a trajectory for the particle in phase space - again, chosing values compatible with the Hamiltonian constraint - it is clear that all other CM must be functionally dependent of this set of $2n-1$ independent CM. For instance, the generators of boosts and rotations, $C^{\alpha\beta} := q^\alpha p^\beta - p^\alpha q^\beta$, can be written as $C^{\alpha\beta} = \frac{1}{p^0}(p^\alpha C^{0\,\beta} - p^\beta C^{0\,\alpha})$, and the generators of special conformal transformations, $C^\alpha := 2q^\alpha S - p^\alpha\eta_{\mu\nu}q^\mu q^\nu$, are related to other CM by

$$C^\alpha - \frac{1}{p^0}(p^\alpha C^0 - 2C^{0\,\alpha}S) = 0\,.$$

One can also verify the relationship among these CM

$$p^\alpha C^\beta - p^\beta C^\alpha + 2C^{\alpha\beta}S = 0\,.$$

This discussion for the AdS background is in contrast with the study of the massive free particle in a Minkowski background, undertaken in [2]. In this case, with the same procedure outlined above, one obtains as CM extracted from the observables, n generators of translations and $n-1$ generators of boosts, totaling $2n-1$ independent CM. With the usual restriction of imposing compatiblility with the Hamiltonian constraint, they will determine a trajectory in phase space. All other CM must then be functionally dependent of the CM derived from the observables.

[9] In fact we only need to care that $\eta^{\mu\nu}p_\mu p_\nu < 0$ because the mass m is arbitrary and it does not appear in the CM mentioned above.

6.1. Constants of motion and Killing vectors

It is easy to show that the CM that happen to be linear in the momenta are associated with the Killing symmetries of the background. Indeed, consider $K^\mu(q)p_\mu$ a candidate to be a CM. Its Poisson bracket with the Hamiltonian constraint is

$$\{K^\rho(q)p_\rho, \mathcal{H}\} = \{K^\rho(q)p_\rho, \frac{1}{2}(g^{\mu\nu}p_\mu p_\nu + m^2)\} = -(\mathcal{L}_K g^{\mu\nu})p_\mu p_\nu$$

where $\mathcal{L}_K g^{\mu\nu}$ is the Lie derivative of the metric under the vector field $K^\rho(q)\partial_\rho$. Thus

$$\{K^\rho(q)p_\rho, \mathcal{H}\} = 0 \quad \Longleftrightarrow \quad \mathcal{L}_K g^{\mu\nu} = 0\,.$$

Applied to our AdS case, we see that our observables account for the Killing vectors associated with translations (α directions), boost (i directions), rotations (i, j directions), and the scale transformation. The only missing Killing vectors are those associated with the special conformal transformations, which have not been obtained within our method. It is perhaps possible that another choice for the gauge fixing could account for these CM, but we do not know for sure. In fact, the choice $\chi = t - r$ as the gauge fixing yields the observable

$$\mathcal{I}_{p_r^2} = p_r^2 + \frac{R^2\,m^2}{r^2} - \frac{R^2\,m^2}{t^2}\,,$$

which does not introduce new CM.

7. Conclusion: a case closed

We have shown that it is possible to construct, albeit in a formal way, observables in general relativity by employing a gauge fixing using a scalar coordinatization. In this way we obtain a new understanding as to why finding observables in generally covariant theories is such a difficult mathematical task. But once these two points - the existence and the difficulty of construction of observables - have been made, a new vision emerges: that constructing these observables through the use of active diffeomorphism-induced symmetry transformations - which are valid for every observer with his/her own coordinatization - is not the most efficient procedure. Indeed, once we have proven that observables can be built for any observer, we can gladly dispose of this construction and just take the passive view of diffeomorphism invariance. We simply instruct each observer, having constructed his or her phase space solutions, to transform them to the intrinsic coordinate system! We have indeed proven that the final result is coincident with the active construction. Of course one benefit of the active construction is that it can be used to prove that the resulting invariants have the property that their Poisson brackets are precisely the Dirac brackets of the invariant associate with their ordinary Poisson brackets. This construction procedure was in fact undertaken in [3] for the case of the relativistic free particle and in [29] for a homogeneous anisotropic cosmological model with a massless scalar field source.

In the gauge orbit picture within the space of on shell field configurations, this agreement has the effect of choosing the gauge fixed description, that is, of moving to the only point in the gauge orbit where the GF constraints are satisfied. Indeed, fixing the gauge by selecting a coordinate system is the standard procedure used for the experimental tests of general relativity [18, 19].

Thus here is the guiding principle: *let everyone adopt the same instrinsic coordinates.* Once this instruction is implemented all geometric onjects becomes observable! All observers attain the same description regardless of the coordinate system with which they begin their construction. In other words, the final description is invariant under alternations in this iniital arbitrary coordinate choice.

In this sense we believe that our analysis of the connection between the gauge fixing procedures and the construction of observables closes the case of the problem of observables in general relativity. The accompanying "mysteries" of "frozen time" and "nothing happens" have faded away. Indeed, the case is closed.

References

[1] J M Pons, D C Salisbury and K A Sundermeyer 2009 "Gravitational observables, intrinsic coordinates, and canonical maps,"*Mod. Phys. Lett. A*,**24**, 725-732 [arXiv:0902.0401 [gr-qc]].
[2] J M Pons, D C Salisbury and K A Sundermeyer 2009 "Revisiting observables in generally covariant theories in the light of gauge fixing methods," *Phys. Rev. D* **80** 084015 [arXiv:0905.4564 [gr-qc]].
[3] J M Pons and D C Salisbury 2005 "The issue of time in generally covariant theories and the Komar-Bergmann approach to observables in general relativity" *Phys. Rev* **D71**, 124012 [arXiv:gr-qc/0503013]
[4] P G Bergmann and A Komar 1972 "The coordinate group symmetries of general relativity" *Int. J. Th. Phys.* **5** 15-28
[5] D C Salisbury and K. Sundermeyer 1983 "The Realization In Phase Space Of General Coordinate Transformations" *Phys. Rev. D* **27** 740
[6] J M Pons, D C Salisbury and L C Shepley 1997 "Gauge transformations in the Lagrangian and Hamiltonian formalisms of generally covariant theories" *Phys. Rev.* **D55** 658-668 [arXiv:gr-qc/9612037]
[7] D C Salisbury 2008 "Rosenfeld, Bergmann, Dirac and the Invention of Constrained Hamiltonian Dynamics," *Proc. of the MG11 Meeting on General Relativity* Berlin, ed H Kleinert, R Jantzen and R Ruffini, World Scientific, New Jersey, 2467 [arXiv:physics/0701299]
[8] D. C. Salisbury, "Léon Rosenfeld's pioneering steps toward a quantum theory of gravity"
[9] R. L. Arnowitt, S. Deser and C. W. Misner, "The dynamics of general relativity," arXiv:gr-qc/0405109.
[10] C Teitelboim 1973 "How commutators of constraints reflect the space-time structure" *Annals Phys.* **79** 542-557
[11] J M Pons 2003 "Generally covariant theories: The Noether obstruction for realizing certain space-time diffeomorphisms in phase space," *Class. Quant. Grav.* **20** 3279 [arXiv:gr-qc/0306035]
[12] J Earman 2002 "Thoroughly Modern McTaggart: Or, What McTaggart Would Have Said if He Had Read the General Theory of Relativity" *Philosopher's Imprint* **2** 3
[13] T Maudlin 2002 "Thoroughly Muddled McTaggart: Or, How to Abuse Gauge Freedom to Create Metaphysical Monostrosities" *Philosopher's Imprint* **2** 4
[14] P A M Dirac 1964 *Lectures on quantum mechanics* Yeshiva Univ. Press, New York.
[15] J M Pons 2005 "On Dirac's incomplete analysis of gauge transformations" *Stud. Hist. Philos. Mod. Phys.* **36** 491 [arXiv:physics/0409076].
[16] K Sundermeyer 1982 "Constrained Dynamics - with Applications To Yang-Mills Theory, General Relativity, Classical Spin, Dual String Model" *Lect. Notes Phys.* **169**
[17] J M Pons and L C Shepley 1995 "Evolutionary laws, initial conditions, and gauge fixing in constrained systems" *Class. Quant. Grav.* **12** 1771 [arXiv:gr-qc/9508052]
[18] C M Will 1993 *Theory and experiment in gravitational physics*, Cambridge University Press, Cambridge, U.K.; New York, U.S.A., 2nd edition.
[19] C M Will 2005 "The confrontation between general relativity and experiment," Living Rev. Rel. **9** 3 [arXiv:gr-qc/0510072].
[20] M Soffel *et al.* 2003 "The IAU 2000 resolutions for astrometry, celestial mechanics and metrology in the relativistic framework: explanatory supplement," Astron. J. **126** 2687 [arXiv:astro-ph/0303376].
[21] C Rovelli 2002 "GPS observables in general relativity," Phys. Rev. D **65** 044017 [arXiv:gr-qc/0110003].
[22] Ch Misner, private communication
[23] M Henneaux and C Teitelboim 1994 *The Quantization of Gauge Systems* Princeton University Press
[24] B Dittrich 2007 "Partial and complete observables for Hamiltonian constrained systems" *Gen. Rel. Grav.* **39**, 1891 [arXiv:gr-qc/0411013]
[25] T Thiemann 2006 "Reduced phase space quantization and Dirac observables" *Class. Quant. Grav.* **23** 1163 [arXiv:gr-qc/0411031]
[26] C Rovelli 1990 "Quantum mechanics without time: A model" *Phys. Rev. D* **42** 2638
[27] C Batlle, J Gomis, X Gracia and J M Pons 1989 "Noether's theorem and gauge transformations: applications to the bosonic string and CP(2)**(n-1) model," *J. Math. Phys.* **30** 1345
[28] C G Torre 1993 "Gravitational observables and local symmetries," *Phys. Rev. D* **48** R2373 [arXiv:gr-qc/9306030]
[29] D C Salisbury, J Helpert and A. Schmitz 2008 "Reparameterization invariants for anisotropic Bianchi I cosmology with a massless scalar source," *Gen. Rel. Grav.* **40** 1475 [arXiv:gr-qc/0503014]

Journal of Physics: Conference Series **222** (2010) 012019 doi:10.1088/1742-6596/222/1/012019

Gravitating Axially-Symmetric Monopole-Antimonopole Configurations

Ya. Shnir

Institute of Physics, Jagiellonian University, Krakow, Poland

E-mail: shnir@maths.tcd.ie

Abstract. We discuss static axially symmetric solutions of $SU(2)$ Einstein-Yang-Mills-Higgs theory. These regular asymptotically flat solutions represent monopole-antimonopole chain and vortex ring solutions, as well as new configurations, present only for larger values of the scalar coupling λ. When gravity is coupled to the Yang-Mills-Higgs system, branches of gravitating solutions emerge from the flat-space solutions, and extend up to critical values of the gravitational coupling constant. For small scalar coupling only two branches of gravitating solutions exist, where the second branch connects to a generalized Bartnik-McKinnon solution. For large scalar coupling, however, a plethora of gravitating branches can be present and indicate the emergence of new flat-space branches.

1. Introduction

The 't Hooft-Polyakov monopole [1] represents but the simplest of a rich variety of regular classical non-perturbative finite energy solutions of $SU(2)$ Yang-Mills-Higgs (YMH) theory. Besides this spherically symmetric solution, axially symmetric multimonopoles [2, 3, 4, 5], monopole-antimonopole pairs (MAPs) [6, 7, 8], monopole-antimonopole chains (MACs) [9], and vortex ring solutions [9, 10] are known, as well as multimonopole solutions with only discrete symmetries [11].

In the Bogomol'nyi-Prasad-Sommerfield (BPS) limit of vanishing Higgs potential, the repulsive and attractive forces between two monopoles exactly compensate for any separation, thus BPS monopoles experience no net interaction [12]. As the scalar field becomes massive, the fine balance of forces between the monopoles is broken, since the attractive Yukawa interaction becomes short-ranged. Consequently non-BPS monopoles experience repulsion [5]. This repulsion can be overcome by the inclusion of gravity, which allows for bound multimonopoles [13].

In monopole and multimonopole solutions, the nodes of the Higgs field are associated with the location of the magnetic charges. This also holds for monopole-antimonopole pair and chain solutions, where m monopoles and antimonopoles are located symmetrically on the symmetry axis in alternating order [6, 7, 8, 9]. When the magnetic charge n of the individual monopoles and antimonopoles becomes large, the balance between the repulsive and attractive interactions of the monopoles and antimonopoles is shifted. For vanishing Higgs potential, the structure of the solutions then changes completely, and the zeros of the Higgs field form one or more rings, centered around the symmetry axis. Thus vortex ring solutions arise [9].

When the scalar coupling λ is small, the transition from MAPs and MACs to vortex ring solutions appears for magnetic charge $n = 3$ [9]. For larger scalar coupling λ, however, bifurcations arise, where new pairs of solutions appear [10]. Thus MAP or MAC solutions and vortex ring solutions may coexist beyond some critical value of λ. Moreover new configurations with mixed node structure arise, possessing isolated nodes on the symmetry axis as well as vortex rings [10]. The strength of the Higgs self-interaction is clearly critical for the existence and the properties of these new equilibrium configurations.

The balance of forces changes again, when gravity is coupled to YMH theory. Depending on the strength of gravity, the structure of the solutions may change again [14]. Also new solutions may arise, which do not possess a flat-space limit [16, 17, 18]. But gravity cannot become too strong, for such regular equilibrium configurations to exist, since beyond a certain gravitational coupling strength, a horizon is expected to form [16].

For gravitating monopoles and multimonopoles a degenerate horizon forms indeed at a critical value of the gravitational coupling [16, 19, 13]. For monopole-antimonopole systems, however, no horizon is seen to form. Instead the flat-space branch of regular graviting monopoles bifurcates with a second branch at a maximal value of the gravitational coupling. This second branch then extends backwards and, in the limit of vanishing coupling constant α, it becomes linked to a generalized Bartnik-McKinnon (BM) solution of the Einstein-Yang-Mills (EYM) theory [20, 21, 22].

We here address the effect of gravity on the monopole-antimonopole systems at large values of the scalar coupling and discuss a rich pattern of bifurcating branches of solutions arising at finite gravitational coupling strength, which encompasses all possible combinations of isolated nodes and vortex rings.

This review is based on results of the work in collaboration with Burkhard Kleihaus, Jutta Kunz and Ulrike Neemann [9, 10]

2. $SU(2)$ EYMH Action and Axially Symmetric Ansatz

We consider the $SU(2)$ Einstein-Yang-Mills-Higgs theory with action

$$S = \int \left\{ \frac{R}{16\pi G} - \frac{1}{2}\mathrm{Tr}\,(F_{\mu\nu}F^{\mu\nu}) - \frac{1}{4}\mathrm{Tr}\,(D_\mu\Phi\, D^\mu\Phi) - \frac{\lambda}{4}\mathrm{Tr}\left[\left(\Phi^2 - \eta^2\right)^2\right] \right\} \sqrt{-g}d^4x \ , \tag{1}$$

with curvature scalar R, $su(2)$ field strength tensor $F_{\mu\nu} = \partial_\mu A_\nu - \partial_\nu A_\mu + ie[A_\mu, A_\nu]$, gauge potential $A_\mu = A^a_\mu \tau^a/2$, and covariant derivative of the Higgs field $\Phi = \Phi^a\tau^a$ in the adjoint representation $D_\mu\Phi = \partial_\mu\Phi + ie[A_\mu, \Phi]$. Here G and e denote the gravitational and gauge coupling constants, respectively, η denotes the vacuum expectation value of the Higgs field, and λ represents the strength of the scalar coupling.

Variation of the action (1) with respect to the metric $g^{\mu\nu}$ leads to the Einstein equations

$$G_{\mu\nu} = R_{\mu\nu} - \frac{1}{2}g_{\mu\nu}R = 8\pi G T_{\mu\nu} \tag{2}$$

Variation with respect to the gauge field A_μ and the Higgs field Φ leads to the matter field equations,

$$\frac{1}{\sqrt{-g}} D_\mu(\sqrt{-g}F^{\mu\nu}) - \frac{1}{4} ie[\Phi, D^\nu \Phi] = 0 \; , \tag{3}$$

$$\frac{1}{\sqrt{-g}} D_\mu(\sqrt{-g}D^\mu \Phi) + \lambda(\Phi^2 - \eta^2)\Phi = 0 \; , \tag{4}$$

respectively.

To obtain gravitating static axially symmetric solutions, we employ isotropic coordinates [13, 21, 22]. In terms of the spherical coordinates r, θ and φ the isotropic metric reads

$$ds^2 = -f dt^2 + \frac{m}{f} dr^2 + \frac{mr^2}{f} d\theta^2 + \frac{lr^2 \sin^2\theta}{f} d\varphi^2 \; , \tag{5}$$

where the metric functions f, m and l are functions of the coordinates r and θ, only. The z-axis ($\theta = 0, \pi$) represents the symmetry axis.

We take a purely magnetic gauge field, $A_0 = 0$, and parametrize the gauge potential and the Higgs field by the Ansatz [9]

$$A_\mu dx^\mu = \left(\frac{K_1}{r} dr + (1 - K_2) d\theta \right) \frac{\tau_\varphi^{(n)}}{2e} - n \sin\theta \left(K_3 \frac{\tau_r^{(n,m)}}{2e} + K_4 \frac{\tau_\theta^{(n,m)}}{2e} \right) d\varphi \; , \tag{6}$$

$$\Phi = \eta \left(\Phi_1 \tau_r^{(n,m)} + \Phi_2 \tau_\theta^{(n,m)} \right) \; . \tag{7}$$

where the $su(2)$ matrices $\tau_r^{(n,m)}$, $\tau_\theta^{(n,m)}$, and $\tau_\varphi^{(n)}$ are defined as products of the spatial unit vectors

$$\begin{aligned} \hat{e}_r^{(n,m)} &= (\sin(m\theta)\cos(n\varphi), \sin(m\theta)\sin(n\varphi), \cos(m\theta)) \; , \\ \hat{e}_\theta^{(n,m)} &= (\cos(m\theta)\cos(n\varphi), \cos(m\theta)\sin(n\varphi), -\sin(m\theta)) \; , \\ \hat{e}_\varphi^{(n)} &= (-\sin(n\varphi), \cos(n\varphi), 0) \; , \end{aligned} \tag{8}$$

with the Pauli matrices $\tau^a = (\tau_x, \tau_y, \tau_z)$. For $m = 2$, $n = 1$ the Ansatz corresponds to the one for the monopole-antimonopole pair solutions [7, 8, 17], while for $m = 1$, $n > 1$ it corresponds to the Ansatz for axially symmetric multimonopoles [3, 5, 13]. The four gauge field functions K_i and two Higgs field functions Φ_i depend on the coordinates r and θ, only. All profile functions are even or odd w.r.t. reflection symmetry, $z \to -z$. We also fix the residual degrees of freedom by choosing the gauge condition [9] $r\partial_r K_1 - \partial_\theta K_2 = 0$. With this Ansatz the equations of motion reduce to a set of 9 coupled partial differential equations, to be solved numerically subject to the usual set of boundary conditions, discussed in [9, 10]. The numerical calculations are then performed with help of the package FIDISOL, based on the Newton-Raphson iterative procedure [24]. Technically, the system of non-linear elliptic partial differential equations of second order is first discretized on a non-equidistant grid in $\bar{x}$ and θ. Typical grids used have sizes 60×40 covering the integration domain $0 \leq \bar{x} \leq 1$ and $0 \leq \theta \leq \pi/2$. The numerical errors are typically of order of 10^{-4}.

3. Results and discussion

Let us introduce the dimensionless coordinate x and the dimensionless coupling constant α [16], $x = \frac{e\alpha}{\sqrt{4\pi G}} r$, $\alpha = \sqrt{4\pi G}\eta$. The limit $\alpha \to 0$ can be approached in two different ways: $G \to 0$ while the Higgs vacuum expectation value η remains finite (flat-space limit), and $\eta \to 0$ while Newton's constant G remains finite. Corresponding to these two limits two branches of solutions may exist.

The dimensionless mass M of the solutions is obtained from the asymptotic expansion of the metric function f,

$$M = \frac{1}{2\alpha^2} \lim_{x\to\infty} x^2 \partial_x f \ . \tag{9}$$

The asymptotic expansion of the gauge fields yields the dimensionless magnetic charge Q of the solutions,

$$Q = \frac{n}{2} \left[1 - (-1)^m\right] \ , \tag{10}$$

i.e., solutions in the topologically trivial sector have no charge, $Q = 0$, whereas solutions in the non-trivial sectors have charge $Q = n$.

3.1. MAPs and vortex rings: $m = 2$, $n = 3$

3.1.1. Flat space solutions Monopole-antimonopole pairs (MAPs) reside in the topologically trivial sector. These solutions represent unstable equilibrium configurations and can be thought of as consisting of two solitons of individual charges $\pm n$. The position of each constituent can be identified according to the location of the nodes of the Higgs field. The separation between the poles depends on the subtle interplay of forces and thus on the strength of the scalar coupling λ [8].

The MAPs with $n = 1$ exist for arbitrary values of λ. They have two isolated nodes on the symmetry axis, indicating the position of the poles, and their energy density possesses two local maxima, associated with the nodes of the Higgs field [17].

As the charge of the poles increases, the interaction between the nonabelian matter fields becomes stronger than in the case of unit charge constituents, leading to new types of solutions. Indeed, for small and vanishing values of λ, MAPs with $n > 2$ no longer exist. Instead solutions of vortex ring type emerge, where the Higgs field vanishes on a ring centred around the symmetry axis in the xy plane [9]. These flat-space vortex ring solutions form the fundamental λ-branches, which persist for arbitrary values of λ. As the scalar coupling strength increases, however, bifurcations are observed, where pairs of additional flat-space solutions emerge at critical values of λ [10].

For $n = 3$, in particular, a pair of MAP solutions arises at the critical value $\lambda_c^1 = 1.382$. The λ-dependence of these flat-space solutions is demonstrated in Fig. 1, where we show their mass and the location of the nodes of their Higgs field configurations. At the bifurcation, the mass of the new MAP solutions is higher than the mass of the vortex ring solution. For large values of λ, however, the MAP solution on the lower mass branch becomes energetically favoured. Note, that the node structure of the solutions on the upper mass branch changes from MAP to vortex ring, as λ increases, but the radius of the ring remains small.

3.1.2. Gravitating solutions When gravity is coupled, branches of gravitating solutions arise from these flat-space configurations. For sufficiently large gravitational coupling α, however, a

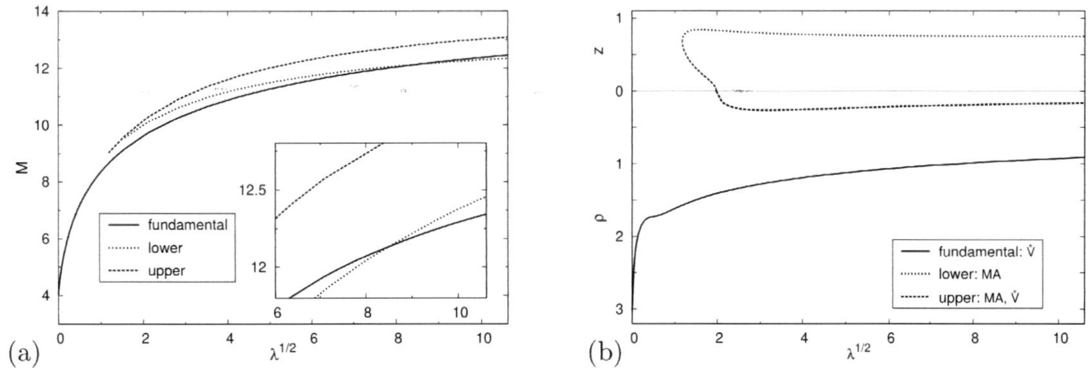

Figure 1. (a) The mass M of the flat-space solutions on the fundamental λ-branch as well as on the two additional λ-branches for $m = 2$, $n = 3$ solutions. (b) The location of the isolated nodes z and the radius ρ of the vortex ring of the Higgs field for the same set of solutions. (See Table 1 for the notation of the node and vortex ring configurations of the Higgs field.)

MA	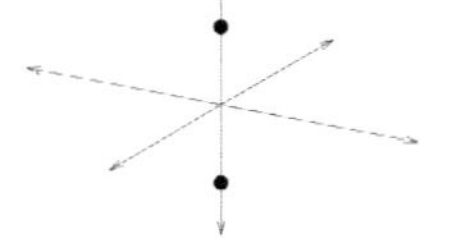	A monopole-antimonopole pair with a monopole above and an antimonopole below the xy-plane on the symmetry axis.
V̊	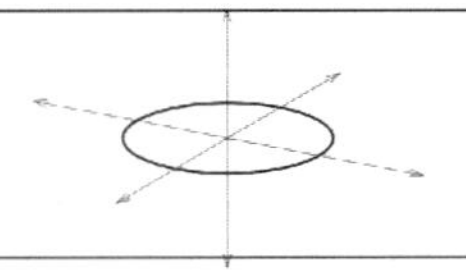	A vortex ring in the xy-plane.

Table 1. Configurations of nodes and vortex rings of the Higgs field for $m = 2$, $n = 3$ solutions.

gravitational instability should develop. Therefore regular gravitating solutions will exist only up to a maximal value of α.

The critical behaviour of these α-branches, emerging from flat-space solutions, is elucidated by numerical investigation. As such an α-branch approaches its maximal value of α, it merges with a second α-branch of regular gravitating solutions, where $\alpha_{\max}$ is the endpoint of both branches. The nature of the second branch, however, depends on the nature of the first branch and on the value of λ.

Fundamental solutions

When considering sets of solutions, emerging from the fundamental flat-space λ-branch, we always find two α-branches, where the solutions on the second α-branches shrink to zero size in the limit $\alpha \to 0$, while their mass diverges. By scaling the coordinates and the Higgs field of the solutions via

$$\hat{x} = \frac{x}{\alpha} \ , \quad \hat{\Phi} = \alpha \Phi \ , \tag{11}$$

one realizes that the second α-branches connect to limiting EYM solutions, when $\alpha \to 0$, solutions with finite scaled size and finite scaled mass $\hat{M}$ [17],

$$\hat{M} = \alpha M \ . \tag{12}$$

Indeed, after the scaling, the field equations do not depend on α. Instead α appears in the asymptotic boundary conditions of the Higgs field, $|\hat{\Phi}| \to \alpha$. The Higgs field then becomes trivial on the second α-branches as $\alpha \to 0$, and the solutions thus approach EYM solutions. Depending on n, the limiting solutions correspond to the lowest BM or a generalized BM solution, i.e., the solutions on these second α-branches possess no flat-space limit.

The two α-branches are illustrated in Fig. 2 for solutions with $m = 2$, $n = 3$ at scalar coupling strength $\lambda = 0$, $\lambda = 2.0$ and $\lambda = 4.5$. The figure exhibits the mass and the scaled mass of these solutions and the location of their nodes. Note, that they always correspond to vortex ring solutions.

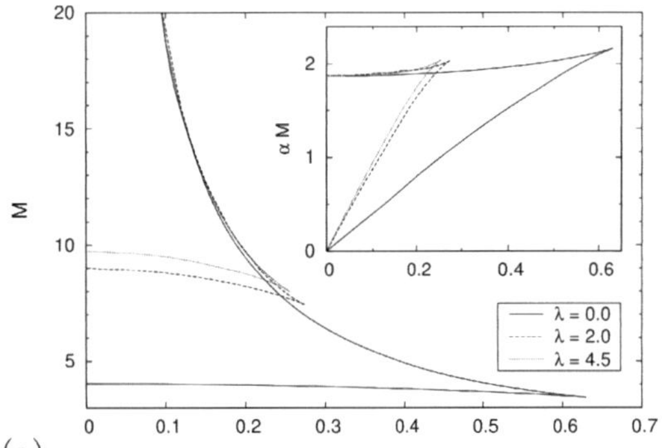

(a)

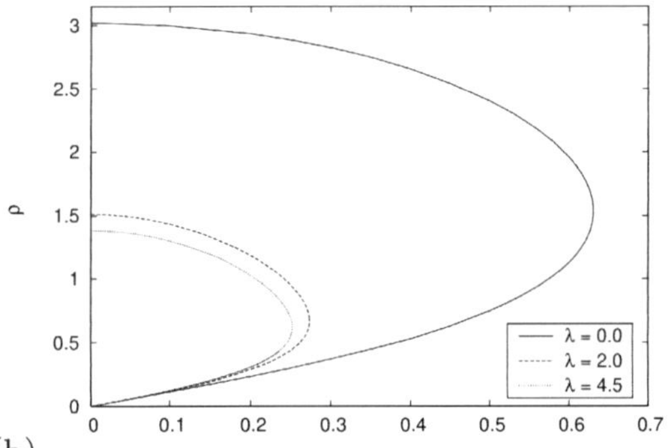

(b)

Figure 2. (a) The mass M and the scaled mass αM of $m = 2$, $n = 3$ solutions, emerging from the fundamental flat-space λ-branch versus α. (b) The radius ρ of the vortex ring of the Higgs field for the same set of solutions.

New solutions: intermediate λ

The second α-branch always connects to an EYM solution, when the first α-branch emerges from a fundamental λ-branch solution. But for the α-branches of the new flat-space solutions, which exist only beyond a critical value of λ, one may anticipate a different pattern, because these solutions always come in pairs.

In the presence of gravity, each of the solutions of such a pair forms an α-branch, extending up to a maximal value of α. But the critical behaviour at $\alpha_{\max}$ now depends on the strength of the scalar coupling λ. In the simplest case, observed for not too large values of λ, the two α-branches simply merge and end at $\alpha_{\max}$. Indeed, this critical behaviour is different from the above, since both branches merging at $\alpha_{\max}$ have flat-space limits.

This is illustrated in Fig. 3, where we exhibit the mass and the location of the nodes for $m = 2$, $n = 3$ configurations at $\lambda = 2.0$, $\lambda = 2.645$, $\lambda = 3.645$, $\lambda = 4.5$, $\lambda = 5.25$ and $\lambda = 12.5$ versus the coupling constant α. Evidently, for the smaller λ values, a first α-branch of solutions originating from the lower mass flat-space solution, extends up to a maximal value $\alpha_{\max}$, where it merges with a second α-branch, originating from the upper mass flat-space solution. The solutions on these branches correspond to MAPs.

However, as λ increases, a new phenomenon arises on the second α-branch, as it evolves back from $\alpha_{\max}$ towards the flat-space solution. As seen in Fig. 3, at a critical value of λ ($\lambda \approx 4.5$) a bifurcation appears on the second α-branch, giving rise to two additional α-branches of solutions.

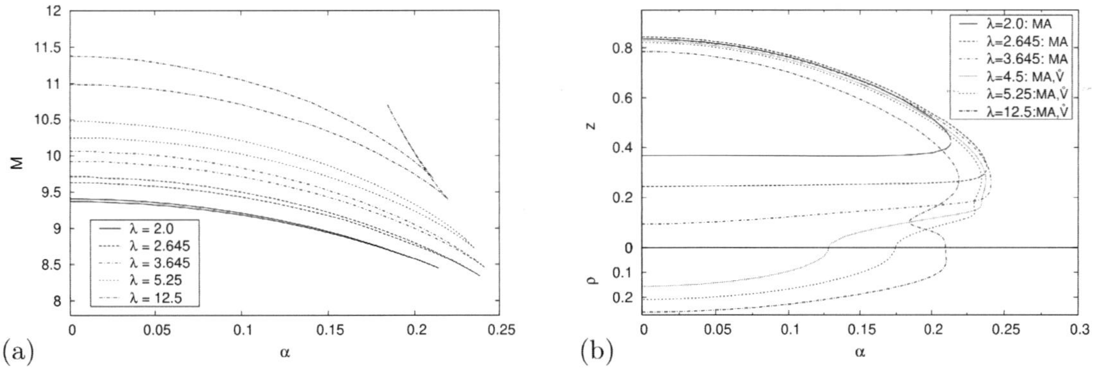

Figure 3. (a) The mass M of $m = 2$, $n = 3$ solutions, related to the new flat-space λ-branches versus α for several values of the scalar coupling λ. (b) The location of the isolated nodes z and radius ρ of the vortex ring of the Higgs field for the same set of solutions. (See Table 1 for the notation of the node and vortex ring configurations of the Higgs field.)

These 3rd and 4th α-branches exist beyond a minimal value of α, $\alpha_{\min}$, and below a respective maximal value of α, where they merge with the first and second α-branch, respectively. The range $\alpha_{\min} \leq \alpha_{\max}^{(i)}$, where these additional branches exist, increases with increasing λ. In Fig. 3 the additional branches manifest in the mass of the $\lambda = 12.5$ solutions as a higher mass spike. Note, that in this range of λ vortex ring solutions arise on a part of the second α-branch.

The emergence and evolution of these new α-branches is illustrated further in Fig. 4, where we exhibit the values of the metric functions f and l at the origin, $f(0)$ and $l(0)$. In particular, $l(0)$ clearly exhibits the various branches and critical points and therefore represents an instructive means to clarify the pattern of solutions.

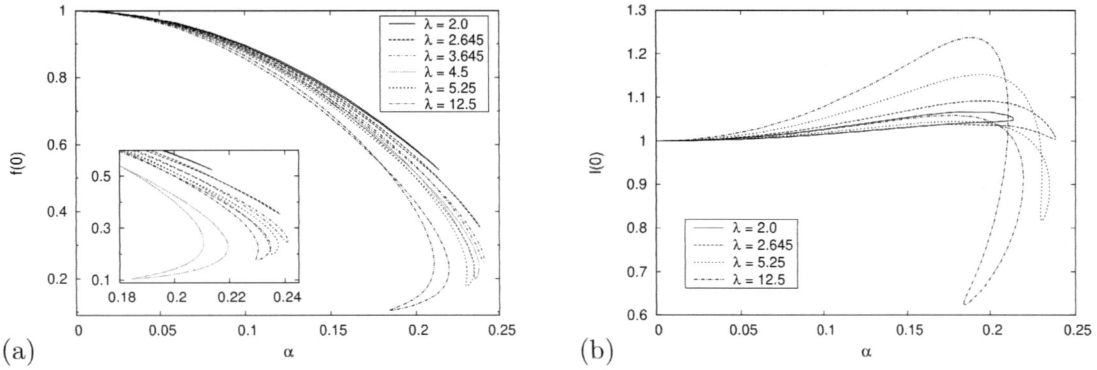

Figure 4. The values of metric functions at the origin, $f(0)$ (a) and $l(0)$ (b) of $m = 2$, $n = 3$ solutions, related to the new flat-space λ-branches versus α for several intermediate values of the scalar coupling λ.

To complement the previous discussion of the new solutions, let us consider the λ-

dependence of the solutions for fixed values of α. In Fig. 5 we exhibit the mass, the location of the nodes, and the values of the metric functions at the origin versus λ for $m = 2$, $n = 3$ configurations at $\alpha = 0.2$, $\alpha = 0.21$, $\alpha = 0.22$, $\alpha = 0.23$ and $\alpha = 0.24$. One clearly recognizes the new solutions, present in a certain region of parameter space.

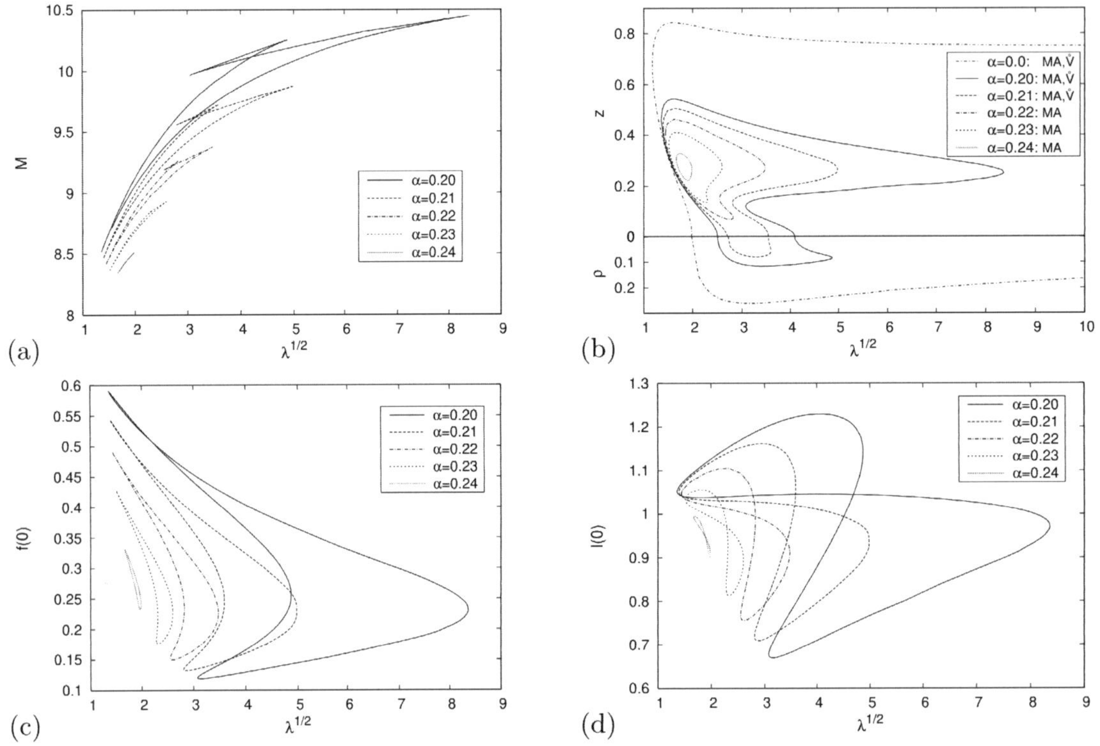

Figure 5. (a) The mass M of $m = 2$, $n = 3$ solutions, related to the new flat-space λ-branches versus λ for several values of the coupling α. (b) The location of the isolated nodes z and the radius ρ of the vortex ring of the Higgs field , and the values of metric functions at the origin, $f(0)$ (c) and $l(0)$ (d) for the same set of solutions. (See Table 1 for the notation of the node and vortex ring configurations of the Higgs field.)

Since both the scalar interaction and gravity are attractive, one can expect a similar effect concerning the changing structure of the nodes, when λ is increased at fixed α as when α is increased at fixed λ. For the new flat-space upper mass λ-branch, a transition from MAPs to vortex rings appears at $\lambda = 3.941$ [10].

As the gravitational coupling constant α increases, the pattern of the nodes of the solutions changes significantly as well, and there are no solutions with vortex rings (except for those connected to the fundamental λ-branch), when $\alpha > 0.215$ for any value of λ. This pattern is illustrated in Figure 5, where the location of the nodes, both isolated zeros and vortex rings, is shown versus the scalar coupling for several values of α. Evidently, the coupling to gravity restricts the range of possible values of λ, where these solutions can exist, and this restriction is the stronger, the larger α, shrinking the domain of existence to zero at a critical value of α.

Journal of Physics: Conference Series **222** (2010) 012019 doi:10.1088/1742-6596/222/1/012019

New solutions: very large λ

As λ increases further, the new α-branches extend further backwards. Along these branches the mass of the solutions increases further as α decreases. At the same time the locations of the nodes of the Higgs field move towards the origin, as illustrated in Fig. 6 for very large values of the scalar coupling λ.

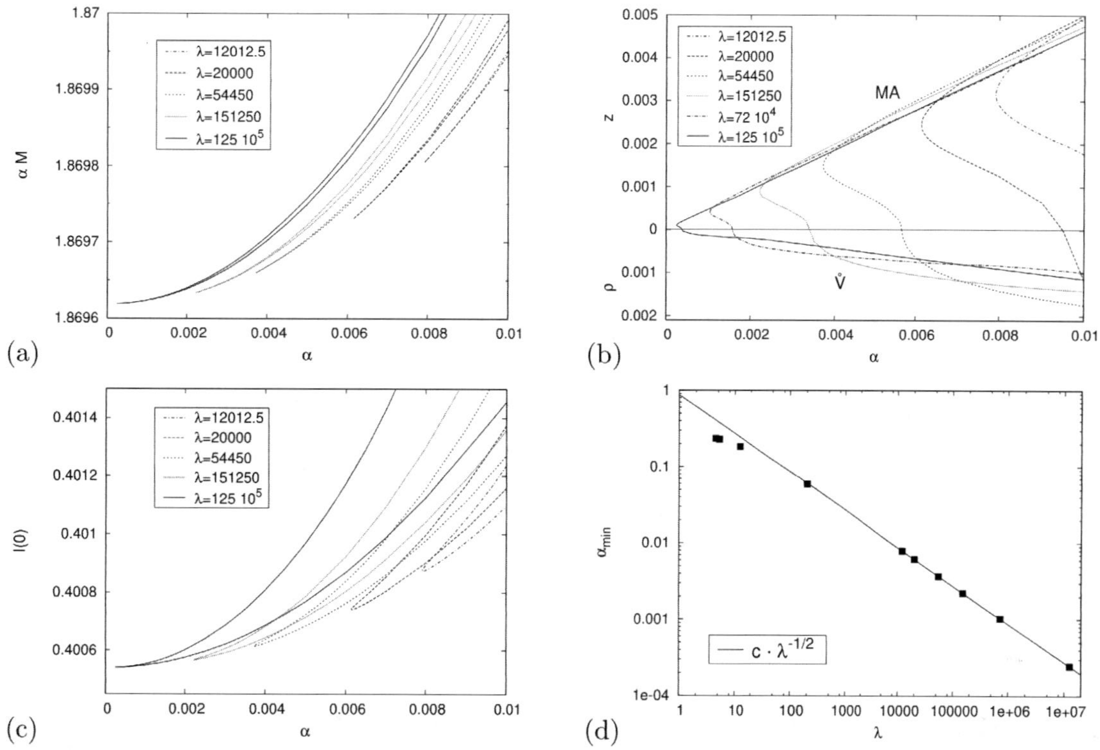

Figure 6. (a) The scaled mass αM of $m = 2$, $n = 3$ solutions, related to the new flat-space λ-branches versus α for very large values of the scalar coupling λ. (b) The location of the isolated nodes z and the radius ρ of the vortex ring of the Higgs field, the value of metric functions at the origin, $l(0)$ (c) for the same set of solutions. (d) The minimal value $\alpha_{\min}$ of the new α-branches of $m = 2$, $n = 3$ solutions versus λ. (See Table 1 for the notation of the node and vortex ring configurations of the Higgs field.)

With help of Fig. 6 we can now address the limit $\lambda \to \infty$ for these new α-branches. As λ increases without bound, the minimal value $\alpha_{\min}$, beyond which the new branches exist, decreases towards zero. (Note, that $\alpha_{\min} \sim 1/\sqrt{\lambda}$ for very large λ.) At the same time the scaled mass of the critical solutions at $\alpha_{\min}$ decreases towards the mass of the generalized BM solution with $n = 3$ [21], while their mass itself diverges. Likewise, the values of their metric functions at the origin approach those of the $n = 3$ generalized BM solution [22]. Their nodes finally move continuously towards the origin, and the solutions shrink to zero size as $\lambda \to \infty$. Scaling the coordinates and Higgs field again reveals the generalized BM solution as limiting solution of finite scaled mass and finite scaled size [17, 18].

This limiting behavior for the critical solutions on the new α-branches is similar to that observed for the solutions on the second α-branches of the fundamental solutions. In both cases the respective generalized BM solution is approached.

One can understand the reason for this behaviour here, by noting that, as the scalar field becomes infinitely heavy, it decouples in these solutions and, in the double limit of infinite λ and zero η, it does not affect the dynamics of the gauge sector anymore. In that sense the EYMH system is getting actually truncated here to an EYM system.

3.2. Topologically nontrivial sector: $\boldsymbol{m = 3}$, $\boldsymbol{n = 3}$

3.2.1. Flat space solutions Let us now turn to solutions in the topologically nontrivial sector with $m = 3$, $n = 3$. In these solutions a triply charged monopole resides at the origin, providing the topological charge of the solutions. Again we first address the flat-space solutions, reviewing and supplementing previous results [10].

For $\lambda \to 0$, the $m = 3$, $n = 3$ solution possesses a triply charged monopole at the origin and two oppositely oriented vortex rings located symmetrically above and below the xy-plane [9]. For larger values of λ one expects a bifurcation to occur, where new branches of solutions appear, which possess a different node structure, with three isolated nodes on the symmetry axis. For high values of λ such a monopole-antimonopole chain (MAC) should then represent the energetically most favourable solution.

Indeed, at $\lambda_{c_1} = 0.8099$ two new branches of solutions appear which possess the node structure of MACs [9]. But as seen in Fig. 7, the vortex ring solutions on the first (fundamental) λ-branch turn already into MACs before λ_{c_1} is reached: As λ increases, two nodes emerge from the origin and separate from each other along the z-axis. The solutions then possess three nodes on the symmetry axis and two vortex rings, located symmetrically above and below the xy-plane. As λ increases further, the new nodes move further apart, while the vortex rings shrink to zero size, merging with the new nodes on the symmetry axis.

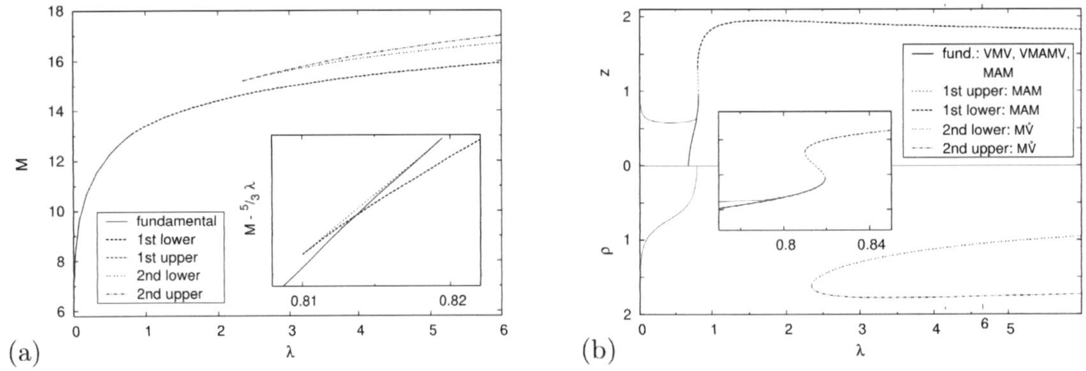

Figure 7. (a) The mass M of the flat-space solutions on the fundamental λ-branch as well as on the additional λ-branches for $m = 3$, $n = 3$ solutions. (b) The location of the isolated nodes z and vortex rings (ρ, z) of the Higgs field for the same set of solutions. (See Table 2 for the notation of the node and vortex ring configurations of the Higgs field.)

Thus beyond a certain value of λ, which is smaller than the first critical value λ_{c_1}, no vortex ring solutions are present any longer [9]. The fundamental branch and the new upper

Journal of Physics: Conference Series **222** (2010) 012019 doi:10.1088/1742-6596/222/1/012019

MAM	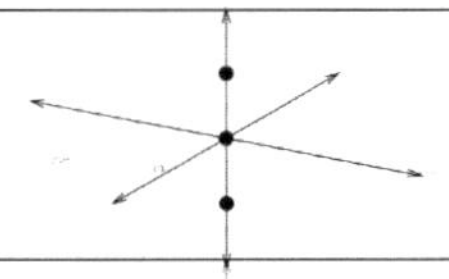	A monopole-antimonopole chain with two monopoles on the symmetry axis, one above and one below the xy-plane, and an antimonopole at the origin.
VMV	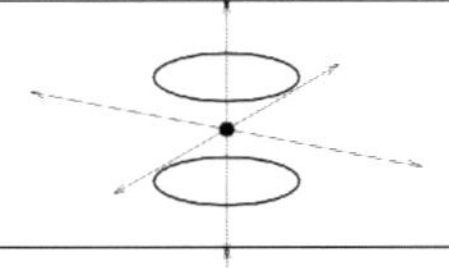	A monopole at the origin and two vortex rings, one above and one below the xy-plane.
M$\mathring{V}$	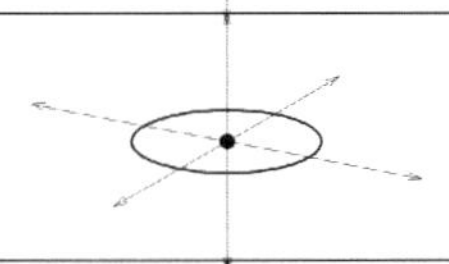	A monopole at the origin and a vortex ring in the xy-plane.
VMAMV	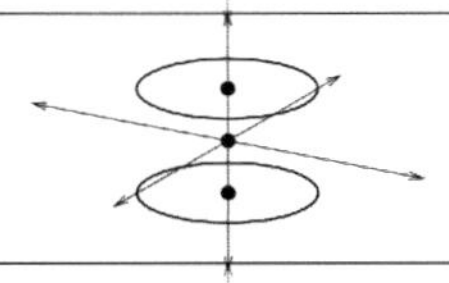	Two vortex rings, one above and one below the xy-plane, two monopoles on the symmetry axis and an antimonopole at the origin.
VM$\mathring{V}$V	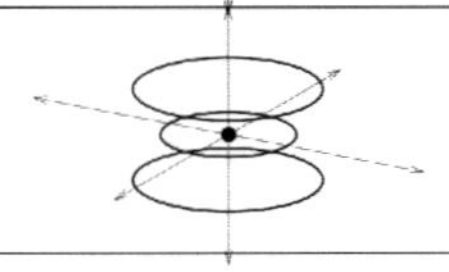	Three vortex rings, one above and one below the xy-plane and one in the xy-plane, and a monopole at the origin.

Table 2. Configurations of the nodes and vortex rings of the Higgs field for $m = 3$, $n = 3$ solutions.

branch then merge at $\lambda_{c_2} = 0.8194$, while the new lower branch continues to large values of λ, representing the fundamental branch there. This is seen in Fig. 7, where the mass and the node structure of the $m = 3$, $n = 3$ λ-branches are exhibited, with emphasis on the three branches of MACs present in the small range of λ, $\lambda_{c_1} \leq \lambda \leq \lambda_{c_2}$.

Interestingly, a further bifurcation arises at $\lambda_{c_3} = 2.3436$, which was missed before. At λ_{c_3} two new branches of vortex ring solutions appear, whose single ring is located in the xy-plane. Thus beyond λ_{c_3}, vortex ring solutions are present again, but possess higher masses than the fundamental MAC solutions. These new λ-branches are also exhibited in Fig. 7.

3.2.2. Gravitating solutions When gravity is coupled, branches of gravitating solutions arise from these $m = 3$, $n = 3$ flat-space configurations. Depending on the value of λ, two or more α-branches are present, whose nature is considered in detail in the following.

The α-dependence of the $\lambda < \lambda_{c_1}$ solutions is simple. A first α-branch emerges from the flat-space solution and merges at some $\alpha_{\max}$ with a second α-branch, which connects to the generalized BM solution in the limit $\alpha \to 0$. Indeed, for small values of α on the second branch, the solutions may be thought of as composed of a scaled $n = 3$ generalized BM solution in an inner region and a flat-space charge-3 monopole configuration in an outer region [18]. In the limit $\alpha \to 0$, the mass of the solutions diverges, while their scaled mass approaches the mass of the generalized $n = 3$ BM solution.

Solutions in the range $\lambda_{c_1} \leq \lambda \leq \lambda_{c_2}$

While for $\lambda < \lambda_{c_1}$ only a single flat-space solution is present, there are three flat-space solutions in the range $\lambda_{c_1} \leq \lambda \leq \lambda_{c_2}$. This complicates the picture considerably. We exhibit in Fig. 8 the node structure of the α-branches, associated with all three flat space solutions, for several values of λ in this critical range.

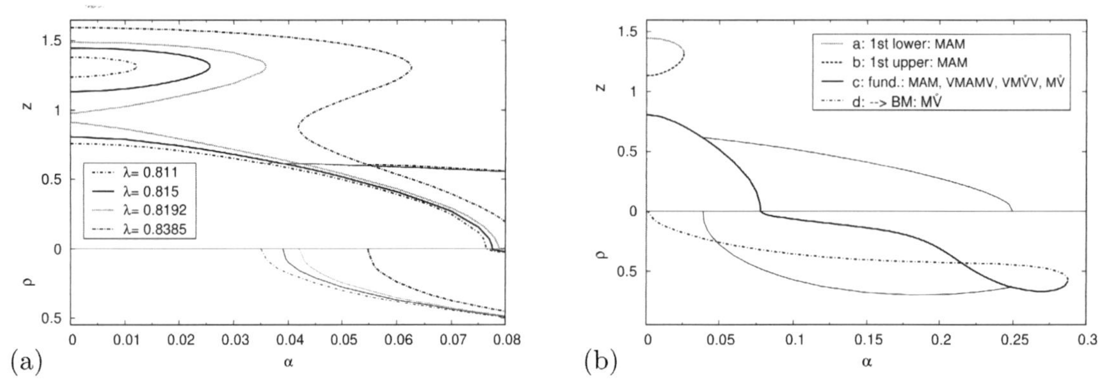

Figure 8. (a) The location of the isolated nodes z and vortex rings (ρ, z) of the Higgs field of $m = 3$, $n = 3$ solutions versus α for several values of the scalar coupling λ in the critical range $\lambda_{c_1} \leq \lambda \leq \lambda_{c_2}$ and beyond. (b) The isolated nodes and vortex rings of the Higgs field for the complete set of solutions for $\lambda = 0.815$. (See Table 2 for the notation of the node and vortex ring configurations of the Higgs field.)

While the left part of the figure shows the evolution of the zeros of the Higgs field with increasing scalar coupling λ, the right part gives for a particular value, $\lambda = 0.815$, the assignment of the branches, to which the zeros belong. The α-branch emerging from the fundamental λ-branch is labeled c, and the α-branches emerging from the new lower λ-branch and the new upper λ-branch are labeled a and b, respectively.

It is intuitively clear that in the complicated pattern of interaction between the monopoles and antimonopoles, the gravitational attraction plays a role similar to the scalar interaction, thus one can expect that coupling to gravity may provide a compensation for the weakening of the scalar interaction. This is indeed, what we here observe regarding the existence of the solutions and their node structure. For increasing α, the branches a and b merge and disappear, while the branch c persists, but changes character, reversing the steps seen for the fundamental flat-space λ-branch in Fig. 7.

Indeed, at $\alpha = 0$, all three branches begin as MAC solutions. As α increases, the two MAC branches a and b merge and end at a small $\alpha_{\max}$. The branch c, however, extends further and changes its node structure accordingly, as α increases. First a vortex ring emerges from each outer node. Next the nodes move towards each other, merging at the origin. Then a central vortex ring appears in the xy-plane, and finally the outer rings merge with the central ring.

The branch c still persists to somewhat larger values of α, and then merges at an $\alpha_{\max}$ with a second branch of vortex ring solutions labeled d. As expected, this branch extends backwards to $\alpha = 0$, where it connects to a generalized BM solution.

The evolution of the solutions with increasing λ can now be understood from the left part of the figure. As λ increases, the branches a and b extend to larger values of α, before they merge. At the same time the nodes of the branches b and c approach each other. At the critical value λ_{c_2} the two flat-space branches merge, and their nodes coincide. As λ increases further, this critical point (where two branches merge) is retained, but it moves to finite values of α. Note, that the pattern of nodes of the gravitating solutions in the interval $[0.03 < \alpha < 0.07]$ for $\lambda = 0.8385$ exhibited in Fig. 8, nicely reflects the (reversed) pattern of nodes of the flat-space solutions in the interval $[0.8 < \lambda < 0.85]$ exhibited in Fig. 7.

Emergence of the new flat-space branches at λ_{c_3}

As seen above, the bifurcations of the flat-space branches at λ_{c_1} and λ_{c_2} are reflected by bifurcations in curved space. In particular, the critical values of the gravitational coupling strength α, where bifurcations arise for a given value of the scalar coupling strength λ, move towards smaller values of α with decreasing λ. As the minimal critical value of α reaches zero, the flat-space bifurcation point λ_{c_2} arises.

It is thus obvious, that the new flat-space λ-branches arising at λ_{c_3}, should also have precursors in curved space. These precursors indeed led to the detection of the new flat-space branches. In the following we analyze the branches of gravitating solutions, obtained with increasing scalar coupling strength λ, to clarify the emergence of the new flat-space branches.

Let us first consider the sets of solutions at $\lambda = 1$ and $\lambda = 2$, exhibited in Fig. 9. While the node structure nicely reveals the types of solutions present at the respective value of λ, and their evolution with α, the values of the metric functions at the origin are instructive to get an overview of the sets of solutions present. For better identification, we again label the various branches of solutions a, b, c, etc.

For $\lambda = 1$ there are four α-branches, just like for the smaller λ values of Fig. 8. When moving continuously along these α-branches, the node structure changes from MAC, present for a and most of b, to MAC plus vortex rings, shortly before the bifurcation with c. The size and the location of the nodes and the rings then evolves along c until a single ring is left in the xy-plane (together with the always present node at the origin). This node structure is then retained on branch d, where the solutions evolve towards the generalized BM solution as $\alpha \to 0$.

As λ increases further, interestingly, another bifurcation arises, and two more α-branches appear. Consequently, there is now an interval of α with six α-branches. The branches labeled a to f are exhibited in the figure for $\lambda = 2$. Moving continuously along the branches, one again observes the structure of the solutions change from MACs to MACs plus vortex rings, and then to a monopole plus vortex ring(s).

Most important for understanding the emergence of the new flat-space branches is the evolution of the branches d and e with λ, with particular emphasis on the critical value of α, where these two branches merge. We therefore consider in Fig. 10 sets of solutions with still higher values of λ.

As λ increases further, the branches d and e extend to increasingly smaller values of α. At the critical value λ_{c_3} of the scalar coupling, the critical value of α, where the branches d and e merge, then precisely reaches zero, giving rise to the bifurcation point of the new flat-space solutions.

For values of the scalar coupling beyond λ_{c_3}, the set of solutions then splits into two disconnected parts. The first part emerges with branch a from the respective fundamental flat-space solution, evolves via branch b and branch c, and reaches along branch d the new upper mass flat-space solution in the limit $\alpha \to 0$. The second part emerges with branch e from the new upper mass flat-space solution, and evolves with branch f towards the limiting generalized BM solutions in the limit $\alpha \to 0$. This is demonstrated in Fig. 10 for the set of

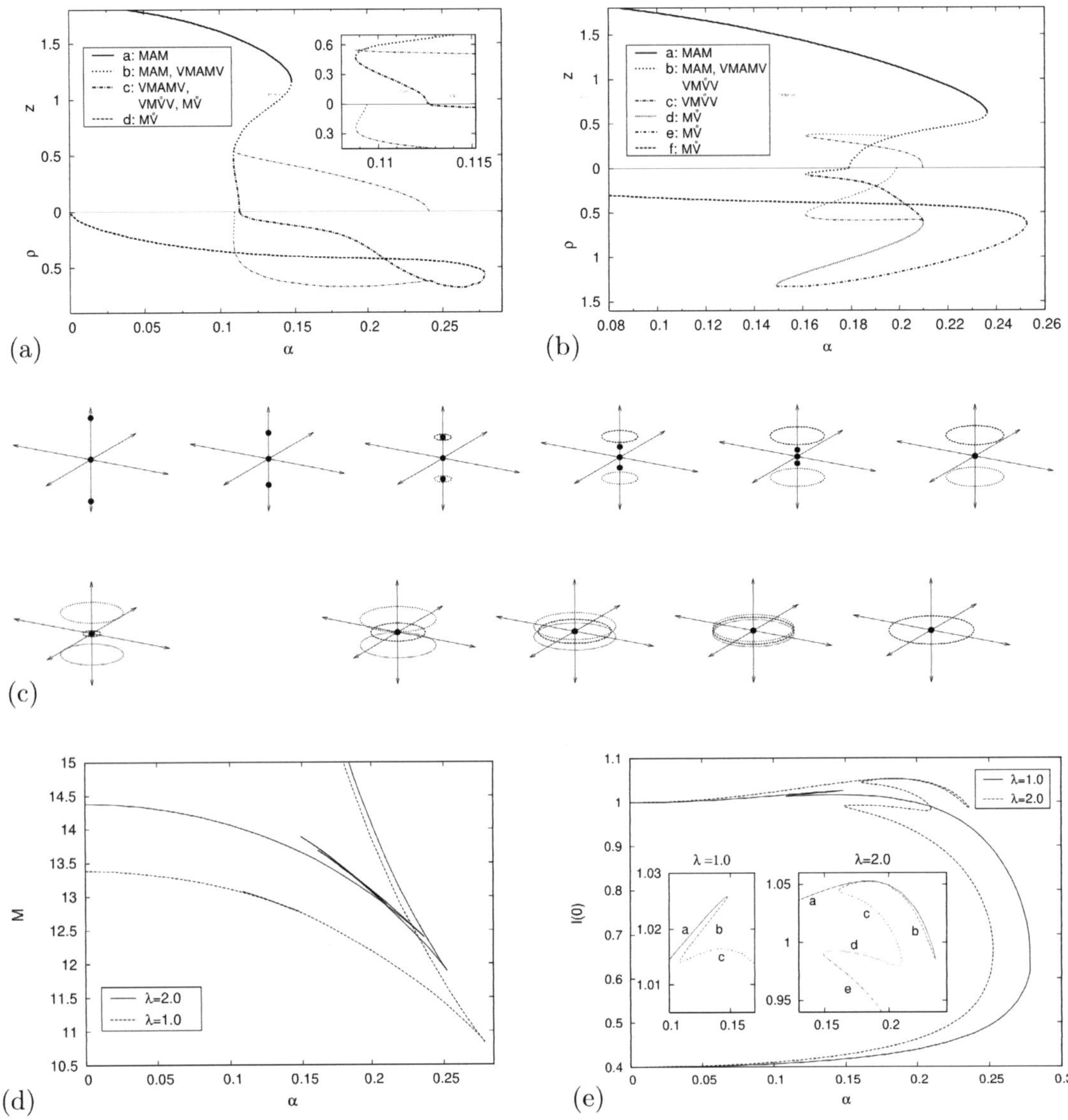

Figure 9. The location of the isolated nodes z and vortex rings (ρ, z) of the Higgs field for the complete set of solutions for $\lambda = 1$ (a) and $\lambda = 2$ (b) versus α. Change of the structure of the nodes of the Higgs field following the evolution of the α-branches (c). The mass M (d) and the value of metric function at the origin $l(0)$ (e) for the same set of solutions.

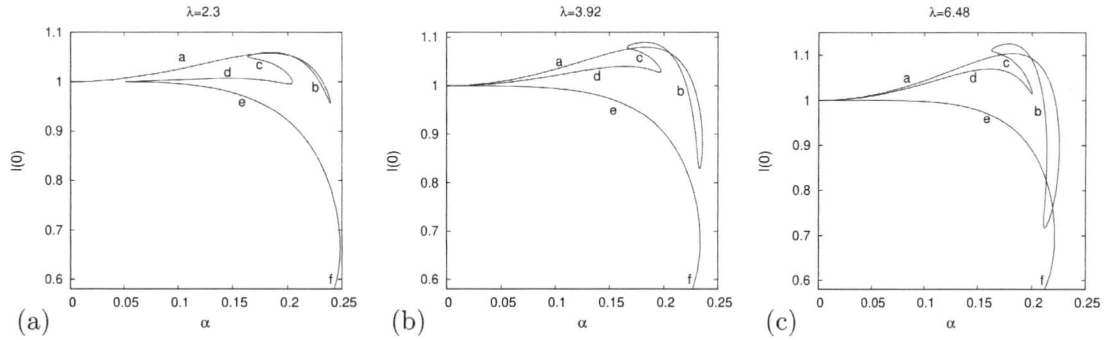

Figure 10. The value of metric function at the origin, $l(0)$, for $\lambda = 2.3$ (a) and $\lambda = 3.92$ (b) and for $\lambda = 6.48$ (c).

solutions with $\lambda = 3.92$.

Solutions at large λ

In the case of the $m = 2$, $n = 3$ solutions we observed the appearance of new α-branches of solutions for large values of λ, exhibiting an intriguing behaviour: In the limit $\lambda \to \infty$ the minimal value $\alpha_{\min}$, beyond which these new branches exist, decreases towards zero, while the critical solutions themselves approach the generalized BM solution as $\alpha_{\min} \to \infty$.

We here observe an analogous pattern for the $m = 3$, $n = 3$ solutions. To see this pattern arising, we reconsider the evolution of the branches a and b with λ, exhibited in Fig. 9 and Fig. 10. As λ increases, the branches a and b first elongate towards larger values of α. Then at a critical value of λ, a new bifurcation arises, and two additional branches appear. The minimal value $\alpha_{\min}$ possible for these new branches now decreases with increasing λ, analogous to the case of the respective branches of the $m = 2$, $n = 3$ solutions.

4. Conclusions

We have investigated static axially symmetric solutions of $SU(2)$ Einstein-Yang-Mills-Higgs theory, representing monopole-antimonopole pairs, chains, vortex rings, and new types of configurations with mixed node structure. Such new configurations appear for larger values of the scalar coupling, when the subtle interplay between repulsive and attractive forces allows for more than one non-trivial equilibrium configuration of these systems.

In flat space, at critical values of the scalar coupling, bifurcations arise, where pairs of new branches of solutions appear, which possess a different node structure than the solutions on the fundamental branches. In particular, new solutions appear, where vortex rings present in the fundamental solutions, are replaced by isolated nodes on the symmetry axis. For high values of λ the new solutions with monopole-antimonopole chain structure have the lowest mass.

While we have studied here in detail only the monopole-antimonopole systems with $m = 2$, 3, 4 and $n = 3$, we conjecture, that this phenomenon is not restricted to these particular systems but that it is of a more general nature, implicating an enormous richness of configuration space for high values of λ and larger m and n. For $m = 4$ and $n = 3$, for instance, we observed already seven distinct flat-space solutions in a certain range of the scalar coupling.

The coupling to gravity leads to additional attraction between the components of these monopole-antimonopole systems, shifting the balance of forces and thus affecting the possible equilibrium configurations. Whereas for small scalar coupling only two gravitating branches of solutions are associated with a flat-space solution, there is a surge of further branches when the scalar coupling becomes large.

Particular of these branches can be shown to give rise to new pairs of flat-space solutions. Thus study of the Einstein-Yang-Mills-Higgs configuration space gives valuable insight into the configuration space of Yang-Mills-Higgs theory. Indeed, by exploiting this interrelationship, we were able to find the new flat-space solutions presented.

But the Einstein-Yang-Mills-Higgs configurations are also related to Einstein-Yang-Mills solutions. First of all, in the limit $\alpha \to 0$, one of the α-branches of a particular set of solutions, characterized by m, n and λ, always connects to a generalized Bartnik-McKinnon solution [20, 21, 22]. But another connection arises in the limit of infinite scalar coupling. In this limit the bifurcation point $\alpha_{\rm min}$ of two of the α-branches is seen to approach $\alpha = 0$, while the solutions at $\alpha_{\rm min}$ approach the respective generalized Bartnik-McKinnon solution.

Acknowledgement This review is entirely based on the work has been done in collaboration with Burkhard Kleihaus, Jutta Kunz and Ulrike Neemann [9, 10]. Most of the diagrams presented here have been prepared by Ulrike Neemann [10, 15].

I would like to acknowledge valuable discussions with Eugen Radu, Tigran Tchrakian, and Michael Volkov. This work is supported by the Marie Curie Host Fellowships Programm MTKD-CT-2006-042360 "Krakow Geometry in Mathematical Physics - KRAGEOMP" I am deeply thankful to the organizers of the First Mediterranean Conference on Classical and Quantum Gravity for invitation to attend a well organized and stimulating meeting and kind hospitality in Columbary.

References

[1] G. 't Hooft, Nucl. Phys. **B79** (1974) 276;
A. M. Polyakov, Pis'ma JETP **20** (1974) 430.
[2] E.J. Weinberg, and A.H. Guth, Phys. Rev. **D14** (1976) 1660.
[3] C. Rebbi, and P. Rossi, Phys. Rev. **D22** (1980) 2010.
[4] R.S. Ward, Comm. Math. Phys. **79** (1981) 317;
P. Forgacs, Z. Horvath, and L. Palla, Phys. Lett. **99B** (1981) 232;
M.K. Prasad, Comm. Math. Phys. **80** (1981) 137;
M.K. Prasad, and P. Rossi, Phys. Rev. **D24** (1981) 2182.
[5] B. Kleihaus, J. Kunz, and D. H. Tchrakian, Mod. Phys. Lett. **A13** (1998) 2523.
[6] C. H. Taubes, Commun. Math. Phys. **86** (1982) 257;
C. H. Taubes, Commun. Math. Phys. **86** (1982) 299;
C. H. Taubes, Commun. Math. Phys. **97** (1985) 473.
[7] W. Nahm, unpublished;
Bernhard Rüber, Diploma Thesis, University of Bonn, 1985.
[8] B. Kleihaus, and J. Kunz, Phys. Rev. **D61** (2000) 025003.
[9] B. Kleihaus, J. Kunz, and Ya. Shnir, Phys. Lett. **B570** (2003) 237;
B. Kleihaus, J. Kunz, and Ya. Shnir, Phys. Rev. **D68** (2003) 101701;
B. Kleihaus, J. Kunz, and Ya. Shnir, Phys. Rev. **D70** (2004) 065010.
[10] J. Kunz, U. Neeman and Ya. Shnir, Phys. Lett. **B640** (2006) 57.
B. Kleihaus, J. Kunz, U. Neemann and Y. Shnir, J. Phys. Conf. Ser. **68** (2007) 012035.
J. Kunz, U. Neemann and Y. Shnir, Phys. Rev. D **75** (2007) 125008.

[11] N. J. Hitchin, N. S. Manton and M. K. Murray, Nonlinearity **8** (1995) 661;
C. J. Houghton and P. M. Sutcliffe, Commun. Math. Phys. **180** (1996)343;
C. J. Houghton and P. M. Sutcliffe, Nonlinearity **9** (1996) 385;
P. M. Sutcliffe, Int. J. Mod. Phys. **A12** (1997) 4663.

[12] N. S. Manton, Nucl. Phys. **B126** (1977) 525;
W. Nahm, Phys. Lett. **85B** (1979) 373;
N. S. Manton, Phys. Lett. **154B** (1985) 397.

[13] B. Hartmann, B. Kleihaus, and J. Kunz, Phys. Rev. Lett. **86** (2001) 1422;
B. Hartmann, B. Kleihaus, and J. Kunz, Phys. Rev. **D65** (2001) 024027.

[14] B. Kleihaus, J. Kunz and U. Neemann, Phys. Lett. B **623** (2005) 171.

[15] U. Neemann, PhD Thesis, University of Oldenburg, 2007.

[16] K. Lee, V.P. Nair, and E.J. Weinberg, Phys. Rev. **D45** (1992) 2751;
P. Breitenlohner, P. Forgacs, and D. Maison, Nucl. Phys. **B383** (1992) 357;
P. Breitenlohner, P. Forgacs, and D. Maison, Nucl. Phys. **B442** (1995) 126.

[17] B. Kleihaus, and J. Kunz, Phys. Rev. Lett. **85** (2000) 2430.

[18] B. Kleihaus, J. Kunz, and Ya. Shnir, Phys. Rev. **D71** (2005) 024013.

[19] A. Lue and E.J. Weinberg, Phys. Rev. **D60** (1999) 084025;
P. Breitenlohner, P. Forgacs, and D. Maison, in preparation.

[20] R. Bartnik, and J. McKinnon, Phys. Rev. Lett. **61** (1988) 141.

[21] B. Kleihaus and J. Kunz, Phys. Rev. Lett. **78** (1997) 2527;
B. Kleihaus and J. Kunz, Phys. Rev. **D57** (1998) 834.

[22] R. Ibadov, B. Kleihaus, J. Kunz, and Ya. Shnir, Phys. Lett. **B609** (2005) 150.

[23] Y. Brihaye, and J. Kunz, Phys. Rev. **D50** (1994) 4175.

[24] W. Schönauer, and R. Weiß, J. Comput. Appl. Math. **27** (1989) 279;
M. Schauder, R. Weiß, and W. Schönauer, The CADSOL Program Package, Universität Karlsruhe, Interner Bericht Nr. 46/92 (1992).

First Mediterranean Conference on Classical and Quantum Gravity (MCCQG 2009) IOP Publishing
Journal of Physics: Conference Series **222** (2010) 012020 doi:10.1088/1742-6596/222/1/012020

Noncommutative effects in astrophysical objects: a survey

Orfeu Bertolami[1,2] **and Carlos A D Zarro**[1,2]

[1] Departamento de Física, Instituto Superior Técnico,
Av. Rovisco Pais, 1049-001 Lisboa, Portugal
[2] Instituto de Plasmas e Fusão Nuclear, Instituto Superior Técnico, Av. Rovisco Pais, 1049-001 Lisboa, Portugal

E-mail: orfeu@cosmos.ist.utl.pt, carlos.zarro@ist.utl.pt

Abstract. The main implications of noncommutativity over astrophysical objects are examined. Noncommutativity is introduced through a deformed dispersion relation $E^2 = p^2c^2(1+\lambda E)^2+m^2c^4$ and the relevant thermodynamical quantities are calculated using the grand canonical ensemble formalism. These results are applied to simple physical models describing main-sequence stars, white-dwarfs and neutron stars. The stability of main-sequence stars and white dwarfs is discussed.

1. Introduction

Noncommutativity is believed to be an important feature of space-time at quantum gravity scales [1, 2]. Interestingly, it also appears in the context of string theory [3, 4].

Noncommutative quantum field theory (NCQFT) models can be implemented by substituting the normal product between fields by a noncommutative (NC) one, the so-called Moyal product [5, 6]. This procedure introduces a minimum length scale, but does not solve, as one could expect, the renormalization problem of the quantum field theory. One also encounters additional difficulties on issues such as causality and unitarity. Related questions associated with noncommutativity involve the violation of translational invariance [7], scalar fields and their stability in curved spaces [8, 9, 10]. NCQFT can also be implemented by generalizing the Heisenberg-Weyl algebra of noncommutative quantum mechanics [11, 12, 13, 14, 15, 16, 17, 18] to field algebra [17]. This approach can be used, for instance, to investigate the inflationary period [19]. Another way to introduce noncommutativity in field theory is through a deformed dispersion relation in a commutative space-time rather than considering a noncommutative space-time over which the fields are defined [20]. Since there is no consistent theory of noncommutative gravity [21, 22] this approach is justified. This is inspired by quantum groups methods [23], by a possible breaking of Lorentz symmetry at high-energies [24, 25, 26] and the modifications of special relativity that take into account the possible existence of a minimum length [27, 28].

One mentions here some possible ways on how noncommutativity can lead to a deformed dispersion relation. The first approach is related to quantum groups methods [23]. Assuming the following commutation relations

$$[x^i, t] = i\lambda x^i \qquad [x^i, x^j] = 0, \tag{1}$$

where $i, j = 1, 2, 3$ and λ is constant, it can be shown that the associated relationship between dynamical quantities is given by [23]

$$\lambda^{-2}(e^{\lambda E} + e^{-\lambda E} - 2) - p^2c^2e^{-\lambda E} = m^2c^4, \tag{2}$$

which is a deformed dispersion relation.

A relationship between noncommutativity and a deformed dispersion relation also arises in the context of special relativity proposal with an invariant length [27]. In this instance, one modifies the boost generator and from the invariant quantity

$$||p||^2 = m^2c^4 = \frac{p^\mu p_\mu}{(1 - L_P E)^2}, \tag{3}$$

where L_P is the invariant length, a deformed dispersion relation follows. The thermodynamics of an ideal fluid following the deformed dispersion relation Eq. (3) was investigated in Ref. [29].

From the previous discussion, it is natural to expect that NC arises at high energies, and hence in the early universe [20, 30]. However, in this contribution one discusses the results of Ref. [31] where NC low-energy effects on astrophysical objects are investigated. Also here the noncommutativity is introduced via a deformed dispersion relation. Using the grand canonical ensemble, one finds the leading noncommutative correction to particle number, energy densities and pressure. This method is applied to radiation, non-relativistic ideal gas and degenerate fermion gas and then used to describe the main features of main-sequence stars, white dwarfs and neutron stars.

2. Deformed Dispersion Relation

One can define a simplified version of Eq. (2) as in Ref. [20]

$$E^2 = p^2c^2(1 + \lambda E)^2 + m^2c^4. \tag{4}$$

Indeed, performing $\lambda \to 2\lambda$ in Eq. (2), Eq. (4) matches Eq. (2) up to the first order in λ [20]. Solving for E one gets

$$E = \frac{\lambda p^2c^2 + \sqrt{p^2c^2 + m^2c^4(1 - \lambda^2p^2c^2)}}{1 - \lambda^2p^2c^2}. \tag{5}$$

The general behaviour of Eq. (5) is depicted in Fig. 1. One considers the particle branch, where all energies can be attained although there is a maximum momentum given by $p_{max} = 1/\lambda c$. Thus the parameter λ is associated with the maximum momentum [30]. Since one aims to investigate energy scales encountered in astrophysical objects and λ is presumably related to the inverse of the quantum gravity energy scale, it suffices to keep only the first correction in λ. Therefore, at first order in λ, Eq. (5) reads

$$E = \lambda p^2c^2 + \sqrt{p^2c^2 + m^2c^4}. \tag{6}$$

Notice that the usual relativistic dispersion relation is recovered for $\lambda \to 0$.

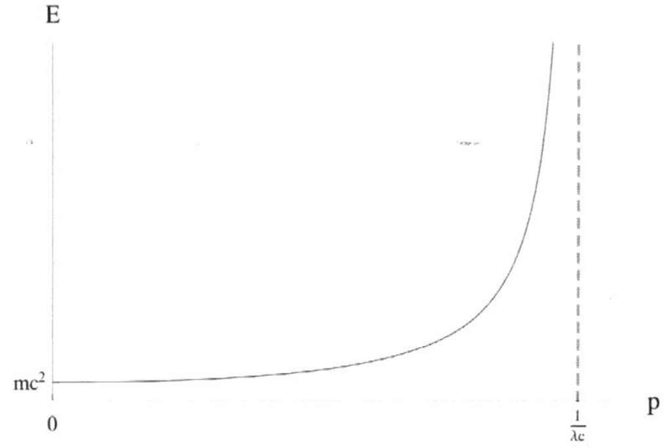

Figure 1. Deformed dispersion relation for the particle branch of Eq. (5).

3. Deformed Statistical Mechanics

The foundations of statistical mechanics are not modified for more general dispersion relations [20, 32, 33]. One considers the grand canonical ensemble, since it is the suitable formalism to describe radiation, ideal gas and degenerate fermion gas. Consider a system with N particles. Each state is labeled by j ($j = 1, 2, \cdots$) and corresponds to n_j particles with energy E_j. One defines the fugacity as $e^{\beta\mu}$, where μ is the chemical potential, $\beta = (k_B T)^{-1}$, k_B is the Boltzmann constant and T the absolute temperature. The grand canonical partition function is defined as

$$\mathcal{Z} = \sum_{n_j}\prod_j \left[ze^{-\beta E_j}\right]^{n_j} = \sum_{n_j}\prod_j \left[e^{\beta(\mu - E_j)}\right]^{n_j}. \tag{7}$$

Through the grand canonical potential, Φ, one obtains the thermodynamics [32]

$$\Phi = -PV = -\frac{1}{\beta}\ln\mathcal{Z} = -\frac{1}{a\beta}\sum_j \ln(1 + aze^{-\beta E_j}) \tag{8}$$

where P is the pressure, V the volume, $a = 1$ for fermions and $a = -1$ for bosons. Using Eq. (8), the pressure is obtained in the large-volume limit $\sum_E \to \int \frac{d^3\vec{x}d^3\vec{p}}{(2\pi\hbar)^3}$,

$$\Phi = -PV = -\frac{\gamma}{a\beta}\int \frac{d^3\vec{x}d^3\vec{p}}{(2\pi\hbar)^3}\ln(1 + aze^{-\beta E(\vec{x},\vec{p})}), \tag{9}$$

where $E = E(\vec{x},\vec{p})$ and γ is the multiplicity of states due to spin. At this point one has to introduce an specific relationship for $E = E(\vec{x},\vec{p})$ and one chooses Eq. (5) for this function. Since the deformed dispersion relation Eq. (5) depends only on the absolute value of the momentum, one can compute the configuration variable integral in Eq. (9), which yields the volume V. The pressure then reads

$$P = \frac{\gamma}{2\pi^2\hbar^3}\int_0^{\frac{1}{\lambda c}} dp \left(\frac{p^2}{z^{-1}e^{\beta E} + a}\right)\left(\frac{p}{3}\frac{dE}{dp}\right). \tag{10}$$

Notice that the upper limit of this integral is $p_{max} = (\lambda c)^{-1}$. The average number of particles and the average energy are given respectively by

$$\langle N\rangle = -\left(\frac{\partial\Phi}{\partial\mu}\right)_\beta = \sum_E n(E), \qquad \langle E\rangle = \sum_E E\, n(E), \tag{11}$$

where $n(E) = \left(z^{-1}e^{\beta E} + a\right)^{-1}$ is the occupation number of particles with energy E. Considering the large-volume limit and integrating, one finds the particle number and energy densities

$$\frac{N}{V} = \frac{\gamma}{2\pi^2\hbar^3}\int_0^{\frac{1}{\lambda c}} dp \left(\frac{p^2}{z^{-1}e^{\beta E}+a}\right), \tag{12}$$

$$u = \frac{E}{V} = \frac{\langle E\rangle}{V} = \frac{\gamma}{2\pi^2\hbar^3}\int_0^{\frac{1}{\lambda c}} dp\, p^2 \left(\frac{E}{z^{-1}e^{\beta E}+a}\right). \tag{13}$$

3.1. Deformed Radiation

A gas of photons is used to describe radiation. For photons, the deformed dispersion relation Eq. (4) reads

$$E = pc(1+\lambda E). \tag{14}$$

To obtain the energy density one substitutes Eq. (14) into Eq. (13) with $a=-1$, $\mu=0$ and $\gamma=2$, that is:

$$u = \frac{1}{\pi^2\hbar^3c^3}\int_0^\infty dE \frac{E^3}{e^{\beta E}-1}\frac{1}{(1+\lambda E)^4}. \tag{15}$$

This integral cannot be solved analytically. The approximation $\lambda k_B T \ll 1$ is introduced. This limit is justified for astrophysical objects since the highest temperatures attained in stars are around $(10^{11} - 10^{12})$ K [34], yielding the condition $\lambda \ll 10$ GeV^{-1} which is satisfied if $\lambda \sim E_{QG}^{-1}$ where E_{QG} is the quantum gravity energy scale. The limit, $\lambda k_B T \gg 1$, presumably relevant at the early universe, was investigated in Ref. [20].

Defining a variable $y = \lambda E$. Eq. (15) can be written as

$$u = \frac{1}{\pi^2\hbar^3c^3\lambda^4}\int_0^\infty dy \frac{y^3}{e^{\frac{y}{\lambda k_B T}}-1}\frac{1}{(1+y)^4}. \tag{16}$$

As $\lambda k_B T \ll 1$, it suffices to expand $y^3/(1+y)^4$ in Taylor series around zero, since away from the origin other values will be exponentially suppressed by $(e^{\frac{y}{\lambda k_B T}}-1)^{-1}$. One gets[1]

$$u = \frac{1}{\pi^2\hbar^3c^3\lambda^4}\left[\int_0^\infty dy \frac{y^3}{e^{\frac{y}{\lambda k_B T}}-1} - 4\int_0^\infty dy \frac{y^4}{e^{\frac{y}{\lambda k_B T}}-1}\right] = \frac{4\sigma}{c}T^4 - \frac{96\zeta(5)}{\pi^2\hbar^3c^3}\lambda k_B^5 T^5, \tag{17}$$

where $\sigma = \frac{\pi^2 k_B^4}{60\hbar^3 c^2}$ is the Stefan-Boltzmann constant. Notice that the Stefan-Boltzmann law is recovered for $\lambda \to 0$. This expression can be rewritten as $u = 4\sigma_{\text{eff}}(T;\lambda)T^4/c$, where the effective Stefan-Boltzmann "constant" is given by:

$$\sigma_{\text{eff}}(T;\lambda) = \sigma\left(1 - \frac{1440\zeta(5)}{\pi^4}\lambda k_B T\right). \tag{18}$$

So the leading noncommutative correction reduces the energy density. Using Eq. (10), the pressure is

$$P = \frac{1}{3\pi^2\hbar^3c^3}\int_0^\infty dE\, \frac{E^3}{e^{\beta E}-1}\frac{1}{(1+\lambda E)^3}. \tag{19}$$

This integral can be solved in the same way as the energy density integral. One obtains

[1] Using the formula $\int_0^\infty \frac{x^{\nu-1}\,dx}{e^{\mu x}-1} = \frac{1}{\mu^\nu}\Gamma(\nu)\zeta(\nu)$ $[\text{Re}\mu > 0, \text{Re}\nu > 0]$, where $\Gamma(\nu)$ and $\zeta(\nu)$ are gamma and zeta functions, respectively [35].

$$P = \frac{1}{\pi^2\hbar^3c^3\lambda^4}\left[\frac{1}{3}\int_0^\infty dy\frac{y^3}{e^{\frac{y}{\lambda k_B T}}-1} - \int_0^\infty dy\frac{y^4}{e^{\frac{y}{\lambda k_B T}}-1}\right] = \frac{4\sigma}{3c}T^4 - \frac{24\zeta(5)}{\pi^2\hbar^3c^3}\lambda k_B^5 T^5. \tag{20}$$

To find an Equation of State (EoS), one must divide Eq. (20) by Eq. (17)

$$\frac{P}{u} = \frac{1}{3} + (\lambda k_B T)\frac{120\zeta(5)}{\pi^4}, \tag{21}$$

Using Eq. (17), one writes T as a power series in λ. Substituting this result into Eq. (21) one finally gets the leading noncommutative correction to the EoS

$$P = \frac{u}{3}\left[1 + \frac{360\zeta(5)}{\pi^4}\left(\frac{15}{\pi^2}\right)^{1/4}\lambda(\hbar c)^{3/4}u^{1/4}\right]. \tag{22}$$

Notice that the usual relationship $u = 3P$ is recovered for $\lambda \to 0$.

3.2. Non-relativistic Ideal Gas

Consider Eq. (6), hence for a non-relativistic ideal gas $p \ll mc$, it reads

$$E = \lambda p^2c^2 + mc^2 + \frac{p^2}{2m} + \mathcal{O}\left(\frac{p}{mc}\right)^4. \tag{23}$$

The condition to be satisfied in order to ensure the relevance of the noncommutative correction is $\lambda \lesssim (2mc^2)^{-1}$. However, for matter found in main-sequence stars, hydrogen for simplicity, one finds $\lambda \lesssim 1\ \text{GeV}^{-1}$, which certainly is a too stringent bound. Hence, one does not consider any noncommutative correction to the non-relativistic ideal gas. The pressure is the well know expression [32] $P = Nk_BT/V$. After introducing the mean molecular weight $\mu_N = \rho/(nm_N)$, where $n = N/V$, ρ is the mass density and m_N is the nucleon mass[2], the pressure reads

$$P = \frac{\rho}{\mu_N m_N}k_B T. \tag{24}$$

3.3. Deformed Degenerate Fermion Gas

A fermion gas becomes degenerate at low temperatures ($T \to 0$). In this case, the fermionic character gives origin to a very simple expression for the occupation number. In terms of the momentum, the occupation number is $n(p) = H(p_F - p)$, where p_F is the Fermi momentum, above which there are no occupied states and $H(x)$ is the Heaviside step function. For stars, the momentum varies between MeV/c – GeV/c, so $p_F \ll (\lambda c)^{-1}$ and one uses Eq. (6) as the deformed dispersion relation.

For spin one-half particles ($\gamma = 2$), the particle number density Eq. (12) is given by

$$n = \frac{N}{V} = \frac{1}{\pi^2\hbar^3}\int_0^{\frac{1}{\lambda c}} dp\, n(p)p^2 = \frac{1}{\pi^2\hbar^3}\int_0^{p_F} dp p^2 = \frac{p_F^3}{3\pi^2\hbar^3} = \frac{(mc)^3x^3}{3\pi^2\hbar^3}, \tag{25}$$

where $x = p_F/mc$. Using Eqs. (6) and (13), one obtains the energy density

$$\begin{aligned} u &= \frac{E}{V} = \frac{1}{\pi^2\hbar^3}\int_0^{p_F} dp\, p^2\sqrt{p^2c^2+m^2c^4} + \frac{\lambda c^2}{\pi^2\hbar^3}\int_0^{p_F} dp\, p^4 \\ &= \frac{(mc^2)^4}{\pi^2(\hbar c)^3}\left(\frac{x^3\sqrt{1+x^2}}{3} - \frac{f(x)}{24} + (\lambda mc^2)\frac{x^5}{5}\right). \end{aligned} \tag{26}$$

[2] $m_N = 1\text{amu} = 931,494\ \text{MeV/c}^2$[36].

where $f(x) = x(2x^2-3)\sqrt{1+x^2} + 3\sinh^{-1}x$ [37]. This shows that the first noncommutative correction is given by λmc^2. The pressure is obtained using Eqs. (6) and (10):

$$P = \frac{c^2}{3\pi^2\hbar^3}\int_0^{p_F} dp\, \frac{p^4}{\sqrt{p^2c^2+m^2c^4}} + \frac{2\lambda c^2}{3\pi^2\hbar^3}\int_0^{p_F} dp\, p^4 = \frac{(mc^2)^4}{\pi^2(\hbar c)^3}\left(\frac{f(x)}{24} + (\lambda mc^2)\frac{2x^5}{15}\right). \quad (27)$$

4. Application to Astrophysical Objects

4.1. Main-sequence Stars: The Sun

The simplest model for a main-sequence star consists in assuming that the star is composed by a non-relativistic ideal gas and radiation that are maintained in equilibrium by gravity [37]. The total pressure P comprises the sum of the radiation pressure $P_{\rm rad}$ (Eq. (20)) and the gas pressure $P_{\rm gas}$ (Eq. (24)). One denotes the ratio between $P_{\rm gas}$ and P by β_S. One then has $P = P_{\rm gas}/\beta_S = P_{\rm rad}/(1-\beta_S)$. Substituting Eqs. (20) and (24) into this relationship one finds

$$\frac{\rho}{\mu_N m_N \beta_S}k_BT = \frac{4\sigma}{3c(1-\beta_S)}T^4 - \lambda\frac{24\zeta(5)k_B^5}{\pi^2\hbar^3c^3(1-\beta_S)}T^5, \quad (28)$$

The aim is to obtain $P = P(\rho)$, so one must solve this equation for T, obtaining an expression $T = T(\rho)$ up to the first order in λ. Inserting this result into $P_{\rm gas} = \beta P$ and using Eq. (24) one finds that $P = K_1\rho^{4/3} + \lambda K_2\rho^{5/3}$ where

$$K_1 = \frac{(1-\beta_S)^{1/3}}{(\beta_S\mu_N)^{4/3}}\left(\frac{3ck_B^4}{4\sigma m_N^4}\right)^{1/3} = 2.67\times10^{10}\left[\frac{(1-\beta_S)^{1/3}}{(\beta_S\mu_N)^{4/3}}\right]\frac{\text{J m}}{\text{kg}^{4/3}} \quad (29)$$

$$K_2 = \frac{(1-\beta_S)^{2/3}}{(\beta_S\mu_N)^{5/3}}\left(\frac{360\zeta(5)}{\pi^4}\right)\left(\frac{3ck_B^4}{4\sigma m_N^{5/2}}\right)^{2/3} = 4.52\times10^{-6}\left[\frac{(1-\beta_S)^{2/3}}{(\beta_S\mu_N)^{5/3}}\right]\frac{\text{J}^2\text{ m}^2}{\text{kg}^{5/3}}. \quad (30)$$

One assumes the Eddington's Standard Model of Stars [34]: β_S is constant inside the star, and μ_N is constant since the chemical composition is not supposed to change. Hence, one finds that the pressure is a contribution of two polytropes: a $n=3$ polytrope for the mixture of gas and radiation and a $n=3/2$ polytrope for the NC contribution, which is considered as a perturbation. The question of stability can be understood as follows. Defining $\Gamma = 1+\frac{1}{n}$, a $n=3$ ($\Gamma = 4/3$) polytrope represents a marginally stable configuration [34]. If one introduces corrections due to general relativity, one shows that, in fact, this polytrope is unstable. One follows here the purely Newtonian analysis [34]. The total energy is written as the sum of the internal energy ($\sim PV$) and the gravitational potential energy ($\sim\frac{GM^2}{R}$). The pressure is given by $P = K_1\rho^{\Gamma_1} + \lambda K_2\rho^{\Gamma_2}$ and to ensure that the second term is a perturbation $\lambda\frac{K_2}{K_1}\rho^{\Gamma_2-\Gamma_1}\ll 1$. The energy can be written as

$$E = k_0PV - k_1\frac{GM^2}{R} = C_1M\rho_c^{\Gamma_1-1} + C_2M\rho_c^{\Gamma_2-1} - k_3M^{5/3}\rho_c^{1/3}, \quad (31)$$

where k_0, k_1, k_3, C_1, C_2 are constants and ρ_c is the central density. The value of M at the critical point $\left(\frac{\partial E}{\partial\rho_c}\right) = 0$ is computed and one performs

$$\frac{d\ln M}{d\ln\rho_c} = \frac{3}{2}\left(\Gamma_1 - \frac{4}{3}\right) + \lambda\frac{3C_2(\Gamma_2-1)}{2C_1(\Gamma_1-1)}(\Gamma_2-\Gamma_1)\rho_c^{\Gamma_2-\Gamma_1}. \quad (32)$$

A necessary, but not sufficient, condition for stability of the configuration is that $\frac{d\ln M}{d\ln\rho_c} > 0$, for M given by $\frac{\partial E}{\partial\rho_c} = 0$ [38]. Here, $\Gamma_1 = 4/3$ and $\Gamma_2 = 5/3 > \Gamma_1$, so $\frac{d\ln M}{d\ln\rho_c} > 0$. Thus, the effect of noncommutativity is to turn this configuration into a stable one.

To quantify the relevance of noncommutativity for these type of stars one chooses the Sun, as it is a very typical main-sequence star. In the model investigated here, at the center, $1 - \beta_S \approx 10^{-3}$, $\rho_c = 1.53 \times 10^5$ kg/m^3 and $\mu_N = 0.829$ [34]. The ratio between the noncommutative and the commutative contributions to the pressure is

$$\frac{P_{NC}}{P_C} = \frac{\lambda K_2 \rho^{5/3}}{K_1 \rho^{4/3}} = 9.66 \times 10^{-16} \lambda, \tag{33}$$

where $P_{NC} = \lambda K_2 \rho^{5/3}$ is the noncommutative correction to the pressure, $P_C = K_1 \rho^{4/3}$ is the usual (commutative) term and λ must be expressed in S.I. units. One computed this value at the center of the star using Eqs. (29) and (30). The value of λ will be discussed later (Section 5), but one sees that this ratio is fairly small.

4.2. White Dwarfs

As one has seen, for main-sequence stars the noncommutative correction is rather small and, even though it plays a role in the stability of the star, one should seek more energetic configurations. White dwarfs are stars that have masses comparable to th Sun and planet sizes [39]. At their center, the range of possible mass densities is 10^8 kg/m$^3 \lesssim \rho_c \lesssim 10^{12}$ kg/m^3. At these densities the matter is completely ionized and the electrons are free. The treatment of matter must necessarily involve quantum mechanics. The simplest model of white dwarfs consists in considering an ideal electronic gas maintained in equilibrium by gravity. Although the central temperature is $T \sim 10^7$ K, it is much smaller than the temperature associated with the Fermi energy ($T \sim 10^{10}$ K). This justifies describing white dwarfs, at first approximation, as a degenerate electron gas.

Defining the electronic molecular weight as $\mu_e = \rho/n_e m_p$, where n_e is the electronic particle number density and m_p is the proton mass. One can rewrite Eq. (25) as

$$x = (3\pi^2)^{1/3} \frac{\hbar}{m_e c} n_e^{1/3} = \left(\frac{3\pi^2}{m_p}\right)^{1/3} \frac{\hbar}{m_e c} \left(\frac{\rho}{\mu_e}\right)^{1/3} = 10^{-3} \left(\frac{\rho}{\mu_e}\right)^{1/3}, \tag{34}$$

where m_e is the mass of the electron and this result is presented in S.I. units. One considers $\mu_e = 2$, as this corresponds to a star composed mainly by Helium. For the possible values of the central density, one chooses a low density value $\rho = 10^8$ kg/m^3, an intermediate density ($p_F \sim m_e c$) value $\rho = 2 \times 10^9$ kg/m^3 and a high one $\rho = 10^{12}$ kg/m^3. One finds $x_8 = 0.37$, $x_9 = 1.01$, $x_{12} = 7.99$, respectively. One now quantifies the relevance of noncommutativity, using Eq. (27):

$$\frac{P_{NC}}{P_C} = \frac{48 x^5}{15 f(x)} \lambda m_e c^2. \tag{35}$$

This function is plotted in Fig. 2 for the usual values of x found in white dwarfs. One sees that the relevance of noncommutativity is just a few $\lambda m_e c^2$. Since $m_e c^2 \sim 0.5$ MeV and $\lambda \sim (E_{QG})^{-1}$ (see Section 5), this correction is much smaller than other improvements to the simplest white dwarf model, such as Coulomb interaction for low densities, and general relativity for high densities. Also neglected are the corrections due to rotation and magnetic fields, which are likely to be greater than the noncommutative correction.

Let us now analyse the problem of stability (for details see e.g. Ref. [34]): the noncommutative correction is proportional to $x^5 \propto \rho^{5/3}$ (Eq. (34)) which represents a $\Gamma = 5/3$ polytrope. This lies in the region of stability and hence noncommutativity does not bring any instability for white dwarfs.

Journal of Physics: Conference Series **222** (2010) 012020 doi:10.1088/1742-6596/222/1/012020

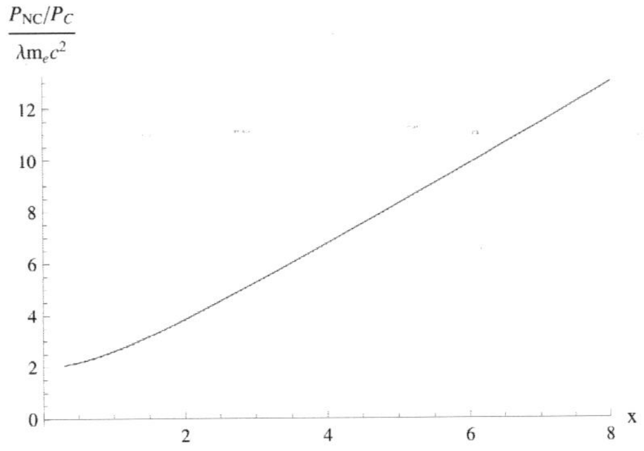

Figure 2. Ratio between the noncommutative and the commutative contributions to the pressure in units of $\lambda m_e c^2$ for white dwarfs and neutron stars.

4.3. Neutron Stars

Consider now that $\rho \gtrsim 10^{12}$ kg/m^3, a density for which the protons and electrons combine and through the weak interaction form neutrons. These objects must be described by general relativity since $\frac{GM}{Rc^2} \sim 0.1$. Their masses are about the solar mass ($M \sim 1M_\odot$) and their radius about $R \sim 10$ km.

These stars are mainly composed by neutrons, hence the simplest model for a neutron star, the Oppenheimer-Volkoff (OV) model [40], considers an ideal neutron gas counterbalanced by gravity. This model is rather unrealistic as one must introduce nuclear interaction effects. However, this simpler formulation has some advantages, the first being that one can compute the thermodynamical quantities in a closed and analytical form. Furthermore, the obtained results do not differ significantly from a more realistic EoS [38]; finally the OV EoS is stiffer than other EoS that include nuclear interaction [41]. This neutron gas is degenerate since the Fermi energy of neutrons (~ 1 GeV) is much greater than the thermal energy associated to the gas ($\sim 10^{-1}$ MeV).

Using Eqs. (25), (26) and (27), one obtains

$$x = \frac{\hbar}{m_n c}\left(3\pi^2 n\right)^{1/3}, \tag{36}$$

$$u = \frac{(m_n c^2)^4}{\pi^2(\hbar c)^3}\left(\frac{x^3\sqrt{1+x^2}}{3} - \frac{f(x)}{24} + (\lambda m_n c^2)\frac{x^5}{5}\right), \tag{37}$$

$$P = \frac{(m_n c^2)^4}{\pi^2(\hbar c)^3}\left(\frac{f(x)}{24} + (\lambda m_n c^2)\frac{2x^5}{15}\right), \tag{38}$$

where m_n is the neutron mass. One cannot obtain the EoS $P = P(u)$ in an analytical form. The relevance of noncommutativity is depicted in Fig. 2, with the modification $m_e \to m_n$. For the usual baryonic matter, $n = 0.15 \times 10^{45}$ m^{-3} [38], one obtains $x = 0.35$ and inserting this value into Eq. (38), one obtains that $\frac{P_{NC}}{P_C} = 2.1(\lambda m_n c^2)$. Notice that as $m_n \sim 10^3 m_e$, the effect of noncommutativity is much more important for this class of stars than for white dwarfs.

For neutron stars the question of stability is more complex since general relativity is required. It involves solving numerically the Oppenheimer-Volkoff equation with the OV EoS.

5. Discussions and conclusions

In this contribution, the most relevant effects of noncommutativity for astrophysical objects were estimated. Noncommutativity is introduced through a deformed dispersion relation and the relevant thermodynamical quantities were calculated using the grand canonical ensemble. These results are applied to physical models describing main-sequence stars, white-dwarfs and neutron stars.

In what concerns the noncommutative parameter, λ, (cf. Eqs. (1) and (6)), an upper bound for its value can be inferred from the breaking of Lorentz invariance applied to the problem of ultra-high energetic cosmic rays. The most stringent limit is $\lambda < 2.5\times10^{-19}$ GeV^{-1} $= 1.6\times10^{-9}$ J^{-1}. For the stars, one is certainly in the low-energy limit, and hence this justifies the expansion of the quantities up to first order in λ.

Using $\lambda \sim 10^{-19}$ GeV^{-1}, one can estimate the ratio between P_{NC} and P_C for the main-sequence stars (as an example one chooses the Sun), white-dwarfs and neutron stars. One obtains $\frac{P_{NC}}{P_C} \sim 10^{-25}$, $\frac{P_{NC}}{P_C} \sim 10^{-22}$ and $\frac{P_{NC}}{P_C} \sim 10^{-19}$ (cf. Eqs. (33), (35) and (38)), respectively. These results indicate that noncommutative correction is fairly small, actually smaller than other usually neglected standard effects in stellar physics. As one expects the relevance of noncommutativity to grow as one considers denser configurations, one is lead to conclude that noncommutative might be relevant for black holes. In fact, full phase-space noncommutativity is shown to be quite important for Schwarzschild black holes [42].

For main sequence stars and white-dwarfs, the question of stability has been analysed and shown that noncommutativity does not introduce any instability. Actually, for main sequence stars, noncommutativity is beneficial for the star stability.

Acknowledgments

The work of C. A. D. Z. is fully supported by the FCT fellowship SFRH/BD/29446/2006.

References

[1] Connes A 1996 *Commun. Math. Phys.* **182** 155 (*Preprint* hep-th/9603053)
[2] Madore J 1999 *Preprint* arXiv:gr-qc/9906059
[3] Witten E 1986 *Nucl. Phys.* B **268** 253
[4] Seiberg N and Witten E 1999 *J. High Energy Phys.* JHEP09(1999)032 (*Preprint* hep-th/9908142)
[5] Szabo R J 2003 *Phys. Rept.* **378** 207 (*Preprint* hep-th/0109162)
[6] Douglas M R and Nekrasov N A 2001 *Rev. Mod. Phys.* **73** 977 (*Preprint* hep-th/0106048)
[7] Bertolami O and Guisado L 2003 *J. High Energy Phys.* JHEP12(2003)013 (*Preprint* hep-th/0306176)
[8] Lizzi F, Mangano G, Miele G and Peloso M 2002 *J. High Energy Phys.* JHEP06(2002)049 (*Preprint* hep-th/0203099)
[9] Bertolami O and Guisado L 2003 *Phys. Rev.* D **67** 025001 (*Preprint* gr-qc/0207124)
[10] Bertolami O and Zarro C A D 2009 *Phys. Lett.* B **673** 83 (*Preprint* 0812.4607 [hep-th])
[11] Zhang J Z 2004 *Phys. Rev. Lett.* **93** 043002 (*Preprint* hep-ph/0405143); Zhang J Z 2004 *Phys. Lett.* B **584** 204 (*Preprint* hep-th/0405135)
[12] Bertolami O, Rosa J G, de Aragão C M L, Castorina P and Zappalà D 2005 *Phys. Rev.* D **72** 025010 (*Preprint* hep-th/0505064)
[13] Acatrinei C S 2005 *Mod. Phys. Lett.* A **20** 1437 (*Preprint* hep-th/0311134)
[14] Bertolami O, Rosa J G, de Aragão C M L, Castorina P and Zappalà D 2006 *Mod. Phys. Lett.* A **21** 795 (*Preprint* hep-th/0509207)
[15] Bastos C and Bertolami O 2008 *Phys. Lett.* A **372** 5556 (*Preprint* gr-qc/0606131)
[16] Bastos C, Bertolami O, Dias N C and Prata J N 2008 *J. Math. Phys.* **49** 072101 (*Preprint* hep-th/0611257)
[17] Gamboa J, Loewe M, Mendez F and Rojas J C 2001 *Mod. Phys. Lett.* A **16** 2075 (*Preprint* hep-th/0104224); Gamboa J, Loewe M and Rojas J C 2001 *Phys. Rev.* D **64** 067901 (*Preprint* hep-th/0010220); Balachandran A P, Queiroz A R, Marques A M and Teotonio-Sobrinho P 2008 *Phys. Rev.* D **77** 105032 (*Preprint* 0706.0021 [hep-th])
[18] Snyder H S 1947 *Phys. Rev.* **71** 38
[19] Barosi L, Brito F A and Queiroz A R . *Cosmol. Astropart. Phys.* JCAP04(2008)005 (*Preprint* 0801.0810 [hep-th])
[20] Alexander S and Magueijo J 2001 *Preprint* hep-th/0104093
[21] Alvarez-Gaumé L, Meyer F and Vazquez-Mozo M A 2006 *Nucl. Phys.* B **753** 92 (*Preprint* hep-th/0605113)
[22] Meyer F 2005 *Preprint* hep-th/0510188.
[23] Amelino-Camelia G and Majid S 2000 *Int. J. Mod. Phys.* A **15** 4301 (*Preprint* hep-th/9907110)

[24] Bertolami O and Carvalho C S 2000 *Phys. Rev.* D **61** 103002 (*Preprint* gr-qc/9912117)
[25] Bertolami O 2003 *Lect. Notes Phys.* **633** 96 (*Preprint* hep-ph/0301191)
[26] Bertolami O 2003 Talk given at 38th Rencontres de Moriond on Gravitational Waves and Experimental Gravity, Les Arcs, Savoie, France, 22-29 Mar 2003, *Preprint* gr-qc/0306093
[27] Magueijo J and Smolin L 2002 *Phys. Rev. Lett.* **88** 190403 (*Preprint* hep-th/0112090)
[28] Amelino-Camelia G 2002 *Int. J. Mod. Phys.* D **11** 35 (*Preprint* gr-qc/0012051)
[29] Das S, Ghosh S and Roychowdhury D 2009 *Preprint* 0908.0413 [hep-th]
[30] Alexander S, Brandenberger R and Magueijo J 2003 *Phys. Rev.* D **67** 081301 (*Preprint* hep-th/0108190)
[31] Bertolami O and Zarro C A D 2009 to appear in *Phys. Rev.* D (*Preprint* 0908.4196 [gr-qc])
[32] Landau L D and Lifshitz E M 1980 *Statistical Physics* (Course of Theoretical Physics vol 5) (Oxford: Butterworth-Heinemann)
[33] Padmanabhan T 2000 *Theoretical Astrophysics Volume I: Astrophysical Processes* (Cambridge: Cambridge University Press 2000)
[34] Padmanabhan T 2001 *Theoretical Astrophysics Volume II: Stars and Stellar Systems* (Cambridge: Cambridge University Press)
[35] Gradshteyn I S and Ryzhik I M 2000 *Table of Integrals, Series and Products* (San Diego: Academic Press)
[36] Amsler C *et al.* [Particle Data Group] 2008 *Phys. Lett.* B **667** 1
[37] S. Chandrasekhar 1967 *An Introduction to the Study of Stellar Structure* (Mineola: Dover)
[38] Glendenning N K 1997 *Compact stars: Nuclear physics, particle physics, and general relativity* (New York: Springer 1997)
[39] Bhatia V B 2001 *Textbook of Astronomy and Astrophysics With Elements of Cosmology* (Pangbourne: Alpha Science International)
[40] Oppenheimer J R and Volkoff G M 1939 *Phys. Rev.* **55** 374
[41] Fraga E S, Pisarski R D and Schaffner-Bielich J 2002 *Nucl. Phys.* A **702** 217 (*Preprint* nucl-th/0110077); Fraga E S, Pisarski R D and Schaffner-Bielich J 2001 *Phys. Rev.* D **63** 121702 (*Preprint* hep-ph/0101143)
[42] Bastos C, Bertolami O, Dias N C and Prata J N 2009 to appear in *Phys. Rev.* D (*Preprint* 0907.1818 [gr-qc])

First Mediterranean Conference on Classical and Quantum Gravity (MCCQG 2009) IOP Publishing
Journal of Physics: Conference Series **222** (2010) 012021 doi:10.1088/1742-6596/222/1/012021

Gravitational waves from cosmological phase transitions

Ruth Durrer

Département de Physique Théorique, Université de Genève
24 Quai E. Ansermet, 1211 Genéve 4, Switzerland

E-mail: ruth.durrer@unige.ch

Abstract. In this talk I discuss the generation of a stochastic background of gravitational waves during a first order phase transition. I present simple general arguments which explain the main features of the gravitational wave spectrum like the k^3 power law growth on large scales and a estimate for the peak amplitude. In the second part I concentrate on the electroweak phase transition and argue that the nucleosynthesis bound on its gravitational wave background seriously limits seed magnetic fields which may have been generated during this transition.

1. Introduction

The Universe has expanded and cooled down from a very hot initial state to (presently) 2.7^oK. It seems likely that it underwent several phase transitions during its evolution of adiabatic expansion. It has already been proposed in the 80ties, that a cosmological phase transition can lead to the generation of a stochastic gravitational wave background [1]. If the phase transitions are of second order or only a crossover as it has been obtained from standard model calculations, we do not expect them to lead to an observable cosmological signal. However, if a phase transition is first order, it proceeds via bubble nucleation which is a highly inhomogeneous process. Some of the kinetic energy of the rapidly expanding bubbles is transferred to the cosmic plasma and leads to turbulence. If small seed magnetic fields are present, they are amplified by the turbulent motion which becomes MHD turbulence. Since the fluid and magnetic Reynolds numbers in the hot cosmic plasma are extremely high, MHD turbulence persists even after the phase transition, where it freely decays. These inhomogeneities int the energy and momentum of the cosmic plasma generate a stochastic background of gravitational waves. Here we want to investigate this background.

1.1. Events in the early Universe

The following important events have most probably taken place in the early Universe and they may, under certain circumstances have induced a background of stochastic gravitational waves.

- *Inflation* is supposed to have generated the scalar fluctuations in the geometry and energy density which have led to the observed anisotropies in the cosmic microwave background (CMB) [2, 3]. Simple inflationary models do not only induce scalar perturbations but the quantum fluctuations in the gravitational spin-2 degrees of freedom also lead to a scale invariant background of gravitational waves of similar amplitude. It is one of the most

Journal of Physics: Conference Series **222** (2010) 012021 doi:10.1088/1742-6596/222/1/012021

important goals of present CMB experiments like Planck [4] to observe the B-polarization which this gravitational wave background should induce in the CMB photons.

- *Pre-heating:* At the end of inflation the inflaton decays into other degrees of freedom which eventually reheat the Universe and lead to a thermal bath of relativistic degrees of freedom containing especially also the standard model particles. The details of this process are complicated and in some cases they generate inhomogeneities leading to a significant stochastic background of gravitational waves [5]. Using the standard relation between temperature and time in a Friedmann Universe filled with relativistic particles [3], we obtain for a reheat temperature T_i

$$T_i \simeq 10^{14}\text{GeV}, \quad t_i = 2.3\text{sec}\left(\frac{1\text{MeV}}{T_i}\right)^2 g_{\text{eff}}(T)^{-1/2} \simeq 10^{-32}\text{sec}, \tag{1}$$

$$\eta_i = \frac{1}{a(t_i)H(t_i)} = 2t_i(1+z_i) \simeq 10^{-7}\text{sec}, \quad \omega_i \gtrsim \frac{1}{\eta_i} \simeq 10^7\text{Hz}, \tag{2}$$

$g_{\text{eff}}(T)$ denotes the effective number of relativistic degrees of freedom at temperature T [3]. To find the conformal time η_i we have used that in a radiation dominated universe

$$\eta = \frac{1}{\mathcal{H}} = \frac{1}{aH} = \frac{1+z}{H} = 2t(1+z),$$

where $\mathcal{H} = aH$ is the comoving Hubble rate and a is the scale factor normalized to one today, $a(t_0) = 1$. Furthermore, since $g_{\text{eff}}(aT)^3$ is constant during adiabatic expansion

$$1+z = \frac{1}{a} = \frac{g_{\text{eff}}^{1/3}T}{g_0^{1/3}T_0} = \frac{10^{10}}{2.35}\left(\frac{T}{1\text{MeV}}\right)\left(\frac{g_{\text{eff}}}{2}\right)^{1/3},$$

so that

$$\eta = 0.67 \times 10^{10}\left(\frac{T}{1\text{MeV}}\right) g_{\text{eff}}^{1/3} t \simeq 1.5 \times 10^{10}\text{sec}\left(\frac{1\text{MeV}}{T}\right) g_{\text{eff}}(T)^{-1/6}. \tag{3}$$

We have set $k_B = 1$ and measure temperature in MeV. In the following we use units with $c = \hbar = 1$ so that length and time have the same units as inverse energy.
For the $\gtrsim$ sign in eq. (2) we have used that the typical frequency generated by such an event must be larger than the inverse of the correlation scale which is always smaller than the horizon size at the given epoch. Note that ω_i denotes the comoving frequency and we have normalized the scale factor to unity today, so that co-moving frequencies, length scales and times correspond to physical scales today.
So far no experiment which can detect gravitational waves with frequencies as high as ω_i has been proposed.

- *The electroweak transition:* In the standard model, for realistic values of the Higgs mass, the electroweak transition is not even second order, but only a crossover [6]. However, small deviations from the standard model e.g. in the Higgs sector or supersymmetric models [7] can can lead to a first order electroweak phase transition at temperature

$$T_{ew} \simeq 10^2\text{GeV}, \qquad t_{ew} \simeq 10^{-10}\text{sec}, \tag{4}$$

$$\eta_{ew} \simeq 10^5\text{sec}, \qquad \omega_{ew} \gtrsim 10^{-5}\text{Hz}. \tag{5}$$

If the electroweak phase transition is strongly first order, the bubbles expand very rapidly and the duration of the phase transition is a small fraction, maybe 1% of the Hubble time. In this case one expects a peak frequency of $\omega_{ew} \sim 100/\eta_{ew} \simeq 10^{-3}$Hz. This is the frequency of best sensitivity for the gravitational wave satellite LISA proposed for launch in 2018 [8].

Journal of Physics: Conference Series **222** (2010) 012021 doi:10.1088/1742-6596/222/1/012021

- *Confinement transition:* Also in this case, standard lattice QCD calculations [9] predict a simple crossover. However, the confinement transition can be first order if the neutrino chemical potentials are sufficiently large (still within the limits allowed by nucleosynthesis) [10]. Models which lead to such high neutrino chemical potentials have been proposed for leptogenesis with sterile neutrini [11]. The temperature of the QCD transition is about

$$T_c \simeq 10^2\text{MeV}, \quad t_c \simeq 10^{-5}\text{sec}, \tag{6}$$
$$\eta_c \simeq 10^7\text{sec}, \quad \omega_c \gtrsim 10^{-7}\text{Hz}. \tag{7}$$

  Gravitational waves in this frequency range could be observed by the pulsar timing array which has been proposed recently [12].
- *Surprises:* Last but not least we may hope to find a gravitational wave background from a phase transition at some other not predicted temperature which then would indicate new physics at the corresponding energy scale, see, e.g [13]. For example, a first order phase transition at a critical temperature $T_* \simeq 10^7$GeV may lead to a gravitational wave background with a peak frequency of about 100Hz which is the best sensitivity of LIGO. As we shall argue below, in a very optimistic scenario it is even conceivable that such a background might be detected by the advanced configuration of the LIGO [14] experiment.

A more comprehensive overview on first order phase transitions as sources for gravitational waves can be found in Ref. [15].

In the following we present a semi-analytical determination of the gravitational wave (GW) signal from a first order phase transition in terms of a few free parameters. We then apply our general results to the electroweak phase transition and discuss some consequences.

2. General results

Before we outline their derivation, let us present the main results. We consider a phase transition at $T = T_*$ which generates (relativistic) anisotropic stresses which source gravitational waves. We assume that the energy density of this source is ρ_X. Since the source is relativistic its anisotropic stresses are of the same order of magnitude. The GW spectrum then has the following main properties:

- *The peak frequency* (the inverse of the correlation scale) is larger than the Hubble rate, $k_* \gtrsim \mathcal{H}_*$. Here we always compare comoving quantities which become physical scales today.
- The GW *energy density* is of the order of

$$\Omega_{gw}(\eta_0) \sim \epsilon\, \Omega_{\rm rad}(\eta_0) \left(\frac{\Omega_X(\eta_*)}{\Omega_{\rm rad}(\eta_*)}\right)^2 \qquad \text{where}$$
$$\epsilon = \begin{cases} (\mathcal{H}_*\Delta\eta_*)^2 & \text{if} \quad \mathcal{H}_*\Delta\eta_* < 1 \\ 1 & \text{if} \quad \mathcal{H}_*\Delta\eta_* \geq 1\,. \end{cases} \tag{8}$$

  Here $\Delta\eta_*$ is the duration over which the source is active, typically the duration of the phase transition. Ω is the density parameter, $\Omega_X = \rho_X/\rho_c$ where ρ_X is the energy density of the component X and ρ_c is the critical density.
- On large scales, $k \ll k_*$ the spectrum is blue,

$$\frac{d\Omega_{gw}(k)}{d\log(k)} \propto k^3\,, \qquad \Omega_{gw} = \int \frac{dk}{k}\frac{d\Omega_{gw}(k)}{d\log(k)}\,.$$

  The second equality defines the the gravitational wave energy per logarithmic frequency interval, $\frac{d\Omega_{gw}(k)}{d\log(k)}$.

- On small scales, $k \gg k_*$ the spectrum decays. The decay law depends on the details of the source. If the anisotropic stress spectrum of the source decays with power $k^{-\gamma}$, and the source is totally coherent in time (see below), the gravitational wave energy density decays like $k^{-(\gamma+1)}$. If the source is coherent for about one wavelength, the gravitational wave energy density decays like $k^{-(\gamma-1)}$ and if the source is incoherent, the gravitational wave energy density decays like $k^{-(\gamma-3)}$. see Ref. [17]. The unequal time correlator of the source is discussed in the next section.

3. Generation of gravitational waves

Gravitational waves are sourced by fluctuations in the energy momentum tensor which have a non-vanishing spin-2 contribution. The perturbed metric is of the form

$$ds^2 = a^2\left(-d\eta^2 + (\gamma_{ij} + 2h_{ij})dx^i dx^j\right) , \tag{9}$$

where γ_{ij} is the metric of a 3–space of constant curvature K and h_{ij} is transverse and traceless. In Fourier space $k^i h_{ij} = h^i_i = 0$.

Einstein's equation to first order in h_{ij} gives (we set $K = 0$ in what follows)

$$\left(\partial_\eta^2 + 2\mathcal{H}\partial_\eta + k^2\right) h_{ij} = 8\pi G a^2 \Pi_{ij} \,. \tag{10}$$

Here $\Pi_{ij}(\mathbf{k}, \eta)$ is the Fourier component of the tensor type (spin-2) anisotropic stress of the source and $\mathcal{H} = \frac{a'}{a}$. A prime denotes the derivative w.r.t. conformal time η.

During a first order phase transition, anisotropic stresses can be generated by colliding bubbles and by inhomogeneities in the cosmic fluid (e.g. turbulence, magnetic fields) [16].

For a duration of the phase transition given by $\Delta\eta_*$, the final bubble size, i.e. the typical size of the bubbles when they start to coalesce and the phase transition terminates, is $R_* = v_b \Delta\eta_*$, where v_b is the bubble velocity. Typically $\Delta\eta_* < \mathcal{H}^{-1}$. As we shall see, depending on the unequal time correlation properties of the anisotropic stress, the peak frequency can be given either by the time scale, $k_* \simeq (\Delta\eta_*)^{-1}$, or by the spatial scale $k_* \simeq R_*^{-1}$. For the most promising case of a strongly first order phase transition, the bubble velocity is however close to the speed of light, $v_b \simeq 1$ so that this distinction is not very relevant.

4. The spectrum

Because of causality, the correlator $\langle \Pi_{ij}(\eta_1, \mathbf{x})]\Pi_{lm}(\eta_2, \mathbf{y})\rangle = \mathcal{M}_{ijlm}(\eta_1, \eta_2, \mathbf{x} - \mathbf{y})$ is a function of compact support. For distances $|\mathbf{x} - \mathbf{y}| > \max(\eta_1, \eta_2)$, $\mathcal{M} \equiv 0$. Therefore, the spatial Fourier transform, $\mathcal{M}_{ijlm}(\eta_1, \eta_2, \mathbf{k})$ is analytic in $\mathbf{k}$ around $\mathbf{k} = 0$. We decompose Π_{ij} into two helicity modes which we assume to be uncorrelated so that the spectrum is even under parity

$$\Pi_{ij}(\eta, \mathbf{k}) = e^+_{ij}\Pi_+(\eta, \mathbf{k}) + e^-_{ij}\Pi_-(\eta, \mathbf{k})$$

$$\langle \Pi_+(\eta, \mathbf{k})\Pi_+^*(\eta', \mathbf{k}')\rangle = \langle \Pi_-(\eta, \mathbf{k})\Pi_-^*(\eta', \mathbf{k}')\rangle = (2\pi)^3\delta^3(\mathbf{k} - \mathbf{k}')\rho_X^2 P_\Pi(\eta, \eta', k)$$

$$\langle \Pi_+(\eta, \mathbf{k})\Pi_-^*(\eta', \mathbf{k}')\rangle = 0 \,.$$

Here ρ_X is the energy density of the component X with anisotropic stress Π which has been factorized in order to keep $k^3 P_\Pi(\eta, \eta', k)$ dimensionless. Causality implies that the function $P_\Pi(\eta, \eta', k)$ is analytic in k. We therefore expect it to start out as white noise and to decay beyond a certain correlation scale $k_*(\eta, \eta') > \min(1/\eta, 1/\eta')$.

If the gravitational wave source is active only for a short duration $\Delta\eta_*$ (less than one Hubble time), we can neglect the damping term $2\mathcal{H}$ in the equation of motion for h during the time when Π is active. Here h denotes either of the two helicities of the gravitational wave and Π is

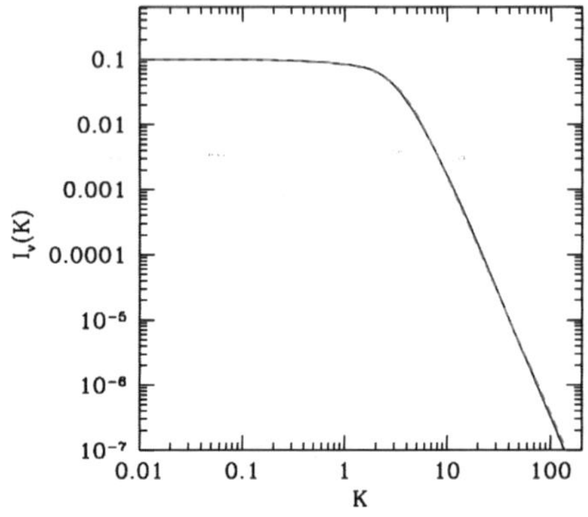

Figure 1. The equal time anisotropic stress power spectrum from turbulence. On large scales the spectrum is scale invariant, on scales much smaller than the correlation scale it falls of like $k^{-11/3}$ which is determined by the Kolmogorov turbulence assumed for the velocity power spectrum. The black line is the numerical result and the red dashed line a simple fit. The normalization is arbitrary. Figure from [17].

the anisotropic stress source of the corresponding helicities. The solution with vanishing initial conditions is then

$$\begin{aligned} h(\mathbf{k},\eta) &= \frac{8i\pi G a_*^3}{6ak}\left[e^{-ik\eta}\int_{\eta_*}^{\eta_*+\Delta\eta_*} d\eta' e^{ik\eta'}\Pi(\eta',\mathbf{k})+\right. \\ &\qquad \left. e^{ik\eta}\int_{\eta_*}^{\eta_*+\Delta\eta_*} d\eta' e^{-ik\eta'}\Pi(\eta',\mathbf{k})\right] \\ &= \frac{8i\pi G a_*^3}{6ak}\left[e^{-ik\eta}\Pi(k,\mathbf{k})+e^{ik\eta}\Pi(-k,\mathbf{k})\right], \qquad \eta>\eta_*+\Delta\eta_* . \end{aligned}$$

The gravitational wave energy density is given by

$$\rho_{gw}(\eta,\mathbf{x}) = \frac{1}{32\pi G a^2}\langle\partial_\eta h_{ij}(\eta,\mathbf{x})\partial_\eta h^*_{ij}(\eta,\mathbf{x})\rangle .$$

If the Universe is radiation dominated at the time η_* when the gravitational waves are generated, this leads to the gravitational wave energy spectrum

$$\frac{d\Omega_{gw}}{d\ln(k)}(\eta_0) = \frac{12\Omega_{\rm rad}(\eta_0)}{\pi^2}\left(\frac{\Omega_X(\eta_*)}{\Omega_{\rm rad}(\eta_*)}\right)^2 \mathcal{H}_*^2 k^3 {\rm Re}[P_\Pi(k,k,k)] . \tag{11}$$

Here

$$P_\Pi(\omega,\omega',k) \equiv \int_{-\infty}^{-\infty} d\eta \int_{-\infty}^{-\infty} d\eta' P_\Pi(\eta,\eta',k)e^{i(\omega\eta-\omega'\eta')} . \tag{12}$$

On large scales, $k<k_*$ the GW energy density from a 'causal' source always scales like k^3. This remains valid also for long duration sources. The only change for long lived sources is that the Green function $e^{ik(\eta-\eta')}$ of the wave operator in Minkowski spacetime has to be replaced by the Green function of the wave operator in an expanding universe which in the case of a radiation dominated universe is given in terms of spherical Bessel functions.

The scale $1/k_*$ is the correlation scale which is smaller than the co-moving Hubble scale $1/k_* < 1/\mathcal{H}_* = \eta_*$. The behavior or the spectrum close to the peak and its decay rate on smaller scales depend on the source characteristics, on its temporal behavior and on its power spectrum.

Let us now briefly discuss the different results which have been obtained for the position of the peak of the gravitational wave spectrum. As we see from eqs. (12) and (11), to determine the the gravitational wave energy density we must know the unequal time correlator of the of the anisotropic stresses, $P_\Pi(\eta, \eta', k)$. This is not well known from numerical simulations but there exist different simple possibilities:

- A *totally coherent source* is one for which the time dependence is deterministic. In this case the randomness is solely in the initial conditions and $\Pi(\eta, \mathbf{k}) = f(\eta, k)\Pi(\eta_*, \mathbf{k})$ for some deterministic function f. For the unequal time correlator P_Π this leads to

$$P_\Pi(\eta, \eta', k) = \sqrt{P_\Pi(\eta', \eta', k)P_\Pi(\eta, \eta, k)}\,.$$

  It has been shown in Refs. [18, 17] that the peak position of the GW spectrum $\propto k^3 P_\Pi(k, k, k)$ for a totally coherent source is given by the peak of the *temporal* Fourier transform of the source, i.e. $k_* = (\Delta\eta_*)^{-1}$.
- For a *source with finite coherence time* we can model the unequal time correlator of the of the anisotropic stresses, by

$$P_\Pi(\eta, \eta', k) = \sqrt{P_\Pi(\eta', \eta', k)P_\Pi(\eta, \eta, k)}\Theta(x_c - |\eta - \eta'|k)\,, \quad x_c \sim 1\,.$$

  This ansatz assumes that the source is totally coherent if the time difference is smaller than about one wavelength and uncorrelated for larger time differences. This ansatz resembles the suggestion by Kraichnan [19] for turbulence and has therefore been used for both turbulence and magnetic fields in Ref. [17]. However, this ansatz, like also the Kraichan decorrelation ansatz still lacks numerical confirmation.
  In this case, the peak of the GW spectrum is determined by the peak of the *spatial* Fourier transform of the source.
- A *totally incoherent source* is uncorrelated for different times. Its unequal time correlator can be modeled as

$$P_\Pi(\eta, \eta', k) = P_\Pi(\eta', \eta', k)\Delta\eta_*\delta(\eta - \eta')\,.$$

  Here the factor $\Delta\eta_*$ is present to make the dimensions right, it can be replaced by some other short correlation time. In this case, the GW spectrum is also determined by the peak of the *spatial* Fourier transform of the source.

The resulting gravitational wave spectra from the different hypotheses for the unequal time correlator are shown in Fig. 2. They are derived in detail in Ref. [17].

5. The electroweak phase transition

According to the standard model, the electroweak transition is not even second order, but only a cross-over. Then, this transition does not lead to the formation of gravitational waves. Even though the Higgs vacuum expectation value does cross-over from zero to a finite value, the corresponding Higgs field fluctuations have a very short correlation length of the order of the inverse temperature and do not invoke significant gravitational field fluctuations.

However, if the standard model is somewhat modified e.g. in the Higgs sector, or in certain regions of the MSSM parameter space, the electroweak phase transition can become first order, even strongly first order [7]. Then it proceeds via the formation of large bubbles of true vacuum. The bubbles themselves are spherically symmetric and do not generate gravitational waves but the collision events do.

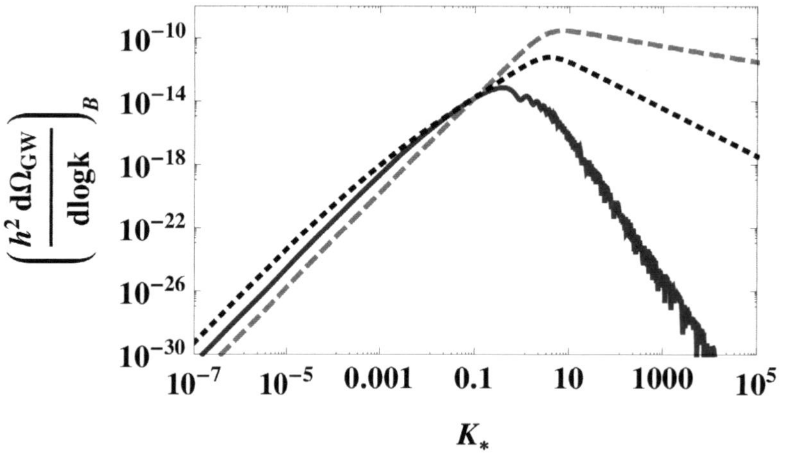

Figure 2. The GW energy density spectrum for a totally incoherent source (red, long-dashed), a totally coherent source (blue solid) and for a source with finite coherence time (black, short-dashed). The parameters are: $T_* = 100$ GeV, $\Delta\eta_*\mathcal{H}_* = 0.01$, $\Omega_X/\Omega_{\rm rad} = 2/9$ $(\langle v^2\rangle = 1/3)$. The horizontal variable is $K_* = kR_*/(2\pi)$. Note that the peak position is roughly at $k = \pi/\Delta\eta_* =$ a few, hence $K_* \simeq R_*/\Delta\eta_* = v_b$ in the totally coherent case and at $k = \pi/R_* =$ a few, hence $K_* \simeq 1$ in the incoherent case and in the case with finite coherence length. In order to clearly separate the peak positions, $v_b \simeq 0.1$ has been chosen. Figure from [17].

In addition, the Reynolds number of the cosmic plasma is very high and the bubbles are expected to induce turbulence and, if there are small magnetic seed fields these are amplified by turbulence leading to an MHD turbulent plasma with turbulent kinetic energy and magnetic fields in equipartition [20].

The correlation scale is expected to be [21] of the order of $k_* = \beta = (\Delta\eta_*)^{-1} \simeq 100/\eta_* \sim 10^{-3}$Hz, which is close to the frequency of the peak sensitivity for the space born gravitational wave antenna LISA, proposed for launch in 2018 by ESA [8].

5.1. Bubble collisions

New numerical simulations of gravitational waves from bubble collisions during the electroweak phase transition [22] have given a slower decay law on small scales, like $\propto k^{-1}$, see Fig. 3.

In Ref. [18] it has been shown that the decays law $\frac{d\Omega_{gw}(k)}{d\log k} \propto k^{-1}$ is generically expected on small scales (large k) if $P_\Pi(\eta,\eta,k)$ is continuous in time but not differentiable. This behavior seems reasonable for bubble collisions which start from vanishing overlap continuously but with a finite slope.

5.2. Turbulent MHD

Since the Reynolds number Re, of the cosmic plasma at $T \sim 100$GeV is very high, the rapidly expanding bubbles of the broken phase provoke the formation of turbulence. Furthermore, in the broken phase the electromagnetic field does generically not vanish. Since the magnetic Reynolds number R_m is even higher than the Reynolds number of the plasma [17], and thus the Prandl number [20] $P_m = R_m/Re \gg 1$, the high conductivity rapidly damps the electric fields and we are left with a magnetic field in a turbulent plasma, *MHD turbulence* [20].

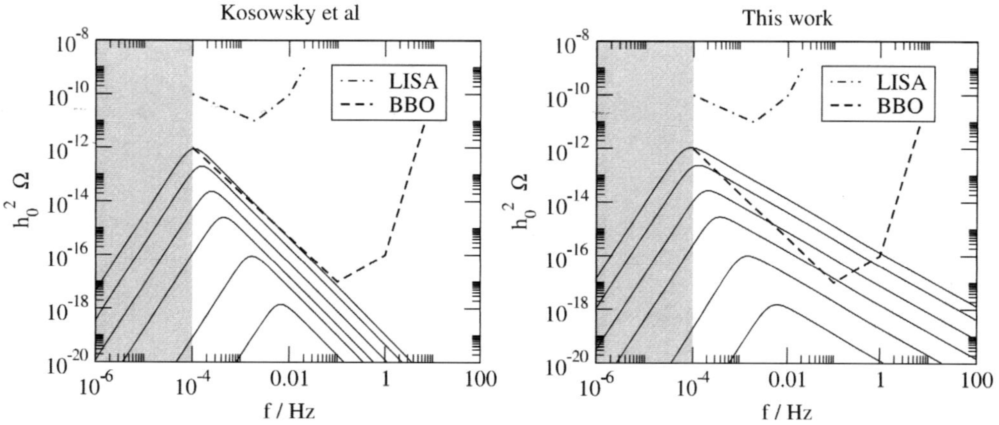

Figure 3. $\frac{d\Omega_{gw}(f)}{d\log f}$ from colliding bubbles, $f = k/2\pi$, numerical results for $\Omega_X/\Omega_{\rm rad} = 0.03$. On the left panel the old results from Ref. [16] are shown while the right panel shows the result of Ref. [22]. Figure from [22].

To determine the anisotropic stress tensors from the turbulent flow and from the magnetic field which is the source of gravitational waves, we first discuss the vorticity and magnetic field spectra. Because both, the vorticity and the magnetic field are divergence free, causality requires that their spectra behave like k^2 for small k:

$$\langle v_i(\mathbf{k})v_j(\mathbf{k}')\rangle = (2\pi)^3\delta^3(\mathbf{k}-\mathbf{k}')(\hat{k}_j\hat{k}_i - \delta_{ij})P_v(k)\,, \quad (13)$$

$$\langle B_i(\mathbf{k})B_j(\mathbf{k}')\rangle = (2\pi)^3\delta^3(\mathbf{k}-\mathbf{k}')(\hat{k}_j\hat{k}_i - \delta_{ij})P_B(k)\,. \quad (14)$$

The functions $(\hat{k}_j\hat{k}_i - \delta_{ij})P_\bullet(k)$ must be analytic at $k = 0$ because of causality (the correlation functions are functions with compact support, they vanish for separations larger than the Hubble scales, thus their Fourier transforms must be analytic) hence $P_v(k)$ and $P_B(k) \propto k^2$ for small k, see [23] . Turbulence is steered during the phase transition and then decays freely with decay time of the order of $\Delta\eta_*$. This source is not truly long-lived, but since it decays only like a power law and not exponentially, it does still contribute significantly for $\eta > \eta_* + \Delta\eta_*$ and we expect a suppression factor ϵ in eq. (8) which is larger than the naive factor $(\mathcal{H}_*\Delta\eta_*)^2$.

The behavior of the spectrum on scales smaller than the correlation scale $k > k_*$ is expected to be a *Kolmogorov spectrum* for the vorticity field, $P_v \propto k^{-11/3}$ and an *Iroshnikov–Kraichnan spectrum* [24] for the magnetic field, $P_B \propto k^{-7/2}$. For the induced GW spectrum this yields

$$\frac{d\Omega_{GW\bullet}(k,\eta_0)}{d\ln(k)} \simeq \epsilon\;\Omega_{\rm rad}(\eta_0)\left(\frac{\Omega_\bullet(\eta_*)}{\Omega_{\rm rad}(\eta_*)}\right)^2 \times \begin{cases} (k/k_*)^3 & \text{for } k \ll k_* \\ (k/k_*)^{-\alpha} & \text{for } k \gg k_*\,. \end{cases}$$

For $\bullet = v$ we have $\alpha = 11/3 - 1 = 8/3$ and for $\bullet = B$ we have $\alpha = 7/2 - 1 = 5/2$, see [25]. To find α we use that on small scale we expect the anisotropic stress spectrum to decay like the velocity, respectively magnetic field spectrum. With $\langle|\dot{h}|^2\rangle \propto P_\Pi/k^2$ and $\frac{d\Omega_{gw}(k)}{d\ln(k)} \propto k^3\langle|\dot{h}|^2\rangle \propto kP_\Pi$ so that $\alpha = \gamma - 1$, where γ denotes the decay exponent of the anisotropic stress spectrum.

In Figs. 4 and 5 the gravitational spectrum obtained from the wlectroweak phase trnasityion and from a first order phase transition at $T_* = 5\times 10^6$ GeV are shown and compared with future gravitational wave experiments like LISA [8], the Bib Bang Observer (BBO) [26] asnd advances LIGO [14].

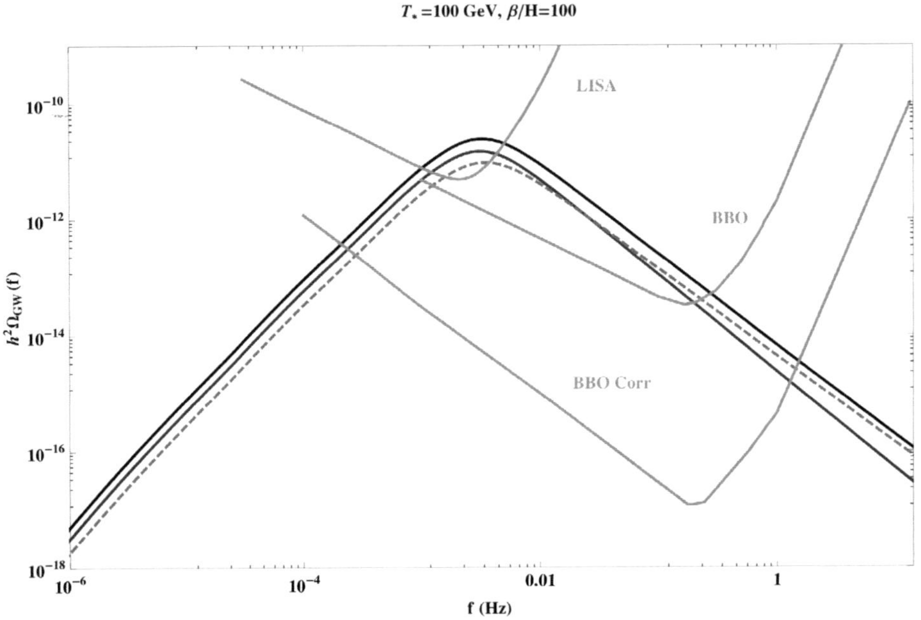

Figure 4. The gravitational wave energy density from magnetic fields (red) and turbulence (blue, the sum is shown in black) generated during the electroweak phase transition at $T_* = 100\text{GeV}$. The time-decorrelation of the source is modeled by the finite correlation approach (Kraichnan decorrelation). The source energy density spectrum is assumed to be maximal, $\Omega_v/\Omega_{\text{rad}} = 2/9$, see [17]. Equipartition in turbulent kinetic energy and magnetic field energy is assumed. The sensitivity curves are from [27]. Figure taken from [17].

5.3. Limits of primordial magnetic fields from the EW transition

It is difficult to estimate $\Omega_B(\eta_*)$ or $\Omega_v(\eta_*)$ directly, but since causality requires the spectra to be so blue, the limit on gravitational waves which comes from small scales $k \simeq k_*$, yields very strong limits on primordial magnetic fields on large scales already from the simple *nucleosynthesis constraint*, $\Omega_{gw} \lesssim 0.1\Omega_{\text{rad}}$. If the magnetic field spectrum behaves like $P_B(k) \propto k^n$, the gravitational wave energy density per log interval must go like $\frac{d\Omega_B(k,\eta_*)}{d\ln(k)} \propto k^{3+n}$ on large scales. For causal magnetic fields we have $n = 2$, hence the magnetic field energy spectrum is very blue,

$$\frac{d\Omega_B(k,\eta_*)}{d\ln(k)} \propto k^5 \quad \text{for} \quad k < k_* \,. \tag{15}$$

The main contribution to the gravitational wave energy density comes from the correlation scale, $\Omega_{gw} \simeq \frac{d\Omega_{gw}(k_*)}{d\ln(k_*)}$, whereas we are interested in the magnetic fields on much larger scales, for example $k_L = (0.1\text{Mpc})^{-1}$. Using our simple formula (8) with $\epsilon \sim 1$ the nucleosynthesis limit yields $k_*^{3/2}B(k_*,\eta_0) \lesssim 10^{-6}$Gauss. The time η_0 denotes today. For the electroweak phase transition $k_* \simeq 100\mathcal{H}_* \simeq 10^{-3}\text{sec}^{-1}$ while we are interested in magnetic fields on scales of 0.1Mpc giving $k_L = 10(\text{Mpc})^{-1} \simeq 10^{-13}\text{sec}^{-1}$. With the above scaling, the limit on a magnetic field on this scale is

$$k_L^{3/2}B(k_L,\eta_0) \lesssim \left(\frac{k_L}{k_*}\right)^{5/2} 10^{-6}\text{Gauss} = 10^{-31}\text{Gauss}\,. \tag{16}$$

Note also that this limit is very stable with respect to the small unknown pre-factor ϵ. Since $\Omega_{gw} \propto B^4$, changing ϵ by 2 orders of magnitude yields a change of only a factor of 3 in the limit

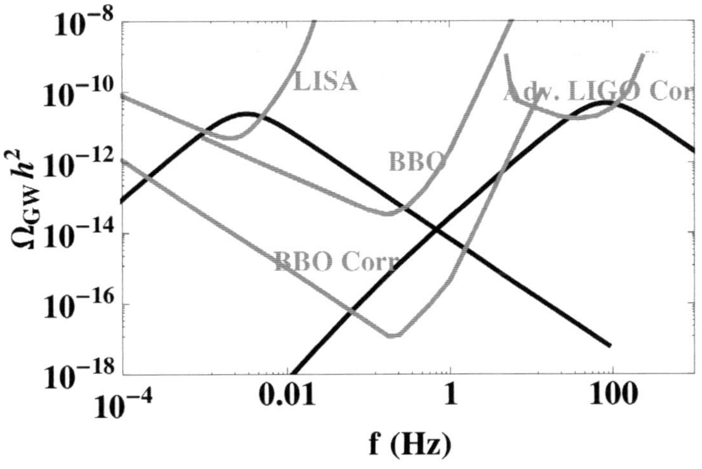

Figure 5. To the GW background from the electroweak phase transition also shown in Fig. 4 (here left black line) we add the result for a phase transition at $T = 5 \times 10^6$ GeV with $\Delta\eta_*\mathcal{H}_* = 0.02$ (black curve to the right). In this case, the gravitational wave background might even lie within the reach of advanced LIGO. The sensitivity curves are from [27].

for B.

This limit is unavoidable if the magnetic field on large scales changes simply by flux conservation so that $B(\eta) = B(\eta_*)(a_*/a)^2$. Scaling this to today yields $B(\eta_0) = B(\eta_*)/(1+z_*)^2$.

The limit is so severe because the magnetic field spectrum is so red. The simple requirement that $\rho_B(t_*) < \rho_{\rm rad}(t_*)$ yields roughly the same limit. However, this limit is not entirely save since at very early time most of the energy density might in principle have been in the form of small scale magnetic fields which later are converted into heat by viscosity damping on small scales. The fact that such magnetic fields would generate a sizable background of gravitational waves, however, forbids this possibility. This has been realized for the first time in Ref. [28]. The inferred limits on magnetic fields as a function of the spectral index are shown in Fig. 6.

5.4. Helicity

The most optimistic dynamo models for the generation of the magnetic fields of about 10^{-6}Gauss which are observed in galaxies and clusters, need seed fields of the order of at least 10^{-21}Gauss. These are excluded by the above limit. There is a possible way out: If the magnetic field has non-vanishing helicity it can develop an inverse cascade and power can be transferred from small to larger scales [29]. In this case the large scale magnetic fields do not simply decay by flux conservation in an expanding universe but they can feed from the small scale fields by inverse cascade.

During the electroweak phase transition parity is broken. Actually, the Chern-Simons winding number of the gauge field, $N_{CS} \propto \int F \wedge A$, which is related to the baryon number, has an electromagnetic part to it which is simply the helicity of the magnetic field [30], $H = V^{-1} \int_V (A \cdot B) d^3x$. It is interesting to note that this relates the baryon number and the magnetic helicity. It has been shown that the parity violation in the stress energy tensor coming from such helical magnetic fields lead to T-B and E-B correlations in the temperature and polarization spectra of the cosmic microwave background [31]. By the same token they also

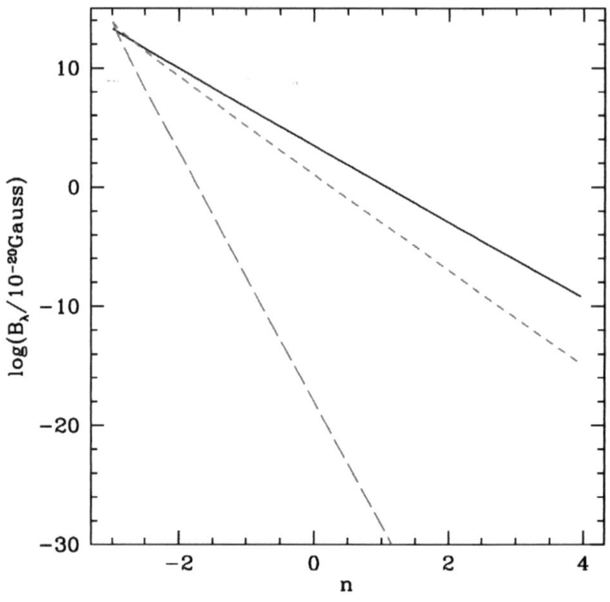

Figure 6. The nucleosynthesis limit on magnetic fields from $\rho_B(t_{\rm nuc}) < 0.1\rho_{\rm rad}(t_{\rm nuc})$ (black solid line), and from the gravitational wave constraint (red lines) as a function of the spectral index n. The limits for magnetic fields from the electroweak phase transition are short-dashed (red), while the long dashed (red) line corresponds to magnetic fields from inflation at a GUT scale. Causality requires that the spectral index be $n = 2$ for magnetic fields from the electroweak phase transition while n is in principle undetermined for inflation. Figure from [28].

generate gravitational waves with non-vanishing helicity [31].

The inverse cascade which is caused by helicity conservation for a helical field and is absent for non-helical magnetic fields has been investigated numerically, e.g., in [32] and is shown in Fig. 7

In Ref. [32] simple power laws are derived for the evolution of the correlation scale in the inverse cascade regime. Using this simple prescription, the mitigation of the magnetic field limits by an inverse cascade of maximally helical magnetic fields has been estimated in Ref. [33]. It has been found, however, that the inverse cascade is not sufficient to present a way out for the magnetic fields from the electroweak phase transition, but it can work for helical magnetic fields generated during the QCD phase transition. The QCD phase transition might be first order if the neutrino chemical potential is large, as discussed in [10].

If the magnetic field is helical, the GW background would not be parity symmetric. There would be more GW's of one helicity than of the other [31].

6. Conclusions

- The rapid expansion and collision of bubbles during a first order phase transition stirs the relativistic cosmic plasma sufficiently to lead to the generation of a (possibly observable) stochastic gravitational wave background.
- Observing such a background would open a new window to the early Universe and to high energy physics.

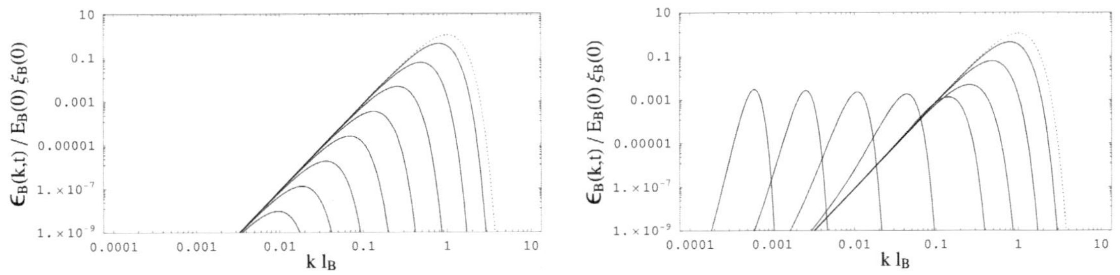

Figure 7. Left: the MHD evolution of the large scale magnetic field energy density spectrum with vanishing helicity. The field is simply damped by viscosity on small scales. Large scales are not effected. The viscosity scale is growing with time.
Right: the case of non-vanishing helicity. First, the helicity is relatively small and the evolution of the helical field is not very different from the non-helical case. But due to helicity conservation, helicity rapidly becomes maximal and the evolution enters the inverse cascade regime.
For simplicity, the correlation scale is set equal to the damping scale. Figure from [32].

- Generically, the density parameter of the GW background is of the order of

$$\Omega_{gw}(t_0) \simeq \epsilon\, \Omega_{\rm rad}(t_0) \left(\frac{\Omega_X(t_*)}{\Omega_{\rm rad}(t_*)}\right)^2 ,$$

  where $\epsilon < 1$ is determined by the duration of the source, $\epsilon \simeq 1$ if the decay time of the source is roughly a Hubble time.

- The spectrum grows like $\frac{d\Omega_{gw}(k,t_0)}{d\ln(k)} \propto k^3$ on large scales and decays on scales smaller than the correlations scale $k_* \sim 1/\Delta\eta_*$. The decay law depends on the physics of the source.
- Within the standard model, the electroweak phase transition is not of first order and does (probably) not generate an appreciable gravitational wave background. However, simple deviations from the standard model can make it first order.
  In this case we expect a GW background which can marginally be detected by the LISA satellite.
- It has been proposed that the magnetic fields generated during the electroweak phase transition, could represent the seeds for the fields observed in galaxies and clusters. Unfortunately this idea does not withstand detailed scrutiny. The magnetic fields cannot have enough power on large scale to represent such seed fields. On small scales the magnetic field power which is initially present is dissipated later on by the viscosity of the cosmic plasma.
- If there is no inverse cascade acting on the magnetic field spectrum, the limits on the large scale fields coming from the generated GW background are too strong to allow significant magnetic fields even for the most optimistic dynamo mechanism.
- If the magnetic field is helical, helicity conservation provokes an inverse cascade which can alleviate these limits. Most probably the relaxed limits are still too stringent for magnetic fields from the electroweak phase transition, but magnetic fields from the QCD phase transition might be sufficient to seed the fields in galaxies and clusters if they are helical.
- Helical magnetic fields generate a parity violating gravitational wave background, $|h_+(k)|^2 \neq |h_-(k)|^2$. This parity violation is in principle observable.

Acknowledgments

I am grateful to my collaborators Chiara Caprini, Elisa Fenu, Tina Kahniashvili, Thomas Konstandin and Geraldine Servant who worked with me on most of the results reported here. I tremendously enjoyed the many hours of animated and stimulating discussions which helped us to gain insight in this subject which touches on so many fascinating areas of physics. I thank Leonardo Campanelli, Chiara Caprini and Geraldine Servant for helpful comments on a first draft of this paper. Finally, it is a pleasure to thank the organizers of the First Mediterranean Conference on Classical and Quantum Gravity (MCCQG) for inviting me to give this talk and to participate in a lively conference on beautiful Crete. My work is supported by the Swiss National Science Foundation.

References

[1] C. J. Hogan, Phys. Rev. Lett. **51**, 1488 (1983);
E. Witten, Phys. Rev. D 30, 272 (1984).
[2] V. Mukhanov, *Physical Foundations of Cosmology*, Cambridge University Press (2005).
[3] Ruth Durrer, *The Cosmic Microwave Background*, Cambridge University Press (2008).
[4] Planck URL: `http://www.esa.int/Planck`
[5] R. Easther, J. T. Giblin and E. A. Lim, Phys. Rev. Lett. **99**, 221301 (2007) [arXiv:astro-ph/0612294];
J. Garcia-Bellido, D. G. Figueroa, Phys. Rev. Lett. **98**, 061302 (2007) [arXiv:astro-ph/0701014];
J. Garcia-Bellido, D. G. Figueroa and A. Sastre, Phys. Rev. D **77**, 043517 (2008) [arXiv:0707.0839 [hep-ph]];
J. F. Dufaux, A. Bergman, G. N. Felder, L. Kofman, J. P. Uzan, Phys. Rev. D **76**, 123517 (2007) [arXiv:0707.0875 [astro-ph]];
A. Diaz-Gil, J. Garcia-Bellido, M. Garcia Perez and A. Gonzalez-Arroyo, Phys. Rev. Lett. **100**, 241301 (2008) [arXiv:0712.4263].
[6] K. Rummukainen et al., Nucl. Phys. **B532**, 283 (1998) [arXiv:hep-lat/9805013];
K. Kajantie et al. arXiv:hep-ph/9809435.
[7] C. Grojean, G. Servant, J. Wells, Phys. Rev. **D71**, 036001 (2005) [arXiv:hep-th/0407019];
C. Delaunay, C. Grojean, J.D. Wells, JHEP0804:029 (2008) [arXiv:0711.2511];
S. J. Huber and T. Konstandin,JCAP0805:017 (2008) [arXiv:0709.2091];
J. R. Espinosa, T. Konstandin, J. M. No, M. Quiros, Phys. Rev. **D78**, 123528 (2008) [arXiv:0809.3215];
A. Ashoorioon and T. Konstandin, JHEP 0907:086 (2009) [arXiv:0904.0353];
T. Stevens and M. B. Johnson, arXiv:0903.2227 [astro-ph.CO], arXiv:1001.3694 [astro-ph.CO];
M. Jarvinen, C. Kouvaris and F. Sannino, arXiv:0911.4096 [hep-ph].
[8] LISA: URL `http://lisa.jpl.nasa.gov/`
[9] Y. Aoki et al., Phys. Lett. **B643**, 46 (2006) [arXiv:hep-lat/0609068];
JHEP **0906**, 088 (2009) [arXiv:0903.4155].
[10] D. J. Schwarz, M. Stuke, JCAP **11** (2009) 025 [arXiv:0906.3434].
[11] A.Boyarsky, O. Ruchayskiy, M. Shaposhnikov, Ann. Rev. Nucl. Part.Sci. **59**, 191 (2009) [arXiv:0901.0011].
[12] G. Hobbs et al. , Proceedings of the 8th Amaldi Conference, Class. Quant. Grav. (2010) [arXiv:0911.5206].
[13] L. Randall and G. Servant, JHEP 0705:054 (2007) [arXiv:hep-ph/0607158].
[14] LIGO: URL `http://www.ligo.caltech.edu`
[15] C. Grojean and G. Servant, Phys. Rev. **D75**, 043507 (2007) [arXiv:hep-ph/0607107].
[16] A. Kosowsky, M. S. Turner and R. Watkins, Phys. Rev. D **45**, 4514 (1992);
A. Kosowsky and M. S. Turner, Phys. Rev. D **47**, 4372 (1993) [arXiv:astro-ph/9211004];
M. Kamionkowski, A. Kosowsky and M. S. Turner, Phys. Rev. D **49**, 2837 (1994) [arXiv:astro-ph/9310044].
[17] C. Caprini, R. Durrer and G. Servant, JCAP **12**, 024 (2009) [arXiv:0909.0622].
[18] C. Caprini, R. Durrer, T. Konstandin and G. Servant, Phys. Rev. D **79**, 083519 (2009) [arXiv:0901.1661 [astro-ph]].
[19] R.H. Kraichnan, Phys. Fluids **7**, 1163 (1964).
[20] D. Biskamp, *Magnetohydrodynamical Turbulence*, Cambridge University Press (Cambridge, 2003).
[21] C. Caprini, R. Durrer and G. Servant, Phys. Rev. D **77**, 124015 (2008) [arXiv:0711.2593 [astro-ph]].
[22] S. J. Huber and T. Konstandin, JCAP **0809**, 022 (2008) [arXiv:0806.1828 [hep-ph]].
[23] R. Durrer and C. Caprini, JCAP **0311**, 010 (2003) [arXiv:astro-ph/0305059].
[24] P. Iroshnikov, AZh. **40**, 742 (1963);
R.H. Kraichnan, Phys. Fluids **8**, 1385 (1965).
[25] C. Caprini and R. Durrer, Phys. Rev. D **74**, 063521 (2006) [arXiv:astro-ph/0603476].

[26] J. Crowder and N.J. Cronish, Phys. Rev. **D72**, 083005 (2005) [ariv:gr-qc/0506015].
[27] A. Buonanno, arXiv:gr-qc/0303085.
[28] C. Caprini and R. Durrer, Phys. Rev. D **65**, 023517 (2001) [arXiv:astro-ph/0106244];
[29] A. Brandenburg and K. Subramanian, Phys. Rep. **417**, 1 (2005) [arXiv:astro-ph/0405052].
[30] T. Vachaspati, Phys. Rev. Lett. **87** 251302 (2001).
[31] C. Caprini, R. Durrer and T. Kahniashvili, Phys. Rev. D **69** (2004) 063006 [arXiv:astro-ph/0304556].
[32] L. Campanelli, Phys. Rev. Lett. **98**, 251302 (2007) [arXiv:0705.2308 [astro-ph]].
[33] C. Caprini, R. Durrer and E. Fenu, JCAP **11**, 001 (2009) [arXiv:0906.4976 [astro-ph.CO]].

First Mediterranean Conference on Classical and Quantum Gravity (MCCQG 2009) IOP Publishing
Journal of Physics: Conference Series **222** (2010) 012022 doi:10.1088/1742-6596/222/1/012022

Generalized particle model: a possible source for Dark Energy

Sudipta Das[a], Subir Ghosh[a], Jan-Willem van Holten[b] and Supratik Pal[a]

[a]Physics and Applied Mathematics Unit, Indian Statistical Institute, 203 B.T.Road, Kolkata 700108, India
[b]NIKHEF, PO Box 41882, 1009 DB Amsterdam, Netherlands

Abstract. We propose a generalized particle dynamics scenario in brane world formalisms, for an asymptotically flat as well as for an asymptotically anti de Sitter background. The generalization implies that the effective mass of the particle becomes dynamical. The present framework results in a new model that accounts for the late acceleration of the universe. An effective Dark Energy equation of state, exhibiting a phantom like behavior, is generated. The model is derived by embedding the physical FRW universe in an effective space-time, induced by the generalized particle dynamics. We corroborate our results with present day observed cosmological parameters.

1. Introduction

In recent times theoretical understanding of Dark Energy (DE) has become the Holy Grail of cosmological investigations. Existence of DE is unavoidable if one wants to explain the (present day) accelerated expansion of the universe. However, if the dynamical laws of motion pertaining to General Relativity are held sacred, it seems inevitable that a paradigm shift in the properties of DE is needed. This is simply because normal matter creates a positive pressure that decelerates the universe expansion whereas DE has to generate a negative pressure to have the accelerated expansion.

Observational vindication [1] of the present-day acceleration of the universe, and the subsequent precise measurements of observable parameters [2, 3, 4] indicate that the entity Dark Energy (DE) which is responsible for the recent acceleration contribute to $\sim 70\%$ of cosmic energy density. This DE density-fraction of total cosmic density ($\Omega_{\rm DE}$) is determined conclusively from several independent probes ($\Omega_{\rm DE} = 0.726 \pm 0.015$ at 95% C.L. from latest WMAP5 data [3]). On the other hand, DE *effective* Equation of State (EOS) ($w_{\rm DE}$) is still inconclusive. But the big surprise is that this agent is even more exotic in nature than imagined before because there is some evidence that the EOS can cross the so-called Phantom Divider (i.e. $w_{\rm DE} = -1$). Though SNLS data show no general behavior for $w_{\rm DE} < -1$, the analysis of the most reliable SNIa Gold dataset show strong indication that $w_{\rm DE} < -1$ [5] (the lower bound being $-1.11 < w_{\rm DE}$ from WMAP5 data [3]), leading to the conclusion that models with phantom divider crossing are preferred over ΛCDM (or quintessential candidates) at 2σ level. This clearly weakens the claims of cosmological constant Λ or dynamical models like quintessence, Kessence, Chaplygin gas etc. [6] as viable DE models. One can resurrect the scalar field models only at the cost of phantom fields (quintom models) [7], with a negative kinetic energy term but they bring in severe instability problems and are better avoided.

In this perspective, instead of looking for contrived and phenomenologically motivated DE models, it seems reasonable to explore modified gravity theories [8, 9] which do not suffer from any such major drawbacks. But one might still feel skeptical since more often than not explicit forms of these modified gravity theories appear to be designer made and/or fine tuned without a deeper dynamical framework based on first principles. In our work, accelerated expansion emerges in a brane world scenario. The modified gravity provides the metric of the higher dimensional spacetime in which the brane is embedded. (This well established formalism is explained below and later as we proceed.) This specific modified gravity metric can be derived from a generalized particle dynamics [10]. We emphasize that in this scheme conventional relativistic dynamical principles are maintained and an extended form of spin-orbit coupling is introduced. We also have extended our model to physical (3+1)-dimensions. We also demonstrate that the phantom-like behavior can be induced without explicitly invoking the phantom field with negative kinetic energy term. It is well-known (for details see [9, 10, 13]) that by embedding techniques one can relate cosmological surface dynamics (Friedmann equations) in lower (e.g., 3+1) dimensions with particle motion in a higher (e.g., 4+1) dimensional black-hole space-time. In fact the latter is taken as asymptotically anti de Sitter space-time, taking Schwarzschild-anti de Sitter (Sch-AdS) as a representative example. In the standard brane world scenario the AdS background induces an effective cosmological constant. However, a more interesting situation occurs in our framework. Here the AdS bulk induces a *dynamical* quantity which is essential in imparting the phantom behavior. We demonstrate that in the subsequent cosmological scenario, the induced effective negative pressure can result in an accelerating universe, even capable of showing phantom behavior. We further establish our model by an analysis of the equation of state and a determination of the relevant parameters describing the evolution of the observable universe.

2. The Generalized Particle Formulation

We start with a generalized relativistic point particle model in a $(3+1)$ dimensional spacetime where the action is given by:

$$S = \int L d\tau = m \int d\tau \left[\left(\frac{1}{2e} g_{\mu\nu} \dot{x}^\mu \dot{x}^\nu - \frac{e}{2} \right) - \lambda g_{\mu\nu} \xi^\mu \dot{x}^\nu + \frac{e\lambda^2}{2} g_{\mu\nu} \xi^\mu \xi^\nu + \frac{e\beta\lambda^2}{2} \right]. \quad (1)$$

Here m is the mass of the particle, τ is the worldline evolution parameter, $x^\mu(\tau)$ are the particle co-ordinates, $e(\tau)$ is the worldline einbein introduced to make the action reparametrization invariant and the Lagrange multiplier $\lambda(\tau)$ is an auxiliary worldline scalar variable. Furthermore β is a numerical constant, whilst the metric $g_{\mu\nu}(x)$ and the vector $\xi^\mu(x)$ are arbitrary functions of the co-ordinates x^μ.

The action is invariant under infinitesimal co-ordinate transformations of the form $\delta x^\mu = \alpha \xi^\mu(x)$, provided the Lie-derivative of the metric with respect to ξ vanishes:

$$\xi^\lambda \partial_\lambda g_{\mu\nu} + g_{\nu\lambda} \partial_\mu \xi^\lambda + g_{\mu\lambda} \partial_\nu \xi^\lambda = 0. \quad (2)$$

This shows that $\xi^\mu(x)$ is the Killing vector associated with the symmetry of the metric.

2.1. Hamiltonian formulation

To analyze the dynamics implied by the action (1), we follow the hamiltonian analysis of constrained systems as formulated by Dirac [15]. The conjugate canonical momenta p_e, p_λ, p_μ are computed and the Hamiltonian follows,

$$H = p_e \dot{e} + p_\lambda \dot{\lambda} + p_\mu \dot{x}^\mu - L = \frac{e}{2m} [p^2 + m^2 + 2m\lambda(\xi \cdot p) - \beta m^2 \lambda^2]. \quad (3)$$

The two significant constraints are:

$$\psi_1 \equiv e(\xi \cdot p - m\beta\lambda) \approx 0; \quad \psi_2 \equiv p^2 + m^2 + \frac{1}{\beta}(\xi \cdot p)^2 \approx 0. \tag{4}$$

The first constraint signifies a conserved quantity associated with Killing vector and the second constraint is nothing but the modified mass-shell condition. It is interesting to note that in the limit $\beta \to \infty$, we get back the normal particle model.

3. Our model in Schwarzschild background

We apply our generalized particle model in the $(3+1)$-dimensional Schwarzschild background where the metric is given by

$$g_{\mu\nu}dx^\mu dx^\nu = -\left(1 - \frac{2M}{r}\right)dt^2 + \frac{dr^2}{1 - \frac{2M}{r}} + r^2 d\theta^2 + r^2 \sin^2\theta d\phi^2. \tag{5}$$

We choose the Killing vector corresponding to rotational symmetry as: $\xi^\mu_{(\phi)} = (0, 0, 0, 1)$. Without loss of generality, we can restrict the particle to move in the equatorial plane and our generalized action in the equatorial plane $\theta = \pi/2$ takes the form:

$$S = m\int \left[\frac{1}{2e}\left(-\left(1 - \frac{2M}{r}\right)\dot{t}^2 + \frac{\dot{r}^2}{1 - \frac{2M}{r}} + r^2\dot{\phi}^2\right) - \frac{e}{2} - \lambda r^2\dot{\phi} + \frac{e\lambda^2}{2}\left(r^2 + \beta\right)\right] d\tau. \tag{6}$$

It should be noted that for our choice of Killing vector, $(\xi \cdot p) = p_\phi$. The Hamiltonian is given by

$$H = \frac{e}{2m}\left[m^2 - \frac{p_t^2}{1 - \frac{2M}{r}} + \left(1 - \frac{2M}{r}\right)p_r^2 + \frac{p_\phi^2}{r^2} + 2m\lambda p_\phi - \beta m^2\lambda^2\right]. \tag{7}$$

Performing the same constraint analysis as before whilst fixing the gauge $e = 1$, we find the modified mass-shell constraint and the relation,

$$2mH = m^2 - \frac{p_t^2}{1 - \frac{2M}{r}} + \left(1 - \frac{2M}{r}\right)p_r^2 + \left(\frac{1}{r^2} + \frac{1}{\beta}\right)p_\phi^2 = 0 \quad ; \quad p_\phi = m\beta\lambda. \tag{8}$$

Consequently the Hamiltonian equations of motion are:

$$\dot{p}_\phi = \{p_\phi, H\} = 0 \Rightarrow p_\phi = m\beta\frac{r^2\dot{\phi}}{r^2 + \beta} = m\beta\lambda = ml, \tag{9}$$

$$\dot{p}_t = \{p_t, H\} = 0 \Rightarrow p_t = m(1 - \frac{2M}{r})\dot{t} = m\varepsilon = E. \tag{10}$$

So we have the two conserved quantities: $p_\phi = ml$ and $p_t = E$, where l and E are two different constants. It is worth mentioning here that ml and E are nothing but the angular momentum and energy respectively, for the particle moving in Schwarzschild background, but now depending on the parameter β as well. In the large β limit the standard results are regained.
The modified mass-shell constraint now can be written as:

$$p^2 + m^2 + \frac{m^2 l^2}{\beta} = 0 \Rightarrow p^2 + \mu^2 m^2 = 0, \tag{11}$$

where μ is a dimensionless parameter defined by the relation

$$\mu^2 = 1 + l^2/\beta$$

and we have taken the velocity of light $c = 1$. Thus for $\mu^2 = 1$, i.e. $\beta \to \infty$, the above equation reduces to the well-known dispersion relation for normal particle model which is

$$p^2 + m^2 = 0.$$

It is interesting to observe that for negative β, the effective mass of the particle is reduced. We proceed further to study this behavior in details.
The radial equation of motion is given by

$$\dot{r}^2 + \frac{V_{eff}(r)}{m^2} = \varepsilon^2, \tag{12}$$

where the effective potential $V_{eff}(r)$ is given by

$$V_{eff} = \left(1 - \frac{2M}{r}\right)\left(\mu^2 + \frac{l^2}{r^2}\right). \tag{13}$$

We observe from the above equation that the radial dynamics in our generalized model is like that in normal Schwarzschild case with a modified velocity of light $c^2 \to \mu^2 c^2$. We get back normal results in the limit $\beta \to \infty$.

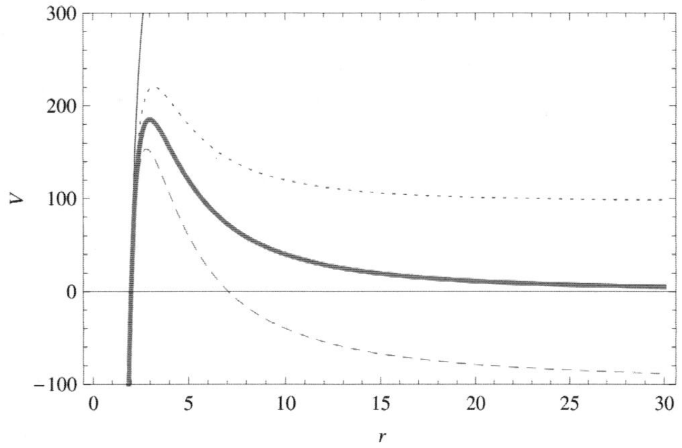

Figure 1. Plot for the variation of the effective potential V with the radial coordinate r for $l = 100$ with different values of β

The effect of different values of β on the dynamics of a particle with mass m can be seen in Figure 1, where we plot $V_{eff}(r)$ *vs.* r for different β at a fixed $l = 100$. In this figure the normal line corresponds to $\beta = 10$ whereas the dotted, thick and dashed lines represent the values $\beta = 50, 10^5, -50$ respectively. Also, we have taken $M = 1$, $m = 0.1$. Thus the thick line (large positive β) represents the same behavior as in normal Schwarzschild case. But for negative β (dashed line), we can clearly see a sign-flip of the effective potential. This negative sign of the effective potential can be interpreted as if there is a repulsive force acting on the particle. We study the cosmological implications of this negative β in the following section.

3.1. Cosmological implications of negative β

In this section we shall point out a crucial implication of our model in the cosmological context. From embedding geometry using the Gauss-Codazzi equation [16] it can be shown that a $(D-1)$-dimensional surface representing a cosmological Friedmann-Robertson-Walker (FRW) metric can be embedded consistently in a D-dimensional black hole spacetime in such a way that the expansion of the FRW surface is realized by the particle trajectory along the radial direction in the gravitational field of the black hole (the so called expanding bubble universe) [13]. In the present article, the particle motion induces a $(2+1)$-dimensional FRW metric and the effective potential contains essential information for the evolution of the embedded cosmological surface. Below we shall show how this can be realized, followed by a discussion of its cosmological implications.

As demonstrated by observations, we consider the physical matter density of the universe ρ to be small compared to brane tension ρ_0. Here we embed $(2+1)$ dimensional spatially flat FRW universe with the line element

$$ds_3^2 = -d\tau^2 + a^2(\tau)[d\sigma^2 + \sigma^2 d\theta^2] \tag{14}$$

into the $(3+1)$-dimensional black hole space-time induced by our generalized particle model whose line element is given by

$$ds_4^2 = -F(r)dt^2 + \frac{dr^2}{F(r)} + r^2 d\Omega_2^2, \tag{15}$$

where $d\Omega_2^2$ is the two-sphere. The radial function $F(r)$ is

$$F(r) = \left(1 - \frac{2M}{r}\right)\left(\frac{\mu^2 r^2 + l^2}{r^2 + l^2}\right). \tag{16}$$

As stated before, we identify the radial coordinate $r(t)$ of the particle with the scale factor $a(t)$ of the universe to arrive at the Friedmann equations:

$$\left(\frac{\dot{a}}{a}\right)^2 = \frac{8\pi G}{3}\rho + \frac{2M}{a^3} - \frac{l^2}{\beta}\left(\frac{1 - 2M/a}{l^2 + a^2}\right) \tag{17}$$

$$\frac{\ddot{a}}{a} = -\frac{4\pi G}{3}(\rho + 3p) - \frac{M}{a^3} - \frac{l^2}{\beta(l^2 + a^2)^2}\left[l^2 - \frac{Ml^2}{a} + Ma\right]. \tag{18}$$

For late time universe, a should be large which is also the physical region in our model as we have $a > 2M$. Thus, the right hand side of (18) becomes positive for negative β and large a implying the late accelerating behavior of our universe. Further, if we consider late time universe and hence taking only the β-dominating terms in the Friedmann equations, the deceleration parameter q turns out to be:

$$q = -\frac{\ddot{a}/a}{(\dot{a}/a)^2} = -\frac{l^2}{l^2 + a^2} - \frac{M}{a - 2M}. \tag{19}$$

This clearly reveals that the deceleration parameter is negative for the cosmologically relevant region $a > 2M$, confirming an accelerated expansion at late time.

Also the behavior for the scale factor can be analyzed to some extent with the above considerations of β-dominance. In the large l limit, an analytic expression for the scale factor $a(t)$ equations can be obtained by solving the first of the Friedmann equation:

$$a(t) \approx M + \frac{1}{2}\left[e^{t/\sqrt{-\beta}} - M^2 e^{-t/\sqrt{-\beta}}\right]. \tag{20}$$

Journal of Physics: Conference Series **222** (2010) 012022 doi:10.1088/1742-6596/222/1/012022

In the above equation, the first term is nothing but a scaling and with $M = 1$ the second term reduces to $\sinh(t/\sqrt{-\beta})$, which interpolates between a decelerating phase at early time and an accelerated expansion at late time. Thus our model behaves pretty close to ΛCDM (which matches very well with observations), with the inverse of the β parameter playing the role of the cosmological constant Λ. Because of this identification, it is worthwhile to emphasize that negative β is consistent with small positive value for Λ that is favored observationally. This makes the scenario further interesting from observational ground since now one can calculate the other observable parameters and compare them with highly accurate observational data. But it should be remembered that here we have represented a toy model of our universe as it is $(2+1)$-dimensional. But this can be easily generalized to the physical $(3+1)$-dimensional universe which we do in the following section.

4. An Application to Schwarzschild-Anti de Sitter geometry

We now study the cosmological applications of our generalized particle model in our physical universe. We start with $(4+1)$-dimensional Schwarzschild-Anti de Sitter (Sch-AdS) background with the line element

$$ds^2_{4+1} = -F(r)dt^2 + \frac{dr^2}{F(r)} + r^2 d\Omega_3^2 \tag{21}$$

where

$$F(r) = \left(k - \frac{2M}{r^2} + \Lambda_5 r^2\right)\left(\frac{\mu^2 r^2 + l^2}{r^2 + l^2}\right)$$

and $k = (0, \pm 1)$ is the curvature parameter whilst Λ_5 is the constant curvature of the spacetime. We embed a $(3+1)$-dimensional FRW spacetime $(T, \sigma, \theta, \varphi)$ with the line element

$$ds^2_{3+1} = -dT^2 + a^2(T)\left[\frac{d\sigma^2}{1 - k\sigma^2} + \sigma^2(d\theta^2 + \sin^2\theta \ \ d\varphi^2)\right], \tag{22}$$

into the $(4+1)$-dimensional effective metric (21). The Friedmann equations take the form

$$\left(\frac{\dot{a}}{a}\right)^2 = \frac{8\pi G_4}{3}\rho + \frac{2M}{a^4} + \alpha - \Lambda_5 + \left[\frac{l^2(2M/a^2 - \Lambda_5 a^2)}{\beta(a^2 + l^2)}\right] \tag{23}$$

$$\frac{\ddot{a}}{a} = -\frac{4\pi G_4}{3}(\rho + 3p) - \frac{2M}{a^4} + \alpha - \Lambda_5 - \left[\frac{l^2(\Lambda_5 a^4 + 2M + 2\Lambda_5 l^2 a^2)}{\beta(a^2 + l^2)^2}\right] \tag{24}$$

where $\dot{a} = da/dT$, $8\pi G_4/3 = 2\rho_0 (8\pi G_5/3)^2$ and $\alpha = (8\pi G_5 \rho_0/3)^2$. We express the Hubble parameter $H = \dot{a}/a$ in terms of the redshift z (where $1 + z = a_0/a = 1/a$). The terms containing M ($\propto a^{-4}$) contribute to the present day radiation energy density of the universe, which is a very small proportion of total radiation energy density [13]. Hence we neglect contributions from any cosmic constituent which redshifts away at the rate of radiation or faster (i. e. the terms of the order of $(1+z)^4$ or higher) for a late time universe. With this assumption, we can express the Friedmann equation (23) in the following convenient form:

$$H^2 = H_0^2\left[\Omega_X\left(1 + b(1+z)^2\right) + \Omega_M(1+z)^3\right]. \tag{25}$$

where $\Omega_M = \rho_M/\rho_c = 8\pi G_4 \rho_M/3H_0^2$ is the density parameter for matter, $\rho_c = 3H_0^2/8\pi G$ is the critical density of our universe and $\Omega_X = (\alpha - \Lambda_5\mu^2)/H_0^2$ is the density parameter for dark energy. The effect of our modified gravity theory appears in Ω_X and in the dimensionless parameter $b = \Lambda_5\beta(\mu^2 - 1)^2/(\alpha - \Lambda_5\mu^2)$. Observations fix $H_0 = 70.5 \pm 1.3$ km/s/Mpc according to the

WMAP5 data [3] and $H_0 = 74.2 \pm 3.6$ km/s/Mpc from HST data [4]. The above equation (25) is the major result of our work the prospects of which we describe below.
Ω_M is the sum-total of the density-fraction for luminous and dark matter which contributes 0.28 ± 0.08 of total cosmic density (fixed independently by the CMB [2] and large scale structure data [17]. For a valid dark energy model, Ω_X accounts for the DE density. But observations indicate a universe close to the ΛCDM ($\Omega_X \simeq .72, \Omega_M \simeq .28$). This forces b to be small and β large in magnitude. Observationally, these values will be restricted by χ^2 fitting, which we discuss in our later issue.

4.1. Estimation of observable parameters

First we estimate the luminosity-redshift relation $d_L(z)$ which determines the dark energy density Ω_X from observations. For our model (25), $d_L(z)$ is given by

$$d_L(z) = (1+z)\int_0^z \frac{dz'}{H(z')} = \frac{(1+z)}{H_0}\int_0^z \frac{dz'}{\left[\Omega_X\left(1+b(1+z')^2\right)+\Omega_M(1+z')^3\right]^{1/2}}. \tag{26}$$

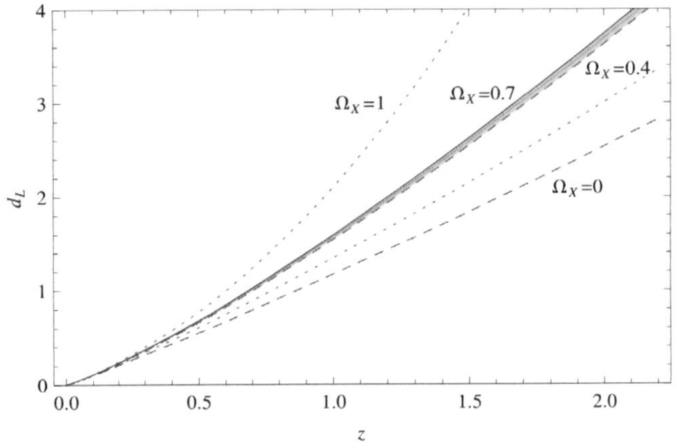

Figure 2. Variation of $d_L(z)$ with redshift z for different Ω_X

In Figure 2 the variation of d_L with redshift in units of H_0^{-1} for different Ω_X (with $\Omega_M + \Omega_X = 1$) is shown using numerical integration. The plots show that a universe with $\Omega_X \sim 0.7$, $\Omega_M \sim 0.3$ is favored by observation, and confirms that Ω_X accounts for the dark energy density in our model. The shaded region gives the bound for b as $-0.07 \le b < 0$. Throughout the rest of the paper, we take a representative small negative value for $b = -0.05$. The age of the universe has the currently accepted value of 13.7 ± 0.02 Gyr [2] which is very close to the estimate of our model.
In our model, the deceleration parameter $q(z)$ takes the form:

$$q(z) = \frac{-\ddot{a}/a}{\dot{a}^2/a^2} = \frac{H'(z)}{H(z)}(1+z) - 1 = \frac{\Omega_M(1+z)^3 - 2\Omega_X}{2(\Omega_X(1+b(1+z)^2)+\Omega_M(1+z)^3)}. \tag{27}$$

In Figure 3, the deceleration parameter has been plotted against redshift for $\Omega_X \sim 0.7, \Omega_M \sim 0.3$. The plot confirms that our model indeed results in an early decelerating and late accelerating universe. Moreover, onset of the recent accelerating phase, when the universe was $\sim 60\%$ of its

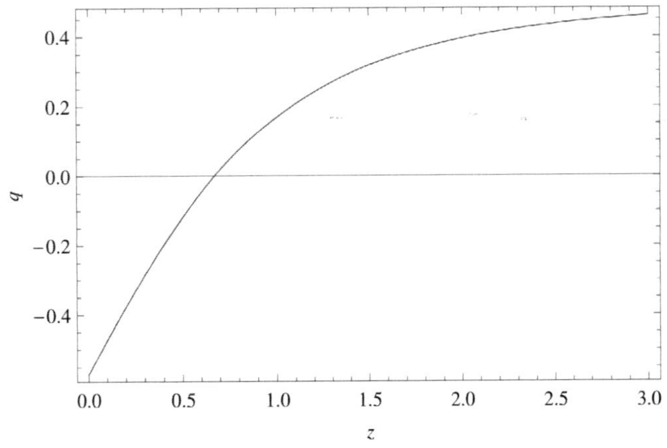

Figure 3. Variation of the deceleration parameter q with redshift

present size ($z = 0.6$), is also confirmed by our model.
The effective equation of state (EOS) of DE in our model is

$$w_X(z) = \frac{2q(z) - 1}{3[1 - \Omega_M(z)]} = -1 + \frac{2b(1+z)^2}{3}. \tag{28}$$

The expression has been obtained by a binomial expansion considering small b and dropping terms of order $(1+z)^4$ or higher, as before. Obviously, since b is negative and non-zero, the effective EOS of dark energy candidate satisfies $w_X < -1$. So it shows a phantom like behaviour that is as claimed by SN1a Gold data set [5].

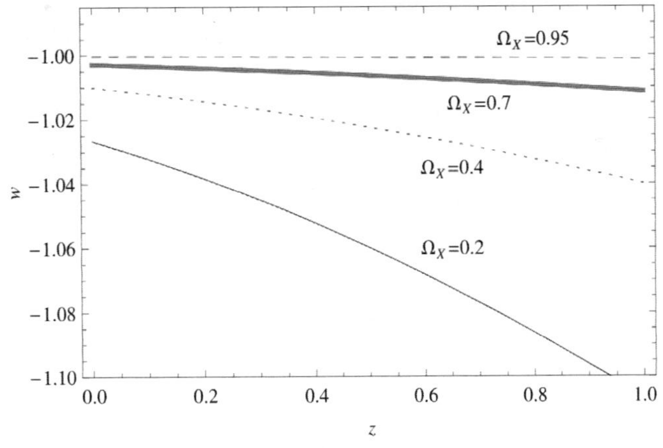

Figure 4. Variation of the effective EOS of dark energy with redshift

In Figure 4, the effective EOS parameter of dark energy with redshift is shown for different values of Ω_X. The latest WMAP5 data constrains the lower bound of the dark energy EOS today to be $-1.11 < w_{DE}$ [3]. This sets the lower bound of b at $(-.15 \leq b)$, which is way below its lower bound $(-.07 \leq b)$ as predicted before from the luminosity-redshift relation. This model with $(-.07 \leq b < 0)$ will thus fit well with a more precise bound for the EOS available in the future.

The statefinder parameters [18]

$$r = \frac{\dddot{a}/a}{(\dot{a}/a)^3} = 1 + \left[\frac{H''}{H} + \left(\frac{H'}{H}\right)^2\right](1+z)^2 - 2\frac{H'}{H}(1+z) \; ; \; s = \frac{2}{3}\frac{r-1}{2q-1} \tag{29}$$

can distinguish dynamical models from ΛCDM. In our model, $\{r, s\}$ pair is

$$r = 1 - \frac{b\Omega_X(1+z)^2}{\Omega_X(1+b(1+z)^2) + \Omega_M(1+z)^3} \; ; \; s = \frac{2}{3}\frac{b(1+z)^2}{[3+b(1+z)^2]}. \tag{30}$$

Thus in our model, as b is non-trivial, the $\{r, s\}$ pair can distinguish our alternative gravity model from ΛCDM for which $r = 1, s = 0$.

5. Summary and Future prospects

We have constructed a specific modified gravity structure, induced by a higher dimensional non-minimal particle dynamics framework. This particular generalized relativistic particle model was introduced by us in [10]. Subsequently we consider brane models embedded in this modified 4+1-dimensional AdS-Schwarzschild spacetime. The induced particle dynamics can simulate cosmological evolution which is our primary interest. We find a late time expanding universe. The most notable feature of our formulation is the possibility of having an effective phantom dark energy model *without the phantom*. This is (quantitatively) manifested in the crossover of the phantom barrier. Our model is qualitatively distinct, but not quantitatively far off, from the ΛCDM model during recent times. Hence all the positive features of ΛCDM along with a phantom behavior (without the problems related to the negative kinetic term) can be accommodated in our model.

Several aspects of the proposed framework can be investigated further. The parameters used in the model can be constrained observationally by using a maximum likelihood method involving the minimization of the function $\chi^2 = \sum_{1=1}^{N} \left[d_L(z)_{\rm obs} - d_L(z)_{\rm th}\right]^2 / \sigma_i^2$, where $d_L(z)_{\rm th}$ contains the parameters used in a specific theory, N is the number of Supernovae (taken as 157 for the most reliable Gold dataset) and σ_i are the 1σ errors from the observational method used [6]. Observationally, this is the most accurate probe of $\Omega_{\rm DE}$; it will further constrain the parameters used in our model. Further, the variable EOS may be reflected in the Integrated Sachs-Wolfe (ISW) effect, which will serve as another test for the model. Studying features related to perturbations in this cosmological framework is another open issue.

[1] S. Perlmutter et al., Astrophys. J. **517** (1999) 565; A. Riess et al., Astrophys. J. **116**, 1009 (1998); P. Astier et. al., Astron. Astrophys. **447** (2006) 31
[2] W. L. Freedman et al., Astrophys. J. 553 (2001) 47; D. N. Spergel et al., Astrophys. J. **148** (2003) 175; *ibid*, Astrophys. J. Suppl. 170, 377 (2007)
[3] E. Komatsu et. al., Astrophys. J. Suppl. 180, 330 (2009)
[4] A. Riess et al., Astrophys. J. 699, 539 (2009) [arXiv:0905.0695]
[5] R. R. Cadwell, Phys. Lett. B 545, 23 (2002)
[6] S. M. Carroll, Living Rev. Relativity **3** (2001) 1; T. Padmanabhan, Phys. Rept. **380**, 235 (2003); E. J. Copeland, M. Sami and S. Tsujikawa, Int. J. Mod. Phys. **D 15**, 1753 (2006); J. Frieman, M. Turner and D. Huterer, Ann. Rev. Astron. Astrophys. 46, 385 (2008); A. Silvestri and M. Trodden, [arXiv:0904.0024]
[7] S. Nesseris and L. Perivolaropoulos, Phys. Rev. **D70** (2004) 043531; *ibid* **D72** (2005) 123519; J. Kujat, R. J. Scherrer and A. A. Sen, Phys. Rev. D74 (2006) 083501; S. Sur and S. Das, JCAP 0901 (2009) 007; C. G. Boehmer and J. Burnett, [arXiv:0906.1351]
[8] L. Perivolaropoulos, JCAP 0510 (2005) 001; R. Gannouji, D. Polarski, A. Ranquet and A. A. Starobinsky, JCAP 0609, 016 (2006); L. P. Chimento, R. Lazkoz, R. Maartens and I. Quiros, JCAP 0609 (2006) 004;

S. Nesseris and L. Perivolaropoulos, JCAP **0701**, 018 (2007); S-F. Wu, A. Chatrabhuti, G-H. Yang and P-M. Zhang, Phys. Lett. B659, 45 (2008); H. Garcia-Compean, G. Garcia-Jimenez, O. Obregon and C. Ramirez, JCAP 0807, 016 (2008)

[9] R. A. Brown, R. Maartens, E. Papantonopoulos and V. Zamarias, JCAP **11** (2005) 008; V. Sahni and Y. Shtanov, JCAP **11** (2003) 014; Y. Shtanov, A. Viznyuk and V. Sahni, Class. Quant. Grav. **24** (2007) 6159; Y. Shtanov, V. Sahni, A. Shafieloo and A. Toporensky, JCAP 0904 (2009) 023

[10] S. Das, S. Ghosh, J. W. van Holten and S. Pal, JHEP **0904** (2009) 115 [arXiv:0902.2304]; S. Das, S. Ghosh, J. W. van Holten and S. Pal, [arXiv:0906.1044]

[11] Lisa Randall, Raman Sundrum, Phys. Rev. Lett. 83:3370 1999 [arXiv:hep-ph/9905221]; Phys. Rev. Lett. 83:4690,1999 [arXiv:hep-th/9906064]

[12] For AdS-CFT correspondence, see for example M.Rangamani, Class. Quant. Grav. 26:224003, 2009 [arXiv:0905.4352]

[13] R. Maartens, Living Rev. Relativity 7 (2004) 7; C. Barcelo and M. Visser, Phys. Lett. B **482**, 183 (2000); J. Garriga and M. Sasaki, Phys. Rev. **D62** (2000) 043523; S. Mukohyama, T. Shiromizu and K. Maeda, Phys. Rev. D **62**, 024028 (2000); P. Bowcock, C. Charmousis and R. Gregory, Class. Quant. Grav. **17**, 4745 (2000); S. Pal, Phys. Rev. **D 74**, 024005 (2006); *ibid* **D 78**, 043517 (2008)

[14] J. W. van Holten, Phys. Rev. D75 (2007) 025027 [hep-th/0612216]; R. H. Rietdijk and J. W. van Holten, Nucl. Phys. B472 (1996) 427 [hep-th/9511166]; R. H. Rietdijk and J. W. van Holten, Nucl. Phys. B404 (1993) 42, [hep-th/9303112]; J. W. van Holten and R. H. Rietdijk, J. Geom. Phys. 11:559,1993 [hep-th/9205074]

[15] P. A. M. Dirac, *Lectures on Quantum Mechanics*, Yeshiva University Press, New York, 1964

[16] E. Poisson, *A Relativist's Toolkit: The Mathematics of Black-Hole Mechanics*, Cambridge University Press, Cambridge (2004)

[17] L. Verde et.al., Mon. Not. Roy. Astron. Soc. 335, 432 (2002)

[18] V. Sahni, T. D. Saini, A. A. Starobinsky and U. Alam, JETP Lett. **77** (2003) 201; U. Alam, V. Sahni, T. D. Saini and A. A. Starobinsky, Mon. Not. Roy. Astron. Soc. **344** (2003) 1057

Journal of Physics: Conference Series **222** (2010) 012023 doi:10.1088/1742-6596/222/1/012023

Cosmology with primordial black holes

A. S. Majumdar

S. N. Bose National Centre for Basic Sciences, Salt Lake, Kolkata 700098, India

E-mail: archan@bose.res.in

Abstract. We study the effect of accretion of radiation in the early universe on primordial black holes in various modified gravity theories. We consider separately, a braneworld model, Brans-Dicke theory, and a more general scalar-tensor model. The rate of growth of a primordial black hole due to accretion of radiation in all these modified gravity models is smaller than the rate of growth of the cosmological horizon, thus making available sufficient radiation density for the black hole to accrete causally. We show that accretion of radiation overrides the effect of Hawking evaporation during the radiation dominated era. The subsequent evaporation of the black holes in later eras could further be modified due to either the modified braneworld geometry, or the variable gravitational coupling in scalar-tensor theories, and they could survive up to much longer times compared to the case of standard cosmology.

1. Introduction

The cosmological evolution of primordial black holes is an interesting subject that has been studied by a number of researchers over the last few decades. Black holes could be formed in the early universe through several mechanisms [1] and in a wide mass range. Their impact on cosmological evolution could be diverse, and is yet to be investigated in full details for different models of black hole formation and cosmological evolution. Among the notable cosmolological consequences of surviving primordial black holes are the possibility of generation of the baryon asymmetry in the universe [2], [3], and their viability as candidates of cold dark matter [4], [5]. Primordial black holes could also have interesting astrophysical consequences, such as the seeding of supermassive black holes [6]. The most relevant issue concerning primordial black holes is their longevity which depends crucially on the effectiveness of various accretion processes.

In this article we provide a brief description of the evolution of primordial black holes in different alternate gravity scenarios [7]. We consider a braneworld model [8], the Brans-Dicke theory [9], and a more general scalar-tensor model [10]. The evolution of primordial black holes in the RS-II braneworld model leads to their growth due to accretion in the early radiation dominated high energy era [11] The possibility of black hole solutions in BD theory was first proposed by Hawking [12]. Due to the variable gravitational constant, the rate of energy loss and gain of primordial black holes through evaporation and accretion of radiation, respectively, gets modified [13]. Using scalar-tensor gravity theories Barrow and Carr [14] have studied PBH evaporation during various eras. It has been recently observed that in the context of generalised Brans-Dicke theory, inclusion of the effect of accretion leads to the prolongation of PBH lifetimes [15]. In all the considered alternate gravity models we show that accretion of radiation in the early radiation dominated universe is a dominant process overcoming the effects of Hawking radiation, enabling the black holes to survive for much longer times. A significant number of

Journal of Physics: Conference Series **222** (2010) 012023 doi:10.1088/1742-6596/222/1/012023

primordial black holes in these scenarios could survive up to present times and contribute to the dark matter density of the universe.

2. Cosmological evolution of primordial black holes

Consider a population of primordial black holes in the radiation dominated era of the universe that could be formed by various mechanisms [1] operative in the post-inflationary stage of the universe. To keep the analysis simple, we do not consider an initial mass distribution of black holes, but assume that all the black holes are formed at time t_0 with mass M_0. The time evolution of the FRW scale factor $a(t)$ of the universe is governed by the Friedmann equation with a dominating radiation component and also a matter component provided by the black holes.

The mass of a black hole will change with time due to two factors. First, the process of Hawking radiation causes a loss of mass. Secondly, accretion of the surrounding matter by the black hole leads to an increase in its mass. The rate of change of mass of a single black hole can be written as

$$\dot{M} = 4\pi R_{BH}^2 \left(-\sigma T_{BH}^4 + f\rho_R \right) \tag{1}$$

where R_{BH} is the Schwarzschild radius, σ is the Stefan-Boltzmann constant, T_{BH} is the Hawking temperature, and f is an $O(1)$ accretion efficiency factor. Relativistic particles and photons accreted by the black hole have density $\rho_R = \frac{\pi^2 g T^4}{30}$ where g is the number of particle species.

The energy density of the relativistic matter ρ_R will evolve both due to the expansion of the universe, as well as because of the absorption or emission by the black holes. The time-variation of ρ_R is thus given by

$$\frac{d}{dt}(\rho_R(t)a^4(t)) = -\dot{M}N_{BH}a(t) \tag{2}$$

where N_{BH} denotes the number of primordial black holes present. When the right-hand side of Eq.(1) vanishes, it yields the usual result $\rho_R \propto a^{-4}$. This happens when either no black holes are formed, i.e., $N_{BH} = 0$, or when the individual black hole masses do not change with the cosmic time.

The equations governing the cosmological evolution of the universe beyond the time t_0 are given by [16] the hubble expansion

$$\frac{\dot{a}^2}{a^2} = \frac{8\pi}{3M_4^2}\left(\rho_R + \frac{M\beta\rho(t_0)}{M_0}\frac{a^3(t_0)}{a^3} \right) \tag{3}$$

the evolution of radiation density

$$\dot{\rho_R} = -4\rho_R\frac{\dot{a}}{a} - \frac{\dot{M}\beta\rho(t_0)}{M_0}\frac{a^3(t_0)}{a^3} \tag{4}$$

and the rate of change of black hole mass Eq.(1). The above coupled equations have to be integrated for the specific gravity model under consideration. The modified geometry for the black holes alters the evaporation and accretion rates. Further, the cosmological evolution of the scale factor could also be altered in a model-dependent way, as we would discuss later in context of a braneworld model, and also in context of scalar-tensor models.

3. Survival of primordial black holes in the braneworld scenario

The cosmology of the RS-II model [8] entails a modified high energy phase in the early radiation dominated era of the universe during which the right hand side of the Einstein equation contains terms that are quadratic in the brane energy momentum tensor [17]. Other modifications include the so-called "dark-radiation" term which is given by the projection of the bulk Weyl tensor. Transition to the standard radiation dominated era takes place when $t \geq t_c \equiv l/2$. Such a modified high energy evolution has rich consequences for the physics of the early universe[17]). In particular, the inflationary scenario is altered, allowing the possibility of steep inflaton potentials to accomplish the desired features [18], [19]. Assuming a radiation dominated equation of state, the solutions for the Friedmann equation lead to an evolution such that for times earlier that t_c, i.e., $t \leq t_c$ (or $\rho \geq \lambda$), one has the non-standard high energy regime during which the radiation density and the scale factor evolve as $\rho_R = \frac{3M_4^2}{32\pi t_c t}$, and $a = a_0(t/t_0)^{1/4}$, respectively. But, for times much later than t_c, i.e., $t >> t_c$ (or $\rho << \lambda$), one gets back the standard radiation dominated cosmological evolution given by $\rho_R = \frac{3M_4^2}{32\pi t^2}$, and $a = a_0(t/(t_0 t_c)^{1/2})^{1/2}$.

The formation and evolution of black holes is an interesting and complex issue of investigation in braneworld cosmology [7]. Horizon sized density perturbations in the high energy phase could lead to the formation of black holes on the brane by the process of gravitational collapse. The geometry of such black holes is inherently 5-dimensional [20] with an altered mass-radius relarionship given by [21]

$$r_{BH} = \left(\frac{8}{3\pi}\right)^{1/2} \left(\frac{l}{l_4}\right)^{1/2} \left(\frac{M}{M_4}\right)^{1/2} l_4 \tag{5}$$

Taking into account these effects of accretion and evaporation together, the rate of change of mass $\dot{M}$ of a braneworld black hole is given by [11]

$$\dot{M} = -\frac{AM_4^2}{Mt_c} + \frac{BM}{t} \tag{6}$$

where $A \simeq (3/(16)^3\pi)$ and $B \simeq (2f/\pi)$. The exact solution for the black hole rate equation was obtained earlier [11]. It was shown that a black hole grows in size by the accretion of radiation during the high energy radiation dominated era, with its mass given by [22], [11]

$$\frac{M(t)}{M_0} \simeq \left(\frac{t}{t_0}\right)^B \tag{7}$$

The growth of mass decreases in the low energy radiation dominated regime, and subsequently, in the matter dominated era there is a lack of sufficient radiation density for the black hole to accrete. Hence, the black hole mass grows to a maximum value M_{max} at time t_t before evaporation starts dominating. The expression for the lifetime t_{end} of a black hole in this scenario is given by [11]

$$\frac{t_{end}}{t_4} \simeq \frac{4}{A}(2\sqrt{2})^B \left(\frac{M_0}{M_4}\right)^{2-B} \frac{t_c}{t_4} \left(\frac{t_t^2}{t_c t_4}\right)^B \tag{8}$$

Depending upon the values of the parameters t_c and l, one could obtain several examples of primordial black holes surviving till cosmologically interesting eras [11]. For example, a black hole formed with an initial mass $M_0 = 10^8 M_4 \simeq 10^3$g, survives up to the present era if one chooses $(l/l_4) \simeq 10^{30}$. If the black holes are produced with an initial mass spectrum, then one would have evaporating black holes at different eras. Observational constraints impacting different cosmological impose restrictions on the initial mass spectrum of braneworld black holes. Clancy *et al* [23] have shown how standard constraints are modified in the case of braneworld cosmology. There have been further studies on the impact of braneworld primordial black holes on the high energy diffuse photon background [24] and cosmic ray antiprotons [25]. It has been also shown how the high energy era dynamics is conducive to the formation of black hole binaries [26].

4. Primordial black holes in scalar-tensor theories

Scalar-tensor models (or Generalyzed Jordan-Brans-Dicke (JBD) models) are obtained in the low energy limit of higher dimensional theories. The JBD [9] theory is one of the earliest and most well-motivated alternatives of the Einstein theory of gravitation. The value of the gravitational "constant" is set by the inverse of a time-dependent classical scalar field with a coupling parameter ω. General relativity is recovered in the limit of $\omega \to \infty$. Solar system observations impose lower bounds on ω [27]. Higher dimensional models after compactification of the extra dimensions yield variants of scalar-tensor models in which the the scalar field coupling ω may become dynamical.

4.1. Brans-Dicke theory

For a spatially flat($k = 0$) FRW universe with scale factor a , the Einstein equations and the equation of motion for the JBD field Φ take the form

$$\frac{\dot{a}^2}{a^2} + \frac{\dot{a}}{a}\frac{\dot{\Phi}}{\Phi} - \frac{\omega}{6}\frac{\dot{\Phi}^2}{\Phi^2} = \frac{8\pi\rho}{3\Phi} \tag{9}$$

$$2\frac{\ddot{a}}{a} + \frac{\dot{a}^2}{a^2} + 2\frac{\dot{a}}{a}\frac{\dot{\Phi}}{\Phi} + \frac{\omega}{2}\frac{\dot{\Phi}^2}{\Phi^2} + \frac{\ddot{\Phi}}{\Phi} = -\frac{8\pi p}{\Phi} \tag{10}$$

$$\frac{\ddot{\Phi}}{8\pi} + 3\frac{\dot{a}}{a}\frac{\dot{\phi}}{8\pi} = \frac{\rho - 3p}{2\omega + 3} \quad . \tag{11}$$

The energy conservation equation is given by

$$\dot{\rho} + 3(\gamma + 1)H\rho = 0 \quad . \tag{12}$$

Here we have used the perfect fluid equation of state $p = \gamma\rho$.

Barrow and Carr [14] have obtained the following solutions for a and G for different eras, as

$$a(t) \propto \begin{cases} t^{(1-\sqrt{n})/3} & (t < t_1) \\ t^{1/2} & (t_1 < t < t_e) \\ t^{(2-n)/3} & (t > t_e) \end{cases} \tag{13}$$

and

$$G(t) = \begin{cases} G_0\left(\frac{t_1}{t}\right)^{\sqrt{n}}\left(\frac{t_0}{t_e}\right)^n & (t < t_1) \\ G_0\left(\frac{t_0}{t_e}\right)^n & (t_1 < t < t_e) \\ G_0\left(\frac{t_0}{t}\right)^n & (t > t_e) \end{cases} \tag{14}$$

where the radiation dominated era lasts from t_1 to t_e, G_0 is the present value of G, t_0 the present time n is a parameter which related to ω by $n = \frac{2}{4+3\omega}$. Since observations[30] require that ω be large ($\omega \geq 10^4$), n is very small ($n \leq 0.00007$).

Barrow and Carr [14] have considered only evaporation of the primodial black holes due to Hawking radiation. If we consider accretion which is effective in radiation dominated era, then the primordial black holes take more time to evaporate. Here we have studied how accretion changes the life time of the primodial black holes. If we consider both evaporation and accretion simultaneously, then the rate at which the primordial black hole mass changes is given by

$$\dot{M}_{BH} = 6fG\left(\frac{\dot{a}}{a}\right)^2 M^2 - \frac{a_H}{256\pi^3}\frac{1}{G^2M^2} \quad . \tag{15}$$

This equation can not be solved exactly. But we can very well approximate it during different regimes when either accretion or evaporation is the dominant process.

Journal of Physics: Conference Series **222** (2010) 012023 doi:10.1088/1742-6596/222/1/012023

For $t < t_1$:
The Hawking evaporation rate for this era is given by

$$\dot{M}_{evap} = -\alpha\Big(\frac{t_e}{t_0}\Big)^{2n}\Big(\frac{1}{t_1}\Big)^{2\sqrt{n}} \quad \Big(\tfrac{t^{2\sqrt{n}}}{M^2}\Big) \tag{16}$$

where $\alpha = \frac{a_H}{256\pi^3 G_0^2}$. Integrating the above equation, we get

$$M^3 = M_i^3 + 3\alpha\Big(1+2\sqrt{n}\Big)^{-1}\Big(\frac{t_e}{t_0}\Big)^{2n}\Big(t_1\Big)^{-2\sqrt{n}} \quad \Big(t_i^{2\sqrt{n}+1} - t^{2\sqrt{n}+1}\Big) \tag{17}$$

where M_i is the black hole mass at its formation time $t_i \simeq G^{-1}(t_i)t_i$. The black hole evaporation time is given by

$$\tau = \Big[(3\alpha)^{-1}(1+2\sqrt{n})\Big(\frac{t_0}{t_e}\Big)^{2n} M_i^3 t_1^{2\sqrt{n}} + t_i^{1+2\sqrt{n}}\Big]^{\frac{1}{1+2\sqrt{n}}} \tag{18}$$

For $t_1 < t < t_e$:
In the radiation dominated era we consider two possibilities like: PBHs created before t_1 and those created after t_1.
CASE-I $(t_i < t_1)$:
Again considering both evaporation and accretion, we can write

$$\dot{M}_{BH} = \Big[-\alpha\Big(\frac{1}{t_1}\Big)^{2\sqrt{n}}\Big(\frac{t_e}{t_0}\Big)^{2n}\Big(\frac{t^{2\sqrt{n}}}{M^2}\Big)\Big] + \Big[\frac{3}{2}fG_0\Big(\frac{t_0}{t_e}\Big)^n\Big(\frac{M^2}{t^2}\Big) - \alpha\Big(\frac{t_e}{t_0}\Big)^{2n}\frac{1}{M^2}\Big] \quad . \tag{19}$$

The first term of r.h.s. is integrated over the time period t_i to t_1 and last two terms of r.h.s. are integrated over the time period t_1 to t. PBHs formed at time $t_i < t_1$ evaporate out in the radiation dominated era [13].

CASE-II $(t_i > t_1)$:

Taking both accretion and evaporation into account, we can write

$$\dot{M}_{BH} = \frac{3}{2}fG_0\Big(\frac{t_0}{t_e}\Big)^n\Big(\frac{M^2}{t^2}\Big) - \alpha\Big(\frac{t_e}{t_0}\Big)^{2n}\frac{1}{M^2} \quad . \tag{20}$$

In the radiation dominated era for a particular primordial black hole, accretion is dominant upto a value of t at which accretion equals evaporation (t_c) and after that evaporation dominates over accretion. Taking this into account we have considered only accretion upto t_c and only evaporation after that. For $t = t_c$, the magnitude of the accretion term is equal to the magnitude of the evaporation term. So we have

$$\frac{3}{2}fG_0\Big(\frac{t_0}{t_e}\Big)^n\Big(\frac{M^2}{t^2}\Big) = \alpha\Big(\frac{t_e}{t_0}\Big)^{2n}\Big(\frac{1}{M^2}\Big) \quad . \tag{21}$$

which gives

$$M_{max} = \Big(\frac{A}{f}\Big)^{\frac{1}{4}} \times \Big(t_c\Big)^{\frac{1}{2}} \tag{22}$$

where $A = \frac{2}{3}G^{-1}\alpha\Big(\frac{t_e}{t_0}\Big)^{3n}$ and $M_{max} = M(t_c)$. But, from the PBH accretion equation, we have

$$M_{max} = M_i\Big[1 + \frac{3}{2}f\Big(\frac{t_i}{t_c} - 1\Big)\Big]^{-1} \quad . \tag{23}$$

Equating the above two equations, we get

$$t_c^{1/2} = \Big(\frac{f}{A}\Big)^{\frac{1}{4}} \times \frac{M_i}{1-\frac{3}{2}f} \quad . \tag{24}$$

and hence

$$M_{max} = \frac{M_i}{1-\frac{3}{2}f} \quad . \tag{25}$$

This leads to the condition $f < \frac{2}{3}$.

Since evaporation dominates over accretion from t_c, we get

$$M = M_{max}\Big[1 + 3\alpha\Big(\frac{t_e}{t_0}\Big)^{2n}\Big(\frac{t_c}{M_{max}^3}\Big)\Big\{1 - \Big(\frac{t}{t_c}\Big)\Big\}\Big]^{\frac{1}{3}} \quad . \tag{26}$$

So, the lifetime for these PBHs is given by

$$t_{evap} = t_c\Big[1 + \Big(3\alpha\Big)^{-1}\Big(\frac{t_0}{t_e}\Big)^{2n}\Big(\frac{M_{max}^3}{t_{crit}}\Big)\Big] \quad . \tag{27}$$

For the PBHs which do not evaporete out till the end of radiation dominated era, one has

$$M = M_e\Big[1 + 3\alpha\Big(2n+1\Big)^{-1}\Big(\frac{t_e}{t_0}\Big)^{2n}\Big(\frac{t_e}{M_e^3}\Big)\Big\{1 - \Big(\frac{t}{t_e}\Big)^{2n+1}\Big\}\Big]^{\frac{1}{3}} \tag{28}$$

where $M_e = M(t_e)$. Hence, the lifetime becomes

$$t_{evap} = t_e\Big[1 + \Big(3\alpha\Big)^{-1}(2n+1)\Big(\frac{t_0}{t_e}\Big)^{2n}\Big(\frac{M_e^3}{t_e}\Big)\Big]^{\frac{1}{2n+1}} \quad . \tag{29}$$

which leads to

$$M_e = M_{max}\Big[1 + 3\alpha\Big(\frac{t_e}{t_0}\Big)^{2n}\Big(\frac{t_c}{M_{max}^3}\Big)\Big\{1 - \Big(\frac{t_e}{t_c}\Big)\Big\}\Big]^{\frac{1}{3}} \quad . \tag{30}$$

From the above expression we can calculate the initial masses of the primordial black holes which are evaporating today for different values of the accretion efficiency f. The results are presented in Ref. [13].

For $t > t_e$:

The primordial black holes which are created before radiation dominated era are completely evaporated during radiation dominated era. So for $t > t_e$, only those primordial black holes, which are created after t_1, exist . Considering both accretion and evaporation into account, we can write

$$\dot{M}_{BH} = \Big\{\frac{3}{2}fG_0\Big(\frac{t_0}{t_e}\Big)^n\Big(\frac{M^2}{t^2}\Big) - \alpha\Big(\frac{t_e}{t_0}\Big)^{2n}\frac{1}{M^2}\Big\} - \alpha\Big(\frac{1}{t_0}\Big)^{2n}\Big(\frac{t^{2n}}{M^2}\Big) \quad . \tag{31}$$

The first two term of r.h.s. are integrated over the period t_i to t_e and the last term is integrated over period t_e to t. The above equation can be integrated numerically to yield the lifetimes of PBHs formed at various initial times.

4.2. Scalar-tensor cosmology

Let us now consider a more general scalar-tensor model of gravity. A generalysed JBD action is of the form

$$S = \frac{1}{16\pi G}\int d^4x\sqrt{-g}[\phi R - \frac{\omega(\phi)}{\phi}(\partial_\mu\phi)^2] + S_{\text{matter}} \tag{32}$$

where S_{matter} corresponds to the action of the relativistic fluid of particles in the early universe. The Friedman equation and the equation of motion for the JBD field ϕ admit solutions of the type [10]

$$a(t) = a_i\left(\frac{t}{t_i}\right)^{\frac{3}{\omega_i+6}} \tag{33}$$

$$\phi(t) = \phi_i\left(\frac{t}{t_i}\right)^{-\frac{3}{\omega_i+6}} \tag{34}$$

$$2\omega(t) + 3 = (2\omega_i + 3)\left(\frac{t}{t_i}\right)^{\frac{2\omega_i+3}{\omega_i+6}} \tag{35}$$

where the sbscript i indicates the initial values of the variables. In order to minimize the departure from the standard radiation dominated evolution, we choose $|\omega_i| \ll 1$. Note that the solution for the JBD field ϕ indicating the strengthening of gravity with the increase of cosmic time is a generic feature of several generalyzed JBD or scalar tensor models [28].

We now consider the evolution of PBHs in the cosmological background governed by the above solutions. The rates of accretion and evaporation by a JBD black hole can be obtained by identifying $r_{BH} = 2M/\phi$ as the black hole radius, and $T = \phi/(8\pi M)$ as the Hawking temperature. The complete evolution for the PBH is described by [15]

$$\dot{M} = -\frac{A}{M^2 t} + \frac{BM^2}{t} \tag{36}$$

For PBHs with initial mass $M_i < (A/B)^{1/4}$, the rate of evaporation exceeds that of accretion. Hence, a PBH with initial mass $M_i < (A/B)^{1/4}$ evaporates out quickly. But PBHs with initial mass $M_i > (A/B)^{1/4}$ experience growth, reaching a maximum mass given by [15]

$$\frac{M_{max}}{M_i} \approx \frac{1}{1 - BM_i \ln(t_{eq}/t_i)} \tag{37}$$

The lifetime for the PBHs t_{evap} is given by

$$t_{evap} = t_{eq}\left[1 - \frac{5}{3}\frac{M_i^3}{(1 - BM_i\ln(t_{eq}/t_i))^3(At_iM_{pl}^4)}\right]^{-3/5} \tag{38}$$

The overall lifetime of a JBD PBH has an interesting comparison to a PBH lifetime in standard cosmology. A JBD PBH with initial mass $M_i < (A/B)^{1/4}$, evaporates out quicker than a PBH in standard cosmology, being unable to accrete in the radiation dominated era. However, when the initial mass exceeds $(A/B)^{1/4}$, accretion proceeds effectively and dominates over evaporation in the radiation dominated era increasing the PBH mass at the time of matter-radiation equality. Certain interesting examples of evolving PBHs in JBD cosmology have been observed [15]). A PBH which forms at time t_{EW} with an initial mass of the order of the cosmological horizon mass at t_{EW} satisfies the condition $M_i > (A/B)^{1/4}$, enabling it to survive beyond the present era ($t_{evap} > t_{now}$), with the PBH thus being a component of cold dark matter.

5. Conclusions

We have discussed the evolution of primordial black holes in separate scenarios of modified gravity leading to different cosmological evolutions in the early universe. The modified high energy behaviour of the expansion of the Hubble volume in the braneworld scenario makes it feasible for accretion by the primordial black holes of the surrounding radiation to take place. This feature is primarily responsible for the mass growth and prolonged survival [11], [22] of the braneworld black holes. In Brans-Dicke theory, the evolution of primordial black holes due to accretion and evaporation is modified by the variable gravitational constant. This leads to the increase of black hole lifetimes, and consequently the modification of several observational constraints from their evaporation [13]. In the context of scalar-tensor cosmology, we have obtained a cut-off value for the initial mass which decides whether the PBHs could survive today. An interesting consequence of this analysis is that no PBHs evaporate during the matter dominated [15] era, and hence there is no impact on the cosmic microwave background radiation due to evaporating PBHs.

References

[1] B. J. Carr, Lect. Notes Phys. **631**, 301 (2003).
[2] A. S. Majumdar, P. Das Gupta, R. P. Saxena Int. J. Mod. Phys. D **4**, 517 (1995).
[3] P. S. Custodio, and J. E. Horvath, Phys. Rev. D **58**, 023504 (1998).
[4] D. Blais, Phys. Lett. B **535**, 11 (2002).
[5] N. Afshordi, ApJ **594**, L71 (2003).
[6] R. Bean, and J. Magueijo, Phys. Rev. D **66**, 063505 (2002).
[7] A. S. Majumdar and N. Mukherjee, Int. J. Mod. Phys. D **14**, 1095 (2005).
[8] L. Randall and R. Sundrum, Phys. Rev. Lett. **83**, 4690 (1999).
[9] C. H. Brans and R. H. Dicke, Phys. Rev. **124**, 925 (1961).
[10] B. K. Sahoo and L. P. Singh, Mod. Phys. Lett. A **17**, 2409 (2002).
[11] A. S. Majumdar, Phys. Rev. Lett. **90**, 031303 (2003).
[12] S. W. Hawking, Comm. Math. Phys. **25**, 167 (1972).
[13] B. Nayak, L. P. Singh and A. S. Majumdar, Phys. Rev. D **80**, 023529 (2009).
[14] J. D. Barrow and B. J. Carr, Phys. Rev. D **54**, 3920 (1996).
[15] A. S. Majumdar, D. Gangopadhyay and L. P. Singh, Mon. Not. R. Astron. Soc. **385**, 1467 (2008).
[16] A. S. Majumdar, in *Focus on black hole research*, (Nova Science Publishers, 2006).
[17] R. Maartens, Living Rev. Relativity **7**, 1 (2004).
[18] E. J. Copeland, *et al* , Phys. Rev. D **64**, 023509 (2001).
[19] A. S. Majumdar, Phys. Rev. D **64**, 083503 (2001).
[20] R. C. Myers and M. J. Perry, Ann. Phys. **172**, 304 (1986).
[21] R. Guedens, *et al* Phys. Rev. D **66**, 043513 (2002).
[22] R. Guedens, D. Clancy and A. R. Liddle, Phys. Rev. D **66**, 083509 (2002).
[23] D. Clancy, R. Guedens and A. R. Liddle, Phys. Rev. D **68**, 023507 (2003).
[24] Y. Sendouda, S. Nagataki and K. Sato, Phys. Rev. D **68**, 103510 (2003).
[25] Y. Sendouda, K. Kohri, S. Nagataki and K. Sato, Phys. Rev. D **71**, 063512 (2005).
[26] A. S. Majumdar, A. Mehta and J.-M. Luck, Phys. Lett. B **607**, 219 (2005).
[27] C. M. Will, Living Rev. Relativity **9**, 3 (2006).
[28] A. S. Majumdar, T. R. Seshadri and S. K. Sethi, Phys. Lett. B **312**, 67 (1993).

Consistency of ΛCDM with Geometric and Dynamical Probes

L. Perivolaropoulos
Department of Physics, University of Ioannina, Greece
E-mail: leandros@uoi.gr

Abstract. The ΛCDM cosmological model assumes the existence of a small cosmological constant in order to explain the observed accelerating cosmic expansion. Despite the dramatic improvement of the quality of cosmological data during the last decade it remains the simplest model that fits remarkably well (almost) all cosmological observations. In this talk I review the increasingly successful fits provided by ΛCDM on recent geometric probe data of the cosmic expansion. I also briefly discuss some emerging shortcomings of the model in attempting to fit specific classes of data (eg cosmic velocity dipole flows and cluster halo profiles). Finally, I summarize recent results on the theoretically predicted matter overdensity ($\delta_m = \frac{\delta\rho_m}{\rho_m}$) evolution (a dynamical probe of the cosmic expansion), emphasizing its scale and gauge dependence on large cosmological scales in the context of general relativity. A new scale dependent parametrization which describes accurately the growth rate of perturbations even on scales larger than $100h^{-1}Mpc$ is shown to be a straightforward generalization of the well known scale independent parametrization $f(a) = \Omega_m(a)^\gamma$ valid on smaller cosmological scales.

1. Introduction

Converging geometrical [1, 2, 3, 4, 5, 6] and dynamical [7, 8, 9, 10, 11] cosmological observations indicate that the universe has entered a phase of accelerating expansion. This expansion could be driven either by the negative pressure of a homogeneous dark energy component [12] or by a modified gravitational force which is repulsive on large cosmological scales. Alternatively it could simply be due to an unusually large underdensity (void) on scales of about 1Gpc which induces an apparent isotropic accelerating expansion to observers located approximately at the center of the void [13]. All of the above three classes of models attempting to explain the observed accelerating expansion suffer from fine tuning problems which are reflected in the unexpected values of the parameters required to fit the cosmological expansion data. For example dark energy originating from an evolving scalar field requires a scalar field mass of order $10^{-42}GeV$ which is several orders of magnitude smaller from expectations based on particle physics theories. Similarly, models assuming the existence of large voids require fine tuning of the location of the earth based observer within about $20Mpc$ from the $1Gpc$ large void while the formation of such a large void is also very unlikely in the context of standard cosmology [13].

The simplest and most successful model consistent with (almost) all cosmological data is the ΛCDM model[14]. This model is based on a modified version of Einstein's equations of the form

$$R_{\mu\nu} - \frac{1}{2}Rg_{\mu\nu} + \Lambda g_{\mu\nu} = 8\pi G T_{\mu\nu} \qquad (1)$$

where Λ is a new free parameter, the cosmological constant. When the new term $\Lambda g_{\mu\nu}$ is placed on the left hand side (lhs) of equation (1) it is interpreted as a modification of the gravitational law corresponding to an effective Newton potential [14]

$$V(r) = -\frac{GM}{r} - \Lambda r^2 \tag{2}$$

where $-\Lambda r^2$ corresponds to a new repulsive gravitational term. When the new term $\Lambda g_{\mu\nu}$ is placed on the right hand side (rhs) of equation (1) it is interpreted as a new contribution to the energy momentum tensor $T_{\mu\nu}$ corresponding to dark energy with constant density $\rho_\Lambda = \frac{\Lambda}{8\pi G}$ and constant negative pressure $p_\Lambda = -\rho_\Lambda$.

Consistency with cosmological data requires a fine tuned value for ρ_Λ

$$\rho_\Lambda^{(obs)} \simeq (10^{-12} GeV)^4 \simeq 2 \times 10^{-10} erg/cm^3 \tag{3}$$

A physically motivated origin of the cosmological constant could be the energy of the quantum field vacuum which is predicted to have a constant diverging energy density and negative pressure. Under the assumption of a proper cutoff the quantum vacuum can be made finite and play the role of a cosmological constant. A natural scale for this cutoff is the Planck scale leading to a vacuum density

$$\rho_\Lambda^{(Pl)} \simeq (10^{18} GeV)^4 \simeq 2 \times 10^{110} erg/cm^3 \tag{4}$$

which is 120 orders of magnitude larger than the observed value.

Despite of the unnaturally small value of the observed $\rho_\Lambda^{(obs)}$ when compared to the anticipated value $\rho_\Lambda^{(Pl)}$ in the context of vacuum energy, the cosmological constant of equation (1) has some unique physically motivated features. In particular the lhs of equation (1) is the most general second rank tensor which is

- local
- coordinate covariant
- divergenceless (needed for energy momentum conservation)
- symmetric

An additional important attractive feature of ΛCDM is simplicity. It is the only model based on General Relativity (GR) which (assuming flatness) involves a single free parameter ($\Omega_\Lambda \equiv \frac{\rho_\Lambda}{\rho_c} = 1 - \Omega_{0\mathrm{m}}$ where ρ_c is the present day critical density required for a flat universe) and its predicted expansion rate $H(z) = \frac{\dot{a}}{a}(z)$ as a function of redshift z

$$H(z)^2 = H_0^2 \left[\Omega_{0\mathrm{m}}(1+z)^3 + (1 - \Omega_{0\mathrm{m}})\right] \tag{5}$$

is currently consistent with all cosmological expansion probes (geometric and dynamical). It is therefore clear that the identification of potential conflicts of ΛCDM with cosmological data is a prerequisite before seriously considering alternative more complicated models unless such models are free from any type of fine tuning. Unfortunately no such model is currently known.

Cosmological observations testing the ΛCDM model include two classes of probes

- Geometric probes which measure directly the cosmic metric (eg Type Ia supernovae as standard candles or Cosmic Microwave Background (CMB) spectrum peaks and Baryon Acoustic Oscillations (BAO) which use the last scattering horizon scale as a standard ruler)

- Dynamical probes which measure simultaneously the cosmic metric and the gravitational law on cosmological scales. The main dynamical probe is the evolution of linear dark matter cosmological perturbations $\delta_m(k,z) = \frac{\delta\rho_m}{\rho_m}(k,z)$ as a function of scale and redshift. This evolution can be probed either directly through weak lensing observations[15] or indirectly though the large scale power spectrum of luminus matter at various redshifts [16].

Geometric probes provide currently the most accurate information about the cosmic expansion rate $H(z)$ as a function of redshift. In the first part of this brief review (section 2) I discuss the following three basic questions

(i) What is the figure of merit [17] (constraining power) of various recent Type Ia supernovae (SnIa) datasets and how does it compare with the corresponding figure of standard ruler (CMB+BAO) geometric probe data?

(ii) What is the consistency level of various geometric probe datasets with ΛCDM?

(iii) What is the level of consistency between standard candle SnIa datasets and standard ruler CMB+BAO geometric probes?

A few cosmological observations which appear to be in some tension [18] with the predictions of ΛCDM will also be discussed in section 2. The most interesting of these observations appears to be the observed dipole velocity flows on scales larger than $50h^{-1}Mpc$ which appear to be a factor of about four larger than the ΛCDM model predictions assuming normalization of the matter fluctuation power spectrum using the WMAP5 CMB data[19, 20, 21]. Other puzzling observations include the high redshift brightness of SnIa [22], the cluster halo profiles [23, 24] and the emptiness of voids [25, 26, 27].

The theoretically predicted linear evolution of cosmological perturbations $\delta_m(k,z)$ as a function of redshift and scale, depends sensitively on both the expansion rate and the gravitational law on cosmological scales. This dual sensitivity makes the linear evolution of density perturbations a particularly useful dynamical cosmological probe. Even though current observational constraints on $\delta_m(k,z)$ are not as powerful as geometric constraints on $H(z)$ this is expected to change dramatically in the next decade [28].

The time evolution of matter density perturbations $\delta_m(k,t)$ is described by the well known equation (see eg [8])

$$\ddot{\delta}_m + 2H\dot{\delta}_m - 4\pi G\rho_m f(k,t)\delta_m = 0 \tag{6}$$

where in the context of general relativity $f(k,t) = 1$ and $\delta_m(k,t) = \delta_m(t)$ becomes independent of the scale k. In fact, this scale independence of δ_m is occasionally considered to be a signature of validity of general relativity[29]. However, the derivation of equation (18) is based on two important assumptions whose validity is questionable on scales ($\gtrsim 200h^{-1}Mpc$). These assumptions are the following:

(i) The scale of the perturbation δ_m is significantly smaller than the Hubble scale *at all* times during the perturbation evolution.

(ii) The choice of gauge plays a minor role in the form of the evolution equation for δ_m.

An important question discussed in the second part of this review (section 3) is the following: *What is the level of validity of these assumptions on large cosmological scales?* It can be shown that the validity of both assumptions breaks down rapidly as the scale increases beyond $200h^{-1}Mpc$.

2. ΛCDM confronts Recent Geometric Cosmological Data

A useful way to test the ΛCDM model is to consider a generalized parametrization of the dark energy equation of state which includes ΛCDM as a special parameter case and find the likelihood of the parameters corresponding to ΛCDM in this context. A commonly used such parametrization is the Chevallier-Polarski, Linder (CPL)[30, 31] ansatz

$$w = w_0 + w_1(1-a) = w_0 + w_1\frac{z}{1+z} \tag{7}$$

which reduces to ΛCDM for $(w_0, w_1) = (-1, 0)$. In this section, I show the ranking of the six latest Type Ia supernova (SnIa) datasets (see Table 1) (Constitution (C), Union (U), ESSENCE (Davis) (E), Gold06 (G), SNLS 1yr (S) and SDSS-II (D)) in the context of the CPL parametrization (7) according to their Figure of Merit (FoM)[17], their consistency with the cosmological constant (ΛCDM) and their consistency with standard rulers (Cosmic Microwave Background (CMB) and Baryon Acoustic Oscillations (BAO)). The datasets considered are shown in Table (1) along with some useful features such as the redshift range or the subsets of each set[32].

Assuming a CPL parametrization for $w(z)$ (equation (7)) it is possible to apply the maximum likelihood method separately for standard rulers (CMB+BAO) and standard candles (SnIa) assuming flatness. The corresponding late time form of $H(z)$ for the CPL parametrization is

$$\begin{aligned} H^2(z) &= H_0^2[\Omega_{0\mathrm{m}}(1+z)^3 + \\ &+ (1-\Omega_{0\mathrm{m}})(1+z)^{3(1+w_0+w_1)}e^{\frac{-3w_1 z}{(1+z)}}]\,. \end{aligned} \tag{8}$$

Table 1. The datasets used in the present analysis. See respective references for details on the sources of the SnIa data points.

Dataset	Date Released	Redshift Range	# of SnIa	Filtered subsets included
SNLS1 [33]	2005	$0.015 \le z \le 1.01$	115	SNLS [33], LR [34]
Gold06 [35]	2006	$0.024 \le z \le 1.76$	182	SNLS1 [33], HST [35], SCP [1], HZSST [2]
ESSENCE [36],[37]	2007	$0.016 \le z \le 1.76$	192	SNLS1 [33], HST [35], ESSENCE[36],[37]
Union [38]	2008	$0.015 \le z \le 1.55$	307	Gold06 [35], ESSENCE[36], [37]
Constitution [3]	2009	$0.015 \le z \le 1.55$	397	Union [38], CfA3[3]
SDSS [39]	2009	$0.022 \le z \le 1.55$	288	Nearby [40], SDSS-II [39], ESSENCE [36], SNLS [33], HST [35]

The SnIa observations use light curve fitters[40] to provide the apparent magnitude $m(z)$ of the supernovae at peak brightness. The resulting apparent magnitude $m(z)$ is related to the dimensionless luminosity distance $D_L(z)$ through [41, 32]

$$m_{th}(z) = \bar{M}(M, H_0) + 5log_{10}(D_L(z))\,, \tag{9}$$

where

$$D_L(z) = (1+z)\int_0^z dz' \frac{H_0}{H(z';\Omega, w_0, w_1)} \tag{10}$$

is the Hubble free luminosity distance ($H_0 d_L$), and $\bar{M}$ is the magnitude zero point offset and depends on the absolute magnitude M and on the present Hubble parameter H_0 as

$$\begin{aligned} \bar{M} &= M + 5log_{10}(\frac{H_0^{-1}}{Mpc}) + 25 = \\ &= M - 5log_{10}h + 42.38. \end{aligned} \quad (11)$$

The parameter M is the absolute magnitude which is assumed to be constant after proper corrections (using light curve fitters) have been implemented in $m(z)$.

The theoretical model parameters are determined by using the maximum likelihood method ie by minimizing the quantity

$$\chi^2_{SnIa}(\Omega, w_0, w_1) = \sum_{i=1}^{N} \frac{(\mu_{obs}(z_i) - \mu_{th}(z_i))^2}{\sigma^2_{\mu\ i}} \quad (12)$$

where N is the number of SnIa of the dataset and $\sigma^2_{\mu\ i}$ are the errors due to flux uncertainties, intrinsic dispersion of SnIa absolute magnitude and peculiar velocity dispersion. These errors are assumed to be Gaussian and uncorrelated. The theoretical distance modulus is defined as

$$\mu_{th}(z_i) \equiv m_{th}(z_i) - M = 5log_{10}(D_L(z)) + \mu_0\,, \quad (13)$$

where

$$\mu_0 = 42.38 - 5log_{10}h\,, \quad (14)$$

The steps followed for the usual minimization of (12) in terms of its parameters are described in detail in [32, 41, 16, 42].

In the context of constraints from standard rulers from CMB spectrum peaks [41, 32], the maximum likelihood method is used and the datapoints $(R, l_a, 100\Omega_b h^2)$ of Ref. [4] (WMAP5) where R, l_a are two shift parameters [41]. In the case of BAO, the maximum likelihood method is also applied [41] using the datapoints of Ref. [43] (SDSS5). For comparison, the more recent data of Ref. [5] (SDSS7) have also been considered with only minor differences in the results (slightly reduced consistency with ΛCDM in the context of the CPL parametrization but no change in the ranking sequences).

The Figure of Merit (FoM) is a useful measure of the effectiveness of a set of data in constraining cosmological parameters. In the case of two parameters (as for the CPL parametrization) it is defined as the reciprocal area of the 95.4% contour (see Fig. 2), in parameter space (w_0, w_1) [17]. Clearly, the larger the FoM the more effective the dataset in constraining the parameters (w_0, w_1). Fig. 1 shows the FoM in terms of the number of the SnIa data for the datasets of Table 1. Clearly, the FoM is an increasing function of the number of SnIa in the datasets. An exception to this rule is the ESSENCE dataset which has a slightly smaller FoM compared to the Gold06 dataset even though it has a larger number of SnIa. A possible origin of this effect is the fact that the FoM does not depend only on the total number of SnIa of the dataset but mainly on the number of SnIa at low and high redshifts (the redshift space distribution of the ESSENCE data includes more data at intermediate redshifts than the Gold06 dataset).

In order to study the consistency of the various SnIa datasets with the cosmological constant and the standard rulers we consider the distance in units of σ (σ-distance d_σ) of the best fit point to a model with parameters (w_0, w_1), where this reference point can be either ΛCDM or some other reference point, (see Fig. 2). The σ-distance can be found by converting $\Delta\chi^2 = \chi^2_{(w_0,w_1)} - \chi^2_{min}$ to d_σ, i.e. solving [44]

$$1 - \Gamma(1, \Delta\chi^2/2)/\Gamma(1) = \mathrm{Erf}(d_\sigma/\sqrt{2}) \quad (15)$$

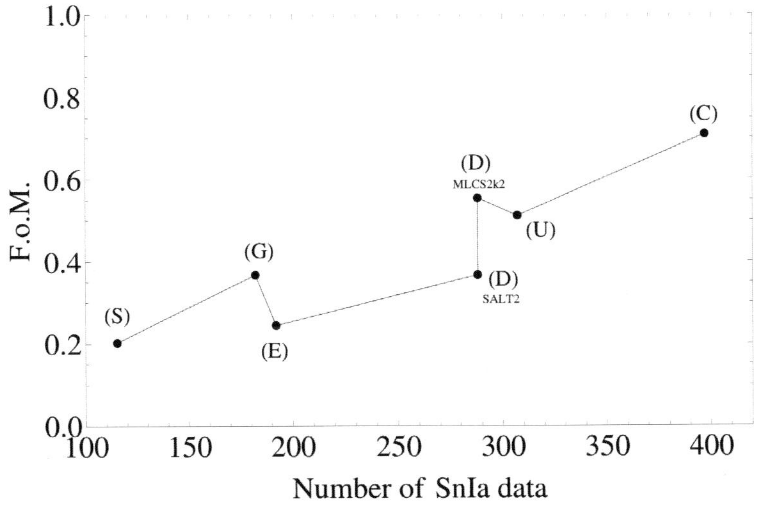

Figure 1. The Figure of Merit (FoM) in terms of the number of the SnIa data for $\Omega_{0m} = 0.28$ using the CPL parametrization.

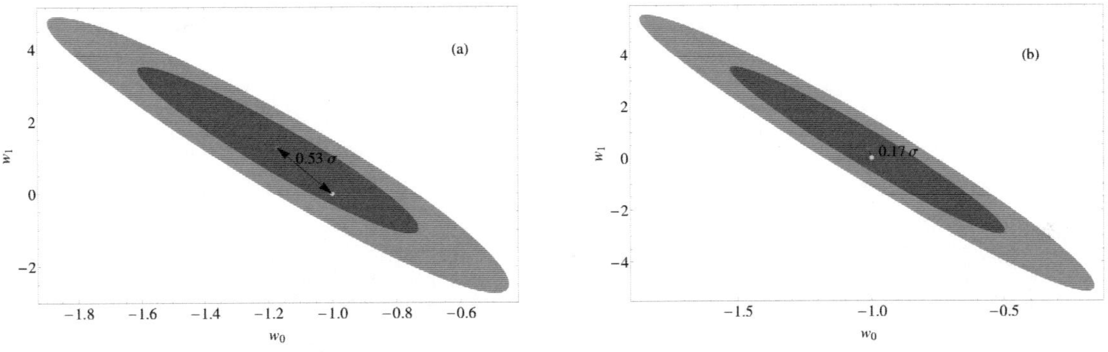

Figure 2. The 68.3%(1σ) – 95.4% (2σ) χ^2 confidence contours in the $w_0 - w_1$ plane based on parametrization (8) for the ESSENCE (left) and SNLS1 datasets (right) for $\Omega_{0m} = 0.24$. The arrows indicate the σ-distance of ΛCDM (green points: $(w_0, w_1) = (-1, 0)$) to the best fit points (red points).

for d_σ (σ-distance), where $\Delta\chi^2$ is the χ^2 difference between the best-fit and the reference point (w_0, w_1) (eg ΛCDM) and Erf() is the error function. The right hand side of Eq. (15) comes from integrating $\int_{-n\sigma}^{n\sigma} \frac{1}{\sigma\sqrt{2\pi}}\, e^{-\frac{x^2}{2\sigma^2}} dx$, where n is the desired number of σs, while the left hand side corresponds to the Cumulative Distribution Function (CDF) of a χ^2 distribution[44] with two degrees of freedom. Note that Eq. (15) is only valid for the two parameters (w_0, w_1) and should be generalized accordingly for more parameters [44]. In the special case of $n = 1$ or $n = 2$ we obtain the well known results $\Delta\chi_{1\sigma} = 2.30$ and $\Delta\chi_{2\sigma} = 6.18$ valid for two parameter parametrizations [44].

Even though the σ-distance is not a commonly used statistic it is quite useful because it can directly give information about probability of a given region in parameter space. The integer values of sigma distance (1σ and 2σ) are commonly used to draw the corresponding contours in parameter space. The extension of this statistic to non-integer values is used to find the specific contours that go through particular reference points of parameter space and thus

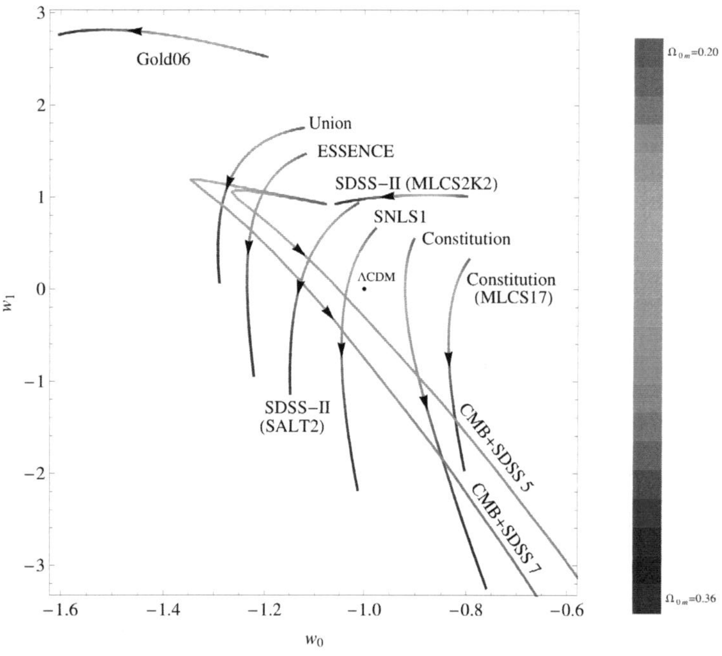

Figure 3. Trajectories of the best fit points (w_0, w_1) obtained for each of the datasets of Table 1 and also for the standard ruler CMB-BAO (WMAP5+SDSS5 and WMAP5+SDSS7) data as Ω_{0m} varies in the range $\Omega_{0m} \in [0.2, 0.36]$. The arrows in the best fit lines indicate the direction of growing Ω_{0m}. Note that for the SDSS5 data the standard ruler best fit parameters stretch out to $(w_0, w_1) \simeq (2, -30)$ for $\Omega_{0m} \simeq 0.36$, whereas for the SDSS7 data $(w_0, w_1) \simeq (0.90, -20)$.

estimate quantitatively the consistency of these points. The advantage of using the σ-distance instead of $\Delta\chi^2$ is the fact that the σ-distance takes into account the number of parameters of the parameterizations and can therefore be directly translated into probability for each point in parameter space. This is not possible for $\Delta\chi^2$ because it does not include information about the number of parameters of the parameterizations considered.

It is straightforward to apply the likelihood method to find the trajectory of the best fit point (w_0, w_1) in parameter space as Ω_{0m} varies in the range $\Omega_{0m} \in [0.2, 0.36]$. These trajectories obtained for each of the datasets of Table 1 and also for the standard ruler CMB-BAO (WMAP5+SDSS5, WMAP5+SDSS7) data are shown in Fig. 3. These trajectories can not be used to directly rank the datasets according to their consistency with any given reference point in parameter space (e.g. ΛCDM) because they contain no information about the 68.3% (1σ) contours. However, they provide useful hints for the trend of the best fit parameters as Ω_{0m} varies. For example, such a trend is the increase of the best fit value of the slope w_1 as the prior of Ω_{0m} decreases towards the value 0.2 or that the best fit value of w_0 remains less than -1 for all datasets except of the Constitution compilation.

The ranking sequence of the datasets of Table 1 with respect to any reference point in parameter space can be studied quantitatively using the σ-distance statistic discussed above. In order to test the sensitivity of the ranking sequence of datasets with respect to the choice of consistency reference point (w_0, w_1) we consider two such reference points: $(w_0, w_1) = (-1, 0)$ (ΛCDM) and $(w_0, w_1) = (-1.4, 2)$ which corresponds to dynamical dark energy with a $w(z)$ that crosses the line $w = -1$. It should be noted that there is nothing special about the parameter

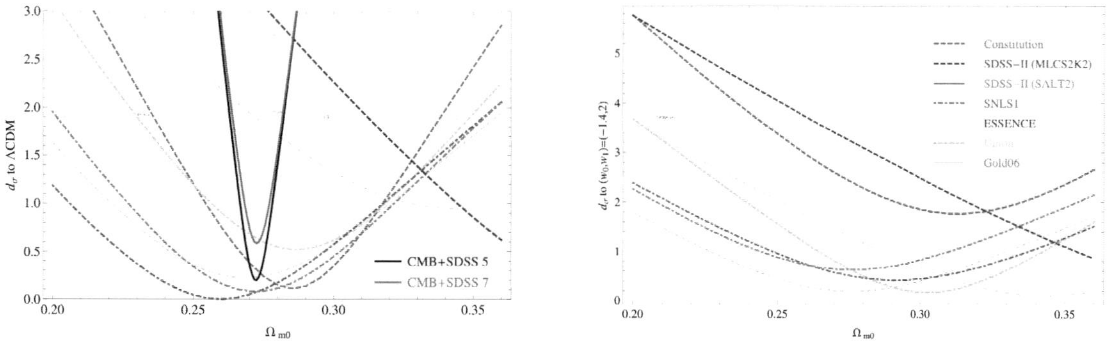

Figure 4. a: σ-distances $d_{\sigma i}(\Omega_{0\mathrm{m}}; -1, 0)$ off the ΛCDM point (reference point) from the best fit of each dataset in Table 1. Notice that the best fit of each dataset, except for the Gold06 and the SDSS-II (MLCS2k2), are minimized at similar values of $\Omega_{0\mathrm{m}}$. The σ-distance $d_\sigma(\Omega_{0\mathrm{m}}; -1, 0)$ between the standard ruler best fits and ΛCDM are also shown as a function of $\Omega_{0\mathrm{m}}$ (black and grey solid lines). b: Similar to (a) for the dynamical dark energy reference point $(w_0, w_1) = (-1.4, 2)$. Notice that the σ-distances $d_{\sigma i}(\Omega_{0\mathrm{m}}; -1.4, 2)$ are minimized at more widely separated values of $\Omega_{0\mathrm{m}}$.

point (-1.4,2). We have selected it as a representative of a wide region in parameter space (upper left from ΛCDM) which corresponds to dynamical dark energy crossing the phantom divide line $w = -1$. Any other point in the same parameter region would lead to similar results and the same ranking of datasets. This particular parameter region is interesting because it is spanned by the best fit trajectories and it also mildly favored by the Gold06 dataset (see Fig. 3).

The resulting $d_\sigma(\Omega_{0\mathrm{m}})$ for each dataset of Table 1 are shown in Figs. 4 in the range $\Omega_{0\mathrm{m}} \in [0.2, 0.36]$.

Table 2. Minimum σ-distance $d_\sigma^{min}(\Omega_{0\mathrm{m}}^{min}; -1, 0)$ from the best fit point for each of the datasets to the ΛCDM point. Also listed are the corresponding values of $\Omega_{0\mathrm{m}}$, and the best fit parameters (w_0, w_1) (see also Fig. 4a). The SDSS-II (MLCS2K2) data showed no minimum of d_σ with respect to $\Omega_{0\mathrm{m}}$ in the range $\Omega_{0\mathrm{m}} \in [0.2, 0.36]$. We thus have simply displayed the lowest value of d_σ in the corresponding range of $\Omega_{0\mathrm{m}}$.

Dataset	d_σ^{min}	Ω_{0m}^{min}	w_0	w_1
SNLS1	0.004	0.260	-1.03	0.16
SDSS-II (SALT2)	0.084	0.270	-1.09	0.51
Constitution	0.114	0.285	-0.91	-0.54
ESSENCE	0.227	0.270	-1.20	1.04
Union	0.525	0.285	-1.25	1.40
SDSS-II (MLCS2K2)	0.623	0.360	-1.06	0.93
Gold06	0.950	0.345	-1.56	2.80
CMB+BAO (SDSS5)	0.200	0.272	-1.15	0.51
CMB+BAO (SDSS7)	0.588	0.272	-1.30	0.97

Clearly, there are values of $\Omega_{0\mathrm{m}}$ that minimize the σ-distance $d_\sigma(\Omega_{0\mathrm{m}})$ between the best fit of each dataset and the reference point. These values of $\Omega_{0\mathrm{m}}$ maximize the consistency of the datasets with the given reference point in this range of $\Omega_{0\mathrm{m}}$. The minima σ-distances $d_\sigma(\Omega_{0\mathrm{m}})$ for each dataset, corresponding to maximum consistency with ΛCDM along with the corresponding values of Ω_{0m} are shown (properly ranked) in Table 2. The corresponding results for the reference point $(w_0, w_1) = (-1.4, 2)$ are shown in Table 3.

Table 3. The minimum σ-distances d_σ^{min} to the reference point $(w_0, w_1) = (-1.4, 2)$, the values of Ω_{0m} at which the minimum distance is attained, and the best fit parameters (w_0, w_1) at Ω_{0m}^{min} are displayed for each dataset (see also Fig. 4b). We omit the rows corresponding to CMB+BAO data as the resulting σ-distance is always $\gg 1$ due to the dominance of the dark energy at early times. The SDSS-II (MLCS2K2) data showed no minimum of d_σ with respect to Ω_{0m} in the range $\Omega_{0m} \in [0.2, 0.36]$. We thus have simply displayed the lowest value of d_σ in the corresponding range of Ω_{0m}.

Dataset	d_σ^{min}	Ω_{0m}^{min}	w_0	w_1
Gold06	0.11	0.345	-1.56	2.80
Union	0.17	0.300	-1.26	1.25
ESSENCE	0.19	0.275	-1.21	0.99
SNLS1	0.42	0.290	-1.04	-0.26
SDSS-II (SALT2)	0.63	0.275	-1.10	0.46
SDSS-II (MLCS2K2)	0.87	0.360	-1.06	0.93
Constitution	1.77	0.315	-0.88	-1.32

The following comments can be made with respect to the results shown in Figs. 4 and in the corresponding Tables 2, 3.

(i) The consistency with ΛCDM of all datasets, except Gold06 and SDSS-II when using the MLCS2K2 method, is maximized in a narrow range of $\Omega_{0m} \in [0.26, 0.29]$ which also includes the value of Ω_{0m} favored by standard rulers. On the other hand, the consistency with the dynamical dark energy point $(w_0, w_1) = (-1.4, 2)$ is maximized over a wider range of Ω_{0m} ($\Omega_{0m} \in [0.27, 0.35]$) thus decreasing the consistency among the datasets in the context of dynamical dark energy.

(ii) The ranking sequence changes dramatically when the consistency with the dynamical dark energy is considered as a reference point instead of ΛCDM (Table 3). Essentially the ranking is reversed! Thus, the choice of the consistency reference point plays an important role in determining the ranking sequence of the datasets (see also Fig. 4).

(iii) The SDSS-II dataset obtained with the MLCS2k2 fitter has some peculiar features compared to other datasets. In particular it favors particularly high values of Ω_{0m} ($\Omega_{0m} \simeq 0.4$) while for $\Omega_{0m} < 0.3$ its consistency with ΛCDM is significantly reduced to a level of 3σ or larger ($d_\sigma > 3$). In addition, the trajectory of its best fit parameter point as Ω_{0m} varies is perpendicular to the corresponding trajectory of most other datasets (see Fig. 4).

It is straightforward to apply the σ-distance statistic to rank the SnIa datasets according to their consistency with standard ruler CMB-BAO data. We simply use as a consistency reference point the best fit point $(w_0, w_1)^{SR}$ for standard rulers obtained as described in section 2 using the WMAP5+SDSS5 data. In this case, the location of the reference point $(w_0, w_1)^{SR}$ in parameter space depends on Ω_{0m} but this does not complicate the analysis. The σ-distance between the reference point $(w_0, w_1)^{SR}$ and the best fit of each dataset is shown in Fig. 5 as a function of Ω_{0m} for the datasets of Table 1. These distances are minimized for values of Ω_{0m} that are different for each dataset but they are all in the narrow range $\Omega_{0m} \in [0.27, 0.3]$.

These minimum distances along with the corresponding value of Ω_{0m} are shown in Table 4 for each dataset (properly ranked according to consistency with standard rulers). Notice that the

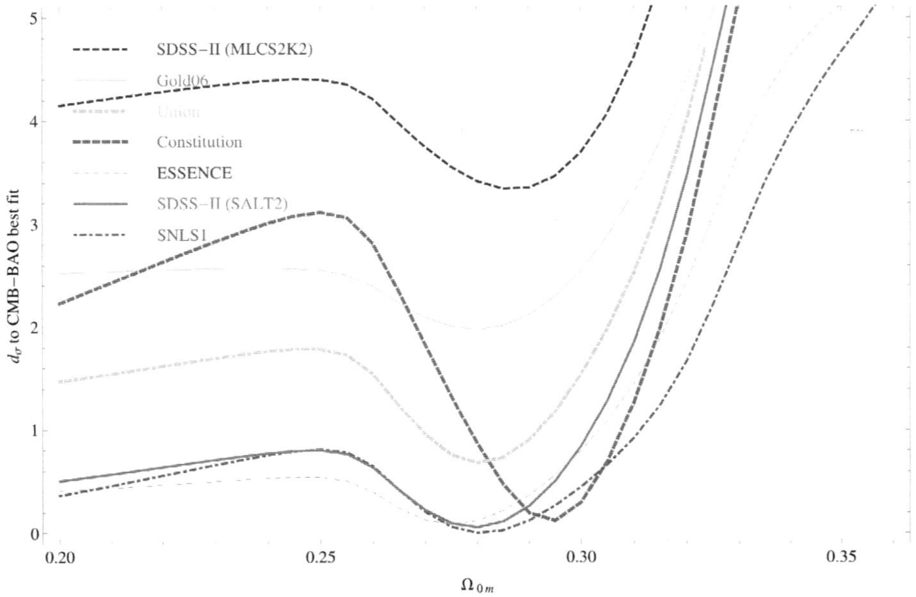

Figure 5. σ-distance $d_{\sigma i}(\Omega_{0\mathrm{m}}; (w_0, w_1)^{SR})$ between the reference point $(w_0, w_1)^{SR}$ of standard rulers (using the WMAP5+SDSS5 data) and the best fit of each dataset as a function of $\Omega_{0\mathrm{m}}$ for the datasets of Table 1. Using the SDSS7 data, the minimum distances are found in the range $\Omega_{0\mathrm{m}} \in [0.29, 0.31]$.

ranking sequence for the consistency with standard rulers is practically identical to the ranking sequence of the consistency with ΛCDM (Table 2) but differs from the ranking sequence of the consistency with dynamical dark energy (Table 3). This is an interesting feature of the data in favor of ΛCDM.

Table 4. Consistency with standard rulers. Minimum σ-distance $d_{\sigma i}^{min}(\Omega_{0\mathrm{m}}; (w_0, w_1)^{SR})$ between best fit parameters for each dataset, (w_0, w_1), and best fit for standard rulers, $(w_0, w_1)^{SR}$. $d_{\sigma i}^{min}(\Omega_{0\mathrm{m}}; (w_0, w_1)^{SR})$ is minimized at $\Omega_{0\mathrm{m}}^{min}$ (see Fig. 5).

Dataset	d_σ^{min}	Ω_{0m}^{min}	w_0	w_1	w_0^{SR}	w_1^{SR}
SNLS1	0.003	0.280	-1.04	-0.10	-1.07	0.08
SDSS-II (SALT2)	0.058	0.280	-1.11	0.40	-1.07	0.08
ESSENCE	0.087	0.275	-1.21	0.99	-1.12	0.38
Constitution	0.121	0.295	-0.90	-0.76	-0.84	-1.28
Union	0.681	0.280	-1.24	1.44	-1.07	0.08
Gold06	1.976	0.280	-1.38	2.75	-1.07	0.08
SDSS-II(MLCS2K2)	3.342	0.285	-0.92	1.02	-1.00	-0.28

It is therefore clear that the ΛCDM cosmological model is a well defined, simple and predictive model which is consistent with the majority of current cosmological observations. Despite of these successes there are specific cosmological observations which differ from the predictions of ΛCDM at a level of 2σ or higher. The observations conflicting the WMAP5 normalized ΛCDM model at a level of 2σ or larger include the following [18]:

- Large Scale Velocity Flows (ΛCDM predicts significantly smaller amplitude and scale of flows than what observations indicate)[19, 20, 21]. The probability of consistency with ΛCDM is about 1%.
- Brightness of Type Ia Supernovae (SnIa) at High Redshift z (ΛCDM predicts fainter SnIa at High z)[22]. The probability of consistency with ΛCDM is about $3-5\%$ for the Union and Gold06 datasets.
- Emptiness of Voids (ΛCDM predicts more dwarf or irregular galaxies in voids than observed)[25, 26, 27].
- Profiles of Cluster Haloes (ΛCDM predicts shallow low concentration and density profiles in contrast to observations which indicate denser high concentration cluster haloes) [23, 24].
- Profiles of Galaxy Haloes[45] (ΛCDM predicts halo mass profiles with cuspy cores and low outer density while lensing and dynamical observations indicate a central core of constant density and a flattish high dark mass density outer profile),
- Sizable Population of Disk Galaxies[46] (ΛCDM predicts a smaller fraction of disk galaxies due to recent mergers expected to disrupt cold rotationally supported disks).

Even though some of the puzzles discussed here may be resolved by more complete observations or astrophysical effects, the possible requirement of more fundamental modifications of the ΛCDM model remains valid.

3. Dynamical Probes: The Growth Function

3.1. Growth of Perturbations in General Relativity: Beyond the sub-Hubble Approximation

As discussed in the introduction, the evolution of the matter overdensity $\delta_m(k,a)$ (the growth function) consists a useful probe of both the expansion rate and the gravitational law on large scales. The standard parametrization of the linear growth function $\delta_m(k,a)$ is scale independent and is obtained by introducing a growth index γ defined through the growth rate $f(a)$ by

$$f_0(a) \equiv \frac{d\ln\delta_m}{d\ln a} = \Omega_m(a)^\gamma \tag{16}$$

where $a = \frac{1}{1+z}$ is the scale factor and

$$\Omega_m(a) \equiv \frac{H_0^2\Omega_{0\mathrm{m}}a^{-3}}{H(a)^2} \tag{17}$$

is the ratio of the matter density to the critical density when the universe has scale-factor a where H_0 is Hubble constant and Ω_{0m} is ratio of mass density to critical density. This parametrization [47] provides an excellent fit to the evolution equation for $\delta_m(a)$ in general relativity and in the small scale (sub-Hubble) approximation

$$\ddot{\delta}_m + 2H\dot{\delta}_m - 4\pi G\rho_m\delta_m = 0 \tag{18}$$

where an overdot denotes the derivative with respect to time and ρ_m is the matter density. Changing variables from t to $\ln a$ we obtain the evolution equation for the growth factor f as

$$f' + f^2 + f(\frac{\dot{H}}{H^2} + 2) = \frac{3}{2}\Omega_m \tag{19}$$

where $' = d/dlna$. For dark energy models in a flat universe with a slowly varying equation of state $w(a) \equiv \frac{p(a)}{\rho(a)} = w_0$, the solution of eq. (19) is well approximated by eq. (16) with [47]

$$\gamma = \frac{3(w_0-1)}{6w_0-5} \tag{20}$$

which reduces to $\gamma = \frac{6}{11}$ for the ΛCDM case ($w_0 = -1$). It is therefore clear that the observational determination of the growth index γ can be used to test ΛCDM [9]. It has been shown [11] that even in the context of dynamical dark energy models consistent with Type Ia supernovae (SnIa) observations the parameter γ does not vary by more than 5% from its ΛCDM value. However, in the context of modified gravity models γ can vary by as much as 30% (e.g. for the DGP model[48] $\gamma_{DGP} \simeq 0.68$ [11]) while scale dependence is also usually introduced[49, 50, 10].

Current observational constraints on γ are based on redshift distortions of galaxy power spectra [51], the rms mass fluctuation $\sigma_8(z)$ inferred from galaxy and $Ly-\alpha$ surveys at various redshifts [52, 53], weak lensing statistics [54], baryon acoustic oscillations [55], X-ray luminous galaxy clusters [56], Integrated Sachs-Wolfe (ISW) effect [57] etc. Unfortunately, the currently available data are limited in number and accuracy and come mainly from the first two categories. They involve significant error bars and non-trivial assumptions that hinder a reliable determination of γ. Thus, the current constraints on γ are fairly weak [9] and are expressed as

$$\gamma = 0.674^{+0.195}_{-0.169} \tag{21}$$

This however is expected to change in the next few years when more detailed weak lensing surveys [15] are anticipated to narrow significantly the above range[28].

A crucial assumption made in the derivation of eq. (18) in the context of general relativistic metric perturbations is the assumption that the scale of the perturbations is significantly smaller than the Hubble scale [47]. This assumption however does not lead to a good approximation on scales larger than about $100h^{-1}Mpc$ [58, 59]. In order to demonstrate this fact consider the perturbed metric of spacetime which takes the form (in the Newtonian gauge):

$$ds^2 = -(1+2\Phi)dt^2 + (1-2\Phi)a^2\gamma_{ij}dx^i dx^j, \tag{22}$$

where γ_{ij} is the metric of the spatial section and we are ignoring anisotropic stresses. The evolution of density perturbations on all scales is dictated by combining the background equations

$$H^2 = \frac{8\pi G}{3}(\rho_m + \rho_{de}) \tag{23}$$

$$\dot{\rho} = -3H(\rho + p) \tag{24}$$

(assuming a flat universe with only pressureless dark matter and (non-clustering) dark energy) with the perturbed linear order Einstein equations in the Newtonian gauge [60] (ρ_m and ρ_{de} are the matter and dark energy densities respectively while $p = w\rho$ is the pressure). The resulting (anisotropic stress-free) equations are of the form

$$\ddot{\Phi} = -4H\dot{\Phi} + 8\pi G\rho_{de}w_{de}\Phi \tag{25}$$

$$\dot{\delta} = 3\dot{\Phi} + \frac{k^2}{a^2}v_f \tag{26}$$

$$\dot{v}_f = -\Phi \tag{27}$$

with constraint equations

$$3H(H\Phi + \dot{\Phi}) + \frac{k^2}{a^2}\Phi = -4\pi G\delta\rho_m \tag{28}$$

$$(H\Phi + \dot{\Phi}) = -4\pi G\rho_m v_f \tag{29}$$

where Φ is the Newtonian potential, $v_f \equiv -va$ (v is the velocity potential for dark matter). Clearly, equations (25)-(27) involve a scale k dependence in contrast to the small scale approximate equation (18) which is scale independent.

The derivation of equation (18) in the context of general relativity is made using the sub-Hubble approximation. The linear matter overdensity $\delta\rho_m$ may be expressed [60] in terms of the gravitational potential Φ and the background variables as follows:

$$-4\pi G\delta\rho_m = \frac{k^2}{a^2}\Phi + 3H^2\Phi + 3H\dot{\Phi} \tag{30}$$

In the sub-Hubble (small scale) approximation ($\frac{k^2}{a^2} \gg H^2$) equation (30) takes the form

$$-4\pi G\delta\rho_m = \frac{k^2}{a^2}\Phi \tag{31}$$

where a slowly varying gravitational potential Φ has also been assumed.

The general relativistic equations (25)-(29) lead to the following equation for the matter overdensity δ_m

$$\ddot{\delta}_m + 2H\dot{\delta}_m + \frac{k^2}{a^2}\Phi = 0 \tag{32}$$

which also expresses the conservation of the perturbed energy momentum tensor for matter. Using the sub-Hubble approximation (31) we obtain the scale independent approximate equation (18). On the other hand, if we avoid this approximation in equation (30), solve for Φ (ignoring the time derivative) and substitute in equation (32) we obtain the following scale dependent evolution equation for δ_m [58, 59]:

$$\ddot{\delta}_m + 2H\dot{\delta}_m - \frac{4\pi G\rho_m\delta_m}{1+\xi(a,k)} = 0 \tag{33}$$

where

$$\xi(a,k) = \frac{3a^2H(a)^2}{k^2} \tag{34}$$

The solution of equation (33) provides a much better approximation to the full linear general relativistic system (25)-(27) up to horizon scales. On scales larger than the horizon even equation (33) breaks down since on these scales, the time derivative of Φ can not be ignored.

Given the successful approximation of the solution of (33) to the exact linear general relativistic solution, it becomes important to construct a scale dependent parametrization that is analogous to (16) and solves (approximately) (33) for all scales k. In order to construct such a parametrization we focus on the matter dominated era when most of the growth occurs and express $\xi(a,k)$ as

$$\xi(a,k) = \frac{3H_0^2\Omega_{0\mathrm{m}}}{ak^2} \tag{35}$$

Equation (33) may be expressed in terms of the growth factor $f = \frac{d\ln\delta_m}{d\ln a}$ in the form

$$f' + f^2 + \left(2 - \frac{3}{2}\Omega_m(a)\right)f = \frac{3}{2}\frac{\Omega_m(a)}{1+\xi(a,k)} \tag{36}$$

where $' \equiv \frac{d}{d\ln a}$, and we have assumed ΛCDM for $H(a)$.

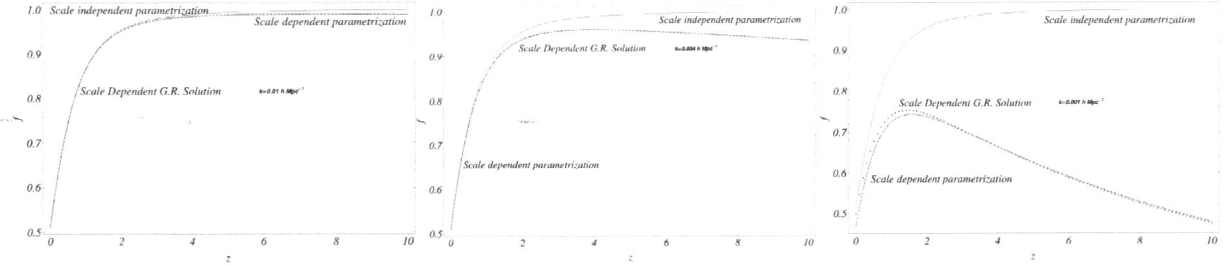

Figure 6. a: The growth rate f obtained from the solution of the general relativistic system ($k = 0.01h^{-1}Mpc$, $\Omega_{0\mathrm{m}} = 0.3$, ΛCDM , dotted line) compared with the scale independent parametrization (continuous line) and the corresponding generalized scale dependent parametrization (thick dashed line) b: Similar to a. for $k = 0.004h^{-1}Mpc$. c: Similar to a. for $k = 0.001h^{-1}Mpc$.

For sub-Hubble scales $\xi(k,a) \to 0$ and equation (36) reduces to (19) whose solution is well approximated by (16) with $\gamma = \frac{6}{11}$. It is straightforward to show using an expansion method [59] that the parametrization

$$f(k,a) = \frac{f_0(a)}{1+\xi(k,a)} = \frac{\Omega_m(a)^\gamma}{1+\frac{3H_0^2\Omega_{0\mathrm{m}}}{ak^2}} \tag{37}$$

is an approximate solution of equation (36) and provides a good approximation to the solution of the general relativistic system (25)-(27) up to horizon scales.

The accuracy of the parameterization (37) is demonstrated in Fig. 6 where we compare the form of the scale dependent parametrization Eq. (37) with the general relativistic numerical solution and with the standard parametrization for three different scales. It is clear that up to approximately the Hubble scale ($k \simeq 0.001hMpc^{-1}$) the scale dependent parametrization (37) is accurate at a level better than 5% at least up a to redshift $z = 10$. It can be shown that this parametrization is also a good approximation to the growth rate not only in the ΛCDM model but also in the case of dynamically evolving dark energy. Thus the parametrization described by equation (37) approximates very well the full linear general relativistic solution up to horizon scales and differs significantly from the standard parametrization on scales larger than about $100h^{-1}Mpc$. Its use may play an important role when comparing cosmological structure data on large scales with theoretical model predictions.

As shown in Fig. 6 the range of scales where the standard scale independent parametrization (16) starts breaking down involves scales larger than $100h^{-1}Mpc$ ($k < 0.01hMpc^{-1}$). This scale is much smaller than the present Hubble scale which is about $3000h^{-1}Mpc$. Therefore, the sub-Hubble approximation breaks down at much smaller scales than the anticipated Hubble scale. This is due to the fact that at early times during the matter era when most growth occurs, the comoving Hubble scale is significantly smaller compared to its present value. In fact, at recombination it is close to $200h^{-1}Mpc$. Therefore, scales that remain at the sub-Hubble level during the whole time when the growth of matter perturbations take place are only scales below $200h^{-1}Mpc$.

The generalized Poisson equation (30) in the Newtonian gauge may be solved for the slowly varying Φ leading to a generalized gravitational potential in Fourier space of the form

$$\Phi^{GR}(k) = \frac{\Phi^{Poisson}(k)}{1+\frac{3a^2H(a)^2}{k^2}} \tag{38}$$

where $\Phi^{Poisson}(k)$ is the usual small scale gravitational potential which emerges as a solution of the usual Poisson equation. In coordinate space the generalized large scale potential potential has a Yukawa form with a Hubble scale cutoff namely

$$\Phi^{GR}(r) = -\frac{GM}{ar}e^{-\sqrt{3}Har} \tag{39}$$

It is this general relativistic potential that should be compared with corresponding modified gravity potentials on scales comparable to the Hubble scale.

In a modified gravity theory the Poisson equation on large scales is modified for two reasons: First due to the modified gravitational law and second due to the Hubble scale effects that are present also in the case of general relativity as discussed above. The result is a Poisson equation of the form

$$\frac{k^2}{a^2}\Phi = -4\pi G\rho_m\delta_m(k,a)f(k,a) \tag{40}$$

where the scale dependent function $f(k,a)$ incorporates the scale dependence due to both Hubble scale effects and scale dependence due to modification of the gravitational law. This function is to be compared with the scale dependence due to pure Hubble scale effects $f_{GR}(k,a) = \frac{1}{1+\xi(a,k)}$ present in general relativity. Therefore a signature of modified gravity would not be a scale dependence of the evolution of matter density perturbations as claimed occasionally in the literature[29], but a deviation from the scale dependence predicted on large scale in the context of general relativity.

3.2. Gauge Dependence

The scale dependence of the matter density growth rate obtained above was based on calculations made in the Newtonian gauge. This is a physically important gauge because it corresponds to a time slicing of isotropic expansion. Nevertheless since the matter density perturbation $\delta_m(k,t)$ is a gauge dependent quantity, it is important to clarify to what extend do the results of section 3.1 persist in a gauge different from the Newtonian gauge.

Another important gauge is the synchronous gauge which corresponds to a time slicing obtained by the matter local rest frame everywhere in space (the free falling observer frame). The synchronous gauge is generally considered to be the most efficient reference system for doing calculations. It is used in many modern cosmology codes for calculating the evolution of cosmological perturbations (eg CMBFAST [61]). The line element of the perturbed spacetime in the synchronous gauge is given by

$$ds^2 = a^2(\tau)(-d\tau^2 + (\delta_{ij} + h_{ij})dx^i dx^j) \tag{41}$$

where τ is the conformal time. It is straightforward to derive the growth equation for $\delta_m \equiv \delta\rho_m/\rho_m$ in a matter dominated universe in the synchronous gauge to obtain [60, 58]

$$\ddot{\delta}_m^{SG} + 2H\dot{\delta}_m^{SG} - 4\pi\rho_m G\delta_m^{SG} = 0 \tag{42}$$

This growth equation is exact in the synchronous gauge in the case of matter domination and involves no scale dependence as in the case of equation (33) of the Newtonian gauge. This scale independence is an artifact of the particular time slicing of the synchronous gauge which is a good approximation on small scales but is unable to capture the horizon scale effects modifying the growth function on large scales.

Nevertheless equations (33) and (42) clearly agree on small scales where $\xi \rightarrow 0$. Therefore, for larger scales ($k < 0.01hMpc$) the question that arises is the following: *What is the proper gauge to use when comparing with observations?* This question was recently addressed in Ref. [62]

where a gauge invariant observable replacement was obtained for δ_m. This observable $\delta_{obs}(k,t)$ involves the matter density perturbation $\delta_m(k,t)$ corrected for redshift distortions due to peculiar velocities and gravitational potential. It also includes volume and position corrections. The final expression however is complicated and makes the theoretical predictions based on it not easy to implement and manipulate.

Alternatively, the gauge-invariant (GI) approach to the cosmological perturbations evolution, pioneered by Bardeen[63] may be used to identify observables (e.g., Refs.[64, 65]). The most general form of the line element for a spatially flat background and scalar metric perturbations can be written as [65]

$$ds^2 = a^2\{(1+2\Phi)d\tau^2 - 2B_{|i}dx^i d\tau - [\delta_{ij} - 2(\Psi\delta_{ij} - E_{|_{ij}})]dx^i dx^j\}, \tag{43}$$

where a and τ are the conformal cosmic expansion scale factor and the conformal cosmic time; "$|$" denotes the background three-dimensional covariant derivative. The corresponding perturbed energy-momentum tensor T^μ_ν has the form

$$\begin{aligned} T^0_{\ 0} &= \rho_m(1+\delta_m) , \\ T_0^{\ i} &= \rho_m U_{|i} , \\ T^0_{\ i} &= -\rho_m(U-B)_{|i} , \\ T^i_{\ j} &= -\rho_m \Sigma_{|ij} , \end{aligned} \tag{44}$$

where ρ_m is the unperturbed pressureless matter density; U and Σ determine velocity perturbation and anisotropic shear perturbation.

A gauge-invariant matter density perturbation may be constructed as [63, 64]

$$\delta_m^{GIS} \equiv \delta_m + 3\frac{\dot{a}}{a}(U-B) , \tag{45}$$

δ_m^{GIS} coincides with the density perturbation $\delta_m^{(CTG)}$ in the Comoving Time-orthogonal Gauge (CTG, in which $U = B = 0$), which denotes the density perturbation relative to the spacelike hypersurface which represents the matter local rest frame everywhere[63]. This quantity also coincides with the density perturbation δ_m^{SG} in the Synchronous Gauge (in which $\Psi = \Phi = B = 0$) for the pressureless matter system. In other words, δ_m^{GIS} denotes the density perturbation relative to the observers everywhere comoving with the matter. These *free falling* observers do not experience the isotropic expanding background of the universe because the peculiar velocity of matter is distinct from the Hubble flow. Thus δ_m^{SGI} has physical significance only for perturbations on scales small compared to the Hubble scale.

An alternative gauge-invariant variable is more closely related to the matter overdensity in the Newtonian gauge. This gauge-invariant perturbation variable is of the form [63, 64, 65],

$$\delta_m^{GIN} \equiv \delta_m + \frac{\dot{\rho}}{\rho}(B - \dot{E}) = \delta_m - 3\frac{\dot{a}}{a}(B - \dot{E}) . \tag{46}$$

and has important advantages over δ_m^{GIS}. δ_m^{GIN} coincides with the density perturbation δ_m^{NG} in the Newtonian Gauge (NG), in which $B = E = 0$.

In addition to the gauge invariant perturbation variable it is straightforward to construct two gauge-invariant scalar potentials ϕ and ψ, both of which become the same as the gravitational potential in the Newtonian limit. These are constructed from metric perturbations as follows [65]:

$$\begin{aligned} \phi &\equiv \Phi - \frac{\dot{a}}{a}(B - \dot{E}) , \\ \psi &\equiv \Psi + \frac{1}{a}\frac{d}{d\tau}[(B - \dot{E})a] . \end{aligned} \tag{47}$$

Journal of Physics: Conference Series **222** (2010) 012024 doi:10.1088/1742-6596/222/1/012024

The relation between δ_m^{GIS} and the general gravitational potential ϕ obeys the Poisson equation[63, 64]:

$$\nabla^2\phi = -k^2\phi = 4\pi G\rho a^2\delta_m^{GIS}. \quad (48)$$

where k is the (comoving) wavenumber of Fourier mode. As discussed above, the Poisson equation is valid only for scales small compared to the Hubble radius $1/H$ while on scales larger than the Hubble scale the growth of matter density perturbations is frozen. Hence, δ_m^{GIS} can not be regarded as the observable matter density perturbation on scales comparable to the Hubble scale. Therefore, the observable density perturbation on both the small-scale and the large-scale modes can not be described *directly* by δ_m^{GIS} even though it is a gauge-invariant quantity.

Contrary to δ_m^{GIS}, the other gauge invariant perturbation δ_m^{GIN} has some important attractive features with respect to observability. These are summarized as follows:

- It reduces to the Newtonian gauge perturbation δ_m^{NG} *ie* it corresponds to a frame which respects the isotropic expansion of the universe and is therefore more appropriate for description of large scale perturbations. This reduction also simplifies the calculation of this perturbation.
- It drives a scale dependent modification of the Poisson equation for the gauge invariant potential ϕ. Indeed, the time-time part of the linearized Einstein equation gives [65, 60]

$$\nabla^2\phi - 3\frac{\dot{a}}{a}(\frac{\dot{a}}{a}\psi + \dot{\phi}) = -k^2\phi - 3\frac{\dot{a}}{a}(\frac{\dot{a}}{a}\psi + \dot{\phi}) = 4\pi G\rho a^2\delta_m^{GIN}. \quad (49)$$

  Thus, the anticipated scale dependence on scales comparable to the Hubble scale is picked up by the perturbation δ_m^{GIN}.
- It is gauge invariant as anticipated for any observable quantity.

These features make the gauge invariant δ_m^{GIN} and the Newtonian gauge δ_m^{NG} to which it reduces, an attractive choice for making theoretical calculations to obtain the gravitational potential and the matter density perturbation that can be directly compared with observations on large scales. However, these theoretically obtained quantities need to also be corrected for bias, redshift distortions (due to gravitational potential and peculiar velocities), lensing magnification and volume distortion[62].

4. Conclusion

I have reviewed the consistency of recent geometric cosmological data with the ΛCDM cosmological model. The standard candle SnIa datasets considered in this study are consistent with ΛCDM and with standard rulers at a level of 95.4% (2σ) or less for certain prior values of the matter density Ω_{0m} in the range $\Omega_{0m} \in [0.25, 0.35]$. It is interesting that the ranking sequence based on consistency with ΛCDM is practically identical with the corresponding ranking based on consistency with standard rulers even though the two criteria are completely independent. Thus, despite the improvement of standard ruler and standard candle data quality during the last decade the consistency of ΛCDM with data has not decreased despite the fact that ΛCDM is a simple, specific and well defined model which appears as a measure-zero point in all generalized models. On the contrary its consistency seems to be improving with time as new and more accurate data appear. For example, the Constitution SnIa dataset which is a very recent compilation with a drastic improvement on the crucial nearby SnIa sample, is also one of the most consistent datasets with both ΛCDM and standard rulers.

Despite of its excellent consistency with both SnIa standard candles and CMB-BAO standard rulers, ΛCDM has to face potential challenges from other cosmological data [18] (e.g. large scale velocity flows, galaxy and cluster halo profiles, peculiar features of CMB maps etc.) which may lead the quest for the properties of dark energy to interesting surprises in the near future.

Such surprises may also come from future standard candle observations or standard ruler CMB experiments (e.g. Planck [66]) which are expected to significantly improve the accuracy of the constraints discussed in the present study.

I have also discussed dynamical probes of the cosmological constant and reviewed a scale dependent parametrization of the growth rate $f = \frac{d\ln\delta_m}{d\ln a}$ that is free from the sub-Hubble approximation of the standard parametrization (16). This parametrization described by equation (37) approximates very well the full linear general relativistic solution up to horizon scales and differs significantly from the standard parametrization on scales larger than about $100h^{-1}Mpc$. This parametrization describes well the growth of matter density perturbations on large scales in the Newtonian gauge.

Even though the matter density perturbation is a gauge dependent quantity, the Newtonian gauge has certain advantages that make it particularly useful when making a direct comparison of theoretical predictions with observations. These advantages are summarized as follows:

- The evolution of perturbations in the Newtonian gauge picks up the anticipated scale dependence when the Hubble scale is approached.
- The time slicing of the Newtonian gauge respects the isotropic expansion of the background and therefore it is more appropriate to describe large scale fluctuations. On the other hand, the synchronous gauge corresponds to free falling observers at all points and therefore it is physically relevant on small scales where the two gauges give identical evolution of cosmological perturbations.
- Gauge invariant generalizations of the density fluctuation δ_m and the gravitational potential reduce to the corresponding Newtonian quantities in the Newtonian gauge. This identification with gauge invariant quantities is a prerequisite for the direct observability of the density perturbation and the gravitational potential in the Newtonian gauge

Even though additional corrections need to be made to the theoretically predicted value of the density perturbation evolution along the lines of Ref. [62], it is clear that the Newtonian gauge offers a good starting point for the direct comparison of theoretical predictions of dark energy models with observations.

References

[1] Perlmutter et al. 1999, ApJ, 517, 565; Knop et al. 2003, ApJ, 598, 102;
[2] Riess et al. 1998 AJ, 116, 1009; Clocchiatti et al. 2003, astro-ph/0310432; Riess et al. 2000, ApJ, 536, 62; Tonry et al. 2003, Astrophys.J.594:1-24,2003.
[3] M. Hicken *et al.*, Astrophys. J. **700**, 1097 (2009) [arXiv:0901.4804 [astro-ph.CO]].
[4] Komatsu E. *et al.* 2009 [WMAP Collaboration], *Ap.J. Suppl.* **180**, 330.
[5] Percival W. J. *et al.* 2009, arXiv:0907.1660.
[6] Nesseris S. and Perivolaropoulos L. 2007 *JCAP* **0701**, 018;
Sanchez J. C. B., Nesseris S. and Perivolaropoulos L. 2009 *JCAP* **0911**, 029.
[7] Tegmark M. *et al.* 2004 [SDSS Collaboration], *Phys. Rev. D* **69**, 103501.
[8] Bertschinger E. 2006 *Ap. J.* **648**, 797;
Nesseris S. and Perivolaropoulos L. 2008 *Phys. Rev. D* **77**, 023504;
Di Porto C. and Amendola L. 2008 *Phys. Rev. D* **77**, 083508;
[9] S. Nesseris and L. Perivolaropoulos, Phys. Rev. D **77**, 023504 (2008) [arXiv:0710.1092 [astro-ph]].
[10] D. Polarski and R. Gannouji, Phys. Lett. B **660**, 439 (2008) [arXiv:0710.1510 [astro-ph]].
[11] E. V. Linder and R. N. Cahn, Astropart. Phys. **28**, 481 (2007) [arXiv:astro-ph/0701317].
[12] Copeland E. J., Sami M. and Tsujikawa S. 2006 *Int. J. Mod. Phys. D* **15**, 1753.
[13] Iguchi H., Nakamura T. and Nakao K. 2002 *Prog. Theor. Phys.* **108**, 809;
Caldwell R. R. and Stebbins A. 2008 *Phys. Rev. Lett.* **100**, 191302;
Vanderveld R. A., Flanagan E. E. and Wasserman I. 2006 *Phys. Rev. D* **74**, 023506.
[14] Peebles P.J.E. and Ratra B. 2003 *Rev. Mod. Phys.* **75**, 559;
Padmanabhan T. 2003 *Phys. Rept.* **380**, 235;
Carroll S. M. 2001 *Living Rev. Rel.* **4**, 1;

Sahni V 2002 *Class. Quant. Grav.* **19**, 3435.
[15] L. Fu *et al.*, arXiv:0712.0884 [astro-ph]; O. Dore *et al.*, arXiv:0712.1599 [astro-ph]; D. Munshi, P. Valageas, L. Van Waerbeke and A. Heavens, Phys. Rept. **462**, 67 (2008) [arXiv:astro-ph/0612667].
[16] S. Nesseris and L. Perivolaropoulos 2004 *Phys. Rev. D* **70**, 043531.
[17] Albrecht A.J. *et al.* 2006 *"Report of the Dark Energy Task Force,"* arXiv:astro-ph/0609591; arXiv: 0901.0721.
[18] L. Perivolaropoulos, arXiv:0811.4684 [astro-ph].
[19] R. Watkins, H. A. Feldman and M. J. Hudson, arXiv:0809.4041 [astro-ph].
[20] A. Kashlinsky, F. Atrio-Barandela, D. Kocevski and H. Ebeling, arXiv:0809.3734 [astro-ph].
[21] H. A. Feldman, R. Watkins and M. J. Hudson, arXiv:0911.5516 [Unknown].
[22] L. Perivolaropoulos and A. Shafieloo, arXiv:0811.2802 [astro-ph].
[23] T. Broadhurst, K. Umetsu, E. Medezinski, M. Oguri and Y. Rephaeli, Astrophys. J. **685**, L9 (2008) [arXiv:0805.2617 [astro-ph]].
[24] T. J. Broadhurst, M. Takada, K. Umetsu, X. Kong, N. Arimoto, M. Chiba and T. Futamase, Astrophys. J. **619**, L143 (2005) [arXiv:astro-ph/0412192].
[25] A. Tikhonov and A. Klypin, arXiv:0807.0924 [astro-ph].
[26] P. J. E. Peebles, Nucl. Phys. Proc. Suppl. **138**, 5 (2005) [arXiv:astro-ph/0311435].
[27] A. A. Klypin, A. V. Kravtsov, O. Valenzuela and F. Prada, Astrophys. J. **522**, 82 (1999) [arXiv:astro-ph/9901240].
[28] Zhan H., Knox L. and Tyson J. A. 2009 *ApJ* **690**, 923;
Joudaki S., Cooray A. and Holz D. E. 2009 *Phys. Rev. D* **80**, 023003;
T. Schrabback *et al.*, arXiv:0911.0053;
A. Cimatti *et al.*, arXiv:0912.0914.
[29] V. Acquaviva, A. Hajian, D. N. Spergel and S. Das, Phys. Rev. D **78**, 043514 (2008) [arXiv:0803.2236 [astro-ph]].
[30] Chevallier M. and Polarski D. 2001 *Int. J. Mod. Phys. D* **10**, 213.
[31] Linder E. V. 2003 *Phys. Rev. Lett.* **90**, 091301.
[32] J. C. B. Sanchez, S. Nesseris and L. Perivolaropoulos 2009 *JCAP* **0911**, 029.
[33] P. Astier *et al.* [The SNLS Collaboration] 2006 *Astron. Astrophys.* **447**, 31.
[34] Hamuy et al. 1996 *Ap.J.* **112**, 2408;
Riess et al. 1999 *Ap.J.* 117, 707;
Jha 2002 (PhD thesis Harvard);
Krisciunas 2001 *Ap.J.* 122, 1616;
Krisciunas, et al. 2004 *Ap.J.* **128**, 3034;
Altavilla et al 2004 *MNRAS*, **349**, 1344.
[35] A. G. Riess *et al.* 2007 *Ap. J* **659**, 98.
[36] W. M. Wood-Vasey *et al.* [ESSENCE Collaboration] 2007 *Ap. J* **666** 694.
[37] T. M. Davis *et al.* 2007 *Ap.J.* **666**, 716.
[38] M. Kowalski *et al.* 2008 *Ap. J.* **686**, 749.
[39] R. Kessler *et al.* 2009 arXiv:0908.4274 [astro-ph.CO].
[40] J. Guy et al. 2005 *A&A* **443**, 781; J. Guy et al., A&A **466**, 11 (2007);
S. Jha, A. G. Riess and R. P. Kirshner 2007 *Ap.J.*, **659**, 122.
[41] R. Lazkoz, S. Nesseris and L. Perivolaropoulos 2008 *JCAP* **0807**, 012.
[42] S. Nesseris and L. Perivolaropoulos 2005 *Phys. Rev. D* **72**, 123519.
[43] W. J. Percival, S. Cole, D. J. Eisenstein, R. C. Nichol, J. A. Peacock, A. C. Pope and A. S. Szalay 2007 *Mon. Not. Roy. Astron. Soc.* **381**, 1053;
D. Eisenstein *et al.* 2005 *Ap. J.* **633**, 560.
[44] W. H. Press *et. al.* 1994 *"Numerical Recipes"* Cambridge University Press.
[45] G. Gentile, C. Tonini and P. Salucci 2007 *Astron. Astrophys.* **467**, 925.
[46] J. S. Bullock, K. R. Stewart and C. W. Purcell 2008 arXiv:0811.0861.
[47] L. M. Wang and P. J. Steinhardt 1998 *Ap. J.* **508**, 483.
[48] C. Deffayet 2001 *Phys. Lett. B* **502**, 199.
[49] S. Nesseris and L. Perivolaropoulos 2007 *JCAP* **0701**, 018.
[50] J. P. Uzan 2007 *Gen. Rel. Grav.* **39**, 307.
[51] E. Hawkins *et al.* 2003 *Mon. Not. Roy. Astron. Soc.* **346**, 78;
E. V. Linder 2007 arXiv:0709.1113 [astro-ph].
[52] M. Viel, M. G. Haehnelt and V. Springel 2004 *Mon. Not. Roy. Astron. Soc.* **354**, 684.
[53] M. Viel and M. G. Haehnelt 2006 *Mon. Not. Roy. Astron. Soc.* **365**, 231.
[54] N. Kaiser 1998 *Ap.J.* **498**, 26;
L. Amendola, M. Kunz and D. Sapone 2007 arXiv:0704.2421 [astro-ph];

H. Hoekstra *et al.* 2006 *Ap. J.* **647**, 116.
[55] H. J. Seo and D. J. Eisenstein 2003 *Ap. J.* **598**, 720 ;
D. Sapone and L. Amendola 2007 arXiv:0709.2792 [astro-ph].
[56] A. Mantz, S. W. Allen, H. Ebeling and D. Rapetti 2007 arXiv:0709.4294 [astro-ph].
[57] M. J.Rees and D. W. Sciama 1968 *Nature* **217** 511;
R. G. Crittenden and N. Turok 1996 *Phys. Rev. Lett.* **76**, 575;
L. Pogosian, P. S. Corasaniti, C. Stephan-Otto, R. Crittenden and R. Nichol 2005 *Phys. Rev. D* **72**, 103519.
[58] J. B. Dent and S. Dutta 2009 *Phys. Rev. D* **79**, 063516.
[59] J. B. Dent, S. Dutta and L. Perivolaropoulos 2009 *Phys. Rev. D* **80**, 023514.
[60] C. P. Ma and E. Bertschinger 1995 *Ap. J.* **455**, 7.
[61] M. Zaldarriaga and U. Seljak 2000 *Ap. J. Suppl.* **129**, 431.
[62] J. Yoo, A. L. Fitzpatrick and M. Zaldarriaga 2009 *Phys. Rev.* D **80**, 083514.
[63] Bardeen J M 1980 *Phys. Rev. D* **22** 1882
[64] Kodama H and Sasaki M 1984 *Prog. Theo. Phys. Suppl.* **78** 1.
[65] Mukhanov V F, Feldman H A and Brandenberger R H 1992 *Phys. Rep.* **215** 203.
[66] Planck Collaboration, The Science Programme of Planck [arXiv:astro-ph/0604069].

Journal of Physics: Conference Series **222** (2010) 012025 doi:10.1088/1742-6596/222/1/012025

The Dark Energy Equation of State using Alternative Cosmic High-z Tracers

M. Plionis[1,2], **R. Terlevich**[2], **S. Basilakos**[3], **F. Bresolin**[4], **E. Terlevich**[2], **J. Melnick**[5], **R. Chavez**[2]

[1] Institute of Astronomy & Astrophysics, National Observatory of Athens, Palaia Penteli 152 36, Athens, Greece.

[2] Instituto Nacional de Astrofísica Optica y Electrónica, AP 51 y 216, 72000, Puebla, México.

[3] Academy of Athens, Research Center for Astronomy and Applied Mathematics, Soranou Efesiou 4, 11527, Athens, Greece

[4] Institute for Astronomy of the University of Hawaii, 2680 Woodlawn Drive, 96822 Honolulu, HI USA

[5] European Southern Observatory, Alonso de Cordova 3107, Santiago, Chile

E-mail: `mplionis@astro.noa.gr`

Abstract. We propose to use alternative cosmic tracers to measure the dark energy equation of state and the matter content of the Universe [w(z) & Ω_m]. Our proposed method consists of two components: (a) tracing the Hubble relation using HII galaxies which can be detected up to very large redshifts, $z \sim 4$, as an alternative to supernovae type Ia, and (b) measuring the clustering pattern of X-ray selected AGN at a median redshift of ~ 1. Each component of the method can in itself provide interesting constraints on the cosmological parameters, especially under our anticipation that we will reduce the corresponding random and systematic errors significantly. However, by joining their likelihood functions we will be able to put stringent cosmological constraints and break the known degeneracies between the *dark energy* equation of state (whether it is constant or variable) and the matter content of the universe and provide a powerful and alternative route to measure the contribution to the global dynamics and the equation of state of *dark energy*. A preliminary joint analysis of X-ray selected AGN clustering (based on the largest to-date XMM survey; the 2XMM) and the currently largest SNIa sample, the *Constitution* set (Hicken et al.), using as priors a flat universe and the WMAP5 normalization of the power-spectrum, provides: $\Omega_{\rm m} = 0.27 \pm 0.02$ and w= -0.96 ± 0.07. Equivalent and consistent results are provided by the joint analysis of X-ray selected AGN clustering and the latest Baryonic Acoustic Oscillation measures, providing: $\Omega_{\rm m} = 0.27 \pm 0.02$ and w= -0.97 ± 0.04.

1. Introduction

We live in a very exciting period for our understanding of the Cosmos. Over the past decade the accumulation and detailed analyses of high quality cosmological data (eg., supernovae type Ia, CMB temperature fluctuations, galaxy clustering, high-z clusters of galaxies, etc.) have strongly suggested that we live in a flat and accelerating universe, which contains at least some sort of cold dark matter to explain the clustering of extragalactic sources, and an extra component which acts as having a negative pressure, as for example the energy of the vacuum (or in a more general setting the so called *dark energy*), to explain the observed accelerated cosmic expansion (eg. Riess, et al. 1998; 2004; 2007, Perlmutter et al. 1999; Spergel et al. 2003, 2007, Tonry et al. 2003; Schuecker et al. 2003; Tegmark et al. 2004; Seljak et al. 2004; Allen et al. 2004;

Basilakos & Plionis 2005; 2006; 2009; Blake et al. 2007; Wood-Vasey et al. 2007, Davis et al. 2007; Kowalski et al. 2008, Komatsu et al. 2008; Hicken et al. 2009, etc).

Due to the absence of a well-motivated fundamental theory, there have been many theoretical speculations regarding the nature of the exotic *dark energy*, on whether it is a cosmological constant, a scalar or vector fields which provide a time varying dark-energy equation of state, usually parametrized by:

$$p_Q = \mathrm{w}(z)\rho_Q \;, \tag{1}$$

with p_Q and ρ_Q the pressure and density of the exotic dark energy fluid and

$$\mathrm{w}(z) = \mathrm{w}_0 + \mathrm{w}_1 f(z) \;, \tag{2}$$

with $\mathrm{w}_0 = \mathrm{w}(0)$ and $f(z)$ an increasing function of redshift [eg., $f(z) = z/(1+z)$] (see Peebles & Ratra 2003 and references therein, Chevalier & Polarski 2001, Linder 2003, Dicus & Repko 2004; Wang & Mukherjee 2006). Of course, the equation of state could be such that w does not evolve cosmologically.

Two very extensive recent reports have identified *dark energy* as a top priority for future research: "Report of the Dark Energy Task Force (advising DOE, NASA and NSF) by Albrecht et al. (2006), and "Report of the ESA/ESO Working Group on Fundamental Cosmology", by Peacock et al. (2006). It is clear that one of the most important questions in Cosmology and cosmic structure formation is related to the nature of *dark energy* (as well as whether it is the sole interpretation of the observed accelerated expansion of the Universe) and its interpretation within a fundamental physical theory. To this end a large number of very expensive experiments are planned and are at various stages of development, among which the *Dark Energy Survey* (DES: `http://www.darkenergysurvey.org/`), the *Joint Dark Energy Mission* (JDEM: `http://jdem.gsfc.nasa.gov/`), *HETDEX* (`http://www.as.utexas.edu/hetdex/`), Pan-STARRS: `http://pan-starrs.ifa.hawaii.edu`, etc.

Therefore, the paramount importance of the detection and quantification of *dark energy* for our understanding of the cosmos and for fundamental theories implies that the results of the different experiments should not only be scrutinized, but alternative, even higher-risk, methods to measure *dark energy* should be developed and applied as well. It is within this paradigm that our current work falls. Indeed, we wish to constrain the *dark energy* equation of state using, individually and in combination, the Hubble relation and large-scale structure (clustering) methods, but utilizing alternative cosmic tracers for both of these components.

From one side we wish to trace the Hubble function using HII galaxies, which can be observed at higher redshifts than those sampled by current SNIa surveys and thus at distances where the Hubble function is more sensitive to the cosmological parameters. The HII galaxies can be used as standard candles (Melnick, Terlevich & Terlevich 2000, Melnick 2003; Siegel et al. 2005; Plionis et al. 2009) due to the correlation between their velocity dispersion, metallicity and H_β luminosity (Melnick 1978, Terlevich & Melnick 1981, Melnick, Terlevich & Moles 1988), once we reduce significantly their distance modulus uncertainties, which at present are unacceptably large for precision cosmology ($\sigma_\mu \simeq 0.52$ mag; Melnick, Terlevich & Terlevich 2000). Furthermore, the use of such an alternative high-z tracer will enable us to check the SNIa based results and lift any doubts that arise from the fact that they are the only tracers of the Hubble relation used to-date (for possible usage of GRBs see for example, Ghirlanda et al. 2006; Basilakos & Perivolaropoulos 2008)[1].

[1] GRBs appear to be anything but standard candles, having a very wide range of isotropic equivalent luminosities and energy outputs. Nevertheless, correlations between various properties of the prompt emission and in some cases also the afterglow emission have been used to determine their distances. A serious problem that hampers a straight forward use of GRBs as Cosmological probes is the intrinsic faintness of the nearby events, a fact which introduces a bias towards low (or high) values of GRB observables and therefore the extrapolation of their

From the other side we wish to use X-ray selected AGN at a median redshift of ~ 1, which is roughly the peak of their redshift distribution (see Basilakos et al. 2004; 2005, Miyaji et al. 2007), in order to determine their clustering pattern and compare it with that predicted by different cosmological models.

Although each of the previously discussed components of our project (Hubble relation using HII galaxies and angular/spatial clustering of X-ray AGN) will provide interesting and relatively stringent constraints on the cosmological parameters, especially under our anticipation that we will reduce significantly the corresponding random and systematic errors, it is the combined likelihood of these two type of analyses that enables us to break the known degeneracies between cosmological parameters and determine with great accuracy the *dark energy* equation of state (see Basilakos & Plionis 2005; 2006; 2009).

Below we present the basic methodology of each of the two main components of our proposal, necessary in order to constrain the *dark energy* equation of state.

2. Cosmological Parameters from the Hubble Relation

It is well known that in the matter dominated epoch and in flat universes, the Hubble relation depends on the cosmological parameters via the following equation:

$$H(z) = H_0 E(z) \quad \text{with} \quad E(z) = \left[\Omega_m (1+z)^3 + \Omega_Q \exp\left(3\int_0^z \frac{1+\mathrm{w}(x)}{1+x}\mathrm{d}x\right)\right]^{\frac{1}{2}} , \tag{3}$$

which is simply derived from Friedman's equation. We remind the reader that Ω_m and $\Omega_Q (\equiv 1 - \Omega_m)$ are the present fractional contributions to the total cosmic mass-energy density of the matter and dark energy source terms, respectively.

Supernovae SNIa are considered standard candles at peak luminosity and therefore they have been used not only to determine the Hubble constant (at relatively low redshifts) but also to trace the curvature of the Hubble relation at high redshifts (see Riess et al. 1998, 2004, 2007; Perlmutter et al. 1998, 1999; Tonry et al. 2003; Astier et al. 2006; Wood-Vasey et al. 2007; Davis et al. 2007; Kowalski et al 2008; Hicken et al. 2009). Practically one relates the distance modulus of the SNIa to its luminosity distance, d_L, through which the cosmological parameters enter:

$$\mu = m - M = 5\log d_L + 25 \quad \text{where} \quad d_L = (1+z)\frac{c}{H_0}\int_0^z \frac{\mathrm{d}z}{E(z)} . \tag{4}$$

The main result of numerous studies using this procedure is that distant SNIa's are dimmer on average by $\sim$0.2 mag than what expected in an Einstein-deSitter model, which translates in them being $\sim 10\%$ further away than expected.

The amazing consequence of these results is that we live in an accelerating phase of the expansion of the Universe, an assertion that needs to be scrutinized on all possible levels, one of which is to verify the accelerated expansion of the Universe using alternative to SNIa's extragalactic standard candles. Furthermore, the cause and rate of the acceleration is of paramount importance, ie., the *dark energy* equation of state is the next fundamental item to search for and to these directions we hope to contribute with our current project.

2.1. Theoretical Expectations:

To appreciate the magnitude of the Hubble relation variations due to the different *dark energy* equations of state, we plot in Figure 1 the relative deviations of the distance modulus, $\Delta\mu$, of

correlations to low-z events is faced with serious problems. One might also expect a significant evolution of the intrinsic properties of GRBs with redshift (also between intermediate and high redshifts) which can be hard to disentangle from cosmological effects. Finally, even if a reliable scaling relation can be identified and used, the scatter in the resulting luminosity and thus distance modulus is still fairly large.

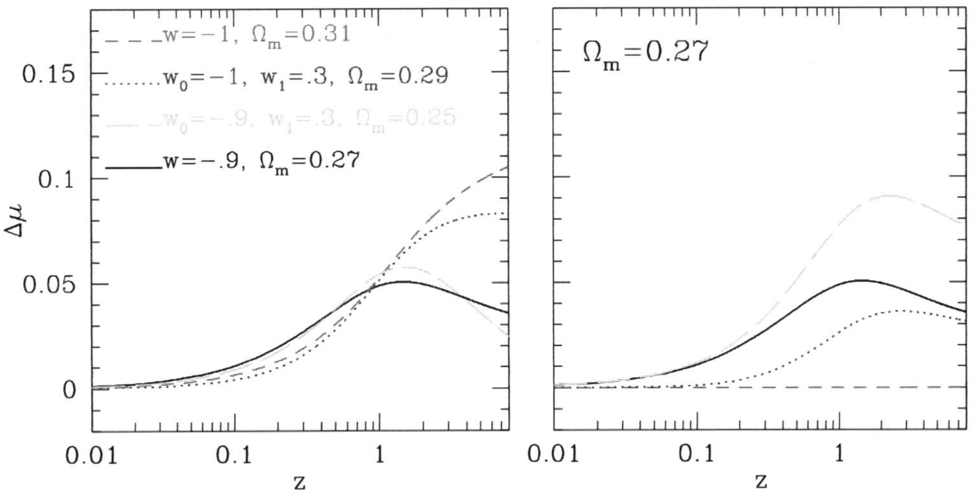

Figure 1. *Left Panel:* The expected distance modulus difference between the *dark-energy* models shown and the reference Λ-model (w= −1) with $\Omega_m = 0.27$. *Right Panel:* The expected distance modulus differences once the Ω_m-w(z) degeneracy is broken (imposing the same Ω_m value as in the comparison model).

different *dark-energy* models from a nominal *standard* (w= −1) Λ-cosmology (with $\Omega_m = 0.27$ and $\Omega_\Lambda = 0.73$), with the relative deviations defined as:

$$\Delta\mu = \mu_\Lambda - \mu_{\rm model} \; . \tag{5}$$

The parameters of the different models used are shown in Figure 1. As far as the *dark-energy* equation of state parameter is concerned, we present the deviations from the *standard* model of two models with a constant w value and of two models with an evolving equation of state parameter, utilizing the form of eq.(2). In the left panel of Figure 1 we present results for selected values of Ω_m, while in the right panel we use the same *dark-energy* equations of state parameters but for the same value of $\Omega_m (= 0.27)$ (ie., we eliminate the $\Omega_m - {\rm w}(z)$ degeneracy).

Three important observations should be made from Figure 1:

(i) The relative magnitude deviations between the different *dark-energy* models are quite small (typically $\lesssim 0.1$ mag), which puts severe pressure on the necessary photometric accuracy of the relevant observations.

(ii) The largest relative deviations of the distance moduli occur at redshifts $z \gtrsim 1.5$, and thus at quite larger redshifts than those currently traced by SN Ia samples, and

(iii) There are strong degeneracies between the different cosmological models at redshifts $z \lesssim 1$, but in some occasions even up to much higher redshifts (one such example is shown in Figure 1 between the models with $(\Omega_m, {\rm w}_0, {\rm w}_1) = (0.31, -1, 0)$ and $(0.29, -1, 0.3)$.

Luckily, such degeneracies can be broken, as discussed already in the introduction, by using other cosmological tests (eg. the clustering of extragalactic sources, the CMB shift parameter, BAO's, etc). Indeed, current evidence overwhelmingly show that the total matter content of the universe is within the range: $0.2 \lesssim \Omega_m \lesssim 0.3$, a fact that reduces significantly the degeneracies between the cosmological parameters.

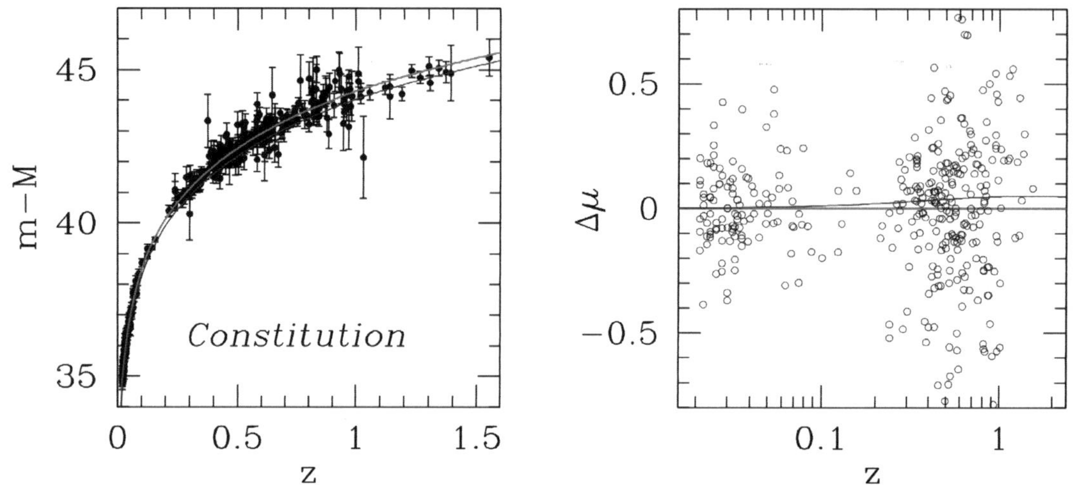

Figure 2. *Left Panel:* SNIa distance moduli as a function of redshift. *Right Panel:* Distance moduli difference between the Λ-model and the SNIa data. The blue line is the corresponding difference between the reference (w= −1) and the w= −0.85 *dark-energy* models.

2.2. Larger numbers or higher redshifts ?

In order to define an efficient strategy to put stringent constraints on the *dark-energy* equation of state, we have decided to re-analyse two recently compiled SNIa samples, the Davis et al. (2007) [hereafter *D07*] compilation of 192 SNIa (based on data from Wood-Vasey et al. 2007, Riess et al. 2007 and Astier et al. 2007) and the *Constitution* compilation of 397 SNIa (Hicken et al. 2009). Note that the two samples are not independent since most of the *D07* is included in the *Constitution* sample.

Firstly, we present in the left panel of Figure 2 the *Constitution* SNIa distance moduli overploted (red-line) with the theoretical expectation of a flat cosmology with $(\Omega_m, \mathrm{w}) = (0.27, -1)$. In the right panel we plot the distance moduli difference between the SNIa data and the previously mentioned model. To appreciate the level of accuracy needed in order to put constraints on the equation of state parameter, we also plot the distance moduli difference between the reference $(\Omega_m, \mathrm{w}) = (0.27, -1)$ and the $(\Omega_m, \mathrm{w}) = (0.27, -0.85)$ models (thin blue line).

We proceed to analyse the SNIa data by defining the usual likelihood estimator[2] as:

$$\mathcal{L}^{\mathrm{SNIa}}(\mathbf{p}) \propto \exp[-\chi^2_{\mathrm{SNIa}}(\mathbf{p})/2] \tag{6}$$

where $\mathbf{p}$ is a vector containing the cosmological parameters that we want to fit for, and

$$\chi^2_{\mathrm{SNIa}}(\mathbf{p}) = \sum_{i=1}^{N} \left[\frac{\mu^{\mathrm{th}}(z_i, \mathbf{p}) - \mu^{\mathrm{obs}}(z_i)}{\sigma_i} \right]^2 , \tag{7}$$

where μ^{th} is given by eq.(3), z_i is the observed redshift and σ_i the observed distance modulus uncertainty. Here we will constrain our analysis within the framework of a flat ($\Omega_{\mathrm{tot}} = 1$) cosmology and therefore the corresponding vector $\mathbf{p}$ is: $\mathbf{p} \equiv (\Omega_m, \mathrm{w}_0, \mathrm{w}_1)$. We will use only SNIa with $z > 0.02$ in order to avoid redshift uncertainties due to the local bulk flow (eg. Hudson et al. 1999 and references therein).

[2] Likelihoods are normalized to their maximum values.

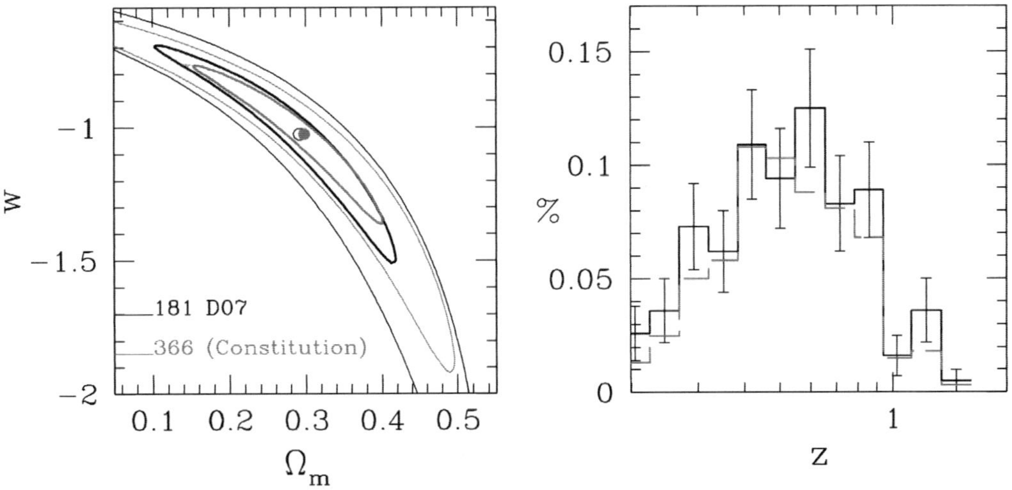

Figure 3. *Left panel:* Cosmological parameter solution space using either of the two SNIa data sets (*Constitution*: red contours and *D07*: black contours). Contours corresponding to the 1 and 3σ confidence levels are shown (ie., plotted where $-2\ln\mathcal{L}/\mathcal{L}_{\max}$ is equal to 2.30 and 11.83, respectively). *Right Panel:* Normalized redshift distributions of the two SNIa data sets.

2.2.1. Larger Numbers? The first issue that we wish to address is how better have we done in imposing cosmological constraints by increasing the available SNIa sample from 181 to 366, ie., increasing the sample by more than 100%. In Table 1 we present various solutions using each of the two previously mentioned samples Note that since only the relative distances of the SNIa are accurate and not their absolute local calibration, we always marginalize with respect to the internally derived Hubble constant (note that fitting procedures exist which do not need to *a priori* marginalize over the internally estimated Hubble constant; eg., Wei 2008). Although the derived cosmological parameters are consistent between the two data sets, possibly indicating the robustness of the method, the corresponding goodness of fit (the reduced χ^2) is significantly larger in the case of the *Constitution* set (1.21 compared to 1.045 of the *D07* set). This appears to be the outcome of the different approaches chosen in order to join the different contributing SNIa sets. According to Hicken 2009 (private communication) in the case of the *D07* the nearby SNIa were imposed to provide a $\chi^2/df \simeq 1$ by hand, while no such fine-tuning was imposed on the UNION set (on which the *Constitution* set is based). A secondary reason could be that the latter set includes distant SNIa which have typically larger distance modulus uncertainties, with respect to those used in *D07*. Overall, the higher χ^2/df value of the *Constitution* set should be attributed to a typically lower uncertainty in μ. As a crude test, we have increased by 20% the distance modulus uncertainty of the *Constitution* nearby SNIa ($z \lesssim 0.4$) and indeed we obtain $\chi^2/df \simeq 1.07$, similar to that of *D07*.

In Figure 3 we can also see that although the SNIa sample has doubled in size, the well-known *banana* shape region of the (Ω_m,w) solution space, indicating the degeneracy between the two cosmological parameters, is reproduced by both data sets. However, there is a reduction of the size of the solution space when using the *Constitution* SNIa compilation (see also Table 1), and at roughly the level expected from Poisson statistics.

A first conclusion is therefore that *the increase by* $\sim$ 100% *of the Constitution sample has not provided significantly more stringent constraints to the cosmological parameters.* We have verified that the larger number of SNIa's in the *Constitution* sample are not preferentially

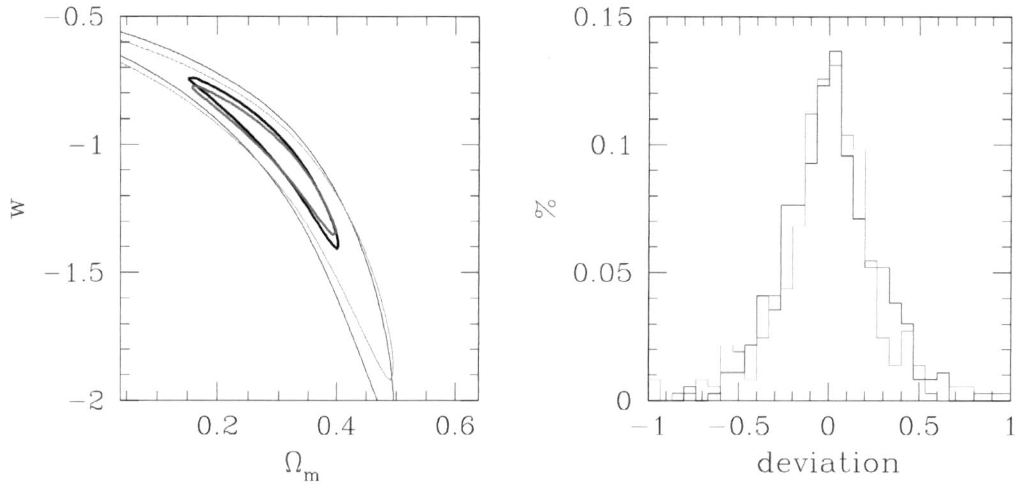

Figure 4. *Left Panel:* Comparison between the *Constitution* SNIa constraints (red contours) and those derived by a Monte-Carlo procedure designed to closely reproduce them (for clarity we show only contours corresponding to 1 and 3 σ confidence levels). *Right Panel:* The *Constitution* SNIa distance modulus deviations from the best fit model $(\Omega_m, \mathrm{w}) \simeq (0.30, -1.01)$; see Table 1 - and a random realization of the model deviations (red histogram).

Table 1. Cosmological parameter fits using the SNIa data within flat cosmologies. Note that for the case where $\mathbf{p} = (\Omega_m, \mathrm{w})$ (ie., last row), the errors shown are estimated after marginalizing with respect to the other fitted parameter.

D07			*Constitution*		
w	Ω_m	$\chi^2_{\rm min}/\mathrm{df}$	w	Ω_m	$\chi^2_{\rm min}/\mathrm{df}$
$\mathbf{-1}$	$0.280^{+0.025}_{-0.015}$	187.03/180	$\mathbf{-1}$	$0.286^{+0.012}_{-0.018}$	439.78/365
$-1.025^{+0.060}_{-0.045}$	0.292 ± 0.018	187.02/179	-1.025 ± 0.030	0.298 ± 0.012	439.79/364

located at low-z's (see right panel of Fig.2) - in which case we should have not expected more stringent cosmological constraints using the latter SNIa sample, but rather have a very similar z-distribution.

We already have a strong hint, from the previously presented comparison between the *D07* and *Constitution* results, that increasing the number of Hubble relation tracers, covering the same redshift range and with the current level of uncertainties as the available SNIa samples, does not appears to be an effective avenue for providing further constraints of the cosmological parameters.

2.2.2. Lower uncertainties or higher-z's: We now resort to a Monte-Carlo procedure which will help us investigate which of the following two directions, which bracket many different possibilities, would provide more stringent cosmological constraints:

- Reduce significantly the distance modulus uncertainties of SNIa, tracing however the same redshift range as the currently available samples, or
- use tracers of the Hubble relation located at redshifts where the models show their largest relative differences ($z \gtrsim 2$), with distance modulus uncertainties comparable to that of the highest redshift SNIa's ($\langle\sigma_\mu\rangle \simeq 0.4$)

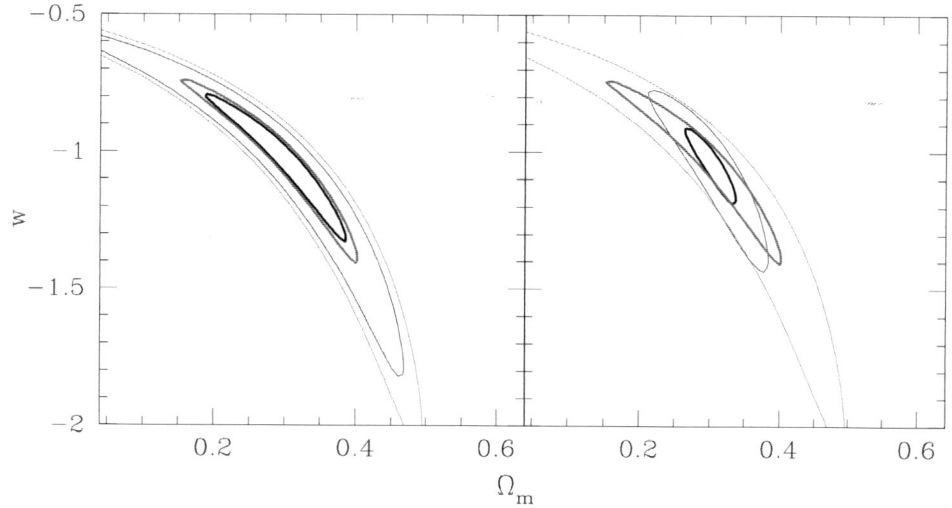

Figure 5. Comparison of the model *Constitution* SNIa constraints (red contours) with those derived by reducing to half their uncertainties (*left panel*, and with those derived by adding a sample of 82 high-z tracers ($2.7 \lesssim z \lesssim 3.5$) with a distance modulus mean uncertainty of $\sigma_\mu \simeq 0.38$ (*right panel*. For clarity we show only contours corresponding to the 1 and 3 σ confidence levels.

The Monte-Carlo procedure is based on using the observed high-z SNIa distance modulus uncertainty distribution (σ_μ) and a model to assign random μ-deviations from a reference $H(z)$ function, that reproduces exactly the original banana-shaped contours of the (Ω_m, w) solution space of Figure 3 (left panel). Indeed, after a trial and error procedure we have found that by assigning to each SNIa (using their true redshift) a distance modulus deviation ($\delta\mu$) from a reference model having a Gaussian distribution with zero mean and variance given by the observed $\langle\sigma_\mu\rangle^2$, and using as the relevant individual distance modulus uncertainty the following: $\sigma_i^2 = \sqrt{(1.2\delta\mu_i)^2 + \phi^2}$ (with ϕ a random Poisson deviate within $[-0.01, 0.01]$) we reproduce exactly the banana-shaped solution range of the reference model. This can be seen clearly in the upper left panel of Figure 4, where we plot the original *Constitution* SNIa solution space (red contours) and the model solution space (black contours). In the right panel we show the distribution of the true SNIa deviations from the best fitted model as well as a random realization of the model deviations.

Armed with the above procedure we can now address the questions posed previously. Firstly, we reduce to half the random deviations of the SNIa distance moduli from the reference model (with the corresponding reduction of the relevant uncertainty, σ_i). The results of the likelihood analysis can be seen in the left panel of Figure 5. There is a reduction of the range of the solution space, but indeed quite a small one. Secondly, we add to the *Constitution* SNIa sample, a mock high-z subsample of 82 objects constructed as follows: For each $z > 0.68$ SNIa we add one mock SNIa having as a redshift $z + \delta z$ where $\delta z = 2$ and the same distance modulus uncertainty. Note that these new 82 SNIa are distributed between $2.68 \lesssim z \lesssim 3.55$, ie., in a range where the largest deviations between the different cosmological models occur (see Figure 1). The deviations from the reference model of these additional SNIa are based on their original μ uncertainty distribution (ie., we have assumed that the new high-z tracers will have similar uncertainties as their $z \gtrsim 0.68$ counterparts, which on average is $\langle\sigma_\mu\rangle \simeq 0.38$). We now find a significantly reduced solution space (right panel of Figure 5), which shows that indeed by increasing the $H(z)$ tracers by a few tens, at those redshifts where the largest deviations between models occur, can have a significant impact on the recovered cosmological parameter solution space. Similarly, if we use a reference

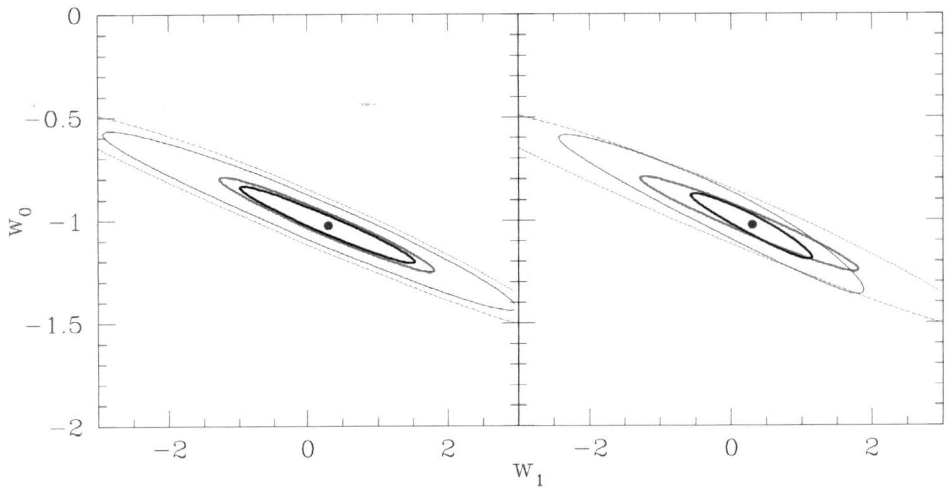

Figure 6. Similar as in Figure 5, but allowing for an evolving DE equation of state according to eq. (2) and after marginalizing with respect to Ω_m. The input cosmological model has (w_0, w_1) = $(-1.025, 0.3)$ and is represented by the red contours. Again we show for clarity only contours corresponding to the 1 and 3 σ confidence levels.

model with an evolving dark-energy equation of state (as that of eq. 2), and after marginalizing with respect to Ω_m, we also find a significantly larger reduction of the (w_0, w_1) solution space when we include the high redshift tracer subsample (Fig.6, right panel) with respect to the distance modulus uncertainty reduction (by 1/2) case (Fig. 6, left panel).

The main conclusion of the previous analysis is that a relatively better strategy to decrease the uncertainties of the cosmological parameters, based on the Hubble relation, is to use standard candles which trace also the redshift range $2 \lesssim z \lesssim 4$. Below we present one such possibility by suggesting an alternative, to the SNIa standard candles, namely HII galaxies (eg. Melnick 2003; Siegel et al. 2005).

2.3. Hubble Relation using HII galaxies

We now reach our suggestion to use an alternative and potentially very powerful technique to estimate cosmological distances, which is the relation between the luminosity of the H_β line and the stellar velocity dispersion, measured from the line-widths, of HII regions and galaxies (Terlevich & Melnick 1981, Melnick, Terlevich & Moles 1988). The cosmological use of this distance indicator has been tested in Melnick, Terlevich & Terlevich (2000) and Siegel et al (2005) (see also the review by Melnick 2003). The presence of O and B-type stars in HII regions causes the strong Balmer line emission, in both H_α and H_β. Furthermore, the fact that the bolometric luminosities of HII galaxies are dominated by the starburst component implies that their luminosity per unit mass is very large, despite the fact that the galaxies are low-mass. Therefore they can be observed at very large redshifts, and this fact makes them cosmologically very interesting objects. Furthermore, it has been shown that the $L(H_\beta) - \sigma$ correlation holds at large redshifts (Koo et al. 1996, Pettini et al. 2001, Erb et al. 2003) and therefore it can be used to trace the Hubble relation at cosmologically interesting distances. One of the most important prerequisites in using such relations, as distance estimators, is the accurate determination of their zero-point. To this end, Melnick et al (1988) used giant HII regions in nearby late-type galaxies and derived the following empirical relation (using a Hubble constant of $H_0 = 71$ km/sec/Mpc):

$$\log L(H_\beta) = \log M_z + 29.60 \text{ with } M_z = \sigma^5/(O/H) \tag{8}$$

where O/H is the metallicity. Based on the above relation and the work of Melnick, Terlevich & Terlevich (2000), the distance modulus of HII galaxies can be written as:

$$\mu = 2.5\log(\sigma^5/F_{H\beta}) - 2.5\log(O/H) - A_{H\beta} - 26.44 \ , \tag{9}$$

with $F_{H\beta}$ and $A_{H\beta}$ the flux and extinction in H_β, respectively. The rms dispersion in distance modulus was found to be ~ 0.52 mag. The analysis of Melnick, Terlevich & Terlevich (2000) has shown that most of this dispersion (~ 0.3 mags) comes from observational errors in the stellar velocity dispersion measurements, from photometric errors and metallicity effects. It is therefore possible to understand and correct the sources of random and systematic errors of the $L(H_\beta) - \sigma$ relation, and indeed with the availability of new observing techniques and instruments, we hope to reduce significantly the previously quoted rms scatter.

A few words are also due to the possible systematic effects of the above relation. Such effects may be related to the age of the HII galaxy (this can be dealt with by putting a limit in the equivalent width of the H_β line, eg. $EW(H_\beta) > 25$ Angs; see Melnick 2003), to extinction, to different metallicities and environments. Also the $EW(H_\beta)$ of HII galaxies at intermediate and high redshifts is smaller than in local galaxies, a fact which should also be taken into account.

We have commenced an investigation of all these effects by using high-resolution spectroscopy of a relatively large number of SDSS low-z HII galaxies, having a range of H_β equivalent widths, luminosities, metal content and local overdensity, in order to understand systematics and reduce the scatter of the HII-galaxy based distance estimator to about half its present value, ie., our target is ~ 0.25 mag. We will also define a medium and high redshift sample (see Pettini et al. 2001; Erb et al. 2003), consisting of a few hundred objects distributed in a range of high-redshifts, which will finally be used to define the high-z Hubble function.

Summarizing, the use of HII galaxies to trace the Hubble relation, as an alternative to the traditionally used SN Ia, is based on the following facts:

(a) *local and high-z HII galaxies and HII regions are physically very similar systems (Melnick et al 1987) providing a phenomenological relation between the luminosity of the H_β line, the velocity dispersion and their metallicity as traced by O/H (Melnick, Terlevich & Moles 1988). Therefore HII galaxies can be used as alternative standard candles (Melnick, Terlevich & Terlevich 2000, Melnick 2003; Siegel et al. 2005)*

(b) *such galaxies can be readily observed at much larger redshifts than those currently probed by SNIa samples, and*

(c) *it is at such higher redshifts that the differences between the predictions of the different cosmological models appear more vividly.*

Already a sample of 15 such high-z HII galaxies have been used by Siegel et al. (2005) in an attempt to constrain cosmological parameters but the constraints, although in the correct direction, are very weak. We have performed our own re-analysis of this data-set, following however the same procedure as that applied to the SNIa data (ie., we factorize out the internally derived Hubble constant using a fit of the distance moduli in the Ω_m, H_0 space). The resulting constraints on the (Ω_m,w) plane can be seen in Figure 7. We find that $\Omega_m \lesssim 0.42$ at a 99.99% level, independent of w. Note that imposing w$= -1$, our analysis of the Siegel et al (2005) data set provides $\Omega_m = 0.19 \pm 0.05$, which is towards the lower side of the generally accepted values.

Comparing these HII-based results to the present constraints of the latest (*Constitution* SNIa data (right panel of Fig. 7) clearly indicates the necessity to:

- re-estimate carefully the local zero-point of the $L(H_\beta) - \sigma$ relation,

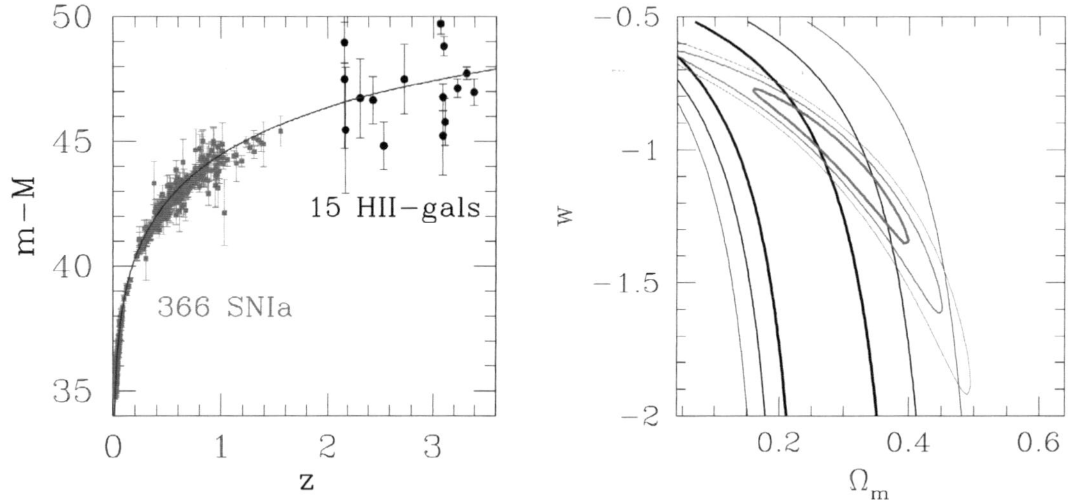

Figure 7. *Left Panel:* Distance moduli comparison between the *Constitution* SNIa's and the 15 high-z HII galaxies of Siegel et al. (2005). *Right Panel:* The corresponding constraints in the (Ω_m, w) plane. Although, the HII-galaxy constraints are weak, leaving completely unconstrained the value of w, they appear to be consistent at a $\sim 2\sigma$ level with the SNIa results. This plot serves only to indicate the potential of using high-z HII galaxies (once of course we have reduced significantly their distance modulus uncertainties).

- suppress the HII-galaxy distance modulus uncertainties,
- increase the high-z HII galaxy sample by a large fraction,
- make sure to select high-z bona-fide HII-galaxies, excluding those that show indications of rotation (Melnick 2003).

3. Cosmological Parameters from the Clustering of X-ray AGN

The method used to put cosmological constraints, based on the angular clustering of some extragalactic mass-tracer (Matsubara 2004, Basilakos & Plionis 2009 and references therein), consists in comparing the observed angular clustering with that predicted by different primordial fluctuations power-spectra, using Limber's integral equation (Limber 1968) to invert from spatial to angular clustering. By minimizing the differences of the observed and predicted angular correlation function, one can constrain the cosmological parameters that enter in the power-spectrum determination as well as in Limber's inversion. Using the latter we can relate the angular and spatial clustering of any extragalactic population under the assumption of power-law correlations and the small angle approximation.

3.1. X-ray surveys: Biases and Systematics

X-ray selected AGN provide a relatively unbiased census of the AGN phenomenon, since obscured AGN, largely missed in optical surveys, are included in such surveys. Furthermore, they can be detected out to high redshifts and thus trace the distant density fluctuations providing important constraints on super-massive black hole formation, the relation between AGN activity and Dark Matter (DM) halo hosts, the cosmic evolution of the AGN phenomenon (eg. Mo & White 1996, Sheth et al. 2001), and on cosmological parameters and the dark-energy equation of state (eg. Basilakos & Plionis 2005; 2006).

The earlier ROSAT-based analyses (eg. Boyle & Mo 1993; Vikhlinin & Forman 1995; Carrera et al. 1998; Akylas, Georgantopoulos, Plionis, 2000; Mullis et al. 2004) provided conflicting results on the nature and amplitude of high-z AGN clustering. With the advent of the XMM and *Chandra* X-ray observatories, many groups have attempted to settle this issue, but in vain. Different surveys have provided again a multitude of conflicting results, intensifying the debate (eg. Yang et al. 2003; Manners et al. 2003; Basilakos et al. 2004; Gilli et al. 2005; Basilakos et al 2005; Yang et al. 2006; Puccetti et al. 2006; Miyaji et al. 2007; Gandhi et al. 2006; Carrera et al. 2007). However, the recent indications of a flux-limit dependent clustering appears to remove most of the above inconsistencies (Plionis et al. 2008).

Furthermore, there are indications for a quite large high-z AGN clustering length, reaching values $\sim 15 - 18\ h^{-1}$ Mpc at the brightest flux-limits (eg., Basilakos et al 2004; 2005, Puccetti et al. 2006, Plionis et al. 2008), a fact which, if verified, has important consequences for the AGN bias evolution and therefore for the evolution of the AGN phenomenon (eg. Miyaji et al. 2007; Basilakos, Plionis & Ragone-Figueroa 2008). An independent test of these results would be to establish that the environment of high-z AGN is associated with large DM haloes, which being massive should be more clustered (work in progress).

It is also important to understand and overcome the shortcomings and problems that one is facing in order to reliably and unambiguously determine the clustering properties of the X-ray selected AGN. Such a list of problems includes the effects of Cosmic Variance, the so-called amplification bias, the reliability of the $\log N - \log S$ distribution of the X-ray AGN luminosity function, etc. (see discussion in Plionis et al. 2009).

Recently, Ebrero et al. (2009) derived the angular correlation function of the soft (0.5-2 keV) X-ray sources using 1063 XMM-*Newton* observations at high galactic latitudes. A full description of the data reduction, source detection and flux estimation are presented in Mateos et al. (2008). Note, that the survey contains $\sim$ 30000 point sources within an effective area of $\sim$ 125.5 deg^2 (for $f_x \geq 1.4 \times 10^{-15}$ erg cm^{-2} s^{-1}). The details regarding the angular correlation function, the various biases that should be taken into account (the amplification bias and integral constraint), the survey luminosity and selection functions as well as issues related to possible non-AGN contamination, which are estimated to be $\lesssim 10\%$, can be found in Ebrero et al (2009).

3.2. Details of the Method

An optimal approach to unambiguously determine the clustering pattern of X-ray selected AGN would be to determine both the angular and spatial clustering pattern. The reason being that various systematic effects or uncertainties enter differently in the two types of analyses. On the one side, using $w(\theta)$ and its Limber inversion, one by-passes the effects of redshift-space distortions and uncertainties related to possible misidentification of the optical counter-parts of X-ray sources. On the other side, using spectroscopic or accurate photometric redshifts to measure $\xi(r)$ or $w_p(\theta)$ one by-passes the inherent necessity, in Limber's inversion of $w(\theta)$, of the source redshift-selection function (for the determination of which one uses the integrated X-ray source luminosity function, different models of which exist).

The basic integral equation relating the angular and sptial correlction functions is:

$$w(\theta) = 2\frac{H_0}{c}\int_0^\infty \left(\frac{1}{N}\frac{\mathrm{d}N}{\mathrm{d}z}\right)^2 E(z)\mathrm{d}z \int_0^\infty \xi(r,z)\mathrm{d}u \quad , \tag{10}$$

where $\mathrm{d}N/\mathrm{d}z$ is the source redshift distribution, estimated by integrating the appropriate source luminosity function (in our case that of Ebrero et al. 2009b), folding in also the area curve of the survey. Note that to derive the spatial correlation length from eq. (10), it is necessary to model the spatial correlation function as a power law, assume the small angle approximation as well as a cosmological background model, the latter given by $E(z)$ (from eq. 3) which for a flat background and a constant *dark energy* equation of state parameter, w, reduces to:

$$E(z) = [\Omega_m(1+z)^3 + (1-\Omega_m)(1+z)^{3(1+\mathrm{w})}]^{1/2} . \tag{11}$$

The AGN spatial correlation function is:

$$\xi(r,z) = (1+z)^{-(3+\epsilon)} b^2(z) \xi_{\rm DM}(r) , \tag{12}$$

where $b(z)$ is the evolution of the linear bias factor (eg. Mo & White 1997; Matarrese et al 1997; Sheth & Tormen 1999; Basilakos & Plionis 2001; 2003, Basilakos et al. 2008 and references therein), ϵ is a parameter related to the model of AGN clustering evolution (eg. de Zotti et al. 1990)[3] and $\xi_{\rm DM}(r)$ is the corresponding correlation function of the underlying dark matter distribution, given by the Fourier transform of the spatial power spectrum $P(k)$ of the matter fluctuations, linearly extrapolated to the present epoch:

$$\xi_{\rm DM}(r) = \frac{1}{2\pi^2}\int_0^\infty k^2 P(k) \frac{\sin(kr)}{kr} {\rm d}k \ . \tag{13}$$

The CDM power spectrum is given by: $P(k) = P_0 k^n T^2(k)$, with $T(k)$ the CDM transfer function (Bardeen et al. 1986; Sugiyama 1995) and $n \simeq 0.96$, following the 5-year WMAP results (Komatsu et al. 2009), and a baryonic density of $\Omega_{\rm b}h^2 = 0.022(\pm 0.002)$. The normalization of the power-spectrum, P_0, can be parametrized by the rms mass fluctuations on $R_8 = 8h^{-1}$Mpc scales (σ_8), according to:

$$P_0 = 2\pi^2\sigma_8^2 \left[\int_0^\infty T^2(k) k^{n+2} W^2(kR_8) {\rm d}k\right]^{-1} , \tag{14}$$

where $W(kR_8) = 3(\sin kR_8 - kR_8 \cos kR_8)/(kR_8)^3$. Regarding the Hubble constant we use $H_0 \simeq 71$ kms^{-1}Mpc^{-1} (Freedman 2001; Komatsu et al. 2009). Note, that we also utilize the non-linear corrections introduced by Peacock & Dodds (1994).

We can now compare the observed AGN clustering with the predicted, for different cosmological models, correlation function of the underlying mass, $\xi_{\rm DM}(r,z)$, and thus constrain the cosmological parameters. To this end it is necessary to use a bias evolution model and although a large number of such models have been proposed in the literature, we use here our own approach, which was described initially in Basilakos & Plionis (2001; 2003) and extended in Basilakos, Plionis & Ragone-Figueroa (2008). This bias model is based on linear perturbation theory and the Friedmann-Lemaitre solutions of the cosmological field equations, while it also allows for interactions and merging of the mass tracers. Considering that each X-ray AGN is hosted by a dark matter halo, we can analytically predict its bias evolution behavior and conversely by comparing with observations we can determine the mass of the DM halo, M_h, within which AGN live.

In Figure 8 we show a comparison between our model predictions and observationally based AGN bias results, estimated at the sample's median redshift by:

$$b(\bar{z}) = \left(\frac{r_0}{r_{0,m}}\right)^{\gamma/2} D^{3+\epsilon}(\bar{z}) \tag{15}$$

where r_0 and $r_{0,m}$ are the measured AGN and the theoretical dark-matter clustering lengths, respectively, while $D(z)$ is the perturbation's linear growing mode (scaled to unity at the present

[3] Following Kúndic (1997) and Basilakos & Plionis (2005; 2006) we use the constant in comoving coordinates clustering model, ie., $\epsilon = -1.2$.

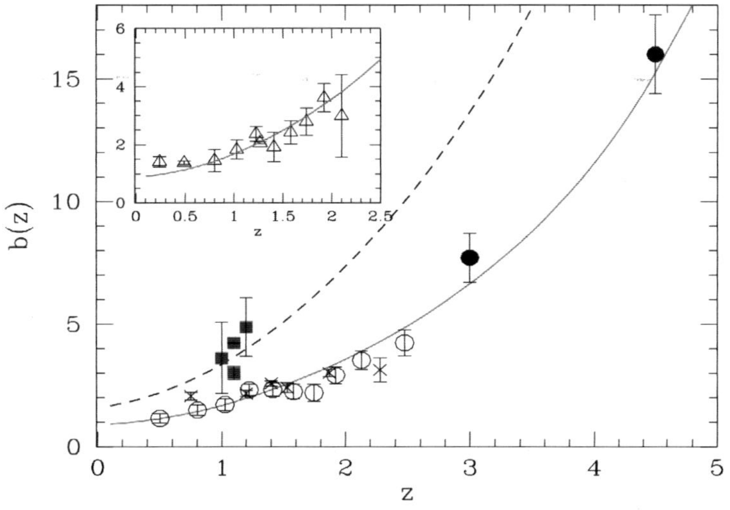

Figure 8. The observed evolution of AGN bias compared with the Basilakos et al. (2008) model. Optically selected (SDSS and 2dF) quasars (Croom et al. 2005; Myers et al. 2007; Shen et al. 2007) correspond to circles (empty or filled) points and crosses while soft-band X-ray selected AGN correspond to filled (blue) squares. In the insert we plot the most recent bias values of Ross et al. (2009), based on the SDSS optical quasar uniform sample. The curves reflect our bias evolution model, with the solid (red) lines corresponding to a DM halo mass of $M \simeq 10^{13}\ h^{-1}\ M_{\odot}$ and the dashed lines to $M \simeq 2.5 \times 10^{13}\ h^{-1}\ M_{\odot}$.

Table 2. The best fit values from the likelihood analysis of the X-ray AGN clustering: Errors of the fitted parameters represent 1σ uncertainties, while the parameters in bold indicate those fixed during the fitting process.

$\Omega_{\rm m}$	w	σ_8	$M_h/10^{13}h^{-1}M_{\odot}$
0.27 ± 0.04	$-0.90^{+0.10}_{-0.16}$	$0.74^{+0.14}_{-0.12}$	$2.50^{+0.50}_{-1.50}$
0.26 ± 0.05	$-0.93^{+0.11}_{-0.19}$	**0.8**	$2.0^{+0.30}_{-0.2}$
0.24 ± 0.06	$\mathbf{-1}$	$0.83^{+0.11}_{-0.16}$	**2.50**

time), useful expressions of which can be found for the dark energy models in Silveira & Waga (1994) and in Basilakos (2003).

The preliminary analysis of Basilakos & Plionis (2005; 2006) compared the measured XMM source angular correlation function based on a shallow XMM survey, covering a small solid angle, with the prediction of different spatially flat and constant-w cosmological models. The recent availability, however, of the highly accurate angular correlation function XMM results of Ebrero et al (2009), based on the largest available X-ray AGN sample ($N \simeq 30000$), provided us the means to put stringent constraints on the cosmological parameters (see Basilakos & Plionis 2009). In Table 2 (and in Figure 9) we present the cosmological constraints provided by the current analysis. The power of our procedure can be appreciated by comparing our constraints, under the prior of a flat Universe, with those provided by most other cosmological test (see Fig.2 of Basilakos & Plionis 2009).

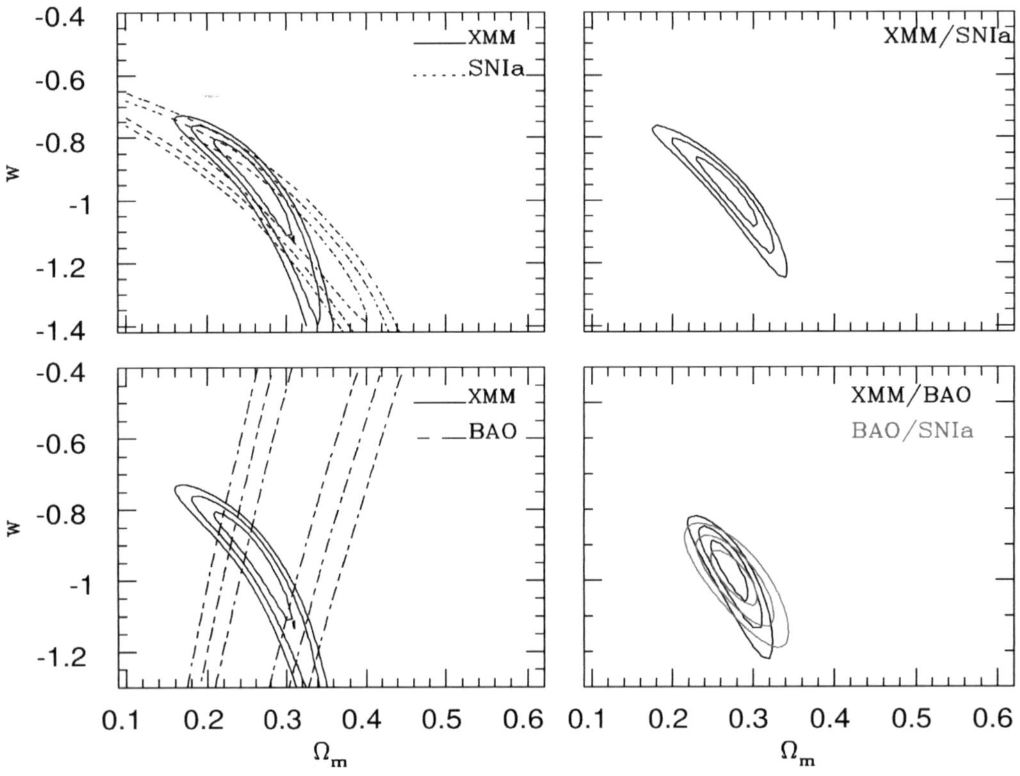

Figure 9. *Upper Left Panel:* Likelihood contours on the Ω_m,w plane from the X-ray AGN clustering analysis of Basilakos & Plionis (2009) (solid contours) and the *Constitution* SNIa analysis (dashed contours). *Upper Right Panel:* The corresponding joint likelihood contours. *Lower Left Panel:* Likelihood contours from the X-ray AGN clustering analysis and the BAO technique (dashed contours). *Lower Right Panel:* Their corresponding joint likelihood contours (while the BAO-SNIa joint contours are shown in red).

4. Joint Hubble-relation and Clustering analysis

It is evident from Figure 9 (left panels) that with the new X-ray AGN clustering analysis we have been able to reduce significantly the degeneracy between w and Ω_m (see also Table 2). We can further break the degeneracy by adding the constraints provided by either the Hubble relation technique, using here the *Constitution* SNIa sample, or the baryonic acoustic oscillation technique (BAO), which was identified by the U.S. Dark Energy Task Force as one of the four most promising techniques to measure the properties of the *dark energy* and the one less likely to be limited by systematic uncertainties. We remind the reader that BAOs are produced by pressure (acoustic) waves in the photon-baryon plasma in the early universe, generated by dark matter (DM) overdensities. At the recombination era ($z \sim 1100$), photons decouple from baryons and free stream while the pressure wave stalls. Its frozen scale, which constitutes a standard ruler, is equal to the sound horizon length, $r_s \sim 100\ h^{-1}$ Mpc (e.g. Eisenstein, Hu & Tegmark 1998). This appears as a small, $\sim 10\%$ excess in the galaxy, cluster or AGN power spectrum (and its Fourier transform, the 2-point correlation function) at a scale corresponding to r_s. First evidences of this excess were recently reported in the clustering of luminous SDSS red-galaxies (Eisenstein et al. 2005, Padmanabhan et al. 2007; Percival et al. 2009).

We therefore perform a joint likelihood analysis, assuming that any two pairs of data sets are independent (which indeed they are) and thus the joint likelihood can be written as the product of the two individual ones. The current joint likelihood analysis, once we impose $h = 0.71$ and $\sigma_8 = 0.8$ (according to Komatsu et al. 2009), provide quite stringent and consistent constraints of the Ω_m and w parameters:

$$\Omega_{\rm m} = 0.27 \pm 0.02 \quad \text{and} \quad {\rm w} = -0.96 \pm 0.07 \quad ({\rm XMM - SNIa})$$

$$\Omega_{\rm m} = 0.27 \pm 0.02 \quad \text{and} \quad {\rm w} = -0.97 \pm 0.04 \quad ({\rm XMM - BAO})$$

These results can be compared with the more *traditional* joint analysis of the SNIa (*Cosntitution* set in this case) and the BAOs, which provide:

$$\Omega_{\rm m} = 0.28^{+0.02}_{-0.03} \quad \text{and} \quad {\rm w} = -0.98 \pm 0.06 \quad ({\rm SNIa - BAO})$$

It is evident that within the errors all the previously presented results are consistent, with the XMM-BAO joint analysis providing the smallest uncertainties of the fitted parameters. This can also be appreciated by inspecting the lower-right panel of Figure 9, which shows that the XMM-BAO joint analysis provides indistinguishable constraints with the corresponding SNIa-BAO analysis.

The necessity, however, to impose constraints on a more general, time-evolving, *dark-energy* equation of state (eq. 3), implies that there is ample space for great improvement and indeed our aim is to complete this project by using a new Hubble relation analysis, based on high-z HII galaxies, as detailed in these proceedings.

Acknowledgments
We thank Dr J. Ebrero for comments and for providing us with an electronic version of the clustering results and their XMM survey area curve. We also thank Dr. N.P. Ross for providing us an electronic version of the QSO uniform sample bias results. MP also acknowledges financial support under Mexican government CONACyT grant 2005-49878.

References

[1] Akylas, A., Georgantopoulos, I., Plionis, M., 2000, MNRAS, 318, 1036
[2] Albrecht, A., et al., 2006, `astro-ph/0609591`
[3] Allen, S. W., et al., 2004, MNRAS, 353, 457
[4] Astier, P., et al., 2006, A&A, 447, 31
[5] Bardeen, J.M., Bond, J.R., Kaiser, N. & Szalay, A.S., 1986, ApJ, 304, 15
[6] Basilakos, S. & Plionis, M., 2001, ApJ, 550, 522
[7] Basilakos, S. & Plionis, M., 2003, ApJ, 593, L61
[8] Basilakos, S., Georgakakis, A., Plionis, M., Georgantopoulos, I., 2004, ApJ, 607, L79
[9] Basilakos, S., Plionis, M., Georgantopoulos, I., Georgakakis, A., 2005, MNRAS, 356, 183
[10] Basilakos, S. & Plionis, M., 2005, MNRAS, 360, L35
[11] Basilakos, S. & Plionis, M., 2006, ApJ, 650, L1
[12] Basilakos, S., Plionis, M. & Ragone-Figueroa, C., 2008, ApJ, 678, 627
[13] Basilakos, S. & Perivolaropoulos, L., 2008, MNRAS, 391, 411
[14] Boyle B.J., Mo H.J., 1993, MNRAS, 260, 925
[15] Blake, C., et al., 2007, MNRAS, 374, 1527
[16] Carrera, F.J., et al. 1998, MNRAS, 299, 229
[17] Carrera, F.J., et al. 2007, A&A, 469, 27
[18] Chevallier M., & Polarski D., 2001, Int. J. Mod. Phys. D, 10, 213
[19] Croom, S.M., et al., 2005, MNRAS, 365, 415
[20] de Zotti, G., et al. 1990, ApJ, 351, 22
[21] Davis, T.M., et al., 2007, ApJ, 666, 716
[22] Dicus, D.A. & Repko, W.W., 2004, Phys.Rev.D, 70, 3527,
[23] Erb, D.K. et al., 2003, ApJ, 591, 101
[24] Eisenstein, Hu, & Tegmark 1998, ApJ, 504, L57

[25] Eisenstein et al. 2005, ApJ, 633, 560
[26] Gandhi, P., et al., 2006, A&A, 457, 393
[27] Gilli, R., et al. 2003, ApJ, 592, 721
[28] Ghirlanda, G., Ghisellini, G., Firmani, C., 2006, NewJPhys, 8, 123
[29] Hudson, M.J., Smith, R.J., Lucey, J.R., Schlegel, D.J., Davies, R.L., 1999, ApJ, 512, L79
[30] Gilli, R., et al. 2005, A&A, 430, 811
[31] Kim, M., et al. 2007, ApJ, 659, 29
[32] Kowalski M. et al., 2008, ApJ, 686, 749
[33] Kundić, T., 1997, ApJ, 482, 631
[34] Linder, V. E., 2003, Phys. Rev. Lett., 90, 091301
[35] Matsubara, T., 2004, ApJ, 615, 573
[36] Manners, J.C., et al., 2003, MNRAS, 343, 293
[37] Matarrese, S., Coles, P., Lucchin, F., Moscardini, L., 1997, MNRAS, 286, 115
[38] Mateos, S., et al., 2008, A&A, 429, 51
[39] Myers, A.D., Brunner. R.J., Nichol, R.C., Richards, G.T., Schneider D.P., Bahcall, N.A., 2007, ApJ, 658, 85
[40] Melnick, J., 1978, A&A, 70, 157
[41] Melnick J., Terlevich, R., Moles, M., 1988, MNRAS, 235, 313;
[42] Melnick, J., Terlevich, R., Terlevich, E., 2000, MNRAS, 311, 629
[43] Melnick, J., 2003, *Star Formation Through Time'*, ASP Conference Proceedings, 297, Eds E.Perez, et al.
[44] Miyaji, T., et al., 2007, ApJS, 172, 396
[45] Mullis C.R., et al., 2004, ApJ, 617, 192
[46] Mo, H.J, & White, S.D.M 1996, MNRAS, 282, 347
[47] Padmanabhan et al. 2007, MNRAS, 378, 852
[48] Peacock et al., 2006, astro-ph/0610906
[49] Percival, W.J., et al., 2007, MNRAS, 381, 1053
[50] Peebles P.J.E., &, Ratra, B., 2003, RvMP, 75, 559
[51] Perlmutter, S., et al., 1998, Nature, 391, 51
[52] Perlmutter, S., et al., 1999, ApJ, 517, 565
[53] Pettini, M., et al., 2001, ApJ, 554, 981
[54] Plionis, M., Rovilos, M., Basilakos, S., Georgantopoulos, I., Bauer, F., 2008, ApJ, 674, L5
[55] Plionis, M., et al., 2009, JoP, 189, 012032
[56] Puccetti, S., et al., 2006, A&A, 457, 501
[57] Riess, A. G., et al., 1998, AJ, 116, 1009
[58] Riess, A. G., et al., 2004, ApJ, 607, 665
[59] Riess A.G., 2007, ApJ, 659, 98
[60] Ross N.P., et al., 2009, ApJ, 697, 1634
[61] Seljak, U., et al., 2004, Phys.Rev.D, 71, 3515
[62] Shen, Y., et al., 2007, AJ, 133, 2222
[63] Sheth, R.K. & Tormen, G., 1999, MNRAS, 308, 119
[64] Sheth, R.K., Mo, H.J., Tormen, G., 2001, MNRAS, 323, 1
[65] Siegel, E.R., et al., 2005, MNRAS, 356, 1117
[66] Spergel, D. N., et al., 2003, ApJS, 148, 175
[67] Spergel, D.N., et al. 2007, ApJS, 170, 377
[68] Sugiyama, N., 1995, ApJS, 100, 281
[69] Tegmark, M., et al., 2004, Phys.Rev.D., 69, 3501
[70] Terlevich, R., Melnick, J., 1981, MNRAS, 195, 839
[71] Tonry, et al. , 2003, ApJ, 594, 1
[72] Vikhlinin, A. & Forman, W., 1995, ApJ, 455, 109
[73] Yang, Y., Mushotzky, R.F., Barger, A.J., Cowie, L.L., Sanders, D.B., Steffen, A.T., 2003, ApJ, 585, L85
[74] Yang, Y., Mushotzky, R.F., Barger, A.J., Cowie, L.L., 2006, ApJ, 645, 68
[75] Wang, Y. & Mukherjee, P., 2006, ApJ, 650, 1
[76] Wei, H., 2008, European Phys.J. C, 60, 449
[77] Wood-Vasey W.M., et al., 2007, ApJ, 666, 694

Cosmology from Gamma Ray Bursts

Athina Pouri[1] **and Spyros Basilakos**[2]

[1] Section of Astronomy & Astrophysics, Department of Physics University of Athens, Zografou, Athens, Greece.
[2]Academy of Athens, Research Center for Astronomy and Applied Mathematics

E-mail: `svasil@academyofathens.gr`

Abstract. In this study we propose to use Gamma Ray Bursts (GRBs) as standard candles in order to constrain the expansion history of the universe up to redshifts of $z \sim 6$. In particular, we utilize the 69 GRB dataset recently compiled by Cardone et al. (2009). Performing a joint likelihood analysis of the recent supernovae type Ia (SNIa) data and the GRBs we can put constraints on the main cosmological parameters (Ω_m, w). However, the use of the current GRBs to trace the Hubble relation, as an alternative to the traditionally used SNIa, can not break the degeneracy between the Ω_m and the dark energy equation of state parameter.

1. Introduction

Several cosmological observations (Tegmark et al. 2004; Davis et al. 2007; Kowalski et al. 2008; Komatsu et al. 2009; Hicken et al. 2009 and references therein) have converged during the last decade towards a cosmic expansion history that involves a recent accelerating expansion of the universe. This expansion has been attributed to an energy component (dark energy) with negative pressure which dominates the universe at late times and causes the observed accelerating expansion. The simplest type of dark energy corresponds to the cosmological constant (Peebles & Ratra 2003 and references therein).

The geometrical probes (Lazkoz, Nesseris & Perivolaropoulos 2008) used to map the cosmic expansion history involve a combination of standard candles [Type Ia supernovae (SNIa) Davis et al. 2007] and standard rulers [clusters, CMB sound horizon detected through Baryon Acoustic Oscillations (BAO; Percival et al. 2007) and through the CMB perturbations angular power spectrum Komatsu et al. (2009). These observations probe the integral of the Hubble expansion rate $H(z)$ either up to redshifts of order $z \simeq 1-2$ (SNIa, BAO, clusters) or up to the redshift of recombination ($z \simeq 1089$). Alternatively, dynamical probes (Bertschinger 2006; Nesseris & Perivolaropoulos 2008) of the expansion history based on measures of the growth rate of cosmological perturbations are also confined to relatively low redshifts up to $z \simeq 1$. It is therefore clear that the redshift range $\sim 2-1000$ is not directly probed by any of the above observations. Even though many models of dark energy predict a decelerating expansion in that redshift range due to matter domination, the possibility of non-trivial expansion properties at higher redshifts can not be excluded. In order to investigate this possibility we need a visible distance indicator at redshifts $z > 2$.

Gamma Ray Bursts (GRBs) are the most violent and bright explosions in the universe. They are produced by a highly relativistic bipolar jet outflow from a compact source (Rhoads 1999; Piran 2004). The fact that GRBs are detected up to very high redshifts (Kawai et al. 2006)

Journal of Physics: Conference Series **222** (2010) 012026 doi:10.1088/1742-6596/222/1/012026

makes it tempting to try and use them as standard candles that could be used to constrain the cosmological expansion history in a similar way as SNIa. The problem is that GRBs appear to be anything but standard candles: they have a very wide range of isotropic equivalent luminosities and energy outputs. Several suggestions have been made to calibrate them as better standard candles by using correlations between various properties of the prompt emission, and in some cases also the afterglow emission. While there is good motivation for such cosmological applications of GRBs, there are many practical difficulties (Basilakos & Perivolaropoulos 2008).

Despite the latter, the potential benefits of obtaining even approximate standard candles at redshifts as high as $z \sim 6$ has prompted a significant activity towards both testing the usefulness of GRBs as standard candles (Amati et al. 2002; Ghirlanda, Ghisellini & Firmani 2006; Firmani et al. 2006; Li 2007; Butler et al. 2007; Zhang & Xie 2007; Hooper & Dodelson 2007; Basilakos & Perivolaropoulos 2008) and eagerly utilizing them to constrain cosmological parameters (Schaefer 2003; Zhang & Meszaros 2004; Dai, Liang & Xu 2004; Di Girolamo, et al. 2005; Schaefer 2007; Bertolami & Tavares Silva 2006; Wang & Dai 2006; Demianski, et al. 2006). This activity has lead to a debate about the usefulness of GRBs as standard candles with both discouraging (Li 2007; Butler et al. 2007) and encouraging (Amati et al. 2002; Ghirlanda et al. 2006; Firmani et al. 2006; Schaefer 2007; Zhang & Xie 2007; Hooper & Dodelson 2007; Capozziello & Izzo 2008; Izzo et al. 2009; Wang et al. 2009; Tsutsui et al. 2009a,b) results.

The aim of the present study is along the above lines. In particular, we use the 69 GRB dataset compiled by Cardone et al. (2009) and using a maximum likelihood method of χ^2 we attempt to put constrains on the cosmological parameters. Finally, we perform a direct statistical comparison of the GRB results with those found using the SNIa data. The structure of the paper is as follows. The basic elements of the cosmological equations are presented in section 2. Section 3 outlines the statistical results of the current analysis. Finally, we draw our conclusions in section 4.

2. The basic Cosmological Equations

For homogeneous, isotropic and flat cosmologies, driven by non relativistic matter and an exotic fluid with equation of state, $p_Q = w(\alpha)\rho_Q$ with $p_Q < 0$, the Einstein field equations can be given by:

$$\left(\frac{\dot{\alpha}}{\alpha}\right)^2 = \frac{8\pi G}{3}(\rho_m + \rho_Q) \tag{1}$$

and

$$\frac{\ddot{\alpha}}{\alpha} = -4\pi G\left[\left(w(\alpha) + \frac{1}{3}\right)\rho_Q + \frac{1}{3}\rho_m\right] , \tag{2}$$

where $\alpha(t)$ is the scale factor, $\rho_m = \rho_{m0}\alpha^{-3}$ is the background matter density and $\rho_Q = \rho_{Q0} f(\alpha)$ is the dark energy density, with:

$$f(\alpha) = \exp\left[3\int_\alpha^1 \left(\frac{1+w(u)}{u}\right) \mathrm{d}u\right] . \tag{3}$$

Thus, one can prove that the Hubble parameter $H \equiv \dot{\alpha}/\alpha$ is given by

$$H(\alpha) = H_0 E(\alpha) \quad E(\alpha) = \left[\Omega_m \alpha^{-3} + \Omega_Q f(\alpha)\right]^{1/2} \tag{4}$$

where H_0 is the Hubble constant [1]. In this context, $\Omega_m = 8\pi G\rho_{m0}/3H_0^2$ is the matter density parameter and $\Omega_Q = 8\pi G\rho_{Q0}/3H_0^2$ is the dark energy parameter at the present time with $\Omega_m + \Omega_Q = 1$.

[1] Note that we always marginalize with respect to the internally derived Hubble constant using the SNIa data. This is $H_0 = 65.4\mathrm{Km/s/Mpc}$.

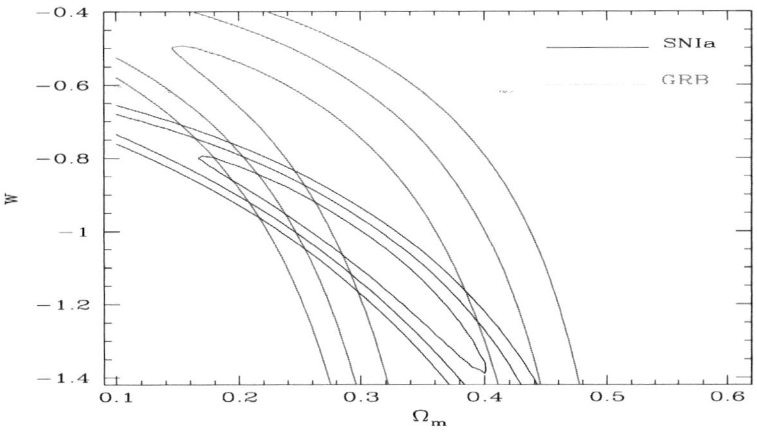

Figure 1. Likelihood contours (1σ, 2σ and 3σ) in the (Ω_m, w) plane (GRBs-red and SNIa-black). The contours are plotted where $-2\ln\mathcal{L}/\mathcal{L}_{\rm max}$ is equal to 2.30, 6.16 and 11.83.

In addition, to $\Omega_m(\alpha)$ also $\Omega_Q(\alpha)$ could evolve with the scale factor as

$$\Omega_m(\alpha) = \frac{\Omega_m \alpha^{-3}}{E^2(\alpha)} \quad \text{and} \quad \Omega_Q(\alpha) = \frac{\Omega_Q f(\alpha)}{E^2(\alpha)} \ . \tag{5}$$

Of course, in order to address the negative pressure term it is essential to define the functional form of the equation parameter $w = w(\alpha)$. In this work, we consider the well known Chevalier-Polarski-Linder (CPL)[Chevallier & Polarski 2001; Linder 2003] parametrization for which

$$w(\alpha) = w_0 + w_1(1 - \alpha) \tag{6}$$

where w_0 and w_1 are constants. Therefore, inserting eq.(6) into eq.(3) we simply derive

$$f(\alpha) = \alpha^{-3(1+w_0+w_1)} e^{3w_1(\alpha-1)} \ . \tag{7}$$

Obviously, if $w_1 = 0$ the current models reduce to the quintessence cosmological models $[f(\alpha) = \alpha^{-3(1+w)}]$.

3. Likelihood Analysis

In this section, we present the statistical results using the Hubble relation provided by the GRB and SNIa data respectively.

- **GRB distance modulii:** We first use $N_{GRB} = 69$ GRBs provided by Cardone et al. (2009) up to 6.6 redshift.
- **SNIa distance modulii:** We additionally utilize the 'Constitution set' of 397 type Ia supernovae of Hicken et al. (2009). In order to avoid possible problems related with the local bulk flow, we use a sub-sample of the overall sample in which we select those SNIa with $z > 0.023$. This sub-sample contains $N_{SNIa} = 351$ entries.

Journal of Physics: Conference Series **222** (2010) 012026 doi:10.1088/1742-6596/222/1/012026

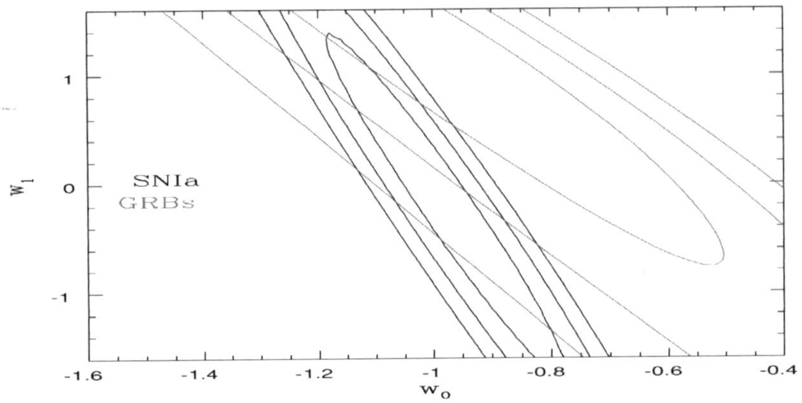

Figure 2. Likelihood contours in the (w_0, w_1) plane after marginalizing with respect to $\Omega_m = 0.28$.

The corresponding χ_j^2 function is:

$$\chi_j^2(\mathbf{p}) = \sum_{i=1}^{N_j} \left[\frac{\mu_{\rm th}(a_i, \mathbf{p}) - \mu_{\rm obs}(a_i)}{\sigma_i}\right]^2 \quad j = GRB, SNIa \tag{8}$$

where $a_i = (1+z_i)^{-1}$ is the observed scale factor of the Universe, z_i is the observed redshift, μ is the distance modulus corresponding to flat space:

$$\mu = m - M = 5\log d_L(a) + 25 \tag{9}$$

and $d_L(a, \mathbf{p})$ is the luminosity distance

$$d_L(\alpha, \mathbf{p}) = \frac{c}{\alpha}\int_a^1 \frac{dy}{y^2 H(y)}, \tag{10}$$

Note, that c is the speed of light. We can combine the above cosmologically tests, using a joint likelihood analysis, in order to put even more stringent constraints on the free-parameter space, according to:

$$\mathcal{L}_{tot}(\mathbf{p}) = \mathcal{L}_{GRB} \times \mathcal{L}_{SNIa}$$

or

$$\chi_{tot}^2(\mathbf{p}) = \chi_{GRB}^2 + \chi_{SNIa}^2, \tag{11}$$

with the likelihood estimator defined as $\mathcal{L}_j \propto \exp[-\chi_j^2/2]$. Note that the overall number of data points used is $N_{tot} = 420$ and the degrees of freedom: *dof*$= N_{tot} - n_{\rm fit}$, with $n_{\rm fit}$ the number of fitted parameters, which vary for the different models.

3.1. Constant equation of state - QP model

In this case the equation of state parameter is constant, $w(a) = w$, (see Peebles & Ratra 2003). Such dark energy models do not have much physical motivation. In particular, a constant equation of state parameter requires a fine tuning of the potential and kinetic energies of the

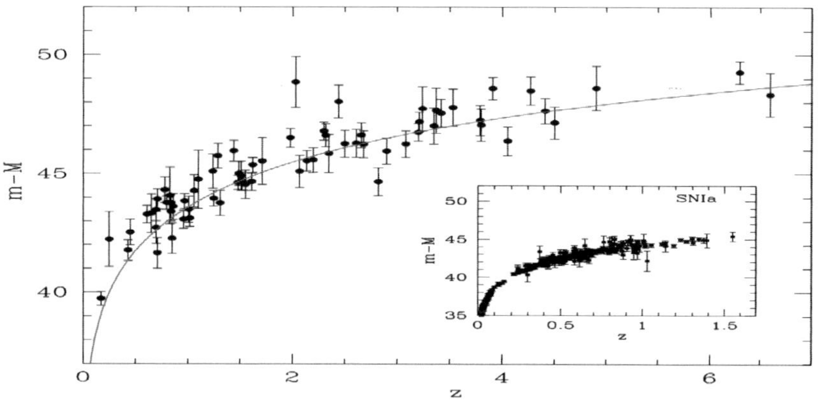

Figure 3. *Insert Panel:* SNIa distance moduli as a function of redshift. Distance moduli comparison difference between the Λ-model ($\Omega_m = 0.28$ solid line) and the GRB data.

real scalar field. Despite the latter problem, these dark energy models have been used in the literature due to their simplicity. Notice, that dark energy models with a canonical kinetic term have $-1 \leq w < -1/3$. On the other hand, models of phantom dark energy ($w < -1$) require exotic nature, such as a scalar field with a negative kinetic energy. Comparing the QP-models with the observational data (we sample $\Omega_m \in [0.1, 1]$ and $w \in [-1.4, -0.6]$ in steps of 0.01) we find that the best fit values are: a) using the GRBs only we find $\Omega_m = 0.40^{+0.03}_{-0.08}$ (while w remains unconstraint due to large errors) and b) for the joint analysis the likelihood function peaks at $\Omega_m = 0.30^{+0.10}_{-0.02}$ and $w = -1.04^{+0.10}_{-0.30}$ with $\chi^2_{tot}(\Omega_m, w)/dof \simeq 540/418$. In figure 1 we present the 1σ, 2σ and 3σ confidence levels (corresponding to where $-2\ln\mathcal{L}/\mathcal{L}_{\max}$ equals 2.30, 6.16 and 11.83) in the (Ω_m, w) plane. We would like to stress that our results are in good agreement with the 5 years WMAP data (Komatsu et al. 2009). Also Davis et al. (2007) using the Essence-SNIa+BAO+CMB and a Bayesian statistics found $\Omega_m = 0.27 \pm 0.04$, while Kowalski et al. (2008) utilizing the Union08-SNIa+BAO+CMB obtained $\Omega_m = 0.285^{+0.02+0.01}_{-0.02-0.01}$. Obviously, our results coincide within 1σ errors. It is worth noting that the concordance Λ-cosmology can be described by QP models with w strictly equal to -1. In this case we find: $\Omega_m = 0.30 \pm 0.30$ with $\chi^2_{tot}(\Omega_m)/dof \simeq 540/419$.

3.2. The parametric Dark Energy model - CPL model

In the CPL (Chevallier & Polarski 2001; Linder 2003) parametrization the dark energy equation of state parameter is defined as a first order Taylor expansion around the present epoch (see eq.6). We sample the unknown parameters as follows: $w_0 \in [-2, -0.4]$ and $w_1 \in [-2.6, 2.6]$ and we find that for the prior of $\Omega_m = 0.28$ the overall likelihood function peaks at $w_0 = -1.15^{+0.22}_{-0.16}$ and $w_1 = 1.12^{+0.62}_{-1.54}$ while the corresponding $\chi^2_{tot}(w_0, w_1)/dof$ is 539/418. In figure 2 we show the GRB constraints in the (w_0, w_1) plane by marginalizing our solution over $\Omega_m = 0.28$. For comparison reasons we also present red contours, which where based on the SNIa data. Obviously, the different shape between the contours is due to the fact that the two data sets cover different redshift ranges. This is a hint that the use of alternative tracers for the Hubble relation at large redshifts ($z \gtrsim 2$), could be a useful avenue for providing further constraints of the cosmological parameters (Plionis et al. 2009). However, in order to use the GRBs properly

Journal of Physics: Conference Series **222** (2010) 012026 doi:10.1088/1742-6596/222/1/012026

as standard candles to constrain the cosmological parameters, we have to significantly reduce the scatter in the distance modulus diagram (see figure 3). This implies that we need a much better calibration of the observed correlations required in order to measure the correct distance modulus (Basilakos & Perivolaropoulos 2008). In fact, a Monte-Carlo analysis have shown that the best strategy to decrease the uncertainties of the cosmological parameters based on the Hubble relation, is to build a sample of $\sim$ 80 standard candles that trace the redshift range $2 \lesssim z \lesssim 6$ as well as to reduce the scatter in the resulting luminosity and thus in the distance modulus diagram by a factor of 1/2.

4. Conclusions

We have utilized the recent GRB data (Cardone et al. 2009) in order to constraint the main cosmological parameters. We find that within the context of flat cosmological models and under of specific conditions the GRB data can place constrains on the main cosmological parameters. However, due to the small number statistics as well as due to the large uncertainties involved in the statistical analysis, the next generation of the GRB catalogs have to (a) contain at least $\sim$ 80 entries at large redshifts ($2 \lesssim z \lesssim 6$) and (b) to reduce the scatter in the diagram of the distance modulus by a factor of 1/2.

References

[1] Amati, L., et al., 2002, A&A, 390, 81
[2] Basilakos, S. & Perivolaropoulos, L., 2008, MNRAS, 391, 411
[3] Bertolami, O., &, Tavares Silva, P., 2006, MNRAS, 365, 1149
[4] Bertschinger, E., 2006, ApJ, 648, 797
[5] Butler, N. R., Kocevski, D., Bloom, J. S., &, Curtis, J. L, 2007, ApJ, 671, 656
[6] Cardone V.F., Capozziello S.,&, Dainotti M.G., 2009, MNRAS, 400, 775
[7] Capozziello S.,&, Izzo L., A&A, 2008, 490, 31
[8] Chevallier M., & Polarski D., 2001, Int. J. Mod. Phys. D, 10, 213
[9] Davis, T.M., et al., 2007, ApJ, 666, 716
[10] Firmani, C., Ghisellini, G., Avila-Reese, V., &, Ghirlanda, G., 2006, MNRAS, 370, 185
[11] Dai, Z. G., Liang, E. W., &, Xu, D., 2004, ApJ, 612, L101
[12] Demianski, M., Piedipalumbo, E., Rubano, C., &, Tortora, C., 2006 A&A, 454, 55
[13] Di Girolamo, T., Catena, R., Vietri, M., &, Di Sciascio, G., 2005, JCAP, 0504, 008
[14] Ghirlanda, G., Ghisellini, G., &, Firmani, C., 2006, New J. Phys., 8, 123
[15] Hicken M., et al., 2009, ApJ, 700, 1097
[16] Hooper, D., &, Dodelson, S., 2007, Astropart. Phys., 27, 113
[17] Izzo, L., Capozziello, S., Covone, G.,&, Capaccioli, M., 2009, A&A, 508, 63
[18] Kawai, N., et al., 2006, Nature, 440, 184
[19] Komatsu E., et al., 2009, ApJS, 180, 330
[20] Kowalski M., et al., 2008, ApJ, 686, 749
[21] Lazkoz, R., Nesseris, S., &, Perivolaropoulos, L., 2007, (arXiv:0712.1232)
[22] Li, L-X., 2007, MNRAS, 379, L55
[23] Linder, V. E., 2003, Phys. Rev. Lett., 90, 091301
[24] Nesseris, S., &, Perivolaropoulos, L., 2008, Phys. Rev. D, 77, 023504
[25] Peebles P.J.E., &, Ratra, B., 2003, RvMP, 75, 559
[26] Percival, W. J., Cole, S., Eisenstein, D. J., Nichol, R. C., Peacock, J. A., Pope, A. C., &, Szalay, A. S., 2007, MNRAS, 381, 1053
[27] Plionis M., et al., 2009, JPhCS, 189, 2032
[28] Piran, T., 2004, Rev. Mod. Phys., 76, 1143
[29] Rhoads, J. E., 1999, ApJ, 525, 737
[30] Schaefer, B. E., 2003, ApJ, 583, L67
[31] Schaefer, B. E., 2007, ApJ, 660,16
[32] Tegmark, M., et al., 2004, Phys.Rev.D., 69, 3501
[33] Tsutsui R. et al., 2009a, MNRAS, 349, L31
[34] Tsutsui R. et al., 2009b, JCAP, 08,015
[35] Wang, F. Y., &, Dai, Z. G., 2006, MNRAS, 368,371
[36] Wang, F. Y., Dai, Z. G., &, Qi Shi, 2009, A&A, 507,53

[37] Zhang, Z. B.&, Xie, G. Z., 2007, (arXiv:0711.1411)
[38] Zhang, B., &, Meszaros, P., 2004, Int. J. Mod. Phys. A, 19, 2385

Journal of Physics: Conference Series **222** (2010) 012027 doi:10.1088/1742-6596/222/1/012027

Phenomenology of loop quantum cosmology

Mairi Sakellariadou

Department of Physics, King's College London, University of London, Strand WC2R 2LS, UK

E-mail: `Mairi.Sakellariadou@kcl.ac.uk`

Abstract. After introducing the basic ingredients of Loop Quantum Cosmology, I will briefly discuss some of its phenomenological aspects. Those can give some useful insight about the full Loop Quantum Gravity theory and provide an answer to some long-standing questions in early universe cosmology.

1. Introduction

The variety of precise astrophysical and cosmological data, available at present, combined with the large number of high energy physics experiments, are expected to give us the necessary ingredients to test fundamental theories and understand the very early phases of the evolution of our universe.

Cosmological inflation [1] remains the most promising candidate to solve the shortcomings of the standard hot big bang model, while it offers a causal explanation for the primordial fluctuations with the correct Cosmic Microwave Background (CMB) features. Despite this success, inflation suffers from a number of drawbacks [2]. In particular, compatibility between theory and measurements often necessitates fine-tuning of the inflationary parameters, inflation remains still a paradigm in search of a model, while it must prove itself generic [3, 4, 5, 6, 7, 8, 9].

There is undoubtfully an additional list of fundamental cosmological questions, still lacking a satisfying answer. One does not know, for instance, how close to the big bang a smooth space-time can be considered as the correct framework. Quantum gravity, a full theory which is supposed to resolve the big bang singularity, is still missing, while it is not known whether a new boundary condition is needed at the big bang, or whether quantum dynamical equations remain well-behaved at singularities. It is nevertheless clear that a smooth space-time background cannot be assumed as the correct description close to the big bang.

In a Hamiltonian formulation to a quantum theory, the absence of background metric indicates that Hamiltonian dynamics is generated by constraints. Physical states are solutions to quantum constraints implying that all physical laws are obtained from these solutions, while there is no external time according which evolution can be studied. One has to define a monotonic variable to play the rôle of emergent, or internal, time, and then build a framework within which the short-distance drawbacks of General Relativity near the big bang can be cured, maintaining however the agreement with General Relativity at large scales.

Loop Quantum Gravity (LQG) [10, 11] is a non-perturbative and background independent canonical quantisation of General Relativity in four space-time dimensions. Loop Quantum Cosmology (LQC) [12] is a cosmological mini-superspace model quantised with methods of LQG. The discreteness of spatial geometry and the simplicity of the setting allow for a complete

Journal of Physics: Conference Series **222** (2010) 012027 doi:10.1088/1742-6596/222/1/012027

study of dynamics. The difference between LQC and other approaches of quantum cosmology, is that the input is motivated by a full quantum gravity theory. The simplicity of the setting, combined with the discreteness of spatial geometry provided by LQG, render feasible the overall study of LQC dynamics.

In what follows, I will briefly discuss some of the phenomenological consequences of LQC, a subject which gains constantly an increasing interest from the scientific community, due in particular to its successes.

2. Elements of LQG/LQC

Loop quantum cosmology is formulated in terms of SU(2) holonomies of the connection and triads. The canonically conjugated variables are defined by the densitised triad E_i^a, and an SU(2) valued connection A_a^i, where i refers to the Lie algebra index, and a is a spatial index with $a, i = 1, 2, 3$. The densitised triad gives information about spatial geometry (*i.e.*, the three metric), while the connection gives information about curvature (spatial and extrinsic one). Let me explain this: Quantum gravity introduces a discreteness to space-time. To quantise quantum gravity, this discreteness manifests itself as quanta of space. Since the quanta are three-dimensional, they can be characterised by a triad of numbers. The connection between adjacent pairs of these *quanta* form a two-dimensional surface, through which the geometric connection is defined. To ensure that local rotations do not induce different geometries, this connection must be SO(3) invariant. By using connection-triad variables, arising from a canonical transformation of Arnowitt-Desner-Misner (variables, one can make an analogy with gauge theories, which is particularly useful when dealing with quantisation issues.

For any quantisation scheme based on a Hamiltonian framework (*e.g.*, LQC) or an action principle (*e.g.*, path integral) for a homogeneous and flat model, divergences appear which need to be regularised. To remove the divergences that occur in non-compact topologies, the spatial homogeneity and Hamiltonian are restricted to a finite fiducial cell, of scale μ_0, with finite volume $V_0 = \int \mathrm{d}^3x\sqrt{|^0q|}$, where ^0q is the determinant of the fiducial background metric.

In the case of a spatially flat background, derived from the Bianchi I model, the isotropic connection can be expressed in terms of the dynamical component of the connection $\tilde{c}(t)$ as

$$A_a^i = \tilde{c}(t)\omega_a^i \ , \tag{1}$$

with ω_a^i a basis of left-invariant one-forms $\omega_a^i = \mathrm{d}x^i$. The densitised triad can be decomposed using the Bianchi I basis vector fields $X_i^a = \delta_i^a$ as

$$E_i^a = \sqrt{^0q}\tilde{p}(t)X_i^a \ , \tag{2}$$

where ^0q stands for the determinant of the fiducial background metric $^0q_{ab} = \omega_a^i\omega_{bi}$, and $\tilde{p}(t)$ denotes the remaining dynamical quantity after symmetry reduction.

In terms of the metric variables with three-metric $q_{ab} = a^2\omega_a^i\omega_{bi}$, the dynamical quantity is just the scale factor $a(t)$. Given that the Bianchi I basis vectors are $X_i^a = \delta_i^a$,

$$|\tilde{p}| = a^2 \ , \tag{3}$$

where the absolute value is taken because the triad has an orientation. Since the basis vector fields are spatially constant in the spatially flat model, the connection component is

$$\tilde{c} = \mathrm{sgn}(\tilde{p})\gamma\frac{\dot{a}}{N} \ ; \tag{4}$$

N is the lapse function and γ the Barbero-Immirzi parameter, a quantum ambiguity parameter approximately equal to 0.2375.

The canonical variables $\tilde{c}, \tilde{p}$ are related through the Poisson bracket

$$\{\tilde{c}, \tilde{p}\} = \frac{\kappa\gamma}{3} V_0 \ , \tag{5}$$

where V_0 the volume of the elementary cell adapted to the fiducial triad and $\kappa \equiv 8\pi G$.

Defining the triad component p, determining the physical volume of the fiducial cell, and the connection component c, determining the rate of change of the physical edge length of the fiducial cell, as

$$p = V_0^{2/3}\tilde{p} \quad , \quad c = V_0^{1/3}\tilde{c} \ , \tag{6}$$

respectively, we obtain

$$\{c, p\} = \frac{\kappa\gamma}{3} \ , \tag{7}$$

independent of the volume V_0 of the fiducial cell.

To quantise the theory one faces the usual difficulty of quantum cosmology. Namely, the metric itself has to be considered as a physical field, thus it must be quantised; it is not a fixed background. Any *standard* quantisation scheme fails: Fock quantisation fails, while even free Hamiltonians are non-quadratic in metric dependence, so there is no simple perturbation analysis. The way out is to use gauge theory variables to define holonomies of the connection along a given edge

$$h_e(A) = \mathcal{P} \exp \int \mathrm{d}s \dot{\gamma}^\mu(s) A^i_\mu(\gamma(s))\tau_i \ , \tag{8}$$

where $\mathcal{P}$ indicates a path ordering of the exponential, γ^μ is a vector tangent to the edge and $\tau_i = -i\sigma_i/2$, with σ_i the Pauli spin matrices, and fluxes of a triad along an S surface

$$E(S, f) = \int_S \epsilon_{abc} E^{ci} f_i \mathrm{d}x^a \mathrm{d}x^b \ , \tag{9}$$

with f_i an SU(2) valued test function. Certainly, at the level of LQC these variables may seem ad-hoc and unnatural, however their motivation follows naturally from the full LQG theory.

Thus, the basic configuration variables in LQC are holonomies of the connection

$$h_i^{(\mu_0)}(A) = \cos\left(\frac{\mu_0 c}{2}\right) \mathbf{1} + 2 \sin\left(\frac{\mu_0 c}{2}\right) \tau_i \ , \tag{10}$$

along a line segment $\mu_0 \, {}^0e_i^a$ and the flux of the triad

$$F_S(E, f) \propto p \ ,$$

where the basic momentum variable is the triad component p, $\mathbf{1}$ is the identity 2×2 matrix and $\tau_i = -i\sigma_i/2$ is a basis in the Lie algebra SU(2) satisfying the relation

$$\tau_i \tau_j = (1/2)\epsilon_{ijk}\tau^k - (1/4)\delta_{ij} \ .$$

In the classical theory, curvature can be expressed as a limit of the holonomies around a loop as the area enclosed by the loop shrinks to zero. In quantum geometry however, the loop cannot be continuously shrunk to zero area and the eigenvalues of the area operator are discrete. Thus, there is a smallest non-zero eigenvalue, the area gap Δ [13]. As a result, the Wheeler-de Witt differential equation gets replaced by a difference equation whose step size is controlled by Δ.

Let me start with the *old* quantisation procedure. Along the lines of LQG, one considers $e^{i\mu_0 c/2}$ (with μ_0 an arbitrary real number) and p, as the elementary classical variables, which have well-defined analogues. Using the Dirac bra-ket notation and setting $e^{i\mu_0 c/2} = \langle c|\mu\rangle$, the action of the operator $\hat{p}$ acting on the basis states $|\mu\rangle$ reads

$$\hat{p}|\mu\rangle = \frac{\kappa\gamma\hbar|\mu|}{6}|\mu\rangle \ , \tag{11}$$

where μ (a real number) stands for the eigenstates of $\hat{p}$, satisfying the orthonormality relation $\langle\mu_1|\mu_2\rangle = \delta_{\mu_1,\mu_2}$. The action of the $\widehat{\exp\left[\frac{i\mu_0}{2}c\right]}$ operator acting on basis states $|\mu\rangle$ is

$$\widehat{\exp\left[\frac{i\mu_0}{2}c\right]}|\mu\rangle = \exp\left[\mu_0\frac{\mathrm{d}}{\mathrm{d}\mu}\right]|\mu\rangle = |\mu+\mu_0\rangle \ , \tag{12}$$

where μ_0 is any real number. Thus, in the *old* quantisation, the operator $e^{i\mu_0 c/2}$ acts as a simple shift operator. As in the full LQG theory, there is no operator corresponding to the connection, however the action of its holonomy is well defined. The action of the holonomies, $\hat{h}_i^{(\mu_0)}$, of the gravitational connection on the basis states is given by [14]

$$\hat{h}_i^{(\mu_0)}|\mu\rangle = \left(\widehat{\mathrm{cs}}\mathbf{1} + 2\widehat{\mathrm{sn}}\tau_i\right)|\mu\rangle \ , \tag{13}$$

where,

$$\begin{aligned} \widehat{\mathrm{cs}}|\mu\rangle \equiv \widehat{\cos(\mu_0 c/2)}|\mu\rangle &= [\ |\mu+\mu_0\rangle + |\mu-\mu_0\rangle\ ]/2 \ , \\ \widehat{\mathrm{sn}}|\mu\rangle \equiv \widehat{\sin(\mu_0 c/2)}|\mu\rangle &= [\ |\mu+\mu_0\rangle - |\mu-\mu_0\rangle\ ]/(2i) \ . \end{aligned} \tag{14}$$

The gravitational part of the Hamiltonian operator in terms of SU(2) holonomies and the triad component, in the irreducible $J = 1/2$ representation, reads [14]

$$\hat{C}_{\mathrm{g}} = \frac{2i}{\kappa^2\hbar\gamma^3\mu_0^3}\mathrm{tr}\sum_{ijk}\epsilon^{ijk}\left(\hat{h}_i^{(\mu_0)}\hat{h}_j^{(\mu_0)}\hat{h}i^{(\mu_0)-1}\hat{h}_j^{(\mu_0)-1}\hat{h}_k^{(\mu_0)}\left[\hat{h}_k^{(\mu_0)-1},\hat{V}\right]\right)\mathrm{sgn}(\hat{p}) \ . \tag{15}$$

The action of the self-adjoint Hamiltonian constraint operator, $\hat{\mathcal{H}}_{\mathrm{g}} = (\hat{C}_{\mathrm{g}} + \hat{C}_{\mathrm{g}}^\dagger)/2$ on the basis states $|\mu\rangle$ is

$$\hat{\mathcal{H}}_{\mathrm{g}}|\mu\rangle = \frac{3}{4\kappa^2\gamma^3\hbar\mu_0^3}\Big\{\big[R(\mu)+R(\mu+4\mu_0)\big]|\mu+4\mu_0\rangle - 4R(\mu)|\mu\rangle + \big[R(\mu)+R(\mu-4\mu_0)\big]|\mu-4\mu_0\rangle\Big\} \ , \tag{16}$$

where

$$R(\mu) = (\kappa\gamma\hbar/6)^{3/2}\left||\mu+\mu_0|^{3/2} - |\mu-\mu_0|^{3/2}\right| \ . \tag{17}$$

The dynamics are then determined by the Hamiltonian constraint

$$(\hat{\mathcal{H}}_{\mathrm{g}} + \hat{\mathcal{H}}_\phi)|\Psi\rangle = 0 \ , \tag{18}$$

where $\hat{\mathcal{H}}_\phi$ stands for the matter Hamiltonian. In the full LQG theory there is an infinite number of constraints, while in LQC there is only one integrated Hamiltonian constraint. Matter is introduced by just adding the actions of matter components to the gravitational action. We can finally obtain difference equations analogous to the differential Wheeler-de Witt equations.

Journal of Physics: Conference Series **222** (2010) 012027 doi:10.1088/1742-6596/222/1/012027

More precisely, the constraint equation on the physical wave-functions $|\Psi\rangle$, which can be expanded using the basis states as $|\Psi\rangle = \sum_\mu \Psi_\mu(\phi)|\mu\rangle$ with summation over values of μ and where the dependence of the coefficients on ϕ represents the matter degrees of freedom, reads [15]

$$\left[\left|V_{\mu+5\mu_0} - V_{\mu+3\mu_0}\right| + \left|V_{\mu+\mu_0} - V_{\mu-\mu_0}\right|\right]\Psi_{\mu+4\mu_0}(\phi) - 4\left|V_{\mu+\mu_0}V_{\mu-\mu_0}\right|\Psi_\mu(\phi)$$
$$+\left[\left|V_{\mu-3\mu_0} - V_{\mu-5\mu_0}\right| + \left|V_{\mu+\mu_0} - V_{\mu-\mu_0}\right|\right]\Psi_{\mu-4\mu_0}(\phi) = -\frac{4\kappa^2\gamma^3\hbar\mu_0^3}{3}\mathcal{H}_\phi(\mu)\Psi_\mu(\phi) \; ; \quad (19)$$

the matter Hamiltonian $\hat{\mathcal{H}}_\phi$ is assumed to act diagonally on the basis states with eigenvalues $\mathcal{H}_\phi(\mu)$. Equation (19) is the quantum evolution (in internal time μ) equation; there is no continuous variable (the scale factor in classical cosmology), but a label μ with discrete steps. The wave-function $\Psi_\mu(\phi)$, depending on internal time μ and matter fields ϕ, determines the dependence of matter fields on the evolution of the universe, with a massless scalar field playing the rôle of emergent time. Thus, in LQC the quantum evolution is governed by a second order difference equation, rather than the second order differential equation of the Wheeler-de Witt quantum cosmology. As the universe becomes large and enters the semi-classical regime, its evolution can be approximated by the differential Wheeler-de Witt equation.

3. Lattice refinement

The quantised holonomies were at first assumed to be shift operators with a fixed magnitude, leading to a quantised Hamiltonian constraint being a difference equation with a constant interval between points on the lattice. While these models can be used to study certain aspects of the quantum regime, they were found to lead to serious instabilities in the continuum, semi-classical limit [16, 17]. Considering the continuum limit of the Hamiltonian constraint operator, we have so far assumed that Ψ does not vary much on scales of the order of $4\mu_0$ (known as *pre-classicality*), so that one can smoothly interpolate between the points on the discrete function $\Psi_\mu(\phi)$ and approximate them by the continuous function $\Psi(\mu, \phi)$. Under this assumption, the difference equation can be very well approximated by a differential equation for a continuous wave-function. However, the form of the wave-function indicates that the period of oscillations can decrease as the scale increases. Thus, the assumption of *pre-classicality* can break down at large scales, leading to deviations from the classical behaviour.

Let me explain this point a bit further: In the underlying full LQG theory, the contributions to the (discrete) Hamiltonian operator depend on the state which describes the universe. As the volume grows (i.e., the universe expands), the number of contributions increases. Thus, the Hamiltonian constraint operator is expected to create new vertices of a lattice state (in addition to changing their edge labels), which in LQC results in a refinement of the discrete lattice [18]. Lattice refinement is also required from phenomenological reasons [17, 19]; for instance, it renders a successful inflationary era more natural [19]. The effect of lattice refinement has been modelled and the elimination of the instabilities in the continuum era has been explicitly shown.

The appropriate lattice refinement model should be obtained from the full LQG theory. In principle, one should use the full Hamiltonian constraint and find the way that its action balances the creation of new vertices as the volume increases. Instead, phenomenological arguments have been used, where the choice of the lattice refinement model is constrained by the form of the matter Hamiltonian [20]. In particular, LQC can generically support inflation, and other matter fields, without the onset of large scale quantum gravity corrections, only for a particular model of lattice refinement [20]. This choice is the only one for which physical quantities are independent of the elementary cell adopted to regulate spatial integrations [21], and moreover, it is exactly the choice required for the uniqueness of the factor ordering of the Wheeler-de Witt equation [22].

Allowing the length scale of the holonomies to vary dynamically, the form of the difference equation, describing the evolution, changes. Since the parameter μ_0 determines the step-size of the difference equation, assuming the lattice size is growing, the step-size of the difference equation is not constant in the original triad variables. Let us consider the particular case of

$$\mu_0 \to \tilde{\mu}\,(\mu) = \mu_0 \mu^{-1/2} \ , \tag{20}$$

suggested by certain intuitive heuristic approaches, such as noting that the minimum area used to regulate the holonomies should be a physical area [23], or that the discrete step size of the difference equation should always be of the order of the Planck volume. Moreover, this choice also results in a significant simplification of the difference equation, compared to more general lattice refinement schemes. The basic operators are given by replacing μ_0 with $\tilde{\mu}$. Upon quantisation [23]

$$\widehat{e^{i\tilde{\mu}c/2}}|\mu\rangle = e^{-i\tilde{\mu}\frac{\mathrm{d}}{\mathrm{d}\mu}}|\mu\rangle \ , \tag{21}$$

which is no longer a simple shift operator since $\tilde{\mu}$ is a function of μ. Changing the basis to

$$\nu = \mu_0 \int \frac{\mathrm{d}\mu}{\tilde{\mu}} = \frac{2}{3}\mu^{3/2} \ , \tag{22}$$

one gets

$$e^{-i\tilde{\mu}\frac{\mathrm{d}}{\mathrm{d}\mu}}|\nu\rangle = e^{-i\mu_0\frac{\mathrm{d}}{\mathrm{d}\nu}}|\nu\rangle = |\nu + \mu_0\rangle \ . \tag{23}$$

Thus, in the new variables the holonomies act as simple shift operators, with parameter length μ_0. In this sense, the basis $|\nu\rangle$ is a much more natural choice than $|\mu\rangle$. One can then proceed as in the previous case of a fixed spatial lattice and write down the Hamiltonian constraint.

3.1. Constraints on inflation

Let me consider first a fixed and then a dynamically varying lattice and solve the second order difference equation in the continuum limit. Two constraints can be imposed on the inflaton potential: the first one so that the continuum approximation is valid, and the second one so that there is agreement with the CMB measurements on large angular scales. A combination of the two constraints in the context of a particular inflationary model, will give us [20] the conditions for natural and successful inflation within LQC.

More precisely, let us separate the wave-function $\Psi(p,\phi)$ into $\Psi(p,\phi) = \Upsilon(p)\Phi(\phi)$ and approximate the dynamics of the inflaton field, ϕ, by setting $V(\phi) = V_\phi p^{\delta-3/2}$, where V_ϕ is a constant and $\delta = 3/2$ in the case of slow-roll, to get [20]

$$p^{-1/2}\frac{\mathrm{d}}{\mathrm{d}p}\left[p^{-1/2}\frac{\mathrm{d}}{\mathrm{d}p}\left(p^{3/2}\Upsilon\,(p)\right)\right] + \beta V_\phi p^\delta \Upsilon(p) = 0 \ , \tag{24}$$

with solutions [20]

$$\Upsilon\,(p) \approx p^{-(9+2\delta)/8}\sqrt{\frac{2\delta+3}{2\sqrt{\beta V_\phi}\pi}}\left[C_1\cos\left(x - \frac{3\pi}{2(2\delta+3)} - \frac{\pi}{4}\right) + C_2\sin\left(x - \frac{3\pi}{2(2\delta+3)} - \frac{\pi}{4}\right)\right] , \tag{25}$$

where $x = 4\sqrt{\beta V_\phi}(2\delta+3)^{-1}p^{(2\delta+3)/4}$ and $\beta = 96/(\kappa\hbar^2)$.

Without lattice refinement, the discrete nature of the underlying lattice would eventually be unable to support the oscillations and the assumption of *pre-classicality* will break down, implying that the discrete nature of space-time becomes significant on large scales. For the end of inflation to be describable using classical General Relativity, it must end before a scale, at which the assumption of *pre-classicality* breaks down and the semi-classical description is no longer valid, is reached. Let me quantify this constraint: The separation between two successive zeros of $\Upsilon(p)$ is

$$\Delta p = \frac{\pi}{\sqrt{\beta V_\phi}} p^{(1-2\delta)/4} \ . \tag{26}$$

For the continuum limit to be valid, the wave-function must vary slowly on scales of the order of $\mu_c = 4\tilde{\mu}$, leading to [20]

$$\Delta p > 4\mu_0 \left(\frac{\kappa\gamma\hbar}{6}\right)^{3/2} p^{-1/2} \ , \tag{27}$$

which implies [20]

$$V_\phi < \frac{27\pi^2}{192\mu_0^2\gamma^3\kappa^2\hbar} p^{(3-2\delta)/2} \ . \tag{28}$$

Assuming slow-roll inflation, we set $\delta \approx 3/2$. Setting $\mu_0 = 3\sqrt{3}/2$, the constraint on the inflationary potential, in units of $\hbar = 1$, reads

$$V(\phi) \lesssim 2.35 \times 10^{-2} l_{\rm Pl}^{-4} \ , \tag{29}$$

which is a weaker constraint than the one imposed for fixed lattices, namely [20]

$$V_\phi \ll 10^{-28} l_{\rm pl}^{-4} \ , \tag{30}$$

assuming that half of the inflationary era takes place during the classical era.

One can further constrain the inflationary potential so that the fractional over-density in Fourier space and at horizon crossing is consistent with the COBE-DMR measurements, namely

$$\frac{[V(\phi)]^{3/2}}{V'(\phi)} \approx 5.2 \times 10^{-4} M_{\rm Pl}^3 \ . \tag{31}$$

To do so, one must however adopt a particular inflationary model. As such, let us select $V(\phi) = m^2\phi^2/2$. Combining the two constraints we obtain [20]

$$m \lesssim 70(e^{-2N_{\rm cl}})\ M_{\rm Pl} \quad \text{and} \quad m \lesssim 10\ M_{\rm Pl} \ , \tag{32}$$

for the fixed and varying lattices, respectively.

In conclusion, the requirement that a significant proportion of a successful (in particular with respect to the CMB measurements) inflationary regime takes place during the classical era, imposes a strong constraint on the inflaton mass. Such a constraint turns out however to be much softer, and therefore rather natural, once lattice refinement is considered. In this sense, we argue [20] that lattice refinement is essential to achieve a successful inflationary era, provided inflation can be generically (*i.e.*, with generic initial conditions) realised.

3.2. Constraints on the matter Hamiltonian

Given a lattice refinement model, we will show [19] that only certain types of matter can be allowed. To do so, we parametrise the lattice refinement by A and the matter Hamiltonian by δ and solve the Hamiltonian constraint. The restrictions on the two-dimensional parameter space will become apparent once physical restrictions to the solutions of the wave-functions are imposed [19].

To be more precise, consider

$$\tilde{\mu} = \mu_0 \mu^A \ , \tag{33}$$

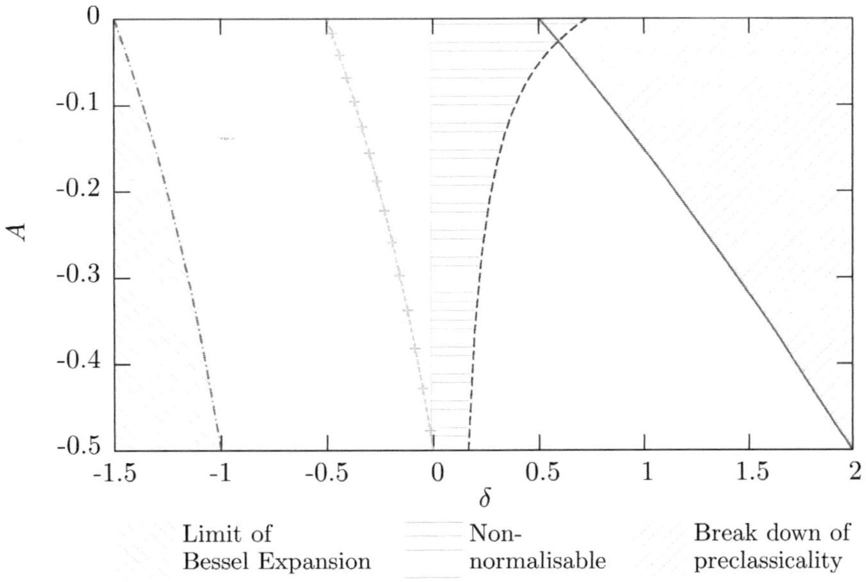

Figure 1. The allowed types of matter content are significantly restricted. For a varying lattice ($A \neq 0$) it is not always possible to treat the large scale behaviour of the wave-functions perturbatively (dashed line with crosses) [19].

leading to

$$\nu = \frac{\tilde{\mu}_0 \mu^{1-A}}{\mu_0(1-A)} \ . \tag{34}$$

Being only interested in the large scale limit, we approximate [19] the matter Hamiltonian by $\hat{\mathcal{H}}_\phi = \hat{\nu}^\delta \hat{\epsilon}\,(\phi)$, implying

$$\hat{\epsilon}\,(\phi)\,\Psi \equiv \epsilon\,(\phi)\,\Psi = -\nu^{-\delta}\hat{\mathcal{H}}_{\rm g}\Psi \ . \tag{35}$$

A necessary condition so that the wave-functions are physical, is that the finite norm of the physical wave-functions, defined by $\int_{\phi=\phi_0} d\nu|\nu|^\delta\overline{\Psi}_1\Psi_2$, is independent of the choice of $\phi = \phi_0$. The solutions of the constraint are renormalisable provided they decay, on large scales, faster than $\nu^{-1/(2\delta)}$.

To solve the constraint equation, we need to specify the from of $\mathcal{H}_\phi$, which has in general two terms with different scale dependence. Being interested in the large scale limit, there is one dominant term, allowing to write [19]

$$\beta\mathcal{H}_\phi = \epsilon_\nu(\phi)\nu^{\delta_\nu} \ , \tag{36}$$

where the function ϵ_ν is constant with respect to ν.

Solving [19] the constraint equation, we only consider the physical solutions. The large scale behaviour of the wave-functions must be normalisable, which is a necessary condition for having physical wave-functions, while the wave-functions should preserve *pre-classicality* at large scales, which is a necessary condition for the validity of the continuum limit. We thus obtain [19] constraints to the two-dimensional parameter space (A, δ), shown in Fig. 1 for A in the range $0 < A < -1/2$, imposed from full LQG theory considerations [17].

In conclusion, the continuum limit of the Hamiltonian constraint equation is sensitive to the choice of the lattice refinement model and only a limited range of matter components can be supported within a particular choice.

3.3. Uniqueness of the factor ordering in the Wheeler-de Witt equation

I will now show that the lattice refinement model $\tilde{\mu} = \mu_0\mu^{-1/2}$, argued [22] to be the only one achieved by both physical considerations of large scale physics and consistency of the quantisation structure, is also the only model which makes the factor ordering ambiguities of LQC to disappear in the continuum limit [22].

Let me be more specific. Writing the gravitational part of the Hamiltonian constraint in terms of the triad and the holonomies of the connection, one realises that there are many ways of doing so. Considering for example,

$$\hat{C}_{\rm g} = \frac{2i}{\kappa^2\hbar\gamma^3 k^3}{\rm tr}\sum_{\rm ijk}\epsilon^{\rm ijk}\left(\hat{h}_{\rm i}\hat{h}_{\rm j}\hat{h}_{\rm i}^{-1}\hat{h}_{\rm j}^{-1}\hat{h}_{\rm k}\left[\hat{h}_{\rm k}^{-1},\hat{V}\right]\right) , \tag{37}$$

there are many possible choices of factor ordering that could have been made at this point; classically, the actions of the holonomies commute. However, each of these factor ordering choices leads to a different factor ordering of the Wheeler-de Witt equation in the continuum limit. The action of the factor ordering chosen in Eq. (37) leads [22] to

$$\epsilon_{\rm ijk}{\rm tr}\left(\hat{h}_{\rm i}\hat{h}_{\rm j}\hat{h}_{\rm i}^{-1}\hat{h}_{\rm j}^{-1}\hat{h}_{\rm k}\left[\hat{h}_{\rm k}^{-1},\hat{V}\right]\right) = -24\widehat{\rm sn}^2\widehat{\rm cs}^2\left(\widehat{\rm cs}\hat{V}\widehat{\rm sn} - \widehat{\rm sn}\hat{V}\widehat{\rm cs}\right) , \tag{38}$$

while other choices have certainly a different action.

Defining $\hat{V}|\nu\rangle = V_\nu|\nu\rangle$, with $\hat{V}$ the volume operator with eigenvalues V_ν, the action of the above factor ordering on a general state $|\Psi\rangle = \sum_\nu \psi_\nu|\nu\rangle$ in the Hilbert space, reads [22]

$$\begin{aligned}\epsilon_{\rm ijk}{\rm tr}\left(\hat{h}_{\rm i}\hat{h}_{\rm j}\hat{h}_{\rm i}^{-1}\hat{h}_{\rm j}^{-1}\hat{h}_{\rm k}\left[\hat{h}_{\rm k}^{-1},\hat{V}\right]\right)|\Psi\rangle &= \frac{-3i}{4}\sum_\nu\Big[\Big(V_{\nu-3k} - V_{\nu-5k}\Big)\psi_{\nu-4k} - 2\Big(V_{\nu+k} - V_{\nu-k}\Big)\psi_\nu \\ &\quad + \Big(V_{\nu+5k} - V_{\nu+3k}\Big)\psi_{\nu+4k}\Big]|\nu\rangle ;\end{aligned} \tag{39}$$

the action of any other factor ordering choice can be obtained in a similar manner. By noting that the volume is given by

$$V_\nu|\nu\rangle \sim [\mu(\nu)]^{3/2}|\nu\rangle , \tag{40}$$

where $\mu(\nu)$ is obtained by

$$\nu = \frac{k\mu^{1-A}}{\mu_0(1-A)} , \tag{41}$$

we find [22]

$$V_{\nu\pm nk} \sim \Big[(\nu \pm nk)\,\alpha\Big]^{3/[2(1-A)]} , \tag{42}$$

where $\alpha = \mu_0\,(1-A)\,/k$.

We then take the continuum limit of these expressions by expanding $\psi_\nu \approx \psi(\nu)$ as a Taylor expansion in small k/ν. For the particular factor ordering chosen above, the large scale continuum limit of the Hamiltonian constraint reads [22]:

$$\begin{aligned}&\lim_{k/\nu\rightarrow 0}\epsilon_{\rm ijk}{\rm tr}\left(\hat{h}_{\rm i}\hat{h}_{\rm j}\hat{h}_{\rm i}^{-1}\hat{h}_{\rm j}^{-1}\hat{h}_{\rm k}\left[\hat{h}_{\rm k}^{-1},\hat{V}\right]\right)|\Psi\rangle \sim \\ &\frac{-36i}{1-A}\alpha^{3/[2(1-A)]}k^3\sum_\nu \nu^{(1+2A)/[2(1-A)]}\left[\frac{{\rm d}^2\psi}{{\rm d}\nu^2} + \frac{1+2A}{1-A}\frac{1}{\nu}\frac{{\rm d}\psi}{{\rm d}\nu} + \frac{(1+2A)(4A-1)}{(1-A)^2}\frac{1}{4\nu^2}\psi(\nu)\right]|\nu\rangle .\end{aligned} \tag{43}$$

One can easily confirm that one obtains the same continuum limit for the Wheeler-de Witt equation, only for the choice $A = -1/2$ [22], in which case the Wheeler-de Witt equation reads [22]

$$\lim_{k/\nu \to 0} \mathcal{C}_{\rm g}|\Psi\rangle = \frac{72}{\kappa^2 \hbar \gamma^3} \left(\frac{\kappa \gamma \hbar}{6} \right)^{3/2} \sum_\nu \frac{{\rm d}^2 \psi}{{\rm d}\nu^2} |\nu\rangle \; . \tag{44}$$

Thus, there is only one lattice refinement model, namely $\tilde{\mu} = \mu_0 \mu^{-1/2}$, with a non-ambiguous continuum limit.

In conclusion, phenomenological and consistency requirements lead to a particular lattice refinement model, implying that LQC predicts a unique factor ordering of the Wheeler-de Witt equation in its continuum limit. Alternatively, demanding that factor ordering ambiguities disappear in LQC at the level of Wheeler-de Witt equation leads to a unique choice for the lattice refinement model.

3.4. Numerical techniques in solving the Hamiltonian constraint

Lattice refinement leads to new dynamical difference equations, which being of a non-uniform step-size, imply technical complications. More precisely, the information needed to calculate the wave-function at a given lattice point is not provided by previous iterations. This becomes clear in the case of two-dimensional wave-functions, as for instance in the study of Bianchi models or black hole interiors. I will present below a method [24] based on Taylor expansions that can be used to perform the necessary interpolations with a well-defined and predictable accuracy.

For a one-dimensional difference equation defined on a varying lattice, the Hamiltonian constraint can be mapped onto a fixed lattice simply by a change of basis [24]. This method is however not useful for the two-dimensional case, where the Hamiltonian constraint is a difference equation on a varying lattice [17]:

$$\begin{aligned} &C_+ (\mu, \tau) \left[\Psi_{\mu+2\delta_\mu, \tau+2\delta_\tau} - \Psi_{\mu-2\delta_\mu, \tau+2\delta_\tau} \right] \\ &+ C_0 (\mu, \tau) \left[(\mu + 2\delta_\mu) \Psi_{\mu+4\delta_\mu, \tau} - 2\left(1 + 2\gamma^2 \delta_\mu^2\right) \mu \Psi_{\mu,\tau} + (\mu - 2\delta_\mu) \Psi_{\mu-4\delta_\mu, \tau} \right] \\ &+ C_- (\mu, \tau) \left[\Psi_{\mu-2\delta_\mu, \tau-2\delta_\tau} - \Psi_{\mu+2\delta_\mu, \tau-2\delta_\tau} \right] = \frac{\delta_\tau \delta_\mu^2}{\delta^3} \mathcal{H}_\phi \Psi_{\mu,\tau} \; , \end{aligned} \tag{45}$$

with

$$C_\pm \equiv 2\delta_\mu \left(\sqrt{|\tau \pm 2\delta_\tau|} + \sqrt{|\tau|} \right) \; , \tag{46}$$

$$C_0 \equiv \sqrt{|\tau + \delta_\tau|} - \sqrt{|\tau - \delta_\tau|} \; , \tag{47}$$

where δ_μ and δ_τ denote the step-sizes along the μ and τ directions, respectively.

In the case of lattice refinement, δ_μ and δ_τ are decreasing functions of μ and τ, respectively, and the data needed to calculate the value of the wave-function at a particular lattice site are not given by previous iterations, as it is illustrated in Fig.2. We propose [24] to use Taylor expansions to calculate the necessary data points. Let us assume that the matter Hamiltonian acts diagonally on the basis states of the wave-function, namely

$$\hat{\mathcal{H}}_\phi |\Psi\rangle \equiv \hat{\mathcal{H}}_\phi \sum_{\mu,\tau} \Psi_{\mu,\tau} |\mu,\tau\rangle = \sum_{\mu,\tau} \mathcal{H}_\phi \Psi_{\mu,\tau} |\mu,\tau\rangle \; . \tag{48}$$

Given a function evaluated at three (non-collinear) coordinates, the Taylor approximation to the value at a fourth position is

$$\begin{aligned} f(x_4, y_4) &= f(x_2, y_2) + \delta^x_{42} \left.\frac{\partial f}{\partial x}\right|_{x_2,y_2} + \delta^y_{42} \left.\frac{\partial f}{\partial y}\right|_{x_2,y_2} \\ &+ \mathcal{O}\left((\delta^x_{42})^2 \left.\frac{\partial^2 f}{\partial x^2}\right|_{x_2,y_2} \right) + \mathcal{O}\left((\delta^y_{42})^2 \left.\frac{\partial^2 f}{\partial y^2}\right|_{x_2,y_2} \right) \; , \end{aligned} \tag{49}$$

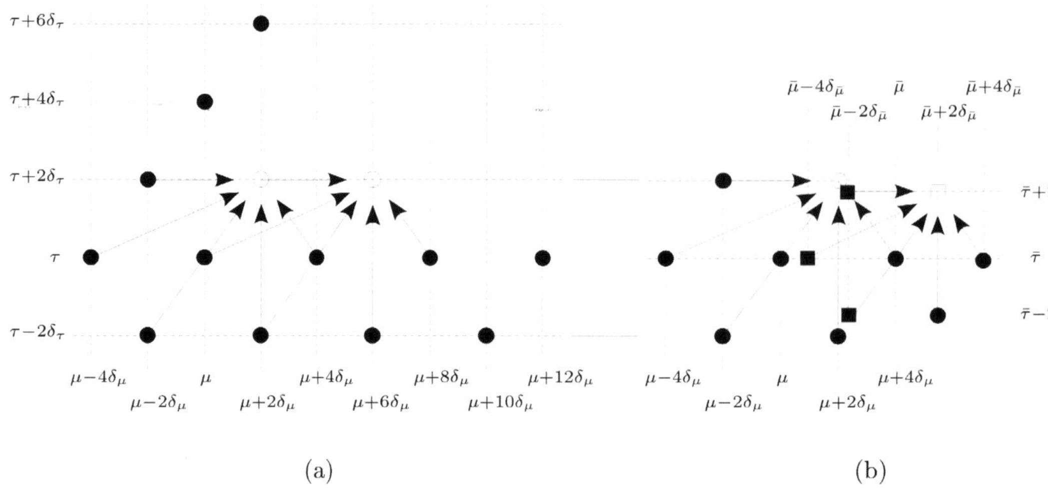

Figure 2. (a) For the fixed lattice case the two-dimensional wave-function can be calculated, given suitable initial conditions (solid circles). (b) In the case of a refining lattice, the data needed to calculate the value of the wave-function at a particular lattice site (open square) is not given by previous iterations (solid squares) [24].

where the Taylor expansion is taken about the position (x_2, y_2), we have defined $\delta^x_{ij} \equiv x_i - x_j$ and $\delta^y_{ij} \equiv y_i - y_j$, and the differentials can be approximated using

$$f(x_1, y_1) = f(x_2, y_2) + \delta^x_{12} \frac{\partial f}{\partial x}\bigg|_{x_2,y_2} + \delta^y_{12} \frac{\partial f}{\partial y}\bigg|_{x_2,y_2} + \cdots , \tag{50}$$

$$f(x_3, y_3) = f(x_2, y_2) + \delta^x_{32} \frac{\partial f}{\partial x}\bigg|_{x_2,y_2} + \delta^y_{32} \frac{\partial f}{\partial y}\bigg|_{x_2,y_2} + \cdots , \tag{51}$$

where the dots indicate higher order terms.

For slowly varying wave-functions, linear approximation is very accurate and higher order terms in Taylor expansion can only improve the accuracy by $10^{-2}\%$ [24]. This method can be applied in any lattice refinement model, while its accuracy can be estimated.

By using this Taylor expansions method, we confirmed [24] numerically the stability criterion of the Schwarzchild interior, found [17] earlier using a von Neumann analysis, and investigated [24] the way that lattice refinement modifies the stability properties of the system.

Let me now show how the underlying discreteness of space-time leads to a twist [24] in the wave-functions, for both a fixed and a varying lattice model. In the case of a constant lattice, an initially centred Gaussian will move to larger μ, as τ is increased [24]. However, for regions where the lattice discreteness is important, this no longer holds, and the value at one lattice point introduces a non-zero component to the value at a lattice point with larger μ coordinate, *i.e.*, the wave-function moves to larger μ. This implies the existence of some induced rotation on the wave-function due to the underlying discreteness of the space-time. Including lattice refinement, this effect persists [24]. In other words, there is no motion for $\tau \gg \delta_\tau$, while in the case of lattice refinement this requirement is reached for lower τ, since δ_τ reduces as τ increases. As the wave-function moves into a region in which the discreteness of the lattice is important, a motion will be induced, as in the constant lattice case.

4. Anisotropic LQC

Various aspects of anisotropic cosmologies have been studied within LQC in the past [17, 25], however the first full and consistent quantisation of a Bianchi I cosmology (the simplest of anisotropic cosmological models) was achieved recently in Ref. [26]. Moreover, the link back to the underlying full LQG theory has been strengthened, by considering the flux of the triads through surfaces consistent with the Bianchi I anisotropic case. With the quantisation of the Bianchi I model under control, it is possible to ask whether LQC features, obtained within the context of isotropic Friedmann-Lemaître-Roberston-Walker cosmology, are robust, at least with respect to this limited extension of the symmetries of the system. Bianchi type I models, apart their simplicity, they present a particular interest for space-like singularities within the full LQG theory. Following the same vein as for the isotropic case, a massless scalar field plays the rôle of an internal time parameter. In the absence of such a field, physical evolution can be found by constructing families of unitarily related partial observables, parametrised by geometry degrees of freedom [27].

4.1. Unstable Bianchi I LQC

Studying stability conditions of the full Hamiltonian constraint equation describing the quantum dynamics of the diagonal Bianchi I model, I will show [28] that there is robust evidence of an instability in the explicit implementation of the difference equation. On the one hand, such a result may question the choice of the quantisation approach, the model of lattice refinement, and/or the rôle of ambiguity parameters. On the other hand, one may argue that such an instability may not be necessarily a problem since it might be that unstable trajectories are explicitly removed by the physical inner product.

Consider the difference equation arising from the loop quantisation of the Bianchi I model [26]:

$$\begin{aligned}\partial_T^2\Psi\left(\lambda_1,\lambda_2,\nu;T\right) &= \frac{\pi G}{2}\sqrt{\nu}\Big[(\nu+2)\sqrt{\nu+4}\Psi_4^+\left(\lambda_1,\lambda_2,\nu;T\right)-(\nu+2)\sqrt{\nu}\Psi_0^+\left(\lambda_1,\lambda_2,\nu;T\right)\\ &\quad -(\nu-2)\sqrt{\nu}\Psi_0^-\left(\lambda_1,\lambda_2,\nu;T\right)+(\nu-2)\sqrt{|\nu-4|}\Psi_4^-\left(\lambda_1,\lambda_2,\nu;T\right)\Big] \quad (52)\end{aligned}$$

where

$$\begin{aligned}\Psi_4^+\left(\lambda_1,\lambda_2,\nu;T\right) &= \sum_{i\neq j=(0,1,2)}\Psi\left(a_i\lambda_1,a_j\lambda_2,\nu+4;T\right)\\ \Psi_4^-\left(\lambda_1,\lambda_2,\nu;T\right) &= \sum_{i\neq j=(-3,-2,0)}\Psi\left(a_i\lambda_1,a_j\lambda_2,\nu-4;T\right)\\ \Psi_0^+\left(\lambda_1,\lambda_2,\nu;T\right) &= \sum_{i\neq j=(-1,0,1)}\Psi\left(a_i\lambda_1,a_j\lambda_2,\nu;T\right)\\ \Psi_0^-\left(\lambda_1,\lambda_2,\nu;T\right) &= \sum_{i\neq j=(-2,0,3)}\Psi\left(a_i\lambda_1,a_j\lambda_2,\nu;T\right)\ ,\end{aligned} \quad (53)$$

$$\begin{aligned}&a_{-3}\equiv\left(\frac{\nu-4}{\nu-2}\right)\ ,\quad a_{-2}\equiv\left(\frac{\nu-2}{\nu}\right)\ ,\quad a_{-1}\equiv\left(\frac{\nu}{\nu+2}\right)\ ,\\ a_0\equiv 1\ ,\quad &a_1\equiv\left(\frac{\nu+4}{\nu+2}\right)\ ,\quad a_2\equiv\left(\frac{\nu+2}{\nu}\right)\ ,\quad a_3\equiv\left(\frac{\nu}{\nu-2}\right)\ .\end{aligned} \quad (54)$$

Note that $\lambda_1, \lambda_2, \lambda_3$ are related to the volume of the fiducial cell $\mathcal{V}$ by

$$\hat{V}\Psi(\lambda_1,\lambda_2,\nu)=2\pi|\gamma|\sqrt{\Delta}|\nu|l_{\rm pl}^3\Psi(\lambda_1,\lambda_2,\nu)\ , \quad (55)$$

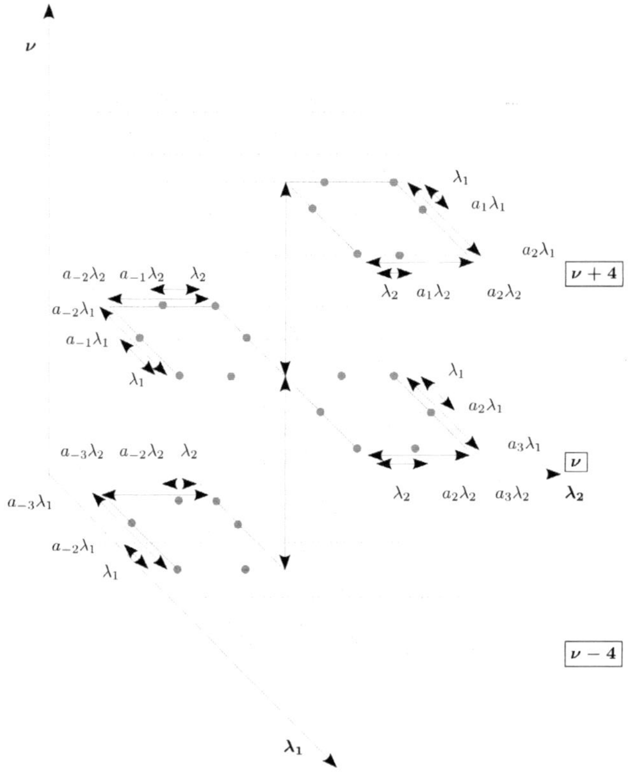

Figure 3. The geometry of the points used in the difference equation that results from the Hamiltonian constraint, for the Bianchi I model [28].

with $\gamma = \text{sgn}(p_1 p_2 p_3)|\gamma|$ and

$$\nu = 2\lambda_1\lambda_2\lambda_3 \ , \tag{56}$$

We will investigate the stability of the vacuum solutions, in which case the solution is static, namely $\Psi(\lambda_1, \lambda_2, \nu; T) = \Psi(\lambda_1, \lambda_2, \nu)$, simplifying considerably Eq. (52), whose geometry is drawn in Fig. 3. To search for growing mode solutions we will perform a von Neumann analysis. In addition to specifying the boundary conditions on the ν and $\nu - 4$ planes, we are also required to specify the value at five of the points given in $\Psi_4^+(\lambda_1, \lambda_2, \nu)$. There are in total 23 values that are required and with such initial data the difference equation, Eq. (52), can be used to evaluate the 24th point. Once this point has been evaluated, it can be used to *move* the central point and evaluate the wave-function at subsequent positions in the $\nu + 4$ plane.

Following the standard von Neumann stability analysis, we decompose the solutions of the difference equation into Fourier modes and look for growing modes. Considering the ansatz [28]

$$\Psi(\lambda_1, \lambda_2, \nu) = T(\lambda_1) \exp\left(i\left(\omega\lambda_2 + \chi\nu\right)\right) \ , \tag{57}$$

where we have chosen the λ_1 direction to be the direction in which the $\nu + 4$ plane is *evolved*, the difference equation becomes

$$\begin{aligned} e^{4\chi i} \sum_{i\neq j=(0,1,2)} T(a_i\lambda_1)\, e^{i(\omega a_j\lambda_2+\chi\nu)} &= \sqrt{\frac{\nu}{\nu+4}} \sum_{i\neq j=(-1,0,1)} T(a_i\lambda_1)\, e^{i(\omega a_j\lambda_2+\chi\nu)} \\ &+ \left(\frac{\nu-2}{\nu+2}\right)\sqrt{\frac{\nu}{\nu+4}} \sum_{i\neq j=(-2,0,3)} T(a_i\lambda_1)\, e^{i(\omega a_j\lambda_2+\chi\nu)} \\ &- \left(\frac{\nu-2}{\nu+2}\right)\sqrt{\frac{|\nu-4|}{\nu+4}} e^{-4\chi i} \sum_{i\neq j=(-3,-2,0)} T(a_i\lambda_1)\, e^{i(\omega a_j\lambda_2+\chi\nu)} \ . \end{aligned} \tag{58}$$

As it has been explicitly shown in Ref. [28], expanding in terms of small $1/\nu$, the difference equation can be written in the form of a vector equation as

$$M_1\overline{T}_3 = M_2\overline{T}_2 \ , \tag{59}$$

where we have defined the vectors

$$\overline{T}_i = \begin{bmatrix} T(a_i\lambda_1) \\ T(a_{i-1}\lambda_1) \\ T(a_{i-2}\lambda_1) \\ T(a_{i-3}\lambda_1) \\ T(a_{i-4}\lambda_1) \\ T(a_{i-5}\lambda_1) \end{bmatrix} \quad \text{for } i = 2,3 \tag{60}$$

and the matrices

$$M_1 = \begin{pmatrix} A & 0 & 0 & 0 & 0 & 0 \\ 0 & 1 & 0 & 0 & 0 & 0 \\ 0 & 0 & 1 & 0 & 0 & 0 \\ 0 & 0 & 0 & 1 & 0 & 0 \\ 0 & 0 & 0 & 0 & 1 & 0 \\ 0 & 0 & 0 & 0 & 0 & 1 \end{pmatrix}, \quad M_2 = \begin{pmatrix} B & C & D & E & F & G \\ 1 & 0 & 0 & 0 & 0 & 0 \\ 0 & 1 & 0 & 0 & 0 & 0 \\ 0 & 0 & 1 & 0 & 0 & 0 \\ 0 & 0 & 0 & 1 & 0 & 0 \\ 0 & 0 & 0 & 0 & 1 & 0 \end{pmatrix} . \tag{61}$$

The condition for stability can be written as follows: If

$$\max|\tilde{\lambda}| \leq 1 \quad \forall \ \omega \text{ and } \chi \ , \tag{62}$$

where $\tilde{\lambda}$ are the eigenvalues of the matrix $(M_1)^{-1} M_2$, the amplitude $T(a_3\lambda_1)$ is less than that of previous points, in other words, the difference equation is stable. Let us define the parameter $\Lambda = \omega\lambda_2/\nu$. The presence of an instability in the difference equation has been demonstrated [28] in several ways: (i) there is a particular set of critical modes, $\Lambda = (2n-1)\pi/2$, with $n \in \mathbb{Z}$, for which the system is unstable; (ii) in the large ν limit, the system is unstable for the modes $\Lambda = \pi/4$ and $\chi = 0$; (iii) the system is unstable for a general ν, for modes that approach the critical value.

In conclusion, the difference equation, Eq. (52), is unconditionally unstable, in the sense that there is no region of $(\lambda_1, \lambda_2, \nu)$ in which Eq. (52) is stable.

4.2. Lattice refinement from isotropic embedding of anisotropic cosmology

Given a consistent quantum anisotropic model, one can find [29] isotropic states for which the discrete step-size of the isotropically embedded Hamiltonian constraint is not necessarily that of the $p^{-1/2}$ (*new quantisation*) step-size. The choice of different embeddings has important consequences for the precise form of discretisation in the isotropic sub-system. In this sense, lattice refinement could be interpreted as being due to the degrees of freedom that are absent in the isotropic model [29].

In the standard approach, isotropic states are taken to be those in which the three scale factors are $\lambda_1 = \lambda_2 = \lambda_3$, or more precisely, defining the volume of the state as $\nu = 2\lambda_1\lambda_2\lambda_3$ to eliminate one of the directions (λ_3), the map

$$|\Psi(\lambda_1, \lambda_2, \nu)\rangle \to \Big| \sum_{\lambda_1,\lambda_2} \Psi(\lambda_1, \lambda_2, \nu)\rangle \equiv |\Psi(\nu)\rangle \ , \tag{63}$$

produces isotropic states. Working with the three scale factors λ_i, one can show [29] that there is an ambiguity in exactly what the volume of such an isotropic state is. Consider the state

$$|\tilde{\Psi}(\lambda_1, \lambda_2, \lambda_3)\rangle \equiv \frac{1}{A}\Big[|\Psi(\lambda_1, \lambda_2, \lambda_3)\rangle + |\Psi(\lambda_3, \lambda_1, \lambda_2)\rangle + |\Psi(\lambda_2, \lambda_3, \lambda_1)\rangle\Big] \ , \tag{64}$$

with the restriction

$$\langle\Psi(\lambda_1, \lambda_2, \lambda_3)|\Psi(\lambda_3, \lambda_1, \lambda_2)\rangle = \langle\Psi(\lambda_1, \lambda_2, \lambda_3)|\Psi(\lambda_2, \lambda_3, \lambda_1)\rangle = \langle\Psi(\lambda_3, \lambda_1, \lambda_2)|\Psi(\lambda_2, \lambda_3, \lambda_1)\rangle \ ,$$

on the anisotropic states. The expectation values of the scale factors along each direction of such a state are

$$\langle\hat{\lambda}_i\rangle = \frac{\lambda_1 + \lambda_2 + \lambda_3}{3} \ . \tag{65}$$

The measured scale factor is equal in each direction and is given by the average of the scale factors of the underlying, anisotropic states. However, the measured volume of such a state is just $\nu = 2\lambda_1\lambda_2\lambda_3$, which is *not* the cube of the measured scale factor. Thus, while it is the eigenvalue of the anisotropically defined volume operator, it is not necessarily what we would measure as the volume. Essentially, the reason for this is that while the scale factors λ_i are measured to be equal in each direction, they are not eigenvalues of the state, *i.e.*, $\hat{\lambda}_i|\tilde{\Psi}(\lambda_1, \lambda_2, \lambda_3)\rangle \neq \lambda_i|\tilde{\Psi}(\lambda_1, \lambda_2, \lambda_3)\rangle$. However, both the average and the product (*i.e.*, the volume) of the scale factors are eigenvalues. It is this ambiguity, that leads to the possibility of deviations from the standard isotropic case.

In conclusion, the difference between the two procedures is essentially due to what one considers to be more fundamental, the volume of the underlying states (ν) or the measured volume of the symmetric state ($\langle\lambda\rangle^3$), which are not necessarily equal. Choosing ν leads to the *new* quantised Hamiltonian of isotropic cosmology, while choosing $\langle\lambda\rangle^3$ results in some kind of different *lattice refinement*. This *lattice refinement* is significantly more complicated that the single power law behaviour, p^A, usually considered.

5. Conclusions

Loop Quantum Gravity proposes a method of quantising gravity in a background independent, non-perturbative way. Quantum gravity is essential when curvature becomes large, as for example in the early stages of the evolution of the universe. Applying LQG in a cosmological context leads to Loop Quantum Cosmology which is a symmetry reduction of the infinite dimensional phase space of the full theory, allowing us to study certain aspects of the theory analytically. The discreteness of spatial geometry, a key element of the full theory, leads to successes in LQC which do not hold in the Wheeler-de Witt quantum cosmology.

By studying phenomenological consequences of LQC we can get some useful insight about the full LQG theory and get an answer to some long-standing questions in early universe cosmology. Here, I have briefly described some of the phenomenological aspects of LQC.

Acknowledgments
It is a pleasure to thank the organisers of the First Mediterranean Conference on Classical and Quantum Gravity (MCCQG) (http://www.phy.olemiss.edu/mccqg/), for inviting me to give this talk, in the beautiful island of Crete in Greece. This work is partially supported by the European Union through the Marie Curie Research and Training Network *UniverseNet* (MRTN-CT-2006-035863).

References
[1] Guth A 1981 *Phys. Rev. D* **23** 347
[2] Sakellariadou M 2008 *Lect. Notes Phys.* **738** 359
[3] Piran T 1986 *Phys. Lett. B* **181** 238
[4] Goldwirth D 1991 *Phys. Rev. D* **43** 3204
[5] Calzetta E and Sakellariadou M 1992 *Phys. Rev. D* **45** 2802
[6] Calzetta E and Sakellariadou M 1993 *Phys. Rev. D* **47** 3184
[7] Germani C, Nelson W and Sakellariadou M 2007 *Phys. Rev. D* **76** 043529
[8] Gibbons G and Turok N 2008 *Phys. Rev. D* **77** 063516
[9] Ashtekar A and Sloan D 2009 Loop quantum cosmology and slow roll inflation (*Preprint* 0912.4093)
[10] Rovelli C 2004 *Quantum Gravity* (Cambridge: Cambridge University Press)
[11] Thiemann T 2007 *Modern Canonical Quantum General Relativity* (Cambridge: Cambridge University Press)
[12] Bojowald M 2005 *Living Rev. Rel.* **8** 11
[13] Ashtekar A and Lewandowski J 1997 *Class. & Quant. Grav.* **14** A55
[14] Ashtekar A, Pawlowski T and Singh P 2006 Phys. Rev. D **74** 084003
[15] Vandersloot K 2005 *Phys. Rev. D* **71** 103506
[16] Rosen J, Jung J H and Khanna G 2006 *Class. Quant. Grav.* **23** 7075
[17] Bojowald M, Cartin D and Khanna G 2007 Phys. Rev. D **76** 064018
[18] Sakellariadou M 2009 *J. Phys. Conf. Ser.* **189** 012035
[19] Nelson W and Sakellariadou M (2007) *Phys. Rev. D* **76** 104003
[20] Nelson W and Sakellariadou M 2007 *Phys. Rev. D* **76** 044015
[21] Corichi A and Singh P 2008 *Phys. Rev. D* **78** 024034
[22] Nelson W and Sakellariadou M 2008 *Phys. Rev. D* **78** 024006
[23] Ashtekar A, Pawlowski T and Singh P 2006 *Phys. Rev. D* **74**, 084003
[24] Nelson W and Sakellariadou M 2008 *Phys. Rev. D* **78** 024030
[25] Chiou D W and Vandersloot K 2007 *Phys. Rev. D* **76** 084015
[26] Ashtekar A and Wilson-Ewing E 2009 *Phys. Rev. D* **79** 083535
[27] Martin-Benito M, Mena Marugan G A and Pawlowski T 2009 *Phys. Rev. D* **80** 084038
[28] Nelson W and Sakellariadou M 2009 *Phys. Rev. D* **80** 063521
[29] Nelson W and Sakellariadou M 2009 Lattice Refining Loop Quantum Cosmology from an Isotropic Embedding of Anisotropic Cosmology (*Preprint* 0906.0292)

First Mediterranean Conference on Classical and Quantum Gravity (MCCQG 2009) IOP Publishing
Journal of Physics: Conference Series **222** (2010) 012028 doi:10.1088/1742-6596/222/1/012028

Space-time as a deformable continuum

A. Tartaglia and N. Radicella

Dipartimento di Fisica, Politecnico, Corso Duca degli Abruzzi 24, I-10129 Torino, Italy and INFN, sezione di Torino

E-mail: `angelo.tartaglia@polito.it`, `ninfa.radicella@polito.it`

Abstract. Space-time may be thought of as a physical continuum endowed with properties similar to the ones of material threedimensional continua. In this view a non-trivial metric tensor can be considered to be the sum of the Minkowski metric plus an appropriate strain tensor. The global symmetry of the universe can be seen as the effect of a spontaneous strained state due to the presence of a texture defect. Consistently with this approach the Lagrangian of space time is obtained adding to the scalar curvature, acting as a kinetic term, a potential term depending on the strain and modeled on the one of the elasticity theory, extended to four dimensions. The theory is applied to the fit of the luminosity dependence of type Ia supernovae on the redshift. A result is obtained slightly better than the one of the ΛCDM theory.

1. Introduction

The vision of the cosmos we have, after one century of General Relativity (GR), is essentially a dual one: the universe is assumed to be made of two basic ingredients, space-time and matter/energy. The situation is perfectly well expressed by the Einstein equations:

$$G_{\mu\nu} = \kappa T_{\mu\nu} \tag{1}$$

The left hand side of the equations (the "marble side", according to Einstein) describes space-time and its properties, the right hand side (the "wooden side", in Einstein's words again) is matter/energy. We think we are familiar with the right hand side, because it contains the stuff our bodies and everything is made of, but, despite its cleaner mathematical description, the left hand side has a rather ambiguous status. Space-time has indeed to be real, since it produces real and fundamental effects, but is different from matter to which it couples giving rise to the gravitational interaction. Actually the attitude of the scientific community towards space-time oscillates between considering it more or less as an useful mathematical device, and treating it as a peculiar field interacting geometrically with any other field and trying to quantize it. Coming to the observations, we see that, despite the tremendous advances of scientific cosmology during the 20th century, recent years have cumulated a number of facts which remain not fully understood or, at least, which can be explained introducing "dark" components of the universe: both dark matter, producing gravitational effects, but not interacting with electromagnetism, and dark energy, producing the accelerated expansion of the universe without any self-gravitational behaviour. The use of "dark" components may just be the anticipation of something real that will soon be discovered but is also the occasion for developing a host of more or less conjectural theories, where the internal mathematical consistency is the main requirement, whereas the link to experiment and observation is some times impossible or at least dubious.

The subject of this paper will be space-time with its nature and properties. The approach we have adopted, in order to avoid an excessive use of conjectures, has been to move from some theory which is already known for situations that could look like space-time with its geometrical properties. Actually space-time is indeed a physical entity, it is continuous (as far as we treat it classically) and displays geometrical properties. The starting point has then been the description of ordinary three-dimensional material continua with their internal strains and stresses. The existing theory of these ordinary continua is not per se relativistic so we had to generalize it to four dimensions and the Lorentzian signature. This generalization has been done in the first part of this work. Then, after introducing also the concept of structural defect, again in analogy to ordinary material continua, our theory has been applied to the cosmic scale and the evolution of the universe. The result, tested on the luminosity curve of type Ia supernovae, has been good and even a little bit better than the one obtained by the mostly used ΛCDM theory.

2. N-dimensional "elastic" continua

Our approach to the description of general N-dimensional continua is summarized in fig.1. We start from an $(N+n)$-dimensional Minkowskian space[1] which will be used to embed the manifolds we want to use and describe; let X^a be the coordinates we use to locate positions in the embedding manifold, with the label a ranging from 1 to N+n. Our embedding space $\{X\}$ contains two different N-dimensional spaces. One, which is called the reference manifold [1], is flat; the second, named natural manifold, is curved. Each manifold is equipped with appropriated coordinates: let us call ξ^α the ones of the reference manifold, $\{\xi\}$, and x^μ the coordinates on the natural manifold, $\{x\}$; Greek indices range from 1 to N. The two N-dimensional Riemannian manifolds are geometrically defined by a set of conditions allowing for the dimensional reduction, such as

$$f_i(X^1, X^2, ...X^{(N+n)}) = 0 \tag{2}$$

for the reference manifold and

$$h_i(X^1, X^2, ...X^{(N+n)}) = 0 \tag{3}$$

for the natural one; in both cases it is $n = 1, 2, ...n$.

We assume that all functional relations are smooth enough for all subsequent purposes; this means that it is always possible to go from one set of coordinates to the other without troubles. Physically we may think to obtain the natural manifold deforming the reference one, which means that the natural manifold will be considered as a strained version of the reference one. In the case of a material support the deformation process will imply the presence of stresses either externally or internally generated. The above ideal process can be formally described introducing a displacement vector field in the flat embedding manifold: the **u**'s in fig. 1. The flatness of the embedding allows for global definitions of vectors. If **r** points at an event in $\{\xi\}$ and **r'** localizes an event in $\{x\}$ it will be

$$\mathbf{u} = \mathbf{r'} - \mathbf{r} \tag{4}$$

The displacement vector field may be expressed either in terms of the ξ's or of the x's since the end points of the arrow are on corresponding points of the two manifolds. An actual strain is present when the **u**-field is non-uniform. This fact is better seen considering corresponding line elements on the two manifolds and using Eqs. (4), (2) and (3). On the natural manifold we have:

$$ds'^2 = \eta_{ab} X'^a X'^b|_{nat} \rightarrow g_{\mu\nu} dx^\mu dx^\nu \tag{5}$$

[1] In what follows it will always be $(N+1)$-dimensional, but there are curved manifolds which require more than 1 extradimensions in order to be embedded in a flat manifold.

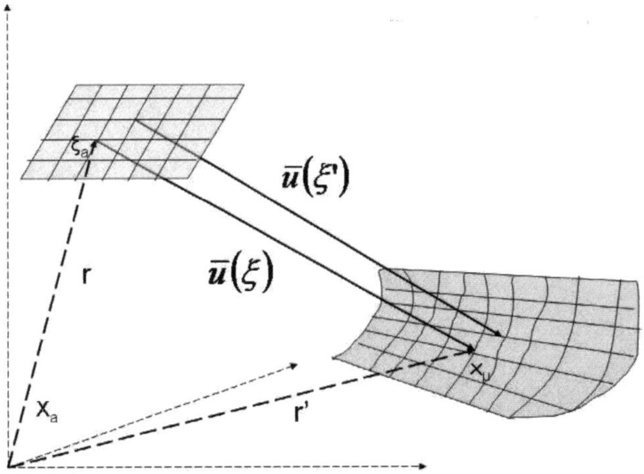

Figure 1. Embedding of the reference and the natural manifolds in an $N+1$ - dimensional flat manifold. X_a are the Cartesian coordinates in the embedding manifold; ξ_μ are the coordinates on the reference (sub)-manifold (N - dimensional), x_ν are the coordinates in the natural (generally curved) (sub)-manifold. **u** represents the displacement vector from points of the reference manifold to points of the natural manifold.

and on the reference manifold it is

$$ds^2 = \eta_{ab} X^a X^b|_{ref} \rightarrow \eta_{\mu\nu} dx^\mu dx^\nu. \tag{6}$$

The symbol η labels the Minkowski metric tensor expressing the flatness both of the embedding $(N+1)$-dimensional and the reference N-dimensional manifold. The natural manifold is in general curved and its metric tensor is g. For convenience everything has been expressed in terms of the x's. Comparing (5) with (6) it is easy to see that

$$g_{\mu\nu} = \eta_{\mu\nu} + 2\varepsilon_{\mu\nu} \tag{7}$$

being

$$\varepsilon_{\mu\nu} = \frac{1}{2}(\eta_{a\mu}\frac{\partial u^a}{\partial x^\nu} + \eta_{b\nu}\frac{\partial u^b}{\partial x^\mu} + \eta_{ab}\frac{\partial u^a}{\partial x^\mu}\frac{\partial u^b}{\partial x^\nu}) \tag{8}$$

the strain tensor of the natural manifold.

3. "Elasticity" and the role of defects

If it is globally possible to write

$$g_{\mu\nu} = \eta_{\alpha\beta}\frac{\partial \xi^\alpha}{\partial x^\mu}\frac{\partial \xi^\beta}{\partial x^\nu} \tag{9}$$

we have a diffeomorphism and we can verify that it is also

$$R^\lambda_{\mu\nu\rho} = 0. \tag{10}$$

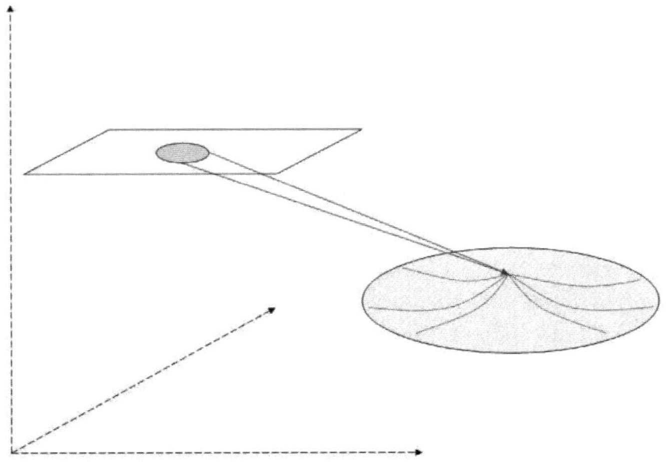

Figure 2. We have a defect whenever an entire region of the reference manifold corresponds to a point (or other lower dimensional variety) in the natural manifold, or viceversa.

$R^{\lambda}_{\mu\nu\rho}$ is the Riemann curvature tensor and eq. (10) corresponds to De Saint Venant's integrability condition for (9). In practice this means that the curvature of the natural manifold cannot be felt from within (using intrinsic coordinates, i.e. the x's); this is the typical situation of a pure and global elastic deformation. In the case of space-time the non-trivial part of the metric tensor, which, in our approach, corresponds to the strain tensor of the manifold, contains the gravitational interaction, so we see that a real gravitational field exists only when the displacement $\mathbf{u}$ field is singular. The presence of some singularity in $\mathbf{u}$ prevents (9) to hold globally and expresses a well known result of general relativity. The relevance of singularities in the displacement field lends the opportunity to introduce in our theory another ingredient of the classical theory of material continua: the defects. It is again convenient to have a look first to a graphic schematization of what a defect is; it can be found in fig.2. We say that a defect is present whenever a whole region of the reference manifold corresponds to a (less than N)-dimensional variety in $\{x\}$, or viceversa. This definition of defects is consistent with the one given by Volterra [2] at the beginning of the 20th century while studying elastic and plastic deformations in solids.

Formally the passage from the coordinates on $\{\xi\}$ to the ones on $\{x\}$ is written

$$dx^{\mu} = \omega^{\mu}_{\alpha} d\xi^{\alpha}. \tag{11}$$

If $\omega^{\mu}_{\alpha} = \frac{\partial x^{\mu}}{\partial \xi^{\alpha}}$ we have a diffeomorphism (even though the coordinates are on different manifolds). Otherwise ω^{μ}_{α} is a general 1-form and

$$\oint \omega^{\mu}_{\alpha} d\xi^{\alpha} \neq 0. \tag{12}$$

Condition (12) is typical of defects known as dislocations; other similar conditions using for instance 2-forms lead to other kinds of defects.

In practice the presence of a defect produces two effects: a spontaneous strained state and a lower symmetry of the natural manifold. Actually in GR the two above effects are caused by matter/energy; here we introduce a new type of source, which is intrinsic to space-time.

4. Generalized Lagrangian for space-time

The concept of strain in the manifold suggests a Lagrangian slightly different from the one commonly used in GR. In fact a typical Lagrangian for classical problems has two additive terms: one contains time derivatives and is known as the kinetic term; the other is a function of the coordinates and represents the potential energy in the system. Taking our view of space-time as a deformable 4-dimensional continuum seriously, we are led to interpret the usual Einstein-Hilbert Lagrangian density, i.e. the scalar curvature R, as the "kinetic" term, since it contains linearly second order derivatives with respect to the Lagrangian coordinates (i.e. the elements of the metric tensor); a potential energy term would be missing. Following our analogy we can easily build such potential energy term borrowing the form valid for ordinary three-dimensional elastic materials. It would be:

$$L_e = \frac{1}{2}\sigma_{\mu\nu}\varepsilon^{\mu\nu} \tag{13}$$

L_e contains the elements of the stress tensor $\sigma_{\mu\nu}$. Stresses are the expression of the causes of the strain; in ordinary conditions they depend on forces and defects, in GR they will depend on matter/energy and again on defects. Stresses and strains are not independent from each other; in the linear theory of elasticity they are mutually proportional. In the case of space-time we cannot a priori say whether the theory is linear or not, however let us assume it is (alternatively we may always think this to be the lowest approximation order), so we shall write:

$$\sigma_{\mu\nu} = C_{\mu\nu\alpha\beta}\varepsilon^{\alpha\beta} \tag{14}$$

$C_{\mu\nu\alpha\beta}$ are the elements of the elastic modulus tensor which expresses the properties of the material continuum under consideration: in our case this is space-time. Eq. 14 is the tensorial version of Hooke's law. If the material continuum is isotropic the elements of the elastic modulus tensor depend on two parameters only. As far as the natural manifold admits a tangent Minkowskian space we may assume local isotropy for space-time too, so we are allowd to write

$$C_{\mu\nu\alpha\beta} = \lambda\eta_{\mu\nu}\eta_{\alpha\beta} + \mu(\eta_{\mu\alpha}\eta_{\nu\beta} + \eta_{\mu\beta}\eta_{\nu\alpha}) \tag{15}$$

The two independent parameters λ and μ are known as the Lamé coefficients. Now introducing Eq. (15) into Eq. (13) we get

$$L_e = \frac{1}{2}(\lambda\epsilon^2 + 2\mu\varepsilon_{\alpha\beta}\varepsilon^{\alpha\beta}). \tag{16}$$

In order to raise and lower indices the full metric tensor (7) is used; $\epsilon = \varepsilon^{\alpha}_{\alpha}$ is the trace of the strain tensor.

Finally the full action integral for space-time in presence of matter will be:

$$S = \int (R + L_e + \kappa L_m)\sqrt{-g}d^4x. \tag{17}$$

L_m is the usual matter term; $\kappa = 16\pi G/c^2$ is the coupling constant between matter and geometry; g is the determinant of the metric tensor. Both R and L_e are geometry, even though applying the usual variational procedure it is possible to write the Einstein's equations in the form

$$G_{\mu\nu} = T_{e\mu\nu} + \kappa T_{\mu\nu}. \tag{18}$$

The tensor $T_{e\mu\nu}$ comes from L_e and appears as a new source of curvature (due to the presence of strain), written on the right of the equations rather than as being part of the geometry of the manifold on the left.

5. Cosmological application

It is commonly accepted that the universe has a global Robertson-Walker (RW) symmetry based on space homogeneity and isotropy. Considering space-time the RW symmetry is not obvious and is not a direct consequence of the matter content, so why is it there? Applying our theory the natural candidate to fix the global symmetry of the manifold is a cosmic defect (CD) corresponding to the initial singularity or big bang. In a sense the CD gives the "container" (i.e. the space-time) a predefined "shape", or, to say better, strain field; matter then couples to the given manifold with its global symmetry, contributing additional and local curvature.

5.1. A closed Robertson-Walker space-time

A useful example of the application of our theory to cosmology may be given studying a closed RW space-time. The situation is sketched in fig.3. The bell-shaped surface is obtained from a plane by cutting a portion out of it, then sewing the rims together. The corresponding defect induces an axial symmetry when seen from the embedding three-dimensional flat space. The most appropriate coordinates are cylindrical: z, r and σ. The picture is three-dimensional, however it represents a four-dimensional situation if we assume σ to be a bi-dimensional surface element. The embedding space is assumed to be Minkowskian and z is a time-like variable. In this way the reference flat manifold is the $z = constant$ space-like Euclidean (hyper)-plane. For the natural curved manifold r and σ are the same as for the flat one, but it is $z = f(r)$ being f some regular (except, possibly, at the origin) function of r.

The line element on the reference manifold is

$$dl^2 = -dr^2 - r^2 d\sigma^2, \tag{19}$$

the corresponding line element on the natural manifold is

$$ds^2 = dz^2 - dr^2 - r^2 d\sigma^2 = (f'^2 - 1)dr^2 - r^2 d\sigma^2. \tag{20}$$

f' is the derivative of f with respect to r. Subtracting (19) from (20) we can read out the strain tensor in the embedding coordinates. The only non-zero component is

$$\varepsilon_{rr} = \frac{f'^2}{2}. \tag{21}$$

In order to convert everything to the natural coordinates we put

$$\sqrt{f'^2 - 1}dr = d\tau \tag{22}$$

so defining the "radial" (time-like) coordinate on the natural manifold, τ (the cosmic time). It will also be $r = a(\tau)$ and the line element on the natural manifold assumes the typical RW form:

$$ds^2 = d\tau^2 - a^2 d\sigma^2 \tag{23}$$

In the natural coordinates the only non-null component of the strain tensor becomes

$$\varepsilon_{\tau\tau} = \frac{1 - \dot{a}^2}{2} \tag{24}$$

being $\dot{a}$ the derivative of the a function with respect to τ. The situation represented in fig.3 corresponds, for reasons of graphical clarity, to a closed space (finite $z = constant$ sections), however the outlined method can equally well be applied to open and critical RW space-times.

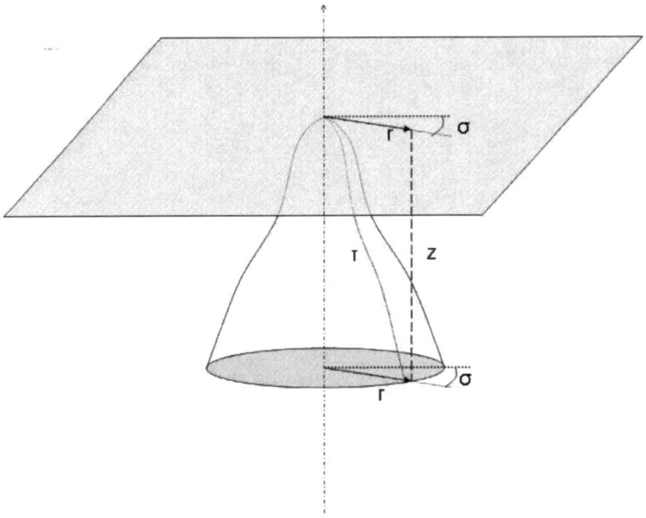

Figure 3. A curved surface with a central symmetry is embedded in a three-dimensional manifold. The reference frame is a plane and the global coordinates are cylindrical. The situation reproduces, in a three-dimensional view, a closed RW space-time

5.2. The accelerated expansion of the universe

If we wish to describe our universe we must start from the fact that it appears to be flat in space. This situation can be dealt with using the same procedure outlined in the previous subsection (details may be read in [3]). Using Cartesian coordinates for space, one obtains three equal non-zero elements of the strain tensor:

$$\varepsilon_{xx} = \varepsilon_{yy} = \varepsilon_{zz} = \frac{1-a^2}{2}. \tag{25}$$

Once the global symmetry has been fixed and the strain tensor is at hands, we are able to compute the L_e to be introduced into (16). The final explicit Lagrangian density turns out to be

$$L = -6(a\ddot{a} + \dot{a}^2) + \frac{9}{8}B\frac{(1-a^2)^2}{a} + \kappa L_m. \tag{26}$$

The Lamé coefficient appear to be combined into the bulk modulus B:

$$B = \lambda + \frac{2}{3}\mu. \tag{27}$$

The only variable is cosmic time and the only unknown function is a. We need to specify the matter Lagrangian. Considering the possibility of having a number of different components, we can deduce from (26) the equation:

$$H^2 = \frac{\dot{a}^2}{a^2} = \frac{9}{16}B\frac{(1-a^2)^2}{a^4} + \kappa\sum_i \rho_{i0}\frac{a_0^{3(1+w_i)}}{a^{3(1+w_i)}}. \tag{28}$$

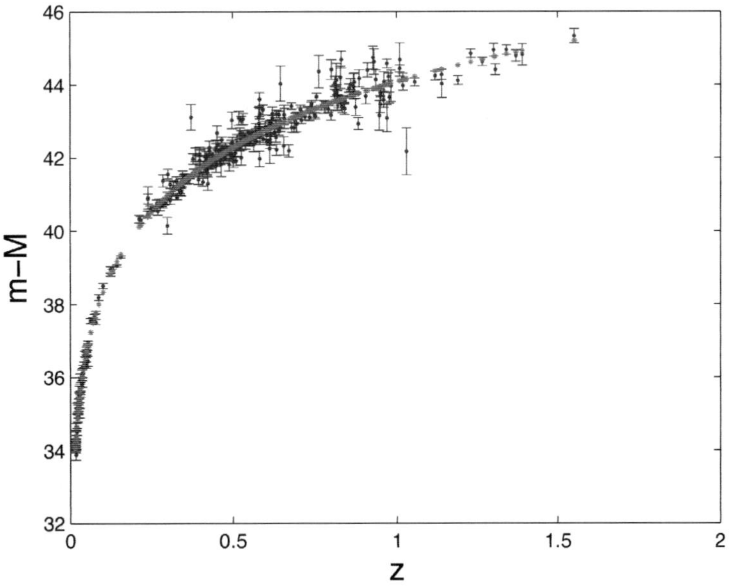

Figure 4. Fit of the luminosity data from 307 type Ia supernovae obtained applying the CD theory. Three optimization parameters have been used. The horizontal error bars are not visible at scale of the graph.

The summation index i runs from 1 to the number of different matter components; the ϱ's are matter/energy densities; w's are equation of state parameters; the $_0$ index labels present day values.

We shall apply (28) to the fit of the luminosity data from type Ia supernovae (SnIa) from which the accelerated expansion of the universe has been discovered [4]. For that purpose we have considered two components only: dust, for which it is $w = 0$, and radiation for which $w = -1/3$. The latter contribution, however, turns out to be negligible since the highest redshift factor z for an SnIa is less than 1.8. The observable quantity related to (28) is the so called distance modulus [5] of the supernova:

$$m - M = 25 + 5\log\left((1+z)\int_0^z \frac{dz'}{H(z')}\right) \tag{29}$$

The observed magnitude is m, the absolute magnitude is M; the link between z and the scale factor a is $a = \frac{a_0}{1+z}$; all distances must be expressed in Mpc.

The fit of the experimental observations has been obtained by means of an optimization method applied to a set of 307 SnIa's [4]. The optimization parameters were a_0, ϱ_0 and B, variously combined. The result is shown in fig. 4.

The reduced χ^2 of the fit is 1.017, slightly better than the value (1.019) obtained, with the same number of parameters, using the ΛCDM theory. The optimal values for the Hubble parameter H_0 and the present matter density ϱ_0 are within the range of commonly accepted ones, with big uncertainties coming from the dispersion of the actual data. The value obtained for the bulk modulus is

$$B = (3 \pm 2) \times 10^{-7}\ \mathtt{Mpc}^{-2} = (3 \pm 2) \times 10^{-52}\ \mathtt{m}^{-2}. \tag{30}$$

6. Conclusion

We have exploited the existing analogy between GR and the theory of ordinary elastic continua. First of all we have generalized the theory to Riemannian arbitrary dimensional manifolds, showing that the gravitational field may be described as a strain in space-time treated as a four-dimensional continuum. The strain can be originated both by a matter/energy distribution and by defects in the texture of space-time. The definition of defect also has been extrapolated from the one valid for ordinary material continua. Going further in our generalization we have hypothesized that the strain in the four-dimensional manifold should show up also in the Lagrangian of empty space-time. The corresponding additional term has been modeled on the classical elastic potential energy. Considering the consequent action integral, in presence of matter, we have succeeded in reproducing the luminosity curve of type Ia supernovae with an accuracy slightly better than the one obtainable from the popular λ-cold-dark-matter theory. The "elastic" parameters of space-time as obtained through the optimization process for the SnIa fit are compound in a bulk modulus of classical vacuum, B. The value obtained by the fit, as reported in (30), is extremely small, so that its effects appear at the cosmic scale only. At the scale of galaxies or even galaxy clusters the possible signatures of the CD theory (the present theory) are negligible. In fact, our theory is, in a sense, isomorphic to GR, since it is a metric theory and space-time always admits a tangent space, then a Newtonian limit; our final equations are in practice Einstein's equations with one more geometric source of curvature originated by the strain of the manifold. Our final description of the geometry of space-time somehow resembles bimetric theories [6], however we actually have only one metric. Our cosmic defect simply fixes the global symmetry of space-time. So far the CD theory proves to be more than a formal analogy and could be a fruitful new paradigm for gravity and space-time. The next steps will be to work out all implications of the theory, for instance concerning gravitational waves and the inhomogeneities in the CMB.

References

[1] Landau L and Lifshitz E 1986 *Theory of Elasticity* 3rd edn (Oxford: Pergamon)
Eshelby J D 1956 *Solid State Physics* (New York: Academic)
[2] Volterra V 1904 *Ann. Sci. de l' É.N.S.* **24** 401–517
[3] Tartaglia A and Radicella N 2010 *Class. Quantum Grav.* **27**, in press.
[4] Perlmutter S *et al* 1999 *Astrophys. J.* **517** 565–86
Riess A G *et al* 1998 *Astron. J.* **116** 1009–38
[5] Weinberg S 1972 *Gravitation and Cosmology: Principles and Applications of the General Theory of Relativity* (New York: Wiley)
[6] Hohmann M and Wohlfarth M 2009 *Phys. Rev.* D **80** 104011

The interaction between dark energy and dark matter

Jian-Hua He, Bin Wang

Department of Physics, Fudan University, 200433 Shanghai, China

E-mail: wangb@fudan.edu.cn

Abstract. In this review we first present a general formalism to study the growth of dark matter perturbations in the presence of interactions between dark matter(DM) and dark energy(DE). We also study the signature of such interaction on the temperature anisotropies of the large scale cosmic microwave background (CMB). We find that the effect of such interaction has significant signature on both the growth of dark matter structure and the late Integrated Sachs Wolfe effect(ISW). We further discuss the potential possibility to detect the coupling by cross-correlating CMB maps with tracers of the large scale structure. We finally confront this interacting model with WMAP 5-year data as well as other data sets. We find that in the 1σ range, the constrained coupling between dark sectors can solve the coincidence problem.

1. Introduction

There has been a great deal of evidence at present indicating that our universe has entered an epoch of accelerated expansion[1, 2, 3]. Theoretical effects to interpret the phenomena in accordance with the observational data has to attribute the acceleration of the universe to a so-called dark energy(DE). The leading interpretation of such dark energy is a cosmological constant(CC), but such model suffers the cosmological constant problem and coincidence problem[4]. It is natural to consider the interaction between DE and DM and such consideration has been widely discussed[5] and it is expected that it can provide a mechanism to alleviate the coincidence problem [5, 6, 7, 8]. However, it was claimed that such interaction may cause serious instabilities of the perturbations among dark sectors[9]. Latter, it was clarified [10] that the instability depends on the choice of the forms of the coupling between dark sectors and there exists rooms to avoid such instability. Furthermore, such interacting models are shown viable in confront with observational data. [11, 12, 13, 14, 15].

Nevertheless, the DE is usually not supposed to clump in the formation of cosmic structures, and the most powerful way to unveil its nature is believed to investigate the expansion history of the universe. However, current observations of cosmic expansion cannot break the degeneracies among different approaches to explain the acceleration of the expansion. Considering the interaction between DE and DM, it was observed that due to the coupling, the DE involves in the growth of the cosmic structure [16, 17, 18].

On the other hand, the WMAP data showed the deficit of large scale power in the temperature map, in particular in the CMB quadrupole. The significant contribution to the fluctuations on these scales is the late time Integrated Sachs Wolfe (ISW) effect which is induced by the shift of CMB photons frequencies through the time evolving gravitational potential when the universe enters a rapid expansion phase once DE dominates. The late time ISW effect has the unique ability to probe the "size" of DE. Much effort has been put into determining the EoS and the speed of sound of DE [19, 20]. Moreover,as found in [11], the couplings also has significant influence on the ISW effect and so does the large scale power in the temperature map. In this short review article we present a short review on these results.

The review is organized as follows: In Section 2, we go over the formalism of the perturbation theory in the presence of interactions between dark sectors. In Section 3, we present the influence of interactions on the dark matter structure formation. In Section 4, we talk about the influence on ISW effect due to the interaction between dark sectors. In Section 5, we report the fitting results from WMAP 5-year and other observational data sets on the coupling. In the last Section, we conclude our main results.

2. Analytical formalism

In this section we review the basic formalism of the perturbation theory when dark sectors are in interactions. The detailed descriptions can be found in [10, 16]. The scalar perturbation of the metric at first order for spatially flat Friedmann-Robertson-Walker (FRW) background is of the form

$$ds^2 = a^2[-(1+2\psi)d\tau^2 + 2\partial_i B d\tau dx^i + (1+2\phi)\delta_{ij}dx^i dx^j + D_{ij}E dx^i dx^j], \tag{1}$$

where ψ, B, ϕ, E are scalar metric perturbations, a is the cosmic scale factor and the operator D_{ij} is defined as

$$D_{ij} = (\partial_i\partial_j - \frac{1}{3}\delta_{ij}\nabla^2)$$

.

The energy-momentum tensor of a general perfect fluid can be represented as

$$T^{\mu\nu} = \rho U^\mu U^\nu + p(g^{\mu\nu} + U^\mu U^\nu), \tag{2}$$

For a two-component system which is composed of DE and DM, the conservation law is satisfied for the whole system, while for each component we have

$$\nabla_\mu T^{\mu\nu}_{(\lambda)} = Q^\nu_{(\lambda)} \tag{3}$$

where $Q^\nu_{(\lambda)}$ describes the interaction and λ denotes either DE or DM.

In the Fourier space the perturbed energy-momentum tensor reads [16]

$$\begin{aligned}
\delta'_\lambda + 3\mathcal{H}(\frac{\delta p_\lambda}{\delta\rho_\lambda} - w_\lambda)\delta_\lambda = -(1 &+ w_\lambda)kv_\lambda - 3(1+w_\lambda)\phi' + (2\psi - \delta_\lambda)\frac{a^2Q^0_\lambda}{\rho_\lambda} + \frac{a^2\delta Q^0_\lambda}{\rho_\lambda}. \\
(v_\lambda + B)' + \mathcal{H}(1-3w_\lambda)(v_\lambda + B) &= \frac{k}{1+w_\lambda}\frac{\delta p_\lambda}{\delta\rho_\lambda}\delta_\lambda - \frac{w'_\lambda}{1+w_\lambda}(v_\lambda + B) + k\psi - \frac{a^2Q^0_\lambda}{\rho_\lambda}v_\lambda \\
&- \frac{w_\lambda a^2 Q^0_\lambda}{(1+w_\lambda)\rho_\lambda}B + \frac{a^2\delta Q_{p\lambda}}{(1+w_\lambda)\rho_\lambda}.
\end{aligned} \tag{4}$$

v is the potential of three velocity and the prime denotes the derivative with respect to the conformal time τ. By constructing the gauge invariant quantities[21],

$$\begin{aligned}
\Psi &= \psi - \frac{1}{k}\mathcal{H}(B + \frac{E'}{2k}) - \frac{1}{k}(B' + \frac{E''}{2k}) \\
\Phi &= \phi + \frac{1}{6}E - \frac{1}{k}\mathcal{H}(B + \frac{E'}{2k})
\end{aligned}$$

$$
\begin{aligned}
\delta\rho^I_\lambda &= \delta\rho_\lambda - \rho'_\lambda \frac{v_\lambda + B}{k} \\
\delta p^I_\lambda &= \delta p_\lambda - p'_\lambda \frac{v_\lambda + B}{k} \\
\Delta_\lambda &= \delta_\lambda - \frac{\rho'_\lambda}{\rho_\lambda}\frac{v_\lambda + B}{k} \\
V_\lambda &= v_\lambda - \frac{E'}{2k} \\
\delta Q^{0I}_\lambda &= \delta Q^0_\lambda - \frac{Q^{0'}_\lambda}{\mathcal{H}}(\phi + \frac{E}{6}) + Q^0_\lambda \left[\frac{1}{\mathcal{H}}(\phi + \frac{E}{6})\right]' \\
\delta Q^I_{p\lambda} &= \delta Q_{p\lambda} - Q^0_\lambda \frac{E'}{2k}.
\end{aligned} \tag{5}
$$

the perturbed Einstein equations can be written as:

$$
\begin{aligned}
&\Phi = 4\pi G \frac{a^2}{k^2} \sum_\lambda \left(\Delta_\lambda + \frac{a^2 Q^0_\lambda}{\rho_\lambda}\frac{V_\lambda}{k}\right)\rho_\lambda \\
&k\left(\mathcal{H}\Psi - \Phi'\right) = 4\pi G a^2 \sum_\lambda (\rho_\lambda + p_\lambda) V_\lambda \\
&\Psi = -\Phi,
\end{aligned} \tag{6}
$$

where we have neglected the anisotropic stress tensor $\Pi^i_j = 0$.

For the coupling, there is no fundamental theory that selects a specific form of the interaction between dark sectors. Here we introduce the phenomenological description of the interaction in the comoving frame[10, 16]

$$
\begin{aligned}
Q^\nu_c &= \left[\frac{3\mathcal{H}}{a^2}(\xi_1\rho_c + \xi_2\rho_d), 0, 0, 0\right]^T \\
Q^\nu_d &= \left[-\frac{3\mathcal{H}}{a^2}(\xi_1\rho_c + \xi_2\rho_d), 0, 0, 0\right]^T,
\end{aligned} \tag{7}
$$

where ξ_1, ξ_2 are small dimensionless constants and T is the transverse of the matrix. The positive sign of ξ_1, ξ_2 indicates that the energy transfers from DE to DM .

3. Structure formation

The growth of structure happens in small scale. In the subhorizon approximation, from the perturbed Einstein equations eq(6) we can get the "Poission equation"

$$
-\frac{k^2}{a^2}\Psi = \frac{3}{2}H^2\left\{\Omega_m\Delta_m + (1-\Omega_m)\Delta_d\right\} \tag{8}
$$

and from the equations of matter perturbations eq(4), we obtain the density evolution equations for DM and DE respectively

$$
\begin{aligned}
\frac{d^2 ln\Delta_m}{dlna^2} &= -\left(\frac{dln\Delta_m}{dlna}\right)^2 - \left[\frac{1}{2} - \frac{3}{2}w(1-\Omega_m)\right]\frac{dln\Delta_m}{dlna} - (3\xi_1 + 6\frac{\xi_2}{r})\frac{dln\Delta_m}{dlna} \\
&+ \frac{3[exp(ln\frac{\Delta_d}{\Delta_m}) - 1]}{r}\left\{\xi_2 + 3\xi_1\xi_2 + 3\xi_2^2/r + \xi_2(\frac{dlnH}{dlna} + 1) - \xi_2\frac{dlnr}{dlna}\right\} \\
&+ 3\frac{\xi_2}{r}\frac{dln\Delta_d}{dlna}exp(ln\frac{\Delta_d}{\Delta_m}) + \frac{3}{2}\left[\Omega_m + (1-\Omega_m)exp(ln\frac{\Delta_d}{\Delta_m})\right] \quad ,
\end{aligned} \tag{9}
$$

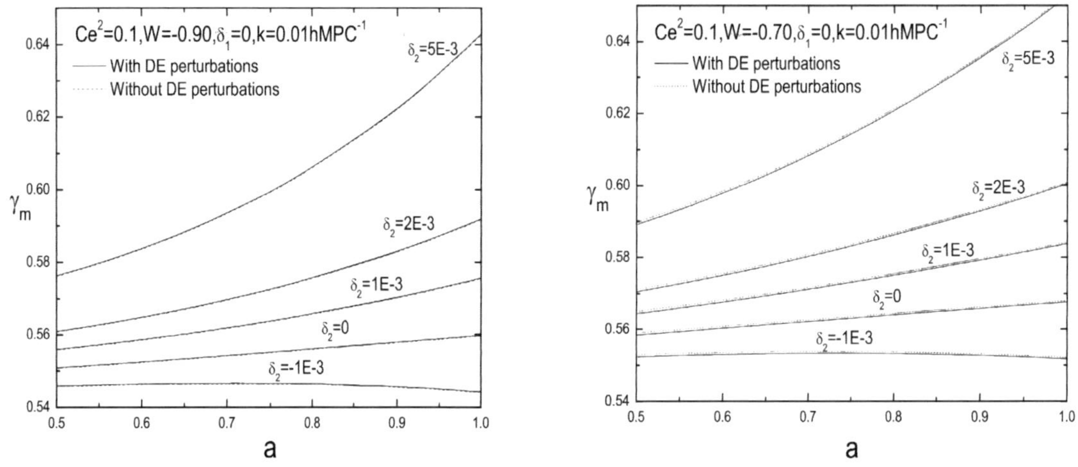

Figure 1. The growth index behavior when the interaction between DE and DM presents. Solid lines are for the result with DE perturbation, while dotted lines are for the result without DE perturbation.

$$\begin{aligned}
\frac{d^2ln\Delta_d}{dlna^2} &= -\left(\frac{dln\Delta_d}{dlna}\right)^2 - \left[\frac{1}{2} - \frac{3}{2}w(1-\Omega_m)\right]\frac{dln\Delta_d}{dlna} - \frac{k^2C_e^2}{a^2H^2} \\
&+ (1+w)\frac{3}{2}\left[\Omega_m exp(ln\frac{\Delta_m}{\Delta_d}) + (1-\Omega_m)\right] - 3\xi_1\left[(\frac{dlnH}{dlna}+1)r + \frac{dr}{dlna}\right]exp(ln\frac{\Delta_m}{\Delta_d}) \\
&+ \left[3\xi_2 + 6\xi_1 r + 6w - 3C_a^2 + 3(C_e^2 - C_a^2)\frac{\xi_1 r + \xi_2}{1+w} + \frac{C_e^2}{1+w}\frac{dln\rho_d}{dlna}\right]\frac{dln\Delta_d}{dlna} \\
&- 3r\xi_1 exp(ln\frac{\Delta_m}{\Delta_d})\frac{dln\Delta_m}{dlna} + 3(\frac{dlnH}{dlna}+1)(w - C_e^2) + 3\xi_1(\frac{dlnH}{dlna}r + r + \frac{dr}{dlna}) \\
&+ 3\left[w - C_e^2 + \xi_1 r(1 - exp(ln\frac{\Delta_m}{\Delta_d}))\right]\left[(1-3w) - 3\frac{C_e^2 - C_a^2}{1+w}(1+w+\xi_1 r+\xi_2)\right] \\
&+ 3\left[w - C_e^2 + \xi_1 r(1 - exp(ln\frac{\Delta_m}{\Delta_d}))\right]\left[-3(\xi_1 r + \xi_2) - \frac{C_e^2}{1+w}\frac{dln\rho_d}{dlna}\right] \quad . \qquad (10)
\end{aligned}$$

We introduce the growth index γ with the definition [22]

$$\gamma_m = (\ln\Omega_m)^{-1}\ln\left(\frac{a}{\Delta_m}\frac{d\Delta_m}{da}\right). \qquad (11)$$

The growth index is generically not constant which was first emphasized and investigated in terms of cosmological parameters in [23]. The growth index has been argued as a useful way in principle to distinguish the modified gravity models from DE models [24, 25]. In Fig 1, we present our numerical results when we incorporate the interaction between DE and DM. The solid lines are for the results with DE perturbation, while the dotted lines are for the results without DE perturbation. It is clearly shown that the growth index got more influenced from the interaction between dark sectors than the DE perturbation. Although the enhancement of the growth index due to the interaction is clearly shown, the available accuracy from the observations such as DUNE[26] etc. is calculated for the ΛCDM model and may not be true for our interacting cosmology, since the current values of Ω_{c0} may not be typically equal in the two cases as argued in [17]. However, our result is interesting, as it opens the possibility that the future measurement of the growth factor may be helpful to reveal the presence of the interaction between DE and DM.

4. ISW effect

The ISW effect can be classified into early and late time effects. The early ISW effect takes place from the time following recombination to the time when radiation is no longer dynamically significant, which gives clues about what has happened in the universe from the radiation domination to matter domination. The late time ISW effect arises when DE becomes dynamically important and has the unique ability to probe the "size" of DE. When DE becomes non-negligible, the gravitational potential decays. When a photon passes through a decaying potential well, it will have a net gain in energy and thus leads to the late time ISW effect. The late time ISW effect is a significant contribution to the large scale power in the temperature map of CMB.

However, such late time ISW effect can not be measured directly from observations. The most efficient way is to investigate the galaxies-ISW correlation. The angular correlation function for galaxies-ISW and galaxies-galaxies correlation are represented as

$$C_l^{gg} = 4\pi \int \frac{dk}{k} \mathcal{P}_\chi(k) I_l^{g*}(k) I_l^g(k) \tag{12}$$

$$C_l^{gI} = 4\pi \int \frac{dk}{k} \mathcal{P}_\chi(k) I_l^{g*}(k) \Delta_l^{ISW}(k), \tag{13}$$

where

$$\Delta_l^{ISW} = \int_{\tau_i}^{\tau_0} d\tau j_l(k[\tau_0 - \tau]) e^{\kappa(\tau_0)-\kappa(\tau)} [\Psi' - \Phi'], \tag{14}$$

is the contribution due to the change of the gravitational potential when photons passing through the universe on their way to earth, namely the ISW effect. The integrand $I_l^g(k)$ for galaxy densities reads,

$$I_l^g(k) = \int dz b_g(z) \Pi(z) (D_{gc} + D_{gb}) j_l[k\chi(z)]. \tag{15}$$

Here $b_g(z)$ is the galaxy bias, $\Pi(z)$is the redshift distribution and $\chi(z)$ is the conformal distance, or equivalently the look-back time from the observer at redshift $z = 0$,

$$\chi(z) = \int_0^z \frac{dz'}{H(z)} = \int_{\tau_i(z)}^{\tau_0} d\tau. \tag{16}$$

We assume $b(z) \sim 1$ for simplicity and adopt the redshift distribution of the form[27],

$$\Pi(z) = \frac{3}{2} \frac{z^2}{z_0^3} \exp\left[-(\frac{z}{z_0})^{3/2}\right] \tag{17}$$

which is normalized to unity and peaks near the median redshift $z_m = 1.4z_0$. For illustrative purpose,we choose $z_m = 0.1$ and $z_m = 0.4$ throughout our analysis. The first choice resembles a shallow survey like 2MASS and the second one resembles a survey similar to SDSS photo-z galaxy samples.

When the coupling is proportional to the energy density of DE and the EoS is larger than $-1(w > -1)$, the cross power spectra and auto correlation power spectra of galaxies are shown

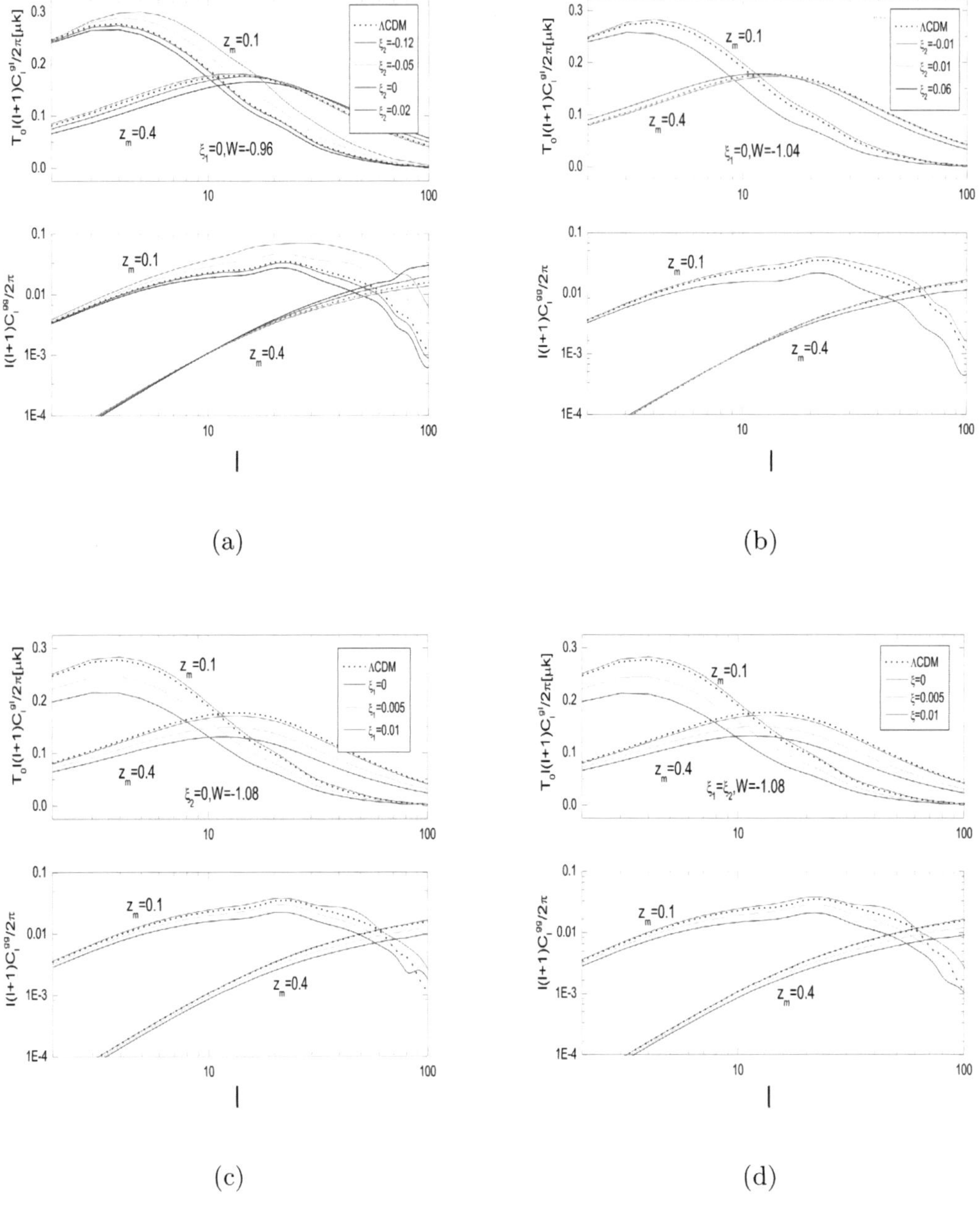

Figure 2. The upper panel shows the cross-spectra and the lower panel shows the galaxies auto power spectra.

in Fig 2a. For lower redshift galaxies survey $z_m = 0.1$, comparing with the LCDM model we see that the couplings significantly change both the cross spectra and the auto spectra. The negative couplings enhance the power of the correlation while the positive couplings hinder such correlation. When the EoS is smaller than -1, we find very similar results as shown in Fig 2b. However, for deeper redshift galaxies survey $z_m = 0.4$, the couplings do not imprint significantly on the cross power spectra and auto correlation power spectra of galaxies as compared with $z_m = 0.1$.

When the coupling is proportional to the energy density of DM or total dark sectors, we find from Fig 2c,Fig 2d that the cross power spectra are more sensitive to the couplings at lower l part than that of higher l part when the median redshift around 0.1. This feature is different from that shown in Fig 2a and Fig 2b when the interaction is proportional to the energy density of DE, where it was found that at small l when ISW effect amplified, the ISW-LSS cross-correlation is not so much different due to the coupling. While in the other way, for higher redshift survey $z_m = 0.4$, we find less such effect on the cross and auto spectra.

The qualitative behaviors presented here show that the 2MASS survey has more possibility in discriminating the interaction between dark sectors than that of SDSS survey. It is expected in the future that galaxy surveys with photo-z measurements or even spec-z measurements, along with better CMB measurements, could provide better ISW-LSS cross-correlation measurements at each redshift bin and provides more precise signature of the interaction between dark sectors.

5. Global fitting and coincidence problem

In this section we confront the interacting model with observational data. We take the parameter space as

$$P = (h, \omega_b, \omega_{cdm}, \tau, \ln[10^{10}A_s], n_s, \xi_1, \xi_2, w)$$

where h is the hubble constant, $\omega_b = \Omega_b h^2, \omega_{cdm} = \Omega_{cdm} h^2$, A_s is the amplitude of the primordial curvature perturbation, n_s is the scalar spectral index, ξ_1 and ξ_2 are coupling constants proportional to the energy density of DM and DE respectively, w is the EoS of DE. We choose the flat universe with $\Omega_k = 0$ and our work is based on CMBEASY code[28].

In the global fitting, we have used CMB data coming from WMAP5 temperature and polarization power spectra. We used Gibbs sampling routine provided by WMAP team for the likelihood calculation. In the small scale CMB measurements, we included BOOMERanG[29], CBI [30], VSA[31] and ACBAR[32] data. In order to get better constraint on the background evolution, we have added SNIa[33] data and marginalized over the nuisance parameters. We also incorporated the data from large scale luminous red galaxies(LRGs) survey, we used SDSS[34] data as powerful constraint on real-space power spectrum $P(k)$ at redshift $z \sim 0.1$.

The global fitting results for different forms of interaction between dark sectors are shown in Table 1. When the coupling proportional to the energy density of DM($\xi_1 \neq 0, \xi_2 = 0$) and total dark sectors ($\xi_1 = \xi_2 = \xi$), we constrain the coupling constant to be positive to avoid the zero crossing problem of the energy density of DE at the early time of the universe. From the fitting results, we find that in 1σ range the coupling naturally is positive. Thus with this kind of interaction the coincidence problem can be overcome(see, Fig 3a,b).

6. Conclusions and discussions

We have reviewed the formalism of the perturbation theory when there is interaction between DE and DM. Based upon this formalism we have studied the signature of the interaction between dark sectors on the structure formation and the large scale temperature CMB fluctuations. We found that the effect of the interaction between dark sectors overwhelms that of the DE perturbation on the growth function of DM perturbation. When the DE EoS w is in the vicinity of -1 somewhere around the best fitted value at the moment, the DE perturbation is

Table 1. The best-fitted results for coupling models

h	$\Omega_b h^2$	$\Omega_{cdm} h^2$	n_s	$\xi_1 = 0, \xi_2 \neq 0$	$1 + w > 0$
$0.687^{+0.013}_{-0.013}$	$0.0225^{+0.0006}_{-0.0005}$	$0.107^{+0.007}_{-0.009}$	$0.963^{+0.013}_{-0.013}$	$-0.028^{+0.023}_{-0.032}$	< 0.052
h	$\Omega_b h^2$	$\Omega_{cdm} h^2$	n_s	$\xi_1 = 0, \xi_2 \neq 0$	$1 + w < 0$
$0.700^{+0.012}_{-0.011}$	$0.0223^{+0.0006}_{-0.0005}$	$0.119^{+0.009}_{-0.006}$	$0.956^{+0.013}_{-0.014}$	$-0.010^{+0.025}_{-0.020}$	$-0.051^{+0.051}_{-0.043}$
h	$\Omega_b h^2$	$\Omega_{cdm} h^2$	n_s	$\xi_1 > 0, \xi_2 = 0$	$1 + w < 0$
$0.690^{+0.015}_{-0.015}$	$0.0224^{+0.0005}_{-0.0005}$	$0.121^{+0.003}_{-0.003}$	$0.953^{+0.013}_{-0.013}$	$0.0007^{+0.0006}_{-0.0006}$	$-0.072^{+0.072}_{-0.053}$
h	$\Omega_b h^2$	$\Omega_{cdm} h^2$	n_s	$\xi = \xi_1 = \xi_2 > 0$	$1 + w < 0$
$0.690^{+0.014}_{-0.014}$	$0.0224^{+0.0006}_{-0.0006}$	$0.121^{+0.004}_{-0.003}$	$0.955^{+0.014}_{-0.014}$	$0.0006^{+0.0006}_{-0.0005}$	$-0.065^{+0.065}_{-0.054}$

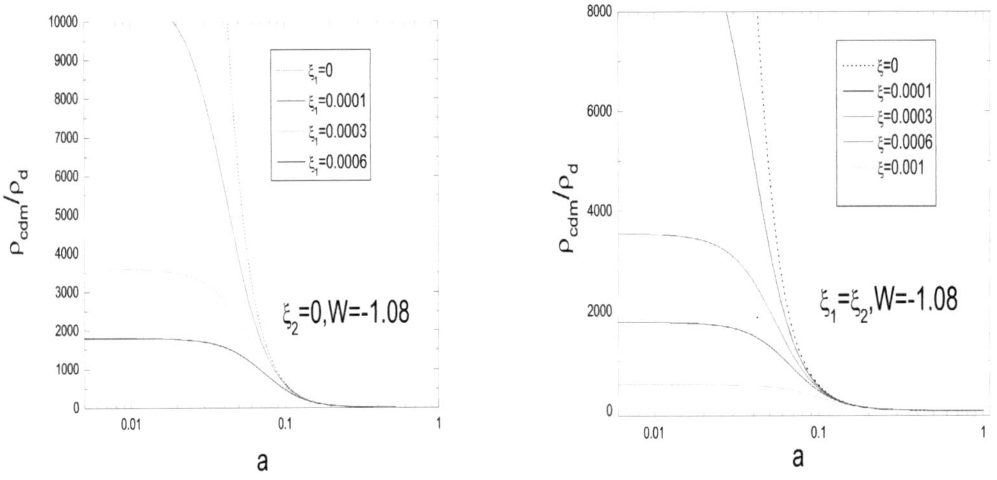

(a) (b)

Figure 3. The evolution of the ratio ρ_{cdm}/ρ_d for different coupling models.

suppressed, however, when the interaction presents, the growth index can differ from the value without interaction by a big amount up to the observational sensibility.

We also found that in addition to disclosing the DE EoS, sound speed, the late ISW effect is a promising tool to measure the coupling between dark sectors. When the interaction between DE and DM takes the form proportional to the energy density of DM or of the total dark sectors, because in these cases DE and DM started to chase each other since early time, the interaction not only presented in the late ISW source term but also left imprint in the SW and early ISW effects. These properties provide a possible way to examine the interaction between DE and DM even from smaller scale in CMB observations.

We have performed the global fitting by using the CMB power spectrum data including WMAP5 data and balloon observational data together with SNIa and SDSS data to constrain the interaction between DE and DM. When the interaction between DE and DM takes the form proportional to the energy density of DM or of the total dark sectors, the coupling can be constrained in a very precise range. In 1σ the coupling is positive indicating that there is energy transfer from DE to DM. This kind of energy transfer can help to alleviate the coincidence problem compared to the noninteracting case.

It is of great interest to extend our study to a field theory description of the interaction between DE and DM and examine its signature in observations. A possible field theory model was proposed in [35] and further investigation in this direction is called for.

7. Acknowledgement

This work was partially supported by NNSF of China and the National Basic Research Program of China under grant No. 2010CB833004.

References

[1] S. J. Perlmutter et al., *Nature* **391**, 51 (1998); A. G. Riess et al., *Astron. J.* **116**, 1009 (1998); S. J. Perlmutter et al., *Astroph. J.* **517**, 565 (1999); J. L. Tonry et al., *Astroph. J.* **594**, 1, (2003); A. G. Riess et al., *Astroph. J.* **607**, 665 (2005); P. Astier et al., *Astron. Astroph. 447*, 31 (2005); A G. Riess et al., *Astroph. J.* **659**, 98 (2007).
[2] M. Kowalski et al., 2008, arXiv:0804.4142.
[3] M. Tegmark et al., *Phys. Rev. D* **74**, 123507 (2006).
[4] S. Weinberg, *Rev. Mod. Phys.* **61**, 1 (1989); I. Zlatev, L. -M. Wang, and P. J. Steinhardt, *Phys. Rev. Lett.* **82**, 896 (1999).
[5] L. Amendola, *Phys. Rev. D* **62**, 043511 (2000); L. Amendola and C. Quercellini, *Phys. Rev. D* **68**, 023514 (2003); L. Amendola, S. Tsujikawa and M. Sami, *Phys. Lett. B* **632**, 155 (2006).
[6] D. Pavon, W. Zimdahl, *Phys. Lett. B* **628**, 206 (2005), S. Campo, R. Herrera, D. Pavon, *Phys. Rev.D* **78**, 021302(R) (2008).
[7] C. G. Boehmer, G. Caldera-Cabral, R. Lazkoz, R. Maartens, *Phys. Rev. D* **78**, 023505 (2008).
[8] S. B. Chen, B. Wang, J. L. Jing, *Phys.Rev.D* **78**, 123503 (2008).
[9] J. Valiviita, E. Majerotto, R. Maartens, *JCAP* **07**, 020 (2008), ArXiv:0804.0232.
[10] J. H. He, B. Wang, E. Abdalla, *Phys. Lett. B* **671**, 139 (2009), ArXiv:0807.3471
[11] H. He, B. Wang, P. Zhang, *Phys. Rev. D* **80**, 063530 (2009) ArXiv:0906.0677.
[12] J. Valiviita, R. Maartens, E. Majerotto, ArXiv:0907.4987
[13] Z. K. Guo, N. Ohta and S. Tsujikawa, *Phys. Rev. D* **76**, 023508 (2007).
[14] J. H. He, B. Wang, *JCAP* **06**, 010 (2008), ArXiv:0801.4233.
[15] C. Feng, B. Wang, E. Abdalla, R. K. Su, *Phys. Lett. B* **665**, 111 (2008), ArXiv:0804.0110.
[16] J. H. He, B. Wang, Y. P. Jing, *JCAP* **07**, 030 (2009) ArXiv:0902.0660.
[17] G. Caldera-Cabral, R. Maartens, B. Schaefer, ArXiv:0905.0492.
[18] E. Abdalla, L.Raul W. Abramo, L. Sodre Jr., B. Wang, Phys. Lett. B673, 107 (2009); E. Abdalla, L. Abramo, J. Souza, arXiv:0910.5236.
[19] J. Weller, A. M. Lewis, *Mon. Not. Roy. Astron. Soc.* **346**, 987 (2003), ArXiv:astro-ph/0307104.
[20] R. Bean, O. Dore, *Phys.Rev. D* **69**, 083503 (2004),ArXiv:astro-ph/0307100.
[21] V. Mukhanov, Physical Foundation of Cosmology, Cambridge *Univ. Press*, 2005; H. Kodama, M. Sasaki, *Prog. Theor. Phys. Suppl.* **78**, 1, (1984).
[22] G. Ballesteros, A. Riotto, *Phys. Lett. B* **668**, 171 (2008).
[23] D. Polarski, R. Gannouji *Phys. Lett. B* **660**, 439 (2008).
[24] E. Bertschinger and P. Zukin, arXiv:0801.2431; S. Tsujikawa, *Phys. Rev. D 76*, 023514, (2007); B. Jain and P. Zhang, ArXiv: 0709.2375; H. Wei and S. N. Zhang, ArXiv: 0803.3292; Y. G. Gong, ArXiv: 0808.1316.
[25] R. Gannouji, D. Polarski *JCAP* **0805**, 018 (2008); R. Gannouji, B. Moraes, D. Polarski arXiv:0809.3374.
[26] A. Refrefier, et al, ArXiv: 0802.2522.
[27] J. Lesgourgues, W. Valkenburg, E. Gaztanaga, *Phys, Rev. D* **77** 063505 (2008).
[28] M. Doran, *JCAP* **0510**, 011 (2005).
[29] C. J. MacTavish, et al., *Astrophys. J.* **647**, 799 (2006).
[30] A. C. S. Readhead, et al., *Astrophys. J. 609*, 498 (2004).
[31] C. Dickinson, et al., *Mon. Not. Roy. Astron. Soc.* **353**, 732 (2004).
[32] C. L. Reichardt, et al., ArXiv:0801.1491.
[33] A. G. Riess, et al., ArXiv:astro-ph/0611572.
[34] P. McDonald, et al., *Astrophys. J. Suppl.* **163**, 80 (2006); P. McDonald, et al., *Astrophys. J.* **635**, 761 (2005).
[35] S. Micheletti, E. Abdalla, B. Wang, *Phys. Rev. D***79**, 123506 (2009) ArXiv:0902.0318.

Journal of Physics: Conference Series **222** (2010) 012030 doi:10.1088/1742-6596/222/1/012030

Laser Interferometers, Gravitational Waves and Echos from the Universe

Laura Cadonati, for the LIGO Scientific Collaboration and the Virgo Collaboration

Physics Department, University of Massachusetts, Amherst, MA 01003

E-mail: cadonati@physics.umass.edu

Abstract. The Laser Interferometer Gravitational-wave Observatory (LIGO) and its sister project Virgo aim for the first direct detections of gravitational waves. Such detections will provide not only a test of general relativity, but also a fundamental, new probe into the Universe. To achieve this goal, LIGO and Virgo use laser interferometers to monitor changes in the relative separation of mirrors at the ends of each of two perpendicular arms of km-scale length, in response to the space-time distortions induced by the passage of gravitational waves. This paper gives an overview of the status and the science of LIGO and Virgo, with selected results from the initial detector phase and predictions for the enhanced and advanced detector configurations.

1. Gravitational Waves and How to Listen

The theory of general relativity predicts that, when large masses accelerate, their gravitational field changes. Such changes propagate as *gravitational waves*, tiny ripples in the fabric of space-time which carry information from the most catastrophic events in the Universe: the death of a star, the collision of black holes, the Big Bang. Gravitational waves are generated by coherent relativistic motions of large masses; they travel through opaque matter and can reveal the dynamics of strongly curved space-time. As such, they will provide information that is complementary to the multi-wavelength electromagnetic spectrum of traditional astronomy and to the neutrinos and cosmic rays of particle astrophysics, with an impact on the understanding of our Universe comparable to the transition from visible to radio, infrared and X-ray astronomy. The existence of gravitational waves was indirectly proven by measurements of the orbital period of binary pulsar P1913+16 (Hulse-Taylor, 1974), which over two decades diminished due to emanation of gravitational radiation, as predicted by general relativity [1]. However, due to their small amplitude when they reach earth, gravitational waves have never been directly measured.

The Laser Interferometer Gravitational-wave Observatory (LIGO) [3] aims to detect gravitational waves using laser interferometers: two in Hanford, WA with 4 and 2 km long arms (H1 and H2, respectively) and one in Livingston, LA with 4 km arms (L1). The interferometers monitor changes in the relative separation of mirrors at the ends of each of two perpendicular arms, in response to the space-time distortions induced by gravitational waves. The goal for the initial phase of LIGO is to measure differences in length to one part in 10^{21}, or 10^{-18} m, which is one thousand times smaller than a typical nuclear diameter. LIGO achieved its design sensitivity in 2005 [2] and carried out a long observing run, S5, which produced one year of

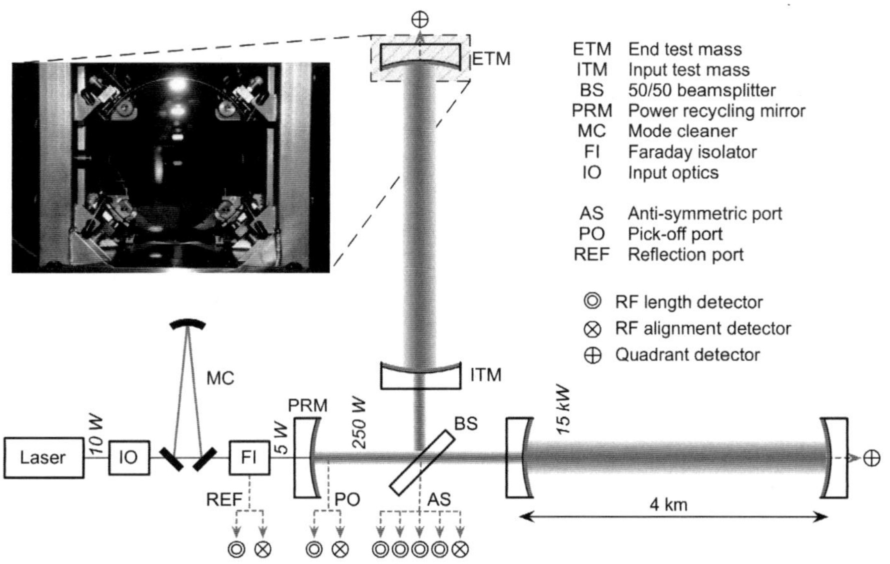

Figure 1. Schematics of the LIGO interferometer, from reference [2]. The laser beam is split in two components that travel along the perpendicular arms, are reflected by mirrors (ITM and ETM) and recombine at the beam splitter (BS). When a gravitational wave modifies the length of the arms, the two beams accumulate phase differences and the photodetector measures a different interference pattern. The inset photo shows a test mass mirror in its pendulum suspension.

integrated Livingston-Hanford coincident data from November 2005 to September 2007, with strain sensitivity $h = \Delta L/L \sim 10^{-21}$ in a 100 Hz band around 150 Hz and broadband coverage between 60 Hz and a few kHz. Previous science runs (S1 through S4) had durations of only one or two months and lower sensitivity. According to the current plans, the S6 LIGO run (Enhanced LIGO) is scheduled to end in late 2010, when the interferometers will undergo invasive hardware upgrades for a ten-fold improvement in sensitivity. Advanced LIGO was approved by NSF in 2008, and design and fabrication of new components are well under way; installation of new hardware will begin in late 2010, for a scheduled start of science data in 2014.

The Virgo interferometer [4], installed in Cascina, Italy, has a 3 km arm length, and it reached comparable sensitivity to that of LIGO in 2007, when it began its first science run, VSR1, in overlap with the S5 LIGO run, beginning May 18, 2007. The two projects now have an agreement for data sharing; all data is analyzed and results are reviewed and published together by the LIGO Scientific Collaboration and the Virgo collaborations. Current and future data runs and instrumental development are coordinated between the two projects: a joint run plan committee is coordinating the ongoing science runs S6 (LIGO) and VSR2 (Virgo). Virgo also has plans for a second generation phase: Advanced Virgo, approved Fall 2009, should achieve comparable sensitivity to Advanced LIGO, on a similar time scale. Figure 2 shows the measured noise spectra for the initial phases of LIGO and Virgo, as well as a projected noise curve for the advanced phases. As the figure shows, the advanced configuration will yield a factor 10 improvement across the frequency spectrum, and even better at low frequencies.

In the remainder of this paper, we will highlight recent results from LIGO and Virgo, based on data from S5 and VSR1, the first long-term operation of a worldwide network of interferometers of similar performance at three different locations.

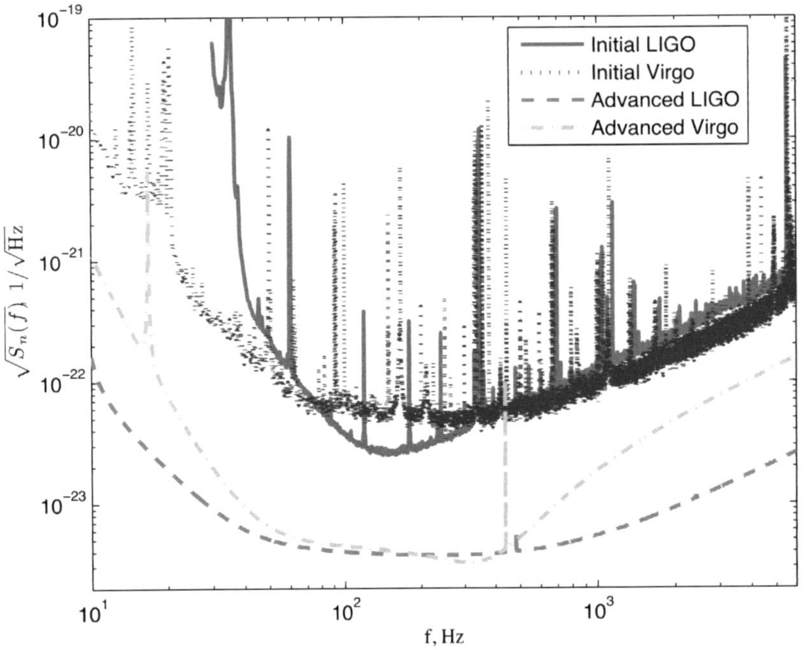

Figure 2. Measured noise amplitude spectra for initial LIGO and initial Virgo, compared to the expected spectra in the advanced configuration [5; 6]. On the vertical axis is the square root of the noise power spectral density, in units of $Hz^{-1/2}$. From reference [7].

2. Recent Results

2.1. Gravitational wave transients

Gravitational wave transients are any short signal, with duration ranging from a few milliseconds to several seconds or minutes. Two approaches have been developed in the search for transient signatures. (1) Templated searches are based on knowledge of the signal being sought, using matched filtering to a class of templates to extract a known, weak signal from noise. These searches have been used by LIGO and Virgo to target the coalescence of compact binary systems, isolated ring-downs and cosmic (super)string cusps, which will be discussed in section 2.5. (2) Un-modeled searches, instead, are designed to be sensitive to gravitational wave signatures with a less accurate model, and serendipitous sources as well. They only make minimal assumptions on the signal morphology and rely on the coherence properties of gravitational waves.

The gravitational wave signature of stellar mass compact binary coalescences is considered the most promising for ground based interferometer, both because of waveform knowledge and predicted rate. LIGO and Virgo data has been match-filtered to a bank of inspiral templates to look for binary neutron stars with total mass M_T=2–35 M$_\odot$ [8; 9]. The inspiral waveforms generated by these systems can be reliably predicted using post-Newtonian (PN) perturbation theory, until the last fraction of a second prior to merger. We used waveforms from non-spinning compact binaries calculated in the frequency domain using the stationary-phase approximation (SPA) [10–12]. The waveforms are calculated to Newtonian order in amplitude and second PN order in phase, and extend until the Schwarzschild innermost stable circular orbit (ISCO). The analysis of data from the first 18 months of S5 yielded no gravitational wave coalescence candidates, so we set upper limits on their rate. The resulting 90% C.L. upper limits on neutron star (1.35 M$_\odot$) and black hole (5.0 M$_\odot$) binary coalescences are [8; 9]:

- neutron star – neutron star rate $< 1.4 \times 10^{-2}/\mathrm{L}_{10}/\mathrm{year}$
- black hole – black hole rate $< 7.3 \times 10^{-4}/\mathrm{L}_{10}/\mathrm{year}$
- black hole – neutron star rate $< 3.6 \times 10^{-3}/\mathrm{L}_{10}/\mathrm{year}$

where $\mathrm{L}_{10} = 10^{10}$ times the blue solar luminosity; the luminosity of the Milky Way is $1.7\,\mathrm{L}_{10}$; these units are used because the rate of binary coalescences in a galaxy follows approximately the star formation rate, or blue light luminosity. These results are still 1 to 2 orders of magnitude above optimistic astrophysical predictions and ~ 3 orders of magnitude above best estimates [7], but we are preparing for better targets in Advanced LIGO/Virgo.

A matched-filter search with ringdown templates is also performed for higher mass systems [13–15]. A perturbed black hole, such as a coalescence remnant, is expected to settle to a stationary configuration through emission of gravitational radiation. The waveform is a superposition of quasi-normal modes, or ringdowns, with fundamental mode $l = m = 2$, a well known waveform that can be used in matched filtering. Using data from the S4 run, we placed a 90% C.L. upper limit on the rate of ringdowns from black holes with mass between 85 $\mathrm{M}_\odot$ and 390 $\mathrm{M}_\odot$ in the local Universe, assuming a uniform source distribution, of $3.2 \times 10^{-5}/yr/Mpc^3 = 1.6 \times 10^{-3}/yr/L_{10}$.

The matched filter technique [16] is now being extended to intermediate mass black holes up to 100 $\mathrm{M}_\odot$, in a regime where only few cycles are in the LIGO band and most of the signal-to-noise ratio comes from the merger and ringdown phase of the coalescence [17; 18], using EOBNR analytical templates (Effective One Body waveforms, tuned to match Numerical Relativity simulations) that follow the full coalescence [19]. At this moment, these templates only describe non-spinning systems, but an intense effort is in progress to extend them to a broad parameter space, including spinning and precessing systems.

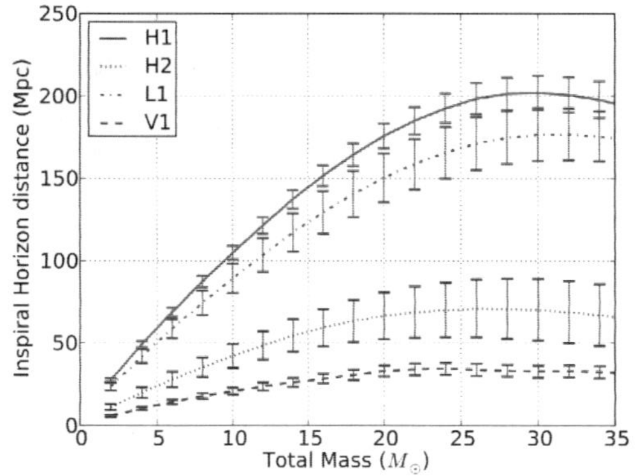

Figure 3. Average inspiral horizon distance during 4.5 months of S5/VSR1, as a function of the total mass of the binary system for each interferometer. The error bars indicate variations over this period. From reference [20].

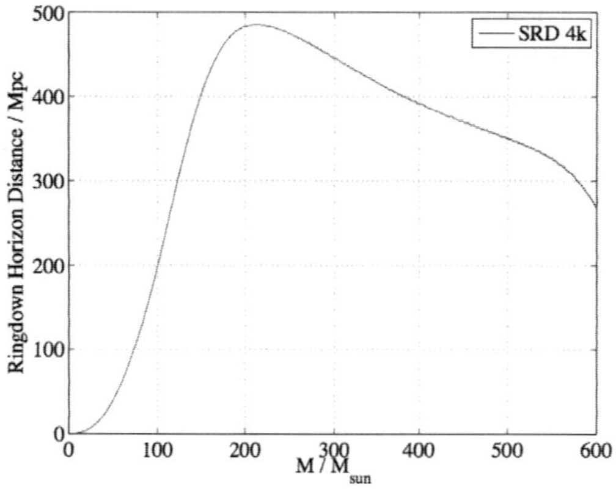

Figure 4. Expected horizon distance in the ringdown matched-filter search for the initial LIGO reference noise (SRD), for a black hole with spin parameter $a = 0.9$ and $\varepsilon = 0.01$, as a function of the black hole mass. For more details, see [15].

The physical reach of templated searches can be quoted as the *horizon*, or the distance at which an optimally oriented source can be detected with signal-to-noise ratio SNR=8. Figure 3 shows the horizon for the detection of compact binary coalescences with SPA matched filtering at the three LIGO detectors and Virgo during the S5 run, function of the system total mass and assuming non-spinning sources. As the mass increases, the inspiral phase of the systems goes out of band and matched filtering to SPA templates becomes less sensitive, while other techniques, such as matched filtering to EOBNR IMR templates and template-less searches, become more effective. Similarly, figure 4 shows the horizon for black hole ringdowns with spin parameter $a = Jc/GM^2 = 0.9$, assuming a fraction $\epsilon = 0.01$ of the black hole's mass is radiated in gravitational waves.

A complementary approach to the detection of gravitational wave transients is to make no assumptions on the signal morphology, but only rely on the properties of gravitational waves. This is the approach followed by LIGO and Virgo burst searches, which look for coherent excess power in multiple detector and effectively match detectors to each other rather than to a template bank. A burst *trigger* is characterized by a time, duration, frequency, bandwidth and root-square-sum amplitude h_{rss}, in units of strain/$\sqrt{\text{Hz}}$:

$$h_{rss} = \sqrt{\int_{-\infty}^{\infty} |h(t)|^2 \, dt} = \sqrt{\int_{-\infty}^{\infty} \left|\tilde{h}(f)\right|^2 df}, \tag{1}$$

where $h(t)$ is the time dependent waveform, in units of strain, and $\tilde{h}(t)$ its Fourier transform. The challenge is to make sure we control the false alarm rate due to noise transients (to ~ 1 false alarm per century) so these searches are particularly sensitive to data quality and to coherent coincidence criteria.

The burst all-sky search of S5/VSR1 [21–23] is a network analysis that combines times when all four detectors (Virgo and the three LIGO) were functioning, times with only the three LIGO detectors and times where only the two colocated Hanford detectors were online; other network configurations were not included, since they had minor contribution to the livetime. The analysis covers a frequency band of 50–6000 Hz.

This analysis yielded no plausible gravitational wave candidate, so we set upper limits on rates of gravitational wave bursts of 2.0 events per year at 90% CL in the 64–2048 Hz band, and 2.2 events per year for higher-frequency bursts up to 6 kHz. The IGEC network of resonant bar detectors set a tighter rate limit, 1.5 events per year at 95% confidence level [24; 25]. However, the IGEC detectors were sensitive only around their resonant frequencies, near 900 Hz, and achieved that rate limit only for signal amplitudes one to two orders of magnitude louder than the LIGO-Virgo burst seach is sensitive to. Figure 5, from reference [23], shows exclusion curves for the rate of gravitational wave bursts of different frequencies, as a function of the burst h_{rss} amplitude, marginalized over the orientation and inclination of the source. Exclusion curves for previous runs (S1, S2, S4) are also shown. Figure 6 shows the rate upper limit per unit volume at 90% CL for a linearly polarized, sine-Gaussian standard candle with $E_{GW} = \mathrm{M}_\odot c^2$ [23].

The astrophysical sensitivity of the S5y2/VSR1 search is quoted as the amount of mass converted into gravitational wave burst energy at a given distance r_0, that would be sufficient to be detected by the search with 50% probability (M_{GW}). A rough estimate for a standard-candle isotropic emission is [23]:

$$M_{\text{GW}} \approx \frac{\pi^2 c}{G} \, r_0^2 \, f_0^2 \, h_{\text{rss}}^2 \,. \tag{2}$$

For instance, for a sine-Gaussian signal with $f_0 = 153$ Hz and $Q = 9$ (oscillations under the Gaussian envelope) in the four-detector network, the root-sum-square amplitude with 50% detection probability is $h_{\text{rss}}^{50\%} = 6.0 \times 10^{-22}\,\text{Hz}^{-1/2}$. Assuming a typical Galactic distance of 10 kpc, that h_{rss} corresponds to $M_{\text{GW}} = 1.8 \times 10^{-8}\,M_\odot$. For a source in the Virgo galaxy cluster, approximately 16 Mpc away, the same h_{rss} would be produced by a mass conversion of $\sim 0.05\,M_\odot$.

For a model-dependent interpretation of the burst result, we quote a detection range for gravitational wave signals from core-collapse supernovae and from neutron star collapse to a black hole. Such signals are expected to be produced at a high frequency, up to a few kHz, and with a relatively small gravitational wave energy output ($10^{-9} - 10^{-5}\, M_{\odot}c^2$). For a possible supernova scenario, we consider a numerical simulation of core collapse by Ott et al. [26]. For the model s25WW, which undergoes an acoustically driven explosion, as much as $8 \times 10^{-5}\, M_{\odot}$ may be converted to gravitational waves, with dominant frequency content at ~ 940 Hz and the duration of order one second. We estimate such signal could be detected out to a distance of around 30 kpc [23]. The axisymmetric neutron star collapse signals D1 and D4 of Baiotti *et al.* [27] have detection ranges (at 50% confidence) of only about 25 pc and 150 pc, due mainly to their lower energy ($M_{\rm GW} < 10^{-8}\, M_{\odot}$) and also to emitting most of that energy at 2–6 kHz, where the detector noise is greater.

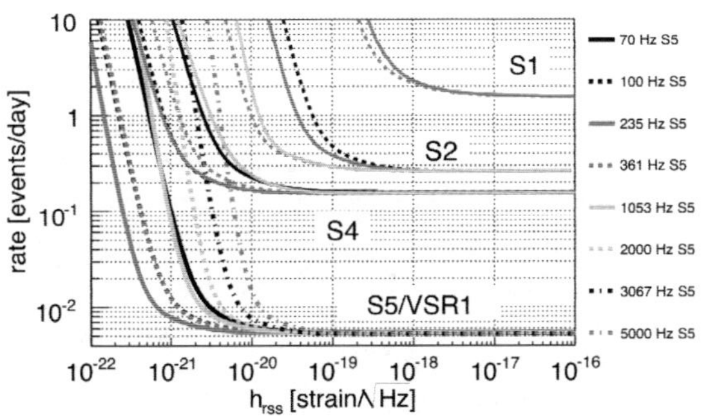

Figure 5. 90% confidence rate limit function of signal amplitude h_{rss} for Q=9 sine-Gaussian pulses and various frequencies. The most recent S5/VSR1 result is compared to the limits from previous LIGO runs [23].

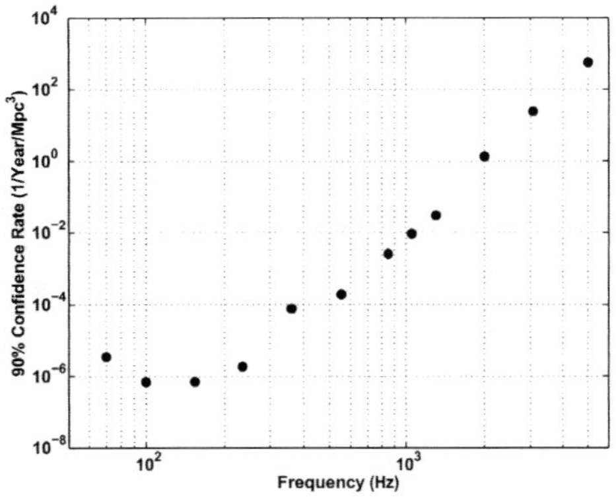

Figure 6. 90% CL rate limit per unit volumetfor a linearly polarized, sine-Gaussian standard candle with $E_{GW} = M_{\odot}c^2$ [23].

2.2. Multi-messenger searches for gravitational wave transients

One of the most exciting prospects of gravitational wave transient searches is the potential for multi-messenger astrophysics, where the information on gravitational wave signatures is combined to that of electromagnetic observations. Of particular interest, in this context, are Gamma Ray Bursts (GRBs), intense flashes of γ-rays which occur approximately once per day and are isotropically distributed over the sky [28]. The variability of the bursts on time scales as short as a millisecond indicates that their progenitors are compact, while the identification of host galaxies and redshift measurements have shown they are extra-galactic. GRBs are grouped into two broad classes by their characteristic duration and spectral hardness [29; 30]. The progenitors of most short GRBs (≤ 2 s, with hard spectra) are widely thought to be mergers of neutron star binaries or neutron star-black hole binaries [31]; a small fraction (up to 15%) of short-duration GRBs are also thought to be due to giant flares from a local distribution of soft-gamma repeaters (SGRs) [32–34]. Long GRBs (≥ 2 s, with soft spectra), on the other hand, are associated with core-collapse supernovae [35–38]. Both the merger and supernova scenarios result in the formation of a stellar-mass black hole with accretion disk [39; 40], and the emission of gravitational radiation is expected in this process.

GRB 070201 was a short, hard gamma-ray burst detected during the S5 run. Its error box was consistent with being located in M31, at a distance of 770 kpc. At that time, only the two Hanford detectors were operating. We searched for gravitational wave signatures in a time window of 180 s around the event, both with the inspiral matched filter and with the unmodeled burst approach, but no plausible candidate was found [41]. Figure 7 shows exclusion regions, at difference confidence levels, for the distance of a compact binary progenitor of GRB 070201, with masses in the range $1\mathrm{M}_\odot < m_1 < 3\mathrm{M}_\odot$ and $1\mathrm{M}_\odot < m_2 < 40\mathrm{M}_\odot$. If GRB 070201 were caused by a binary neutron star merger, we find that a location in M31 is excluded at more than 99% confidence, while $D < 3.5$ Mpc is excluded at 90% confidence, assuming random inclination. The analysis also implies that an unmodeled gravitational wave burst from GRB 070201 emitted less than $4.4 \times 10^{-4}\mathrm{M}_\odot c^2$ ($7.9 \times 10^{50} ergs$) in any 100 ms long period within the signal region, if the source was in M31 and radiated isotropically at the same frequency as LIGO's peak sensitivity ($f \sim 150$ Hz). This upper limit does not exclude current models of SGRs at the M31 distance. This result prompted a review of the properties of this GRB, leading to the conclusion that it was not a binary merger in M31, but possibly an SGR giant flare instead [42; 43].

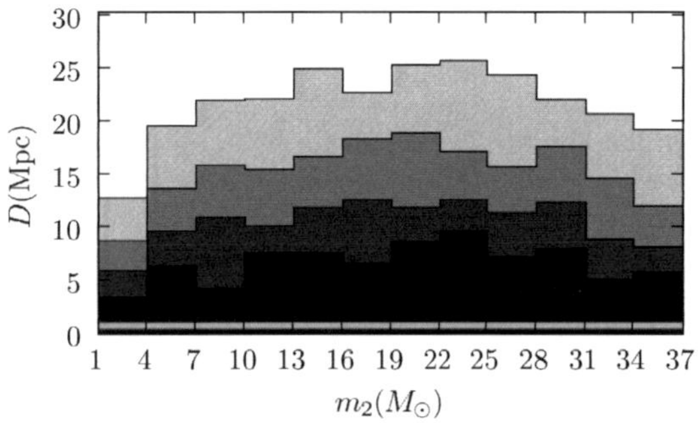

Figure 7. The shaded regions represent 90%, 75%, 50%, and 25% exclusion regions, from darkest to lightest, respectively. The distance to M31 is indicated by the horizontal line at D=0.77 Mpc. Both the amplitude calibration uncertainty and Monte Carlo statistics are included in this result; apparent uctuations as a function of mass are due to Monte Carlo uncertainty. [41]

In a recent paper [44], we published the analysis of 137 GRBs that were detected by satellite-based gamma-ray experiments during the S5/VSR1 run, mostly by the *Swift* satellite, searching for gravitational wave bursts from both short and long duration GRBs. Since the precise nature of the radiation depends on the progenitor model, we looked for unmodelled burst signals with duration < 1 s in LIGO/Virgo band, 60–2000 Hz.

The search implemented a coherent network analysis method that accounts for the different locations and orientations of the interferometers at the three LIGO-Virgo sites. We found no evidence for gravitational-wave burst signals associated with this sample of GRBs; using simulated short-duration (< 1 s) waveforms, we set upper limits on the amplitude of gravitational waves associated with each GRB. We also placed lower bounds on the distance to each GRB under the assumption of a isotropic gravitational wave energy emission $E_{GW}^{iso} = 0.01\mathrm{M}_\odot c^2$, at frequencies around 150 Hz, where the LIGO-Virgo detector network has best sensitivity, and found typical limits of $D \sim 15$ Mpc, as shown in figure 8. $E_{GW}^{iso} = 0.01\mathrm{M}_\odot c^2$ is a reasonable expectation for emission in the LIGO-Virgo band by various progenitor models, such as mergers of neutron star binaries or neutron star – black hole binaries for short GRBs, or fragmentation of the accretion disk for long GRBs. For other values of E_{GW}^{iso}, the distance limit scales as $D \propto (E_{GW}^{iso})^{1/2}$.

We also performed a search with matched filtering to templates for compact binary coalescences [44], in temporal and directional coincidence with 22 GRBs during S5/VSR1. This analysis was motivated by progenitor scenarios for short gamma-ray bursts (short GRBs) that include coalescences of two neutron stars or a neutron star and black hole. We found no statistically significant gravitational wave candidates within a $[-5, +1)$ s window around the time of any GRB. We constrained the distance to each GRB assuming it was caused by a compact binary coalescence with a neutron star with $1 < m_1 < 3$ $\mathrm{M}_\odot$ and a companion of mass $1 < m_2 < 3$ $\mathrm{M}_\odot$ (neutron star–neutron star system, or NS–NS) or $7 < m_2 < 10$ $\mathrm{M}_\odot$ (neutron star–black hole system, or NS-BH): the median exclusion distance for a NS–NS system is 3.3

Mpc and for a NS–BH system is 6.7 Mpc, assuming no beaming. Lower limits on the distance of individual GRBs are shown in figure 9.

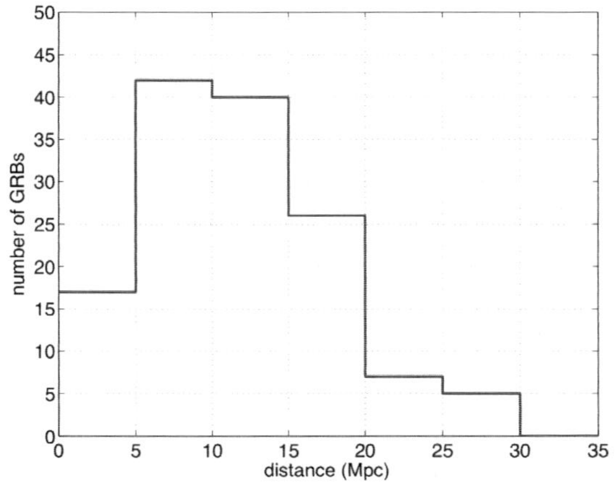

Figure 8. Distribution of lower limits on the distance to each of the 137 GRBs studied, assuming the GRB progenitors emit $0.01\mathrm{M}_{\odot}c^2 = 1.8 \times 10^{52}$ erg of energy in circularly polarized gravitational waves at 150Hz. [44]

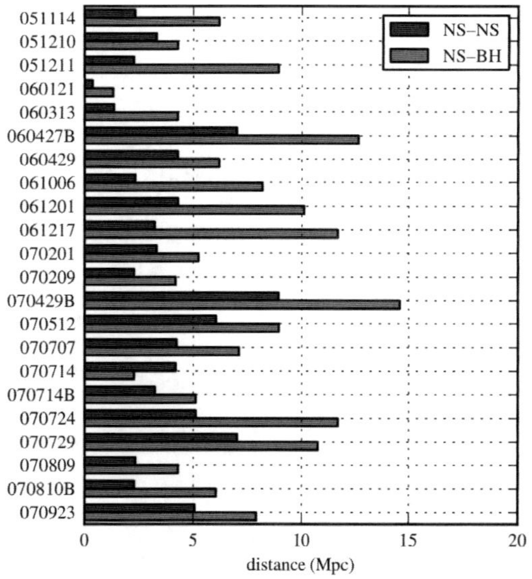

Figure 9. Lower limits on distances at 90% CL to NS–NS and NS–BH progenitor systems for 22 GRBs during the S5/VSR1 run [45].

2.3. Continuous sources: gravitational waves from known pulsars

Continuous waves represent a different class of signatures searched for in LIGO–Virgo data. Millisecond and young pulsars are some of the best galactic targets for gravitational wave searches in the 40–2000 Hz frequency band of current interferometric detectors; there are currently over 200 known pulsars with spin frequencies greater than 20 Hz. A search was conducted [46] for 116 known millisecond and young pulsars using LIGO S5 data, using for each pulsar ephemerides from radio and X-ray observations. Knowing the position and spin evolution of a pulsar allows a fully coherent analysis of long time spans, and dig deeply into the detector noise to search for evidence of gravitational wave emission by the pulsar.

This search assumes that the pulsars are triaxial stars emitting gravitational waves at twice their observed spin frequencies, i.e. the emission mechanism is an $l = m = 2$ quadrupole, and that gravitational waves are phase-locked with the electromagnetic signal. For each pulsar, the target senstivity is the spin-down limit on strain tensor amplitude h_0^{sd}, which can be calculated by assuming that the observed spin-down rate of a pulsar is entirely due to energy loss through gravitational radiation from an $l = m = 2$ quadrupole, as:

$$h_0^{sd} = 8.06 \times 10^{-19} \frac{I_{38}}{r_{kpc}} \sqrt{\frac{|\dot{\nu}|}{\nu}} \tag{3}$$

where I_{38} is the pulsars principal moment of inertia I_{zz} in units of 10^{38} kg m; r_{kpc} is the pulsar distance in kpc, ν is the spin-frequency in Hz, and $\dot{\nu}$ is the spin-down rate in Hz s^{-1}.

We did not detect signals from any of the targets; therefore we interpret our results as upper limits on the gravitational wave signal strength, as shown in figure 10.

Journal of Physics: Conference Series **222** (2010) 012030 doi:10.1088/1742-6596/222/1/012030

Our best upper limit on gravitational wave amplitude is 2.3×10^{-26} for pulsar J1603-7202 and our best limit on the inferred pulsar ellipticity is 7.0×10^{-8} for pulsar J2124-3358. Of the recycled millisecond pulsars several of the measured upper limits are only about an order of magnitude above their spin-down limits. For the young pulsars J1913+1011 and J1952+3252 we are only a factor of a few above the spin-down limit, and for the X-ray pulsar J0537-6910 we reach the spin-down limit under the assumption that any gravitational wave signal from it stays phase locked to the X-ray pulses over timing glitches.

Of particular interest are the limits on gravitational radiation from the Crab pulsar [47]. The Crab pulsar has a high spin-down rate, $\dot{\nu} \approx -3.7 \times 10^{-10}$ Hz/s, corresponding to a kinetic energy loss rate of $\dot{E} \approx 4.4 \times 10^{31}$ W. This loss is due to a variety of mechanisms, including magnetic dipole radiation and particle acceleration in the magnetosphere, but a significant fraction of this energy could be shed in the form of gravitational radiation. We searched for signal in first 9 months of LIGO S5 data and found no evidence for gravitational radiation. The measured upper limit for gravitational emission from the Crab pulsar is now a factor of seven below the spin-down limit. Assuming the gravitational signal is locked to electromagnetic pulses, this null search result implies that no more than ~2% of the available spin-down power is radiated in the form of gravitational waves.

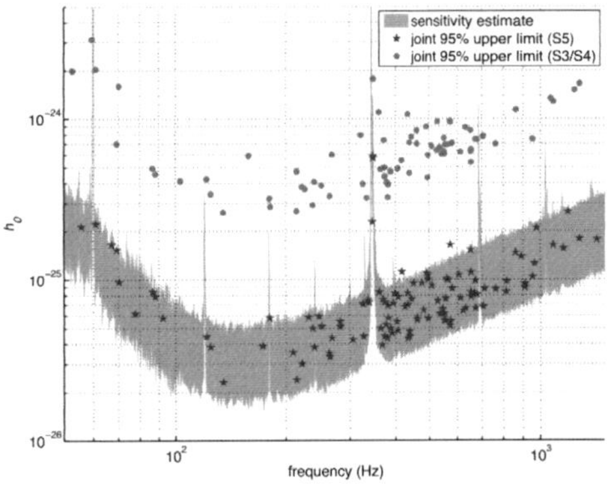

Figure 10. Upper limits for gavitational wave amplitude from known pulsars, superimposed to the estimated sensitivity of the search (grey band). The results of the S3/S4 search are also shown, for comparison [46].

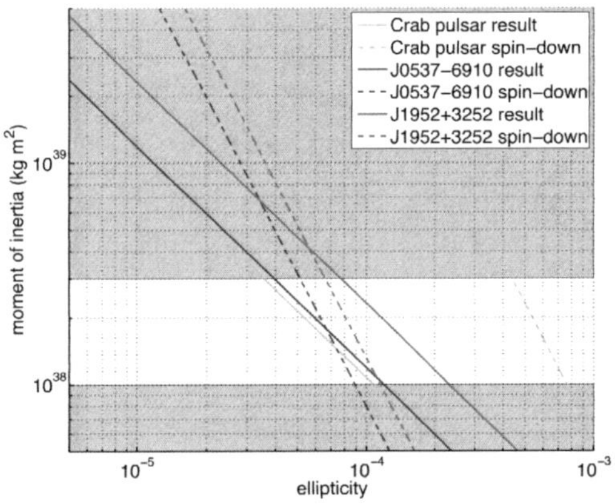

Figure 11. Results for the Crab pulsar, J0537–6910 and J1952+3252 analyses, compared to the spin-down limits, in the moment of inertia–ellipticity plane. Areas to the right of the diagonal lines are excluded. The shaded regions are outside the theoretically predicted range of moments of inertia $I_{38} = 1 - 3$. [46]

2.4. Stochastic Background

A stochastic background of gravitational waves is expected to arise from a superposition of a large number of unresolved gravitational wave sources of astrophysical and cosmological origin. Such signal is expected to carry unique signatures from the earliest epochs in the evolution of the Universe, inaccessible to standard astrophysical observations. Direct measurements of the amplitude of this background would provide unaltered information on the physical processes that generated it, and understanding of the evolution of the Universe in its first minute of life.

We quantify the stochastic gravitational wave background by its spectrum:

$$\Omega_{GW}(f) = \frac{f}{\rho_c}\frac{\mathrm{d}\rho_{GW}}{\mathrm{d}f} \tag{4}$$

where $\mathrm{d}\rho_{GW}$ is the energy density of gravitational radiation contained in the frequency range f to $f + \mathrm{d}f$ and ρ_c is the critical energy density of the Universe. To measure Ω_{GW}, we rely on cross correlation of data from pairs of interferometers. The cross correlation is computed in frequency domain and multiplied by a filter factor that is inversely proportional to the noise in each detector, and also depends on the frequency-dependent overlap in the antenna patterns of the two interferometers.

We analyzed data from the LIGO S5 run, and constrained the energy density of the stochastic gravitational-wave background normalized by the critical energy density of the Universe, in the 41.5–169.25 Hz frequency band, to be $< 6.9 \times 10^{-6}$ at 95% confidence. This result improves on the indirect limits from Big Bang Nucleosynthesis (BBN) and Cosmic Microwave Background (CMB) at 100 Hz [48].

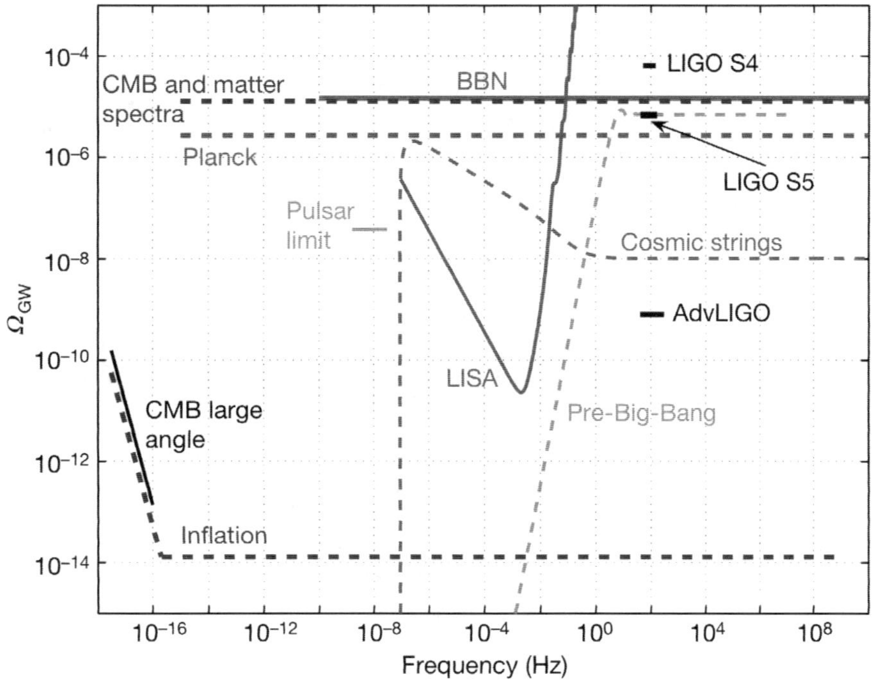

Figure 12. The landscape of sotchastic gravitational wave background upper limits (see text for explanation). From reference [48].

Figure 12 compares the upper limit on gravitational wave stochastic background set by LIGO to previous measurements, other experiments and models [48]. The 95% upper limit $\Omega_0 < 6.9 \times 10^{-6}$, in the 41.5–169.25 Hz band, is compared to the previous LIGO S4 result and to the projected Advanced LIGO sensitivity, using the two co-located instruments at the Hanford site. The corresponding 95% upper bound on the total gravitational-wave energy density in this band, assuming frequency independent spectrum, is 9.7×10^{-6} . The indirect bound due to BBN acts on $\Omega_{BBN} = \int \Omega_{GW} \mathrm{d}(\ln f)$, rather than on the density Ω_{GW}, over the $10^{-13} - 10^{10}$Hz band. A similar integral bound over $10^{-15} - 10^{10}$Hz can be placed using CMB and matter power spectra.

The pulsar limit is based on the fluctuations in the pulse arrival times of millisecond pulsars, at $\sim 10^{-8}$Hz frequencies. Measurements of the CMB at large angular scales constrain the possible redshift of CMB photons due to the stochastic gravitational wave background, and therefore limit the amplitude of the stochastic gravitational wave background at largest wavelengths. In addition, the figure shows projected sensitivities of the Planck and LISA experiments, as well as examples of inflationary cosmic strings and pre-Big-Bang models.

Although the evolution of the Universe following the BBN is well understood, there is little observational data probing the evolution before BBN, when the Universe was less than one minute old. The gravitational wave spectrum Ω_{GW} is related to the parameters that regulate the evolution of the Universe. As such the upper limit on Ω_{GW} allows to constrain parameter space, and to rule out models of early Universe evolution with relatively large equation of state parameter; we refer to reference [48] for details.

2.5. Cosmic Strings and SuperStrings

Cosmic strings were originally proposed as topological defects formed during phase transitions in the early Universe. More recently, it was realized that fundamental strings may also be expanded to cosmological scales. Hence, searching for cosmic strings may provide a unique and powerful window into string theory and into particle physics at the highest energy scales.

The network of cosmic strings is parametrized by the string tension μ (multiplied by the Newton constant G), and reconnection probability p. The CMB observations limit Gμ to 10^{-6}, while the size of cosmic string loops, parametrized by ε, is essentially unconstrained. The mechanism for production of gravitational waves relies on cosmic string cusps: regions of string that move at speeds close to the speed of light. If the cusp motion points towards Earth, a detectable burst of gravitational radiation may be produced. The superposition of gravitational waves from all string cusps in the cosmic string network would produce a stochastic gravitational wave background, thus the limits described in section 2.4 also rule out cosmic (super)string models with relatively small string tension, favoured in some string theory models.

Figure 13 shows how different experiments probe the $\varepsilon - \text{G}\mu$ plane for a typical value of the reconnection probability $p = 10^{-3}$. The region to the right of this curve is expected to produce at least one cosmic string burst event detectable by LIGO during the S5 run. Note that this search is complementary to the search for the stochastic background, as it probes a different part of the parameter space. Also shown is the region that will be probed by the Planck satellite measurements of the CMB. The entire plane shown here will be accessible to the stochastic gravitational background search of Advanced LIGO.

We also performed a matched filter search for bursts from cosmic string cusps performed on 14.9 days of LIGO S4 data, using times when all three LIGO detectors were operating [49]. No gravitational waves were detected in 14.9 days of data from. This non-detection can be interpreted in terms of a frequentist upper limit on the rate of gravitational wave bursts and use the limits on the rate to constrain the parameter space (string tension, reconnection probability, and loop sizes) of cosmic string models. The sensitivity of the LIGO instruments during the S4 run does not allow us to place constraints as tight as the indirect bounds from Big Bang Nucleosynthesis. In the future, however we expect our sensitivity to surpass these limits for large areas of cosmic string model parameter space.

3. Conclusion

This paper presented a survey of recent results from the analysis of LIGO and Virgo data from the S5/VSR1 run. No gravitational wave detection is reported, but these null results are triggering the interest of the astrophysics community, as they constrain source populations, with limits on the rate of compact binary systems close to coalescence or transient signatures

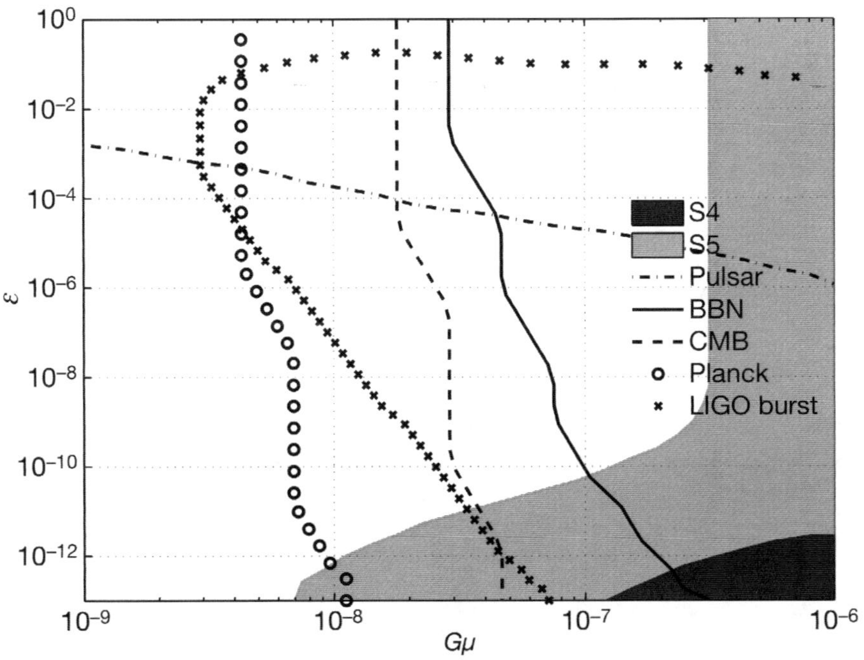

Figure 13. Exclusion regions in the $\varepsilon - G\mu$ plane for reconnection probability $p = 10^{-3}$ from the LIGO S5 stochastic gravitational wave background search, compared to the previous limits from the LIGO S4 analysis and limits from the BBN and CMB. Also shown are the region that will be probed by the Planck satellite and the expected limits from a search of LIGO data for cosmic string cusps with gravitational wave transient techniques. The region to the right of these curves is expected to produce at least one consmic string burst event detectable by LIGO during the S5 run. For details, see [48] and references therein. For results of a cosmic string cusp search in S4, with different values of the reconnection probability, see [49].

from nearby galaxies, as well as specific source models (e.g. the nature of GRB070201, Crab spindown, stochastic gravitational background spectrum and cosmic string parameters).

At the time of this writing, the S6/VSR2 science run is in progress, part of an *enhanced* program, where some of the advanced technology is being tested on existing interferometers. By 2011, LIGO will halt data acquisition and begin installation of Advanced LIGO, a tunable, quantum-noise-limited interferometer, with a predicted ten-fold sensitivity improvement down to 10 Hz, due to new fused silica multi-stage suspension, higher laser power, active seismic isolation, signal recycling a careful quantum engineering to balance radiation pressure and shot noise [5]. In parallel, Advanced Virgo will install new monolithic suspensions and better mirrors for lower thermal noise, and higher laser power and signal recycling for an improved sensitivity at intermediate and high frequencies [6].

Within the next few years, Advanced LIGO and Advanced Virgo will probe ~1000 more galaxies, and the science from the first three hours will be comparable to a whole year of Initial LIGO, and the numerical relativity, astrophysics and data analysis communities are now preparing for the predicted first detections, and the inauguration of a new gravitational wave astronomy.

Acknowledgments
The authors gratefully acknowledge the support of the United States National Science Foundation for the construction and operation of the LIGO Laboratory, the Science and Technology Facilities Council of the United Kingdom, the Max-Planck-Society and the State of Niedersachsen/Germany for support of the construction and operation of the GEO 600 detector, and the Italian Istituto Nazionale di Fisica Nucleare and the French Centre National de la Recherche Scientifique for the construction and operation of the Virgo detector. The authors also gratefully acknowledge the support of the research by these agencies and by the Australian Research Council, the Council of Scientific and Industrial Research of India, the Istituto Nazionale di Fisica Nucleare of Italy, the Spanish Ministerio de Educación y Ciencia, the Conselleria d'Economia Hisenda i Innovació of the Govern de les Illes Balears, the Foundation for Fundamental Research on Matter supported by the Netherlands Organisation for Scientific Research, the Polish Ministry of Science and Higher Education, the FOCUS Programme of Foundation for Polish Science, the Royal Society, the Scottish Funding Council, the Scottish Universities Physics Alliance, the National Aeronautics and Space Administration, the Carnegie Trust, the Leverhulme Trust, the David and Lucile Packard Foundation, the Research Corporation, and the Alfred P. Sloan Foundation.
This paper was assigned LIGO Document Control Number LIGO-P1000016.

References

[1] Weisberg J M and Taylor J H 2005 *ASP Conf. Ser.* **328** 25 (*Preprint* astro-ph/0407149)
[2] Abbott B *et al.* (LIGO Scientific Collaboration) 2009 *Rep. Prog. Phys.* **72** 076901 (*Preprint* 0711.3041)
[3] LIGO URL http://www.ligo.caltech.edu
[4] Acernese F *et al.* 2008 *Class. Quant. Grav.* **25** 114045
[5] Advanced LIGO anticipated sensitivity curves URL https://dcc.ligo.org/cgi-bin/DocDB/ShowDocument?docid=T0900288
[6] Advanced Virgo Baseline Design, The Virgo Collaboration, note VIR027A09 May 16, 2009 URL https://pub3.ego-gw.it/codifier/includes/showTmpFile.php?doc=2219&calledFile=VIR-027A-09.pdf
[7] ? 2010 *in preparation (will be in arXiv by end of 02/10)* (*Preprint* LIGO DCC P0900125)
[8] Abbott B P *et al.* (LIGO Scientific Collaboration) 2009 *Phys. Rev. D* **79** 122001 (*Preprint* 0901.0302)
[9] Abbott B P *et al.* (LIGO Scientific Collaboration) 2009 *Phys. Rev. D* **80** 047101 (*Preprint* 0905.3710)
[10] Thorne K 1987 *Cambridge University Press, Cambridge* **9** 330–458
[11] Sathyaprakash B S and Dhurandhar S V 1991 *Phys. Rev. D* **44** 3819–3834
[12] Droz S, Knapp D J, Poisson E and Owen B J 1999 *Phys. Rev. D* **59** 124016 (*Preprint* gr-qc/9901076)
[13] Creighton J D E 1999 *Phys. Rev. D* **60** 022001 (*Preprint* gr-qc/9901084)
[14] Goggin L M 2008 *A search for gravitational waves from perturbed black hole ringdowns in LIGO data* Ph.D. thesis Caltech
[15] Abbott B P *et al.* (LIGO Scientific Collaboration) 2009 *Phys. Rev. D* **80** 062001 (*Preprint* 0905.1654)
[16] Wainstein L A and Zubakov V D 1962 *Extraction of signals from noise* (Englewood Cliffs, NJ: Prentice-Hall)
[17] Flanagan E E and Hughes S A 1998 *Phys. Rev. D* **57** 4535

[18] Flanagan E E and Hughes S A 1998 *Phys. Rev. D* **57** 4566

[19] Buonanno A *et al.* 2007 *Phys. Rev. D* **76** 104049 (*Preprint* 0706.3732)

[20] Abadie J *et al.* (LIGO Scientific Collaboration and Virgo Collaboration) 2010 *in preparation*

[21] Abbott B *et al.* (LIGO Scientific Collaboration) 2009 *Phys. Rev. D* **80** 102001 (*Preprint* arXiv:0905.0020)

[22] Abbott B *et al.* (LIGO Scientific Collaboration) 2009 *Phys. Rev. D* **80** 102002 (*Preprint* arXiv:0904.4910)

[23] Abadie J *et al.* (LIGO Scientific Collaboration and Virgo Collaboration) 2010 (*Preprint* arXiv:1002.1036)

[24] Astone P *et al.* 2003 *Phys. Rev. D* **68** 022001

[25] Astone P *et al.* 2007 *Phys. Rev. D* **76** 102001

[26] Ott C D, Burrows A, Dessart L and Livne E 2006 *Phys. Rev. Lett.* **96** 201102

[27] Baiotti L *et al.* 2007 *Class. Quantum Grav.* **24** S187

[28] Meszaros P 2006 *Rept. Prog. Phys.* **69** 2259–2322 (*Preprint* astro-ph/0605208)

[29] Kouveliotou C, Meegan C A, Fishman G J, Bhat N P, Briggs M S, Koshut T M, Paciesas W S and Pendleton G N 1993 *Astrophys. J.* **413** L101–L104

[30] Gehrels N *et al.* 2006 *Nature* **444** 1044 (*Preprint* astro-ph/0610635)

[31] Nakar E 2007 *Physics Reports* **442** 166–236

[32] Duncan R C and Thompson C 1992 *Astrophys. J.* **392** L9–L13

[33] Chapman R, Priddey R S and Tanvir N R 2008 *AIP Conf. Proc.* **983** 304–306 (*Preprint* 0709.4640v1)

[34] Nakar E, Gal-Yam A and Fox D B 2006 *Astrophys. J.* **650** 281–290 (*Preprint* astro-ph/0511254)

[35] Galama T J *et al.* 1998 *Nature* **395** 670 (*Preprint* astro-ph/9806175)

[36] Hjorth J *et al.* 2003 *Nature.* **423** 847–850 (*Preprint* astro-ph/0306347)

[37] Malesani D *et al.* 2004 *Astrophys. J.* **609** L5 (*Preprint* astro-ph/0405449)

[38] Campana S *et al.* 2006 *Nature* **442** 1008–1010 (*Preprint* astro-ph/0603279)

[39] Fryer C L, Woosley S E and Hartmann D H 1999 *Astrophys. J.* **526** 152 (*Preprint* astro-ph/9904122)

[40] Cannizzo J K and Gehrels N 2009 *Astrophys. J.* **700** 1047–1058 (*Preprint* 0901.3564)

[41] Abbott B *et al.* (LIGO Scientific Collaboration) 2008 *Astrophys. J.* **681** 1419 (*Preprint* 0711.1163)

[42] Ofek E O *et al.* 2007 (*Preprint* 0712.3585)

[43] Mazets E P, Aptekar R L, Cline T L, Frederiks D D, Goldsten J O, Golenetskii S V, Hurley K, von Kienlin A, and Pal'shin V D 2008 *The Astrophysical Journal* **680** 545–549 URL http://stacks.iop.org/0004-637X/680/545

[44] Abbott B *et al.* (LIGO Scientific Collaboration and Virgo Collaboration) 2009 (*Preprint* 0908.3824)

[45] Abadie J *et al.* (LIGO Scientific Collaboration and Virgo Collaboration) 2010 (*Preprint* 1001.0165)

[46] Abbott B P *et al.* (LIGO Scientific Collaboration and Virgo Collaboration) 2009 (*Preprint* 0909.3583)

[47] Abbott B *et al.* (LIGO Scientific Collaboration) 2008 *Astrophys. J.* **683** L45–L50 (*Preprint* 0805.4758)

[48] Abbott B P *et al.* (LIGO Scientific Collaboration and Virgo Collaboration) 2009 *Nature* **460** 990

[49] Abbott B P *et al.* (LIGO Scientific Collaboration) 2009 *Phys. Rev.* **D80** 062002 (*Preprint* 0904.4718)

Journal of Physics: Conference Series **222** (2010) 012031 doi:10.1088/1742-6596/222/1/012031

Neutron Star Dynamics and Gravitational Waves

Kostas D. Kokkotas[1,2], **Erich Gaertig**[1] **and Antonella Colaiuda**[1]

[1]Theoretical Astrophysics, Eberhard-Karls University of Tübingen, Tübingen 72076, Germany
[2]Department of Physics, Aristotle University of Thessaloniki, Thessaloniki 54124, Greece

Abstract. As several large scale interferometers are beginning to take data at sensitivities where astrophysical sources are predicted, the direct detection of gravitational waves may well be imminent. This would open the gravitational-wave window to our Universe and should lead to a much improved understanding of the most violent processes imaginable; the formation of black holes and neutron stars following core collapse supernovae and the merger of compact objects at the end of binary inspiral. Among the best candidate sources for gravitational waves are the oscillations, but mainly the rotational instabilities of neutron stars which can emit quite strong gravitational wave signals via which one may reveal the details of their structure. Magnetars also are neutron stars with ultra strong magnetic field whose periodic flaring activity is associated with starquakes. They are also a potential source of gravitational waves while even the quasi-periodic oscillations (QPOs) observed in the electromagnetic spectrum can reveal the details of their structure.

1. Introduction

Gravitational waves are ripples of spacetime generated as masses are accelerated. It is one of the central predictions of Einstein's general theory of relativity but despite decades of effort, these ripples in spacetime have still not been observed directly. Yet we have strong indirect evidence for their existence from the excellent agreement between the observed inspiral rate of the binary pulsar PSR1913+16 and the theoretical prediction (better than 1% in the phase evolution). This provides confidence in the theory and suggests that "gravitational-wave astronomy" should be viewed as a serious proposition. This new window into the universe will complement our view of the cosmos and will help us unveil the fabric of spacetime around black-holes, observe directly the formation of black holes or the merging of binary systems consisting of black holes or neutron stars, search for rapidly spinning neutron stars, dig deep into the very early moments of the origin of the universe and look at the very center of the galaxies where supermassive black holes weighing millions of solar masses are hidden. Secondly, detecting gravitational waves is important for our understanding of the fundamental laws of physics; the proof that gravitational waves exist will verify a fundamental 90-year-old prediction of general relativity. Also, by comparing the arrival times of light and gravitational waves from e.g. supernovae, Einstein's prediction that light and gravitational waves travel at the same speed could be checked. Finally, we could verify that they have the polarization predicted by general relativity. For detailed reviews on gravitational wave astronomy one can refer to [1, 2]

Neutron stars are cosmic laboratories for extreme physics. With densities reaching values beyond nuclear saturation they represent the densest states of cosmic matter. These extreme densities cannot be achieved in terrestrial laboratories and thus the astrophysical observations remain as the only tool of testing our theoretical models. During the last 50 years we managed via radio, X-ray and gamma ray observations to improve our understanding of the features observed on these extreme objects. They are observed as radio pulsars, as magnetars or as millisecond pulsars, isolated or in binary systems. Still our models remain quite basic, despite of the efforts into modeling the violent dynamics of supernovae and gamma-ray bursts, or the radio pulsar emission mechanism, the glitches etcetera.

During their evolution, neutron stars may undergo oscillations which can become unstable under certain conditions. Newly born neutron stars are expected to oscillate wildly during their creation shortly after the supernovae collapse [3, 4]. They are also expected to oscillate if they are members of binary systems and there is tidal interaction [5, 6] or mass and angular momentum transfer from a companion star and also when they undergo phase transitions [7, 8], which might be responsible for the observed glitches in pulsars. Rotation strongly affects these oscillations and perturbed stars can become unstable if they rotate faster than some critical velocities. During these oscillatory phases of their lifes, compact stars emit copious amounts of gravitational waves which together with viscosity tend to suppress the oscillations. The oscillations are divided into different families according to the restoring force [9, 10]. If pressure is the main restoring force, then these modes are called (pressure) *p-modes*, buoyancy is the restoring force of another class of oscillation modes, the *g-modes* while Coriolis force is the restoring force for the *inertial modes*. Spacetime induces another family of oscillations which couple only weakly to the fluid, these are the so-called *w-modes* [11]. There are more families of modes if one assumes the presence of crust [12, 13, 9, 14] or magnetic fields [15]. For a complete description of perturbation theory for relativistic stars one may refer to [10, 16, 17].

During the last two decades, the study of neutron star dynamics has become even more important due to the relations of the oscillations and instabilities to the emission of gravitational waves and the possibility of getting information about the stellar parameters (mass, radius, equation of state) by the proper analysis of the oscillation spectrum [18, 19, 20, 21, 22, 23]. Still, all these studies were mainly dealing with non-rotating stars, because the combination of rotation and general relativity made both the analytic and numerical studies extremely involved. This led to certain approximations in studying rotating stars in GR. The most obvious of them include the so called "slow-rotation" and the "Cowling" approximation. In the *slow rotation approximation*, one expands the perturbation equations in terms of a small parameter $\varepsilon = \Omega/\Omega_K$, where Ω is the angular velocity of the star and Ω_K stands for the "Kepler angular velocity" which is the maximum velocity that can be attained before the star splits apart due to centrifugal forces. In the *Cowling approximation*, one typically neglects the perturbations of the Newtonian potential or the spacetime in the case of GR. This is a quite good approximation, both qualitatively and quantitatively, for high order *p-modes*, for the *g-modes* and the inertial modes while it is only qualitatively good for the *f-modes*.

Although, most of our understanding about the oscillations of relativistic stars is due to perturbative studies, recently, the study of stellar oscillations using evolutions of the non-linear equations of motion for the fluid became possible [24, 25, 26, 27, 28]. Finally, differential rotation is another key issue that is believed to play an important role in the dynamics of nascent neutron stars. Actually, it is associated with dynamical instabilities both for fast and slowly rotating neutron stars [29, 30] and affects the onset of secular instabilities [31, 32, 33, 34].

Since rotational instabilities are typically connected with fast rotating stars, it is of great importance to study oscillations of this type of objects. As a first step one can drop the "slow rotation" approximation but still maintain the Cowling approximation i.e. to freeze the spacetime perturbations or in the best case to freeze the radiative part of these perturbations. This was the approach used up to now for most of the studies of the oscillations of fast rotating relativistic stars either in perturbative approaches [35, 36] or in non-linear (but axisymmetric) approaches [37].

If neutron stars rotate rapidly, nonaxisymmetric *dynamical* or *secular* instabilities can develop. These arise from non-axisymmetric perturbations which may grow if appropriate conditions met

Journal of Physics: Conference Series **222** (2010) 012031 doi:10.1088/1742-6596/222/1/012031

(temperature, rotation rate etc.) [16]. The amplitude of the emitted gravitational waves can be estimated as $h \sim MR^2\Omega^2/d$, where M is the mass of the body, R its size, Ω the rotation rate and d the distance to the source. This leads to an estimation of the maximum GW amplitude

$$h \approx 10^{-22} f^2_{1-3\mathrm{kHz}} d^{-1}_{15\mathrm{Mpc}} M_{1.4M_\odot} R^2_{10\mathrm{km}}. \quad (1)$$

If the instability persists for n rotation periods then the *effective amplitude* will be $h_{\mathrm{eff}} = h\sqrt{n}$ which means that the source can be detected from distances $d_{\mathrm{eff}} = d\sqrt{n}$. Earlier studies for the efficiency of the secular rotational instabilities [38, 39, 40] suggested, that the previous back of the envelope calculation is correct within an order of magnitude and this prompts for a detailed study of the problem using proper general relativistic hydrodynamics. The ingredients that one needs for a proper estimation are: i) the critical rotational value for the onset of the instability, ii) the growth time of the instability, iii) the viscous damping times associated with the specific instability, iv) the maximum amplitude that the instability may reach before it saturates due to non-linear couplings and v) the waveform of the emitted gravitational wave which is vital for the detection.

In the next section we review the most recent efforts in estimating the critical rotational values for the onset of the secular instabilities for uniform and differentially rotating neutron stars [41, 42, 34] while in the last section we present the recent efforts in understanding the dynamics of magnetars and to reveal their structure via the observed quasi-periodic oscillations.

2. Oscillations and Instabilities of Fast Rotating Compact Stars

It has already been known from the Newtonian case, that rotating stars are secularly unstable to non-axisymmetric perturbations once a dissipative effect extracts energy from the system [43]. For constant-density MacLaurin spheroids, this leads to the well-known transition from the axisymmetric MacLaurin-sequence to the triaxial Jacobi-sequence for a ratio of rotational to gravitational energy of $T/|W| = 0.14$. The precise mechanism of this type of instability is not limited to the Newtonian case though. Is is a generic behaviour of a new class of instabilities which are called rotational dragging instabilities, since everything that is needed for them in order to work is a coupling to some radiation field (which will dissipate energy) and a rotating background that leads to a splitting between pro- and retrograde travelling, non-axisymmetric modes. In GR, the emission of gravitational waves will extract energy from the star and for sufficiently high rotation rates, certain modes may destabilize the star due to this so-called CFS-instability [44, 45], which will then emit a copious amount of gravitational radiation. For more realistic equations of state (EoS) than the constant-density MacLaurin-spheroids, the onset of the gravitational wave-driven instability is actually shifted towards lower values of $T/|W|$ as we will see.

Early Newtonian calculations [46] also suggested, that polytropes with a polytropic index of $N > 0.808$ may never be secularly unstable because for these models, mass-shedding at the equator would set in before the instability regime is reached. General relativistic effects will also slightly change this result. Finally, differential rotation will dramatically change this picture, because depending on the precise rotation law one is using and the actual degree of differential rotation, the redistribution of angular momentum inside the star will lead to strongly deformed stars which can attain much higher rotation rates and an even earlier onset of the CFS-instability in terms of $T/|W|$ when compared to uniform rotation [47, 35, 34].

As already mentioned in Section 1, there are various ways to study neutron star oscillations either via direct time-integration of the full, non-linear equations of general relativity or by using certain simplifying approximations. Neglecting non-linear effects and possible mode-couplings, linear perturbation theory is a powerful tool to obtain mode-frequencies and eigenfunctions of small-scale oscillations around the static equilibrium. In addition, freezing the space-time degrees of freedom, i.e. working in the Cowling-approximation, has proven to be a quantitatively good approach for high-order pressure modes, g- and inertial modes while it is of limited accuracy for predicting fundamental mode frequencies [48, 49].

In the following discussion, we will mostly focus on the quadrupolar fundamental pressure mode although it should be noted, that other mode classes (g- and inertial modes) may also become unstable at considerably lower rotation rates e.g. [50, 42]. The current understanding is, that viscous effects most likely suppress these instabilities though e.g. [51].

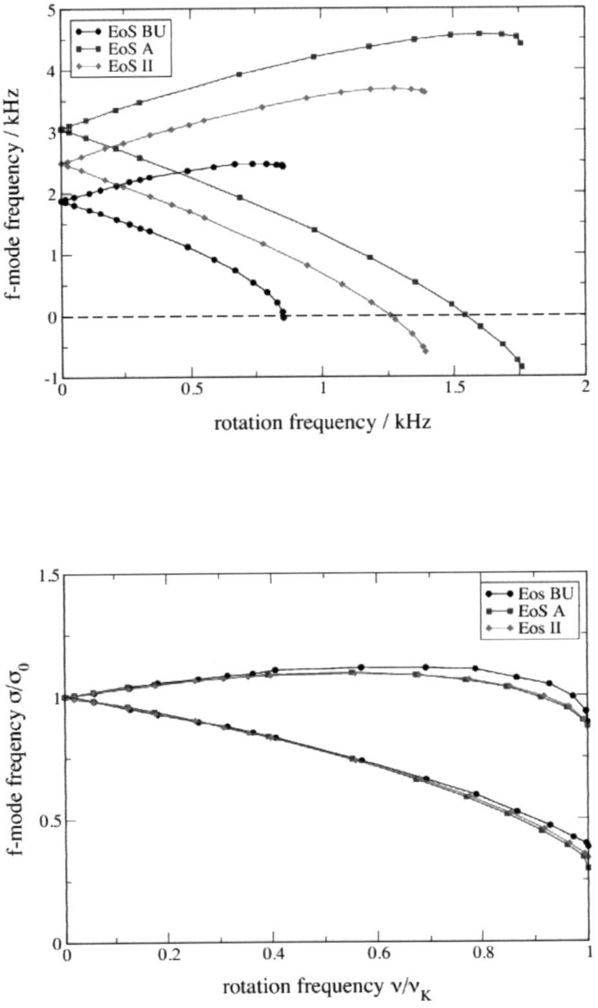

Figure 1. Splitting of the ^2f-mode for three different EoS. *Top panel:* Frequencies of co- and counterrotating branches for an inertial observer. *Bottom panel:* Normalized frequencies in the comoving frame.

Figure 1 shows the splitting of the fundamental quadrupolar mode for three different EoS; a detailed description of the equilibrium models can be found in [41]. The BU sequence is a set of constant energy-density models with a polytropic EoS and a polytropic index of $N = 1$, while EoS A and EoS II are polytropic fits to realistic equations of state with smaller polytropic indices. As one can see, actually all three sequences become CFS-unstable at a certain rotation rate; for the $N = 1$-polytrope this happens virtually at its mass-shedding limit while for the other, more compact models, the instability occurs well below the Kepler-frequency.

As it turns out, despite the obvious differences in the mode-frequencies for the various EoS in the inertial frame (top panel of Figure 1), when normalized in the comoving frame of reference, the frequencies change nearly in the same manner when rotation is increased (bottom panel of Figure 1). If σ_0 is the f-mode frequency in the nonrotating limit and ν_K the Kepler-frequency, for all rotation rates we can write

$$\frac{\sigma}{\sigma_0} = 1.0 + C^{(1)}_{lm}\left(\frac{\nu}{\nu_K}\right) + C^{(2)}_{lm}\left(\frac{\nu}{\nu_K}\right)^2 \tag{2}$$

and perform a quadratic fitting of the curves in the right panel of Figure 1. We find $C^{(1)}_{22} \approx -0.25$ and $C^{(2)}_{22} \approx -0.38$ for the $m = 2$ solution and $C^{(1)}_{2-2} \approx 0.48$ and $C^{(2)}_{2-2} \approx -0.55$ for the $m = -2$ branch. This can potentially be used in order to determine stellar parameters like mass, radius and angular velocity once one can actually detect oscillation frequencies of neutron stars.

Differential rotation adds a new degree of freedom to the problem which is of great importance for young hot nascent neutron stars. Depending on the exact rotation law for differential rotation, which is unknown currently, it usually allows for much higher rotation rates than for uniformly rotating stars since the total angular momentum can be disposed in a variety of possible configurations. Here, we focus on the commonly used *j-constant law*

$$A^2\left(\Omega_c - \Omega\right) = \frac{(\Omega-\omega)r^2\sin^2\theta e^{2(\psi-\nu)}}{1-(\Omega-\omega)^2r^2\sin^2\theta e^{2(\psi-\nu)}}, \tag{3}$$

where Ω_c is the angular velocity along the rotation axis and the parameter A controls the degree of differential rotation, see also [52].

The splitting of the fundamental mode for various degrees of differential rotation can be seen in Figure 2, where the dashed line shows the limit of uniform rotation. Since there is no unambiguous definition of a global angular velocity any more, we plot the mode frequencies against the ratio of rotational to gravitational energy $T/|W|$.

As one can see, a higher degree of differential rotation, which is indicated by a larger value of the dimensionless parameter $\hat{A}^{-1}$, leads to equilibrium models that can store a lot more rotational energy than the uniformly rotating case and second, the CFS-instability typically sets in at a smaller fraction of the maximum allowed $T/|W|$, which is indicated by the endpoints of the curves in Figure 2. This behaviour is generic in the sense that it not only applies for the fundamental quadrupolar mode but for high order non-axisymmetric modes as well. To make this point more clear, we can define the dimensionless quantity $\tilde{\beta}_c$ to be the value of $T/|W|$ at the onset of the CFS-instability, normalized by its corresponding value at the mass-shedding limit. This is somewhat similar to the normalization we did for the uniform rotation rate in equation (2). It then turns out, that increasing the degree of differential rotation leads to a continuous decrease in the critical value $\tilde{\beta}_c$ not only for the quadrupolar mode but for high order modes as well, for which the secular instability sets in even earlier when compared to the quadrupolar mode; see also Figure 3.

So, in general the onset of the CFS-instability is eased by differential rotation and for stiffer equations of state, these effects are even more pronounced, see [34].

Up to now, we were only considering the f-mode and its instability for uniformly and differentially rotating stars. There is another class of barotropic oscillation modes which only works for rotating stars and which we shortly want to mention here. They are called inertial

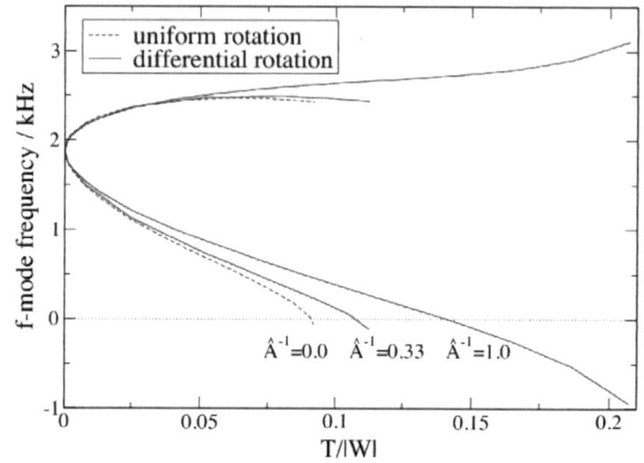

Figure 2. Splitting of the ${}^2f$-mode for different degrees of differential rotation for the B sequence. All frequencies are computed in the inertial frame.

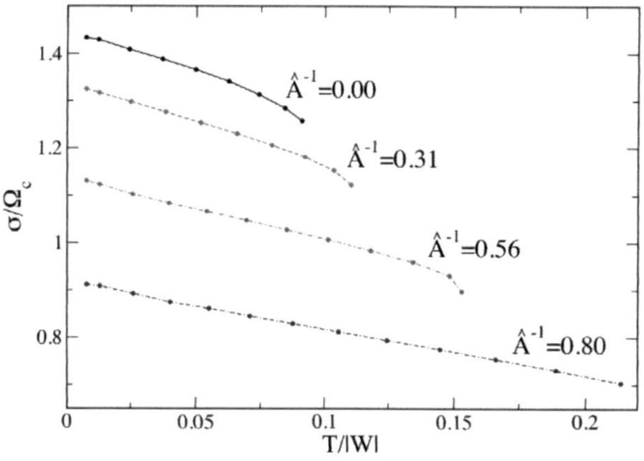

Figure 3. The critical value for the onset of the instability for different degrees of differential rotation and $m = 2, 3, 4$.

modes and are restored by the Coriolis force. Due to their low frequency, they are CFS-unstable already at very low rotation rates, see [54].

In Figure 4 we show the frequencies of the $l = 2, m = 2$ inertial mode, the so-called r-mode, along sequences of different degrees of differential rotation. We normalize them by the central angular velocity Ω_c of the corresponding background model, as it has already been done in [55]. This kind of normalization is suggested by the relation $\sigma = 4/3\,\Omega$ which is valid in Newtonian theory, where σ is the r-mode's angular frequency measured in the inertial frame and Ω the angular velocity of the star.

First, consider the solid curve representing the rigidly rotation case, i. e. $\hat{A}^{-1} = 0$. Concerning the value of σ/Ω_c in the non-rotating limit, which is approximately 1.43, we observe a small deviation from Newtonian theory. However, this is consistent with general relativistic corrections to the Newtonian result. As it has been shown in [56], the Newtonian value of 4/3 increases proportional to the frame-dragging potential ω in the metric.

The curves corresponding to the differentially rotating sequences have lower values, because the central angular velocity increases compared to the uniformly rotating ones, and this effect becomes even stronger the higher the degree of differential rotation is. On the other hand, the angular velocity at the equator, Ω_e, is lowered so that a normalization by this quantity would move the lines in the other direction. Finally, we chose to normalize by the central angular velocity in order to compare qualitatively with Newtonian findings in [55]. We find that the qualitative behaviour is the same.

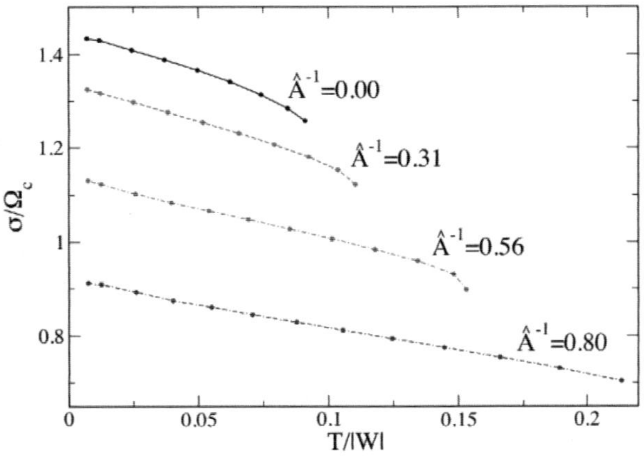

Figure 4. Frequencies of the r-mode for several sequences with different degrees of differential rotation. The frequencies are normalized by the central angular velocity Ω_c.

3. Magnetar Dynamics

Magnetars are neutron stars with a magnetic field, greatly exceeding $B > 10^{14}$ Gauss. The Soft Gamma Repeaters (SGRs) are astronomical objects with a giant flare activity: they release a huge amount of energy ($L \simeq 10^{44} - 10^{46}$ ergs/s) in few seconds. These giant flares are thought of being associated with starquakes, due to the huge magnetic field and its rearrangement. For this reason, the SGRs are considered to be magnetars.

Up to now, three SGRs have been detected: SGR 05026-66 in 1979, SGR 1900+14 in 1998 and SGR 1860-20 in 2004. In two of those objects (SGR 1900+14 and SGR 1860-20), a tail lasting several hundred seconds has been observed. The study of this part of the spectrum has revealed the presence of quasi-periodic oscillations QPOs (see [60], [64] and [65]), whose frequencies are approximately 18, 26, 30, 92, 150, 625 and 1840 Hz for SGR 1860-20 (with possible additional frequencies at 720 Hz and 2384 Hz) and 28, 53, 84 and 155 Hz for SGR 1900+14.

Since the range in frequencies is so wide, many types of modes have been excluded: for example, the interface modes generated in the layer between core and crust (they have frequencies ranging 10-30 Hz) and the shear modes in the solid crust of neutron stars since their frequencies start from around 100 Hz.

It is believed that, after fracturing of the crust, seismic oscillations could be excited: such oscillations are known as torsional oscillations (t-modes) and depend just on the shear velocity

Journal of Physics: Conference Series **222** (2010) 012031 doi:10.1088/1742-6596/222/1/012031

$v_s = \mu/\rho$, where μ is the shear modulus which takes into account the elastic response of the crust to the oscillations. The t-modes are axial-type oscillation with a typical frequency of $\sigma = v_s/R$ with R being the radius of the star.

Since the detection of QPOs, many models were constructed which tried to fit those frquencies. The first works focused purely on crustal torsional oscillations, as for example in [62] where a dipole field and a realistic EoS were used. However, not all the frequencies could be explained with such a model: [59] suggested that the observed spectra could be explained via global magneto-elastic oscillations while [61] pointed out that the spectruum could be a continuum. This idea was confirmed in [63], where a 2-dimensional time evolution equation of linearized Alfvén oscillations was used, together with a dipole magnetic field and realistic EoS. However, the two-dimensional equation that has been used there has a pathological behaviour that drives the code unstable and in order to stabilize the problem, a certain amount of artificial viscosity was needed.

In two recent papers, [57] and [58], the use of artificial viscosity was avoided: in the first case, the authors used a simple coordinate transformation which was based on the distribution of the magnetic field lines and which reduces the two-dimensional time-evolution equation of [63] to a one-dimensional equation, also avoiding the use of artificial viscosity while in the second case, the authors used a full non-linear code. In all these papers, the coupling between core and crust was not taken into account: the star was assumed to be a pure fluid. The coupling between crust and core is one of the main open issue in the study of QPOs: it may be able to explain the small gap between the lower frequencies 18 Hz, 26 Hz and/or 30 Hz in the data for SGR 1860-20. In fact, the crust modes are related to the lower frequencies by the following rule [66]:

$$t_n = \left[1 + a_n \left(B/B_\mu\right)^2\right]^{1/2} t_0\,, \tag{4}$$

where t_0 is the fundamental torsional frequency, t_n indicates the overtones of order n, a_n is a certain coefficient and $B_\mu = 4 \times 10^{15}$ Gauss is the magnetic field, measured at the pole. On the other hand, the Alfvén waves in the core seem to form a continuum and they are linked by

$$f_n^{\text{even}} = (2n+1) f_0\,, \quad f_n^{\text{odd}} = (n+1) f_0\,, \quad f_n^{\text{close}} = (n+1) f_0\,, \tag{5}$$

where *even* and *odd* indicates whether the boundary condition on the equator is of even or odd parity while the label *close* indicates the closed magnetic field lines that close inside the star without reaching the surface and which match with the dipole magnetic field outside the star. In Figures 5 and 6 we show a plot of the frequencies for open and closed lines respectively. If both core and crust frequencies are taken into account, then the complete QPO spectrum could easily be explained. A proper combined study of the coupled core and crust oscillations is still underway. Some preliminary results are shown in Table 1: the frequencies labelled as f_{crust} are due to the presence of the crust, while the frequencies labelled as $f_{\text{crust+core}}$ are present in the interior of the star. The frequencies in bold are approaching the 22, 58 and 116 Hz found in SGRs 1806-20 by [67]. The results in Table (1) seem to confirm the hypothesis, that the coupling between crust and core should lead to extra features in the spectrum [69]. In figure (7) we plot the frequencies in the core (green-dot line) and the extra frequencies given by the presence of the crust (red-continue line).

The study of QPOs in magnetars can constrain the mass and the radius of the neutron star and gives us information about the topology and the strenght of the magnetic field. In fact, taking a model with a given equation of state and performing our time-evolution analysis, we can see if the frequencies that we obtain can fit or no the observed frequencies. If this is the case, then we know the equation of state, the mass and the radius of our star. In our study, we find that the model with a mass $M = 1.4 - 1.5\ M_\odot$ and an equation of state not too stiff, may fit better the data.

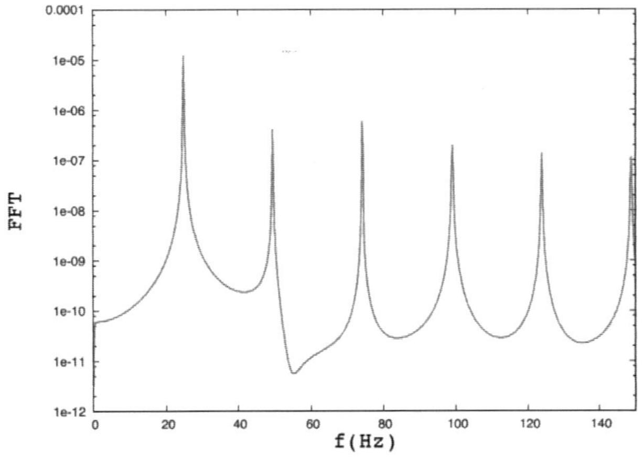

Figure 5. The QPOs value in the case of a fluid star. The mass of the star is $M = 1.4M_{\odot}$ and the model is APR2. The frequencies follow the pattern given in (5) for open lines.

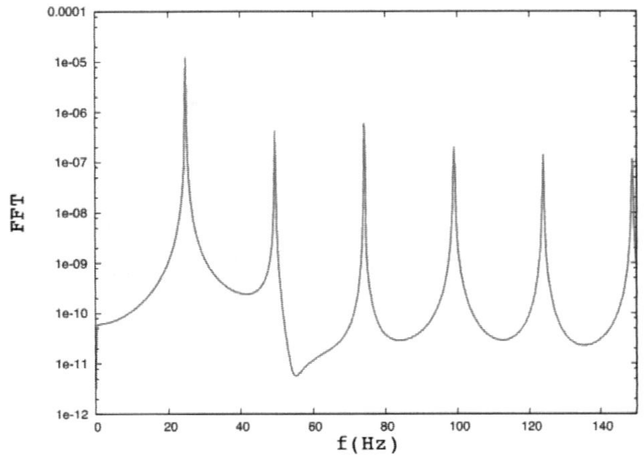

Figure 6. The QPOs value in the case of a fluid star. The mass of the star is $M = 1.4M_{\odot}$ and the model is APR2. The frequencies follow the pattern given in (5) for closed lines .

In the study of magnetar dynamics, one must not forget the importance of magnetic field configurations. It is known that different magnetic field configurations could influence the surface temperature of the star and induce anisotropies, see [70]. This effect is more relevant if the magnetic field is confined in the crust: such configurations are called crustal fields and they occur when the core is a type I superconductor. In this case, because of the Meissner effect, the magnetic field is expelled from the core in about $\simeq 100$ years.

In a paper by [71], different magnetic field configurations were studied in order to examine if and how they can fit the QPOs observations. The authors consider both a toroidal and a poloidal component of the magnetic field. The study included a configuration that owns a

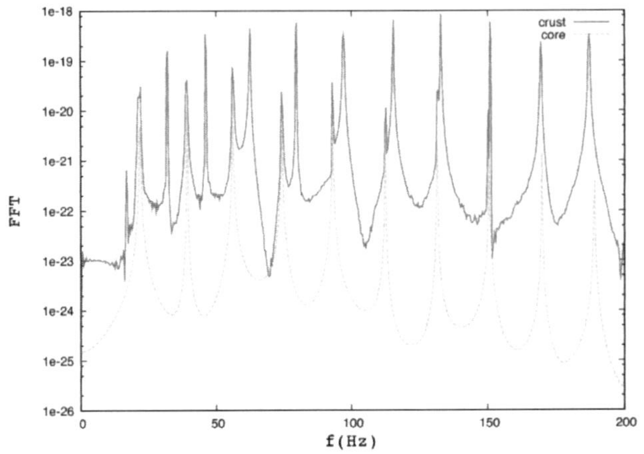

Figure 7. The QPOs value in presence of the coupling between crust and core. The mass of the star is $M = 1.4M_{\odot}$ and the model is APR2. It is possible to see extra frequencies showing.

Model	$f_{\text{crust+core}}$(Hz)	$f_{\text{crust}}(Hz)$
		16.93
	22.28	**22.28**
		31.57
APR_{14}	39.74	39.74
		45.80
	56.08	**56.08**
		62.53
	75.91	75.91
		80.09
	93.32	93.32
		96.95
	112.8	112.8
		115.3

Table 1. Frequencies present only in the crust (f_{crust}) and frequencies present in the core and in the crust ($f_{\text{crust+core}}$) for stellar model APR2 ([68]) and a mass of $M = 1.4M_{\odot}$

type I superconductor core (with both components of the magnetic field confined to the crust) while another configuration was constructed with the assumption that the magnetic field could permeate the star. They found that only if the magnetic field was permeating the whole star, the observed QPO data can be explained. So the case of a magnetic field solely confined in the crust should be excluded in QPO-studies; actually this result was confirmed in [72].

The most crucial missing points for the gravitational wave research is the absence of proper simulations of the way and the degree of excitation of the density variations. They are ultimately linked with the emission of gravitational waves which, if it takes place, will be a unique observation for astronomy. The event, a flaring magnetar, could potentially be observed in various parts of the electromagnetic and gravitational wave spectrum [73]. Such an observation will provide us with unique constraints on the structure and dynamics of magnetars and it could be a tremendous boost in the attempt to understand the structure of magnetized neutron stars.

The analysis of data taken during the super-flare event in SGR 1806-20 that took place in December 2004, when LIGO was in operation, revealed no evidence of gravitational wave emission [74, 75]. This is not a surpizing result since the actual sensitivity of LIGO at that time was three to four orders of magnitude below the expected sensitivity of the next generation of gravitational wave detectors such as the "Einstein telescope" (ET). Furthermore, a detailed analysis should not be performed only for signals at the observed QPO frequencies alone, since the accompanying gravitational wave signal might be emitted at completely different frequencies.

Acknowledgments

This work was supported by the German Science Council (DFG) via SFB/TR7, E.G. was funded by EGO via the VESF fellowship program.

References

[1] Maggiore, M. 2007, Gravitational Waves: Theory and Experiments, Oxford Univ. Press
[2] Kokkotas, K.D. 2007, Acta Phys. Polonica B, 38, 3891
[3] Ott, C.D. 2007, Class. Quant. Grav., 24, S139
[4] Ott, C.D. 2009, Class. Quant. Grav., 26, 204015
[5] Kokkotas, K.D & Schäfer, G. 1995, M.N.R.A.S., 275, 301
[6] Flanagan, E.E & Racine, E. 2007, Phys. Rev. D, 75, 044001
[7] Marranghello, G.F., Vasconcellos, C. A. & de Freitas Pacheco J. A. 2002, Phys. Rev. D, 66, 064027
[8] Ferrari, V., Gualtieri, L., & Pons, J.A. 2007, Class. Quant. Grav., 24, 5093
[9] McDermott, P.N., van Horn, H. M. & Hansen, C. J. 1988, ApJ, 325, 725
[10] Kokkotas, K.D. & Schmidt, B.G. 1999, Living Reviews in Relativity, 2, 2
[11] Kokkotas, K. D. & Schutz, B. F. 1992 M.N.R.A.S., 255, 119
[12] Schumaker, B.L. & Thorne, K.S. 1983 M.N.R.A.S., 203, 457
[13] McDermott, P. N., van Horn, H. M., Hansen, C. J. & Buland, R. 1985, ApJ, 297, L37
[14] Vavoulidis, M., Kokkotas, K.D. & Stavridis, A. 2008 M.N.R.A.S. 384, 1711
[15] Carroll, B. W. , Zweibel, E. G., Hansen, C. J., McDermott, P. N., Savedo, M. P., Thomas, J. H. & van Horn, H. M. 1986, ApJ, 305, 767
[16] Andersson, N. & Kokkotas, K.D. 2001, Int. J. Modern Phys. D, 10, 381
[17] Stergioulas, N. 2003, Living Reviews in Relativity, 6 , 3
[18] Andersson, N. & Kokkotas, K.D. 1996, Phys. Rev. Lett., 77, 4134
[19] Andersson, N. & Kokkotas, K.D. 1998 M.N.R.A.S., 299, 1059
[20] Benhar, O., Berti, E., & Ferrari, V. 1999, M.N.R.A.S., 310, 797
[21] Kokkotas, K.D., Apostolatos, T. & Andersson, N. 2001, M.N.R.A.S., 320, 307
[22] Sotani, H., Kohri, K. & Harada, T. 2004, Phys. Rev. D, 69, 084008
[23] Sotani H. & Kokkotas, K.D. 2004, Phys. Rev. D, 70, 084026
[24] Font, J.A., Stergioulas, N. & Kokkotas, K.D. 2000, M.N.R.A.S., 313, 678
[25] Font, J.A., Dimmelmeier, H., Gupta, A. & Stergioulas, N. 2001, M.N.R.A.S., 325, 1463
[26] Stergioulas, N., Apostolatos, T.A. & and Font, J.A. 2004, M.N.R.A.S., 352, 1089
[27] Dimmelmeier, H., Stergioulas, N. & Font, J.A. 2006, M.N.R.A.S., 368, 1609
[28] Kastaun, W. 2006, Phys. Rev. D, 74, 124024
[29] Centrella, J.M., New, K.C.B., Lowe, L.L. & Brown, J. D. 2001, ApJ, 550, L193
[30] Shibata, M., Karino, S. & Eriguchi, Y. 2002, M.N.R.A.S. 334, L27
[31] Dimmelmeier, H., Font, J. A. & Müller, E. 2002, A&A, 393, 523
[32] Stavridis, A., Passamonti, A. & Kokkotas, K.D. 2007, Phys. Rev. D, 75, 064019
[33] Passamonti, A., Stavridis, A. & Kokkotas, K.D. 2008, Phys. Rev. D, 77, 024029
[34] Krüger, C., Gaertig, E. & Kokkotas, K.D. 2009, arXiv:0911.2764
[35] Yoshida, S., Rezzolla, L., Karino, S. & Eriguchi, Y. 2002, ApJ, 568, L41
[36] Boutloukos, S. & Nollert, H.-P. 2007, Phys. Rev. D, 75, 043007
[37] Dimmelmeier, H., Stergioulas, N. & Font, J. A. 2006, M.N.R.A.S., 368, 1609
[38] Lai, D. & Shapiro, S.L. 1995, ApJ, 442, 259
[39] Shibata, M. & Karino, S. 2004 Phys. Rev. D, 70, 084022
[40] Ou, S., Tohline, J.E. & Lindblom, L. 2004, ApJ, 617, 490
[41] Gaertig, E. & Kokkotas, K.D. 2008, Phys. Rev. D, 78, 064083

[42] Gaertig, E. & Kokkotas, K.D. 2009, Phys. Rev. D, 80, 064026
[43] Chandrasekhar, S. 1969, Yale University Press
[44] Chandrasekhar, S. 1970, Phys. Rev. Lett., 24, 611
[45] Friedman, J. L. & Schutz, B. F. 1978, ApJ, 22, 281
[46] James, R. A. & Schutz, B.F. 1964, ApJ, 140, 552
[47] Karino, S. and Eriguchi, Y. 2002, ApJ, 578, 413
[48] Lindblom, L. and Splinter, R. J. 1990, ApJ, 348, 198
[49] Yoshida, S. and Kojima, Y. 1997, MNRAS, 289, 117
[50] Passamonti et al. 2009, MNRAS, 394, 730
[51] Chatterjee, D. and Bandyopadhyay, D. 2008, Journal of Physics G, 35, 10, 104078
[52] Komatsu, H. and Eriguchi, Y. and Hachisu, I. 1989, MNRAS, 237, 355
[53] Komatsu, H. and Eriguchi, Y. and Hachisu, I. 1989, MNRAS, 239, 153
[54] Andersson, N. 1998, ApJ, 502, 708
[55] Karino, S. and Yoshida, S. and Eriguchi, Y. 2001, Phys. Rev. D, 64, 024003
[56] Yoshida, S. and Yoshida, S. and Eriguchi, Y. 2005, MNRAS 356, 217
[57] Colaiuda A., Beyer H., Kokkotas K.D., 2009, MNRAS, 396,1441
[58] Cerda-Duran P., Stergioulas N., Font J.A., 2009, MNRAS, 397, 1607C
[59] Glampedakis K., Samuelsson L., Andersson N., 2006, MNRAS, 371, L74
[60] Israel G. et al., 2005 ApJ, 628, L53
[61] Levin Y., 2006, MNRAS, 368, L35
[62] Samuelsson L., Andersson N., 2007, MNRAS, 374, 256
[63] Sotani H., Kokkotas K.D., Stergioulas N. 2008a, MNRAS, 385, L5
[64] Watts A.,Strohmayer T.E., 2006 ApJ, 637, L117
[65] Watts A.L., Strohmayer T.E., 2006, Advances in Space Research, 2007, 308, 625W
[66] Sotani H., Kokkotas K.D., Stergioulas N. 2007, MNRAS, 375, 261
[67] Hambaran V., Neuhaeuser R., 2009, to be published
[68] Akmal A.,Pandharipande V.R., Ravenhall D.G., Phys. Rev. C58, 1804 (1988)
[69] Colaiuda A., Kokkotas K.D., 2009, preprint
[70] Pons J.A., Geppert U., 2007, A&A, 407, 303P
[71] Sotani H., Colaiuda A.,Kokkotas K.D. 2008b, MNRAS, 385, 2161
[72] Van Hoven M., Levin Y., 2008, MNRAS, 391, 283
[73] Stella, L., Dall' Osso, S., Israel, G.L. 2009, Mem. S.A.It., 80, 186
[74] Abbott B., et al 2007, Phys. Rev. D, 76, 062003
[75] B. Abbott et al. 2008 Phys. Rev. Lett., 101, 211102

Journal of Physics: Conference Series **222** (2010) 012032 doi:10.1088/1742-6596/222/1/012032

Gravitational-wave signature of a thin-shell gravastar

Paolo Pani

E-mail: paolo.pani@ca.infn.it
Dipartimento di Fisica, Università di Cagliari, and INFN sezione di Cagliari, Cittadella Universitaria 09042 Monserrato, Italy

Emanuele Berti

E-mail: berti@phy.olemiss.edu
Department of Physics and Astronomy, The University of Mississippi, University, MS 38677-1848, USA
Theoretical Astrophysics 350-17, California Institute of Technology, Pasadena, CA 91125, USA

Vitor Cardoso

E-mail: vitor.cardoso@ist.utl.pt
Department of Physics and Astronomy, The University of Mississippi, University, MS 38677-1848, USA
Centro Multidisciplinar de Astrofísica - CENTRA, Dept. de Física, Instituto Superior Técnico, Av. Rovisco Pais 1, 1049-001 Lisboa, Portugal

Yanbei Chen, Richard Norte

E-mail: yanbei@tapir.caltech.edu, norte@caltech.edu
Theoretical Astrophysics 350-17, California Institute of Technology, Pasadena, CA 91125, USA

Abstract. We discuss how the emission of gravitational waves by ultra-compact objects can be qualitatively di erent depending on the presence or absence of an event horizon. Our case study are nonrotating thin-shell gravastars. The model has an in nitely thin shell with nite tension separating a de Sitter interior and a Schwarzschild exterior. The shell can be located arbitrarily close to the would-be event horizon, so a gravastar might seem indistinguishable from a black hole when tests are only performed on its external metric. We discuss gravitational perturbations of thin-shell gravastars and show that the spectrum of axial and polar quasinormal modes is completely di erent from that of a Schwarzschild black hole, even in the limit when the surface redshift becomes in nite. Furthermore, we study gravitational wave emission from the quasi-circular extreme mass ratio inspiral of compact objects of mass m_0 into massive thin-shell gravastars of mass $M \gg m_0$. The power radiated in gravitational waves during the inspiral shows distinctive peaks corresponding to the excitation of the polar oscillation modes of the gravastar. The frequency of these peaks usually depends on the equation of state, but for ultra-compact gravastars it typically lies within the optimal sensitivity bandwidth of LISA, providing a very speci c signature of the horizonless nature of the central object.

1. Introduction

One of the most elusive properties characterizing black holes (BHs) in general relativity is the presence of an event horizon. Evidence supporting the astrophysical reality of BHs necessarily depends on direct or indirect observations of an event horizon in astrophysical ultra-compact objects [1, 2]. Astrophysical observations usually probe the weak-gravity region, i.e. the space-time far away from the event horizon. The most precise measurements so far indicates the presence of a "dark object" of mass $M \simeq (4.1 \quad 0.6) \quad 10^6 M_\odot$ in our own galaxy [3]. Recent millimeter and infrared observations of Sagittarius A , the compact source of radio, infrared and X-ray emission at the center of the Milky Way, infer an intrinsic diameter of $37^{+16}_{\ 10}$ microarcseconds, which is even *smaller* than the expected apparent size of the event horizon of the presumed BH [4]. Traditional electromagnetic astronomy can at best yield lower limits on the gravitational redshift corresponding to hypothetical surfaces replacing the event horizon, and indeed some hold the view that an observational proof of the existence of event horizons based on electromagnetic observations is basically impossible (but see [1, 2, 5] for di erent viewpoints on this delicate issue). For this reason studies of the gravitational radiation signatures of event horizons, possibly probing the strong-gravity region, are necessary to con rm or disprove the BH paradigm [5, 6].

Gravitational-wave detectors o er a new way of observing BHs, complementing the wealth of information from present electromagnetic observations [7, 8]. A promising approach consists in probing the structure of compact objects which is encoded in their free oscillation spectrum, i.e. in their quasinormal modes (QNMs) [9]. Measuring several free oscillation frequencies, and comparing them with the QNM spectrum of BHs, provides a promising method to discern between BHs and horizonless objects [10]. These tests are one of the main goals of the Laser Interferometer Space Antenna (LISA) and they require a signal-to-noise ratio which may be even achieved by advanced Earth-based gravitational-wave interferometers [11, 12, 13].

In this paper we review some recent results in testing possible alternatives to the BH paradigm, following Refs. [14, 15]. We discuss compact objects whose external metric is virtually identical to that of a Schwarzschild BH, but which do not possess an event horizon. We focus on one of the simplest exemplars of such ultracompact horizonless objects: nonrotating thin-shell gravastars [16, 17]. We investigate their complete linearized dynamics, generalizing previous results [17, 18] by considering the thin-shell as a dynamical entity. Gravitational perturbations are generically discontinuous across the shell, due to a nite tension, and suitable matching conditions between interior and exterior perturbations have to be imposed [14]. By applying this general formalism, we discuss the QNM spectrum of a thin-shell gravastar. The spectrum of both axial and polar modes is complex and profoundly di erent from that of a BH, mainly because of the di erent boundary conditions at the surface replacing the event horizon. In particular polar modes depend on the equation of state (EOS) of matter on the shell and they can be used to discern between di erent gravastar models.

Furthermore polar modes can be excited during the inspiral of compact objects in quasi-circular orbits around stars [19, 20]. Thus di erences between the QNM spectrum of an horizonless object and that of a BH correspond to a di erent gravitational signal emitted by the orbiting object because of the resonant scattering of gravitational radiation. We discuss gravitational-wave emission from the quasi-circular extreme mass ratio inspiral of compact objects of mass m_0 into massive thin-shell gravastars of mass $M \gg m_0$ [15]. As expected, the power radiated in gravitational waves during the inspiral shows distinctive peaks corresponding to the excitation of the polar oscillation modes of the gravastar.

The plan of the paper is as follows. In Sec. 2 we study free gravitational perturbations of a thin-shell gravastar, discussing matching conditions across the shell and reviewing the QNM spectrum. In Sec. 3 we discuss how the perturbation functions outside the shell (as obtained by the matching conditions) can be used to compute the gravitational-wave emission from the quasi-circular extreme mass ratio inspiral of a point-like mass into a massive thin-shell gravastar. We conclude by discussing results and possible extensions of our work.

2. Gravitational perturbations of thin-shell gravastars

Let us start with the equilibrium model, which is a static thin-shell gravastar described by the metric [16, 17]

$$ds_0^2 = \quad f(r)dt^2 + \frac{1}{h(r)}dr^2 + r^2(d\theta^2 + \sin^2\theta d\varphi^2)\,, \tag{1}$$

with

$$f(r) = \begin{cases} h = 1 \quad \frac{2M}{r}\,, & r > a\,, \\ \alpha\, h = \alpha \quad 1 \quad \frac{8\ \rho}{3} r^2 \quad , & r < a\,, \end{cases} \tag{2}$$

where M is the gravastar mass and $\rho \quad 3M/(4 \quad a^3)$ is the "energy density" of the interior region. Junction conditions require the induced metric to be continuous across the shell, at $r = a$. This implies that also $f(r)$ has to be continuous at $r = a$, i.e. $\alpha = 1$. The metric above reduces to the de Sitter metric in the region $r < a$ and to the Schwarzschild metric for $r > a$. On the shell we must impose Israel's junction conditions [21] between the extrinsic curvature K_{ij} and the surface-energy tensor S_{ij}, i.e. $[[K_{ij}]] = 8 \quad [[S_{ij} \quad \gamma_{ij}S/2]]$. Here and in the following we adopt the same index notation as in Ref. [14]. The symbol "[[...]]" denotes the "jump" of a given quantity across the shell, i.e. the di erence between the limits of the corresponding quantity as $r \to a$. The surface energy and surface tension Θ are then related to discontinuities in the metric coefficients as [16]

$$[[\sqrt{h}]] = \quad 4 \quad a \quad , \quad \left[\left[f'\sqrt{h}/f\right]\right] = 8 \ (\qquad 2\Theta)\,. \tag{3}$$

The QNM spectrum of thin-shell gravastars has been studied in great detail in Ref. [14] using the Regge-Wheeler (RW) gauge [22] (see also [17] for the axial QNM spectrum of ve-layer gravastar models). In the RW gauge, perturbations split into two independent sets: two *a ial* functions, usually called h_0 and h_1 and four *olar* functions, $H_0\,, H_1\,, H_2\,, K$. Axial and polar perturbations can be recast into two single functions, the so-called RW and Zerilli functions respectively. Here we brie y review the method developed in [14], which consists in matching perturbations in the de Sitter interior with perturbations in the Schwarzschild exterior. The matching procedure is performed by using Israel's junction conditions.

In the de Sitter interior, *both* the RW and Zerilli functions are described by the following analytical solution of the linearized equations [14]

$$\Psi^{\rm in} = r^{l+1} \quad 1 \quad C \quad \frac{r}{2M} \quad ^2 \quad ^{i\frac{2M\omega}{2\sqrt{C}}} F \quad \frac{l+2 \quad i\frac{2M\omega}{\sqrt{C}}}{2}, \frac{1+l \quad i\frac{2M\omega}{\sqrt{C}}}{2}, l+\frac{3}{2}, C \quad \frac{r}{2M} \quad ^2 \Big)\,, \tag{4}$$

where $C \quad (2M/a)^3 = 8\mu^3$ and $F(a,b,c,z)$ is the hypergeometric function. In obtaining the solution above, one must require regularity at the center ($r = 0$). For a thin-shell gravastar the surface stress-energy tensor Θ is, in general, nonvanishing. This implies that the perturbation functions are discontinuous across the shell. For axial perturbations these matching conditions read [14]

$$[[h_0]] = 0\,, \quad \left[\left[\sqrt{h}h_1\right]\right] = 0\,. \tag{5}$$

For a thin-shell gravastar Eqs. (5) just imply continuity of the RW function across the shell. In the polar case, the matching of interior and exterior perturbations at the gravastar shell requires a more careful analysis because (unlike axial perturbations) polar perturbations of spherical objects actually induce motions of matter, which in turn couples back to gravitational perturbations. In Ref. [14] the following relations for the jump of the polar metric functions across the shell have been derived:

$$[[K]] = 0\,, \quad [[K']] = \quad 8 \frac{\quad}{\sqrt{f(a)}}\,,$$

$$\frac{2M}{a^2}[[H]] \quad [[H\, f']] \quad 2f(a)[[H']] + 4i\omega[[H_1]] = 16 \quad \sqrt{f(a)}(1+2v_s^2) \quad . \qquad (6)$$

The parameter v_s depends on the EOS on the thin shell, $\Theta = \Theta(\;)$:

$$v_s^2 \quad \frac{\partial\Theta}{\partial}\Big|_{\Sigma=0}\,, \qquad (7)$$

and it has the dimensions of a velocity. Roughly speaking, this parameter is related to the sound speed on the shell, although it is not necessarly bounded by $0 < v_s^2 < 1$ [23].

From Eq. (6) it is clear that polar QNMs (unlike axial QNMs) should depend on v_s, i.e. on the equation of state on the shell. This is a new feature that does not arise in the case of axial perturbations [17]. The situation closely parallels the ordinary stellar perturbation problem [24, 25]. The role played by the equation of state in the dynamical stability of gravastars against *s herically symmetric* perturbations was discussed in Ref. [16]. Our calculations extend that investigation to nonradial oscillations (see also Ref. [26] for a discussion on stability of thin-shells surrounding a BH).

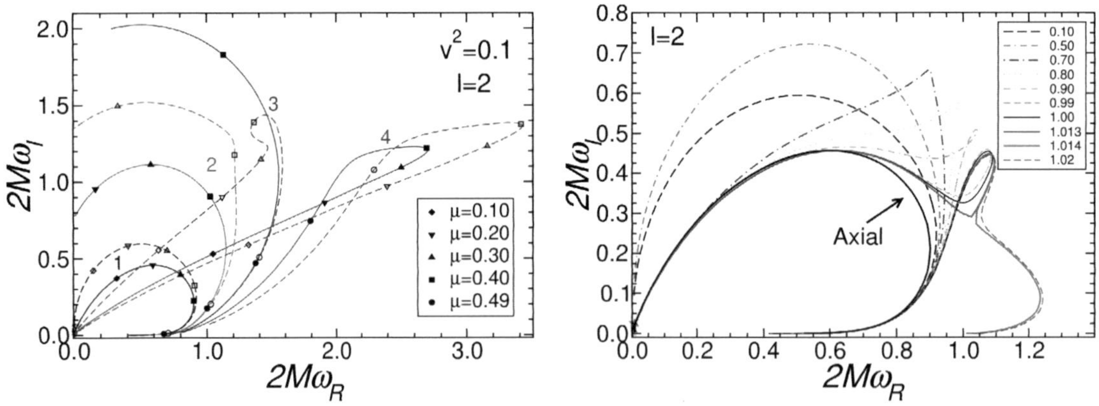

Figure 1. Left: Some axial (continuous lines) and polar (dashed lines) QNMs of a thin-shell gravastar with $v_s^2 = 0.1$ and $l = 2$ as the compactness μ varies. Numbers next to the polar and axial modes refer to the overtone index N ($N = 1$ being the fundamental mode). Right: Tracks of the fundamental polar and axial modes for di erent values of the "sound speed" parameter v_s when $v_s^2 > 0$. Figures are taken from Ref. [14].

2.1. N ectrum

Here we brie y review results obtained in Ref. [14] for the QNM spectrum of thin-shell gravastars. We focus of $l = 2$ modes, although qualitatively similar results hold for $l > 2$. Axial and polar QNM frequencies have been computed by using the junction conditions (5) and (6) and the continued fraction method [27]. Some QNMs of a static thin-shell gravastar are shown in Fig. 1.

The QNM spectrum is qualitatively similar to that of 'ordinary" ultracompact stars (compare Fig. 1 with those in Refs. [28, 29]). For numerical calculations we rst choose to keep v_s xed and to vary μ from some starting value. In the left panel of Fig. 1 we chose $v_s^2 = 0.1$, and we computed the QNM spectrum for $\mu = 0.4$ by using the continued fraction method. We then follow each QNM as $\mu \to 0$ and as $\mu \to 1/2$ to produce the tracks displayed in the gure. The imaginary part of both axial modes (continuous lines) and polar modes (dashed lines) becomes very small in the "Schwarzschild limit", i.e. as $\mu \to 1/2$. Of course this method does not allow to nd QNMs which do not exist when $\mu = 0.4$, but otherwise exist for smaller (or larger) compactness. Thus we choose di erent starting values of μ, in order to cover the entire spectrum.

Numerical results are in excellent agreement with previous results [17, 18] for axial modes. A new feature emerging in our case is that, unlike axial modes, polar modes depend on the EOS. In the right panel of Fig. 1 we show the tracks described in the complex plane by the fundamental polar and axial mode as we vary the compactness parameter μ and for di erent positive values of v_s^2. Although not shown, when $v_s^2 < 0$ polar modes are qualitatively similar to the axial ones.

One of the most important conclusions is that *neither a ial nor olar modes of a gravastar reduce to the N s of a chwar schild H when* $\mu \to 1/2$. In this limit, the real part of most modes is extremely small (much smaller than the Schwarzschild result, $2M\omega_R \simeq 0.74734$ for the fundamental mode with $l = 2$ [7]). Indeed, the QNM spectrum is drastically di erent from the QNM spectrum of a Schwarzschild BH: when $\mu \to 1/2$ the entire spectrum seems to collapse towards the origin. This is in sharp contrast with the Schwarzschild BH case and it can be used to tell very compact gravastars from BHs [17].

The dependence of polar QNMs on the EOS as shown in the right panel of Fig. 1 is quite interesting. For small values of the sound speed parameter (v_s^2 0.5 or so) the modes show an ordinary behavior, which is similar to that (not shown) for $v_s^2 < 0$ and to that for axial modes. However for larger values modes have a peculiar behavior, until a critical value $v_{\rm crit} \simeq 1.007$ is reached and, for $v_s > v_{\rm crit}$, the QNM behavior changes quite drastically.

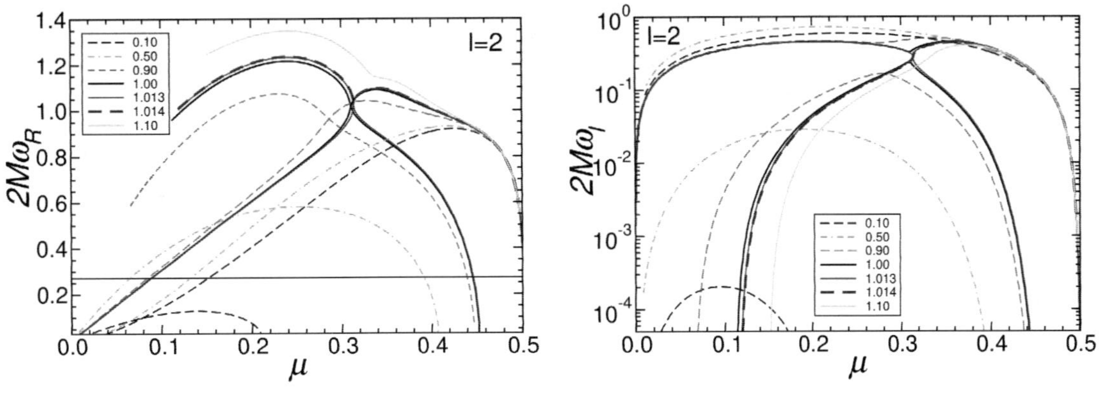

Figure 2. Real (left) and imaginary parts (right) of the fundamental polar mode for di erent values of the equation of state parameter v_s. For each (μ,v_s) a couple of mode exists, which is shown with the same linestyle. The horizontal line in the left panel corresponds to twice the orbital frequency of a particle in circular orbit at the innermost stable circular orbit (ISCO): only modes below the line can be excited during a quasicircular inspiral (cfr. Sect. 3). Figures are taken from Ref. [14].

Even more interestingly, when $v_s^2 > 0$ the QNM spectrum has a very rich and complex structure. A *second* family of QNMs with very small imaginary part appears. In the left (right) panel of Fig. 2 we show the real (imaginary) part of both the rst and the second family of modes, for some selected values of v_s. The second family of QNMs has very long damping, and in this sense it is similar to the s-modes of "ordinary" ultra-compact stars discussed by Chandrasekhar and Ferrari [25], although for gravastars these modes appear for *small* compactness $\mu < \mu_{\rm crit}(v_s, l)$. These mode are suitable to be computed also by the resonant method [25, 30] whose results are in excellent agreement with the continued fraction method. The imaginary part of modes with $v_s^2 \gtrsim 0.84$ rapidly approaches zero at some nite compactness μ while the real part of the modes stays nite. This suggests an instability at high v_s, as shown by the critical threshold in Fig. 3. In the region below the solid line, thin-shell gravastars may undergo an instability to nonradial perturbations with $l = 2$. Moreover the dashed line in Fig. 3 corresponds to the critical line above which the second family of modes disappears (both the real and imaginary part of modes tend to zero).

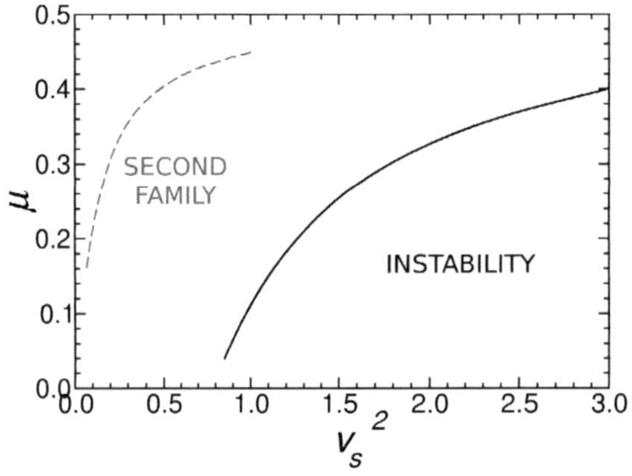

Figure 3. Signi cant thresholds in the (μ, v_s^2) plane. Numerical results suggest that the region below the black solid line corresponds to unstable thin-shell gravastar models. The second family of modes exists in the region of parameter below the red dashed line (see the text for a detailed discussion). Figures are taken from Ref. [14].

3. Gravitational-wav emission from extreme mass ratio inspirals

Here we consider a physical process where the GW energy spectrum has distinct signatures depending on whether the dynamics involve a thin-shell gravastar or a BH. We review recent results in Ref. [15], in which the gravitational emission by a particle with mass m_0 orbiting a horizonless object with mass $M \gg m_0$ has been computed in detail (see Refs. [19, 20, 31] for related works considering neutron stars). We consider a particle orbiting a thin-shell gravastar in a circular orbit with radius R_0. This should be sufficient to show that gravitational radiation from extreme mass ratio inspirals around horizonless objects is drastically di erent from the BH case. Similar arguments can presumably be made for more complex (or contrived) gravastar models and for eccentric orbits.

The time-averaged energy- ux can be shown to be [19, 15]

$$\dot{E}^R(m\omega_K) \quad \left\langle \frac{dE_{GW}}{dt} \right\rangle = \sum_{lm} \frac{m_0^2}{4 \ (m\omega_K)^2} |\hat{A}_{lm}(m\omega_K)|^2 \quad \sum_{lm} \dot{E}^R_{lm}. \tag{8}$$

where the dot indicates di erentiation with respect to proper time, E is the energy per unit mass of the particle, $\omega_K = \sqrt{M/R_0^3}$ denotes the Keplerian orbital frequency and (l, m) are the usual quantum numbers. As discussed in detail in Ref. [19], the function $\hat{A}_{lm}(\omega)$ is related to the amplitude of the wave at radial in nity apart a Dirac delta contribution, $(\omega \quad m\omega_K)$, and it can be computed in terms of two independent solutions Ψ^0_{lm} and Ψ^1_{lm} of the following inhomogeneous Bardeen-Press-Teukolsky (BPT) equation [32, 33]

$$\left\{\Delta^2 \frac{d}{dr}\left[\frac{1}{\Delta}\frac{d}{dr}\right] + \left[\frac{r^4\omega^2 + 4i(r-M)r^2\omega}{\Delta} - 8i\omega r - 2n\right]\right\}\Psi_{lm}(\omega, r) = -T_{lm}(\omega, r), \quad (9)$$

satisfying suitable boundary conditions [31]. In the equation above $\Delta = r^2 - 2Mr$ and the source term $T_{lm}(\omega, r)$ describes the point mass m_0 moving on a given orbit around the gravastar. In order to evaluate Ψ^0_{lm} and Ψ^1_{lm}, the BPT equation is integrated with an adaptive Runge-Kutta method starting with initial conditions on the gravastar shell obtained from the Zerilli and RW functions (as well as their derivatives), which can be constructed from the interior perturbations, Eq. (4), by using the junction conditions Eqs. (5)-(6) (see [15] for details).

The energy flux emitted in gravitational waves $\dot{E}^R$, normalized by the Newtonian quadrupole energy flux $\dot{E}^N$, reads

$$P(v) \equiv \frac{\dot{E}^R}{\dot{E}^N} = \sum_{lm} \frac{5}{128}\frac{M^2}{(m\omega_K)^2 v^{10}}|\hat{A}_{lm}(m\omega_K)|^2 . \quad (10)$$

where $\dot{E}^N = 32 m_0^2 v^{10}/(5M^2)$, $v = (M\omega_K)^{1/3} = p^{-1/2}$ being the orbital velocity and $p = R_0/M$ the semilatus rectum of the circular orbit. The normalized energy flux (10) can be computed up to $v \le 1/\sqrt{6} \simeq 0.408$, which corresponds to the innermost stable circular orbit (ISCO) at $R_0 = 6M$. The instability of circular orbits with $R_0 < 6M$ sets an upper bound on the velocity of the point mass. If the radius of the gravastar is larger than the ISCO, i.e. $\mu < 0.1666$, the upper limit in v is smaller. From the analytical form of the stress-energy tensor in Eq. (9) (see also [19]) it is easy to see that, for each assigned l, a mode of the star is excited when the orbital frequency satisfies the resonant condition

$$m\omega_{\rm K} = \omega_{\rm QNM} , \quad (11)$$

where $\omega_{\rm QNM}$ is the QNM frequency. Thus we expect sharp peaks to appear at the values of v corresponding to the excitation of the gravastar QNMs for different values of the angular momentum parameter l. This offers an intriguing signature of the absence of event horizons, since the emitted power for a Schwarzschild BH does not show any peak. In fact one can easily check that the frequency of the fundamental QNM of a Schwarzschild BH is *higher* than the critical value $m\omega_{\rm K}$ corresponding to a particle at the ISCO [31]. In other words, Schwarzschild QNMs can only be excited by particles plunging into the BH, while the QNMs of a gravastar can be excited *during the inspiral*.

3.1. Comparing the gravitational flux of a thin-shell gravastar and of a black hole

Our purpose here is to compare and contrast the energy flux from particles orbiting Schwarzschild BHs to the energy flux from particles orbiting thin-shell gravastars. The gravitational emission of a Schwarzschild BH perturbed by a particle has been studied analytically and numerically in great detail for both circular and eccentric orbits [34, 35, 36]. We consider thin-shell gravastars as a case study for generic compact horizonless objects, but several different models can be explored (see e.g. [16, 17]). However we expect the qualitative results of our analysis to apply quite in general, since the main difference in the gravitational-wave emission of a compact object comes from the different boundary conditions at the "surface" replacing the BH event horizon, rather than from the specific nature of this surface.

The numerical study in Ref. [15] covers the whole range in compactness ($0 < \mu < 0.5$) and it mainly focuses on the most physical range of the EOS parameter ($0 < v_s^2 < 1$), although the superluminal case ($v_s^2 > 1$) and models with $v_s^2 < 0$ were also investigated. For each value of the gravastar parameters (μ, v_s^2) we integrate the perturbations equations for a point-like object of mass m_0 moving on a circular orbit of radius R_0 with orbital velocity v and we compute the energy ux (10) by using the BPT formalism, as previously discussed.

As shown in Section 2, in the Schwarzschild limit $\mu \to 0.5$ the real part of the QNM frequency tends to zero and to a very good approximation it is independent of v_s [14]. In order for a QNM to be excited by particles in circular orbits, the QNM frequency must be small enough to allow for the resonant condition (11).

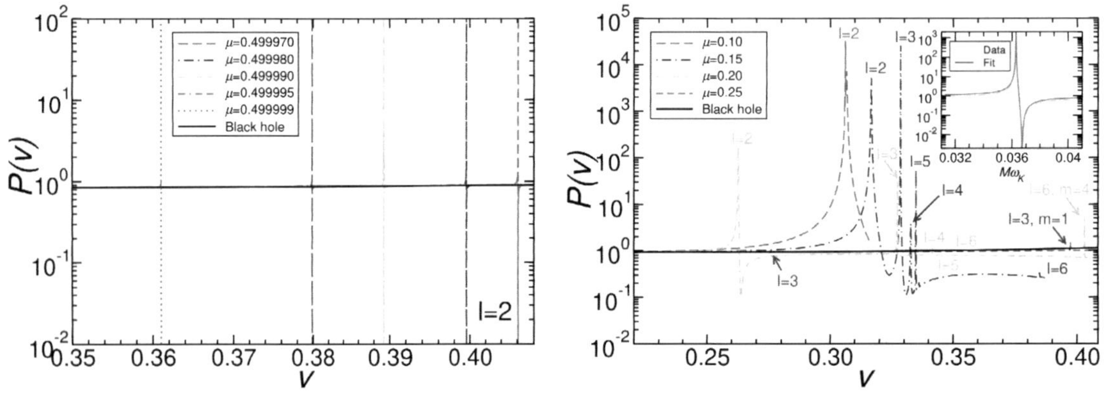

Figure 4. Left: Dominant ($l = 2$) contribution to the energy flux for very high compactness and $v_s^2 = 0.1$ (but when $\mu \sim 0.5$ resonances are almost independent on v_s^2). Right: Energy flux (summed up to $l = 6$) of the gravitational radiation emitted by a small mass orbiting thin-shell gravastars with $v_s^2 = 0.1$ and di erent values of μ, compared with the flux for a Schwarzschild BH, as a function of the particle orbital velocity v. In the inset a typical resonance is plotted.The comparison between numerical data and a simple harmonic oscillator [20] is shown. Figures are taken from Ref. [15].

In Fig. 4 we show the energy ux for a gravastar. In the left panel the dominant ($l = 2$) contribution at very high compactness is shown. The frequencies of the lowest QNMs of a Schwarzschild BH are higher than those of an ultra-compact gravastar, and cannot be excited by particles in stable circular orbits. For this reason the power emitted by a BH (on the scale of the plot) appear almost as a at line. Resonance peaks do appear for gravastars, as expected, when $\omega_{\rm QNM} = 2\omega_{\rm K}$. Notice that these resonances are extremely narrow and they would get even narrower for $l > 2$. This is because the imaginary part of the excited modes is extremely small ($2M\omega_I$ $10^{\ 7}$, $10^{\ 10}$ for $l = 2$ and $l = 3$, respectively) in the high-compactness limit $\mu \to 0.5$, leading to a corresponding decrease in the quality factor of the oscillations. Whether these resonances are actually detectable is an interesting question for LISA data analysis [31].

In Table 1 we list the expected excited modes, as extracted from the QNM spectrum for di erent values of μ corresponding to ultra-compact gravastars. Both for $l = 2$ and for $l = 3$ the typical frequencies lay within the optimal sensitivity bandwidth of LISA, providing a very speci c signature of the horizonless nature of the central object. A simple t of the resonant frequency as a function of the compactness suggests that LISA has the potential to reveal solid surfaces replacing horizons even when these solid surfaces are extremely close to the location of the Schwarzschild horizon [15].

The rich structure of the QNM spectrum allows gravastars which are as compact as neutron stars to leave a signature on the gravitational signal emitted by small, inspiralling compact objects. In the right panel of Fig. 4 we plot the normalized energy ux $P(v)$ as a function of the

Journal of Physics: Conference Series **222** (2010) 012032 doi:10.1088/1742-6596/222/1/012032

Table 1. Values of the compactness μ, angular momentum quantum number l, QNM frequency, orbital velocity v and gravitational-wave frequency $\nu_{\rm GW}$ of the circular orbits which correspond to the excitation of the fundamental QNM of the gravastar for the given multipole. The Keplerian frequency is given in mHz and rescaled to the typical gravastar mass $M_6 = 10^6 M_\odot$ that would be targeted by LISA [15].

	l	$M\omega_{\rm QNM}$	v	(M_6/M) GW (mHz)
0.49997	2	0.1339	0.4061	4.328
	3	0.1508	0.3691	4.873
0.49998	2	0.1276	0.3996	4.123
	3	0.1429	0.3625	4.616
0.49999	2	0.1180	0.3893	3.812
	3	0.1310	0.3521	4.232
0.499995	2	0.1096	0.3799	3.543
0.499999	2	0.0941	0.3610	3.041

orbital velocity for gravastar models with $v_s^2 = 0.1$ and compactness in the range $0.1 \lesssim \mu \lesssim 0.25$, as well as for a Schwarzschild BH. The total ux was computed by adding all multipoles ($|m|$ l) and by truncating the multipolar expansion at $l = 6$. Roughly speaking, a truncation at $l = 6$ produces a relative error (in the non-resonant regime) of order p ^{5} $= v^{10}$ [19, 36]. When $\mu \gtrsim 0.166$ the ISCO is located outside the gravastar and the plots extend up to the ISCO velocity $v_{\rm ISCO} \simeq 0.408$ (corresponding to $R_0 = 6M$). For less compact gravastars the energy ux is truncated at the velocity corresponding to the location of the shell. When $l = 2$ and $v_s^2 = 0.1$ no QNMs can be excited for 0.21 μ 0.49997. In this compactness range the energy ux emitted by either the gravastar or the BH is mostly due to the orbital motion and it only depends on the compactness of the central object.

In the inset of Fig. 4 we zoom around a typical resonance as function of the Keplerian orbital frequency $M\omega_{\rm K}$ of the particle for $\mu = 0.2$ and $l = 2$ (dashed green line). A t using the simple harmonic oscillator model of Ref. [20] (red line) reproduces the qualitative features of both resonance and antiresonance.

The complex structure of the spectrum for values of μ smaller than about 0.2 is best understood by considering the real and imaginary parts of the weakly damped QNM frequencies of a gravastar as plotted in the left and right panels of Fig. 2, respectively. Only QNMs whose real part lies *below* the horizontal line in the left panel (corresponding to twice the ISCO orbital frequency for a particle in circular orbit) can be excited. The range of μ over which QNMs can be excited depends on the speed of sound parameter v_s and on l.

4. Conclusions

We have discussed in detail gravitational perturbations of a thin-shell gravastar, a model which is representative of generic horizonless ultra-compact objects. We have studied the QNM spectrum and showed that both axial and polar modes are drastically di erent from Schwarzschild QNMs, even when the compactness approaches the Schwarzschild value, $\mu \to 0.5$. In particular polar modes depend on the equation of state of matter on the thin-shell and they may be used to discern between di erent gravastar models. More importantly, polar modes can be excited by small masses orbiting the thin-shell gravastar. We have discussed point-like particles in circular orbit and we have shown that for some orbital frequencies the energy ux in GWs shows resonances. The existence of resonant frequencies depends both on the compactness and on the equation of state on the shell. However, in the most interesting case of ultra-compact gravastars ($\mu \to 0.5$) resonant frequencies are independent on the EOS and above a certain critical compactness resonances always exist. Interestingly enough, typical resonant frequencies lay within the optimal sensitivity bandwidth of LISA, providing a very speci c signature of the horizonless nature of the central object. It would be interesting to generalize our method to more realistic models for horizonless objects [16, 17], possibly in connection with both circular and eccentric inspirals. Another interesting and nontrivial development is the generalization to the case of rotating gravastars [37, 38].

Acknowledgements
This work was partially supported by FCT - Portugal through projects PTDC/FIS/64175/2006, PTDC/FIS/098025/2008 and PTDC/FIS/098032/2008. Y.C. was supported by NSF grants PHY-0653653 and PHY-0601459, and the David and Barbara Groce Start-up Fund at Caltech. E.B.'s research was supported by NSF grant PHY-0900735.

References

[1] R. Narayan, New J. Phys. **7**, 199 (2005)
[2] M. A. Abramowicz, W. Kluzniak and J. P. Lasota, Astron. Astrophys. **396**, L31 (2002)
[3] A. M. Ghez *et al.*, Astrophys. J. **689**, 1044 (2008)
[4] S. Doeleman *et al.*, Nature **455**, 78 (2008)
[5] M. Visser,
[6] W. Kundt, Gen. Rel. Grav. **41** (2009) 1967
[7] E. Berti, V. Cardoso and A. O. Starinets, Class. Quant. Grav. **26**, 163001 (2009)
[8] F. D. Ryan, Phys. Rev. D **52**, 5707 (1995).
[9] K. D. Kokkotas and B. G. Schmidt, Living Rev. Rel. **2**, 2 (1999)
[10] O. Dreyer et *al.*, Class. Quant. Grav. **21**, 787 (2004)
[11] E. Berti, V. Cardoso and C. M. Will, Phys. Rev. D **73**, 064030 (2006)
[12] E. Berti, J. Cardoso, V. Cardoso and M. Cavaglia, Phys. Rev. D **76**, 104044 (2007)
[13] B. S. Sathyaprakash and B. F. Schutz, Living Rev. Rel. **12**, 2 (2009)
[14] P. Pani, E. Berti, V. Cardoso, Y. Chen and R. Norte, arXiv:0909.0287 [gr-qc].
[15] P. Pani, E. Berti, V. Cardoso, Y. Chen and R. Norte, in preparation
[16] M. Visser and D. L. Wiltshire, Class. Quant. Grav. **21**, 1135 (2004)
[17] C. B. M. Chirenti and L. Rezzolla, Class. Quant. Grav. **24**, 4191 (2007)
[18] P. P. Fiziev, Class. Quant. Grav. **23**, 2447 (2006)
[19] L. Gualtieri, E. Berti, J. A. Pons, G. Miniutti and V. Ferrari, Phys. Rev. D **64**, 104007 (2001)
[20] J. A. Pons, E. Berti, L. Gualtieri, G. Miniutti and V. Ferrari, Phys. Rev. D **65**, 104021 (2002)
[21] W. Israel, Nuovo Cim. B **44S10**, 1 (1966) [Erratum-ibid. B **48**, 463 (1967 NUCIA,B44,1.1966)].
[22] T. Regge and J.A, Wheeler, Phys.Rev. **108**, 1063 (1957).
[23] M. Visser, 'Lorentzian wormholes: From Einstein to Hawking," *Woodbury, USA: AIP (1995) 412 p*
[24] S. Chandrasekhar and V. Ferrari, Proc. Roy. Soc. Lond. A **432**, 247 (1991).
[25] S. Chandrasekhar and V. Ferrari, Proc. Roy. Soc. Lond. A **434**, 449 (1991).
[26] P. R. Brady, J. Louko and E. Poisson, Phys. Rev. D **44** (1991) 1891.
[27] E. W. Leaver, Proc. Roy. Soc. Lond. A **402** (1985) 285.
[28] N. Andersson, Y. Kojima and K. D. Kokkotas, Astrophys. J. **462**, 855 (1996)
[29] N. Andersson, Gen. Rel. Grav. **28**, 1433 (1996).
[30] E. Berti, V. Cardoso and P. Pani, Phys. Rev. D **79** (2009) 101501
[31] E. Berti, J. A. Pons, G. Miniutti, L. Gualtieri and V. Ferrari, Phys. Rev. D **66** (2002) 064013
[32] J. M. Bardeen and W. H. Press, J. Math. Phys. **14**, 7 (1973).
[33] S. A. Teukolsky, Astrophys. J. **185**, 635 (1973).
[34] C. Cutler, D. Kenne ck and E. Poisson, Phys. Rev. D **50**, 3816 (1994).
[35] E. Poisson, Phys. Rev. D **52**, 5719 (1995) [Addendum-ibid. D **55**, 7980 (1997)]
[36] N. Yunes and E. Berti, Phys. Rev. D **77**, 124006 (2008)
[37] V. Cardoso, P. Pani, M. Cadoni and M. Cavaglia, Phys. Rev. D **77** (2008) 124044
[38] C. B. M. Chirenti and L. Rezzolla, Phys. Rev. D **78**, 084011 (2008)

Journal of Physics: Conference Series **222** (2010) 012033 doi:10.1088/1742-6596/222/1/012033

Noncommutative Black Holes [1]

C Bastos[1,2], O Bertolami[1,2], N C Dias[3,4] and J N Prata[3,4]

[1] Departamento de Física, Instituto Superior Técnico, Avenida Rovisco Pais 1, 1049-001 Lisboa, Portugal

[2] Instituto de Plasmas e Fusão Nuclear, Instituto Superior Técnico, Avenida Rovisco Pais 1, 1049-001 Lisboa, Portugal

[3] Departamento de Matemática, Universidade Lusófona de Humanidades e Tecnologias, Avenida Campo Grande, 376, 1749-024 Lisboa, Portugal

[4] Grupo de Física Matemática, Universidade de Lisboa, Avenida Prof. Gama Pinto 2, 1649-003, Lisboa, Portugal

E-mail: cbastos@fisica.ist.utl.pt, orfeu@cosmos.ist.utl.pt, ncdias@mail.telepac.pt, joao.prata@mail.telepac.pt

Abstract.
One considers phase-space noncommutativity in the context of a Kantowski-Sachs cosmological model to study the interior of a Schwarzschild black hole. It is shown that the potential function of the corresponding quantum cosmology problem has a local minimum. One deduces the thermodynamics and show that the Hawking temperature and entropy exhibit an explicit dependence on the momentum noncommutativity parameter, η. Furthermore, the $t = r = 0$ singularity is analysed in the noncommutative regime and it is shown that the wave function vanishes in this limit.

1. Introduction

The microscopic properties of black holes (BHs) and the singularity problem most likely require a quantum theory of gravity to be properly understood. Given that this theory is beyond reach, a quantum cosmology approach based on the minisuperspace approximation might be a helpful guideline. In this context noncommutative features should not be discarded. If so, one should seek for a noncommutative version of the Wheeler-DeWitt (WDW) equation for BHs.

In this contribution one presents a study of the interior of a Schwarzschild BH [1] using a phase-space noncommutativity of the Kantowski-Sachs (KS) cosmological model developed in Ref. [2]. One obtains the temperature and the entropy of this noncommutative BH and examine the singularity $t = r = 0$ in the context of this noncommutative model.

A Schwarzschild BH is described by the metric

$$ds^2 = -\left(1 - \frac{2M}{r}\right) dt^2 + \left(1 - \frac{2M}{r}\right)^{-1} dr^2 + r^2(d\theta^2 + \sin^2\theta d\varphi^2) , \tag{1}$$

where r is the radial coordinate. For $r < 2M$, within the horizon of events, the time and radial

[1] Based on a talk presented by CB at 1st Mediterranean Conference on Classical and Quantum Gravity 2009, Kolymabari, Crete, 14th - 18th September 2009, Greece.

coordinates are interchanged ($r \leftrightarrow t$) and space-time is described by the metric

$$ds^2 = -\left(\frac{2M}{t} - 1\right)^{-1} dt^2 + \left(\frac{2M}{t} - 1\right) dr^2 + t^2(d\theta^2 + \sin^2\theta d\varphi^2) \ . \tag{2}$$

That is, an isotropic metric turns into an anisotropic one, implying that the interior of a Schwarzschild BH can be described by an anisotropic cosmological space-time. Indeed, the metric (2) can be mapped into the KS cosmological model [3], which, in the Misner parametrization, can be written as

$$ds^2 = -N^2 dt^2 + e^{2\sqrt{3}\beta} dr^2 + e^{-2\sqrt{3}(\beta+\Omega)}(d\theta^2 + \sin^2\theta d\varphi^2) \ , \tag{3}$$

where Ω and β are scale factors, and N is the lapse function. The following identification for $t < 2M$,

$$N^2 = \left(\frac{2M}{t} - 1\right)^{-1} \quad , \quad e^{2\sqrt{3}\beta} = \left(\frac{2M}{t} - 1\right) \quad , \quad e^{-2\sqrt{3}\beta} e^{-2\sqrt{3}\Omega} = t^2 \ , \tag{4}$$

allows for mapping the metric Eq. (3) into the metric Eq. (2).

Thus, it is assumed that, at the quantum level, the interior of the BH can be described by the phase-space noncommutative extension of the quantum KS cosmological model. Recently, the noncommutative Wheeler-DeWitt (NCWDW) equation of this problem was obtained and its solutions calculated explicitly. From this new NCWDW equation one computes the partition function for the Schwarzschild BH through the Feynman-Hibbs procedure [4], in order to get the temperature and entropy of the BH. These quantities are explicitly dependent on the noncommuatative parameter, η [1]. Furthermore, the solutions of the NCWDW equation provide some insight about the BH singularity. It is shown that these solutions vanish in the neighbourhood of $t = r = 0$, but that this does not necessarily imply a vanishing probability of finding the system at the singularity [1]. One concludes that canonical phase-space noncommutativity is insufficient to ensure the avoidance of singularities. Interestingly, the obtained results suggest that the singularity may be removed through other forms of phase-space noncommutativity.

2. Phase-Space Noncommutative Quantum Cosmology

The ultimate structure of space-time should be determined by quantum gravity, and at this scale, presumably Planck scale, space-time might be noncommutative [5]. Before addressing the problem at hand, one discusses the mathematical background of the model.

Noncommutative quantum mechanics has been extensively discussed in the last few years [6]. A canonical extension of the Heisenberg-Weyl algebra is considered and time is assumed as being a commutative parameter. The theory is set in a $2d$-dimensional phase-space of operators with non-commuting position and momentum variables. Thus, the noncommutative algebra reads,[2]

$$[\hat{q}_i, \hat{q}_j] = i\theta_{ij} \ , \ [\hat{q}_i, \hat{p}_j] = i\delta_{ij} \ , \ [\hat{p}_i, \hat{p}_j] = i\eta_{ij} \ , \ i, j = 1, ..., d \tag{5}$$

where η_{ij} and θ_{ij} are antisymmetric real constant $(d \times d)$ matrices and δ_{ij} is the identity matrix. The extended algebra is related to the standard Heisenberg-Weyl algebra:

$$\left[\hat{R}_i, \hat{R}_j\right] = 0 \ , \ \left[\hat{R}_i, \hat{\Pi}_j\right] = i\hbar\delta_{ij} \ , \ \left[\hat{\Pi}_i, \hat{\Pi}_j\right] = 0 \ , \ i, j = 1, ..., d \ , \tag{6}$$

[2] The units $c = \hbar = k_B = G = 1$ are used.

by a class of linear (non-canonical) set of transformations:

$$\hat{q}_i = \hat{q}_i\left(\hat{R}_j, \hat{\Pi}_j\right) \ , \ \hat{p}_i = \hat{p}_i\left(\hat{R}_j, \hat{\Pi}_j\right) \tag{7}$$

the so-called Seiberg-Witten (SW) maps[3].

One reviews now some features of the phase-space noncommutative extension of the KS minisuperspace model [2]. At least, two scale factors are needed in order to impose the noncommutative relations. Classically, one has the noncommutative Poisson algebra imposed on the scale factors β, Ω and on their conjugate momenta P_β, P_Ω:

$$\{\Omega, P_\Omega\} = 1 \ , \ \{\beta, P_\beta\} = 1 \ , \ \{\Omega, \beta\} = \theta \ , \ \{P_\Omega, P_\beta\} = \eta \ . \tag{8}$$

Through the ADM formalism and taking Ω, β as configuration variables, one derives the Hamiltonian for this system,

$$H = N\mathcal{H} = Ne^{\sqrt{3}\beta+2\sqrt{3}\Omega}\left[-\frac{P_\Omega^2}{24} + \frac{P_\beta^2}{24} - 2e^{-2\sqrt{3}\Omega}\right] \ . \tag{9}$$

Choosing the lapse function associated to the KS metric, Eq. (3), as $N = 24e^{-\sqrt{3}\beta-2\sqrt{3}\Omega}$ one obtains a system of equations of motion for this problem [2]. For the noncommutative Poisson algebra (8), these read:

$$\begin{aligned} &\dot{\Omega} = -2P_\Omega \ , \quad (a) \\ &\dot{P}_\Omega = 2\eta P_\beta - 96\sqrt{3}e^{-2\sqrt{3}\Omega} \ , \quad (b) \\ &\dot{\beta} = 2P_\beta - 96\sqrt{3}\theta e^{-2\sqrt{3}\Omega} \ , \quad (c) \\ &\dot{P}_\beta = 2\eta P_\Omega \ . \quad (d) \end{aligned} \tag{10}$$

An analytical solution of this system is difficult to obtain given the entanglement of the variables. However, one can get a numerical solution that yields some predictions for the relevant physical quantities [2]. Moreover, one finds that Eqs. (10a) and (10d) yield a constant of motion

$$\dot{P}_\beta = -\eta(-2P_\Omega) = -\eta\dot{\Omega} \Rightarrow P_\beta + \eta\Omega = C \ , \tag{11}$$

which is crucial to solve the phase-space NCWDW equation.

The canonical quantization of the classical Hamiltonian constraint, $\mathcal{H} \approx 0$, based on the ordinary Heisenberg-Weyl algebra, yields the commutative WDW equation for the wave function of the universe. For the simplest factor ordering of operators this equation reads

$$\left[-\hat{P}_\Omega^2 + \hat{P}_\beta^2 - 48e^{-2\sqrt{3}\hat{\Omega}}\right]\psi(\Omega, \beta) = 0 \ . \tag{12}$$

where $\hat{P}_\Omega = -i\frac{\partial}{\partial\Omega}$, $\hat{P}_\beta = -i\frac{\partial}{\partial\beta}$ are the fundamental momentum operators conjugate to $\hat{\Omega} = \Omega$ and $\hat{\beta} = \beta$, respectively. The quantization of the classical algebra (8), is achieved through the following noncommutative algebra:

$$\left[\hat{\Omega}, \hat{P}_\Omega\right] = 1 \ , \ \left[\hat{\beta}, \hat{P}_\beta\right] = i \ , \ \left[\hat{\Omega}, \hat{\beta}\right] = i\theta \ , \ \left[\hat{P}_\Omega, \hat{P}_\beta\right] = i\eta \ . \tag{13}$$

[3] In the mathematics literature this mapping is usually referred to as Darboux map.

One can relate this noncommutative algebra with the Heisenberg-Weyl algebra via a SW map [2]:

$$\hat{\Omega} = \lambda\hat{\Omega}_c - \frac{\theta}{2\lambda}\hat{P}_{\beta_c} \ , \quad \hat{\beta} = \lambda\hat{\beta}_c + \frac{\theta}{2\lambda}\hat{P}_{\Omega_c} \ ,$$
$$\hat{P}_\Omega = \mu\hat{P}_{\Omega_c} + \frac{\eta}{2\mu}\hat{\beta}_c \ , \quad \hat{P}_\beta = \mu\hat{P}_{\beta_c} - \frac{\eta}{2\mu}\hat{\Omega}_c \ , \tag{14}$$

where the index c denotes commutative variables, i.e. variables for which $\left[\hat{\Omega}_c, \hat{\beta}_c\right] = \left[\hat{P}_{\Omega_c}, \hat{P}_{\beta_c}\right] = 0$ and $\left[\hat{\Omega}_c, \hat{P}_{\Omega_c}\right] = \left[\hat{\beta}_c, \hat{P}_{\beta_c}\right] = i$. It is possible invert the transformations Eqs. (14) as long as

$$\xi \equiv \theta\eta < 1. \tag{15}$$

The dimensionless constants λ and μ satisfy the relationship [2]

$$(\lambda\mu)^2 - \lambda\mu + \frac{\xi}{4} = 0 \Leftrightarrow \lambda\mu = \frac{1+\sqrt{1-\xi}}{2} \ . \tag{16}$$

Thus, Eqs. (14) allow for a representation of the operators (13) as self-adjoint operators acting on the Hilbert space $L^2(\mathbb{R}^2)$. In this representation the WDW Eq. (12) is deformed into a modified second order partial differential equation, which exhibits an explicit dependence on the noncommutative parameters:

$$\left[-\left(-i\mu\frac{\partial}{\partial\Omega_c} + \frac{\eta}{2\mu}\beta_c\right)^2 + \left(-i\mu\frac{\partial}{\partial\beta_c} - \frac{\eta}{2\mu}\Omega_c\right)^2 - 48\exp\left[-2\sqrt{3}\left(\lambda\Omega_c + i\frac{\theta}{2\lambda}\frac{\partial}{\partial\beta_c}\right)\right]\right]\psi(\Omega_c,\beta_c) = 0 \ . \tag{17}$$

From the constant of motion, Eq. (11), one defines a new constant operator $\hat{A} = \frac{\hat{C}}{\sqrt{1-\xi}}$, from which follows that:

$$\mu\hat{P}_{\beta_c} + \frac{\eta}{2\mu}\hat{\Omega}_c = \hat{A} \ . \tag{18}$$

As this operator commutes with the noncommutative Hamiltonian of Eq. (17), one looks for solutions of Eq. (17) that are also eigenstates of $\hat{A}$. A generic eigenstate obeys the eigenvalue problem

$$\left(-i\mu\frac{\partial}{\partial\beta_c} + \frac{\eta}{2\mu}\Omega_c\right)\psi_a(\Omega_c,\beta_c) = a\psi_a(\Omega_c,\beta_c) \ . \tag{19}$$

where $a \in \mathbb{R}$ [2]. Thus,

$$\psi_a(\Omega_c,\beta_c) = \Re_a(\Omega_c)\exp\left[\frac{i}{\mu}\left(a - \frac{\eta}{2\mu}\Omega_c\right)\beta_c\right] \ . \tag{20}$$

Substituting this solution into Eq. (17) implies that $\phi_a(z) \equiv \Re_a(\Omega_c(z))$ satisfies:

$$\phi_a''(z) + (\eta z - a)^2\phi_a(z) - 48\exp[-2\sqrt{3}z + \frac{\sqrt{3}\theta}{\lambda\mu}a]\phi_a(z) = 0 \ , \tag{21}$$

where one has introduced the new variable

$$z = \frac{\Omega_c}{\mu} \quad \rightarrow \quad \frac{d}{dz} = \mu\frac{d}{d\Omega_c} \ . \tag{22}$$

So, one has found a second order ordinary differential equation which actually can be solved numerically. This equation depends on the eigenvalue a and on the noncommutative parameters θ and η.

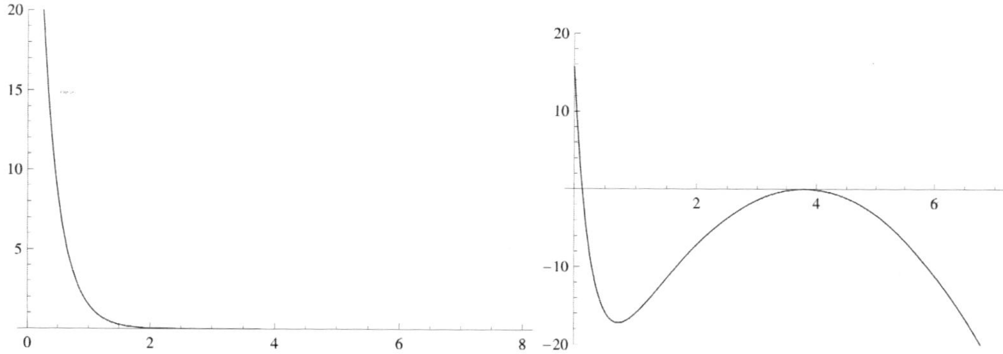

(a) $\eta = 0$ and $c = 0.01$ (b) $\eta = 1.5$ and $c = 5.68$

Figure 1. Potential function for some typical values of η and c: (a) The potential function for the noncommutative case, $\theta \neq 0$ and $\eta = 0$ and (b) The potential function with $\eta \neq 0$.

3. Thermodynamics of Phase-Space Noncommutative Black Hole

In order to get the temperature and the entropy for the phase-space noncommutative BH, one computes the partition function through the Feynman-Hibbs procedure [7]. This method is based on the minisuperspace potential function obtained from Eq. (21):

$$V(z) = 48 \exp\left[-2\sqrt{3}z + \frac{\sqrt{3}\theta}{\lambda\mu}a\right] - (\eta z - a)^2 \ . \tag{23}$$

After a change of variable, $x = z - \frac{\theta}{2\lambda\mu}a$, the potential becomes

$$V(x) = 48 \exp\left(-2\sqrt{3}x\right) - (\eta x - c)^2 \ , \tag{24}$$

where $c = P_\beta(0) + \eta\Omega(0)$ is a constant from the classical constraint.

In Fig. 1 one presents the potential function for the noncommutative case, $\theta \neq 0$ and $\eta = 0$, and the noncommutative case, where $\eta = 1.5$. One sees that, qualitatively, there is no difference between the commutative case and the noncommutativity one for the configuration variables ($\eta = 0$, $\theta \neq 0$) [8], that is, there is no local minimum. Thus, only when the noncommutativity in the momentum sector is introduced that one obtains a local minimum of the potential. The values of η used in Fig. 1 are fairly typical and the qualitative behaviour of the potential function is similar for other values. The constant c is obtained using $P_\beta(0) = 0.01$. Notice that the wave function, solution of the reduced NCWDW Eq. (21), is well-defined for the chosen values of η [2].

In order to use the Feynman-Hibbs method one expands the exponential term in the potential function in Eq. (24) to second order in $(x - x_0)$, where the minimum x_0 is given by [1]

$$\frac{dV}{dx}|_{x_0} = -96\sqrt{3}\exp\left(-2\sqrt{3}x_0\right) - 2\eta(\eta x_0 - c) = 0 \ . \tag{25}$$

The relationship,

$$\exp\left(-2\sqrt{3}x_0\right) = \zeta D - \frac{\zeta^2}{\sqrt{3}}x_0 \ , \tag{26}$$

and the fact that the second derivative of the potential is positive, define the minimum of the potential function,

$$6\exp\left(-2\sqrt{3}x_0\right) - \zeta^2 > 0 \Leftrightarrow x_0 < -\frac{1}{\sqrt{3}}\ln\left(\frac{\zeta}{\sqrt{6}}\right) \ , \tag{27}$$

where $\zeta = \eta/4\sqrt{3}$ and $D = c/12$. Thus, the NCWDW equation becomes [1]

$$-\frac{1}{2}\frac{d^2\phi}{dx^2} + 24(6e^{-2\sqrt{3}x_0} - \zeta^2)(x - x_0)^2\phi + [24e^{-2\sqrt{3}x_0} - \frac{1}{2}(\eta x_0 - c)^2]\phi = 0 \ . \tag{28}$$

This equation is analogous to the Schrödinger equation of the harmonic oscillator. So, one can identify the potential function as

$$V_{NC}(y) = 24(6e^{-2\sqrt{3}x_0} - \zeta^2)y^2 \ , \tag{29}$$

where $y = x - x_0$. The Feynman-Hibbs procedure allows for introducing quantum corrections to the partition function through the potential, i.e. the quantum correction to the potential is given by [4],

$$\frac{\beta_{BH}}{24}V''_{NC}(y) = 2\beta_{BH}(6e^{-2\sqrt{3}x_0} - \zeta^2) \ , \tag{30}$$

where β_{BH} is the inverse of the BH temperature. The noncommutative potential used to compute the partition function is the following,

$$U_{NC}(y) = 24(6e^{-2\sqrt{3}x_0} - \zeta^2)\left(y^2 + \frac{\beta_{BH}}{12}\right) \ . \tag{31}$$

Thus, the noncommutative partition function is given by

$$Z_{NC} = \sqrt{\frac{1}{48(6e^{-2\sqrt{3}x_0} - \zeta^2)}\frac{1}{\beta_{BH}}} \exp\left[-2\beta_{BH}^2\left(6e^{-2\sqrt{3}x_0} - \zeta^2\right)\right] \ . \tag{32}$$

The noncommutative internal energy of the BH, $\bar{E}_{NC} = -\frac{\partial}{\partial\beta_{BH}}\ln Z_{NC}$, is given by

$$\bar{E}_{NC} = \frac{1}{\beta_{BH}} + 4(6e^{-2\sqrt{3}x_0} - \zeta^2)\beta_{BH} \ . \tag{33}$$

Setting that $\bar{E}_{NC} = M$, one derives the BH temperature

$$\beta_{BH} = \frac{M}{8(6e^{-2\sqrt{3}x_0} - \zeta^2)}\left[1 \pm \left(1 - \frac{16}{M^2}(6e^{-2\sqrt{3}x_0} - \zeta^2)\right)^{1/2}\right] \ . \tag{34}$$

The BH temperature is given by inverting, Eq. (34). Assuming that $M >> 1$ and taking the positive root, then:

$$T_{BH} = \frac{4}{M}(6e^{-2\sqrt{3}x_0} - \zeta^2) \ . \tag{35}$$

Notice that this quantity must be positive (cf. Eq. (27)). To compare this result with the Hawking temperature $T_{BH} = \frac{1}{8\pi M}$, one should be cautious as the limit $\eta \to 0$ is ill-defined. Notice that Eq. (35) has the same mass dependence as the Hawking temperature

$$T_{BH} = \frac{b(\zeta)}{M} \ , \tag{36}$$

where $b(\zeta) = 4(6e^{-2\sqrt{3}x_0} - \zeta^2)$.

One can recover the Hawking temperature even in the presence of the momentum noncommutativity for a specific value of η. Indeed, equating (35) with the Hawking temperature and using the stationarity condition Eq. (26), one gets for $c = 12D = 5.68$:

$$x_0 = 1.8478 \qquad \eta = 0.025 \ . \tag{37}$$

Journal of Physics: Conference Series **222** (2010) 012033 doi:10.1088/1742-6596/222/1/012033

Since η cannot be exactly equal to zero, one uses $\eta_0 = 0.025$ as a reference value which yields the Hawking temperature. As η increases one gets a gradual noncommutative deformation of the Hawking temperature.

Finally, through the relationship, $S_{NC} = \ln Z_{NC} + \beta_{BH}\bar{E}_{NC}$ one obtains the phase-space noncommutative BH entropy:

$$\begin{aligned} S_{BH} &= \ln\frac{1}{\sqrt{12b(\zeta)}} + \frac{M^2}{2b(\zeta)}\left(1+\sqrt{1-\frac{4b(\zeta)}{M^2}}\right) - \frac{M^2}{8b(\zeta)}\left(1+\sqrt{1-\frac{4b(\zeta)}{M^2}}\right)^2 + \\ &- \ln\frac{M}{2b(\zeta)}\left(1+\sqrt{1-\frac{4b(\zeta)}{M^2}}\right) . \end{aligned} \tag{38}$$

Once again, as $M >> 1$, neglecting terms proportional to M^{-2} yields

$$S_{BH} \simeq \frac{M^2}{2b(\zeta)} + \ln\frac{\sqrt{b(\zeta)}}{M\sqrt{3}} . \tag{39}$$

For $\eta = \eta_0$, $b(\zeta_0) = 1/(8\pi)$, one recovers the Hawking entropy and some "stringy" corrections

$$S_{BH} = 4\pi M^2 + \ln\sqrt{\frac{2\pi}{3}} - \ln(8\pi M) . \tag{40}$$

4. Singularity

One examines now the $r = t = 0$ singularity. Notice that one employs the KS metric to describe the interior of the Schwarzschild BH through the identification Eq. (4). Thus, $t = 0$ corresponds to $\Omega \to +\infty$ and $\beta \to +\infty$. Therefore, one is interested in studying the limit

$$\lim_{\Omega_c,\beta_c\to+\infty} \psi(\Omega_c,\beta_c) , \tag{41}$$

where $\psi(\Omega_c,\beta_c)$ is a generic solution of Eq. (17). In terms of the eigenstates of $\hat{A}$ Eq. (18), the solutions of Eq. (17) can be represented by

$$\psi(\Omega_c,\beta_c) = \int da\, C(a)\psi_a(\Omega_c,\beta_c) \tag{42}$$

where $C(a) \in \mathbf{C}$ and $\psi_a(\Omega_c,\beta_c)$ is of the form

$$\psi_a(\Omega_c,\beta_c) = \phi_a\left(\frac{\Omega_c}{\mu}\right)\exp\left[\frac{i}{\mu}\left(a - \frac{\eta}{2\mu}\Omega_c\right)\beta_c\right] \tag{43}$$

and $\phi_a(z)$, $z = \Omega_c/\mu$ satisfies Eq. (21). For $\Omega \to \infty$, one keeps only the dominant terms, such as

$$\phi_a''(z) + (\eta z - a)^2\phi_a(z) = 0 . \tag{44}$$

This equation can be rewritten for $\eta \neq 0$ as

$$\left\{-\frac{\partial^2}{\partial\tilde{z}^2} - \eta^2\tilde{z}^2\right\}\tilde{\phi}_a(\tilde{z}) = 0 , \tag{45}$$

Journal of Physics: Conference Series **222** (2010) 012033 doi:10.1088/1742-6596/222/1/012033

where one has performed a change of variables, $\tilde{z} = z - \frac{a}{\eta}$ and $\tilde{\phi}_a(x) = \phi_a(x + \frac{a}{\eta})$. As can be clearly seen, this equation is similar to the eigenvalue equation of an inverted harmonic oscillator. This Hamiltonian is self-adjoint in $L^2(\mathbb{R}^2)$, its spectrum is continuous and its zero eigenfunction (the solution of Eq. (45)) displays the asymptotic form (for $\eta \neq 0$) [10]

$$\tilde{\phi}_a(\tilde{z}) \sim \frac{1}{\tilde{z}^{1/2}} \exp\left[\pm i\frac{\eta}{2}\tilde{z}^2\right] \tag{46}$$

and so, for all a,

$$\lim_{z\to+\infty} \phi_a(z) = \lim_{z\to+\infty} \tilde{\phi}_a(z - \frac{a}{\eta}) = 0 \quad \Longrightarrow \quad \lim_{\Omega_c,\beta_c\to+\infty} \psi_a(\Omega_c, \beta_c) = 0 \tag{47}$$

Therefore, considering a fairly general choice of coefficients $C(a)$, it is expected that

$$\lim_{\Omega_c,\beta_c\to+\infty} \psi(\Omega_c, \beta_c) = 0 \ , \tag{48}$$

which is a necessary condition to provide a quantum regularization of the classical singularity of the Schwarzschild BH. However, one should be cautious before concluding that the probability of finding the BH at the singularity is zero. Although the calculation of probabilities for general covariant systems is a delicate issue, in here, given that the wave function is oscillatory in β_c, it is natural to fix a β_c-hypersurface, corresponding to the introduction of the measure $\delta(\beta - \beta_c)d\beta d\Omega_c$ in the probability distribution. The probability $P(r = 0, t = 0)$ of finding the BH at the singularity is then given by

$$P(r = 0, t = 0) = \lim_{\Omega_c,\beta_c\to+\infty} \int_{\Omega_c}^{+\infty} |\psi(\Omega'_c, \beta_c)|^2 d\Omega'_c \simeq \lim_{\Omega_c\to+\infty} \int_{\Omega_c}^{+\infty} |\phi_a(\frac{\Omega'_c}{\mu})|^2 d\Omega'_c \tag{49}$$

which, unfortunately, is divergent. This follows from the conclusion (which can be derived from the asymptotic expression) that the inverted harmonic oscillator displays non-normalizable eigenstates. Hence, the noncommutativity of the form (13) cannot be regarded as the final answer for the singularity problem of the Schwarzschild BH.

However, if one considers instead a Hamiltonian like

$$H = -\frac{\partial^2}{\partial x^2} + V(x) \quad , \quad V(x) \sim -\eta^2 x^{2+2\epsilon} \tag{50}$$

for some $\epsilon > 0$, one concludes that it displays zero energy eigenstates solutions of the [9]

$$\psi(x) \sim \frac{1}{x^{(1+\epsilon)/2}} \exp\left[\pm i\tfrac{\eta}{2+\epsilon}\tilde{x}^{2+\epsilon}\right] \tag{51}$$

which are normalizable. One can then conjecture that a suitable deformation of the noncommutative structure (13) may lead to a NCWDW equation associated to a potential of the form $V(x) \sim -\eta^2 x^{2(1+\epsilon)}$ for some arbitrarily small $\epsilon > 0$. The solutions of this new NCWDW equation would then display vanishing probability (in the sense of Eq. (49)) at the singularity, thus solving this problem for the Schwarzschild BH [11].

5. Conclusions

In this contribution the temperature and the entropy of a phase-space noncommutative Schwarzschild BH have been computed. This has been performed through a KS cosmological model and the NCWDW equation of the problem.

The $t = r = 0$ singularity was examined in the same context and found that for $t = 0$, or $\Omega, \beta \to +\infty$, one encounters the Schrödinger problem of the inverted harmonic oscillator. The wave function vanishes in this limit, but given that it is not square integrable, it is not possible to conclude that the probability vanishes at the singularity. However, the discussed approach suggests that this feature may be achieved through more general noncommutativity relations [11].

Acknowledgments

The work of CB is supported by Fundação para a Ciência e a Tecnologia (FCT) under the fellowship SFRH/BD/24058/2005. The work of NCD and JNP was partially supported by Grant No. POCTI/0208/2003 and PTDC/MAT/69635/2006 of the FCT.

References

[1] C. Bastos, O. Bertolami, N.C. Dias and J.N. Prata, Phys. Rev. D **80** (2009) 124038.
[2] C. Bastos, O. Bertolami, N.C. Dias and J.N. Prata, Phys. Rev. D **78** (2008) 023516.
[3] P.R. Anderson, W.A. Hiscock and S. L. Larson, Phys. Rev. **D 56** (1997) 3571; R. Kantowski and R.K. Sachs, J. Math. Phys. **7** (1966) 443.
[4] R.P. Feynman and A.R. Hibbs, *Quantum Mechanics and Path Integrals* (1965).
[5] A. Connes, M.R. Douglas and A. Schwarz, JHEP **9802** (1998) 003; N. Seiberg and E. Witten, JHEP **9909** (1999) 032.
[6] O. Bertolami, J. G. Rosa, C. Aragão, P. Castorina and D. Zappalà, Phys. Rev. **D 72** (2005) 025010; O. Bertolami, J. G. Rosa, C. Aragão, P. Castorina and D. Zappalà, Mod. Phys. Lett. **A 21** (2006) 795; Jian-zu Zhang, Phys. Rev. Lett. **93** (2004) 043002; Jian-zu Zhang, Phys. Lett. **B 584** (2004) 204; C. Acatrinei, Mod. Phys. Lett. **A 20** (2005) 1437; C. Bastos, O. Bertolami, N.C. Dias and J.N. Prata, J. Math. Phys. **49** (2008) 072101; C. Bastos, O. Bertolami, N.C. Dias and J.N. Prata, Int. J. Mod. Phys. A **24** (2009) 2741.
[7] O. Obregon, M. Sabido and V.I. Tkach, Gen. Rel. Grav. **33** (2001) 913.
[8] J.C. López-Dominguez, O. Obregón, M. Sabido and C. Ramirez, Phys. Rev. D **74** (2006) 084024.
[9] B.L. Voronov, D.M. Gitman and I.V. Tyutin, *Self-adjoint differential operator associated with self-adjoint differential expressions*, quant-ph/0603187.
[10] D.M. Gitman, Private communication.
[11] C. Bastos, O. Bertolami, N. Dias and J. Prata, *The singularity problem and phase-space noncanonical noncommuativity*, 0912.4027 [hep-th].

First Mediterranean Conference on Classical and Quantum Gravity (MCCQG 2009) IOP Publishing
Journal of Physics: Conference Series **222** (2010) 012034 doi:10.1088/1742-6596/222/1/012034

Vacuum polarization by topological defects in de Sitter spacetime

Eugênio R. Bezerra de Mello

Departamento de Física-CCEN, Universidade Federal da Paraíba. 58.059-970, J. Pessoa, PB. Brazil

E-mail: emello@fisica.ufpb.br

Abstract. In this paper we investigate the vacuum polarization effects associated with a massive quantum scalar field in de Sitter spacetime in the presence of gravitational topological defects. Specifically we calculate the vacuum expectation value of the field square, $\langle \Phi^2 \rangle$. Because this investigation has been developed in a pure de Sitter space, here we are mainly interested on the effects induced by the presence of the defects.

1. Introduction

De Sitter (dS) space is one of the curved spacetime which has attracted more attention in quantum field theory during the past two decades. One of the main reason resides in the fact that it is maximally symmetric and several physical problems can be exactly solvable on this background; moreover in great number of inflationary models, approximated dS spacetime is employed to solve relevant problems in standard cosmology. During an inflationary epoch, quantum fluctuations in the inflaton field introduce inhomogeneities and may affect the transition toward the true vacuum. These fluctuations play important role in the generation of cosmic structures from inflation.

It is well known that different types of topological defects may have been formed in the early universe after Planck time by the vacuum phase transition [1, 2]. Depending on the topology of the vacuum manifold, $\mathcal{M}$, these are domain wall, strings, monopole and texture. Cosmic strings and monopoles seem to be the most probably to be observed. Cosmic strings are linear topologically stable gravitational. The gravitational field produced by a cosmic string may be approximated by a planar angle deficit in the two-dimensional sub-space perpendicular to the string. Although recent observations data on the cosmic microwave background have ruled out cosmic strings as the primary source for primordial density perturbation, they are still candidate for the generation of a number of interesting physical effects such as gamma ray burst [3], gravitational waves [4] and high energy cosmic rays [5]. Recently, cosmic strings have attracted renewed interest partly because a variant of their formation mechanism is proposed in the framework of brane inflation [6]-[8]. Global monopoles are spherically symmetric topological stable gravitational defects which appear in the framework of grand unified theories. These objects could be produced as a consequence of spontaneous breakdown of a global $O(3)$ gauge symmetry to $U(1)$ [9]. The gravitational field produced by global monopoles may be approximated by a solid angle deficit in the three-dimensional space.

The analysis of quantum vacuum fluctuations associated with matter fields in curved spacetime are relevant in a semiclassical theory of gravity [10]. The validity conditions under which the vacuum expectation values of the energy-momentum tensor, $\langle T_{\mu\nu}\rangle$, can act as sources for a semiclassical gravitational field are discussed in [11, 12]. In this context, as the first steps in this direction, here we shall investigate the vacuum expectation value (VEV) of the square of quantum fields, $\langle\Phi^2\rangle$, on the background of four-dimensional de Sitter spacetime considering the presence of a cosmic string and global monopole in it.

This paper is organized as follow. In section 2 we present backgrounds associated with the geometries under considerations and the solutions of the Klein-Gordon equations admitting an arbitrary curvature coupling. Also we present the corresponding Wightman functions. In section 3 we calculate the vacuum expectation values of the field squared. Finally we leave for the section 4 our more relevant conclusions about this paper.

2. Topological defects in de Sitter

Although gravitational topological defects have been first introduced in a Minkowski bulk, they can also considered in dS background:

- For a cosmic string in dS spacetime, the geometry can be given by the following line element [13]:

$$ds^2 = dt^2 - e^{2t/\alpha}\left(dr^2 + r^2 d\phi^2 + dz^2\right) , \tag{1}$$

where $r \geq 0$ and $\phi \in [0,\ 2\pi/q]$ define the coordinates on the conical geometry. The parameter $q \geq 1$, is related with the linear mass density of the string, and the parameter α is related with the cosmological constant and Ricci scalar by the formulae

$$\Lambda = \frac{3}{2\alpha^2} \text{ and } R = \frac{12}{\alpha^2} . \tag{2}$$

- For a global monopole in dS spacetime, the geometry can be given by the following line element [14]:

$$ds^2 = dt^2 - e^{2t/\alpha}\left(dr^2 + \beta^2 r^2 d\Omega^2\right) , \tag{3}$$

where the coordinates, $x^\mu = (t,\ r,\ \theta, \phi)$, are defined in the intervals $t \in (-\infty, \infty)$, $\theta \in [0, \pi]$, $\phi \in [0,\ 2\pi]$ and $r \geq 0$. The parameter $\beta \leq 1$, which codifies the presence of the global monopole, is associated with the energy scale where the defect is formed. For this geometry the curvature scalar reads

$$R = \frac{12}{\alpha^2} + \frac{2e^{-2t/\alpha}(1-\beta^2)}{\beta^2 r^2} . \tag{4}$$

In addition to the synchronous time coordinates t, we may use the conformal time τ defined according to

$$\tau = -\alpha e^{-t/\alpha} , \ -\infty < \tau < 0 , \tag{5}$$

to express the above line elements. In this new coordinate system the line elements (1) and (3) read:

- For the cosmic string,

$$ds^2 = (\alpha/\tau)^2\left(d\tau^2 - dr^2 - r^2 d\phi^2 - dz^2\right) . \tag{6}$$

- For global monopole,

$$ds^2 = (\alpha/\tau)^2\left(d\tau^2 - dr^2 - \beta^2 r^2 d\Omega^2\right) . \tag{7}$$

As we can see, both line elements above are conformally related with the corresponding cosmic string and global monopole geometries by the function factor $(\alpha/\tau)^2$.

Journal of Physics: Conference Series **222** (2010) 012034 doi:10.1088/1742-6596/222/1/012034

2.1. Wightman function

An important quantity to calculate vacuum polarization effects is the positive frequency Wightman function. Because in this paper we are more interested to analyse the vacuum polarization effects due to a cosmic string and a global monopole in dS space, the main objective of this subsection is to calculate the positive frequency Wightman functions associated with a quantum massive scalar field in a four-dimensional dS spacetime taking into account the presence of the defects. In order to do that we first obtain the complete set of eigenfunctions for the Klein-Gordon equation below, admitting an arbitrary curvature coupling:

$$(\nabla_\mu \nabla^\mu + m^2 + \xi R)\Phi(x) = 0 \ , \tag{8}$$

where ξ is the arbitrary curvature coupling constant.

2.1.1. Cosmic string case The complete set of solutions of (8) in the coordinate system defined by (6) is:

$$\Phi_\sigma(x) = C_\sigma \eta^{3/2} H_\nu^{(1)}(\lambda\eta) J_{q|n|}(pr) e^{ikz+in\varphi} \ , \ \eta = \alpha e^{-t/\alpha} \ , \tag{9}$$

where $\lambda = \sqrt{p^2+k^2}$, $H_\mu^{(1)}$ and J_ν represent the Hankel and Bessel functions, respectively, and

$$\nu = \sqrt{9/4 - 12\xi - m^2\alpha^2}, \quad n = 0, \pm 1, \pm 2, \ \dots \ . \tag{10}$$

This solution is characterized by the set of quantum numbers $\sigma \equiv (k, \ p, \ n)$, where $k \in (-\infty, \ \infty)$ and $p \in [0, \ \infty)$. The coefficient C_σ can be found by the orthonormalization condition

$$-i \int d^{D-1}x \sqrt{|g|} g^{00} [\Phi_\sigma(x) \partial_t \Phi^*_{\sigma'}(x) - \Phi^*_{\sigma'}(x) \partial_t \Phi^*_\sigma(x)] = \delta_{\sigma,\sigma'} \ , \tag{11}$$

where the integral is evaluated over the spatial hypersurface $\tau =$ const, and $\delta_{\sigma,\sigma'}$ represents the Kronecker-delta for discrete index and Dirac-delta function for continuous ones. This leads to the result

$$C_\sigma^2 = \frac{q\, p}{16\alpha^2\pi} \tag{12}$$

for the normalization coefficient.

We shall employ the mode-sum formula to calculate the positive frequency Wightman function:

$$G(x,x') = \sum_\sigma \Phi_\sigma(x) \Phi^*_\sigma(x') \ . \tag{13}$$

Substituting (9), with respective coefficient (12), into (13) we obtain

$$\begin{aligned} G(x,x') \ &= \ \frac{q(\eta\eta')^{3/2}}{16\alpha^2\pi} \int_0^\infty dp \, p \int_{-\infty}^\infty dk \ e^{ik\Delta z} \ \times \\ &\sum_n e^{inq\Delta\phi} J_{q|n|}(pr) J_{q|n|}(pr') H_\nu^{(1)}(\lambda\eta) [H_\nu^{(1)}(\lambda\eta')]^* \ , \end{aligned} \tag{14}$$

with $\Delta z = z - z'$, $\Delta\phi = \phi - \phi'$.

The analysis of vacuum polarizations in dS spacetime, have been developed by many authors in literature. Here we are mainly interested in quantum effects induced by the presence of the cosmic string. In order to investigate these effects we introduce below the subtracted Wightman function

$$G_s(x,x') = G(x,x') - G(x,x')|_{q=1} \ . \tag{15}$$

As the presence of the string does not change the curvature for the background manifold for $r \neq 0$, this function does not present additional divergences in the calculations of vacuum polarizations. Hence, for these points the function $G_s(x, x')$ is finite in the coincidence limit.

In this paper we shall consider only the particular case where the parameter q is an integer number.[1] For this case the Wightman function given in (14) can be expressed by the sum of q images of dS Wightman function [13]

$$G(x', x) = \sum_{k=0}^{q-1} G_k(x', x) , \tag{16}$$

with

$$G_k(x', x) = \frac{\Gamma\left(\frac{3}{2}+\nu\right)\Gamma\left(\frac{3}{2}-\nu\right)}{8\alpha^2\pi^2(1-u_k^2)^{1/2}} P^{-1}_{\nu-1/2}(u_k) , \tag{17}$$

being

$$u_k = \frac{\Delta x_k^2}{2\eta\eta'} = -1 + \frac{\Delta z + r^2 + r'^2 - 2rr'\cos(\Delta\phi + 2\pi k/q) - (\Delta\eta)^2}{2\eta\eta'} , \tag{18}$$

where P^ν_μ represents the associated Legendre function of first kind.

We can see that only the $k = 0$ component of (16) is divergent in the coincidence limit, $x' \to x$. Finally we have that $G_s(x, x')$ is given by

$$G_s(x, x') = \sum_{k=1}^{q-1} G_k(x, x') , \tag{19}$$

which is finite at the coincidence limit.

2.1.2. Global monopole case The complete set of solutions of (8) in the coordinate system defined by (7) is:

$$\Phi_\sigma(x) = C_\sigma \frac{\eta^{3/2}}{\sqrt{r}} H^{(1)}_\mu(\beta\omega\eta) J_{\nu_l}(\beta\omega r) Y^m_l(\theta,\ \phi) \ ,\ \eta = \alpha e^{-t/\alpha} , \tag{20}$$

where

$$\mu = \sqrt{9/4 - 12\xi - m^2\alpha^2} \ ,\ \nu_l = \frac{1}{\beta}\sqrt{(l+1/2)^2 + 2(1-\beta^2)(\xi - 1/8)} . \tag{21}$$

Y^m_l represents the spherical harmonics of degree l. In (20), $\sigma \equiv (\omega,\ l,\ m)$ denotes the set of quantum numbers, being $\omega \in [0,\ \infty)$. The coefficient C_σ can be found by the orthonormalization condition (11). It reads

$$C^2_\sigma = \frac{\pi\omega}{4\alpha^2} . \tag{22}$$

Employing the mode-sum formula (13) to calculate the positive frequency Wightman function, after many intermediate steps, we obtain [14]:

$$G(x, x') = \frac{1}{4\pi^2(\alpha\beta)^2}\left(\frac{\eta\eta'}{rr'}\right)^{3/2} \int_0^\infty dy \frac{\cosh(2\mu y)}{\sinh(v)} \sum_l (2l+1) P_l(\cos\gamma) e^{-\nu_l v} , \tag{23}$$

where the variable v is defined by

$$\cosh(v) = \frac{1}{2rr'}(r^2 + r'^2 - \Delta\eta^2 + 4\eta\eta'\sinh^2(y)) . \tag{24}$$

[1] Although being a very special situation, the results found can be used to provide the behavior on the vacuum polarization effects induced by a cosmic string for a general case.

Journal of Physics: Conference Series **222** (2010) 012034 doi:10.1088/1742-6596/222/1/012034

3. The computation of $\langle\Phi^2\rangle$

Having obtained the positive frequency Wightman functions the next steps is the calculation of the vacuum expectation value of the field square. Formally this value is obtained by computing the Wightman function at the coincidence limit:

$$\langle\Phi^2(x)\rangle = \lim_{x'\to x} G(x,x') \ . \tag{25}$$

However this procedure provides a divergent result. To obtain a finite and well defined result we shall apply in this calculation the point-splitting renormalization method. This procedure is based upon a divergence subtraction scheme in the coincidence limit of the Wightman function. This will be done in a manifest form, by subtracting from the Wightman function the singular part of the Hadamard expansion, $G_H(x,x')$, before applying the coincidence limit, as shown below:

$$\langle\Phi^2(x)\rangle_{Ren.} = \lim_{x'\to x}\left[G(x,x') - G_H(x,x')\right] \ . \tag{26}$$

This calculation will be developed separately in this section specifying the physical situation under consideration.

3.1. Cosmic string case

According to our previous analysis we may write the complete Wightman function as the sum of a pure dS part plus the subtracted one as shown below:

$$G(x,x') = G_{dS}(x,x') + G_s(x,x') \ . \tag{27}$$

Consequently

$$\langle\Phi^2(x)\rangle = \langle\Phi^2(x)\rangle_{dS} + \langle\Phi^2(x)\rangle_s \ . \tag{28}$$

The divergence is contained only in the part corresponding to the pure dS spacetime. The part induced by the cosmic string is finite for points outside the string.

Due to the maximal symmetry of the dS spacetime, $\langle\Phi^2(x)\rangle_{dS}$ does not depend on the spacetime point. In the special case of q being an integer number $\langle\Phi^2(x)\rangle_s$ is essentially simplified and can be given by (19):

$$\langle\Phi^2(x)\rangle_s = \frac{\Gamma\left(\frac{3}{2}+\nu\right)\Gamma\left(\frac{3}{2}-\nu\right)}{8\alpha^2\pi^2}\sum_{k=1}^{q-1}\frac{P^{-1}_{\nu-1/2}(u^0_k)}{[1-(u^0_k)^2]^{1/2}} \tag{29}$$

with

$$u^0_k = -1 + 2(r/\eta)^2\sin^2(\pi k/q) \ . \tag{30}$$

Note that the VEV (29) is a function of the ratio r/η which is the proper distance from the string, $\alpha r/\eta$, measured in the units of the dS curvature radius α.

In figure 1 we have plotted the string induced part in the VEV of the field squared versus the ratio r/η for minimally coupled scalar field ($\xi = 0$) with $m\alpha = 1$ (left panel) and $m\alpha = 2$ (right panel). The numbers near the curves correspond to the values of the parameter q. For the left panel the parameter ν is real and for the right one this parameter is imaginary. In the latter case the oscillatory behavior of the VEV is seen at large distances from the string.

For a conformally coupled massless scalar field, $\nu = 1/2$, so (29) becomes

$$\langle\Phi^2\rangle_s = \frac{1}{16\pi^2}\left(\frac{\eta}{\alpha r}\right)^2\sum_{k=1}^{q-1}\frac{1}{\sin^2(\pi k/q)} \ . \tag{31}$$

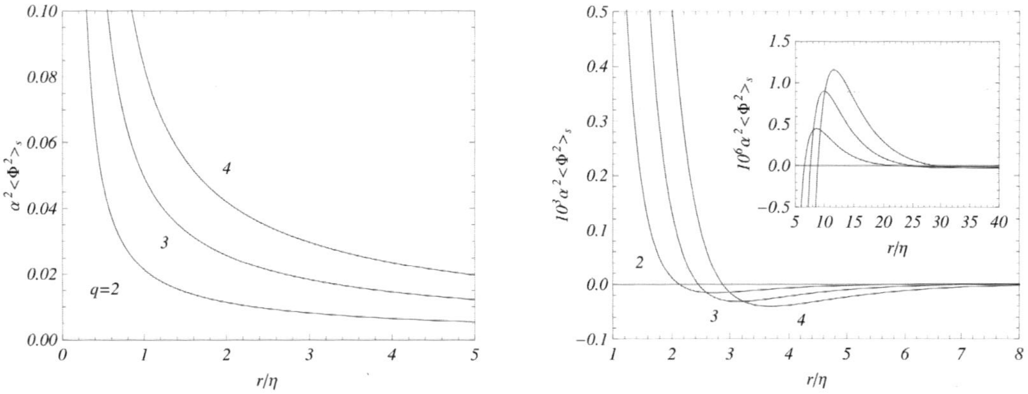

Figure 1. The string induced part in the VEV of the field squared as a function of the ratio r/η for various values of the parameter q for a minimally coupled scalar field with $m\alpha = 1$ (left panel) and $m\alpha = 2$ (right panel) in dS spacetime.

The summation on the right-hand side of the above result can be obtained in a closed form [13]:

$$I(x) = \sum_{k=1}^{q-1} \sin^{-2}(x + k\pi/q) = \frac{q^2}{\sin^2(qx)} - \frac{1}{\sin^2(x)} . \tag{32}$$

Taking the limit $x \to 0$ we have $I(0) = (q^2 - 1)/3$; consequently

$$\langle\Phi^2\rangle_s = \frac{q^2 - 1}{48\pi^2}\left(\frac{\eta}{\alpha r}\right)^2 . \tag{33}$$

The above result is an analytical function of q and by analytical continuation is valid for arbitrary value of this parameter.

3.2. *Global monopole case*

In general it is not possible to provide a closed expression for the Wightman function given in (23). This fact is consequence of the non trivial dependence of the order of the Bessel function, ν_l, given in (21), with the angular quantum number l. With the objective to provide a more workable Wightman function which allows us to develop the calculation of the VEV of the field squared, in [14] is presented an approximated expression for (23) considering that the parameter β is close to the unity.[2] In this way, taking $\gamma = 0$, and developing an expansion in the parameter $\Delta^2 = 1 - \beta^2 << 1$ in ν_l, up to the first order, the summation on the angular quantum number l can be developed, providing for the Wightman function the expression below:

$$G(x,x') = \frac{1}{16\pi^2\alpha^2\beta^2}\left(\frac{\eta\eta'}{rr'}\right)^{3/2}\int_0^\infty dy \frac{\cosh(2\mu y)}{\sinh^3(v/2)}\left[1 - \frac{v\Delta^2}{\sinh(v)}\left(1 + 4\xi\sinh^2(v/2)\right)\right] . \tag{34}$$

Moreover, expanding the factor $1/\beta^2$ up to the first power in Δ^2, (34) can be expressed as the sum of two terms, the first one corresponds to the Wightman function in a pure dS spacetime, and the second term, corresponds to the contribution coming from the presence of the monopole, as shown below:

$$G(x,x') = G_{dS}(x,x') + G_{gm}(x,x') , \tag{35}$$

[2] In fact for a typical grand unified theories the parameter $\beta^2 \approx 1 - 10^{-5}$.

with

$$G_{gm}(x,x') = \frac{1}{16\pi^2\alpha^2}\frac{\Delta^2}{(\rho\rho')^{3/2}}\int_0^\infty dy\frac{\cosh(2\mu y)}{\sinh^3(v/2)}\left[1 - \frac{v}{\sinh(v)}\left(1 + 4\xi\sinh^2(v/2)\right)\right] , \tag{36}$$

being

$$\cosh(v) = \frac{\rho^2 + \rho'^2}{2\rho\rho'} + \frac{2}{\rho\rho'}\sinh^2(y) . \tag{37}$$

In the above expressions we have introduced the dimensionless coordinate $\rho = r/\eta$, which corresponds to the proper distance from the monopole, $\alpha r/\eta$, measured in units of dS curvature radius, α.

As consequence of this procedure, the VEV of the field squared, (26), is given as the sum of two distinct contributions. The first one corresponds the value in a pure dS spacetime, and the second contribution is consequence of the presence of the global monopole:

$$\langle\Phi^2(x)\rangle_{Ren} = \langle\Phi^2(x)\rangle_{dS} + \langle\Phi^2(x)\rangle_{gm} . \tag{38}$$

Because the global monopole modifies the curvature of the background geometry introducing new divergences at the radial coordinate coincidence limit, $r' = r$, in the Wightman function, a finite and well defined contribution to the last term of (38) is given by:

$$\langle\Phi^2(x)\rangle_{gm} = \lim_{x'\to x}\left[G_{gm}(x,x') - \tilde{G}_H(x,x')\right] , \tag{39}$$

where $\tilde{G}_H(x,x')$ corresponds to the expansion in the Hadamard function due only to the presence of the monopole.

In [15] is provided a general expression for the singular expansion of the Hadamard function. It is given in terms of the one-half of the geodesic distance between two points x and x'. It reads:

$$G_H(x,x') = \frac{1}{8\pi^2\sigma(x,x')} + \frac{1}{16\pi^2}\left[m^2 + \left(\xi - \frac{1}{6}\right)R\right]\left[2\gamma + \ln\left(\frac{m_{DS}^2\sigma(x,x')}{2}\right)\right] , \tag{40}$$

where m_{DS} is equal to m for massive scalar field, and equal to μ, an arbitrary infrared cutoff energy scale, for a massless field. The curvature scalar R corresponds to dS one, $R_{dS} = \frac{12}{\alpha^2}$, plus the correction due to the presence of the monopole, $R_{gm} = 2\frac{(1-\beta^2)\eta^2}{\alpha^2\beta^2 r^2}$. In our approximation we may use $R_{gm} \approx 2\frac{\Delta^2\eta^2}{\alpha^2 r^2}$. Consequently the term in the Hadamard function needed to renormalize (39) is:

$$\tilde{G}_H(x,x') = \frac{\Delta^2}{8\pi^2\alpha^2}\frac{1}{\rho^2}\left(\xi - \frac{1}{6}\right)\left[2\gamma + \ln\left(\frac{m^2\sigma(x,x')}{2}\right)\right] . \tag{41}$$

Substituting into the above expression, the radial one-half of the geodesic distance given by $\sigma(x,x') = \frac{\alpha^2\Delta r^2}{2\eta^2}$, we obtain

$$\tilde{G}_H(x,x') = \frac{\Delta^2}{4\pi^2\alpha^2}\frac{1}{\rho^2}\left(\xi - \frac{1}{6}\right)\ln\left(\frac{m\alpha\Delta\rho}{2}\right) + \frac{\Delta^2}{4\pi^2\alpha^2}\frac{\gamma}{\rho^2}\left(\xi - \frac{1}{6}\right) . \tag{42}$$

At this point it is convenient to express the logarithmic singular contribution in (42) by an integral form. This can be done by using the Legendre function of second kind Q_0, in an integral representation [16]:

$$Q_0(\cosh(u)) = \frac{1}{\sqrt{2}}\int_u^\infty \frac{dy\ e^{-y/2}}{\sqrt{\cosh(y) - \cosh(u)}} , \tag{43}$$

being $\cosh(u) = \frac{\rho^2+\rho'^2}{2\rho\rho'}$. Doing this we find

$$\tilde{G}_H(x,x') = \frac{\Delta^2}{4\pi^2\alpha^2}\frac{1}{\rho^2}\left(\xi-\frac{1}{6}\right)\ln\left(\frac{m\alpha(\rho+\rho')}{2}\right)+\frac{\Delta^2}{4\pi^2\alpha^2}\frac{\gamma}{\rho^2}\left(\xi-\frac{1}{6}\right) - \frac{\Delta^2}{4\pi^2\alpha^2}\frac{1}{\rho^2}\left(\xi-\frac{1}{6}\right)Q_0(\cosh(u)) \ . \quad (44)$$

Now substituting (36) and (44) into (39), and after some intermediate steps, we obtain:

$$\langle\Phi^2(x)\rangle_{gm} = \frac{\Delta^2}{16\pi^2\alpha^2}\int_0^\infty dy\left\{\frac{\cosh(2\mu y)}{\sinh^3(y)}\left[1-\frac{\tilde{v}}{2\sinh(y)\sqrt{\rho^2+\sinh^2(y)}}\left(\rho^2+4\xi\sinh^2(y)\right)\right] + \frac{2(\xi-1/6)}{\rho^2}\frac{e^{-y/2}}{\sinh(y/2)}\right\} - \frac{\Delta^2}{4\pi^2\alpha^2}\frac{1}{\rho^2}\left(\xi-\frac{1}{6}\right)\ln(m\alpha\rho) - \frac{\Delta^2}{4\pi^2\alpha^2}\frac{\gamma}{\rho^2}\left(\xi-\frac{1}{6}\right) \quad (45)$$

where

$$\tilde{v} = 2\,\mathrm{arcsinh}\left(\frac{\sinh(y)}{\rho}\right) \ . \quad (46)$$

As we can see, the global monopole induced part in the VEV of the field squared, presents divergences at the monopole's position, given explicitly by the terms proportional to $\frac{1}{\rho^2}$ and $\frac{\ln(m\alpha\rho)}{\rho^2}$. Unfortunately it is not clear the behavior in (45) given by the integral contribution. So, in order to exhibit this behavior, in figure 2 we have plotted this contribution as function of the ratio $\rho = r/\eta$ (proper distance from the monopole measured in units of α) for minimally coupled scalar field ($\xi = 0$), with $m\alpha = 1$, represented by points, and $m\alpha = \sqrt{2}$, by solid line, in the left panel; and with $m\alpha = \sqrt{14}/2$, by points, and $m\alpha = \sqrt{10}/2$, by solid line, in the right panel. For the left panel the correspondent values for μ are reals and for the right panel they are imaginaries. Also by this figure we can observe that for $\rho \rightarrow 0$, all the integrals become large and negative, indicating a possible divergence. In fact analysing the integrand of (45) for small values of ρ we observe a divergence $\frac{1}{\rho^2}$, which vanishes for $\xi = 1/6$, plus a logarithmic divergent term, $\ln(\rho)$.

4. Conclusions

In this paper we have revisited the quantum effects associated with massive scalar fields in four-dimensional dS space in presence of two specific gravitational topological defects: *i*) an idealized cosmic string and *ii*) point-like global monopole. In fact in these analysis we were more interested to calculate the contributions induced by the gravitational defects themselves on the VEV of the field square. The reason is because the vacuum polarization effects associated with a quantum scalar quantum fields in dS spacetime have been investigated by many authors in literature, and the corresponding results do not depend on the spacetime point. In this way we can compare the influence due to the defects on the vacuum polarization effects with the dS contribution.

In order to develop these analysis we have presented the complete Wightman functions as the sum of two contributions: *i*) The first ones correspond to the Wightman function on the bulk in the absence of the defects and *ii*) the second ones are induced by the presence of the defects. As consequence of this formalism, we have obtained, for the VEV of the field square, the following result:

$$\langle\phi^2\rangle = \langle\phi^2\rangle_{dS} + \langle\phi^2\rangle_d \ , \quad (47)$$

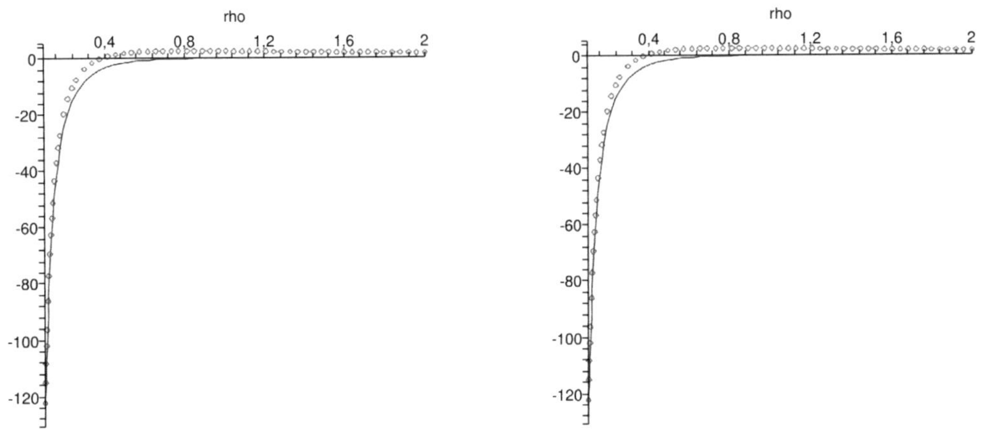

Figure 2. The behavior of the integral contribution induced by the global monopole in the VEV of the field squared, Eq. (45), as a function of the ratio $\rho = r/\eta$ for a minimally coupled scalar field with $m\alpha = 1$, $\sqrt{2}$ (left panel), and $m\alpha = \sqrt{14}/2$, $\sqrt{10}/2$ (right panel). For the first values, the behavior are represented by points, and for the seconds ones by solid lines.

where

$$\begin{aligned}\langle \Phi^2(x) \rangle_{dS} &= \frac{1}{8\pi^2\alpha^2}\left\{\left(\frac{m^2\alpha^2}{2}+6\xi-1\right)\left[\Psi\left(\frac{3}{2}+\nu\right)+\Psi\left(\frac{3}{2}-\nu\right)-2\ln(m\alpha)\right]\right. \\ &- \left.\frac{(6\xi-1)^2}{m^2\alpha^2}+\frac{1}{30m^2\alpha^2}-6\xi+\frac{2}{3}\right\}, \qquad (48)\end{aligned}$$

corresponds to the dS spacetime contribution,[3] and the second term on the right hand side of (47), $\langle \phi^2 \rangle_d$, corresponds to the induced gravitational defects contributions. By the results obtained in (29) and (45), for the latter, the corresponding results become more relevant than (48) near the defects. At large distance the situation is the opposite, $\langle \phi^2 \rangle_{dS}$ becomes dominant.

The results found for $\langle \phi^2 \rangle_d$ are divergent at the defects' position, i.e. $r \approx 0$. These facts are consequence of the idealized models adopted for them. In a realistic point of view, cosmic strings and global monopoles have a non-vanishing radius. Taking into account a nontrivial inner structure for these defects, the VEV associated with quantum fields should be considered in both, external and internal regions, separately. Regarding to this fact, considering these defects in a Minkowski bulk, the VEV associated with scalar field in global monopole spacetime have been analysed in [17, 18] and [19] for scalar and fermionic fields, respectively. In these analysis we have obtained that the vacuum polarization effects in the region inside the defect are regular at $r \approx 0$. We expect that assuming a more realistic model for cosmic string and global monopole in dS spacetime, similar results will also be found.

Acknowledgments

E.R.B.M. thanks Conselho Nacional de Desenvolvimento Científico e Tecnológico (CNPq) for partial financial support, FAPESQ-PB/CNPq (PRONEX) and FAPES-ES/CNPq (PRONEX).

[3] The $\Psi(x)$ in (48) corresponds the logarithmic derivative of the gamma-function.

References

[1] Kibble T W 1976 *J. Phys.* A **9** 1387.
[2] Vilenkin A and Shellard E P S 1994 *Cosmic Strings and Other Topological Defects* (England: Cambridge University Press, Cambridge).
[3] Berezinski V, Hnatyk V B, and Vilenkin A 2001 *Phys. Rev.* D **64** 043004.
[4] Damour T and Vilenkin A 2000 *Phys. Rev. Lett.* **85** 3761.
[5] Bhattacharjee P and Sigl G 2000 *Phys. Rep.* **327** 109.
[6] Sarangi S and Henry Tye S-H 2002 *Phys. Lett.* B **536** 185.
[7] Copeland E J, Myers R C and Polchinski J *J. High Energy Phys.* JHEP06(2004)013.
[8] Dvali G and Vilenkin A *J. Cosmol. Astropart. Phys.* JCAP03(2004)010.
[9] Barriola M and Vilenkin A 1989 *Phys. Rev. Lett.* **63** 341.
[10] Birrell N D and Davies P C W 1982 *Quantum Fields in Curved Space* (England: Cambridge University Press, Cambridge).
[11] Ford L H 1982 *Ann. Phys.* NY **144** 238.
[12] Singh T P and Padmanabhan T 1989 *Ann. Phys.* NY **196** 296.
[13] Bezerra de Mello E R and Saharian A A *J. High Energy Phys.* JHEP04(2009)046.
[14] Bezerra de Mello E R, *Vacuum polarization by a scalar field in de Sitter spacetime in the presence of global monopole*, arXiv:0905.1615 [hep-th].
[15] Christensen S M 1976 *Phys. Rev.* D **14** 2490.
[16] Gradshteyn I S and Ryzhik I M 1980 *Table of Integrals, Series and Products* (New York: Academic Press).
[17] Spinelly J and Bezerra de Mello E R 2005 *Class. Quantum Grav.* **22** 3247.
[18] Bezerra de Mello E R and Saharian A A 2006 *J. High Energy Phys.* JHEP10(2006)049.
[19] Bezerra de Mello E R and Saharian A A 2007 *Phys. Rev.* D **75** 065019.

First Mediterranean Conference on Classical and Quantum Gravity (MCCQG 2009) IOP Publishing
Journal of Physics: Conference Series **222** (2010) 012035 doi:10.1088/1742-6596/222/1/012035

Extensions of Birkhoff's Theorem in 6D Gauss-Bonnet Gravity

Charalampos Bogdanos

Laboratoire de Physique Theorique
Université de Paris-Sud 11
Bâtiment 210, 91405 Orsay CEDEX, France

E-mail: Charalampos.Bogdanos@th.u-psud.fr

Abstract. We present a generalization of Birkhoff's theorem in the context of six-dimensional Gauss-Bonnet theory. Contrary to what we encounter in ordinary General Relativity, the presence of the higher curvature terms leads to novel constraints on the geometry of the admissible black hole horizons. As a result, we can have static solutions which depart from spherical symmetry, but must satisfy specific constraints involving the Weyl tensor of the four-dimensional horizon. After obtaining the general profile of the static solutions in the transverse space, we give explicit examples of such black hole horizons, which are in general of an anisotropic nature.

1. Introduction

Higher dimensional theories of gravity have become a very prominent research field in theoretical physics over the last decade[1]. With the introduction of novel ways to incorporate extra dimensions to explain puzzling phenomena such as the hierarchy problem[2, 3, 4, 5], a considerable effort has been devoted to the study and clarification of the rich phenomenology of these models. Braneworlds and other higher dimensional modifications of ordinary General Relativity, like Gauss-Bonnet and Lovelock[6] theory have been considered as possible extensions in the quest for a more complete theory of gravity. The absence of a consistent theory of quantum gravity, as well as mounting evidence from the front of observational cosmology seem to suggest that General Relativity may indeed need some non-trivial modification along these lines. In particular, the observed accelerated expansion of the universe[7, 8], as well as the problem of a very small cosmological constant[9] responsible for this expansion cannot be addressed in a satisfying manner within the context of pure GR. Higher dimensional theories like Gauss-Bonnet gravity thus seem as a likely candidate for an infrared completion of GR, accounting for accelerated expansion in geometrical terms (for a review see[10]).

One of the most well known results in ordinary GR is Birkhoff's theorem: A spherically symmetric solution to Einstein's equations is necessarily static. As a result, the Schwarzschild solution is the unique static and spherically solution in four dimensions and any other static solution will be necessarily (locally) isomorphic to Schwarzschild[11, 12], with the possibility of a non-trivial overall topological difference. This theorem is in fact amplified in strength when we consider higher dimensional GR with just the ordinary Einstein-Hilbert action[13, 14]. As it turns out in this case, spherical symmetry is not necessary in order to ensure the staticity of the

Journal of Physics: Conference Series **222** (2010) 012035 doi:10.1088/1742-6596/222/1/012035

solutions. In fact, the horizon of the higher dimensional black hole, and consequently any other hypersurface in the transverse direction can be any kind of Einstein space. We are thus allowed to take the horizon geometry of a higher dimensional Schwarzschild black hole, substitute it with any Einstein space of our liking and the result will be a new, static solution to the full Einstein's equations.

When new terms are introduced in the gravitational action of a higher dimensional theory, not surprisingly, non-trivial modifications to Birkhoff's theorem occur. This has already been studied in five dimensions for the Gauss-Bonnet theory[15], as well as in arbitrary dimensions for a more general Lovelock theory in the context of spherically symmetric solutions[16]. As it turns out, for the Gauss-Bonnet case, there are two classes of solutions emerging, with only one of them leading to static solutions. Within the static category, we encounter two distinct branches, characterized by the sign of the Gauss-Bonnet coupling. Only one of them is perturbatively connected to pure GR, while the other is found to be unstable. Similar results also emerge in the Lovelock case.

Here we consider the case of six-dimensional Gauss-Bonnet theory, where the assumption of spherical symmetry is relaxed. Our goal is to study whether a generalization of Birkhoff's theorem is possible and what are the restrictions on the admissible horizon geometry of a six-dimensional black hole. As we will see, the presence of the Gauss-Bonnet term leads to a number of non-trivial constraints on the geometry of the horizon, in order to have a static solution. We can thus prove a version of Birkhoff's theorem in six-dimensional Gauss-Bonnet theory, which is stronger than in four-dimensional GR (we can also have static solutions which are not spherically symmetric), but more constrained than its higher dimensional GR counterpart (not every Einstein space is a good horizon solution). The square of the Weyl tensor of the horizon, $C^{abcd}C_{abcd}$, figures prominently in the constraint. This is also one of the reasons why the six-dimensional model is qualitatively different from the five-dimensional one. In five dimensions, the horizon of the black hole is three-dimensional and the Weyl tensor is automatically zero. In six dimensions the horizon becomes four-dimensional and the Weyl tensor (and its square) is in general non-vanishing. So six dimensions is the least we need to have a non-trivial contribution from the horizon geometry. When considered in the context of Lovelock gravity, this is also the last case for which we can safely neglect any higher curvature corrections in the action. For $d = 7$ and above, we should also include the next Lovelock density $\mathcal{L}_3$ and even more densities as we keep increasing the dimensionality of the construction.

That the Weyl tensor of the horizon enters the solution should come as no surprise. Contrary to ordinary GR, the Gauss-Bonnet term exposes the entire Riemann tensor to the equations of motion, hence the dependence on C_{abcd}. As it was recently shown, these higher curvature terms can constrain the form of the solution for a priori static configurations[17]. We show that this is a general characteristic which can be used to classify solutions obeying Birkhoff's theorem. The static solutions have as an additional parameter the square of the Weyl tensor. We also give some explicit examples of horizon geometries compatible with the constraints. In general, these are topologically non-trivial and of an anisotropic nature.

2. Equations of Motion and Staticity Conditions

We begin by considering the Einstein-Gauss-Bonnet action with a cosmological constant in six dimensions

$$S^{(6)} = \frac{{M^{(6)}}^4}{2} \int d^6x\sqrt{-g^{(6)}}\left[R - 2\Lambda + \alpha\hat{G}\right], \tag{1}$$

where $M^{(6)}$ is the fundamental mass scale in six-dimensional spacetime, $\hat{G}$ the Gauss-Bonnet density defined as

$$\hat{G} = R_{ABCD}R^{ADCB} - 4R_{AB}R^{AB} + R^2\,. \tag{2}$$

and the cosmological constant $\Lambda = -10k^2 < 0$ and $\Lambda = 10a^2 > 0$. Using these conventions we can vary the action with respect to the field metric to derive the equations of motion

$$\mathcal{E}_{AB} = G_{AB} + \Lambda g_{AB} + \alpha H_{AB} = 0\,, \tag{3}$$

where G_{AB} stands for the Einstein tensor. Uppercase indices refer to six-dimensional coordinates. We have also introduced the Gauss-Bonnet tensor,

$$H_{AB} = \frac{g_{AB}}{2}\hat{G} - 2RR_{AB} + 4R_{AC}R^{C}_{B} + 4R_{CD}R^{C\ D}_{\ A\ B} - 2R_{ACDE}R^{CDE}_{B}\,. \tag{4}$$

We now choose an appropriate ansatz to look for static solutions of a black hole profile,

$$ds^2 = e^{2\nu(t,z)}B\left(t,z\right)^{-3/4}\left(-dt^2 + dz^2\right) + B\left(t,z\right)^{1/2}h^{(4)}_{\mu\nu}\left(x\right)dx^{\mu}dx^{\nu}\,. \tag{5}$$

The coordinates t, z parametrize the "transverse space". Lowercase greek indices correspond to internal coordinates of the 4-space, which also characterizes the black hole horizon. The two functions $B(t,z)$ and $\nu(t,z)$ determine the profile of the transverse geometry. We then switch the coordinates of the transverse space to light-cone coordinates,

$$u = \frac{t-z}{\sqrt{2}}\,, \quad v = \frac{t+z}{\sqrt{2}}\,, \tag{6}$$

in terms of which the metric reads

$$ds^2 = -2e^{2\nu(u,v)}B\left(u,v\right)^{-3/4}dudv + B\left(u,v\right)^{1/2}h^{(4)}_{\mu\nu}\left(x\right)dx^{\mu}dx^{\nu}\,. \tag{7}$$

Using the above prescription, we are now able to write down the equations of motion. The uu and vv equations yield

$$\mathcal{E}_{uu} = \frac{2\nu_{,u}B_{,u} - B_{,uu}}{B}\left(1 + \alpha\left(B^{-1/2}R^{(4)} + \frac{3}{2}e^{-2\nu}B^{-5/4}B_{,u}B_{,v}\right)\right)\,, \tag{8}$$

$$\mathcal{E}_{vv} = \frac{2\nu_{,v}B_{,v} - B_{,vv}}{B}\left(1 + \alpha\left(B^{-1/2}R^{(4)} + \frac{3}{2}e^{-2\nu}B^{-5/4}B_{,u}B_{,v}\right)\right)\,. \tag{9}$$

The off-diagonal equation reads

$$\begin{aligned}
\mathcal{E}_{uv} &= \frac{B_{,uv}}{B} - \Lambda e^{2\nu}B^{-3/4} + \frac{\alpha}{2}e^{2\nu}B^{-7/4}\hat{G}^{(4)} \\
&+ R^{(4)}\left(\frac{1}{2}e^{2\nu}B^{-5/4} - \alpha B^{-3/2}\left(\frac{1}{2}\frac{B_{,u}B_{,v}}{B} - B_{,uv}\right)\right) \\
&+ \alpha e^{-2\nu}B^{-5/4}\left(-\frac{15}{16}\left(\frac{B_{,u}B_{,v}}{B}\right)^2 + \frac{3}{2}\frac{B_{,u}B_{,v}}{B}B_{,uv}\right)\,.
\end{aligned} \tag{10}$$

We also have the $\mu\nu$ equations, which can be brought into the form

$$\begin{aligned}
\mathcal{E}_{\mu\nu} &= G^{(4)}_{\mu\nu} - e^{-2\nu}B^{1/4}\left(\frac{3}{4}B_{,uv} + 2B\nu_{,uv}\right)h^{(4)}_{\mu\nu} + \Lambda B^{1/2}h^{(4)}_{\mu\nu} \\
&+ \frac{3}{2}\alpha e^{-4\nu}\left(B_{,uu} - 2\nu_{,u}B_{,u}\right)\left(B_{,vv} - 2\nu_{,v}B_{,v}\right)h^{(4)}_{\mu\nu} \\
&- \alpha e^{-4\nu}\left(\frac{45}{32}\left(\frac{B_{,u}B_{,v}}{B}\right)^2 - \frac{21}{8}\frac{B_{,u}B_{,v}}{B}B_{,uv} + \frac{3}{2}B^2_{,uv} + 3B_{,u}B_{,v}\nu_{,uv}\right)h^{(4)}_{\mu\nu} \\
&- \alpha e^{-2\nu}B^{-1/4}\left(\frac{3}{4}\frac{B_{,u}B_{,v}}{B} - \frac{1}{2}B_{,uv} + 4B\nu_{,uv}\right)\left(R^{(4)}h^{(4)}_{\mu\nu} - 2R^{(4)}_{\mu\nu}\right)\,.
\end{aligned} \tag{11}$$

The classification of solutions proceeds by inspecting the so-called *integrability conditions*, (8) and (9). We see that we can set these equal to zero provided that we either choose

$$1+\alpha\left(B^{-1/2}R^{(4)}+\frac{3}{2}e^{-2\nu}B^{-5/4}B_{,u}B_{,v}\right)=0\,, \tag{12}$$

or

$$2\nu_{,u}B_{,u}-B_{,uu}=2\nu_{,v}B_{,v}-B_{,vv}=0\,. \tag{13}$$

Choosing the former leads to the class of solutions which we call *class I*. These are new solutions specific to the Gauss-Bonnet case (since for $\alpha = 0$ this equation cannot be satisfied). The later choice gives rise to the *class II* solutions. These are essentially the same integrability conditions that we also encounter in ordinary GR. As we will see, class I solutions are generally time-dependent and thus do not obey Birkhoff's theorem, however their physical relevance is ambiguous. Class II solutions on the other hand are the ones obeying Birkhoff under certain restrictions on the internal geometry of the horizon.

We deal with class I solutions first. From the integrability condition (12) we can solve for $\nu(u,v)$ as

$$\nu\left(u,v\right)=\frac{1}{2}\ln\left(-\frac{3\alpha}{2}\frac{B_{,u}B_{,v}}{B^{5/4}\left(1+\alpha B^{-1/2}R^{(4)}\right)}\right)\,. \tag{14}$$

Restrictions are already imposed on the four-dimensional geometry of the internal space, as only spaces of constant scalar curvature $R^{(4)}$ allow for separation of variables. Plugging the above expression for $\nu(u,v)$ into the uv equation leads to the following constraints

$$5+12\alpha\Lambda=0,\;\hat{G}^{(4)}=\frac{1}{6}{R^{(4)}}^2\,, \tag{15}$$

and the $\mu\nu$ equation reduces to

$$\mathcal{E}_{\mu\nu}=\left(R^{(4)}_{\mu\nu}-\frac{1}{4}R^{(4)}h^{(4)}_{\mu\nu}\right)\left[1+2\alpha e^{-2\nu}B^{-1/4}\left(\frac{3}{4}\frac{B_{,u}B_{,v}}{B}-\frac{1}{2}B_{,uv}+4B\nu_{,uv}\right)\right]. \tag{16}$$

There are obviously two ways to satisfy this final equation and get a solution. Setting the first parenthesis equal to zero is practically equivalent to the definition of an Einstein space for the internal geometry. Coupled with (15), this leads to the additional condition $C^{(4)}_{\alpha\beta\mu\nu}C^{(4)\alpha\beta\mu\nu}=0$, so the Weyl-squared of the internal space must also be zero. Under these conditions, we do have a valid solution, with no other equation left to satisfy. But the function $B(u,v)$ is undetermined up to this point! We thus realise that the system of equations is in this case underdetermined and since $B(u,v)$ is arbitrary, it can also be time-dependent and Birkhoff's theorem will not hold in general for this kind of solutions.

On the other hand, by requiring the second parenthesis in (16) to be zero, we end up with a third order differential equation for $B(u,v)$, which can in principle be solved and thus completely determine the transverse geometry. Again, these solution can in general be time-dependent, so Birkhoff's theorem is again violated. Still, there is now no additional equation constraining the internal space, which may now not be of the Einstein variety. The underdeterminacy of the system is in this sense shifted from the transverse to the internal space. The presence of an underdetermined system is a sign of pathology, so although these solutions are not static, they may be in the end physically irrelevant. As a result, we will now focus on the class II solutions, for which no such problems emerge.

For the class II solutions the integrability conditions promptly lead to an equation of the form

$$e^{2\nu}=B_{,u}=B_{,v}\,, \tag{17}$$

Journal of Physics: Conference Series **222** (2010) 012035 doi:10.1088/1742-6596/222/1/012035

which, in turn, yields $B = B(U+V)$, with $U = U(u)$ and $V = V(v)$. In this way, under a change of variables of the form of (6) the function B does not depend on time and thus Birkhoff's theorem is satisfied as well. Additionally, rewriting (17), $\nu(u,v)$ is now defined as

$$e^{2\nu} = B'U'V' , \tag{18}$$

where primes denote differentiation with respect to the single argument of each function. The uu and vv equations thus determine the staticity of the metric, as well as the relation between B and ν. We can then determine $B(u,v)$, or equivalently the form of the black hole potential, from the uv equation. Taking advantage of the already deduced staticity, we can express this as

$$\begin{aligned} B'' &+ \frac{1}{2}R^{(4)}B^{-1/4}B' - \frac{15}{16}\alpha B^{-9/4}B'^3 + \frac{3}{2}\alpha B^{-5/4}B''B' \\ &- \frac{1}{2}\alpha R^{(4)}B^{-3/2}B'^2 + \alpha B^{-1/2}R^{(4)}B'' + \frac{1}{2}\alpha B^{-3/4}B'\hat{G}^{(4)} - \Lambda B^{1/4}B' = 0 . \end{aligned} \tag{19}$$

Inspection of the above expression leads to the conclusion that a priori only solutions with a constant Ricci scalar and Gauss-Bonnet density for the internal space are permissible [1]. Upon integration, this leads to a quadratic equation for B'. We can then solve for B' and determine the black hole potential $V(r)$

$$ds^2 = -V(r)\,dt^2 + \frac{dr^2}{V(r)} + r^2 h^{(4)}_{\mu\nu}(x)\,dx^\mu dx^\nu , \tag{20}$$

using the change of variables $r = B^{1/4}$. The corresponding potential turns out to be

$$V(r) = \frac{R^{(4)}}{12} + \frac{r^2}{12\alpha}\left[1 \pm \sqrt{1 + \frac{12\alpha\Lambda}{5} + \frac{\alpha^2\left(R^{(4)2} - 6\hat{G}^{(4)}\right)}{r^4} + 24\frac{\alpha M}{r^5}}\right] , \tag{21}$$

where M is an integration constant independent of x, related to the mass of the six-dimensional black hole .

As in the five-dimensional case, we encounter within the static class II solutions two branches, corresponding to the two possible signs in the potential (21). The plus sign corresponds to the so-called *Gauss-Bonnet branch* and is not perturbatively connected to the usual six-dimensional black hole solutions of general relativity, as can be easily inferred. This branch has been shown to be unstable[19]. The minus sign gives rise to the *Einstein branch*, which has a well-defined limit for $\alpha \to 0$. We will consider this to be the physical branch from now on.

We will now clarify the significance of the constraints on the internal geometry. We have already seen that the horizon should first have constant Ricci scalar. Furthermore, the combination $R^{(4)2} - 6\hat{G}^{(4)}$ appearing in the expression for the potential is indeed related to the square of the four-dimensional Weyl tensor and must also be a constant. To illustrate this better, we consider a horizon geometry of the Einstein kind with a constant curvature $R^{(4)}_{\mu\nu} = 3\kappa h_{\mu\nu}$. Coupled to the requirement of constant Gauss-Bonnet scalar, this leads to the relation $C^{\alpha\beta\gamma\mu}C_{\alpha\beta\gamma\mu} = \Theta$ where Θ is a positive constant. Thus we get the condition

$$C^{\alpha\beta\gamma\mu}C_{\alpha\beta\gamma\nu} = \Theta\delta^\mu_\nu , \tag{22}$$

[1] There are, however, special cases which should be treated separately. For a more detailed discussion and connection with black strings, see [18].

i.e. the square of the Weyl tensor of the internal space must be constant. The potential in this case reduces to

$$V(r) = \kappa + \frac{r^2}{12\alpha}\left(1 \pm \sqrt{1 - 24k^2\alpha - 24\frac{\alpha^2\Theta}{r^4} + 24\frac{\alpha M}{r^5}}\right) , \tag{23}$$

and in general the parameter Θ acts now as a classifier for the admissible horizons. Static solutions that obey Birkhoff's theorem have to satisfy the constraints of constant $R^{(4)}$ and Θ, which is a more restrictive situation compared to higher dimensional GR. Still, the theorem remains stronger than in four-dimensional GR, since spherical symmetry is not a necessary condition for staticity. In fact, as we will discuss in the following section, we can have horizon geometries which significantly depart from spherical symmetry.

3. Horizon Geometries

As we have already discussed, the presence of the Gauss-Bonnet term leads to non-trivial constraint on the Weyl tensor of the internal geometry. We will now present some explicit examples of such admissible geometries. In looking for such four-dimensional spaces of constant Θ, we immediately realize that these spaces cannot be asymptotically flat, as this would lead us to the admissible but trivial case of $\Theta = 0$. In general, we have to consider constructions of a non-trivial topology and a shift from asymptotically flat to asymptotically *locally* flat. A first example is the case of an $S^2 \times S^2$ geometry, the product of two spheres, bearing the ansatz

$$ds^2 = \rho_1^2\left(d\theta_1^2 + \sin^2\theta_1 d\phi_1^2\right) + \rho_2^2\left(d\theta_2^2 + \sin^2\theta_2 d\phi_2^2\right) , \tag{24}$$

where we consider the (dimensionless) radii ρ_1 and ρ_2 of the spheres to be constant. The entire six-dimensional space has the form

$$ds^2 = -V(r)\,dt^2 + \frac{dr^2}{V(r)} + r^2\rho_1^2\left(d\theta_1^2 + \sin^2\theta_1 d\phi_1^2\right) + r^2\rho_2^2\left(d\theta_2^2 + \sin^2\theta_2 d\phi_2^2\right) , \tag{25}$$

with the potential

$$V(r) = \frac{R^{(4)}}{12} + \frac{r^2}{12\alpha}\left(1 \pm \sqrt{1 - 24k^2\alpha - 24\Theta\frac{\alpha^2}{r^4} + 24\alpha\frac{M}{r^5}}\right) . \tag{26}$$

In order for (25) to be a solution to the Gauss-Bonnet equations of motion, we are led to the condition of equal sphere radii, $\rho_1 = \rho_2$. The Ricci scalar and Θ are in this case $R^{(4)} = \frac{4}{\rho_1^2}$, $\Theta = \frac{4}{3\rho_1^2}$. Clearly, this is an anisotropic, non-asymptotically flat space.

Another example of topologically non-trivial space is the Taub-NUT/Bolt geometry. This is a homogeneous but anisotropic space, which is asymptotically locally flat. It is not globally flat because of the non-trivial topology. The general metric describing Taub-NUT/Bolt spaces of Euclidean signature is

$$ds^2 = V(r)\left(d\tau^2 + 2n\cos\theta d\phi\right)^2 + \frac{dr^2}{V(r)} + \left(r^2 - n^2\right)\left(d\theta^2 + \sin^2\theta d\phi^2\right) . \tag{27}$$

Here, n is the nut parameter. Depending on the form of the potential $V(r)$, we can either have a nut, i.e. a zero-dimensional fixed point set of the Euclidean time $U(1)$ isometry, or a bolt, which is a two-dimensional fixed point set. Still, this is not yet a good candidate for our horizon.

What we need is the AdS variant of the Taub-NUT/Bolt space[20]. Written in the Pedersen form[21], it is

$$ds^2 = \frac{4}{(1-k^2r^2)^2}\left[\frac{1-\mu^2r^2}{1-k^2\mu^2r^4}dr^2 + r^2\left(1-\mu^2r^2\right)\left(\sigma_1^2+\sigma_2^2\right) + r^2\frac{1-k^2\mu^2r^4}{1-\mu^2r^2}\sigma_3^2\right], \tag{28}$$

where the mass parameter μ is defined in terms of k, the curvature of AdS space and the nut parameter by $\mu^2 = k^2 - \frac{1}{4n^2}$. By calculating Θ for this space we find that $\Theta = 6\mu^4\frac{(1-k^2r^2)^6}{(1-\mu^2r^2)^6}$, which only becomes constant at infinity. But we notice that taking the $n \to \infty$ limit, or $\mu = k$, Θ becomes constant everywhere. Surprisingly, there is already a name for this geometry, known as *Bergman space*. In these coordinates it can be written as

$$ds^2 = \frac{4}{(1-k^2r^2)^2}\left[\frac{1}{1+k^2r^2}dr^2 + r^2\left(1-k^2r^2\right)\left(\sigma_1^2+\sigma_2^2\right) + r^2\left(1+k^2r^2\right)\sigma_3^2\right]. \tag{29}$$

It describes the coset space $SU(2,1)/U(2)$, which is a Kähler-Einstein manifold with Kähler potential

$$K(z_1,\bar{z}_1,z_2,\bar{z}_2) = 1 - z_1\bar{z}_1 - z_2\bar{z}_2\,, \qquad \text{for } z_1\bar{z}_1 + z_2\bar{z}_2 < 1, \tag{30}$$

and the topology of the open ball in C^2. Setting $z_1 = k\rho\cos(\theta/2)e^{i(\phi+\psi)/2}$ and $z_2 = k\rho\sin(\theta/2)e^{i(\phi-\psi)/2}$ the metric $g_{\alpha\bar{\beta}} = -\partial_\alpha\partial_{\bar{\beta}}\ln K^{1/k^2}$ reproduces exactly (29) after a change of coordinate $\rho^2 = 2r^2/(1+k^2r^2)$. The Bergman metric (29) has an isometry group of $SU(2,1)$. In practice, the choice $\mu = k$ corresponds to infinite "squashing" of the 3-sphere at the boundary $r \to 1/k$, such that only a one-dimensional circle remains intact. By comparing the terms multiplying $\sigma_1^2 + \sigma_2^2$ (2-sphere) and σ_3^2, we see that as we approach the boundary, the σ_3^2 part blows up faster and becomes dominant. The space has this circle as its conformal boundary. This space is asymptotically locally flat. It is now possible to see from the expression for Θ in AdS Taub-NUT that the Bergman space has $\Theta = 6k^4$ and is thus a suitable horizon solution. Substituting (29) as the metric of the internal space $h^{(4)}_{\mu\nu}$, we verify that it is a solution to the equations of motion. To do so, we first rescale the radial coordinate as $\rho = r/l$, with l having dimensions of $mass^{-1}$ in order to make the metric dimensionless. As a result, we identify the dimensionless curvature scale $\kappa = kl$. The resulting black hole potential is then

$$V(r) = -\kappa^2 + \frac{r^2}{12\alpha}\left(1 \pm \sqrt{1 - 24k^2\alpha - 144\kappa^4\frac{\alpha^2}{r^4} + 24\alpha\frac{M}{r^5}}\right). \tag{31}$$

Bergman space is known to be unstable in ordinary GR[22], but here we consider its Euclidean version, for which the same conclusion doesn't necessarily hold. It is also not known how the presence of the Gauss-Bonnet corrections modifies the stability properties, so it is hard to argue on the physical relevance of this construction.

4. Conclusions

We studied the staticity properties of six-dimensional black hole solutions in Gauss-Bonnet theory and proved a version of Birkhoff's theorem for such setups. As we saw, we can have four-dimensional spaces of constant Ricci scalar and Θ as possible horizon geometries. Non-spherically symmetric solutions are admissible and we gave explicit examples in the form of $S^2 \times S^2$ and the Bergman space. In general, the internal space has a non-trivial topology and should not be asymptotically flat in order to obtain novel solutions. Geometries which exhibit asymptotic flatness only locally are good candidates. Intuitively, we also expect similar conclusions to hold for even higher dimensional scenarios, in the context of the full Lovelock theory. There, the presence of higher curvature terms in the action will potentially lead to further constraints, involving quantities similar to Θ. It is an open question whether a more systematic classification of such spaces can be established.

5. Acknowledgments

The author is indebted to the organisers of the First Mediterranean Conference in Classical and Quantum Gravity. This presentation is based on work done in collaboration with C. Charmousis, B. Gouteraux at LPT Orsay and R. Zegers at Durham University. The author is supported by the CNRS and the Université de Paris-Sud XI.

References

[1] Rubakov V A and Shaposhnikov M E 1983 *Phys. Lett.* **B125** 136–138
[2] Antoniadis I 1990 *Phys. Lett.* **B246** 377–384
[3] Arkani-Hamed N, Dimopoulos S and Dvali G R 1998 *Phys. Lett.* **B429** 263–272 (*Preprint* hep-ph/9803315)
[4] Randall L and Sundrum R 1999 *Phys. Rev. Lett.* **83** 3370–3373 (*Preprint* hep-ph/9905221)
[5] Randall L and Sundrum R 1999 *Phys. Rev. Lett.* **83** 4690–4693 (*Preprint* hep-th/9906064)
[6] Lovelock D 1971 *J. Math. Phys.* **12** 498–501
[7] Riess e A G *et al.* (Supernova Search Team) 1998 *Astron. J.* **116** 1009–1038 (*Preprint* astro-ph/9805201)
[8] Astier P *et al.* (The SNLS) 2006 *Astron. Astrophys.* **447** 31–48 (*Preprint* astro-ph/0510447)
[9] Weinberg S 1989 *Rev. Mod. Phys.* **61** 1–23
[10] Charmousis C 2009 *Lect. Notes Phys.* **769** 299–346 (*Preprint* 0805.0568)
[11] Israel W 1968 *Commun. Math. Phys.* **8** 245–260
[12] Israel W 1967 *Phys. Rev.* **164** 1776–1779
[13] Gibbons G and Hartnoll S A 2002 *Phys. Rev.* **D66** 064024 (*Preprint* hep-th/0206202)
[14] Gibbons G W, Hartnoll S A and Pope C N 2003 *Phys. Rev.* **D67** 084024 (*Preprint* hep-th/0208031)
[15] Charmousis C and Dufaux J F 2002 *Class. Quant. Grav.* **19** 4671–4682 (*Preprint* hep-th/0202107)
[16] Zegers R 2005 *J. Math. Phys.* **46** 072502 (*Preprint* gr-qc/0505016)
[17] Dotti G and Gleiser R J 2005 *Phys. Lett.* **B627** 174–179 (*Preprint* hep-th/0508118)
[18] Bogdanos C, Charmousis C, Gouteraux B and Zegers R 2009 (*Preprint* 0906.4953)
[19] Charmousis C and Padilla A 2008 *JHEP* **0812** 038 (*Preprint* 0807.2864)
[20] Chamblin A, Emparan R, Johnson C V and Myers R C 1999 *Phys. Rev.* **D59** 064010 (*Preprint* hep-th/9808177)
[21] Zoubos K 2002 *JHEP* **12** 037 (*Preprint* hep-th/0209235)
[22] Kleban M, Porrati M and Rabadan R 2005 *JHEP* **08** 016 (*Preprint* hep-th/0409242)

Journal of Physics: Conference Series **222** (2010) 012036 doi:10.1088/1742-6596/222/1/012036

Correlations of Hawking radiation in acoustic black holes

Serena Fagnocchi

SISSA-ISAS International School of Advanced Studies, and INFN sezione Trieste, I-34151 Trieste, Italy

E-mail: serena.fagnocchi@sissa.it

Abstract. I will review the main motivations and results regarding the proposal to use non-local correlations measurements to get evidence of the Hawking effect in acoustic black holes, focussing in particular on those built up with weakly interacting Bose Einstein condensates and rings of ions.

1. Introduction

Hawking's prediction of black holes evaporation is generally regarded as a milestone of modern theoretical physics. Combining Einstein's General Relativity and Quantum Mechanics, Hawking was able to show that black holes are not "black", but emit thermal radiation at a temperature inversely proportional to their mass [1]. This quantum mechanical process is triggered by the formation of a horizon and proceeds via the conversion of vacuum fluctuations into on-shell particles. Unfortunately, due to the extreme faintness of the effect, so far there is no experimental support for this amazing theoretical prediction.

In a remarkable work Unruh [2] showed that Hawking radiation is not peculiar to gravity, but is rather a purely kinematical effect of quantum field theory which only depends on field propagation on a black hole-type curved space-time background. This opens the concrete possibility to study the Hawking radiation process in completely different physical systems. As an example, the propagation of sound waves in Eulerian fluids can be described in terms of the same equation describing a massless scalar field on a curved spacetime characterized by an *acoustic metric* $G_{\mu\nu}$ which is function of the background flow: the curvature of the acoustic geometry is induced by the inhomogeneity of the fluid flow, while flat Minkowskii spacetime is recovered in the case of a homogeneous system. In particular, an *acoustic black hole* (or *dumb hole*) configuration is obtained whenever a subsonic flow turns supersonic: sound waves in the supersonic region are in fact dragged away by the flow and can not propagate back towards the *acoustic horizon* separating the supersonic and subsonic regions. Upon quantization, Hawking radiation is expected to appear as a flux of thermal phonons emitted from the horizon at a temperature proportional to its surface gravity. Even though a substantial effort has been spent on a variety of *analog models* [3] (e.g. superfluid Helium [4], phonons in atomic Bose-Einstein condensates [5], degenerate Fermi gases [6], slow light in moving media [7], or traveling refractive index interfaces in nonlinear media [8], ion rings [9]), the small Hawking temperature predicted even in these systems, and the actual impossibility to distinguish the Hawking contribution

Journal of Physics: Conference Series **222** (2010) 012036 doi:10.1088/1742-6596/222/1/012036

from the thermal background in the detection of the phonon content, has so far prevented any experimental verification of these predictions.

To overcome these difficulties, the detection of Hawking radiation via non-local correlation measurements has been first proposed in [10]. The idea is to get the quantum signature of the correlated Hawking quanta as a proof of the Hawking effect. Differently from the case of astrophysical black holes, both the external and the internal region of the acoustic black hole are in fact accessible to experiments: the quantum correlations between the Hawking radiation emitted inside and outside the acoustic black hole are responsible for a significant long-range correlation at points respectively inside and outside the acoustic black hole. This unique signature can be exploited to isolate Hawking radiation from the background of competing processes and experimental noise.

The main prediction contained in [10] has been numerically validated in [11]. Later correlations are calculated for black holes built with ion rings [9]. Then semi-analytical works followed [12, 13] explaining features that the first proposal paper didn't address.

2. Correlations of Hawking radiation in hydrodynamical systems

Eulerian fluids are described by the continuity and Bernoulli equations:

$$\dot{n} + \vec{\nabla} \cdot (n\vec{v}) = 0 \qquad \dot{\theta} + \frac{1}{2}\vec{v}^2 + \mu(n) = 0 \tag{1}$$

where n is the density (the mass of the fluid constituents is taken $m = 1$), $\vec{v} = \vec{\nabla}\theta$ the flow velocity, $\mu(n)$ is the specific enthalpy from which the state equation derives. The phonons propagation in these fluids can be obtained by the linearization of around the solution of Eqs. (1): $n + \hat{n}_1$ and $\theta + \hat{\theta}_1$. The core of the analogy is in the fact that it is possible to show that the fluctuation of the phase $\hat{\theta}_1$ satisfies an equation which is formally equivalent to a curved spacetime equation for a massless scalar field [2, 3]

$$\Box\hat{\theta}_1 = \frac{1}{\sqrt{-G}}\partial_\mu(\sqrt{-G}\,G^{\mu\nu}\partial_\nu)\,\hat{\theta}_1 = 0\,, \tag{2}$$

where $\Box$ is the curved d'Alembertian for the acoustic metric $G_{\mu\nu}$ of line element

$$ds^2 = G_{\mu\nu}dx^\mu dx^\nu = \frac{n}{c}[-c^2dt^2 + (d\vec{x} - \vec{v}dt)\cdot(d\vec{x} - \vec{v}dt)]\,, \tag{3}$$

G is the metric determinant, and c, defined by $c^2 = nd\mu/dn$, is the local sound speed. The density fluctuation $\hat{n}_1$ is algebraically related to the phase fluctuation $\hat{\theta}_1$:

$$\hat{n}_1 = -\frac{n}{c^2}(\partial_t\hat{\theta}_1 + \vec{v}\cdot\vec{\nabla}\hat{\theta}_1) \tag{4}$$

An acoustic horizon appears when the fluid turns supersonic, and, after its formation, Hawking radiation is expected as a thermal emission of phonons [2].

The Hawking process consists of the creation of correlated pairs of outgoing quanta triggered by the horizon formation [14, 15]: one member is emitted into an escaping mode, and corresponds to what is usually called Hawking particle; the other member, the so-called *partner*, is emitted into a trapped mode inside the horizon. Of course, this latter particle is definitively lost and can not be detected in the case of astrophysical black holes.

To get the correlations associated to the Hawking process the two-point function $H_2(x, x') = \langle\hat{\theta}_1(\vec{x}, t)\hat{\theta}_1(\vec{x}', t')\rangle$ is the relevant one, as it will be always possible to derive the density-density $\langle\hat{n}_1(\vec{x}, t)\hat{n}_1(\vec{x}', t')\rangle$ or the mixed one $\langle\hat{\theta}_1(\vec{x}, t)\hat{n}_1(\vec{x}', t')\rangle$ by simply acting the derivative operator

$\mathcal{D}_x$, defined as $\mathcal{D}_x = -\frac{n(\vec{x},t)}{c^2(\vec{x},t)}\left(\partial_t + v(\vec{x})\vec{\nabla}_{\vec{x}}\right)$ on H_2 once or twice, and then taking the $t' \to t$ limit if interested in the same-time correlations.

To obtain workable analytical expressions, the simplest case of an effective $1+1$ dimensional geometry is considered. This will be an excellent approximation whenever one can assume that the mode propagation in the transverse dimensions is negligible or actually frozen. Performing a dimensional reduction along the transverse direction [18], the two-point function for the field $\hat{\theta}_1$ can be approximated as:

$$\langle\hat{\theta}_1(x,t)\hat{\theta}_1(x',t')\rangle \simeq -\frac{\hbar}{4\pi\sqrt{C(x,t)C(x',t')}}\ln[\Delta\tilde{x}^-\Delta\tilde{x}^+] \tag{5}$$

where $C = \frac{n}{c}$ is the conformal factor for the metric (3), and $\tilde{x}^\pm$ are coordinates related to the light (sound) cone coordinates $x^\pm = t \pm \int\frac{dx}{(c\mp v)}$ by a conformal transformation, defining the proper quantum state. The expectation value entering the two-point function (5) is to be taken in the vacuum state of the positive frequency modes with respect to the proper $\tilde{x}^\pm$ coordinates [16]. After dimensional reduction, an effective potential term appears in the $1+1$ dimensional wave equation (2), which is responsible for backscattering of the modes. In what follows this term is neglected, and the final result is expected to be just slightly overestimated [17].

After the horizon formation, while the ingoing modes remain positive frequency modes with respect to the x^+ light cone coordinate, the outgoing modes are no longer positive frequency with respect to the x^- light cone coordinate. Indeed they emerge as positive frequency with respect to the generalized outgoing coordinate $\tilde{x}^- = \pm\frac{e^{-kx^-}}{k}$ [14, 16, 18], where the surface gravity at the horizon (located at $x^- = \infty$ or $\tilde{x}^- = 0$) is defined as

$$k = \frac{1}{2c}\frac{d}{dx}(c^2 - v^2)|_H \tag{6}$$

and fully determines the Hawking temperature $T_H = \hbar k/(2\pi\kappa_B)$ [16]. The $\pm$ signs in the definition of $\tilde{x}^-$ refer to outgoing modes which are respectively trapped inside the horizon or able to escape towards the asymptotic subsonic region.

To get the correlator associated to late times after the horizon has formed one must use $\ln(\Delta\tilde{x}^-\Delta x^+)$ as the master function defining the correct quantum state in Eq. (5).

3. Density-density correlations in weakly interacting Bose Einstein condensates

Let's move to the analog model of black hole based on a flowing atomic Bose-Einstein condensate (BEC) [5]. A weakly interacting BEC is described by an order parameter satisfying the Gross-Pitaevskii equations [19]. One can consider phonons by considering small fluctuations around a stationary and fully condensed state. One can write the Bose field operator as $\hat{\Psi} = e^{i\hat{\theta}}\sqrt{\hat{n}}$ in terms of the number density $\hat{n}$ and the phase $\hat{\theta}$ and then expand the density $\hat{n} = n + \hat{n}_1$ and phase $\hat{\theta} = \theta + \hat{\theta}_1$ operators around the classical background values n and θ fixed by the mean-field Gross-Pitaevskii equation.

We limit our attention to the so-called hydrodynamic limit, where perturbations are considered with wavelengths much longer than the healing length $\xi = \hbar/\sqrt{gn}$ (g the atom-atom interaction constant in the Gross-Pitaevskii equation). In this limit the Gross-Pitaevskii equations assume exactly the hydrodynamical form of Eqs. (1).

As the density operator $\hat{n}_1$ is algebraically related to $\hat{\theta}_1$ by $\hat{n}_1 = -\frac{1}{g}\left(\partial_t\hat{\theta}_1 + \vec{\nabla}\theta\cdot\vec{\nabla}\hat{\theta}_1\right)$, the one-time density-density correlation function

$$G_2(x,x') = \langle\hat{n}(x)\hat{n}(x')\rangle - \langle\hat{n}(x)\rangle\langle\hat{n}(x')\rangle = \langle\hat{n}_1(x)\hat{n}_1(x')\rangle \tag{7}$$

can be simply expressed in term of the two-points function for the field $\hat{\theta}_1$:

$$G_2(x, x') = \lim_{t' \to t} \mathcal{D}_x \mathcal{D}_{x'} \langle \hat{\theta}_1(x,t) \hat{\theta}_1(x',t') \rangle \,. \tag{8}$$

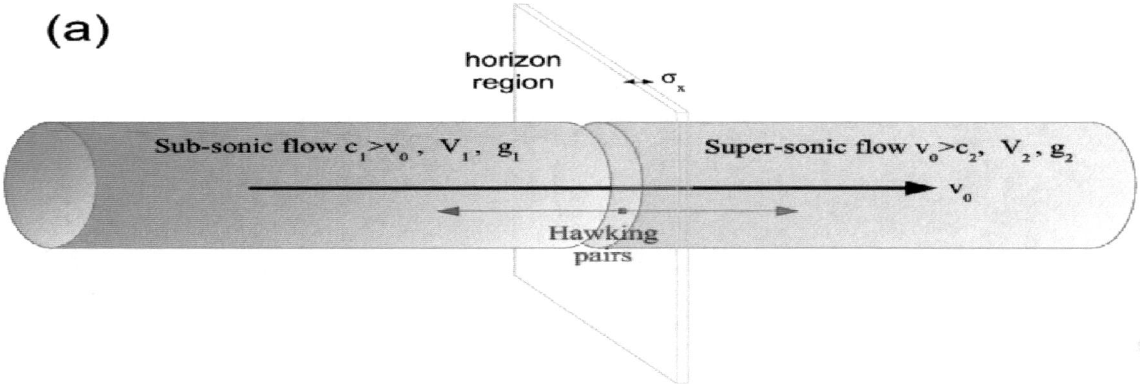

Figure 1. Sketch of the proposed set-up: the density n and the flow velocity v are kept constant, while the sound velocity c is linearly modulated in the region $[-x_0, x_0]$. The sonic horizon is at $x = 0$.

Even if the basic ingredients to get the general form of the G_2 corralator are all present, the calculation is done for a realistic experimental proposal to form an acoustic black hole. A one-dimensional BEC of uniform density n_{1D} is flowing with constant speed v along the x direction. For $t < 0$ the sound velocity $c(x,t)$ is constant in space and equal to $c(x,t) = c_1 > v$. At $t = 0$, the sound velocity in the $x < 0$ region is rapidly switched to a lower $c_2 < v$ value in order to create a horizon that separates a region of sub-sonic $v < c_1$ flow for $x > 0$ from a super-sonic $v > c_2$ one for $x < 0$. In the crossover region $x \in [-x_0, x_0]$, the sound velocity is assumed to vary linearly in x. In order for the hydrodynamic approximation to be valid x_0 has to be larger than ξ [1]. The sonic horizon (i.e. the locus where $c = v$) is taken at $x = 0$ (see Fig. 1).

A compact formula can then be obtained from Eqs.(5, 8) for the most interesting case when the x, x' points are located on opposite sides with respect to the horizon and lie well outside the modulation region i.e. $x > x_0$ and $x' < -x_0$:

$$\frac{G_2(x,x')}{n_{1D}^2} \simeq -\frac{k^2\,\xi_2\xi_1}{16\pi\,c_2 c_1}\,\frac{1}{\sqrt{(n_{1D}\xi_1)(n_{1D}\xi_2)}}\,\frac{c_1\,c_2}{(c_1-v)(v-c_2)}\,\frac{1}{\cosh^2\left[\frac{k}{2}\left(\frac{x}{c_1-v}+\frac{x'}{v-c_2}\right)\right]} \tag{9}$$

As shown in Fig. 2, $G_2(x,x')$ has a quite narrow, valley-shaped feature centered on the $x'/(v-c_2) = -x/(c_1-v)$ straight line that describes correlations between pair of points located

[1] Such a modulation can be obtained by spatially modulating the transverse confinement of the waveguide and/or the nonlinear interaction constant $g(x)$ by means of a spatially varying magnetic field in the vicinity of a Feshbach resonance.

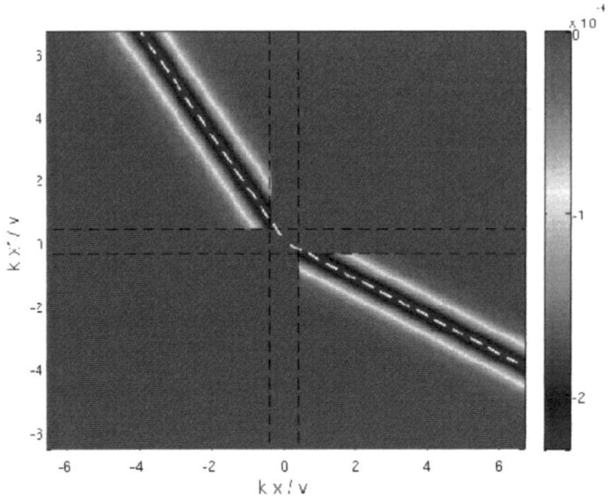

Figure 2. Density-density correlation pattern $|G_2(x, x')| / n_{1D}^2$. The valley-shaped feature indicated as a white dashed line is the signature of Hawking radiation. The black dashed lines indicate the $[-x_0, x_0]$ horizon region.

respectively inside and outside the black hole. The bottom value of the valley goes as the squared surface gravity and it is inversely proportional to the diluteness parameter $n_{1D}\xi$ of the gas. The neglected subleading term $O[|x - x'|^{-2}]$ contains the vacuum contribution of the ingoing modes and it is a consequence of the short-range repulsion of bosons [20]. On the other hand, no specific signal appears for pair of points on the same side of the horizon.

The position of the valley has a transparent physical interpretation: at all times, pairs of Hawking phonons almost simultaneously emerge from the horizon region and propagate into the sub- and super-sonic regions at speeds respectively $c_1 - v$ and $v - c_2$. After a propagation time Δt, the initial quantum correlation then reflects into a density correlation between $x = (c_1 - v)\Delta t$ and $x' = -(v - c_2)\Delta t$. The valley-shaped feature is immediately recovered once we integrate over all possible values of Δt. The width of the valley is proportional to the inverse of the surface gravity k^{-1} and it provides a simple way of estimating the Hawking temperature of the emission. Note that under the hydrodynamic assumption the width is much larger than the healing length ξ. The contribution of mode back-scattering, here neglected, quickly vanishes in the hydrodynamic limit: the corresponding terms are in fact of higher order in k. Although all the predictions have been obtained for the simplest configuration, the behavior of the correlation function at late ($t \gg k^{-1}$) times can be shown to be generic and independent on the details of the horizon formation process [16].

4. Numerical simulations

In [11] the analytical result of Eq. (9) has been validated by numerically simulating the dynamics of fluctuations around the mean field solution of the Gross-Pitaevskii equations during and after the formation of the horizon, by means of the so-called truncated Wigner method for the interacting Bose field [21].

At $t = 0$ well before the formation of the black-hole horizon, the condensate is assumed to be uniform and at thermal equilibrium in the moving frame at v_0 at a temperature T_0.

The numerical study of the Hawking radiation as appeared in [11] is based on a vast campaign of Wigner simulations of the condensate evolution after the formation of the black-hole horizon. Specific attention has been paid to the normal-ordered density-density correlation function

Journal of Physics: Conference Series **222** (2010) 012036 doi:10.1088/1742-6596/222/1/012036

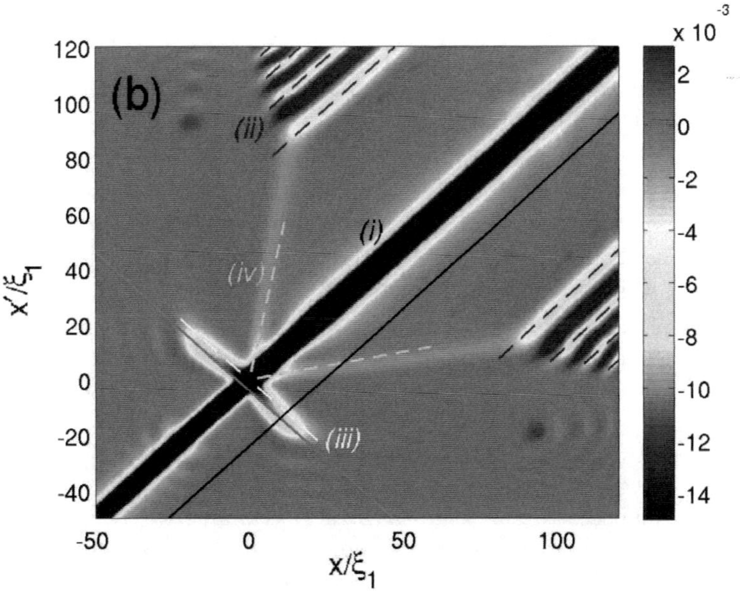

Figure 3. Density plot of the rescaled density correlation $(n\xi_1)\times[G_2(x,x')]$ well after the switch-on of the horizon. The dashed lines and the (i), (ii), (iii), (iv) labels identify the main features discussed in the text. The horizon has a spatial width $\sigma_x/\xi_1 = 0.5$. The initial temperature is $T_0 = 0$.

$G_2(x,x')$. One snapshots of $G_2(x,x')$ well after the horizon formation is shown in Fig.3.

The main features that are observed in the figure can be classified as follows:

(i) A strong, negative correlation strip is always present along the main diagonal $x = x'$ and is almost unaffected by the horizon formation process.

(ii) A system of fringes parallel to the main diagonal appears inside the black hole as soon as the horizon is formed. As time goes on, these fringes move away from the main diagonal at an approximately constant speed and eventually disappear from the region of sight.

(iii) Symmetric pairs of negative correlation tongues extend from the horizon point almost orthogonally to the main diagonal. While their maximum height remains almost constant in time, their length linearly grows with time. These tongues involve pairs of points located on opposite sides of the horizon.

(iv) A second pair of tongues appears for pairs of points located inside the black hole. Both their height and length scale in the same way as for feature (iii).

Apart from feature (i) which is the usual antibunching due to the repulsive atom-atom interactions [20], these observations illustrate a variety of effects of quantum field theory in a spatially and temporally varying background [16].

Feature (ii) is a transient effect that originates in the bulk of the internal $x > 0$ region and has no relation with the presence of a horizon. Detailed analysis is given in [22]: as a consequence of the time-modulation of the interaction constant g, correlated pairs of Bogoliubov phonons are generated during the short modulation time by a phonon analog of the dynamical Casimir effect [23]. As usual, the longer the switching time σ_t, the weaker the dynamical Casimir signal which eventually disappears in the limit of a very long switch.

On the other hand, features (*iii*) and (*iv*) do not depend on σ_t, but only on the possible presence of a horizon at long times. In particular, they completely disappear if the flow remains everywhere sub-sonic $v_0 < c_{1,2}$ [11]. This fact, together with their shape in the (x, x') plane and their persistence for indefinite times after the horizon formation suggests a strict link with the Hawking effect.

The predictions made in [10] are confirmed, by showing the quantum correlations within a pair of Hawking phonons emitted in respectively the inward and outward direction translate into a correlation between the density fluctuations at distant points located on opposite sides from the horizon. Once the horizon is formed, correlated pairs of Hawking phonons are continuously emitted at all times $t > t_0$. These phonons then propagate from the horizon in respectively the outward and inward direction at speeds $v_0 - c_1 < 0$ and $v_0 - c_2 > 0$. A generic time τ after their emission they are located at $x = (v_0 - c_1)\tau < 0$ and $x' = (v_0 - c_2)\tau > 0$, which defines a straight line of slope. As one can see in Fig.3, this line (indicated as a white dashed line) almost coincides with the axis of the numerically observed tongue (*iii*).

Feature (*iv*) originates from the (partial) elastic back-scattering of the Hawking particles in the horizon region: in this case both the Hawking and the partner phonons eventually propagate in the inward direction at speeds $v_0 \pm c_2$. Again, the analytically predicted slope $(v_0 - c_2)/(v_0 + c_2)$ (green dashed lines) well agrees with the axis of the numerically observed tongue (*iv*). Extended semi-analytically analysis of the process have appeared in [12, 13].

The identification of features (*iii*) and (*iv*) as signatures of Hawking radiation is further confirmed by a quantitative comparison of the numerical data with the predictions in [10]. The peak value G_2(peak) and the transverse full width at half maximum (FWHM) Δx_{FWHM} of the tongue (*iii*) confirms, as expected, that the Hawking emission is in fact stationary in both space and time, and that the agreement with the gravitational prediction is quantitatively excellent in the hydrodynamic limit $\sigma_x/\xi \gg 1$ where the physics of the many-body system is dominated by the hydrodynamic modes [11].

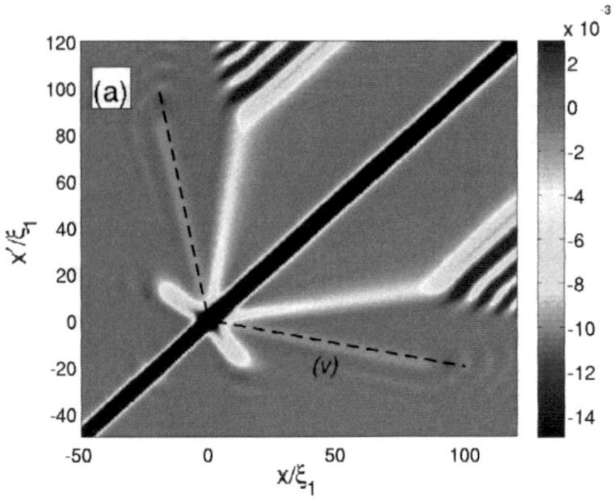

Figure 4. Panel (a): plot of the density correlation $(n\xi_1)G_2(x, x')$ for the same parameters as in Fig.3 but a finite initial temperature $k_B T_0/\mu_1 = 0.1$.

Even for an initial temperature $(k_B T/\mu_1 = 0.1)$ higher than the Hawking temperature $k_B T_H/\mu_1 \simeq 0.05$, the (*iii*) and (*iv*) Hawking tongues remain perfectly visible and are even a bit strengthened by stimulation effects [12, 24]. The new feature (*v*) that originates from thermal effects is located between the tongues (*iii*) and (*iv*) and is well distinct from them. Its attribution to partial reflection of thermal phonons is confirmed by the value of its slope, which is in very good agreement with the analytical prediction $(v_0-c_1)/(v_0+c_2)$ indicated as a dashed black line in Fig.4.

As the slope of the thermal tongue is completely different from the one $(v_0-c_1)/(v_0-c_2)$ of the Hawking tongue, the Hawking signal can be easily isolated in a correlation image from spurious thermal effects also at temperatures higher than T_H. This is even more remarkable as in this regime it appears difficult to separate the Hawking phonons from the thermal ones by looking just at the total phonon flux.

5. Acoustic black holes in ion rings

It has been recently proposed in [9] to get experimental evidence of Hawking radiation by using ion rings, and measuring the peculiar correlations associated to the Hawking process. This is a discrete analog of a hydrodynamical system that can show super- and subsonic regions, built up with ions trapped in a quadrupole ring trap [25] as schematically depicted in Fig. 5. Ions rotate on a ring with circumference L with an inhomogeneous, but stationary velocity profile $v(\theta)$. Thus the ions are inhomogeneously spaced. The necessary (de-)acceleration of the ions is guaranteed by additional electrodes exerting a force $F^e(\theta)$ on the ions. The harmonic oscillations of the ions around this equilibrium motion are phonons with velocities $c(\theta) \propto v(\theta)^{-1/2}$. Regions with large enough ion spacings and low enough velocities are supersonic; here phonons can only move into the direction of the ion flow and are trapped like light inside a black hole. The considered system shows a super- and a subsonic region. The border between these regions is analogous to a black hole horizon and is shown to emit Hawking radiation [2]. The details of the systems are given in [9].

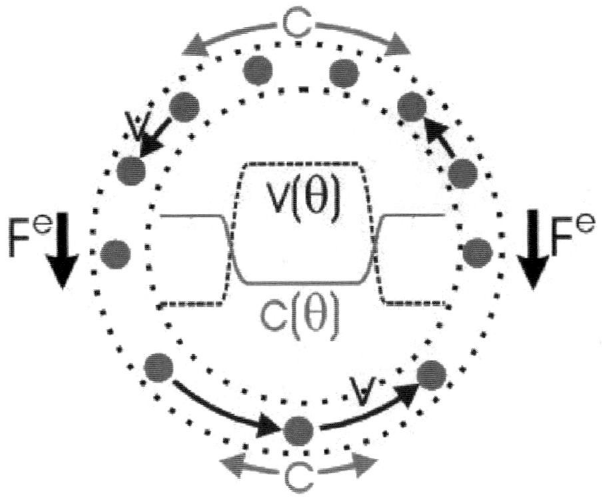

Figure 5. Schematic depiction of the ions rotating with velocity $v(\theta)$ and the phononic excitations with velocity $\pm c(\theta)$, that depend on the varying ion spacing. The external force $F^e(\theta)$ localized at the transition between the super- and subsonic region (de-)accelerates the ions. The inset shows a typical velocity profile with a black hole like region for $v(\theta) > c(\theta)$.

Small perturbations around the equilibrium motion positions $\theta_i(t) = \theta_i^0(t) + \delta\theta_i(t)$ are considered, and the canonical operators $\delta\theta_i$ and $-i\hbar\partial_{\delta\theta_i}$ describe the phononic oscillations of the ions. The quasi-free quantum dynamics of this harmonic system are governed by the classical linear equations of motion for the first and second moments. The stability of this equilibrium motion has been studied numerically. The limit of an infinite number of ions and only the Coulomb interaction between neighboring ions are considered, and an analogy with the standard Hawking effect is obtained. Then for a slowly varying $v(\theta)$ the system Lagrangian for the scalar

[2] Being a periodic system, it will show also the analog of the white hole horizon. It's presence will give constraints on the time-scale to perform the experiment.

field $\Phi(\theta_i^0(t), t) = \delta\theta_i(t)$ becomes

$$\mathcal{L} = \int d\theta \frac{\rho(\theta)}{2} \left[(\partial_t \Phi + v(\theta)\partial_\theta \Phi)^2 - [iD(\theta, -i\partial_\theta)\Phi]^2 \right] \tag{10}$$

with the dispersion relation $D(\theta, k) = c(\theta)k + \mathcal{O}(k^3)$, the conformal factor $\rho(\theta) = n(\theta) \cdot mL^2/(2\pi)^2$, and the density $n(\theta) = N/(v(\theta)T)$. This scalar field satisfying a linear dispersion relation at low wavenumbers, with sound velocity $c(\theta) = \sqrt{2(2\pi)^3 n(\theta)e^2/(4\pi\epsilon_0 \cdot mL^3)}$ is analogous to a massless scalar field in a black hole spacetime [2, 3]. Its quanta cannot escape a supersonic region with $v(\theta) > c(\theta)$, like photons trapped inside a black hole. The horizon of this analog model is located at $c(\theta_H) = v(\theta_H)$. Pairs of Hawking particles are emitted close to the black hole horizon with a black body distribution at the Hawking temperature given in Eq. (6).

For an experimental proof for Hawking radiation on the ion ring, correlations between the different regions on the ion ring after the creation of a black hole are studied and discussed. Starting from a thermal state of the excitations around homogeneously spaced rotating ions, a supersonic region is created in a small time interval τ. This is done by reducing the subsonic fluid velocity v_{min} in a Gaussian way, while leaving the average rotation velocity constant.

The normalized angle-angle correlations is analyzed

$$C_{ij} = \langle \delta\theta_i \delta\theta_j \rangle / \sqrt{\langle \delta\theta_i^2 \rangle \langle \delta\theta_j^2 \rangle}. \tag{11}$$

Fig. 6 shows the results at a fixed time starting from the ground state. Correlations between the super- and subsonic regions are created close to the black hole horizon and are moving away from it as expected. These correlations correspond to the pair creation of Hawking particles. In

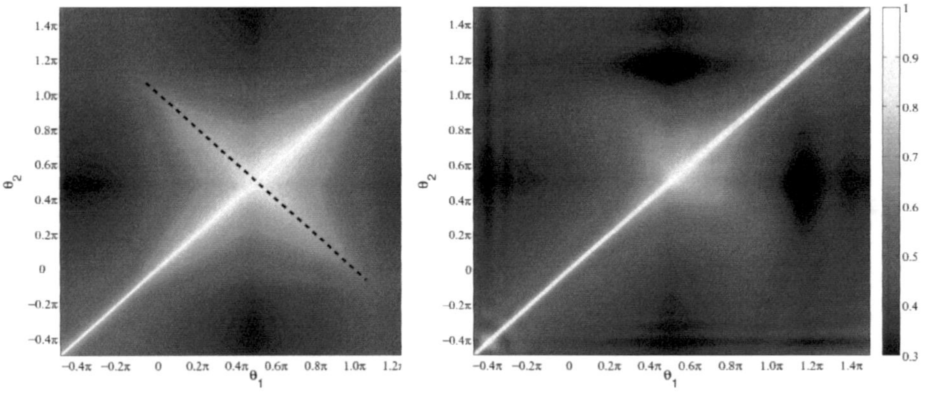

Figure 6. Correlations $C_{ij}(t)$ at time $t = T$ with real space lab frame positions. System parameters: $N = 500$ ions, $\sigma v_{min}T = 2\pi \cdot 0.25$, $e^2/4\pi\epsilon_0 = \frac{1.1}{2N}\frac{mL^3}{T^2}$, and $\tau = 0.05T$. Only nearest-neighbor interactions are considered (left panel). Full Coulomb interactions considered (right panel).

Fig. 6(left) the simulation is performed with interactions between neighboring ions only. The angle and the propagation velocity of the correlation features are in good agreement with the predictions based on the sound velocities and the ion velocities. For the long range Coulomb interactions the correlations behave similarly, and the slope agrees with the prediction one gets from the real dispersion relation [9]. This gives the first evidence of the robustness of Hawking radiation in a discrete system. The Hawking signature in the correlation pattern is estimated to be visible in a set of parameter usually used in everyday experiments with ions, and therefore the prospect to observe it in near future experiment is realistic.

6. Conclusions

In this paper the proposal to use correlation measurements to reveal the Hawking radiation in analog black holes is reviewed. The original proposal for weakly interacting BECs and the following application to ion rings, show that the presence of Hawking radiation is clear and peculiar in the correlation patterns.. It is therefore fair to conclude that so far correlation measurements are the most promising tool to get evidence of such an elusive effect in the near future experiments.

Acknowledgements

This paper is based on the talk given by the author at the 1st Mediterranean Conference of Classical and Quantum Gravity, Kolimbary, Greece, September 14-18 2009. It is based on the papers: R.Balbinot, A. Fabbri, S. Fagnocchi, A. Recati, I. Carusotto *"Non-local density correlations as signal of Hawking radiation in BEC acoustic black holes"*, Phys.Rev.A 78,02160 (2008); I. Carusotto, S. Fagnocchi, A. Recati, R. Balbinot, A. Fabbri *"Numerical observation of Hawking radiation from acoustic black holes in atomic Bose-Einstein condensates"*, New.J.Phys.10, 103001 (2008) regarding the density correlations in weakly interacting BECs; and on B. Horstmann, B. Reznik, S. Fagnocchi, J. I. Cirac, *"Hawking Radiation from an Acoustic Black Hole on an Ion Ring"*, arXiv:0904.4801 (2009) about Hawking correlations in ion rings. I acknowledge all the coauthors.

References

[1] S.W. Hawking, Nature (London) **248**, 30 (1974); S. W. Hawking, Commun. Math. Phys. **43**, 199 (1975)
[2] W.G. Unruh, Phys. Rev. Lett. **46**, 1351 (1981)
[3] C. Barcelo, S. Liberati, and M. Visser, Living Rev. Rel. **8** (2005)
[4] T. A. Jacobson and G. E. Volovik, Phys. Rev. D **58**, 064021 (1998)
[5] L.J. Garay, J.R. Anglin, J.I. Cirac and P. Zoller, Phys. Rev. Lett. **85**, 4643 (2000); P. O. Fedichev and U. R. Fischer, Phys. Rev. Lett. **91**, 240407 (2003)
[6] S. Giovanazzi, Phys. Rev. Lett. **94**, 061302 (2005)
[7] U. Leonhardt and P. Piwnicki, Phys. Rev. Lett. **84**, 822 (2000); W. G. Unruh and R. Schützhold, Phys. Rev. D **68**, 024008 (2003).
[8] R. Schützhold and W. G. Unruh, Phys. Rev. Lett. **95**, 031301 (2005); T. G. Philbin, *et al.*, Science **319**, 1367 (2008)
[9] B. Horstmann, B. Reznik, S. Fagnocchi and J. I. Cirac, arXiv:0904.4801 [quant-ph]
[10] R. Balbinot, A. Fabbri, S. Fagnocchi, A. Recati and I. Carusotto, Phys. Rev. A **78**, 021603 (2008)
[11] I. Carusotto, S. Fagnocchi, A. Recati, R. Balbinot and A. Fabbri, New J. Phys. **10**, 103001 (2008)
[12] A. Recati, N. Pavloff and I. Carusotto, arXiv:0907.4305 [cond-mat.quant-gas]
[13] J. Macher and R. Parentani, arXiv:0905.3634 [cond-mat.quant-gas]
[14] R.D. Carlitz, R.S. Willey, Phys. Rev. D **36**, 2327 (1987)
[15] S. Massar, R. Parentani, Phys. Rev. D **54**, 7444 (1996); R. Brout *et al*, Phys. Rept. **260**, 328 (1995)
[16] N.D. Birrell and P.C.W. Davies, *Quantum fields in curved space*, Cambridge University Press, Cambridge, UK, (1982)
[17] R. Balbinot *at al.*, Phys. Rev. **D** 63, 084029 (2001)
[18] R. Balbinot, A. Fabbri, S. Fagnocchi, and R. Parentani, Riv.Nuovo Cim.28, 1 (2005)
[19] L.P. Pitaevskii, S. Stringari, *Bose-Einstein condensation*, Clarendon Press, Oxford (2003)
[20] C. Mora, Y. Castin, Phys. Rev. **A 67**, 053615 (2003)
[21] A. Sinatra, C. Lobo, Y. Castin, J. Phys. **B 35**, 3599 (2002)
[22] I. Carusotto, R. Balbinot, A. Fabbri and A. Recati, arXiv:0907.2314 [cond-mat.quant-gas].
[23] A. Lambrecht, J. Opt. B: Quantum Semiclass. Opt. 7, S3 (2005); C. K. Law, Phys. Rev. A 49, 433 (1994)
[24] J. D. Bekenstein and A. Meisels, Phys. Rev. **D** 15, 2775 (1977)
[25] G. Birkl *et al.*, Nature 357, 310 (1992); T. Schätz *et al.*, Nature 412, 717 (2001)

First Mediterranean Conference on Classical and Quantum Gravity (MCCQG 2009) IOP Publishing
Journal of Physics: Conference Series **222** (2010) 012037 doi:10.1088/1742-6596/222/1/012037

The black hole information paradox and macroscopic superpositions

Stephen D.H. Hsu

Institute for Theoretical Science, University of Oregon, Eugene OR 97403-5203, USA

E-mail: hsu@uoregon.edu

Abstract. We investigate the experimental capabilities required to test whether black holes destroy information. We show that an experiment capable of illuminating the information puzzle must necessarily be able to detect or manipulate macroscopic superpositions (i.e., Everett branches). Hence, it could also address the fundamental question of decoherence versus wavefunction collapse.

1. Introduction: black hole information

In 1976 Hawking proposed that black holes destroy information: pure states which collapse to form black holes evaporate into mixed states described by density matrices [1]. The argument in favor of information destruction can be pared to a few essential components; for reviews, see [2, 3, 4, 5, 6]. Hawking radiation, into which the hole evaporates, originates from outside the horizon and is causally disconnected from the interior: a spacelike slice can be constructed which intersects both the infalling matter and the outgoing Hawking radiation. The no-cloning theorem [7] in quantum mechanics prevents information from residing in two places on the same slice, so the outgoing radiation must be independent of the initial state. (See Figure 1.)

It is safe to say that, over 30 years after Hawking's paper, theoreticians remain divided as to whether Hawking was originally correct, or whether some locality-violating mechanism somehow allows the information to escape.

In this discussion [8], we investigate the following questions: Is the black hole information puzzle simply philosophy, or is it subject to experimental test? If the latter, then what capabilities are required? We find connections to a different question, from the foundations of quantum mechanics: do wavefunctions collapse, or is quantum evolution strictly unitary, leading inevitably to macroscopic superposition states? An experiment which sheds light on the black hole information puzzle would also be capable of addressing fundamental issues in quantum mechanics. See Zeh [9] and also [10] for related discussion.

To highlight the importance of macroscopic superpositions, we emphasize that the very formulation of the information puzzle relies on the use of the semiclassical black hole spacetime – e.g., in the construction of the spacelike slice used in the no-cloning argument, or in the original Hawking calculation. But, in any fully quantum mechanical treatment of the black hole formation process there exist branches of the wavefunction, possibly of very small amplitude, in which no apparent horizon is formed and the initial state particles all escape to infinity. In other words, in which the spacetime is radically different. This is most easily seen if one considers black holes formed in the collision of two particles [11] (see Figure 2): there is always a non-zero

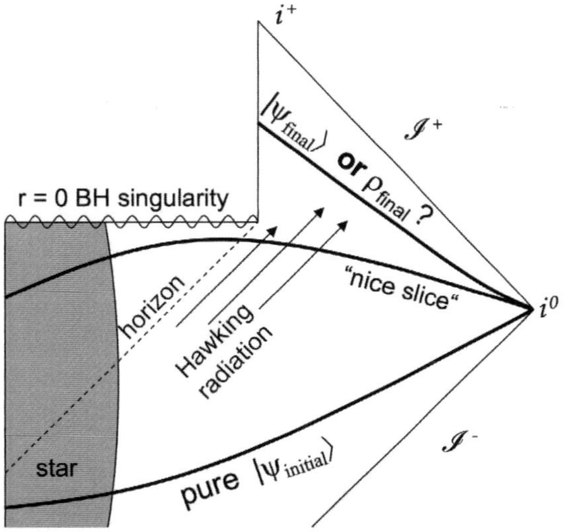

Figure 1. Formulation of the black hole information puzzle on a semiclassical spacetime. The no-cloning theorem prevents the same quantum information from existing in two different places on the nice slice.

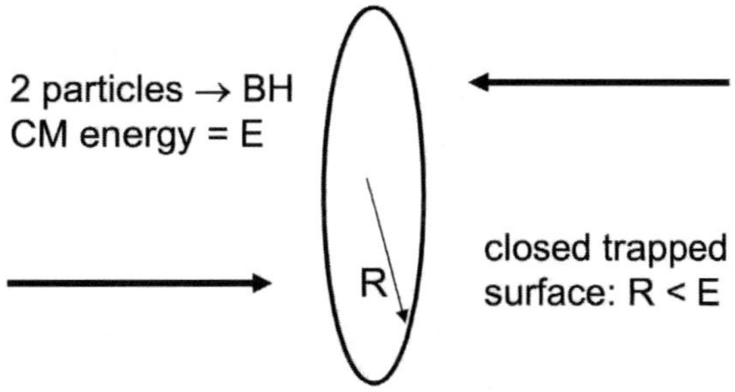

Figure 2. Two particle scattering with sufficiently small impact parameter leads to black hole formation.

amplitude for no scattering – i.e., no black hole creation – in which the particles simply pass each other. This remains the case even as the number of particles in the initial state becomes large, although for certain semiclassical initial data one can make the no-formation amplitude arbitrarily small. Nevertheless, the information puzzle cannot avoid the issue of macroscopic superpositions: could such small amplitude branches restore purity or unitarity? [12]

2. Decoherence: pure to mixed evolution

Quantum mechanics as conventionally formulated (the Copenhagen interpretation) allows for two kinds of time evolution: the usual Schrödinger evolution, which is unitary, and measurement collapse, which is non-unitary and leads to von Neumann projection onto a particular eigenstate of the operator associated with the measurement. It is appealing to think that wavefunction collapse might only be an apparent phenomenon, which results from unitary evolution. This idea dates to Everett [13], but has been developed substantially in recent decades as the theory of decoherence [14].

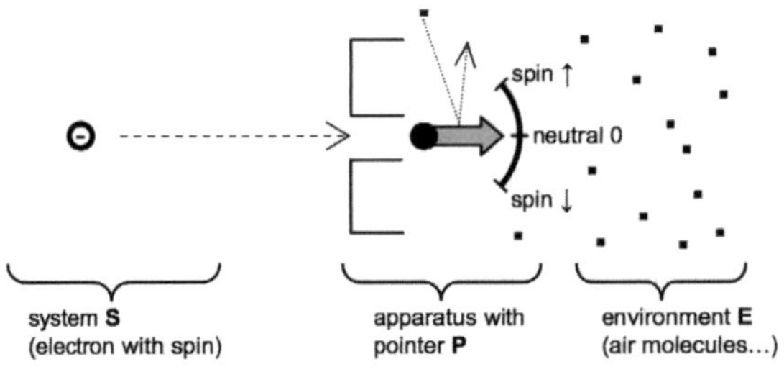

Figure 3. Measurement of a single spin leads to entangled state in Eqn. (2).

Consider a system prepared in a superposition state (see Figure 3)

$$|\psi\rangle = c_1|1\rangle + c_2|2\rangle \quad . \tag{1}$$

Suppose that, due to interactions between the system and its environment (or, equivalently, a measuring apparatus), the two evolve into an entangled state

$$|\psi\rangle \otimes |E\rangle \quad \longrightarrow \quad |\Psi\rangle = c_1|1\rangle \otimes |E_1\rangle + c_2|2\rangle \otimes |E_2\rangle \quad . \tag{2}$$

The states $E_{1,2}$ are referred to as pointer states of the environment or measuring device. These pointer states are determined by the dynamics – that is, the $1, 2$ bases along which the device makes measurements is determined by its specific properties. A Stern-Gerlach machine measures spin along a particular axis given by its magnetic field; it will not decohere the system into spin states along any other directions. After the measurement interaction, the state of the system is encoded redundantly in the environment states $E_{1,2}$ and can be read out by making simple measurements on subsets of the degrees of freedom – i.e., did the red light flash (look for red photons), or did the green light flash?

The environment is assumed to have a large number of degrees of freedom, so that after a relatively short time (determined by the specific dynamics) the states E_1 and E_2 are nearly orthogonal. The dimensionality of a Hilbert space describing N degrees of freedom is exponential in N: for qubits, $d = 2^N$. Two randomly chosen vectors from this space will have overlap $\langle E_1|E_2\rangle \sim 1/d$, which is tremendously small for any macroscopic environment or measuring device (e.g., $N \sim$ Avogadro's number). Consequently, interference phenomena between the two "branches" of system plus environment are highly suppressed (see below).

Absent the capability to measure the environmental degrees of freedom, it is appropriate to trace over the degrees of freedom of E:

$$\hat{\rho} \;=\; \mathrm{Tr}_E|\Psi\rangle\langle\Psi| \;=\; |c_1|^2\,|1\rangle\langle 1| + |c_2|^2\,|2\rangle\langle 2| \quad ; \tag{3}$$

this density matrix, from which the outcome of all subsequent measurements on the system alone can be predicted just as well as from the knowledge of the complete state $|\Psi\rangle$, is diagonal if one neglects the exponentially small overlap $\langle E_1|E_2\rangle$. This process then has the appearance of measurement with fundamental wavefunction collapse and probabilistic outcomes, despite purely unitary Schrödinger evolution. For All Practical Purposes – FAPP, as formulated by Bell [15] – decoherence leads to the usual Copenhagen interpretation.

But a nagging issue remains – the presence of the other "branches" in the pure state $|\Psi\rangle$. Are they real? Can they ever be detected experimentally? The off-diagonal elements of the reduced density matrix are suppressed by the small overlap of typical environmental states E_1, E_2, but can their effects be measured?

Journal of Physics: Conference Series **222** (2010) 012037 doi:10.1088/1742-6596/222/1/012037

Omnès [16] made detailed estimates of the capabilities required to distinguish between decoherence and fundamental collapse, which we will examine in the following section. He found that for macroscopic objects, e.g., containing Avogadro's number of degrees of freedom, any beyond-FAPP device would have to be larger than the visible universe. Hence, Omnès asserted, any distinction between the Copenhagen interpretation and unitary wavefunction evolution (leading to Everett branches) for macroscopic objects is untestable, and beyond the realm of scientific inquiry.

The foregoing discussion, in particular Omnès' estimate, assumes big environments (e.g., $N \sim$ Avogadro's number). However, decoherence mechanisms are also at work if the number of degrees of freedom N of the environment is smaller, or if the interaction between system and environment is weaker; in these cases, however, decoherence effects are either not as strong (the off-diagonal elements in $\hat{\rho}$ cannot be completely neglected as in (3), i.e. FAPP does not hold in this case), or happen only over time scales longer than in the strongly-interacting case, respectively. In recent years, this gradual onset of decoherence in controlled environments (also called the "Quantum-to-Classical transition") has been the focus of quite a number of laboratory experiments. For example, in [17] the gradual loss of spatial coherence (interference pattern) of fullerene molecules in a slit experiment was observed with increasing pressure of the gas, i.e., with increasing interaction strength (cross section) between fullerenes and gas molecules (environment). Other experiments, e.g. [18], have verified the gradual loss of coherence in a system (superposition of two states of a Rydberg atom) with increasing number of degrees of freedom of the interacting environment ($N \sim 10$ photons in a cavity). On the other hand, one of the big challenges in achieving useful quantum computing [19] is to build and control large and scalable quantum systems ($N \sim 100$ or more) in which coherence is maintained (possibly via quantum error correction) over the time of the computation.

In the decoherence approach to measurement an initially pure state is later described by a reduced density matrix which, FAPP, represents a mixed state. The black hole information puzzle is often described in similar terms: infalling matter in a pure state is somehow transformed into a mixed state of Hawking radiation. Or, equivalently, if the quantum information in the black hole precursor is not to be found in the outgoing radiation, the radiation is surmised to be in a mixed state. It has been claimed that pure to mixed evolution implies, necessarily, catastrophic consequences, such as violation of energy conservation [20] (see [21] for additional arguments, for and against this point of view). Potential resolutions of the puzzle in which the information ends up somewhere outside our universe (e.g., involving baby universes or spacetime topology change [22]) must still have an effective description in our universe in terms of pure to mixed state evolution. Decoherence provides an example of *effective* pure to mixed evolution without catastrophic consequences. Tracing over the environmental states, one loses track of the total energy of the system, but without any resulting catastrophe.

3. Black hole information experiments

Below we describe two categories of experiments which test different aspects of theoretical ideas about black hole information. Despite their differences, both require the ability to detect or manipulate macroscopic superpositions.

A. Test purity vs decohered mixed state. A basic goal would be to differentiate between pure and mixed states of the type produced by decoherence. Without this minimal capability one can hardly investigate whether pure states evolve to mixed states, as proposed originally by Hawking. (Was the initial state pure? Is the final state pure or mixed?)

In the black hole context, one could imagine forming the hole from an initial state with at least some degrees of freedom in a superposition (e.g., two spin states). The remaining degrees of freedom can be considered the environment E from our earlier discussion, assuming the dynamics are such that the environment evolves into two different pointer states corresponding to the superposition. This would be the case if, for example, the two spin states had slightly different energy due to a magnetic field provided by the other degrees of freedom.

If, for each $i = 1, 2$, the initial state $|i\rangle$ of the system leads to the final state $|\Psi_i\rangle = |i\rangle \otimes |E_i\rangle$ for system plus environment, then, starting from the initial state $c_1|1\rangle + c_2|2\rangle$, two candidates for the final state would be: on the one hand a pure state superposition

$$|\Psi\rangle = c_1|\Psi_1\rangle + c_2|\Psi_2\rangle \tag{4}$$

as predicted by unitary evolution, corresponding to the pure density matrix

$$\hat{\rho}_P = |\Psi\rangle \langle\Psi| \ , \tag{5}$$

and, on the other hand, a mixed state density matrix

$$\hat{\rho}_M \ = \ |c_1|^2 \, |\Psi_1\rangle \, \langle\Psi_1| \ + \ |c_2|^2 \, |\Psi_2\rangle \, \langle\Psi_2| \ , \tag{6}$$

predicted by wavefunction collapse. What is required to differentiate between these alternatives? Consider a measurement operator M, which, without loss of generality, we might take to be a projector onto some subspace of the Hilbert space. The pure and mixed states can be distinguished if we can detect the difference

$$\mathrm{Tr}[\hat{\rho}_P M] - \mathrm{Tr}[\hat{\rho}_M M] \ = \ c_1^* c_2 \langle\Psi_1|M|\Psi_2\rangle + c_2^* c_1 \langle\Psi_2|M|\Psi_1\rangle \ . \tag{7}$$

Let us consider two possibilities. If M is a generic operator – for example, only couples to a limited subset of the degrees of freedom – then the matrix elements in (7) will be exponentially small. In particular,

$$\langle\Psi_1|M|\Psi_2\rangle \sim \frac{1}{\dim \mathcal{H}_{\mathrm{S}\otimes\mathrm{E}}} \sim \exp(-N) \ , \tag{8}$$

where N is the number of degrees of freedom of system plus environment. For a black hole, N is of order its entropy, or area in Planck units. Omnès [16] has argued that when N is of order Avogadro's number (e.g., for one gram of ordinary matter), a measurement of this accuracy is impossible *in principle*. Based on this, Omnès concludes that questions about decoherent branches other than the one observed as an outcome are not scientific. In effect, his calculations purport to extend Bell's FAPP to a statement of principle. Roughly speaking, he argues that the sensitivity of a classical device with N' degrees of freedom only improves as a power of N' (not exponentially with N'). Since the precision needed to detect a decohered macroscopic superposition involving N degrees is $\sim \exp(-N)$, as in (8), the required N' grows exponentially with N. To detect a macroscopic superposition with $N \sim 10^{24}$, Omnès concludes, would require N' larger than the number of particles in the visible universe.

A concrete proposal [23] has been made of how to experimentally decide between unitary (pure to pure) and non-unitary (pure to mixed) evaporation of a black hole (in AdS space), namely by measuring whether a certain correlation function drops, over time, either to zero or to the finite value $\exp(-N)$. Interestingly, the required measurement accuracy is essentially that necessary to detect Everett branches of a system with a similar number N of degrees of freedom, as described in the previous paragraph.

On the other hand, for a carefully engineered operator M, the amplitude $\langle\Psi_1|M|\Psi_2\rangle$ in (7) can be of order $1/2$, even if $\langle\Psi_1|\Psi_2\rangle = 0$. Indeed, the maximum value is obtained when M is a projector onto a macroscopic superposition of the type $\sim |\Psi_1\rangle + |\Psi_2\rangle$, like $|\Psi\rangle$ itself. It is a formidable challenge to perform a measurement of such an operator M, presumably even harder than preparing $|\Psi\rangle$ itself, and we give a concrete, albeit idealized, example below in the context of the Coleman-Hepp model of measurement [24] to illustrate the difficulties arising even in simple cases.

Coleman and Hepp proposed an explicit model of a quantum measurement which results from the interaction of spin states. In their model the interaction between the system (itself a spin) and measuring device causes evolution as in (2), with

$$\begin{aligned} |\Psi_1\rangle &= |1\rangle \otimes |E_1\rangle = |+\rangle \otimes |+++\cdots+\rangle \\ |\Psi_2\rangle &= |2\rangle \otimes |E_2\rangle = |-\rangle \otimes |---\cdots-\rangle \ , \end{aligned} \tag{9}$$

where $\pm$ are spin-up and spin-down states along the z axis, and the measuring device has N degrees of freedom (qubits). That is, the interaction between system and measuring device leads to a correlation between the initial spin state and the (macroscopic, when N is large) state of the device. The state of the spin can be read out by measuring some subset of the N degrees of freedom in the device.

In this context it is straightforward to design an operator M for which $\langle\Psi_1|M|\Psi_2\rangle$ is large: we simply take the tensor product operator

$$M = \bigotimes_i \sigma_x^i \ , \tag{10}$$

where $\sigma_x^i|\pm\rangle = |\mp\rangle$ measures the spin of environmental qubit i in the x-direction. An experimental realization of M would be able to distinguish the two considered post-measurement states of the system plus device: pure $\hat{\rho}_P$ vs. mixed $\hat{\rho}_M$. Note, though, that the simple M considered in this toy example is not a projection (in particular, not the projector into the one-dimensional subspace $|\Psi\rangle$) and so cannot, e.g., distinguish the considered macroscopic superposition pure state (4) from the (seemingly simpler) pure state $|+_x\rangle \otimes |+_x +_x \cdots +_x\rangle$ in which all spins are aligned in the $+x$-direction.

One might object that in this simple example the pointer states $E_{1,2}$ of the device in (9) are mutually orthogonal, but not thermalized. In this sense the model does not represent a realistic measurement (the "environment" is highly constrained). This could easily be remedied by allowing some interactions between the spins in the device, which leads to some (presumably) ergodic but unitary evolution. If one keeps the spins isolated from the rest of the universe, these interactions evolve $\Psi_{1,2}$ into something more random, at least in appearance:

$$|\Psi\rangle \rightarrow |\Psi'\rangle = U\left(c_1|\Psi_1\rangle + c_2|\Psi_2\rangle\right) \tag{11}$$

or, for the proposed mixed state (6) after the measurement,

$$\hat{\rho}_M \rightarrow \hat{\rho}'_M = U\hat{\rho}_M U^\dagger \ . \tag{12}$$

It would still be the case that the operator $M' = UMU^\dagger$ can distinguish between pure and mixed states of the post-measurement device. However, if the N spins are separated in space (e.g., correspond to isolated qubits in a quantum device), then the operator M' would itself have to be realized out of macroscopic superpositions of spatially separated objects, unlike the original M which acted independently on each of the spins.

To summarize, a measurement which can differentiate between pure and mixed states either has to rely on extreme precision to detect very small matrix elements as in (8), or on the ability to prepare a very special, typically non-local, operator like M'.

B. Test Hawking mixed state vs typical pure state. In this scenario we compare Hawking radiation in a mixed state ρ_* to radiation in a pure state ψ.

It is widely believed that large scale violation of locality at the black hole horizon is necessary for unitary evolution. This might lead to significant deviations from Hawking's results describing what is emitted from the hole. If such deviations were observed, they would undermine the usual semiclassical reasoning which leads to the information puzzle, although deviations of the radiation from Hawking's mixed state description do not by themselves imply that evolution is unitary or purity-conserving.

Perhaps a more plausible scenario, assumed in what follows, is that the radiation, although described by a pure state ψ, only deviates in subtle ways from the Hawking mixed state. That is, information is encoded in correlations (phases or superpositions) between particle states in the radiation, but the overall distribution appears to be thermal and the temperature evolution is as predicted by Hawking.

It is extremely difficult to differentiate between a pure state ψ of this type and the Hawking mixed state ρ_*. Local measurements on the radiation (i.e., over length scales much smaller than its full spatial extent) can only exclude tiny subsets of the ψ Hilbert space. Moreover, it can be shown that for almost all states ψ these local measurements are governed by the same (thermal) probability distribution as the one obtained from ρ_*.

A simple way to understand this is to recall that maximizing the entropy of a system subject to an energy constraint leads to a thermal distribution. Pure states which conform, at least macroscopically, to the Hawking predictions are constrained to describe the same total energy emission over any particular interval of time. To be specific, consider a time interval Δt_i over which the Hawking temperature is close to constant, $T = T_i$, but during which many particles are emitted. Let the Hawking calculation predict that a total amount of energy $\mathcal{E}_i$ be emitted during the interval (see Figure 4). Ordinary statistical mechanics tells us that the overwhelming majority of states (of the system in the volume corresponding to Δt_i) with energies close to $\mathcal{E}_i$ will be approximately thermal – i.e., maximizing the entropy leads to a Boltzmann distribution for energy occupation numbers, with temperature equal to the Hawking temperature T_i. The probability distribution governing measurements of the energy distribution of individual emitted quanta over the interval Δt_i will then coincide with that given by ρ_*, except for an exponentially rare subset of states satisfying the constraint (i.e., configurations with much lower entropies than the Boltzmannian, or thermal, ones).

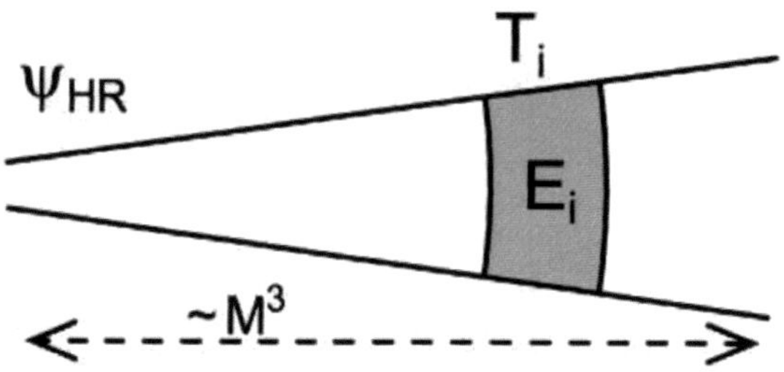

Figure 4. Hawking radiation emitted in a pure state ψ_{HR} which conforms to the semiclassical prediction for temperature and energy radiated in each time interval i.

A more explicit computation follows. Consider the subset of pure states ψ which conform to the Hawking predictions governing the amount of energy radiated in each time interval Δt_i. Specifically, require that, for every i, the reduction (by taking the partial trace) of ψ to the degrees of freedom emitted in the interval Δt_i be a mixture of superpositions of the energy eigenstates (of the theory in that volume) with eigenvalues close to $\mathcal{E}_i$. That is, consider region i and only the degrees of freedom within it (neglecting boundary effects, which are negligible for large regions). The reduction of ψ to these degrees of freedom, when expanded in an energy eigenstate basis for the region Δt_i, must have support only on states with energies close to $\mathcal{E}_i$. The superposition of two pure states ψ_1 and ψ_2 satisfying this condition will also satisfy the condition as superposition does not extend the region of support; therefore the condition defines a subspace of the larger Hilbert space. Denote by $\mathcal{H}_R$ this restricted ("energetically conforming") subspace of the overall radiation Hilbert space $\mathcal{H}$.

Further, divide the radiation into a subsystem S, to be measured, and the remaining degrees of freedom which constitute an environment E, so $\mathcal{H} = \mathcal{H}_S \otimes \mathcal{H}_E$ and

$$\rho_S \equiv \rho_S(\psi) = \mathrm{Tr}_E|\psi\rangle\langle\psi| \tag{13}$$

is the density matrix which governs measurements on S for a given pure state ψ. Note the assumption that these measurements are local to S, hence the trace over E.

Then, a recent theorem [25] on entangled states, which exploits properties of Hilbert spaces of very high dimension, shows that for almost all $\psi \in \mathcal{H}_R$, $\rho_S(\psi) \approx \mathrm{Tr}_E\,(\rho_*)$ as long as $d_E \gg d_S$, where $d_{E,S}$ are the dimensionalities of the $\mathcal{H}_E$ and $\mathcal{H}_S$ Hilbert spaces. In the theorem, $\rho_* = 1_R/d_R$ is the equiprobable mixed state on the restricted Hilbert space $\mathcal{H}_R$ (1_R is the identity projection on $\mathcal{H}_R$ and d_R the dimensionality of $\mathcal{H}_R$), so $\mathrm{Tr}_E\,(\rho_*)$ is the corresponding canonical state of the subsystem S. In other words, ρ_* describes a perfectly thermalized radiation system with temperature profile equal to that of Hawking radiation from an evaporating black hole.

To state the theorem in [25] more precisely, the (measurement-theoretic) notion of the *trace-norm* is required, which can be used to characterize the distance between two mixed states ρ_S and Ω_S:

$$\|\rho_S - \Omega_S\|_1 \equiv \mathrm{Tr}\sqrt{(\rho_S - \Omega_S)^2} \ . \tag{14}$$

This sensibly quantifies how easily the two states can be distinguished by measurements, according to the identity

$$\|\rho_S - \Omega_S\|_1 = \sup_{\|O\|\leq 1} \mathrm{Tr}\,(\rho_S O - \Omega_S O) \ , \tag{15}$$

where the supremum runs over all observables O with operator norm $\|O\|$ smaller than 1 (projectors $P = O$ are in some sense the best observables, all other observables can be composed out of them, and they have $\|P\| = 1$). Note that the trace on the right-hand side of (15) is the difference of the observable averages $\langle O\rangle$ evaluated on the two states ρ_S and Ω_S, and therefore specifies the experimental accuracy necessary to distinguish these states in measurements of O. A special form of the theorem then states that the probability that

$$\|\rho_S\,(\psi) - \mathrm{Tr}_E\,(\rho_*)\,\|_1 \ \geq \ d_R^{-1/3} + \sqrt{\frac{d_S^2}{d_R}} \tag{16}$$

is less than $2\exp(-d_R^{1/3}/18\pi^3)$. In words: let ψ be chosen randomly (according to the natural Hilbert space measure) out of the space of allowed states $\mathcal{H}_R$; the probability that a measurement on the subsystem S *only*, with measurement accuracy of $d_R^{-1/3} + \sqrt{d_S^2/d_R}$, will be able to tell the pure state ψ (of the entire system) apart from the mixed state ρ_* is exponentially small ($\sim \exp -d_R^{1/3}$) in the dimension of the space $\mathcal{H}_R$ of allowed states. That is, the overwhelming majority of energetically conforming pure Hawking evaporation states $\psi \in \mathcal{H}_R$ cannot be distinguished from Hawking's predicted density matrix ρ_* by measurements on a small subsystem S of the radiation, even if the experimental error in measurements of projectors is only $d_R^{-1/3} + \sqrt{d_S^2/d_R}$. This is an incredible precision, considering the estimates $d_R \sim \exp S_{BH} \sim \exp M^2$ and $d_S \sim \exp S_S$, where the entropy $S_S \sim VT^3 \ll S_{BH}$ of the (energetically conforming) subsystem S can be computed from its volume V and the Hawking temperature T of the particle excitation distribution inside.

Thus, as long as individual measurements are localized in spacetime, so that S is small relative to E, one cannot distinguish a typical state $\psi \in \mathcal{H}_R$ from ρ_* without exponential sensitivity in the measurement on S. This is true even if one performs many measurements on distinct subsystems S_i, in particular even if the union of S_i covers all of the radiation. This is because, even if each subsystem were in a pure state that could in principle be measured exactly (as opposed to merely measuring all the single-particle excitations in it separately), at the very least the phase relations between the states of the different subsystems S_i are lost. Only complete measurements (including phase relations) on very big subsystems S ($\sim$ half of the degrees of freedom of E) have a non-infinitesimal chance of distinguishing ψ from ρ_*.

In analogy to what we learned in the previous cases, measurements that can distinguish generic $\psi \in \mathcal{H}_R$ from ρ_* with reasonable probability, or without exquisite sensitivity, must be highly non-local, covering a spacetime region which includes most of the radiation emitted by the black hole. They must, in a sense, measure it all "at once". But measurements of this sophistication could, again, also differentiate between macroscopic superpositions and mixed states.

4. Conclusions

Black hole information experiments are at least as hard as experiments which test decoherence vs fundamental collapse. One has to create a semiclassical black hole (in fact, many of them in *identical* states) and then make very challenging measurements on the relativistic decay products, which include gravitons. (Note, it appears difficult to determine the quantum state of gravitons with physically realizable detectors that obey the positive energy condition [26].) In particular, the experiment must be sensitive to the phase relations in coherent superpositions of Fock states rather than simply counting occupation numbers as ordinary particle detectors do. By comparison, the most accessible tests of decoherence would be in the context of an artificial toy system like the Coleman-Hepp model, or other controlled quantum computing environment.

Our results can be summarized rather simply. Hawking suggested black holes cause pure states to evolve to mixed states. But, for all practical purposes (FAPP), decoherence does the same thing – *or at least appears to.* In order to test Hawking's proposal one therefore has to go beyond FAPP and beyond decoherence. Such capability allows fundamental tests of quantum measurement.

If fundamental questions about measurement, decoherence and wavefunction collapse are philosophy rather than science – i.e., cannot be tested by experiments – then so is the black hole information puzzle.

Acknowledgments — The author thanks the organizers of the MCCQG for a stimulating and pleasant meeting. This work was supported by the Department of Energy under DE-FG02-96ER40969.

References

[1] S. W. Hawking, *Phys. Rev.* D **14**, 2460 (1976).
[2] A. Strominger, *Preprint* arXiv:hep-th/9501071.

[3] T. Banks, *Nucl. Phys. Proc. Suppl.* **41**, 21 (1995) (*Preprint* arXiv:hep-th/9412131).
[4] S. B. Giddings, *Preprint* arXiv:hep-th/9412138.
[5] T. Jacobson, *Preprint* arXiv:gr-qc/9908031.
[6] L. Susskind and J. Lindesay, An introduction to black holes, information and the string theory revolution: The holographic universe, *World Scientific, Hackensack, USA (2005).*
[7] D. Dieks, *Phys. Lett.* A **92**, 271 (1982); W. K. Wootters, W. H. Zurek, *Nature* **299**, 802 (1982); H. Barnum, C. M. Caves, C. A. Fuchs, R. Jozsa and B. Schumacher, *Phys. Rev. Lett.* **76**, 2818 (1996) (*Preprint* arXiv:quant-ph/9511010).
[8] This contribution is based on a talk presented by the author at the first Mediterranean Conference on Classical and Quantum Gravity, and on earlier work in S. D. H. Hsu and D. Reeb, *Phys. Rev.* D **79** (2009) 124037 (*Preprint* arXiv:0903.2258 [gr-qc]).
[9] H. D. Zeh, *Phys. Lett.* A **347**, 1 (2005) (*Preprint* arXiv:gr-qc/0507051).
[10] C. Kiefer, *Class. Quant. Grav.* **18**, L151 (2001) (*Preprint* arXiv:gr-qc/0110070); C. Kiefer, *Lect. Notes Phys.* **633**, 84 (2003) (*Preprint* arXiv:gr-qc/0304102); H. Nikolic, *Preprint* arXiv:0805.2555 [gr-qc].
[11] D. M. Eardley and S. B. Giddings, *Phys. Rev.* D **66**, 044011 (2002) (*Preprint* arXiv:gr-qc/0201034); S. D. H. Hsu, *Phys. Lett.* B **555**, 92 (2003) (*Preprint* arXiv:hep-ph/0203154).
[12] In analogy with ordinary scattering theory we could write the quantum state at future infinity as $\psi' = \psi_0 + \psi_1$ where ψ_0 is the future-evolved unscattered state (i.e., no black hole formed; initial state particles can propagate to infinity) and ψ_1 the Hawking radiation state (black hole was formed and evaporated). For suitable initial data (e.g., a large ball of dust) one can ensure that the magnitude of ψ_0 is arbitrarily small (a black hole is almost certainly formed) and that ψ_1 is largely independent of the internal state of the hole (the no-cloning argument applies to the semiclassical black hole spacetime). This suggests a unitarity problem: if varying the internal state does not change ψ_1, then many distinct states are mapped to almost the same final state $\psi' = \psi_0 + \psi_1 \approx \psi_1$. This argument does not, however, require the final state to be a mixture.
[13] H. Everett, III, *Rev. Mod. Phys.* **29**, 454 (1957); H. Everett, III, "The Theory of the Universal Wavefunction" (PhD thesis), reprinted in B. S. DeWitt and R. N. Graham (eds.), The Many-Worlds Interpretation of Quantum Mechanics, *Princeton Series in Physics, Princeton University Press, Princeton, USA (1973).*
[14] H. D. Zeh, *Found. Phys.* **1**, 69 (1970); R. Omnès, *Rev. Mod. Phys.* **64**, 339 (1992); W. H. Zurek, *Rev. Mod. Phys.* **75**, 715 (2003) (*Preprint* arXiv:quant-ph/0105127); E. Joos, H. D. Zeh, C. Kiefer, D. Giulini, J. Kupsch, I.-O. Stamatescu, Decoherence and the Appearance of a Classical World in Quantum Theory (2nd edition), *Springer, New York, USA (2003).*
[15] J. S. Bell, Helv. Phys. Acta **48**, 93 (1975); J. S. Bell, Phys. World **3**, 33 (1990).
[16] R. Omnès, The Interpretation of Quantum Mechanics, *Princeton Series in Physics, Princeton University Press, Princeton, USA (1994).*
[17] K. Hornberger, S. Uttenthaler, B. Brezger, L. Hackermüller, M. Arndt and A. Zeilinger, *Phys. Rev. Lett.* **90**, 160401 (2003) (*Preprint* arXiv:quant-ph/0303093).
[18] M. Brune, E. Hagley, J. Dreyer, X. Maitre, A. Maali, C. Wunderlich, J. M. Raimond and S. Haroche, *Phys. Rev. Lett.* **77**, 4887 (1996); P. Bertet, S. Osnaghi, A. Rauschenbeutel, G. Nogues, A. Auffeves, M. Brune, J. M. Raimond and S. Haroche, *Nature* **411**, 166 (2001).
[19] R. P. Feynman, *Int. J. Theor. Phys.* **21**, 467 (1982); D. P. DiVincenzo, *Fortschr. Phys.* **48**, 771 (2000) (*Preprint* arXiv:quant-ph/0002077).
[20] T. Banks, M. E. Peskin and L. Susskind, *Nucl. Phys.* B **244**, 125 (1984).
[21] M. Srednicki, *Nucl. Phys. B* **410**, 143 (1993) (*Preprint* arXiv:hep-th/9206056); S. B. Giddings, *Phys. Rev.* D **49**, 4078 (1994) (*Preprint* arXiv:hep-th/9310101); W. G. Unruh and R. M. Wald, *Phys. Rev.* D **52**, 2176 (1995) (*Preprint* arXiv:hep-th/9503024); J. Oppenheim and B. Reznik, *Preprint* arXiv:0902.2361 [hep-th].
[22] S. D. H. Hsu, "Spacetime topology change and black hole information," *Phys. Lett.* B **644**, 67 (2007) (*Preprint* hep-th/0608175); S. Hossenfelder and L. Smolin, (*Preprint* arXiv:0901.3156 [gr-qc]).
[23] J. M. Maldacena, *JHEP* **0304**, 021 (2003) (*Preprint* arXiv:hep-th/0106112).
[24] K. Hepp, *Helv. Phys. Acta* **45**, 237 (1972).
[25] S. Popescu, A. J. Short, A. Winter, *Nature Physics* **2**, 754 (2006); S. Popescu, A. J. Short, A. Winter, *Preprint* arXiv:quant-ph/0511225.
[26] L. Smolin, *Gen. Rel. Grav.* **17**, 417 (1985); L. Smolin, Gen. Rel. Grav. **16**, 205 (1984).

First Mediterranean Conference on Classical and Quantum Gravity (MCCQG 2009) IOP Publishing
Journal of Physics: Conference Series **222** (2010) 012038 doi:10.1088/1742-6596/222/1/012038

Geon black holes and quantum field theory

Jorma Louko
School of Mathematical Sciences, University of Nottingham, Nottingham NG7 2RD, UK
E-mail: jorma.louko@nottingham.ac.uk

Abstract. Black hole spacetimes that are topological geons in the sense of Sorkin can be constructed by taking a quotient of a stationary black hole that has a bifurcate Killing horizon. We discuss the geometric properties of these geon black holes and the Hawking-Unruh effect on them. We in particular show how correlations in the Hawking-Unruh effect reveal to an exterior observer features of the geometry that are classically confined to the regions behind the horizons.

1. Introduction

A geon, short for "gravitational-electromagnetic entity", was introduced in 1955 by John Archibald Wheeler [1] as a configuration of the gravitational field, possibly coupled to other zero-mass fields such as massless neutrinos [2] or the electromagnetic field [3], such that a distant observer sees the curvature to be concentrated in a central region with persistent large scale features. The configuration was required to be asymptotically flat [4], allowing the mass to be defined by what are now known as Arnowitt-Deser-Misner methods. The examples discussed in [5] indicate that the configuration was also understood to have spatial topology $\mathbb{R}^3$, excluding black holes. These geons are however believed to be unstable, owing to the tendency of massless fields either to disperse to infinity or to collapse into a black hole [6].

In 1985, Sorkin [7] generalised Wheeler's geon into a *topological geon* by allowing the spatial topology to be nontrivial, as well as dropping the a priori requirement of persistent large scale features. The central regions may thus have complicated topology, as illustrated in Figure 1, and the time evolution of the initial data may lead into singularities and black holes [8, 9]. In particular, topological geons include quotients of stationary black hole spacetimes with a bifurcate Killing horizon, such that the two exteriors separated by the Killing horizon become identified [10, 11]. Such geon black holes are an intermediate case between conventional

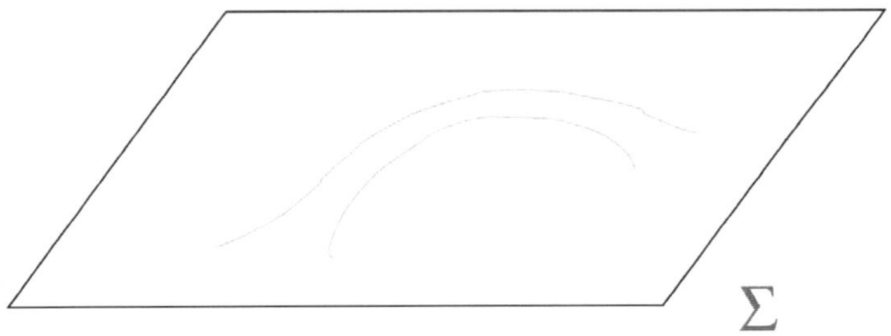

Figure 1. A spatial slice Σ of a topological geon. The geometry of Σ is asymptotically Euclidean, but the topology of Σ does not need to be $\mathbb{R}^3$, as illustrated by the handle drawn.

Journal of Physics: Conference Series **222** (2010) 012038 doi:10.1088/1742-6596/222/1/012038

stationary black holes and dynamical black holes, in the sense that the nonstationary features are confined behind the horizons.

As the nontrivial spatial topology in a geon black hole is present since arbitrarily early times, one does not expect an astrophysical star collapse to result into a geon black hole. Instead, the interest of geon black holes is in their quantum mechanical properties. Because the Killing horizon now does not separate two causally disconnected exteriors, it is not possible to arrive at a thermal density matrix for a quantum field by the usual procedure of tracing over the second exterior. Nevertheless, as we shall discuss, a quantum field on a geon black hole does exhibit thermality in the standard Hawking temperature, although only for a restricted set of observations. We shall in particular see how the quantum correlations in the Hawking-Unruh effect reveal to an exterior observer features of the geometry that are confined to the regions behind the horizons.

We start by introducing in Section 2 the showcase example, the $\mathbb{RP}^3$ geon, formed as a quotient of Kruskal. Section 3 discusses a number of generalisations, including geons with angular momenta and gauge charges. Thermal effects in quantum field theory on geons are addressed in Section 4. Section 5 concludes by discussing the prospects of computing the entropy of a geon from a quantum theory of gravity.

2. Showcase example: $\mathbb{RP}^3$ geon

Before the Kruskal-Szekeres extension of the Schwarzschild solution was known, it had been observed that the time-symmetric initial data for exterior Schwarzschild can be extended into a time-symmetric wormhole initial data, connecting two asymptotically flat infinities, illustrated in Figure 2. Misner and Wheeler [5] noted that this wormhole admits an involutive freely-acting isometry that consists of the radial reflection about the wormhole throat composed with the antipodal map on the S^2 of spherical symmetry. The $\mathbb{Z}_2$ quotient of the wormhole by this isometry, illustrated in Figure 3, is hence a new time-symmetric spherically symmetric initial data for an Einstein spacetime.

The new time-symmetric initial data has topology $\mathbb{RP}^3 \setminus \{\text{point}\}$, the omitted point being at the asymptotically flat infinity [12]. The time evolution is a topological geon spacetime, called the $\mathbb{RP}^3$ geon [10], and it can be obtained as a $\mathbb{Z}_2$ quotient of Kruskal as shown in Figures 4 and 5. A general discussion of quotients of Kruskal can be found in [13].

The $\mathbb{RP}^3$ geon is a black and white hole spacetime, it is spherically symmetric, and it is time and space orientable. It inherits from Kruskal a black hole singularity and a white hole singularity, but the quotient has introduced no new singularities. The distinctive feature is that is has only one exterior region, isometric to exterior Schwarzschild.

As is well known, Kruskal admits a one-parameter group of isometries that coincide with Schwarzschild time translations in the two exteriors. These isometries do however not induce globally-defined isometries on the $\mathbb{RP}^3$ geon, because the quotienting map changes the sign of the corresponding Killing vector, as shown in Figures 6 and 7. As a consequence the geon exterior has a distinguished surface of constant Schwarzschild time, shown in Figure 5, even though one needs to probe the geon geometry beyond the exterior to identify this surface. The existence of this distinguished spacelike surface will turn out significant with the Hawking-Unruh effect in Section 4.

3. Other geons

3.1. Static

The quotienting of Kruskal into the $\mathbb{RP}^3$ geon generalises immediately to the higher-dimensional spherically symmetric generalisation of Schwarzschild [14]. As the antipodal map on a sphere preserves the sphere's orientation in odd dimensions but reverses the sphere's orientation in even dimensions, the geon is space orientable precisely when the spacetime dimension is even.

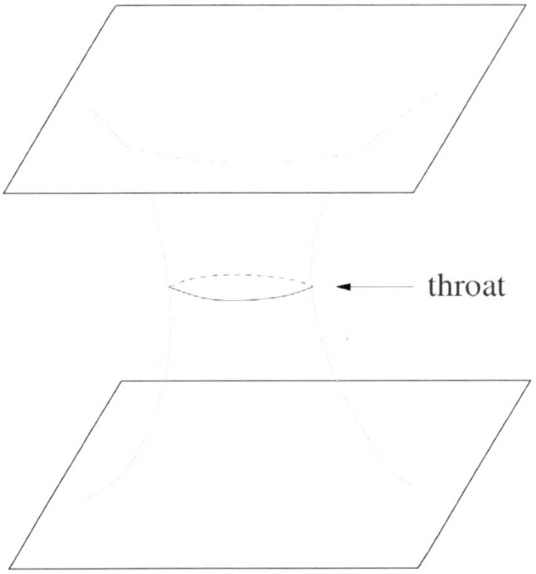

Figure 2. Time-symmetric wormhole initial data for Kruskal manifold, with the radial dimension drawn vertical and the two-spheres of spherical symmetry drawn as horizontal circles. The wormhole connects two asymptotically flat infinities. It admits a freely-acting involutive isometry that consists of the radial reflection about the wormhole throat composed with the antipodal map on the S^2.

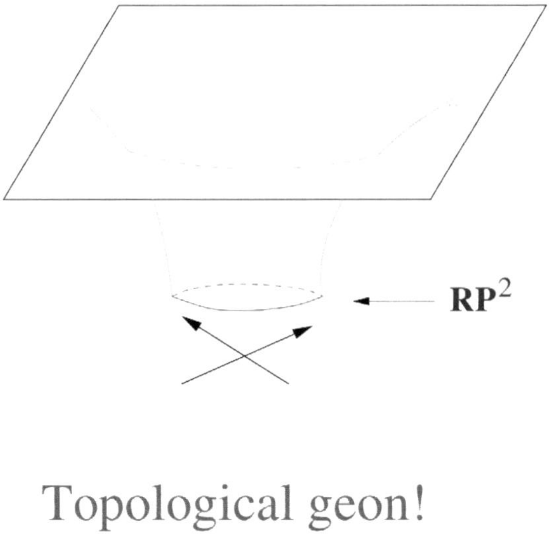

Figure 3. Time-symmetric initial data for the $\mathbb{RP}^3$ geon, obtained as the $\mathbb{Z}_2$ quotient of the wormhole by the freely-acting involutive isometry. The topology of the wormhole is $S^2 \times \mathbb{R} \simeq S^3 \setminus \{2 \text{ points}\}$ and that of the quotient is $\big(S^3 \setminus \{2 \text{ points}\}\big)/\mathbb{Z}_2 \simeq \mathbb{RP}^3 \setminus \{\text{point}\}$, each of the omitted points being at an asymptotically flat infinity.

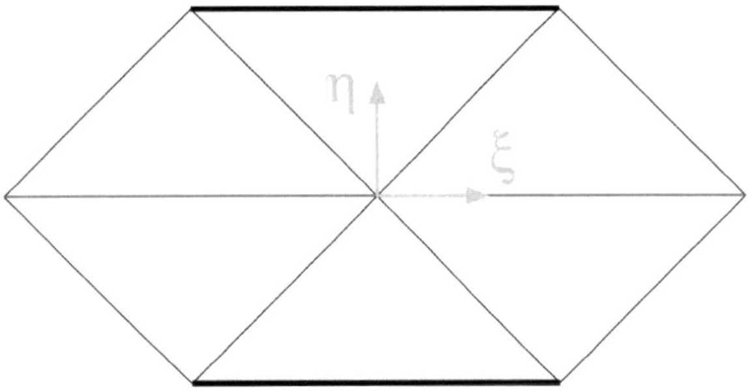

Figure 4. Conformal diagram of Kruskal manifold, with the S^2 orbits of spherical symmetry suppressed and the Kruskal time and space coordinates (η, ξ) shown. The wormhole of Figure 2 is the horizontal line $\eta = 0$. The freely-acting involutive isometry that preserves time orientation and restricts to that of the wormhole is $J_K : (\eta, \xi, \theta, \varphi) \mapsto \big(\eta, -\xi, P(\theta, \varphi)\big)$, where P is the S^2 antipodal map.

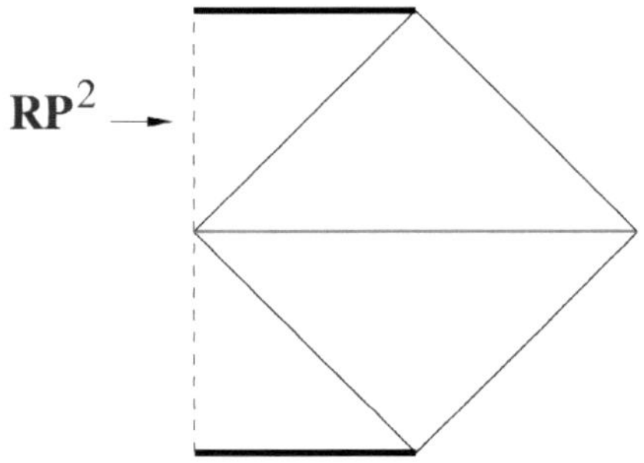

Figure 5. Conformal diagram of the $\mathbb{RP}^3$ geon, obtained as the $\mathbb{Z}_2$ quotient of Kruskal by the freely-acting involutive isometry J_K. The suppressed orbits of spherical symmetry are S^2 except on the dashed line where they are $\mathbb{RP}^2$. The initial data of Figure 3 is shown as the horizontal line.

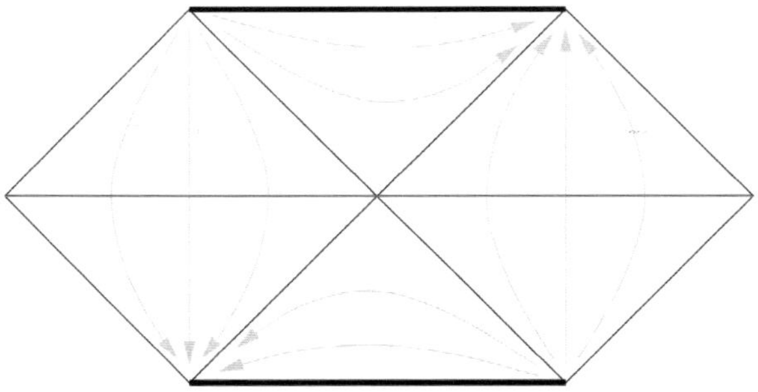

Figure 6. Orbits of Killing time translations on Kruskal. In the exteriors these isometries are translations in Schwarzschild time.

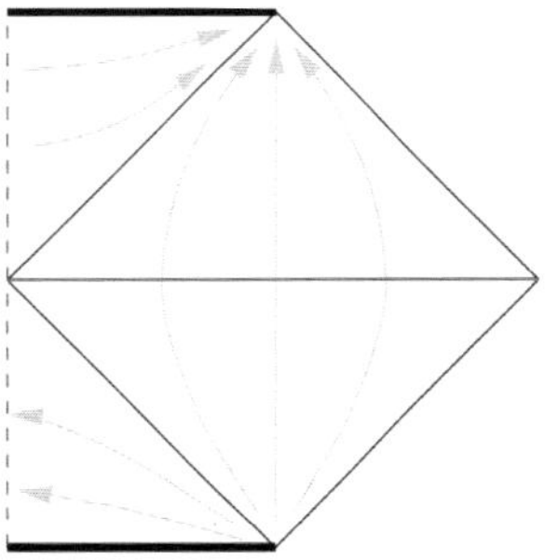

Figure 7. Orbits of the local $\mathbb{RP}^3$ geon isometry inherited from the Killing time translations on Kruskal. The isometry is not globally defined because the corresponding Killing vector has a sign ambiguity in the black and white hole interiors.

A wider class of generalisations occurs when the sphere is replaced by a (not necessarily symmetric) Einstein space. In $D \geq 4$ spacetime dimensions, the vacuum solution can be written in Schwarzschild-type coordinates as [15]

$$ds^2 = -\Delta dt^2 + \frac{dr^2}{\Delta} + r^2 d\bar{\Omega}^2_{D-2} \quad , \quad \Delta = \tilde{\lambda} - \frac{\mu}{r^{D-3}} - \tilde{\Lambda} r^2, \tag{3.1}$$

where $d\bar{\Omega}^2_{D-2}$ is the metric on the $(D-2)$-dimensional Einstein space $\mathcal{M}_{D-2}$, $\tilde{\lambda}$ is proportional to the Ricci scalar of $\mathcal{M}_{D-2}$, $\tilde{\Lambda}$ is proportional to the cosmological constant and μ is the mass parameter. Suppose that Δ has a simple zero at $r = r_+ > 0$ and $\Delta > 0$ for $r > r_+$. The solution has then a Kruskal-type extension across a bifurcate Killing horizon at $r = r_+$, separating two causally disconnected exteriors. If now $\mathcal{M}_{D-2}$ admits a freely-acting involutive isometry, a geon quotient can be formed just as in Kruskal [11]. For a negative cosmological constant these geons are asymptotically locally anti-de Sitter.

For $D = 3$ and a negative cosmological constant, the above $\mathbb{Z}_2$ quotient generalises immediately to the nonrotating Bañados-Teitelboim-Zanelli (BTZ) hole [16]. The BTZ hole admits however also geon versions that arise as other quotients of anti-de Sitter space, even with angular momentum [17, 18, 19].

If a spacelike asymptopia at $r \to \infty$ is not assumed, further quotient possibilities arise, for example by quotienting de Sitter space or Schwarzschild-de Sitter space [20, 21, 22]. We shall not consider these quotients here.

3.2. Spin

Including spin is delicate. To see the problem, consider the Kerr black hole, shown in Figure 8. The two exteriors are isometric by an isometry that preserves the global time orientation, but this isometry inverts the direction of the rotational angle and has hence fixed points in the black hole region and in the white hole region. Kerr does therefore not admit a geon quotient. A way to think about this is that if the angular momentum at one infinity is J, the angular momentum at the opposing infinity is $-J$.

The generalisation of Kerr(-AdS) to $D \geq 5$ dimensions [23, 24, 25, 26] has $[(D-1)/2]$ independent angular momenta. Each of these angular momenta has a different sign at the opposing infinities. This suggests that if the angular momenta are arranged into multiplets, such that within each multiplet the magnitude is constant but the signs alternate, there might

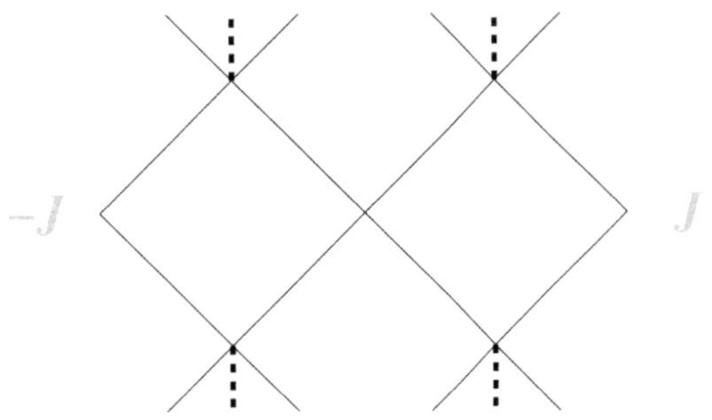

Figure 8. Conformal diagram of Kerr. The angular momenta at the opposing infinities are equal in absolute value but opposite in sign.

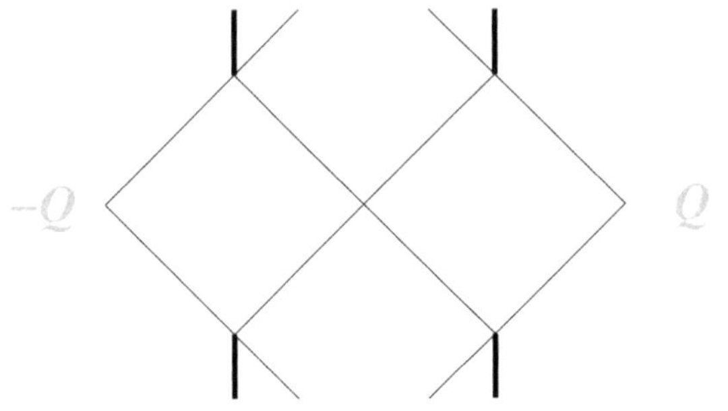

Figure 9. Conformal diagram of Reissner-Nordström. The electric charges at the opposing infinities are equal in absolute value but opposite in sign.

exist a geon quotient by a map that suitably permutes the rotation planes within each multiplet. This turns out to work when D is odd and not equal to 7, but not for other values of D at least in a way that would be continuously deformable to the nonrotating case [11].

For a negative cosmological constant it is possible to define angular momenta also on a torus, rather than on a sphere. The geon quotient must then again involve a permutation of angular momenta with equal magnitude and opposite signs [11].

Finally, the five-dimensional Emparan-Reall black ring [27] has a geon quotient [28], despite having just one nonvanishing angular momentum.

3.3. U(1) *charge*

Including the electromagnetic field is also delicate. To see the problem, consider electrically-charged Reissner-Nordström, shown in Figure 9. Given the global time orientation, Gauss's law says that if the charge at one infinity equals Q, the charge at the opposing infinity equals $-Q$. While there is an involutive isometry by which to take a geon quotient of the spacetime manifold, this isometry reverses the sign the field strength tensor, and a quotient is thus not possible with the usual understanding of the electromagnetic field as a U(1) gauge field [5]. The situation is the same for the electrically-charged generalisations of (3.1) [15].

There is however a way around this minus sign problem. Maxwell's theory has a discrete involutive symmetry that reverses the sign of the field strength tensor as well as the signs of all charges: charge conjugation. The usual formulation of electromagnetism as a U(1) gauge theory treats charge conjugation as a global symmetry rather than as a gauge symmetry. However, it is possible to promote charge conjugation into a gauge symmetry: the gauge group is then no longer $\mathrm{U}(1) \simeq \mathrm{SO}(2)$ but $\mathrm{O}(2) \simeq \mathbb{Z}_2 \ltimes \mathrm{U}(1)$, where in the semidirect product presentation the nontrivial element of $\mathbb{Z}_2$ acts on U(1) by complex conjugation [29]. The disconnected component of the enlarged gauge group then reverses the sign of the field strength tensor. A geon quotient is now possible provided the quotienting map includes a gauge transformation in the disconnected component of the enlarged gauge group [11]. The geon's charge is well-defined only up to the overall sign, but detecting the sign ambiguity would require observations behind the horizons.

Within the magnetically-charged generalisations of (3.1) [15], the options depend on the $(D-2)$-dimensional Einstein space $\mathcal{M}_{D-2}$ [11]. In some cases (including magnetic Reissner-Nordström in even dimensions) the situation is just as with electric charge. In other cases (including magnetic Reissner-Nordström in odd dimensions) the field strength tensor is invariant under the relevant involutive isometry and the usual U(1) formulation of Maxwell's theory suffices. Further options arise if the spacetime has both electric and magnetic charge [11].

3.4. SU(n) *charge*

Black holes with a non-Abelian gauge field also admit a geon quotient. For generic static, spherically symmetric SU(n) black holes with $n > 2$ [30, 31, 32] this requires enlarging the gauge group to $\mathbb{Z}_2 \ltimes \mathrm{SU}(n)$, where the nontrivial element of $\mathbb{Z}_2$ acts on SU(n) by complex conjugation and the quotienting map includes a a gauge transformation in the disconnected component [33, 34]. By contrast, static SU(2) black holes with spherical [35, 36, 37, 38, 39, 40, 41] or axial [42, 43] symmetry admit a geon quotient without enlarging the gauge group [11, 33, 34].

4. Quantum field theory

The geon black holes that we have discussed are genuine black and white holes, and the exterior region is isometric to that in an ordinary stationary black hole with a bifurcate Killing horizon. Does the geon also have a Hawking temperature, and if so, in what sense?

4.1. Real scalar field on the $\mathbb{RP}^3$ *geon*

As the prototype, consider a real scalar field on the $\mathbb{RP}^3$ geon.

Recall first the usual setting for a quantised scalar field on Kruskal. There is a distinguished vacuum state $|0_\mathrm{K}\rangle$, the Hartle-Hawking-Israel (HHI) vacuum [44, 45], which is the unique regular state that is invariant under all the continuous isometries [46]. $|0_\mathrm{K}\rangle$ does *not* coincide with the Boulware vacuum state $|0_{\mathrm{K,B}}\rangle$ [47], which static observers in the exteriors see as a no-particle state. Instead, we have the expansion [44, 45]

$$|0_\mathrm{K}\rangle = \sum_{ij\ldots} f_{ij\ldots}\, \underbrace{a^\dagger_{R,i} a^\dagger_{L,i}}_{\text{corr}}\, \underbrace{a^\dagger_{R,j} a^\dagger_{L,j}}_{\text{corr}} \cdots |0_{\mathrm{K,B}}\rangle\,, \tag{4.1}$$

where $a^\dagger_{R,i}$ and $a^\dagger_{L,i}$ are creation operators of particles seen by the static observers in respectively the right exterior R and the left exterior L, and i stands for the quantum numbers labelling these particles. The creation operators occur in correlated pairs, with one particle in each exterior, as shown in Figure 10. From the properties of the expansion coefficients $f_{ij\ldots}$ it follows that measurements in (say) R see the pure state $|0_\mathrm{K}\rangle$ as a Planckian density matrix, in a temperature that at asymptotically large radii equals the Hawking temperature

$$T_H = \frac{1}{8\pi M}, \tag{4.2}$$

where M is the mass, and at finite radii is corrected by the gravitational redshift factor.

To summarise, observers in an exterior region of Kruskal see the HHI vacuum $|0_\mathrm{K}\rangle$ as a thermal equilibrium state in the Hawking temperature (4.2).

Consider now a scalar field on the $\mathbb{RP}^3$ geon. By the quotient from Kruskal, the HHI vacuum $|0_\mathrm{K}\rangle$ induces on the geon a vacuum state $|0_\mathrm{G}\rangle$, which we call the geon vacuum. Again, $|0_\mathrm{G}\rangle$ does not coincide with the geon's Boulware state $|0_{\mathrm{G,B}}\rangle$, which static observers in the exterior see as a no-particle state. Instead, $|0_\mathrm{G}\rangle$ can be expanded as in (4.1), with creation operators of exterior particles acting on $|0_{\mathrm{G,B}}\rangle$, and these creation operators come again in correlated pairs, as illustrated in Figure 11. For generic observations in the exterior region, the geon vacuum state does hence not appear as a mixed state, let alone a thermal one.

However, the correlated particle pairs in the geon exterior have a special structure: if one particle is localised at asymptotically late times, then the correlated particle is localised at asymptotically early times, and vice versa. Here, "early" and "late" are defined with respect to the distinguished constant Schwarzschild time hypersurface shown in Figure 5. This means that $|0_\mathrm{G}\rangle$ appears as a mixed state when probed by observers at asymptotically late (or early) times, and this mixed state is again Planckian in the Hawking temperature (4.2) [48].

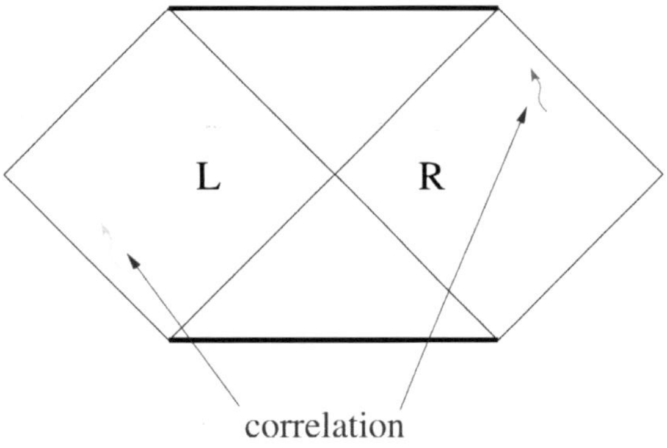

Figure 10. A correlated pair of exterior particles in the expansion (4.1) of the Kruskal HHI vacuum $|0_K\rangle$ as excitations on the Boulware vacuum. The two particles are in the opposite exteriors. From the invariance of $|0_K\rangle$ under Killing time translations it follows that if the particle in R is localised in the far future, the correlated particle in L is localised in the far past, and vice versa, regardless the precise sense of the localisation.

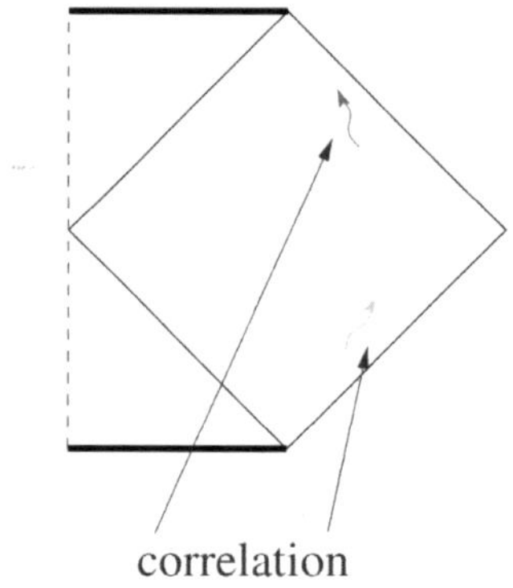

Figure 11. A correlated pair of exterior particles in the expansion of the $\mathbb{RP}^3$ geon HHI vacuum $|0_G\rangle$ as excitations on the Boulware vacuum. Both particles are in the one and only exterior. If one particle is localised in the far future, the correlated particle is localised in the far past, and vice versa.

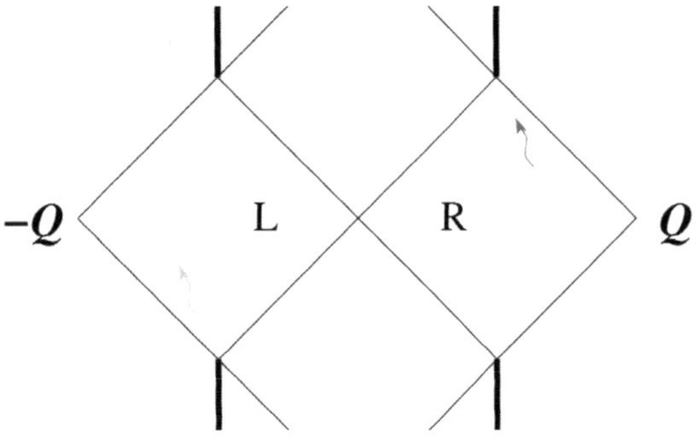

Figure 12. A correlated pair of exterior particles in the expansion of the electrically-charged Reissner-Nordström HHI vacuum as excitations on the Boulware vacuum. The two particles have opposite charge.

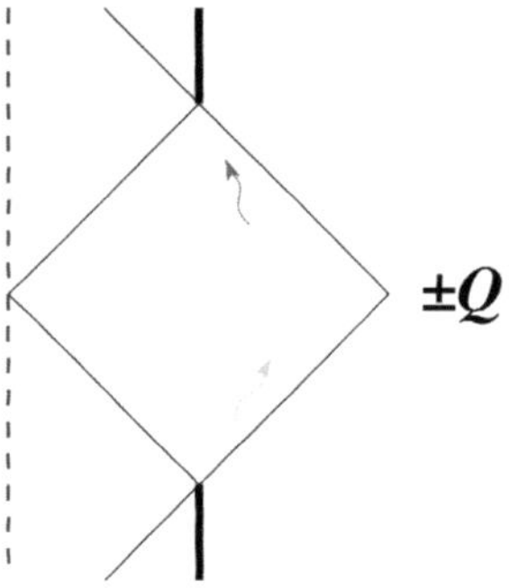

Figure 13. A correlated pair of exterior particles in the expansion of the electrically-charged Reissner-Nordström geon HHI vacuum as excitations on the Boulware vacuum. The two particles have the *same* charge.

Journal of Physics: Conference Series **222** (2010) 012038 doi:10.1088/1742-6596/222/1/012038

To summarise, observers in the geon exterior see the geon vacuum $|0_G\rangle$ as thermal in the Hawking temperature (4.2) at asymptotically late (or early) times. At intermediate times the non-thermal correlations in $|0_G\rangle$ can be observed by local measurements.

4.2. Generalisations

When the scalar field is replaced by a fermion field, a new issue arises in the choice of the spin structure [49]. As Kruskal has only one spin structure, there is only one HHI vacuum on Kruskal. The $\mathbb{RP}^3$ geon has however two spin structures, and each of them inherits a HHI vacuum from that on Kruskal. Observations made at asymptotically late (or early) exterior times see each of these vacua as a thermal state in the usual Hawking temperature, and are in particular unable to distinguish the two vacua. The vacua, and their spin structures, can however be distinguished by observations of the non-thermal correlations at intermediate times.

When the geon has a gauge field, a new issue arises from its coupling to the quantised matter field. For concreteness, consider a complex scalar field on electrically-charged Reissner-Nordström [50, 51]. The HHI vacua on the two-exterior hole and on the geon again contain exterior particles in correlated pairs as shown in Figures 12 and 13, and the sense of thermality is as in the uncoupled case but with an appropriate chemical potential term [52]. However, while on the two-exterior hole the correlated pairs consist of particles with opposite charge, on the geon the pairs consist of particles with the *same* charge. This is a direct consequence of the charge conjugation that appears in the quotienting map. A geon exterior observer can hence infer from the non-thermal correlations at intermediate times that charges on the spacetime are not globally defined, even though this phenomenon only arises from the geometry of the geon's gauge bundle behind the horizons.

When the geon has anti-de Sitter asymptotics, the geometry induces a state for a quantum field that lives on the geon's conformal boundary, via gauge-gravity correspondence [53, 54]. The boundary state exhibits thermality for appropriately restricted observations, and the non-thermal correlations in this state bear an imprint of the geometry behind the geon's horizons [16, 55, 56, 57].

5. Concluding remarks: geon entropy?

We have seen that a geon black hole, formed as a $\mathbb{Z}_2$ quotient of a stationary black hole with a bifurcate Killing horizon, has a Hartle-Hawking-Israel vacuum that appears thermal in the usual Hawking temperature at asymptotically late (and early) times. Does the geon have also an entropy?

For a late time exterior observer, laws of classical black hole mechanics do not distinguish between a geon, a conventional eternal black hole with a bifurcate Killing horizon, or indeed a black hole formed in a star collapse. Combining the laws of classical black hole mechanics with the late time Hawking temperature, the late time observer will hence conclude that the geon has the usual Bekenstein-Hawking entropy.

Could the geon entropy be obtained directly from a quantum theory of gravity? If the theory is formulated in a way that allows focussing on the quantum spacetime in the distant future, state-counting computations for the geon should proceed essentially as for the hole with two exteriors. This is in particular the case in loop quantum gravity, where the counting of black hole microstates is expressly set up on the future horizon [58, 59, 60, 61]. By contrast, it is not clear whether Euclidean path-integral methods allow a localisation at late times. While the geon has a regular Euclidean-signature section, a naïve application of Euclidean path-integral methods to this section yields for the entropy a result that is only half of the late time Bekenstein-Hawking entropy [48, 62].

It would be of interest to develop a geon entropy computation within the gauge-gravity correspondence. Case studies [16, 55, 56, 57] suggest that an appropriate formalism for this could be late time entanglement entropy in the boundary field theory. The challenge would be to give a concrete implementation in which the gauge theory side of the correspondence is well understood.

Acknowledgments

I am indebted to many colleagues, including John Barrett, David Bruschi, John Friedman, Gary Gibbons, Nico Giulini, Bernard Kay, George Kottanattu, Kirill Krasnov, Paul Langlois, Robb Mann, Don Marolf, Simon Ross, Kristin Schleich, Bernard Whiting and Jacek Wiśniewski, for discussions and collaboration on geons. I am grateful to the MCCQG organisers for the invitation to participate and to present this talk. This work was supported in part by STFC (UK) grant PP/D507358/1.

This contribution is dedicated to Rafael Sorkin in celebration of his influence, inspiration and friendship.

References

[1] Wheeler J A 1955 Geons *Phys. Rev.* **97** 511–36
[2] Brill D R and Wheeler J A 1957 Interaction of neutrinos and gravitational fields *Rev. Mod. Phys.* **29** 465–79
[3] Brill D R and Hartle J B 1964 Method of the self-consistent field in general relativity and its application to the gravitational geon *Phys. Rev.* **135** B271–8
[4] Melvin M A 1965 Dynamics of cylindrical electromagnetic universes *Phys. Rev.* **139** B225–43
[5] Misner C W and Wheeler J A 1957 Classical physics as geometry: Gravitation, electromagnetism, unquantized charge, and mass as properties of curved empty space *Ann. Phys.* (N.Y.) **2** 525–603 Reprinted in Wheeler J A 1962 *Geometrodynamics* (Academic: New York)
[6] Gundlach G 2003 Critical phenomena in gravitational collapse *Phys. Rept.* **376** 339–405 (*Preprint* gr-qc/0210101)
[7] Sorkin R D 1986 Introduction to topological geons *Topological properties and global structure of space-time*, Proceedings of the NATO Advanced Study Institute on Topological Properties and Global Structure of Space-Time, Erice, Italy, May 12–22 1985, ed P G Bergmann and V De Sabbata (New York: Plenum) pp 249–70
[8] Gannon D 1975 Singularities in nonsimply connected spacetimes *J. Math. Phys.* **16** 2364–7
[9] Gannon D 1976 On the topology of spacelike hypersurfaces, singularities, and black holes *Gen. Rel. Grav.* **7** 219–32
[10] Friedman J L, Schleich K and Witt D M 1993 Topological censorship *Phys. Rev. Lett.* **71** 1486–9; Erratum 1995 *Phys. Rev. Lett.* **75** 1872 (*Preprint* gr-qc/9305017)
[11] Louko J, Mann R B and Marolf D 2005 Geons with spin and charge *Class. Quant. Grav.* **22** 1451–67 (*Preprint* gr-qc/0412012)
[12] Giulini D 1990 3-manifolds in canonical quantum gravity (PhD Thesis, University of Cambridge)
[13] Chamblin A and Gibbons G W Nucleating black holes via nonorientable instantons *Phys. Rev.* D **55** 2177–85 (*Preprint* gr-qc/9607079)
[14] Tangherlini F R 1963 Schwarzschild field in n dimensions and the dimensionality of space problem *Nuovo Cim.* **27** 636–51
[15] Gibbons G W and Wiltshire D L 1987 Space-time as a membrane in higher dimensions *Nucl. Phys.* **B287** 717–42 (*Preprint* hep-th/0109093)
[16] Louko J and Marolf D 1999 Single-exterior black holes and the AdS-CFT conjecture *Phys. Rev.* D **59** 066002 (*Preprint* hep-th/9808081)
[17] Åminneborg S, Bengtsson I and Holst S 1999 A spinning anti-de Sitter wormhole *Class. Quant. Grav.* **16** 363–82 (*Preprint* gr-qc/9805028)
[18] Brill D 2000 Black holes and wormholes in $2+1$ dimensions *Lect. Notes Phys.* **537** 143–79 (*Preprint* gr-qc/9904083)
[19] Brill D 2000 (2+1)-dimensional black holes with momentum and angular momentum *Annalen Phys.* (Leipzig) **9** 217–26 (*Preprint* gr-qc/9912079)
[20] Louko J and Schleich K 1999 The exponential law: Monopole detectors, Bogolubov transformations, and the thermal nature of the Euclidean vacuum in $\mathbb{RP}^3$ de Sitter spacetime *Class. Quant. Grav.* **16** 2005–21 (*Preprint* gr-qc/9812056)

[21] Schleich K and Witt D M 1999 The generalized Hartle-Hawking initial state: Quantum field theory on Einstein conifolds *Phys. Rev.* D **60** 064013 (*Preprint* gr-qc/9903062)
[22] McInnes B 2003 de Sitter and Schwarzschild-de Sitter according to Schwarzschild and de Sitter *JHEP* **0309** 009 (*Preprint* hep-th/0308022)
[23] Myers R C and Perry M J 1986 Black holes in higher-dimensional spacetimes *Ann. Phys.* (N.Y.) **172** 304–47
[24] Gibbons G W, Lü H, Page D N and Pope C N 2004 Rotating black holes in higher dimensions with a cosmological constant *Phys. Rev. Lett.* **93** 171102 (*Preprint* hep-th/0409155)
[25] Gibbons G W, Lü H, Page D N and Pope C N 2005 The general Kerr-de Sitter metrics in all dimensions *J. Geom. Phys.* **53** 49–73 (*Preprint* hep-th/0404008)
[26] Gibbons G W, Perry M J and Pope C N 2005 The first law of thermodynamics for Kerr-Anti-de Sitter black holes *Class. Quant. Grav.* **22** 1503–26 (*Preprint* hep-th/0408217)
[27] Emparan R and Reall H S 2002 A rotating black ring in five dimensions *Phys. Rev. Lett.* **88** 101101 (*Preprint* hep-th/0110260)
[28] Chruściel P T and Cortier J 2008 On the geometry of Emparan-Reall black rings *Preprint* 0807.2309 [gr-qc]
[29] Kiskis J E 1978 Disconnected gauge groups and the global violation of charge conservation *Phys. Rev.* D **17** 3196–202
[30] Künzle H P 1994 Analysis of the static spherically symmetric SU(n) Einstein Yang-Mills equations *Commun. Math. Phys.* **162** 371–97
[31] Baxter J E, Helbling M and Winstanley E 2007 Soliton and black hole solutions of SU(N) Einstein-Yang-Mills theory in anti-de Sitter space *Phys. Rev.* D **76** 104017 (*Preprint* 0708.2357 [gr-qc])
[32] Baxter J E, Helbling M and Winstanley E 2008 Abundant stable gauge field hair for black holes in anti-de Sitter space *Phys. Rev. Lett.* **100** 011301 (*Preprint* 0708.2356 [gr-qc])
[33] Kottanattu G T 2008 Geon-type black holes in Einstein-Yang-Mills theory (PhD thesis, University of Nottingham)
[34] Kottanattu G T and Louko J (in preparation)
[35] Bizon P 1990 Colored black holes *Phys. Rev. Lett.* **64** 2844–7
[36] Künzle H P and Masood-ul-Alam A K M 1990 Spherically symmetric static SU(2) Einstein-Yang-Mills fields *J. Math. Phys.* **31** 928–35
[37] Winstanley E 1999 Existence of stable hairy black holes in SU(2) Einstein-Yang-Mills theory with a negative cosmological constant *Class. Quant. Grav.* **16** 1963–78 (*Preprint* gr-qc/9812064)
[38] Bjoraker J and Hosotani Y 2000 Stable monopole and dyon solutions in the Einstein-Yang-Mills theory in asymptotically anti-de Sitter space *Phys. Rev. Lett.* **84** 1853–6 (*Preprint* gr-qc/9906091)
[39] Bjoraker J and Hosotani Y 2000 Monopoles, dyons and black holes in the four-dimensional Einstein-Yang-Mills theory *Phys. Rev.* D **62** 043513 (*Preprint* hep-th/0002098)
[40] Winstanley E and Sarbach O 2002 On the linear stability of solitons and hairy black holes with a negative cosmological constant: The even parity sector *Class. Quant. Grav.* **19** 689–724 (*Preprint* gr-qc/0111039)
[41] Winstanley E and Sarbach O 2001 On the linear stability of solitons and hairy black holes with a negative cosmological constant: The odd parity sector *Class. Quant. Grav.* **18** 2125–46 (*Preprint* gr-qc/0102033)
[42] Kleihaus B and Kunz J 1997 Static black hole solutions with axial symmetry *Phys. Rev. Lett.* **79** 1595–8 (*Preprint* gr-qc/9704060)
[43] Kleihaus B and Kunz J 1998 Static axially symmetric Einstein-Yang-Mills-dilaton solutions. II: Black hole solutions *Phys. Rev.* D **57** 6138–57 (*Preprint* gr-qc/9712086)
[44] Hartle J B and Hawking S W 1976 Path integral derivation of black hole radiance *Phys. Rev.* D **13** 2188–203
[45] Israel W 1976 Thermo field dynamics of black holes *Phys. Lett.* A **57** 107–10
[46] Kay B S and Wald R M 1991 Theorems on the uniqueness and thermal properties of stationary, nonsingular, quasifree states on space-times with a bifurcate Killing horizon *Phys. Rept.* **207** 49–136
[47] Boulware D G 1975 Quantum field theory in Schwarzschild and Rindler spaces *Phys. Rev.* D **11** 1404–23
[48] Louko J and Marolf D 1998 Inextendible Schwarzschild black hole with a single exterior: How thermal is the Hawking radiation? *Phys. Rev.* D **58** 024007 (*Preprint* gr-qc/9802068)
[49] Langlois P 2004 Hawking radiation for Dirac spinors on the $\mathbb{RP}^3$ geon *Phys. Rev.* D **70** 104008 (*Preprint* gr-qc/0403011)
[50] Bruschi D E 2009 Talk given at the 12th Marcel Grossmann meeting, Paris, France, 12–18 July 2009.
[51] Bruschi D E and Louko J (in preparation)
[52] Gibbons G W 1975 Vacuum polarization and the spontaneous loss of charge by black holes *Commun. Math. Phys.* **44** 245–64
[53] Maldacena J 1998 The large N limit of superconformal field theories and supergravity *Adv. Theor. Math. Phys.* **2** 231–52 (*Preprint* hep-th/9711200)
[54] Witten E 1998 Anti-de Sitter space and holography *Adv. Theor. Math. Phys.* **2** 253–91 (*Preprint* hep-th/9802150)

[55] Louko J, Marolf D and Ross S F 2000 Geodesic propagators and black hole holography *Phys. Rev.* D **62** 044041 (*Preprint* hep-th/0002111)

[56] Maldacena J M 2003 Eternal black holes in anti-de Sitter *JHEP* **0304** 021 (*Preprint* hep-th/0106112)

[57] Louko J and Wiśniewski J 2004 Einstein black holes, free scalars and AdS/CFT correspondence *Phys. Rev.* D **70** 084024 (*Preprint* hep-th/0406140)

[58] Ashtekar A, Baez J, Corichi A and Krasnov K 1998 Quantum geometry and black hole entropy *Phys. Rev. Lett.* **80** 904–7 (*Preprint* gr-qc/9710007)

[59] Ashtekar A, Baez J and Krasnov K 2000 Quantum geometry of isolated horizons and black hole entropy *Adv. Theor. Math. Phys.* **4** 1–94 (*Preprint* gr-qc/0005126)

[60] Meissner K A 2004 Black hole entropy in loop quantum gravity *Class. Quant. Grav.* **21** 5245–52 (*Preprint* gr-qc/0407052)

[61] Domagala M and Lewandowski J 2004 Black hole entropy from quantum geometry *Class. Quant. Grav.* **21** 5233–44 (*Preprint* gr-qc/0407051)

[62] Louko J 2000 Single-exterior black holes *Lect. Notes Phys.* **541** 188–202 (*Preprint* gr-qc/9906031)

First Mediterranean Conference on Classical and Quantum Gravity (MCCQG 2009) IOP Publishing
Journal of Physics: Conference Series **222** (2010) 012039 doi:10.1088/1742-6596/222/1/012039

Mass bound and thermodynamical behaviour of the charged BTZ Black Hole

M Cadoni, C Monni[1]

E-mail: mariano.cadoni@ca.infn.it

E-mail: cristina.monni@ca.infn.it

Abstract. The charged Bañados-Teitelboim-Zanelli (BTZ) black hole displays divergent boundary terms in the action, which lead to a divergent black hole mass. Once a finite, regularized, mass M is defined black hole states exist for arbitrarily negative values of M, moreover there is no upper bound on the U(1) charge Q. We show that these pathologies can be completely removed by using an alternative renormalization procedure for the black hole mass and defining a new mass M_0, which is the total energy inside the horizon. The new mass satisfies a BPS-like bound $M_0 \geq \frac{\pi}{2}Q^2$ and the heat capacity of the hole is positive. We also discuss the black hole thermodynamics that arises when M_0 is interpreted as the internal energy of the system. We show, using three independent approaches (black hole thermodynamics, Einstein equations, Euclidean action formulation) that M_0 satisfies the first law if a term describing the mechanical work done by the electrostatic pressure is introduced.

1. Introduction

It is well known that three-dimensional (3d) pure gravity has no propagating degrees of freedom. This feature makes it a suitable tool for analyzing the issue of quantum gravity and gauge theory in a rather simplified context. Moreover, 3d gravity with a negative cosmological constant can be expressed classically as a gauge theory, the action can be rewritten as a Chern-Simons action for an $SO(2,2)$ field composed by a spin connection and the vierbein. The discovery in 1992 of the existence of black hole solutions in 3d gravity with negative cosmological constant by Bañados, Teitelboim and Zanelli (BTZ) [1, 2] made the subject of 3D gravity even more interesting. This black hole is asymptotically Anti-de Sitter and unlike higher dimensional counterparts, it has not a curvature singularity: this unusual feature is actually what one would expect, because AdS space has constant negative curvature everywhere. The black hole is obtained by orbifolding the AdS spacetime with a discrete subgroup of its $SO(2,2)$ isometry group. In this way the absence of curvature singularities is manifest and the resulting spacetime can be seen as a black hole because if one extends the metric past $r=0$ to $r<0$ the resulting manifold contains closed timelike lines [2], so that the surface $r=0$ is a singularity in the causal structure.

3d AdS gravity plays also a crucial role in the AdS/CFT correspondence. Much before Maldacena's conjecture, it was pointed out by Brown and Henneaux [4] that the diffeomeorphisms preserving the asymptotic structure of AdS spacetime are generated by vector fields, which span a pair of Virasoro algebras. The conformal symmetry can be realized as

[1] Dipartimento di Fisica, Università di Cagliari and INFN, Sezione di Cagliari
Cittadella Universitaria, 09042 Monserrato, Italy

canonical transformations and for an asymptotically AdS space one is forced to introduce boundary terms to the Hamiltonian and momentum constraints. In this way the Virasoro algebras acquire a central charge.

The BTZ black hole solution allows us to find the simplest realization of a 3d bulk gravity configurations that are holographically dual to thermal CFT states [4, 5]. One of the most striking successes of the AdS_3/CFT_2 correspondence is the exact computation of the Bekenstein-Hawking entropy of the BTZ black hole using the dual two-dimensional CFT [6]. Black holes in the 3d bulk can be viewed as excitations of the AdS background and these excitations are dual to thermal excitations of the boudary conformal field theory.
Another important issue concerns the investigation of supersymmetric solutions of 3d gravity. As shown in ref. [18], extreme dilaton black holes are solutions of $\mathcal{N} = 4$ supergravity, and the cosmic censorship condition on the mass is equivalent to the BPS bound deriving from supersymmetry, requiring that the mass of the asymptotically flat spacetime be larger than or equal to the absolute value of all central charges. Coussaert and Henneaux have shown [19] that this is also true for AdS black holes, in particular the massless solution has 2 supersymmetries, the maximal number, and that it is the ground state of (1,1) AdS supergravity with Ramond boundary conditions on the spinors. The extremal black hole has one supersymmetry, while generically there are no supersymmetries at all.

The BTZ solution has a U(1) charged generalization [1, 7] with a different singularity structure. The 3d charged AdS black hole shows indeed a power-law curvature singularity, sharing this feature with its higher-dimensional counterparts, like the charged Reissner-Nordstrom solution in 4d AdS space. However, several pathologies make this black hole rather different from higher-dimensional ones. These pathologies would spoil any hope of exploiting the analogy between 3d and higher-dimensional charged black holes to investigate in a simplified context peculiar features of higher dimensional charged black holes, such as the existence of extremal black holes states of zero temperature and non-vanishing entropy.

The origin of these pathologies is related to the logarithmic divergence of the electrostatic potential of the hole, which can be traced back to the logarithmic behaviour of harmonic functions in two dimensions. This has some important consequences: a) When we vary the action we get divergent boundary terms, i.e we have a divergent black hole mass; b) Using a suitable renormalization procedure, we can define a mass M of the solution, but we find that black hole state exist for arbitrarily negative values of M; c) At M fixed there is no upper bound on the charge Q. Another, recently discovered, manifestation of this problematic behaviour, is the fact that the entropy function approach do not work for the extremal charged BTZ black hole [14].

These features make the charged BTZ black hole drastically different from the charged solutions in four and higher dimensions. In these latter cases the black hole mass satisfies a BPS bound $M \geq a^2Q^2$, which guarantees that the mass is positive definite and that for a given mass the charge is bounded from above. The existence of these bounds is usually a consequence of the supersymmetry of the extremal black hole.

Recently, an alternative renormalization procedure leading to a finite value M_0 for the mass of the charged BTZ black hole, has been proposed [8, 9]. Physically, M_0 is the total energy (gravitational and electromagnetic) inside the black hole outer horizon. The identification of M_0 with the conserved charge associated with time-translations allows to reproduce microscopically the Bekenstein-Hawking entropy of the hole and to consider the charged BTZ geometry as a bridge between two AdS_2 geometries, a near horizon one ($AdS_2 \times S^1$), and an asymptotic one (linear dilaton AdS_2) [8]. In both cases, these can be realized as the chiral half of a 2D CFT. For the asymptotic CFT, the central charge is generated by the breaking of the $SL(2, \mathbb{R})$ isometry of the AdS_2 background due to the non constant dilaton, while the electromagnetic field only enters in the renormalization of the virasoro operator L_0. For the near horizon CFT, the central charge is generated by the boundary conditions for the electromagnetic vector potential. For the asymptotic CFT one can compute the black hole entropy through Cardy's formula, which matches the Bakenstein Hawking entropy.

In view of these results, one is led to ask if the use of the alternative renormalization procedure for the mass of Ref. [8, 9], allows also to cure the pathologies of the charged BTZ black hole. In this note we address this issue by reviewing the results of Ref. [24]. We will argue that all the problematic features of the charged BTZ black hole can be removed if one uses M_0 as the black hole mass. We will demonstrate that M_0 satisfies a BPS-like bound $M_0 \geq \frac{\pi}{2}Q^2$ and that for a black hole above extremality the heat capacity is positive and becomes zero in the extremal case. We also discuss the formulation of black hole thermodynamics when M_0 is interpreted has the internal energy of the thermodynamical system. We show, using three independent methods (black hole thermodynamics, Einstein equations, Euclidean action formulation) that M_0 satisfies the first law if a term describing the mechanical work done by the electrostatic pressure is introduced.

Recently, some new advances on this topics have been achieved in Ref. [23], through a computation of the microscopic entropy for the charged, rotating BTZ black hole by using the Cardy - Verlinde formula.

The structure of this proceeding is as follows. In Sect. 2 we briefly review the main features and pathologies of the charged BTZ black hole. In Sect. 3 we discuss in detail the two renormalization schemes, which can be used to get a finite black hole mass and show that the mass M_0 satisfies a BPS-like bound. In sect. 4 we discuss the thermodynamics of the charged BTZ black hole in the two cases when the internal energy of the system is identified with M or M_0. In Sect. 5 we present our conclusions.

2. The charged BTZ black hole

The charged BTZ black hole is a U(1) generalization of the uncharged BTZ black hole solution [1]. In this paper we consider the presence of a nonvanishing electric charge Q and zero angular momentum J.

The action is

$$I = \int d^3x\sqrt{-g}\left(\frac{R+2\Lambda}{2\pi} - \frac{1}{4}F^{\mu\nu}F_{\mu\nu}\right), \tag{1}$$

where $F_{\mu\nu}$ is Maxwell tensor, $\Lambda = 1/l^2$ is the cosmological constant related to the AdS length l and we are using units such that three-dimensional (3d) Newton constant G is dimensionless, $G = \frac{1}{8}$.

The solution for the electrically charged, non rotating black hole is given by [7, 9]

$$ds^2 = -f(r)dt^2 + f^{-1}dr^2 + r^2 d\varphi^2, \tag{2}$$
$$0 \leq r < \infty, \quad 0 \leq \varphi < 2\pi, \quad ,$$

with metric function and Maxwell field

$$f = \frac{r^2}{l^2} - M - \pi Q^2 \ln\frac{r}{w}, \quad F_{tr} = \frac{Q}{r}. \tag{3}$$

M, Q, w are integration constants. We choose to keep the explicit dependence of the metric on the arbitrary length scale w. However, since the solution depends only on two integration constants (which we will identify with the mass and the charge), this parameter w could be absorbed in a redefinition of M. The above solution represents a 3d asymptotically AdS black hole, with a power-law singularity at $r = 0$, where $R \sim \pi Q^2/r^2$ and, generically, with an inner (r_-) and outer r_+ horizon. M, Q could be naively seen as the black hole mass and charge, respectively. M can be easily expressed as a function of the charge and of the outer horizon radius,

$$M(r_+, Q, w) = \frac{r_+^2}{l^2} - \pi Q^2 \ln\left(\frac{r_+}{w}\right). \tag{4}$$

Whereas the interpretation of Q as the black hole electric charge is straightforward, the same is not true for M.

In fact, as suggested by the same authors in Ref. [1], by varying the action (1), one finds a surface term which diverges in the limit $r \to \infty$:

$$\left(-\delta M - \pi \delta Q^2 \ln r\right) N(r), \tag{5}$$

where N is the lapse function. The presence of the logarithmic divergent boundary term makes the black hole mass a poorly defined concept. This is the first pathology of the charged solution.

A second unpleasant feature emerges when one imposes a *cosmic censorship* condition, i.e the absence of naked singularities. The requirement that the singularity at $r = 0$ is shielded by an event horizon is equivalent to requiring that the metric function $f(r)$ evaluated on its minimum value, is equal or less than zero (corresponding respectively to one or two horizons). Introducing a function $\Delta(M, Q)$, the condition for the existence of the horizon(s) can be cast in the form

$$\Delta(M, Q) = f\left(r = r_{min} = \sqrt{\frac{\pi}{2}} Ql\right) = -M + \frac{\pi Q^2}{2}\left(1 - \ln\frac{\pi Q^2}{2}\right) \leq 0, \tag{6}$$

Eq. (6) can be satisfied by arbitrarily negative values of the mass M, for Q above a critical value Q_0. This can be immediately seen considering the $M - Q$ phase diagram shown in Fig (1).

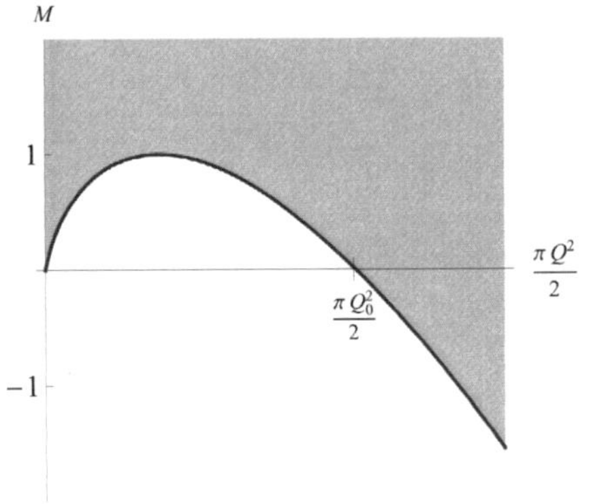

Figure 1. Region in the phase space $M - Q$ where the black hole exists. The region of existence is is the shaded region. Extremal black holes are in the boundary line between the shaded and the unshaded regions.

The presence of black hole states with mass values unbounded from below makes the system intrinsically unstable and the definition of thermodynamical ensembles problematic. In next section we show that the presence of black hole states with arbitrarily negative mass is not an intrinsic physical feature of the system, but instead it is an artifact due to the divergence of the boundary term.

3. Mass renormalization

We address the pathology related to the presence of a divergent boundary term through a renormalization procedure. First, we enclose the system in a circle of radius r_0. This allows to

Journal of Physics: Conference Series **222** (2010) 012039 doi:10.1088/1742-6596/222/1/012039

rewrite the metric function in the form

$$f(r) = -M_0(r_0, w) + \frac{r^2}{l^2} - \pi Q^2 \ln\left(\frac{r}{r_0}\right), \tag{7}$$

and to define a regularized mass as a function of r_0. This was also suggested by the authors of Ref. [7], but now we also have a dependence on the scale w, which we will interpret as a running scale:

$$M_0(r_0, w) = M + \pi Q^2 \ln\left(\frac{r_0}{w}\right). \tag{8}$$

$M_0(r_0)$ is the total energy (gravitational and electromagnetic) inside a circle of radius r_0. In the limit $r \to \infty$ one takes also $r_0 \to \infty$, keeping the ratio $r/r_0 = 1$ [8]. One is left with two possible options: a) M is held fixed and the space-time metric is scale-dependent; b) The metric is w-invariant and M runs with w.

Option a) is the renormalization procedure proposed in [7]. Eq. (8) is used to identify M as the total mass of the solution, an infinite constant is absorbed in M_0 and the logarithmic divergent term in Eq. (5) is removed. Apart from the drawback of having the metric (hence the positions of the horizons and the black hole entropy) running with w, this procedure does not solve the stability problem, hence does not allow for a consistent interpretation of the charged BTZ black hole as a thermodynamical system.

In this paper we use the w-invariant renormalization prescription b) first proposed in Ref. [8]. To keep the metric function (7) unchanged as w runs, we hold $M_0(r_0, w)$ fixed, whereas M changes with w: $w \to \lambda w, \quad M \to M + \pi Q^2 \ln \lambda$. The boundary term (5) becomes now

$$\left(-\delta M_0 - \pi \delta Q^2 \ln \frac{r}{r_0}\right) N(r), \tag{9}$$

and in the limit $r, r_0 \to \infty$ the divergent part is removed.

As a consequence of the w-invariance of f, M_0, we can arbitrarily choose w and M_0. Following [9] we choose to fix in terms of the AdS length $w = l$, and r_0 to the horizon position $r_0 = r_+$. Thus we associate to every charged BTZ black hole solution (3) a finite mass given by

$$M_0(r_+) = M + \pi Q^2 \ln\left(\frac{r_+}{l}\right). \tag{10}$$

The metric function becomes

$$f(r, M_0) = -M_0 + \frac{r^2}{l^2} - \pi Q^2 \ln \frac{r}{r_+}. \tag{11}$$

Thus we can consider the w-invariant mass $M_0(r_+)$ as the conserved charge associated with time translation invariance, instead of the mass M. The renormalization prescription b) has further nice features. The renormalized mass M_0 depends only on the horizon position, is always positive-definite and shares with the uncharged BTZ black hole the mass/horizon-position dependence:

$$M_0(r_+) = \frac{r_+^2}{l^2}. \tag{12}$$

Moreover, the identification of M_0 with the conserved charge associated with time-translation allows to reproduce exactly the Bekenstein-Hawking entropy of the charged BTZ black hole using a Cardy formula for the 2D dual CFT [9].

Journal of Physics: Conference Series **222** (2010) 012039 doi:10.1088/1742-6596/222/1/012039

Let us now show that the use of M_0 as the physical mass of the system allows, at least in principle, to solve the instability problem. The new mass spectrum can be found by using Eq. (12) and expressing Δ of Eq. (6) as a function of M_0,

$$\Delta(M_0, Q) = -M_0 + \frac{\pi Q^2}{2}\left(1 - \ln\frac{\pi Q^2}{2M_0}\right), \tag{13}$$

Setting $\alpha = 2M_0/\pi Q^2$ the condition for the existence of the horizons, $\Delta \leq 0$ together with the relation $r_+ \geq \sqrt{\pi/2}Ql$, implies

$$M_0 \geq \frac{\pi}{2}Q^2, \tag{14}$$

and takes $\alpha - 1 \geq \ln\alpha$, which is always true.

Eq. (14) represents a BPS-like bound for the black hole mass. When the black hole is extremal, the bound is saturated and in Eq. (14) the equality holds. The extremal solution has mass $M_0 = (\pi/2)Q^2$, zero temperature and nonvanishing entropy $S = 2\sqrt{2\pi}\,lQ$ (see Eq. (15) below). All these features are shared by higher dimensional charged black holes. The quadratic form of the $M_0 - Q$ phase diagram, which results from Eq. (14), eliminates the negative, unbounded from below, tail present in Fig. 1. This result implies that the presence of black hole states with arbitrary negative mass is a consequence of identifying the energy of the system with the mass M. Moreover, they give a strong hint that the $M_0 = \pi Q^2/2$ extremal black hole could be a stable configuration. Obviously, in the context of our discussion stability is just a consequence of the validity of the cosmic censorship conjecture. A formal proof of the stability of this configuration would require a detailed analysis of the perturbation spectrum around the extremal black hole solution. Alternatively, stability can be proved by showing that the extremal background is supersymmetric, i.e. it allows for the existence of Killing spinors [18]. A detailed analysis of the stability of the extremal black hole is outside the aim of this paper. In the next sections we will show that using M_0 as black hole mass allows for a consistent formulation of the thermodynamics of the charged BTZ black hole.

4. The first law of thermodynamics for the charged BTZ black hole

The thermodynamical behavior of our charged BTZ black hole will depend on the identification of the black hole parameters in terms of thermodynamical variables. For the temperature T, the entropy S, the electric potential Φ (thought of as chemical potential) there is no ambiguity. T, S, Φ are given as usual in terms of respectively surface gravity, horizon area and time component of the vector potential,

$$\begin{aligned} T &= \frac{1}{4\pi}\left(\frac{2r_+}{l^2} - \frac{\pi Q^2}{r_+}\right), \\ S &= 4\pi r_+ = 4\pi l\sqrt{\pi Q^2 \ln\frac{r_+}{l} + M}, \\ \Phi &= A_0(r_+) = -2\pi Q \ln\frac{r_+}{l}. \end{aligned} \tag{15}$$

On the other hand, for the internal energy E of the thermodynamical system we have two possible choices: we can identify E either with M or with M_0. For $E = M$ the internal energy of the system is the total energy (gravitational and electrostatic) of the black hole and we expect the first principle to take the usual form, as was previously suggested in ref. [1]. In fact differentiating $M(r_+, Q)$ in Eq. (4) and making use of Eqs. (15) one easily obtains the first principle in the form

$$dM = TdS + \Phi dQ. \tag{16}$$

The exact form $M(S,Q)$ can be easily determined. We have

$$M(S,Q) = \frac{S^2}{16\pi^2 l^2} - \pi Q^2 \ln\left(\frac{S}{4\pi l}\right) \tag{17}$$

Conversely, when $E = M_0$ the internal energy is identified with the energy of the black hole inside the radius r_+. In this case we expect that the presence of radial pressure gradients will give rise to additional terms in Eq. (16). The new form of the first principle can be obtained differentiating M_0 given in equation (10) and using Eq. (16). One has

$$dM_0 = TdS + \Phi dQ + dK, \tag{18}$$

where $K = \pi Q^2 \ln\left(\frac{r_+}{l}\right)$ is minus the electrostatic energy outside the horizon. The variation of K cannot change the black hole entropy but represents mechanical work done by electrostatic pressure. We can compute dK keeping the electrostatic potential Φ constant: this allows to express charge variation in terms of displacement of the horizon,

$$2\pi \ln \frac{r_+}{l} dQ = -\frac{2\pi Q}{r_+} dr_+. \tag{19}$$

Using Eq. (18) one finds

$$dM_0 = TdS + \Phi dQ - \frac{\pi Q^2}{r_+} dr_+. \tag{20}$$

The last term in the previous equation is the work done by the radial pressure

$$P_r = T_r^r \tag{21}$$

generated by the electrostatic field ($T_{\mu\nu}$ is the stress-energy tensor for the Maxwell field). Explicit computation of the T_r^r gives

$$P_r(r_+) = -\frac{Q^2}{2r_+^2}. \tag{22}$$

Using this equation in (18) one obtains the first principle in the final form

$$dM_0 = TdS + \Phi dQ + P_r d\mathcal{A}, \tag{23}$$

where $\mathcal{A} = \pi r_+^2$ is the area inside the radius r_+. Notice that when $d\mathcal{A} > 0$, the mechanical work $P_r d\mathcal{A}$ in Eq. (23) is negative, i.e it is done by the thermodynamical system. The pressure $P_r(r_+)$ goes to zero when $r_+ \to \infty$, in this situation $M = M_0$ and the two thermodynamical descriptions are equivalent.

The internal energy M_0 appearing in the first principle (23) at first sight seems to be a function of three independent extensive thermodynamical parameters $S, Q, \mathcal{A}$; a simple calculation gives

$$M_0(S,Q,\mathcal{A}) = \frac{S^2}{16\pi^2 l^2} - \pi Q^2 \ln\left(\frac{S}{4\pi l}\right) + \frac{\pi Q^2}{2} \ln\left(\frac{\mathcal{A}}{\pi l^2}\right). \tag{24}$$

However, there are only two independent parameters because of the presence of a constraint. This constraint takes a different form for thermodynamical transformations at constant Φ or constant Q. In the first case the constraint takes the form (19), which can be also written in the form

$$\Phi dQ = -2P_r d\mathcal{A}. \tag{25}$$

Conversely, keeping the charge Q constant the constraint takes the form

$$Qd\Phi = -2\mathcal{A}dP_r. \tag{26}$$

It is also interesting to compute the thermal capacity of the black hole at constant charge as a function of M_0. We have

$$C = T\frac{\partial S}{\partial T}|_Q = 4\pi l\sqrt{M_0}\frac{2M_0 - \pi Q^2}{2M_0 + \pi Q^2}. \tag{27}$$

The heat capacity is always positive when the black hole is above extremality, $M_0 \geq \pi Q^2/2$, and becomes zero in the extremal case.

4.1. Derivation of the first law from Einstein's equations

It is well known that black hole thermodynamics can be derived from the laws of black hole mechanics, i.e it is codified in Einstein Equations. The first principle of thermodynamics for the charged BTZ black hole (23) can be derived from the rr component of Einstein's equation,

$$G_r^r - \Lambda g_r^r = \pi T_r^r, \tag{28}$$

where $G_{\mu\nu} = R_{\mu\nu} - \frac{1}{2}g_{\mu\nu}R$ is the Einstein tensor. Evaluating the equations (28) on the horizon and using Eq. (21) one has

$$\frac{f'(r_+)}{2r_+} - \frac{1}{l^2} = \pi P_r \tag{29}$$

Multiplying both terms of this equation by $d(r_+^2)$ and using Eqs. (15) and (12) we obtain

$$dM_0 = TdS - P_r d\mathcal{A}. \tag{30}$$

Using Eq. (25) we easily find that equation (30) is equivalent to the first law (23).

4.2. Euclidean action formulation

In this section we will show that the thermodynamics of the charged BTZ black hole described in the previous sections can be also derived using the Euclidean action formalism. In the Euclidean action approach to black hole thermodynamics, one can get Gibbs and Helmoltz free energies through analytic continuation of the action, with suitable boundary terms, in Euclidean space.

To compute the Euclidean action I_E we will follow the method of Bañados, Teitelboim e Zanelli [2], which uses the Hamiltonian version of the action (1). The bulk contribution is equal to zero and the Euclidean action is completely given by three surface terms. The first surface term must be added at infinity and is given by the mass of the solution times the periodicity of Killing time $\beta = 1/T$. The other two surface terms make sure that the variational derivative of the action vanishes on the horizon.

If we use the regularization scheme a) of section 3 the boundary term at infinity is given by M and all together one has, for the Euclidean action:

$$I_E = \beta M - 4\pi r_+ - \beta A_0(r_+)Q. \tag{31}$$

The Gibbs free energy, $G(T, \Phi)$ describing the system in the grand canonical ensemble, is given by $G = TI_E$ and using Eq. (31) turns out to be, as expected, the Legendre transform of M with respect to S and Q:

$$G(T, \Phi) = M - TS - \Phi Q. \tag{32}$$

The description of the thermodynamical system through $G(T,\Phi)$ corresponds to the choice of M as the internal energy of the system. One can easily reproduce the entropy S and the charge Q as $S = -(\partial G)/(\partial T)$ and $Q = -(\partial G)/(\partial \Phi)$.

If we use instead the renormalization scheme b) of section 3, the Euclidean action (31), hence Gibbs free energy do not change. The mass of the solution is now M_0 but the boundary term at infinity is still given by M, the logarithmic term in Eq. (10) being an horizon contribution. From Eq. (10) it follows that M_0 can be written as

$$M_0 = M - \frac{1}{2}Q\Phi(r_+). \tag{33}$$

Thus, the term needed to cancel the variational derivatives of the action on the horizon is now given by $-4\pi r_+ - (1/2)\beta Q\Phi$. All these contributions sum up to the same result given in Eq. (31).

Corresponding to the choice of M_0 as the internal energy of the system, Gibbs free energy can be now expressed as a function of T, Φ, P_r. This can be done by first making use of Eq. (33) to write G of equation (32) in terms of M_0: $G = M_0 - TS - \Phi Q + \frac{1}{2}\Phi dQ$, then differentiating, using the first law (23) and the constraints (25), (26). One obtains

$$dG(T,\Phi,P) = -SdT - Qd\Phi - \mathcal{A}dP_r. \tag{34}$$

5. Conclusions

We have shown that the renormalization of the mass of the charged BTZ black hole, in the prescription introduced by Cadoni, Melis, Setare [8, 9, 10], leads to a consistent thermodynamical description, once we introduce a pressure term in the first law generated by the electric field. Physically, the new mass is the total energy inside the horizon. Moreover, the new renormalized mass automatically cures the pathologies of the black hole, since it is positive definite for every value of the U(1) charge Q. This suggests the presence of a bound for the new mass which resembles the BPS bound for a supersymmetric system; in particular this bound is saturated in the extremal case, a state with zero temperature and nonvanishing entropy, as one would expect. This bound also provides an upper bound for the charge. The thermal capacity of the hole is always positive and becomes zero at extremality. All these features make this black hole very similar to higher dimensional solutions such as Reissner Nordstrom in AdS_4.

Our results improve our understanding of charged black hole solution and could be also very useful in the AdS/CFT context. Similarly to the higher dimensional cases the BTZ black hole is the bridge between AdS_2 near-extremal, near-horizon and asymptotic AdS_3 geometries [9]. This feature could be very useful for understanding the nature of AdS_2 quantum gravity and in particular the microscopic entropy of extremal charged black holes [13, 12, 11, 10, 15, 16, 17].

The presence of the BPS-like bound (14), the vanishing of the temperature and of the thermal capacity in the extremal configuration strongly indicate that the extremal charged BTZ black hole is stable. An interesting application could be the investigation of the existence of a supersymmetric Killing spinor, which would automatically ensure that the extremal state is stable. Alternatively, a deeper understanding could be reached through a detailed analysis of the perturbation spectrum.

References

[1] M. Banados, C. Teitelboim and J. Zanelli, Phys. Rev. Lett. **69** (1992) 1849 [arXiv:hep-th/9204099].
[2] M. Banados, M. Henneaux, C. Teitelboim and J. Zanelli, Phys. Rev. D **48** (1993) 1506 [arXiv:gr-qc/9302012].
[3] S. Carlip, Class. Quant. Grav. **12** (1995) 2853 [arXiv:gr-qc/9506079].
[4] J. D. Brown and M. Henneaux, Symmetries: An Commun. Math. Phys. **104**, 207 (1986).
[5] E. Witten, arXiv:0706.3359 [hep-th].
[6] A. Strominger, JHEP **9802** (1998) 009 [arXiv:hep-th/9712251].

[7] C. Martinez, C. Teitelboim and J. Zanelli, Phys. Rev. D **61** (2000) 104013 [arXiv:hep-th/9912259].
[8] M. Cadoni, M. Melis and M. R. Setare, Class. Quant. Grav. **25** (2008) 195022 [arXiv:0710.3009 [hep-th]].
[9] M. Cadoni and M. R. Setare, quantum arXiv:0806.2754 [hep-th].
[10] M. Cadoni, M. Melis and P. Pani, arXiv:0812.3362 [hep-th].
[11] T. Hartman and A. Strominger, arXiv:0803.3621 [hep-th].
[12] A. Sen, arXiv:0805.0095 [hep-th].
[13] R. K. Gupta and A. Sen, arXiv:0806.0053 [hep-th].
[14] Y. S. Myung, Y. W. Kim and Y. J. Park, arXiv:0903.2109 [hep-th].
[15] A. Castro, D. Grumiller, F. Larsen and R. McNees, JHEP **0811** (2008) 052 [arXiv:0809.4264 [hep-th]].
[16] K. Hotta, Y. Hyakutake, T. Kubota, T. Nishinaka and H. Tanida, JHEP **0901** (2009) 010 [arXiv:0811.0910 [hep-th]].
[17] Y. S. Myung, Y. W. Kim and Y. J. Park, arXiv:0901.2141 [hep-th].
[18] R. Kallosh, A. D. Linde, T. Ortin, A. W. Peet and A. Van Proeyen, Phys. Rev. D **46** (1992) 5278 [arXiv:hep-th/9205027].
[19] O. Coussaert and M. Henneaux, Phys. Rev. Lett. **72** (1994) 183 [arXiv:hep-th/9310194].
[20] M. M. Caldarelli, G. Cognola and D. Klemm, Class. Quant. Grav. **17** (2000) 399 [arXiv:hep-th/9908022].
[21] M. Akbar, BTZ Black Chin. Phys. Lett. **24** (2007) 1158 [arXiv:hep-th/0702029].
[22] A. Paranjape, S. Sarkar and T. Padmanabhan, gravity," Phys. Rev. D **74** (2006) 104015 [arXiv:hep-th/0607240].
[23] M. R. Setare and M. Jamil, BTZ black Phys. Lett. B **681** (2009) 469 [arXiv:0912.0861 [hep-th]].
[24] M. Cadoni and C. Monni, Phys. Rev. D **80** (2009) 024034 [arXiv:0905.3517 [gr-qc]].

Journal of Physics: Conference Series **222** (2010) 012040 doi:10.1088/1742-6596/222/1/012040

On the thermodynamics of moving bodies

Jorge G. Russo[a] **and Paul K. Townsend**[b]

[a] Institució Catalana de Recerca i Estudis Avançats (ICREA), Departament ECM, Facultat de Física, Universitat de Barcelona, Diagonal 647, E-08028 Barcelona, Spain.

[b] Department of Applied Mathematics and Theoretical Physics Centre for Mathematical Sciences, University of Cambridge Wilberforce Road, Cambridge, CB3 0WA, UK.

Abstract. We model the situation in which a photon detector moves with constant velocity in a Minkowski heat-bath by considering an Unruh-DeWitt detector, with variable energy gap $\hbar\omega$, moving on a brane, at a non-zero (Unruh) temperature T due to acceleration in an orthogonal direction. We compute the angular response for a 2-brane. At low velocity we find agreement with the standard Doppler shift formula in the limit $\hbar\omega \gg T$ in which the photon gas is classical; otherwise there is a discrepancy. At relativistic velocities our result disagrees with the standard formula even when $\hbar\omega \gg T$, and above a critical velocity there is no response from the detector within a 'backward' cone. We discuss potential implications for observations of the cosmic microwave background radiation. [Talk given by JGR at this 1st Mediterranean Conference in Classical and Quantum Gravity].

1. Introduction

One might suppose that the formulation of thermodynamics for systems in uniform relative motion should follow in a straightforward way from its formulation in the rest-frame of a heat-bath at temperature T and the kinematics of special relativity. Planck [1] and Einstein [2] evidently thought so in 1907 when they proposed that a body in uniform motion with velocity v should have temperature $T\sqrt{1-v^2}$. The history up to 1921 is summarized by Pauli in his book of that year on relativity [3], and the logic is presented in detail in Tolman's 1934 book [4]. Nevertheless, the conclusion of Planck and Einstein was disputed in the 1960s and the current consensus, which is sometimes traced back to a 1976 paper of Israel [5], appears to be that the term "temperature" should be reserved to mean the temperature in the rest frame because the thermal properties of a body in motion cannot be characterized by a temperature understood in any other way (see e.g. [6], but also e.g. [7] for a different view).

The frame-dependence of thermodynamic quantities acquired importance following the 1965 discovery of the cosmic microwave background radiation (CMBR) because one needs to take into account the (constant) velocity $\mathbf{v}$ of a photon detector relative to the rest-frame of this radiation. Fortunately, there is no need to use any dubious temperature Lorentz transformation: given a gas of particles at temperature T in its rest frame, the distribution of these particles in any other inertial frame may be unambiguously determined. The result for massless particles, which is especially simple, was derived in the context of the CMBR by Peebles and Wilkinson (PW) [8] (see also [9]): the distribution is anisotropic in a moving frame but is still thermal for fixed direction, specified by unit vector $\mathbf{n}$, with an effective temperature

$$T_{\text{eff}} = \frac{T/\gamma}{[1 - \mathbf{v} \cdot \mathbf{n}]}, \qquad \gamma = \frac{1}{\sqrt{1 - v^2}}. \tag{1}$$

Applied to photons of the CMBR, this implies a direction-dependent photon density

$$n_{\text{PW}}(\omega, \mathbf{n})\, d\omega\, d\Omega = \frac{\omega^2}{8\pi^3 \left(e^{\hbar\omega/T_{\text{eff}}} - 1\right)} d\omega\, d\Omega\,, \tag{2}$$

where we have set Boltzmann's constant and the speed of light to unity. This result, which may also be deduced by an application of the relativistic Doppler effect, is implicit in Pauli's book although he did not express it in terms of an effective temperature. An integration over solid angle yields the 'integrated' number density

$$n_{\text{PW}}(\omega)\, d\omega = \frac{\omega T}{4\pi^2 \hbar \gamma v} \log \left[\frac{1 - e^{-\hbar\omega\gamma(1-v)/T}}{1 - e^{-\hbar\omega\gamma(1+v)/T}} \right] d\omega\,. \tag{3}$$

Note the factor of v in the denominator, which ensures that the $v \to 0$ limit is non-zero; naturally, this limit yields the usual Planckian distribution.

Any practical application of these results also depends on an understanding of the behaviour of photon detectors. The purpose of this paper is to point out a feature of this behaviour that leads to an apparent correction to the formula (2). We say "apparent" because the effect may be attributable to the detector, but it is no less real for that and must be taken into account when interpreting observational results. As photon polarization will play no part in our analysis, apart from doubling the number of states, we shall replace photons by 'scalar photons', i.e. particle excitations of a massless scalar field; these have only one polarization state so there will now be an extra factor of $1/2$ on the right-hand side of (2). Also, the term "photon gas" can be understood to refer to any gas of neutral massless particles at zero chemical potential.

The behaviour of a photon detector moving inertially in a photon heat bath is relevant in the context of motion near the horizon of a black brane. Consider first a Schwarzschild black hole in thermal equilibrium with radiation at the Hawking temperature. A static photon detector will register a thermal spectrum at the local temperature determined by the Tolman law for thermal equilibrium. As shown by Unruh [10], the local temperature near the horizon has a purely kinematic origin that is due to the constant proper acceleration needed to keep the detector at a fixed distance r from the horizon, which is locally indistinguishable from the acceleration horizon of the 2-dimensional Rindler spacetime. By 'promoting' the initial black hole spacetime to a $(4 + p)$-dimensional black p-brane spacetime, we arrive instead at a detector in a flat $(2 + p)$-dimensional spacetime with metric

$$ds^2 = -r^2 d\eta^2 + dr^2 + \sum_{i=1}^{p} dy_i^2\,. \tag{4}$$

The temperature still varies with r in the way required by the Tolman law: $T = \hbar/(2\pi r)$, which equals the Unruh temperature $T_U = \hbar a/(2\pi)$ because the proper acceleration of a static observer in this spacetime is $a = 1/r$. This is true of all static observers, so each flat p-surface at fixed r, which we may view for kinematic purposes as a p-brane, is accelerating uniformly with proper acceleration $a = 1/r$ [11]. Although the r-dependence of the temperature implies an anisotropy in the thermal distribution of photons arriving at a static detector, the distribution of photons arriving from within the subspace at fixed r, which is a $(p+1)$-dimensional Minkowski spacetime, will be isotropic.

Journal of Physics: Conference Series **222** (2010) 012040 doi:10.1088/1742-6596/222/1/012040

What this shows is that a flat p-brane undergoing constant uniform proper acceleration in a direction orthogonal to its Minkowski worldvolume can be used to model a $(p+1)$-dimensional Minkowski heat bath. The same general idea underlies the GEMS approach to black hole thermodynamics: the black hole spacetime is globally and isometrically embedded in a higher-dimensional flat spacetime; it is then found that a static observer in the embedded spacetime is accelerating in the higher-dimensional flat spacetime, just such that the Unruh formula yields the local temperature in thermal equilibrium [12]. At spatial infinity, we have essentially the accelerating brane set up described above. It should be appreciated that there is no necessity to ascribe physical reality to the extra dimension(s): we may view the brane set-up as just a physically possible model in which to investigate the behaviour of photon detectors in a heat bath. It may be that other physical phenomena can distinguish between a 'real' extra dimension (as in the case of a detector restricted to a fixed distance from a black p-brane horizon) and a 'fictitious' extra dimension (as in the GEMS program); e.g. the Coulomb force between point charges (although we are free to replace point sources by line sources in the extra dimension, if this is fictitious, such that the Coulomb force is the same as in the lower dimension). However, we consider only photons in a photon gas; if we restrict the photon detector to detect photons arriving isotropically, in the rest-frame of the gas, from directions 'on the brane' to which we confine the motion of the detector, then we expect the behaviour of the detector to be independent of other factors that are unobservable from its perspective.

Thus, *we have a model in which a detector that is moving with constant velocity* $\mathbf{v}$ *with respect to the rest-frame of a heat bath, at temperature* T*, is equivalent to a 'braneworld' detector that is 'drifting' with constant velocity* $\mathbf{v}$ *on a brane that is itself accelerating in an additional dimension with proper acceleration* $a = 2\pi T/\hbar$. The motion of the detector in the higher-dimensional spacetime is therefore one of 'acceleration with drift', which is a *stationary* motion. Stationary motions are those in which the worldline is an orbit of a timelike Killing vector field, which has the consequence that the detector's excitation spectrum is time independent. We wish to compare the known thermal response of a 'braneworld' detector undergoing pure constant proper acceleration in the embedding spacetime with the time-independent response of a similar detector that is undergoing an 'acceleration with drift'. We certainly expect to find a difference because the stationary motions of 'pure' acceleration and 'acceleration with drift' are distinct stationary motions according to the classification (for 4-dimensional Minkowski spacetime) of [13], which is based on the local curvature, torsion and hypertorsion of the detector's worldline. There are six distinct stationary motions in all; the others are inertial motion, circular motion, a 'null' case and a 'generic' case. A Lorentz covariant version of this classification follows from consideration of the detector's relativistic acceleration, jerk and snap [14].

We shall consider a simple two-state Unruh-DeWitt 'monopole' detector [10, 15] with variable energy gap $\hbar\omega$. The response of an Unruh-DeWitt detector to acceleration is known to have an 'artificial' dimension dependence: the thermal distribution of 'photons' seen by a detector undergoing constant proper acceleration in an odd-dimensional spacetime is apparently of Fermi-Dirac type rather than Bose-Einstein type [16]. Although this is indeed 'artificial', for reasons explained in [17, 18], it is relevant here because we would need to embed the 4-dimensional Minkowski braneworld of a 3-brane in a 5-dimensional Rindler spacetime: the $p = 3$ case of (4). We avoid this problem by going down one dimension: we consider instead the 3-dimensional Minkowski braneworld embedded in 4-dimensional Rindler spacetime; i.e. we consider the $p = 2$ case of (4). This means that we can make use of results obtained previously for the response of an Unruh-DeWitt detector in a 4-dimensional Minkowski spacetime.

The integrated response, i.e. the result after integration over solid angle, has been extensively investigated for all types of stationary motion in a 4-dimensional Minkowski spacetime [13, 19], although only numerical results are known for 'acceleration with drift'. Here, however, we need the *angular-dependence* of the excitation spectrum. We undertake this computation for 'acceleration with drift' in section 2, and our results are potentially of interest in the wider context of the problem of understanding the behaviour of particle detectors undergoing stationary motion; we comment further on this in the conclusions. Our result must be compared to the analog of (2) for a three-dimensional (braneworld) spacetime (i.e. $p = 2$), which is

$$n_{\mathrm{PW}}(\omega, \varphi) d\omega\, d\varphi = \frac{\omega}{4\pi^2 \left(e^{\hbar\omega/T_{\mathrm{eff}}} - 1\right)} d\omega\, d\varphi\,, \qquad T_{\mathrm{eff}} = T\frac{\sqrt{1-v^2}}{1 - v\cos\varphi}\,, \tag{5}$$

where φ is the angle to the direction defined by the detector's velocity. This assumes a single polarization state, as appropriate to 'scalar photons'. Experimentally, one would deduce the 'photon' number density from the excitation rate of the detector. These quantities have the same dimension for $D = 3$ and we shall see that they are proportional for $v = 0$; we use this fact to fix the constant of proportionality. We then find a result that agrees with (5) for small drift velocity in the ultra-violet (UV) regime $\hbar\omega \gg T$, which is where we may think of the photons in the photon gas as classical particles. This is encouraging because it shows that the classical Doppler-shift effect is indeed reproduced by our accelerating braneworld set-up.

Despite this agreement in the UV regime, we find a discrepancy with (5) in the infra-red (IR) regime $\hbar\omega \ll T$. In the context of our brane set-up, there is a simple likely explanation for this discrepancy. As shown in [21, 22], there is an ergo-region outside the Rindler horizon at non-zero drift velocity and the analogy with rotating black holes leads us to expect spontaneous emission of radiation from this ergo-region, in addition to the thermal radiation from the horizon. We discuss this in section 3. There is also a more obvious reason for the discrepancy that does not rely on the accelerating brane picture. Induced emission is important to the detector's response in the IR limit, and the angular dependence of the photon distribution due to the motion of the detector implies an angular dependence in the induced emission rate that is not captured by the classical Doppler shift. This observation applies as much to motion in a 4-dimensional Minkowski heat bath as to a 3-dimensional one, which is one indication that our results in $2+1$ dimensions are likely to have implications for the detection of photons in the CMBR (for which $v \sim 0.002$); we discuss this in section 4.

For a relativistic drift velocity, the disagreement between the formula (5) and our result is significant even in the UV limit. In fact, above a critical velocity $v_c \sim 0.22$ we find that there is *no response* from the detector to photons arriving within some angle to the half-line $\varphi = \pi$; in other words, there is a 'cone of silence' in place of a wake as the detector moves through the photon heat bath.

2. Detector response

In this section we compute the response of an Unruh DeWitt detector undergoing the stationary motion that we refer to as "acceleration with drift" in a four-dimensional Minkowski spacetime with metric

$$d\mathbb{X} \cdot d\mathbb{X} = -dX_0^2 + dX^2 + dY^2 + dZ^2\,. \tag{6}$$

The worldline of the detector in this spacetime is given by

$$X_0 = a^{-1}\sinh\left[a\gamma\tau\right]\,, \qquad X = a^{-1}\cosh\left[a\gamma\tau\right]\,, \qquad Y = \gamma v\tau\,, \qquad Z = 0\,, \tag{7}$$

where $\gamma = 1/\sqrt{1-v^2}$. The 4-acceleration A is non-zero for any v. For $v = 0$ we have the well-known trajectory of a detector undergoing constant proper acceleration $|A| = a$. When $v \neq 0$ we have "acceleration with drift" in which uniform motion with velocity v in the Y direction is superposed on constant proper acceleration $|A| = \gamma^2 a$ in the X direction. For pure acceleration, the worldline has non-zero curvature but zero torsion, whereas the torsion is also non-zero for acceleration with drift [13]. Equivalently, the relativistic jerk [11, 14] is zero for pure acceleration but not for acceleration with drift. We therefore expect a v-dependence of the detector's response to the motion.

As explained in the introduction, we will interpret a photon heat-bath in a 3-dimensional Minkowski spacetime as a 2-brane at the Unruh temperature associated with an orthogonal acceleration. This set-up is achieved by the following family of embeddings of a 3-dimensional worldvolume with coordinates (τ, y, z) [14]:

$$\begin{aligned} X_0 &= a^{-1}\sinh\left[a\gamma(\tau+vy)\right], \qquad X = a^{-1}\cosh\left[a\gamma(\tau+vy)\right], \\ Y &= \gamma(y+v\tau), \qquad Z = z, \qquad \left(\gamma = 1/\sqrt{1-v^2}\right) \end{aligned} \tag{8}$$

A computation of the induced metric shows that the embedded spacetime is indeed Minkowski and also that the coordinates (τ, y, z) are cartesian. This Minkowski worldvolume can be viewed as the congruence of worldlines parametrized by position (y, z) on the brane. Each such worldline defines a stationary motion of acceleration with drift, with proper time τ. The embedding with non-zero v is obtained from that with $v = 0$ by a worldvolume Lorentz boost in the y direction. Although the induced metric is not affected by this boost, the *extrinsic* geometry of the worldlines of the congruence is affected by it. In particular, the torsion (equivalently, relativistic jerk) of the worldlines of the congruence is zero only for $v = 0$. When account is taken of the Unruh temperature on the brane due to the acceleration, this fact can be seen as the *kinematical* equivalent of the fact that the rest frame of a heat bath provides a 'preferred' reference frame.

We begin with some generalities. We assume a 'monopole' detector [15] with (variable) energy gap $\hbar\omega$ coupled to a free quantum scalar field $\Phi(\mathbb{X})$. The interaction is

$$S_{int} = g\int d\tau\, m(\tau)\Phi\left(\mathbb{X}(\tau)\right), \tag{9}$$

where $m(\tau)$ is the detector's monopole moment operator, and $\mathbb{X}(\tau)$ is the detector's worldline, parametrized by proper time τ. The coupling constant g is dimensionless (this is a simplifying feature of four spacetime dimensions) and may be assumed small. The probability of excitation and accompanying emission of a massless particle of 4-momentum $\mathbb{P} = \hbar(k, \mathbf{k})$ into solid angle element $d\Omega$ may then be computed using first-order perturbation theory. Integrating over photon energy, and omitting a small dimensionless constant of proportionality, one has [23, 20]

$$\frac{d\mathcal{P}}{d\Omega} = \int_0^\infty dk\, k\int_{-\infty}^\infty d\tau_+\int_{-\infty}^\infty d\tau_-\ \exp\left(2i\hbar\omega\tau_- - i\,\mathbb{P}\cdot\Delta\mathbb{X}(\tau,\tau')\right), \tag{10}$$

where

$$\Delta\mathbb{X}(\tau,\tau') = \mathbb{X}(\tau) - \mathbb{X}(\tau'), \qquad \tau_\pm = \frac{1}{2}\left(\tau \pm \tau'\right). \tag{11}$$

Henceforth, we choose units such that $\hbar = 1$, which means that all dimensionful quantities have dimensions of mass to some power.

2.1. Integrated response

From the brane perspective that we will eventually adopt, it would make no sense to integrate the detector's response over solid angle. However, for the general purpose of understanding the physics of detector response it is instructive to consider this integral. From (10) we have ($\hbar = 1$)

$$\int d\Omega\,\frac{d\mathcal{P}}{d\Omega} = \int d\tau_+\,\dot{\mathcal{P}}, \qquad \dot{\mathcal{P}} = 16\pi^3\int_{-\infty}^\infty d\tau_-\ e^{2i\omega\tau_-}G^+, \tag{12}$$

where $\dot{\mathcal{P}}$ is the integrated excitation rate, determined by

$$G^+ = \left[4\pi^2\left|\Delta\mathbb{X}(\tau,\tau')\right|^2\right]^{-1}, \tag{13}$$

which is the Wightman function evaluated for points on the detector's worldline specified by proper times τ and τ'. The integral should vanish for inertial motion, for which $G^+ = -1/(16\pi^2\tau_-^2)$. The integrand has a pole at $\tau_- = 0$ but the usual '$i\varepsilon$' prescription will push this pole below the real axis so that it does not lie within the contour completed in the upper half of the complex τ_--plane (we assume $\omega > 0$). With this prescription, we recover the expected result that the detector remains in its ground state if it is moving inertially.

It has been proposed [24] that the response of an inertial detector in a heat bath at temperature T is given by

$$\dot{\mathcal{P}}_T \propto \int_{-\infty}^{\infty} d\tau_-\, e^{2i\omega\tau_-}\, G_T^+(\tau_-)\,, \tag{14}$$

where G_T^+ is the thermal Wightman function at temperature T, again evaluated for points on an inertial worldline specified by proper times τ and τ':

$$G_T^+ = -\frac{1}{16\pi^2}\sum_{n=-\infty}^{\infty}\frac{1}{\left[(\tau_- - in\gamma/2T)^2 + (n\gamma v/2T)^2\right]}\,. \tag{15}$$

Now there are poles inside the contour even for inertial motion and one gets a non-zero result that depends on the detector's velocity with respect to the heat bath. The 'scalar photon' number density may be read off from this result, and this turns out to be half the photon number density (3) [24, 6]. This computation appears to support the proposition that a photon detector moving with constant velocity through a heat-bath will register an angular spectrum given (for a 4-dimensional Minkowski spacetime) by (2). However, the formula (14) has not been derived from first principles.

We return to the conventional formula (12) for the integrated excitation rate, which *is* derived from first principles. This is time independent for any stationary motion since G^+ is then a function only of τ_-. For "acceleration with drift" one finds that [13]

$$G^+ = -\frac{a^2}{16\pi^2\left[\sinh^2\xi - v^2\xi^2\right]}\,, \qquad \xi = a\gamma\tau_-\,, \tag{16}$$

as can be shown directly using (7). This yields a non-zero integrated excitation rate. One can arrive at the same result by first performing the k integral in (10). To this end, we choose spherical angles, $0 \le \theta \le \pi$, $0 \le \varphi \le 2\pi$, so that

$$\mathbb{P}_0 = k\,, \qquad \mathbb{P}_X = k\,\cos\theta\,, \qquad \mathbb{P}_Y = k\,\sin\theta\cos\varphi\,, \qquad \mathbb{P}_Z = k\,\sin\theta\sin\varphi\,. \tag{17}$$

We then find that

$$\frac{d\mathcal{P}}{d\Omega} = -\frac{a}{2\gamma}\int d\tau_+ \int_{-\infty}^{\infty} d\xi\,\frac{e^{2i(\omega/a\gamma)\xi}}{\left[\Gamma_\theta\sinh\xi - v\xi\cos\varphi\sin\theta\right]^2} \tag{18}$$

where

$$\Gamma_\theta = \cosh\left(a\gamma\tau_+\right) - \cos\theta\sinh\left(a\gamma\tau_+\right)\,. \tag{19}$$

Integrating over solid angle, we recover the integrated excitation rate in terms of G^+.

2.2. Angular response

We now focus on the formula (18), which gives the angular response of the detector integrated over time. Consider first the $v = 0$ case. We have, from (18) and (19),

$$\left.\frac{d\mathcal{P}}{d\Omega}\right|_{v=0} = \int_{-\infty}^{\infty} d\tau_+\,\left[\Gamma_\theta(\tau_+)\right]^{-2}\left.\frac{d\dot{\mathcal{P}}}{d\Omega}\right|_{v=0} \tag{20}$$

where

$$\left.\frac{d\dot{\mathcal{P}}}{d\Omega}\right|_{v=0} = -\frac{a}{2\gamma}\int_{-\infty}^{\infty} d\xi\,\frac{e^{2i(\omega/a\gamma)\xi}}{\sinh^2\xi}\,. \tag{21}$$

The integrand of the τ_+ integral is to be identified with an excitation rate but this is τ_+-dependent because of the $[\Gamma_\theta]^{-2}$ factor. As shown in [20, 25], this factor arises because the detector has a time-dependent velocity $dX/dT = \tanh(a\gamma\tau_+)$ relative to an inertial 'laboratory' frame. The two frames coincide at $\tau_+ = 0$, so the time-independent excitation rate in the detector's frame is just $(d\dot{\mathcal{P}}/d\Omega)_{v=0}$. The same logic applied for $v \neq 0$ yields the excitation rate

$$\frac{d\dot{\mathcal{P}}}{d\Omega} = -\frac{a}{2\gamma}\int_{-\infty}^{\infty} d\xi\,\frac{e^{2i(\omega/a\gamma)\xi}}{\left[\sinh\xi - v\xi\cos\varphi\sin\theta + i\varepsilon\right]^2}\,, \tag{22}$$

where we have now made explicit the $i\varepsilon$ prescription.

To compute the integral in (22), we complete the contour in the upper-half complex ξ-plane. For non-zero $u \equiv v\cos\varphi\sin\theta$ it encloses a finite number of poles at

$$\xi = ix\,, \qquad \sin x = ux\,, \quad x > 0\,, \quad (u \equiv v\cos\varphi\sin\theta)\,. \tag{23}$$

There is a unique pole at $x = x_0(u)$ when $u > u_1 \approx 0.13$, and thus the residue calculus yields

$$\frac{d\dot{\mathcal{P}}}{d\Omega} = \frac{\pi a}{\gamma} c\left[(x_0(u))\right] e^{-\frac{2\omega x_0(u)}{a\gamma}}\,, \tag{24}$$

where

$$c[x] \equiv \frac{2\omega u - 2\omega\cos x + a\gamma\sin x}{a\gamma\left[u - \cos x\right]^3}\,. \tag{25}$$

At $u = u_1$ there is an extra double pole, which splits into two poles at $x = x_1 > x_0$ and $x = x_1' > x_1$ for $u < u_1$. As long as $u > u_2$, where $u_2 < u_1$ is a second critical value of u, there will be at most three poles. As u decreases further, new pairs of poles appear at a sequence of critical value values u_ℓ of u, and for any non-zero v one finds that

$$\frac{d\dot{\mathcal{P}}}{d\Omega} = \frac{\pi a}{\gamma}\left\{c(x_0)\,e^{-\frac{2\omega x_0}{a\gamma}} + \sum_{\ell=1}^{\ell_{\max}}\theta(u_\ell - u)\left[c(x_\ell)\,e^{-\frac{2\omega x_\ell}{a\gamma}} + c(x_\ell')\,e^{-\frac{2\omega x_\ell'}{a\gamma}}\right]\right\}\,, \tag{26}$$

where $2\ell_{\max}+1$ is the number of roots of $\sin x = ux$. For $u \to 0$ (e.g. as a result of $v \to 0$) $\ell_{\max} \to \infty$ and the poles in the contour move to $\xi = \pi k i$, thus reproducing the $v = 0$ result. Despite the step functions, $d\dot{\mathcal{P}}/d\Omega$ is a smooth function of u because the pairs of poles that that appear as u decreases though a critical value u_ℓ give contributions of opposite signs that cancel at $u = u_\ell$. In the ultraviolet (UV) limit, the first term in (26) dominates, exponentially, so that (24) gives the asymptotic behaviour.

So far we have considered only positive u, but u becomes negative when $\frac{3\pi}{2} > \varphi > \frac{\pi}{2}$. The equation $f(u) \equiv \sin x - ux = 0$ then has no solution for u below a certain critical value u_*, determined as follows. For u slightly above the critical value u_*, there are two roots, which become a double root at $u = u_*$. At this point $u = u_*$ also solves $f'(u) = \cos x - u = 0$. Combining both equations, we find the critical value at

$$\tan x_* = x_*\,, \qquad x_* \approx 4.4934\,, \qquad u_* = \cos x_* \approx -0.2172\,. \tag{27}$$

Since $u = v\cos\varphi\sin\theta$, this critical point exists only at velocities $v > v_c = |u_*| \approx 0.2172$. At any velocity $v > v_c$, there is no pole of the integrand of (22) within the contour whenever

$$-\sin\theta\,\cos\varphi > \frac{v_c}{v}\,. \tag{28}$$

In other words, there is no response from the detector in a 'backward' cone, with forward axis defined by the detector velocity, of angle α such that

$$\cos\alpha = \frac{v_c}{v} . \tag{29}$$

The angle α goes to zero as $v \to v_c$, while $\alpha \to \arccos(v_c) \approx 0.57\pi \approx 77.5^\circ$ as $v \to 1$. The cone closes at $v = v_c$, and for $v < v_c$ the detector has a response in all directions. For u slightly above u_*, the equation $\sin x - ux = 0$ has two roots x_1, x_2 and and the detector response is of the form

$$\frac{d\dot{\mathcal{P}}}{d\Omega} = \frac{\pi a}{\gamma}\left(c(x_1)\, e^{-\frac{2\omega x_1}{a\gamma}} + c(x_2)\, e^{-\frac{2\omega x_2}{a\gamma}}\right) . \tag{30}$$

Again it should be noted that there is no discontinuous behaviour: the two terms give contributions of opposite signs which cancel as $u \to u_*$.

2.3. Braneworld detector response

We now want to apply our results to the braneworld approach to motion in a Minkowski heat-bath. We therefore consider a detector restricted to detect photons arriving from directions within the brane. Recalling that θ is the angle that a photon's 3-momentum makes with the X-axis, we see that we must set $\theta = \pi/2$. For convenience, we define

$$\frac{d\dot{\mathcal{F}}}{d\varphi} \equiv \left.\frac{d\dot{\mathcal{P}}}{d\Omega}\right|_{\theta=\pi/2} = -\frac{a}{2\gamma}\int_{-\infty}^{\infty} d\xi\, \frac{e^{2i(\omega/a\gamma)\xi}}{\left[\sinh\xi - v\xi\cos\varphi + i\varepsilon\right]^2} . \tag{31}$$

The result of this integration is obtained from the previous formulas (24) and (26) by setting $\theta = \pi/2$, so that $u = v\cos\varphi$. From these formulas we may deduce the detector response for relativistic velocities, but we postpone this analysis until we have considered the low velocity limit.

For $v \ll 1$, it is convenient to return to (31) and expand the integrand in powers of v. This yields

$$\frac{d\dot{\mathcal{F}}}{d\varphi} = -\frac{a}{2\gamma}\sum_{n=0}^{\infty}(n+1)\int d\xi\, e^{\frac{2i\omega\xi}{a\gamma}}\, \frac{(v\,\xi\cos\varphi)^n}{\left[\sinh\xi + i\varepsilon\right]^{n+2}} . \tag{32}$$

Again completing the contour in the upper half complex ξ- plane, we find poles within the contour at $\xi = \pi k i$ for positive integer k, and the residue calculus yields

$$\frac{d\dot{\mathcal{F}}}{d\varphi} = \frac{2\pi\omega}{\gamma^2}\sum_{\ell=1}^{\infty}\sum_{n=0}^{\infty}\frac{p_n}{n!}\left(\frac{(-1)^{\ell+1}\, 2\pi\ell\omega u}{a\gamma}\right)^n e^{-\frac{2\pi\ell\omega}{a\gamma}} , \tag{33}$$

where p_n is a polynomial in a/ω, of degree $n+1$ for $n > 0$, with $p_n(0) = 1$; in particular,

$$p_0 = 1 , \qquad p_1 = 1 - \frac{a\gamma}{k\pi\omega} + \frac{a^2\gamma^2}{4\omega^2} , \tag{34}$$

We can compute exactly the linear terms in the velocity by keeping only the $n = 0, 1$ terms in the sum over n in (33). Using the Unruh formula $2\pi T = |A| = \gamma^2 a$, we find that

$$\frac{d\dot{\mathcal{F}}}{d\varphi} = \frac{2\pi\omega}{e^{\omega/T} - 1} + \frac{2\pi\omega^2 v\cos\varphi\, e^{-\omega/T}}{T\left(1 + e^{-\omega/T}\right)^2}\left[1 - \frac{2T}{\omega} + \frac{\pi^2T^2}{\omega^2} - \frac{2T}{\omega}e^{-\omega/T}\right] + \mathcal{O}\left(v^2\right) . \tag{35}$$

As anticipated in the introduction, the excitation rate for $v = 0$ reproduces the thermal number density (5) up to an overall factor (which includes the coupling constant of the interaction of the detector with the field). This allows us to fix the factor relating the excitation rate and the number density:

$$n(\omega,\varphi) = \frac{1}{8\pi^3}\frac{d\dot{\mathcal{F}}}{d\varphi}$$
$$= \frac{\omega}{4\pi^2(e^{\omega/T}-1)} + \frac{\omega^2 v\cos\varphi\, e^{-\omega/T}}{4\pi^2 T\left(1+e^{-\omega/T}\right)^2}\left[1-\frac{2T}{\omega}+\frac{\pi^2T^2}{\omega^2}-\frac{2T}{\omega}e^{-\omega/T}\right]+\mathcal{O}\left(v^2\right). \tag{36}$$

In the same approximation, the excitation rate implied by (5) is

$$n_{\rm PW}(\omega,\varphi) = \frac{\omega}{4\pi^2(e^{\omega/T}-1)} + \frac{\omega^2 v\cos\varphi\, e^{-\omega/T}}{4\pi^2 T\left(1-e^{-\omega/T}\right)^2}+\mathcal{O}\left(v^2\right), \tag{37}$$

Thus, we find agreement in the UV part of the spectrum where $\omega \gg T$. Otherwise there is a discrepancy already at linear order in v, and this discrepancy is significant in the IR limit, as we anticipated in the introduction on the grounds that an angular-dependent induced emission rate becomes important in this limit.

For relativistic velocities we return to formulas (24) and (26) evaluated at $\theta = \pi/2$. For $v > |u_*| \approx 0.2172$ we find, for negative u, the same "cone of silence" within which the detector has no response because the integrand of (31) has no pole within the contour. This cone is now defined by $\varphi > \varphi_0$, with $\cos\varphi_0 = u_*/v$. The "cone of silence" is obviously not a feature of the PW formula. Consider now $u \to 1$, which implies both $\varphi \to 0$ and $v \to 1$. In this case there is a single pole which is located at $x_0 = \sqrt{6}\sqrt{1-v\cos\varphi} \to 0$; the leading behaviour is given by

$$n(\omega,\varphi) = \frac{1}{8\pi^3}\frac{d\dot{\mathcal{F}}}{d\varphi} \approx \frac{T}{16\pi\gamma^3}\frac{\sqrt{3}}{\sqrt{2}}\frac{\exp\left(-\frac{\sqrt{6}\gamma\omega}{\pi T}\sqrt{1-v\cos\varphi}\right)}{(1-v\cos\varphi)^{\frac{5}{2}}}. \tag{38}$$

This is significantly different from the ultra relativistic limit of the PW-type formula (5).

3. Vacuum for acceleration with drift

It is well-known that the thermal spectrum registered by an accelerating detector can be explained, *in a detector independent way*, as a result of a difference of the Minkowski quantum vacuum to the (Fulling) vacuum of the Rindler spacetime. Specifically, the Bogoliubov transformation that relates the particle creation and annihilation operators in the two spacetimes is non-trivial, so that the Minkowski vacuum is a thermal state in the Rindler spacetime. It would be natural to suppose that our result for 'acceleration with drift' has a similar explanation; after all, there *is* an Unruh-type effect associated with motion through a medium with refractive index $n > 1$ [25]. However, the Bogoliubov transformation that connects the Fulling-Rindler vacuum with the analogous vacuum in a moving frame is trivial [26], so it might seem that the photons available for detection at non-zero drift velocity are just the same photons that were available for detection at zero velocity, but Doppler shifted. If so, this would justify the PW formula based on the Doppler shift. However, there is more to this problem than just the Bogoliubov transformation; in particular, it was shown in [21, 22] that the stationary motion we call "acceleration with drift" leads a geometry in which there is an ergo-region outside the acceleration horizon. We present here a summary of these results and then discuss the implications. This whole analysis can be carried out for arbitrary spacetime dimension D, in particular for the $D = 4$ case as well as for the $D = 3$ case that we have analyzed from the perspective of detector response.

We begin with the $(D+1)$-dimensional Minkowski metric in the form

$$ds^2 = -dt^2 + dx^2 + dy^2 + |d\vec{z}|^2\,. \tag{39}$$

Journal of Physics: Conference Series **222** (2010) 012040 doi:10.1088/1742-6596/222/1/012040

where $\vec{z} = (z_1, \ldots z_{D-2})$. The change of coordinates

$$t = r \sinh \eta \ , \qquad x = r \ \cosh \eta \ , \tag{40}$$

gives the Rindler metric

$$ds^2 = -r^2 d\eta^2 + dr^2 + dy^2 + |d\vec{z}|^2 \ . \tag{41}$$

The Rindler coordinates cover only two portions of the Minkowski spacetime, the positive ("right") and negative ("left") Rindler wedges,

$$R_+ = \{x \mid x > |t|\} \ , \qquad R_- = \{x \mid x < -|t|\} \ ,$$

for positive and negative r, respectively. In these regions the Killing vector field $\partial_\eta = x\partial_t + t\partial_x$ is timelike. The worldlines with constant r correspond to constant proper acceleration $a = 1/r$.

Consider now a complex massless scalar field ϕ obeying the wave equation $\partial^2 \phi = 0$. The norm of any solution ϕ is defined as (we follow the review article [18])

$$||\phi|| = -i \int d\Sigma^\mu \left(\phi^* \partial_\mu \phi - \phi \partial_\mu \phi^*\right) \tag{42}$$

where Σ is the volume element on an arbitrary spacelike hypersurface. A basis of solutions of positive norm is provided by the positive frequency plane parallel waves

$$U_k = \frac{1}{[2\omega_k (2\pi)^{n-1}]^{1/2}} \ e^{i\left(-\omega_k t + k_x x + k_y y + \vec{k}\cdot\vec{z}\right)} \ , \qquad \omega_k = \sqrt{k_x^2 + k_y^2 + |\mathbf{k}|^2} \ . \tag{43}$$

In Rindler coordinates, the solutions with positive norm are again those of positive frequency, but where this is now defined with respect to the Rindler time η; these modes are

$$u_k^{(\sigma)} = \frac{\theta(\sigma r)}{[2\Omega (2\pi)^{n-2}]^{1/2}} \ h_k^{(\sigma)}(r) \ \exp\left[i\left(-\sigma\Omega\eta + k_y y + \vec{k}\cdot\vec{z}\right)\right] \ , \tag{44}$$

where $\sigma = \pm$ for the positive and negative Rindler wedges, respectively. Here we are using the fact that the timelike Killing vector ∂_η is future directed on R_+ and past directed on R_-. Introducing the coordinate $r_* = \log|r|$, the equation of motion for the radial functions $h_k^{(\sigma)}(r)$ takes the form of a one-dimensional Schrodinger equation:

$$\left(-\frac{d^2}{dr_*^2} + \mu_k^2 e^{2r_*}\right) h_k^{(\sigma)} = \Omega^2 h_k^{(\sigma)} \ , \qquad \mu_k = \sqrt{k_y^2 + \left|\vec{k}\right|^2} \ . \tag{45}$$

At this point it is important to note that Ω is an arbitrary parameter that ranges from 0 to ∞, independently of the values of the momenta; the reason is that for any $\Omega > 0$ the solution is always oscillatory for sufficiently small r_*.

The scalar field ϕ can be expanded either in terms of U_k, U_k^* or in terms of $u_k^{(+)}, u_k^{(+)*}, u_k^{(-)}, u_k^{(-)*}$, as both sets constitute complete sets of modes:

$$\phi = \int d^{d-1}k \ (a_k U_k + \bar{a}_k^\dagger U_k^*) = \int_0^\infty d\Omega \int d^{d-2}k \ \sum_{\sigma=\pm} (b_k^{(\sigma)} u_k^{(\sigma)} + \bar{b}_k^{(\sigma)\dagger} u_k^{(\sigma)*}) \ . \tag{46}$$

The linear transformation from $a_k, a_k^\dagger$ to $b_k^{(\sigma)}, \bar{b}_k^{(\sigma)\dagger}$ is a Bogoliubov transformation, which mixes modes of positive and negative norm; it takes the form

$$b_k^{(\sigma)} = \int d^{d-1}k' (\alpha_{kk'}^{(\sigma)} \ a_{k'} + \beta_{kk'}^{(\sigma)} \ \bar{a}_{k'}^\dagger) \ . \tag{47}$$

The important implication is that the Minkowski vacuum state defined by

$$a_k|0\rangle_M = \bar{a}_k|0\rangle_M = 0 \qquad \forall\ k \tag{48}$$

is not equivalent to the Fulling-Rindler vacuum state defined by

$$b_k^{(\sigma)}|0\rangle_R = \bar{b}_k^{(\sigma)}|0\rangle_R = 0 \qquad \forall\ \sigma,\ k\,. \tag{49}$$

In particular, the expectation value of the Rindler particle number operator $N_k = b_k^{(\sigma)\dagger} b_k^{(\sigma)}$ in the Minkowski vacuum is

$$\int d^{d-1}k'\,\beta_{kk'}\beta^*_{k'k}\,, \tag{50}$$

and as a result the Rindler observer is immersed in a thermal distribution of particles.

We now consider the effect of a drift combined with acceleration on the vacuum. The appropriate metric is obtained by a Lorentz boost in the y direction applied to the metric (41). For an observer standing at $r_0 = a^{-1}$, the proper time is $\tau = r_0\eta$. We introduce new coordinates

$$y = \gamma(\tilde{y} - vr_0\tilde{\eta})\,, \qquad r_0\eta = \gamma(r_0\tilde{\eta} - v\tilde{y})\,, \qquad r_0 = a^{-1}\,, \tag{51}$$

so that

$$\begin{aligned} ds^2 &= -r^2\gamma^2(d\tilde{\eta} - \frac{v}{r_0}d\tilde{y})^2 + dr^2 + \gamma^2(d\tilde{y} - vr_0 d\tilde{\eta})^2 + |d\vec{z}|^2 \\ &= -\gamma^2(r^2 - v^2r_0^2)d\tilde{\eta}^2 + 2\gamma^2\frac{v}{r_0}(r^2 - r_0^2)d\tilde{y}d\tilde{\eta} + \gamma^2\left(1 - \frac{r^2v^2}{r_0^2}\right)d\tilde{y}^2 + dr^2 + |d\vec{z}|^2\,. \end{aligned} \tag{52}$$

This metric is stationary, with respect to the new time variable $\tilde{\eta}$, but it is not static. There is still a coordinate singularity at $r = 0$, which is again the acceleration horizon, but there is also a 'static limit' surface at $r = vr_0$ [22]; the region between the horizon and this static limit surface is analogous to the ergo-region of a rotating black hole. The hypersurface at $r = r_0$ is Minkowski spacetime in standard coordinates

$$ds^2\Big|_{r=1} = -r_0^2 d\tilde{\eta}^2 + d\tilde{y}^2 + |d\vec{z}|^2\,. \tag{53}$$

A static detector in this subspace has proper acceleration $\gamma^2 a$ in the larger space, where the γ^2 factor is due to time dilation as a consequence of the velocity v in the y direction.

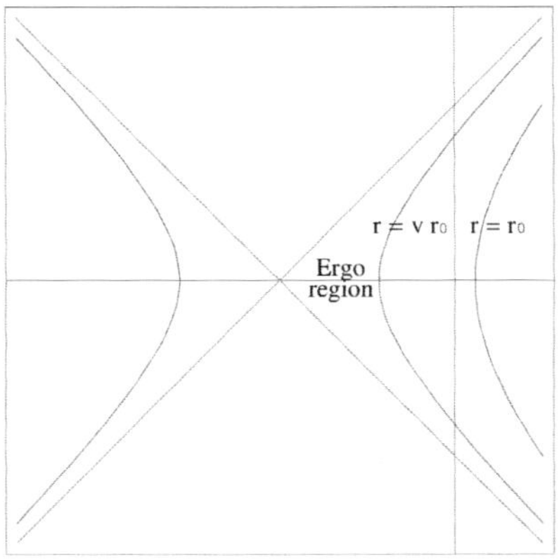

Figure 1. Spacetime diagram for a detector at $r = r_0$ on a stationary trajectory of acceleration with drift velocity v. There is an ergo-region for the natural time-like Killing vector field between the acceleration horizon and the static-limit surface at $r = vr_0$.

As ϕ is a scalar field, we can obtain the solutions of the wave equation in the new coordinates from those in the original Rindler coordinates by simply making the substitution (51):

$$\tilde{u}_k^{(\sigma)} = \frac{\theta(\sigma r)}{[2\Omega(2\pi)^{n-2}]^{1/2}}\, \tilde{h}_k^{(\sigma)}(r)\ \exp\left(i(-\sigma\tilde{\Omega}\tilde{\eta} + \tilde{k}_y\tilde{y} + \vec{k}\cdot\vec{z})\right) \tag{54}$$

where

$$a\tilde{\Omega} = \gamma(a\Omega + \sigma\ vk_y)\ , \qquad \tilde{k}_y = \gamma(k_y + \sigma\ va\Omega)\ , \tag{55}$$

As pointed out above, Ω and k_y take independent values, $0 < \Omega < \infty$ and $-\infty < k_y < \infty$. This has the implication that a mode of positive (negative) frequency before the Lorentz boost, becomes negative (positive) frequency after the boost if $vk_y < -\Omega$. However, the modes that were originally of positive (negative) norm still have positive (negative) norm, so the definition of the vacuum state is not affected by the boost [26]. Nevertheless, we now have a situation in which some states of positive norm have negative energy, and this implies an instability of the vacuum [22].

We can now see, within our braneworld set-up, why the detector should not detect a spectrum of particles that is simply a Doppler-shift of the thermal spectrum at zero velocity. In this extra-dimensional picture, the thermal spectrum is a consequence of the acceleration horizon, but there is also an ergo-region. In the context of black hole physics, ergo-regions are classically stable but quantum-mechanically unstable. An ergo-region is a source of spontaneous emission of radiation that decreases the angular momentum of a rotating black hole. Here we expect an angular-dependent spontaneous emission of radiation from the ergo-region that would decrease the momentum in the y-direction if this direction were to be periodically identified. The detector at $r = a^{-1}$ should detect the photons in this radiation in addition to the photons of the thermal radiation from the horizon. This presumably explains why we find a departure from the Doppler-shift result at non-relativistic velocities, although it is hard to see how it could explain the "cone of silence" that we find above the critical, relativistic, velocity.

4. Possible implications for the CMBR

Our local group of galaxies appears to be moving at $v \approx 0.002$ relative to CMBR rest frame. Using this velocity in our result for detector response in a three-dimensional Minkowski heat bath, we get a deviation from the PW-type spectrum (5) such that

$$\frac{|n - n_{\mathrm{PW}}|}{n} \sim 10^{-4}\ . \tag{56}$$

The angular distribution of our spectrum as compared to (5) at $v = 0.002$ is shown in fig. 1, where one can see that the angular dependence is remarkably similar; the scale has been amplified to make the difference visible.

The energy spectrum at $\varphi = 0$ as compared to $\omega\, n_{\mathrm{PW}}(\omega)$ is shown in fig. 2a for $v = 0.04$; in this case there are seven poles contributing to the sum (26). We see that the spectra are remarkably close at this velocity, differing slightly near the peak. The energy spectra continue to be close even for much higher velocities such as $v = 0.15$ where only a single pole contributes to the integral and the spectrum $n(\omega)$ is exactly given by (24) divided by $8\pi^3$ (see eq.(36)). For this value of v the greatest deviation is at the peak of the spectrum and is of order 8 %. At higher velocities the difference becomes substantial; for example, for $v = 0.8$ the peak of the PW energy spectrum is smaller by a factor of 5 and the PW temperature (defined by the exponent of the tail) is higher by a factor of 1.8, as shown in fig 2b.

Spacetime diagram for a detector at $r = r_0$ on a stationary trajectory of acceleration with drift velocity v. There is an ergo-region for the natural time-like Killing vector field between the acceleration horizon and the static-limit surface at $r = vr_0$.

Journal of Physics: Conference Series **222** (2010) 012040 doi:10.1088/1742-6596/222/1/012040

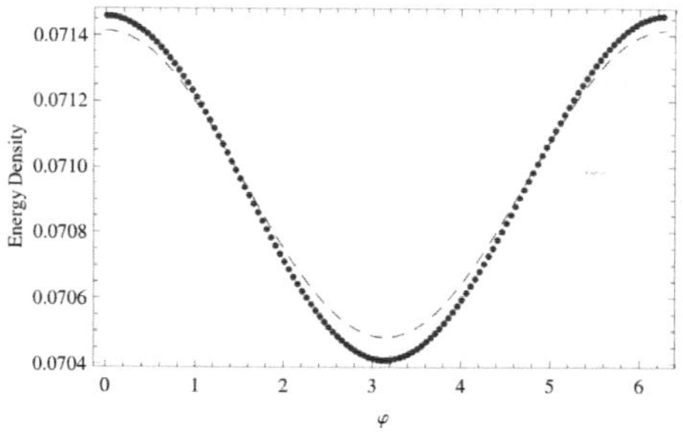

Figure 2. Angular distribution at $v = 0.002$, $\omega = A/2 = 0.5$. The dashed line represents the Peebles-Wilkinson spectrum. The greatest deviations are observed at $\varphi = \pi$, with a normalized value of order 8×10^{-4}, and at $\varphi \sim 0$, of order 5×10^{-4}.

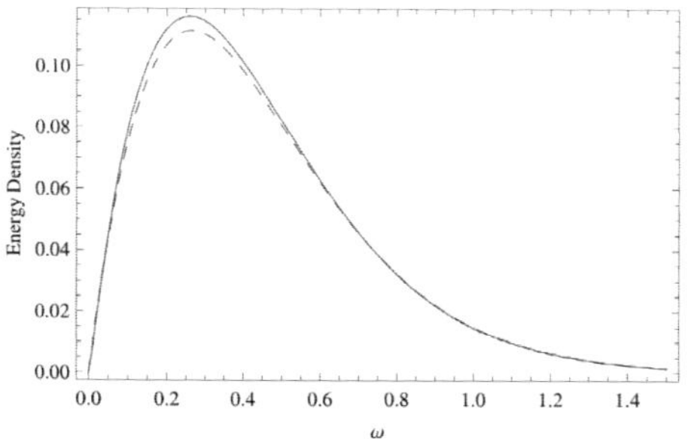

Figure 3. Spectrum at $\varphi = 0$ with $v = 0.04$ (conventions as above).

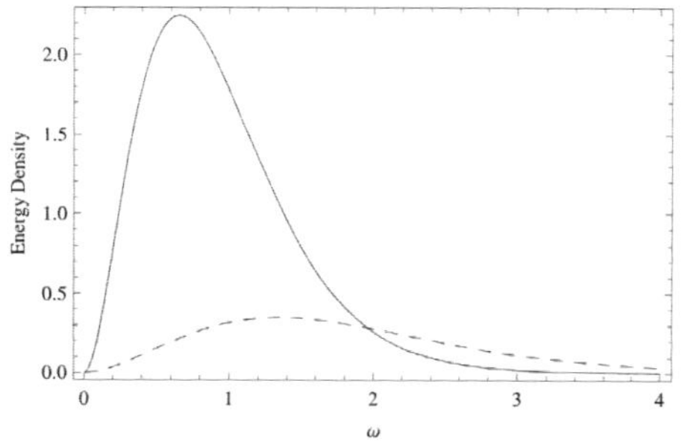

Figure 4. Spectrum at $\varphi = 0$ with $v = 0.8$ (conventions as above).

Although the deviation from PW spectrum is very small at small velocities, the difference becomes appreciable if one focuses on the dipole term proportional to the detector velocity. From (5) one sees that the Doppler-shift approach predicts an angular variability of the effective temperature:

$$T_{\text{eff}}(\varphi) = T\, v\, \cos\varphi + \mathcal{O}(v^2) \tag{57}$$

In contrast, our result (36) can be interpreted as predicting a frequency-dependent angular variability:

$$T_{\text{eff}}(\varphi, \omega) = [T\, v\, \cos\varphi]\, F(\omega)\,, \tag{58}$$

where

$$F(\omega) = \frac{\left(1 - e^{-\omega/T}\right)^2}{\left(1 + e^{-\omega/T}\right)^2} \left[1 - \frac{2T}{\omega} + \frac{\pi^2 T^2}{\omega^2} - \frac{2T}{\omega} e^{-\omega/T}\right] \tag{59}$$

The function $F(\omega)$ represents the influence of frequency-dependent quantum effects neglected in the Doppler-shift calculation. For the experimentally available frequency range (roughly from 50 to 500 GHz), this implies a change from the highest to the lowest frequencies by a factor of order 2 (see fig. 3).

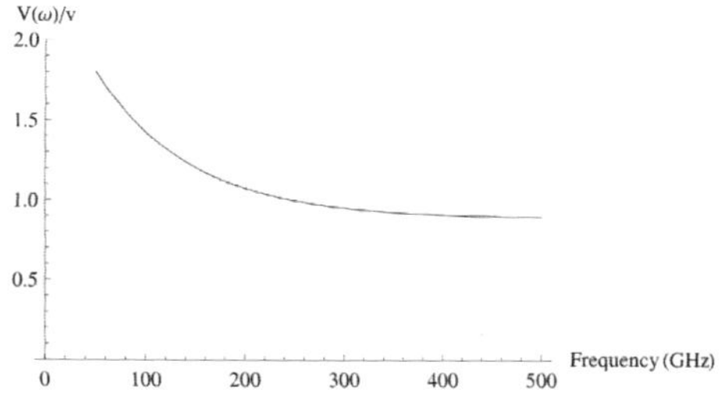

Figure 5. Dipole relative to CMBR in the experimentally available frequency range.

Journal of Physics: Conference Series **222** (2010) 012040 doi:10.1088/1742-6596/222/1/012040

It is worth noting that the different terms appearing in the expression for $F(\omega)$ contain poles at $\omega = 0$ that cancel out thanks to the specific combination appearing in (59): a generic change in the coefficients in the terms appearing inside the square brackets would lead to a much higher factor. Of course, a factor of 2 variation is still large, but this variation occurs only in low frequency part of the available range; otherwise the spectrum is nearly flat. Since our results have been derived for motion in a $(2+1)$-dimensional Minkowski heat bath, it is possible that the analogous spectrum for $(3+1)$ dimensions will not be in conflict with experiment. Also, the CMBR experiment is calibrated on the dipole assuming a frequency independent dipole factor, so a small frequency dependence might show up as an anomaly elsewhere.

5. Discussion

We have studied the behaviour of a photon detector moving in a photon heat bath at constant velocity by exploiting the fact that the heat bath can be generated on a brane by acceleration in an additional direction, by virtue of the Unruh effect. The same basic idea is used in the GEMS approach to black hole thermodynamics, which gives correctly the local temperature T, as would be measured by a static particle detector, of various black hole spacetimes in various dimensions: a static detector undergoes constant proper acceleration $2\pi T/\hbar$ in the embedding spacetime, so the Unruh formula gives the local temperature [12]. We have exploited the fact that inertial motion on an accelerating Minkowski brane corresponds to "acceleration with drift" in the higher dimensional spacetime, this being another of the six possible distinct stationary motions according to the classification of Letaw [13]. Thus, the behaviour of a photon detector moving inertially in a heat bath should be the same as that of a detector undergoing acceleration with drift, assuming that it is constrained to detect particles that move within the brane.

The response of an Unruh-DeWitt detector undergoing stationary motion in a Minkowski vacuum has been the subject of many previous papers, but most have concentrated on what we have referred to as the 'integrated response' obtained by integration over angles. This integration has the feature that the response is then time-independent irrespective of the frame from which it is viewed. Here we have computed the angular response for a detector undergoing acceleration with drift; this is time-independent in its own frame (necessarily for stationary motion) but not in a 'laboratory' frame. By restricting the angles to those "accessible" to a detector 'on the brane', our result can be compared with the analog of the PW formula for a $(2+1)$-dimensional spacetime. There is no agreement for relativistic velocities but let us postpone discussion of this fact and concentrate for the moment on non-relativistic velocities. In this case we may expand to first order in the velocity and then make the comparison. We found agreement in the UV limit $\hbar\omega \gg T$ in which the photon gas is essentially classical, and this shows at least that the Doppler-shift effect is incorporated in our approach. However, as anticipated, we found a disagreement in the IR limit.

As mentioned, there are *a priori* grounds for thinking that quantum effects could effectively modify the PW formula in the IR limit, but if so it should be possible to understand the physical origin of any such modification in the context of the brane set-up described above. One might suspect that it is due to some velocity dependence of the vacuum, analogous to the Unruh effect itself, which would imply the existence of photons at non-zero velocity that were absent at zero velocity (or vice versa). However, it is known that the Bogoliubov transformation from the vacuum of an accelerating detector to the vacuum of a detector accelerating with drift is trivial [26]. Nevertheless, there is an important sense in which the two vacua are not equivalent[1]: strictly speaking, for a detector undergoing acceleration with drift the Fulling-Rindler vacuum is not a true vacuum state; it is unstable, due to negative energy states present in the right Rindler wedge [22]. This is due to the fact that for such a detector there is an ergo-region

[1] The non-equivalence of the two vacua is not surprising. In general, a detector moving at constant velocity in a non-Lorentz invariant medium is expected to see a non-trivial vacuum, and another example can be found in [20].

outside the acceleration horizon. In the context of black holes, ergo-regions are unstable against spontaneous emission. We have proposed that our result for detector response could be explained by superposing on the thermal emission from the horizon an emission of low-frequency photons from the ergo-region.

Of course, these results can be applied to the CMBR only in a fictitious universe of $(2+1)$ dimensions. This restriction has allowed us to make use of many standard results, and acquired intuition, concerning the behaviour of Unruh-DeWitt detectors, and to avoid known difficulties concerning motion in odd-dimensional (ambient) spacetimes. We cannot say what the analogous result should be in $(3+1)$ dimensions, so we can make no definite statement concerning the validity of the Peebles-Wilkinson formula as applied to the CMBR. However, we may get some idea of what to expect by substituting the observed velocity $v \approx 0.002$ into our formula. Although this leads to a spectrum that differs little from the Doppler-shift based formula, comparison of the dipole term linear in the velocity leads to an apparent frequency dependence of the effective temperature that varies by a factor of 2 over the experimentally relevant frequency range. This indicates that it might be worthwhile to look for such frequency dependence in the CMBR data, or to analyze the implications for calibrations of assuming frequency independence.

Our results also have implications for the general understanding of particle detectors. The thermal response of an accelerating particle detector is often attributed to the different global features of Minkowski and Rindler spacetimes, and of the resulting difference between the Minkowski and Fulling-Rindler vacua. However, this cannot be the whole story because there are more distinct stationary motions than there are distinct quantum vacua, at least if two vacua are considered equivalent if the Bogoliubov transformation connecting them is trivial. We have already mentioned that the vacuum for acceleration with drift is equivalent, in this sense, to the Fulling-Rindler vacuum. It was further shown in [26] that for *any* stationary motion the quantum vacuum is equivalent, in this sense, to *either* the Minkowski vacuum *or* the Fulling-Rindler vacuum. In contrast, the local response of particle detectors on stationary worldlines is far more variable, being essentially different for each distinct stationary motion, and it is a challenge to understand why this is so (see e.g. [27, 28, 22]).

This issue was addressed for circular motion in [27]. The quantum vacuum for a detector undergoing uniform motion in a circle is actually equivalent to the Minkowski vacuum, despite the centripetal acceleration, so why does the detector detect particles? One may guess that the answer must have to do with the coordinate singularity at a critical radius in static co-rotating cylindrical polar coordinates, and it was shown in [27] that a co-rotating detector in the Minkowski vacuum does *not* detect particles if one assumes boundary conditions that remove the region of spacetime beyond the critical radius. For acceleration with drift, the resolution appears to be different. As mentioned, the appearance of an ergo-region outside the Rindler horizon is also expected to have an effect on a particle detector. However, it is not clear how this could explain the "cone of silence" that we have found above the critical velocity $v_c \sim 0.2172$.

Acknowledgements
We thank an anonymous referee of an earlier version of this work for emphasizing the importance of the frequency dependence of the CMBR dipole moment and D. Kothawala for a useful discussion. JGR acknowledges support by research grants MCYT FPA 2007-66665, 2005SGR00564. PKT thanks the EPSRC for financial support, and the University of Barcelona for hospitality.

References

[1] M. Planck, S.B. Preuss. Akad. Wiss., p.542 (1907); Ann. der Phys. **76**, 1 (1908)
[2] A. Einstein, Jahrb. der Radioaktivität und Elektronik **4** (1907) 411.

[3] W. Pauli, *Relativitats Theorie*, Teubner, Berlin (1921); 1958 English translation available as *Theory of Relativity* (Dover 1981).
[4] R.C. Tolman, *Relativity, Thermodynamics and Cosmology*, Clarendon Press, Oxford, 1934; republished by Dover Publications Inc. 1987.
[5] W. Israel, "Nonstationary irreversible thermodynamics: A causal relativistic theory", Ann. Phys. (N.Y.) **100**, (1976) 310.
[6] P. T. Landsberg and G. E. A. Matsas, "Laying the ghost of the relativistic temperature transformation," Phys. Lett. A **223** (1996) 401 [arXiv:physics/9610016].
[7] Z. C. Wu, "Inverse Temperature 4-vector in Special Relativity," Europhys. Lett. **88** (2009) 20005 [arXiv:0804.3827 [gr-qc]].
[8] P. J. E. Peebles and D. T. Wilkinson, "Comment on the Anisotropy of the Primeval Fireball," Phys. Rev. **174** (1968) 2168.
[9] G. R. Henry, R. B. Feduniak, J. E. Silver and M. A. Peterson, "Distribution of black body cavity radiation in a moving frame of reference," Phys. Rev. **176** (1968) 1451.
[10] W. G. Unruh, "Notes on black hole evaporation," Phys. Rev. D **14**, 870 (1976).
[11] J. G. Russo and P. K. Townsend, "Accelerating Branes and Brane Temperature," Class. Quant. Grav. **25** (2008) 175017 [arXiv:0805.3488].
[12] S. Deser and O. Levin, "Mapping Hawking into Unruh thermal properties," Phys. Rev. D **59** (1999) 064004 [arXiv:hep-th/9809159].
[13] J. R. Letaw, "Vacuum Excitation Of Noninertial Detectors On Stationary World Lines," Phys. Rev. D **23** (1981) 1709.
[14] J. G. Russo and P. K. Townsend, "Relativistic Kinematics and Stationary Motions," J. Phys. A **42** (2009) 445402 [arXiv:0902.4243 [hep-th]].
[15] B.S. DeWitt, "Quantum Gravity: the new synthesis", in *General Relativity: An Einstein Centenary Survey*, edited by S. W. Hawking and W. Israel (Cambridge University Press, New York, 1979), p. 680.
[16] S. Takagi, Prog. Theor. Phys. **72** (1984) 505; *Ibid.* **74** (1985) 142; **74** (1985) 501.
[17] W. G. Unruh, "Accelerated monopole detector in odd spacetime dimensions" Phys. Rev. D **34** (1986) 1222.
[18] S. Takagi, "Vacuum Noise And Stress Induced By Uniform Accelerator: Hawking-Unruh Effect In Rindler Manifold Of Arbitrary Dimensions," Prog. Theor. Phys. Suppl. **88**, 1 (1986).
[19] J. Louko and A. Satz, "How often does the Unruh-DeWitt detector click? Regularisation by a spatial profile," Class. Quant. Grav. **23** (2006) 6321 [arXiv:gr-qc/0606067].
[20] H. Kolbenstvedt, "Anisotropy of thermal radiation by accelerated detectors; an aberration effect" , Phys. Lett. A **122** (1987) 292; "Inertial interpretation of the Unruh effect" , Phys. Rev. D **38** (1988) 1118.
[21] J. R. Letaw and J. D. Pfautsch, "The Stationary Coordinate Systems In Flat Space-Time," J. Math. Phys. **23** (1982) 425.
[22] J. I. Korsbakken and J. M. Leinaas, "The Fulling-Unruh effect in general stationary accelerated frames," Phys. Rev. D **70** (2004) 084016 [arXiv:hep-th/0406080].
[23] N. D. Birrell and P. C. W. Davies, "Quantum Fields In Curved Space," Cambridge University Press (1982).
[24] S. S. Costa and G. E. A. Matsas, "Temperature And Relativity," arXiv:gr-qc/9505045.
[25] I. H. Brevik and H. Kolbenstvedt, "Quantum Point Detector Moving Through A Dielectric Medium. 1. Constant Velocity," Nuovo Cim. B **102** (1988) 139; "Quantum Point Detector Moving Through A Dielectric Medium. 2. Constant Acceleration," Nuovo Cim. B **103** (1989) 45.
[26] J. R. Letaw and J. D. Pfautsch, "The Quantized Scalar Field In The Stationary Coordinate Systems Of Flat Space-Time," Phys. Rev. D **24** (1981) 1491.
[27] P. C. W. Davies, T. Dray and C. A. Manogue, "The Rotating Quantum Vacuum," Phys. Rev. D **53**, 4382 (1996) [arXiv:gr-qc/9601034].
[28] L. Sriramkumar and T. Padmanabhan, "Probes of the vacuum structure of quantum fields in classical backgrounds," Int. J. Mod. Phys. D **11**, 1 (2002) [arXiv:gr-qc/9903054].

Journal of Physics: Conference Series **222** (2010) 012041 doi:10.1088/1742-6596/222/1/012041

Destroying black holes with test bodies

Ted Jacobson* and Thomas P. Sotiriou†

*Center for Fundamental Physics, University of Maryland, College Park, MD 20742-4111, USA
†Department of Applied Mathematics and Theoretical Physics, Centre for Mathematical Sciences, University of Cambridge, Wilberforce Road, Cambridge, CB3 0WA, UK

E-mail: jacobson@umd.edu; T.Sotiriou@damtp.cam.ac.uk

Abstract.
If a black hole can accrete a body whose spin or charge would send the black hole parameters over the extremal limit, then a naked singularity would presumably form, in violation of the cosmic censorship conjecture. We review some previous results on testing cosmic censorship in this way using the test body approximation, focusing mostly on the case of neutral black holes. Under certain conditions a black hole can indeed be over-spun or over-charged in this approximation, hence radiative and self-force effects must be taken into account to further test cosmic censorship.

1. Black holes and naked singularities

Penrose and Hawking have shown that singularities are inevitable in gravitational collapse [1, 2]. These results do not rely on specific solutions of Einstein's equations or special symmetry requirements. Therefore, spacetime singularities are physically relevant and not simply mathematical peculiarities of special solutions. While the results do not say anything about the precise nature of the singularities, they do indicate the breakdown of general relativity. Einstein's theory is unable to predict the outcome of events in the vicinity of the singularity, and unusual phenomena could take place there. Visible singularities could provide exciting access to the so far unobserved physics of quantum gravity.

In the case of a black hole, however, the event horizon hides the singularity. Although this is a lost opportunity for quantum gravity phenomenology, it is also a blessing, if all we wish to understand is what can be seen from outside without having to grapple with the physics of a singularity. But does an event horizon always hide any singularities that form from a collapsing object? As Roger Penrose first put it in 1969 [3], "does there exist a cosmic censor who forbids the appearance of naked singularities, clothing each one in an absolute event horizon?" The conjecture that such a censor indeed exists is called the "cosmic censorship conjecture".

The physics of event horizons, unlike that of singularities, should be well described by classical general relativity, so cosmic censorship can be tested and possibly proven without going beyond known physics. Evidence from several directions suggests that any singularity arising to the future of generic (not infinitely finely tuned) non-singular initial conditions, may indeed always be hidden behind a black hole event horizon [4, 5]. However, we are far from having a definitive proof.

Much of the evidence in favor of cosmic censorship comes from the failure of attempts to violate it in thought experiments. Given the difficulty of a direct attack on the general question, this strategy continues to offer an attractive approach to the problem. Even in the context of simplifying approximations, such thought experiments can uncover mechanisms tending to uphold censorship, and they can produce scenarios where censorship would be violated if the approximations were valid. In the latter case, they focus our attention on those "dangerous" scenarios and on the limits of validity of the approximation scheme. This type of approach is what will be discussed here[1].

2. "Destroying" a black hole

In general relativity, the spacetime in the vicinity of the black hole is described by the Kerr–Newman (K-N) metric which contains three parameters: the mass M, the spin angular momentum J, and the electric charge Q. The K-N metric describes a black hole as long as the mass is sufficiently large compared to a combination of the charge and angular momentum, $M^2 \geq a^2 + Q^2$, where $a = J/M$. (We adopt units with Newton's constant G and the speed of light c both set equal to unity.) The case where $M^2 = a^2 + Q^2$ is called an extremal black hole, while for $M^2 < a^2 + Q^2$ there is no event horizon and the K-N metric actually describes a naked singularity. Therefore, it would naively seem that to create a naked singularity all one need do is to start with a black hole and toss in matter with enough angular momentum or charge so as to drive its parameters beyond the extremal limit, leaving it no option but to expose the singularity.

On further thought there are some subtleties involved in such a scenario:

(i) The K-N metric describes stationary configurations, so the proposed strategy can work as stated only if, after having absorbed the matter, the system settles down to a stationary configuration containing all the mass, angular momentum and charge, i.e. without having shed the excess angular momentum or charge in the settling down process. Such an outcome is by no means guaranteed. Indeed, it seems rather unlikely, given the evidence that the trans-extremal K-N metric is unstable [8, 9]. If such instability occurs (the uncertainty lies in the proper boundary conditions at the singularity), the system may well shed sufficient angular momentum and/or charge, or not settle down to such a metric, and at present nobody knows what it would do. What this means is that to demonstrate the creation of a naked singularity one would have to follow the evolution further than the initial "absorption" of the extra matter. So the possibility of initially overspinning or overcharging an initial black hole configuration can only be taken as an indication that cosmic censorship *might* fail.

(ii) The notion of "exposing" the singularity may be inappropriate, since the singularity inside a perturbed charged or rotating stationary black hole cannot send signals to any point, even those interior to the horizon. (Here we assume that the Cauchy horizon inside the black hole is indeed unstable, as evidence indicates [10].) That is, it has no nonsingular future. Hence it is not so clear that, if a horizon could be "destroyed", the result would be to expose the singularity that would have been there had the horizon *not* been destroyed. It might produce a wholly different singularity. Nevertheless, either way the process would violate cosmic censorship.

Setting these issues aside for the present, we focus here just on the question of whether a black hole can initially absorb sufficient angular momentum or charge to send its parameters over the extremal limit.

Given the non-linear nature of Einstein's equations, following the evolution of a black hole exactly is a very difficult problem that presumably requires numerical solution of the Einstein

[1] This paper is based partially on a talk given by T.P.S. at the 1st Mediterranean conference on Classical and Quantum gravity following the lines of Ref. [6] and partially on an essay prepared for the "FQXi essay contest: What is Ultimately Possible in Physics?" [7].

equation. Therefore, most studies of cosmic censorship to date either imposed symmetry conditions or have been carried out in the simple framework of a test-body moving on the black hole spacetime. The test-body approximation imposes the conditions

$$\delta E \ll M, \quad \delta J \ll M^2, \quad \delta Q \ll M, \tag{1}$$

where δE, δJ, and δQ denote the energy, angular momentum and charge of the body. These conditions seem sufficient to ensure that the influence of the test body can be treated as a small perturbation on the background spacetime. (In particular, we need not assume that $\delta J \ll J$ or $\delta Q \ll Q$, since addition of angular momentum to a nonspinning black hole, or charge to a neutral black hole, can perfectly well be a small perturbation.) However, they certainly do not guarantee that the effects due to the gravity of the body, such as gravitational radiation and self-force, can be neglected when studying the motion of the body. Issues that arise when including these effects are are discussed briefly at the end of this paper.

Provided the body can be tossed into the black hole, the final composite object would have mass $M + \delta E$, angular momentum $J + \delta J$ and charge $Q + \delta Q$. In order for the K-N metric with these parameters to be a naked singularity they would have to satisfy the inequality

$$(M + \delta E)^2 < \left(\frac{J + \delta J}{M + \delta E}\right)^2 + (Q + \delta Q)^2. \tag{2}$$

Various special cases of such test body experiments have been considered in the literature. Wald focused on an exactly extremal black hole and showed that it cannot be overcharged or over-spun using a particle with charge and/or orbital angular momentum. He also showed that an extremal neutral rotating black hole cannot be overspun using a particle with spin angular momentum falling along the spin axis [11]. However, de Felice and Yu have shown that an extremal charged black hole *can* be sent over the extremal limit by accretion of a neutral spinning test body [14]. If one starts with a non-extremal black hole, the results differ from what Wald found for the extremal case. In particular, Hubeny showed that one can overcharge a *near*-extremal Reisser–Nördstom black hole by tossing in a test body [12], and Hod showed that a near-extremal rotating black hole can be overspun in the limiting case where a particle carrying angular momentum is lowered all the way to the horizon of a black hole and dropped from there [13]. We will return to these results later on and comment on them more extensively. A case we will not consider further is that of dyonic black holes which carry both electric and magnetic charge [15, 16]. We will also not consider here test field experiments via wave scattering, classical or quantum, as discussed for example in [17] and references therein.

Astrophysical black holes tend to increase their angular momentum by accreting matter, whereas they tend to decrease their charge by attracting opposite and repelling same charges. For a point of principle, one may ignore these facts, but nevertheless it would be much more provocative and promising if arguments showed that a naked singularity could be created using only angular momentum, since that might then actually occur in nature. We therefore focus mostly on the case were the both the black hole and the test body have only angular momentum and no charge.

We shall demonstrate that in the test body approximation a trans-extremal condition can be attained, even when taking into account constraints on the size and structure of the body. The limitations of the test-body approximation will then be addressed.

2.1. Over-spinning a neutral black hole

With $Q = \delta Q = 0$, the inequality in eq. (2) takes the form

$$J + \delta J > (M + \delta E)^2. \tag{3}$$

This yields a lower bound on the required angular momentum carried by the body, for a given energy δE:

$$\delta J > \delta J_{min} = (M^2 - J) + 2M\delta E + \delta E^2. \tag{4}$$

Since we are assuming $\delta E \ll M$, it might seem that the δE^2 term may as well just be neglected at this stage. However, as we will see shortly, the presence of that term imposes an upper bound on δE and δJ and, therefore, should not be neglected.

We can already extract a useful piece of information from eq. (4). Dividing both sides by δE^2, and observing that each term on the right hand side is positive and therefore should by itself be smaller that the left hand side, we get

$$\delta J/\delta E^2 > 2M/\delta E \gg 1 \tag{5}$$

(where the last inequality follows from (1). If δE is comparable to the rest mass of the body (it can be much less if the body is deeply bound by the gravitational field of the black hole), and if δJ comes from spin (rather than orbital angular momentum), this would imply that the body has angular momentum far over the extremal ratio. In that case the body could not be a black hole. This does not mean that it would have to be a naked singularity itself, as there is no a priori upper limit to this ratio for bodies other than black holes. Stars for instance can easily have ratios much larger than 1.

The requirement that the composite object be a naked singularity gave us the lower bound (4) on the angular momentum of the body. An upper bound is obtained from the requirement that the body does indeed cross the horizon. One can use the equations of motion for the body in order to derive the bound [11]. These are the Papapetrou equation if the body's angular momentum includes spin, and the geodesic equation if it is purely orbital angular momentum. But a simpler and more transparent method is to just consider the flux of energy and angular momentum into the black hole when the body falls across the horizon. The requirement that the energy momentum tensor satisfies the null energy condition (which follows for example if the energy density is positive in all local reference frames) yields (see for example [6])

$$\delta E \geq \Omega_H \delta J, \tag{6}$$

where $\Omega_H = a/2Mr_+$ is the angular velocity of the horizon and $r_+ = M + (M^2 - a^2)^{1/2}$ is the horizon radius in Boyer-Lindquist coordinates. This condition can be written as

$$\delta J \leq \delta J_{max} = \frac{2Mr_+}{a}\delta E. \tag{7}$$

It guarantees that the body can fall across the horizon starting from *some* point outside, although in general the body is in a bound orbit that does not come from spatial infinity.

We now have both an upper and a lower bound for the angular momentum of the body, for a given energy. As long as $\delta J_{min} < \delta J_{max}$ for some δE, there will be values of δJ and δE satisfying both inequalities (4) and (7).

First let us suppose the black hole starts exactly extremal, i.e. $J = M^2$. Then $a = M = r_+$, so one has $\delta J_{min} = 2M\delta E + \delta E^2$ and $\delta J_{max} = 2M\delta E$. This implies that $\delta J_{min} > \delta J_{max}$ so it is impossible to over-spin the black hole. Thus we recover the result of Wald [11] mentioned earlier. (The analysis here is significantly simpler than that of Ref. [11].)

The physical interpretation of this result is the following: In the case of the spinning body, the spin-spin interaction with the spin of the black hole is sufficiently repulsive to prevent the body from falling in if it would have overspun the black hole. In the orbital angular momentum case, If the body has the angular momentum required to overspin then the impact parameter of the body is too large for it to hit the horizon.

Journal of Physics: Conference Series **222** (2010) 012041 doi:10.1088/1742-6596/222/1/012041

In the sub-extremal case, however, the inequalities *can* be satisfied, as was shown in [6]. As mentioned earlier, the limiting case where the body is dropped from a point on the horizon had been considered previously by Hod [13] (it corresponds to $\delta J = \delta J_{max}$). To understand the range of overspinning parameters, it is helpful to visualize the inequalities graphically. If δJ_{max} and δJ_{min} are plotted vs. δE, the former is a straight line through the origin, with slope $2Mr_+/a$, while the latter is a parabola with positive intercept, slope $2M$ at the intercept, and curved upwards. For a sub-extremal black hole we have $r_+ > M > a$, so the slope of the δJ_{max} line is greater than the initial slope of the δJ_{min} parabola. Some algebra reveals that the parabola always intersects the straight line in two points. The allowed values of δE and δJ are those in the compact region above the parabola and on or below the straight line. Note that if the δE^2 term is neglected in (4), the parabola is replaced by a straight line, and the allowed region becomes an infinite wedge, with no upper bound.

To determine whether the overspinning can be accomplished with a small perturbation satisfying (1) we can expand in the dimensionless quantity ϵ defined by

$$J/M^2 = a/M = 1 - 2\epsilon^2. \tag{8}$$

(Hubeny [12] used the same parameter to analyze the charged case, see below.) The parameter ϵ measures how close to extremality the black hole is to begin with, . and we must have $\epsilon \ll 1$ if the overspinning perturbation is to be small. It is now useful to adopt units with $M = 1$, to keep the expressions simpler. Then we have, from (4) and (7),

$$\delta J_{min} = 2\epsilon^2 + 2\delta E + \delta E^2 \tag{9}$$

$$\delta J_{max} = (2 + 4\epsilon)\delta E, \tag{10}$$

where terms of order $O(\epsilon^2 \delta E)$ have been dropped in (10). The allowed range of δE lies where the difference

$$\delta J_{max} - \delta J_{min} = -2\epsilon^2 + 4\epsilon\delta E - \delta E^2 \tag{11}$$

is positive, i.e.

$$(2 - \sqrt{2})\epsilon < \delta E < (2 + \sqrt{2})\epsilon. \tag{12}$$

In particular, δE must be of order ϵ. For a given δE, the allowed values of δJ are near $2\delta E$, so we must have

$$\delta J \sim \delta E. \tag{13}$$

We conclude that if $\epsilon \ll 1$ the overspinning values of δE and δJ can indeed be consistent with the perturbative requirement.

Note that the *width* (11) of the allowed range of δJ is only of order $\epsilon^2 \ll \epsilon$. Note also that $a - 1 = 2\epsilon^2$ is parametrically smaller than ϵ. For example, if $\epsilon = 10^{-2}$, then the initial black hole must have $a = 0.9998$. For a thought experiment, we can imagine even smaller values of ϵ.

2.2. Over-charging or over-spinning a charged black hole

Let us return to similar conclusions reached in the cases mentioned earlier, where different assumptions for the quantities characterizing the black hole were made. Hubeny considered the case of adding charge to a charged black hole ($J = \delta J = 0$) [12]. In analogy to the spinning case, in the charged case the two constraints are

$$\delta Q > M - Q + \delta E, \tag{14}$$

$$\delta Q \leq \frac{r_+}{Q}\delta E, \tag{15}$$

Journal of Physics: Conference Series **222** (2010) 012041 doi:10.1088/1742-6596/222/1/012041

where now $r_+ = M + \sqrt{M^2 - Q^2}$. If the black hole starts out extremal, $M = Q = r_+$, then overcharging is impossible. However if $M > Q$, then $r_+ > Q$, and one easily sees by visualizing the inequalities graphically (now they are both straight lines) that there is an infinite range of solutions, once δE is greater than a certain minimum value.

de Felice and Yu considered the case of adding angular momentum to an extremally charged black hole ($Q = M$, $J = \delta Q = 0$) [14]. . In this case the minimum δJ to overspin is given by

$$\delta J > (M + \delta E)\sqrt{2M\delta E + \delta E^2}, \tag{16}$$

and there is no maximum δJ, since the only requirement for the body to fall across the horizon is $\delta E \geq 0$, which does not involve δJ. Note that to lowest order in $\delta E/M$ the minimum overspinning δJ is $\delta J_{\min} = M\sqrt{2M\delta E}$.

3. Size and structure requirements

So far we have characterized the body only by its energy δE, angular momentum δJ and charge δQ. We did not consider restrictions placed on its size and structure: It should be sufficiently small to justify use of a test particle approximation, and it should be composed of matter having positive energy density and no unphysically large stresses. The next step is to take into account these requirements.

3.1. Adding spin angular momentum to a neutral black hole

We begin with the case of a spinning test body. For simplicity we assume that the body is dropped along the rotational axis of the black hole. We first consider the case where the body has $\delta E \sim m$, and is not spinning relativistically, so its spin angular momentum is given by $\delta J \sim mvR \sim \delta E\, vR$, where v is the surface velocity and R is the equatorial radius. The condition $v < 1$ then implies $R > \delta J/\delta E$. We saw above that the ratio $\delta J/\delta E$ must be of order unity (13), that is of order M. In this case the body must be larger than the black hole, so it simply will not "fit" in the transverse direction, and in any case treating it as a point particle with spin would be unjustified, since that rests on the smallness of the size of the body compared to the ambient radius of curvature. Moreover, one can show [6] that the radial tidal stress required to hold the body together would be larger than the energy density, violating energy conditions. It cannot help to allow ultra-relativistic tangential velocity: as a simple Newtonian estimate shows, that would require unphysical stresses holding the body together. The conclusion is that it is impossible to over-spin the black hole if the body's energy is close to its rest mass, $\delta E \sim m$.

Since the angular momentum involves the rest mass m, not the energy δE, it might be possible to achieve a large enough δJ with a small enough size R, without requiring unphysical matter, by dropping the body from a position where it is deeply bound, $\delta E \ll m$. This might be achieved by slowly lowering the body on a tether, down to the near the black hole horizon, before dropping it in. Now we reconsider whether the size restrictions can be met in this setting.

We begin with the restrictions on the rest mass m. If m is much greater than δE, then the test body approximation requires that we impose not only $\delta E \ll 1\,(= M)$ (1), but also $m \ll 1$. There is also a lower bound on m, coming from an upper bound on R: the angular momentum is $\delta J \sim mvR$, hence (restricting to nonrelativistic spin $v < 1$ as required by the previously mentioned result) $R > \delta J/m \simeq 4\epsilon/m$. The requirement $R \ll 1$ then yields $m \gg \epsilon$. The mass and size must therefore fall within the ranges

$$\epsilon \ll m \ll 1, \qquad 4\epsilon/m \lesssim R \ll 1. \tag{17}$$

To these conditions we must add the requirement $R \gtrsim m$ that the body is not a black hole, as explained above.

Journal of Physics: Conference Series **222** (2010) 012041 doi:10.1088/1742-6596/222/1/012041

The inequality (6) guarantees that the body can cross the horizon with the chosen values of energy and angular momentum, but since the deeply bound drop point lies at a finite distance from the horizon it is necessary to check that (a) the spinning body would actually fall into the black hole rather than being repelled, and (b) it is possible to choose the polar radius of the body $R_{\rm polar}$ to be smaller than the proper distance d from the horizon to the drop point

$$R_{\rm polar} < d, \tag{18}$$

so that it can fully "fit" outside the black hole and be localized at the drop point. Now it turns out [6] that, in order to fall in, the maximum value that d can have, given the allowed values of δE and δJ, is

$$d_{max} \simeq \epsilon/m. \tag{19}$$

Thus we arrive at the bound

$$R_{\rm polar} < \epsilon/m. \tag{20}$$

Together with (17), this means that the body must be at least somewhat oblate, $R_{\rm polar} \lesssim R/4$.[2] We conclude that the body can be large enough to possess the requisite angular momentum and also have a physically acceptable stress and fit outside the black hole at the drop point[3].

3.2. Adding orbital angular momentum to a neutral black hole

We turn now to the case of orbital angular momentum in the equatorial plane. Here the issue is that in order to have the required values of δE and δJ, the body might have to be in a bound orbit, which would have a turning point at a maximum radius. In that case we would need to require that the body be small enough to fit outside the horizon at this radius. Since the body can be no smaller than a black hole with the same rest mass, it is not clear in advance whether this requirement could be met. However, as has been shown in [6] this size constraint is not an issue, since in fact there are suitable orbits that come in from infinity with no turning point. This can be shown numerically, but also analytically by the use of the effective potential governing the motion of a test particle in a Kerr spacetime (Kerr-Newman with no charge).

3.3. Cases involving a charged black hole

Let us briefly consider the size and structure requirements when attempting to overcharge or overspin a charged black hole. In the case with no angular momentum Hubeny [12] showed that the body can have the required charge and mass, with low internal stresses and size much smaller than the black hole. Also, she demonstrated that there are charged test particle trajectories that fall from infinity into the black hole. Therefore, much like the orbital angular momentum case, size constraints are not an issue. On the contrary, for spinning particles dropped radially with radial spin into an extremal charged black hole, de Felice and Yu [14] found that the test body must be bound very close to the horizon. The same is true for a particle carrying orbital angular momentum but no spin, as we now show. The radial motion is governed by the equation $\dot{r}^2 + (1 - 1/r)^2(1 + \tilde{L}^2/r^2) = \tilde{E}^2$. Here $\dot{r}$ is the derivative of the Reissner-Nordstrom radial coordinate with respect to the particle proper time, and $\tilde{E} = (\delta E)/m$ and $\tilde{L} = (\delta J)/m$ are the energy and and angular momentum divided by the particle rest mass m, and we have again set $M = 1$. As mentioned after (16), the overspinning requirement is $\delta J/\delta E \gtrsim \sqrt{2/\delta E} \gg 1$, hence

[2] In [6] the possibility that $R_{\rm polar} \neq R_{\rm equator}$ was overlooked, so it was erroneously concluded that no value of R could meet all requirements.

[3] de Felice and Yu made a similar analysis for the case of dropping a spinning body into an extremal charged black hole, but they computed the coordinate radius corresponding to $d_{\rm max}$, rather than the proper distance. In the extremal case, the proper distance to the horizon is infinite in the direction orthogonal to the Killing vector, so there is apparently no requirement that the body be disk-shaped, contrary to what was stated in [14].

$\tilde{L}/\tilde{E} \gg 1$. For an unbound orbit $\tilde{E} \geq 1$, so we infer that $\tilde{L} \gg 1$. There are therefore two turning points where $\dot{r} = 0$. To fall into the hole the particle must lie inside the inner turning point, which lies at a radius r_{inner} very close to the horizon, where $r_{inner} - 1 \simeq \tilde{E}/\tilde{L} \lesssim \sqrt{\delta E/2} \ll 1$. However, although the radial coordinate must be very close to that of the horizon, the proper distance to the horizon, measured in the static frame, is infinite for an exactly extremal black hole. Hence, in both the spin and orbital cases, no further size constraints appear to be imposed by the location of the turning point (see also footnote 3).

4. Beyond the test body approximation

Our considerations thus far have been based on an approximation which neglects loss of energy and angular momentum in gravitational radiation and does not take into account self-force effects. Recall that our purely kinematic considerations above yielded a finely tuned relation between the energy and angular momentum of the dropped body for over-spinning to occur. Both quantities have to be of order ϵ in magnitude, but the allowed window for angular momentum, given the energy, is only of order ϵ^2. In the light of this delicate balance, it is certainly possible that, although small, gravitational radiation and/or self-force effects may always manage to preclude the over-spinning.

Given that the inequalities (4) and (7) need only hold on the horizon, one could imagine that the loss of energy and angular momentum in gravitational radiation might be compensated by adjusting the initial conditions. In the case of an axially symmetric spinning body falling along the black hole spin axis there is no radiation of angular momentum, so it should be possible to simply compensate for the energy radiated. To determine whether compensation is actually possible in the orbital case requires further investigation. Perhaps more worrisome than radiation are the self force effects. Indeed Hubeny found strong indications that for the charged case the electromagnetic self-force might prevent the overcharging, although her calculations were not conclusive [12].

Another distinct effect that might prevent the creation of a naked singularity is the tides raised on the black hole horizon by the falling body. These would be irrelevant for the orbital angular momentum case since the body falls in from spatial infinity. In the spinning body case, however, in which the body is lowered to the horizon and then dropped, the tidal bulge of the horizon might perhaps make it impossible for the body to start out in the exterior while still satisfying the size constraints.

Given the existing evidence for cosmic censorship, it seems indeed likely that neglected gravitational effects will come to its rescue. The examples discussed here suggest dynamical regimes in which it may be interesting to study these neglected effects.

Acknowledgements

This research was supported in part by the NSF under Grant Nos. PHY-0601800 and PHY-0903572, and by STFC.

References

[1] Penrose R 1965 "Gravitational collapse and space-time singularities," *Phys. Rev. Lett.* **14** 57
[2] Hawking S W and Penrose R 1970 "The Singularities of gravitational collapse and cosmology," *Proc. Roy. Soc. Lond. A* **314** 529
[3] Penrose R 1969 "Gravitational collapse: The role of general relativity," *Riv. Nuovo Cim.* **1** 252 [2002 *Gen. Rel. Grav.* **34**, 1141]
[4] Wald R M 1997 *Preprint* "Gravitational collapse and cosmic censorship," gr-qc/9710068
[5] Penrose R 1999 "The Question Of Cosmic Censorship," *J. Astrophys. Astron.* **20** 233
[6] Jacobson T and Sotiriou T S 2009 "Overspinning a Black Hole with a Test Body," *Phys. Rev. Lett.* **103** 141101 [Erratum-ibid. **103**, 209903 (2009)]

[7] Jacobson T and Sotiriou T S 2009 "Might black holes reveal their inner secrets?" http://fqxi.org/community/forum/topic/594
[8] Dotti G, Gleiser R J and Pullin J 2007 "Instability of charged and rotating naked singularities," *Phys. Lett. B* **644** 289
[9] Dotti G, Gleiser R J, Pullin J, Ranea-Sandoval I F and Vucetich H, "Instabilities of naked singularities and black hole interiors in General Relativity," *Int. J. Mod. Phys. A* **24** 1578
[10] Poisson E and Israel W 1990 "Internal structure of black holes," *Phys. Rev. D* **41** 1796; Ori A 1992 "Structure Of The Singularity Inside A Realistic Rotating Black Hole," *Phys. Rev. Lett.* **68** 2117
[11] Wald R M 1974 "Gedanken experiments to destroy a black hole," *Ann. Phys.* **82** 548
[12] Hubeny V E 1999 "Overcharging a Black Hole and Cosmic Censorship," *Phys. Rev. D* **59** 064013
[13] Hod S 2002 "Cosmic Censorship, Area Theorem, and Self-Energy of Particles," *Phys. Rev. D* **66** 024016
[14] de Felice F and Yu Y Q 2001 "Turning a black hole into a naked singularity," *Class. Quant. Grav.* **18** 1235
[15] Hiscock W A 1981 "Magnetic charge, black holes, and cosmic censorship," *Ann. Phys.* **131** 245
[16] Semiz I 1990 "Dyon black holes do not violate cosmic censorship," *Class. Quant. Grav.* **7** 353
[17] Matsas G E A, Richartz M, Saa A, da Silva A R R and Vanzella DAT 2009 "Can quantum mechanics fool the cosmic censor?" *Phys. Rev. D* **79** 101502

Journal of Physics: Conference Series **222** (2010) 012042 doi:10.1088/1742-6596/222/1/012042

Non-uniqueness of the Dirac theory in a curved spacetime

Mayeul Arminjon[1] **and Frank Reifler**[2]

[1] CNRS (Section of Theoretical Physics), Laboratory "Soils, Solids, Structures, Risks", Grenoble, France

[2] Lockheed Martin Corporation, MS2 Division, Moorestown, New Jersey, USA

E-mail: `arminjon@hmg.inpg.fr`

Abstract. We summarize a recent work on the subject title. The Dirac equation in a curved spacetime depends on a field of coefficients (essentially the Dirac matrices), for which a continuum of different choices are possible. We study the conditions under which a change of the coefficient fields leads to an equivalent Hamiltonian operator H, or to an equivalent energy operator E. In this paper, we focus on the standard version of the gravitational Dirac equation, but the non-uniqueness applies also to our alternative versions. We find that the changes which lead to an equivalent operator H, or respectively to an equivalent operator E, are determined by initial data, or respectively have to make some point-dependent antihermitian matrix vanish. Thus, the vast majority of the possible coefficient changes lead neither to an equivalent operator H, nor to an equivalent operator E, whence a lack of uniqueness. We show that even the Dirac energy spectrum is not unique.

1. Introduction

This paper summarizes a recent work [1] on the subject title. Thus, rather than to give detailed arguments and to present all relevant results, our aim here is to make the origin of this surprising result more easily visible.

Quantum effects in the classical gravitational field *are observed,* e.g. on neutrons [2, 3, 4], which are spin $\frac{1}{2}$ particles. This motivates work on the curved spacetime Dirac equation. The standard version of the latter (see e.g. Refs. [5, 6]) is due to Fock and to Weyl, and will be referred to as *the Dirac-Fock-Weyl* (**DFW**) *equation.* On the other hand, two alternative Dirac equations in a curved spacetime were derived recently [7], by applying directly the "classical-quantum correspondence", i.e., the correspondence between a classical Hamiltonian and a quantum wave operator. Thus, together with the standard (DFW) version, we have three distinct versions of the Dirac equation in a curved spacetime. The basic quantum mechanics was studied for each of those three equations [8] (see Ref. [9] for a summary):

- the Dirac probability current was defined for any possible field (γ^μ) of Dirac matrices. We derived the condition which has to be satisfied by (γ^μ) in order that the current conservation be true for any solution of the Dirac equation (for DFW, this condition is always satisfied);
- the precise form of the relevant scalar product was shown to be imposed by the axioms of quantum mechanics (it turns out that, for DFW, this is the form used by Leclerc [10]);

- the Hamiltonian operator was written and the condition of its hermiticity w.r.t. the relevant scalar product was derived in a general setting (for DFW, this condition coincides, in the particular case envisaged by Leclerc [10], with that derived by him).

That foregoing work showed, in particular, that the hermiticity of the Hamiltonian is not stable under all admissible changes of the field of Dirac matrices. This implies that there is a non-uniqueness problem for the curved-spacetime Dirac equation.

The *aim of the present work [1]*, therefore, was to study the (non-)uniqueness of the Hamiltonian and energy operators, including the (non-)uniqueness of the energy spectrum. The qualitative conclusions are the same for the three versions, i.e., the non-uniqueness applies to the alternative equations of Ref. [7], too [1]. Thus, finding this non-uniqueness was disappointing for us as well. In this paper, we will focus on the *standard equation* (DFW) and the non-uniqueness results for it, because these results could be more directly of interest to many physicists.

2. Three Dirac equations in a curved spacetime

The three versions of the gravitational Dirac equation have the same form:

$$\gamma^\mu D_\mu \psi = -im\psi, \tag{1}$$

where $\gamma^\mu = \gamma^\mu(X)$ ($\mu = 0, ..., 3$) is a field of 4×4 complex matrices, depending on the point $X \in \mathrm{V}$ in the curved spacetime $(\mathrm{V}, g_{\mu\nu})$, such that

$$\gamma^\mu \gamma^\nu + \gamma^\nu \gamma^\mu = 2g^{\mu\nu}\, \mathbf{1}_4, \quad \mu, \nu \in \{0, ..., 3\} \tag{2}$$

[here $(g^{\mu\nu})(X)$ is the inverse of the matrix $(g_{\mu\nu})(X)$ and $\mathbf{1}_4 \equiv \mathrm{diag}(1, 1, 1, 1)$];
where ψ is a *bispinor* field for the standard equation (DFW), but is a *4-vector* field for the two alternative equations, based on the *tensor representation of the Dirac fields* (**TRD**);
and where D_μ is a covariant derivative, associated with a specific *connection*. For DFW, this is the "spin connection", which depends on the (γ^μ) field [5, 6].

3. Definition of the field of Dirac matrices

For DFW, one defines

$$\gamma^\mu(X) = a^\mu{}_\alpha(X)\, \gamma^{\sharp\alpha}, \tag{3}$$

with $u_\alpha = a^\mu{}_\alpha(X)\, \partial_\mu$ an orthonormal tetrad field (∂_μ is the natural basis associated with the coordinates x^μ) and $(\gamma^{\sharp\alpha})$ a set of "flat" Dirac matrices, i.e., a constant solution of the anticommutation relation (2) with $(g^{\mu\nu}) = (\eta^{\mu\nu}) \equiv \mathrm{diag}(1, -1, -1, -1)$. One should be able to use *any* possible choice of $(\gamma^{\sharp\alpha})$. One should study the influence of both choices: $(\gamma^{\sharp\alpha})$ and (u_α).

For TRD, a tetrad field can also be used, though other possibilities also exist, e.g. parallel transport [7].

In order to be able to use any possible field (γ^μ) of Dirac matrices, we must have recourse to the *hermitizing matrix* A of Bargmann and Pauli [12]. This is a 4×4 complex matrix such that

$$A^\dagger = A, \qquad (A\gamma^\mu)^\dagger = A\gamma^\mu \quad \mu = 0, ..., 3, \tag{4}$$

with $M^\dagger \equiv M^{*T}$ = Hermitian conjugate of matrix M. We proved [11] the existence of A, and also that of B: the latter is a positive-definite hermitizing matrix for the α^μ matrices which are defined by Eq. (10) below. In general, $A = A(X)$ is also a field. However, when we use the definition from a tetrad field (3), we may take [8]

$$A \equiv A^\sharp, \tag{5}$$

where $A^\sharp$ is a (constant) hermitizing matrix for the constant set $(\gamma^{\sharp\alpha})$. Moreover, in practice for DFW, one considers only sets $(\gamma^{\sharp\alpha})$ for which $\gamma^{\sharp 0}$ is a hermitizing matrix. (This includes the usual sets: Dirac's, "chiral", Majorana's, and any set deduced therefrom by a unitary transformation [13].) Thus, in practice,

$$\underline{A = \gamma^{\sharp 0} \text{ for DFW.}} \tag{6}$$

4. Local similarities

In a curved spacetime $(\mathrm{V}, g_{\mu\nu})$, the Dirac matrices γ^μ and the hermitizing matrix A are fields: they depend on $X \in \mathrm{V}$. If one changes from one field (γ^μ) to another one $(\tilde{\gamma}^\mu)$, also satisfying the anticommutation relation (2) and with the *same* metric $g_{\mu\nu}$, then the new field obtains by a *local similarity transformation* [12], which is unique up to a complex factor [11]:

$$\exists S = S(X) \in \mathsf{GL}(4, \mathsf{C}) : \qquad \tilde{\gamma}^\mu(X) = S^{-1}\gamma^\mu(X)S, \quad \mu = 0, ..., 3. \tag{7}$$

For the standard Dirac equation (DFW), the similarities are restricted to the spin group $\mathsf{Spin}(1,3)$, i.e., they result from a change of the tetrad field by a (proper) local Lorentz transform $L(X) \in \mathsf{SO}(1,3)$, depending arbitrarily on $X \in \mathrm{V}$ [5]. The corresponding local similarity transformation $S(X) \in \mathsf{Spin}(1,3)$ is deduced [8] from $L(X)$ through the spinor representation (defined only up to a sign), $L \mapsto S = \pm \mathsf{S}(L)$.

5. The general Dirac Hamiltonian

Rewriting the Dirac equation (1) in the "Schrödinger" form:

$$i\frac{\partial \psi}{\partial t} = \mathrm{H}\psi, \qquad (t \equiv x^0), \tag{8}$$

gives the Hamiltonian operator:

$$\mathrm{H} \equiv m\alpha^0 - i\alpha^j D_j - i(D_0 - \partial_0), \tag{9}$$

with

$$\alpha^0 \equiv \gamma^0/g^{00}, \qquad \alpha^j \equiv \gamma^0\gamma^j/g^{00} \quad (j = 1, 2, 3). \tag{10}$$

One should realize that, in a given spacetime $(\mathrm{V}, g_{\mu\nu})$ and with a given field of Dirac matrices (γ^μ), this operator still depends on the coordinate system, or, more exactly, on the *reference frame* [8]—the latter being understood here as an equivalence class F of local coordinate systems (charts) on the spacetime V, modulo the purely spatial transformations

$$x'^0 = x^0, \ x'^j = f^j((x^k)). \tag{11}$$

In the present work, we fix the reference frame F and even the chart (the latter being, however, an arbitrary one), and we study the dependence of the Hamiltonian and energy operators on the field (γ^μ).

6. Scalar product and equivalent operators

The Hilbert scalar product is fixed by the following result [8]:

Theorem. A necessary *condition for the scalar product of time-independent wave functions to be time independent and for the Hamiltonian* H *to be a Hermitian operator, is that the scalar product should be*

$$(\psi \mid \varphi) \equiv \int \psi^\dagger A \gamma^0 \varphi \ \sqrt{-g}\ \mathrm{d}^3\mathbf{x}. \tag{12}$$

For DFW, we have in practice $A = \gamma^{\sharp 0}$, Eq. (6). Then (12) is the scalar product used by Leclerc [10].

With each of any two possible "coefficient fields": (γ^μ, A) and $(\tilde{\gamma}^\mu, \tilde{A})$, corresponds thus a unique scalar product. [For DFW, we have always $\tilde{A} = A = A^\sharp$, Eq. (5), so the coefficient fields reduce to the field of Dirac matrices γ^μ.] These two scalar products are *isometrically equivalent* through the mapping $\psi \mapsto \tilde{\psi} \equiv S^{-1}\psi$: one shows easily [1] that

$$(\tilde{\psi} \,\tilde{|}\, \tilde{\varphi}) = (\psi \mid \varphi). \tag{13}$$

Moreover, H is fully determined by the set of the products $(\mathrm{H}\psi \mid \varphi)$, for $\psi, \varphi \in \mathcal{D} \equiv \mathrm{Dom(H)}$. The same is true for the Hamiltonian $\tilde{\mathrm{H}}$ after the similarity (7), $\tilde{\mathrm{H}}$ being obtained by substituting $\tilde{\gamma}^\mu$ for γ^μ in Eq. (10), and by using the covariant derivatives $\tilde{D}_\mu \equiv \partial_\mu + \tilde{\Gamma}_\mu$ in (9), with $\tilde{\Gamma}_\mu$ the new connection matrices [see Eq. (15) below]. It follows [1] that, in order that H and $\tilde{\mathrm{H}}$ be equivalent operators, it is necessary and sufficient that

$$\tilde{\mathrm{H}} = S^{-1}\,\mathrm{H}\,S, \tag{14}$$

where S is the relevant similarity in Eq. (7). This, of course, is no surprise. Of course also, the same condition defines the equivalence of the energy operators E and $\tilde{\mathrm{E}}$, E being defined by Eq. (17) below.

7. Invariance condition of the Hamiltonian under a local similarity (DFW)

When does a local similarity $S(X)$, applied to the field of Dirac matrices (γ^μ), leave the Hamiltonian operator H [equation (9)] invariant? I.e., when do we have Eq. (14)?

The Hamiltonian (9) involves the connection matrices, $\Gamma_\mu \equiv D_\mu - \partial_\mu$. For DFW, these matrices change after a similarity according to [6]:

$$\tilde{\Gamma}_\mu = S^{-1}\Gamma_\mu S + S^{-1}(\partial_\mu S). \tag{15}$$

A straightforward calculation shows then that, for DFW, we have (14) iff $S(X)$ is time-independent,

$$\partial_0 S = 0. \tag{16}$$

But, in the general case, we have $g_{\mu\nu,0} \neq 0$, hence any possible field (γ^μ) depends on $t \equiv x^0$. Thus, there is no way of finding a class of fields (γ^μ) exchanging with $\partial_0 S = 0$. I.e.: *The Dirac Hamiltonian is not unique* [1].

8. Invariance condition of the energy operator (DFW)

When the Hamiltonian H is not Hermitian, one should use the Hermitian part of H:

$$\mathrm{E} = \mathrm{H} + \frac{i}{2\sqrt{-g}} B^{-1}\partial_0\left(\sqrt{-g}\,B\right) = \frac{1}{2}\left(\mathrm{H} + \mathrm{H}^\ddagger\right),\ B \equiv A\gamma^0, \tag{17}$$

where the Hermitian conjugate operator $\mathrm{H}^\ddagger$ is w.r.t. the unique scalar product (12). This Hermitian-symmetrized operator E coincides, in the particular case envisaged by Leclerc [10], with what is called the *energy operator,* which is derived from the field Lagrangian. In the general case (thus for TRD and for DFW as well, and with any possible (γ^μ) field), the Dirac equation (1) also derives from a Lagrangian [1]. We can show generally that the expected value of the energy operator E equals the classical field energy as derived from that Lagrangian.

Again a straightforward calculation gives the invariance condition of E (for DFW):

$$B(\partial_0 S)S^{-1} - \left[B(\partial_0 S)S^{-1}\right]^\dagger \equiv 2\left[B(\partial_0 S)S^{-1}\right]^a = 0. \tag{18}$$

Only very particular local similarities $S(X)$ do verify (18). Thus, there is a serious non-uniqueness problem for DFW (and for the alternative, TRD equations as well). Could even the *spectrum* of E be non-unique? In what follows, we investigate this question.

9. Explicit expression of the energy operator (DFW)

The general expression of the change of E after a local similarity $S(X)$ is easily found to be [1]:

$$\delta\mathrm{E} \equiv S\tilde{\mathrm{E}}S^{-1} - \mathrm{E} = -iB^{-1}\left[B(\partial_0 S)S^{-1}\right]^a. \tag{19}$$

In very general coordinates, the matrix $(a^\mu{}_\alpha)$ of that tetrad which defines the starting (γ^μ) field may be chosen to satisfy $a^0{}_j = 0$ [1]. Then $a^0{}_0 = \sqrt{g^{00}}$ from the orthonormality of the tetrad. Let us take for "flat" matrices $\gamma^{\sharp\alpha}$ any usual Dirac matrices, for which $A = \gamma^{\sharp 0}$ [Eq. (6)]. Thus

$$B \equiv A\gamma^0 = \gamma^{\sharp 0}(a^0{}_0\gamma^{\sharp 0}) = \sqrt{g^{00}}\,\mathbf{1}_4, \tag{20}$$

whence from (19):

$$\delta\mathrm{E} = -i\left[(\partial_0 S)S^{-1}\right]^a. \tag{21}$$

For DFW, $S(X)$ is an arbitrary $\mathsf{Spin}(1,3)$ transformation. Thus, $(\partial_0 S)S^{-1}$ is the generic element of $\mathcal{G}$, the Lie algebra of $\mathsf{Spin}(1,3)$ – whose the matrices

$$s^{\alpha\beta} \equiv \left[\gamma^{\sharp\alpha}, \gamma^{\sharp\beta}\right]\ (\alpha < \beta) \tag{22}$$

make a basis. Hence

$$\delta\mathrm{E} = -i\left[\omega_{\alpha\beta}s^{\alpha\beta}\right]^a, \tag{23}$$

where, depending on the local Lorentz $L(X)$ that defines $S(X) = \mathsf{S}(L(X))$, the six coefficients $\omega_{\alpha\beta} = -\omega_{\beta\alpha}$ depend arbitrarily on $X \in \mathrm{V}$. With usual Dirac matrices verifying (6), we have

$$(s^{\alpha\beta})^\dagger = \gamma^{\sharp 0}\, s^{\beta\alpha}\, \gamma^{\sharp 0}, \tag{24}$$

whence from (23) and the anticommutation formula:

$$\delta\mathrm{E} = -i\sum_{j,k=1}^{3}\omega_{jk}s^{jk}. \tag{25}$$

10. The case with the "chiral" Dirac matrices

If the "flat" Dirac matrices $\gamma^{\sharp\alpha}$ are the "chiral" ones {see e.g. Schulten [14], Eqs. (10.257) and (10.260)}, we get from (25):

$$\delta\mathrm{E} = -i\sum_{j,k=1}^{3}\omega_{jk}s^{jk} = \begin{pmatrix} N & 0 \\ 0 & N \end{pmatrix}, \qquad N \equiv -\frac{1}{2}\vec{\theta}.\vec{\sigma} \tag{26}$$

where $\vec{\theta} \equiv (\theta^k)$ with $\theta^1 \equiv \omega_{23}$ (circular), and where $\vec{\sigma} \equiv (\sigma_k)$ with σ_k the Pauli matrices.

Depending on the three real numbers ω_{jk}, $1 \leq j < k \leq 3$, the matrix N can be *any* Hermitian matrix 2×2 with zero trace {see e.g. Schulten [14], Eq. (5.226)}. Any such matrix has two eigenvalues $\mu \in \mathsf{R}$ and $-\mu$, and has an orthonormal basis of eigenvectors: respectively $u \in \mathsf{C}^2$ for μ, and v for $-\mu$.

11. Non-uniqueness of the energy spectrum (DFW)

A small perturbation of the starting Dirac matrices γ^μ is defined by a similarity close to the identity:

$$S(\varepsilon, X) = I + \varepsilon\,(\delta S)(X) + O(\varepsilon^2). \tag{27}$$

It modifies each eigenvalue of the energy operator E according to the well-known formula

$$\delta\lambda = (\psi \mid \delta \mathrm{E}(\varepsilon)\psi) + O(\varepsilon^2), \tag{28}$$

where ψ is the eigenfunction for the unperturbed state. Inserting δE (26) and the expression (12) of the scalar product into Eq. (28), and decomposing the bispinor into two two-components spinors: $\psi = (\phi, \chi)$, we get:

$$\delta\lambda = \int \psi^\dagger \delta \mathrm{E}\, \psi\ \underline{\sqrt{-g\, g^{00}}\ \mathrm{d}^3\mathbf{x}} = \int (\phi^\dagger N \phi + \chi^\dagger N \chi)\ \underline{\mathrm{dV}}. \tag{29}$$

Let us fix $\mu > 0$ and t. For any spatial position $\mathbf{x} \equiv (x^j)$ in our fixed chart, let us select the 2×2 Hermitian matrix with zero trace, $N = N(\mathbf{x})$, in such a way that $\phi(\mathbf{x})$ be the eigenvector of $N(\mathbf{x})$ for the eigenvalue μ. (The 3-vector $\mathbf{x}$ is the coordinate expression of the position x in the space manifold M associated with the arbitrary given reference frame F [8].) That is:

$$\phi^\dagger N \phi = \mu\ \phi^\dagger \phi. \tag{30}$$

Since N has the eigenvalues μ and $-\mu$, we have then necessarily

$$\chi^\dagger N \chi \geq -\mu\ \chi^\dagger \chi. \tag{31}$$

In Eq. (31), the inequality becomes an equality only if $\chi(\mathbf{x})$ is an eigenvector of $N(\mathbf{x})$ with eigenvalue $-\mu$. In that case, $\phi(\mathbf{x})$ is orthogonal to $\chi(\mathbf{x})$ in C^2, $\chi(\mathbf{x}) \perp \phi(\mathbf{x})$. Returning to Eq. (29), we see that $\delta\lambda > 0$ unless if i) $\chi(\mathbf{x}) \perp \phi(\mathbf{x})$ for almost every spatial position $\mathbf{x}$, *and* ii) $\int \phi^\dagger \phi\ \mathrm{dV} = \int \chi^\dagger \chi\ \mathrm{dV}$. This could occur only in extremely particular situations. In fact, i) alone implies [1] that the probability current is light-like for almost every $\mathbf{x}$, and this *is impossible* if $m > 0$ [15]. This proves the non-uniqueness of the DFW energy spectrum, at least for a massive particle.

12. Conclusion

Three distinct gravitational Dirac equations were envisaged: the standard equation ("Dirac-Fock-Weyl" or DFW), plus two alternative equations [7] based on the tensor representation of the Dirac field (TRD). Using the hermitizing matrix A, they were studied together, to a large extent [8, 1]. Non-uniqueness results have been proved for each of the three versions [1]. In this paper, we summarized the latter work, but focussed on the results for DFW.

The Hamiltonian operator H is not unique: it depends on the admissible choice of the field of Dirac matrices. The same is true for the energy operator E. This is true for DFW and for TRD.

The *spectrum* of E is itself non-unique. All of these results apply already to a *flat* spacetime in a non-inertial frame.

Appendix A. The classical energy and its frame dependence

One might ask if the non-uniqueness of the energy spectrum for a spin $\frac{1}{2}$ particle in a curved spacetime, reported here, might have something to do with the well-known fact that, in general relativity (GR), there is no covariant concept of local energy—already in the classical context (as opposed to the quantum context envisaged in this paper). The latter fact may be formulated by saying that there is no covariant conservation law for the (total) energy-momentum—the energy-momentum of the gravitational field giving rise only to a *pseudo*-tensor. As is well known, the usual "conservation law" for the energy-momentum tensor of matter and non-gravitational fields "does not express the conservation of anything" [16]. However, for a classical *test particle,* in any arbitrary reference frame F, there is a well-defined *Hamiltonian energy* [7, 17, 18], though it depends on F and, in general, on the time:

- Geodesic motion in the Lorentzian manifold $(\mathrm{V}, g_{\mu\nu})$ derives from the ("super-")Hamiltonian over the 8-dimensional phase space associated with V:

$$\tilde{H}[(p_\mu),(x^\nu)] \equiv \frac{1}{2} g^{\mu\nu}((x^\rho))p_\mu p_\nu \qquad (c=1). \tag{A.1}$$

  This is because, in a (pseudo-)Riemannian manifold, geodesic motion is identical to free Hamiltonian motion, i.e., $\tilde{H} \equiv T$ with $T[(p_\mu),(x^\nu)] \equiv \frac{1}{2} g^{\mu\nu} p_\mu p_\nu$ the "kinetic energy": Sect. 45 D in Arnold [19]. Note that $\tilde{H}$ does not depend on the (proper) time τ.

- It follows from that "time"-independence that, in any given coordinate system, we deduce a "normal" Hamiltonian H over the usual, 6-dimensional phase space, by *dimensional reduction* (Sect. 45 B in Ref. [19]):

$$H[(p_j),(x^j), t \equiv x^0] \equiv p_0 \tag{A.2}$$

  is extracted from

$$g^{\mu\nu} p_\mu p_\nu - m^2 = 0. \tag{A.3}$$

  {With the $(-+++)$ signature, we have $H \equiv -p_0$ instead, and the last equation writes $g^{\mu\nu} p_\mu p_\nu + m^2 = 0$ [18].} One may note that $E = H \equiv p_0$ is invariant under purely spatial coordinate changes (11), thus it depends only on the reference frame—which is arbitrary, but *fixed* in the present study.

Therefore, in a fixed reference frame, a classical test particle in a given external gravitational field has unique energy, in spite of the absence of a covariant concept of local energy in GR, where gravitational dynamics are also included in the model. Similarly, the answer to the question posed at the beginning of this Appendix is no. The non-uniqueness of the energy spectrum for a spin-half particle in a given external curved spacetime, as found in this study, has nothing to do with the absence of a covariant concept of local energy in GR.

[1] Arminjon M and Reifler F 2009 A non-uniqueness problem of the Dirac theory in a curved spacetime *Preprint* arXiv:0905.3686 (gr-qc)
[2] Colella R Overhauser A W and Werner S A 1975 Observation of gravitationally induced quantum interference *Phys. Rev. Lett.* **34** 1472–1474
[3] Werner S A Staudenmann J A and Colella R 1979 Effect of Earth's rotation on the quantum mechanical phase of the neutron *Phys. Rev. Lett.* **42** 1103–1107
[4] Nesvizhevsky V V *et al.* 2002 Quantum states of neutrons in the Earth's gravitational field *Nature* **415** 297–299
[5] Brill D R and Wheeler J A 1957 Interaction of neutrinos and gravitational fields *Rev. Modern Phys.* **29** 465–479. Erratum: *Rev. Modern Phys.* **33** 623–624 (1961)
[6] Chapman T C and Leiter D L 1977 On the generally covariant Dirac equation *Am. J. Phys.* **44**, No. 9, 858–862.

Journal of Physics: Conference Series **222** (2010) 012042 doi:10.1088/1742-6596/222/1/012042

[7] Arminjon M 2008 Dirac-type equations in a gravitational field, with vector wave function *Found. Phys.* **38** 1020–1045 (*Preprint* arXiv:gr-qc/0702048)
[8] Arminjon M and Reifler F 2008 Basic quantum mechanics for three Dirac equations in a curved spacetime *Preprint* arXiv:0807.0570 (gr-qc)
[9] Arminjon M and Reifler F 2008 Quantum mechanics for three versions of the Dirac equation in a curved spacetime *Preprint* arXiv:0810.0671 (gr-qc)
[10] Leclerc M 2006 Hermitian Dirac Hamiltonian in the time-dependent gravitational field *Class. Quant. Grav.* **23** 4013–4020 (*Preprint* arXiv:gr-qc/0511060)
[11] Arminjon M and Reifler F 2008 Dirac equation: Representation independence and tensor transformation *Braz. J. Phys.* **38** 248–258 (*Preprint* arXiv:0707.1829 (quant-ph))
[12] Pauli W 1933 Über die Formulierung der Naturgesetze mit fünf homogenen Koordinaten, Teil II: Die Diracschen Gleichungen für die Materiewellen *Ann. der Phys.* (5) **18** 337–354
[13] Pal P. B. 2007 Representation-independent manipulations with Dirac spinors *Preprint* arXiv:physics/0703214
[14] Schulten K 1999 Relativistic quantum mechanics, in *Notes on Quantum Mechanics*, online course of the University of Illinois at Urbana-Champaign by the same author
[15] Reifler F and Morris R 2005 Hestenes' tetrad and spin connections *Int. J. Theor. Phys.* **44** 1307–1324 (*Preprint* arXiv:0706.1258 (gr-qc))
[16] Landau L and Lifchitz E 1989 *Théorie des Champs* (Moscow: Mir)
Lifshitz E M and Landau L D 1980 *The Classical Theory of Fields* (Oxford: Butterworth Heinemann)
[17] Arminjon M 1998 Remarks on the mathematical origin of wave mechanics and consequences for a quantum mechanics in a gravitational field *Proc. Sixth Int. Conf. Physical Interpretations of Relativity Theory* ed M C Duffy (Sunderland: University of Sunderland) pp 1–17 (*Preprint* arXiv:gr-qc/0203104)
[18] Bertschinger E 1999 Hamiltonian dynamics of particle motion, in *General Relativity*, online course of the Massachussets Institute of Technology by the same author web.mit.edu/edbert/GR/gr3.pdf
[19] Arnold V 1976 *Méthodes Mathématiques de la Mécanique Classique* (Moscow: Mir)
Arnold V I 1989 *Mathematical Methods of Classical Mechanics* (New York: Springer)

First Mediterranean Conference on Classical and Quantum Gravity (MCCQG 2009) IOP Publishing
Journal of Physics: Conference Series **222** (2010) 012043 doi:10.1088/1742-6596/222/1/012043

The effect of particle creation on space

R Blake

Senior Consultant, Readify Pty Ltd, Suite 408 LifeLabs Building, 198 Harbour Esplanade Docklands, Melbourne, Victoria, Australia 3008

E-mail: `russ.blake@readify.net`

Abstract. General Relativity and Quantum Mechanics have been successful at describing their respective realms, but the two theories remain disjoint. We finesse this difficulty with a classical model of the Universe that unites gravitational, nuclear and electromagnetic forces. This model is derived by examining the nucleus and its atomic quanta. Newton's Law of Gravitation evolves into a formula for the gravitational field within an atom's quantum layers which has the form of Hooke's Law for the potential energy of a spring. The inward force—a reaction to the strain caused by introducing two particles into the space—drives particles towards each other. Gravity is not a mysterious attractive force acting at a distance with no mechanism, but a force acting locally with a well-defined mechanism. The force on the nucleons from the spring stress in the nucleus provides a physical basis for the strong and weak forces of quantum mechanics. The Fundamental Law of Nature gives the relationship between the strong force stress on space, gravity and the wavelength of a photon. The new paradigm offers an alternative to the Standard Model that is easy to integrate with gravitation.

1. Introduction

We present a new model of the physical Universe. The observations of atomic behavior are presently modeled by Quantum Mechanics. The observations of very large bodies in space are modeled by General Relativity. Both of these models have been successful at predicting observed phenomena in their respective domains. But they remain intractably separate theories.

The attempt to knit together diverse mathematical models built to describe what appear at the outset unrelated physical phenomena is understandably fraught with difficulty. Our admittedly radical approach is to finesse the problem by presenting a single physical model for gravitational and nuclear forces. Building unified mathematical models on top of this unified physical model should then be easier.

How can a return to a physical model be a radical solution in physics? Isn't physics, after all, about physical reality? Yes, but Quantum Mechanics was founded on the assumption that classical physical laws do not apply on the subatomic scale. Suggesting at this point that an answer to the logjam in physics might be to utilize classical laws to explore subatomic physical reality is swimming against a tide running strongly in the opposite direction for 90 years.

2. Results

2.1. Quantum law of gravitation

Newton presented his Law of Universal Gravitation in 1687, where the force of gravity ${}_GF$ is given by

$$ {}_GF = G\frac{m_1 m_2}{r^2} \tag{1} $$

where m_1 and m_2 are the masses of two objects, r is the separation of their centres, and G is the Gravitational Constant, not measured until 71 years after Newton's death. Until now G has only been known by measurement. We show how G relates to other physical constants.

Newton's Law of Gravitation shows the force of gravity is dependent on two masses. We isolate a single mass by considering its gravitational field ${}_GU$ at distance r from the centre of its mass m:

$$_GU = G\frac{m}{r^2} \tag{2}$$

This expression has units m/s^2, and is the acceleration imparted towards mass m on another object in its gravitational field. (We use italics for variables like mass (m) and normal case for units like meters (m).) Equation (2) implies that the gravitational field extends indefinitely from the object across the Universe.

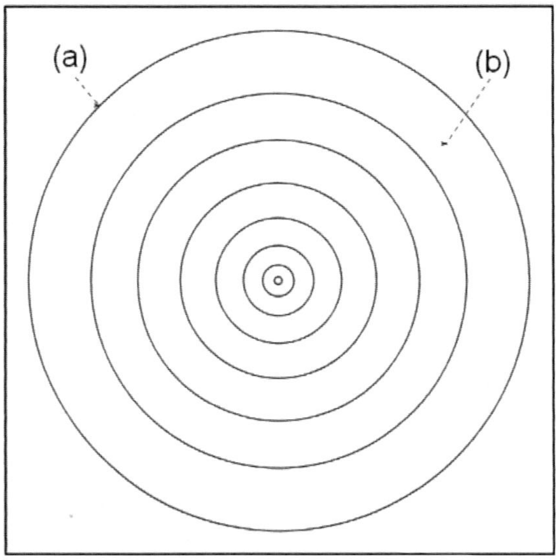

Figure 1. Bohr model of an atom. This figure illustrates the first 8 quantum levels. (a) Quantum level 8; (b) Quantum layer 8. Drawn to scale: radius of each quantum level is proportional to the square of its quantum number. Nucleus is 10^4 or 10^5 times smaller than quantum level 1, depending on the element. The Bohr model was introduced in 1913. Since replaced by more accurate models, it's simplicity and clarity still make it a useful starting place for a new model of atomic phenomena.

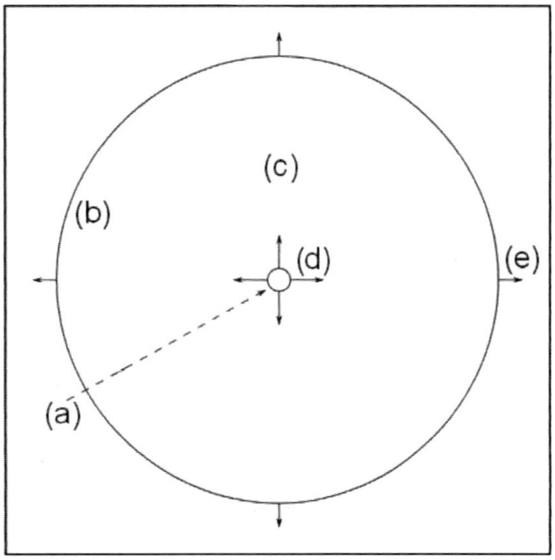

Figure 2. Particle creation. (a) Atomic nucleus. (b) Quantum level 1. (c) Quantum layer 1. (d) Stress of introducing nucleus into space causes strain in quantum layer 1: it acts like a spherical, double-ended, compressed spring. (e) Stress transmitted to quantum layer 2 in proportion to surface area of quantum level 1. This sets the stage for—but does not itself induce—gravitation. A second particle is required to interact with this one before gravitation arises. (Not drawn to scale.)

Figure 1. illustrates the Bohr model of the atom, in which electrons circulate around the nucleus in disjont, spherical shells. The quantum levels of an atom do not stop with the level of an outmost trapped electron, but instead extend throughout space. In our model we imagine that each elementary particle or nuclear group thereof has such quantum levels.

Applying (2) to an atom with A nucleons at distance of quantum level n, n an integer > 0, gives

$$ {}^{A}_{G}U_n = G\frac{{}^{A}m}{r_n^2} \tag{3}$$

where ${}^{A}m$ is the mass of the nucleus, r_n is the distance of the nth quantum level from the centre of the nucleus and G is the gravitational constant introduced by (1). An important implication of (3) is the consistency of the gravitational field throughout each quantum layer (Figure 1(b).) This is at root our notion of quantum gravity.

We start with the formula for the radius of the nth quantum of the nucleus with A nucleons,

$$ {}^{A}r_n = {}^{A}r_{1}n^2 \tag{4}$$

where ${}^{A}r_{1}$ is the radius of the first quantum level. (Constants like ${}^{1}r_{1}$ are listed in Table 1.)

Table 1. Table of Physical Constants.

Symbol	Value	Units	Meaning
G	6.93×10^{-11}	$m^3/kg{\cdot}s^2$	Gravitational constant [4]
${}^{1}r_{1}$	5.29×10^{-11}	m	Radius quantum level 1 of ${}^{1}H$[5]
m_p	1.67×10^{-27}	kg	Mass of proton [2]
${}^{1}\aleph$	8.00×10^{-17}	kg/m	Quantum gravitational constant (8)
${}^{1}r_{0}$	1.58×10^{-15}	m	Radius of ${}^{1}H$ nucleus[5]
r_p	0.875×10^{-15}	m	Radius of a proton[2]
Ω	8.00×10^{-17}	kg/m	Hardness of space (26)

$$ {}^{A}_{G}U_n = G\frac{{}^{A}m}{{}^{A}r_1^2 n^4} \tag{5}$$

We notice that ${}^{A}m$ is just Am_p where m_p is the mass of the proton. (The proton and the neutron have almost exactly the same mass, so for simplicity we will treat them as equivalent here.) This and (4) permit us to write (3) as

$$ {}^{A}_{G}U_n = \frac{Gm_p}{{}^{A}r_1^2}\frac{A}{n^4} \tag{6}$$

To improve our understanding and simplify this expression we define

$$ {}^{A}x_n^2 = \frac{A}{n^4} \tag{7}$$

and define ${}^{A}\aleph$ (using aleph, the first letter of the Hebrew alphabet) as

$$ {}^{A}\aleph = \frac{2Gm_p}{{}^{A}r_1^2} \tag{8}$$

which enables us to express (6) as

$$ {}^{A}_{G}U_n = \frac{1}{2}\ {}^{A}\aleph\ {}^{A}x_n^2 \tag{9}$$

Journal of Physics: Conference Series **222** (2010) 012043 doi:10.1088/1742-6596/222/1/012043

This has the same form as the potential energy of a spring as given by Hooke's Law: ${}^{A}\aleph$ is the spring constant, and ${}^{A}x_n^2$ is the strain on (i.e. compression of) the spring. So (3)—the gravitational field of the nucleus of an element with A nucleons at quantum level n—is analogous to the potential energy of an elastic spring at quantum level n with spring constant ${}^{A}\aleph$ and displacement ${}^{A}x_n^2$. But ${}^{A}_{G}U_n$ is not potential energy: it has units of acceleration. This is because unlike a normal spring, gravity accelerates all objects at the same rate.

We turn our attention first to understanding ${}^{A}x_n^2$ better. As an acceleration ${}^{A}_{G}U_n$ has units of m/s^2, and all the units are in ${}^{A}\aleph$. This is expected since deformation of springs is usually expressed as a unit-less ratio.

Now we need the usual model of the radius ${}^{A}r_0$ of the atomic nucleus of an element with A nucleons[1] (adopting a convention that quantum level "0" is the edge of the nucleus):

$$ {}^{A}r_0 = {}^{1}r_0 \, A^{\frac{1}{3}} \tag{10} $$

This heuristic together with (4) yields an expression for the displacement of the "gravitational spring":

$$ {}^{A}x_n^2 = \frac{{}^{A}S_1}{{}^{A}S_n} \frac{{}^{A}V_0}{{}^{1}V_0} \tag{11} $$

The displacement of the "gravitational spring" in the nth quantum level depends inversely on the ratio of the surface area of the nth quantum level to the surface area of the first quantum level, and directly on the ratio of the volume of the nucleus of an element with A nucleons to the volume of the ^{1}Hydrogen nucleus. As the product of two ratios, it is without dimension. Strain is determined entirely by the geometry of the atom and its quanta.

This observation has implications on the nature of space itself. Our understanding of space is growing. We know the speed of light in space depends on the permittivity and permeability of space; just as does the speed of light in water or glass depend on those characteristics of those materials. Furthermore the conclusion of Einstein's General Theory of Relativity—that space is curved in the presence of a gravitational mass—also implies that space is more than a totally empty void cavity.

One way to interpret the direct dependence of displacement ${}^{A}x_n^2$ upon the volume of the nucleus is that when an element is created, it does not replace space: it *displaces* the space that it now occupies. This is analogous to the insertion of a ball bearing (the nucleus) into a foam rubber block (the Universe.) But this analogy only goes so far. For one thing our model assumes the Universe, from the viewpoint of a nucleus, is divided into quantum levels, whereas in the foam rubber block compression is continuous.

Think of the insertion of the nucleus into the Universe as creating a pressure outward on the space within the first quantum layer (Figure 2(d).) This pressure is distributed equally around the surface of the nucleus. Using the spring analogy, space is "compressed". Think of this space as staying inside the quantum level, but being placed under stress by the strain of insertion. This stress creates an outward pressure on the quantum level 1, distributed around the surface area of the quantum, just as a compressed spring applies equal pressure to both ends. The pressure is the same at every quantum level, so as the surface area increases, the force applied to each surface decreases in proportion. Thus the pressure is the same on the inner surface and the outer surface of each quantum layer. The pressure exerted outward at quantum level 1 induces a stress on the 2nd quantum layer. And so on throughout the Universe.

This model asserts that gravitational acceleration is a side effect of the strain on space. A more elaborate model of this effect will be discussed when we have characterized a few more details about the nucleus.

Journal of Physics: Conference Series **222** (2010) 012043 doi:10.1088/1742-6596/222/1/012043

Now let A = 1 and n = 1, the first quantum level of ^{1}H. In this case we see the gravitational acceleration at the first quantum of ^{1}H is 1/2 $^1\aleph$. The spring constant $^1\aleph$ is easy to evaluate from the Table 1:

$$^1\aleph = 8.00 \times 10^{-17} \tag{12}$$

Combining (7) and (9) gives a "quantum law of gravitation":

$$^A_G U_n = \frac{1}{2}\ {}^A\aleph\ \frac{A}{n^4} \tag{13}$$

The quantum law of gravitation replicates applying Newton's Law of Gravitation to the nucleus of an atom.

2.2. *The fundamental law of nuclear gravitation*

We want to look more closely at the insertion of a particle into space and the effect it has on the Universe. Consider formation of a single proton.

The energy necessary to create the proton is 938 MeV, but the energy needed to create its components, two up quarks and a down quark, amounts to only about 10 MeV [2]. By our model the remaining 928 MeV is used to set the gravitational spring. This is placed into the volume of the proton; dividing the former by the latter gives the pressure from the proton to the gravitational spring, i.e. between the proton and the inner surface of the roughly spherical shell surrounding it (Figure 3(d).) This pressure in N/m^2 is

$$^1P = \frac{^1E}{^1V_p} = 5.30 \times 10^{34} \tag{14}$$

where ^1P is the pressure the proton is putting on the space between itself and the inner edge of the nucleus, ^1E is the formation energy left over after quark formation in Nm, and ^1V_p is the volume of the proton. This is distributed around the surface of the proton (Figure 3(e).) The equal and opposite force on the proton (in Newtons, Figure 3(g)) is the pressure times the surface area:

$$^1_S F_p = {}^1P\,{}^1S_p = 5.10 \times 10^5 \tag{15}$$

where ^1S_p is the surface area of the proton. $^1_S F_p$ is equivalent to what in the Standard Model is called the strong force holding the quarks together inside the proton. As we generalize to multiple particle nuclei $^A_S F_p$ becomes equivalent to the residual strong force, holding nucleons together in the nucleus. In our model this is equal and opposite to the stress force placed on the inner edge of the nucleus: our single comnpressive force performs both functions.

Neutrons outside a nucleus normally decay into a proton and electron (and an anti-neutrino) in 14 minutes and 46 seconds [2]. But deuterium, ^2H, is stable with one neutron and one proton in the nucleus. This nucleus has the shape of a capsule (roughly two hemispheres joined by a cylinder). Using this surface area gives $^2_S F_p = 6.11 \times 10^5$ N, or about 20% higher than for the single proton (15). This we reason is more than enough extra force to hold the neutron together, whereas subject only to the pressure of (15) on its own, the neutron decays.

This is a strong force indeed: the opposing force of two positively charged protons in the spherical nucleus of ^{4}Helium is only 75.3N, and the strong force $^4_S F_p$ on the ^{4}He nucleons is 9.46×10^5N. This is 1.86 times (15), because 4 times the energy and 4.64 times the surface area are divided by only 10.0 times the roughly tetrahedral volume.

By definition the formation energy the strong force of the gravitational spring is enough to hold the quarks together inside protons and neutrons. Holding the protons together against their opposing positive charges is then a small matter by comparison. According to our model the

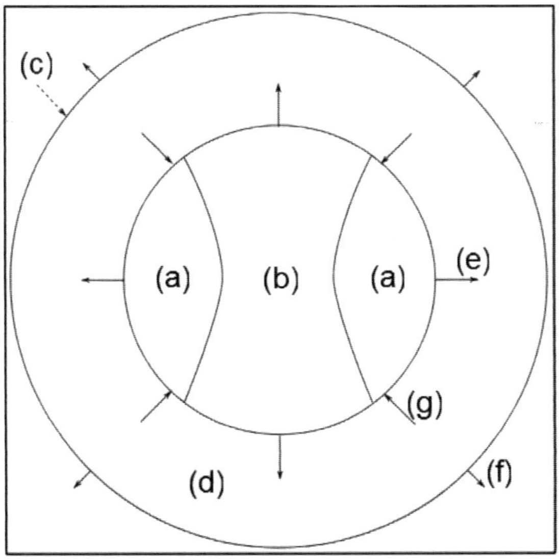

Figure 3. The nuclear spring. The proton is composed of two up quarks (a) and one down quark (b). The outer circle is the edge of the nucleus (c): quantum level "0"; (d) is quantum layer "0". (e) Placing the proton into the space compresses it. (f) The stress of compression is transmitted to the first quantum layer. (g) The equal and opposite force to (e) is the strong force holding the quarks together in the proton. Drawn to scale.

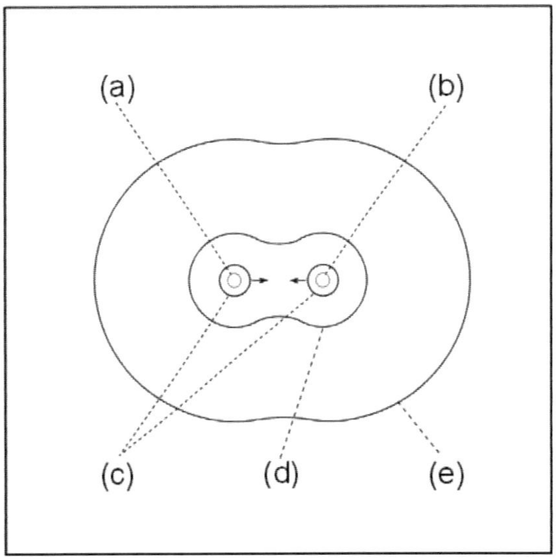

Figure 4. Gravitation. Two particles: (a) Proton. (b) Neutron. (c) Quatum level "0": edge of single particle "nucleus". (d) Quantum level 1. (e) Quantum level 2. The lower energy state achieved as the infinite quantum layers coalesce gives rise to the acceleration called gravitation (arrows.) The pressure on the evolving capsules requires less force to maintain because the capsules have less surface area than two combining spheres. Not drawn to scale.

strong force of the Sandard Model now has a physical basis: the force of setting the gravitational spring within the nucleus.

The strong force in the case of a nucleus with A nucleons is

$$ {}^{A}_{S}F_p = {}^{A}P \, {}^{A}S_p \tag{16} $$

Applying (14) to the element with A nucleons, where ${}^{A}E = A\,{}^{1}E$ gives us

$$ {}^{A}_{S}F_p = \frac{A\,{}^{1}E}{{}^{A}V_p}\,{}^{A}S_p \tag{17} $$

Expanding the volume and surface area in terms of the radius ${}^{A}r_p$ and canceling, (17) reduces to

$$ {}^{A}_{S}F_p = \frac{3\,A\,{}^{1}E}{{}^{A}r_p} \tag{18} $$

As the nucleon count rises, the strong force initially increases rapidly, but then tapers off. Our model would suggest that there will be a point where the incremental strong force would be insufficient to hold the nucleus together, and radioactive decay ensues. In contrast the Standard Model tells us the residual strong force retains particles in the nucleus. Beta radioactivity due to decay of a neutron in the nucleus is then due to the weak force. But our model argues that

Journal of Physics: Conference Series **222** (2010) 012043 doi:10.1088/1742-6596/222/1/012043

the compressive force of space is the only force we need consider, and that the weak force of quantum mechanics is really the stress on space becoming too weak to do its job.

The strong force is transmitted to the edge of the nucleus in inverse proportion to its surface area:

$$ {}^{A}_{S}F_{0} = {}^{A}_{S}F_{p} \, \frac{{}^{A}S_{p}}{{}^{A}S_{0}} \tag{19}$$

which likewise applies to the nth quantum level

$$ {}^{A}_{S}F_{n} = {}^{A}_{S}F_{0} \, \frac{{}^{A}S_{0}}{{}^{A}S_{n}} \tag{20}$$

Substituting Eqs.(19), then (18) and finally (3) rewards us with

$$ {}^{A}_{S}F_{n} = \frac{3\,{}^{1}E\,{}^{A}r_{p}}{{}^{A}r_{1}^{2}}\,\frac{A}{n^{4}} \tag{21}$$

Restating (9) using (8)

$$ {}^{A}_{G}U_{n} = \frac{1}{2}\,{}^{A}\aleph\,\frac{A}{n^{4}} = \frac{1}{2}\frac{2Gm_{p}}{{}^{A}r_{1}^{2}}\,\frac{A}{n^{4}} \tag{22}$$

The ratio of Eqs.(21) and (22) is

$$ \frac{{}^{A}_{S}F_{n}}{{}^{A}_{G}U_{n}} = \frac{3\,{}^{1}E\,{}^{A}r_{p}}{Gm_{p}} \tag{23}$$

Applying the approximation from Special Relativity

$$ {}^{1}E = m_{p}c^{2} \tag{24}$$

we obtain

$$ \frac{{}^{A}_{S}F_{n}}{{}^{A}_{G}U_{n}} = \frac{3\,{}^{A}r_{p}c^{2}}{G} \tag{25}$$

To distill this relationship let us define the "hardness of space", Ω, as the constant

$$ \Omega = \frac{3c^{2}}{G} \tag{26}$$

Then (25) becomes the "fundamental law of nuclear gravitation":

$$ {}^{A}_{S}F_{n} = \Omega\,{}^{A}r_{p}\,{}^{A}_{G}U_{n} \tag{27}$$

where in kg/m

$$ \Omega = 4.30 \times 10^{27} \tag{28}$$

(27) is an expression for the amount of force (originating from the strong force radiating outward) that must be applied to the nth quantum level to create a unit gravitational acceleration. The quantity $\Omega\,{}^{A}r_{p}$ is on order of 10^{13} kg.

Notice that ${}^{A}r_{p}$ is the absolute displacement of the spring. The geometry of the initial condition of the insertion of the mass into space has a fundamental effect on all the outgoing quanta. This is an unexpected geometrical component of more massive objects having a stronger gravitational field, and complements our hypothesis that space is strained by the insertion of a particle.

The only non-zero eigenvalue of the Cauchy stress tensor is (21), and of the strain tensor is

$$ {}^{A}\epsilon_n = \frac{{}^{A}r_p}{{}^{A}r_n} \tag{29} $$

The non-zero term of the coefficient of elasticity tensor is the quotient of the strain and the stress:

$$ {}^{A}C_n = \frac{3\ {}^{1}E\ A}{{}^{A}r_1\ n^2} \tag{30} $$

Let us return now to the assertion that gravitation is a side effect of the strain placed on space by insertion of the particle. At first it might seem gravity is simply the effect of the compressive forces in the displaced quantum layers, but that neglects the equal and opposite force of the stress on space at each quantum level. Based on force alone, everything is in equilibrium, and there is no reason for particles to move towards rather than away from one another: the forces at the two ends of a spring—and at each quantum level—are equal and opposite, consistent with Newton's Third Law. But then, what causes gravitational acceleration?

Consider the two particles in the nucleus of ^{2}H (deuterium): a proton and a neutron. The nuclear boundary forms a capsule shape around the two particles (analogous to the roughly spherical shell at Figure 3(d) for a single particle). What about the first quantum level? Due to its distance from the nucleus we think of it as a single shell rather than two overlapping shells. Whether this is true or not, certainly there is some quantum level far from the nucleus where we can assume the quantum levels are coincident. Imagine forcing these two particles apart. The nuclear capsule starts to collapse, and finally snaps into the form of two separate particles. To do this one must overcome the inward force of the spring, ${}^{2}_{S}F_{p}$. But the outer coincident quantum levels are still acting like a capsule-shaped spring and must be forced apart just like the nuclear capsule, albiet at much lower resistance. Gravitation results from the lower energy state achieved as quantum levels of the two particles become aligned; lower, because a capsule has less surface area than the two separate spheres, permitting less force to sustain greater pressure (Figure(4).) Kinetic energy is released in the form of acceleration as the alignment is achieved, and is required to undo the alignment. By this model the infinite quantum levels of at least two particles are required to create acceleration due to gravity.

Reversing (26) gives us an elegant expression for Newton's gravitational constant, G:

$$ G = \frac{3c^2}{\Omega} \tag{31} $$

2.3. The Fundamental Law of Nature

We turn our attention to that other quantum phenomenon: the production of a photon as an electron transitions down a quantum level. The wavelength of the light emitted is related to the change (Δ) in energy required for the transition:

$$ \Delta\nu = \Delta\frac{1}{\lambda} = \frac{\Delta E}{hc} = \frac{R_M Z^2}{hc}\left(\frac{1}{n^2} - \frac{1}{(n-1)^2}\right) \tag{32} $$

where ν is frequency, λ is wavelength, h is the Planck Constant, c is the speed of light in a vacuum, and R_M is Rydberg's constant for hydrogen-like (i.e. single electron) atoms of atomic mass M with Z nuclear protons[3].

Our model suggests the stress on space within quantum layers is applying a force on electrons that is keeping them in their quantum levels. An indication that this may be the case is the way electrons are repelled from the heavily compressed space in the nucleus between the proton and quantum level 0 (Figure 3(d).) In fact it seems likely that permittivity and/or permeability are increased as a result of the strain. Call the resulting force on the electron the "quantum force". As an outward force it would have to overcome the Coulomb force of the nuclear protons attracting the electron, and the gravitational force between the electron and the nucleus. As an inward force it would have to retain the electron in its quantum level until it had enough energy to get to a higher level, and suppress radiation from the moving electron.

The energy change from quantum level n to quantum level $n-1$ would be the sum of these forces times the distance between the levels (assuming the electron follows a path in line with the centre of the nucleus):

$$\Delta E = ({}^{Z}_{Q}F_n + {}^{Z}_{C}F_n + m_e\, {}^{A}_{G}U_n)(r_n - r_{n-1}) \tag{33}$$

where ${}^{Z}_{Q}F_n$ is the quantum force, ${}^{Z}_{C}F_n$ is the Coulomb force of the Z protons in the nucleus, and m_e is the mass of the electron. Using (32) and solving for quantum force gives

$${}^{Z}_{Q}F_n = \frac{R_M Z^2}{(r_n - r_{n-1})}\left(\frac{1}{n^2} - \frac{1}{(n-1)^2}\right) - {}^{Z}_{C}F_n - m_e\, {}^{A}_{G}U_n \tag{34}$$

To put this in perspective, depending on the size of the nucleus the Coulomb force is a billion to a trillion times smaller than the quantum force; the gravitational force is even smaller. As a first approximation we can safely ignore these terms.

Consider first the ratio of the quantum force in layer n to the same force in quantum layer 2, the lowest quantum across which a photon can be generated.

$$\frac{{}^{Z}_{Q}F_n}{{}^{Z}_{Q}F_2} = \frac{\frac{R_M Z^2}{(r_n - r_{n-1})}\left(\frac{1}{n^2} - \frac{1}{(n-1)^2}\right)}{\frac{R_M Z^2}{(r_2 - r_1)}\left(\frac{1}{2^2} - \frac{1}{1^2}\right)} \tag{35}$$

Using (4) this becomes simply

$$\frac{{}^{Z}_{Q}F_n}{{}^{Z}_{Q}F_2} = \frac{4}{n^2(n-1)^2} \tag{36}$$

Similarly comparing the strong force (27) on quantum level n to the 2nd level gives

$$\frac{{}^{A}_{S}F_n}{{}^{A}_{S}F_2} = \frac{16}{n^4} \tag{37}$$

The quantum force can be compared to the strong force in each quantum layer by taking a ratio of the (36) and (37):

$${}^{Z}_{Q}F_n = \frac{n^2}{4(n-1)^2}\frac{{}^{Z}_{Q}F_2}{{}^{A}_{S}F_2}\, {}^{A}_{S}F_n \tag{38}$$

The model asserts that the strain on the quantum layer created by the insertion of the particle into space creates stress in that layer which in turn exerts the force an electron must overcome to leap up to the next highest quantum level. In effect the strong force stress on space keeps electrons at their quantum levels. In particular it likely keeps the electron at quantum level 1 from every approaching the nucleus, no matter how many protons it contains. It seems to prevent electromagnetic radiation from the electrons as they accelerate around the nucleus. The most fundamental electromagnetic phenomena are thus tied to the nuclear strong force.

Combining (38) with (27) yields the "fundamental law of nature":

$$ {}^{Z}_{Q}F_n = \frac{\Omega\ {}^{A}r_p\ n^2\ {}^{Z}_{Q}F_2}{4(n-1)^2\ {}^{A}_{S}F_2}\ {}^{A}_{G}U_n \tag{39} $$

For any atom this is the relationship between the quantum force on the electron (which determines the wavelength of the light photon produced) and the acceleration due to gravity, mediated by the strong force.

2.4. Discussion

No model is causal: it is only descriptive, and its usefulness ultimately depends on its simplicity, elegance and ability to describe and predict reality. Furthermore it cannot be refuted merely by comparison with another model. Many models can be constructed to describe the same phenomena; only experimental evidence can limit a model's domain.

By our model space displacement creates the strong and weak forces; if correct this finding will have the greatest impact on current theories. The full consequences for quantum mechanics of a compressive force acting on elementary particles and the nucleus must be evaluated.

The scattering of electrons by the strained space in the nucleus is the clearest evidence for this model. And the Standard Model shows no way for neutrons to congregate, while this model explains the existence of tetraneutrons: clusters of 4 neutrons in a nucleus[6]. Dark matter could simply be gravitationally assembled neutron neuclei. The model holds promise for solving other mysteries. Being only 1% of the mass, quarks may be structures of struts supporting the proton void in space. This implies space has surface tension: light photons may be small voids supported merely by the surface tension of space without the need for supporting struts, leading to a simple explanation of the wave-particle duality.

Our model is intuitively pleasing compared to mysterious gravitational attraction acting at a distance. Finally we can understand why things fall.

2.5. Acknowledgements

The author is indebted to Ray Trattles for his insightful comments which led to a much improved derivation of the Atomic Law of Gravitation. The author would also like to thank Rick Sebastian for his thoughtful consideration of these ideas as they emerged. Thanks also to Karla Blake, Graeme Strange, Andrew Cohen, Steven Nagy, Paul Glavich, Philip Beadle, Andrew Cohen, Paul Kilian and Shawn Blake for their encouragement and comments, and Shane Wright for his erudition.

3. References

[1] Krane, K. S., 1988 *Introductory Nuclear Physics*, (USA & Canada, John Wiley & Sons, Inc)
[2] C. Amsler et al., (Particle Data Group) 2008/9/10 *Phys. Lett. B667*, 1 Available: http://pdglive.lbl.gov. Accessed: 15 July 2009
[3] *The NIST Reference on Constants, Units, and Uncertainty* http://physics.nist.gov/cuu/Units/meter.html. Accessed: 24 September 2008
[4] Fixler J B, Foster G T, McGuirk J M and Kasevich M A 2007 Atom interferometer measurement of the Newtonian constant of gravity, *Science* 5 Jan
[5] Nave C R 2005 Nuclear size and density *HyperPhysics* http://hyperphysics.phy-astr.gsu.edu/hbase/nuclear/nucuni.html#c4. Accessed: 20 September 2008
[6] Marqus F M 2002 Detection of neutron clusters *Phys. Rev. C* **65**

First Mediterranean Conference on Classical and Quantum Gravity (MCCQG 2009) IOP Publishing
Journal of Physics: Conference Series **222** (2010) 012044 doi:10.1088/1742-6596/222/1/012044

Fluctuating Twistor-Beam Solutions and Pre-Quantum Kerr-Schild Geometry.

Alexander Burinskii

Laboratory of Theoretical Physics, NSI Russian Academy of Sciences,
B.Tulskaya 52, Moscow, 115191, Russia

E-mail: `bur@ibrae.ac.ru`

Abstract. Kerr-Schild (KS) geometry is based on a congruence of twistors which is determined by the Kerr theorem. We describe time-dependent KS solutions for electromagnetic excitations of black-holes taking into account the consistent back-reaction to metric. The exact solutions have the form of singular beam-like pulses supported on twistor null lines of the Kerr congruence. The twistor-beams have very strong back reaction to metric and BH horizon and produce a fluctuating KS geometry which takes an intermediate position between the Classical and Quantum gravity.

1. Introduction

Singular pp-waves [1], playing the role of plane waves in gravity, differ drastically from plane waves in flat spacetime, which is one of the origin of incompatibility of the Quantum theory and Gravity. Quantum theory works basically in momentum space, while Gravity demands explicit representation in configuration space-time. Twistor theory forms a bridge between them. Geometrically, twistor is a null line formed by a pair $(x^\mu,\ \theta^\alpha)$, where θ^α is a two-component spinor joined to the point $x^\mu \in M^4$. This spinor fixes a null direction $\sigma^\mu_{\dot\alpha\alpha}\bar\theta^{\dot\alpha}\theta^\alpha$ corresponding to momentum p^μ of a massless particle. A plane wave in twistor coordinates $T^I = \{\theta^\alpha,\ \mu_{\dot\alpha}\}$, $\mu_{\dot\alpha} = x_\nu\sigma^\nu_{\dot\alpha\alpha}\theta^\alpha$ has the form $\exp\{ix_\mu\sigma^\mu_{\dot\alpha\alpha}\bar\theta^{\dot\alpha}\theta^\alpha\}$ and may be transformed to twistor space by a 'twistor' Fourier transform. The result, [2](p.2.5), corresponds to a singular beam in direction p^μ supported on the twistor null line with coordinates $T^I = \{\theta^\alpha,\ \mu_{\dot\alpha}\}$. Such twistor-beams are the sources of singular pp-wave solutions in gravity, and similar twistor-beams are obtained for typical exact electromagnetic (em) excitations of black-holes (BH) [3] presented in the KS class of the algebraically special solutions [4] which cover a wide range of the rotating and non-rotating BH's and cosmological solutions. Appearance of the twistor-beams in the exact solutions is not accidental, since twistors form a skeleton of the KS space-time.

In this work we give a detailed description of integration of the Debney-Kerr-Schild equations [4] which describe the exact em excitations of the Kerr black-hole and corresponding back reaction to metric[1] consistent with the Einstein-Maxwell equations with averaged stress-energy tensor.

The KS geometry has twosheeted holographic twistor structure which turns out to be perfectly adapted to quantum treatment suggested in [5, 6], and we show that the resulting set of the

[1] In this treatment we neglect by the recoil of the excitations on the position of the BH.

solutions forms a fluctuating pre-quantum geometry taking intermediate position between the Classical and Quantum gravity.

2. Twistor structure of the KS geometry

The KS solutions are based on the KS form of metric

$$g_{\mu\nu} = \eta_{\mu\nu} + 2H k_\mu k_\nu, \tag{1}$$

in which $\eta_{\mu\nu}$ is metric of auxiliary Minkowski space-time M^4 and k_μ is a field of null directions, forming a principal null congruence (PNC) $\mathcal{K}$. The BH solutions may be represented in two different forms $g^{\pm}_{\mu\nu} = \eta_{\mu\nu} + 2Hk^{\pm}_\mu k^{\pm}_\nu$, were $k^{\mu\pm}(x)$, $(x = x^\mu \in M^4)$ correspond to two different vector fields tangent to different congruences $\mathcal{K}^{\pm}$. The directions $k^{\pm}_\mu$ are determined in the Cartesian null coordinates $u = (z-t)/\sqrt{2}$, $v = (z+t)/\sqrt{2}$, $\zeta = (x+iy)/\sqrt{2}$, $\bar\zeta = (x-iy)/\sqrt{2}$ by the form

$$k^{(\pm)}_\mu dx^\mu = P^{-1}(du + \bar{Y}^{\pm} d\zeta + Y^{\pm} d\bar\zeta - Y^{\pm}\bar{Y}^{(\pm)} dv) \tag{2}$$

which depends on the complex angular coordinate $Y = e^{i\phi}\tan\frac{\theta}{2}$. Two solutions $Y^{\pm}(x)$ are determined by the **Kerr Theorem** [1, 4, 7, 8] which states that the (necessary for solutions of type D) geodesic and shear-free null congruences in M^4 are generated by algebraic equation $F = 0$, where $F(Z^p)$ is an arbitrary holomorphic function of the projective twistor coordinates.[2]

The Kerr congruence represents a vortex of null lines, which covers the spacetime twice: in the form of ingoing ($k^{\mu-} \in \mathcal{K}^-$) and outgoing vector fields ($k^{\mu+} \in \mathcal{K}^+$). It forms two sheets of the KS geometry with two different metrics $g^{\pm}_{\mu\nu}$ on the same spacetime M^4.

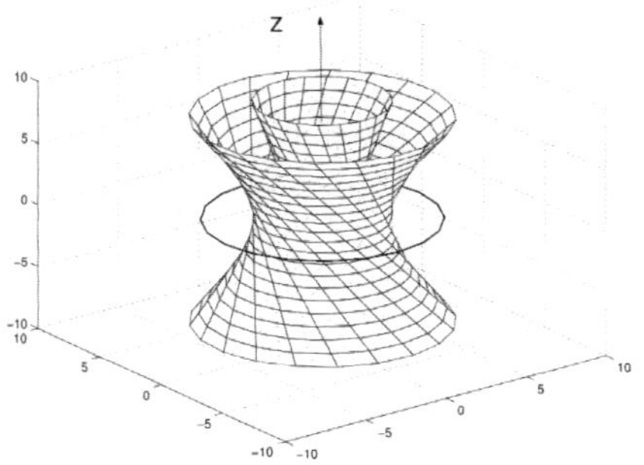

Figure 1. The Kerr singular ring and the Kerr congruence of twistor null lines.

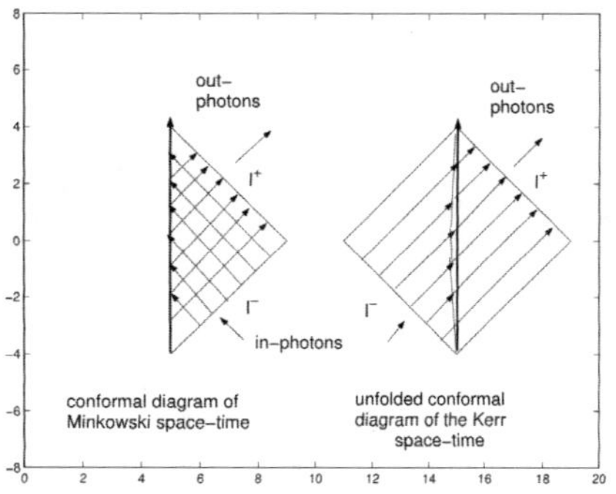

Figure 2. Penrose conformal diagrams. Unfolding of the auxiliary M^4 space of the Kerr spacetime yields twosheeted structure of a pre-quantum BH spacetime.

The em field has to be aligned with the rays of PNC and turns out to be different on the in- and out- sheets, and these two fields should not be mixed, which is ignored usually in the perturbative approach, leading to drastic discrepancy in the form of the fundamental solutions. The typical exact em solutions on the Kerr background have the form of singular twistor-beams propagating along the twistor null rays of the Kerr PNC, contrary to the smooth angular dependence of the typical wave solutions used perturbatively!

[2] Projective twistor coordinates Z^p, $p = 1,2,3$ are related with twistor coordinates $\{\theta^\alpha, \mu_{\dot\alpha}\}$ as follows $T^I = \theta^1(1, Z^p)$.

3. Holographic KS structure

For a long time this twosheetedness was a mystery of the Kerr BH (see refs in [9]). It was suggested by Israel (1968) to truncate the second sheet and replace it by a rotating disk-like bubble covering the Kerr singular ring. Alternatively, the Kerr singular ring was considered as a closed 'Alice' string forming a gate to 'Alice' mirror world of the advanced fields [10]. Holographic approach resolves this problem unifying the both points of view: the source of Kerr solution has to be considered as a membrane separating the in- and out- parts of the KS space. It is supported by the membrane paradigm and by the obtained recently structure of the consistent Kerr-Newman source [11]. The both sheets of the KS space are necessary for description of quantum fields in gravity. In particular, Gibbons states in [5] that the curved spacetime $\mathcal{M}$ should be separated into two time-ordered regions $\mathcal{M}_-$ and $\mathcal{M}_+$ which are associated with ingoing and outgoing vacuum states $|0_- >$ and $|0_+ >$. If the source is absent, the KS basic solutions may be extended analytically from in- to out-sheet and vice versa. Presence of the source breaks this analyticity, separating the retarded and advanced fields, which allows one to consider the BH evaporation as a scattering of the in-vacuum on the BH-source. Similarly, in [6] authors consider a pre-quantum BH spacetime with separated the in- and out- sheets which correspond to a holographic correspondence: the source forms a holographically dual boundary (membrane) separating the in- and out- regions.[3] The usual Penrose conformal diagram, containing the in- and out-fields on the same of M^4, has to be unfolded in the holographic interpretation to split twosheetedness, as it is shown on the Fig.2. The null rays of Kerr congruence perform the lightlike projection of the past null infinity I^-, so the Kerr source appears as a holographic image of the data on the past null infinity I^-.

Conformal structure of the KN solution is determined by complex function $Y(x)$. Tetrad derivatives $\partial_a = e^\mu_a \partial_\mu$ of the function $Y(x)$ determine principal parameters of the KS holographic projection. In particular, function $Y(x)$ for the shear-free and geodesic congruences of the Kerr theorem has to satisfy the conditions

$$Y_{,2} = Y_{,4} = 0, \tag{3}$$

which show that the expression $dY = Y_{,a} e^a = Y_{,1} e^1 + Y_{,3} e^3$ is gradient of the complex null surfaces $Y = const.$, which form a fiber bundle of the KS space-time with fibers spanned by the tetrad forms e^1 and e^3, providing conformal properties of the KS holographic projection.

4. Stationary twistor-beam KS solutions

For the Kerr congruence function $F(Y, x)$ of the Kerr theorem is to be quadratic in Y, and the eq. $F = 0$ may resolved in explicit form. The corresponding solutions $Y(x)$ determine the stationary Kerr congruence via (2) and the adapted (oblate spheroidal) coordinate system r, θ, ϕ in which

$$Y = e^{i\phi} \tan \frac{\theta}{2}. \tag{4}$$

Function $F(Y, x)$ determines also the function H up to an arbitrary function $\psi(Y)$, and therefore, the KS metric (1). In agreement with [4]

$$H = \frac{1}{2}[m(Z + \bar{Z})P^{-1} - |\psi|^2 Z\bar{Z}P^{-2}], \tag{5}$$

where $Z = Y_{,1}$ and $(Z/P)^{-1} = -dF/dY$. Complex function $Z = \rho + i\omega$ characterizes the conformal properties of the congruence: expansion ρ and rotation ω of the projected image. For the Kerr-Newman BH solution at rest, Z is inversely proportional to a complex radial distance

[3] In the model [11] it is a domain wall boundary of a rotating disk-like source forming a superconducting bubble.

$Z = -P/(r + ia\cos\theta)$ where function P is an extra conformal factor determined by Killing vector of the solution [4].

In particular, for the Kerr and Kerr-Newman (KN) geometries at rest

$$H = \frac{mr - |\psi|^2/2}{r^2 + a^2\cos^2\theta} \ . \tag{6}$$

The KN electromagnetic field is determined by the vector potential

$$\alpha = \alpha_\mu dx^\mu = -\frac{1}{2} Re\ [(\frac{\psi}{r + ia\cos\theta})e^3 + \chi d\bar{Y}], \tag{7}$$

where $\chi = 2\int(1+Y\bar{Y})^{-2}\psi dY$, which obeys the alignment condition $\alpha_\mu k^\mu = 0$.

For the Kerr solution m is mass of BH and $\psi = 0$. For the KN solution $\psi = q = const.$. However, any nonconstant holomorphic function $\psi(Y)$ yields also an exact KS solution, [4]. On the other hand, any nonconstant holomorphic functions on sphere acquire at least one pole. A single pole at $Y = Y_i$, $\psi_i(Y) = q_i/(Y - Y_i)$ produces the beam in angular directions $Y_i = e^{i\phi_i}\tan\frac{\theta_i}{2}$. The function $\psi(Y)$ acts immediately on the function H which determines the metric and position of the horizon. The given in [12] analysis showed that electromagnetic beams have very strong back reaction to metric and deform topologically the horizon, forming in the horizon the holes which allow matter to escape interior. For $\psi(Y) = \sum_i \frac{q_i}{Y - Y_i}$, the exact solutions have several beams in angular directions $Y_i = e^{i\phi_i}\tan\frac{\theta_i}{2}$, leading to the horizon with many holes. In far zone the twistor beams tend to the known exact singular pp-wave solutions [1].

5. Pre-quantum fluctuating KS geometry

The stationary KS beamlike solutions may be generalized to time-dependent wave pulses, [3]. Since the horizon is extra sensitive to electromagnetic excitations, it may also be sensitive to the vacuum electromagnetic field, and the vacuum beam pulses may produce a fine-grained structure of fluctuating microholes in the horizon, allowing radiation to escape interior of black-hole.

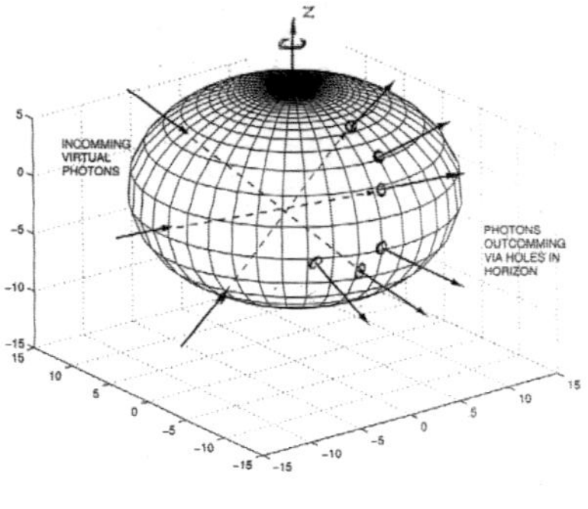

Figure 3. Twistor-beam pulses perforate the BH horizon forming its fluctuating fine-grained structure.

Twistor-beam pulses depend on a retarded time τ and have to obey to the non-stationary Debney-Kerr-Schild (DKS) equations [4]. The corresponding solutions acquire an extra radiative term $\gamma(Y, \tau)$. The long-term attack on the DKS equations [3], has led to the obtained time-dependent solutions of the fluctuating Kerr-Schild space-times. Derivation of the time-dependent solutions will be given in the next section. Here we give a brief preliminary description of the general structure of the DKS solutions obtained by integration in [4].

The general em field is described by the self-dual tetrad components,

$$\mathcal{F}_{12} = AZ^2, \quad \mathcal{F}_{31} = \gamma Z - (AZ)_{,1}\,, \tag{8}$$

where $\mathcal{F}_{ab} = e^\mu_a e^\nu_b \mathcal{F}_{\mu\nu}$. We will assume that the mass of BH is much greater then the energy of the BH excitation. It allows us to neglect by recoil and consider the Kerr congruence as stationary. Fluctuation of the metric will be related with fluctuation of the em field. By these assumptions we obtain, [3], that the exact time-dependent solutions describe the em radiation from BH which contains two components:

a) a set of the singular beam pulses (determined by function $\psi(Y,\tau)$) propagating along the Kerr PNC and breaking the topology and impenetrability of the horizon; and

b)the regularized radiative component (determined by $\gamma_{reg}(Y,\tau)$) which is smooth and determines evaporation of the black-hole.

The obtained solutions describe excitations of singular electromagnetic beams $A = \psi(Y,\tau)/P^2$ on the Kerr-Schild background and the back reaction of the singular field $\psi(Y,\tau)$ on the metric leading to fine-grained fluctuations of the metric and black-hole horizon.

The most essential difference from the stationary case is the appearance of an extra field $\gamma(Y,\tau)$. This term is absent in the expression for (6), and therefore, it does not have immediate action on the KS metric. One sees also from (8) that it generates the null EM radiation along the Kerr congruence which fall off asymptotically as $\sim Z = P/(r + ia\cos\theta) \sim 1/r$, and so, it is the leading term at infinity. The corresponding components of the stress-energy tensor

$$T_{\mu\nu} = T_{33} e^3_\mu e^3_\nu \sim = \frac{1}{2}\bar{\gamma}\gamma k_\mu k_\nu \tag{9}$$

describe a flow of energy along the Kerr congruence and are known as radiation of the Vaidya 'shining star' solution [1]. This term appears also in the Kinnersley 'photon rocket' solution [1], where it is related with acceleration of the source.

It shows explicitly that the horizon turns out to be covered by fluctuating micro-holes which make it penetrable for outgoing radiation. In the same time, the BH radiation is determined by the smooth field $\gamma_{reg}(Y,\tau)$, corresponding to the Vaidya "shining star" radiation. The holographic space-time forms a fluctuating twosheeted pre-geometry which reflects the dynamics of singular beam pulses. This pre-geometry is classical, but has to be still regularized to get the usual smooth classical space-time. In this sense, it takes an intermediate position between the classical and quantum gravity.

Note also that this structure may be generalized by using the Kerr theorem with the higher degree in Y functions $F(Y)$, which leads to the multi-particle KS solutions [13] having the complicate networks of twistor-beams.

6. Time-dependent Kerr-Schild solutions

The nearest time-dependent generalization of (4) is given by the form

$$\psi(Y,\tau) = \sum_i c_i(\tau)(Y - Y_i)^{-1}, \tag{10}$$

where one assumes that an elementary beam has $c_i(\tau) = q_i(\tau)e^{i\omega_i\tau}$ where $q_i(\tau)$ is amplitude and ω_i is a carrier frequency. Similar to stationary case, the non-stationary Kerr-Schild solutions contain the singular beams which change the structure of black hole horizon, but the beams may acquire the form of time-dependent pulses. Since the mass of a black hole is much more than the energy of its excitation, we will neglect by recoil and take the assumption that the black hole is at rest and the Kerr congruence remains undisturbed. It corresponds to $P = 2^{-1/2}(1 + Y\bar{Y})$ and $\dot{P} = 0$.

The obtained in [4] DKS general equations are the equations for functions A and γ which determine electromagnetic field

$$A,_2 - 2Z^{-1}\bar{Z}Y,_3 A = 0, \quad A,_4 = 0, \tag{11}$$

$$\mathcal{D}A + \bar{Z}^{-1}\gamma,_2 - Z^{-1}Y,_3 \gamma = 0, \tag{12}$$

and the equations for function M which determine gravitational field taking into account the back reaction caused by electromagnetic field

$$M,_2 - 3Z^{-1}\bar{Z}Y,_3 M = A\bar{\gamma}\bar{Z}, \tag{13}$$

$$\mathcal{D}M = \frac{1}{2}\gamma\bar{\gamma}. \tag{14}$$

The used here operator $\mathcal{D}$ is

$$\mathcal{D} = \partial_3 - Z^{-1}Y,_3 \partial_1 - \bar{Z}^{-1}\bar{Y},_3 \partial_2 \ . \tag{15}$$

General equations for the nonstationary EM KS solutions were obtained by by Debney, Kerr and Schild [4] in 1968. However, in [4] the equations were fully integrated out only for the particular case $\gamma = 0$ which corresponds to stationary solutions.[4]

The first non-stationary wave EM solutions on the KS background were obtained in [14] and contained singular beams along the $\pm z-$ half-axis. There appeared the conjecture that there is no the exact regular time-dependent solutions at all, and for very seldom exclusions the exact solutions should have the singular beams.

The DKS equations are simplified by using the DKS relation

$$Y,_3 = -ZP_{\bar{Y}}/P. \tag{16}$$

The equation (11) for function A takes the same form as in stationary case,

$$(AP^2),_2 = 0, \ \ A,_4 = 0, \tag{17}$$

where the general solution is $A = \psi/P^2$, and function ψ has to obey

$$\psi,_2 = \psi,_4 = 0. \tag{18}$$

To obtain the non-stationary KS solutions we introduce a complex retarded-time parameter τ which satisfies the relations

$$(\tau),_2 = (\tau),_4 = 0 \ . \tag{19}$$

It allows us to represent the equation (11) in the form $(AP^2),_2 = 0$, and to get solution which has the same form

$$A = \psi(Y, \tau)/P^2 \tag{20}$$

[4] The degree of difficulty of this problem is described in the excerptions from the given by Roy Kerr in [15] (see Part 3 on the page 2481) review on the history of the Kerr solution. In this review Roy calls the eqs. (8), (17) and (18) as "easy" ones which *"... showed that the Einstein-Maxwell field depends on three functions, M, A, and γ ... restricted by "hard" field equations..."* [The "hard" equations are the EM eq. (12) and gravitational equation (13). AB.] ... *"I could not solve the later unless $\gamma = 0$ so I temporarily took this as an additional assumption and continued. This led to the complete charged Kerr-Schild metric metrics including, of course, charged Kerr. The null congruence for the later is the same as the Kerr congruence but the EM field depends on an arbitrary function of a complex variable...If it is a more complicated function then the EM field will probably be singular.* [This is exactly the case of stationary beam-like KS solutions. AB.] *At that point I turned the problem over to G.C. Debney ... to see if he could solve the problem when $\gamma \neq 0$. ... In the end it became clear that the tree of us could not solve the general problem."*

with the exception of the extra dependence of function ψ from the retarded-time parameter τ.

The principal difference from the stationary case is contained in the second electromagnetic DKS equation. Action of operator $\mathcal{D}$ on the variables $Y, \bar{Y}$ and τ is

$$\mathcal{D}Y = \mathcal{D}\bar{Y} = 0. \tag{21}$$

For the considered here case $P = 2^{-1/2}(1 + Y\bar{Y})$, and $\dot{P} = 0$, which yields

$$\mathcal{D}\tau = -1/P \; . \tag{22}$$

Using (16), (21) and (22), we reduce the equation (12) to the form

$$\dot{A} = -(\gamma P),_{\bar{Y}} \, , \tag{23}$$

where $\dot{(\ )} \equiv \partial_\tau$. Integrating this equation we obtain

$$\gamma = -P^{-1}\int \dot{A}d\bar{Y} = \frac{2^{1/2}\dot{\psi}}{P^2 Y} + \phi(Y,\tau)/P, \tag{24}$$

where ϕ is an arbitrary analytic function of Y and τ.

This solution shows that *any non-stationarity in electromagnetic field ($\dot{A} \neq 0$) generates an extra function* γ which, in accord with (8), generates also the lightlike electromagnetic radiation along the Kerr congruence. Such a radiation is well-known for the Vaidya 'shining star' solution [1], in which the field $A = \psi P^{-2}$ is absent and γ is incoherent, being related with the loss of total mass into radiation. The KS analog of this Vaidya relation is obtained from (14) by substitution

$$M = m/P^3, \tag{25}$$

and using (16), (21) and (22), and also the fact that $t = \frac{1}{2}(\tau + \bar{\tau})$,

$$\dot{m} = -\frac{1}{2}P^4\gamma^{(reg)}\bar{\gamma}^{(reg)}. \tag{26}$$

It is one of two gravitational equations determining self-consistency of the Kerr-Schild solution [4]. The most important consequence following from DKS equations in the non-stationary case is the fact that the field γ appears inevitable, however it does not contribute to deformation of the horizon, since it is absent in the function H of (6). Its back reaction on the metric is smooth and circumstantial, acting only via the slowly decreasing mass parameter m. Therefore, in the non-stationary Kerr-Schild case we obtain that the *lightlike fields, determined by functions* ψ *and* γ*, have essentially different impact on the horizon.*

Function $\psi = \psi(\tau, Y)$ obeys the equation (18) which shows that the retarded time τ has to satisfy the conditions similar to (3), and therefore, gradient of τ is to be aligned to congruence,

$$k^\mu \tau,_\mu = 0. \tag{27}$$

It was obtained in [14] that the corresponding retarded-time parameter has the form

$$\tau = t - r - ia\cos\theta. \tag{28}$$

Since function ϕ contributes only to γ it does not impact on the form of the horizon too. Important role of this function is obtained from the analysis of the gravitational Kerr-Schild equation (13) which is reduced by using (16), (21) and (25) to the form

$$m,_{\bar{Y}} = \psi\bar{\gamma}P. \tag{29}$$

If we note that $\gamma \sim \dot{\psi} \approx i\omega\psi$, we obtain that the r.h.s. of this equation tends to zero in the low-frequency limit, as well as the r.h.s of the equation for $\dot{m}$. So, the full solution will tend to consistent with gravity at least in the low-frequency limit [9]. For the simplicity we consider first the case of a single time-dependent beam generated by a single pole at $Y = Y_i$. In this case

$$\dot{\psi}_i = \dot{c}_i(\tau)/(Y - Y_i) \tag{30}$$

and the solution (24) looks singular. However, we have still the free function ϕ, form of which and parameters (time-dependence and position of its pole) may be tuned to cancel the pole of function $\dot{\psi}_i$, and thus, regularize the solution. We define the analytic in Y function

$$P_i = P(Y, \bar{Y}_i) = 2^{-1/2}(1 + Y\bar{Y}_i) \tag{31}$$

and set

$$\phi_i^{(tun)}(Y,\tau) = -\frac{2^{1/2}\dot{c}_i(\tau)}{Y(Y - Y_i)P_i}. \tag{32}$$

One sees that the required analyticity of function $\phi_i^{(tun)}(Y,\tau)$ in Y is ensured. Using the equality

$$(P_i - P)/(Y_i - Y) = \frac{Y}{\sqrt{2}}\frac{(\bar{Y}_i - \bar{Y})}{(Y_i - Y)}, \tag{33}$$

we obtain that the regularized solution $\gamma_i^{(reg)} = \frac{2^{1/2}\dot{\psi}_i}{P^2 Y} + \phi_i^{(tun)}(Y,\tau)/P$, takes the form

$$\gamma_i^{(reg)} = \frac{1}{P^2}\frac{\dot{c}_i}{P_i}[\frac{\bar{Y}_i - \bar{Y}}{Y_i - Y}]. \tag{34}$$

The r.h.s. of the equation (29) for this regular solution takes the form

$$\psi_i\bar{\gamma}_i^{(reg)}P = -\frac{c_i\dot{\bar{c}}_i}{P\bar{P}_i(\bar{Y} - \bar{Y}_i)}. \tag{35}$$

The equation (29) takes the form

$$m = -\int d\bar{Y}\frac{c_i\dot{\bar{c}}_i}{P\bar{P}_i(\bar{Y} - \bar{Y}_i)} \tag{36}$$

and may now be integrated by using the Cauchy integral formula, $\oint \frac{f(z)dz}{z-z'} = 2\pi i f(z')$, and we obtain the expression

$$m = m_0(Y) - 2\pi i\frac{c_i\dot{\bar{c}}_i}{P_i P_{ii}}, \tag{37}$$

containing free function $m_0(Y)$, and also the depending on Y contribution from the residue at singular point $\bar{Y}_i$, in which

$$P_{ii} = \frac{1}{\sqrt{2}}(1 + Y_i\bar{Y}_i) \tag{38}$$

is a constant. On the real slice P is real, $P_i \to P_{ii}$, and the real part of m takes the form

$$\Re e\ m = m_0 - i\pi\frac{c_i\dot{\bar{c}}_i - \dot{c}_i\bar{c}_i}{P_{ii}^2}, \tag{39}$$

where m_0 is a real constant. Imaginary part is

$$\Im m\ m = -i\pi \frac{c_i \dot{\bar{c}}_i + \dot{c}_i \bar{c}_i}{P_{ii}^2} \ . \qquad (40)$$

If $c_i(\tau)$ is expressed via slowly varying amplitude $q_i(\tau)$ and the carrier frequency ω_i, $c_i(\tau) = q_i(\tau)e^{-i\omega_i \tau}$, the impact of the carrier frequency disappears and we obtain

$$\Re e\ m = m_0 - 2\pi\omega_i \frac{|q_i|^2}{P_{ii}^2} - i\pi \frac{q_i \dot{\bar{q}}_i - \dot{q}_i \bar{q}_i}{P_{ii}^2} \ , \qquad (41)$$

So, the result of integration yields the time-dependence of m caused by the amplitude $q_i(\tau)$ and the weak and slow imaginary contribution

$$\Im m\ m = -i\pi \frac{q_i \dot{\bar{q}}_i + \dot{q}_i \bar{q}_i}{P_{ii}^2} \ ., \qquad (42)$$

meaning of which is not clear at this stage. We have

$$m = m_0 - 2\pi\omega_i \frac{|q_i|^2}{P_{ii}^2} - 2\pi i \frac{q_i \dot{\bar{q}}_i}{P_{ii}^2} = m_0 - 2\pi i \frac{q_i}{P_{ii}^2} (\dot{\bar{q}}_i - i\omega_i \bar{q}_i) \ . \qquad (43)$$

Influence of the extra contributions from $\dot{q}_i(\tau)$ falls off as $1/T$, where T is effective time of action of the beam. This contribution is negligible in the limit of the very long beams. Although these relations may be interpreted physically as a reaction of the mass parameter to single pulse, they cannot be considered literally, since the imaginary mass is inconsistent with the KS class of metrics. To provide consistency with gravitational sector we enforced to perform an averaging of the mass term over time, which is equivalent to averaging of the stress-energy tensor.

Further, it is known [1] that the gravitational equation (26) for radiative solutions is not equation really, but is only a definition of the loss of mass in radiation $\dot{m} = -\frac{1}{2} \frac{\dot{c}_i \dot{\bar{c}}_i}{|P_i|^2}$. On the real slice $P_i \to P_{ii}$, and in terms of the amplitudes of beams we obtain

$$\dot{m} = -\frac{1}{2} \frac{\omega_i^2 |q_i|^2 + |\dot{q}_i|^2}{P_{ii}^2} = -\frac{1}{2P_{ii}^2} (\dot{q}_i - i\omega_i q_i)(\dot{\bar{q}}_i - i\omega_i \bar{q}_i). \qquad (44)$$

Similar to the Vaidya 'shining star' solution [1], for the case of the many stochastic pulses, we assume their frequencies as uncorrelated. There appears a double sum over the interacting beams and extra beating between them with fluctuations which drop out after averaging. The surviving radiation contains only slow fluctuations of the mass term caused by amplitudes of the self-correlated beams, i.e. the sum of the partial solutions pulses. Therefore, the obtained solutions turn out to be consistent with respect to the Einstein equations with averaged r.h.s., which for the DKS gravitational equations is equivalent to

$$m_{,\bar{Y}} =< P\psi \bar{\gamma}^{(reg)} >, \quad \dot{m} = -\frac{1}{2} < P^2 \gamma^{(reg)} \bar{\gamma}^{(reg)} > . \qquad (45)$$

The averaging removes the beating of the incoherent beams, however, it does not remove the sharp back-reaction of the beams to metric and horizon, caused by the poles in function $\psi(Y, \tau)$ in agreement with (6).

Therefore, the obtained solutions are exact and consistent with the Einstein-Maxwell system of equations with averaged stress-energy tensor. Recall also that we have used the restriction on the absence of recoil. The case of the non-zero recoil is very hard and was considered earlier by Kinnersley only for the non-rotating case [1]. It may be very important for the problem of scattering of spinning particles in twistor theory.

Acknowledgments

Author thanks the Organizers and especially Elias Vagenas for invitation to this very interesting conference and and the RFBR (grants 07-08-00234-a, 09-02-08353-) for financial support.

[1] D.Kramer, H.Stephani, E. Herlt, M.MacCallum 1980 *Exact Solutions of Einstein's Field Equations* , Cambridge Univ. Press.

[2] E. Witten 2004 *Comm. Math. Phys.* **252** 189.

[3] A. Burinskii, First Award of Gravity Research Foundation 2009 *Gen. Rel. Grav.* **41** 2281, arXiv: 0903.3162[gr-qc]; details in 0903.2365[hep-th].

[4] G.C. Debney, R.P. Kerr and A.Schild 2009 *J. Math. Phys.* **10**, 1842.

[5] G. W. Gibbons 1975 *Comm. Math. Phys.* **45** 191.

[6] C.R. Stephens, G. 't Hooft and B.F. Whiting 994 *Class. Quant. Grav.* **11** 621.

[7] R. Penrose 1967 *J. Math. Phys.* **8** 345;

[8] A. Burinskii 2004 *Phys. Rev.* **D 70**, 086006.

[9] A. Burinskii, E.Elizalde, S. R. Hildebrandt and G. Magli 2002 *Phys. Rev.* **D65** 064039, arXiv:gr-qc/0109085.

[10] A. Burinskii 1974 *Sov.Phys.JETP*, **39** 193; 1974 *Russian Phys.J.* **17**, 1068, DOI:10.1007/BF00901591; 2008 *Grav. Cosmol.* **14** 109.

[11] A. Burinskii, 2009 Superconducting Source of the Kerr-Newman Electron, arXiv: 0910.5388[hep-th].

[12] A. Burinskii, E. Elizalde, S.R. Hildebrandt and G. Magli 2006 *Phys. Rev.* **D 74** 021502; 2006 *Grav. Cosm.* 12 119.

[13] A. Burinskii 2005 *Grav. Cosmol.* **11**, 301; 2007 *IJGMMP*, **4**, n.3, 437.

[14] A. Burinskii, Grav. Cosmol.**10**, (2004) 50.

[15] Roy P. Kerr, Gen Relativ Gravit **41** 2479 (2009); DOI 10.1007/s10714-009-0856-0.

[16] A. Burinskii, *Phys.Rev.***D 67**, (2003) 124024.

Journal of Physics: Conference Series **222** (2010) 012045 doi:10.1088/1742-6596/222/1/012045

Canonical Quantization of Some Midi-Superspace Models in 3+1 Dimensions

T Christodoulakis[1], G Doulis[1], Petros A Terzis[1], E Melas[2], Th Grammenos[3], G O Papadopoulos[4], and A Spanou[5]

[1] Nuclear and Particle Physics Section, Physics Department, University of Athens, GR 157–71 Athens

[2] Technological Educational Institution of Lamia, Electrical Engineering Department, GR 35–100, Lamia

[3] Department of Civil Engineering, University of Thessaly, GR 383–34 Volos

[4] Department of Mathematics and Statistics, Dalhousie University, Halifax, Nova Scotia, Canada B3H 3J5

[5] School of Applied Mathematics and Physical Sciences, National Technical University of Athens, GR 157–80, Athens

E-mail: tchris@phys.uoa.gr, gdoulis@phys.uoa.gr, pterzis@phys.uoa.gr, evangelosmelas@yahoo.co.uk, thgramme@uth.gr, gopapad@mathstat.dal.ca, aspanou@central.ntua.gr

Abstract. A proposal is put forward which enables the canonical quantization of a family of spherically symmetric geometries in 3+1 dimensions. The proposal consists of a particular renormalization **Assumption** and an accompanying **Requirement** and results in a Wheeler–DeWitt equation which is based on a renormalized manifold parametrized by three smooth scalar functionals. The aforementioned equation is analytically solved for the 3+1 case.

1. Introduction

After Dirac's pioneering work on the treatment of systems with constraints [1], [2], [3], [4] the way was open for a systematic treatment of constrained dynamics. Some of the landmarks in the study of constrained systems have been the connection between constraints and invariances [5], the extension of the formalism to describe fields with half-integer spin through the algebra of Grassmann variables [6] and the introduction of the BRST formalism [7]. All these results have helped the quantization of gauge theories. There exist several excellent reviews studying constraint systems with a finite number of degrees of freedom [8] or constraint field theories [9], as well as more general presentations [10], [11], [12], [13], [14], [15]. In particular, the conventional canonical analysis approach of quantum gravity has been initiated by B.S. DeWitt [16] based on earlier work of P.G. Bergmann [17].

Since a full theory of quantum gravity does not exist today, it is reasonably important to address the quantization of (classes of) simplified geometries. An elegant way to achieve a degree of simplification is to impose some symmetry. For example, the assumption of a G_3 symmetry group acting simply transitively on the surfaces of simultaneity, i.e. the existence of three independent space-like Killing vector fields, leads to classical and subsequently quantum homogeneous cosmology (see, e.g., [18], [19]). The imposition of lesser symmetry, e.g. fewer Killing vector fields, results in the various inhomogeneous cosmological models [20]. The canonical analysis under the assumption of spherical symmetry, which is a G_3 group acting multiply transitively on two-dimensional space-like subsurfaces of the three-slices, has been first considered in [21], [22].

In this work the canonical quantization of all geometries admitting two-dimensional surfaces of maximal symmetry, i.e. spheres (constant positive curvature), planes (zero curvature) and Gauss-Bolyai-Lobachevsky (henceforth GBL) spaces (constant negative curvature) is considered. The structure of the paper is as follows: In Section 2 we give the reduced metrics, the space of classical solutions and the Hamiltonian formulation of the reduced Einstein-Hilbert action principle, resulting in one (quadratic) Hamiltonian and two (linear) momentum constraints, all being first class. In Section 3, we consider the quantization of this constraint system following Dirac's proposal of implementing the quantum operator constraints as conditions annihilating the wave-function [4]. Our guide-line is a conceptual generalization of the quantization scheme developed in [24], [25] for the case of constraint systems with finite degrees of freedom, to the present case. Even though after the symmetry reduction the system still represents a field theory (all remaining metric components depend on time and the radial coordinate), we manage to extract and subsequently completely solve a Wheeler-DeWitt equation in terms of three unique smooth scalar functionals of the appropriate components of the reduced spatial metric. This is achieved through an appropriate renormalization assumption we adopt. Finally, some concluding remarks are included in the discussion.

2. Possible Metrics, Classical Solutions and Hamiltonian Formulation

Our starting point is the two-dimensional spaces of positive, zero or negative constant curvature. Their line elements are respectively:

$$ds^2 = d\theta^2 + \sin^2\theta\, d\phi^2, \qquad ds^2 = d\theta^2 + \theta^2\, d\phi^2, \qquad ds^2 = d\theta^2 + \sinh^2\theta\, d\phi^2 \tag{2.1}$$

with an obvious range of the coordinates for each case. The corresponding (maximal) symmetry groups are generated by the following Killing vector fields (KVF):

$$\xi_1 = \frac{\partial}{\partial\phi}, \quad \xi_2 = -\cos\phi\frac{\partial}{\partial\theta} + \cot\theta\sin\phi\frac{\partial}{\partial\phi}, \qquad \xi_3 = \sin\phi\frac{\partial}{\partial\theta} + \cot\theta\cos\phi\frac{\partial}{\partial\phi} \tag{2.2}$$

$$\xi_1 = \frac{\partial}{\partial\phi}, \quad \xi_2 = -\cos\phi\frac{\partial}{\partial\theta} + \frac{\sin\phi}{\theta}\frac{\partial}{\partial\phi}, \qquad \xi_3 = \sin\phi\frac{\partial}{\partial\theta} + \frac{\cos\phi}{\theta}\frac{\partial}{\partial\phi} \tag{2.3}$$

$$\xi_1 = \frac{\partial}{\partial\phi}, \quad \xi_2 = -\cos\phi\frac{\partial}{\partial\theta} + \coth\theta\sin\phi\frac{\partial}{\partial\phi}, \quad \xi_3 = \sin\phi\frac{\partial}{\partial\theta} + \coth\theta\cos\phi\frac{\partial}{\partial\phi} \tag{2.4}$$

Trivially prolonging these KVF's to four-dimensional space-time, we arrive at the three classes of metrics described by the following line element:

$$\begin{aligned} ds^2 = &\left(-\alpha(t,r)^2 + \frac{\beta(t,r)^2}{\gamma(t,r)^2}\right) dt^2 + 2\,\beta(t,r)\,dt\,dr + \gamma(t,r)^2\,dr^2 \\ &+ \psi(t,r)^2\,d\theta^2 + \psi(t,r)^2\,f(\theta)^2\,d\phi^2 \end{aligned} \tag{2.5}$$

where $f(\theta) = \sin\theta$ in the case of spherical symmetry, $f(\theta) = \theta$ in the case of plane symmetry, and $f(\theta) = \sinh\theta$ in the case of the GBL symmetry.

In order to attain the classical space of solution for these geometries one can, exploiting the freedom to change coordinates in the (t, r) subspace, bring the upper left block of the metric in conformally flat form and readily solve the vacuum Einstein's field equations. The result is given, in the light-cone coordinates $u = \frac{t+r}{2}$, $v = \frac{-t+r}{2}$, by the following two line elements:

$$ds^2 = 2\,\epsilon\,\frac{A + 2\,\psi(u,v)}{4\,\psi(u,v)}\,du\,dv + \psi(u,v)^2\,d\theta^2 + \psi(u,v)^2\,f(\theta)^2\,d\phi^2, \tag{2.6}$$

where

$$\psi(u,v) = -\frac{A}{2}\left(1 + \mathrm{ProductLog}(-\frac{\exp^{-\frac{\lambda(A+u+\epsilon\,v)}{A}}}{A})\right),$$

$\epsilon = \pm 1$, $\lambda = \pm 1$ for $f(\theta) = \sin\theta$, $\lambda = \pm i$, for $f(\theta) = \sinh\theta$ and $\mathrm{ProductLog}(z)$ is the principal solution for w to the equation $z = w\,\exp^w$, and

$$ds^2 = 2\,\epsilon\,\frac{1}{\psi(u,v)}du\,dv + \psi(u,v)^2 d\theta^2 + \psi(u,v)^2\theta^2 d\phi^2, \tag{2.7}$$

where

$$\psi(u,v) = \pm\sqrt{2u + 2\epsilon\,v}.$$

The canonical analysis proceeds as follows: Utilizing the Gauss-Codazzi equation (see, e.g., [26]), we eliminate all second time-derivatives from the Einstein-Hilbert action, by subtracting an appropriate divergence, and arrive at an action quadratic in the velocities,

$$I = \int d^4x\sqrt{-g}(R - 2\Lambda - 2\,F^I_{;I})$$

The application of the Dirac algorithm results in the primary constraints

$$P_\alpha \equiv \frac{\delta L}{\delta\dot{\alpha}} \approx 0, \qquad P^\beta \equiv \frac{\delta L}{\delta\dot{\beta}} \approx 0$$

and the reduced Hamiltonian reads

$$H = \int \left(N^o\,\mathcal{H}_o + N^i\,\mathcal{H}_i\right) dr, \tag{2.8}$$

where

$$N^o = \alpha(t,r), \quad N^1 = \frac{\beta(t,r)}{\gamma(t,r)^2}, \quad N^2 = 0, \quad N^3 = 0$$

and $\mathcal{H}_o$, $\mathcal{H}_i$ are given by

$$\mathcal{H}_o = \frac{1}{2}G^{\alpha\beta}\,\pi_\alpha\,\pi_\beta + V \tag{2.9a}$$

$$\mathcal{H}_1 = -\gamma\,\pi'_\gamma + \psi'\,\pi_\psi, \qquad \mathcal{H}_2 = 0, \qquad \mathcal{H}_3 = 0, \tag{2.9b}$$

Journal of Physics: Conference Series **222** (2010) 012045 doi:10.1088/1742-6596/222/1/012045

while the indices $\{\alpha, \beta\}$ take the values $\{\gamma, \psi\}$ and $' = \frac{d}{dr}$. The reduced Wheeler-DeWitt super-metric $G^{\alpha\beta}$ reads

$$G^{\alpha\beta} = \begin{pmatrix} \frac{\gamma}{4\psi^2} & -\frac{1}{4\psi} \\ -\frac{1}{4\psi} & 0 \end{pmatrix}, \tag{2.10}$$

while the potential V is

$$V = -2\epsilon\gamma + 2\Lambda\gamma\psi^2 - 2\frac{\psi'^2}{\gamma} + 4\left(\frac{\psi\psi'}{\gamma}\right)' \tag{2.11}$$

with $\epsilon = \{1, 0, -1\}$ for the families of two-dimensional subspaces with positive, zero or negative constant curvature, respectively. The requirement for preservation, in time, of the primary constraints leads to the secondary constraints

$$\mathcal{H}_o \approx 0, \qquad \mathcal{H}_1 \approx 0 \tag{2.12}$$

The "open" Poisson bracket algebra satisfied by these constraints is:

$$\begin{aligned} \{\mathcal{H}_o(r), \mathcal{H}_o(\tilde{r})\} &= \left[\frac{1}{\gamma^2(r)}\mathcal{H}_1(r) + \frac{1}{\gamma^2(\tilde{r})}\mathcal{H}_1(\tilde{r})\right]\delta'(r, \tilde{r}), \\ \{\mathcal{H}_1(r), \mathcal{H}_o(\tilde{r})\} &= \mathcal{H}_o(r)\delta'(r, \tilde{r}), \\ \{\mathcal{H}_1(r), \mathcal{H}_1(\tilde{r})\} &= \mathcal{H}_1(r)\delta'(r, \tilde{r}) - \mathcal{H}_1(\tilde{r})\delta(r, \tilde{r})' \end{aligned} \tag{2.13}$$

indicating that they are first class and also signaling the termination of the algorithm. Thus, our system is described by (2.12) and the dynamical Hamilton-Jacobi equations $\frac{d\pi_\gamma}{dt} = \{\pi_\gamma, H\}$, $\frac{d\pi_\psi}{dt} = \{\pi_\psi, H\}$. These equations, when expressed in the velocity phase-space with the help of the definitions $\frac{d\gamma}{dt} = \{\gamma, H\}$, $\frac{d\psi}{dt} = \{\psi, H\}$, are completely equivalent to the independent Einstein's field equations satisfied by (2.5).

Under changes of the radial variable r of the form $r \to \tilde{r} = h(r)$, it can easily be inferred from (2.5) that

$$\tilde{\gamma}(\tilde{r}) = \gamma(r)\frac{dr}{d\tilde{r}}, \qquad \tilde{\psi}(\tilde{r}) = \psi(r), \qquad \frac{d\tilde{\psi}(\tilde{r})}{d\tilde{r}} = \frac{d\psi(r)}{dr}\frac{dr}{d\tilde{r}}, \tag{2.14}$$

where the t-dependence has again been omitted. Thus, under the above coordinate transformations, ψ is a scalar, while γ, ψ' are covariant rank 1 tensors (one-forms), or, equivalently in one dimension, scalar densities of weight -1. Therefore, the scalar derivative is not $\frac{d}{dr}$ but rather $\frac{d}{\gamma\, dr}$. Finally, if we consider an infinitesimal transformation $r \to \tilde{r} = r - \eta(r)$, it is easily seen that the corresponding changes induced on the basic fields are:

$$\delta\gamma(r) = (\gamma(r)\,\eta(r))', \qquad \delta\psi(r) = \psi'(r)\,\eta(r) \tag{2.15}$$

i.e., nothing but the one-dimensional analogue of the appropriate Lie derivatives.
With the use of (2.15), we can reveal the nature of the action of $\mathcal{H}_1$ on the basic configuration space variables as that of the generator of spatial diffeomorphisms:

$$\begin{aligned} \left\{\gamma(r)\,,\,\int d\tilde{r}\,\eta(\tilde{r})\,\mathcal{H}_1(\tilde{r})\right\} &= (\gamma(r)\,\eta(r))', \\ \left\{\psi(r)\,,\,\int d\tilde{r}\,\eta(\tilde{r})\,\mathcal{H}_1(\tilde{r})\right\} &= \psi'(r)\,\eta(r). \end{aligned} \tag{2.16}$$

Thus, we are justified to consider $\mathcal{H}_1$ as the representative, in phase-space, of an arbitrary infinitesimal reparametrization of the radial coordinate.

3. Quantization

We are now interested in attempting to quantize this Hamiltonian system following Dirac's general spirit of realizing the classical first class constraints as quantum operator constraint conditions annihilating the wave function. The main motivation behind such an approach is the justified desire to construct a quantum theory manifestly invariant under the "gauge" generated by the constraints. To begin with, let us first note that the system is still a field theory in the sense that all configuration variables and canonical conjugate momenta depend not only on time (as is the case in homogeneous cosmology), but also on the radial coordinate r. Thus, to canonically quantize the system in the Schrödinger representation, we first realize the classical momenta as functional derivatives with respect to their corresponding conjugate fields

$$\hat{\pi}_\gamma(r) = -i\,\frac{\delta}{\delta\,\gamma(r)}, \qquad \hat{\pi}_\psi(r) = -i\,\frac{\delta}{\delta\,\psi(r)}.$$

We next have to decide on the initial space of state vectors. To elucidate our choice, let us consider the action of a momentum operator on some function of the configuration field variables, say

$$\hat{\pi}_\gamma(r)\gamma(\tilde{r})^2 = -2i\gamma(\tilde{r})\delta(\tilde{r},r).$$

The Dirac delta-function renders the outcome of this action a distribution rather than a function. Also, if the momentum operator were to act at the point at which the function is evaluated, i.e. if $\tilde{r} = r$, then its action would produce a $\delta(0)$ and would therefore be ill-defined. Both of these unwanted features are rectified, as far as expressions linear in momentum operators are concerned, if we choose as our initial collection of states all *smooth functionals* (i.e., integrals over r) of the configuration variables $\gamma(r)$, $\psi(r)$ and their derivatives of any order. Indeed, as we infer from the previous example,

$$\hat{\pi}_\gamma(r)\int d\tilde{r}\gamma(\tilde{r})^2 = -2i\int d\tilde{r}\gamma(\tilde{r})\delta(\tilde{r},r) = -2i\gamma(r)$$

thus the action of the momentum operators on all such states will be well-defined (no $\delta(0)$'s) and will also produce only local functions and not distributions. However, even so, $\delta(0)$'s will appear as soon as local expressions quadratic in momenta are considered, e.g.,

$$\hat{\pi}_\gamma(r)\,\hat{\pi}_\gamma(r)\int d\tilde{r}\gamma(\tilde{r})^2 = \hat{\pi}_\gamma(r)(-2i\int d\tilde{r}\gamma(\tilde{r})\delta(\tilde{r},r)) = \hat{\pi}_\gamma(r)(-2i\gamma(r)) = -2\delta(0)$$

Another problem of equal, if not greater, importance has to do with the number of derivatives (with respect to r) considered: A momentum operator acting on a smooth functional of degree n in derivatives of $\gamma(r), \psi(r)$ will, in general, produce a function of degree $2n$, e.g.,

$$\hat{\pi}_\gamma(r) \int d\tilde{r}\gamma''(\tilde{r})^2 = -2i \int d\tilde{r}\gamma''(\tilde{r})\delta''(\tilde{r}, r) = -2i\gamma^{(4)}(r).$$

Thus, clearly, more and more derivatives must be included if we desire the action of momentum operators to keep us inside the space of integrands corresponding to the initial collection of smooth functionals; eventually, we have to consider $n \to \infty$. This, in a sense, can be considered as the reflection to the canonical approach, of the non-renormalizability results existing in the so-called covariant approach. The way to deal with these problems is, loosely speaking, to regularize (i.e., render finite) the infinite distribution limits, and renormalize the theory by, somehow, enforcing n to terminate at some finite value.

In the following, we are going to present a quantization scheme of our systems which: (a) avoids the occurrence of $\delta(0)$'s, (b) reveals the value $n = 1$, as the only possibility to obtain a closed space of state vectors, and (c) extracts a finite-dimensional Wheeler-DeWitt equation governing the quantum dynamics. The scheme closely parallels, conceptually, the quantization developed in [24],[25] for finite systems with one quadratic and a number of linear first class constraints. Therefore, we deem it appropriate and instructive to present a brief account of the essentials of this construction.

To this end, let us consider a system described by a Hamiltonian of the form

$$\begin{aligned} H &\equiv \mu X + \mu^i \chi_i \\ &= \mu \left(\frac{1}{2}G^{AB}(Q^\Gamma)P_A P_B + U^A(Q^\Gamma)P_A + V(Q^\Gamma)\right) + \mu^i \phi_i^A(Q^\Gamma)P_A, \end{aligned} \tag{3.1}$$

where $A, B, \Gamma \ldots = 1, 2 \ldots, M$ count the configuration space variables and $i = 1, 2, \ldots, N < (M-1)$ numbers the super-momenta constraints $\chi_i \approx 0$, which along with the super-Hamiltonian constraint $X \approx 0$ are assumed to be first class:

$$\{X, X\} = 0, \qquad \{X, \chi_i\} = XC_i + C_i^j \chi_j, \qquad \{\chi_i, \chi_j\} = C_{ij}^k \chi_k, \tag{3.2}$$

where the first (trivial) Poisson bracket has been included only to emphasize the difference from the first of (2.13).

The physical state of the system is unaffected by the "gauge" transformations generated by (X, χ_i), **but also** under the following three changes:

(I) Mixing of the super-momenta with a non-singular matrix

$$\bar{\chi}_i = \lambda_i^j(Q^\Gamma)\chi_j$$

(II) Gauging of the super-Hamiltonian with the super-momenta

$$\bar{X} = X + \kappa^{(Ai}(Q^\Gamma)\phi_i^{B)}(Q^\Gamma)P_A P_B + \sigma^i(Q^\Gamma)\phi_i^A(Q^\Gamma)P_A$$

(III) Scaling of the super-Hamiltonian

$$\bar{X} = \tau^2(Q^\Gamma)X$$

Therefore, the geometrical structures on the configuration space that can be inferred from the super-Hamiltonian are really equivalence classes under actions (I), (II) and (III); for example (II), (III) imply that the super-metric G^{AB} is known only up to conformal scalings and additions of the super-momenta coefficients $\bar{G}^{AB} = \tau^2(G^{AB} + \kappa^{(Ai}\phi_i^{B)})$. It is thus mandatory that, when we Dirac-quantize the system, we realize the quantum operator constraint conditions on the wave-function in such a way as to secure that the whole scheme is independent of actions (I), (II), (III). This is achieved by the following steps:

(1) Realize the linear operator constraint conditions with the momentum operators to the right

$$\hat{\chi}_i \Psi = 0 \leftrightarrow \phi_i^A(Q^\Gamma) \frac{\partial\, \Psi(Q^\Gamma)}{\partial\, Q^A} = 0,$$

which maintains the geometrical meaning of the linear constraints and produces the $M-N$ independent solutions to the above equations $q^\alpha(Q^\Gamma)$, $\alpha = 1, 2, \ldots, M-N$ called physical variables, since they are invariant under the transformations generated by χ_i.

(2) Define the induced structure $g^{\alpha\beta} \equiv G^{AB} \dfrac{\partial\, q^\alpha}{\partial\, Q^A} \dfrac{\partial\, q^\beta}{\partial\, Q^B}$ and realize the quadratic in momenta part of X as the conformal Laplace-Beltrami operator based on $g_{\alpha\beta}$.

We are now ready to proceed with the quantization of our system (2.8-2.12) in close analogy to the scheme above outlined. In order to realize the equivalent to step 1, we first define the quantum analogue of $\mathcal{H}_1(r) \approx 0$ as

$$\hat{\mathcal{H}}_1(r)\Phi = 0 \leftrightarrow -\gamma(r)\,(\frac{\delta\, \Phi}{\delta\, \gamma(r)})' + \psi'(r)\,\frac{\delta\, \Phi}{\delta\, \psi(r)} = 0. \tag{3.3}$$

As explained, the action of $\hat{\mathcal{H}}_1(r)$ on all smooth functionals is well defined, i.e., produces no $\delta(0)$'s. It can be proven that, in order for such a functional to be annihilated by this linear quantum operator, it must be scalar, i.e. have the form

$$\Phi \;=\; \int \gamma(\tilde{r})\, L\left(\Psi^{(0)}, \Psi^{(1)}, \ldots, \Psi^{(n)}\right) d\tilde{r} \tag{3.4a}$$

$$\Psi^{(0)} \;\equiv\; \psi(\tilde{r}), \quad \Psi^{(1)} \equiv \frac{\psi'(\tilde{r})}{\gamma(\tilde{r})}, \ldots, \Psi^{(n)} \equiv \frac{1}{\gamma(\tilde{r})} \frac{d}{d\tilde{r}} \left(\underbrace{\cdots}_{n-1} \psi(\tilde{r}) \right) \tag{3.4b}$$

where L is any function of its arguments. We note that $\frac{\psi'}{\gamma}$ is the only scalar first derivative of ψ, and likewise for the higher derivatives. The proof of this statement is analogous to the corresponding result concerning full gravity [27]: consider an infinitesimal r-reparametrization $\tilde{r} = r - \eta(r)$. Under such a change, the left-hand side of (3.4a), being a number, must remain unaltered. If we calculate the change induced on the right-hand side we arrive at

$$0 = \int \left[L\delta\gamma + \gamma \frac{\delta L}{\delta\gamma}\delta\gamma + \gamma \frac{\delta L}{\delta\psi}\delta\psi \right] dr = \int [\gamma\, \hat{\mathcal{H}}_1(L)]\eta(r) dr, \tag{3.5}$$

where use of (2.15) and a partial integration has been made. Since this must hold for any $\eta(r)$, the result sought for is obtained.

In trying to realize step 2 of the programm, we have to define the equivalent of Kuchař's induced metric on the so far space of "physical" states Φ described by (3.4a) which are the analogues, in our case, of Kuchař's physical variables q^α. Let us start our investigation by considering **one** initial candidate of the above form. Then, generalizing the partial to functional derivatives, the induced metric will be given by

$$g^{\Phi\Phi} = G^{\alpha\beta}\,\frac{\delta\Phi}{\delta x^\alpha}\,\frac{\delta\Phi}{\delta x^\beta}, \quad \text{where} \quad x^\alpha = \{\gamma, \psi\} \tag{3.6}$$

and $G^{\alpha\beta}$ is given by (2.10). Note that this metric is well defined since it contains only first functional derivatives of the state vectors, as opposed to any second order functional derivative operator that might be considered as a quantum analogue of the kinetic part of $\mathcal{H}_o$. Nevertheless, $g^{\Phi\Phi}$ *is* a local function and not a smooth functional. It is thus clear that, if we want the induced metric $g^{\Phi\Phi}$ to be composed out of the "physical" states annihilated by $\hat{H}_1$, we must establish a correspondence between local functions and smooth functionals. A way to achieve this is to adopt the following ansatz:

Assumption: *We assume that, as part of the renormalization procedure, we are permitted to map local functions to their corresponding smeared expressions e.g.,* $\psi(r) \leftrightarrow \int d\tilde{r}\psi(\tilde{r})$.

Let us be more specific, concerning the meaning of the above Assumption. Let $\mathcal{F}$ be the space which contains all local functions, and define the equivalence relations

$$\sim: \{f_1(r) \sim f_2(\tilde{r}), \tilde{r} = g(r)\}, \quad \approx: \{h_1(r) \approx h_2(\tilde{r})\,\frac{d\,\tilde{r}}{d\,r}, \tilde{r} = g(r)\} \tag{3.7}$$

for scalars and densities respectively.

Now let $\mathcal{F}_o = \{f \in \mathcal{F}, \ \bmod\ (\sim, \approx)\}$ and $\mathcal{F}_\mathcal{I}$ the space of the smeared functionals. We define the one to one maps $\mathfrak{G}$, $\mathfrak{G}^{-1}$

$$\mathfrak{G} : \mathcal{F}_o \mapsto \mathcal{F}_I : \quad \psi(r) \mapsto \int \psi(\tilde{r})\,d\tilde{r}, \quad \mathfrak{G}^{-1} : \mathcal{F}_I \mapsto \mathcal{F}_o : \quad \int \psi(\tilde{r})\,d\tilde{r} \mapsto \psi(r) \tag{3.8}$$

The necessity to define the maps $\mathfrak{G}, \mathfrak{G}^{-1}$ on the equivalence classes and not on the individual functions, stems out of the fact that we are trying to develop a quantum theory of the geometries (2.5) and not of their coordinate representations. If we had tried to define the map $\mathfrak{G}$ from the original space $\mathcal{F}$ to $\mathcal{F}_I$ we would end up with states which would not be invariant under spatial coordinate transformations (**r**-reparameterizations). Indeed, one can make a correspondence between local functions and smeared expressions, but smeared expressions *must* contain another arbitrary smearing function, say $s(r)$. Then the map between functions and smeared expressions is one to one (as is also the above map) and is given by multiplying by $s(r)$ and integrating over r; while the inverse map is given by varying w.r.t. $s(r)$. However, this would be in the opposite direction from that which led us to the states (3.4a) by imposition of the linear operator constraint. As an example consider the action of this operator on one particular case of the states (3.4a), containing the structure $s(r)$:

$$\hat{\mathcal{H}}_1(r)\int s(\tilde{r})\,\gamma(\tilde{r})\,\psi(\tilde{r})\,d\tilde{r} = -s'(r)\,\gamma(r)\,\psi(r) \neq 0 \quad \text{for arbitrary}\, s(r) \tag{3.9}$$

Thus, every foreign to the geometry structure $s(r)$ is not allowed to enter the physical states.

Now, after the correspondence has been established, we can come to the basic property the induced metric must have. In the case of finite degrees of freedom the induced metric depends,

up to a conformal scaling, on the physical coordinates q^α by virtue of (3.2). In our case, due to the dependence of the configuration variables on the radial coordinate r, the above property is not automatically satisfied; e.g., the functional derivative $\frac{\delta}{\delta\psi(r)}$ acting on $\Psi^{(n)}$ will produce, upon partial integration of the nth derivative of the Dirac delta function, a term proportional to $\Psi^{(2n)}$. Therefore, since L in (3.4a) contains derivatives of $\psi(r)$ up to $\Psi^{(n)}$, the above mentioned property must be *enforced*. The need for this can also be traced to the substantially different first Poisson bracket in (2.13), which signals a non trivial mixing between the dynamical evolution generator $\mathcal{H}_o$ and the linear generator $\mathcal{H}_1$.
Thus, according to the above reasoning, in order to proceed with the generalization of Kuchař's method, we have to demand that:

Requirement: $L\left(\Psi^{(0)},\ldots,\Psi^{(n)}\right)$ *must be such that* $g^{\Phi\Phi}$ *becomes a general function, say* $F\left(\gamma(r)\,L(\Psi^{(0)},\ldots,\Psi^{(n)})\right)$ *of the integrand of* Φ*, so that it can be considered a function of this state:* $g^{\Phi\Phi} \overset{Assumption}{\equiv} F\left(\int\gamma(\tilde{r})\,L(\Psi^{(0)},\ldots,\Psi^{(n)})d\tilde{r}\right) = F(\Phi)$.

At this point, we must emphasize that the application of the **Requirement** in the subsequent development of our quantum theory will result in very severe restrictions on the form of (3.4a). Essentially, all higher derivatives of $\psi(r)$ (i.e $\Psi^{(2)}\ldots\Psi^{(n)}$)) are eliminated from Φ.This might, at first sight, strike as odd; indeed, the common belief is that all the derivatives of the configuration variables should enter the physical states. However, before the imposition of both the linear *and* the quadratic constraints there are no truly physical states. Thus, no physical states are lost by the imposition of the **Requirement**; ultimately the only true physical states are the solutions to the Wheeler–DeWitt equation, which will be constructed at a later stage.

Having clarified the way in which we view the **Assumption** and **Requirement** above, we proceed to the restrictions implied by their use.

We now turn to the degree of derivatives (n) of $\psi(r)$. As we argued before, the functional derivatives $\frac{\delta}{\delta\psi(r)}$ and $\frac{\delta}{\delta\gamma(r)}$ acting on $\Psi^{(n)}$ will produce, upon partial integration of the nth derivative of the Dirac delta function, a term proportional to $\Psi^{(2n)}$ and $\Psi^{(2n-1)}$ respectively. More precisely

$$g^{\Phi\Phi} = \ldots + 2\,G^{12}\frac{\delta\Phi}{\delta\gamma(r)}\frac{\delta\Phi}{\delta\psi(r)}\,.$$

where the functional derivatives are:

$$\begin{aligned}\frac{\delta\Phi}{\delta\psi} &= \ldots + \int\gamma\,\frac{\partial L}{\partial\Psi^{(n)}}\frac{\delta\Psi^{(n)}}{\delta\psi}\,d\tilde{r} = \ldots + \int\gamma\,\frac{\partial L}{\partial\Psi^{(n)}}\,\frac{1}{\gamma}\frac{d}{d\tilde{r}}\left(\underbrace{\ldots}_{n-1}\delta(r,\tilde{r})\right)d\tilde{r} = \\ &= \ldots - \int\frac{d}{d\tilde{r}}\left(\frac{\partial L}{\partial\Psi^{(n)}}\right)\frac{1}{\gamma}\frac{d}{d\tilde{r}}\left(\underbrace{\ldots}_{n-2}\delta(r,\tilde{r})\right)d\tilde{r} =\end{aligned}$$

$$\begin{aligned}
&= \ldots - \int \gamma \frac{\partial^2 L}{\partial \left(\Psi^{(n)}\right)^2} \Psi^{(n+1)} \frac{1}{\gamma} \frac{d}{d\tilde{r}} \left(\underbrace{\ldots}_{n-2} \delta(r,\tilde{r}) \right) d\tilde{r} = \\
&\qquad \vdots \\
&= \ldots + (-1)^n \int \gamma(\tilde{r}) \frac{\partial^2 L}{\partial \left(\Psi^{(n)}\right)^2} \Psi^{(2n)} \delta(r,\tilde{r})\, d\tilde{r} = \\
&= \ldots + (-1)^n \gamma \frac{\partial^2 L}{\partial \left(\Psi^{(n)}\right)^2} \Psi^{(2n)}
\end{aligned}$$

and

$$\begin{aligned}
\frac{\delta\Phi}{\delta\gamma} &= \ldots + \int \gamma \frac{\partial L}{\partial \Psi^{(n)}} \frac{\delta \Psi^{(n)}}{\delta\gamma} d\tilde{r} = \ldots + \int \gamma \frac{\partial L}{\partial \Psi^{(n)}} \frac{1}{\gamma} \frac{d}{d\tilde{r}} \left(\underbrace{\ldots}_{n-2} - \frac{\delta(r,\tilde{r})}{\gamma(\tilde{r})^2} \psi'(\tilde{r}) \right) d\tilde{r} = \\
&= \ldots + \int \gamma \frac{\partial L}{\partial \Psi^{(n)}} \frac{1}{\gamma} \frac{d}{d\tilde{r}} \left(\underbrace{\ldots}_{n-2} - \frac{\delta(r,\tilde{r})}{\gamma(\tilde{r})} \Psi^{(1)} \right) d\tilde{r} \\
&= \ldots - \int \frac{d}{d\tilde{r}} \left(\frac{\partial L}{\partial \Psi^{(n)}} \right) \frac{1}{\gamma} \frac{d}{d\tilde{r}} \left(\underbrace{\ldots}_{n-3} - \frac{\delta(r,\tilde{r})}{\gamma(\tilde{r})} \Psi^{(1)} \right) d\tilde{r} = \\
&= \ldots - \int \gamma \frac{\partial^2 L}{\partial \left(\Psi^{(n)}\right)^2} \Psi^{(n+1)} \frac{1}{\gamma} \frac{d}{d\tilde{r}} \left(\underbrace{\ldots}_{n-3} - \frac{\delta(r,\tilde{r})}{\gamma(\tilde{r})} \Psi^{(1)} \right) d\tilde{r} = \\
&\qquad \vdots \\
&= \ldots + (-1)^{n-1} \int \frac{\partial^2 L}{\partial \left(\Psi^{(n)}\right)^2} \Psi^{(2n-1)} \Psi^{(1)} \delta(r,\tilde{r})\, d\tilde{r} = \\
&= \ldots + (-1)^{n-1} \frac{\partial^2 L}{\partial \left(\Psi^{(n)}\right)^2} \Psi^{(2n-1)} \Psi^{(1)} .
\end{aligned}$$

Therefore

$$g^{\Phi\Phi} = \ldots - \frac{\gamma}{2\psi} (-1)^{2n-1} \left(\frac{\partial^2 L}{\partial \left(\Psi^{(n)}\right)^2} \right)^2 \Psi^{(1)} \Psi^{(2n-1)} \Psi^{(2n)},$$

where the $\ldots$ stand for all other terms, not involving $\Psi^{(2n)}$. Now, according to the aforementioned **Requirement** we need this to be a general function, say $F(\gamma L)$, and for this to happen the coefficient of $\Psi^{(2n)}$ must vanish, i.e.

$$\frac{\partial^2 L}{\partial \left(\Psi^{(n)}\right)^2} = 0 \Leftrightarrow L = L_1 \left(\Psi^{(0)}, \ldots, \Psi^{(n-1)} \right) \Psi^{(n)} + L_2 \left(\Psi^{(0)}, \ldots, \Psi^{(n-1)} \right).$$

Now, the term in Φ corresponding to L_1 is, up to a surface term, equivalent to a general term depending on $\Psi^{(0)}, \ldots, \Psi^{(n-1)}$. The argument can be repeated successively for $n-1$, $n-$

$2,\ldots,2$. The case $n=1$ needs separate consideration since, upon elimination of the linear in $\Psi^{(2)}$ term we are left with a local function of $\Psi^{(1)}$, and thus the possibility arises to meet the **Requirement** by solving a differential equation for L. In more detail, if

$$\Phi \equiv \int \gamma(\tilde{r}) L\left(\psi, \Psi^{(1)}\right) d\tilde{r}, \tag{3.10}$$

$g^{\Phi\Phi}$ reads

$$\begin{aligned} g^{\Phi\Phi} &= \frac{\gamma}{4\psi^2}\left(L - \Psi^{(1)}\frac{\partial L}{\partial \Psi^{(1)}}\right)\left[L - \Psi^{(1)}\frac{\partial L}{\partial \Psi^{(1)}} - 2\,\psi\left(\frac{\partial L}{\partial \psi} - \Psi^{(1)}\frac{\partial^2 L}{\partial \psi\,\partial \Psi^{(1)}}\right)\right] + \\ &+\frac{\gamma}{2\psi}\left(L - \Psi^{(1)}\frac{\partial L}{\partial \Psi^{(1)}}\right)\frac{\partial^2 L}{\partial (\Psi^{(1)})^2}\,\Psi^{(2)}. \end{aligned} \tag{3.11}$$

Through the definition

$$H \equiv L - \Psi^{(1)}\frac{\partial L}{\partial \Psi^{(1)}} \tag{3.12}$$

we obtain

$$\begin{aligned} \frac{\partial H}{\partial \psi} &= \frac{\partial L}{\partial \psi} - \Psi^{(1)}\frac{\partial^2 L}{\partial \psi\,\partial \Psi^{(1)}}, \\ \frac{\partial H}{\partial \Psi^{(1)}} &= -\Psi^{(1)}\frac{\partial^2 L}{\partial\left(\Psi^{(1)}\right)^2}. \end{aligned}$$

Thus (3.11) assumes the form

$$g^{\Phi\Phi} = \frac{\gamma}{4\psi^2}\left(H^2 - 2\,\psi\, H\frac{\partial H}{\partial \psi} - \frac{2\,\psi}{\Psi^{(1)}}\,H\frac{\partial H}{\partial \Psi^{(1)}}\,\Psi^{(2)}\right),$$

which upon addition, by virtue of the **Assumption**, of the surface term

$$A = \frac{d}{dr}\left(\int \frac{1}{2\psi\Psi^{(1)}}\,H\frac{\partial H}{\partial \Psi^{(1)}}\,d\Psi^{(1)}\right)$$

gives

$$g^{\Phi\Phi} = \frac{\gamma}{4\psi^2}\left(H^2 - 2\,\psi\, H\frac{\partial H}{\partial \psi} + 4\psi^2\Psi^{(1)}\frac{\partial}{\partial \psi}\int \frac{1}{2\psi\Psi^{(1)}}\,H\frac{\partial H}{\partial \Psi^{(1)}}\,d\Psi^{(1)}\right). \tag{3.13}$$

Since in the last expression we have only a multiplicative $\gamma(r)$, it is obvious that the **Requirement**

$$g^{\Phi\Phi} = F(\gamma\, L)$$

can be satisfied only by

$$g^{\Phi\Phi} = \kappa\,\gamma\, L, \tag{3.14}$$

with $g^{\Phi\Phi}$ given by (3.13). Upon differentiation of this equation with respect to $\Psi^{(1)}$ we get

$$\frac{\partial}{\partial \psi}\int \frac{1}{2\psi\Psi^{(1)}}\,H\frac{\partial H}{\partial \Psi^{(1)}}\,d\Psi^{(1)} = \kappa\,\frac{\partial L}{\partial \Psi^{(1)}}.$$

Multiplying the last expression by $\Psi^{(1)}$ and subtracting it from (3.14) we end up with the autonomous necessary condition for $H(\psi,\, \Psi^{(1)})$:

$$H\left(\frac{1}{4\psi^2}H - \frac{1}{2\psi}\frac{\partial H}{\partial \psi} - \kappa\right) = 0,$$

where (3.12) was also used. The above equation can be readily integrated giving

$$\begin{aligned} H &= 0, \\ H &= -\frac{4\kappa\psi^2}{3} + \sqrt{\psi}\, a(\Psi^{(1)}), \end{aligned}$$

where $a(\Psi^{(1)})$ is an arbitrary function of its argument. The first possibility gives according to (3.12) $L = \lambda\, \Psi^{(1)}$ which, however, contributes to Φ a surface term, and can thus be ignored. Inserting the second solution into (3.12) we construct a partial differential equation for L, namely

$$L - \Psi^{(1)}\frac{\partial L}{\partial \Psi^{(1)}} = -\frac{4\kappa\psi^2}{3} + \sqrt{\psi}\, a(\Psi^{(1)}),$$

which upon integration gives

$$L = -\frac{4\kappa\psi^2}{3} - \sqrt{\psi}\, \Psi^{(1)} \int \frac{a(\Psi^{(1)})}{\Psi^{(1)2}}\, d\Psi^{(1)} + c_1(\psi)\, \Psi^{(1)}\,.$$

Since this form of L emerged as a necessary condition, it must be inserted (along with H) in (3.14). The result is that $c_1(\psi) = 0$. Thus L reads

$$L = -\frac{4\kappa\psi^2}{3} - \sqrt{\psi}\, \Psi^{(1)} \int \frac{a(\Psi^{(1)})}{\Psi^{(1)2}}\, d\Psi^{(1)}\,. \tag{3.15}$$

By assuming that the $\Psi^{(1)}$–dependent part of L equals $b(\Psi^{(1)})$, i.e.

$$-\Psi^{(1)} \int \frac{a(\Psi^{(1)})}{\Psi^{(1)2}}\, d\Psi^{(1)} = b(\Psi^{(1)}),$$

we get, upon a double differentiation with respect to $\Psi^{(1)}$, the ordinary differential equation

$$-\frac{a'(\Psi^{(1)})}{\Psi^{(1)}} = b''(\Psi^{(1)})$$

with solution

$$a(\Psi^{(1)}) = b(\Psi^{(1)}) + \kappa_1 - \Psi^{(1)}\, b'(\Psi^{(1)}),$$

where κ_1 is a constant. Substituting this equation into (3.15) and performing a partial integration we end up with

$$L = -\frac{4\kappa\psi^2}{3} + \kappa_1\sqrt{\psi} + \sqrt{\psi}\, b(\Psi^{(1)})\,. \tag{3.16}$$

Journal of Physics: Conference Series **222** (2010) 012045 doi:10.1088/1742-6596/222/1/012045

κ, κ_1 and $b(\Psi^{(1)})$ being completely arbitrary and to our disposal; the two simplest choices $\kappa = 0$, $b(\Psi^{(1)}) = 0$ and $\kappa_1 = 0$, $b(\Psi^{(1)}) = 0$ lead respectively to the following two basic local smooth functionals:

$$q^1 = \int d\tilde{r}\gamma(\tilde{r})\sqrt{\psi(\tilde{r})}, \qquad q^2 = \int d\tilde{r}\gamma(\tilde{r})\psi(\tilde{r})^2 .$$

The next simpler choice $\kappa = 0$, $\kappa_1 = 0$ and $b(\Psi^{(1)})$ arbitrary leads to a generic $q^3 = \int d\tilde{r}\gamma(\tilde{r})\sqrt{\psi}\, b(\Psi^{(1)})$. However, it can be proven that, for any choice of $b(\Psi^{(1)})$, the corresponding renormalized induced metric

$$g^{AB} = G^{\alpha\beta}\frac{\delta q^A}{\delta x^\alpha}\frac{\delta q^B}{\delta x^\beta} \qquad where \quad A, B = 1, 2, 3$$

is singular. The calculation of g^{AB} gives:

$$\begin{aligned}
g^{11} &= G^{\alpha\beta}\frac{\delta q^1}{\delta x^\alpha}\frac{\delta q^1}{\delta x^\beta} = 0 \overset{Assumption}{\Longleftrightarrow} g^{11}_{ren} = 0, \\
g^{12} &= G^{\alpha\beta}\frac{\delta q^1}{\delta x^\alpha}\frac{\delta q^2}{\delta x^\beta} = -\frac{3}{8}\gamma\sqrt{\psi} \overset{Assumption}{\Longleftrightarrow} g^{12}_{ren} = -\frac{3\,q^1}{8}, \\
g^{22} &= G^{\alpha\beta}\frac{\delta q^2}{\delta x^\alpha}\frac{\delta q^2}{\delta x^\beta} = -\frac{3}{4}\gamma\,\psi^2 \overset{Assumption}{\Longleftrightarrow} g^{22}_{ren} = -\frac{3}{4}q^2, \\
g^{13} &= G^{\alpha\beta}\frac{\delta q^1}{\delta x^\alpha}\frac{\delta q^3}{\delta x^\beta} = \frac{1}{4}\gamma\,\Psi^{(2)}b'' = \frac{d}{dr}\left(\frac{1}{4}b'\right) \overset{Assumption}{\Longleftrightarrow} g^{13}_{ren} = 0,
\end{aligned}$$

$$\begin{aligned}
g^{23} &= G^{\alpha\beta}\frac{\delta q^2}{\delta x^\alpha}\frac{\delta q^3}{\delta x^\beta} = \frac{1}{8}\gamma\sqrt{\psi}\left(-3\,b + 3\Psi^{(1)}b' + 2\,\psi\,\Psi^{(2)}\,b''\right) \overset{Assumption}{\Longleftrightarrow} \\
g^{23}_{ren} &= \frac{1}{8}\int dr\gamma\sqrt{\psi}\left(-3\,b + 3\Psi^{(1)}b' + 2\,\psi\,\Psi^{(2)}\,b''\right) - \int dr\frac{d}{dr}\left(\frac{1}{4}\int d\Psi^{(1)}\psi^{3/2}\,b''\right) = \\
&= -\frac{3}{8}\int dr\gamma\sqrt{\psi}\,b = -\frac{3\,q^3}{8}, \\
g^{33} &= G^{\alpha\beta}\frac{\delta q^3}{\delta x^\alpha}\frac{\delta q^3}{\delta x^\beta} = \frac{1}{2}\gamma\left(b - \Psi^{(1)}\,b'\right)\Psi^{(2)}\,b'' \overset{Assumption}{\Longleftrightarrow} \\
g^{33}_{ren} &= \frac{1}{2}\int dr\gamma\left(b - \Psi^{(1)}\,b'\right)\Psi^{(2)}\,b'' - \int dr\frac{d}{dr}\left[\frac{1}{2}\int d\Psi^{(1)}\left(b - \Psi^{(1)}\,b'\right)b''\right] = 0,
\end{aligned}$$

where by $'$ we denote differentiation with respect to $\Psi^{(1)}$. Thus the renormalized induced metric reads

$$g^{AB}_{ren}(q^1, q^2, q^3) = -\frac{3}{8}\begin{pmatrix} 0 & q^1 & 0 \\ q^1 & 2\,q^2 & q^3 \\ 0 & q^3 & 0 \end{pmatrix}.$$

Effecting the transformation $(\tilde{q}^1, \tilde{q}^2, \tilde{q}^3) = \left(q^1, q^2, f\left(\frac{q^3}{q^1}\right)\right)$ we bring g^{AB}_{ren} into a manifestly degenerate form:

$$g^{AB}_{ren}(q^1,q^2) = -\frac{3}{8}\begin{pmatrix} 0 & q^1 & 0 \\ q^1 & 2\,q^2 & 0 \\ 0 & 0 & 0 \end{pmatrix}.$$

So, it seems that, as far as the ultra local part of the functionals is concerned, the renormalized metric is given by the upper left 2×2 block of the above expression.

It is interesting to observe that the integrands of q^1, q^2 form a base in the space spanned by γ, ψ. It is convenient to change these two integrands (namely $\gamma\sqrt{\psi}$ and $\gamma\,\psi^2$) to γ and $\gamma\psi^2$ respectively, since the latter choice complies with the ultra local parts of the potential (2.11). It can be proved that the new renormalized metric, resulting from the choice of the new functionals,

$$y^1 = \int \gamma(\tilde{r})d\tilde{r}, \qquad y^2 = \int \gamma(\tilde{r})\psi(\tilde{r})^2 d\tilde{r}$$

is equivalent to the previous one.

However, this is not the end of our investigation for a suitable space of state vectors: the argument leading to y^1, y^2 crucially depends upon the original choice of one initial candidate smooth scalar functional (3.10). Therefore, to complete the search we must close the circle by starting with the two already secured smooth functionals (y^1, y^2), and a third of the general form

$$y^3 = \int d\tilde{r}\,\gamma(\tilde{r})\,L(\Psi^{(1)})$$

(since the ψ dependence has already been fixed to either 1 or ψ^2). The reduced renormalized manifold is thus parameterized by the following three smooth scalar functionals:

$$y^1 = \int \gamma(\tilde{r})d\tilde{r}, \qquad y^2 = \int \gamma(\tilde{r})\psi(\tilde{r})^2 d\tilde{r}, \qquad y^3 = \int \gamma(\tilde{r})L(\Psi^{(1)})d\tilde{r}. \tag{3.17}$$

Any other functional, say $y^4 = \int d\tilde{r}\,\gamma(\tilde{r})\,K\left[\psi(\tilde{r}),\Psi^{(1)}(\tilde{r})\right]$, can be considered as a function of y^1, y^2, y^3; indeed, since the scalar functions appearing in the integrands of y^2, y^3 form a base in the space spanned by $\psi, \Psi^{(1)}(r)$, we can express the generic K in y^4 as $K\left[\sqrt{\frac{\gamma\psi^2}{\gamma}},\Psi^{(1)}\right]$, which (through the **Assumption**) gives $y^4 = y^1\,K\left[\sqrt{\frac{y^2}{y^1}}, L^{-1}\left(\frac{y^3}{y^1}\right)\right]$.

The geometry of this space is described by the induced renormalized metric

$$g^{AB}_{ren}(y^1,y^2,y^3) = -\frac{1}{4}\begin{pmatrix} -\frac{(y^1)^2}{y^2} & y^1 & -\frac{y^1 y^3}{y^2} \\ y^1 & 3\,y^2 & y^3 \\ -\frac{y^1 y^3}{y^2} & y^3 & -\frac{(y^3)^2}{y^2} + \frac{4\,(y^1)^2 F\left(\frac{y^3}{y^1}\right)^2}{3\,y^2 F'\left[F\left(\frac{y^3}{y^1}\right)\right]^2} \end{pmatrix}, \tag{3.18}$$

$$g_{AB\,ren}(y^1, y^2, y^3) = \begin{pmatrix} \frac{3\,y^2}{(y^1)^4}\left((y^1)^2 - \frac{(y^3)^2 F' \;\; F \;\; \frac{y^3}{y^1} \;\; ^2}{F \;\; \frac{y^3}{y^1} \;\; ^2}\right) & -\frac{1}{y^1} & \frac{3\,y^2 y^3 F' \;\; F \;\; \frac{y^3}{y^1} \;\; ^2}{(y^1)^3 F \;\; \frac{y^3}{y^1} \;\; ^2} \\ -\frac{1}{y^1} & -\frac{1}{y^2} & 0 \\ \frac{3\,y^2 y^3 F' \;\; F \;\; \frac{y^3}{y^1} \;\; ^2}{(y^1)^3 F \;\; \frac{y^3}{y^1} \;\; ^2} & 0 & -\frac{3\,y^2 F' \;\; F \;\; \frac{y^3}{y^1} \;\; ^2}{(y^1)^2 F \;\; \frac{y^3}{y^1} \;\; ^2} \end{pmatrix}. \tag{3.19}$$

Any function $\Psi(y^1, y^2, y^3)$ on this manifold is of course annihilated by the quantum linear constraint, i.e.

$$\hat{\mathcal{H}}_1 \Psi(y^1, y^2, y^3) = \frac{\partial \Psi(y^1, y^2, y^3)}{\partial y^1}\,\hat{\mathcal{H}}_1\, y^1 + \frac{\partial \Psi(y^1, y^2, y^3)}{\partial y^2}\,\hat{\mathcal{H}}_1\, y^2 + \frac{\partial \Psi(y^1, y^2, y^3)}{\partial y^3}\,\hat{\mathcal{H}}_1\, y^3 = 0$$

since the derivatives with respect to r are transparent to the partial derivatives of Ψ (which are, just like the y^A's, r-numbers).

The covariant metric (3.18) describes a three dimensional conformally flat geometry, since the Cotton-York tensor vanishes. The Ricci scalar is $R = \frac{3}{8\,y^2}$, indicating that the arbitrariness in F (and thus also in L) is a *pure gauge*. The change of coordinates

$$(y^1, y^2, y^3) = (e^{-\frac{1}{8}(5\,Y^1 + 3\,Y^3)}, e^{Y^1 + Y^2 + Y^3}, e^{-\frac{1}{8}(5\,Y^1 + 3\,Y^3)} F^{-1}(e^{\frac{1}{24}(-9\,Y^1 + 8\,Y^2 - 15\,Y^3)})) \tag{3.20}$$

(where F^{-1} denotes the function inverse to F, i.e $F^{-1}(F(x)) = x$) brings the metric to the manifestly conformally flat form:

$$g_{AB\,ren}(Y^1, Y^2, Y^3) = \begin{pmatrix} e^{Y^1+Y^2+Y^3} & 0 & 0 \\ 0 & -\frac{4}{3}\,e^{Y^1+Y^2+Y^3} & 0 \\ 0 & 0 & -e^{Y^1+Y^2+Y^3} \end{pmatrix}, \tag{3.21}$$

in which all the F dependence has indeed disappeared.

The final restriction on the form of Ψ will be obtained by the imposition of the quantum analog of the quadratic constraint $\mathcal{H}_o$. According to the above exposition we postulate that the quantum gravity of the geometries given by (2.5) will be described by the following partial differential equation (in terms of the Y^A's)

$$\hat{\mathcal{H}}_o \Psi \equiv [-\frac{1}{2}\,\Box_c + V_{ren}]\,\Psi(Y^1, Y^2, Y^3) = 0 \tag{3.22}$$

with

$$\Box_c = \Box + \frac{d-2}{4\,(d-1)}\,R \tag{3.23}$$

being the conformal Laplacian based on $g_{AB\,ren}(Y^1, Y^2, Y^3)$, R the Ricci scalar, and d the dimensions of $g_{AB\,ren}$. The metric (3.21) is conformally flat with the Ricci scalar $R =$

$\frac{3}{8}\,e^{-Y^1-Y^2-Y^3}$, and its dimension is $d = 3$. The renormalized form of the potential (2.11) offers us the possibility to introduce, in a dynamical way, topological effects into our wave functional: Indeed, under our **Assumption**, the first two terms become $-2\,\epsilon\,y^1$ and $2\,\Lambda\,y^2$, respectively, while the last, being a total derivative, becomes $A_T \equiv 4\,\frac{\psi\,\psi'}{\gamma}\,|_\alpha^\beta$ $(if\,\alpha < r < \beta)$. In the spirit previously explained we should drop this term, however one could also keep it. The renormalized form of the remaining, third, term of the potential can be obtained as follows

$$y^3 \;=\; \gamma\,L(\Psi^{(1)}) \Leftrightarrow L(\Psi^{(1)}) = \frac{y^3}{\gamma} \overset{Assumption}{\Longleftrightarrow} L(\Psi^{(1)}) = \frac{y^3}{y^1} \Leftrightarrow \Psi^{(1)} = L^{-1}\left(\frac{y^3}{y^1}\right),$$

thus finally

$$\frac{\psi'}{\gamma} = L^{-1}\left(\frac{y^3}{y^1}\right)$$

and the third term becomes $-2\,y^1\left[L^{-1}\left(\frac{y^3}{y^1}\right)\right]^2$. Finally, effecting the transformation (3.20) the form of the renormalized potential is

$$\begin{aligned} V_{ren} \;=\; & -2\,\epsilon\,e^{-\frac{1}{8}(5\,Y^1+3\,Y^3)} - 2\,e^{-\frac{1}{8}(5\,Y^1+3\,Y^3)}\left[L^{-1}\left(F^{-1}(e^{\frac{1}{24}(-9\,Y^1+8\,Y^2-15\,Y^3)})\right)\right]^2 + \\ & 2\,\Lambda\,e^{Y^1+Y^2+Y^3} + A_T \end{aligned} \tag{3.24}$$

and the Wheeler-DeWitt equation is given as

$$\begin{aligned} & -2\,\epsilon\,e^{\frac{1}{8}(3\,Y^1+5\,Y^3)+Y^2}\,\Psi(Y^1,Y^2,Y^3) + 2\,\Lambda\,e^{2(Y^1+Y^2+Y^3)}\,\Psi(Y^1,Y^2,Y^3) - \\ & 2\,e^{\frac{1}{8}(3\,Y^1+5\,Y^3)+Y^2}\left[L^{-1}\left(F^{-1}(e^{\frac{1}{24}(-9\,Y^1+8\,Y^2-15\,Y^3)})\right)\right]^2\Psi(Y^1,Y^2,Y^3) + \\ & A_T\,e^{Y^1+Y^2+Y^3}\,\Psi(Y^1,Y^2,Y^3) - \frac{3}{128}\Psi(Y^1,Y^2,Y^3) - \frac{1}{4}\frac{\partial\Psi(Y^1,Y^2,Y^3)}{\partial Y^1} + \\ & \frac{3}{16}\frac{\partial\Psi(Y^1,Y^2,Y^3)}{\partial Y^2} + \frac{1}{4}\frac{\partial\Psi(Y^1,Y^2,Y^3)}{\partial Y^3} - \frac{1}{2}\frac{\partial^2\Psi(Y^1,Y^2,Y^3)}{\partial\,(Y^1)^2} + \frac{3}{8}\frac{\partial^2\Psi(Y^1,Y^2,Y^3)}{\partial\,(Y^2)^2} + \\ & \frac{1}{2}\frac{\partial^2\Psi(Y^1,Y^2,Y^3)}{\partial\,(Y^3)^2} = 0. \end{aligned}$$

Since F^{-1} is an arbitrary function of its arguments, we may contemplate the choice:

$$F^{-1}\left(e^{\frac{1}{24}(-9\,Y^1+8\,Y^2-15\,Y^3)}\right) = L\left(\sqrt{e^{\frac{1}{24}(-9\,Y^1+8\,Y^2-15\,Y^3)} - \epsilon}\right)$$

Of course, there is a question of existence for such a choice: any demand that F has a specified form (much more in terms of L) constitutes a restriction on the form of L. Consequently, one has to prove that such an L indeed exists. The appropriate form of L is implicitly given by

$$L(\omega) = m + \int \frac{(\epsilon+\omega^2)^{1/4}}{\omega^{5/8}}\,e^{k-\frac{3\,\epsilon}{16\,\omega^2}}\,d\omega$$

where $\omega \equiv \sqrt{\Psi^{(1)2} - \epsilon}$ and m, k are related by

$$c_1 m + c_2 + c_3 e^k = 0$$

For a detailed elaboration of the results, see [28].

Anyway, the above choice of F reduces the Wheeler–DeWitt equation to the final separable form:

$$\left[2\,\Lambda\, e^{2(Y^1+Y^2+Y^3)} - 2\, e^{\frac{4}{3}Y^2} + A_T\, e^{Y^1+Y^2+Y^3} - \frac{3}{128}\right]\Psi(Y^1,Y^2,Y^3) -$$
$$\frac{1}{4}\frac{\partial\Psi(Y^1,Y^2,Y^3)}{\partial Y^1} + \frac{3}{16}\frac{\partial\Psi(Y^1,Y^2,Y^3)}{\partial Y^2} + \frac{1}{4}\frac{\partial\Psi(Y^1,Y^2,Y^3)}{\partial Y^3} -$$
$$\frac{1}{2}\frac{\partial^2\Psi(Y^1,Y^2,Y^3)}{\partial\,(Y^1)^2} + \frac{3}{8}\frac{\partial^2\Psi(Y^1,Y^2,Y^3)}{\partial\,(Y^2)^2} + \frac{1}{2}\frac{\partial^2\Psi(Y^1,Y^2,Y^3)}{\partial\,(Y^3)^2} = 0. \qquad (3.25)$$

This equation is readily solved, for $\Lambda = 0$ and $A_T = 0$, by the method of separation of variables: assuming $\Psi(Y^1,Y^2,Y^3) = \Psi^1(Y^1)\,\Psi^2(Y^2)\,\Psi^3(Y^3)$ and dividing (3.25) by Ψ we get the three ordinary differential equations:

$$\frac{1}{4\,\Psi^1(Y^1)}\frac{d\Psi^1(Y^1)}{dY^1} + \frac{1}{2\,\Psi^1(Y^1)}\frac{d^2\Psi^1(Y^1)}{d\,(Y^1)^2} = m,$$
$$\frac{3}{16\,\Psi^2(Y^2)}\frac{d\Psi^2(Y^2)}{dY^2} + \frac{3}{8\,\Psi^2(Y^2)}\frac{d^2\Psi^2(Y^2)}{d\,(Y^2)^2} - 2\,e^{\frac{4}{3}Y^2} = n,$$
$$\frac{1}{4\,\Psi^3(Y^3)}\frac{d\Psi^3(Y^3)}{dY^3} + \frac{1}{2\,\Psi^3(Y^3)}\frac{d^2\Psi^3(Y^3)}{d\,(Y^3)^2} - \frac{3}{128} = m - n,$$

where m and n are separation constants. Their solutions are:

$$\begin{aligned}
\Psi^1(Y^1) &= c_1\, e^{\frac{1}{4}\left(-1-\sqrt{1+32\,m}\right)Y^1} + c_2\, e^{\frac{1}{4}\left(-1+\sqrt{1+32\,m}\right)Y^1},\\
\Psi^2(Y^2) &= c_3\, e^{-Y^2/4}\, I_{-\frac{\sqrt{3}}{8}\sqrt{3+128\,n}}\left(2\sqrt{3}\,e^{2\,Y^2/3}\right) + c_4\, e^{-Y^2/4}\, I_{\frac{\sqrt{3}}{8}\sqrt{3+128\,n}}\left(2\sqrt{3}\,e^{2\,Y^2/3}\right)\\
\Psi^3(Y^3) &= c_5\, e^{\frac{1}{8}\left(-2-\sqrt{7+128\,m-128\,n}\right)Y^3} + c_6\, e^{\frac{1}{8}\left(-2+\sqrt{7+128\,m-128\,n}\right)Y^3},
\end{aligned}$$

where $I_{\pm\frac{\sqrt{3}}{8}\sqrt{3+128\,n}}\left(2\sqrt{3}\,e^{2\,Y^2/3}\right)$ are modified Bessel functions of the first kind and of non-integer order.

4. Discussion

We have considered the canonical analysis and subsequent quantization of a midi-superspace model in 3+1 dimensions. The simplifying assumption was the existence of appropriate Killing vector fields, which reduced the components of the metrics to functions of time and a radial coordinate. Thus, the emanating field theory is effectively one-dimensional. The quantum linear constraints reduced the initial Hilbert space consisting of all smooth functionals to the space of all smooth scalar functionals. The appropriate imposition of the quadratic constraint leads us to the the **Assumption** and the **Requirement**, which strongly further restrict the space of functionals to the (3.17) for the 3+1 system. In addition, the quantum analogue of the Hamiltonian constraint produces a Wheeler-DeWitt equation based on the reduced manifold of states, which is completely integrated. If the above quantization scheme truly achieves general coordinate invariant characterization of the wave function, then there must be a way to classify classical geometries using only **first derivatives** of the metrics. The current state of knowledge for the subject is the Cartan-Karlhede equivalence classification scheme, which requires up to seven derivatives of the Riemann tensor.

4.1. Acknowledgments
One of the authors (G. O. Papadopoulos) is a Killam Postdoctoral Fellow and acknowledges the relevant support from the Killam Foundation.

References

[1] Dirac P A M 1950 *Can. J. Math.* **2** 129
[2] Dirac P A M 1951 *Can. J. Math.* **3** 1
[3] Dirac P A M 1958 *Proc. R. Soc.* (London) A **246** 326
[4] Dirac P A M 1964 *Lectures on Quantum Mechanics* (New York: Yeshiva University, Academic Press)
[5] Bergmann P G 1961 *Rev. Mod. Phys.* **33** 510
[6] Berezin F A and Marinov M S 1977 *Ann. Phys.* (N.Y.) **104** 336
[7] Becchi C, Rouet C and Stora R 1976 *Ann. Phys.* (N.Y.) **98** 287
[8] Sudarshan E C G and Mukunda N 1974 *Classical Dynamics: A Modern Perspective* (New York: Wiley)
[9] Hanson A J, Regge T and Teitelboim C 1976 *Constrained Hamiltonian Systems* (Vatican: Accademia Nazionale dei Lincei)
[10] Sundermeyer K 1982 *Lecture Notes in Physics* **169** *Constrained Dynamics* (Heidelberg: Springer-Verlag)
[11] Gitman D M and Tyutin I V 1991 *Quantization of Fields with Constraints* Springer Series in Nuclear and Particle Physics
[12] Govaers J 1991 *Hamiltonian Quantization and Constrained Dynamics* (Leuven: Leuven University Press)
[13] Henneaux M and Teitelboim C 1992 *Quantization of Gauge Systems* (Princeton:Princeton University Press)
[14] Wipf A *Hamilton's Formalism for Systems with Constraints* (*Preprint* hep-th/9312078)
[15] Thiemann Th 2007 *Modern Canonical Quantum General Relativity* (Cambridge: Cambridge University Press)
[16] DeWitt B S 1967 *Phys. Rev.* **160**(5) 1113
[17] Bergmann P G 1966 *Phys. Rev.* **144** 1078
[18] Ryan M P Jr and Shepley L C 1975 *Homogeneous Relativistic Cosmologies* (Princeton: Princeton University Press)
[19] Christodoulakis T 2002 *Quantum Cosmology Lect.Notes Phys.* **592** Springer Verlag 318-350
[20] Krasinski A 1997 *Inhomogeneous Cosmological Models* (Cambridge: Cambridge University Press)
[21] Thomi P, Isaak B and Hajicek P 1984 *Phys. Rev.* D **30** 1168
[22] Hajicek P 1984 *Phys. Rev.* D **30** 1178
[23] Kiefer C, Müller-Hill J and Vaz C 2006 *Phys. Rev.* D **73** 044025
[24] Hajicek P and Kuchař K V 1990 *Phys. Rev.* D **41** 1091
[25] Hajicek P and Kuchař K V 1990 *J. Math.Phys.* **31** 1723
[26] Eisenhart L P 1964 *Riemannian Geometry*, (Princeton: Princeton University Press, New Jersey) 5th printing pp.146
[27] Thomas T Y 1991 *The Differential Invariants of Generalized Spaces*, (New York: Chelsea Publishing Company) pp.140
[28] Christodoulakis T et al. *Towards a Canonical Quantum Gravity for Geometries Admitting Maximally Symmetric Two-Dimensional Surfaces*, submitted

First Mediterranean Conference on Classical and Quantum Gravity (MCCQG 2009) IOP Publishing
Journal of Physics: Conference Series **222** (2010) 012046 doi:10.1088/1742-6596/222/1/012046

Superluminal warp drives are semiclassically unstable

S Finazzi[1], S Liberati[1] and C Barceló[2]

[1] SISSA, via Beirut 2-4, Trieste 34151, Italy and INFN sezione di Trieste
[2] Instituto de Astrofísica de Andalucía, CSIC, Camino Bajo de Huétor 50, 18008 Granada, Spain

E-mail: finazzi@sissa.it, liberati@sissa.it and carlos@iaa.es

Abstract. Warp drives are very interesting configurations of General Relativity: they provide a way to travel at superluminal speeds, albeit at the cost of requiring exotic matter to build them. Even if one succeeded in providing the necessary exotic matter, it would still be necessary to check whether they would survive to the switching on of quantum effects. Semiclassical corrections to warp-drive geometries created out of an initially flat spacetime have been analyzed in a previous work by the present authors in special locations, close to the wall of the bubble and in its center. Here, we present an exact numerical analysis of the renormalized stress-energy tensor (RSET) in the whole bubble. We find that the the RSET will exponentially grow in time close to the front wall of the superluminal bubble, after some transient terms have disappeared, hence strongly supporting our previous conclusion that the warp-drive geometries are unstable against semiclassical back-reaction. This result seems to implement the chronology protection conjecture, forbiddig the set up of a structure potentially dangerous for causality.

1. Introduction

Since they were introduced by Alcubierre [1], warp drives have been certainly one of the most studied spacetime geometries among those requiring exotic matter (*i.e.* energy-conditions-violating matter) for their existence in the context of general relativity (see [2] for a recent review). They are not only an exciting theoretical test-field for our comprehension of general relativity and quantum field theory in curved spacetimes, but they also provide, at least theoretically, a way to travel at superluminal speeds.

A warp-drive metric describes a bubble containing an almost flat region, moving at arbitrary speeds within an asymptotically flat spacetime:

$$ds^2 = -c^2dt^2 + [dx - v(r)dt]^2 + dy^2 + dz^2, \tag{1}$$

where $r \equiv \sqrt{(x - v_0t)^2 + y^2 + z^2}$ is the distance from the center of the bubble and v_0 is the warp-drive velocity. Here $v = v_0 f(r)$, with f a suitable smooth function satisfying $f(0) = 1$ and $f(r) \to 0$ for $r \to \infty$.

After the warp-drive spacetimes were proposed, their most investigated aspect has been the amount of exotic matter that would be required to support them [3, 4, 5, 6]. It was found that, if the exoticity was provided by the quantum nature of a field, satisfying therefore the so-called *Quantum Inequalities* (QI) [7], then the violations of the energy conditions would have to be confined to Planck-size regions, making the bubble-wall thickness to be of Planck size. This bound turns into lower limits on the required amount of exotic matter ($\gtrsim 1\,M_\odot$).

Less effort has been devoted to another important issue regarding the feasibility of these spacetimes: Their stability under the switching on of quantum effects. It was noticed [8] that the backward and forward walls (along the direction of motion) look respectively as the future (black) and past (white) event horizon of a black hole, for an observer inside the warp-drive bubble. In the Heisenberg representation, by imposing over the spacetime a state which is vacuum at the null infinities, it was found that the renormalized stress-energy tensor (RSET) had to diverge on the horizons. However, from equivalent analysis in eternal-black-hole geometries we know that such divergences are related to an inappropriate vacuum choice (the Boulware vacuum) and, therefore, they are not physical.

We consider here the more realistic case of a warp drive created out of Minkowski spacetime and then accelerated up to some superluminal speed in a finite amount of time. Accordingly, the globally defined quantum state is chosen to reproduce the Minkowski vacuum at early times. Using this vacuum state, in a recent previous work [9], we approximately calculated the salient features of the energy density as seen by inertial observers inside the warp-drive bubble, in three specific locations: the center of the bubble, the black and the white horizons. In this work we show exact numerical calculations for the energy density in the entire interior of the bubble (using a specific model), which confirm our previous partial results. We find that some transient effects are generated when the horizons are created. However, these effects disappear in a short time and the behaviors found in our analytical work are recovered. We confirm the presence of a thermal flux of radiation with a Hawking temperature proportional to the surface gravity of the black horizon, which is inversely proportional to the wall thickness. As mention, if the supporting matter is subjected to the QI, then this wall thickness would be of the order of the Planck length, thus leading to a temperature of the order of the Planck temperature. More importantly, we find that the energy density measured by free-falling observers is negative on the white and black horizons (as expected for a vacuum polarization). While on the black one it is exponentially damped, on the white one it diverges exponentially to minus infinity. In [9], in addition to the black and white horizons already envisaged in [8], we discovered the presence of a Cauchy horizon. Rightgoing null rays approach this horizon at late times. As one should expect by analogy with other situation in which Cauchy horizons are present, as in Kerr-Newman black holes, we find that the energy carried by such rays diverges, this time contributing with a positive sign, when approaching this horizon, thus representing an additional source of instability.

In Sect. 2 we summarize the results of [9], using the specific model used for the numerical treatment. Our numerical results are presented in Sect. 3. Conclusions and final remarks follow in Sect. 4.

2. Dynamical warp-drives

2.1. Causal structure of a dynamical warp drive

We restrict our attention to the 1+1 dimensional case and write the metric (1) using coordinates t and $r \equiv x - v_0 t$:

$$ds^2 = -c^2 dt^2 + [dr - \bar{v}(r)dt]^2 \quad , \qquad (2)$$

where $\bar{v} = v - v_0$. As above, $v = v_0 f(r)$, with f defined now on the whole real axis. We require $f(r)$ to be a smooth bell-shaped function with maximum $f(0) = 1$ and $f(r) \to 0$ for $r \to \pm\infty$. It is easy to check that $\bar{v}$ has a maximum in $r = 0$, $\bar{v}(0) = \bar{v}'(0) = 0$, and that $\bar{v}(r) \to -v_0 < -c$ for $r \to \pm\infty$. As a consequence, there are two positions r_1 and r_2 at which $c + \bar{v}(r) = 0$. As seen from an observer inside the bubble, these points will correspond to a black and a white horizons, $\mathscr{H}_1^+$ and $\mathscr{H}_2^-$. A dynamical warp-drive geometry is obtained by replacing $\bar{v}$ with a time dependent velocity $\hat{v}(t, r)$, such that the new metric interpolates between an initial Minkowski spacetime [$\hat{v}(t, r) \to 0$, for $t \to -\infty$] and a final stationary superluminal ($v_0 > c$) bubble [$\hat{v}(t, r) \to \bar{v}(r)$, for $t \to +\infty$]. In Fig. 1 we plot the Penrose diagram of such a geometry (see Ref. [9] for its construction). One can note that a timelike observer can reach $\mathscr{H}_C^+$ and $\mathscr{H}_R^+$

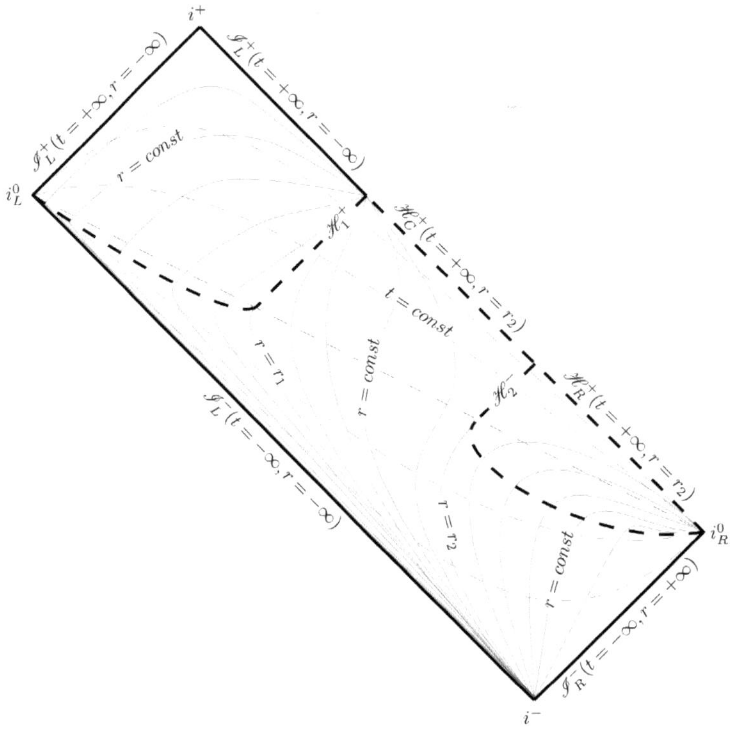

Figure 1. Penrose diagram of a superluminal warp-drive creation. Lines of constant r (*solid lines*) and of constant t (*dashed lines*). The lines of constant r become null at the apparent horizon (*heavy dashed lines*). The warp-drive bubble is located in the central region $(r_1 < r < r_2)$.

in a finite proper time. As a consequence the spacetime can be extended in the future, beyond these lines. However, $\mathscr{H}_C^+$ and $\mathscr{H}_R^+$ are on the boundary of the Cauchy development of $\mathscr{I}^-$. In this sense, they are Cauchy horizons given initial data assigned only on $\mathscr{I}^-$.

2.2. Light-ray propagation

We now consider light-ray propagation in the above described geometry, specializing to the particular model used for numerical calculation. Adapting the method presented in [10, 11] for stellar collapses, in our framework, a dynamical superluminal warp-drive geometry is determined by a late time profile $\bar{v}(r)$ and a prescription for the transition between the early time Minkowski spacetime and the late time regime. We choose $\bar{v}$ as:

$$\bar{v}(r) = \alpha c\,[f(r) - 1] = \alpha c \left[\frac{1}{\cosh(r/a)} - 1\right] , \tag{3}$$

with $\alpha > 0$ and we define a dynamical profile $\hat{v}$

$$\hat{v}(r,t) = \begin{cases} \bar{v}(\xi(t)) & \text{if} \quad |r| \geq \xi(t) , \\ \bar{v}(r) & \text{if} \quad |r| < \xi(t) , \end{cases} \tag{4}$$

where $\xi(t)$ is a monotonically increasing function of t, such that $\xi(t) \to 0$, for $t \to -\infty$, and $\xi(t_{\rm H}) = r_2 = -r_1$, which we called the kink function.[1] The horizons are located at

$$r_{1,2} = \mp a \ln\left(\beta + \sqrt{\beta^2 - 1}\right) , \qquad \beta \equiv \frac{\alpha}{\alpha - 1} > 1 . \tag{5}$$

We also define a signed surface gravity[2] on these horizons:

$$\kappa_{1,2} \equiv \left.\frac{dv(r)}{dr}\right|_{r=r_{1,2}} = \pm \frac{c\,(\alpha-1)\sqrt{\beta^2-1}}{a\beta} \equiv \pm\kappa . \tag{6}$$

In [9] it was argued that the universal features of the RSET are determined only by right-going rays, just like outgoing modes do in the case of a black hole collapse [10, 12]. This result is confirmed by the numerical computation where also the effects of left-going rays are considered, but they turn out to be negligible at late times (see Sect. 3). For the sake of conciseness, in this section we present only the features due to rightgoing rays, since treatment of left going rays is analogous.

First of all, we need the relation between the past null coordinate U relevant on $\mathscr{I}_L^-$ and future null coordinate u relevant on $\mathscr{H}_C^+$. Following [10], we write the differential equation describing right-going rays:

$$\frac{dr}{dt} = c + \hat{v}(r,t) . \tag{7}$$

Integrating this equation at early times or close to $\mathscr{H}_C^+$ ($t \to +\infty$ and $r \to r_2$), one can define U and u:

$$U = \lim_{t_i \to -\infty} \left(t_i - \frac{r_i}{c}\right) , \tag{8}$$

$$u = \lim_{t_f \to +\infty} \left[t_f + \frac{1}{\kappa}\ln\left(r_2 - r_f\right)\right] . \tag{9}$$

We identify initial events $P \equiv (r_i, t_i)$, with $r_i \sim ct_i$, final events $Q \equiv (r_f, t_f)$ with $r_f \sim r_2 - e^{-\kappa t_f}$, and intermediate events $O \equiv (r_0, t_0)$. Integrating Eq. (7) between P and O in the in region of this spacetime (where the velocity profile depends only on t) and, respectively, between O now in the out region (where the velocity profile depends only on r) and Q, one obtains for $r_1 < r_0 < r_2$:

$$U = t_0 - \frac{r_0}{c} + \frac{1}{c}\int_{-\infty}^{t_0} dt\,\bar{v}(\xi(t)) , \tag{10}$$

$$\begin{aligned} u = t_0 &+ \frac{1}{\kappa}\ln\left[r_2 - r_0\right] - \frac{1}{\kappa}\ln\left[r_0 - r_1\right] + \frac{1}{\kappa}\ln\left[r_2 - r_1\right] \\ &+ \int_{r_0}^{r_2} dr \left[\frac{1}{c + \bar{v}(r)} - \frac{1}{\kappa\,(r_2 - r)} - \frac{1}{\kappa\,(r - r_1)}\right] , \end{aligned} \tag{11}$$

where the last integral is finite for each value of r_0. To find the relation between U and u, one must choose a particular form of the kink function ξ. It is particularly useful to choose ξ such that all light rays cross the kink $r = \mp\xi(t)$ only once. In this fashion, the relation $U = p(u)$ can be found by eliminating t_0 and r_0 between Eqs. (10) and (11), with O on the kink, $r_0 = \mp\xi(t_0)$. We show below how the analytical results of [9] can be recovered for the kink model. To this

[1] One may wonder whether defining a velocity profile with a kink, as in Eq. (4), may lead to unphysical phenomena. Indeed this computational trick induces some spurious effects, but these features are just transients and do not affect the results at late times.

[2] In general, κ_2 could be different from κ_1, although it is expected to be comparable with κ_1.

Journal of Physics: Conference Series **222** (2010) 012046 doi:10.1088/1742-6596/222/1/012046

aim, one needs the limit for large and positive values of u (close to $\mathscr{H}_2^-$, for $r \to r_1$, $u \to +\infty$) and for large and negative values of u (close to $\mathscr{H}_1^+$, for $r \to r_2$, $u \to -\infty$). Let us focus to rays peeling off $\mathscr{H}_2^-$,*i.e.*, those rays which cross the left kink at $r_0 = -\xi(t_0)$, when r_0 is very close to r_1.

When $t \to t_{\rm H}$ (the time of the appearance of the trapping horizons) ξ can be expanded as:

$$-\xi(t) = r_1 - \lambda\,(t - t_{\rm H}) + \mathcal{O}\left[(t - t_{\rm H})^2\right] \,, \tag{12}$$

with λ a positive constant. From Eq. (10) one sees that

$$U = U_{\rm BH} - \frac{\lambda}{c}\,(t_{\rm H} - t_0) + \mathcal{O}\left[(t_{\rm H} - t_0)^2\right] \,, \tag{13}$$

where $U_{\rm BH} = U(t_{\rm H}, -\xi(t_{\rm H}))$. Expanding Eq. (11) in the same limit, $t_0 \to t_{\rm H}$, one has

$$u \simeq -\frac{1}{\kappa}\ln\left[\lambda\,(t_{\rm H} - t_0)\right] \,, \tag{14}$$

Putting together the last two equations, one obtains

$$U(u \to +\infty) \simeq U_{\rm BH} + A_1 e^{-\kappa u} + \frac{A_2}{2} e^{-2\kappa u} + \ldots \,, \tag{15}$$

where it can be shown that $A_1 < 0$. This relation is the standard result for the formation of a black hole through gravitational collapse. As a consequence, the quantum state which is vacuum on $\mathscr{I}^-$ will show, inside the bubble, Hawking radiation with temperature $T_H = \kappa_1/2\pi$.

By performing the same steps for light rays passing close to the white horizon one can find that

$$U(u \to -\infty) \simeq U_{\rm WH} + D_1 e^{\kappa u} + \frac{D_2}{2} e^{2\kappa u} + \ldots \,, \tag{16}$$

where $A_2 > 0$.

Of course, this identification is not complete as this radiation does not travel towards an asymptotically flat region but towards the white horizon. However, for large-enough bubbles, the radiation at the center of the bubble will look approximately as Hawking free particles.

2.3. Renormalized stress-energy tensor

For the calculation of the RSET inside the warp-drive bubble we use the method proposed in [12]. In past null coordinates U and W the metric can be written as

$$ds^2 = -C(U, W)dUdW \,. \tag{17}$$

In the asymptotic out region, it is convenient to use a different set of coordinates: u, Eq. (9), and $\tilde{w}$, defined as

$$\tilde{w}(t, r) = t + \int_0^r \frac{dr}{c - \bar{v}(r)} \,. \tag{18}$$

In these coordinates the metric is expressed as

$$ds^2 = -\bar{C}(u, \tilde{w})du d\tilde{w} \,, \quad C(U, W) = \frac{\bar{C}(u, \tilde{w})}{\dot{p}(u)\dot{q}(\tilde{w})} \,, \tag{19}$$

where $U = p(u)$ and $W = q(\tilde{w})$.

For concreteness, we refer to the RSET associated with a quantum massless scalar field living on the spacetime. The RSET components have the following form [13]:

$$T_{UU} = -\frac{1}{12\pi} C^{1/2} \partial_U^2 C^{-1/2} , \tag{20}$$

$$T_{WW} = -\frac{1}{12\pi} C^{1/2} \partial_W^2 C^{-1/2} , \tag{21}$$

$$T_{UW} = T_{WU} = \frac{1}{96\pi} C\ R . \tag{22}$$

If there were other fields present in the theory, the previous expressions would be multiplied by a specific numerical factor. Using the relationships $U = p(u)$, $W = q(\tilde{w})$ and the time-independence of u and $\tilde{w}$, one can calculate [9] the RSET components in the stationary region.

To extract significant information from the RSET we study the energy ρ as measured by a set of free-falling observers, whose four velocity is $u_c^\mu = (1, \bar{v})$ in (t, r) components. For these observers we obtain

$$\rho = T_{\mu\nu} u_c^\mu u_c^\nu = \rho_{\rm st} + \rho_{\text{dyn}-u} + \rho_{\text{dyn}-\tilde{w}} , \tag{23}$$

where we define a static term $\rho_{\rm st}$, depending only on the r coordinate through $\bar{v}(r)$,

$$\rho_{\rm st} \equiv -\frac{1}{24\pi} \left[\frac{(\bar{v}^4 - \bar{v}^2 + 2)}{(1-\bar{v}^2)^2} \bar{v}'^{\,2} + \frac{2\bar{v}}{1-\bar{v}^2} \bar{v}'' \right] , \tag{24}$$

and two dynamic terms $\rho_{\text{dyn}-u}$ and $\rho_{\text{dyn}-\tilde{w}}$

$$\rho_{\text{dyn}-u} \equiv \frac{1}{48\pi} \frac{f(u)}{(1+\bar{v})^2} . \tag{25}$$

$$\rho_{\text{dyn}-\tilde{w}} \equiv \frac{1}{48\pi} \frac{g(w)}{(1-\bar{v})^2} , \tag{26}$$

where we have put $c = 1$ and we have defined

$$f(u) \equiv \frac{3\ddot{p}^2(u) - 2\dot{p}(u)\dddot{p}(u)}{\dot{p}^2(u)} , \tag{27}$$

$$g(\tilde{w}) \equiv \frac{3\ddot{q}^2(\tilde{w}) - 2\dot{q}(\tilde{w})\dddot{q}(\tilde{w})}{\dot{q}^2(\tilde{w})} . \tag{28}$$

These dynamic terms, depending also on u ($\tilde{w}$), correspond to energy travelling on right-going (left-going) rays, red/blue-shifted by a term depending on r.

As said above, $\dot{q}$ contains solely information associated with the dynamical details of the transition region. Moreover, for simple forms of this transition, as the one chosen in this work, $\dot{q}(\tilde{w})$ goes to a constant at late times, such that $g(\tilde{w}) \to 0$. Thus, we briefly report the result of an analytical approximate analysis, by neglecting this term, even if in the numerics we retain the full expression of ρ.

In the center of the bubble, at late times $\rho_{\rm st} = 0$, because $\bar{v}(r=0) = \bar{v}'(r=0) = 0$. Moreover $u(t,0) \to +\infty$ and one can evaluate ρ using a late-time expansion for $f(u)$, which gives $f(u) \approx \kappa_1^2$, so that $\rho(r=0) \approx \kappa_1^2/(48\pi) = \pi T_{\rm H}^2/12$, where $T_{\rm H} \equiv \kappa_1/(2\pi)$ is the usual Hawking temperature. This result confirms the presence of a thermal flux of radiation at temperature $T_{\rm H}$, given that the above expression is the energy density of a scalar field in $1+1$ dimension at finite temperature $T_{\rm H}$.

On the horizons $\mathscr{H}_1^+$ and $\mathscr{H}_2^-$, both $\rho_{\rm st}$ and $\rho_{\text{dyn}-u}$ are divergent because of the $(1+\bar{v})$ factors in the denominators. Using the late time expansion of $f(u)$ in the proximity of the horizons, one obtains that the diverging terms in $\rho_{\rm st}$ and $\rho_{\rm dyn-u}$ exactly cancel each other [9]. An analogous cancellation is found when studying the formation of a black hole through gravitational collapse [12]. It is now clear that the total ρ is $\mathcal{O}(1)$ on the horizon and does not diverge at any finite time (as expected from Fulling-Sweeny-Wald theorem [14]). By looking at the subleading terms for the black horizon, one sees that the contribution of the transient radiation (different from Hawking radiation) dies off with time exponentially, with a time scale $\sim 1/\kappa_1$ [3], while something distinctive occurs close to the white horizon. In fact, these subleading terms now becomes

$$\rho = \frac{e^{2\kappa_2 t}}{48\pi}\left[3\left(\frac{D_2}{D_1}\right)^2 - 2\frac{D_3}{D_1}\right] + D + \mathcal{O}\left(r - r_1\right) \ . \qquad (29)$$

This expression shows an exponential increase of the energy density with time. This means that, moving along $\mathscr{H}_2^-$, ρ grows exponentially, diverging at the crossing point between $\mathscr{H}_2^-$ and $\mathscr{H}_C^+$.

In a completely analogous way, performing an expansion at late times ($t \to +\infty$) for $r \to r_2$, one finds that the RSET diverges on the whole $\mathscr{H}_C^+$. This result does not contradict the Fulling-Sweeny-Wald theorem [14] because $\mathscr{H}_C^+$ is a Cauchy horizon.

3. Numerical results

In this section we present the result of the numerical calculation of ρ inside the bubble. The velocity profile is given in Eq. (3) with $c = 1$, $\alpha = 2$ and $a = 1$. With this choice, the surface gravity is $\kappa = \sqrt{3}/2$, the kink function is $\xi(t) = \operatorname{arccosh}(t+1)$ and the horizons appear at $t_{\rm H} = 1$.

In Fig. 2, the energy density ρ (thick solid line) of Eq. (23) is plotted as a function of r at different times ($t = 0.5, 1, 2, 3$) and r varies between r_1 and r_2 the locations of the $\mathscr{H}_2^-$ and $\mathscr{H}_1^+$. The three terms in the right-hand-side of Eq. (23), $\rho_{\rm st}$, $\rho_{\rm dyn-u}$ and $\rho_{\rm dyn-w}$ are plotted, respectively with thin-solid, dashed and dash-dotted lines. In the first plot, all the components of ρ vanish, for $-0.96 \lesssim r \lesssim 0.96$ because at time $t = 0.5$ the kink is located at $r = \pm 0.96$, and the space is flat for $|r| > \xi(t)$.

The figure shows that the dynamic term $\rho_{\rm dyn-w}$ is transient, confirming the correctness of the approximation done in the analytical analysis, when this term had been neglected to study the behavior of ρ at late times. Note, however, that $\rho_{\rm dyn-w}$ gives an important contribution during the creation process of the warp drive.

In the center of the bubble ($r = 0$) at late times, the energy density is $\rho \approx \kappa^2/48\pi \approx 0.005$ and it is due only to the dynamic term $\rho_{\rm dyn-u}$, which can be naturally interpreted as Hawking radiation. For $t \lesssim 2$ the transient term $\rho_{\rm dyn-w}$ cannot be neglected. For instance, at $t = t_{\rm H} = 1$, it causes the energy density inside the bubble to be about twice its late-time value.

On the horizons $r = r_{1,2}$, both $\rho_{\rm st}$ and $\rho_{\rm dyn-u}$ are found to be divergent, but they cancel each other, leaving only a finite term, which vanishes with time on the black horizon ($r = r_1$), while it diverges to $-\infty$ on the white horizon ($r = r_2$), as shown in Eq. (29). The claimed divergence on the Cauchy horizon is also found, even it is difficult to recognize it from such plots. Looking at the Penrose diagram in Fig. 1, one indeed realizes that the location ($r = r_2, t = +\infty$) does not represent a single point but a whole segment of the diagram. This means that one should disentangle the two effects (one at the Cauchy horizon, the other at the crossing point between $\mathscr{H}_2^-$ and $\mathscr{H}_C^+$) which both take place at r close to r_2. This can be done with a careful analysis: Exactly at $r = r_2$, as $t \to +\infty$, the value of ρ goes towards $-\infty$; this contribution is superposed with a positive energy pulse whose peak-value grows towards $+\infty$ with time and whose center is located progressively closer and closer to r_2, approaching it as $r - r_2 \propto e^{-\kappa t}$. This can be better appreciated in Fig. 3, where the same quantities of Fig. 2 are plotted at time $t = 2, 3$, with different scales, proportional to $e^{2\kappa t}$ to compensate the exponential divergence of ρ.

[3] However, in analogy to the conclusions of [12], a slow approach to the black-horizon formation might lead to large values of the RSET and hence to a large back-reaction.

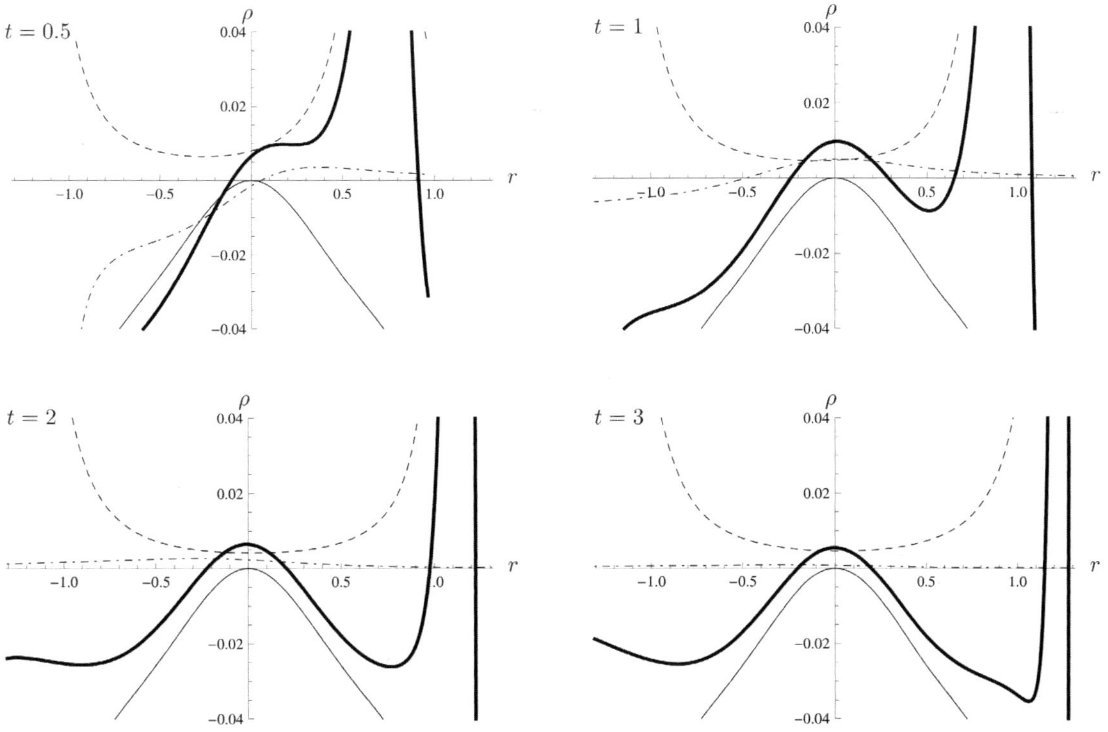

Figure 2. Energy density ρ (*thick solid line*), ρ_{st} (*solid line*), $\rho_{\mathrm{dyn}-u}$ (*dashed line*) and $\rho_{\mathrm{dyn}-w}$ (*dash-dotted line*) as functions of r, $r_1 < r < r_2$ at time $t = 0.5, 1, 2, 3$. Apparent horizons form at $t_{\mathrm{H}} = 1$.

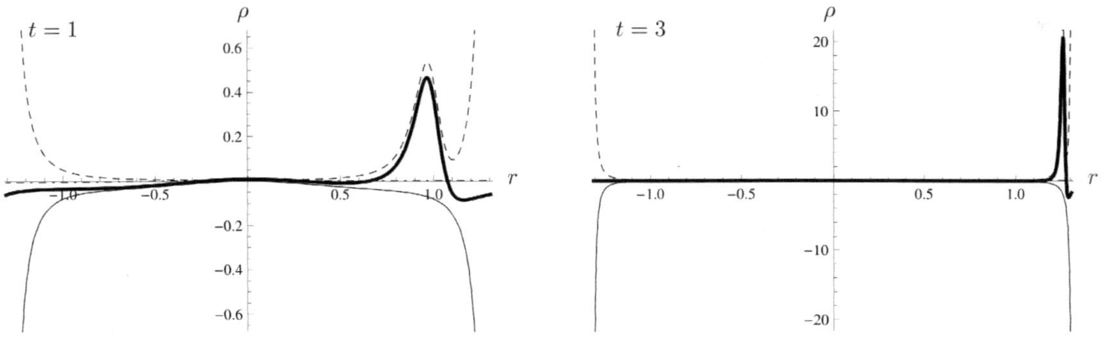

Figure 3. See caption of Fig. 2 for notation. Time slices at $t = 1, 3$. The scale of the ρ-axis is proportional to $e^{2\kappa t}$.

Since this degeneracy is due only to a bad choice of coordinates, it can be removed by using, for instance, the null coordinates u and $\tilde{w}$. In Fig. 4, ρ is plotted as a function of $-\tanh(\kappa u)$ (in order to put the horizon $u = \pm\infty$ to finite positions) and $\tilde{w}$. Here, $\rho \to -\infty$ for $u \to -\infty$ ($-\tanh(\kappa u) = 1$) and $\tilde{w} \to +\infty$, while $\rho \to +\infty$ for u not too much negative and $\tilde{w} \to +\infty$. The former is the divergence on the white horizon, the latter is the divergence close to the Cauchy horizon: Whenever some energy travels on a u-ray, it is exponentially blue-shifted at late times $\tilde{w}$ (late times t), when approaching this horizon.

It is important to stress that these two divergences are of very different nature. In fact, the

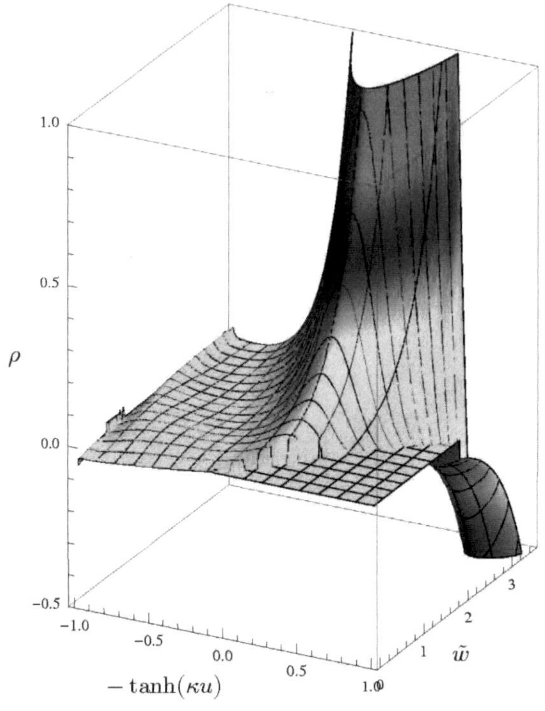

Figure 4. ρ as a function of $-\tanh(\kappa u)$ and $\tilde{w}$.

divergence at the crossing point between $\mathscr{H}_2^-$ and $\mathscr{H}_C^+$ is intrinsically due to the inevitable transient disturbances produced by the formation of the white horizon. In this sense it is a new and very effective instability. On the contrary, the divergence on $\mathscr{H}_C^+$ is due to the well known infinite blue-shift suffered by light rays as they approach a Cauchy horizon. It is analogous to the often claimed instability of inner horizons in Kerr-Newman black holes [15, 16, 17]. In any case the backreaction of the RSET will doom the warp drive to be semiclassically unstable.

4. Conclusions

In this work a $1+1$ calculation was performed. Generally in spherically symmetric spacetimes this could be seen as a s-wave approximation to the exact results. However, this is not the case for the axisymmetric warp-drive configuration. Nonetheless, we do expect that the salient features of our results would be maintained in a full 3+1 calculation, given that they will still be valid in a suitable open set of the horizons centered around the axis aligned with the direction of motion.

Hence, we think that this work is convincingly ruling out the semiclassical stability of superluminal warp drives, confirming and extending the approximate analytic results found in our previous work [9], on the base of the following evidence.

(1) The central region of the warp drive behaves like the asymptotic region of a black hole: In both of these regions the static term $\rho_{\rm st}$ vanishes and the whole energy density is due to the Hawking radiation generated at the black horizon. If one trusts the QI [3, 4], the wall thickness for a warp drive with $v_0 \approx c$ would be $\Delta \lesssim 10^2\, L_{\rm P}$, and its surface gravity $\kappa_1 \gtrsim 10^{-2}\, t_{\rm P}^{-1}$, where $t_{\rm P}$ is the Planck time. Hence, the Hawking temperature of this radiation would be unacceptably large: $T_{\rm H} \sim \kappa_1 \gtrsim 10^{-2}\, T_{\rm P}$, where $T_{\rm P}$ is the Planck temperature, about 10^{32} K.

(2) The formation of a white horizon produces a transient radiation which accumulates on the white horizon itself. This causes the energy density ρ seeing by a free-falling observer to grow unboundedly with time on this horizon. The semiclassical backreaction of the RSET will make the superluminal warp drive to become rapidly unstable, in a time scale of the order of $1/\kappa_2$, the inverse of the surface gravity of the white horizon. In fact, in order to get even a time scale $\tau \sim 1$ s for the growing rate of the RSET, one would need a wall as large as 3×10^8 m. Thus, most probably, one would be able to maintain a superluminal speed for just a very short interval of time.

(3) The formation of a Cauchy horizon gives rise to an instability, similar to inner horizon instability in black holes, due to the blue-shift of Hawking radiation produced by the black horizon.

A suggestive interpretation of these results can be argued in connection with the so called chronology protection conjecture [18]. In fact, a time machine [19] could be build through a couple of superluminal warp drives traveling in opposite directions. Thus, a protection mechanism seems to act at an early stage, forbidding the creation of a system which could be dangerous for causality.

Of course, all the aforementioned problems disappear when the bubble remains subluminal. In that case no horizon forms, no Hawking radiation is created, and neither strong temperature nor white horizon instability is found. Traveling at just 99% of the speed of light could be not that bad, after all.

Acknowledgments

The authors wish to thank S. Sonego and M. Visser for illuminating discussions. S.F. acknowledge the support provided by a INFN/MICINN collaboration. C.B. has been supported by the Spanish MICINN under project FIS2008-06078-C03-01/FIS and Junta de Andalucia under project FQM2288.

References

[1] Alcubierre M 1994 *Class. Quant. Grav.* **11** L73–L77
[2] Lobo F S N *Preprint* gr–qc/0710.4474
[3] Pfenning M J and Ford L H 1997 *Class. Quant. Grav.* **14** 1743–1751
[4] Van Den Broeck C 1999 *Class. Quant. Grav.* **16** 3973–3979
[5] Lobo F S N and Visser M 2004 *Class. Quant. Grav.* **21** 5871–5892
[6] Lobo F S N and Visser M *Preprint* gr–qc/0412065
[7] Roman T A 2005 *The Tenth Marcel Grossmann Meeting* ed Novello M, Perez Bergliaffa S and Ruffini R p 1909
[8] Hiscock W A 1997 *Class. Quant. Grav.* **14** L183–L188
[9] Finazzi S, Liberati S and Barcelo C 2009 *Phys. Rev.* D **79** 124017
[10] Barceló C, Liberati S, Sonego S and Visser M 2006 *Class. Quant. Grav.* **23** 5341–5366
[11] Barceló C, Liberati S, Sonego S and Visser M 2006 *Phys. Rev. Lett.* **97** 171301
[12] Barcelo C, Liberati S, Sonego S and Visser M 2008 *Phys. Rev.* D **77** 044032
[13] Birrell N D and Davies P C W 1984 *Quantum Fields in Curved Space* (Cambridge: Cambridge University Press)
[14] Fulling S A, Sweeny M and Wald R M 1978 *Comm. Math. Phys.* **63** 257–264
[15] Simpson M and Penrose R 1973 *Int. J. Theor. Phys.* **7** 183–197
[16] Poisson E and Israel W 1990 *Phys. Rev.* D **41** 1796–1809
[17] Marković D and Poisson E 1995 *Phys. Rev. Lett.* **74** 1280–1283
[18] Hawking S W 1992 *Phys. Rev.* D **46** 603–611
[19] Everett A E 1996 *Phys. Rev.* D **53** 7365–7368

First Mediterranean Conference on Classical and Quantum Gravity (MCCQG 2009) IOP Publishing
Journal of Physics: Conference Series **222** (2010) 012047 doi:10.1088/1742-6596/222/1/012047

Gravity duals for logarithmic conformal field theories

Daniel Grumiller and Niklas Johansson

Institute for Theoretical Physics, Vienna University of Technology
Wiedner Hauptstrasse 8-10/136, A-1040 Vienna, Austria

E-mail: grumil@hep.itp.tuwien.ac.at, niklasj@hep.itp.tuwien.ac.at

Abstract. Logarithmic conformal field theories with vanishing central charge describe systems with quenched disorder, percolation or dilute self-avoiding polymers. In these theories the energy momentum tensor acquires a logarithmic partner. In this talk we address the construction of possible gravity duals for these logarithmic conformal field theories and present two viable candidates for such duals, namely theories of massive gravity in three dimensions at a chiral point.

Outline

This talk is organized as follows. In section 1 we recall salient features of 2-dimensional conformal field theories. In section 2 we review a specific class of logarithmic conformal field theories where the energy momentum tensor acquires a logarithmic partner. In section 3 we present a wish-list for gravity duals to logarithmic conformal field theories. In section 4 we discuss two examples of massive gravity theories that comply with all the items on that list. In section 5 we address possible applications of an Anti-deSitter/logarithmic conformal field theory correspondence in condensed matter physics.

1. Conformal field theory distillate

Conformal field theories (CFTs) are quantum field theories that exhibit invariance under angle preserving transformations: translations, rotations, boosts, dilatations and special conformal transformations. In two dimensions the conformal algebra is infinite dimensional, and thus two-dimensional CFTs exhibit a particularly rich structure. They arise in various contexts in physics, including string theory, statistical mechanics and condensed matter physics, see e.g. [1].

The main observables in any field theory are correlation functions between gauge invariant operators. There exist powerful tools to calculate these correlators in a CFT. The operator content of various CFTs may differ, but all CFTs contain at least an energy momentum tensor $T_{\mu\nu}$. Conformal invariance requires the energy momentum tensor to be traceless, $T^\mu_\mu = 0$, in addition to its conservation, $\partial_\mu T^{\mu\nu} = 0$. In lightcone gauge for the Minkowski metric, $\mathrm{d}s^2 = 2\,\mathrm{d}z\,\mathrm{d}\bar{z}$, these equations take a particularly simple form: $T_{z\bar{z}} = 0$, $T_{zz} = T_{zz}(z) := \mathcal{O}^L(z)$ and $T_{\bar{z}\bar{z}} = T_{\bar{z}\bar{z}}(\bar{z}) := \mathcal{O}^R(\bar{z})$. Conformal Ward identities determine essentially uniquely the form of 2- and 3-point correlators between the flux components $\mathcal{O}^{L/R}$ of the energy momentum tensor:

$$\langle \mathcal{O}^R(\bar z)\,\mathcal{O}^R(0)\rangle = \frac{c_R}{2\bar z^4} \tag{1a}$$
$$\langle \mathcal{O}^L(z)\,\mathcal{O}^L(0)\rangle = \frac{c_L}{2z^4} \tag{1b}$$
$$\langle \mathcal{O}^L(z)\,\mathcal{O}^R(0)\rangle = 0 \tag{1c}$$
$$\langle \mathcal{O}^R(\bar z)\,\mathcal{O}^R(\bar z')\,\mathcal{O}^R(0)\rangle = \frac{c_R}{\bar z^2 \bar z'^2(\bar z-\bar z')^2} \tag{1d}$$
$$\langle \mathcal{O}^L(z)\,\mathcal{O}^L(z')\,\mathcal{O}^L(0)\rangle = \frac{c_L}{z^2 z'^2(z-z')^2} \tag{1e}$$
$$\langle \mathcal{O}^L(z)\,\mathcal{O}^R(\bar z')\,\mathcal{O}^R(0)\rangle = 0 \tag{1f}$$
$$\langle \mathcal{O}^L(z)\,\mathcal{O}^L(z')\,\mathcal{O}^R(0)\rangle = 0 \tag{1g}$$

The real numbers c_L, c_R are the left and right central charges, which determine key properties of the CFT. We have omitted terms that are less divergent in the near coincidence limit $z, \bar z \to 0$ as well as contact terms, i.e., contributions that are localized (δ-functions and derivatives thereof).

If someone provides us with a traceless energy momentum tensor and gives us a prescription how to calculate correlators,[1] but does not reveal whether the underlying field theory is a CFT, then we can perform the following check. We calculate all 2- and 3-point correlators of the energy momentum tensor with itself, and if at least one of the correlators does not match precisely with the corresponding correlator in (1) then we know that the field theory in question cannot be a CFT. On the other hand, if all the correlators match with corresponding ones in (1) we have non-trivial evidence that the field theory in question might be a CFT. Let us keep this stringent check in mind for later purposes, but switch gears now and consider a specific class of CFTs, namely logarithmic CFTs (LCFTs).

2. Logarithmic CFTs with an energetic partner

LCFTs were introduced in physics by Gurarie [2]. We focus now on some properties of LCFTs and postpone a physics discussion until the end of the talk, see [3,4] for reviews. There are two conceptually different, but mathematically equivalent, ways to define LCFTs. In both versions there exists at least one operator that acquires a logarithmic partner, which we denote by $\mathcal{O}^{\log}$. We focus in this talk exclusively on theories where one (or both) of the energy momentum tensor flux components is the operator acquiring such a partner, for instance $\mathcal{O}^L$. We discuss now briefly both ways of defining LCFTs.

According to the first definition "acquiring a logarithmic partner" means that the Hamiltonian H cannot be diagonalized. For example

$$H\begin{pmatrix}\mathcal{O}^{\log}\\ \mathcal{O}^L\end{pmatrix} = \begin{pmatrix}2 & 1\\ 0 & 2\end{pmatrix}\begin{pmatrix}\mathcal{O}^{\log}\\ \mathcal{O}^L\end{pmatrix} \tag{2}$$

The angular momentum operator J may or may not be diagonalizable. We consider only theories where J is diagonalizable:

$$J\begin{pmatrix}\mathcal{O}^{\log}\\ \mathcal{O}^L\end{pmatrix} = \begin{pmatrix}2 & 0\\ 0 & 2\end{pmatrix}\begin{pmatrix}\mathcal{O}^{\log}\\ \mathcal{O}^L\end{pmatrix} \tag{3}$$

The eigenvalues 2 arise because the energy momentum tensor and its logarithmic partner both correspond to spin-2 excitations.

[1] This is exactly what the AdS/CFT correspondence does: given a gravity dual we can calculate the energy momentum tensor and correlators.

The second definition makes it more transparent why these CFTs are called "logarithmic" in the first place. Suppose that in addition to $\mathcal{O}^{L/R}$ we have an operator $\mathcal{O}^M$ with conformal weights $h = 2 + \varepsilon$, $\bar{h} = \varepsilon$, meaning that its 2-point correlator with itself is given by

$$\langle \mathcal{O}^M(z,\bar{z})\, \mathcal{O}^M(0,0)\rangle = \frac{\hat{B}}{z^{4+2\varepsilon}\,\bar{z}^{2\varepsilon}} \tag{4}$$

The correlator of $\mathcal{O}^M$ with $\mathcal{O}^L$ vanishes since the latter has conformal weights $h = 2$, $\bar{h} = 0$, and operators whose conformal weights do not match lead to vanishing correlators. Suppose now that we send the central charge c_L and the parameter ε to zero, and simultaneously send $\hat{B}$ to infinity, such that the following limits exist:

$$b_L := \lim_{c_L\to 0} -\frac{c_L}{\varepsilon} \neq 0 \qquad B := \lim_{c_L\to 0} \big(\hat{B} + \frac{2}{c_L}\big) \tag{5}$$

Then we can define a new operator $\mathcal{O}^{\log}$ that linearly combines $\mathcal{O}^{L/M}$.

$$\mathcal{O}^{\log} = b_L\, \frac{\mathcal{O}^L}{c_L} + \frac{b_L}{2}\, \mathcal{O}^M \tag{6}$$

Taking the limit $c_L \to 0$ leads to the following 2-point correlators:

$$\langle \mathcal{O}^L(z)\mathcal{O}^L(0,0)\rangle = 0 \tag{7a}$$

$$\langle \mathcal{O}^L(z)\mathcal{O}^{\log}(0,0)\rangle = \frac{b_L}{2z^4} \tag{7b}$$

$$\langle \mathcal{O}^{\log}(z,\bar{z})\mathcal{O}^{\log}(0,0)\rangle = -\frac{b_L \ln{(m_L^2|z|^2)}}{z^4} \tag{7c}$$

These 2-point correlators exhibit several remarkable features. The flux component $\mathcal{O}^L$ of the energy momentum tensor becomes a zero norm state (7a). Nevertheless, the theory does not become chiral, because the left-moving sector is not trivial: $\mathcal{O}^L$ has a non-vanishing correlator (7b) with its logarithmic partner $\mathcal{O}^{\log}$. The 2-point correlator (7c) between two logarithmic operators $\mathcal{O}^{\log}$ makes it clear why such CFTs have the attribute "logarithmic". The constant b_L, sometimes called "new anomaly", defines crucial properties of the LCFT, much like the central charges do in ordinary CFTs. The mass scale m_L appearing in the last correlator above has no significance, and is determined by the value of B in (5). It can be changed to any finite value by the redefinition $\mathcal{O}^{\log} \to \mathcal{O}^{\log} + \gamma\, \mathcal{O}^L$ with some finite γ. We set $m_L = 1$ for convenience.

Conformal Ward identities determine again essentially uniquely the form of 2- and 3-point correlators in a LCFT. For the specific case where the energy momentum tensor acquires a logarithmic partner the 3-point correlators were calculated in [5]. The non-vanishing ones are given by

$$\langle \mathcal{O}^L(z,\bar{z})\mathcal{O}^L(z',\bar{z}')\mathcal{O}^{\log}(0,0)\rangle = \frac{b_L}{z^2 z'^2 (z-z')^2} \tag{8a}$$

$$\langle \mathcal{O}^L(z,\bar{z})\mathcal{O}^{\log}(z',\bar{z}')\mathcal{O}^{\log}(0,0)\rangle = -\frac{2b_L \ln|z'|^2 + \frac{b_L}{2}}{z^2 z'^2 (z-z')^2} \tag{8b}$$

$$\langle \mathcal{O}^{\log}(z,\bar{z})\mathcal{O}^{\log}(z',\bar{z}')\mathcal{O}^{\log}(0,0)\rangle = \frac{\text{lengthy}}{z^2 z'^2 (z-z')^2} \tag{8c}$$

If also $\mathcal{O}^R$ acquires a logarithmic partner $\mathcal{O}^{\widetilde{\log}}$ then the construction above can be repeated, changing everywhere $L \to R$, $z \to \bar{z}$ etc. In that case we have a LCFT with $c_L = c_R = 0$ and

Journal of Physics: Conference Series **222** (2010) 012047 doi:10.1088/1742-6596/222/1/012047

$b_L, b_R \neq 0$. Alternatively, it may happen that only $\mathcal{O}^L$ has a logarithmic partner $\mathcal{O}^{\log}$. In that case we have a LCFT with $c_L = b_R = 0$ and $b_L, c_R \neq 0$. This concludes our brief excursion into the realm of LCFTs.

Given that LCFTs are interesting in physics (see section 5) and that a powerful way to describe strongly coupled CFTs is to exploit the AdS/CFT correspondence [6] it is natural to inquire whether there are any gravity duals to LCFTs.

3. Wish-list for gravity duals to LCFTs

In this section we establish necessary properties required for gravity duals to LCFTs. We formulate them as a wish-list and explain afterwards each item on this list.

(i) We wish for a 3-dimensional action S that depends on the metric $g_{\mu\nu}$ and possibly on further fields that we summarily denote by ϕ.

(ii) We wish for the existence of AdS_3 vacua with finite AdS radius ℓ.

(iii) We wish for a finite, conserved and traceless Brown–York stress tensor, given by the first variation of the full on-shell action (including boundary terms) with respect to the metric.

(iv) We wish that the 2- and 3-point correlators of the Brown–York stress tensor with itself are given by (1).

(v) We wish for central charges (a la Brown–Henneaux [7]) that can be tuned to zero, without requiring a singular limit of the AdS radius or of Newton's constant. For concreteness we assume $c_L = 0$ (in addition c_R may also vanish, but it need not).

(vi) We wish for a logarithmic partner to the Brown–York stress tensor, so that we obtain a Jordan-block structure like in (2) and (3).

(vii) We wish that the 2- and non-vanishing 3-point correlators of the Brown–York stress tensor with its logarithmic partner are given by (7) and (8) (and the right-handed analog thereof).

We explain now why each of these items is necessary. (i) is required since the AdS/CFT correspondence relates a gravity theory in $d+1$ dimensions to a CFT in d dimensions, and we chose $d = 2$ on the CFT side. (ii) is required since we are not merely looking for a gauge/gravity duality, but really for an AdS/CFT correspondence, which requires the existence of AdS solutions on the gravity side. (iii) is required since we desire consistency with the AdS dictionary, which relates the vacuum expectation value of the renormalized energy momentum tensor in the CFT $\langle T_{ij}\rangle$ to the Brown–York stress tensor T_{ij}^{BY}:

$$\langle T_{ij}\rangle = T_{ij}^{\mathrm{BY}} = \frac{2}{\sqrt{-g}}\,\frac{\delta S}{\delta g^{ij}}\Big|_{\mathrm{EOM}} \tag{9}$$

The right hand side of this equation contains the first variation of the full on-shell action with respect to the metric, which by definition yields the Brown–York stress tensor. (iv) is required since the 2- and 3-point correlators of a CFT are fixed by conformal Ward identities to take the form (1). (v) is required because of the construction presented in section 2, where a LCFT emerges from taking an appropriate limit of vanishing central charge, so we need to be able to tune the central charge without generating parametric singularities. Actually, there are two cases: either left and right central charge vanish and both energy momentum tensor flux components acquire a logarithmic partner, or only one of them acquires a logarithmic partner, which for sake of specificity we always choose to be left. (vi) is required, since we consider exclusively LCFTs where the energy momentum tensor acquires a logarithmic partner. (vii) is required since the 2- and 3-point correlators of a LCFT are fixed by conformal Ward identities to take the form (7), (8). If any of the items on the wish-list above is not fulfilled it is impossible that the gravitational theory under consideration is a gravity dual to a LCFT of the type discussed in section 2.[2] On the other hand, if all the wishes are granted by a given gravitational theory there are excellent chances that this theory is dual to a LCFT. Until recently no good gravity duals for LCFTs were known [8–12].

Before addressing candidate theories that may comply with all wishes we review briefly how to calculate correlators on the gravity side [6], since we shall need such calculations for checking several items on the wish-list. The basic identity of the AdS/CFT dictionary is

$$\langle \mathcal{O}_1(z_1)\, \mathcal{O}_2(z_2) \dots \mathcal{O}_n(z_n) \rangle = \frac{\delta^{(n)} S}{\delta j_1(z_1) \delta j_2(z_2) \dots \delta j_n(z_n)} \Big|_{j_i=0} \tag{10}$$

The left hand side is the CFT correlator between n operators $\mathcal{O}_i$, where $\mathcal{O}_i$ in our case comprise the left- and right-moving flux components of the energy momentum tensor and their logarithmic partners. The right hand side contains the gravitational action S differentiated with respect to appropriate sources j_i for the corresponding operators. According to the AdS/CFT dictionary "appropriate sources" refers to non-normalizable solutions of the linearized equations of motion. We shall be more concrete about the operators, actions, sources and non-normalizable solutions to the linearized equations of motion in the next section. For now we address possible candidate theories of gravity duals to LCFTs.

The simplest candidate, pure 3-dimensional Einstein gravity with a cosmological constant described by the action

$$S_{\text{EH}} = -\frac{1}{8\pi\, G_N} \int_{\mathcal{M}} \mathrm{d}^3x \sqrt{-g}\, \Big[R + \frac{2}{\ell^2} \Big] - \frac{1}{4\pi\, G_N} \int_{\partial\mathcal{M}} \mathrm{d}^2x \sqrt{-\gamma}\, \Big[K - \frac{1}{\ell} \Big] \tag{11}$$

does not comply with the whole wish list. Only the first four wishes are granted: The 3-dimensional action (12) depends on the metric. The equations of motion are solved by AdS_3.

$$\mathrm{d}s^2_{\text{AdS}_3} = g_{\mu\nu}^{\text{AdS3}}\, \mathrm{d}x^\mu\, \mathrm{d}x^\nu = \ell^2 \left(\mathrm{d}\rho^2 - \frac{1}{4} \cosh^2\!\rho\, (\mathrm{d}u + \mathrm{d}v)^2 + \frac{1}{4} \sinh^2\!\rho\, (\mathrm{d}u - \mathrm{d}v)^2 \right) \tag{12}$$

The Brown–York stress tensor (9) is finite, conserved and traceless. The 2- and 3-point correlators on the gravity side match precisely with (1). However, the central charges are given by [7]

$$c_L = c_R = \frac{3\ell}{2G_N} \tag{13}$$

and therefore allow no tuning to $c_L = 0$ without taking a singular limit. Moreover, there is no candidate for a logarithmic partner to the Brown–York stress tensor. Thus, pure 3-dimensional Einstein gravity cannot be dual to a LCFT.

Adding matter fields to Einstein gravity does not help neither. While this may lead to other kinds of LCFTs, it cannot produce a logarithmic partner for the energy momentum tensor. This is so, because the energy momentum tensor corresponds to graviton (spin-2) excitations in the bulk, and the only field producing such excitations is the metric.

Therefore, what we need is a way to provide additional degrees of freedom in the gravity sector. The most natural way to do this is by considering higher derivative interactions of the metric. The first gravity model of this type was constructed by Deser, Jackiw and Templeton [13] who introduced a Chern–Simons term for the Christoffel connection.

$$S_{\text{CS}} = -\frac{1}{16\pi G_N\, \mu} \int \mathrm{d}^3x\, \epsilon^{\lambda\mu\nu} \Gamma^\rho{}_{\sigma\lambda} \Big[\partial_\mu \Gamma^\sigma{}_{\rho\nu} + \frac{2}{3} \Gamma^\sigma{}_{\kappa\mu} \Gamma^\kappa{}_{\sigma\nu} \Big] \tag{14}$$

[2] Other types of LCFTs exist, e.g. with non-vanishing central charge or with logarithmic partners to operators other than the energy momentum tensor. The gravity duals for such LCFTs need not comply with all the items on our wish list.

Here μ is a real coupling constant. Adding this action to the Einstein–Hilbert action (11) generates massive graviton excitations in the bulk, which is encouraging for our wish list since we need these extra degrees of freedom. The model that arises when summing the actions (11) and (14),

$$S_{\rm CTMG} = S_{\rm EH} + S_{\rm CS} \tag{15}$$

is known as "cosmological topologically massive gravity" (CTMG) [14]. It was demonstrated by Kraus and Larsen [15] that the central charges in CTMG are shifted from their Brown–Henneaux values:

$$c_L = \frac{3\ell}{2G_N}\big(1-\frac{1}{\mu\ell}\big) \qquad c_R = \frac{3\ell}{2G_N}\big(1+\frac{1}{\mu\ell}\big) \tag{16}$$

This is again good news concerning our wish list, since c_L can be made vanishing by a (non-singular) tuning of parameters in the action.

$$\mu\ell = 1 \tag{17}$$

CTMG (15) with the tuning above (17) is known as "cosmological topologically massive gravity at the chiral point" (CCTMG). It complies with the first five items on our wish list, but we still have to prove that also the last two wishes are granted. To this end we need to find a suitable partner for the graviton.

4. Keeping logs in massive gravity

4.1. Login

In this section we discuss the evidence for the existence of specific gravity duals to LCFTs that has accumulated over the past two years. We start with the theory introduced above, CCTMG, and we end with a relatively new theory, new massive gravity [16].

4.2. Seeds of logs

Given that we want a partner for the graviton we consider now graviton excitations ψ around the AdS background (12) in CCTMG.

$$g_{\mu\nu} = g_{\mu\nu}^{\rm AdS_3} + \psi_{\mu\nu} \tag{18}$$

Li, Song and Strominger [17] found a nice way to construct them, and we follow their construction here. Imposing transverse gauge $\nabla_\mu\psi^{\mu\nu} = 0$ and defining the mutually commuting first order operators

$$(\mathcal{D}^M)_\mu^\beta = \delta_\mu^\beta + \frac{1}{\mu}\,\varepsilon_\mu{}^{\alpha\beta}\nabla_\alpha \qquad (\mathcal{D}^{L/R})_\mu^\beta = \delta_\mu^\beta \pm \ell\,\varepsilon_\mu{}^{\alpha\beta}\nabla_\alpha \tag{19}$$

allows to write the linearized equations of motion around the AdS background (12) as follows.

$$(\mathcal{D}^M\mathcal{D}^L\mathcal{D}^R\psi)_{\mu\nu} = 0 \tag{20}$$

A mode annihilated by $\mathcal{D}^M$ ($\mathcal{D}^L$) [$\mathcal{D}^R$] {$(\mathcal{D}^L)^2$ but not by $\mathcal{D}^L$} is called massive (left-moving) [right-moving] {logarithmic} and is denoted by ψ^M (ψ^L) [ψ^R] {$\psi^{\log}$}. Away from the chiral point, $\mu\ell \neq 1$, the general solution to the linearized equations of motion (20) is obtained from linearly combining left, right and massive modes [17]. At the chiral point $\mathcal{D}^M$ degenerates with $\mathcal{D}^L$ and the general solution to the linearized equations of motion (20) is obtained from linearly combining left, right and logarithmic modes [18]. Interestingly, we discovered in [18] that the modes $\psi^{\log}$ and ψ^L behave as follows:

$$(L_0 + \bar{L}_0)\begin{pmatrix}\psi^{\log}\\ \psi^L\end{pmatrix} = \begin{pmatrix}2 & 1\\ 0 & 2\end{pmatrix}\begin{pmatrix}\psi^{\log}\\ \psi^L\end{pmatrix} \tag{21}$$

where $L_0 = i\partial_u$, $\bar{L}_0 = i\partial_v$ and

$$(L_0 - \bar{L}_0)\begin{pmatrix} \psi^{\log} \\ \psi^L \end{pmatrix} = \begin{pmatrix} 2 & 0 \\ 0 & 2 \end{pmatrix}\begin{pmatrix} \psi^{\log} \\ \psi^L \end{pmatrix} \tag{22}$$

If we define naturally the Hamiltonian by $H = L_0 + \bar{L}_0$ and the angular momentum by $J = L_0 - \bar{L}_0$ we recover exactly (2) and (3), which suggests that the CFT dual to CCTMG (if it exists) is logarithmic, as conjectured in [18]. It was further shown with Jackiw that the existence of the logarithmic excitations $\psi^{\log}$ is not an artifact of the linearized approach, but persists in the full theory [19].

Thus, also the sixth wish is granted in CCTMG. The rest of this section discusses the last wish.

4.3. Growing logs

We assume now that there is a standard AdS/CFT dictionary [6] available for LCFTs and check if CCTMG indeed leads to the correct 2- and 3-point correlators. To this end we have to identify the sources j_i that appear on the right hand side of the correlator equation (10). Following the standard AdS/CFT prescription the sources for the operators $\mathcal{O}^L$ ($\mathcal{O}^R$) [$\mathcal{O}^{\log}$] are given by left (right) [logarithmic] non-normalizable solutions to the linearized equations of motion (20). Thus, our first task is to find all solutions of the linearized equations of motion and to classify them into normalizable and non-normalizable ones, where "normalizable" refers to asymptotic (large ρ) behavior that is exponentially suppressed as compared to the AdS background (12).

A construction of all normalizable left and right solutions was provided in [17], and the normalizable logarithmic solutions were constructed in [18].[3] The non-normalizable solutions were constructed in [25]. It turned out to be convenient to work in momentum space

$$\psi_{\mu\nu}^{L/R/\log}(h,\bar{h}) = e^{-ih(t+\phi)-i\bar{h}(t-\phi)}\, F_{\mu\nu}^{L/R/\log}(\rho) \tag{23}$$

The momenta $h, \bar{h}$ are called "weights". All components of the tensor $F_{\mu\nu}$ are determined algebraically, except for one that is determined from a second order (hypergeometric) differential equation. In general one of the linear combinations of the solutions is singular at the origin $\rho = 0$, while the other is regular there. We keep only regular solutions. For each given set of weights $h, \bar{h}$ the regular solution is either normalizable or non-normalizable. It turns out that normalizable solutions exist for integer weights $h \geq 2$, $\bar{h} \geq 0$ (or $h \leq -2$, $\bar{h} \leq 0$). All other solutions are non-normalizable.

An example for a normalizable left mode is given by the primary with weights $h = 2$, $\bar{h} = 0$

$$\psi_{\mu\nu}^L(2,0) = \frac{e^{-2iu}}{\cosh^4\rho}\begin{pmatrix} \frac{1}{4}\sinh^2(2\rho) & 0 & \frac{i}{2}\sinh(2\rho) \\ 0 & 0 & 0 \\ \frac{i}{2}\sinh(2\rho) & 0 & -1 \end{pmatrix}_{\mu\nu} \tag{24}$$

Note that all components of this mode behave asymptotically ($\rho \to \infty$) at most like a constant. The corresponding logarithmic mode is given by

$$\psi_{\mu\nu}^{\log}(2,0) = -\frac{1}{2}\left(i(u+v) + \ln\cosh^2\rho\right)\psi_{\mu\nu}^L(2,0) \tag{25}$$

Evidently, it behaves asymptotically like its left partner (24), except for overall linear growth in ρ. It is also worthwhile emphasizing that the logarithmic mode (25) depends linearly on time

[3] All these modes are compatible with asymptotic AdS behavior [20,21], and they appear in vacuum expectation values of 1-point functions. Indeed, the 1-point function $\langle T^{ij}\rangle$ involves both $\psi^{\log}$ and ψ^R [21–24].

$t = (u+v)/2$. Both features are inherent to all logarithmic modes. All other normalizable modes can be constructed from the primaries (24), (25) algebraically.

An example for a non-normalizable left mode is given by the mode with weights $h = 1$, $\bar{h} = -1$

$$\psi^L_{\mu\nu}(1,-1) = \frac{1}{4}\, e^{-iu+iv} \begin{pmatrix} 0 & 0 & 0 \\ 0 & \cosh(2\rho)-1 & -2i\sqrt{\frac{\cosh(2\rho)-1}{\cosh(2\rho)+1}} \\ 0 & -2i\sqrt{\frac{\cosh(2\rho)-1}{\cosh(2\rho)+1}} & -\frac{4}{\cosh(2\rho)+1} \end{pmatrix}_{\mu\nu} \tag{26}$$

Note that all components of this mode behave asymptotically ($\rho \to \infty$) at most like a constant, except for the vv-component, which grows like $e^{2\rho}$. The corresponding logarithmic mode grows again faster than its left partner (26) by a factor of ρ and depends again linearly on time.

Given a non-normalizable solution ψ^L obviously also $\alpha\,\psi^L$ is a non-normalizable solution, with some constant α. To fix this normalization ambiguity we demand standard coupling of the metric to the stress tensor:

$$S(\psi_v^{u\,L}, T_u^v) = \frac{1}{2}\int \mathrm{d}t\,\mathrm{d}\phi\sqrt{-g^{(0)}}\,\psi_L^{uu} T_{uu} = \int \mathrm{d}t\,\mathrm{d}\phi\, e^{-ihu-i\bar{h}v}\, T_{uu} \tag{27}$$

Here S is either some CFT action with background metric $g^{(0)}$ or a dual gravitational action with boundary metric $g^{(0)}$. The non-normalizable mode ψ^L is the source for the energy-momentum flux component T_{uu}. The requirement (27) fixes the normalization. The discussion above focussed on left modes. For the right modes essentially the same discussion applies, but with the substitutions $L \leftrightarrow R$, $h \leftrightarrow \bar{h}$ and $u \leftrightarrow v$.

4.4. Logging correlators

Generically the 2-point correlators on the gravity side between two modes $\psi^1(h,\bar{h})$ and $\psi^2(h',\bar{h}')$ in momentum space are determined by

$$\langle \psi^1(h,\bar{h})\,\psi^2(h',\bar{h}')\rangle = \frac{1}{2}\left(\delta^{(2)} S_{\text{CCTMG}}(\psi^1,\psi^2) + \delta^{(2)} S_{\text{CCTMG}}(\psi^2,\psi^1)\right) \tag{28}$$

where $\langle \psi^1\,\psi^2\rangle$ stands for the correlation function of the CFT operators dual to the graviton modes ψ^1 and ψ^2. On the right hand side one has to plug the non-normalizable modes ψ^1 and ψ^2 into the second variation of the on-shell action and symmetrize with respect to the two modes. The second variation of the on-shell action of CCTMG

$$\delta^{(2)} S_{\text{CCTMG}} = -\frac{1}{16\pi\, G_N}\int \mathrm{d}^3x\sqrt{-g}\,(\mathcal{D}^L \psi^{1\,*})^{\mu\nu}\,\delta G_{\mu\nu}(\psi^2) + \text{boundary terms} \tag{29}$$

turns out to be very similar to the second variation of the on-shell Einstein–Hilbert action

$$\delta^{(2)} S_{\text{EH}} = -\frac{1}{16\pi\, G_N}\int \mathrm{d}^3x\sqrt{-g}\,\psi^{1\,\mu\nu\,*}\,\delta G_{\mu\nu}(\psi^2) + \text{boundary terms} \tag{30}$$

This similarity allows us to exploit results from Einstein gravity for CCTMG, as we now explain.[4] The bulk term in CCTMG (29) has the same form as in Einstein theory (30) with ψ^1 replaced by $\mathcal{D}^L\psi^1$. Now, consider boundary terms. Possible obstructions to a well-defined Dirichlet boundary value problem can come only from the variation $\delta G_{\mu\nu}(\psi^2)$, since $\mathcal{D}^L$ is a first order operator. Thus any boundary terms appearing in (29) containing normal derivatives must be

[4] Alternatively, one can follow the program of holographic renormalization, as it was done by Skenderis, Taylor and van Rees [23]. Their results for 2-point correlators agree with the results presented here.

Journal of Physics: Conference Series **222** (2010) 012047 doi:10.1088/1742-6596/222/1/012047

identical with those in Einstein gravity upon substituting $\psi^1 \to \mathcal{D}^L\psi^1$. In addition there can be boundary terms which do not contain normal derivatives of the metric. However, it turns out that such terms can at most lead to contact terms in the holographic computation of 2-point functions. The upshot of this discussion is that we can reduce the calculation of all possible 2-point functions in CCTMG to the equivalent calculation in Einstein gravity with suitable source terms. To continue we go on-shell.[5]

$$\mathcal{D}^L\psi^L = 0 \qquad \mathcal{D}^L\psi^R = 2\psi^R \qquad \mathcal{D}^L\psi^{\log} = -2\psi^L \tag{31}$$

These relations together with the comparison between CCTMG (29) and Einstein gravity (30) then establish

$$\langle\psi^R(h,\bar h)\,\psi^R(h',\bar h')\rangle_{\text{CCTMG}} \sim 2\langle\psi^R(h,\bar h)\,\psi^R(h',\bar h')\rangle_{\text{EH}} \tag{32a}$$
$$\langle\psi^L(h,\bar h)\,\psi^L(h',\bar h')\rangle_{\text{CCTMG}} \sim 0 \tag{32b}$$
$$\langle\psi^L(h,\bar h)\,\psi^R(h',\bar h')\rangle_{\text{CCTMG}} \sim 0 \tag{32c}$$
$$\langle\psi^R(h,\bar h)\,\psi^{\log}(h',\bar h')\rangle_{\text{CCTMG}} \sim 0 \tag{32d}$$
$$\langle\psi^L(h,\bar h)\,\psi^{\log}(h',\bar h')\rangle_{\text{CCTMG}} \sim -2\,\langle\psi^L(h,\bar h)\,\psi^L(h',\bar h')\rangle_{\text{EH}} \tag{32e}$$

Here the sign $\sim$ means equality up to contact terms. Evaluating the right hand sides in Einstein gravity yields

$$\langle\psi^L(h,\bar h)\,\psi^L(h',\bar h')\rangle_{\text{EH}} = \delta_{h,h'}\,\delta_{\bar h,\bar h'}\,\frac{c_{BH}}{24}\,\frac{h}{\bar h}(h^2-1)\int\limits_{t_0}^{t_1}\mathrm{d}t \tag{33}$$

and similarly for the right modes, with $h \leftrightarrow \bar h$. The quantity c_{BH} is the Brown–Henneaux central charge (13). The calculation of the 2-point correlator between two logarithmic modes cannot be reduced to a correlator known from Einstein gravity. The result is given by [25]

$$\langle\psi^{\log}(h,\bar h)\,\psi^{\log}(h',\bar h')\rangle_{\text{CCTMG}} \sim -\delta_{h,h'}\,\delta_{\bar h,\bar h'}\,\frac{\ell}{4G_N}\,\frac{h}{\bar h}\,(h^2-1)\,\big(\psi(h-1)+\psi(-\bar h)\big)\int\limits_{t_0}^{t_1}\mathrm{d}t \tag{34}$$

where ψ is the digamma function. An ambiguity in defining $\psi^{\log}$, viz., $\psi^{\log} \to \psi^{\log} + \gamma\,\psi^L$, was fixed conveniently in the result (34). This ambiguity corresponds precisely to the ambiguity of the LCFT mass scale m_L in (7c) (see also the discussion below that equation).

To compare the results (32)-(34) with the Euclidean 2-point correlators in the short-distance limit (1), (7) we take the limit of large weights $h, -\bar h \to \infty$ (e.g. $\lim_{h\to\infty}\psi(h) = \ln h + \mathcal{O}(1/h)$) and Fourier-transform back to coordinate space (e.g. $h^3/\bar h$ is Fourier-transformed into $\partial_z^4/(\partial_z\partial_{\bar z})\,\delta^{(2)}(z,\bar z) \propto \partial_z^4 \ln|z| \propto 1/z^4$). Straightforward calculation establishes perfect agreement with the LCFT correlators (1), (7), provided we use the values

$$c_L = 0 \qquad c_R = \frac{3\ell}{G_N} \qquad b_L = -\frac{3\ell}{G_N} \tag{35}$$

These are exactly the values for central charges c_L, c_R [15] and new anomaly b_L [23, 25] found before. Thus, at the level of 2-point correlators CCTMG is indeed a gravity dual for a LCFT.

[5] Above by "on-shell" we meant that the background metric is AdS_3 (12) and therefore a solution of the classical equations of motion. Here by "on-shell" we mean additionally that the linearized equations of motion (20) hold.

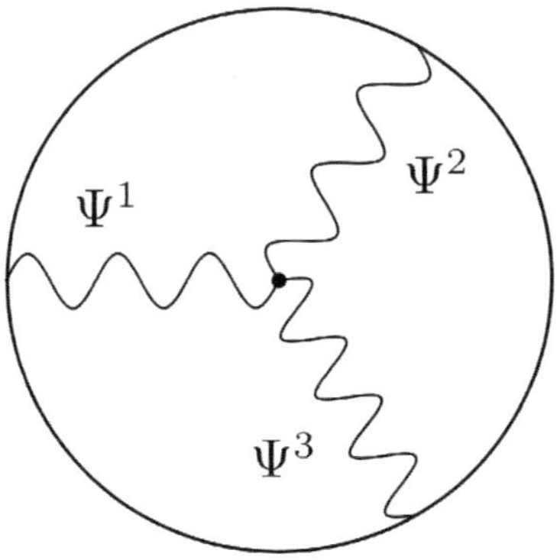

Figure 1. Witten diagram for three graviton correlator

We evaluate now the Witten diagram in Fig. 1, which yields the 3-point correlator on the gravity side between three modes $\psi^1(h,\bar{h})$, $\psi^2(h',\bar{h}')$ and $\psi^3(h'',\bar{h}'')$ in momentum space.

$$\langle \psi^1(h,\bar{h})\,\psi^2(h',\bar{h}')\,\psi^3(h'',\bar{h}'')\rangle = \frac{1}{6}\left(\delta^{(3)} S_{\text{CCTMG}}(\psi^1,\psi^2,\psi^3) + 5 \text{ permutations}\right) \tag{36}$$

On the right hand side one has to plug the non-normalizable modes ψ^1, ψ^2 and ψ^3 into the third variation of the on-shell action and symmetrize with respect to all three modes.

$$\delta^{(3)} S_{\text{CCTMG}} \sim -\frac{1}{16\pi\, G_N}\int \mathrm{d}^3x\sqrt{-g}\left[\left(\mathcal{D}^L\psi^1\right)^{\mu\nu}\delta^{(2)}R_{\mu\nu}(\psi^2,\psi^3) + \psi^{1\,\mu\nu}\,\Delta_{\mu\nu}(\psi^2,\psi^3)\right] \tag{37}$$

The quantity $\delta^{(2)}R_{\mu\nu}(\psi^2,\psi^3)$ denotes the second variation of the Ricci-tensor and the tensor $\Delta_{\mu\nu}(\psi^2,\psi^3)$ vanishes if evaluated on left- and/or right-moving solutions. All boundary terms turn out to be contact terms, which is why only bulk terms are present in the result (37) for the third variation of the on-shell action. We compare again with Einstein gravity.

$$\delta^{(3)} S_{\text{EH}} \sim -\frac{1}{16\pi\, G_N}\int \mathrm{d}^3x\sqrt{-g}\,\psi^{1\,\mu\nu}\,\delta^{(2)}R_{\mu\nu}(\psi^2,\psi^3) \tag{38}$$

Once more we can exploit some results from Einstein gravity for CCTMG, and we find the following results [25] for 3-point correlators without log-insertions:

$$\langle \psi^R(h,\bar{h})\,\psi^R(h',\bar{h}')\,\psi^R(h'',\bar{h}'')\rangle_{\text{CCTMG}} \sim 2\,\langle \psi^R(h,\bar{h})\,\psi^R(h',\bar{h}')\,\psi^R(h'',\bar{h}'')\rangle_{\text{EH}} \tag{39a}$$
$$\langle \psi^L(h,\bar{h})\,\psi^R(h',\bar{h}')\,\psi^R(h'',\bar{h}'')\rangle_{\text{CCTMG}} \sim 0 \tag{39b}$$
$$\langle \psi^L(h,\bar{h})\,\psi^L(h',\bar{h}')\,\psi^R(h'',\bar{h}'')\rangle_{\text{CCTMG}} \sim 0 \tag{39c}$$
$$\langle \psi^L(h,\bar{h})\,\psi^L(h',\bar{h}')\,\psi^L(h'',\bar{h}'')\rangle_{\text{CCTMG}} \sim 0 \tag{39d}$$

with one log-insertion:

$$\langle \psi^R(h,\bar{h})\,\psi^R(h',\bar{h}')\,\psi^{\log}(h'',\bar{h}'')\rangle_{\text{CCTMG}} \sim 0 \tag{40a}$$
$$\langle \psi^L(h,\bar{h})\,\psi^R(h',\bar{h}')\,\psi^{\log}(h'',\bar{h}'')\rangle_{\text{CCTMG}} \sim 0 \tag{40b}$$
$$\langle \psi^L(h,\bar{h})\,\psi^L(h',\bar{h}')\,\psi^{\log}(h'',\bar{h}'')\rangle_{\text{CCTMG}} \sim -2\,\langle \psi^L(h,\bar{h})\,\psi^L(h',\bar{h}')\,\psi^L(h'',\bar{h}'')\rangle_{\text{EH}} \tag{40c}$$

Journal of Physics: Conference Series **222** (2010) 012047 doi:10.1088/1742-6596/222/1/012047

and with two or more log-insertions:

$$\lim_{|\text{weights}|\to\infty}\langle\psi^R(h,\bar h)\,\psi^{\log}(h',\bar h')\,\psi^{\log}(h'',\bar h'')\rangle_{\text{CCTMG}}\sim 0 \tag{41a}$$

$$\lim_{|\text{weights}|\to\infty}\langle\psi^L(h,\bar h)\,\psi^{\log}(h',\bar h')\,\psi^{\log}(h'',\bar h'')\rangle_{\text{CCTMG}}\sim \delta_{h'',-h-h'}\delta_{\bar h'',-\bar h-\bar h'}\,\frac{P^{\log}(h,h',\bar h,\bar h')}{\bar h\bar h'(\bar h+\bar h')} \tag{41b}$$

$$\lim_{|\text{weights}|\to\infty}\langle\psi^{\log}(h,\bar h)\,\psi^{\log}(h',\bar h')\,\psi^{\log}(h'',\bar h'')\rangle_{\text{CCTMG}}\sim \delta_{h'',-h-h'}\delta_{\bar h'',-\bar h-\bar h'}\,\frac{\text{lengthy}}{\bar h\bar h'(\bar h+\bar h')} \tag{41c}$$

The last two correlators so far could be calculated qualitatively only ($P^{\log}$ is a known polynomial in the weights and also contains logarithms in the weights, as expected on general grounds), and it would be interesting to calculate them exactly. They are in qualitative agreement with corresponding LCFT correlators. All other correlators have been calculated exactly [25], and they are in precise agreement with the LCFT correlators (1), (8), provided we use again the values (35) for central charges and new anomaly.

In conclusion, also the seventh wish is granted for CCTMG.[6] Thus, there are excellent chances that CCTMG is dual to a LCFT with values for central charges and new anomaly given by (35).

4.5. Logs don't grow on trees

From the discussion above it is clear that possible gravity duals for LCFTs are sparse in theory space: Einstein gravity (11) does not provide a gravity dual for any tuning of parameters and CTMG (15) does potentially provide a gravity dual only for a specific tuning of parameters (17). Any candidate for a novel gravity dual to a LCFT is therefore welcomed as a rare entity.

Very recently another plausible candidate for such a gravitational theory was found [26]. That theory is known as "new massive gravity" [16].

$$S_{\text{NMG}}=\frac{1}{16\pi G_N}\int d^3x\sqrt{-g}\left[\sigma R+\frac{1}{m^2}\left(R^{\mu\nu}R_{\mu\nu}-\frac{3}{8}R^2\right)-2\lambda m^2\right] \tag{42}$$

Here m is a mass parameter, λ a dimensionless cosmological parameter and $\sigma=\pm1$ the sign of the Einstein-Hilbert term. If they are tuned as follows

$$\lambda = 3 \qquad\Rightarrow\qquad m^2 = -\frac{\sigma}{2\ell^2} \tag{43}$$

then essentially the same story unfolds as for CTMG at the chiral point. The main difference to CCTMG is that both central charges vanish in new massive gravity at the chiral point (CNMG) [27,28].

$$c_L=c_R=\frac{3\ell}{2G_N}\left(\sigma+\frac{1}{2\ell^2m^2}\right)=0 \tag{44}$$

Therefore, both left and right flux component of the energy momentum tensor acquire a logarithmic partner. It is easy to check that CNMG grants us the first six wishes from section 3. The seventh wish requires again the calculation of correlators. The 3-point correlators have not been calculated so far, but at the level of 2-point correlators again perfect agreement with a LCFT was found, provided we use the values [26]

$$c_L=c_R=0 \qquad b_L=b_R=-\sigma\,\frac{12\ell}{G_N} \tag{45}$$

[6] The sole caveat is that two of the ten 3-point correlators were calculated only qualitatively. It would be particularly interesting to calculate the correlator between three logarithmic modes (41c), since it contains an additional parameter independent from the central charges and new anomaly that determines LCFT properties.

It is likely that a similar story can be repeated for general massive gravity [16], which combines new massive gravity (42) with a gravitational Chern–Simons term (14). Thus, even though they are sparse in theory space we have found a few good candidates for gravity duals to LCFTs: cosmological topologically massive gravity, new massive gravity and general massive gravity. In all cases we have to tune parameters in such a way that a "chiral point" emerges where at least one of the central charges vanishes.

4.6. Chopping logs?

So far we were exclusively concerned with finding gravitational theories where logarithmic modes can arise. In this subsection we try to get rid of them. The rationale behind the desire to eliminate the logarithmic modes is unitarity of quantum gravity. Gravity in 2+1 dimensions is simple and yet relevant, as it contains black holes [29], possibly gravity waves [13] and solutions that are asymptotically AdS. Thus, it could provide an excellent arena to study quantum gravity in depth provided one is able to come up with a consistent (unitary) theory of quantum gravity, for instance by constructing its dual (unitary) CFT. Indeed, two years ago Witten suggested a specific CFT dual to 3-dimensional quantum gravity in AdS [30]. This proposal engendered a lot of further research (see [31–37] for some early references), including the suggestion by Li, Song and Strominger [17] to construct a quantum theory of gravity that is purely right-moving, dubbed "chiral gravity". To make a long story [18,19,24,38–81] short, "chiral gravity" is nothing but CCTMG with the logarithmic modes truncated in some consistent way.

We discuss now two conceptually different possibilities of implementing such a truncation. The first option was proposed in [18]. If one imposes periodicity in time for all modes, $t \to t+\beta$, then only the left- and right-moving modes are allowed, while the logarithmic modes are eliminated since they grow linearly in time, see e.g. (25). The other possibility was pursued in [22]. It is based upon the observation that logarithmic modes grow logarithmically faster in $e^{2\rho}$ than their left partners, see e.g. (25). Thus, imposing boundary conditions that prohibit this logarithmic growth eliminates all logarithmic modes.

Currently it is not known whether chiral gravity has its own dual CFT or if it exists merely as a zero-charge superselection sector of the logarithmic CFT. In the latter case it is unclear whether or not the zero-charge superselection sector is a fully-fledged CFT. Another alternative is that neither the LCFT nor its chiral truncation dual to chiral gravity exists. In that case CTMG is unlikely to exist as a consistent quantum theory on its own. Rather, it would require a UV completion, such as string theory.

4.7. Logout

We summarize now the key results reviewed in this section as well as some open issues. Cosmological topologically massive gravity (15) at the chiral point (17) is likely to be dual to a LCFT with a logarithmic partner for one flux component of the energy momentum tensor since 2- [23] and 3-point correlators [25] match. The values of central charges and new anomaly are given by (35). The detailed calculation of the correlator with three log-insertions (41c) still needs to be performed and will determine another parameter of the LCFT. New massive gravity (42) at the chiral point (43) is likely to be dual to a LCFT with a logarithmic partner for both flux components of the energy momentum tensor since 2-point correlators match [26]. The central charges vanish and the new anomalies are given by (45). The calculation of 3-point correlators still needs to be performed and will provide a more stringent test of the conjectured duality to a LCFT. A similar story is likely to repeat for general massive gravity (the combination of topologically and new massive gravity) at a chiral point, and it could be rewarding to investigate this issue. Finally we addressed possibilities to eliminate the logarithmic modes and their partners, since such an elimination might lead to a chiral theory of quantum gravity [17], called "chiral gravity". The issue of whether chiral gravity exists still remains open.

Journal of Physics: Conference Series **222** (2010) 012047 doi:10.1088/1742-6596/222/1/012047

5. Towards condensed matter applications

In this final section we review briefly some condensed matter systems where LCFTs do arise, see [3, 4] for more comprehensive reviews. We focus on LCFTs where the energy-momentum tensor acquires a logarithmic partner, i.e., the class of LCFTs for which we have found possible gravity duals.[7] Condensed matter systems described by such LCFTs are for instance systems at (or near) a critical point with quenched disorder, like spin glasses [83]/quenched random magnets [84, 85], dilute self-avoiding polymers or percolation [86]. "Quenched disorder" arises in a condensed matter system with random variables that do not evolve with time. If the amount of disorder is sufficiently large one cannot study the effects of disorder by perturbing around a critical point without disorder — standard mean field methods break down. The system is then driven towards a random critical point, and it is a challenge to understand its precise nature. Mathematically, the essence of the problem lies in the infamous denominator arising in correlation functions of some operator $\mathcal{O}$ averaged over disordered configurations (see e.g. chapter VI.7 in [87])

$$\overline{\langle \mathcal{O}(z)\,\mathcal{O}(0)\rangle} = \int \mathcal{D}V P[V]\,\frac{\int \mathcal{D}\phi\,\exp\big(-S[\phi]-\int \mathrm{d}^2z'V(z')\mathcal{O}(z')\big)\,\mathcal{O}(z)\,\mathcal{O}(0)}{\int \mathcal{D}\phi\,\exp\big(-S[\phi]-\int \mathrm{d}^2z'V(z')\mathcal{O}(z')\big)} \tag{46}$$

Here $S[\phi]$ is some 2-dimensional[8] quantum field theory action for some field(s) ϕ and $V(z)$ is a random potential with some probability distribution. For white noise one takes the Gaussian probability distribution $P[V] \propto \exp\big(-\int d^2zV^2(z)/(2g^2)\big)$, where g is a coupling constant that measures the strength of the impurities. If it were not for the denominator appearing on the right hand side of the averaged correlator (46) we could simply perform the Gaussian integral over the impurities encoded in the random potential $V(z)$. This denominator is therefore the source of all complications and to deal with it requires suitable methods, see e.g. [88]. One possibility is to eliminate the denominator by introducing ghosts. This so-called "supersymmetric method" works well if the original quantum field theory described by the action $S[\phi]$ is very simple, like a free field theory. Another option is the so-called replica trick, where one introduces n copies of the original quantum field theory, calculates correlators in this setup and takes the limit $n \to 0$ in the end, which formally reproduces the denominator in (46). Recently, Fujita, Hikida, Ryu and Takayanagi combined the replica method with the AdS/CFT correspondence to describe disordered systems [89] (see [90,91] for related work), essentially by taking n copies of the CFT, exploiting AdS/CFT to calculate correlators and taking formally the limit $n \to 0$ in the end. Like other replica tricks their approach relies on the existence of the limit $n \to 0$.

One of the results obtained by the supersymmetric method or replica trick is that correlators like the one in (46) develop a logarithmic behavior, exactly as in a LCFT [84]. In fact, in the $n \to 0$ limit prescribed by the replica trick, the conformal dimensions of certain operators degenerate. This produces a Jordan block structure for the Hamiltonian in precise parallel to the $\mu\ell \to 1$ limit of CTMG. More concretely, LCFTs can be used to compute correlators of quenched random systems!

This suggests yet-another route to describe systems with quenched disorder, and our present results add to this toolbox. Namely, instead of taking n copies of an ordinary CFT we may start directly with a LCFT. If this LCFT is weakly coupled we can work on the LCFT side perturbatively, using the results mentioned above [3,4,84–86]. On the other hand, if the LCFT becomes strongly coupled, perturbative methods fail. To get a handle on these situations we can exploit the AdS/LCFT correspondence and work on the gravity side. Of course, to this end

[7] A well-studied alternative case is a LCFT with $c = -2$ [2,82]. There is no obvious way to construct a gravity dual for such LCFTs, even when considering CTMG or new massive gravity away from the chiral point. We thank Ivo Sachs for discussions on this issue.

[8] Analog constructions work in higher dimensions, but we focus here on two dimensions.

Journal of Physics: Conference Series **222** (2010) 012047 doi:10.1088/1742-6596/222/1/012047

one needs to construct gravity duals for LCFTs. The models reviewed in this talk are simple and natural examples of such constructions.

Acknowledgments

We thank Matthias Gaberdiel, Gaston Giribet, Olaf Hohm, Roman Jackiw, David Lowe, Hong Liu, Alex Maloney, John McGreevy, Ivo Sachs, Kostas Skenderis, Wei Song, Andy Strominger and Marika Taylor for discussions. DG thanks the organizers of the "First Mediterranean Conference on Classical and Quantum Gravity" for the kind invitation and for all their efforts to make the meeting very enjoyable. DG and NJ are supported by the START project Y435-N16 of the Austrian Science Foundation (FWF). During the final stage NJ has been supported by project P21927-N16 of FWF. NJ acknowledges financial support from the Erwin-Schrödinger-Institute (ESI) during the workshop "Gravity in three dimensions".

References

[1] Di Francesco P, Mathieu P and Senechal D 1997 *Conformal Field Theory* (Springer)
[2] Gurarie V 1993 *Nucl. Phys.* **B410** 535–549 (*Preprint* hep-th/9303160)
[3] Flohr M 2003 *Int. J. Mod. Phys.* **A18** 4497–4592 (*Preprint* hep-th/0111228)
[4] Gaberdiel M R 2003 *Int. J. Mod. Phys.* **A18** 4593–4638 (*Preprint* hep-th/0111260)
[5] Kogan I I and Nichols A 2002 *JHEP* **01** 029 (*Preprint* hep-th/0112008)
[6] Aharony O, Gubser S S, Maldacena J M, Ooguri H and Oz Y 2000 *Phys. Rept.* **323** 183–386 (*Preprint* hep-th/9905111); for a review focussed on the condensed matter perspective see McGreevy J, (*Preprint* 0909.0518)
[7] Brown J D and Henneaux M 1986 *Commun. Math. Phys.* **104** 207–226
[8] Ghezelbash A M, Khorrami M and Aghamohammadi A 1999 *Int. J. Mod. Phys.* **A14** 2581–2592 (*Preprint* hep-th/9807034)
[9] Myung Y S and Lee H W 1999 *JHEP* **10** 009 (*Preprint* hep-th/9904056)
[10] Kogan I I 1999 *Phys. Lett.* **B458** 66–72 (*Preprint* hep-th/9903162)
[11] Lewis A 2000 *Phys. Lett.* **B480** 348–354 (*Preprint* hep-th/9911163)
[12] Moghimi-Araghi S, Rouhani S and Saadat M 2004 *Nucl. Phys.* **B696** 492–502 (*Preprint* hep-th/0403150)
[13] Deser S, Jackiw R and Templeton S 1982 *Phys. Rev. Lett.* **48** 975–978
Deser S, Jackiw R and Templeton S 1982 *Ann. Phys.* **140** 372–411; 1988 *Erratum-ibid.* **185** 406
[14] Deser S 1984 *Quantum Theory Of Gravity* ed Christensen S M (Bristol: Adam Hilger) pp 374–381 (*Preprint* Print-82-0692 (Brandeis))
[15] Kraus P and Larsen F 2006 *JHEP* **01** 022 (*Preprint* hep-th/0508218)
[16] Bergshoeff E A, Hohm O and Townsend P K 2009 *Phys. Rev. Lett.* **102** 201301 (*Preprint* 0901.1766)
[17] Li W, Song W and Strominger A 2008 *JHEP* **04** 082 (*Preprint* 0801.4566)
[18] Grumiller D and Johansson N 2008 *JHEP* **07** 134 (*Preprint* 0805.2610)
[19] Grumiller D, Jackiw R and Johansson N 2009 *Fundamental Interactions - A Memorial Volume for Wolfgang Kummer* (World Scientific) (*Preprint* 0806.4185)
[20] Grumiller D and Johansson N 2009 *Int. J. Mod. Phys.* **D17** 2367–2372 (*Preprint* 0808.2575)
[21] Henneaux M, Martinez C and Troncoso R 2009 *Phys. Rev.* **D79** 081502R (*Preprint* 0901.2874)
[22] Maloney A, Song W and Strominger A 2009 (*Preprint* 0903.4573)
[23] Skenderis K, Taylor M and van Rees B C 2009 *JHEP* **09** 045 (*Preprint* 0906.4926)
[24] Ertl S, Grumiller D and Johansson N 2009 (*Preprint* 0910.1706)
[25] Grumiller D and Sachs I 2009 (*Preprint* 0910.5241)
[26] Grumiller D and Hohm O 2009 (*Preprint* 0911.4274)
[27] Liu Y and Sun Y W 2009 *JHEP* **04** 106 (*Preprint* 0903.0536)
[28] Bergshoeff E A, Hohm O and Townsend P K 2009 *Phys. Rev.* **D79** 124042 (*Preprint* 0905.1259)
[29] Banados M, Teitelboim C and Zanelli J, 1992 *Phys. Rev. Lett.* **69** 1849 (*Preprint* hep-th/9204099)
[30] Witten E 2007 (*Preprint* 0706.3359)
[31] Manschot J 2007 *JHEP* **10** 103 (*Preprint* 0707.1159)
[32] Gaiotto D and Yin X 2007 *JHEP* **08** 029 (*Preprint* 0707.3437)
[33] Gaberdiel M R 2007 *JHEP* **11** 087 (*Preprint* 0707.4073)
[34] Yin X 2007 (*Preprint* arXiv:0710.2129 [hep-th])
[35] Yin X 2008 *JHEP* **09** 120 (*Preprint* 0711.2803)
[36] Maloney A and Witten E 2007 (*Preprint* 0712.0155)

[37] Manschot J and Moore G W 2007 (*Preprint* 0712.0573)
[38] Carlip S, Deser S, Waldron A and Wise D K 2009 *Class. Quant. Grav.* **26** 075008 (*Preprint* 0803.3998)
[39] Carlip S 2008 *JHEP* **10** 078 (*Preprint* 0807.4152)
[40] Hotta K, Hyakutake Y, Kubota T and Tanida H 2008 *JHEP* **07** 066 (*Preprint* 0805.2005)
[41] Li W, Song W and Strominger A 2008 (*Preprint* 0805.3101)
[42] Park M I 2008 *JHEP* **09** 084 (*Preprint* 0805.4328)
[43] Sachs I and Solodukhin S N 2008 *JHEP* **08** 003 (*Preprint* 0806.1788)
[44] Lowe D A and Roy S 2008 *Phys. Lett.* **B668** 159–162 (*Preprint* 0806.3070)
[45] Myung Y S, Lee H W and Kim Y W 2008 (*Preprint* 0806.3794)
[46] Carlip S, Deser S, Waldron A and Wise D K 2008 *Phys. Lett.* **B666** 272–276 (*Preprint* 0807.0486)
[47] Lee H W, Kim Y W and Myung Y S 2008 (*Preprint* 0807.1371)
[48] Sachs I 2008 *JHEP* **09** 073 (*Preprint* 0807.1844)
[49] Gibbons G W, Pope C N and Sezgin E 2008 *Class. Quant. Grav.* **25** 205005 (*Preprint* 0807.2613)
[50] Anninos D, Li W, Padi M, Song W and Strominger A 2009 *JHEP* **03** 130 (*Preprint* 0807.3040)
[51] Giribet G, Kleban M and Porrati M 2008 *JHEP* **10** 045 (*Preprint* 0807.4703)
[52] Strominger A 2008 (*Preprint* 0808.0506)
[53] Compere G and Detournay S 2009 *Class. Quant. Grav.* **26** 012001 (*Preprint* 0808.1911)
[54] Myung Y S 2008 *Phys. Lett.* **B670** 220–223 (*Preprint* 0808.1942)
[55] de Haro S 2009 *JHEP* **01** 042 (*Preprint* 0808.2054)
[56] Stevens K A, Schleich K and Witt D M 2009 *Class. Quant. Grav.* **26** 075012 (*Preprint* 0809.3022)
[57] Deser S 2009 *Fundamental Interactions - A Memorial Volume for Wolfgang Kummer* (World Scientific) (*Preprint* 0810.5384)
[58] Hotta K, Hyakutake Y, Kubota T, Nishinaka T and Tanida H 2009 *JHEP* **01** 010 (*Preprint* 0811.0910)
[59] Quevedo H and Sanchez A 2009 *Phys. Rev.* **D79** 024012 (*Preprint* 0811.2524)
[60] Oh J J and Kim W 2009 *JHEP* **01** 067 (*Preprint* 0811.2632)
[61] Garbarz A, Giribet G and Vasquez Y 2009 *Phys. Rev.* **D79** 044036 (*Preprint* 0811.4464)
[62] Kim W and Son E J 2009 *Phys. Lett.* **B673** 90–94 (*Preprint* 0812.0876)
[63] Mann R B, Oh J J and Park M I 2009 *Phys. Rev.* **D79** 064005 (*Preprint* 0812.2297)
[64] Blagojevic M and Cvetkovic B 2009 *JHEP* **05** 073 (*Preprint* 0812.4742)
[65] Nam S and Park J D 2009 *Class. Quant. Grav.* **26** 145015 (*Preprint* 0902.0982)
[66] Hellerman S 2009 (*Preprint* 0902.2790)
[67] Sezgin E and Tanii Y 2009 *Class. Quant. Grav.* **26** 235005 (*Preprint* 0903.3779)
[68] Anninos D, Esole M and Guica M 2009 *JHEP* **10** 083 (*Preprint* 0905.2612)
[69] Compere G and Detournay S 2009 *JHEP* **08** 092 (*Preprint* 0906.1243)
[70] Hotta K, Hyakutake Y, Kubota T, Nishinaka T and Tanida H 2009 *Phys. Lett.* **B680** 279–285 (*Preprint* 0906.1255)
[71] Anninos D 2009 (*Preprint* 0906.1819)
[72] Carlip S 2009 *JHEP* **09** 083 (*Preprint* 0906.2384)
[73] Chow D D K, Pope C N and Sezgin E 2009 (*Preprint* 0906.3559)
[74] Becker M, Bruillard P and Downes S 2009 *JHEP* **10** 004 (*Preprint* 0906.4822)
[75] Blagojevic M and Cvetkovic B 2009 *JHEP* **09** 006 (*Preprint* 0907.0950)
[76] Vasquez Y 2009 (*Preprint* 0907.4165)
[77] Duncan J F and Frenkel I B 2009 (*Preprint* 0907.4529)
[78] Andrade T and Marolf D 2009 (*Preprint* 0909.0727)
[79] Miskovic O and Olea R 2009 *JHEP* **12** 046 (*Preprint* 0909.2275)
[80] Skenderis K, Taylor M and van Rees B C 2009 (*Preprint* 0909.5617)
[81] Afshar H R, Alishahiha M and Naseh A 2009 (*Preprint* 0910.4350)
[82] Gaberdiel M R and Kausch H G 1996 *Phys. Lett.* **B386** 131–137 (*Preprint* hep-th/9606050)
[83] Binder K and Young A P 1986 *Rev. Mod. Phys.* **58** 801–976
[84] Cardy J L 1999 (*Preprint* cond-mat/9911024)
[85] Reza Rahimi Tabar M 2000 *Nucl. Phys.* **B588** 630–637 (*Preprint* cond-mat/0002309)
[86] Gurarie V and Ludwig A W W 2002 *J. Phys.* **A35** L377–L384 (*Preprint* cond-mat/9911392)
[87] Zee A 2003 *Quantum field theory in a nutshell* (Princeton University Press)
[88] Bernard D 1995 (*Preprint* hep-th/9509137)
[89] Fujita M, Hikida Y, Ryu S and Takayanagi T 2008 *JHEP* **12** 065 (*Preprint* 0810.5394)
[90] Kiritsis E and Niarchos V 2009 *Nucl. Phys.* **B812** 488–524 (*Preprint* 0808.3410)
[91] Myers R C and Wapler M C 2008 *JHEP* **12** 115 (*Preprint* 0811.0480)

First Mediterranean Conference on Classical and Quantum Gravity (MCCQG 2009) IOP Publishing
Journal of Physics: Conference Series **222** (2010) 012048 doi:10.1088/1742-6596/222/1/012048

Casimir Energy of the Universe and New Regularization of Higher Dimensional Quantum Field Theories

Shoichi Ichinose

Laboratory of Physics, School of Food and Nutritional Sciences, University of Shizuoka, Yada 52-1, Shizuoka 422-8526, Japan

E-mail: `ichinose@u-shizuoka-ken.ac.jp`

Abstract. Casimir energy is calculated for the 5D electromagnetism and 5D scalar theory in the *warped* geometry. It is compared with the flat case. A new regularization, called *sphere lattice regularization*, is taken. In the integration over the 5D space, we introduce two boundary curves (IR-surface and UV-surface) based on the *minimal area principle*. It is a *direct* realization of the geometrical approach to the *renormalization group*. The regularized configuration is *closed-string like*. We do *not* take the KK-expansion approach. Instead, the position/momentum propagator is exploited, combined with the *heat-kernel method*. All expressions are closed-form (not KK-expanded form). The *generalized* P/M propagators are introduced. We numerically evaluate Λ(4D UV-cutoff), ω(5D bulk curvature, warp parameter) and T(extra space IR parameter) dependence of the Casimir energy. We present two *new ideas* in order to define the 5D QFT: 1) the summation (integral) region over the 5D space is *restricted* by two minimal surfaces (IR-surface, UV-surface) ; or 2) we introduce a *weight function* and require the dominant contribution, in the summation, is given by the *minimal surface*. Based on these, 5D Casimir energy is *finitely* obtained after the *proper renormalization procedure*. The *warp parameter* ω suffers from the *renormalization effect*. The IR parameter T does not. We examine the meaning of the weight function and finally reach a *new definition* of the Casimir energy where *the 4D momenta(or coordinates) are quantized* with the extra coordinate as the Euclidean time (inverse temperature). We examine the cosmological constant problem and present an answer at the end. Dirac's large number naturally appears.

1. Introduction

In the dawn of the quantum theory, the *divergence* problem of the specific heat of the radiation cavity was the biggest one (the problem of the blackbody radiation). It is historically so famous that the difficulty was solved by Planck's idea that the energy is quantized. In other words, the phase space of the photon field dynamics is not continuous but has the "cell" or "lattice" structure with the unit area $(\Delta x \cdot \Delta p)$ of the size $2\pi\hbar$ (Planck constant). The radiation energy is composed of two parts, E_{Cas} and E_β:

$$E_{4dEM} = E_{Cas} + E_\beta \ , \ E_{Cas} = \sum_{m_x, m_y, n \in \mathbf{Z}} \tilde{\omega}_{m_x m_y n} \ , \ E_\beta = 2 \sum_{m_x, m_y, n \in \mathbf{Z}} \frac{\tilde{\omega}_{m_x m_y n}}{e^{\beta \tilde{\omega}_{m_x m_y n}} - 1},$$

$$\tilde{\omega}^2_{m_x m_y n} = (m_x \frac{\pi}{L})^2 + (m_y \frac{\pi}{L})^2 + (n \frac{\pi}{l})^2 \quad , \quad l \ll L \quad , \qquad (1)$$

where the parameter β is the inverse temperature, l is the separation length between two perfectly-conducting plates, and L is the IR regularization parameter of the plate-size. The second part E_β is, essentially, Planck's radiation formula. The first one E_{Cas} is the vaccuum energy of the radiation field, that is, the Casimir energy. It is a very delicate quantity. The quantity is formally *divergent*, hence it must be defined with careful *regularization*. $E_{Cas}/(2L)^2$ does depend only on the *boudary* parameter l. The quantity is a quantum effect and , at the same time, depends on the *global* (macro) parameter l.

$$\frac{E_{Cas}}{(2L)^2} = \frac{\pi^2}{(2l)^3}\frac{B_4}{4!} = -\frac{\pi^2}{720}\frac{1}{(2l)^3} \quad , \quad B_4(\text{the fourth Bernoulli number}) = -\frac{1}{30} \quad , \tag{2}$$

In Fig.1, Planck's radiation spectrum distribution is shown. Introducing the axis of the

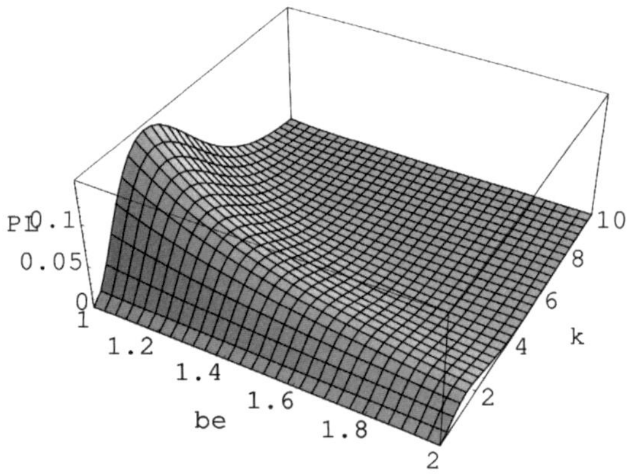

Figure 1. Graph of Planck's radiation formula. $\mathcal{P}(\beta, k) = \frac{1}{(c\hbar)^3}\frac{1}{\pi^2}k^3/(e^{\beta k} - 1)$ $(1 \leq \beta \leq 2,\ 0.01 \leq k \leq 10)$.

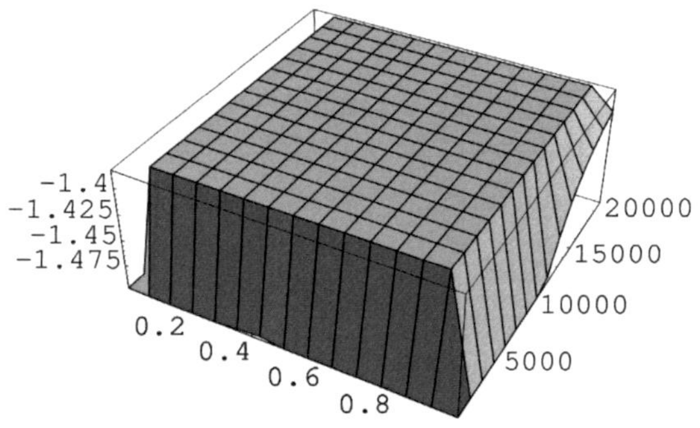

Figure 2. Behavior of $\ln|\frac{1}{2}\mathcal{F}^-(\tilde{k}, z)| = \ln|\tilde{k}\ G^-_{\tilde{k}}(z, z)/(\omega z)^3|$. $\omega = 10^4, T = 1, \Lambda = 2 \times 10^4$. $1.0001/\omega \leq z \leq 0.9999/T$. $\Lambda T/\omega \leq \tilde{k} \leq \Lambda$. Note $\ln|(1/2) \times (1/2)| \approx -1.39$.

inverse temperature(β), besides the photon energy or frequency (k), it is shown stereographically. Although we will examine the 5D version of the zero-point part (the Casimir energy), the calculated quantities in this paper are much more related to this Planck's formula. We see, near the β-axis, a sharply-rising surface, which is the Rayleigh-Jeans region (the energy density is proportional to the *square* of the photon frequency). The *damping* region in high k is the Wien's region. [1] The ridge (the line of peaks at each β) forms the *hyperbolic* curve (Wien's displacement law). When we will, in this paper, deal with the energy distribution over the 4D momentum and the extra-coordinate, we will see the similar behavior (although top and bottom appear in the opposite way).

In the quest for the unified theory, the higher dimensional (HD) approach is a fascinating one from the geometrical point. Historically the initial successful one is the Kaluza-Klein model[1, 2], which unifies the photon, graviton and dilaton from the 5D space-time approach. The HD theories , however, generally have the serious defect as the quantum field theory(QFT) : un-renormalizability. The HD quantum field theories, at present, are not defined within the QFT. One can take the standpoint that the more fundamental formulation, such as the string theory

[1] We recall that the old problem of the *divergent* specific heat was solved by the Wien's formula. This fact strongly supports the present idea of introducing the *weight function* (see Sec.6).

and D-brane theory, can solve the problem. In the present paper, we have the *new standpoint* that the HD theories should be defined by themselves within the QFT. In order to escape the dimension requirement D=10 or 26 from the quantum consistency (anomaly cancellation)[3], we treat the gravitational (metric) field only as the background one. This does *not* mean the space-time is not quantized. See later discussions (Sec.7). We present a way to define 5D quantum field theory through the analysis of the Casimir energy of 5D electromagnetism.

In 1983, the Casimir energy in the Kaluza-Klein theory was calculated by Appelquist and Chodos[4]. They took the cut-off (Λ) regularization and found the quintic (Λ^5) divergence and the finite term. The divergent term shows the *unrenormalizability* of the 5D theory, but the finite term looks meaningful [2] and, in fact, is widely regarded as the right vacuum energy which shows *contraction* of the extra axis.

In the development of the string and D-brane theories, a new approach to the renormalization group was found. It is called *holographic renormalization* [6, 7, 8, 9, 10, 11]. We regard the renormalization flow as a curve in the bulk (HD space). The flow goes along the extra axis. The curve is derived as a dynamical equation such as Hamilton-Jacobi equation. It originated from the AdS/CFT correspondence[12, 13, 14]. Spiritually the present basic idea overlaps with this approach. The characteristic points of this paper are: a) We do *not* rely on the 5D supergravity; b) We do *not* quantize the gravitational(metric) field; c) The divergence problem is solved by reducing the degree of freedom of the system, where we require, not higher symmetries, but some restriction based on the *minimal area principle*; d) No local counterterms are necessary.

In the previous paper[15], we investigated the 5D electromagnetism in the *flat* geometry. The results show the *renormalization of the compactification size l.*

$$E^W_{Cas}/\Lambda l = -\frac{\alpha}{l^4}(1-4c\ln(l\Lambda)) = -\frac{\alpha}{l'^4} \quad , \quad \beta = \frac{\partial}{\partial(\ln\Lambda)}\ln\frac{l'}{l} = c \quad , \tag{3}$$

where α and c are some numbers. They are, at present, not fixed, but are numerically obtained depending on the weight function W. The aim of this paper is to examine how the above results change for the 5D warped geometry case. One additional massive parameter, that is, the warp (bulk curvature) parameter ω appears. This introduction of the "thickness" $1/\omega$ comes from the expectation that it softens the UV-singularity, which is the same situation as in the string theory.

2. Kaluza-Klein expansion approach

In order to analyze the 5D EM-theory, we start with 5D *massive* vector theory.

$$S_{5dV} = \int d^4x dz \sqrt{-G}(-\frac{1}{4}F_{MN}F^{MN} - \frac{1}{2}m^2A^MA_M) \ , \ F_{MN} = \partial_M A_N - \partial_N A_M \ ,$$

$$ds^2 = \frac{1}{\omega^2 z^2}(\eta_{\mu\nu}dx^\mu dx^\nu + dz^2) = G_{MN}dX^MdX^N \quad , \quad G \equiv \det G_{AB} \quad . \tag{4}$$

The 5D vector mass, m, is regarded as a IR-regularization parameter. In the limit, $m = 0$, the above one has the 5D local-gauge symmetry. Casimir energy is given by some integral where the (modified) Bessel functions, with the index $\nu = \sqrt{1+\frac{m^2}{\omega^2}}$, appear. Hence the 5D EM limit is given by $\nu = 1$ ($m = 0$). We consider, however, the *imaginary* mass case $m = i\omega$ ($m^2 = -\omega^2$, $\nu = 0$) mainly for the simplicity. We can simplify the model furthermore. Instead of analyzing the $m^2 = -\omega^2$ of the massive vector (4), we take the 5D massive *scalar* theory on AdS_5 with $m^2 = -4\omega^2$, $\nu = \sqrt{4+m^2/\omega^2} = 0$.

$$\mathcal{L} = \sqrt{-G}(-\frac{1}{2}\nabla^A\Phi\nabla_A\Phi - \frac{1}{2}m^2\Phi^2), \ ds^2 = G_{AB}dX^AdX^B, \ \nabla^A\nabla_A\Phi - m^2\Phi + J = 0, \tag{5}$$

[2] The gauge independence was confirmed in Ref.[5].

where $\Phi(X)=\Phi(x^a,z)$ is the 5D scalar field. The integral region is given by

$$-\frac{1}{T}\leq z\leq -\frac{1}{\omega}\quad\text{or}\quad \frac{1}{\omega}\leq z\leq \frac{1}{T}\quad (-l\leq y\leq l\ ,\ |z|=\frac{1}{\omega}\mathrm{e}^{\omega|y|})\quad ,\quad \frac{1}{T}\equiv\frac{1}{\omega}\mathrm{e}^{\omega l}\quad , \tag{6}$$

where we take into account Z_2 symmetry: $z\leftrightarrow -z$. ω is the bulk curvature (AdS_5 parameter) and T^{-1} is the size of the extra space (Infrared parameter). In this section, we present the standard expressions, that is, those obtained by the Kaluza-Klein expansion. The Casimir energy E_{Cas} is given by

$$\mathrm{e}^{-T^{-4}E_{Cas}}=\int\mathcal{D}\Phi\exp\{i\int d^5X\mathcal{L}\}$$

$$=\int\mathcal{D}\Phi_p(z)\exp\left[i\int\frac{d^4p}{(2\pi)^4}2\int_{1/\omega}^{1/T}dz\left\{\frac{1}{2}\Phi_p(z)s(z)(s(z)^{-1}\hat{L}_z-p^2)\Phi_p(z)\right\}\right]$$

$$=\int\prod_n dc_n(p)\exp\left[\int\frac{d^4p_E}{(2\pi)^4}\sum_n\{-\frac{1}{2}c_n(p)^2(p_E^2+M_n^2)\}\right]=\exp\sum_{n,p}\{-\frac{1}{2}\ln(p_E^2+M_n^2)\}\quad , \tag{7}$$

where $\Phi_p(z)$ is the partially(4D world only)-Fourier-transformed one of $\Phi(X)$. $\Phi_p(z)$ is expressed in the expansion form using the eigen functions, $\psi_n(z)$, of this AdS_5 system.

$$\Phi_p(z)=\sum_n c_n(p)\psi_n(z),\{s(z)^{-1}\hat{L}_z+{M_n}^2\}\psi_n(z)=0,\hat{L}_z\equiv\frac{d}{dz}\frac{1}{(\omega z)^3}\frac{d}{dz}-\frac{m^2}{(\omega z)^5},$$
$$\psi_n(z)=-\psi_n(-z)\quad\text{for}\quad P=-\quad ;\quad \psi_n(z)=\psi_n(-z)\quad\text{for}\quad P=+\quad . \tag{8}$$

where $s(z)=\frac{1}{(\omega z)^3}$. The expression (7) is the familiar one of the Casimir energy.

3. Heat-Kernel Approach and Position/Momentum Propagator

Eq.(7) is the expression of E_{Cas} by the KK-expansion. In this section, the same quantity is re-expressed in a *closed* form using the heat-kernel method and the P/M propagator. First we can express it, using the heat equation solution, as follows.

$$\mathrm{e}^{-T^{-4}E_{Cas}}=(\text{const})\times\exp\left[T^{-4}\int\frac{d^4p}{(2\pi)^4}2\int_0^\infty\frac{1}{2}\frac{dt}{t}\mathrm{Tr}\ H_p(z,z';t)\right]\quad ,$$
$$\mathrm{Tr}\ H_p(z,z';t)=\int_{1/\omega}^{1/T}s(z)H_p(z,z;t)dz\quad ,\quad \{\frac{\partial}{\partial t}-(s^{-1}\hat{L}_z-p^2)\}H_p(z,z';t)=0\quad . \tag{9}$$

The heat kernel $H_p(z,z';t)$ is formally solved, using the Dirac's bra and ket vectors $(z|,|z)$, as

$$H_p(z,z';t)=(z|\mathrm{e}^{-(-s^{-1}\hat{L}_z+p^2)t}|z')\quad . \tag{10}$$

We here introduce the position/momentum propagators $G_p^{\mp}$ as follows.

$$G_p^{\mp}(z,z')\equiv\int_0^\infty dt\ H_p(z,z';t)=\sum_{n\in\mathbf{Z}}\frac{1}{M_n^2+p^2}\frac{1}{2}\{\psi_n(z)\psi_n(z')\mp\psi_n(z)\psi_n(-z')\}\quad . \tag{11}$$

They satisfy the following differential equations of *propagators*.

$$(\hat{L}_z-p^2s(z))G_p^{\mp}(z,z')=\sum_{n\in\mathbf{Z}}\frac{\{\psi_n(z)\psi_n(z')\mp\psi_n(z)\psi_n(-z')\}}{2}=\left\{\begin{array}{ll}\epsilon(z)\epsilon(z')\hat{\delta}(|z|-|z'|) & \text{P}=-1\\ \hat{\delta}(|z|-|z'|) & \text{P}=1\end{array}\right. \tag{12}$$

Therefore the Casimir energy E_{Cas} is given by

$$-E^-_{Cas}(\omega,T)=\int\frac{d^4p_E}{(2\pi)^4}2\int_0^\infty\frac{dt}{t}2\int_{1/\omega}^{1/T}dz\ s(z)H_{p_E}(z,z;t)$$
$$=\int\frac{d^4p_E}{(2\pi)^4}2\int_0^\infty\frac{dt}{t}2\int_{1/\omega}^{1/T}dz\ s(z)\left\{\sum_{n\in\mathbf{Z}}\mathrm{e}^{-(M_n^2+p_E^2)t}\psi_n(z)^2\right\}\quad, \tag{13}$$

where $s(z)=1/(\omega z)^3$. The momentum symbol p_E indicates Euclideanization. This expression leads to the same treatment as the previous section. Note that the above expression shows the *negative definiteness* of E^-_{Cas}. Finally we obtain the following useful expression of the Casimir energy for $P=\mp$.

$$-E^\mp_{Cas}(\omega,T)=\int\frac{d^4p_E}{(2\pi)^4}\int_{1/\omega}^{1/T}dz\ s(z)\int_{p_E^2}^\infty\{G^\mp_k(z,z)\}dk^2\quad. \tag{14}$$

The P/M propagators $G^\mp_p$ in (11) and (14) can be expressed in a *closed* form. Taking the *Dirichlet* condition at all fixed points, the expression for the fundamental region ($1/\omega\leq z\leq z'\leq 1/T$) is given by

$$G^\mp_p(z,z')=\mp\frac{\omega^3}{2}z^2z'^2\frac{\{\mathbf{I}_0(\frac{\tilde p}{\omega})\mathbf{K}_0(\tilde pz)\mp\mathbf{K}_0(\frac{\tilde p}{\omega})\mathbf{I}_0(\tilde pz)\}\{\mathbf{I}_0(\frac{\tilde p}{T})\mathbf{K}_0(\tilde pz')\mp\mathbf{K}_0(\frac{\tilde p}{T})\mathbf{I}_0(\tilde pz')\}}{\mathbf{I}_0(\frac{\tilde p}{T})\mathbf{K}_0(\frac{\tilde p}{\omega})-\mathbf{K}_0(\frac{\tilde p}{T})\mathbf{I}_0(\frac{\tilde p}{\omega})}, \tag{15}$$

where $\tilde p\equiv\sqrt{p^2},p^2\geq 0$. We can express Casimir energy in terms of the following functions $F^\mp(\tilde p,z)$.

$$-E^{\Lambda,\mp}_{Cas}(\omega,T)=\int\frac{d^4p_E}{(2\pi)^4}\bigg|_{\tilde p\leq\Lambda}\int_{1/\omega}^{1/T}dz\ F^\mp(\tilde p,z)\quad,$$
$$F^\mp(\tilde p,z)\equiv s(z)\int_{p_E^2}^{\Lambda^2}\{G^\mp_k(z,z)\}dk^2=\frac{2}{(\omega z)^3}\int_{\tilde p}^{\Lambda}\tilde k\ G^\mp_k(z,z)d\tilde k\equiv\int_{\tilde p}^{\Lambda}\mathcal{F}^\mp(\tilde k,z)d\tilde k\quad, \tag{16}$$

where $\mathcal{F}^\mp(\tilde k,z)$ are the integrands of $F^\mp(\tilde p,z)$ and $\tilde p=\sqrt{p_E^2}$. Here we introduce the UV cut-off parameter Λ for the 4D momentum space. In Fig.2, we show the behavior of $\mathcal{F}^-(\tilde k,z)$. The table-shape graph says the "Rayleigh-Jeans" dominance. That is, for the wide-range region $(\tilde p,z)$ satisfying both $\tilde p(z-\frac{1}{\omega})\gg 1$ and $\tilde p(\frac{1}{T}-z)\gg 1$,

$$\mathcal{F}^-(\tilde p,z)\approx\frac{1}{2}\quad,\quad\mathcal{F}^+(\tilde p,z)\approx\frac{1}{2}\quad,\quad(\tilde p,z)\in\{(\tilde p,z)|\tilde p(z-\frac{1}{\omega})\gg 1\text{ and }\tilde p(\frac{1}{T}-z)\gg 1\}\quad. \tag{17}$$

4. UV and IR Regularization Parameters and Evaluation of Casimir Energy

The integral region of the above equation (16) is displayed in Fig.3. In the figure, we introduce the regularization cut-offs for the 4D-momentum integral, $\mu\leq\tilde p\leq\Lambda$. As for the extra-coordinate integral, it is the finite interval, $1/\omega\leq z\leq 1/T=\mathrm{e}^{\omega l}/\omega$, hence we need not introduce further regularization parameters. For simplicity, we take the following IR cutoff of 4D momentum : $\mu=\Lambda\cdot\frac{T}{\omega}=\Lambda\mathrm{e}^{-\omega l}$. Hence the new regularization parameter is Λ only.

Let us evaluate the (Λ,T)-*regularized* value of (16).

$$-E^{\Lambda,\mp}_{Cas}(\omega,T)=\frac{2\pi^2}{(2\pi)^4}\int_\mu^\Lambda d\tilde p\int_{1/\omega}^{1/T}dz\ \tilde p^3F^\mp(\tilde p,z)\quad,\quad F^\mp(\tilde p,z)=\frac{2}{(\omega z)^3}\int_{\tilde p}^{\Lambda}\tilde k\ G^\mp_k(z,z)d\tilde k\quad. \tag{18}$$

Journal of Physics: Conference Series **222** (2010) 012048 doi:10.1088/1742-6596/222/1/012048

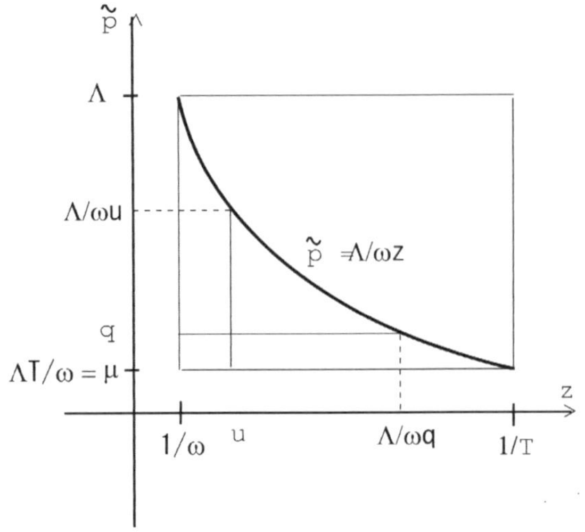

Figure 3. Space of (z,$\tilde{p}$) for the integration. The hyperbolic curve will be used in Sec.5.

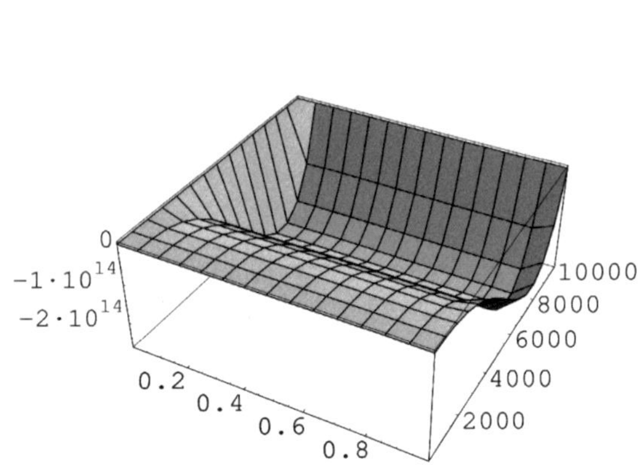

Figure 4. Behaviour of $(-1/2)\tilde{p}^3 F^-(\tilde{p}, z)$ (18). $T = 1, \omega = 10^4, \Lambda = 10^4$. $1.0001/\omega \leq z < 0.9999/T$, $\Lambda T/\omega \leq \tilde{p} \leq \Lambda$.

The integral region of $(\tilde{p}, z)$ is the *rectangle* shown in Fig.3. Note that eq.(18) is the *rigorous* expression of the (Λ, T)-regularized Casimir energy. We show the behavior of $(-1/2)\tilde{p}^3 F^-(\tilde{p}, z)$ taking the values $\omega = 10^4, T = 1$ in Fig.4($\Lambda = 10^4$). [3] Behavior along $\tilde{p}$-axis does not so much depend on z. A valley runs parallel to the z-axis with the bottom line at the fixed ratio of $\tilde{p}/\Lambda \sim 0.75$. The depth of the valley is proportional to Λ^4. Because E_{Cas} is the $(\tilde{p}, z)$ 'flat-plane' integral of $\tilde{p}^3 F(\tilde{p}, z)$, the *volume* inside the valley is the quantity E_{Cas} . Hence it is easy to see E_{Cas} is proportional to Λ^5. This is the same situation as the flat case. Importantly, (18) shows the *scaling* behavior for large values of Λ and $1/T$. From a *close* numerical analysis of $(\tilde{p}, z)$-integral (18), we have confirmed : Eq.(4A) $E^{\Lambda,-}_{Cas}(\omega, T) = \frac{2\pi^2}{(2\pi)^4} \times \left[-0.0250\frac{\Lambda^5}{T}\right]$. It does *not* depend on ω and has no $\ln\frac{\Lambda}{T}$-term. (Note: $0.025 = 1/40$.) Compared with the flat case, we see the factor T^{-1} plays the role of *IR parameter* of the extra space. We note that the behavior of Fig.4 is similar to the Rayleigh-Jeans's region (small momentum region) of the Planck's radiation formula (Fig.1) in the sense that $\tilde{p}^3 F(\tilde{p}, z) \propto \tilde{p}^3$ for $\tilde{p} \ll \Lambda$.

Finally we notice, from the Fig.4, the approximate form of $F(\tilde{p}, z)$ for the large Λ and $1/T$ is given by : Eq.(4B) $F^{\mp}(\tilde{p}, z) \approx \frac{f}{2}\Lambda(1 - \frac{\tilde{p}}{\Lambda})$, $f = 1$. It does *not* depend on z, ω and T. f is the degree of freedom. The above result is consistent with (17).

5. UV and IR Regularization Surfaces, Principle of Minimal Area and Renormalization Flow

The advantage of the new approach is that the KK-expansion is replaced by the integral of the extra dimensional coordinate z and all expressions are written in the *closed* (not expanded) form. The Λ^5-divergence, (4A), shows the notorious problem of the higher dimensional theories, as in the flat case. In spite of all efforts of the past literature, we have not succeeded in defining the higher-dimensional theories. (The divergence causes problems. The famous example is the divergent *cosmological constant* in the gravity-involving theories. [4]) Here we notice that the divergence problem can be solved if we find a way to *legitimately restrict the integral region in* $(\tilde{p}, z)$*-space.*

[3] The requirement for the three parameters ω, T, Λ is $\Lambda \gg \omega \gg T$. See ref.[18] for the discussion about the hierarchy Λ, ω, T.

One proposal of this was presented by Randall and Schwartz[16]. They introduced the *position-dependent cut-off*, $\mu < \tilde{p} < \Lambda/\omega u$, $u \in [1/\omega, 1/T]$, for the 4D-momentum integral in the "brane" located at $z = u$. See Fig.3. The total integral region is the lower part of the *hyperbolic* curve $\tilde{p} = \Lambda/\omega z$. They succeeded in obtaining the *finite* β-function of the 5D warped vector model. We have confirmed that the value E_{Cas} of (18), when the Randall-Schwartz integral region (Fig.3) is taken, is proportional to Λ^5. The close numerical analysis says

$$E_{Cas}^{-RS}(\omega,T) = \frac{2\pi^2}{(2\pi)^4}\int_\mu^\Lambda dq \int_{1/\omega}^{\Lambda/\omega q} dz\, q^3 F^-(q,z) = \frac{2\pi^2}{(2\pi)^4}\int_{1/\omega}^{1/T} du \int_\mu^{\Lambda/\omega u} d\tilde{p}\, \tilde{p}^3 F^-(\tilde{p},u)$$
$$= \frac{2\pi^2}{(2\pi)^4}\frac{\Lambda^5}{\omega}\left\{-1.58\times 10^{-2} - 1.69\times 10^{-4}\ln\frac{\Lambda}{\omega}\right\} \quad , \quad (19)$$

which is *independent* of T . This shows the divergence situation does *not* improve compared with the non-restricted case of (4A). *T of (4A) is replaced by the warp parameter ω.* This is contrasting with the flat case where $E_{Cas}^{RS} \propto -\Lambda^4$. The UV-behavior, however, *does improve* if we can choose the parameter Λ in the way: $\Lambda \propto \omega$. This fact shows the parameter ω "smoothes" the UV-singularity to some extent.

Although they claim the holography is behind the procedure, the legitimateness of the restriction looks less obvious. We have proposed an alternate approach and given a legitimate explanation within the 5D QFT[17, 19, 15, 20]. Here we closely examine the *new regularization*. On the "3-brane" at $z = 1/\omega$, we introduce the IR-cutoff $\mu = \Lambda \cdot \frac{T}{\omega}$ and the UV-cutoff Λ

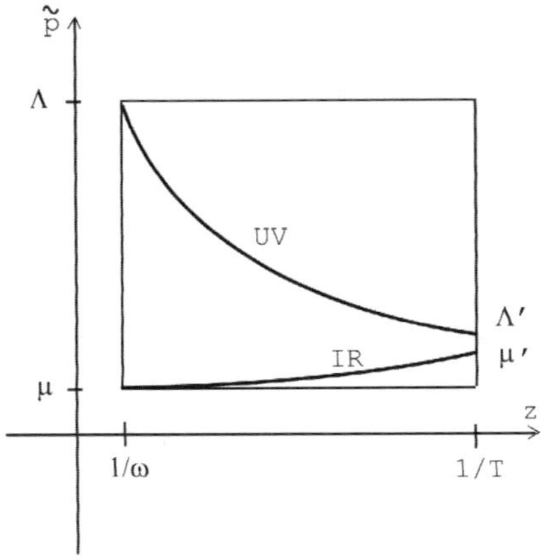

Figure 5. Space of ($\tilde{p}$,z) for the integration (present proposal).

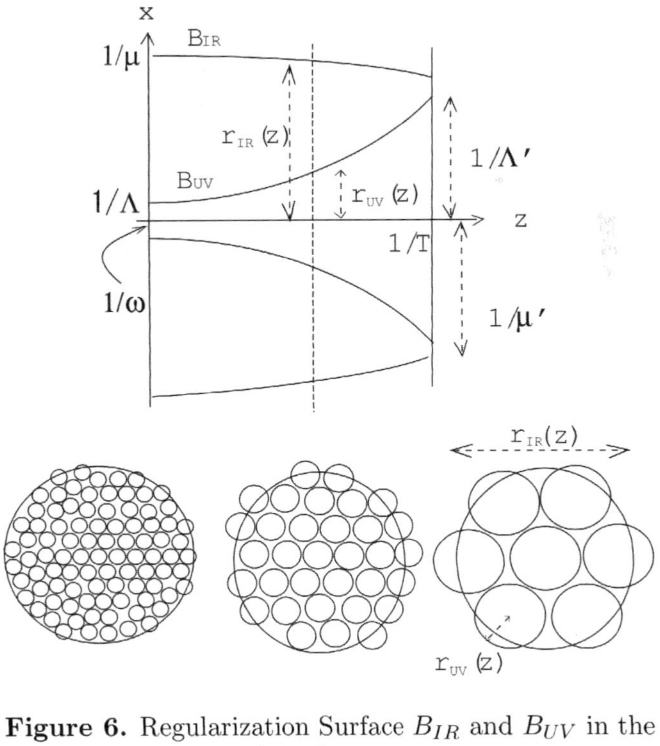

Figure 6. Regularization Surface B_{IR} and B_{UV} in the 5D coordinate space (x^μ, z).

($\mu \ll \Lambda$) in the way : Eq.(5A) $\mu \ll \Lambda$ ($T \ll \omega$). See Fig.5. This is legitimate in the sense that we generally do this procedure in the 4D *renormalizable* thoeries. (Here we are considering those 5D theories that are *renormalizable* in "3-branes". Examples are 5D free theories (present model), 5D electromagnetism[15], 5D Φ^4-theory, 5D Yang-Mills theory, e.t.c..) In the same reason, on the "3-brane" at $z = 1/T$, we may have another set of IR and UV-cutoffs, μ' and Λ'. We consider the case: Eq.(5B) $\mu' \leq \Lambda'$, $\Lambda' \ll \Lambda$, $\mu \sim \mu'$. This case will lead us to introduce the *renormalization flow.* (See the later discussion.) We claim here, as for the regularization treatment of the "3-brane" located at other points z $(1/\omega < z \leq 1/T)$, the regularization parameters are determined by the *minimal area principle.* To explain it, we move to the 5D coordinate space (x^μ, z). See Fig.6. The $\tilde{p}$-expression can be replaced by $\sqrt{x_\mu x^\mu}$-expression by the *reciprocal relation* : Eq.(5C) $\sqrt{x_\mu(z)x^\mu(z)} \equiv r(z) \leftrightarrow \frac{1}{\tilde{p}(z)}$. The UV and IR cutoffs change their values along z-axis and their trajectories make *surfaces* in the 5D bulk space (x^μ, z). We *require* the two surfaces do *not cross* for the purpose of the renormalization group interpretation (discussed later). We call them UV and IR regularization (or boundary) surfaces(B_{UV}, B_{IR}). The cross sections of the regularization surfaces at z are the spheres S^3 with the radii $r_{UV}(z)$ and $r_{IR}(z)$. Here we consider the Euclidean space for simplicity. The UV-surface is stereographically shown in Fig.7 and reminds us of the *closed string* propagation. Note that the boundary surface B_{UV} (and B_{IR}) is the 4 dimensional manifold.

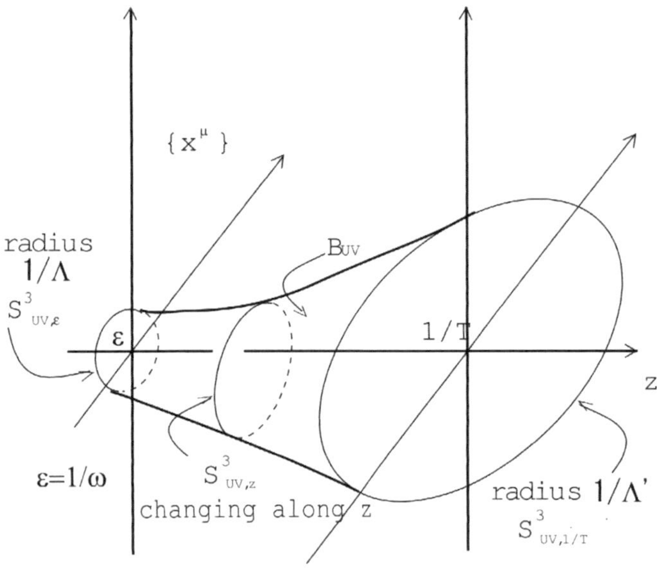

Figure 7. UV regularization surface (B_{UV}) in 5D coordinate space.

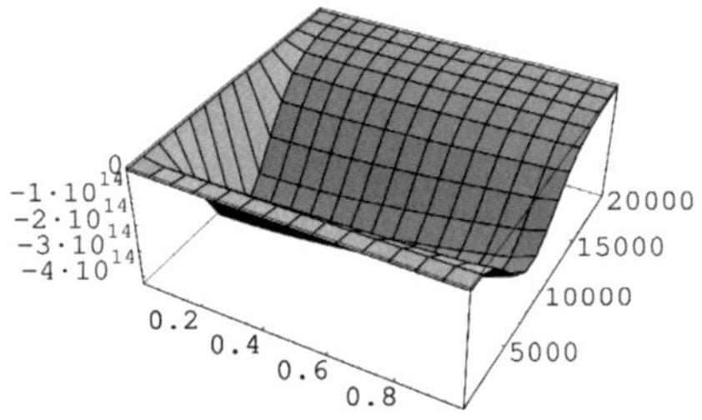

Figure 8. Behavior of $(-N_1/2)\tilde{p}^3 W_1(\tilde{p}, z)F^-(\tilde{p}, z)$(elliptic suppression). $\Lambda = 20000$, $\omega = 5000$, $T = 1$. $1.0001/\omega \leq z \leq 0.9999/T$, $\mu = \Lambda T/\omega \leq \tilde{p} \leq \Lambda$.

The 5D volume region bounded by B_{UV} and B_{IR} is the integral region of the Casimir energy E_{Cas}. The forms of $r_{UV}(z)$ and $r_{IR}(z)$ can be determined by the *minimal area principle.*

$$3 + \frac{4}{z}r'r - \frac{r''r}{r'^2 + 1} = 0 \quad , \quad r' \equiv \frac{dr}{dz} \quad , \quad r'' \equiv \frac{d^2r}{dz^2} \quad , \quad 1/\omega \leq z \leq 1/T \quad . \tag{20}$$

We have confirmed, by numerically solving the above differential eqation (Runge-Kutta), those curves that show the flow of renormalization really occur. The results imply the *boundary conditions* determine the property of the renormalization flow.

The present regularization scheme gives the *renormalization group* interpretation to the change of physical quantities along the extra axis. See Fig.6. [4] In the "3-brane" located

[4] This part is contrasting with AdS/CFT approach where the renormalization flow comes from the Einstein equation of 5D supergravity.

Journal of Physics: Conference Series **222** (2010) 012048 doi:10.1088/1742-6596/222/1/012048

at z, the UV-cutoff is $r_{UV}(z)$ and the regularization surface is the sphere S^3 with the radius $r_{UV}(z)$. The IR-cutoff is $r_{IR}(z)$ and the regularization surface is the another sphere S^3 with the radius $r_{IR}(z)$. We can regard the regularization integral region as the *sphere lattice* of the following properties: a) A unit lattice (cell) is the sphere S^3 with radius $r_{UV}(z)$ and its inside; b) Total lattice is the sphere S^3 with radius $r_{IR}(z)$ and its inside; c) The integration region of this regularization is made of many cells and the total number of cells is const.$\times\left(\frac{r_{IR}(z)}{r_{UV}(z)}\right)^4$. The total number of cells changes from $(\frac{\Lambda}{\mu})^4$ at $z=1/\omega$ to $(\frac{\Lambda'}{\mu'})^4$ at $z=1/T$. Along the z-axis, the number increases or decreases as : Eq.(5D) $\left(\frac{r_{IR}(z)}{r_{UV}(z)}\right)^4 \equiv N(z)$. For the "scale" change $z \rightarrow z+\Delta z$, N changes as : Eq.(5E) $\Delta(\ln N) = 4\frac{\partial}{\partial z}\{\ln(\frac{r_{IR}(z)}{r_{UV}(z)})\}\cdot\Delta z$. When the system has some coupling $g(z)$, the renormalization group $\tilde{\beta}(g)$-function (along the extra axis) is expressed as

$$\tilde{\beta} = \frac{\Delta(\ln g)}{\Delta(\ln N)} = \frac{1}{\Delta(\ln N)}\frac{\Delta g}{g} = \frac{1}{4}\frac{1}{\frac{\partial}{\partial z}\ln(\frac{r_{IR}(z)}{r_{UV}(z)})}\frac{1}{g}\frac{\partial g}{\partial z} \quad , \tag{21}$$

where $g(z)$ is a renormalized coupling at z. [5]

We have explained, in this section, that the *minimal area principle* determines the flow of the regularization surfaces.

6. Weight Function and Casimir Energy Evaluation

In the expression (14), the Casimir energy is written by the integral in the $(\tilde{p},z)$-space over the range: $1/\omega \le z \le 1/T,\ 0 \le \tilde{p} \le \infty$. In Sec.5, we have seen *the integral region should be properly restricted* because the cut-off region in the 4D world *changes along the extra-axis* obeying the bulk (warped) geometry (*minimal area principle*). We can expect the singular behavior (UV divergences) reduces by the integral-region restriction, but the concrete evaluation along the proposed prescription is practically not easy. In this section, we consider an alternate approach which respects the *minimal area principle* and evaluate the Casimir energy.

We introduce, instead of restricting the integral region, a *weight function* $W(\tilde{p},z)$ in the $(\tilde{p},z)$-space for the purpose of suppressing UV and IR divergences of the Casimir Energy.

$$-E^{\mp\,W}_{Cas}(\omega,T) \equiv \int\frac{d^4p_E}{(2\pi)^4}\int_{1/\omega}^{1/T} dz\, W(\tilde{p},z)F^{\mp}(\tilde{p},z) \quad , \quad \tilde{p}=\sqrt{p_4^2+p_1^2+p_2^2+p_3^2} \quad ,$$

$$\text{Examples of } W(\tilde{p},z):\quad W(\tilde{p},z) = \left\{\begin{array}{lr} (N_1)^{-1}\mathrm{e}^{-(1/2)\tilde{p}^2/\omega^2-(1/2)z^2T^2} \equiv W_1(\tilde{p},z),\ N_1 = 1.711/8\pi^2 & \text{elliptic suppr.} \\ (N_2)^{-1}\mathrm{e}^{-\tilde{p}zT/\omega} \equiv W_2(\tilde{p},z),\ N_2 = 2\frac{\omega^3}{T^3}/8\pi^2 & \text{hyperbolic suppr.1} \\ (N_8)^{-1}\mathrm{e}^{-1/2(\tilde{p}^2/\omega^2+1/z^2T^2)} \equiv W_8(\tilde{p},z),\ N_8 = 0.4177/8\pi^2 & \text{reciprocal suppr.1} \end{array}\right. \tag{22}$$

where $F^{\mp}(\tilde{p},z)$ are defined in (16). In the above, we list some examples expected for the weight function $W(\tilde{p},z)$. W_2 is regarded to correspond to the regularization taken by Randall-Schwartz. How to specify the form of W is the subject of the next section. We show the shape of the energy integrand $(-1/2)\tilde{p}^3W_1(\tilde{p},z)F^-(\tilde{p},z)$ in Fig.8. We notice the valley-bottom line $\tilde{p}\approx 0.75\Lambda$, which appeared in the un-weighted case (Fig.4), is replaced by a new line: $\tilde{p}^2+z^2\times\omega^2T^2 \approx$ const. It is located *away from* the original Λ-effected line ($\tilde{p}\sim 0.75\Lambda$).

[5] Here we consider an interacting theory, such as 5D Yang-Mills theory and 5D Φ^4 theory, where the coupling $g(z)$ is the renormalized one in the '3-brane' at z.

We can check the divergence (scaling) behavior of $E_{Cas}^{\mp W}$ by *numerically* evaluating the $(\tilde{p}, z)$-integral (22) for the rectangle region of Fig.3.

$$-E_{Cas}^{W} = \begin{cases} \frac{\omega^4}{T}\Lambda \times 1.2\left\{1 + 0.11 \ln\frac{\Lambda}{\omega} - 0.10 \ln\frac{\Lambda}{T}\right\} & \text{for } W_1 \\ \frac{T^2}{\omega^2}\Lambda^4 \times 0.062\left\{1 + 0.03 \ln\frac{\Lambda}{\omega} - 0.08 \ln\frac{\Lambda}{T}\right\} & \text{for } W_2 \\ \frac{\omega^4}{T}\Lambda \times 1.6\left\{1 + 0.09 \ln\frac{\Lambda}{\omega} - 0.10 \ln\frac{\Lambda}{T}\right\} & \text{for } W_8 \end{cases} \tag{23}$$

The suppression behavior of W_2 improves, compared with (19) by Randall-Schwartz. The quintic divergence of (19) reduces to the quartic divergence in the present approach of W_2. The hyperbolic suppressions, however, are still insufficient for the renormalizability. After dividing by the normalization factor, ΛT^{-1}, the cubic divergence remains. The desired cases are others. The Casimir energy for each case consists of three terms. The first terms give *finite* values after dividing by the overall normalization factor ΛT^{-1}. The last two terms are proportional to $\log \Lambda$ and show the *anomalous scaling*. Their contributions are order of 10^{-1} to the first leading terms. The second ones ($\ln \frac{\Lambda}{\omega}$) contribute positively while the third ones ($\ln \frac{\Lambda}{T}$) negatively. They give, after normalizing the factor Λ/T, *only* the *log-divergence*.

$$E_{Cas}^{W}/\Lambda T^{-1} = -\alpha\omega^4\left(1 - 4c\ln(\Lambda/\omega) - 4c'\ln(\Lambda/T)\right) \quad , \tag{24}$$

where α, c and c' can be read from (23) depending on the choice of W. This means the 5D Casimir energy is *finitely* obtained by the ordinary renormalization of the warp factor ω. (See the final section.) In the above result of the warped case, the IR parameter l in the flat result (3) is replaced by the inverse of the warp factor ω.

So far as the legitimate reason of the introduction of $W(\tilde{p}, y)$ is not clear, we should regard this procedure as a *regularization* to define the higher dimensional theories. We give a clear definition of $W(\tilde{p}, y)$ and a legitimate explanation in the next section. It should be done, in principle, in a consistent way with the bulk geometry and the gauge principle.

7. Meaning of Weight Function and Quantum Fluctuation of Coordinates and Momenta

In the previous work[15], we have presented the following idea to define the weight function $W(\tilde{p}, z)$. In the evaluation (22):

$$-E_{Cas}^{W}(\omega, T) = \int \frac{d^4p_E}{(2\pi)^4}\int_{1/\omega}^{1/T} dz\ W(\tilde{p}, z)F^{\mp}(\tilde{p}, z) = \frac{2\pi^2}{(2\pi)^4}\int d\tilde{p}\int_{1/\omega}^{1/T} dz\ \tilde{p}^3 W(\tilde{p}, z)F^{\mp}(\tilde{p}, z)\ , \tag{25}$$

the $(\tilde{p}, z)$-integral is over the rectangle region shown in Fig.5 (with $\Lambda \to \infty$ and $\mu \to 0$). $F^{\mp}(\tilde{p}, z)$ is explicitly given in (16). Following Feynman[21], we can replace the integral by the summation over all possible pathes $\tilde{p}(z)$.

$$-E_{Cas}^{W}(\omega, T) = \int \mathcal{D}\tilde{p}(z)\int_{1/\omega}^{1/T} dz\ S[\tilde{p}(z), z]\ ,\ S[\tilde{p}(z), z] = \frac{2\pi^2}{(2\pi)^4}\tilde{p}(z)^3 W(\tilde{p}(z), z)F^{\mp}(\tilde{p}(z), z). \tag{26}$$

There exists the *dominant path* $\tilde{p}_W(z)$ which is determined by the minimal principle : $\delta S = 0$.

$$\text{Dominant Path } \tilde{p}_W(z)\ :\quad \frac{d\tilde{p}}{dz} = \frac{-\frac{\partial \ln(WF)}{\partial z}}{\frac{3}{\tilde{p}} + \frac{\partial \ln(WF)}{\partial \tilde{p}}} \quad . \tag{27}$$

Hence it is fixed by $W(\tilde{p}, z)$. An example is the valley-bottom line in Fig.8. On the other hand, there exists another independent path: the minimal surface curve $r_g(z)$.

$$\text{Minimal Surface Curve } r_g(z)\ :\quad 3 + \frac{4}{z}r'r - \frac{r''r}{r'^2 + 1} = 0 \quad , \quad \frac{1}{\omega} \le z \le \frac{1}{T} \quad , \tag{28}$$

which is obtained by the *minimal area principle*: $\delta A = 0$ where

$$ds^2 = (\delta_{ab} + \frac{x^a x^b}{(rr')^2})\frac{dx^a dx^b}{\omega^2 z^2} \equiv g_{ab}(x)dx^a dx^b, A = \int \sqrt{\det g_{ab}}\ d^4x = \int_{1/\omega}^{1/T} \frac{1}{\omega^4 z^4}\sqrt{r'^2 + 1}\ r^3 dz. (29)$$

Hence $r_g(z)$ is fixed by the *induced geometry* $g_{ab}(x)$. Here we put the *requirement*[15]: Eq.(7A) $\tilde{p}_W(z) = \tilde{p}_g(z)$, where $\tilde{p}_g \equiv 1/r_g$. This means the following things. We *require* the dominant path coincides with the minimal surface line $\tilde{p}_g(z) = 1/r_g(z)$ which is defined independently of $W(\tilde{p}, z)$. In other words, $W(\tilde{p}, z)$ is defined here by the induced geometry $g_{ab}(x)$. In this way, we can connect the integral-measure over the 5D-space with the (bulk) geometry. We have confirmed the (approximate) coincidence by the numerical method.

In order to most naturally accomplish the above requirement, we can go to a *new step*. Namely, we *propose* to *replace* the 5D space integral with the weight W, (25), by the following *path-integral*. We *newly define* the Casimir energy in the higher-dimensional theory as follows.

$$-\mathcal{E}_{Cas}(\omega, T, \Lambda) \equiv \int_{1/\Lambda}^{1/\mu} d\rho \int_{\substack{r(1/\omega) \\ = r(1/T) \\ = \rho}} \prod_{a,z} \mathcal{D}x^a(z) F(\frac{1}{r}, z)\ \exp\left[-\frac{1}{2\alpha'}\int_{1/\omega}^{1/T} \frac{1}{\omega^4 z^4}\sqrt{r'^2 + 1}\ r^3 dz \right], (30)$$

where $\mu = \Lambda T/\omega$ and the limit $\Lambda T^{-1} \to \infty$ is taken. The string (surface) tension parameter $1/2\alpha'$ is introduced. (Note: Dimension of α' is $[\text{Length}]^4$.) The square-bracket ($[\cdots]$)-parts of (30) are $-\frac{1}{2\alpha'}\text{Area} = -\frac{1}{2\alpha'}\int \sqrt{\det g_{ab}} d^4x$ where g_{ab} is the induced metric on the 4D surface. $F(\tilde{p}, z)$ is defined in (22) or (16) and shows the *field-quantization* of the bulk scalar (EM) fields. In the above expression, we have followed the path-integral formulation of the *density matrix* (See Feynman's text[21]). The validity of the above definition is based on the following points: a) When the weight part (exp $[\cdots]$-part) is 1, the proposed quantity $\mathcal{E}_{Cas}$ is equal to E^W_{Cas}, (25), with $W = 1$; b) The leading path is given by $r_g(z) = 1/p_g(z)$, (28); c) The proposed definition, (30), clearly shows the 4D space-coordinates x^a or the 4D momentum-coordinates p^a are *quantized* (quantum-statistically, not field-theoretically) with the Euclidean time z and the "*area* Hamiltonian" $A = \int \sqrt{\det g_{ab}}\ d^4x$. Note that $F(\tilde{p}, z)$ or $F(1/r, z)$ appears, in (30), as the *energy density operator* in the quantum statistical system of $\{p^a(z)\}$ or $\{x^a(z)\}$.

In the view of the previous paragraph, the treatment of Sec.6 is an *effective* action approach using the (trial) weight function $W(\tilde{p}, z)$. Note that the integral over (p^μ, z)-space, appearing in (16), is the summation over all degrees of freedom of the 5D space(-time) points using the "naive" measure d^4pdz. An important point is that we have the possibility to take another measure for the summation in the case of the higher dimensional QFT. We have adopted, in Sec.6, the new measure $W(p^\mu, z)d^4pdz$ in such a way that the Casimir energy *does not show physical divergences*. We expect the *direct* evaluation of (30), numerically or analytically, leads to the similar result.

8. Discussion and Conclusion

The log-divergence in (24) is the familiar one in the ordinary QFT. It can be *renormalized* in the following way.

$$\frac{E^W_{Cas}}{\Lambda T^{-1}} = -\alpha\omega^4 \left(1 - 4c\ln(\frac{\Lambda}{\omega}) - 4c'\ln(\frac{\Lambda}{T})\right) = -\alpha(\omega_r)^4, \omega_r = \omega\sqrt[4]{1 - 4c\ln(\frac{\Lambda}{\omega}) - 4c'\ln(\frac{\Lambda}{T})}, \quad (31)$$

where ω_r is the *renormalized* warp factor and ω is the *bare* one. No local counterterms are necessary. Note that this renormalization relation is *exact* (not a perturbative result). In the familiar case of the 4D renormalizable theories, the coefficients c and c' depend on the coupling, but, in the present case, they are *pure numbers*. It reflects the *interaction between (EM) fields and the boundaries*. When c and c' are sufficiently small we find the renormalization group function for the warp factor ω as

$$|c| \ll 1 \ , \ |c'| \ll 1 \quad , \quad \omega_r = \omega(1 - c\ln(\Lambda/\omega) - c'\ln(\Lambda/T)) \ , \ \beta \equiv \frac{\partial}{\partial(\ln\Lambda)}\ln\frac{\omega_r}{\omega} = -c - c' \quad . (32)$$

We should notice that, in the flat geometry case, the IR parameter (extra-space size) l is renormalized. In the present warped case, however, the corresponding parameter T is *not renormalized*, but the *warp parameter* ω *is renormalized.* Depending on the sign of $c+c'$, the 5D bulk curvature ω *flows* as follows. When $c+c' > 0$, the bulk curvature ω decreases (increases) as the the measurement energy scale Λ increases (decreases). When $c+c' < 0$, the flow goes in the opposite way. When $c+c' = 0$, ω does not flow ($\beta = 0$) and is given by $\omega_r = \omega(1 + c\ln(\omega/T))$.

The final result (31) is the *new* type Casimir energy, $-\omega^4$. ω appears as a boundary parameter like T. The familiar one is $-T^4$ in the present context. In ref.[22], another type $T^2\omega^2$ was predicted using a "quasi" Warped model (bulk-boundary theory).

Through the Casimir energy calculation, in the higher dimension, we find a way to quantize the higher dimensional theories within the QFT framework. The quantization *with respect to the fields* (except the gravitational fields $G_{AB}(X)$) is done in the standard way. After this step, the expression has the summation *over the 5D space(-time) coordinates or momenta* $\int dz \prod_a dp^a$. We have proposed that this summation should be replaced by the *path-integral* $\int \prod_{a,z} \mathcal{D}p^a(z)$ with the *area* action (Hamiltonian) $A = \int \sqrt{\det g_{ab}}d^4x$ where g_{ab} is the *induced* metric on the 4D surface. This procedure says the 4D momenta p^a (or coordinates x^a) are *quantum statistical* operators and the extra-coordinate z is the inverse temperature (Euclidean time). We recall the similar situation occurs in the standard string approach. The space-time coordinates obey some uncertainty principle[23].

Recently the dark energy (as well as the dark matter) in the universe is a hot subject. It is well-known that the dominant candidate is the cosmological term. We also know the prototype higher-dimensional theory, that is, the 5D KK theory, has predicted so far the *divergent* cosmological constant[4]. This unpleasant situation has been annoying us for a long time. If we apply the present result, the situation drastically improves. The cosmological constant λ appears as: Eq.(8A) $R_{\mu\nu} - \frac{1}{2}g_{\mu\nu}R - \lambda g_{\mu\nu} = T^{matter}_{\mu\nu}, S = \int d^4x\sqrt{-g}\{\frac{1}{G_N}(R+\lambda)\} + \int d^4x\sqrt{-g}\{\mathcal{L}_{matter}\}, g =$ det $g_{\mu\nu}$, where G_N is the Newton's gravitational constant, R is the Riemann scalar curvature. We consider here the 3+1 dim Lorentzian space-time ($\mu,\nu = 0,1,2,3$). The constant λ observationally takes the value : Eq.(8B) $\frac{1}{G_N}\lambda_{obs} \sim \frac{1}{G_N {R_{cos}}^2} \sim m_\nu^4 \sim (10^{-3}eV)^4, \lambda_{obs} \sim \frac{1}{R_{cos}^2} \sim 4 \times 10^{-66}(eV)^2$, where $R_{cos} \sim 5 \times 10^{32}\mathrm{eV}^{-1}$ is the cosmological size (Hubble length), m_ν is the neutrino mass. [6] On the other hand, we have theoretically so far : Eq.(8C) $\frac{1}{G_N}\lambda_{th} \sim \frac{1}{{G_N}^2} = {M_{pl}}^4 \sim (10^{28}eV)^4$. This is because the mass scale usually comes from the quantum gravity. (See ref.[26] for the derivation using the Coleman-Weinberg mechanism.) We have the famous huge discrepancy factor : Eq.(8D) $\frac{\lambda_{th}}{\lambda_{obs}} \sim N_{DL}^2, N_{DL} \equiv M_{pl}R_{cos} \sim 6 \times 10^{60}$, where N_{DL} is the Dirac's large number[27]. If we use the present result (31), we can obtain a natural choice of T, ω and Λ as follows. By identifying $T^{-4}E_{Cas} = -\alpha_1 \Lambda T^{-1}\omega^4/T^4$ with $\int d^4x\sqrt{-g}(1/G_N)\lambda_{ob} = R_{cos}^2(1/G_N)$, we obtain the following relation: Eq.(8E) $N_{DL}^2 = R_{cos}^{\ 2}\frac{1}{G_N} = -\alpha_1\frac{\omega^4\Lambda}{T^5}, \alpha_1$: some coefficient. The warped (AdS_5) model predicts the cosmological constant *negative*, hence we have interest only in its absolute value. We take the following choice for Λ and ω : Eq.(8F) $\Lambda = M_{pl} \sim 10^{19}GeV, \omega \sim \frac{1}{\sqrt[4]{G_N {R_{cos}}^2}} = \sqrt{\frac{M_{pl}}{R_{cos}}} \sim m_\nu \sim 10^{-3}$eV. The choice

[6] The relation $m_\nu \sim \sqrt{M_{pl}/R_{cos}} = \sqrt{1/R_{cos}\sqrt{G_N}}$, which appears in some extra dimension model[24, 25], is used. The neutrino mass is, at least empirically, located at the *geometrical average* of two extreme ends of the mass scales in the universe.

Journal of Physics: Conference Series **222** (2010) 012048 doi:10.1088/1742-6596/222/1/012048

for Λ is accepted in that the largest known energy scale is the Planck energy. The choice for ω comes from the experimental bound for the Newton's gravitational force.

As shown above, we have the standpoint that the cosmological constant is mainly made from the Casimir energy. We do not yet succeed in obtaining the value α_1 negatively, but succeed in obtaining the finiteness of the cosmological constant and its gross absolute value. The smallness of the value is naturally explained by the renormalization flow as follows. Because we already know the warp parameter ω *flows* (32), the $\lambda_{obs} \sim 1/R_{cos}^2$ expression (8F), $\lambda_{obs} \propto \omega^4$, says that the *smallness of the cosmological constant comes from the renormalization flow* for the non asymptotic-free case ($c + c' < 0$ in (32)).

The IR parameter T, the normalization factor Λ/T in (24) and the IR cutoff $\mu = \Lambda\frac{T}{\omega}$ are given by : Eq.(8G) $T = R_{cos}^{-1}(N_{DL})^{1/5} \sim 10^{-20}eV, \frac{\Lambda}{T} = (N_{DL})^{4/5} \sim 10^{50}, \mu = M_{pl}N_{DL}^{-3/10} \sim 1GeV \sim m_N$, where m_N is the nucleon mass. The Fig.6 strongly suggests that the degree of freedom of the universe (space-time) is given by : Eq.(8H) $\frac{\Lambda^4}{\mu^4} = \frac{\omega^4}{T^4} = N_{DL}^{6/5} \sim 10^{74} \sim (\frac{M_{pl}}{m_N})^4$.

References

[1] Th. Kaluza, Sitzungsberichte der K.Preussischen Akademite der Wissenschaften zu Berlin. p966 (1921)

[2] O. Klein, *Z. Physik* **37** 895 (1926)

[3] M. B. Green, J. H. Schwartz and E. Witten, *Superstring theory, Vol.I and II*, Cambridge Univ. Press, c1987, Cambridge
J. Polchinski, *STRING THEORY, Vol.I and II*, Cambridge Univ. Press, c1998, Cambridge

[4] T. Appelquist and A. Chodos, *Phys. Rev.* **D28**(1983)772
T. Appelquist and A. Chodos, *Phys. Rev. Lett.* **50**(1983)141

[5] S. Ichinose, *Phys. Lett.* **152B**(1985),56

[6] L. Suskind and E. Witten, "The Holographic Bound in Anti-de Sitter Space", arXiv:hep-th/9805114

[7] M. Henningson and K. Skenderis, *JHEP* **9807**(1998)023, arXiv:hep-th/9806087
M. Henningson and K. Skenderis, *Fortsch.Phys.* **48**(2000)125, arXiv:hep-th/9812032

[8] K. Skenderis and P.K. Townsend, *Phys. Lett.* **B468**(1999)46, arXiv:hep-th/9909070

[9] O. DeWolfe, D.Z. Freedman, S.S. Gubser and A. Karch, *Phys. Rev.* **D62**(2000) 046008, arXiv:hep-th/9909134

[10] D.Z. Freedman, S.S. Gubser, K. Pilch and N.P. Warner, *Adv.Theor.Math.Phys.***3**(1999)363

[11] J. de Boer, E. Verlinde and H. Verlinde, *JHEP* **0008**(2000)003, arXiv:hep-th/9912012

[12] J.M. Maldacena, *Adv.Theor.Math.Phys.***2**(1998)231 [*Int. J. Theor. Phys.***38**(1999)1113]

[13] S. S. Gubser, I. R. Klebanov and A. M. Polyakov, *Phys. Lett.* **B428**(1998)105, arXiv:hep-th/9802109

[14] E. Witten, *Adv. Theor. Math. Phys.***2**(1998)253, arXiv:hep-th/9802150

[15] S. Ichinose, *Prog.Theor.Phys.***121**(2009)727, ArXiv:0801.3064v8[hep-th].

[16] L. Randall and M.D. Schwartz, *JHEP* **0111** (2001) 003, hep-th/0108114

[17] S. Ichinose and A. Murayama, *Phys. Rev.* **D76**(2007)065008, hep-th/0703228

[18] S. Ichinose, *Class. Quantum Grav.* **18**(2001)421, hep-th/0003275

[19] S. Ichinose, Proc. of VIII Asia-Pacific Int. Conf. on Gravitation and Astrophysics (ICGA8,Aug.29-Sep.1,2007,Nara Women's Univ.,Japan),Press Section p36-39, arXiv:/0712.4043

[20] S. Ichinose, *Int.Jour.Mod.Phys.***23A**(2008)2245-2248, Proc. of Int. Conf. on Prog. of String Theory and Quantum Field Theory (Dec.7-10,2007,Osaka City Univ.,Japan), arXiv:/0804.0945

[21] R.P. Feynman, *Statistical Mechanics*, W.A.Benjamin,Inc., Massachusetts, 1972

[22] S. Ichinose and A. Murayama, *Nucl. Phys.* **B710**(2005)255, hep-th/0401011

[23] T. Yoneya, *Duality and Indeterminacy Principle in String Theory* in "Wandering in the Fields", eds. K. Kawarabayashi and A. Ukawa (World Scientific,1987), p.419
T. Yoneya, *String Theory and Quantum Gravity* in "Quantum String Theory", eds. N. Kawamoto and T. Kugo (Springer,1988), p.23
T. Yoneya, *Prog.Theor.Phys.***103**(2000)1081

[24] S. Ichinose, hep-th/0012255, US-00-11, 2000, "Pole Solution in Six Dimensions and Mass Hierarchy"

[25] S. Ichinose, Proc. of 10th Tohwa Int. Symp. on String Thery (Jul.3-7, 2001, Tohwa Univ., Minerva Hall, Fukuoka, Japan), ed. H. Aoki and T. Tada, C2002, AIP Conf. Proc. 607, American Inst. Phys., Melville, New York, p307

[26] S. Ichinose, *Nucl. Phys.* **B231**(1984)335

[27] P.A.M. Dirac, *Nature* **139**(1937)323; *Proc.Roy.Soc.***A165**(1938)199; "Directions in Physics", John Wiley & Sons, Inc., New York, 1978

On the Avoidance of Classical Singularities in Quantum Cosmology

Claus Kiefer

Institut für Theoretische Physik, Universität zu Köln, Zülpicher Straße 77, 50937 Köln, Germany.

E-mail: kiefer@thp.Uni-Koeln.DE

Abstract.
Many cosmological models based on general relativity contain singularities. I address here the question whether consistent models without singularities can exist in quantum cosmology. The discussion is based on the Wheeler–DeWitt equation of quantum geometrodynamics. The models under consideration are partly motivated by recent discussions of dark energy. Natural criteria of singularity avoidance in the quantum theory are the vanishing of the wave function at the classical singularity or the breakdown of the semiclassical approximation upon approaching the region of the singularity. I show that singularity avoidance in this sense can happen in the models under consideration.

1. Introduction

The search for a quantum theory of gravity belongs to the most important tasks in fundamental physics [1]. Whilst such a theory is not yet available in final form, it is more or less obvious which expectations it should fulfill. Among them is the fate of the classical singularities in quantum gravity: a reasonable theory of quantum gravity should predict their avoidance.

Given the incomplete understanding of quantum gravity, it is obvious that general conclusions cannot yet be drawn. However, the issue of singularity avoidance may be addressed within the context of simple models. This is the topic of my contribution. I shall restrict myself to singularities for homogeneous and isotropic cosmological models. For such models, the quantization can be performed and the question can be investigated whether singularity-free solutions exist and whether they are natural.

My paper is organized as follows. In section 2, I shall quote one of the famous singularity theorems and give an overview of cosmological singularities in Friedmann models; here, one does not only have the standard big-bang and big-crunch singularities, but also more exotic ones such as a big rip or a big brake. Section 3 presents the general framework of my discussion, quantum geometrodynamics and the Wheeler–DeWitt equation, as well as arguments why this framework is an appropriate one for the issue under investigation. In section 4, then, I review the quantum cosmology of some of these models and show that there, indeed, singularities can be avoided in a natural way. One could imagine that singularity avoidance is similarly implementable in full quantum gravity. I end with some brief conclusions.

My contribution here is closely related to a similar contribution for another conference [2].

2. Classical singularities in cosmology

We call a spacetime singular if it is incomplete with respect to a timelike or null geodesic and if it cannot be embedded in a bigger spacetime. There exists a variety of theorems which postulate the existence of singularities in general relativity [3]. Typically, they involve an energy condition, a condition on the global structure, and the condition that gravitation be strong enough to lead to the existence of a closed trapped surface. Let me quote one of the famous theorems proved by Hawking and Penrose in 1970 [4]:

> A spacetime M cannot satisfy causal geodesic completeness if, together with Einstein's equations, the following four conditions hold:
>
> (i) M contains no closed timelike curves.
> (ii) The strong energy condition is satisfied at every point.
> (iii) The generality condition (...) is satisfied for every causal geodesic.
> (iv) M contains either a trapped surface, or a point P for which the convergence of all the null geodesics through P changes sign somewhere to the past of P, or a compact spacelike hypersurface

If we take an ideal fluid as a model for matter, with ρ as the energy density and p_i $(i = 1, 2, 3)$ as the principal pressures, the strong energy condition used in this theorem reads as

$$\rho + \sum_i p_i \geq 0, \quad \rho + p_i \geq 0, \quad i = 1, 2, 3\,. \tag{1}$$

In fact, as was already shown in [4] and discussed in [3], the observation of the Cosmic Microwave Background Radiation indicates that there is enough matter on the past light-cone of our present location P to imply that the divergence of this cone changes somewhere to the past of P. Thus, if in addition the strong energy condition is fulfilled, the conditions stated in the above theorem will apply and the origin of our Universe cannot be described within general relativity.

Modern cosmology entertains the idea that a period of quasi-exponential acceleration ("inflation") took place in the very early Universe. Since such an inflationary phase may violate the strong energy conditions, it has been speculated that the singularity can thereby be avoided. Is this true? There are two aspects of this. On the one hand, it was shown by Borde *et al.* in [5] that singularities are not avoided by an inflationary phase if the universe has open spatial sections or if the Hubble expansion rate is bounded away from zero in the past. On the other hand, Ellis *et al.* have presented an example of a singularity-free inflationary model with closed spatial sections [6]. Thus, independent of what describes more precisely the real conditions in the early Universe, one should at least envisage the possibility that the classical solution describing it is singular.

Cosmological singularities cannot only happen in the past. There is the option (presently not supported by observations) that the Universe will recollapse in the future and encounter a big-crunch singularity. But even if the Universe continues to expand, it can encounter singularities in the future for non-vanishing scale factor. This can happen for various equations of state which can describe a form of dark energy. (Such singularities may also occur at a finite value $a > 0$ in the past). In fact, present observations still allow the possibility of an equation of state leading to such "exotic" singularities, cf. [7] and the references therein. In the light of the above-quoted singularity theorem it is of interest to remark that these exotic singularities may or may not violate energy conditions, so this theorem is not directly applicable and one has to explore additional options for relevant energy conditions.

One can classify cosmological singularities for homogeneous and isotropic spacetimes following the schemes proposed in [7, 9]. Different singularities can be distinguished according to the behaviour of scale factor a, energy density ρ, and pressure p. The classification is

Big Bang/Crunch $a = 0$ at finite proper time, but ρ diverges.

Type I (Big Rip) a diverges in finite proper time, and both ρ and p diverge. A big rip is obtained from phantom dark energy, that is, matter with an equation of state $p = w\rho$, $w < -1$. Phantoms not only violate the strong energy condition (cf. the theorem quoted above), but also all the other energy conditions.

Type II (Sudden singularity) a and ρ remain finite, p diverges, but the Hubble parameter stays finite ($\dot{H}$ diverges). Sudden singularities only violate the dominant energy condition. Special cases

Journal of Physics: Conference Series **222** (2010) 012049 doi:10.1088/1742-6596/222/1/012049

are the big-brake singularity in the future and the big-démarrage singularity in the past; these are characterized as sudden singularities arising from a generalized Chaplygin equation of state, $p = -A/\rho^\beta$. For sudden singularities, geodesics can be continued through these singularities, but tidal forces diverge. Sudden singularities can appear as close as 8.7 million years in the future [7]. There exist also generalized sudden singularities, which are singularities of pressure derivatives and which fulfill all energy conditions.

Type III (Finite scale factor singularity) a remains finite, but both ρ and p diverge (as well as H and $\dot{H}$); particular examples are the "big-freeze" singularities (both in the past and the future), which are characterized by a generalized Chaplygin equation of state.

Type IV (Big Separation) a, ρ, and p remain finite, but the rth derivative of a and the $(r-1)$th derivative of H diverge ($r \geq 3$).

w**-singularity** The only singularity here is in the barotropic index w [8].

In extension of the Type I to IV classification, one could call big bang/big crunch Type 0 and the w-singularity Type V.

It is now of interest to see whether some or all of these singularities can be avoided in quantum gravity. We have investigated the quantum theory for some of these models [10, 11, 12], and some of the results will be discussed below. Before that, however, I shall briefly introduce the framework on which the investigation is based.

3. Quantum geometrodynamics

A full quantum theory of gravity remains elusive [1]. Can one nevertheless say something reliable about quantum gravity without knowing the exact theory? In [13] I have made the point that this is indeed possible. The situation is analogous to the role of the quantum mechanical Schrödinger equation. Although this equation is not fundamental (it is non-relativistic, it is not field-theoretic, and so on), important insights can be drawn from it. For example, in the case of the hydrogen atom, one has to impose boundary conditions for the wave function at the origin $r \to 0$, that is, at the centre of the atom. This is certainly not a region where one would expect non-relativistic quantum mechanics to be exactly valid, but its consequences, in particular the resulting spectrum, are empirically correct to an excellent approximation.

Erwin Schrödinger has found his equation by "guessing" a wave equation from which the Hamilton–Jacobi equation of classical mechanics can be recovered in the limit of small wavelengths, in analogy to the limit of geometric optics from wave optics. The same approach can be applied to general relativity. One can start from the Hamilton–Jacobi version of Einstein's equations and "guess" a wave equation from which it can be recovered in the classical limit. The only assumption that is required is the universal validity of quantum theory, that is, its linear structure. It is not yet needed for this step to impose a Hilbert-space structure. Such a structure is employed in quantum mechanics because of the probability interpretation for which one needs a scalar product and its conservation in time (unitarity). Its status in quantum gravity remains open.

The Hamilton–Jacobi equation for general relativity was formulated by Peres [14]. In the vacuum case, it reads (we set $c = 1$ in the following)

$$16\pi G\, G_{abcd} \frac{\delta S}{\delta h_{ab}} \frac{\delta S}{\delta h_{cd}} - \frac{\sqrt{h}}{16\pi G}\left({}^{(3)}R - 2\Lambda\right) = 0\ . \qquad (2)$$

The functional $S[h_{ab}]$ depends on the three-dimensional metric, h_{ab}. It also obeys the equation

$$D_a \frac{\delta S}{\delta h_{ab}} = 0\ , \qquad (3)$$

which guarantees that $S[h_{ab}]$ is invariant under three-dimensional coordinate transformations.

The task is now to find a wave equation which yields the Hamilton–Jacobi equation (2) as well as Eq. (3) in the semiclassical limit ("WKB approximation") where the wave functional reads

$$\Psi[h_{ab}] = C[h_{ab}] \exp\left(\frac{\mathrm{i}}{\hbar} S[h_{ab}]\right) \tag{4}$$

with a rapidly varying phase and a slowly varying amplitude. In the vacuum case, one then finds from (2) and (3) the equations

$$\hat{H}\Psi \equiv \left(-16\pi G\hbar^2 G_{abcd}\frac{\delta^2}{\delta h_{ab}\delta h_{cd}} - (16\pi G)^{-1}\sqrt{h}({}^{(3)}R - 2\Lambda)\right)\Psi = 0\,, \tag{5}$$

and

$$\hat{D}^a\Psi \equiv -2\nabla_b \frac{\hbar}{\mathrm{i}}\frac{\delta\Psi}{\delta h_{ab}} = 0\,. \tag{6}$$

The first equation is called the Wheeler–DeWitt equation, the second one – which guarantees the invariance of the wave functional under three-dimensional coordinate transformations – is called the quantum diffeomorphism (momentum) constraint. Whether these equations hold at the most fundamental level or not, they should approximately be valid away from the Planck scale, provided that quantum theory is universally valid.

We recognize from (5) und (6) that no external time is present anymore – spacetime has *disappeared* and only space remains [1]. Nevertheless, a local *intrinsic time* can be defined through the local hyperbolic structure of the Wheeler–DeWitt equation. This intrinsic time is solely constructed from the three-geometry (and matter degrees of freedom, if present).

If non-gravitational degrees of freedom are present, the equations (5) and (6) have to be augmented by the corresponding terms. For the semiclassical approximation one then makes the following "Born–Oppenheimer"-type of ansatz [1]

$$|\Psi[h_{ab}]\rangle = C[h_{ab}]\mathrm{e}^{\mathrm{i}m_{\mathrm{P}}^2 S[h_{ab}]}|\psi[h_{ab}]\rangle\,, \tag{7}$$

where the *bra* and *ket* notation refers to non-gravitational fields, for which we can assume that a standard Hilbert-space is at our disposal (m_{P} denotes the Planck length).

One then evaluates the wave functional $|\psi[h_{ab}]\rangle$ along a solution of the classical Einstein equations, $h_{ab}(\mathbf{x},t)$, corresponding to a solution, $S[h_{ab}]$, of Equations (2) and (3); this solution is obtained from

$$\dot{h}_{ab} = NG_{abcd}\frac{\delta S}{\delta h_{cd}} + 2D_{(a}N_{b)}\,, \tag{8}$$

where N and N_b denote the lapse function and the shift vector, respectively [1]. Employing in the semiclassical limit the following definition of an *approximate* time parameter t,

$$\frac{\partial}{\partial t}|\psi(t)\rangle = \int \mathrm{d}^3x\, \dot{h}_{ab}(\mathbf{x},t)\,\frac{\delta}{\delta h_{ab}(\mathbf{x})}|\psi[h_{ab}]\rangle\,,$$

one obtains a functional Schrödinger equation for quantized matter fields in the chosen external classical gravitational field:

$$\begin{aligned} \mathrm{i}\hbar\frac{\partial}{\partial t}|\psi(t)\rangle &= \hat{H}^{\mathrm{m}}|\psi(t)\rangle\,, \\ \hat{H}^{\mathrm{m}} &\equiv \int \mathrm{d}^3x\left\{N(\mathbf{x})\hat{\mathcal{H}}^{\mathrm{m}}_{\perp}(\mathbf{x}) + N^a(\mathbf{x})\hat{\mathcal{H}}^{\mathrm{m}}_a(\mathbf{x})\right\}\,. \end{aligned} \tag{9}$$

Journal of Physics: Conference Series **222** (2010) 012049 doi:10.1088/1742-6596/222/1/012049

Here, $\hat{H}^{\mathrm{m}}$ denotes the matter-field Hamiltonian in the Schrödinger picture, parametrically depending on the (generally non-static) metric coefficients of the curved space–time background; the "WKB time" t controls the dynamics in this approximation.

The special ansatz (7) can be justified through the process of decoherence [15]. Unobserved degrees of freedom such as tiny gravitational waves or small density fluctuations interact with the relevant degrees of freedom (e.g. the volume of the three-dimensional space) in such a way that, for example, the superposition of (7) with its complex conjugate cannot be distinguished from a corresponding mixture; the state (7) can thus be treated separately. This is analogous to the emergence of the chiral structure for certain molecules [15].

The semiclassical approximation and the ensuing time parameter exist, of course, only under special conditions. Only then can the notions of spacetime and geodesics, needed for the application of the classical singularity theorems, be employed. In the quantum case we have to think first about the meaning of singularities and singularity avoidance.

4. Quantum avoidance of singularities

There does not yet exist any general agreement on the necessary criteria for quantum avoidance of singularities. I shall not attempt here to provide such necessary criteria either, but shall focus instead on two sufficient criteria which have turned out to be useful in the discussion of the models presented below. The first criterium dates back to the pioneering work of Bryce DeWitt on canonical quantum gravity [16]. He has suggested that the wave function should vanish at the point of the classical singularity. We shall thus interpret the vanishing of the wave function in the region of the classical singularity as singularity avoidance. As a second criterium we shall adopt the spreading of wave packets when approaching the region of the classical singularity: the semiclassical approximation discussed above breaks down in this case and no classical time parameter is available. The classical singularity theorems can then no longer be applied.

It must be emphasized that $\Psi \longrightarrow 0$ is really only a sufficient, but by no means a necessary criterium for singularity avoidance. Consider, for example, the solution of the Dirac equation for the ground state of hydrogen-like atoms (in standard notation). This solution diverges at the origin,

$$\psi_0(r) \propto (2mZ\alpha r)^{\sqrt{1-Z^2\alpha^2}-1}\mathrm{e}^{-mZ\alpha r} \stackrel{r\to 0}{\longrightarrow} \infty \,,$$

but $\int \mathrm{d}r\ r^2|\psi_0|^2$ remains finite. So the important quantity is, in fact, the inner product. As already mentioned, there is no general consensus about the role of the inner product in quantum cosmology, but the situation could be analogous to the hydrogen atom. In the case of the Wheeler–DeWitt equation for a Friedmann universe with a massless scalar field, for example, the simplest solution diverges for scale factor $a \longrightarrow 0$,

$$\Psi \propto K_0(a^2/2) \stackrel{a\to 0}{\longrightarrow} c \ln a \,,$$

but nevertheless the integral $\int \mathrm{d}a\mathrm{d}\phi\sqrt{|G|}|\psi(a,\phi)|^2$, where G here denotes the determinant of the DeWitt metric, may be finite.

For a closed Friedmann–Lemaître universe with scale factor a, containing a homogeneous scalar field ϕ with potential $V(\phi)$, that is, a two-dimensional *minisuperspace*, the classical line element is

$$\mathrm{d}s^2 = -N^2(t)\mathrm{d}t^2 + a^2(t)\mathrm{d}\Omega_3^2 \,.$$

The Wheeler–DeWitt equation reads (with units $2G/3\pi = 1$)

$$\frac{1}{2}\left(\frac{\hbar^2}{a^2}\frac{\partial}{\partial a}\left(a\frac{\partial}{\partial a}\right) - \frac{\hbar^2}{a^3}\frac{\partial^2}{\partial\phi^2} - a + \frac{\Lambda a^3}{3} + 2a^3V(\phi)\right)\psi(a,\phi) = 0 \,,$$

where the factor ordering has been chosen in order to achieve covariance in minisuperspace; this is the Laplace–Beltrami factor ordering. (For a general overview of quantum cosmology, see, for example, [1, 17] and the references therein.)

Let us consider now various models with classical singularities. The first one is a model with "phantoms". These are fields with negative kinetic energy, which are certainly very exotic but which cannot yet be excluded on the basis of supernova data. Classically, the ensuing dynamics develops a "big-rip singularity", that is, ρ and p diverge as a goes to infinity at a *finite time* (see above). The corresponding quantization was discussed in [10]. It was found that wave-packet solutions of the Wheeler–DeWitt equation necessarily disperse in the region of the classical big-rip singularity; time and the classical evolution thus come to an end and only a stationary quantum state is left. Quantum effects thus become important for a big universe, not only for a small Planck-size universe. Interestingly, it was shown that an alternative quantization, employing the Bohm interpretation of quantum theory, cannot avoid the big-rip singularity [18].

Another model is cosmology with a big brake [11]. We take an equation of state of the form $p = A/\rho$, $A > 0$, that is, an "Anti-Chaplygin gas". For a Friedmann universe with scale factor $a(t)$ and a scalar field $\phi(t)$, this equation of state can be realized by the potential

$$V(\phi) = V_0 \left(\sinh\left(\sqrt{3\kappa^2}|\phi|\right) - \frac{1}{\sinh\left(\sqrt{3\kappa^2}|\phi|\right)} \right) \; ; \; V_0 = \sqrt{A/4}\,,$$

where $\kappa^2 = 8\pi G$. The classical dynamics develops a pressure singularity (only $\ddot{a}(t)$ becomes singular) and comes to an abrupt halt in the future.

The Wheeler–DeWitt equation for this model reads

$$\frac{\hbar^2}{2}\left(\frac{\kappa^2}{6}\frac{\partial^2}{\partial\alpha^2} - \frac{\partial^2}{\partial\phi^2}\right)\Psi(\alpha,\phi) + V_0 e^{6\alpha}\left(\sinh\left(\sqrt{3\kappa^2}|\phi|\right) - \frac{1}{\sinh\left(\sqrt{3\kappa^2}|\phi|\right)} \right)\Psi(\alpha,\phi) = 0\,, \quad (10)$$

where $\alpha = \ln a$, and Laplace–Beltrami factor ordering has again been used. The vicinity of the big-brake singularity is the region of small ϕ; we can therefore use the approximation

$$\frac{\hbar^2}{2}\left(\frac{\kappa^2}{6}\frac{\partial^2}{\partial\alpha^2} - \frac{\partial^2}{\partial\phi^2}\right)\Psi(\alpha,\phi) - \frac{\tilde{V_0}}{|\phi|}e^{6\alpha}\Psi(\alpha,\phi) = 0\,,$$

where $\tilde{V_0} = V_0/3\kappa^2$.

It was shown in [11] that all normalizable solutions read

$$\Psi(\alpha,\phi) = \sum_{k=1}^{\infty} A(k) k^{-3/2} \mathrm{K}_0\left(\frac{1}{\sqrt{6}}\frac{V_\alpha}{\hbar^2 k\kappa}\right) \times \left(2\frac{V_\alpha}{k}|\phi|\right) e^{-\frac{V_\alpha}{k|\phi|}} \mathrm{L}^1_{k-1}\left(2\frac{V_\alpha}{k}|\phi|\right)\,, \quad (11)$$

where K_0 is a Bessel function, L^1_{k-1} are Laguerre polynoms, and $V_\alpha \equiv \tilde{V_0} e^{6\alpha}$. They all *vanish* at the classical singularity. This model therefore implements our other criterium above: the vanishing of the wave function in the spirit of DeWitt. A similar result holds for the corresponding loop quantum cosmology. Interestingly, the solutions also implement the avoidance of the big-bang singularity, $\Psi \to 0$ for $\alpha \to -\infty$.

The big-brake singularity is an example of type-II singularities. A more general class of type-II and type-III singularities was discussed in [12]. The equation of state in this class is chosen to be that of a generalized Chaplygin gas:

$$p = -\frac{A}{\rho^\beta}$$

with general real parameters A and β. For example, in the big-freeze (type III) singularity, both H and $\dot{H}$ blow up at a finite value of the scale factor in the past. The big-freeze singularity occurs in the past at a minimal scale factor $a_{\min} > 0$; there are thus no classical solutions in the limit $\alpha \to -\infty$. For this reason we have to demand that the wave function go to zero in the classically forbidden region, $\Psi \stackrel{\alpha\to-\infty}{\longrightarrow} 0$, because otherwise one would not obtain the correct classical limit.

The class of solutions then reads

$$\Psi_k(\alpha,\phi) \propto \sqrt{|\phi|}\mathrm{J}_\nu(k|\phi|)\left[b_1\mathrm{e}^{\mathrm{i}\frac{\sqrt{6}k}{\kappa}\alpha}+b_2\mathrm{e}^{-\mathrm{i}\frac{\sqrt{6}k}{\kappa}\alpha}\right] ,$$

with ν as a function of α. These solutions obey, in fact, DeWitt's boundary condition at the singularity, $\Psi_k(0,0)=0$. The same also holds for the other cases discussed in [12].

The vanishing of the wave function at the classical singularity plays also a key role in other approaches. An especially interesting example are the supersymmetric quantum cosmological billiards discussed in [19]. In $D=11$ supergravity, one can employ near a spacelike singularity a cosmological billiard description based on the Kac–Moody group E_{10} and address the corresponding Wheeler–DeWitt equation. It was found there, too, that $\Psi \to 0$ near the singularity. Such a behaviour also occurs in a model with null dust shells where the classical black-hole and white-hole singularities are thereby avoided in the quantum theory [20]. DeWitt's criterium of singularity avoidance may thus be a viable criterium implementable in a wide range of quantum cosmological models.

5. Conclusion

I have shown that the classical cosmological singularities of various models can be avoided in quantum cosmology. Two sufficient (but by no means necessary) criteria have been employed for this purpose: unavoidable spreading of the wave function when approaching the singular region and vanishing of the wave function at the singularity itself.

The models discussed here can have (in the classical theory) singularities at large values of the scale factor, that is, far away from the Planck length. The corresponding quantum avoidance thus means that quantum gravitational effects can occur for large universes – an intriguing thought. Such macroscopic quantum effects have hitherto only been envisaged near the turning point of a classically recollapsing universe [21, 22]. It is clear that these scenarios have also consequences for the arrow of time in the universe [21, 23].

Singularity avoidance also occurs in the framework of loop quantum cosmology, although in a somewhat different way, see [24]. The big-bang singularity can be avoided by solutions of the difference equation which there replaces the Wheeler–DeWitt equation. The avoidance can also be achieved by the occurrence of a bounce in the effective Friedmann equations. For the singularities at large scale factors of the big-rip or big-brake type, the mechanism of avoidance is similar to the one discussed above for the Wheeler–DeWitt equation, but there are other sudden singularities that do not seem to be avoided [25].

Acknowledgments

I am grateful to the organizers of the First Mediterranean Meeting on Quantum Gravity for inviting me to a stimulating meeting at a beautiful place. I am grateful to my co-authors Mariam Bouhmadi-López, Mariusz Dąbrowski, Alexander Kamenshchik, Barbara Sandhöfer, and Paulo Vargas Moniz for our collaboration on this faszinating subject.

[1] Kiefer C 2007 *Quantum Gravity* second edition (Oxford: Oxford University Press)
[2] Kiefer C 2009 Can singularities be avoided in quantum cosmology? To appear in *Annalen der Physik*
[3] Hawking S W and Ellis G F R 1973 *The Large Scale Structure of Space–Time* (Cambridge: Cambridge University Press)
[4] Hawking S W and Penrose R 1970 *Proc Roy Soc Lond A* **314** 529
[5] Borde A, Guth A H, and Vilenkin A 2003 *Phys Rev Lett* **90** 151301
[6] Ellis G F R and Maartens R 2004 *Class Quantum Grav* **21** 223
[7] Dąbrowski M P and Denkiewicz T 2009 Exotic-singularity-driven dark energy `arXiv:0910.0023v1 [gr-qc]`
[8] Dąbrowski M P and Denkiewicz T 2009 *Phys. Rev. D* **79** 063521
[9] Nojiri S, Odintsov S D, and Tsujikawa S 2005 *Phys Rev D* **71** 063004

[10] Dąbrowski M P, Kiefer C, and Sandhöfer B 2006 *Phys Rev D* **74** 044022
[11] Kamenshchik A, Kiefer C, and Sandhöfer B 2007 *Phys Rev D* **76** 064032
[12] Bouhmadi-López M, Kiefer C, Sandhöfer B, and Moniz P V 2009 *Phys Rev D* **79** 124035
[13] Kiefer C 2009 *Gen Rel Grav* **41** 877
[14] Peres A 1962 *Nuovo Cimento* **XXVI** 53
[15] Joos E, Zeh H D, Kiefer C, Giulini D, Kupsch J, and Stamatescu I O 2003 *Decoherence and the appearance of a classical world in quantum theory* second edition (Berlin: Springer)
[16] DeWitt B S 1967 *Phys. Rev.* **160** 1113
[17] Kiefer C 2006 *Ann Phys (Berlin)* **15** 316
[18] Pinto-Neto N and Moraes Pantoja D 2009 *Phys Rev D* **80** 083509
[19] Kleinschmidt A, Koehn M, and Nicolai N 2009 *Phys Rev D* **80** 061701
[20] Hájíček P and Kiefer C 2001 *Int J Mod Phys D* **10** 775
[21] Zeh H D 2007 *The physical basis of the direction of time* fifth edition (Berlin: Springer)
[22] Kiefer C and Zeh H D 1995 *Phys Rev D* **51** 4145
[23] Kiefer C 2009 Can the Arrow of Time be understood from Quantum Cosmology? `arXiv:0910.5836v1 [gr-qc]`
[24] Bojowald M 2007 Singularities and Quantum Gravity `arXiv:gr-qc/0702144v1`
[25] Cailleteau T, Cardoso A, Vandersloot K, and Wands D 2008 *Phys. Rev. Lett.* **101** 251302

First Mediterranean Conference on Classical and Quantum Gravity (MCCQG 2009) IOP Publishing
Journal of Physics: Conference Series **222** (2010) 012050 doi:10.1088/1742-6596/222/1/012050

Routes towards Emergent Gravity

Stefano Liberati
SISSA and INFN, Trieste
E-mail: `liberati@sissa.it`

Florian Girelli
School of Physics, Sydney University
E-mail: `girelli@physics.usyd.edu.au`

Lorenzo Sindoni
Albert Einstein Institute, Golm, Germany
E-mail: `sindoni@aei.mpg.de`

Abstract. We discuss a toy model for an emergent non-relativistic gravitational theory. Within a certain class of Bose–Einstein condensates, it is possible to show that, in a suitable regime, a modified version of non-relativistic Newtonian gravity does effectively describes the low energy dynamics of the coupled system condensate/quasi-particles. Furthermore, we study the role of Lorentz and diffeomorphism invariance in emergent gravity scenarios by developing a toy model showing an emergent Lorentzian signature from a Euclidean setting and simultaneously and emergent gravitational dynamics of the Nördstrom type (scalar gravity). Some lessons about the crucial challenges awaiting in the future the emergent gravity proposal are finally drawn.

1. Introduction: BEC dynamics

In recent years the emergent gravity approach has constantly gained momentum [1]. In this particular perspective, the gravitational field, encoded in the spacetime geometry and its dynamics, is seen as a kind of large number/thermodynamical limit of some more fundamental theory.

While there is no proof that this is the case, there are evidences to support this point of view. First of all, let us remind the striking correspondence existing between the laws of black hole mechanics in General Relativity (GR) and the laws of thermodynamics [2]. This correspondence has inspired a deeper study of the thermodynamical aspects of gravitation. In particular, it has been shown that Einstein's equation can be seen as some sort of equation of state of a thermodynamical system at equilibrium [3, 4, 5, 6]. Furthermore, from the study of the properties of gravitational collapse and the resulting black hole formation, very specific patterns have been recognized in the parameter space of the initial conditions which resemble the behavior of critical phenomena (phase transitions) [7, 8].

In addition, it has been realized that within several condensed matter systems it is possible to distinguish some peculiar regimes in which the effective degrees of freedom are represented by fields propagating over effective pseudo-Riemannian structures [9].

For instance, in the case of perfect, irrotational and barotropic fluids, it can be proved that the perturbations in the velocity potential (*i.e.* the scalar function θ whose gradient gives the velocity of the fluid, $\vec{v} \propto \vec{\nabla}\theta$) do obey a massless Klein–Gordon equation in a curved effective spacetime whose metric tensor is given by the so-called acoustic metric,

$$g_{\mu\nu} = \frac{\rho}{c_s^2} \left(\begin{array}{ccc} -(c_s^2 - v^2) & \vdots & v^i \\ \cdots & \cdot & \cdots \\ v^i & \vdots & \delta_{ij} \end{array} \right), \tag{1}$$

where ρ is the local density of the fluid, c_s it the (local) speed of sound and v_i is the velocity field of the fluid flow.

Analogue models have been used to understand (and possibly to test in a laboratory) some peculiar aspects of physics in curved spacetimes, otherwise inaccessible (*e.g.* Hawking radiation). For a review of the subject see [9]. For the large majority, these analogue models for gravity do offer the possibility of studying some kinematical aspects of physics of curved spacetimes, leaving aside the issue of dynamics.

Despite the fact that, to date, there are no completely successful models of emergent gravity, *i.e.* non-gravitational system possessing an effective GR like limit, it is interesting to study simple toy models in which some aspects of the gravitational interaction can be mimicked. Here we present a toy model [10], based on a well-known condensed matter system, a Bose–Einstein condensate of dilute, weakly interacting gas of bosons [11]. While this toy model is very far from representing an analogue of a realistic theory of gravity, it gives nonetheless some insights on some interesting features of emergent theories.

The behaviour of dilute, weakly interacting Bose gases can be described with the formalism of second quantization, by introducing the field operators describing the gas atoms (V is the volume of the box in which we consider the gas)

$$\hat{\Psi}(\mathrm{x}) = \frac{1}{\sqrt{V}} \sum_{\mathrm{k}} \hat{a}_{\mathrm{k}} e^{i\mathrm{k}\cdot\mathrm{x}}, \tag{2}$$

and a suitable Hamiltonian:

$$\hat{H}_0 = \int d^3x\, \hat{\Psi}^\dagger(\mathrm{x}) \left(\frac{-\hbar^2\nabla^2}{2m} - \mu + \frac{\kappa}{2} |\hat{\Psi}(\mathrm{x})|^2 \right) \hat{\Psi}(\mathrm{x}). \tag{3}$$

As it is well known, for μ, κ positive, this Hamiltonian leads to a condensation, *i.e.* the ground state of the system is characterized by a macroscopic occupation number of a single particle state. Within the formalism, this corresponds to the field operator $\hat{\Psi}$ having a non-vanishing vacuum expectation value,

$$\hat{\Psi}(\mathrm{x}) \approx \psi(\mathrm{x})\mathbb{I} + \hat{\chi}(\mathrm{x}), \qquad \langle\Omega|\hat{\Psi}(\mathrm{x})|\Omega\rangle = \psi(\mathrm{x}), \tag{4}$$

where ψ is a classical complex scalar field, the condensate wavefunction, describing the mean field, and $\hat{\chi}$ is a quantum operator describing the residual quantum fluctuations around the condensate, or, in more physical terms, the atoms which are out of the condensate.

Obviously, the nonlinearity in the Hamiltonian (3) makes the analysis of the physical properties of this system rather involved. As a first approximation, it is useful to assume that the non-condensed fraction is small compared to the condensed fraction.

The lowest order equation describes the condensate wave-function alone, without the backreaction terms induced by the $\hat{\chi}$ field. It is called the Gross–Pitaevski equation and it is giving the first approximation to the condensate wavefunction dynamics. It is obtained from the equations of motion for the system by the replacement $\hat{\Psi} \to \psi$.

The most simple solutions to this equations are the constant ones, describing homogeneous condensates, which, in the analogue model perspective, correspond to a flat acoustic metric, *i.e.* to Minkowski spacetime, and therefore will play a particular role in the following. The weak field limit of GR will correspond to the weak field limit around this constant configuration. In fact, as a direct calculation show, the correspondence between the acoustic metric $g_{\mu\nu}$ and the complex scalar field ψ is such that:

$$\left.\begin{array}{c} g_{\mu\nu} \approx \eta_{\mu\nu} + h_{\mu\nu} \\ \psi \approx (\kappa/\mu)^{1/2}(1 + u(x) + iv(x)) \end{array}\right\} \Rightarrow h_{00} \propto u(x) \tag{5}$$

so that the perturbations in the number density encoded in $u(x)$ are the natural candidate to represent some sort of Newtonian gravitational field.

The next order equation is an equation for the noncondensed fraction $\hat{\chi}$ in the external field ψ. The equation of motion has the shape:

$$i\hbar\frac{\partial}{\partial t}\hat{\chi} = -\frac{\hbar^2}{2m}\nabla^2\hat{\chi} + \mu\hat{\chi} + \mu\hat{\chi}^\dagger. \tag{6}$$

The Hamiltonian (3) does possess a global $U(1)$ symmetry, $\hat{\Psi} \to e^{i\alpha}\hat{\Psi}$, which is spontaneously broken by the condensation mechanism (leading to the $\psi \neq 0$ ground state). As a consequence of the Goldstone theorem, the excitations will be gapless. (This can also be proved directly by diagonalizing the equation (6) using the Bogoliubov transformations.) Finally, it can be showed that, in some low energy regime, quasiparticles are described by a massless scalar field propagating on an effective acoustic geometry of the same shape of (1).

1.1. The coupled system

In order to see how some sort of gravitational dynamics is encoded in the BEC, a suitable dynamical framework must be set up in order to see how the quasiparticles backreact over the condensate. This formalism consists in considering an improved version of the Gross–Pitaevski equation which consistently takes into account the effect of the particles out of the condensate. The Gross–Pitaevski (GP) equation is replaced by the so-called Bogoliubov–de Gennes (BdG) equation

$$i\hbar\frac{\partial}{\partial t}\psi = -\frac{\hbar^2}{2m}\nabla^2\psi - \mu\psi + \kappa|\psi|^2\psi + 2\kappa\mathfrak{n}\psi + \kappa\mathfrak{m}\psi^*, \tag{7}$$

where $\mathfrak{n}, \mathfrak{m}$ are given by the expectation values:

$$\mathfrak{n} = \langle\Xi|\hat{\chi}(x)^\dagger\hat{\chi}(x)|\Xi\rangle, \qquad \mathfrak{m} = \langle\Xi|\hat{\chi}(x)^2|\Xi\rangle, \tag{8}$$

where the state $|\Xi\rangle$ is the particular state one is considering. Notice that, if this state were the Fock vacuum state for particles, these expectation values would be identically zero. Notice also that one is implicitly taking a normal ordering in the particle operator, so that an unphysical (divergent) zero point energy is removed automatically.

This equation, suitably modified, will give rise to the dynamics for the field $u(x)$, and hence to an analogue of the Poisson equation.

To construct some analogue of Newtonian gravity, we need massive particles as sources of the gravitational field (massless particles do not gravitate in Newtonian gravity). Therefore, the quasiparticles must not be Goldstone bosons, but instead pseudo-Goldstone: the $U(1)$ symmetry has to be broken explicitly at the level of the Hamiltonian. This is achieved by adding a term of the form $-\lambda\hat{\Psi}^\dagger\hat{\Psi}^\dagger + \text{h.c.}$ to the Hamiltonian (3), which in turn implies an extra $-\lambda\psi^*$ term on the right hand side of Eq. (7).[1]

The analysis of the properties of the quasiparticles in the case of homogeneous background (see [10] for details) leads to the conclusion that the quasiparticles dispersion relation is

$$E = \left(\mathcal{M}^2 c_s^4 + p^2 c_s^2 + \frac{p^4}{4m^2}\right)^{1/2}, \tag{9}$$

where c_s and $\mathcal{M}$ are respectively the speed of sound and the mass of the quasiparticles (which are function of the microscopic parameters λ, μ, m).

In the case of small momenta, and when the condensate wavefunction is not exactly homogeneous (*i.e.* when $u(\mathrm{x}) \neq 0$), the Hamiltonian for the quasiparticles takes the shape

$$\hat{H}_{q.p.} \approx \mathcal{M}c_s^2 - \frac{\hbar^2\nabla^2}{2\mathcal{M}} + \frac{C(\lambda,\mu,m)}{\mathcal{M}c_s^2}u(\mathrm{x}), \tag{10}$$

(with $C(\lambda,\mu,m)$ denotes a certain function of the various coupling constants, see [10] for the details) which leads to the identification of a "gravitational potential":

$$\Phi_{grav}(\mathrm{x}) = D(\lambda,\mu,m)u(\mathrm{x}), \tag{11}$$

with $D(\lambda,\mu,m)$ being a constant depending on the microphysics of the system.

1.1.1. The emergent gravitational system Having presented the main ideas and required tools, we pass to the results. Consider the Hamiltonian with the $U(1)$ breaking term. In the limit in which the backreaction of the condensate is small, *i.e.* in the limit in which there are few quasiparticles, when the condensate is almost homogeneous, the Bogoliubov–de Gennes equation can be rewritten as:

$$\left(\nabla^2 - \frac{1}{L^2}\right)\Phi_{grav} = 4\pi G_N \rho_{\text{matter}} + \Lambda, \tag{12}$$

where G_N and L, the analogue of the Newton's constant and the healing length, are function of the various coupling constants, and

$$\Lambda \equiv B(\lambda,\mu,m)\left(\mathfrak{n}_\Omega + \frac{1}{2}\mathfrak{m}_\Omega\right), \tag{13}$$

were again $B(\lambda,\mu,m)$ is a function of the microscopic parameters characterizing the model [10].

Notice the peculiar splitting of the source term. A detailed analysis [10] shows that the expectation values (8) always split into two contributions, one nonlocal term due to the quasiparticles, ρ_{matter}, and an unavoidable vacuum contribution, Λ, due to the inequivalence between the Fock vacuum for particles and the Fock vacuum for quasiparticles.

The reader will immediately realize that this interaction is very far away from a realistic Newtonian gravitational interaction: indeed, the would-be Poisson equation includes a term which makes the interaction short ranged. In particular, this range is set by the healing length L, which is an UV scale for the physics of the BEC (very much like the Planck scale in quantum gravity). This might have been guessed from the beginning, given that the healing length represents the typical scale for the dynamics of the condensate. Despite the fact that this system is not an analogue for a realistic form of gravitational interaction, it does offer some intriguing inspirations which we develop now.

[1] The breaking of the $U(1)$ symmetry has an obvious interpretation: the number of bosons is no longer a conserved charge (although it is conserved on average). This could lead to some issues about the physical realizability of such a system. For a critical discussion on this point see [10].

2. Beyond BECs

The discussion of the BEC model has shown that by limiting the analysis to condensed matter systems there are rather strong constraints on the kind of gravitational models which is reasonable to simulate. For example, in the above investigation we had a single scalar field: it would be interesting to see what happens if several different species are present. In that case, besides the issue of having a short range rather than a long range interaction, also the coupling to the gravitational field must be carefully discussed. Indeed, in order to have some sort of equivalence principle, all the fields must be coupled to the gravitational field in the same way.

The natural setup to discuss these issues is the 2-BEC model [12, 13]: in fact in this case one could treat a multi-particle system whose richness could allow a closer mimicking of Newtonian gravity with a long range potential. However, the fact that emergent gravity has to be Newtonian in a BEC-based analogue model seems to be unavoidable since the gravitational potential depends on the condensate, which is typically described by non-relativistic equations. A possible way to avoid this issue is either to consider relativistic BEC [14, 15] (however in this case we would still expect to get only some type of scalar gravity), or to change completely paradigm and identify gravity not as the condensate but as linked, together with matter, to the perturbations around the condensate. We will consider later this second point of view in a different model.

Furthermore, there is another important issue that requires attention. In our treatment we neglected the quantum potential, i.e. we have deliberately worked in the hydrodynamic limit of the theory, carefully avoiding the issue of the breakdown of acoustic Lorentz invariance in the system at suitably high energies of the quasi-particles. Presumably the breakdown of this emergent spacetime symmetry, namely local Lorentz invariance, will be linked also to some relevant regime change in the gravitational dynamics (which is anyway affected by the presence of a Lorentz symmetry breaking scale, the healing length, which ends up setting the graviton mass scale). Should we take Lorentz symmetry breaking as a crucial ingredient of the emerging gravity paradigm or as an accident of the condensed matter analogue models? In the first case, how the breakdown of such spacetime symmetry affects the symmetries of the gravitational dynamics and in particular diffeomorphism invariance? Furthermore, does this imply that an emergent gravity scenario should give up the relativity principle and bring us back to Newton's absolute space and time?

In order to explore these issues we can start investigating the role of Lorentz invariance in emergent gravity scenarios by considering the most well known "no-go theorem" against them, i.e. the so called Weinberg–Witten theorem [16].

2.1. Lorentz invariance and emergent gravity: the Weinberg–Witten theorem

The idea of having the graviton as a composite particle/emergent field is certainly a fascinating idea. However, there are limitations to what it is possible to do. In particular, there is a theorem, due to Weinberg and Witten [16], which is often presented as a crucial (fatal, in fact) obstruction for a successful emergent gravity program.

The theorem states precise limits for the existence of consistent theories with massless particles. It has two parts, and it says that (quoting from [16]):

(i) A theory that allows the construction of a Lorentz-covariantly conserved four-vector current J^μ cannot contain massless particles of spin $j > 1/2$ with nonvanishing values of the conserved charge $\int J^0 d^3x$.

(ii) A theory that allows for the construction of a conserved Lorentz covariant energy-momentum tensor $\theta^{\mu\nu}$ for which $\int \theta^{0\nu} d^3x$ is the energy-momentum four-vector cannot contain massless particles of spin $j > 1$.

For a careful discussion of the proof of the theorem, and for references, see [17]. For additional comments, see [18, 19]

Journal of Physics: Conference Series **222** (2010) 012050 doi:10.1088/1742-6596/222/1/012050

Crucial ingredients for the proof of this theorem are Lorentz invariance and the nonvanishing of the charges obtained from Lorentz covariant vectors and tensors. Interestingly, the gauge bosons like the gluons and the graviton are not forbidden since the current for the gluons is not Lorentz-covariant conserved, and the graviton does not possess a covariant stress-energy tensor (but rather a pseudo-tensor).

This theorem, then, poses rather strong constraints on the possible theories that can be built in Minkowski spacetime. Of course, gravity is not just the theory of a spin-2 particle in Minkowski spacetime. Nevertheless, it surely makes sense to consider the linearized theory in sufficiently small neighborhoods. In this limit, then, the theorem does apply.

With this caveat in mind, we can say that in an emergent gravity program this theorem must be taken appropriately into account and appropriately evaded. There are (at least) two "obvious" way out:

- allow for Lorentz symmetry breaking, or
- make the spacetime manifold to emerge as well.

The first option is rather straightforward, and it is essentially what could be pursued within scenarios like the one considered in analogue models, in which a preferred time function is specified.

The possibility of Lorentz Violation (LV) is not at all an exotic one. Specific hints of LV arose from various approaches to Quantum Gravity (QG). Examples include string theory tensor VEVs [20], spacetime foam [21], semiclassical spin-network calculations in Loop QG [22], non-commutative geometry [23, 24, 25], some brane-world backgrounds [26] and condensed matter analogues of "emergent gravity" [27]. Although none of these calculations proves that Lorentz symmetry breaking is a necessary feature of Planck scale physics, they did stimulate research aimed at understanding the possible measurable consequences of LV [28, 29, 30]. Furthermore, recent investigations strongly suggested that an high energy breakdown of Lorentz invariance might strongly improve the renormalizability of field theories [31, 32] including gravitation [33].

Therefore, the option of introducing Lorentz symmetry violation deserves at least some consideration. However, there is apparently also a (conceptually high) price to pay: a step back from Minkowski spacetime to the notions of absolute space and time. Moreover, and most importantly, there is the issue of recovering a low energy approximate Lorentz invariance. We shall come back later on this point.

The second option is probably the most viable, conceptually appealing, but most demanding in terms of new concepts to be introduced. If no reference is made to a background Minkowski spacetime, but rather the graviton emerges in the same limit in which the manifold emerges, then there is no obvious conflict with the Weinberg-Witten theorem. Simply, what is called the gauge symmetry in terms of fields living of spacetime is the manifestation of an underlying symmetry acting on the fundamental degrees of freedom in the limit when they are reorganized in terms of a spacetime manifold and fields (gauge fields and gravitons in particular).

There are already two examples of this possibility, namely matrix models and quantum graphity models. In both cases, the very notion of spacetime manifold is immaterial for the foundations of the theory. The manifold and the metric are derived concepts, obtained in precise dynamical regimes of the theory. The interested reader can find additional comments and references in [34, 35].

3. A concrete example: emergent Lorentzian signature and Nordström gravity

To give support to the ideas presented so far, let us discuss some results concerning a toy model which we have used to get some further insight. We are not going to reproduce the full calculations, which can be found in [36]. Rather, we will discuss the main outcomes.

Journal of Physics: Conference Series **222** (2010) 012050 doi:10.1088/1742-6596/222/1/012050

Initially, we have considered fields ψ_i that live in a Euclidean space, and showed that there exists a class of Lagrangians (essentially purely kinetic K-essence ones) such that the perturbations φ_i around some classical solutions $\bar{\psi}$ propagate in a Minkowski spacetime. In this case $\bar{\psi}$ is essentially picking up a preferred direction, so that we have a spontaneous symmetry breaking of the Euclidean symmetry. The apparent change of signature is free of the problems usually met in signature change frameworks since the theory is fundamentally Euclidean. Lorentz symmetry is only approximate, and in this sense it is *emergent*.

The main lesson we want to emphasize here is that *Lorentzian signature can emerge from a fundamental Euclidean theory* and this process can in principle be reconstructed by observers living in the emergent system. In fact, while from the perturbations point of view it is a priori difficult to see the fundamental Euclidean nature of the world, this could be guessed from the fact that some Lorentz symmetry breaking would appear at high energy (in our case in the form of a non-dynamical ether field). In this sense, we have a toy-model for the emergence of the Poincaré symmetries. This construction can be seen as a generalization of the typical situation in analogue models of gravity [9] where one has Poincaré symmetries emerging from fundamental Galilean symmetries [9]. However, let us stress that in our case no preferred system of reference is present in the underling field theory given that the fundamental Lagrangian is endowed with a full Euclidean group $ISO(4)$.

In identifying an emergent gravitational dynamics our starting point is then the first order truncated Lagrangian for the perturbations φ_i. This comes out to be of the simple form

$$\mathcal{L}_{\text{eff}}(\varphi_1, ...\varphi_N) = \sum_i \eta^{\mu\nu}\partial_\mu\varphi_i\partial_\nu\varphi_i. \tag{14}$$

which can simply be rewritten in terms of the (real) multiplet $\varphi = (\varphi_1, ..., \varphi_N)$ as

$$\mathcal{L}_{\text{eff}}(\varphi) = \eta^{\mu\nu}(\partial_\mu\varphi)^T(\partial_\nu\varphi). \tag{15}$$

This system has a global $O(N)$ symmetry which has emerged as well from the initial Lagrangian [36]. It is hence quite natural to rewrite the multiplet φ by introducing an amplitude characterized by a scalar field $\Phi(x)$ and a multiplet $\phi(x)$ with N components such that[2]

$$\begin{pmatrix} \varphi_1 \\ \vdots \\ \varphi_N \end{pmatrix} = \Phi \begin{pmatrix} \phi_1 \\ \vdots \\ \phi_N \end{pmatrix}, \qquad \text{with } |\phi|^2 \equiv \sum_i \phi_i^2 = \ell^2. \tag{16}$$

ℓ is an arbitrary length parameter to keep the dimension right. In particular, Φ is dimensionless and ϕ has the dimension of a length. Φ is the field invariant under $O(N)$ transformations, whereas ϕ does transform under $O(N)$. As we shall see, this field redefinition will provide us the means to identify gravity and matter degrees of freedom. The Lagrangian for the perturbations (15) reads now as[3]

$$\mathcal{L}_{\text{eff}}(\varphi_1, ...\varphi_N) \to \mathcal{L}_{\text{eff}}(\Phi, \phi_1, ...\phi_N) = \ell^2\eta^{\mu\nu}\partial_\mu\Phi\partial_\nu\Phi + \sum_i \Phi^2\eta^{\mu\nu}\partial_\mu\phi_i\partial_\nu\phi_i + \lambda(|\phi|^2 - \ell^2), \tag{17}$$

where λ is a Lagrange multiplier. We recognize in particular the action for a non-linear sigma model given in terms of the fields ϕ_i. The associated equations of motion are

$$\eta^{\mu\nu}(\ell^2\partial_\mu\partial_\nu\Phi - \Phi\sum_i \partial_\mu\phi_i\partial_\nu\phi_i) = 0, \tag{18}$$

$$\eta^{\mu\nu}(2\partial_\mu\Phi\partial_\nu\phi_i + \Phi^2\partial_\mu\partial_\nu\phi_i + \frac{1}{\ell^2}\partial_\mu\phi_j\partial_\nu\phi_k\delta^{jk}\phi_i) = 0, \tag{19}$$

$$|\phi|^2 - \ell^2 = 0. \tag{20}$$

[2] Our field redefinition is the generalization of the so-called Madelung representation [9].
[3] We use the normalization condition $|\phi|^2 = \ell^2$, which implies in particular $\sum_i \phi_i\partial_\mu\phi_i = 0$.

If we introduce the (conformally flat) metric

$$g_{\mu\nu}(x) = \Phi^2(x)\eta_{\mu\nu}, \tag{21}$$

the equations of motion (19) can be simply rewritten as

$$(\sqrt{-g})^{-1}\partial_\mu(\sqrt{-g}g^{\mu\nu}\partial_\nu\phi_i) + \frac{1}{\ell^2}g^{\mu\nu}\partial_\mu\phi_j\partial_\nu\phi_k\delta^{jk}\phi_i = \tag{22}$$

$$\Box_g\phi_i + \frac{1}{\ell^2}g^{\mu\nu}\partial_\mu\phi_j\partial_\nu\phi_k\delta^{jk}\phi_i = 0, \tag{23}$$

where we have introduced the d'Alembertian $\Box_g$ for the metric g and used that $\sqrt{-g} = \Phi^4$ and $g^{\mu\nu} = \Phi^{-2}\eta^{\mu\nu}$. Notice that equation (19) can be rewritten in the form (23) using the metric redefinition (21) only in four dimensions. To be consistent, the change of variable $\Phi \to g_{\mu\nu}$ should be completed with the constraint that $g_{\mu\nu}$ is conformally flat, that is

$$C_{\alpha\beta\gamma\delta}(g) = 0, \tag{24}$$

where $C_{\alpha\beta\gamma\delta}$ is the Weyl tensor.

Eq. (23) suggests that *the gravitational degree of freedom should be encoded in the scalar field* Φ, whereas *matter should be encoded in the* ϕ_i. We are therefore aiming at a scalar theory of gravity with actions:

$$S_{\text{eff}} = \int dx^4\sqrt{-\eta}\,\mathcal{L}_{\text{eff}} = S_{\text{grav}} + S_{\text{matter}}, \tag{25}$$

$$S_{\text{grav}} = \ell^2 \int dx^4\sqrt{-\eta}\,\eta^{\mu\nu}\partial_\mu\Phi\partial_\nu\Phi, \tag{26}$$

$$S_{\text{matter}} = \int dx^4\sqrt{-\eta}\left(\sum_i \Phi^2\eta^{\mu\nu}\partial_\mu\phi_i\partial_\nu\phi_i + \lambda(|\phi|^2 - \ell^2)\right), \tag{27}$$

where we have explicitly written the volume element $\sqrt{-\eta} = 1$ so to make clear that these actions are given in flat spacetime.

It is easy to see that the very same actions can be recast in the form of actions in a curved spacetime endowed with the metric (21). In particular for the matter action in (27) one has

$$S_{\text{matter}} = \int dx^4\left(\sum_i \Phi^2\eta^{\mu\nu}\partial_\mu\phi_i\partial_\nu\phi_i + \lambda(|\phi|^2 - \ell^2)\right) = \tag{28}$$

$$\int \sqrt{-g}dx^4\left(\sum_i g^{\mu\nu}\partial_\mu\phi_i\partial_\nu\phi_i + \lambda'(|\phi|^2 - \ell^2)\right), \tag{29}$$

where we have suitably rescaled the Lagrange multiplier to λ'. This allows to construct the stress-energy tensor $T_{\mu\nu}$ for the non-linear sigma model, and its trace $\mathbf{T}$ with respect to the metric g:

$$T_{\mu\nu} = \frac{2}{\sqrt{-g}}\frac{\delta S_{\text{matter}}}{\delta g^{\mu\nu}} = \sum_i\left(\partial_\mu\phi_i\partial_\nu\phi_i - \frac{1}{2}g_{\mu\nu}(g^{\alpha\beta}\partial_\alpha\phi_i\partial_\beta\phi_i)\right), \tag{30}$$

$$\mathbf{T} = g^{\mu\nu}T_{\mu\nu} = -\Phi^{-2}\sum_i \eta^{\mu\nu}\partial_\mu\phi_i\partial_\nu\phi_i. \tag{31}$$

Finally, the above result, together with the recognition that the Ricci scalar $\mathbf{R}$, associated to the metric $g_{\mu\nu}$, can be written as $\mathbf{R} = -6\Box_\eta\Phi/\Phi^3$, allows us to rewrite Eq. (18) as the Einstein–Fokker equation

$$\Box_\eta\Phi = \frac{1}{\ell^2}\eta^{\mu\nu}\Phi\sum_i \partial_\mu\phi_i\partial_\nu\phi_i \quad \Leftrightarrow \quad \mathbf{R} = \frac{6}{\ell^2}\mathbf{T}. \tag{32}$$

In summary, we can gather together the equations of motion (23, 24, 32), obtained by introducing the metric (21), we have

$$\mathbf{R} = \frac{6}{\ell^2}\mathbf{T}, \qquad C_{\alpha\beta\gamma\delta} = 0. \tag{33}$$

$$\Box_g\phi_i + \frac{1}{\ell^2}g^{\mu\nu}\partial_\mu\phi_j\partial_\nu\phi_k\delta^{jk}\phi_i = 0, \qquad |\phi|^2 - \ell^2 = 0. \tag{34}$$

We recognize the equations of motion as those for Nordström gravity

$$\mathbf{R} = 24\pi G_\mathrm{N}\,\mathbf{T}, \qquad C_{\alpha\beta\gamma\delta} = 0, \tag{35}$$

coupled to a non-linear sigma model. Indeed, the rewriting of (18)-(20) into the form (33)-(34), is a special case of the procedure suggested by Einstein and Fokker so to cast Nordström gravity in a geometrical form [37].

We see from the above equation that the Newton constant G_N in our model has to be proportional to ℓ^{-2}. However, in identifying the exact relation between the two quantities, some care has to be given to the fact that the stress-energy tensors appearing respectively in equation (33) and equation (35) do not share the same dimensions. This is due to the fact that the fields ϕ_i have the dimension of a length rather than the usual one of an energy. This implies that in order to really compare the expressions one has to suitably rescale our fields with a dimensional factor, Ξ, which in the end would combine with ℓ so to produce an energy, $\dim[\ell\,\Xi] =$ energy. In particular, is easy to check that one has to assume $4\pi\ell^2\Xi^2 \equiv E^2_\mathrm{Planck}$ in order to recover the standard value of G_N (assuming c as the observed speed of signals and $\hbar$ as the quantum of action). As a final remark, we should stress that the scale ℓ is completely arbitrary within the emergent system and in principle should be derived from the physics of the "atoms of spacetime" whose large N limit gives rise to the initial fundamental Lagrangian (the one for the initial fields ψ_i).

Accidentally, the above discussion also shows that, once the fields are suitably rescaled so to have the right dimensions, the constraint appearing in Eq. (34) is fixing the norm of the multiplet to be equal to the square of the Planck energy. This implies that the interaction terms in the aforementioned equation are indeed Planck-suppressed and hence negligible at low energy. This should not be a surprise, given that in the end $\ell\Xi$ is the only energy scale present in our model. It is conceivable that more complicate frameworks, possibly endowed with many dimensional constants, will introduce a hierarchy of energy scales and hence break the degeneracy between the scale of gravity and the scale of matter interactions.

In the second part of the discussion of this toy model, using a natural field redefinition adapted to the symmetries of the system, we have identified from the perturbations φ_i, a scalar field Φ encoding gravitational degrees of freedom and a set of scalar fields ϕ_i (a non-linear sigma model) encoding matter fields. In this sense, gravity and matter are both emergent at the same level. This approach is then rather different from the one of analogue models of gravity where one usually identifies the analogue of the gravitational degrees of freedom with the "background" fields, *i.e.* the condensate or the solution ψ of the equations of motion. Indeed, following this line of thought in looking for a theory of gravitational dynamics, we would be led to require that the fundamental field theory must be endowed with diffeomorphisms invariance from the very start — the symmetries of the background are identical by construction to the ones of the fundamental theory. This would imply that one would have to obtain gravity from a theory which is already diffeomorphisms invariant and hence most probably with a form very close to some known theory of gravitation.

For these reasons, we do expect that if an emergent picture is indeed appropriate for gravitation, then it should be of the sort presented here, with both matter and gravity emerging at the same level. Of course, it is not possible to exclude that a full fledged theory of gravity could emerge, together with the notion of manifold, in a single step from the eventual semiclassical/large number limit of the fundamental objects. In this case, however, we would still have a very different picture from the one envisaged in analogue models of gravity.

In particular, this allows not only for an emergent local Lorentz invariance for the perturbations dynamics but it leads as well to an emergent diffeomorphisms invariance. In fact, the (lowest order) equations of motion for the perturbations can be naturally rewritten in an evidently diffeomorphisms invariant form, from the point of view of "matter fields observers". In fact, following the standard hole argument (see [38] for a careful discussion of the various issues related to diffeomorphism invariance), this also implies that the coordinates x^μ, used to parameterized our theory, do not have any physical meaning from the point of view of the ϕ_i "matter observers". They are merely parameters. In agreement with the fact that diffeomorphisms invariance is emergent in our system, it can be noted that the higher order contributions contributions ends up breaking it at the same level it breaks Lorentz invariance.

Furthermore, Nordström gravity is also a nice framework for discussing the subtle distinction between background independence and diffeomorphisms invariance [39]. We call background some geometrical degrees of freedom that are not dynamical. For example, in GR the topology of the manifold and its dimension, or the signature of the metric, can be considered as (trivial) background quantities. We can therefore have some specific background structures while still having diffeomorphisms invariance. Nordström gravity is encoded in conformally flat metrics. If one considers fields which are conformally coupled to the metric (such as the electromagnetic field), these fields only see the metric $\eta_{\mu\nu}$ which is of course not dynamical. The Minkowski metric can be see then as a background structure, this is what one may call a "prior geometry" (*e.g.* see [40]). One may hence say that diffeomorphism invariance is somewhat of a weaker form in Nordström gravity with respect the one present in GR.

In particular, while the essence of diffeomorphism invariance in GR is encoded in the associated Hamiltonian constraints, these are not defined in the present formulation of Nordström gravity. Furthermore, in the most general implementations of Norström theory, quantities can be built which manifestly include the background structure $\eta_{\mu\nu}$ and hence are not diffeomorphism invariant. However, within our model, the prior geometry cannot be detected. Indeed, in order to detect the Minkowski background, one should be able to propose a method to pinpoint the conformal factor Φ^2 in the relation $g_{\mu\nu} = \Phi^2\eta_{\mu\nu}$. However, a careful analysis shows that this is actually impossible. Let us elaborate on this point. If we perform a conformal transformation, $x^\mu \to \bar{x}^\mu(x)$, the (lowest order) equations of motions for the perturbations (in their original parametrization) are transforming like

$$\Box_\eta \varphi_i = 0 \to \Box_{\bar{\eta}} \varphi_i = 0, \tag{36}$$

where η and $\bar{\eta}$ are two different Minkowski metrics related by some conformal factor $\lambda(x)$. Therefore, η and $\bar{\eta}$ are indistinguishable, due to conformal invariance the equations of motion for φ_i. Hence, what appears to be a background structure, namely $\eta_{\mu\nu}$, is ambiguously defined, and the coordinates x^μ in which the equations of motion for the fields φ_i are written have no operational meaning, they are mere labels. Furthermore, this ambiguity in the definition of what would be called a background structure implies an ambiguity on the definition of the conformal factor relating the physical metric to the would-be background structure. In this sense, within this very specific implementation of the model which has conformal invariance, there is no Minkowski geometry as a background. There is a background structure, which is the conformal structure of Minkowski spacetime. This is a mild limitation of our simple toy model as a diffeomorphism invariant, background independent system.

Of course, the above discussion holds only at the lowest order in the fields φ_i. As previously discussed, higher orders in perturbation theory will generate terms leading to a breaking of the conformal symmetry and hence the appearance of the background structures, *i.e.* the Euclidean space and the $\partial_\mu\bar{\psi}$ which have selected the timelike direction.

Finally, it is interesting to discuss in details the features that allowed the construction of such a toy model. In particular, it is important to stress the role of symmetries, in order to make clear the way in which they enter at the various levels. As in the case of selecting Riemannian geometry out of Finsler geometry (see e.g. [41] for a pedagogical introduction), here there are some symmetries which are absolutely essential: it is only due to their presence that we do have an emergent gravitational system possessing a geometrical nature.

We have seen that in order to produce a working model, a number of properties must be assumed. First of all, there is an underlying $ISO(4)$ symmetry which allows us to use particularly simple affine solutions. This $ISO(4)$, when spontaneously broken, can lead to an approximate Poincaré invariance. Moreover, the masslessness of the resulting modes is promoting this Poincaré invariance to a full conformal invariance, which is approximate as well. This conformal invariance is the key symmetry which hides the background structure, forbidding a low energy observer to detect a background metric structure (there is only a background conformal structure).

Conformal invariance seems to be deeply intertwined with the possibility of writing down the resulting equations of motion in the form of a system of diffeomorphism invariant equations, as we have seen. However, in order for the Lagrangian (14) to be conformal invariant, there must be an overall $O(N)$ symmetry between the fields. This symmetry is just the other side of the coin of the mechanism leading to the monometricity. If two fields move in different metrics, clearly this $O(N)$ is broken and the entire model fails to provide a geometric picture, let alone a diffeo-invariant one.

In general, one should expect that in any situation in which the metric is an emergent structure, there should be a mechanism taking care of the fact that different matter fields propagate over the same geometry. In this picture, where a manifold is given from the beginning, the role of internal and spacetime symmetries is crucial. The behavior we have described is not general at all. Of course, one could conclude that this kind of models is somehow contrived and unnatural.

However, there is also a positive side: given that symmetries (both of the equations of motion and of the ground state) play a crucial role in the emergence mechanism, the fact that our universe seems to be ruled, at large scales, by GR and locally by special relativity, suggests that not all the pre-geometric scenarios are viable, and that there are rather strong constraints on what are the possible mechanism of emergence. In particular we want to conclude this analysis with a discussion about the constraints related to the breakdown of Lorentz invariance.

4. The naturalness problem

Should we take these results as a strong hint that Lorentz symmetry breaking should be a part of any working emergent gravity scenario? It is at this stage unclear if we can be that bold. Surely one open issue is the naturalness of theories endowed with Lorentz symmetry breaking. In general, radiative corrections lead to a dangerous "percolation" in the infrared regimes of the Lorentz breaking [42, 43, 44], something strongly constrained by current observations [28, 29, 30].

More precisely, it was found that it is generic that even starting with an effective field theory with only Lorentz breaking operators in the Lagrangian of mass dimension 5 and 6 for free particles, radiative corrections due to particle interactions will generate lower-dimension LV

Journal of Physics: Conference Series **222** (2010) 012050 doi:10.1088/1742-6596/222/1/012050

terms that will then become dominant [42], as their dimensionless coefficients are of the same order as the higher dimension ones ($O(1)$, given our previous assumption, see [45]). Thus, either a symmetry (or some other mechanism) protects the lower dimension operators from large LV, or the suppression of the non-renormalizable operators indeed will always be greater than that of the renormalizable ones.

SuperSymmetry (SUSY) was proposed as a possible candidate for a custodial symmetry doing this job [46, 47]. SUSY is closely related to Poincaré invariance: the composition of two SUSY generators is proportional to the momenta, the generators of space-time translations. However, the idea is that SUSY could still be an exact symmetry even in the presence of LV and it can serve as a custodial symmetry, preventing certain operators from appearing in LV field theories.

The effect of SUSY on LV is to prevent dimension ≤ 4, renormalizable LV operators to be present in the Lagrangian. Moreover, it has been demonstrated [46, 47] that the renormalization group equations for Supersymmetric QED plus the addition of dimension 5 LV operators à la Myers & Pospelov do not generate lower dimensional operators, if SUSY is unbroken. However, this is not the case for our low energy world, of which SUSY is definitely not a symmetry. The effect of soft SUSY breaking was again investigated in [46, 47]. As expected, it was found that, when SUSY is broken, the renormalizable operators appear in the Lagrangian. In particular, dimension κ operators arise from the percolation of dimension $\kappa + 2$ LV operators[4]. The effect of SUSY soft-breaking is, however, to introduce a suppression of order m_s^2/M_{pl} ($\kappa = 3$) or $(m_s/M_{pl})^2$ ($\kappa = 4$), where m_s is the scale of soft SUSY breaking. Given the present constraints, the theory in which $\kappa = 3$ must be fine-tuned to be viable, because the SUSY-breaking-induced suppression is not powerful enough to eliminate linear modifications in the dispersion relation of electrons. However, if $\kappa = 4$, the induced dimension 4 terms are sufficiently suppressed, provided that $m_s < 100$ TeV.

Let us stress, however, that all this studies have implemented SUSY in the standard (Lorentz invariant) form and it is not clear if and how a SUSSY group can be found in a Lorentz breaking context. In this sense the naturalness problem of Lorentz breaking theories is probably the most important issues to be solved not only for the quantum gravity phenomenology field but also for the whole emergent gravity proposal. We hope that studies like the one reported here will help in putting this issue under the spot of future research in the field.

5. Conclusions

In summary, we hope that the two examples of emergent gravitational dynamics presented here have suitably illustrated the potentialities of emergent gravity models inspired by the analogue gravity perspective. The first case, the one of a BEC system, has shown us that a gravitational-like dynamics (with a small cosmological constant) seems to be a natural by product of a condensation mechanism. However, the analogy with the real world was not only limited by the Lorentz breaking scale but also from the fact that gravity and matter seem to be living at rather different levels (gravity is the condensate while matter is associated to the quasi-particle states).

There are important conceptual and practical obstructions which forbid the BEC system to be used to discuss analogue of the gravitational field. In this sense, it is an attempt doomed to fail from the very beginning. First of all, the model is non-relativistic. Despite some encouraging results towards the extension to a relativistic theory made in [36], it is still not clear how to evade the Weinberg–Witten theorem [16, 17, 19] in order to produce a viable model of emergent spin-2 gravity. Nevertheless, the brief discussion of the BEC-based toy model offers the possibility of showing very nice properties which could be of help in understanding some of the puzzles we find in the study of the gravitational field.

[4] We consider only $\kappa = 3, 4$, for which these relationships have been demonstrated.

For instance, the naturalness of the small cosmological constant term present in a BEC. As we have seen, the cosmological constant term is proportional to the depletion factor, *i.e.* by the ratio between the number of particles on excited states and the number of particles in the ground state (the condensate). Whenever this ratio is small, the cosmological constant is similarly suppressed.

The toy model discussed in the second part of this proceedings is aimed at overcoming some of the problems encountered in the case of BECs. The model shows Lorentz violation but only in the limit of large fluctuations of the fundamental fields φ_i. This implies that the gravitational dynamics is no more endowed with a massive graviton whose mass scale is set by the UV Lorentz breaking scale of the system. Furthermore, this toy models shows how time and diffeomorphism invariance might emerge. In particular the latter is allowed by the special symmetries of the system. As we have said previously, similar symmetries are probably needed anyway to protect the IR limit of the theory from large violations of Lorentz invariance of the equations. Given that the presence of such violations seems to be a very natural way around to the Weinberg-Witten theorem obstruction, it might be that the next step towards a satisfactory emergent gravity scenario might have to consist in finding which sort of mechanism, possibly a custodial symmetry, could simultaneously guarantee the background independence of the emerging dynamics as well as a very accurate Local Lorentz invariance of the emergent spacetime. We hope to address these questions in future work.

6. Acknowledgments

[1] B. L. Hu, J. Phys. Conf. Ser. **174**, 012015 (2009) [arXiv:0903.0878 [gr-qc]].
[2] J. M. Bardeen, B. Carter and S. W. Hawking, Commun. Math. Phys. **31**, 161 (1973).
[3] T. Jacobson, Phys. Rev. Lett. **75**, 1260 (1995) [arXiv:gr-qc/9504004];
[4] C. Eling, R. Guedens and T. Jacobson, Phys. Rev. Lett. **96**, 121301 (2006) [arXiv:gr-qc/0602001].
[5] G. Chirco and S. Liberati, arXiv:0909.4194 [gr-qc].
[6] T. Padmanabhan, arXiv:0911.5004 [gr-qc].
[7] M. W. Choptuik, Phys. Rev. Lett. **70**, 9 (1993).
[8] C. Gundlach and J. M. Martin-Garcia, Living Rev. Rel. **10** 5 (2007) [arXiv:0711.4620 [gr-qc]].
[9] C. Barceló, S. Liberati and M. Visser, Living Rev. Rel. **8**, 12 (2005) [arXiv:gr-qc/0505065].
[10] F. Girelli, S. Liberati and L. Sindoni, Phys. Rev. D **78**, 084013 (2008) [arXiv:0807.4910 [gr-qc]].
[11] C. J. Pethick and H. Smith, *Bose-Einstein Condensation in Dilute Gases,* Ed. Cambridge University Press, Cambridge, U.K. 2002
[12] S. Liberati, M. Visser and S. Weinfurtner, Class. Quant. Grav. **23**, 3129 (2006) [arXiv:gr-qc/0510125]; Phys. Rev. Lett. **96**, 151301 (2006) [arXiv:gr-qc/0512139].
[13] S. Weinfurtner, S. Liberati and M. Visser, Lect. Notes Phys. **718**, 115 (2007) [arXiv:gr-qc/0605121].
[14] J. Bernstein and S. Dodelson. Phys. Rev. Lett. **66**, 683 (1991).
[15] E. Witkowska, P. Zin and M. Gajda, Phys. Rev. D **79** 025003 (2009) [arXiv:0812.0260 [hep-th]].
[16] S. Weinberg and E. Witten, Phys. Lett. B **96**, 59 (1980).
[17] F. Loebbert, Annalen Phys. **17**, 803 (2008).
[18] A. Jenkins, *Topics in particle physics and cosmology beyond the standard model,* PhD thesis, arXiv:hep-th/0607239.
[19] A. Jenkins, e-print: arXiv:0904.0453 [gr-qc].
[20] V. A. Kostelecky and S. Samuel, Phys. Rev. D **39**, 683 (1989).
[21] G. Amelino-Camelia, J. R. Ellis, N. E. Mavromatos and D. V. Nanopoulos, Int. J. Mod. Phys. A **12**, 607 (1997) [arXiv:hep-th/9605211];
G. Amelino-Camelia, J. R. Ellis, N. E. Mavromatos, D. V. Nanopoulos and S. Sarkar, Nature **393**, 763 (1998).
[22] R. Gambini and J. Pullin, Phys. Rev. D **59**, 124021 (1999). [arXiv:gr-qc/9809038]
[23] S. M. Carroll, J. A. Harvey, V. A. Kostelecky, C. D. Lane and T. Okamoto, Phys. Rev. Lett. **87**, 141601 (2001). [arXiv:hep-th/0105082].
[24] J. Lukierski, H. Ruegg and W. J. Zakrzewski, Annals Phys. **243** (1995) 90 [arXiv:hep-th/9312153].
[25] G. Amelino-Camelia and S. Majid, Int. J. Mod. Phys. A **15** (2000) 4301 [arXiv:hep-th/9907110].
[26] C. P. Burgess, J. Cline, E. Filotas, J. Matias and G. D. Moore, JHEP **0203**, 043 (2002). [arXiv:hep-ph/0201082].

[27] *Artificial Black Holes*, Ed. M. Novello, M. Visser and G. Volovik, World Scientific, Singapore;
G. E. Volovik, Phys. Rept. **351**, 195 (2001) [arXiv:gr-qc/0005091];
The Universe in a Helium Droplet (Oxford University Press, 2003);
C. Barcelo, S. Liberati and M. Visser, Living Rev. Rel. **8**, 12 (2005). [arXiv:gr-qc/0505065].
[28] D. Mattingly, Living Rev. Rel. **8**, 5 (2005) [arXiv:gr-qc/0502097].
[29] G. Amelino-Camelia, arXiv:0806.0339 [gr-qc].
[30] S. Liberati and L. Maccione, arXiv:0906.0681 [astro-ph.HE]. To appear in Annual Review of Particle Physics 2009.
[31] D. Anselmi and M. Halat, Phys. Rev. D **76**, 125011 (2007) [arXiv:0707.2480 [hep-th]].
[32] M. Visser, Phys. Rev. D **80**, 025011 (2009) [arXiv:0902.0590 [hep-th]].
[33] P. Horava, Phys. Rev. D **79**, 084008 (2009) [arXiv:0901.3775 [hep-th]];
Phys. Rev. Lett. **102**, 161301 (2009) [arXiv:0902.3657 [hep-th]].
[34] T. Konopka, arXiv:0805.2283 [hep-th]; T. Konopka, F. Markopoulou and S. Severini, Phys. Rev. D **77**, 104029 (2008) arXiv:0801.0861 [hep-th]; T. Konopka, F. Markopoulou and L. Smolin, arXiv:hep-th/0611197.
[35] H. Steinacker, J. Phys. Conf. Ser. **174** (2009) 012044 [arXiv:0903.1015 [hep-th]].
[36] F. Girelli, S. Liberati and L. Sindoni, Phys. Rev. D **79**, 044019 (2009) [arXiv:0806.4239 [gr-qc]].
[37] A. Einstein and A. D. Fokker, *Nordström's Theory of Gravitation from the Point of View of the Absolute Differential Calculus*, Annalen Phys. **44**, 321 (1914). Annalen Phys. **14**, 500 (2005).
[38] H. Westman and S. Sonego, arXiv:0711.2651 [gr-qc].
[39] D. Giulini, Lect. Notes Phys. **721** 105 (2007) [arXiv:gr-qc/0603087].
[40] C. W. Misner, K. S. Thorne and J. A. Wheeler, *Gravitation*, Freeman, San Francisco, 1973.
[41] L. Sindoni, "Emergent Gravity: The Analogue Models Perspective," SPIRES entry
[42] J. Collins, A. Perez, D. Sudarsky, L. Urrutia and H. Vucetich, Phys. Rev. Lett. **93**, 191301 (2004) [arXiv:gr-qc/0403053].
[43] T. Jacobson, S. Liberati and D. Mattingly, Annals Phys. **321**, 150 (2006) [arXiv:astro-ph/0505267].
[44] R. Iengo, J. G. Russo and M. Serone, arXiv:0906.3477 [hep-th].
[45] D. Mattingly, PoS **QG-PH**, 026 (2007). arXiv:0802.1561 [gr-qc].
[46] S. Groot Nibbelink and M. Pospelov, Phys. Rev. Lett. **94**, 081601 (2005) [arXiv:hep-ph/0404271].
[47] P. A. Bolokhov, S. G. Nibbelink and M. Pospelov, Phys. Rev. D **72**, 015013 (2005) [arXiv:hep-ph/0505029].

Journal of Physics: Conference Series **222** (2010) 012051 doi:10.1088/1742-6596/222/1/012051

Reduced Hamiltonian for intersecting shells and Hawking radiation

Pietro Menotti

Department of Physics, University of Pisa, Italy and INFN Sezione di Pisa, Italy

E-mail: menotti@df.unipi.it

Abstract. We consider the dynamics of one or more self gravitating shells of matter in a centrally symmetric gravitational field in the Painlevé family of gauges. We give the reduced hamiltonian for two intersecting shells, both massless and massive. Such a formulation is applied to the computation of the semiclassical action of two intersecting shells. The relation of the imaginary part of the space-part of the action to the computation of the Bogoliubov coefficients is revisited.

1. Introduction

In this paper we shall study the dynamics of one or more self-gravitating spherical shells of matter subject to a centrally symmetric gravitational field and the application of the ensuing formalism to the semiclassical treatment of Hawking radiation. Such field of research was started by the papers by Kraus and Wilczek [1, 2]. The main results we shall report here are the extension of the treatment to one massive shell of matter and also the extension to two or more massive shells of matter which in the time development can also intersect [3]. The main appeal of the approach of [1, 2] is that energy conservation is taken exactly into account and thus the back reaction effects can be computed.

We shall keep the formalism for more that one shell as close as possible to the original formalism of [1] which can be summarized in words that we shall adopt a Painlevé gauge.

The extraction of the reduced hamiltonian is often classified as a very complicated procedure. We shall show that by introducing a "generating function" the derivation can be drastically simplified and also by the same token it can be extended to deal not only with massive shells but also with a finite number of massive shells. The treatment of massless shells is just a particular case.

In Sec.2 we shall give the derivation of the reduced action in the case of one shell while proving the independence of the canonical momentum within the Painlevé class of gauges. In Sec.3 we shall discuss the equations of motion. In Sec.4 we work out the analytic properties of the conjugate momentum p_c which appears in the reduced action and in Sec.5 we shall briefly describe the extension of the treatment to more than one shell. An application of the results is given in Sec.6 giving a derivation within such a formalism of the Dray-'t Hooft- Redmount exchange relations. An important integrability relation is reported in Sec.7 which allows to compute the imaginary part of the space integral of the canonical momenta for two shells. In Sec.8 we revisit the role of the imaginary part of the canonical momentum in determining the

Bogoliubov coefficients and thus the most important features of Hawking radiation. In Sec.9 we give some concluding remarks.

2. The reduced action

As usual we write the metric for a spherically symmetric configuration in the ADM form

$$ds^2 = -N^2dt^2 + L^2(dr + N^r dt)^2 + R^2 d\Omega^2. \tag{1}$$

were following [1, 2, 4, 5] we shall choose the functions N, N^r, L, R as continuous functions of the coordinates. We shall work on a finite region of space time $(t_i, t_f) \times (r_0, r_m)$. On the two initial and final surfaces we give the intrinsic metric by specifying $R(r, t_i)$ and $L(r, t_i)$ and similarly $R(r, t_f)$ and $L(r, t_f)$.

The complete action in hamiltonian form, boundary terms included is [1, 2, 6, 7]

$$\begin{aligned} S = S_{shell} + \int_{t_i}^{t_f} dt \int_{r_0}^{r_m} dr(\pi_L \dot{L} + \pi_R \dot{R} - N\mathcal{H}_t - N^r\mathcal{H}_r) + \\ \int_{t_i}^{t_f} dt\,(-N^r\pi_L L + \frac{NRR'}{L})\Big|_{r_0}^{r_m} \end{aligned} \tag{2}$$

where

$$S_{shell} = \int_{t_i}^{t_f} dt\ \hat{p}\ \dot{\hat{r}}. \tag{3}$$

$\hat{r}$ is the shell position and $\hat{p}$ its canonical conjugate momentum. $\mathcal{H}_r$ and $\mathcal{H}_t$ are the constraints

$$\mathcal{H}_r = \pi_R R' - \pi'_L L - \hat{p}\ \delta(r - \hat{r}), \tag{4}$$

$$\mathcal{H}_t = \frac{RR''}{L} + \frac{R'^2}{2L} + \frac{L\pi_L^2}{2R^2} - \frac{RR'L'}{L^2} - \frac{\pi_L\pi_R}{R} - \frac{L}{2} + \sqrt{\hat{p}^2L^{-2} + m^2}\ \delta(r - \hat{r}). \tag{5}$$

Action (2) is immediately generalized to a finite number of shells. The shell action as given by eqs.(3,4,5) refers to a dust shell even though generalizations to more complicated equations of state have been considered [7, 8, 9]. The boundary terms are those given in the paper by Hawking and Hunter [6] and will play a very important role in the following. The function F [3]

$$F = RL\sqrt{\left(\frac{R'}{L}\right)^2 - 1 + \frac{2\mathcal{M}}{R}} + RR' \log\left(\frac{R'}{L} - \sqrt{\left(\frac{R'}{L}\right)^2 - 1 + \frac{2\mathcal{M}}{R}}\right) \tag{6}$$

has the remarkable property of generating the conjugate momenta as solutions of the constraints as follows

$$\pi_L = \frac{\delta F}{\delta L} = \frac{\partial F}{\partial L} \tag{7}$$

$$\pi_R = \frac{\delta F}{\delta R} = \frac{\partial F}{\partial R} - \frac{\partial}{\partial r}\frac{\partial F}{\partial R'} \tag{8}$$

where $\mathcal{M}$ is a mass which is constant in r except at the shell position $\hat{r}$ and that we shall denote by H for r above all the shell positions and by M for r below all the shell positions. We shall adopt a Painlevé gauge defined by $L = 1$ everywhere. With regard to the remaining freedom in the choice of the gauge we shall choose $R = r$ except for a deformation region near the shell positions. Such deformation is unavoidable because the constraints impose a discontinuity in the derivative $R'(r)$ at $r = \hat{r}$ as we shall see in the following. In the Painlevé gauges F becomes

$$F = RW(R, R', \mathcal{M}) + RR'(\mathcal{L}(R, R', \mathcal{M}) - \mathcal{B}(R, \mathcal{M})) \tag{9}$$

Journal of Physics: Conference Series **222** (2010) 012051 doi:10.1088/1742-6596/222/1/012051

where

$$W(R,R',\mathcal{M}) = \sqrt{R'^2 - 1 + \frac{2\mathcal{M}}{R}}; \quad \mathcal{L}(R,R',\mathcal{M}) = \log(R' - W(R,R',\mathcal{M})) \tag{10}$$

and

$$\mathcal{B}(R,\mathcal{M}) = \sqrt{\frac{2\mathcal{M}}{R}} + \log\left(1 - \sqrt{\frac{2\mathcal{M}}{R}}\right) \tag{11}$$

where we exploited the freedom of adding to (6) a total derivative, thus gaining for F the useful property of vanishing wherever $R' = 1$. For $L = 1$ eqs.(7,8) become

$$\pi_L = R\sqrt{R'^2 - 1 + \frac{2\mathcal{M}}{R}} \equiv RW; \quad \pi_R = \frac{[RR'' + R'^2 - 1 + \mathcal{M}/R]}{W}. \tag{12}$$

With regard to $R(r,t)$ one can choose several gauges, within the Painlevé family.

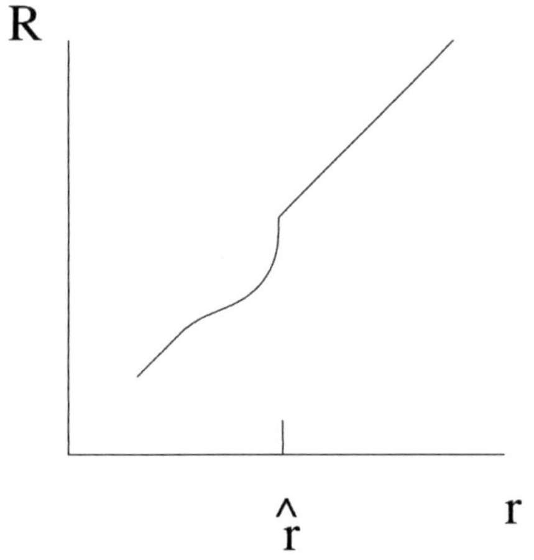

Figure 1. Outer gauge

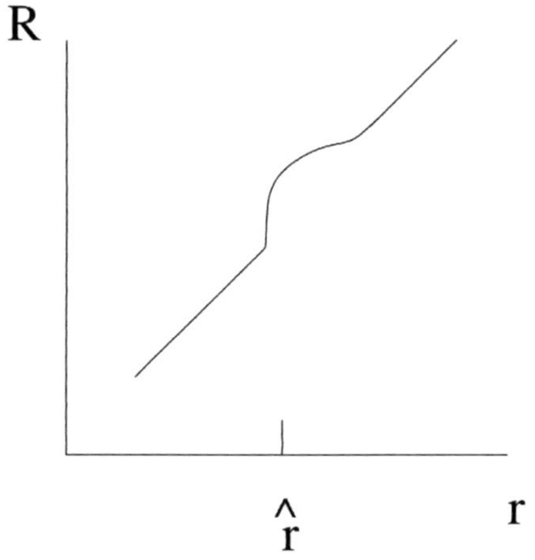

Figure 2. Inner gauge

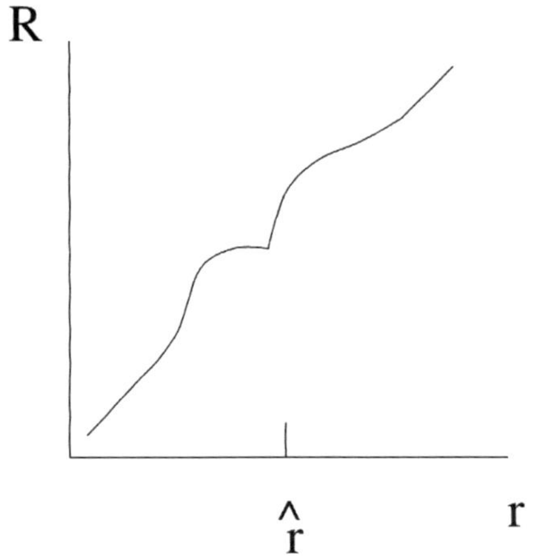

Figure 3. Generic gauge

Typical are the "outer gauge" characterized by $R(r,t) = r$ for $r \geq \hat{r}(t)$ and shown in Fig.1 and the inner gauge characterized by $R(r,t) = r$ for $r \leq \hat{r}(t)$ shown in Fig.2, but there are also more general choices as shown in Fig.3. We stress that such deformation g is not related to the thickness of the shell which is always zero.

Using the solutions (7,8) of the constraints one obtains in the outer gauge the following reduced [3] action where only the shell position $\hat{r}$ appears as degree of freedom

$$\int_{t_i}^{t_f}\left(p_c\,\dot{\hat{r}} - \dot{M}(t)\int_{r_0}^{\hat{r}(t)}\frac{\partial F}{\partial M}dr + (-N^r\pi_L + NRR')|_{r_0}^{r_m}\right)dt. \tag{13}$$

The p_c is easily computed

$$p_c = \hat{r}(\Delta\mathcal{L} - \Delta\mathcal{B}) \tag{14}$$

where $\Delta\mathcal{L} = \mathcal{L}(\hat{r}+\varepsilon) - \mathcal{L}(\hat{r}-\varepsilon)$ and similarly for $\Delta\mathcal{B}$. In deriving eq.(14) we used the consequences of the constraints (4,5)

$$\Delta R' = -\frac{V}{R}; \quad \text{where } V = \sqrt{\hat{p}^2 + m^2}; \qquad \Delta\pi_L = -\hat{p}. \tag{15}$$

In the outer gauge we find

$$p_c = \sqrt{2M\,\hat{r}} - \sqrt{2H\,\hat{r}} - \hat{r}\log\left(\frac{\hat{r} + \sqrt{\hat{p}^2 + m^2} - \hat{p} - \sqrt{2H\,\hat{r}}}{\hat{r} - \sqrt{2M\,\hat{r}}}\right) \tag{16}$$

with $\hat{p}$ given implicitly by

$$H - M = V + \frac{m^2}{2\hat{r}} - \hat{p}\sqrt{\frac{2H}{\hat{r}}}; \qquad V = \sqrt{\hat{p}^2 + m^2}. \tag{17}$$

This is the result obtained by Friedman, Louko, Winters-Hilt [4] through a limit procedure in which the support of the deformation function goes to zero but actually it is independent of the deformation. For working out the dynamics of one shell the limit procedure is all right, but in dealing with two or more shell it is important to have a deformation on a finite range because otherwise a limit procedure would give a result dependent on the order in which the two limits, deformation and $\hat{r}_1 \to \hat{r}_2$, are taken.

Similarly one can compute using the general formula the reduced canonical momentum of the system in the inner gauge

$$p_c^i = \sqrt{2M\,\hat{r}} - \sqrt{2H\,\hat{r}} - \hat{r}\log\left(\frac{\hat{r} - \sqrt{2H\,\hat{r}}}{\hat{r} - V + \hat{p} - \sqrt{2M\hat{r}}}\right) \tag{18}$$

and $\hat{p}$, again determined by the discontinuity equation (15), is given now by the implicit equation

$$H - M = V - \frac{m^2}{2\hat{r}} - \hat{p}\sqrt{\frac{2M}{\hat{r}}}; \qquad V = \sqrt{\hat{p}^2 + m^2}. \tag{19}$$

The two reduced canonical momenta p_c, p_c^i appear to be completely different but they can be proven to be the same function of $\hat{r}$. One can consider also the more general gauges depicted in Fig.3 and it can be proven [10] that p_c is always the same. However in inner gauge a term $\dot{H}(t)$ appears in the reduced action while in the more general gauges both a term $\dot{M}$ and a term $\dot{H}$ appear in the action.

The boundary term given in eq.(2) is equivalent to

$$-HN(r_m) + MN(r_0). \tag{20}$$

3. Equations of motion

In deriving the equation of motion from action (13) one can consider $M(t) = M$, the interior mass, as a datum of the problem and vary H to obtain

$$\dot{\hat{r}}\frac{\partial p_c}{\partial H} - N(r_m) = 0 \tag{21}$$

which using the expression of p_c and the relation between N and N^r imposed by the gravitational equations [3, 4] can be written as

$$\dot{\hat{r}} = \frac{\hat{p}}{V}N(\hat{r}) - N^r(\hat{r}) = \left(\frac{\hat{p}}{V} - \sqrt{\frac{2H}{\hat{r}}}\right)N(\hat{r}). \tag{22}$$

Alternatively one can consider H, the total energy as a datum of the problem and vary $M(t)$. The calculation is far more complicated due to the presence of $\dot{M}$ in the action (13), but using the equations for the gravitational field one reaches [3] the same equation of motion (22). In addition a consequence of the gravitational equations of motion is the constancy in time of $M(t)$ and $H(t)$.

4. Analytic properties of p_c

We saw that in the outer gauge p_c is given by (16,17). The solution of eq.(17) for $\hat{p}$ is

$$\frac{\hat{p}}{\hat{r}} = \frac{A\sqrt{\frac{2H}{\hat{r}}} \pm \sqrt{A^2 - (1 - \frac{2H}{\hat{r}})\frac{m^2}{\hat{r}^2}}}{1 - \frac{2H}{\hat{r}}} \tag{23}$$

where

$$A = \frac{H - M}{\hat{r}} - \frac{m^2}{2\hat{r}^2}. \tag{24}$$

If we want $\hat{p}$ to describe an outgoing shell we must choose the plus sign in front of the square root. Moreover the shell reaches $r = +\infty$ if and only if $H - M > m$ as expected.

The logarithm in p_c, eq.(16), has branch points at zero and infinity and thus we must investigate for which values or $\hat{r}$ such values are reached. At $\hat{r} = 2H$, $\hat{p}$ has a simple pole with positive residue. Thus the numerator in the argument of the logarithm in p_c goes to zero and below $2H$ it becomes

$$\hat{r} - V - \hat{p} - \sqrt{2H\hat{r}} \tag{25}$$

where here V is the absolute value of the square root. Expression (25) is negative irrespective of the sign of $\hat{p}$ and stays so for $\hat{r} < 2H$ because $\hat{p}$ is no longer singular. As a consequence p_c below $2H$ acquires the imaginary part $i\pi\hat{r}$. Below $\hat{r} = 2M$ the denominator of the argument of the logarithm in eq.(16) becomes negative so that the argument of the logarithm reverts to positive values. Thus the so called classically forbidden region where p_c becomes complex is $2M < \hat{r} < 2H$ independent of m and of the deformation g and the integral of the imaginary part of p_c for any deformation g and for any mass m of the shell is

$$\text{Im} \int p_c dr = \pi \int_{2M}^{2H} r dr = 2\pi(H^2 - M^2) = \frac{\Delta S}{2} \tag{26}$$

which is the original result derived in [1, 11] in the zero mass case. Parikh and Wilczek [11] gave to

$$\exp(-2\text{Im} \int p_c dr) \tag{27}$$

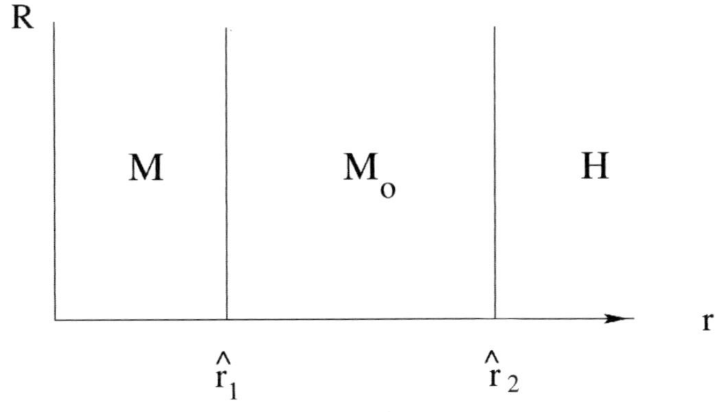

Figure 4. Two shell dynamics

the interpretation of the tunneling probability for the emission of a quantum of energy ω. Criticism and alternative proposals for the emission probability like

$$\exp(-\mathrm{Im}\oint p_c dr) \tag{28}$$

$$\exp\left(-2\mathrm{Im}(\int p_c dr + \text{temporal contribution})\right) \tag{29}$$

followed [12, 13, 14, 15, 16, 17, 18, 19, 20, 29, 30, 31]. Here instead we shall discuss the role of eq.(16) in the framework of mode analysis. This will be done in Sec.(8).

5. Two shell reduced action

Now we have three characteristic masses, M, H and an intermediate mass M_0, see Fig.4, which can change only if the two shells cross. Working as before with M = const considered as a datum of the problem we reach the reduced action [3]

$$\begin{aligned}&\dot{\hat{r}}_1 p_{c1} + \dot{\hat{r}}_2 p_{c2} + \dot{H}(R(\hat{r}_1) - \hat{r}_1)\frac{\partial T}{\partial H}\mathcal{D} + \dot{M}_0(R(\hat{r}_1) - \hat{r}_1)\frac{\partial T}{\partial M_0}\mathcal{D} + \\ &+\frac{d}{dt}\int_{r_0}^{\hat{r}_2} F dr - \dot{M}_0 \int_{\hat{r}_1}^{\hat{r}_2} \frac{\partial F}{\partial M_0} dr + (-N^r \pi_L + NRR')|_{r_0}^{r_m}\end{aligned} \tag{30}$$

where

$$T = \log\frac{V_2}{R(\hat{r}_2)}; \quad \mathcal{D} = R(\Delta\mathcal{L} - \Delta\mathcal{B})|_{\hat{r}_1}; \tag{31}$$

$$p_{c1} = R'(\hat{r}_1 + \varepsilon)\mathcal{D}; \quad p_{c2} = p_{c2}^0 + \frac{d}{d\hat{r}_2}(R(\hat{r}_1) - \hat{r}_1)\mathcal{D} \tag{32}$$

and p_{c2}^0 is given by eq.(16) with M replaced by M_0. The novelty is that now even in the outer gauge the time derivative of H intervenes in addition to $\dot{M}_0$ and p_{c1} and p_{c2} depend both on $\hat{r}_1$ and $\hat{r}_2$. We can vary $\hat{r}_1$, $\hat{r}_2$, H and M_0 independently obtaining the correct equations of motion [3]. As expected one finds that the exterior shell moves irrespective of the dynamics which develops at lower values of r until a crossing occurs.

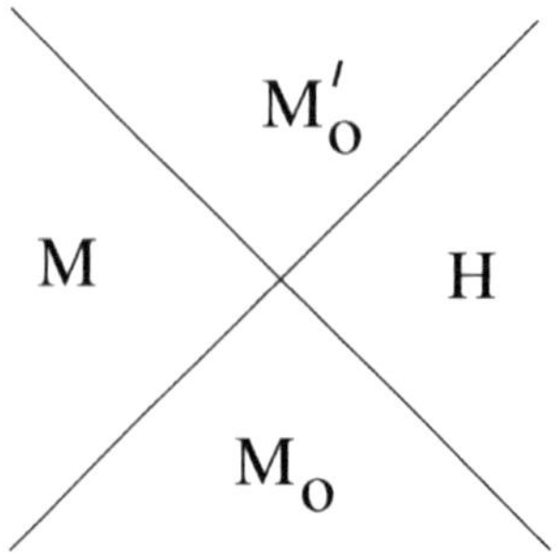

Figure 5. The exchange diagram

6. The exchange relations

In case of crossing of the two shells from the equations of Sec.(5) we can obtain relations between $M, M_0, H, \hat{r}_e$ and M'_0 being $\hat{r}_e$ the the shell position at the crossing and M'_0 the intermediate mass after the crossing. During the crossing the masses of the shells can change, provided they satisfy a relation analogous to the energy-momentum conservation is special relativity

$$\hat{p}_1 + \hat{p}_2 = \hat{p}'_1 + \hat{p}'_2 \tag{33}$$
$$V_1 + V_2 = V'_1 + V'_2; \quad V_n = \sqrt{\hat{p}_n^2 + m_n^2}; \quad V'_n = \sqrt{\hat{p'}_n^2 + m'^2_n}.$$

This is an outcome of the constraints. It is not possible to predict the final masses of the two shells as it depends on the details of the interaction which has to be specified. A relatively simple case is when the masses are unchanged during the crossing (transparent crossing) [3] and the simplest case is the crossing of two massless shells which remain massless, thus re-obtaining the well known Dray-'t Hooft-Redmount relations [21, 22] which we report here below

$$H\hat{r}_e + M\hat{r}_e - 2HM = M_0\hat{r}_e - 2M_0M'_0 + M'_0\hat{r}_e \tag{34}$$

being $\hat{r}_e$ the crossing radius.

7. Integrability of the form $p_{c1}d\hat{r}_1 + p_{c2}d\hat{r}_2$ and imaginary part of the space-component of the action

The space-part of the on-shell action for two massive shells of matter is given from eq.(30) by

$$\int_{t_i}^{t_f} (p_{c1}\ \dot{\hat{r}}_1 + p_{c2}\ \dot{\hat{r}}_2)\ dt \tag{35}$$

as on the equations of motion $\dot{H}(t) = \dot{M}(t) = 0$. It is possible to prove a theorem analogous to the one found in the books of Whittaker and Arnold [23, 24] i.e. that in presence of a constant of motion, in addition to the hamiltonian, the form $p_{c1}d\hat{r}_1 + p_{c2}d\hat{r}_2$ is closed even though the proof is somewhat different [3]. The intermediate mass M_0 plays the role of the additional constant of motion. The above result allows to deform the integration path as to bring $\hat{r}_1$ immediately below or above $\hat{r}_2$. This is allowed by the absence of discontinuities in our scheme. But, in words, two coaleshed shell have the same properties of a single shell with energy $H - M$. The final result is that

$$\text{Im} \int_{t_i}^{t_f} dt(p_{c1}\ \dot{\hat{r}}_1 + p_{c2}\ \dot{\hat{r}}_{c2}) = \text{Im} \int_{r_{1i},r_{2i}}^{r_{1f},r_{2f}} (p_{c1}\ d\hat{r}_1 + p_{c2}\ d\hat{r}_2) = 2\pi(H^2 - M^2). \tag{36}$$

The result (36) holds also for in the case when at the crossing the two shells can change their mass. For details we refer to [3].

8. Mode analysis

The original way to extract information on the spectrum of the radiation is mode analysis [26, 1, 27]. In the massless case

$$\int^{\hat{r}} p_c d\hat{r}' = f(\hat{r}, M) - f(\hat{r}, H) \tag{37}$$

can be computed exactly and its expansion to first order in $\omega = H - M$ is

$$\int^{\hat{r}} p_c d\hat{r}' = 4M\omega \log(\hat{r} - 2M) + \text{ regular terms.} \tag{38}$$

Given the semiclassical mode

$$\phi(r, t) = e^{i\int^{\hat{r}} p_c d\hat{r}' - i\omega t} \tag{39}$$

we can perform the analysis of the mode regular at the horizon in terms of the above mode. This analysis can be performed either by scalar product [25] i.e. space integration where one has to keep into account that the background metric is Painlevé or by time Fourier analysis. In this way one obtains the well known formulas for the Bogoliubov coefficients. The regular outgoing modes near the horizon are given by

$$\psi(\hat{r}, t) = e^{ik(\hat{r}-2M)e^{-\frac{t}{4M}}}. \tag{40}$$

Computing the scalar product, and taking into account that the background metric is the Painlevé metric, we have

$$-i\int (\psi^* \partial_\rho \phi - \phi \partial_\rho \psi^*) g^{\rho 0} \varepsilon_{0r\theta\phi} \sqrt{-g}\; d\hat{r} d\theta d\phi \tag{41}$$

with

$$g^{rt} = N^r = \sqrt{\frac{2M}{\hat{r}}} \quad \text{and} \quad \sqrt{-g} = 1 \tag{42}$$

and the integration region is outside the horizon. Keeping only the most singular terms at the horizon eq.(41) reduces, with $\tau = \exp(-t/4M)$, to

$$\int_0^\infty e^{-ikx\tau} e^{4iM\omega \log x - i\omega t} \frac{dx}{x}. \tag{43}$$

We can compute such scalar products at $t = 0$ obtaining the standard Hawking integral

$$\int_0^\infty e^{-ikx} e^{4iM\omega \log x} \frac{dx}{x} = e^{2\pi\omega M} (k)^{-4i\omega M} \Gamma(4i\omega M) = \text{const } \alpha^*_{k\omega} \tag{44}$$

which gives the dominant contribution for large k. The coefficient $\beta_{k\omega}$ is obtained by changing in (44) ω into $-\omega$. Alternatively we can extract the Bogoliubov coefficients by performing a time Fourier transform i.e.

$$\int_{-\infty}^{+\infty} e^{-ikx\tau} e^{4iM\omega \log x - i\omega t} dt = \int_0^{+\infty} e^{-ikx\tau} e^{4iM\omega \log(x\tau)} \frac{d\tau}{\tau} \tag{45}$$

which is, as it should be, a result independent of r (i.e. x) and reproduces eq.(44). The derivation is valid to first order in ω which is the realm of validity of the external field approximation. For

finite ω it is problematic to perform a space integration on modes, because we have not a well defined background metric.
Thus Kraus and Wilczek [1] followed the time Fourier analysis method. In order to do so one has to construct the non-perturbative modes regular at the horizon. This is not a completely trivial task. The regular modes are given by

$$e^{iS} = e^{ik\hat{r}(0)+i\int_{\hat{r}(0)}^{r} p_c d\hat{r} - i(H-M)t} \tag{46}$$

where S is the on shell action computed with the following boundary conditions: 1) At time t the shell position is $\hat{r}$, outside the horizon. 2) At time $t=0$ the conjugate momentum is k, i.e. $S(0,\hat{r}) = k\hat{r}$. Due to these boundary conditions $\hat{r}(0)$, H and as a consequence p_c depend on $k, t, \hat{r}$ even if along each trajectory H is always a constant of motion. Using the saddle point approximation [1, 27] one obtains that the absolute value of the Bogoliubov coefficient $\alpha_{k\omega}$ is given by

$$\left| e^{i\int_{r(0)}^{r} p_c(r',H,M)dr'} \right| = e^{-\mathrm{Im}\int_{r(0)}^{r} p_c(r',H,M)dr'} \tag{47}$$

computed for $H = M + \omega$. What is important here is that only the space part of the action appears, as the time part $(H-M)t$ cancels with ωt at the saddle point. $\hat{r}(0)$ is given by the condition

$$k = \hat{r}(0)\log\frac{\sqrt{\hat{r}(0)}-\sqrt{2M}}{\sqrt{\hat{r}(0)}-\sqrt{2H}} \tag{48}$$

which for $H = M + \omega > M$ is solved by

$$2M < 2H < \hat{r}(0) \tag{49}$$

and thus there is no imaginary contribution to the integral from the gap $2M, 2H$ as we discussed in Sect.(4). The Bogoliubov coefficient $\beta_{k\omega}$ is obtained by changing in (47) ω in $-\omega$ (always $\omega > 0$). Then we have

$$\hat{r}(0) < 2H < 2M \tag{50}$$

The value of the saddle point t, now complex, is given by

$$t = 4H\log\frac{\sqrt{\hat{r}}-\sqrt{2H}}{\sqrt{\hat{r}(0)}-\sqrt{2H}} \tag{51}$$

and the r.h.s. term of eq.(47) becomes

$$e^{-4\pi M\omega(1-\omega/2M)} \tag{52}$$

where the last passage is due to eq.(26) for which we gave a general proof within the family of Painlevé gauges even in the massive case. An explicit derivation of the result (52) has been given in [10] by working out the late time expansion of the time development of the action. The main point is that in the mode analysis what comes in is not the total action of a model particle crossing the horizon, but only the "space part" of it. The analysis and the results depend on the interpretation of the expression

$$\exp(iS) \tag{53}$$

as modes of the field dressed by the gravitational interaction with non vanishing quanta of energy. The result eq.(36) proven for the emission of two shells [3] was interpreted [28] as the absence of correlations among quanta in the emitted radiation. It would be of interest to give a similar "mode interpretation" of the result (36) derived for the emission of two shells which during the time evolution can also interact [3]. To this end one should compute the two shell modes which are regular at the horizon, and perform a time Fourier analysis of them. This has not yet been accomplished.

9. Concluding remarks

In this paper we gave a general treatment of the dynamics of one or more self- gravitating spherical shell of matter in a spherical gravitational field. We extended the treatment of [1] to more than one shell [3] proving on the way the universality of the reduced conjugate momentum within the family of Painlevé gauges [10].

This allows to give a general derivation of the exchange relations and also, exploiting an integrability result, to extend the result on the imaginary part of the space part of the action to more than one shell. Instead of following the tunneling picture here we have revisited the treatment which follows by interpreting the exponential of the action, properly subtracted, as the dressed modes of the Hawking radiation. The main point here is that according such a mode interpretation only the space part of the action and in particular its imaginary part comes in determining the Bogoliubov coefficient and thus in determining the Hawking spectrum corrected for the back reaction effects.

References

[1] P. Kraus, F. Wilczek, *Nucl. Phys.* **B 433** (1995) 403, e-Print: arXiv:gr-qc/9408003
[2] P. Kraus, F. Wilczek, *Nucl. Phys.* **B 437** (1995) 231, e-Print: arXiv:hep-th/9411219
[3] F. Fiamberti, P. Menotti, *Nucl. Phys.* **B 794** (2008) 512, e-Print: arXiv:0708.2868 [hep-th]
[4] J. L. Friedman , J. Louko S. Winters-Hilt, *Phys. Rev.* **D 56** (1997) 7674, e-Print: arXiv:gr-qc/9706051
[5] J. Louko, B. F. Whiting, J. L. Friedman, *Phys. Rev.* **D 57** (1998) 2279, e-Print: arXiv:gr-qc/9708012
[6] S. W. Hawking, C. J. Hunter, *Class. Quantum Grav.* **13** (1996) 2735, e-Print: arXiv:gr-qc/9603050
[7] W. Fischler, D. Morgan, J. Polchinski, *Phys. Rev.* **D 41** (1990) 2638; *Phys. Rev.* **D 42** (1990) 4042
[8] S. Goncalves, *Phys. Rev.* **D 66** (2002) 084021, e-Print: arXiv:gr-qc/0212124
[9] D. Nunez, H. P. Oliveira, J. Salim, *Class. Quantum Grav.* **10** (1993) 1117, e-Print: arXiv:gr-qc/9302003
[10] P. Menotti e-Print: arXiv:0911.4358 [hep-th]
[11] M. Parikh, F. Wilczek, *Phys. Rev. Lett.* **85** (2000) 5042, e-Print: arXiv:hep-th/9907001
[12] B. Chowdhury, *Pramana* **70** (2008) 593; *Pramana* **70** (2008) 3, e-Print: arXix:hep-th/0605197; P. Mitra, *Phys. Lett.* **B 648** (2007) 240, e-Print: arXiv:hep-th/0611265
[13] M. Angheben, M. Nadalini, L. Vanzo, S. Zerbini, JHEP 0505:014 (2005), e-Print: arXiv:hep-th/0503081; A.J.M. Medved, E.C. Vagenas, *Mod. Phys. Lett.* **A 20** (2005) 2449, e-Print: arXiv:gr-qc/0504113
[14] E. T. Akhmedov, V. Akhmedova, D. Singleton, *Phys. Lett.* **B 642** (2006) 124, e-Print: arXiv:hep-th/0608098
[15] V. Akhmedova, T. Pilling, A. de Gill, D. Singleton, *Phys. Lett.* **B 666** (2008) 269, e-Print: arXiv:0804.2289 [hep-th]
[16] S. A. Hayward , R. Di Criscienzo, L. Vanzo, M. Nadalini, S. Zerbini, *Class. Quant. Grav.* **26** (2009) 062001, e-Print: arXiv:0806.0014 [gr-qc]
[17] R. Di Criscienzo, S. A. Hayward, M. Nadalini, L. Vanzo, S. Zerbini, e-Print: arXiv:0906.1725 [gr-qc]; e-Print: arXiv:0911.4417
[18] S.A. Hayward, R. Di Criscienzo, M. Nadalini, L. Vanzo, S. Zerbini,e-Print: arXiv:0909.2956 [gr-qc];
[19] M. Pizzi, e-Print: arXiv:0904.4572 [gr-qc]; e-Print: arXiv:0907.2020 [gr-qc]; e-Print: arXiv:0909.3800 [gr-qc]
[20] V.A. Belinski, e-Print: arXiv:0910.3934 [gr-qc]
[21] T. Dray, G. 't Hooft, *Comm. Math. Phys.* **99** (1985) 613
[22] I. H. Redmount, *Prog. Theor. Phys.* **73** (1985) 1401
[23] E. T. Whittaker, *A treatise on the analytical mechanics of particles and rigid bodies*, Chapt. X (Cambridge University Press, Cambridge U.K., 1947)
[24] V. I. Arnol'd, *Mathematical Methods of Classical Mechanics*, Chapt. IX (Springer, Berlin, 1989)
[25] N.D. Birrel and P.C.W. Davies, *Quantum field theory in curved space* Cambridge Monographs on Mathematical Physics, (Cambridge University Press, Cambridge U.K. 1982); R. Wald, *Quantum field theory in curved spacetime and black- hole thermodynamics*, (The University of Chicago Press, Chicago USA 1994)
[26] S. Hawking, *Comm. Math. Phys.* **43** (1975) 199
[27] E. Keski-Vakkuri, P. Kraus, *Nucl. Phys.* **B 491** (1997) 249, e-Print: arXiv:hep-th/9610045
[28] M. Parikh, *Rio de Janeiro 2003, Recent developments in theoretical and experimental general relativity, gravitation, and relativistic field theories*, B 1585, e-Print: arXiv:hep-th/0402166
[29] K. Srinivasan, T. Padmanabhan, *Phys. Rev.* **D 60** (1999) 024007, e-Print: arXiv:gr-qc/9812028
[30] R. Banerjee, B. R. Majhi, *Phys. Lett.* **B 675** (2009) 243, e-Print:arXiv:0903.0250 [hep-th]
[31] R. Kerner and R.B. Mann, *Phys. Rev.* **D 73** (2006) 104010 e-Print: arXiv:gr-qc/0603019

First Mediterranean Conference on Classical and Quantum Gravity (MCCQG 2009) IOP Publishing
Journal of Physics: Conference Series **222** (2010) 012052 doi:10.1088/1742-6596/222/1/012052

Léon Rosenfeld's pioneering steps toward a quantum theory of gravity

Donald C. Salisbury

Department of Physics, Austin College, Sherman, Texas 75090-4440, USA
Max-Planck-Institut für Wissenschaftsgeschichte, Boltzmannstrasse 22, 14195 Berlin, Germany

E-mail: dsalisbury@austincollege.edu

Abstract. In an article published in 1930 Léon Rosenfeld invented a general Hamiltonian formalism that purported to realize general coordinate, local Lorentz, and U(1) symmetries as canonical phase space transformations. He applied the formalism to a q-number version of tetrad gravity in interaction with both the electromagnetic field and a spinorial Dirac electron matter field. His procedure predated by almost two decades the algorithms of Dirac and Bergmann, and with regard to internal (non-spacetime) symmetries is fully equivalent to them. Dirac was in fact already in 1932 familiar with Rosenfelds work, although as far as I can tell he never acknowledged in print his perhaps unconscious debt to Rosenfeld. I will review the general formalism, comparing and contrasting with the work of Dirac, Bergmann and his associates. Although Rosenfeld formulated a correct prescription for constructing the vanishing Hamiltonian generator of time evolution, he evidently did not succeed in carrying out the construction. Nor did he have the correct phase space generators of diffeomorphism-induced symmetry variations. He did not take into account that some of the Lagrangian symmetries are not projectable under the Legendre transformation to phase space.

1. Introduction

The Belgian physicist Léon Rosenfeld is probably best known for his 1933 analysis with Niels Bohr of the measurability of quantized electric and magnetic fields.[5] But his earlier largely unrecognized and unappreciated development of a Hamiltonian formalism for dealing with gauge symmetries could arguably rank among his most significant achievements. Rosenfeld joined Wolfgang Pauli as his young assistant at the Eidgenössische Technische Hochschule in Zurich in 1929. Pauli and Heisenberg had just published their groundbreaking treatise on relativistic quantum electrodynamics.[21] Perturbative expansions already led to infinite self energies. Pauli was hopeful that these infinities could be eliminated if gravitation, the remaining known fundamental interaction, were included in the theory. Thus in 1930 Rosenfeld became the first researcher to attempt to quantize gravity, albeit in linearized form.[28] Questionable methods had been employed in both papers for dealing with gauge symmetry. At Pauli's instigation and with his active encouragement, Rosenfeld set himself the task of justifying these mathematical techniques. In fact, the resulting constrained Hamiltonian formalism, also published in 1930, purported to include within its domain of applicability the full general coordinate transformation symmetry of general relativity.[29] We shall show that this formalism was deficient with regard to general covariance.But it does provide a comprehensive method for incorporating local gauge symmetries, such as those of linearized gravity and electromagnetism, within a Hamiltonian framework. However, neither Rosenfeld nor his contemporaries seemed to have recognized nor utilized his achievement. Much of the machinery was reinvented by Peter Bergmann and his collaborators almost twenty years later.[3] Paul Dirac also initiated his ostensibly independent investigations at roughly the same time as Bergmann.[13] But surprisingly, although Dirac never acknowledged having been stimulated by Rosenfeld's work, we have documentary evidence that he was aware of it already in 1932.[33] To this day the algorithm for incorporating gauge symmetry into the Hamiltonian framework is, as we shall show, somewhat unjustifiably known as the Dirac-Bergmann procedure. It should more accurately be described as the Rosenfeld-Dirac-Bergmann method.

In this paper we shall first briefly discuss the Heisenberg-Pauli method for dealing with gauge symmetry in quantum electrodynamics, identifying the source of their unease. The circumstances surrounding Rosenfeld's later involvement have been exhaustively discussed elsewhere, as has the legitimate application of Rosenfeld's method to the Heisenberg-Pauli model.[33] It was in the course of his attempts at quantizing gravity that Rosenfeld detected what he surmised were universal features of gauge-invariant Lagrangian systems, and he subsequently turned his attention to developing a general formalism. In Section 3 we shall present the full symmetry analysis of Rosenfeld's 1930 *Annalen der Physik* article, and we will show the degree to which it ultimately fails in implementing general covariance. On the other hand the method is applicable to Rosenfeld's model of linearized gravity in interaction with the electromagnetic field, completed after the publication of his general analysis. We address this model in Section 4. We will also show that the method could have been profitably employed in 1936 by Matvei Bronstein in demonstrating the recovery of the Newtonian gravitational potential energy in his own analysis of quantized linearized gravity in interaction with material sources.[7] Finally, in Section 5 we will discuss the relation between Rosenfeld's work, and that of Bergmann and Dirac. We will conclude in Section 6 with an assessment of the significance of Rosenfeld's contribution to quantum gravity and with some speculation on the reasons for the lack of impact of Rosenfeld's work.

2. Quantum electrodynamics

Quantum electrodynamics can be understood in a sense as having "grown up together" with quantum mechanics. Pascual Jordan had famously introduced quantum field theoretic considerations into the 1926 foundational *Dreimännerarbeit* with Max Born and Werner Heisenberg.[6] Pauli's collaboration with Heisenberg followed shortly thereafter in 1927, resulting ultimately in 1929 in their first paper *On the quantum dynamics of wave fields.*[21] Perhaps less appreciated is the fact that quantum gravity is itself a younger sibling of its still maturing brothers. Rosenfeld's procedure for handling gauge symmetries in linearized gravity mirrored those of Heisenberg and Pauli, and they were employed in the hope that gravity could cure potentially fatal flaws in quantum electrodynamics.

The problem that had for several months taxed Heisenberg and Pauli was that in the fully relativistic and gauge covariant classical Hamiltonian formulation of electrodynamics one component of the momentum conjugate to the electromagnetic potential vanishes identically. They took as their Lagrangian density

$$\mathcal{L}_{em} = -\frac{1}{4}F^{\mu\nu}F_{\mu\nu} + eA_\mu\bar{\psi}\gamma^\mu\psi + i\hbar c\bar{\psi}\gamma^\mu\psi_{,\mu} - mc\bar{\psi}\psi,$$

in where $F_{\mu\nu} := \partial_\mu A_\nu - \partial_\nu A_\mu$ is the electromagnetic tensor and $A_\mu = (V, -\vec{A})$ the electromagnetic 4-potential. (We employ Rosenfeld's metric signature of -2.) ψ is the Dirac spinor electron field. The momentum conjugate to A_μ is $p^\mu = \frac{\partial \mathcal{L}}{\partial \dot{A}_\mu} = F^{0\mu}$, so $p^0 \equiv 0$. But a vanishing p^0 is inconsistent with the canonical commutation relations $[A_0(x), p^0(x')] = i\hbar\delta^3(x - x')$. Pauli and Heisenberg were of course aware that their Lagrangian was invariant under the gauge transformations $\delta A_\mu = \xi_{,\mu}$ and $\delta\psi = \frac{ie}{\hbar c}\psi\xi$, where ξ is an arbitrary spacetime function, and that consequently the gauge potential was not an observable field. Their initial attempt at overcoming the vanishing momentum obstacle was to add a gauge symmetry-breaking term to the Lagrangian, of the form $\frac{\epsilon}{2}\left(A^\mu_{,\mu}\right)^2$, and then take the limit $\epsilon \to 0$ at the end of their calculations. Heisenberg and Pauli worked effectively with the quantized *transverse* components of the electromagnetic vector potential. They lacked a theoretical justification for this strategy, and Rosenfeld was assigned the task of investigating the coupling between transverse and longitudinal components in the $\epsilon \to 0$ limit. He succeeded in showing that the longitudinal modes did not emerge from initial transverse states.[27] In the meantime Enrico Fermi published a new approach in which the Coulomb interaction between electrons was assumed and exclusively transverse electromagnetic modes were admitted.[15] Pauli and Heisenberg then showed that Fermi's method was equivalent to the addition of gauge symmetry-breaking term $\frac{1}{2}\left(A^\mu_{,\mu}\right)^2$, with the imposition of the Lorenz gauge $A^\mu_{,\mu} = 0$ as a condition on initial states.[22] In this same paper the authors described a new approach in which manifest Lorentz covariance was broken in setting A_0 equal to zero. Besides their questionable treatment of gauge symmetry, all of these approaches yielded infinite electron self-energies in second order of perturbation. The latter problem apparently weighed most heavily on Pauli, leading to his suggestion that Rosenfeld incorporate gravitational interaction into quantum field theory.[1]

3. Rosenfeld's constrained Hamiltonian dynamics formalism

Rosenfeld observed in the introduction to his 1930 *Annalen der Physik* paper that identities among configuration and conjugate momentum variables always arise as a consequence of gauge invariance, and that as he " was investigating these relations in the especially instructive example of gravitation theory, Professor Pauli helpfully indicated to me the principles of a simpler and more natural manner of applying the Hamiltonian procedure in the presence of identities. This procedure is not subject to the disadvantages of the earlier methods."[2] Rosenfeld did not specify precisely what was Pauli's conclusion. But judging from Pauli's remark in a 1955 letter to Oskar Klein, the formalism that we shall now describe was his own invention. Pauli writes "I would like to bring to your attention the work by Rosenfeld in 1930. He was known here at the time as the man who quantised the Vierbein (sounds like the title of a Grimms fairy tale doesnt it?). See part II of his work where the Vierbein appears. Much importance was given at that time to the identities among the ps and qs (that is the canonically conjugate fields) that arise from the existence of the group of general coordinate transformations. I still remember that I was not happy with every aspect of his work since he had to introduce certain additional assumptions that no one was satisfied with."[3]

[1] In a 1963 interview, deposited in the Archive for the History of Quantum Physics (AHQP, 7/19/1963, p. 8), Rosenfeld remarked that "I got provoked by Pauli to tackle this problem of the quantization of gravitation and the gravitation effects of light quanta, which at that time were more interesting. When I explained to Pauli what I wanted to work out, I think it was the Kerr effect or some optical effect, he said Well, you may do that, and I am glad beforehand for any result you may find. That was a way of saying that this was a problem that was not instructive, that any result might come out, whereas at that time, the calculation of the self energy of the light quantum arising from its gravitational field was done with a very definite purpose.

[2] A translation and commentary of this article is available online [31]

[3] "Gerne möchte ich Dich in dieser Verbindung auf die lange Arbeit von Rosenfeld, Annalen der Physik (4), 5, 113, 1930 aufmerksam machen. Er hat sie seinerzeit bei mir in Zürich gemacht und hiess hier dementsprechend der Mann, der das Vierbein quantelt (klingt wie der Titel eines Grimmschen Märchens, nicht?). Siehe dazu Teil II seiner Arbeit, wo das Vierbein daran kommt. Auf die Identitäten zwischen den 'p' und 'q' - d.h. kanonisch konjugierten Feldern - die eben aus der Existenz der Gruppe der Allgemeinen Relativittätstheorie (Koordinaten-Transformationen mit 4 willkürlichen Funktionen) entspringen, wurde damals besonderer Wert gelegt. Ich erinnere

Rosenfeld considers infinitesimal coordinate transformations of the form $x'^\mu = x^\mu + \delta x^\nu$, where

$$\delta x^\nu = a_r^{\nu,0}(x)\epsilon^r(x) + a_r^{\nu,\sigma}(x)\frac{\partial\epsilon^r}{\partial x^\sigma} + a_r^{\nu,\sigma\cdots\tau}(x)\frac{\partial^k\epsilon^r}{\partial x^\sigma\cdots\partial x^\tau},$$

where the ϵ^r are arbitrary spacetime functions. (For the moment we will be concerned exclusively with the special case $\delta x^\nu = -\epsilon^\mu(x)$. Later we will include additional internal gauge freedom with corresponding arbitrary spacetime functions ϵ^r, where $r > 3$.) Rosenfeld then assumes that the configuration variables Q_α are subject to infinitesimal gauge symmetry transformations of the form

$$\delta Q_\alpha = c^0_{\alpha r}(x,Q)\xi^r(x) + c^\sigma_{\alpha r}(x,Q)\frac{\partial\xi^r}{\partial x^\sigma} + c^{\sigma\cdots\tau}_{\alpha r}(x,Q)\frac{\partial^j\xi^r}{\partial x^\sigma\cdots\partial x^\tau}.$$

Rosenfeld applied his formalism to the tetrad version of general relativity, but we shall illustrate with the simpler conventional metric form of general relativity in which the metric components

$$g_{\mu\nu} = \begin{pmatrix} -N^2 + N^cN^dg_{cd} & g_{ac}N^c \\ g_{bd}N^d & g_{ab}\end{pmatrix}$$

constitute the configuration variables, where N is the lapse and N^a is the shift vector. We choose this example in part to show the extent to which Rosenfeld could have succeeded using his formalism had he been able to use the Lagrangian that was much later discovered by Arnowitt, Deser, and Misner (ADM).[2][4] As we shall see shortly, this Lagrangian is quadratic in derivatives of the metric components. Now under the infinitesimal coordinate transformation $x'^\mu = x^\mu - \epsilon^\nu$ the resulting variations are $\delta g_{\mu\nu} := g'_{\mu\nu}(x') - g_{\mu\nu}(x) = g_{\mu\alpha}\epsilon^\alpha_{,\nu} + g_{\alpha\nu}\epsilon^\alpha_{,\mu}$.

Rosenfeld considers Lagrangians that are quadratic in field derivatives, i.e., of the form

$$\mathcal{L} = \frac{1}{2}\left(Q_{\alpha,\nu}\mathcal{A}^{\alpha\nu,\beta\mu}(Q)Q_{\beta,\mu} + Q_{\alpha,\nu}\mathcal{B}^{\alpha\nu}(Q) + \mathcal{B}^{\alpha\nu}(Q)Q_{\alpha,\nu} + \mathcal{C}(Q)\right).$$

He then supposes that the δQ_α are Noether symmetry transformations so that the Lagrangian transforms as a scalar density of weight one:

$$\delta\mathcal{L} + \mathcal{L}\frac{\partial\delta x^\mu}{\partial x^\mu} \equiv 0. \tag{1}$$

(Later he relaxes this assumption, admitting the possibility that the variation differs from a density by a total divergence.)

In our example we employ the invariant Hilbert action with Lagrangian $\mathcal{L} = \sqrt{-g}$ where the Ricci scalar takes the form

$$R = {}^{(3)}\!R + {}^{(3)}\!g^{ac(3)}g^{bd}K_{ab}K_{cd} - {}^{(3)}\!g^{ab(3)}g^{cd}K_{ab}K_{cd} + (n^\mu n^\nu_{;\nu})_{;\mu} - (n^\nu n^\mu_{;\nu})_{;\nu}.$$

In this expression the superscript (3) refers to objects constructed from the spatial metric g_{ab}. Also $n^\mu = (N^{-1}, -N^{-1}N^a)$ is the normal to the constant coordinate time hypersurfaces and K_{ab} is the extrinsic curvature of these surfaces. We will use the ADM Lagrangian, $\mathcal{L}_G$, obtained by subtracting the final two total derivative terms,

$$\mathcal{L}_G = N\sqrt{{}^{(3)}g}\left({}^{(3)}\!R + {}^{(3)}\!g^{ac(3)}g^{bd}K_{ab}K_{cd} - {}^{(3)}\!g^{ab(3)}g^{cd}K_{ab}K_{cd}\right).$$

mich noch, dass Rosenfelds Arbeit nicht in jeder Hinsicht befriedigend war, da er gewisse zusätzliche Bedingungen einführen musste, die niemand richtig verstehen konnte ([34], p. 64.

[4] See also [35], p. 464.

3.1. The Hamiltonian

Rosenfeld first constructs the canonical momentum $\mathcal{P}^\alpha = \frac{\partial \mathcal{L}}{\partial \dot{Q}_\alpha} = \mathcal{A}^{\alpha\nu,\mu 0} Q_{\alpha,\nu}$. The key observation is that in the identity (1) the coefficients of each order of time derivative of ϵ^μ must vanish identically. Thus focusing on the second time derivative term we deduce from $\delta\mathcal{L} = p^\mu c^0_{\mu r} \ddot{\xi}^r + \cdots \equiv 0$ that there are primary constraints $p^\mu c^0_{\mu r} \equiv 0$. We note in addition that the primary constraints give us null vectors of the Legendre matrix $\frac{\partial^2 \mathcal{L}}{\partial \dot{Q}_\alpha \partial \dot{Q}_\alpha}$, i.e.,

$$\mathcal{A}^{\alpha 0,\mu 0} c^0_{\mu r} \equiv 0.$$

Consequently, since

$$p^\alpha = \mathcal{A}^{\alpha 0,\mu 0} \dot{Q}_\mu + \cdots, \tag{2}$$

the velocities are not fixed uniquely in terms of the momenta. Rather,

$$\dot{Q}_\mu = \frac{\partial\, {}^0\mathcal{H}}{\partial p^\mu} + \lambda^r c^0_{\mu r} = \frac{\partial \left({}^0\mathcal{H} + \lambda^r p^\nu c^0_{\nu r} \right)}{\partial p^\mu} = \frac{\partial \mathcal{H}}{\partial p^\mu}, \tag{3}$$

where the λ^r are arbitrary spacetime functions. The Hamiltonian ${}^0\mathcal{H}$ is constructed using any particular solution ${}^0\dot{Q}_\mu(Q, \mathcal{P})$ of the defining relation (2), so ${}^0\mathcal{H} := p^{\alpha\, 0}\dot{Q}_\alpha - \mathcal{L}(Q, {}^0\dot{Q})$ and the total Hamiltonian is defined to be $\mathcal{H} = {}^0\mathcal{H} + \lambda^r p^\nu c^0_{\nu r}$.

Continuing with our example, the primary constraints are $p_\mu = \frac{\partial \mathcal{L}_G}{\partial \dot{N}^\mu} = 0$ and the corresponding tangent space null vectors are $\frac{\partial}{\partial \dot{N}^\mu}$. The total Hamiltonian is $\mathcal{H} = N^\mu \mathcal{H}_\mu + \lambda_\mu p^\mu$, where $\mathcal{H}_0$ is the usual scalar constraint:

$$\mathcal{H}_0 = \frac{1}{\sqrt{{}^{(3)}g}} \left(p_{ab} p^{ab} - (p^a_a)^2 \right) - \sqrt{g}\, {}^{(3)}R,$$

where the p^{ab} are the momenta conjugate to g_{ab}. In following through later on with Rosenfeld's construction of gauge symmetry generators we find that we must take as our vector constraints

$$\mathcal{H}_a = 2 D_b p^b{}_a,$$

(and not the conventional expression that differs by a total spatial derivative). We note that as usual the equations of motion for the lapse and shift yield $\dot{N}^\mu = \lambda^\mu$.

3.2. The gauge generators and higher-order constraints

Rosenfeld next constructs the phase space generators of active gauge transformations. It is significant that he, as well as Bergmann and his collaborators, insisted from the start that phase space symmetry transformations should always faithfully realize the configuration-velocity space transformations - in contradistinction to Dirac. We will use the notation that was apparently first introduced by Noether, and later taken over by Bergmann. We define the active transformation $\bar{\delta} Q_\alpha(x) = Q\alpha'(x) - Q\alpha(x)$ and $\bar{\delta}\mathcal{P}^\alpha(x) = \mathcal{P}'^\alpha(x) - \mathcal{P}^\alpha(x)$. (These are of course the Lie derivatives along the vector field $\epsilon^\mu = -\delta x^\mu$.)

Rosenfeld proved that the following integral generates the correct active gauge variations of Q and P:

$$\overline{\mathcal{M}} := \int d^3x\, p^\alpha \delta Q_\alpha - \mathcal{H}\delta x^0 - \mathcal{P}^a Q_{,a} \delta x^a. \tag{4}$$

He then showed that this generator is a constant of the motion, i.e., $\frac{d\overline{\mathcal{M}}}{dt} = 0$. Consequently, the coefficients of the time derivatives of ϵ of each order must vanish. Rosenfeld then proved that this generator could always be written as the sum of time derivatives of the primary constraints multiplying time derivatives of the arbitrary function ϵ^r,

$$\overline{\mathcal{M}} = \int d^3x \left(\frac{d\epsilon^r}{dt} p^\mu c^0_{\mu r} - \epsilon^r \frac{d}{dt} \left(p^\mu c^0_{\mu r} \right) \right). \tag{5}$$

Thus Rosenfeld showed that the preservation in time of primary constraints leads to secondary (and tertiary) constraints. (In (5) we have assumed that this expansion terminates with secondary constraints.) This result has always been attributed to Anderson and Bergmann.[1] The terminology is due to them and was employed later by Dirac.

The generating density of our example is

$$\begin{aligned} \mathcal{M} &= \frac{1}{2}p^{ab}\delta g_{ab} + p_\mu \delta N^\mu - \mathcal{H}\delta x^0 - \frac{1}{2}p^{ab} g_{ab,c}\delta x^c - p_\mu N^\mu_{,c}\delta x^c \\ &= p^{ab}\left(g_{ca}\epsilon^c_{,b} + N^c g_{ca}\epsilon^0_{,b} \right) + p_0 \left(N\dot{\epsilon}^0 - NN^a\epsilon^0_{,a} \right) + p_a \left(N^a\dot{\epsilon}^0 - (N^2 e^{ab} + N^a N^b)\epsilon^0_{,b} + \dot{\epsilon}^a - N^b \epsilon^a_{,b} \right) \\ &+ \left(N\mathcal{H}_0 + N^a\mathcal{H}_a + \dot{N}^\mu p_\mu \right)\epsilon^0 + \frac{1}{2}p^{ab} g_{ab,c}\epsilon^c + p_0 N_{,c}\epsilon^c + p_a N^a_{,c}\epsilon^c \end{aligned} \tag{6}$$

3.3. Problems with projectability and the contemporary resolution

It is a this stage that we encounter a failure in Rosenfeld's formalism when it is applied to generally covariant models. The problem is evident in our example generator (6); a legitimate generator must consist of a sum phase space functions multiplying the arbitrary time derivatives of ϵ^r. However, there is no phase space function corresponding to the time derivative $\dot{N}$. In other words, functions of $\dot{N}$ are not projectable under the Legendre transformation from configuration-velocity space to phase space. Bergmann and Brunings were apparently the first to note this requirement in print.[4] Lee and Wald were the first to begin a systematic exploration of this condition.[24]

It was shown by Pons, Salisbury, and Shepley in 1997 that projectability is attained in generally covariant metric theories through a unique compulsory dependence of the infinitesimal transformations on the lapse and shift. In fact, the infinitesimal functions ϵ^μ must be of the form

$$\epsilon^\mu = \delta^\mu_a \xi^a + n^\mu \xi^0, \tag{7}$$

where the ξ^μ are arbitrary functions of the spacetime coordinates and of the three-metric.[25] In fact, if this ϵ^μ is substituted into the example Rosenfeld form (6), one obtains the form first displayed in 1997 [25], namely

$$\mathcal{M} = p_\mu \dot{\xi}^\mu + \left(\mathcal{H}_\mu + N^\rho C^\nu_{\mu\rho} p_\nu \right) \xi^\mu, \tag{8}$$

where the $C^\nu_{\mu\rho}$ are the structure functions that appear in the Poisson bracket algebra of the secondary constraints, $\{\mathcal{H}_\mu, \mathcal{H}_\nu\} = C^\alpha_{\mu\nu}\mathcal{H}_\alpha$.

Continuing with the general Rosenfeld formalism, it turns out that further modifications of his gauge generator are required if additional (internal) symmetries are present beyond general coordinate symmetry. It turns out that in this case pure diffeomorphism symmetries cannot be realized at all as canonical transformations. An internal symmetry must be added. This fact has been understood in various guises since the 1970's. The first group theoretical explanation was given in 1983 for Einstein-Yang-Mills theory [32]. A projectability analysis followed in 2000 [26]. Surprising, taking this requirement into account, Rosenfeld's 1930 expression for the gauge symmetry generator once again delivers the correct modern form. We will examine some relevant details for Einstein-Yang-Mills theory. So let us consider the variation of the temporal component of the Yang-Mills potential under the infinitesimal coordinate transformation (7),

$$\begin{aligned} \delta A_0^i &= A_\nu^i (n^\nu \xi^0)_{,0} \\ &= A_0^i \left(-N^{-2} \dot{N} \xi^0 + N^{-1} \dot{\xi}^0 \right) + A_b^i \left(-N^{-2} \dot{N} N^b \xi^0 + N^{-1} \dot{N}^b \xi^0 + N^{-1} N^b \dot{\xi}^0 \right). \end{aligned} \tag{9}$$

The $\dot{N}^\mu$ terms are not projectable, and neither are the $\dot{A}_0^i$. They can be uniquely eliminated by supplementing the infinitesimal diffeomorphims (7) with a Yang-Mills internal gauge transformation with descriptor $\Lambda^i = A_\mu^i n^\mu \xi^0$, where the correspond internal gauge transformation $\delta_G A_0^i = -\Lambda_{,0}^i - C_{jk}^i \Lambda^j A_0^k$. The C_{jk}^i are the structure constants of the Yang-Mills group. The resulting composite variation of A_0^i is then

$$\delta A_0^i + \delta_G A_0^i = A_b^i \dot{\xi}^0 + N^{-1} F_{0b}^i N^b \xi^0, \tag{10}$$

where F_{0b}^i are components of the field tensor derived from the Yang-Mills potential. If this and corresponding variations of A_a^i are substituted into the Rosenfeld expression (4), then the modern result is again obtained,

$$\mathcal{M} = p_A \dot{\xi}^A + \left(\mathcal{H}_A + N^C C_{AC}^B p_B \right) \xi^A, \tag{11}$$

where the index A now ranges over the four spacetime coordinate indices μ and the internal indices i.

Rosenfeld's ambition in his 1930 paper was no less than a quantum unification of all of the forces known at that time, including Einstein's generally covariant curved spacetime gravitational theory. Several pieces of the fully interacting model had only recently been invented, starting with Dirac's relativistic wave equation in 1928.[10] The coupling of the Dirac electron matter field to gravity required a tetrad formulation of Einstein's theory. We will represent the tetrad field as E_I^μ, where the Minkowski index I ranges from 0 to 3. The spinor connection $\Omega_\mu = \frac{1}{4} \gamma^I \gamma^J \omega_{\mu IJ}$ was written down independently in 1929 by Weyl[36] and Fock.[19], and they each were apparently unaware of Cartan's introduction of tetrads in 1928.[8] In this expression $\omega_{\mu IJ}$ is the Ricci rotation coefficient and the γ^I are the flat Dirac γ matrices. Instead of the Hilbert gravitational action Rosenfeld removed second derivatives by subtracting a total covariant divergence from the Ricci scalar density,

$${}^4\mathcal{R} - \nabla_\mu \left(2(-g)^{\frac{1}{2}} E_I^\mu E_J^\nu \omega_\nu^{IJ} \right) = -(-g)^{\frac{1}{2}} E_I^\mu E_J^\nu \left(\omega_\mu{}^I{}_L \omega_\nu{}^{LJ} - \omega_\nu{}^I{}_L \omega_\mu{}^{LJ} \right) =: -\mathcal{G}. \tag{12}$$

For the matter contribution he employed the expressions of Fock and Weyl,

$$\mathcal{L}_M = i\hbar c(-g)^{1/2} \overline{\psi} E_L^\mu \Gamma^L \left(\frac{\partial}{\partial x^\mu} + \Omega_\mu - i \frac{e}{\hbar c} \phi_\mu \right) \psi + mc^2 \overline{\psi} \psi (-g)^{1/2}, \tag{13}$$

where $\overline{\psi} := \psi^\dagger \Gamma^0$. Rosenfeld did not obtain an explicit phase space expression for either the Hamiltonian or the gauge symmetry generators for this model. Indeed, as we have discussed above, such an expression does not exist without taking Legendre projectability conditions into account.

4. Rosenfeld, Bronstein, and linearized quantum gravity

4.1. Calculation of the gravitational self-energy of photons

After completing this formal analysis of constrained Hamiltonian dynamics Rosenfeld turned to the problem that had originally been suggested to him by his host. Pauli wished to know whether by bringing gravity into his nascent quantum electrodynamics it might be possible to avoid the unpleasant infinities that plagued the theory. For this purpose it was sufficient to consider a linearized version of general relativity. Rosenfeld then became the first to undertake its quantization, though not in complete generality.[28] He considered the electromagnetic radiation field in interaction with gravity. Linearized gravity was first treated by Einstein himself in 1916[14]. Assuming a small perturbation of the metric from flatness,

$$g_{\mu\nu} = \eta_{\mu\nu} + \epsilon h_{\mu\nu}, \tag{14}$$

where $\epsilon := \sqrt{8\pi G/c^4}$, and keeping along first order in ϵ terms in the gravitational action, Rosenfeld obtained the following flat space Lagrangian:

$$L = -\frac{1}{4}F_{\mu\nu}F^{\mu\nu} - \frac{1}{8}\left(\bar{h}_{\mu\nu,\alpha}\bar{h}^{\mu\nu,\alpha} - \frac{1}{2}\bar{h}_{,\alpha}\bar{h}^{,\alpha}\right) + \frac{\epsilon}{2}\bar{h}_{\mu\nu}T_{\mu\nu}. \tag{15}$$

Indices are raised with the Minkowski metric. The "barred" perturbation is defined as $\bar{h}_{\mu\nu} := h_{\mu\nu} - \frac{1}{2}\eta_{\mu\nu}h$, where h is the trace $h^{\mu}{}_{\mu}$. $T_{\mu\nu}$ is the electromagnetic stress energy tensor. Rosenfeld's expression for the gravitational Lagrangian, $L_R := -\frac{1}{8}\left(\bar{h}_{\mu\nu,\alpha}\bar{h}^{\mu\nu,\alpha} - \frac{1}{2}\bar{h}_{,\alpha}\bar{h}^{,\alpha}\right) = -\frac{1}{8}\left(h_{\mu\nu,\alpha}h^{\mu\nu,\alpha} - \frac{1}{2}h_{,\alpha}h^{,\alpha}\right)$, differs from the linearized Einstein-Hilbert Lagrangian,

$$L_{EH} := -\frac{1}{4}h_{\alpha\gamma,\beta}h^{\alpha\gamma,\beta} + \frac{1}{2}h_{\alpha\gamma,\beta}h^{\alpha\beta,\gamma} + \frac{1}{4}h_{,\beta}h^{,\beta} - \frac{1}{2}h_{,\beta}h^{\alpha\beta}{}_{,\alpha}. \tag{16}$$

This Lagrangian differs by a total divergence from the Lagrangian L_F employed in 1962 by Feynman,

$$L_F := -\frac{1}{4}h_{\alpha\gamma,\beta}h^{\alpha\gamma,\beta} + \frac{1}{2}h_{\alpha\beta}{}^{,\beta}h^{\alpha\gamma}{}_{,\gamma} + \frac{1}{4}h_{,\beta}h^{,\beta} - \frac{1}{2}h_{,\beta}h^{\alpha\beta}{}_{,\alpha}, \tag{17}$$

and is therefore equivalent to L_E. Having been derived from the Einstein action, these latter two actions are invariant under the under the residual coordinate gauge symmetry $\delta h_{\mu\nu} = \xi_{\mu,\nu} + \xi_{\nu,\mu}$.[5] Rosenfeld's action is not. He did not make this observation in the paper, though it is clear that this choice was made purposely. Curiously, he chooses a quantization procedure that parallels the method that had been employed by Fermi in quantum electrodynamics[15], a method that he had in that context rigorously validated, and that he would continue to employ in subsequent publications.[6] The method makes use of a non-singular, in this case, a gauge-fixed Lagrangian. He selected as a "coordinate condition" $\bar{h}_{\mu\nu}{}^{,\nu} = 0$, requiring in addition that it's time derivative also vanishes at the initial time. He then confirmed that these two conditions were preserved under time evolution. Following the Fermi method, the two relations were then imposed as conditions on physically admissible quantum states.

In 1936 Bronstein used a gravitational Lagrangian, L_B, that differed from Rosenfeld's by a total divergence. Contrary to Rosenfeld, Bronstein stated explicitly not only that he was employing the Fermi method, but that his action was not gauge invariant. Feynman was apparently the first to use the gauge-invariant linearized Lagrangian, describing his quantum gravitational results in lectures at the Californian Institute of Technology in 1962.[17] In that same year he also reported his preliminary results to the relativity community at the Conference on Relativistic Theories of Gravitation in Jablonna, Poland.[16] In neither of these papers is there a mention of the earlier work of Rosenfeld and Bronstein. Rosenfeld was present at the Jablonna

[5] The quickest way to demonstrate the invariance of the linearized action under these gauge transformations is to note that $L_F = -\frac{1}{4}h_{\alpha\gamma,\beta}\bar{h}^{\alpha\gamma,\beta} + \frac{1}{2}\bar{h}_{\alpha\beta}{}^{,\beta}\bar{h}^{\alpha\gamma}{}_{,\gamma}$ and therefore $\delta L_F = -\frac{1}{2}\delta h_{\alpha\gamma,\beta}\bar{h}^{\alpha\gamma,\beta} + \delta\bar{h}_{\alpha\beta}{}^{,\beta}\bar{h}^{\alpha\gamma}{}_{,\gamma}$ Substitution of the gauge transformation and an integration by parts yields the desire result.

[6] See [33] for details.

meeting, but there is no record of his having participated in the discussion following Feynman's presentation. Rather, in another session in which Feynman was present, Rosenfeld called into question the need to quantize the gravitational field.[23]

As we shall discuss in more detail below, Rosenfeld's non-gauge-invariant approach to gravitation is surprising. Because he had chosen this route he was not in position to avail himself of the sophisticated formalism that he had completed six months before this paper was submitted for publication in September of 1930. This is even more surprising since the formalism when applied to internal symmetries is correct and perfectly suited for dealing with the residual internal gauge freedom. It could have been employed to justify on a group-theoretical basis the gauge choices he subsequently made. Nor did he refer to this foundational paper.

The equation of motion for metric perturbation in the gauge $\bar{h}_{\mu\nu}{}^{,\nu} = 0$ is

$$\eta^{\mu\nu} \frac{\partial^2}{\partial x^\mu \partial x^\nu} h_{\mu\nu} = -2\epsilon T_{\mu\nu}. \tag{18}$$

Expressing the electromagnetic stress energy in terms of photon creation and annihilation operators, Rosenfeld could then solve (18) for the gravitational field $h_{\mu\nu}$ ingendered by the electromagnetic source. He then substituted this photon field operator into the pure gravitational field contribution to the Hamiltonian. It turned out, to the dismay (but perhaps not the surprise) of all, that this operator exhibited an ultraviolet divergence, independently of the electromagnetic state on which it acted.

Rosenfeld also considered first order transition amplitudes. For this purpose he for the first time invented graviton creation and annihilation operators, though he did not use this terminology. He called the exitations "gravitational quanta". He showed that the pure gravitational Hamiltonian could be written as $\Sigma_{\vec{k}_n,i} \left(\hat{N}_{\vec{k}_n,i} + 1\right) h\nu_{\vec{k}_n,i}$, where i ranges over two helicities, $\hat{N}_{\vec{k}_n,i}$ is the graviton number operator for wavenumber $\vec{k}_n$ and the frequency is $\nu_{\vec{k}_n,i}$. He recognized that the interaction term in the Lagrangian yielded the two possibilities, each with its inverse: the annihilation of a graviton and creation of two photons or the annihilation of a photon with the production of a photon (of lower frequency) and a graviton.

4.2. Bronstein and the quantum recovery of the Newtonian potential

As mentioned above, Bronstein in 1936 used the same gauge-fixing procedure as Rosenfeld, proceeding *a la* Fermi from a non-gauge-invariant Lagrangian. By this time, and for many years after, the second quantized approach of Heisenberg and Pauli was out of favor. Most researchers, evidently including Rosenfeld, preferred the conceptually and computationally simpler multi-particle approach of Dirac in which electromagnetic radiation was quantized, but individual electrons were described by Dirac wave functions.[11] Rosenfeld had himself demonstrated that Dirac's procedure was equivalent to the second quantization approach of Heisenberg and Pauli.[30] But the explicit realization of gauge symmetry was far simpler to carry out when both radiation and matter were quantized - using the Rosenfeld formalism we have examined above. This is particularly pertinent to Bronstein's work. Dirac had showed how the Coulomb static potential energy could be recovered in his program.[11] I have shown elsewhere how the Rosenfeld symmetry group formalism could be employed to gain the much more satisfactory modern quantum field theoretical derivation of the electromagnetic interaction Hamiltonian.[33] The method employs the group to construct operators that are invariant under the action of the group. We do know that Bronstein was aware of Rosenfeld's formalism, even though he did not cite the 1930 paper. The two met in Kharkov, Russia, in 1934.[7][20] It is perhaps pertinent to note here that, as far as I can tell, the first reference to the massless spin 2 nature of the

[7] available online at http://people.bu.edu/gorelik

graviton appeared in the work of Fierz and Pauli in 1939.[18] Thus the nature of the internal symmetry group of linearized gravity may not have been fully appreciated before this date.

5. Dirac, Bergmann and Rosenfeld

Most relativists associate Paul Dirac and Peter Bergmann with the development of constrained Hamiltonian dynamics, developments that occurred almost twenty years after Rosenfeld's groundbreaking paper. It is natural to ask to what extent they were aware of Rosenfeld's work. The relation to Bergmann and his group is clear.[8] The Syracuse group did not learn of Rosenfeld's work until 1951. The work was consistently cited thereafter. The relation with Dirac is, however, problematical. The two first encountered each other in Göttingen as early as 1928. Dirac had already in 1926 struggled to incorporate special relativity first into Heisenberg's matrix mechanics, and then shortly afterwards into Schrödinger's new wave mechanics. He was attempting a relativistic description of Compton scattering. The first effort is especially noteworthy since it signals apparently the first appearance of Hamiltonian constraints in quantum mechanics.[9] He promoted the time and energy to quantum operators, imposing the relationship between the Hamiltonian and the energy as a constraint. We have a correspondence between Dirac and Rosenfeld in 1932 in which Rosenfeld brings to Dirac's attention his 1930 paper.[9] Dirac followed up with a question referring explicitly to the origin and significance of Rosenfeld's arbitrary functions in his constrained Hamiltonian formalism. Rosenfeld provided a detailed explanation. It is thus surprising that, as far as I can tell, Dirac never in any of his publications cited Rosenfeld's work. One might suppose that he had simply forgotten about it in the seventeen years between this interchange and Dirac's constrained Hamiltonian publications from 1949 and onwards. But in fact, Dirac published related material much earlier. In 1933 Dirac published a paper on "Homogeneous variables in classical mechanics".[12] In this paper he dealt with singular Langrangians that were homogeneous of first degree in the velocities. The Rosenfeld procedure was not immediately suited for dealing with this type of gauge symmetry - so Dirac could perhaps be excused for not mentioning a possible connection to Rosenfeld's work.

6. Conclusions

I have shown in this paper that Rosenfeld developed in 1930 a powerful technique for handling local gauge symmetries within a Hamiltonian framework. The method is strictly correct, and still employed today, when applied to internal gauge symmetries. However, as we have noted, it has the names of two other illustrious physicists attached to it. It can for example be applied without modification to linearized gravity. Yet neither Rosenfeld in 1930 or Bronstein in 1936 made explicit use of it in their pioneering work in linearized quantum gravity. On the other hand, we have seen the method requires some modification when applied to generally covariant dynamical systems like general relativity. The class of admissible projectable coordinate transformations must be correctly delineated, and the gauge symmetry generators appropriately modified. We have seen that when these modifications are carried through both in conventional classical vacuum relativity and Einstein-Yang-Mills theory, the Rosenfeld expressions do transform into the correct generators.

References

[1] James L. Anderson and Peter G. Bergmann. Constraints in covariant field theories. *Physical Review*, 83:1018 – 1025, 1951.

[2] Richard Arnowitt, Stanley Deser, and Charles Misner. The dynamics of general relativity. In L. Witten, editor, *Gravitation: an introduction to current research*, pages 227–264, 1962.

[3] Peter G. Bergmann. Non-linear field theories. *Physical Review*, 75:680 – 685, 1949.

[8] See [?] for a discussion of Bergmann's early contributions.

[9] See [33] for more details.

[4] Peter G. Bergmann and J. H. M. Brunings. Non-linear field theories II. Canonical equations and quantization. *Reviews of Modern Physics*, 21:480 – 487, 1949.

[5] Niels Bohr and Leon Rosenfeld. Zur Frage der Messbarkeit der electromagnetischen Feldgrössen. *Kgl. Danske Vidensk*, 78:94 – 798, 1933.

[6] Max Born, Pascual Jordan, and Werner Heisenberg. Zur Quantenmechanik II. *Zeitschrift für Physik*, 35:557–615, 1926.

[7] Matvei Petrovich Bronstein. Quantentheorie schwacher Gravitationsfelder. *Physikalische Zeitschrift der Sowjetunion*, 9:140–157, 1936.

[8] Élie Cartan. *Leçons sur la Géométrie des Espaces de Riemann*. Gauthiers-Villars, 1928.

[9] P. A. M. Dirac. Relativity quantum mechanics with an application to compton scattering. *Proceedings of the Royal Society of London*, A111:405–423, 1926.

[10] P. A. M. Dirac. The quantum theory of the electron. *Proceedings of the Royal Society of London*, A117:610–624, 1928.

[11] P. A. M. Dirac. Relativistic quantum mechanics. *Proceedings of the Royal Society of London*, A136:453–464, 1932.

[12] P. A. M. Dirac. Homogeneous variables in classical dynamics. *Proceedings of the Cambridge Philosphical Society*, 29:389–401, 1933.

[13] P. A. M. Dirac. Generalized Hamiltonian dynamics. *Canadian Journal of Mathematics*, 2:129 – 148, 1950.

[14] Albert Einstein. Näherungsweise Integration der Feldgleichungen der Gravitation. *Königlich Preussische Akademie der Wissenschaft. Sitzungsberichte (Berlin)*, pages 688–696, 1916.

[15] Enrico Fermi. Sopra l'elettrodinamica quantistica. *Rend. Lincei*, 9:881–887, 1929.

[16] Richard P. Feynman. Quantum theory of gravitation. *Acta Physica Polonica*, 24:697–722, 1963.

[17] Richard P. Feynman, Fernando B. Morinigo, and William J. Wagner. *Feynman Lectures on Gravitation*. Addison-Wesley, 1995.

[18] Markus Fierz and Wolfgang Pauli. Über relativistische Feldgleichungen von Teilchen mit beliebigem Spin im electromagnetischen Feld. *Helvetica Physica Acta*, 12:297–300, 1939.

[19] V. Fock. Geometrisierung der Diracschen Theorie des Elektrons. *Zeitschrift für Physik*, 57:261–277, 1929.

[20] Gennady Gorelik. Matvei bronstein and quantum gravity: 70th anniversary of the unsolved problem. *Physics-Uspekhi*, 48:1039–1053, 2005.

[21] Werner Heisenberg and Wolfgang Pauli. Zur Quantendynamik der Wellenfelder. *Zeitschrift für Physik*, 56(1-61), 1929.

[22] Werner Heisenberg and Wolfgang Pauli. Zur Quantentheorie der Wellenfelder.II. *Zeitschrift für Physik*, 59(168-190), 1930.

[23] L. Infeld, editor. *Relativistic Theories of Gravitation: Proceedings of a conference held in Warsaw and Jablonna, July 1962*. Pergamon Press, 1964.

[24] Joohan Lee and Robert M. Wald. Local symmetries and constraints. *Journal of Mathematical Physics*, 31:725–743, 1990.

[25] Josep Pons, Donald Salisbury, and Lawrence Shepley. Gauge transformations in the Lagrangian and Hamiltonian formalisms of generally covariant theories. *Physical Review D*, 55:658–668, 1997.

[26] Josep Pons, Donald Salisbury, and Lawrence Shepley. Gauge transformations in Einstein-Yang-Mills theories. *Journal of Mathematical Physics*, 41(8):5557–5571, 2000.

[27] Léon Rosenfeld. Über die longitudinalen Eigenlösungen der Heisenberg-Paulischen electromagnetischen Gleichungen. *Zeitschrift für Physik*, 58:540–555, 1929.

[28] Léon Rosenfeld. Über die Gravitationswirkungen des Lichtes. *Zeitschrift für Physik*, 65:589–599, 1930.

[29] Léon Rosenfeld. Zur Quantelung der Wellenfelder. *Annalen der Physik*, 5:113 – 152, 1930.

[30] Léon Rosenfeld. Über eine mögliche fassung des Diracschen Programms zur Quantenelektrodynamik und deren formalen Zusammenhang mit der Heisenberg-Paulischen Theorie. *Zeitschrift für Physik*, 76:729–734, 1932.

[31] Donald Salisbury. Translation and commentary of Léon Rosenfeld's "Zur Quantelung der Wellenfelder", Annalen der Physik 397, 113 (1930), Max Planck Institute for the History of Science Preprint 381, http://www.mpiwg-berlin.mpg.de/en/resources/preprints.html, November 2009.

[32] Donald Salisbury and Kurt Sundermeyer. Local symmetries of the Einstein-Yang-Mills theory as phase space transformations. *Physical Review D*, 27(4):757–763, 1983.

[33] Donald C. Salisbury. Léon Rosenfeld and the challenge of the vanishing momentum in quantum electrodynamics. *Studies in History and Philosophy of Modern Physics*, 40:363–373, 2009.

[34] Karl von Meyenn. *Wolfgang Pauli (Scientific Correspondence With Bohr, Einstein, Heisenberg, a.O./Volume IV : 1955-1956)*, volume 4. Springer, 2001.

[35] Robert M. Wald. *General Relativity*. University of Chicago Press, 1984.

[36] Hermann Weyl. Elektron und Gravitation. I. *Zeitschrift für Physik*, 56:330–352, 1929.

First Mediterranean Conference on Classical and Quantum Gravity (MCCQG 2009) IOP Publishing
Journal of Physics: Conference Series **222** (2010) 012053 doi:10.1088/1742-6596/222/1/012053

Causal Sets: Quantum gravity from a fundamentally discrete spacetime

Petros Wallden

1. University of Athens, Physics Department, Nuclear & Particle Physics Section, Panepistimiopolis 157-71, Ilissia Athens, Greece
2. Chalkis Institute of Technology, Department of Technological Applications, Psahna-34400, Greece

E-mail: petros.wallden@gmail.com

Abstract.
In order to construct a quantum theory of gravity, we may have to abandon certain assumptions we were making. In particular, the concept of spacetime as a continuum substratum is questioned. Causal Sets is an attempt to construct a quantum theory of gravity starting with a fundamentally discrete spacetime. In this contribution we review the whole approach, focusing on some recent developments in the kinematics and dynamics of the approach.

1. Motivation

The challenge to construct a quantum theory of gravity is great. There exist many approaches, each of them having their merits, however there is no widely accepted and confirmed theory. The difficulty of the problem suggests, similarly with other paradigms in the history of physics, that we may need to abandon some of our a-priori assumptions about reality. In particular the continuity of spacetime that we assume, is questioned in several places. Approaches to quantum gravity that start from a continuum theory, observe a (possibly effective) discreteness[1]. Another field that we may get hints about quantum gravity, is the black holes thermodynamics. Infinities that arise in this case, could be avoided if space was discrete as it is observed in[1][2]. The infinities we face in quantum field theory (though we have learnt to live with them), and singularities in general relativity are two other reasons why discreteness could simplify our physical picture. All the above, would be resolved, if one was willing to abandon spacetime

[1] Discreteness of volume in Loop Quantum Gravity, dualities between very small and large scales in string theory, etc.

[2] Research in causal sets, involves some treatment of black hole thermodynamics. In this contribution we will not deal with it, but we give the following references for the interested reader [2].

Journal of Physics: Conference Series **222** (2010) 012053 doi:10.1088/1742-6596/222/1/012053
continuity[3]. It is therefore a natural starting point, for building a new quantum theory of gravity, to take spacetime fundamentally discrete.

The second element in causal sets, is *causality*. Here again, it seems more intuitive to speak of causal relations (that have the natural interpretation of cause and effect), rather than some standard spacetime where one needs to assume topology, differential structure and metric. After all, as we will see below causal relations encode most of the information of the metric.

We therefore, come to conclude that it may be better starting a quantum gravity theory from a discrete spacetime that comes along with the causal relations of the elements (points) of this spacetime. This is precisely the starting point of the causal set approach to quantum gravity. It was originally introduced in [3].

2. Definition

Mathematically a causal set (or causet) is a set C endowed with a partial order relation $\prec$ which is irreflexive ($x \not\prec x$), transitive ($x \prec y \prec z \Rightarrow x \prec z$), and locally finite $[x,y] \equiv (|\{y|x \prec y \prec z\}| < \infty \,\forall \quad x,z \in C)$ (where $|A|$ indicates cardinality of the set A). The local finiteness condition imposes the requirement of discreteness, because it requires that there is only a finite number of elements between every pair in the causet. The partial order relation, represent the causal relation between elements of C, so if $x \prec y$ it means that x is at the past of y.

The next thing to do is to see why a causal set (as defined above) could possibly replace continuum spacetime. This is based on a theorem by Malament [4]:

> The metric of a globally hyperbolic spacetime can be reconstructed uniquely from its causal relations up to a conformal factor.

The theorem suggests that the causal structure encodes most of the information of the metric. What we lack to obtain the full metric is a way to fix the conformal factor, in other words to fix the volume. This is precisely done by considering a discrete spacetime. In causal set we make the correspondence Volume= Number of elements[4]. By having only a finite number of causal set elements between any pair of causally related spacetime points, we can use that number to fix the volume. We thus have the "slogan" Order+Number=Geometry. We are now in position to make the central conjecture of causal sets (also called the Hauptvermutung):

> **Central Conjecture**: Two distinct, non-isometric spacetimes cannot arise from a single causal set.

The conjecture, while it still remains a conjecture has very solid theoretical and numerical evidence in its support[5]. The central conjecture guarantees that there can be only one (essentially) spacetime approximating a given causal set. Here we must stress the fact that a causal set behaves as a continuum manifold, under several conditions, and in particular, when the number of causal set elements is very big. In this case we speak of the *continuum approximation*[6].

[3] There are also models for modifying gravity that would resolve the cosmological constant problem, that favor discreteness.

[4] In particular each element of the causal set corresponds to 1 Plank unit Volume.

[5] Most of the work on the kinematics of causal sets, regarding the dimension (e.g. [5]), timelike and spacelike distances ([6, 7]), topology ([8]), provide support to the conjecture.

[6] It is the analogue of continuum limit of other approaches, however it is not really a limit, since we never take the discreteness scale going to zero, as it is done for example in CDT [9].

Journal of Physics: Conference Series **222** (2010) 012053 doi:10.1088/1742-6596/222/1/012053

We define the concept of *faithful embedding* to make precise when a continuum spacetime approximates a causal set:

A faithful embedding is a map ϕ from a causal set $\mathcal{P}$ to a spacetime $\mathcal{M}$ that:

(i) preserves the causal relation (*i.e.*, $x \prec y \iff \phi(x) \prec \phi(y)$) and

(ii) is "volume preserving", meaning that the number of elements mapped to every spacetime region is Poisson distributed, with mean the volume of the spacetime region in fundamental units, and

(iii) $\mathcal{M}$ does not possess curvature at scales smaller than that defined by the "intermolecular spacing" of the embedding (discreteness scale).

The central conjecture therefore reads as "a causal set cannot be faithfully embedded in two non isometric spacetimes". Taking a closer look at the definition of faithful embedding, we see that a regular lattice cannot correspond to a faithful embedding of Minkowski spacetime. It fails because, in a regular lattice, considering a very boosted frame, we will end up with big volumes having no elements at all and fail to satisfy the condition (ii) of faithful embedding. Instead, a *random* lattice would do this job. For example one generated by sprinkling elements in spacetime randomly with probability $P(n) = \frac{(\rho v)^n \exp^{-\rho V}}{n!}$, in other words a Poisson sprinkling[7].

Causal sets are constructed in such a way, that they respect Lorentz invariance at the kinematic level already [8]. These two observations are analyzed in detail in [10].

3. Kinematics

In all discrete/combinatoric approaches to quantum gravity one important step is the so called "inverse problem" [11]. It consists of two parts, the first is how do continuum like structure arise from the fundamentally discrete base. The second part, is how shall we recognize that we do have something continuum like (and what) when we actually have it.

In our case, we want to address this issue for a causal set. In particular we want to obtain continuum notions from the partial order. A typical causal set (if taken by pure counting and attributing equal weights to all causal sets), it is a Kleitman-Rothschild order [12]. These orders have only three layers (i.e. longest chain is a 3-element chain) and 1/4 of elements sit at the first layer, 1/2 in the second and 1/4 in the third. This is certainly not manifold-like (it contains only 3-moments of time). We thus need the dynamics to select for us a causal set that looks like a continuum manifold. For now we will deal with properties of causal sets that in the case there exist a manifold that faithfully embeds to it, correspond to continuum properties such as lengths of curves.

Some definitions are in order:

Definition 1: A pair of elements x, y is a *link* if $x \prec y$ and $\nexists \quad z | x \prec z \prec y$. In other words is the most basic relation, that cannot be implied by transitivity.

Definition 2: Chain C is a collection of elements such that $\forall x, y \in C$ either $x \prec y$ or $y \prec x$. In other words it is linearly ordered [9].

Definition 3: We define a *path* to be a chain, where each pair of consecutive elements are links.

Definition 4: We define a *2-link*, given a pair of unrelated elements x, y, an element z that x, z and y, z are both links.

[7] And the expected variance is of course $\sqrt{V}$.

[8] Unlike for example spin networks, where Lorentz invariance is expected to arise due to some properties of the superimposed Lorentz-violating spin networks

[9] this intuitively corresponds to a timelike curve at the continuum

Journal of Physics: Conference Series **222** (2010) 012053 doi:10.1088/1742-6596/222/1/012053

The concept of timelike curve is represented by a chain. Moreover, it can be shown that we can define the timelike distance between two related elements $x \prec y$ of the causal set. In the continuum it corresponds to the maximum length of timelike curve that joins x and y. In the causal set this is simply the chain of maximum cardinality starting from x and ending at y [6][10].

$$d_t(x,y) = \max |C| \text{ where } C \text{ start at x and ends at y} \tag{1}$$

It has been tested theoretically but also numerically [6, 7], that in the case that the causal set is faithfully embedded at a manifold [11] the above distance agrees with the continuum spacelike distance, in the limit of a sufficiently big causal set.

To get spacelike distance for a causal set that corresponds to Minkowski spacetime is considerably more difficult. Lee Smolin for example has claimed that what lacks from causal sets is some spatial information. The first naive attempt to get the spatial distance would be the following. Consider x, y two unrelated (spacelike) elements of the causal set. Then let $u \in J^+(x) \cap J^+(y)$ and $v \in J^-(x) \cap J^-(y)$.

$$d_s(x,y) = \min d_t(v,u) \tag{2}$$

Unfortunately this fails as was already noted in [6] and confirmed numerically in [7]. A more refined treatment is required where $u \in J^+(x) \cap J^+(y)$ but also u is a 2-link (w.r.t. x, y). The reader is referred too [7] for further details.

It can be also shown, that using this notion of spacelike distance, one can define closest spatial neighbors (timelike closest neighbors are simply the links). Using this concept we can measure the length of any curve in the causal set (not only on chains). More importantly this generalizes in curved spacetimes, provided that there is no curvature at very small scales. This is a reasonable assumption made also by Gibbons in [13].

The dimension of a causal set is also examined in [5]. Assume we have two related elements $x \prec y$, with $d_t(x,y) = n$ and the number of elements between them being N ($|V_{xy}| = N$ where $z \in V_{xy}$ if $x \prec z \prec y$). We then see how the volume scale as we double the timelike distance [12]. This and other dimension measures that work for causal sets embedable in continuum spacetimes as well as extensions of the notion of dimension for non-manifold like causal sets can be found in [5].

Finally, the spatial topology has been analyzed (e.g. in [8]). The concept of spatial slice is replaced by the following: We consider a *maximal anti-chain* a collection of elements $\{e_i\}$ such that $e_i \not\prec e_j$ and does not exist any other element being unrelated to all the elements of our anti-chain[13]. However, this is not a good analogue for a spatial slice. We need to "thicken" by considering all elements with distance a number n from the anti-chain. There is a nice way to recover the topology of this slice, using simplicial compleces. The results are in very good agreement with continuum results for causal sets that embed in continuum spacetimes [8].

4. Dynamics

While a lot can be said from the kinematic sector of a theory (and as we will see later, some phenomenological effects), the true physical content of a theory lies in the dynamics. There are two ways to try and adopt the dynamics. The first[14] takes as fundamental ingredient the

[10] To be more precise it is proportional and the proportionality constant in general depends on the dimension
[11] Basically, it was tested that causal sets that arose from sprinkling to some continuum manifold
[12] in other words take an element z such that $d_t(x,z) = 2n$ and find $|V_{xz}|$
[13] Adding any other element makes our set stop being an anti-chain
[14] called initially by Sorkin as the "principled" approach

Journal of Physics: Conference Series **222** (2010) 012053 doi:10.1088/1742-6596/222/1/012053
causal set itself and attempts to generate the full causal set from some basic principles that are in some sense natural to the causal set approach, in other words simply related with the partial order. The second approach [15], is an attempt to guess some dynamics (possibly by rephrasing the continuum Einstein action, in terms of the order relation) and get "correct" result[16]. Finally, of interest is to consider, quantum matter (or fields) on one classical causal set (an analogue of QFT in curved spacetime). There has been considerable progress in the latter recently. Here we will first consider some classical (alas stochastic dynamics). Then we will discuss about quantum dynamics. This task has not been fully accomplished, and we will only mention some toy models, attempts and directions for future research. Finally we will deal with quantum matter on a causal set.

4.1. Classical Dynamics for Causal Sets

The stochastic classical dynamics have been analyzed in depth (see [14]). One can construct a partial order that consists of all the causal sets (a partial order of partial orders), where two elements are related if the one is subset (when viewed as a partial order) of the other. It is called poscau (see Fig. 4.1). This is the arena for causal set dynamics and is an analogue of the superspace. It is the set of all possible spacetimes. We can imagine a process that generates all the causal sets, by growing the causal set element by element. This process is characterized by some probability, that each new-born element will be at the future or spacelike (unrelated) from each of the already existing elements. These processes are know as *(classical) growth dynamics* (CGD)[17]. We impose some physical conditions that these transitions probabilities must obey. In particular the requirements are:

(i) **Internal temporality**. Each new element is born to the future or spacelike to all existing elements (never at the past).

(ii) **Discrete general covariance**. The probability of arriving at a particular causal set does not depend on the order the elements are born. In other words the probability does not depend on the path chosen on the Figure.

(iii) **"Bell's causality"**. This condition states that the probability of new-born elements depends on the past of the new-born element and thus is not affected by births in spacelike regions. The name cames from the fact that it is related with the paradoxes that arise in quantum theory due to non-locality. It is believed that this condition may need to be relaxed (or transformed) if we were to consider a quantum theory of causal sets[18].

(iv) **The Markov Sum rule**. This means that the sum of all probabilities starting from a particular causal set (element of the poscau), sum up to one.

These conditions turn out to be very restrictive, and there is only one free parameter left for each level[19]. One particular choice of these parameters results in the transitive percolation that has been studied in depth (see references in [14]). The different choices give rise to very different causal sets. While some of these have striking similarities with cosmological models [15], the resulting causal sets are not in general manifold like and thus we would need to resort to quantum dynamics to explain the continuum like behavior we see.

[15] called the "opportunistic" approach

[16] It is opportunistic, since the Einstein action has a very good motivation in terms of continuum spacetime, but the analogue on a causal set is not something intuitive.

[17] They can be thought as generalizations of random walks

[18] However, it is not necessary since non-local effects on a causal set, may arise even if the dynamics of the causal set itself are in some sense local.

[19] By level, we mean the number of total elements of the causal set.

First Mediterranean Conference on Classical and Quantum Gravity (MCCQG 2009) IOP Publishing
Journal of Physics: Conference Series **222** (2010) 012053 doi:10.1088/1742-6596/222/1/012053

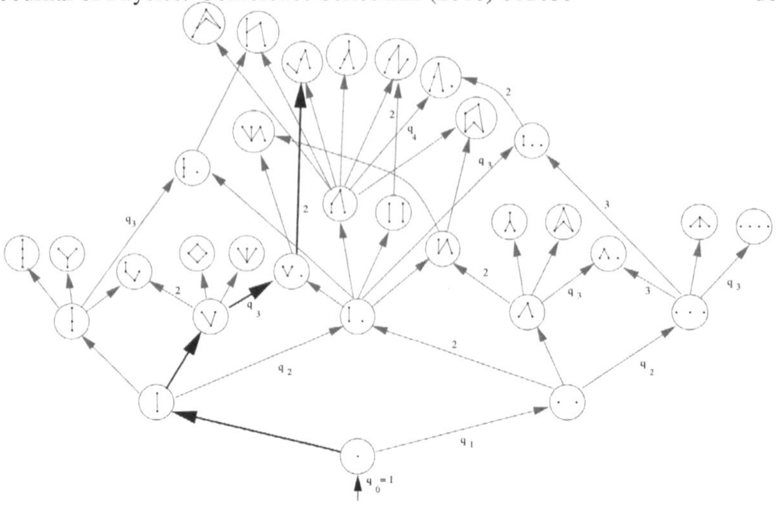

The poset of finite causal sets

4.2. Quantum Causal Sets

A first attempt to quantize the above dynamics, would be to use quantum amplitudes instead of probabilities. In other words to allow the transition probabilities to be complex numbers[20].This would be a generalization of what is known as quantum random walk (QRW). The whole causal sets program is based on spacetime rather than space. It is not possible to do a canonical quantization, since there is no well defined way to break up the full causal set to spatial slices with an external time parameter[21]. It is generally believed that in order to deal with quantum causal sets, one is forced to resort to sum over histories quantization. The mathematical ingredient one needs to use is the *quantum measure* [16][22]. To interpret it properly, one needs to address the basic problems of the foundations of quantum theory. Consistent (or decoherent) histories is one approach, while there is considerable recent activity in a novel interpretation the "Pimbino" interpretation (also called the co-event or anhomomorphic logic interpretation) [17].

The most interesting result however comes from considering 2-dimensional partial orders [18]. Here the dimension is not to be understood as spacetime dimension, it is rather a technical term for partial orders. In particular it means the following: Assume that we have a set $\mathcal{P} = \{e_1, e_2, \cdots\}$ and we have two linear orderings of this set . Then we take the intersection of this linear orderings, in other words an element $e_1 \prec e_2$ iff e_1 is before e_2 in both linear order we have. The resulting partial order is called 2-dimensional, if it can be generated from the intersection of 2 linear orderings[23]. For 2-dimensional partial orders it turns out, that they do correspond to 2-dimensional spacetimes, however this analogy breaks down for higher dimensions.

Brightwell et al in [18] considered 2-d orders and atttempted to find what the typical 2-

[20] Tentative results show that the typical causal set still is not one manifold-like

[21] Despite the fact that as we have seen there exist an analogue of a spacelike surface for causal sets that correspond to spacetimes that allow such a foliation. However is not fundamental for the approach.

[22] In classical stochastic theories such as the Brownian system, one uses again histories, and a measure (the Wiener measure) on the set of possible histories. In a sense we believe that quantum theory comes as a natural generalization of classical stochastic theories rather than deterministic theories.

[23] By analogy it is n-dimensional if it is the intersection of n linear orderings.

Journal of Physics: Conference Series **222** (2010) 012053 doi:10.1088/1742-6596/222/1/012053
d order would look like. Note, that as we have already mentioned a typical causal set is not manifold-like (it is a Kleitman-Rothschild order). This problem is also known as the entropy problem and is faced in other approaches such as the DT (dynamical triangulation) and a possible solution is to restrict to sum over sum particular class rather than all possible configurations (as was done in CDT (causal dynamical triangulations) [9]). Interestingly, in 2-d orders the major contribution comes from causal sets that are manifold-like and moreover there are faithfully embeddable in 2-d Minkowski. This can be seen as a toy model to see how manifold-like behavior (and indeed flat!) arises from simple considerations on a causal set.

Finally, recently Bombelli and Svedrlov in [19] have taken the "opportunist" approach. They have written the Einstein action in terms of causal set elements, and have used this weight to make a sum over all possible causal sets. However, this is work is at the beginning. One has to see what is a typical causal set, if we were to use this weight and also what consequences that would have. The reader is referred to the original papers for further details.

4.3. Quantum Matter on Causal Set

One thing of great interest, would be to see how matter and fields behave, on a single fixed (and thus classical) causal set. This question corresponds to fixing the background to some (generally) curved spacetime and allowing quantum fields on it. For causal sets, we need to firstly see how we recover standard results, and then see wether the assumption of discrete background may have some effect that could be possible to test and thus confirm or falsify the causal set assumption[24].

The first attempt would be to consider point (massive) particles moving along a chain of the causal set that faithfully embeds at Minkowski spacetime. There are several models considered [20, 21], that in all of those, when taking the continuum approximation, the particle follows approximately timelike geodesic, but deviating (swerving) slightly. In other words it is like having some drift, and all models result in a diffusion equation depending on a single parameter the diffusion strength k. Let us here briefly describe the first model ([20]). The particles trajectory is a chain $\{e_1, e_2, \cdots\}$ in the causal set. we define a forgetting time t_f, which is the time below which the causal set may behave non-locally, but above which it behaves normally. If the particle has position e_n with four momentum p_n the next element is chosen

- e_{n+1} is at the causal future of e_n and at proper time t_f
- the momentum change $|p_n - p_{n+1}|$ is minimized (where p_{n+1} is on the mass shell and proportional to the vector between e_n and e_{n+1}).

The resulting motion, heuristically, stays as close to straight as possible, having however some random fluctuations in the particles momentum. The diffusion equation resulting in the continuum approximation was (see [20])

$$\frac{\partial\rho}{\partial\tau} = k\nabla^2\rho - \frac{1}{m}p^\mu\partial_\mu\rho \tag{3}$$

Following similar lines, one may attempt to see what happens for massless particles. The added difficulty being that the trajectory is no longer a chain (which by definition is a timelike curve). The resulting equation has two parameters (diffusion and drift parameters) k_1, k_2 and there is again some non-locality scale. The reader is referred to [21].

[24] Of course the observed effect, typically would depend on some parameters and perhaps on the full quantum dynamics.

This particular feature, of some non-locality scale, appears also from consideration of particles propagators. In order to do this one needs to define the D'Alembertian operator $\Box$. In [22], the d'Alembertian is introduced. We will follow briefly [23], to obtain the retarded propagator for a particle on a causal set.

To get the propagator $K(x, y)$ from an element x to y in its future, we need to sum over all the chains (or paths) on a causal set from x to y with a particular amplitude (c.f. quantum mechanics where one adds the quantum amplitudes to go from one point to another and sums over all the possible trajectories). We need to make a choice of which trajectories to sum over (either all chains or all paths), and what weight to put on each trajectory. These choices have to be made in such a way, that when a causal set is faithfully embedable in Minkoweski spacetime, to give us back a propagator of the Klein-Gordon equation (in particular, with the choices made we shall get the retarded propagator). In particular the amplitude depends on two parameters a being the probability that the particle would 'hop' once along the trajectory from one element to another, and b being the probability that would stop at an element of the trajectory. For a chain of length n (where we have n hopes and $n-1$ intermediate stops) the amplitude would be $a^n b^{n-1}$. For causal set in 1+1 Minkowski and 3+1 Minkowski we fix the values of a and b to be respectively[25]:

$$a = \frac{1}{2}, \quad b = -\frac{m^2}{\rho} \tag{4}$$

$$a = \frac{\sqrt{\rho}}{2\pi\sqrt{6}}, \quad b = -\frac{m^2}{\rho} \tag{5}$$

With these choices the retarded Klein-Gordon propagator is obtained as shown in [23].

Building up on this work, using the retarded Klein-Gordon propagator, Johnston in [24] proceeded and considered scalar quantum field on a causal set, and he computed the Feynman propagator[26].

5. A Solution to the Cosmological Constant Problem

The biggest probably success of the causal sets approach to date, in terms of concrete result, is the prediction of the correct order of magnitude of the cosmological constant [26]. The argument is heuristic, but conclusive, in the sense that a zero cosmological constant (or of different magnitude) is ruled out by causal sets (or else a different result rules out causal sets). Another interesting point of this prediction is that it was first made in 1990, when at the time it was believed that its value is zero, and only much later was measured to be of that order of magnitude. Here we will present briefly the argument, and the reader is referred to [26] where it first appeared and [27] where it is analyzed in some depth along with possible problems and open questions.

The cosmological constant problem arises from the following observation. The dark energy is the 70% of effective energy density of the universe. Moreover it has negative pressure. This is usually explained by the use of a cosmological constant Λ. However, it appears to be very small, since the natural value we would expect from the vacuum expectation value is $\rho_{vac} \simeq m_{pl}^4$ which is $\simeq 10^{120}\rho_{obs}$ the observed value. Most mechanism that makes it small brings it to identically

[25] The values depend only on the mass of the particle and the volume corrseponding to each causal set element, i.e. the density.

[26] Of crucial importance was the use of Pauli-Jordan function [25] $\Delta(x) := G_R(x) - G_A(x)$ and its analogue for a causal set. Note that this could also generalize for causal sets embedable in curved spacetimes.

Journal of Physics: Conference Series **222** (2010) 012053 doi:10.1088/1742-6596/222/1/012053

zero (e.g. supersymmetry). The other strange feature is that the cosmological constant is of the order of magnitude of the ambient density, but only *now* (i.e. in our epoch). This also seems as an unnatural fine-tuning. The causal set proposal will suggest why the latter fine tuning appears and explain how the cosmological constant is non-zero (and of the desired magnitude) given a mechanism that would otherwise bring it to zero.

We need to introduce here an alternative formulation of general relativity the so-called *unimodular* gravity. It is a path integral formulation where the total volume V is kept fixed. The equation of motions are identically the same as in ordinary GR. The cosmological constant term $\Lambda = \Lambda_0 + \lambda$ appears in the action along with V in a $-\Lambda V$ term:

$$\delta(\int(\frac{1}{2\kappa}R - \Lambda_0)dV - \lambda V) = 0 \tag{6}$$

In unimodular gravity, the cosmological constant and the volume are conjugate pairs very much like Energy-Time. It obeys a similar uncertainty principle where $\Delta V \cdot \Delta\Lambda \sim \hbar$. In standard unimodular gravity, the volume is fixed, and the cosmological constant is completely undetermined from the parameters of the theory as expected.

However, in causal set things change. The role of time is replaced by the number of elements N (e.g. growth dynamics) and in QT we do not sum over different times. It is natural to causal sets to consider fixed N, which is *almost* unimodular gravity. The difference being that we have fixed N instead of V. N is proportional to V *up to poisson fluctuations*, i.e. $\pm\sqrt{V}$. We thus have some "kinematic fluctuations". Taking the conjugate of the volume we get $\Delta\Lambda \sim 1/\Delta V \sim 1/\sqrt{V}$. Assuming that there is a mechanism that brings the cosmological constant to zero, then applying this on a causal set we get $\Lambda \sim V^{-1/2} \sim H^2 \sim \rho_{\text{critical}}$. This cosmological constant is thus of the correct magnitude and is *always* of the order of magnitude of the critical density. This happens by simple considerations of the kinematics of the causal set (being a random partial order). The V is the volume of the past and thus result in an everpresent Λ of varying magnitude and sign. Particular models/implimantations of dynamics are needed to check whether this assumption change the standard picture of cosmology (as e.g. the Big Bang Nucleosynthesis). In [27] such a model is considered.

6. Summary and Conclusion

We have briefly reviewed the causal sets approach to quantum gravity, giving the original references for deeper study. In particular we have seen what a causal set is and how it succeeds in being both discrete and Lorentz invariant. The continuum concept arise from solely the partial (causal) order, and it has been shown that these do agree with the continuum notions for large enough causal set that have a continuum spacetime that faithfully embeds in it. Further we have seen how with some very simple and natural requirements for the (classical) dynamics we get a family of solutions that is quite restrictive. Getting the full quantum dynamics of causal sets, is the major open issue for causal sets, even though there are considerable recent developments. Firstly a 2-dimensional full gravity model that gives as the typical causal set Minkowski spacetime (which is certainly manifold-like!). Secondly, transforming the Einstein action in terms of causal relations has been done and it remains to be examined. Thirdly, there is progress in the conceptual part, and the development of a histories quantum theory of closed systems, the Piombino interpretation. A semi-classical description, where spacetime is a single causal set, but matter (and or fields) lying on it are quantum has been developed recently in a satisfactory degree and what remains now, is to explore possible deviations from standard QFT on curved background physics, in order to test the assumption of causal sets. Finally, we reviewed the proposed explanation, using causal sets, for the small non-zero cosmological constant observed.

Acknowledgments
The author would like to thank all the people that he interacted and discussed about causal sets and in particular Rafael Sorkin, Fay Dowker, Sumati Surya, David Rideout and Yousef Ghazi-Tabatabai. This work is partly supported by IKY. Finally, the author would like to thank the organizers of the conference.

References

[1] L. Bombelli, R. Koul, J. Lee and R. Sorkin, *Phys. Rev. D* **34** (1986) 373; M. Srednicki, *Phys. Rev. Lett.* **71** (1993) 666.
[2] D. Dou and R. Sorkin, *Found. Phys.* **33** (2003) 279; D. Rideout and S. Zohren, *Class. Quant. Grav.* **23** (2006) 6195.
[3] L. Bombelli, J.H. Lee, D. Meyer, and R. Sorkin, *Phys. Rev. Lett.* **59** (1987) 521.
[4] David. Malament, J. Math. Phys. **18**, 1399 (1977).
[5] D.A. Meyer, *Order* **10** 227–237 (1993); Jan Myrheim, "Statistical Geometry", CERN preprint Ref.TH.2538-CERN (1978).
[6] G. Brightwell and R. Gregory, Phys. Rev. Lett. 66: 260-263 (1991).
[7] D. Rideout and P. Wallden, *Class. Quant. Grav.* **26** (2009) 155013.
[8] S. Major, D. Rideout, and S. Surya, *J. Math. Phys.* **48**, 032501, (2007); Sumati Surya, *Theoretical Computer Science* **405**, 1–2, pp. 188–197 (2008).
[9] J. Ambjorn, J. Jurkiewicz, R. Loll, To appear as part of a Springer Lecture Notes in Physics publication: "Quantum Gravity - New Paths Towards Unification", [arXiv:gr-qc/0906.3947].
[10] L. Bombelli, J. Henson and R. D. Sorkin, *Mod. Phys. Lett.* **A24** (2009), 2579; J. Henson *J. Phys. Conf. Ser.* **174** (2009) 012020.
[11] Lee Smolin, in Rickles, D. (ed.) et al.: *The structural foundation of quantum gravity* 196–239. (2005).
[12] D. Kleitman and B. L. Rothschild, *Trans. Am. Math. Soc.* **205**, (1975) 205.
[13] G. W. Gibbons and S. N. Solodukhin, Phys. Lett. B 649: 317 (2007).
[14] D. Rideout and R. Sorkin, *Phys.Rev.* **D61** (2000) 024002 ; D. Rideout and R. Sorkin, *Phys.Rev.* **D63** (2001) 104011.
[15] M. Ahmed and D. Rideout, preprint [arXiv:gr-qc/0909.4771].
[16] R. D. Sorkin, Mod. Phys. Lett. A **9**, 3119 (1994); R. D. Sorkin, *Quantum Classical Correspondence: Proceedings of the 4th Drexel Symposium on Quantum Nonintegrability*, Philadelphia, September 8-11, 1994, pages 229-251, (International Press, Cambridge Mass. 1997) D.H. Feng and B-L Hu, editors. [arXiv:gr-qc/9507057]; S. Surya and P. Wallden to appear in Foundation of Physics [arXiv:quant-ph/0809.1951].
[17] R. D. Sorkin, J. Phys. A **40**, 3207 (2007); R. D. Sorkin, J. Phys. Conf. Ser. **67**, 012018 (2007); F. Dowker and Y. Ghazi-Tabatabai, J. Phys. A **41**, 105301 (2008); Y. Ghazi-Tabatabai and P. Wallden, J. Phys. A: Math. Theor. **42** 235303 (2009).
[18] G. Brightwell, J. Henson and S. Surya, *Class. Quant. Grav.* **25** (2008) 105025.
[19] Roman Sverdlov and Luca Bombelli, *Class. Quant. Grav.* **26** (2009) 075011.
[20] F. Dowker, J. Henson and R. Sorkin, *Mod. Phys. Lett.* **A19** (2004) 1829.
[21] F. Doeker, L. Philpott and R. Sorkin, *Phys. Rev. D* **79** (2009) 124047.
[22] R. D. Sorkin, in *Approaches to Quantum Gravity*, D. Oriti (ed.), Cambridge University Press (2009).
[23] Steven Johnston, *Class. Quant. Grav.* **25** (2008) 202001.
[24] Steven Johnston, *Phys. Rev. Lett.* **103** (2009) 180401.
[25] N. N. Bogoliubov and D. V. Shirkov, Introduction To The Theory Of Quantized Fields Interscience Monographs and Texts in Physics and Astronomy (Wiley Interscience, London, 1959), Vol 3.
[26] Rafael Sorkin, in Relativity and Gravitation: Classical and Quantum, pp. 150-173. World Scientific, Singapore, 1991.
[27] M. Ahmed, S. Dodelson, P. Greene and R. Sorkin, *Phys. Rev. D* **69** (2004) 103523.

First Mediterranean Conference on Classical and Quantum Gravity (MCCQG 2009) IOP Publishing
Journal of Physics: Conference Series **222** (2010) 012054 doi:10.1088/1742-6596/222/1/012054

Projectable Hořava–Lifshitz gravity in a nutshell

Silke Weinfurtner[1], **Thomas P. Sotiriou**[2] **and Matt Visser**[3]

[1] Department of Physics and Astronomy, University of British Columbia, 6224 Agricultural Road, Vancouver, B.C. V6T 1Z1 Canada;
SISSA-International School for Advanced Studies, Via Beirut 2-4, 34014 Trieste, Italy
[2] Department of Applied Mathematics and Theoretical Physics, Centre for Mathematical Sciences, University of Cambridge, Wilberforce Road, Cambridge, CB3 0WA, UK
[3] School of Mathematics, Statistics, and Operations Research, Victoria University of Wellington, New Zealand

E-mail: [1]`silkiest@gmail.com`, [2]`T.Sotiriou@damtp.cam.ac.uk`,
[3]`Matt.Visser@msor.vuw.ac.nz`

Abstract. Approximately one year ago Hořava proposed a power-counting renormalizable theory of gravity which abandons local Lorentz invariance. The proposal has been received with growing interest and resulted in various different versions of Hořava–Lifshitz gravity theories, involving a colourful potpourri of new terminology. In this proceedings contribution we first motivate and briefly overview the various different approaches, clarifying their differences and similarities. We then focus on a model referred to as projectable Hořava–Lifshitz gravity and summarize the key results regarding its viability.

1. Gravity as quantum field theory

Even though general relativity has been a very successful classical field theory of gravity, unlike other field theories, *e.g.* electrodynamics, one cannot quantize it in the most straightforward manner, *i.e.* using the canonical quantization or path integral formalism. A perturbative loop expansion for gravity results in infinitely many ultraviolet (UV) divergent Feynman diagrams, and at each order the theory requires counterterms of ever-increasing degree in curvature. Consequently, it is necessary to fix an infinite number of free parameters to have a well-defined ultraviolet behavior, and we are left with a quantum theory that has no predictive power at small distance scales, where the counterterms are crucial. In this sense gravity is said to be non-renormalizable.

There are in principle two different ways to deal with the situation at hand. The first one is to abandon the whole idea of renormalizability, by assuming that gravity is an effective field theory only valid below the cutoff scale. Here the cutoff is a finite, physically meaningful quantity that acts as a short-distance regulator fixing the divergences in the Feynman diagrams. Beyond that scale the theory requires an ultraviolet completion. A typical example of such a theory is the Fermi theory of the weak nuclear force. The other way to proceed is to insist on the concept of a self-sufficient gravitational theory and further investigate perturbative or non-perturbative tools resulting in a renormalizable (or even finite) quantum theory of gravity.

Hořava–Lifshitz gravity [1, 2] is grouped within the second category and investigates extensions to Einstein–Hilbert gravity that explicitly break local Lorentz invariance resulting in a power counting renormalizable $d+1$ dimensional theory of quantum gravity. In the manuscript at hand key features of Hořava–Lifshitz gravity are reviewed. We then focus on the projectable version of Hořava–Lifshitz gravity [2, 3, 4].

The big appeal of Hořava–Lifshitz gravity as a quantum gravity candidate lies in the fact that — once the theory is set up — one has the chance to perform concrete computations and explicitly check its viability. As usual, it would be very difficult to prove the overall working success of the theory. Instead, one tends to focus on the various shortcomings in order to eventually establish whether they can be overcome or whether the theory is ruled out. Our intention here is to overview projectable Hořava–Lifshitz gravity and summarize its shortcomings.

2. Emergence of local Lorentz symmetry

The key idea behind Hořava–Lifshitz gravity is to sacrifice local Lorentz symmetry in exchange for renormalizability: Local Lorentz symmetry is violated at the non-perturbative level of our field theory, but will hopefully be recovered at long distance / low energy scales. The particular kind of Lorentz violations we will encounter here have been discussed to some extent within quantum gravity phenomenology [5, 6, 7]. Let us consider the following class of theories that break Lorentz invariance fundamentally, and that at the kinematical level lead to a dispersion relation which is some function of momentum and mass, $E^2 = m^2 + p^2 \to E^2 = F(p, m)$. Specifically, the dispersion relations encountered by Hořava–Lifshitz theories are truncated at finite order $2z$,

$$E^2 = m^2 + p^2 + (\tilde{\eta}_1)_i\, p^i + (\tilde{\eta}_2)_{ij}\, p^i p^j + \ldots + (\tilde{\eta}_{2z})_{i \ldots j}\, p^i \ldots p^j, \tag{1}$$

The truncation coefficient z is also referred to as the degree of anisotropy, adopted from condensed-matter jargon referring to the anisotropic scaling between space and time.

As will be demonstrated shortly, local Lorentz symmetry breaking of the above type can act as a quantum field theory regulator in the ultraviolet for appropriate choices of z. Next we focus on two particular examples and discuss their (power-counting) renormalizability.

3. Renormalizability of anisotropic field theories

We will lay out in brief a road map to construct field theories in which Lorentz symmetry breaking acts as a quantum field theory regulator. The first example will be a polynomial interacting scalar field, and the second will be gravity. The study of the scalar field case (Lifshitz scalar) is mainly of pedagogical interest, as any detailed renormalization group calculations for the case of gravity are still pending and all assumptions are based on power-counting [8, 9, 10, 11, 12, 13]. This is in contrast to scalar field theories in this setting, which have been investigated more closely over the last few months [8, 9, 10, 11, 12, 14, 13].

3.1. Lifshitz scalar field theories

The simplest yet insightful field theory to demonstrate the above concepts is a (massive) scalar field with polynomial interactions on a flat background,

$$S = \int \left\{ \dot{\phi}^2 - \phi \tilde{D} \phi - m^2 \phi^2 + \sum_{n=1}^{N} g_n \phi^n \right\} dt\, d^d x, \tag{2}$$

where the differential operator $\tilde{D}$ is given by

$$\tilde{D} = -c^2 \triangle + \ldots + (-\triangle)^z. \tag{3}$$

For a relativistic field theory $z = 1$ and $\tilde{D}$ is the standard Laplacian, $\triangle$. These kind of theories are well understood, and we know that $\phi^n \,/\, \phi^6 \,/\, \phi^4 \,/\, \phi^3$ are renormalizable in $1 + 1 \,/\, 2 + 1 \,/\, 3 + 1 \,/\, 5 + 1$ dimensions respectively.

Journal of Physics: Conference Series **222** (2010) 012054 doi:10.1088/1742-6596/222/1/012054

Suppose now that z is left arbitrary and we impose anisotropic scaling between space and time, using the engineering dimensions

$$[dx] = [\kappa]^{-1} \quad \text{and} \quad [dt] = [\kappa]^{-z}\,, \tag{4}$$

where κ is a placeholder symbol for some object with dimensions of momentum (this scaling is the most convenient for power counting in this case). Choosing the dimensions of ϕ such that $[\phi]^2[dt\,dx^d] = [1]$, $\triangle^z$ becomes a marginal operator in eq. (3), and the lower-order $\triangle^n$ operators become relevant. The coupling constants g_n in action (2) have dimensions $[g_n] = [\kappa]^{d+z-n(d-z)/2} = [m]^{[d+z-n(d-z)/2]/z}$, which are non-negative as long as $d+z-n(d-z)/2 \geq 0$. For $n \to \infty$ this would require simply $z \geq d$. If we apply our experiences from studies of Lorentz invariant scalar theories, non-negative momentum dimensions are a good indication for renormalizability. In reference [12] the superficial degree of divergence was derived and it was shown that for $z = d$ ($z > d$) and with (without) normal ordering the model at hand is perturbatively ultraviolet finite.

Before we move on to gravity we would like to stress that although the ultraviolet behaviour appears to be significantly improved by the inclusion of suitable higher order operators in (2) it has already been shown that the one-loop renormalization and evolution of the coupling constants, for two interacting anisotropic scalar fields, is facing the naturalness problem, see reference [15, 14]: The difference in the speed of light between the two fields runs logarithmically to zero, and thus persists at low energy scales. This is in contradiction to our experimental constraints on Lorentz symmetry breaking, unless we impose some unnatural fine tuning on the bare coupling constants.

3.2. Hořava–Lifshitz gravity

The natural setting in which to apply the concepts discussed above to gravity is within the Hamiltonian formulation of general relativity based on the Arnowitt–Deser–Misner (ADM) decomposition of spacetime,

$$\mathrm{d}s^2 = -N^2c^2\mathrm{d}t^2 + g_{ij}(\mathrm{d}x^i - N^i\mathrm{d}t)(\mathrm{d}x^j - N^j\mathrm{d}t). \tag{5}$$

The ADM formalism foliates spacetime into a family of non-intersecting spacelike hypersurfaces, where the metric field g_{ij} is the induced geometry on the spatial slice, the scalar function N is the lapse, and the three-vector N^i is the shift-vector. The extrinsic curvature is defined as

$$K_{ij} = \frac{1}{2N}\left\{-\dot{g}_{ij} + \nabla_i N_j + \nabla_j N_i\right\}. \tag{6}$$

Within this setting one can write down an action of the form

$$S = \int [T - V]\,\sqrt{g}\,N\,d^3x\,dt, \tag{7}$$

where T denotes the part that contains time derivatives and is given by

$$T(K) = g_K\left\{(K^{ij}K_{ij} - K^2) + \xi K^2\right\}. \tag{8}$$

The potential V on the other hand does not contain time derivative but it can in principle contain any 3-covariant term constructed with the pieces of the ADM metric. We are abandoning Lorentz invariance and therefore we cannot automatically require ξ to vanish, as is the case in general relativity. Similarly, V can now include higher order spatial derivatives instead of just being equal to the 3-dimensional Ricci scalar, as in general relativity.

Journal of Physics: Conference Series **222** (2010) 012054 doi:10.1088/1742-6596/222/1/012054

An action of the above form is clearly not invariant under general diffeomorphisms. It is, however, invariant under the restricted set of diffeomorphisms that preserve the foliation, *i.e.* $t \to t + \chi^0(t)$, $x^i \to x^i + \chi^i(t, \mathbf{x})$. This symmetry is enough to prevent N^i from appearing in V. In the original proposal from Hořava N was also assumed not to enter in V [2], however, this constitutes an ad hoc assumption as noticed by [16] (we will return later to the various variants of the theory).

The way to proceed now is to invoke anisotropic engineering dimensions, as in equation (4), and to evaluate the dimensionality of coupling constants of the leading order operators in the UV. Starting from the ADM decomposition (5), we need to have $[N^i] = [\kappa]^{z-1}$ and we are free to choose $[g_{ij}] = [N] = [1]$, resulting in the dimensionality for the line-element $[ds] = [\kappa]^{-1}$, for the speed of light $[c] = [\kappa]^{z-1}$, the volume-element $[dV_{d+1}] = [\kappa]^{-d-z}$, and the extrinsic curvature $[K] = [\kappa]^z$. Under the constraint that the action has to be dimensionless, we obtain for g_K and for the coupling(s) g_{2z} of the operator(s) of order $2z$ in V

$$[g_K] = [g_{2z}] = [\kappa]^{(d-z)}. \tag{9}$$

These operators will dominate at the UV as they are the leading ones in temporal and spatial derivatives respectively. Choosing $z = d$ will make these couplings dimensionless. (A more detailed treatment of the power-counting and the derivation in general can be found in references [3, 4].) This implies that by including in V operators of order $2d$, we would end up with a power-counting renormalizable theory, along the lines of the Lifshitz scalar case discussed previously.

What remains to be discussed is which can be the exact form of V. As already mentioned, in his initial proposal Hořava assumed that V depends only on g_{ij} and its spatial derivatives. To simplify things further he proposed a specific combination of terms, dubbed detailed balance (see [2] for more details on the exact form of V in this case), which is inspired by quantum critical systems and non-equilibrium critical phenomena. The hope seemed to be that such a rigid construction would be imposed as a symmetry, simplifying renormalization group calculations. There is, however, no clear physical motivation for detailed balance apart from simplicity, and instead detailed balance appears to be disfavored by the first consistency tests, see for example [3, 17, 18].

Once detailed balance is abandoned there is no reason to exclude any possible term. With or without detailed balance, however, Hořava's model propagates an extra scalar mode with respect to general relativity and appears to be burdened with serious shortcomings , such as instabilities, overconstrained evolution and strong coupling at low energies [17, 19, 20, 21]. In Ref. [16] an extension was proposed, after noticing that terms involving N and its spatial derivatives can be included in V without violating the symmetry of the action. The scalar mode in this model does exhibit improved behavior. However, it still possesses strong coupling at energies orders of magnitude lower than the Planck scale [22]. One could do away with the strong coupling if the higher order operators become important at energies below this strong coupling scale, but this requires tuning of their coupling constants (or the introduction of a new scale) [23, 22].

We will not review the literature any further here. On what comes next, we will focus instead on a specific version of Hořava–Lifshitz gravity, the so-called projectable version, for reasons that will become clear shortly.

4. Projectable Hořava–Lifshitz gravity

We have argued that it is indeed possible to construct a power-counting renormalizable theory of gravity by adding ad-hoc spatial derivatives with twice the order of the spatial dimension of the theory. Therefore, in 3 spatial dimensions there is a plethora of terms which, when added to the Einstein–Hilbert action could lead to a theory with the desirable characteristics. As discussed above, there is no real guiding principle that would allow us to exclude any lower or 6th order operators. On the other hand, taking into account all terms is hardly an easy task from a practical perspective.

Another restriction proposed by Hořava was to impose that the lapse N be just a function of time t. The motivation behind this restriction came from the desire to match the reduced symmetry of the action mentioned earlier. There seems to be nothing fundamental in this restriction. On the contrary, it is bound to cause several problems, for instance, it will not allow us to access any gauge choice for which N is not a function of time only.[1] Most importantly, when $N = N(t)$ the Hamiltonian constraint will not be local, as one cannot get rid of the spatial integral after the variation. Taking this shortcomings into account, assuming $N = N(t)$ has also a serious advantage: it allows us to do away with several terms in the action, as all spatial derivatives of N vanish and at the same times many of the possible terms one can write end up differing only by a total divergence. This makes calculations tractable even for the most general action one could write (see also [3, 4]). In this sense, it is worth exploring this version of the theory, called projectable Hořava–Lifshitz gravity.

Within this setup the most general action can be put together following the lines of Refs. [3, 4]: applying integration by parts (discarding surface terms), commutator and Bianchi identities, and special relations arising in 3 dimensions (vanishing of the Weyl tensor), one obtains, besides the standard Einstein–Hilbert terms,

$$S_{\rm EH} = \int \left\{ (K^{ij}K_{ij} - K^2) + \zeta^4 R - g_0\, \zeta^6 \right\} \sqrt{g}\; N\; {\rm d}^3x\; {\rm d}t, \tag{10}$$

new Lorentz-violating terms involving in total eight dimensionless coupling constants $(\xi, g_2, \dots, g_8)$

$$\begin{aligned} S_{\rm LIV} \;=\; & \int \Big\{ \xi\, K^2 - g_2\, \zeta^2\, R^2 - g_3\, \zeta^2\, R_{ij}R^{ij} - g_4\, R^3 - g_5\, R(R_{ij}R^{ij}) - g_6\, R^i{}_j R^j{}_k R^k{}_i \\ & - g_7\, R\nabla^2 R - g_8\, \nabla_i R_{jk}\, \nabla^i R^{jk} \Big\} \sqrt{g}\; N\; {\rm d}^d x\; {\rm d}t. \end{aligned} \tag{11}$$

Here we have introduced a scale ζ in order to make all couplings dimensionless.

Instead of explicitly deriving the constraint and field equations by variation with respect to the lapse (Hamiltonian constraint) and the shift vector (super-momentum constraint), and the induced metric (dynamical equations of motion), we refer the reader to Refs. [3, 4]. However, we would like to point out the key difference between the projectable and the non-projectable case. As mentioned previously, in the former the Hamiltonian constraint is not a local one, implying that there is no local notion of energy conservation. At best we can say that the integral of the energy at each spatial slice is conserved.

It is worth noting that so far we have been working in units that impose anisotropic scaling between space and time simply because this is convenient for power-counting renormalizability arguments. If instead one needs to work in units where the speed of light c is equal to 1, then this can be easily achieved by setting $dt \to \zeta^{-2} dt$. After writing the action in these new units, see [3, 4], one can read off the Newton and cosmological constant,

$$\frac{1}{16\pi G_{\rm Newton}} = \zeta^2 \qquad \text{and} \qquad \Lambda = \frac{g_0\, \zeta^2}{2}, \tag{12}$$

so that ζ is identified as the Planck scale.

[1] On the other hand, it should be noted that the most physically interesting solutions of the Einstein equations, the Schwarzschild, Reissner–Nordström, Kerr, and Friedmann–Lemaitre–Robertson–Walker spacetimes can by suitable choice of coordinates all be cast into projectable form, in fact with $N = 1$ [24].

The main result up to now is that it is possible to include all possible terms and to explore — within the restricted but complete model — the key features of Hořava–Lifshitz gravity. It should be noted that this result can, in principle, be extended to the non-projectable case through a challenging but straightforward calculation.

5. Excitations in projectable Hořava–Lifshitz gravity

One of the most surprising features of general relativity is that, out of the possible fundamental excitations — spin-0, spin-1 and spin-2 — only the spin-2 excitation, the graviton, is physical. The spin-1 and spin-0 modes can be gauged away using general covariance. This feature of general relativity is in perfect agreement with observations, which have not so far revealed extra fields mediating gravity, see for example [25]. Therefore, a crucial viability test for any gravity theory is whether it exhibits other physical degrees of freedom which would have already been detected by observations.[2] In $3+1$ dimensions it is difficult to answer this question directly. It is much easier to study the degrees of freedom at each order perturbing around some specific background solution.

Within the framework of projectable Hořava–Lifshitz gravity one can assume that $g_0 = 0$ for simplicity and perturb around flat spacetime geometry, along the lines of Ref. [4]. After imposing suitable gauge conditions one can show that there exist a (suitably defined) transverse traceless mode $\widetilde{H}_{ij}$ satisfying the equation

$$\ddot{\widetilde{H}}_{ij} = -\left[-\partial^2 + g_3\zeta^{-2}\partial^4 + g_8\zeta^{-4}\partial^6\right]\widetilde{H}_{ij}. \tag{13}$$

Clearly, the spin-2 particle has the dispersion relation discussed previously and shown in eq. (1).

However, there exist an additional scalar mode: the trace h of the perturbation to the metric h_{ij}, which satisfies the following equation

$$\left(1-\frac{3}{2}\xi\right)\ddot{h} = -\xi\left\{-\frac{1}{2}\partial^2 + \left(-4g_2 - \frac{3}{2}g_3\right)\zeta^{-2}\partial^4 + \left(4g_7 - \frac{3}{2}g_8\right)\zeta^{-4}\partial^6\right\}h. \tag{14}$$

This is a wave equation with a trans-Bogoliubov dispersion relation. At low momentum the group and phase velocity approach the same limiting propagation speed,

$$c^2_{\text{spin}-0} = \frac{\xi}{2-3\xi}, \tag{15}$$

The low momentum linearized action for this scalar mode is

$$S_{\text{spin}-0} = -\int d^3x dt\left[\frac{1}{c^2_{\text{spin}-0}}\dot{h}^2 - (\partial h)^2\right]. \tag{16}$$

From the last two equations one can infer the following: for $\xi < 0$ and $\xi > 2/3$ the scalar modes is classically unstable. On the other hand, for $\xi \in (0, 2/3)$, the mode is classically stable but then the kinetic term has the wrong sign in the action. Therefore, for these values the mode is quantum mechanically unstable (ghost). This leaves us with only two special values allowed, $\xi = 2/3$ and $\xi = 0$. The latter is the value ξ has in general relativity, whereas the former is an order 1 deviation from this value and we will not consider it futher. For $\xi = 0$ the gauge used to derived the propagator presented above is no longer accessible and we have to do a separate special case study. Using a slightly different gauge led to a peculiar mode growing

[2] It is possible to construct condensed matter systems that exhibit spin-2 excitations, but in these systems all the lower-spin excitations are also physical due to the lack of general coordinate covariance. See for example Ref. [26].

quadratically with time in Ref. [4]. As it was later demonstrated in Ref. [27] this mode was actually a gauge artifact and the scalar appears to be frozen around flat background (see also Ref. [28] for Friedmann–Lemaître–Robertson–Walker spacetimes).

It appears that there is no scalar mode for $\xi = 0$ whereas instabilities plagued nearby values. However, there are two different reasons for which setting $\xi = 0$ by fiat will not work. Firstly, ξ is a running coupling and we are lacking a symmetry that would allow us to set it to any specific value. Secondly, appearances deceive and the $\xi = 0$ choice is not as healthy as it seems. In fact, even though for this value the mode appears to freeze around maximally symmetric backgrounds, this is only due to the fact that it actually gets strongly coupled at all scales [17, 29]. It seems that we have run out of available options for the value of ξ. On the other hand we know that we would have to drive ξ to zero eventually if general relativity is to be recovered at the infrared.

6. Discussion

After discussing the motivation and giving a brief overview of the various versions of Hořava–Lifshitz gravity, we focussed on one of them, the projectable version where the lapse is assumed to be only a function of time and not space. The main advantage of this version is that it drastically reduces the number of terms that one must consider in the action and it, therefore, makes it possible to perform explicit calculations using the most general action which respects the symmetries of the theory.

Unfortunately, projectable Hořava–Lifshitz gravity appears to be plagued with serious problems related to the existence of a scalar degree of freedom. This scalar mode turns out to be classically or quantum mechanically unstable in general, and when parameters of the theory are chosen such that the low energy limit corresponds to general relativity, the scalar exhibits strong coupling at all scales. One could still hope for a non-perturbative restoration of the limit to general relativity à la Vainshtein in massive gravity [30]. This is an avenue worth exploring but so far such an effect has not been shown to be present.

In the light of this we could choose to view the projectable version of Hořava–Lifshitz gravity as a useful tool in order to gain insight in the more complete versions of the theory. Indeed, some of the characteristics of this more limited version, such as the strong coupling, are still shared by other versions, but some are not (*e.g.* the fact that the Hamiltonian constraint is not a local one). Therefore, its study can provide some insight in the other versions, but only once the subtle differences are carefully weighed as well.

To conclude, Hořava–Lifshitz gravity constitutes an interesting quantum gravity theory. It appears challenging to construct a viable model within this framework. In fact there are numerous issues that have not even been systematically considered yet, such as matter coupling, renormalization group calculations, *etc.* Given the fact that sensible renormalizable gravity theories are not easy to construct, Hořava–Lifshitz gravity seems to deserve further study.

References

[1] Petr Hořava. Membranes at Quantum Criticality. *JHEP*, 03:020, 2009, arXiv:0812.4287.
[2] Petr Hořava. Quantum Gravity at a Lifshitz Point. *Phys. Rev.*, D79:084008, 2009, arXiv:0901.3775.
[3] Thomas Sotiriou, Matt Visser, and Silke Weinfurtner. Phenomenologically viable Lorentz-violating quantum gravity. *Phys. Rev. Lett.*, 102:251601, 2009, arXiv:0904.4464.
[4] Thomas P Sotiriou, Matt Visser, and Silke Weinfurtner. Quantum gravity without Lorentz invariance. *JHEP*, 10:033, 2009, arXiv:0905.2798.
[5] David Mattingly. Modern tests of Lorentz invariance. *Living Reviews in Relativity*, 8(5), 2005.
[6] Ted Jacobson, Stefano Liberati, and David Mattingly. Lorentz violation at high energy: Concepts, phenomena and astrophysical constraints. *Annals Phys*, 321:150, 2006, arXiv:astro-ph/0505267.
[7] Stefano Liberati, Matt Visser, and Silke Weinfurtner. Naturalness in emergent spacetime. *Phys. Rev. Lett.*, 96:151301, 2006, arXiv:gr-qc/0512139.

[8] Damiano Anselmi. Weighted power counting and Lorentz violating gauge theories. I: General properties. *Annals Phys.*, 324:874–896, 2009, arXiv:0808.3470.
[9] Damiano Anselmi. Weighted power counting and Lorentz violating gauge theories. II: Classification. *Annals Phys.*, 324:1058–1077, 2009, arXiv:0808.3474.
[10] Damiano Anselmi. Weighted power counting, neutrino masses and Lorentz violating extensions of the Standard Model. *Phys. Rev.*, D79:025017, 2009, arXiv:0808.3475.
[11] Damiano Anselmi. Weighted scale invariant quantum field theories. *JHEP*, 02:051, 2008, arXiv:0801.1216.
[12] Matt Visser. Lorentz symmetry breaking as a quantum field theory regulator. *Phys. Rev.*, D80:025011, 2009, arXiv:0902.0590.
[13] Matt Visser. Power-counting renormalizability of generalized Horava gravity. 2009, arXiv:0912.4757.
[14] Roberto Iengo, Jorge G. Russo, and Marco Serone. Renormalization group in Lifshitz-type theories. *JHEP*, 11:020, 2009, arXiv:0906.3477.
[15] John Collins, Alejandro Perez, Daniel Sudarsky, Luis Urrutia, and Hector Vucetich. Lorentz invariance and quantum gravity: An additional fine-tuning problem? *Phys. Rev. Lett.*, 93(19):191301, 2004, arXiv:hep-th/0603002.
[16] D. Blas, O. Pujolas, and S. Sibiryakov. A healthy extension of Horava gravity. 2009, arXiv:0909.3525.
[17] Christos Charmousis, Gustavo Niz, Antonio Padilla, and Paul M. Saffin. Strong coupling in Hořava gravity. *JHEP*, 08:070, 2009, arXiv:0905.2579.
[18] H. Lu, Jianwei Mei, and C. N. Pope. Solutions to Hořava Gravity. *Phys. Rev. Lett.*, 103:091301, 2009, arXiv:0904.1595.
[19] D. Blas, O. Pujolas, and S. Sibiryakov. On the Extra Mode and Inconsistency of Horava Gravity. *JHEP*, 10:029, 2009, arXiv:0906.3046.
[20] Miao Li and Yi Pang. A Trouble with Hořava-Lifshitz Gravity. *JHEP*, 08:015, 2009, 0905.2751.
[21] Marc Henneaux, Axel Kleinschmidt, and Gustavo Lucena Gomez. A dynamical inconsistency of Horava gravity. 2009, arXiv:0912.0399.
[22] Antonios Papazoglou and Thomas P. Sotiriou. Strong coupling in extended Horava-Lifshitz gravity. accepted in *Phys. Lett.* B, 2009, arXiv:0911.1299.
[23] D. Blas, O. Pujolas, and S. Sibiryakov. Comment on 'Strong coupling in extended Horava-Lifshitz gravity'. 2009, arXiv:0912.0550.
[24] M. Visser. *The Kerr spacetime: A brief introduction.* Published in *The Kerr spacetime: Rotating black holes in general relativity*, edited by D. Wiltshire, M. Visser, and S. Scott, (Cambridge University Press, 2009). arXiv:0706.0622.
[25] Clifford M. Will. The confrontation between general relativity and experiment. *Living Rev. Rel.*, 9:3, 2005, arXiv:gr-qc/0510072.
[26] Zheng-Cheng Gu and Xiao-Gang Wen. A lattice bosonic model as a quantum theory of gravity. 2006, arXivg:r-qc/0606100.
[27] Anzhong Wang and Roy Maartens. Linear perturbations of cosmological models in the Hořava-Lifshitz theory of gravity without detailed balance. 2009, arXiv:0907.1748.
[28] Xian Gao, Yi Wang, R. Brandenberger, and A. Riotto. Cosmological Perturbations in Hořava-Lifshitz Gravity. 2009, arXiv:0905.3821.
[29] Kazuya Koyama and Frederico Arroja. Pathological behaviour of the scalar graviton in Hořava-Lifshitz gravity. 2009, arXiv:0910.1998.
[30] A. I. Vainshtein. To the problem of nonvanishing gravitation mass. *Phys. Lett.*, B39:393–394, 1972.

AUTHOR INDEX